# 2022 China Semiconductor Technology International Conference (CSTIC 2022)

Shanghai, China
20 – 21 June 2022

IEEE Catalog Number: CFP2260Y-POD
ISBN: 978-1-6654-9759-6

**Copyright © 2022 by the Institute of Electrical and Electronics Engineers, Inc.
All Rights Reserved**

*Copyright and Reprint Permissions*: Abstracting is permitted with credit to the source. Libraries are permitted to photocopy beyond the limit of U.S. copyright law for private use of patrons those articles in this volume that carry a code at the bottom of the first page, provided the per-copy fee indicated in the code is paid through Copyright Clearance Center, 222 Rosewood Drive, Danvers, MA 01923.

For other copying, reprint or republication permission, write to IEEE Copyrights Manager, IEEE Service Center, 445 Hoes Lane, Piscataway, NJ 08854. All rights reserved.

***\*\*\* This is a print representation of what appears in the IEEE Digital Library. Some format issues inherent in the e-media version may also appear in this print version.***

IEEE Catalog Number:     CFP2260Y-POD
ISBN (Print-On-Demand):  978-1-6654-9759-6
ISBN (Online):           978-1-6654-9758-9

**Additional Copies of This Publication Are Available From:**

Curran Associates, Inc
57 Morehouse Lane
Red Hook, NY  12571 USA
Phone:      (845) 758-0400
Fax:        (845) 758-2633
E-mail:     curran@proceedings.com
Web:        www.proceedings.com

# 2022 China Semiconductor Technology International Conference (CSTIC 2022)

Shanghai, China
20 – 21 June 2022

**IEEE Catalog Number:** CFP2260Y-POD
**ISBN:** 978-1-6654-9759-6

# Table of Contents

**Preface**

## Chapter I - Device Engineering and Memory Technology

**System Technology Co-Optimization for 3D Monolithic Memory-Centric Computing**  1
Qi Dang, Yijun Li, Jianshi Tang, He Qian and Bin Gao
*School of Integrated Circuits, Beijing Innovation Center for Future Chips (ICFC), Tsinghua University, Beijing, China*

**Zinc-alloyed HfO$_2$ Synaptic RRAM with Operating Voltage and Switching Energy Enhancement**  5
Jun Lan[1], Quanzhou Zhu[1], Yiyang Zhang[1], Wenhui Wang[1], Muhammad Zaheer[1], Jinxuan Liang[1], Mei Shen[1], Zhixiong Li[2], Zhen Chen[2], Hongxue Wei[2], Guobiao Zhang[1], and Yida Li[1]
*[1]Southern University of Science and Technology, Shenzhen, China*
*[2]Shenzhen Longsys Electronics Co., Ltd, Shenzhen China*

**Novel Negative-Feedback Method for Writing Variation Suppression in FEFET-Based Computing-in-Memory Macro**  8
Weikai Xu[1], Jin Luo[1], Yide Du[1], Qianqian Huang[1,2] and Ru Huang[1,2]
*[1]Key Laboratory of Microelectronic Devices and Circuits (MOE), School of Integrated Circuits, Peking University, Beijing, China*
*[2]Chinese Institute for Brain Research (CIBR), Beijing, China*

**A Study on the Hf$_{0.5}$Zr$_{0.5}$O$_2$ Ferroelectric Capacitors Fabricated with Hf and Zr Chlorides**  12
Yujin Kim, Zhan Liu, Hao Jiang, T.P. Ma, Jun-Fei Zheng, Phil Chen, Eric Condo, Bryan Hendrix and James A.O'Neill
*[1]Department of Electrical Engineering, Yale University, 15 Prospect St, New Haven, CT 06511*
*[2]Entegris Inc., 7 Commerce Drive, Danbury, CT 06810*

**A Novel Anti-Ferroelectric Negative Capacitance Tunneling FET with Mitigated Subthreshold Swing Degradation Issue**  15
Shaodi Xu[1], Chang Su[1], Yimei Li[1], Qianqian Huang[1] and Ru Huang[1,2]
*[1]Key Laboratory of Microelectronic Devices and Circuits (MOE), School of Integrated Circuits, Peking University, Beijing, China*
*[2]Chinese Institute for Brain Research (CIBR), Beijing, China*

**Air Stable High Mobility ALD ZnO TFT with HfO$_2$ Passivation Layer Suitable for CMOS-BEOL Integration**  18
Wenhui Wang[1], Jiqing Lu[1,2], Jun Lan[1], Bing Zhou[1], Yiyang Zhang[1], Jinxuan Liang[1], Muhammad Zaheer[1], Mei Shen[1] and Yida Li[1]
*[1]Southern University of Science and Technology, Shenzhen, China*
*[2]Harbin Institute of Technology, Harbin, China*

**The Low-Frequency Noise Behavior Of Advanced Logic and Memory Devices**      22
Eddy Simoen[1,2], Romain Ritzenthaler[1], Hans Mertens[1], Eugenio Dentoni Litta[1], Naoto Horiguchi[1], Adrian Vaisman[1], Nouredine Rassoul[1], Gouri Sankar Kar[1] and Cor Claeys[3]
[1]*Imec,Kapeldreef 75, B-3001 Leuven, Belgium*
[2]*also at Solid-State Sciences Depart., Ghent University, Gent, Belgium*
[3]*ESAT-INSYS Depart.KU Leuven, Kasteelpark Arenberg 10, B-3001 Leuven, Belgium*

**Effects of gate Metal and Channel Shape on the Variability of Junctionless Field-Effect Transistors**      26
Xinhe Wang[1], Feng Xu[1,2], Zhigang Zhang[1], Xinyi Li[1], Jianshi Tang[1,2], Bin Gao[1,2], Huaqiang Wu[1,2] and He Qian[1,2]
[1]*School of Integrated Circuits, Beijing Innovation Center for Future Chips (ICFC), Beijing National Research Center for Information Science and Technology (BNRist), Tsinghua University, Beijing, China*
[2]*Beijing Innovation Center for Future Chips (ICFC), Tsinghua University, Beijing, China*

**CFET 6T HD SRAM Designs with 3nm Design Rule**      29
Xiaona Zhu[1], RongZheng Ding[1], Yanli Li[1,2], Qiang Wu[1,2] and Shaofeng Yu[1,2]
[1]*School of Microelectronics, Fudan University, Shanghai, China*
[2]*National Integrated Circuit Innovation Center, Shanghai, China*

**IC Technologies and Systems for Green Future**      33
Min-hwa Chi[1,2]
[1]*SiEn (Qindao) Integrated Circuits Cor., Shandong, China*
[2]*Micro-Nano Technology College, Qindao University, Qindao, Shandong, China*

**Influence of Ion Implantation on Silicon Pits Defect Formation in Oxide Etch Process**      38
Zhiqiang Xiao, Long Feng, Jiaxing Xiao, Haitao Yan, Zhenchao Sui and Xin Zhang
[1]*Semiconductor Manufacturing North China (Beijing) Corp., Beijing, China*

**Wafer to Wafer Bonding to Increase Memory Density**      41
Dube Belinda Langelihe Yolanda
*Systems Plus Consulting, Nantes, France*

**Pathfinding By Process Window Modeling: Advanced Dram Capacitor Patterning Process Window Evaluation Using Virtual Fabrication**      45
Qingpeng Wang, Yu De Chen, Jacky Huang, Benjamin Vincent and Joseph Ervin
*Coventor, Inc., A Lam Research Company, Shanghai, China*

**Systematic Study of Temperature and VDD Impact to Read Current and Standby Leakage of SRAM in FINFET Technology**      49
Yijun Zhang, Yu Li, Yuan Wang
*Technology R&D Center, SMIC, 18 Zhangjiang Road, Pudong New Area, Shanghai, China*

**T2000 ISS FT Solution Using CP Light Source**      53
Hiroyuki Kobayashi, Bin Gong, Senter Liu and Shenqi Cai
*Advantest (China) Co., Ltd, Shanghai, China*

**Impact of Topology of Trench Gate Bottom Corner For Power MOSFET and IGBT**     **56**

Min-hwa Chi[1,2], Janifer Liu[1] and Perry Li[1]

[1]*Micro-Nano Technology College, Qingdao University, Qingdao, Shandong, China*
[2]*SiEn (Qingdao) Integrated Circuits Cor., Shandong, China*

**Investigation of Removing Standing Wave Effect During Litho Process**     **60**

Xueqiang Liu, Biqiu Liu, Xiaobo Guo and Shirui Yu

*Shanghai Huali Integrated Circuit Corporation, Shanghai, China*

**A Novel Method Using Barc EB Process to Improve Split Gate Flash Erase Capability**     **63**

Song Zhang, Yaohui Zhou, Qun Liu and DeJin Wang

*Technology Development, Central Semiconductor Manufacturing Corporation (CSMC), Wuxi, China*

**Optimization of Metal Line Thickness & CD and Effect on RC Delay**     **66**

Pengfei Lyu, Qingpeng Wang and Lifei Sun

*Lam Research Service Co., Ltd, Shanghai, China*

**Novel SOI Based Pseudo-Inverter: Experimental and Simulation Research**     **69**

Sherzod Khaydarov[1], Kai Xiao[1], Haihua Wang[1], Bo Li[2], Fanyu Liu[2] and Jing Wan[1]

[1]*School of Information Science and Technology, Fudan University, Shanghai , China*
[2]*Institute of Microelectronics, Chinese Academy of Sciences, Beijing, China*

**SiGe Epitaxial Growth on SiGe Substrate for Advanced FDSOI Technology**     **72**

Yongyue Chen

*Research and Development Dept., Shanghai Huali Integrated Circuit Corporation, Shanghai, China*

**A PIN Photodetector Based on a RF-SOI Substrate**     **75**

JJ. Chou and J. Wan

*State key lab of ASIC and System, School of Information Science and Engineering, Fudan University, Shanghai, China*

**55nm Ultra-Low-Power Platform 3.3V I/O Device Reliability Improvement**     **78**

Zhao Guo

*Shanghai Huali Microelectronics Corporation (HLMC), Shanghai, China*

**MBIST Repair Mechanism and Implementation**     **81**

Haijing Wu

*Global Application and Development Center, Advantest (China), Co., Ltd., Shanghai, China*

**The Research of Decreasing SiGe Loss in Fully Depleted Silicon on Insulator (FDSOI) Devices**     **84**

Siyuan Che, Xiangguo Meng, Lian Lu, Quanbo Li, Jun Huang and Yu Zhang

*Shanghai Huali Integrated Circuit Corporation, Shanghai, China*

**Investigation of Epoxy Molding Compound on SOP Device**     **87**

Hongjie Liu, Wei Tan, Xingming Cheng, Lanxia Li, Yangyang Duan, Dandan Fan, Ming Gu Jianglong Han

*JiangSu HHCK Advanced materials Co., Ltd., Lianyungang, China*

**Improvement of CMOS Device Performance by a Combination of Spike and Flash Annealing** .......... 90
Kecheng Chen, Lan Jiang and Yaoting Shen
*Research and Development Dept., Shanghai Huali Integrated Circuit Corporation, Shanghai, China*

**Negative Capacitance Double-Gate Vertical Tunnel FET with Improved Subthreshold Characteristics** .......... 93
Guoliang Tian[1,2], Gaobo Xu[1,2], Huaxiang Yin[1,2], Lianlian Li[1,2], Gangping Yan[1,2], Yanna Luo[1,2] and Xiaoting Sun[1]
*[1]Institute of Microelectronics, Chinese Academy of Sciences, Beijing, China*
*[2]University of Chinese Academy of Sciences, Beijing, China*

**Research on Vt Window Improvement Process of 55nm SONOS eFlash Cell** .......... 96
Shipu Li, Jun Qian and Yufei Peng
*Shanghai Huali Microelectronics Corporation, Shanghai, China*

**Suppression of TSV Leakage of Stacking CIS by Optimizing TSV Profile and Uniformity** .......... 98
Zherui Cao, Weiwei Yu, Chunshan Zhao and Wuzhi Zhang
*Shanghai Huali Integrated Circuit Corporation, Shanghai, China*

**ALD Characteristic Study of $Al_2O_3$ Film Deposited by a Dual Single-Wafer Process Chamber** .......... 100
Ge Zhang, Qihang Chen, Gi Kim and Sean Chang
*Piotech Inc., No. 900 ,Shuijia, Hunnan District, Shenyang, China*

**A Novel SiGe Heterojunction Phototransistor Applicable in Wide Spectral Range** .......... 104
Yin Sha, Hongyun Xie, Yan Wang, Yang Xiang, Fu Zhu, Ruilang Ji, Weicong Na and Wanrong Zhang
*Micro-Electronics Department, Beijing University of Technology, Beijing, China*

**Defect Reduction with Advanced Lithographic Filter** .......... 107
Robb Fang[1], Alexander Zhu[1] and Yoshiaki Yamada[2]
*[1]Hangzhou Cobetter Filtration Equipment Co Ltd., Hangzhou, China*
*[2]Nippon Cobetter Co,Ltd., Osaka, Japan*

**FCVD Anneal Process for FIN CD Loss Reduction Study** .......... 112
Jun Yin, Yanxia Hao and Yinshuai Wang
*Research and Development Dept., Shanghai Huali Integrated Circuit Corporation, Shanghai, China*

**The Influence of the Film Adjustment for Deep Trench Isolation (DTI) Filling on the Thermal Quality in CMOS Image Sensor** .......... 114
Yaguo Cai, Chunshan Zhao and Wuzhi Zhang
*Huali Microelectronics Corp., Shanghai, China*

**FDC High Order Analysis** .......... 117
He Yiwen, Gong Danli and Shao Xiong
*School of Electronic Information, Shanghai University of Engineering Science, China*
*Engineering Section 1, Huali IC Manufacturing Co., Ltd., Shanghai, China*

**A Study of LDMOS with High Breakdown Voltage and Low On-Resistance in 22nm Technology**     120

Zhenchao Sui[1,2], Yongqiang Che[1], Xiaoxi Liu[1], Yongjin Tang[1] and Yongsheng Yang[1]

[1]*Semiconductor Manufacturing International Corporation(SMIC), Beijing, China*
[2]*School of Software and Microelectronic, Peking University, Beijing, China*

**A Study of the Effect of SiGe on the Inverse Narrow Width Effect in 28nm Process**     123

Yang Li, Weiwei Ma, Ran Huang, Yamin Cao and Wei Zhou
*Shanghai Huali Integrated Circuit Corporation, Shanghai, China*

**Improving the Device Performance of LDMOS Through the Optimization of Structure**     126

Shuang Jiao, Chenchen Qiu, Jun Qian and Chang Sun
[1]*Shanghai Huali Microelectronics Corporation, Shanghai, China*

**Bottom Dielectric Isolation to Suppress Sub-Fin Parasitic Channel of Vertically-Stacked Horizontal Gate-All-Around Si Nanosheets Devices**     128

Lei Cao[1,2], Yang Liu[1], Qingzhu Zhang[1,2], Zhenhua Wu[1,2], Jiaxin Yao[1,2], Yanna Luo[1,2], Haoqing Xu[1,2], Peng Zhao[1,2], Kun Luo[1] Yongqin Wu[3], Weihai Bu[3] and Huaxiang Yin[1,2]

[1]*Key Laboratory of Microelectronics Devices and Integrated Technology, Institute of Microelectronics of Chinese Academy of Sciences, Beijing, China*
[2]*University of Chinese Academy of Sciences, Beijing, China*

**Influence of Parasitic Capacitance and Resistance on Performance of 6T-SRAM for Advanced CMOS Circuits Design**     131

Yanna Luo[1,2], Gangping Yan[1,2], Lei Cao[1,2], Jiali Huo[1,2], Xuexiang Zhang[1,2], Yanzhao Wei[1,2], Guoliang Tian[1,2], Qingzhu Zhang[1], Zhenhua Wu[1] and Huaxiang Yin[1,2]

[1]*Key Laboratory of Microelectronics Devices and Integrated Technology, Institute of Microelectronics of Chinese Academy of Sciences, Beijing, China*
[2]*University of Chinese Academy of Sciences, Beijing, China*

**Optimization and Demonstration of Low-Voltage NMOS for RF Switch Application in 65nm SOI Technology**     134

Zhaozhao Xu[1], Tianyu Han[1], Wan Song[1], Fei Meng[2] and Wensheng Qian[2]

[1]*Huahong Semiconductor (Wuxi) Limited, Wuxi, China*
[2]*Shanghai Huahong Grace Semiconductor Manufacturing Corporation, Shanghai, China*

**The Optimization of Breakdown Voltage and Specific On-Resistance of 30V N-VDMOS with Short Channel**     137

Xiaoqing Cai, Donghua Liu and Wensheng Qian
*HuaHong Grace Semiconductor Manufacturing Corporation, Shanghai, China*

**Study of Breakdown Voltage Improvement of High-Voltage NLDMOS in Width Direction**     140

*Wenting Duan, Wensheng Qian, Donghua Liu, Haiyang Ling, Feng Jin and Ying Cai*
*HuaHong Grace Semiconductor Manufacturing Corporation, Shanghai, China*

**Improve the On-Resistance for 80V NLDMOS with STI Technology in 0.18 μm BCD Process**      143

Xiaoming Zhang[1,2], Donghua Liu[2], Haiyang Ling[2] and Wensheng Qian[2]
[1]*School of Information Science and Technology, ShanghaiTech University,Shanghai, China*
[2] *HuaHong Grace Semiconductor Manufacturing Corporation, Shanghai, China*

**Low Crosstalk and High Q-Factor Inductor based on Multiple Guard-Ring Design**      146

Xiaodong Wang, Xining Wang, Weihong Qian, Xi Huang, Muyi Liu and Byung Sup Shim
*Semiconductor Manufacturing International Corporation, shanghai, China*

**Flicker Noise Characterization of LDMOS in FINFET Technology**      149

Junwei Gu, Yuning Guo, Yali Liu and Jing Tong
*Semiconductor Manufacturing International Corporation (SMIC), Shanghai, China*

**Low On-Resistance LDMOS with Stepped Field Plates from 12V to 40V in 300-mm 90-nm BCD Technology**      152

Hualun Chen[1], Zhaozhao Xu[1], Yu Chen[1], Mingxu Fang[1], Li Wang[1], Li Xiao[1], Yuanyuan Qian[1], Wan Song[1], Tian Tian[1], Ziquan Fang[2], Donghua Liu[2], Wensheng Qian[2] and Weiran Kong[1,2]
[1]*Huahong Semiconductor (Wuxi) Limited, Wuxi, China*
[2]*Shanghai Huahong Grace Semiconductor Manufacturing Corporation, Shanghai, China*

**Suppression of Poly Depletion Effect by Optimizing Ion Implantation and Rapid Thermal Annealing Process in 55nm Nor Flash Technology**      156

Hualun Chen, Jiahui Jiao, Hu Wang, Haochun Xiao, Zhaozhao Xu and Lin Gu
*Huahong Semiconductor (WUXI) Limited, Wuxi, China*

**Data Retention Enhancement of Modern 55nm Nor Flash Memory**      159

Hualun Chen, Botong Liu and Lin Gu
*Huahong Semiconductor (Wuxi) Limited, Wuxi, China*

**Optimization of Integrated Process By Reducing Dual Gate and Cell Lightly Doped Drain Photo Layer in ETOX NOR Flash Technology**      162

Hualun Chen, Yihang Du, Lin Gu, Zhuangzhuang Wang, Hongjie Shen, Hu Wang, Jiuli Hu and Jiaming Pu
*Wuxi Huahong Grace Semiconductor Manufacturing Corporation, Wuxi, China*

**Improvement of Disturb And Endurance in Nor Flash Memory**      166

Hualun Chen, Ran Xu, Hui Wang, Xiaobing Ren, Xiaojun Xu and Jian Zhang
*Huahong Semiconductor (WUXI) Limited, Wuxi, , China*

**Process Optimization and Yield Improvement for Erase Failure in Embedded Flash Memory**      169

Hualun Chen, Yang Dang, Hui Wang, Xiaojun Xu, Jian Zhang and Yunlong Huang
*Wuxi Huahong Grace Semiconductor Manufacturing Corporation, Wuxi, China*

**A Wafer-On-Wafer Non-Uniform High Power Thermal Model for 3D Chip Package** — 172
Song Wang[1,3], Xiping Jiang[1,2,4], Fengguo Zuo[1], Huimei Wang[1], Xiaofeng Zhou[1], Jingrui Chai[1], Yubing Wang[1], Bin Hou[1], Liang Zhong[1], Wenxin Li[1] and Yi Kang[3]
[1]Xi'an UniIC Semiconductors, Xi'an, China
[2]Institute of Microelectronics of the Chinese Academy of Sciences, Beijing, China
[3]University of Science and Technology of China, Hefei, China
[4]University of Chinese Academy of Sciences, Beijing, China

**Evaluation of Different Cut Approaches on Advanced BEOL Self-aligned Double Patterning Scheme** — 176
Chia Lin Lu[1], Peng Fei Lyu, Qing Peng Wang and Yu Shan Chi
Lam Research Service Co., Ltd, Shanghai 201203, China

**Addressable WAT Test of Domestic Semitronix Tester** — 179
Yangyang Xing, Yanping Zhu and Xiong Shao
Shanghai Huali Integrated Circuit Manufacturing Co., LTD

**40nm Backside Optimization and Improvement** — 181
Lu Xu, Wei Lu and Li Ding
Shanghai Huali Integrated Circuit Manufacturing Co., Ltd, Shanghai, China.

**Methods of Reducing Etching Defects in Back End of Line for Semiconductor in 28nm Technology** — 184
Shanshan Chen, Hunglin Chen, Yin Long and Kai Wang
Shanghai Huali Integrated Circuit Corporation, Shanghai, China

**Hard Mask Open Tilting Improvement with Advanced Source Coil** — 186
Xiantao Luo, Qiang Ge and Ying Huang
Applied Materials (China), Inc. China

## Chapter II – Lithography and Patterning

**Considerations in the Setting Up of Industry Standards for Photolithography Process, Historical Perspectives, Methodologies, and Outlook** — 189
Qiang Wu, Yanli Li, Xianhe Liu, Xiaona Zhu, Shaofeng Yu and Wei Zhang
School of Microelectronics, Fudan University, Shanghai, China

**The Status of Stochastic Issues of EUV Lithography Photon Stochastic and Chemical Stochastic** — 196
Toru Fujimori
Electronic Materials Research Laboratories, FUJIFILM Corporation 4000 Kawashiri, Yoshida-cho, Haibara-Gun, Shizuoka, 421-0396, Japan

**Process Window, and Process Optimization in both Low and High NA EUV Processes for Advanced Logic Technologies Nodes** — 199
Yanli Li, Qiang Wu, Xianhe Liu, Xiaona Zhu and Shaofeng Yu
School of Microelectronics, Fudan University, Shanghai, China

**A Study of Improved Design Rules Through Allowing 45-Degree Metal Lines** 205
Xianhe Liu, Yanli Li and Qiang Wu
*School of Microelectronics, Fudan University, Shanghai, China*

**GaN-Based Fast Mask Near-Field Calculation** 209
Yijiang Shen, Zhijie Jiao and Jiaxiang Zhuo
*School of Automation, Guangdong University of Technology, Mega Education Center South, Guangzhou, China*

**Dose Control Strategy Using Random Logic Device Patterns and Massive Metrology in** 213
**a Foundry High Volume Manufacturing Environment**
Kan Zhou, Xin Guo, Yu Yang Bian, Wen Zhan Zhou and Yu Zhang
*Advanced Module Technology Development, Shanghai Huali Integrated Circuit Corporation, Shanghai, China*

**Patterning Capability of Surface Plasmon Imaging**
Lihong Liu[1], Huwen Ding[1,2], Libin Zhang[1], Lisong Dong[1], Le Ma[1,2] and Yayi Wei[1,2] 221
[1]*Institute of Microelectronics, Chinese Academy of Sciences, Beijing, China*
[2]*University of Chinese Academy of Sciences, Beijing, China*

**Utilizing Bossung Plot To Calibrate OPC Optical Model** 224
Jian Wang
*Semiconductor Manufacturing International Corporation, Shanghai, China*

**CD-SEM Contour Extraction for Complex Features Measurement** 227
Ting He, Yingchun Zhang and Jian Wang
*Semiconductor Manufacturing International Corporation, Shanghai, China*

**New Model-Rules-Hybrid SBARS Placement Strategy for 2D Pattern** 232
Ge Zhang, Yibin Huang and Zhongwen Yana
*Semiconductor Manufacturing International Corporation, Shanghai, China*

**Simulation-Based Source and Mask Optimization for Advanced Nodes** 236
Lianbo Luo[1] and Yu Yan[2]
[1]*Semiconductor Manufacturing International Corporation (SMIC), Shanghai, China,*
[2]*Mentor Graphics, Wilsonville, Oregon, USA*

**Process Window Integration Rule Check for BEOLIN Advanced Tech Node** 239
Xiaoyan Wang, Qing Yang, Honghong Ren, Xuanbo Chen and Qingwei Liu
*Semiconductor Manufacturing International Corporation, Shanghai, China*

**Effect of High Energy Implantation on the Photoresist for Smaller Size CMOS Image** 243
**Sensor**
Hui Chen, Xiaoyu Li, Zhengying Wei, Chang Sun, Jiong Xu and Yufei Peng
*Shanghai Huali Microelectronics Corporation, Shanghai, China*

**Stitching Process Development on 300mm Wafer CMOS BEOL for High Performance** 246
**Chip Application**
Ming Li, Xiaoxu Kang and Kaiyan Zang
*Process Technology Department, Shanghai IC R&D Center, Shanghai, China*

# Chapter III – Dry & Wet Etch and Cleaning

**Pitch Walking Simulation and Process Window Evaluation of Self Aligned Quadruple Patterning for Advanced Nodes** — 249
Xing Ke, Changcheng Jiang, Shiliang Ji, Zhenyang Zhao, Bo Su, Fengmei Li, Chongchong Zheng and Haiyang Zhang
*R&D Infrastructure, Semiconductor Manufacturing International Corp. Pudong New Area, Shanghai, China*

**Bitline Etch Process Development for Advanced Process DRAM Manufacturing** — 252
Xing-Jun Yao[1], Li-Tian Xu[1], Jan-Kun Zhang[1], Zun-Hua Zhao[1], Guang Yang[1], Chen Chen[1], Jing-Lun Ma[2], Zheng-Qing Sun[2], You-Quan Yu[2], Ming Cheng[2], Fei Li[2], Bing-Hui Lin[2], Xin-Wen Huang[2], Ying-Yi Chen[2] and Xian-Wen Su[2]
*[1]Beijing NAURA Microelectronics Equipment Co. Ltd, Beijing, China*
*[2]ChangXin Memory Technologies, Inc. Ltd, Hefei, China*

**SADP Etch Process Development Using PR Core for Sub 17nm DRAM** — 254
Li Tian Xu[1], Shuai Zhang[2], Ling Feng Li[1], Hao Liu[1], Xin Wen Huang[2], Ying Yi Chen[2], Xian Wen Su[2], Zhong Ning Guo[2], Chun Yu Xiu[2], Tian Lei Mu[2], Bing Hui Lin[2], Zhong Yi He[1] and Qing Jun Zhou[1]
*[1] Beijing NAURA Microelectronics Equipment Co. Ltd, Beijing, China*
*[2]ChangXin Memory Technologies, Inc. Ltd, Hefei City, China*

**Beyond 20nm DRAM Capacitor Etch Challenge and Process Solution** — 257
Jianqiu Hou[1], Jun Xia[2], Ya Zhou[1], Kangshu Zhan and Zengwen Hu
*[1]Advanced Micro-Fabrication Equipment Inc. China, Shanghai, China*
*[2]Chang Xin Memory Technologies Corporation, Hefei, China*

**Localized Pattern Loading Improvement of SOC Recess Process for Airgap Spacer** — 265
Bo Su, Abraham Yoo and Hansu Oh
*Semiconductor Manufacturing International Corporation, Shanghai, China*

**A Study of Wafer Center Via Etch Arcing Mechanism and Solution** — 268
Jun Wang, Yuan Li and Xinruo Su
*Semiconductor Manufacturing North China Corporation, China*

**Trench Etch for SiC Power Devices** — 271
Xiaoyu Tan and Qiushi Xie
*Department of Semiconductor Etching, NAURA Technology Group Co., Ltd., Beijing, China*

**Removing (Sub)Surface Defects Induced by Si Wafer Thinning Processes Enables High-Performance Backscattered Electron Detector** — 274
Zhu Chen[1], Lilei Hu[1,2], Li Zhang[2], Jingxuan Shen[1], and Chang Chen[1,2,3,4,5]
*[1]School of Microelectronics, Shanghai University, China*
*[2]Shanghai Industrial μTechnology Research Institute, China*
*[3]State Key Laboratory of Transducer Technology, Shanghai Institute of Microsystem and Information Technology Chinese Academy of Sciences, Shanghai, China*
*[4]Shanghai Academy of Experimental Medicine, Shanghai, China*
*[5]Shanghai Si-Gene Biotech Co., Ltd, Shanghai, China*

**Advanced Ruthenium Selective Etch for MEMS and Sub-3nm Applications** 278
Chien-Pin Sherman Hsu
*Avantor, No. 38-1, Tai-Yuan St., Chu-Bei, Hsinchu, Taiwan, China*

**Tri-Layer Mask Dry Etch Process Optimizing and Wet Effect For Straight Profile** 280
Xiaobing Liu, Shaojian Hu, Haihua Chen, Haojun Huang and YuShu Yang
*Shanghai IC R&D Center, Shanghai, China*

**Effects of Ion Incident Angles on Etching Morphology of Blazed Grating by IBE** 283
Jie Yuan[1], Xingyu Li[1], Zhongyuan Jiang[2], Yuxin Yang[2], Jiahe Li[2], Kaidong Xu[1,2] and Shiwei Zhuang[1]
*[1]School of Physics and Electronic Engineering, Jiangsu Normal University, Xuzhou, China*
*[2]Jiangsu Leuven Instruments Co. Ltd, Xuzhou, China*

**Investigation of FIN Bowing Formation Mechanism During STI Etching by Virtual Fabrication** 286
Lifei Sun, Pengfei Lyu, Qingpeng Wang, Qinghua Zhong, Kui Wang and Yushan Chi
*Lam Research Service Co., Ltd., Shanghai, China*

**PSR Silicon Trench Profile Optimization in FinFET Fabrication** 289
Zhengning Li, Xing Ke, Changcheng Jiang, Fengmei Li, Yangkui Lin and Haiyang Zhang
*Semiconductor Manufacturing International Corporation, Shanghai, China*

**Simulation Study on Different Integration Schemes to Form Single Diffusion Break** 292
Pengfei Lyu, Jian Huang, Minxiang Wang, Tianhao Zhang, Qingpeng Wang, Kui Wang and YuShan Chi
*Lam Research Service Co., Ltd, Shanghai, China*

**Study on Process Improvement and Yield Enhancement of 40nm E-Flash AIO Wet Strip** 295
Zhiyuan Xu, Youfeng Xu, Jin Chen and Zhenghong Liu
*Shanghai Huali Integrated Circuit Manufacturing Corporation, No. 6 Liang Teng Rd., Pudong New Area,Shanghai, China*

**Study and Optimization of Photo Resistor Etch Back Loop in HK Metal Gate** 298
Yajie Li, Baichun Zhang, Jianguo Yang, Lei Sun, Quanbo Li and Yu Zhang
*Shanghai Huali Integrated Circuit Corporation, Shanghai, China*

**Study on the Optimization of FinFET Ultra-Shallow Junction Ion Implantation Process** 301
Li Wenqiang, Yang Jianguo, Sun Lei, Quanbo Li, Jun Huang and Zhang Yu
***SHANGHAI HUALI INTEGRATED CIRCUIT CORPORATION, SHANGHAI, CHINA***

**N/P Split Boundary Profile Improvement in High K Metal Gate Dummy Poly Remove Process** 304
Huang Shan, Qu Xiaofeng, Sun Lei, Li Quanbo and Zhang Yu
*Shanghai Huali Integrated Circuit Corporation, Shanghai, China*

**Exploration and Optimization of Metal Gate Etch Back Process in Advanced Technology Node** 307
Shaoxiong Liu, Linpeng Jiang, Junjie Pan, Lian Lu, Quanbo Li, Jun Huang and Yu Zhang
*Shanghai Huali Integrated Circuit Corporation, Shanghai, China*

**Statistic Big Data Analysis Method Used for Tilting Mismatch Problem Solving**    310
Qingpeng Wang, Rui Bao, Shaw Zhang, Lisa Wu and Cheng Li
*Lam Research Inc., Shanghai, China*

**Selective Wet-Etching of GeSi in Multi-Layer GeSi/Si Stacks**    314
Jiajia Tian, Zhijun Cao, Qingzhu Zhang, Cinan Wu, Zhaohao Zhang and Huaxiang Yin
*[1]College of Big Data and Information Engineering, Guizhou University Guiyang, China*
*[2]Key Laboratory of Microelectronic Devices and Integration Technology, Institute of Microelectronics, Chinese Academy of Sciences, Beijing, China*

**Optimization of Approach for Metal Contamination Reduction**    317
Meng-Yu Xie, Jian-Kun Zhang[1], Zun-Hua Zhao[1], Zhong-Hai Tang[1], Jian-Zhu Zhuang[2], Wei Wan[2], Bing-Hui Lin[2], Da-Wei Tao[2], Mei-Hua Liu[2] and Pan Wang[2]
*[1]Beijing NAURA Microelectronics Equipment Co. Ltd, Beijing. Beijing, China*
*[2]ChangXin Memory Technologies, Inc. Ltd, Hefei, China*

**Optimization of Shallow Trench Isolation CD Micro Loading in Advance CMOS**    320
Guang Yang, Zhong-Wei Jiang and Jian-Kun Zhang
*[1] Beijing NAURA Microelectronics Equipment CO., Ltd, Beijing, Beijing, China*
*[2] ChangXin Memory Technologies, Inc. Ltd, Economic and Technological Development Area, Hefei, China*

**Research of Ultra High Aspect Ratio Silicon Etching in the Advanced Process**    323
Zheng Ji[1], Jian-kun Zhang[1], Guang Yang[1], Zun-hua Zhao[1], Jing Wang[1], Xue-Sheng Wang[2], Xia-Yu Shi[2], Song-Yu Li[2], Bing-Hui Lin[2], Xin-Wen Huang[2], Ying-Yi Chen[2], Zhong-Ning Guo[2] and Xian-Wen Su[2]
*[1] Beijing NAURA Microelectronics Equipment Co. Ltd, Beijing, China*
*[2] ChangXin Memory Technologies, Inc. Ltd, Hefei, China*

**SNC SADP Spacer Etch Process Development Using Carbon Hard Mask Mandrel for Sub Advanced Process DRAM**    326
Hao Liu[1], Jian-kun Zhang[1], Zun-hua Zhao[1], Li-Tian Xu[1], Chen Chen[1], Yang Chen[2], Xin-Ru Han[2], Shi-Ran Zhang[2], Jing-Lun Ma[2], Bing-Hui Lin[2], Xin-Wen Huang[2], Ying-Yi Chen[2] and Xian-Wen Su[2]
*[1] Beijing NAURA Microelectronics Equipment Co. Ltd, Beijing, China*
*[2] ChangXin Memory Technologies, Inc. Ltd, Hefei, China*

**Silicon Partial Etch Defect Researches in BSI CMOS Image Sensor Process Product**    329
Hebao Liu[1], Xiaoyu Li[2], Qixin Wu[1], Ji Feng[2], Zhe Wang[2], Dongmei Zhai[2] and Fanyan Meng[1]
*[1] School of Mathematics and Physics, University of Science and Technology Beijing, China*
*[2] Semiconductor Manufacturing International Corporation, Beijing, China*

**Hard Mask Etch Process Development For Patterning 60nm Magnetic Tunnel Junction**    332
Xiaohui Li, Hailong Liu, Ruiping Zhu, Jipeng Liu, Zhongyi He and Qingjun Zhou
*Beijing NAURA Microelectronics Equipment Co. Ltd, Beijing, China*

## Chapter IV – Thin Film, Plating and Process Integration

**Increasing the Post Halo Implantation Anneal Temperature for the Effective Improvement of Threshold Voltage Roll-off Induced by the Unique Non-uniform Boron Diffusion from the Embedded Source/Drain**          335
Run-Ling Li and Yu-Long Jiang
*State Key Laboratory of ASIC and System, School of Microelectronics, Fudan University, Shanghai, China*

**Investigation of Nanosheet Deformation During Channel-Release in Gate-All-Around Nanosheet Transistors**          338
Jingwen Yang[1], Kun Chen[1,2], Xin Sun[1], Dawei Wang[1], Qiang Wang[1], Tao Liu[1], Ziqiang Huang[1], Zhecheng Pan[1], Saisheng Xu[1], Chunlei Wu[1], Min Xu[1,2] and David Wei Zhang[1,2]
[1]*School of Microelectronics, Fudan University, Shanghai, China*
[2] *Shanghai Integrated Circuit Manufacturing Innovation Center Co., Ltd, Shanghai, China*

**The Application of Lau's Schottky-Poole-Frenkel Theory to Distinguish Leakage Current Mechanisms in High-K MIM Capacitors by Pattern Recognition**          341
W.S. Lau
*Nanyang Technological University (Retired), School of EEE, Singapore*

**Minimizing Residual Stress of Aluminum Nitride (AlN) Thin Films Using Multi-Step Deposition of DC Pulsed Sputtering**          344
Wei-Lun Chen[1], Shang Shian Yang[1], Ning Hsiu Yuan[1], Wei Yu Zhou[1], Yu-Pu Yang[1], Hsiao-Han Lo[2], Peter J. Wang[2], Walter Lai[2], Yiin-kuen Fuh[1] and Tomi T. Li[1]
[1]*Department of Mechanical Engineering, National Central University*
[2]*Delta Electronics, Inc., Taoyuan City, Taiwan, China*

**Silicon Phosphorus Process Uniformity Improvement Study in Advanced Node**          348
Huojin Tu[1], Hui Wang[1,2], Li Peng[1], Jiaqi Hong[1], Wangxin Nie[1], Jun Tan[1], Qin Sun[1] and Xinhua Cheng[1]
[1]*Research and Development Dept., Shanghai Huali Integrated Circuit Corporation, Shanghai, China*
[2]*School of Information Science and Technology, ShanghaiTech University, Shanghai, China*

**The Study of SiGe Channel Formation for FDSOI**          351
Lan Jiang
*Shanghai Huali Integrated Circuit Corporation, 6 Liangteng Rd., Pudong district, Shanghai, China*

**Low-Temperature Die to Glass Wafer Bonding Based on Au-Au Atomic Diffusion**          353
Shuchao Bao[1,2], Yi Zhong[1,2], Yimin He[2,3], Ke Li[1,2], and Daquan Yu[1,2]
[1]*School of Electronic Science and Engineering, Xiamen University, Xiamen, China*
[2]*Xiamen Sky Semiconductor Technology Co., Ltd., Xiamen, China*
[3]*School of Materials Science and Engineering, Xiamen University of Technology, Xiamen, China*

**The Effect of Interfacial and Bulk Free Energies on the Leakage Current Vs Voltage Characteristics of High-K MIM Capacitors Prepared by Atomic Layer Deposition**          356

W.S. Lau
*Nanyang Technological University (Retired), School of EEE, Singapore*

**A Hybrid Modelling Approach for the Digital Twin of Device Fabrication**     359
Ze Zheng and Dong Ni
*College of Control Science and Engineering, Zhejiang University, Hangzhou, China*

**Effects of Coil Surface Current Density on Plasma Characteristics in a PECVD Chamber**     362
Jiangjie Zeng, Qinrui Zhang, Jiahong Yu, Yongjie Hu, Xingyu Li and Shiwei Zhuang
*School of Physics and Electronic Engineering, Jiangsu Normal University, Xuzhou, China*

**Aluminum Gap Fill Improvement for 28 HKMG Process**     366
Shihao Wang, Zhaoqin Zeng, Yu Bao, Shasha Wang, Haifeng Zhou, Jingxun Fang and Yu Zhang
*Shanghai Huali Integrated Circuit Corporation, Shanghai, China*

**Research on the Oxidation Behavior of Titanium Nitride Thin Films and Resistance Simulation**     370
Xiaotong Zhang, Zhaoqin Zeng, Yanpeng Cao, Yanyan Zhang, Yu Bao, Haifeng Zhou, Jingxun Fang and Yu Zhang
*Shanghai Huali Integrated Circuit Corporation, Shanghai, China*

**Improvement of Aluminum Diffusion in HKMG Process**     374
Xiaoyang Xi, Zhaoqin Zeng, Yu Bao, Haifeng Zhou, Jingxun Fang and Yu Zhang
*Shanghai Huali Integrated Circuit Corporation, Shanghai, China*

**Investigation of 14 nm Contact Tungsten Gap-Filling Performance**     378
Xiaofang Wang, Jiazhang Xu, Zhiqi Yuan, Yingbo Cheng, Zhaoqin Zeng, Yu Bao, Haifeng Zhou, Jingxun Fang and Yu Zhang
*Shanghai Huali Integrated Circuit Corporation, Shanghai, China*

**Review and Thermo-Fluids Numerical Modeling on Electrostatic Chuck**     381
Peng Feng[1,2], Taide Tan[2], Yujie Ji[1], Yaxin Zhang[2] and Toshihisa Nozawa[2]
[1]*Department of Mechanical Engineering, Shenyang Ligong University, Shenyang, China*
[2]*Piotech, Inc, Shenyang, China*

**Film Property Analysis by FTIR on ULK Film Deposition and UV Curing Process**     384
Xinchen Cai[1], Tianjiao Teng[1], Xuanyu Lin[1], Peipei Li[1], Xinyi Chen[1], Zhuo Wang[1], Huaqiang Tan[1], Qing Mi[2], Bo Zhang[2] and Jintao Liu[2]
[1]*piotech Technology Co., Ltd., Shenyang, Liaoning Province, China*
[2]*SMNC Co., Ltd., , Beijing Economic and Technological Development Zone, China*

**Analysis of Influence to Gate Stack of EEPROM Using Cl Doped Re-Oxidation**     388
Hong Zhang[1] and Po-Yu Huang[2]
[1]*School of Microelectronics, Fudan University, Shanghai, China*
[2]*Technical Marketing, Zing Semiconductor Corporation, Shanghai, China*

**The Effect of STI Divot on Planner Logic Device Performance Study**     392
Zhenchao Sui[1,2], Yongqiang Che[1], Wenrong Hou[1], Weichi Cheng[3] and Jingang Wang[1]

[1]*SemiconductorManufacturing North China (Beijing) Corporation, Beijing, China*
[2]*School of Software and Microelectronic, Peking University, Beijing, China*
[3]*Semiconductor Technology Innovation Center (Beijing) Corporation, Beijing, China*

**300 mm Wafer Laser Anneal Process Development for Applications of Multiple Process Condition in Different Zones on Single Wafer**     395
Xiaoxu Kang[1], Y uanhao Huang[2], Wen Luo[2], Jianrui Liu[2], Zhenghui Chu[1], Xiaolan Zhong[1], Min Zhang[1], Xiaoqiang Zhou[1], Kaiyan Zang[1], Ming Li[1], Limin Zhu[3], Hanwei Lu[3] and Bo Zhang[3]
[1]*Process Technology Department, Shanghai IC R&D Center, Shanghai, China*
[2]*Shanghai Micro Electronics Equipment (Group) Co., Ltd., Shanghai, China*
[3]*PIE Department III, Shanghai Huahong Grace Semiconductor Manufacturing Corporation, Shanghai, China*

**A Strategy of Eliminating Salicide Block Film Bubbles**     398
Hualun Chen, Junwen Liu and Chao Bao
*Huahong Semiconductor (WUXI) Limited, Wuxi, China*

**TaN Based Metal-Insulator-Metal Capacitor with Excellent Long-Term Reliability**     401
Hualun Chen, Linlin Zhang, Hongxu Yang and Junwen Liu
*Huahong Semiconductor (Wuxi) Limited, Wuxi, China*

**A Strategy of Eliminating Silicon Dislocation in 55 nm Node Technology**     405
Hualun Chen, Chao Bao, Jiawei Gu, Junwen Liu and Zhongcai Niu
*Huahong Semiconductor (Wuxi) Limited, Wuxi, China*

**12-inch 90 nm BCD Process Optimization: Reducing Wafer Edge LDMOS Leakage**     409
Hualun Chen, Li Wang, Li Xiao, Yong Chen, Tian Chen, Guangyuan Lu, Jian Wang and Fan Ji
*Huahong Semiconductor (Wuxi) Limited, Wuxi, China*

**Enhanced Fill of Tungsten in 3D NAND Wordline: A View of Molecular Diffusion**     412
Xin Gan
*Lam Research (Shanghai) Co., Ltd. Shanghai, China*

**Effects of Stacked $Al_2O_3$ / $HFO_2$ Films Deposited in Deep Trench Isolation on White Pixels Noise**     414
Qixin Wu, Hebao Liu, Haitao Jin, Felix Li, Enjoy Yin, Horse Ma and Xinhe Zheng
*University of Science and Technology Beijing*
*Semiconductor Manufacturing International Corporation Beijing, China*

**An Achievement of Low Resistance Non-Salicide CT on Active Area**     417
Bingquan Wang, Simmons Zhang, Zhigao Wang
*Semiconductor Manufacturing International Corp., Beijing, China*

## Chapter V – CMP and Post-Polish Cleaning

**Developing of CMP Head-To-Head Compensation Function For Gate Height Uniformity Control** 420

Yurong Que[1], Xing Ma[1], Shuxiang Wang[1], Jian Zhang[1], Haifeng Zhou[1], Jingxu Fang[2] and Yu Zhang[2]
*Advanced Module Technology Dept., Shanghai Huali Integrated Circuit Corp., Shanghai, China*

**Several Strategies for Aluminum Metal Gate Chemical Mechanical Planarization Scratch Reduction** 424

Q. X. Hong, P. L. Zhu, F. Luo, D. C. Liu, J. Zhang, H. F. Zhou, J. X. Fang and Y. Zhang
*Shanghai Huali Integrated Circuit Corporation, Shanghai, China*

**Mark Damage Phenomenon Caused by Superimposed CMP Dishing on Large-Area STI Regions** 428

Xu WenSheng[1,2], Ba You[1], Chen YongBo[1], Zhu YuJie[1], Li RunLing[1], Ding ShiJin[2] and Ye JiongHan[1]
[1]*Shanghai Huali Microelectronics Corporation, Shanghai, China*
[2]*School of Microelectronics, Fudan University, Shanghai, China*

**Component Optimization of Sapphire Slurry Based on Response Surface Methodology for Chemical Mechanical Polishing** 431

Minghui Qu[1,2], Xinhuan Niu[1,2], Yanan Lu[1,2], Ziyang Hou[1,2], Han Yan[1,2] and Fu Luo[1,2]
[1]*School of Electronics and Information Engineering, Hebei University of Technology, Tianjin, China*
[2]*Tianjin Key Laboratory of Electronic Materials and Devices, Tianjin, China*

**Defect Law of Cu/Co Patterned Wafers after Using a Novel Bulk/Barrier Slurry and Cleaning Solution** 435

Lifei Zhang, Tongqing Wang and Xinchun Lu
*State Key Laboratory of Tribology, Tsinghua University, Beijing, China*

**Improvement of Cu-CMP EDP Curves for Different Pattern Density** 443

Yi Xian, Yuanyuan Meng, Lei Zhang, Jian Zhang, Haifeng Zhou and Jingxun Fang
*Advanced Module Technology Development, Shanghai Huali Integrated Circuit Corporation, Shanghai, China*

**Effect of TAZ as An Inhibitor on Electrochemical and CMP Characteristics of Molybdenum** 445

Pengfei Wu [1,2], Baoguo Zhang[1,2], Ye Wang[1,2], Mengchen Xie[1,2], Ye Li[1,2] and Haoran Li[1,2]
[1]*School of Electronics and Information Engineering, Hebei University of Technology Tianjin, China*
[2]*Tianjin Key Laboratory of Electronic Materials and Devices, Tianjin, China*

**Effect of OA and JFCE as Surfactants on the Stability of Copper Interconnection CMP Slurry** 448

Han Yan[1,2], Xinhuan Niu[1,2], Fu Luo[12], Minghui Qu[1,2] and Yinchan Zhang[1,2]
[1]*School of Electronics and Information Engineering, Hebei University of Technology, Tianjin, China*
[2] *Tianjin Key Laboratory of Electronic Materials and Devices, Tianjin 300130, China*

**Effect of Different Complexing Agents on Chemical Mechanical Polishing of Copper Film**　　451

Fu Luo[1,2], Xinhuan Niu[1,2], Han Yan[1,2] and Minghui Qu[1,2]

[1]*School of Electronics and Information Engineering, Hebei University of Technology, Tianjin, China*
[2]*Tianjin Key Laboratory of Electronic Materials and Devices, Tianjin, China*

**Improvement of Sensitivity of Eddy Current Thickness Sensors with Ferrite Core for CMP Process**　　454

Chengxin Wang, Tongqin Wang, Fangxin Tian, Bangxu Liu and Xinchun Lu
*State Key Laboratory of Tribology, Tsinghua University, Beijing, China*

**Dishing Improve in Advanced Technology Nodes**　　457

Yu Yang, Wei Zhao, Runcai Xiao, Hu Li, Jian Zhang, Haifeng Zhou, Jingxun Fang and Yu Zhang
*Advanced Module Technology Development, Shanghai Huali Integrated Circuit Corporation, Shanghai, China*

**An Optimized Method for Cu CMP Dishing Improvement**　　460

Lei Zhang, Yuanyuan Meng, Yi Xian, Jian Zhang, Haifeng Zhou, Jingxun Fang and Yu Zhang
*Advanced Module Technology Development, Shanghai Huali Integrated Circuit Corporation, Shanghai, China*

**Study on the Correlation Between CMP Cu Loading and EDP Curve**　　463

Yuanyuan Meng, Yixian, Lei Zhang, Jian Zhang, Haifeng Zhou and Jingxun Fang
*Advanced Module Technology Development, Shanghai Huali Integrated Circuit Corporation, Shanghai, China*

**Tungsten CMP Consumable Localization study at 28nm Technology Node**　　466

Shaojia Zhu, Mingfei Yu, Hongming Pan, Yibin Li, Lei Zhang, Jian Zhang, Haifeng Zhou, Jingxun Fang and Yu Zhang
*Advanced Module Technology Development, Shanghai Huali Integrated Circuit Corporation, Shanghai, China*

**An Optimized Monitoring Method for 28HK ILDCMP**　　469

Jingjing Li, Hongming Pan, Mingfei Yu, Shaojia Zhu, Feng Shi, Jian Zhang, Haifeng Zhou, Jingxun Fang and Yu Zhang
*Advanced Module Technology Development, Shanghai Huali Integrated Circuit Corporation, Shanghai, China*

**CMP Scratch Improve in Advanced Technology Nodes**　　472

Weiran Sun, Yu Yang, Hu Li, Wei Zhao, Runcai Xiao, Jian Zhang, Haifeng Zhou, Jingxun Fang and Yu Zhang
*Advanced Module Technology Development, Shanghai Huali Integrated Circuit Corporation, Shanghai, China*

**Study on Preparation and Polishing Performance of Ceria Slurry**　　475

Ye Wang[1,2], Baoguo Zhang[1,2], Pengfei Wu[1,2], Mengchen Xie[1,2], Ye Li[1,2] and Haoran Li[1,2]
*[1]School of Electronics and Information Engineering, Hebei University of Technology Tianjin, China*
*[2]Tianjin Key Laboratory of Electronic Materials and Devices, Tianjin, China*

**Galvanic Corrosion Caused by Device Structure in Chemical Mechanical Planarization**     478
Lei Wang[1,2], Guilin Chen[1], Luping Liu[1], Xinchun Lu[2]
*[1]Zhejiang Hikstor Technology Co. Ltd., Hangzhou, China*
*[2]State Key Laboratory of Tribology, Tsinghua University, Beijing, China*

**Post CMP Cleaning Study of Ceria Slurry**     480
Lei Wang[1,2], Luping Liu[1], Zhen Li[1], Changzhen Jia[3], Shoutian Li[3], Xinchun Lu[2]
*[1]Zhejiang Hikstor Technology Co. Ltd., Hangzhou, China*
*[2]State Key Laboratory of Tribology, Tsinghua University, Beijing, China*
*[3]Anji Microelectronics Technology (Shanghai) Co. Ltd., Shanghai, China*

## Chapter VI – Metrology, Reliability and Testing

**Study on Weighted Binary Classifier with Imbalanced SEM Defect Data**     484
Hairong Lei, Cho Teh, Lingling Pu and Qian Dong
*ASML, San Jose, USA*

**Advanced Modeling Techniques Expand the Applications of Picosecond Laser Acoustics for RF Process Monitoring**     487
Johnny Dai[1], Cheolkyu Kim[2], Robin Mair[1] and Priya Mukundhan[1]
*[1]Onto Innovation, Budd Lake, New Yersey, USA*
*[2]Onto Innovation, Bundang-gu, Sungnam-si,Gyunggi-do, Korea*

**All Side Optical Chipping Inspection of Dice in A Die Attach Machine at High Throughput**     490
Norbert Ackerl, Anton Rigner, Andreas Wiedmer, Rudolf Grüter, and Katharina Schmeing
*Besi Switzerland AG, Steinhausen, Switzerland*

**Review of Micro- And Nanoprobe Metrology for Direct Electrical Measurements on Product Wafers**     493
Benny Guralnik[1,2], Peter F. Nielsen[1], Dirch H. Petersen[2], Ole Hansen[2], Lior Shiv[1], Wilson Wei[1], Thomas A. Marangoni[2], Jonas D. Buron[1], Frederik W. Østerberg[1], Rong Lin[1], Henrik H. Henrichsen[1] and Mikkel F. Hansen[1]
*[1]CAPRES – a KLA company, Lyngby, Denmark*
*[2]Technical University of Denmark, Lyngby, Denmark*

**An Industry Example to Reduce the Test Time by Optimizing Data Extraction Method**     497
Xiaofeng Liang and Deguang Zheng
*Wafer Test Department, NXP semiconductor (China) Ltd., Tianjin, China*

**High Inheritability and Flexibility Smarttest8 Testmethod Library on Ultra-High-**     501

**Speed SERDES Measurement**
Jiaying Xiang, Yichen Xiao, Juyang Sun, Qingqing Xia and Yanfen Fang
*Advantest, Shanghai, China*

**The Flexible and Inheritable SMT7 Solution for High-Speed DDR PHY IP With IJTAG**    505
Yichen Xiao[1], Hao Wu[2] and Xin Song[3]
*[1]Advantest (China) Co. Ltd, Shanghai, China*

**An ATE Solution for High Resolution ADC/DAC**    508
Feng Qin and Mitch Royals
*Advantest (China) Co., Ltd*

**Towards Lean Front End IC Manufacturing (with AMHS implants)**    511
George W Horn
*Middlesex Industries SA*

**Study on Stress Migration of FCQFN Package with Unbalanced Arrangement of Bumps**    514
Shuanshe Chao[2], Xinyi Lin[2], Dan Yang[2], Na Mei[2], Tuobei Sun[2] and Keqing Ouyan[1]
*[1]State Key Laboratory of Mobile Network and Mobile Multimedia Technology, ZTE Corporation, Shenzhen, China*
*[2]Department of Packaging and Testing, ZTE Corporation, Shenzhen, China*

**Machine Learning Based Aging Critical Path Selection**    518
Keqing Ouyang[1,2], Jiadong Yao[2], Qi Wei[1,2], Chenfei Wu[1,2] and Guohua Zhou[1,2]
*[1]State Key Laboratory of Mobile Network and Mobile Multimedia Technology, China*
*[2]Sanechips Technology Co., Ltd, Shenzhen, Guangdong, China*

**Digital Defect Traceability across Sapphire Processing: Case Study on Micro-LED Chain**    521
Ivan Orlov[1], Gourav Sen[2] and Frédéric Falise[1]
*[1]Scientific Visual, Lausanne, Switzerland*
*[2]Fametec-Ebner GmbH, Leonding, Austria*

**Research on Reliability Optimization Mechanism of 28HKMG Technology**    524
Weiwei Ma, Ran Huang, Yamin Cao and Wei Zhou
*Shanghai Huali Integrated Circuit Corporation, Shanghai, China*

**Effective Usage of Attenuators on Production Test**    527
Hao Chen, XueQuan Chen and Qin Feng
*Advantest (China) Co., Ltd, Shanghai, China*

**MCU+ Test Solution on V93000 SmartTest8**    530
Tianyu Zhang, Lei Zhu and Wenyu Zhang
*Advantest, Shanghai, China*

**Conditionally Executed Tests, Branching and Algorithmic Binning – Getting It Right**    533
Lai-Choon Chan
*Applications Development Centre, Teradyne Asia Pte Ltd, Singapore*

**Study of Negative Charge Accumulation Mechanism and Removal Method on the Wafer Surface in Via Photo Development Process** ........ 536

Haihua Chen, Shaojian Hu, Xiaobing Liu, Heguang Shi, Manhua Shen, and Yushu Yang
*Shanghai IC R&D Center, Shanghai, China*

**14 Bits Current DAC Testing with 16 Bits Instruments** ........ 539

Pio Marcozzi[1] and Luffy Jin[2]
*[1] Teradyne SEG, Milan, Italy*
*[2] Teradyne SEG, Hefei, China*

**A Neural Network Approach to Analyze FDC Data in Semiconductor Manufacture** ........ 542

Wei Yu, Xu Chen, Guiyun Mao, Yong Wang and Zhengying Wei
*E1, HLMC, Zhangjiang Hi-Tech Park, Shanghai, China*

**A Voltage Screen Model and Method For Early Failure Screening** ........ 545

Wen Ying, Canny Chen, Kelly Yang
*Q&R, Semiconductor Manufacturing International Corporation (SMIC) Shanghai, China*

**Study and Improvement on Measurement Accuracy of Image BSED Overlay** ........ 548

Zhi-Feng Gan and Xiao-Chi Xu
*Technology R&D, Semiconductor Manufacturing International Corp., Pudong New Area, Shanghai, China*

**Tiny SADP Defect Detection and Reduction for 19nm NAND Flash Technology Semiconductor Manufacturing Engineering** ........ 551

QiuFeng Cao, Qiliang Ni and Jianye Song
*Shanghai Huali Microelectronics Corporation, Shanghai,China*

**Optimization of ESD Diode Design for RF Applications in FinFET Technology** ........ 554

Mingxing Zhou, TongQing Zhu and Lifei Zhang
*Semiconductor Manufacturing International (Shanghai) Corporation, Shanghai, China*

**Different Module's Process Affect to Poly Pattern Etch Stick Particle** ........ 557

Jiayi Fu, Qiliang Ni and Chao Han
*Shanghai Huali Microelectronics Corporation, Shanghai, China*

**Classification of Wafer Backside Images Via Faster-RCNN Based Neural Network** ........ 560

Junjun Zhuang[1], Guiyun Mao[1], Yong Wang[1], Xu Chen[1], Yansheng Wang[2] and Zhengying Wei[1]
*[1]Shanghai Huali Microelectronics Corporation, Shanghai, China*
*[2]Shanghai Huali Integrated Circuit Corporation, Shanghai, China*

**WAT Throughput Improved by Algorithm Optimization** ........ 564

Tong Chen, Liang Wang, Guiyun Mao and Xun Chen
*Shanghai Huali Microelectronics Corporation, Shanghai,* China

**Anomaly Detection of Semiconductor Processing Data Based On DTW-LOF Algorithm** ........ 569

Wang Yong, Mao Guiyun, Chen Xu and Wei Zhengying
*Shanghai Huali Microelectronics Corporation, Shanghai, China*

**Reason Forecast with BERT on Test Result Alarm in Semiconductor Fab** 572
Meng Xue, Xu Chen, Guiyun Mao, Yong Wang and Zhengying Wei
*Shanghai Huali Microelectronics Corporation, Shanghai, China*

**Virtual Metrology of WAT Value with Machine Learning Based Method** 574
Tao Zhou[1], Xuling Diao[1], Yiyi Jiang[2], Shiyuan Wen[2], Xuelong Shi[1], Quan Jing[2] and Chen Li[1]
[1]*Shanghai Integrated Circuits R&D Center Co., Ltd., Shanghai, China*
[2]*Shanghai Huali Microelectronics Corporation, Shanghai, China*

**Hybrid Solutions for Root Cause Tracing of Random Alarm on Etch Tool** 576
Xuling Diao[1], Yiyi Jiang[2], Shiyuan Wen[2], Xuelong Shi[1], Quan Jing[2], Chen Li and Tao Zhou[1]
[1]*Shanghai Integrated Circuits R&D Center Co., Ltd., Shanghai, China*
[2]*Shanghai Huali Microelectronics Corporation, Shanghai, China*

**The Type and Solution of Inline Poly Residue Defect for 28 HK Process Improvement** 579
Min Wan, Hunglin Chen, Kai Wang, Yin Long and Hao Guo
*Shanghai Huali Microelectronics Corporation, Shanghai, China*

**Study of the Degradation in LDMOS with STI Technology and Improve the Reliability with Several Methods** 582
Xiaoming Zhang[1,2], Donghua Liu[2] and Wensheng Qian[2]
[1]*School of Information Science and Technology, ShanghaiTech University, Shanghai, China*
[2]*HuaHong Grace Semiconductor Manufacturing Corporation, Shanghai, China*

## Chapter VII – Packaging and Assembly

**Integration Strategy on Low-Cost Chip-First Fan-Out Panel Level Packaging** 585
Cheng-Tar Wu, Junbo Jiang, Mengqiang Li, Chen Xian and Minghao Shen
*Chengdu ESWIN System IC Co., Ltd., Chengdu, Sichuan Province, China*

**Development and Application of Electroplating Copper Products with Low or Zero Internal Stress** 589
Yun Zhang[1], Jing Wang[1], Peipei Dong[1], Xingxing Zhang[1], Wei Zhao[1], Michael Herkommer[2], Klaus Leyendecker[2] and Volker Wohlfarth[2]
[1]*Suzhou Shinhao Materials LLC, Suzhou, Jiangsu, China*
[2]*Umicore Galvanotechnik GmbH, Schwaebisch Gmuend, Germany*

**Design and Fabrication of High-Q IPDS for Process Design Kits on Glass Substrate** 593
Haozhe Ma[1,2], Zhihui Hu[1,2], Qing Zhou[1,2], Jiabin Chen[2] and Daquan Yu[1,2]
[1]*School of Electronic science and Engineering, Xiamen University, Xiamen, China*
[2]*Xiamen Sky Semiconductor Co., Ltd, Xiamen, China*

**Investigation on Yield Improvement of Fan-Out Wafer-Level Packaging** 596
Kai Zhu[2], Shuanshe Chao[2], Yang Chen[2], Na Mei[2], Jianmin Fang[2], Tuobei Sun[2] and Keqing Ouyang[1]
*[1] State Key Laboratory of Mobile Network and Mobile Multimedia Technology*
*[2] Department of Packaging and Testing, ZTE Corporation, Shenzhen, China*

**Assembly Reflow Process Effect on Interlayer Dielectric Crack and New Finding of Failure Analysis** 599
Suming Wang, Zhidan He, Wenyuan Chen and Honghu Ji
*TF-AMD, Suzhou, China*

**The Study on Warpage of Epoxy Molding Compound** 603
Yangyang Duan, Wei Tan, Hongjie Liu, Lanxia Li, Dandan Fan, Lingling Li and Xingming Cheng
*Jiangsu Hua Hai Cheng Ke Advanced Material Co. Ltd., Lianyungang, China*

**State of the Art Metal Deposition System for Advanced UBM, RDL and Fan-Out Wafer Level Packaging** 607
Clinton Goh, Junqi Wei, Tuck Foong Koh, Kang Zhang, Kelvin Boh, Hannah Tang and Bridger Hoerner
*Applied Materials Singapore Technology Pte Ltd.*

**Glass Carriers for Advanced Packaging** 610
James Li and Jay Zhang
*Corning Incorporated, New York, USA*

**Dielectric Property Design Based on $BaTi_2O_5$ Nanorods and $BaTiO_3$ Nanoparticles Couple and its Application in Embedded Capacitor** 613
Wenzhong Zou[1], Guoyun Zhou[1,2], Wei He[1,2], Shouxu Wang[1,2], Baoliang Ren[3], Zesheng Wen[4], Quanyong Wang[4] and Yongqiang Xu[4]
*[1] School of Materials and Energy, University of Electronic Science and Technology of China, Chengdu, China*
*[2] Jiangxi Electronic Circuit Research Center of UESTC, Pingxiang, China*
*[3] Pingxiang Fengdaxing Circuits Co., Ltd., Pingxiang, China*
*[4] Ganzhou Sun & Lynn Circuits Co., Ltd., Ganzhou, China*

**Optimization of Wafer Dicing-Saw to Reduce the Chipping Defect by Using the Response Surface Methodology** 616
Hong Zhang[1], WeiFeng Wang[2] and Po-Yu Huang
*[1] School of Microelectronics, Fudan University, Shanghai, China*
*[2] Zing Semiconductor Corporation, Shanghai, China*

**The Research on Small "Dead Zones" Packaging Technology for Mass Production of Silicon Photomultiplier** 620
Yuxiao Liu, Xingan Zhang, Yang Shao, Yongqiang Yan, Heng Yi, Ru Yang, Kun Liang and Dejun Han
*College of Nuclear Science and Technology, Beijing Normal University, Beijing, China*

**Fine - Pitch Cu - Sn Transient - Liquid - Phase Bonding Based on Reflow and Pre -** 624

**Bonding**

Yunfan Shi[1], Zilin Wang[1] and Zheyao Wang[1,2]

[1] *Institute of Microelectronics, Tsinghua University, Beijing, China*
2 *Beijing Innovation Center for Future Chips, Tsinghua University, Beijing, China*

**A Simple Method for Fine Vertical Interconnection by Stencil Printed Vias on Flexible Printed Circuit Board with Low Temperature Sintering Nano-Silver Paste**    627

Xun Xiang[1], Zibai Li[1], Hongjian Zeng[2], Wei Zheng[1], Chuan Hu[1] and Yao Wang[1]

[1] *Institute of Semiconductors, Guangdong Academy of Sciences, Guangzhou, China.*
[2] *Electrical and Electronic Engineering, School of Engineering, Merz Court, Newcastle University, Newcastle Upon Tyne, United Kingdom*

## Chapter VIII – MEMS, Sensors and Emerging Semiconductor Technologies

**Highly Processable WTSV Modular Manufacturing for Next Generation 3D MEMS/NEMS Integrated System**    630

Simian Zhang[1], Xiaonan Deng[1], Yuqi Wang[1], Yifei Wu[1], Shengxian Ke[1], Linsen Li[2], Chaoyang Xing[3], Zhengcao Li[1] and Chen Wang[1]

[1] *State Key Laboratory of New Ceramics and Fine Processing, Key Laboratory of Advanced Materials of Ministry of Education, School of Materials Science and Engineering, Tsinghua University, Beijing, China*
[2] *Anhui Province Key Laboratory of Microsystem, The 43rd Research Institute of CETC, Hefei, Anhui, China*
[3] *Beijing Institute of Aerospace Control Devices, Beijing, China*

**Evaluation of Capacitive Humidity Sensor with Polyimide as Sensing Material**    634

Xiaoxu Kang[1], Xiaolan Zhong[1], Zhenghui Chu[1], Min Zhang[1], Xiaoqiang Zhou[1], Jiming Qi[1], Qi Jia[2], Weifa Zhong[2], Huajian Liang[2], Xiaozhi Kang[3] and Qingqing Sun[3]

[1] *Process Technology Department, Shanghai IC R&D Center, Shanghai, China*
[2] *Zhuhai SiSensor Technology Co. Ltd., Guangdong, China*
[3] *State Key Laboratory of ASIC and System, School of Microelectronics, Fudan University, Shanghai, China*

**Quantum Efficiency Enhancement by Optimizing BST Design for NIR CMOS Sensor**    637

Chunshan Zhao, Wuzhi Zhang, Yamin Cao and Wei Zhou
*Shanghai Huali Integrated Circuit Corporation, Shanghai, China*

**Device Modeling and Simulation of Ferroelectric Tunnel Junction for Computing-In-Memory Application**    639

Yuyao Lu, Linpu Zhai, Bin Gao, Jianshi Tang, Feng Xu, Yue Xi, Qingtian Zhang, Zhigang Zhang, He Qian and Huaqiang Wu
*School of Integrated Circuits (SIC), Beijing Innovation Center for Future Chips (ICFC), Tsinghua University, Beijing, China*

**Polarity-Controllable WSe$_2$ Transistor Enabled by Inserting a H-BN Layer**    642

Zheng Bian[1,2], Jialei Miao[1,2,3], Tianjiao Zhang[1,2] and Yuda Zhao[1,2]
[1]*School of Micro-Nano Electronics, Zhejiang University, Hang Zhou, China*
[2]*ZJU-Hangzhou Global Scientific and Technological Innovation Center, Hang Zhou, China*
[3]*Faculty of Electrical Engineering and Computer Science, Ningbo University, Ningbo, China*

**Study on Small Dead Area High-Voltage Silicon PIN Detectors**      645
Tiesong Li, Xin He and Min Yu
[1]*National Key Laboratory of Nano/Micro Fabrication Technology, Institute of Microelectronics, Peking University, Beijing, China*

**Ultra-Fast On-Chip Sensing for PCR Detection with Machine Learning-Based Image Processing Algorithm**      648
Jingxuan Shen[1], Lilei Hu[1,2], Yichen Zhang[2], Zhu Chen[1], Liying Liu[5], Yang Liu[1], Yuqian Ma[1] and Chang Chen[1, 2, 3, 4, 5]
[1]*School of Microelectronics, Shanghai University, Shanghai, China*
[2]*Shanghai Industrial μTechnology Research Institute, Shanghai, China*
[3]*State Key Laboratory of Transducer Technology, Shanghai Institute of Microsystem and Information Technology Chinese Academy of Sciences, Shanghai, China*
[4]*Shanghai Academy of Experimental Medicine, Shanghai, China*
[5]*Shanghai Si-Gene Biotech Co., Ltd, Shanghai, China*

**Based on Deep Learning CD-SEM Image Defect Detection System**      652
Shijia Yan, Shenglan Ding, Sen Wang, Cong Luo, Lei Li, Juan Ai, Qiang Shen, Qing Xia, Zhi Li, Qilin Cheng, Shilin Li, Hongwei Dai and Xiangang Hu
*Wuhan Xinxin Semiconductor Manufacturing Co., Ltd.(XMC), Wuhan, China*

**Improving White Pixel Through the Optimization of Structure and Implantation in CIS Device**      656
Lu Wang, Cuiyu Mei, Jiong Xu and Chang Sun
*Shanghai Huali Microelectronics Corporation, Shanghai, China*

**Fabrication and Performance Research of Silicon Nanosheet Field Effect Transistor Biosensor**      659
Enyi Xiong[1], Zhaohao Zhang[2], Qingzhu Zhang[2], Shuhua Wei[1], Qianhui Wei[3], Jing Zhang[1], and Jiang Yan[1]
[1]*School of Information Science and Technology, North China University of Technology, Beijing, China*
[2]*Key Laboratory of Microelectronic Devices and Integration Technology, Institute of Microelectronics, Chinese Academy of Sciences, Beijing, China*
[3]*State Key Laboratory of Advanced Materials for Smart Sensing GRINM Group Co. Ltd., Beijing, China*

**Study on Performance of $Si_3N_4$ Enhanced Microbolometer with Salicided Polysilicon Thermistor in CMOS Technology**      663
Haolan Ma, Yaozu Guo, Ke Wang, Haoyu Zhu, Jiabing Liu and Xiaoli Ji
*School of the Electronic Science and Engineering, Nanjing University, Nanjing, China*

**Influence of B Ions Doping on the Performance of *P*-Type Silicon Nanowire Field Effect**      666

**Transistor Biosensor**

Jiawei Hu[1,2], Qingzhu Zhang[2], Shuhua Wei[1], Jing Zhang[1], Zhaohao Zhang[2], Jin biao Liu[2] and Jiang Yan[1]

*[1]School of Information Science and Technology, North China University of Technology, Beijing, China*

*[2]Key Laboratory of Microelectronic Devices and Integration Technology，Institute of Microelectronics，Chinese Academy of Sciences，Beijing, China*

**An SOI-Based SiGe Heterojunction Phototransistor with Large Operation Current Range**     669

Fu Zhu, Hong-Yun Xie, Zi-Hang Wang, Rui-Lang Ji, Yang Xiang, Yin Sha, Dong-Yue Jin and Wan-Rong Zhang

*Faculty of Information Technology, Beijing University of Technology, Beijing, China*

## Chapter IX – Design and Automation of Circuit and Systems

**Using Mixed Logic Synthesis Tools in Open-Source FPGA Design Framework**     672

Liangtao Shi[1], Yong Xiaoy[2], Yun Shaoy[2] and Zhufei Chu[1,2]

*[1]EECS, Ningbo University, Ningbo, China*
*[2]giga Design Automation Co. Ltd., Shenzhen, China*

**Design-for-Recovery Techniques for Combating Chip Aging Issues**     675

Xinfei Guo

*University of Michigan – Shanghai Jiao Tong University Joint Institute*
*Shanghai Jiao Tong University, Shanghai, China*

**Analytical Optimization Method for VLSI Global Placement**     679

Weijie Chen[1], Haishan Huang[1], Zhipeng Huang[2] and Jianli Chen[2,3]

*[1]College of Computer and Data Science, Fuzhou University, Fuzhou, China*
*[2]Center for Discrete Mathematics and Theoretical Computer Science, Fuzhou University, Fuzhou, China*
*[3]State Key Lab of ASIC & System, Fudan University, Shanghai, China*

**Area-aware Optimization of XOR-AND Graph Based on Reed-Muller Logic Expansion**     683

Hongwei Zhou[1], Yong Xiao[2], Yun Shao[2] and Zhufei Chu[1,2]

*[1]EECS, Ningbo University, Ningbo, China*
*[2]Giga Design Automation Co. Ltd., Shenzhen, China*

**Polynomial Formal Verification of General Tree-Like Circuits**     686

Alireza Mahzoon[1] and Rolf Drechsler[1,2]

*[1]Institute of Computer Science, University of Bremen, Bremen, Germany*
*[2]Cyber-Physical Systems, DFKI GmbH, Bremen, Germany*

**AN FPGA-Based Verification Platform for High-Speed Interface IPs**     690

C.-Z. Chen[1,2], Xuhui Liu[1] and Hanming Wu[1,2]
*[1]Peng Cheng Laboratory, Shenzhen, China*
*[2]School of Micro-Nano Electronics, Zhejiang University, Hangzhou, China*

**Automatic Placement Algorithm of Integrated Circuits for Wire Bond Packaging Application**
693
J. Wan and HC. Wang
*School of Information Science and Engineering, Fudan University, Shanghai, China*

**Integrated Superconducting Isolator-Circulator-Isolator Device**
696
Rutian Huang[1], Genting Dai[1], Jianshe Liu[1] and Wei Chen[1,2]
*[1]School of Integrated Circuits, Tsinghua University, Beijing, China*
*[2]Beijing Innovation Center for Future Chips, Tsinghua University, Beijing, China*

**A Gaussian Process And Multi-Swarm Optimizer Assisted Optimization Approach for Analog Circuit Design**
699
Xu Fu[1], Changhao Yan[1], Dian Zhou[2] and Xuan Zeng[1]
*[1]State Key Lab of ASIC & System, Fudan University, Shanghai, China*
*[2]Department of Electrical Engineering, University of Texas at Dallas, Texas, U.S.A*

**An Approximating Twiddle Factor Coefficient Based Multiplier for Fixed-Point FFT**
702
Songyu Sun[1], Zhicheng Xu[1], Xi Chen[1] and Xunzhao Yin[1,2]
*[1]College of Information Science & Electronic Engineering, Zhejiang University, Hangzhou, China*
*[2]Zhejiang Lab, China*

**A Fast Timing Analysis and Optimization for Latch-Based Circuits**
706
Kaixiang Zhu, Xiao Di, Wai-Shing Luk, Lingli Wang and Jun Tao
*School of Microelectronics, Fudan University, Shanghai, China*

**A Transient-Improved Spike Time Reduction Circuit for LDO**
709
Zongyuan Zheng, Chen Zhang, Bo Wang and Xinan Wang
*The Key lab of IMS, School of ECE Peking University Shenzhen Graduate School, GuangDong, China*

**Rule Check of Pad Placement in IC Layout With Yolo V3**
713
Chunxi Lin and Tao Su
*School of Electronics and Information Technology, Sun Yat-sen University, Guangzhou, Guangdong, China*

# China Semiconductor Technology International Conference 2022 (CSTIC 2022)

**Editors:**

**Cor Claeys**
KU Leuven
Leuven, Belgium

**Hanming Wu**
Zhejiang University
Hangzhou, China

**Beichao Zhang**
HFC Semiconductors
Shanghai, China

**Steve X. Liang**
JCET Semiconductor (Shaoxing) Co. Ltd
Shaoxing, China

**Qinghuang Lin**
LAM Research Corporation
Fremont, CA, USA

**Ru Huang**
Peking University
Beijing, China

**Peilin Song**
IBM Thomas J. Watson Research Center
Yorktown Heights, NY, USA

**Linyong (Leo) Pang**
D2S
San Jose, California, USA

**Ying Zhang**
Naura
Santa Clara, CA, USA

**Xinping Qu**
Fudan University
Shanghai, China

**Cheng Zhuo**
Zhejiang University, China
Hangzhou, China

**Hsiang-Lan Lung**
Macronix International, Ltd.
Hsinchu, Taiwan, China

# PREFACE

This issue contains a selection of the accepted papers presented at *China Semiconductor Technology International Conference 2022 (CSTIC 2022)*, June 19-20, 2022 in Shanghai, China. Due to the COVID-19 pandemic the conference has been change into a hybrid conference. After reviewing a selection of the presentations have been considered for publication in IEEE Xplore.

CSTIC is the largest and the most comprehensive annual industrial semiconductor technology conference in China. It aims to provide a platform for executives, managers, engineers and researchers from around the world to exchange the latest developments in semiconductor technology and manufacturing and related fields. It also offers an opportunity for those who are interested in investing and collaboration opportunities in the semiconductor industry in Asia, particularly in China.

CSTIC covers all the aspects of semiconductor technology and manufacturing, including circuit design, system integration, devices, materials, patterning (lithography and etching), processes, integration, testing, reliability, device physics and manufacturing as well as emerging semiconductor technologies, including clean energy such as light emitting diodes (LEDs), III-V semiconductors, sensors and micro-electromechanical systems (MEMS).

CSTIC 2022, organized by Semiconductor Equipment and Material International (SEMI), imec, and The Integrated circuit Materials Industry Technology Innovation Alliance (ICMTIA) and technically sponsored by the IEEE Electron Devices Society relies on a long time tradition, which started in 2001. The original International Semiconductor Technology Conference (ISTC) merged in 2009 to become CSTIC, aiming for a broad international representation and increased paper submissions from around the world. For CSTIC 2022 the papers came from all major semiconductor manufacturing regions in the world, including Austria, Belgium, China, Denmark, Germany, France, Italy, Japan, Korea, Sweden, Switzerland, Taiwan, United Kingdom and the United States of America. About 189 papers have been selected for oral presentations and approximate 247 papers for poster presentations after careful reviews by the conference organizing committee.

In total 216 papers are included in these Proceedings after peer reviews. They represent a snapshot of the recent developments in semiconductor technology and manufacturing in the world. In particular, they offer a glimpse into the state-of-the-art of semiconductor technology and manufacturing in China. These papers are divided into nine (9) chapters according to the nine symposia of CSTIC 2022:

- Device Engineering and Memory Technology

- Lithography and Patterning

- Dry & Wet Etch and Cleaning

- Thin Film, Plating and Process Integration

- Chemical-Mechanical Polishing (CMP) and Post-Polish Cleaning

- Metrology, Reliability and Testing

- Packaging and Assembly

- MEMS, Sensors and Emerging Semiconductor Technologies

- Design and Automation of Circuits and Systems

These Proceedings are very valuable to engineers and researchers in the fast-moving and growing semiconductor industry. It will give readers a clear understanding of the status of semiconductor technology and manufacturing in China. Furthermore, it will also serve as a useful reference for those who are interested in nanofabrication, micro- and nano-fluidics, micro- and nano-photonics, organic electronics, bio-chips, light emitting diodes (LEDs) and other clean energy technologies.

We thank the invited speakers and the authors, particularly the conference plenary speakers, Dr. Martin van den Brink, President and CEO, ASML, The Netherlands, Dr. Gerard Yin, Chairman and CEO, AMEC, Dr. Marvin Liao, Vice President, APTS, TSMC, China, and Dr. Feng Hong, CEO, ICLeague Technology, China for their valuable contributions to CSTIC 2022. We also thank the more than 120 organizing committee members, particularly the symposium chairs, for their dedication and hard work to help improve the quality and to broaden the reach of CSTIC. These committee members are experts in their respective fields of semiconductor technology and are from well-known companies or prestigious institutions. They all have demanding day jobs, yet they have volunteered to help organizing this conference and to critically review papers presented in these Proceedings. Their contributions were crucial for the success of the conference. We are also indebted to the financial support from the sponsors of CSTIC 2022. Finally, we extend our sincere thanks to SEMI for their tireless efforts and their meticulous organizational skills to help organize CSTIC 2022 and to assemble and publish these CSTIC 2022 proceedings.

**Hanming Wu, General Chair CSTIC 2022**
Zhejiang University, Hangzhou, China

**Cor Claeys, Co-Chair, CSTIC 2022**
KU Leuven, Leuven, Belgium

### CSTIC 2022 Organizing Committee

June 2022, Shanghai, China

# SYSTEM-TECHNOLOGY CO-OPTIMIZATION FOR 3D MONOLITHIC MEMORY-CENTRIC COMPUTING

*Qi Dang, Yijun Li, Jianshi Tang, He Qian, and Bin Gao\**

School of Integrated Circuits, Beijing Innovation Center for Future Chips (ICFC),
Tsinghua University, Beijing, China
\*Corresponding Author's Email: gaob1@tsinghua.edu.cn

## ABSTRACT

In post-Moore era, further enhancement of computing efficiency in conventional computing system has become progressively slower due to the memory-wall bottleneck. Monolithic 3D (M3D) integration technology provides ultra-high inter-layer bandwidth, and thus becomes a promising solution for the memory-wall issue. This work will first discuss the principles of M3D memory-centric computing based on back-end-of-line (BEOL) process integration of Resistive Random Access Memory (RRAM) and nano-devices. Next, the recent progresses on system-technology co-optimization (STCO) for the M3D system, including the fabrication process of RRAM devices, computation-in-memory (CIM) circuit and computing system, are reviewed. Furthermore, a cross-layer simulator for M3D STCO is developed. Also, benchmark and design guidelines for the future M3D systems are provided based on the simulator outputs.

## INTRODUCTION

Conventional computing system is facing great challenges due to the slowdown of conventional Moore's law scaling and the intrinsic inefficiency of data-shuttling under the classic von-Neumann architecture [1]. To solve this issues, Computation-in-memory architecture deploys computation onto memory unit and avoid massive amount of data movement, which can empower extremely efficient artificial intelligence applications [2].

On the other hand, Monolithic 3D (M3D) integration technology provides ultra-high inter-layer bandwidth, and thus becomes a promising solution for the memory-wall issue. Monolithic 3D integration, whereby each circuit layer is fabricated directly over the last circuit layers on the same substrate, can employ inter-layer vias (ILVs) to connect between stacked layers [3]. Compared with through-silicon vias (TSVs), the use of conventional ILVs significantly increases vertical interconnection density, potentially maximizing the benefits of 3D integrated circuits (ICs) [4].

RRAM exhibits unique advantages, such as low temperature substrate-independent fabrication processes and fab-friendly materials, which makes it promising for the M3D integration and solve the memory wall issue. However, research on RRAM based memory-centric

computing system still faces some challenges: 1) the variation of multilevel states [5]; 2) CIM accelerators require area-hungry peripheral circuits such as the high-precision ADC to quantize the output results [6]; 3) the scalability of the RRAM based CIM system for the performance boosts [5]. It is clearly that the future design should focus on the cross-layer level, instead of some specific level studies. In this paper, we demonstrated a cross-layer simulator for M3D STCO. The simulator output can be the benchmark and design guidelines for the future M3D systems.

## M3D INTEGRATION OF RRAM
### RRAM integrated with CMOS logic

The RRAM can be fabricated using back-end-of-line (BEOL) process under a low temperature ($\leq$ 300 °C), enabling its integration with Si transistors as one-transistor-one-resistor (1T1R) cells. Figure 1 (a) shows a cross-sectional transmission electron microscopy (TEM) image of the fabricated $HfO_x$-based RRAM. The RRAM stacks are TiN/TEL/$HfO_x$/TiN, where TEL means thermal enhanced layer. The RRAM cell can be fabricated directly over the logic layers on the same substrate and connected through ILV. The fabricated array with 1024 1T1R cells is shown in Figure 1 (b). The 1T1R array exhibits a high operation speed of <10 ns, a high yield (99.9%) and good endurance performance [7]. The transistor gate ports facilitate fine conductance modulation by controlling the compliance current with a

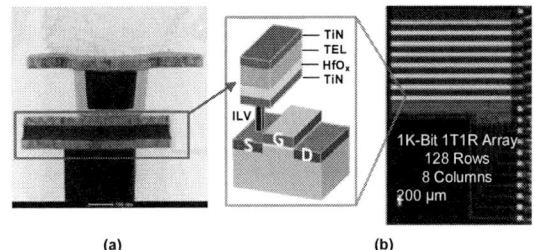

(a)                                    (b)

*Figure 1: Cross-sectional transmission electron microscopy (TEM) image of the RRAM integrated at different layers and connected using interlayer vias. (b) The schematic of 1T1R RRAM cell with TiN/TEL /HfOx /TiN structure and 128×8 array.*

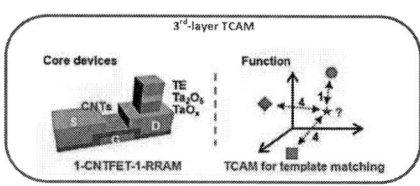

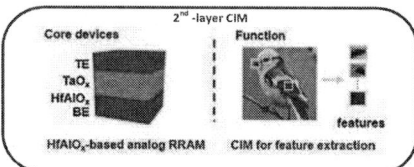

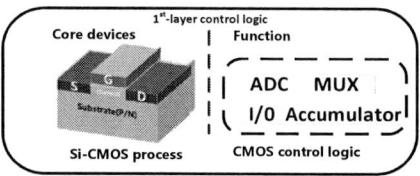

Figure 2: The architecture of monolithic 3D integrated chip, which is consisted of Si MOSFET-based CMOS logic (1st layer), HfOx RRAM-based CIM (2nd layer) and $Ta_2O_5$ RRAM and CNTFET-based TCAM (3rd layer). Adapted from Ref. [8].

specific gate voltage.

### M3D integration with BEOL transistor

Carbon Nanotube Field Effect Transistor (CNTFET), where is BEOL-compatible and can be used for high-performance logic, is a promising candidates for the M3D chip. The CNTFET promises improvement on both performance and energy efficiency (~10x benefit in energy-delay product (EDP) compared to silicon-CMOS) [3]. As illustrated in Figure 2, a M3D chip (M3D-LIME) was successfully demonstrated, which consists of a Si MOSFET-based CMOS logic (1st layer), a $HfO_X$ RRAM-based CIM (2nd layer) and a $Ta_2O_5$ RRAM and CNTFET-based TCAM (3rd layer) [8]. The latter two layers were fabricated using the low-temperature BEOL process without affecting the performance of prior layers. Meanwhile, GPU-equivalent classification accuracy up to 97.8% was achieved in the one-shot/few-shot learning task on the Omniglot dataset with 162× lower energy consumption compared with GPU.

## COMPUTING-IN-MEMORY WITH RRAM

The RRAM array forms a natural processing elements (PE) to implement parallel in-memory matrix-vecter-multiplication (MVM) operations, by directly using Ohm's law for multiplication and Kirchhoff's law for accumulation. As shown in Figure 3, in a practical RRAM-based neuromorphic computing system, the input

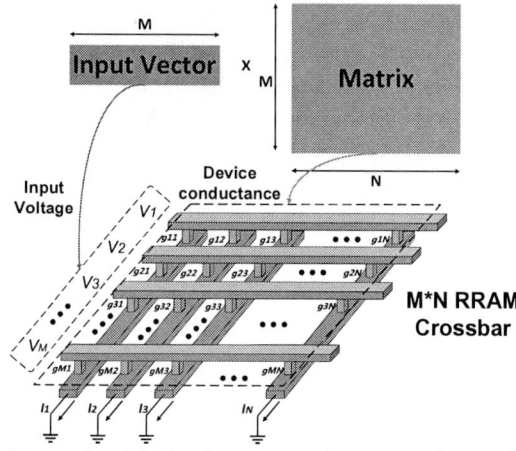

Figure 3: The basic concept of a massively parallel analog vector–matrix multiplication within a RRAM crossbar. The currents accumulated by the columns are given by Eq. (1).

vector is represented by a voltage vector $V$ applied to the word-line (WL) of the RRAM array, and the weight matrix is represented by the conductance of the RRAM array. The MVM results are represented by the output current vector I from each bit line (BL):

$$I_k = \sum_{j=1}^{M} g_{k,j} V_j \qquad (1)$$

where $V_j$ represent the element of the input voltage vector V and $I_k$ represent the output current vector; $g_{k,j}$ is the conductance of the $(i, j)$ cell.

A small-scale analog CNN accelerator has been fully implemented in hardware. Using eight 128*16 TEL/HfO$_x$ RRAM crossbars, hardware-based two convolutional layers and one fully connected layer achieved 96% accuracy on MNIST. Meanwhile, more than two orders of magnitude higher power efficiency and one order of magnitude higher performance density were obtained compared with Tesla V100 GPU [7].

Furthermore, a full-system-integrated CIM chip with 160Kb analog RRAM cell was designed and fabricated, which achieved high recognition accuracy on MNIST database, high inference speed, and 78.4 TOPS/W peak energy efficiency [1].

## STCO FOR RRAM BASED CIM SYSTEM

While M3D integration of logic and memory in arbitrary stacking order is an attractive technology, mixing different technologies increases the complexity of the system design and fabrication. For this purpose, there is an urgent of STCO platform where we can conduct cross-layer codesign from system level to device level.

978-1-6654-9759-6/22 $31.00 © 2022 IEEE

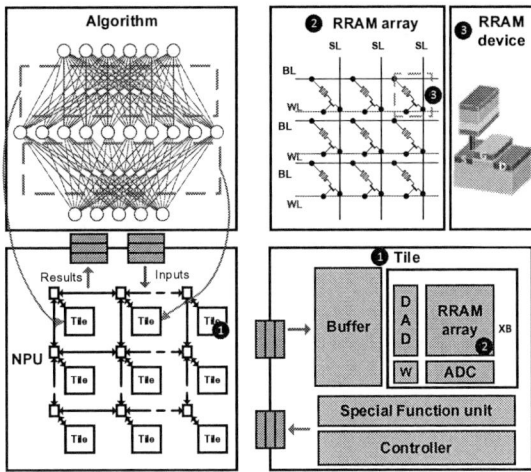

*Figure 4: Left: Schematic workflow of the simulator from DNN algorithm framework to the RRAM based NPU chip. Right: Details of the microarchitecture of the Tile, the modules of Crossbar, and the module of*

TABLE I. ARCHITECTURE CONFIGURATION PARAMETERS OF M3D SIMULATOR

| | **Parameter** | | **Parameter** |
|---|---|---|---|
| Device Level | RRAM Precision | Tile Level | Crossbar number |
| | RRAM Variation | | Input precision |
| | Range of resistance | | Buffer size |
| Crossbar Level | Crossbar Size | System level | Tile number |
| | Cell Type | | NoC Configuration |
| | DAC precision | | Memory Configuration |
| | ADC precision | | Intra Tile Bandwidth |
| | ADC Number | | Intra layer Bandwidth |

In our previous work, XPESim, an end-to-end simulator, was developed to explore the design space of RRAM based neural-processing unit (NPU) [9]. As shown in Figure 5, the simulator can provide detailed circuit-level evaluations from the aspects of device technologies and non-ideal parameters. Furthermore the simulator support end-to-end training and evaluation of computing accuracy with different scales of neural networks.

In this paper, we developed a cross-layer simulator for the M3D chip. This simulator help researchers to fast evaluate the accuracy of the algorithm model and the performance of their hardware architecture.

The features of the simulator include: M3D integration by placing circuit unit on the four layers corresponding to different functions; different neural network types and tasks from VGG [4] to one-shot learning algorithms [10]; fast benchmark of the M3D system, including latency, area, and power. As shown in Figure 5, the simulator consists of model parameter, memory-centric compiler and simulation output analysis.

*a) Configuration parameter:* The Configuration

parameter contains the DNN model that integrates some device variations in the training process and the configurations of architecture. The detailed architecture configuration parameters of different levels are listed in Table I. Users can configure their architecture design from device level to system level.

*b) Memory-centric compiler:* the compiler analyzes the model automatically and splits the model into the crossbar-level weights to fit the crossbar size and mapping the weight to the conductance of RRAM crossbar. Meanwhile, the M3D chip architecture will be defined in the input configurations and manually partition the architecture across multiple layers.

*c) Simulation output:* The simulator is used to calculate on-chip computing accuracy in the inference phase and evaluate the latency and energy consumption of the M3D memory-centric computing system.

As illustrated in Figure 6, in our simulation, we defined a three layers M3D chip (M3D-LIME) as benchmarks to show the system-level execution time and the classification accuracy. The results in Figure 7(a) shows the classification accuracy of one-shot/few-shot learning on the Omniglot dataset and VGG8 Network on the CIFAR-10 dataset using GPU and the M3D platform. On the other hand, Figure 7(b) shows that inference time benchmark on the M3D system and the 2D-chip baseline (in which CIM and RRAM memory are not monolithically integrated) with VGG8 Network on the CIFAR-10 dataset.

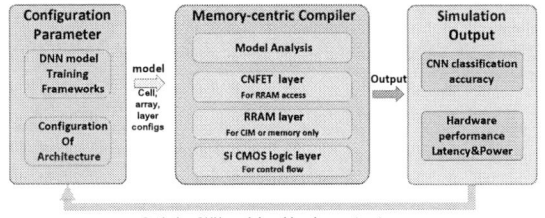

*Figure 5: The simulation framework of the cross-layer simulator for M3D STCO*

## ACKNOWLEDGEMENTS

This work was supported in part by and Natural Science Foundation of China (61874169), Beijing Municipal

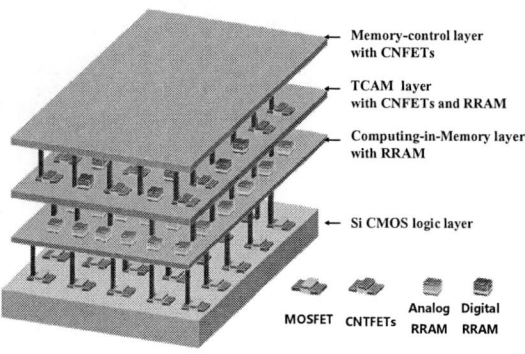

Figure 6: The M3D chip architecture

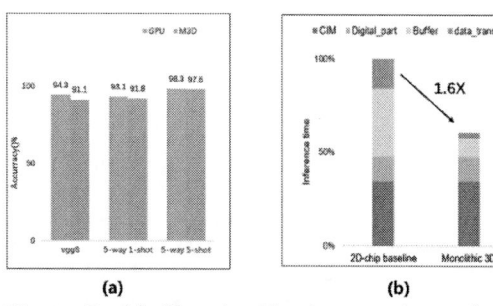

Figure 7: (a) The classification accuracy of one-shot/few-shot learning on the Omniglot dataset and VGG8 Network on the CIFAR-10 dataset using GPU and the M3D platform. (b) Inference time benchmark on the M3D system and 2D-chip baseline (in which CIM and RRAM memory are not monolithically integrated) with VGG8 Network on the CIFAR-10 dataset.

Science and Technology Project (Z191100007519008), and Beijing National Research Center for Information Science and Technology (BNRist).

## REFERENCES

[1] Q. Liu *et al.*, "33.2 A Fully Integrated Analog ReRAM Based 78.4TOPS/W Compute-In-Memory Chip with Fully Parallel MAC Computing," in *2020 IEEE International Solid-State Circuits Conference - (ISSCC)*, 16-20 Feb. 2020 2020, pp. 500-502.

[?] H. M. Chen *et al.*, "On Reconfiguring Memory-Centric AI Edge Devices for CIM," in *2021 18th International SoC Design Conference (ISOCC)*, 6-9 Oct. 2021 2021, pp. 262-263.

[3] M. M. Shulaker *et al.*, "Monolithic 3D integration of logic and memory: Carbon nanotube FETs, resistive RAM, and silicon FETs," in *2014 IEEE International Electron*

*Devices Meeting*, 15-17 Dec. 2014 2014, pp. 27.4.1-27.4.4.

[4] M. M. S. Aly *et al.*, "The N3XT Approach to Energy-Efficient Abundant-Data Computing," *Proceedings of the IEEE,* vol. 107, no. 1, pp. 19-48, 2019.

[5] S. Yu, W. Shim, X. Peng, and Y. Luo, "RRAM for Compute-in-Memory: From Inference to Training," *IEEE Transactions on Circuits and Systems I: Regular Papers,* vol. 68, no. 7, pp. 2753-2765, 2021.

[6] X. Peng *et al.*, "Benchmarking Monolithic 3D Integration for Compute-in-Memory Accelerators: Overcoming ADC Bottlenecks and Maintaining Scalability to 7nm or Beyond," in *2020 IEEE International Electron Devices Meeting (IEDM)*, 12-18 Dec. 2020 2020, pp. 30.4.1-30.4.4.

[7] P. Yao *et al.*, "Fully hardware-implemented memristor convolutional neural network," *Nature,* vol. 577, no. 7792, pp. 641-646, 2020/01/01 2020.

[8] Y. Li *et al.*, "Monolithic 3D Integration of Logic, Memory and Computing-In-Memory for One-Shot Learning," in *2021 IEEE International Electron Devices Meeting (IEDM)*, 2021.

[9] W. Zhang *et al.*, "Design Guidelines of RRAM based Neural-Processing-Unit: A Joint Device-Circuit-Algorithm Analysis," in *2019 56th ACM/IEEE Design Automation Conference (DAC)*, 2-6 June 2019 2019, pp. 1-6.

[10] F.-F. Li, R. Fergus, and P. Perona, "One-shot learning of object categories," *IEEE Transactions on Pattern Analysis and Machine Intelligence,* vol. 28, no. 4, pp. 594-611, 2006.

# ZINC-ALLOYED HFO$_2$ SYNAPTIC RRAM WITH OPERATING VOLTAGE AND SWITCHING ENERGY ENHANCEMENT

*Jun Lan[1], Quanzhou Zhu[1], Yiyang Zhang[1], Wenhui Wang[1], Muhammad Zaheer[1], Jinxuan Liang[1], Mei Shen[1], Zhixiong Li[2], Zhen Chen[2], Hongxue Wei[2], Guobiao Zhang[1], and Yida Li[1,*]*
[1]Southern University of Science and Technology, Shenzhen 518055, China
[2]Shenzhen Longsys Electronics Co., Ltd, Shenzhen 518000, China
*Corresponding Author's Email: liyd3@sustech.edu.cn

## ABSTRACT

In this work, we demonstrate the performance engineering of a HfO$_2$-based analog RRAM device based on a Zn alloying approach using ALD. The Zn-alloyed HfO$_2$ RRAM results in a significant smaller operating voltages and lower switching energy due to the increase in oxygen vacancies elucidated via XPS measurements. Meanwhile, its retention at 85 ℃ can exceed that of $10^5$ s. The LTP and LTD of the Zn-alloyed RRAM shows good linearity for the implementation of artificial neural network in handwriting recognition. Our results provide a pathway for RRAM to meet the requirements of emerging embedded memory applications for analog computing.

## INTRODUCTION

To facilitate more extensive implementation of next generation memory and computing systems, memristor is widely considered to be the most promising building blocks. Metal oxides-based resistive random-access memory (RRAM) are intensively studied and can be proposed as the most implementable candidates due to its simple device structure, fast operating speed and low energy consumption [1-3]. At the same time, RRAM has also shown great potential as an artificial synapse. Its low-power and high-speed programming characteristics, as well as the synaptic characteristics such as long-term potentiation (LTP) and depression (LTD), satisfy the exact requirements in neuromorphic computing architecture [4-6]. However, many kinds of materials applied in RRAM and its fabrication process have disadvantages such as being costly and not fully compatible with current state-of-art complementary metal oxide semiconductor (CMOS) platform, thus limiting its adoption commercially.

In a number of reported works, HfO$_2$ as switching layer is commonly used due to its CMOS compatibility nature[7]. However, as the requirements of RRAM technology continue to evolve, superior performance of the RRAM such as operating voltage, speed and energy are expected. A number of methods have been proposed to improve the performance of RRAM, such as doping, layering, and process tuning[8-14], but the overall improvement still falls short. It is clear that the performance improvement of single transition metal oxides-based RRAM is limited by the material. Instead, controlled approaches to introduce new elements to form binary or higher order oxides could provide a potential way to engineer the RRAM for a performance boost.

In this work, we propose a Zn-alloyed HfO$_2$ (HZnO) RRAM deposited using atomic layer deposition (ALD) process, providing us with high degree of composition and thickness controllability. As compared to HfO$_2$-based devices prepared using ALD process, the HZnO RRAM exhibits significantly lower forming/set voltages (1.7 V/1 V), and >1 order lower switching energy of 3 pJ. In addition, it shows a good retention at relatively high temperatures. The performance enhancement is credited to the ability to modulate the oxygen vacancies (V$_o$) creation as indicated from detailed XPS characterization for both types of films. Finally, we evaluated the handwriting recognition accuracy (MNIST database) using the measured synaptic characteristic.

## EXPERIMENT

The process flow, schematic of the RRAM device with the corresponding microscopic and SEM images are shown in Figure 1.

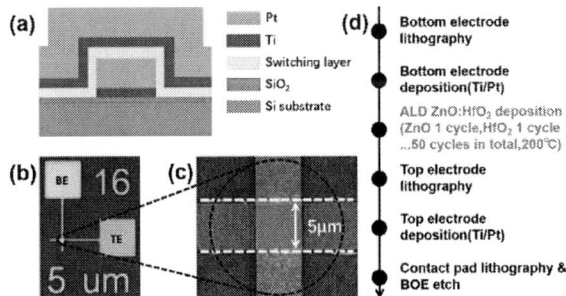

*Figure 1: (a) Schematic diagram, (b) optical microscope image, (c) SEM image, and (d) process flow of the fabricated RRAM.*

The RRAM is of a typical cross-point structure where the switching layer (SL) is sandwiched between an inert BE - Pt and reactive top electrode (TE) – Ti. The SL(s) (HfO$_2$ and HZnO) were deposited using an ALD process at 250 ℃ and 200 ℃, respectively. For the HZnO alloyed film, it was deposited by alternating single-layer of HfO$_2$

and ZnO. In all cases, a total of 50 ALD cycles was performed using $H_2O$ as the oxygen source. Ellipsometer measurements confirmed the deposited alloyed film to be ~5.5 nm. The fabricated RRAM has a cell size of 5 x 5 $\mu m^2$. The electrical characteristics were measured using a Keithley 4200-SCS semiconductor parameter analyzer. Positive voltage bias was applied to the TE while the BE was grounded during the measurements.

## RESULTS AND DISCUSSION

Figure 2(a) and (b) show the I-V characteristic curves of both the $HfO_2$ and HZnO RRAM respectively. The I-V curves include 60 consecutive normal switching cycles and the average cycles are indicated. It can be observed that although the resistance of the fresh $HfO_2$ RRAM is larger than that of the HZnO RRAM, the HZnO RRAM shows lower set voltage (0.9 V vs 1.1 V) and better cycle to cycle uniformity. Figure 2(c) and Figure 2(d) show the device-to-device distribution of the forming, operating voltages, and resistances for both sets of RRAMs, as indicated in the legend.

It can be observed that the HZnO RRAM has a lower forming/set voltage as compared to the $HfO_2$ RRAM (1.7 V/0.9 V vs 3.3 V/1.1 V), instrumental in operational power consumption reduction. In addition, while having a smaller off-resistance (~2x), the HZnO RRAM shows a smaller voltage/resistance variation as compared to the $HfO_2$ RRAM, indicating the advantages of controlled alloying approach in improving devices' uniformity.

Furthermore, AC pulse switching of both type of devices was measured as shown in Figure 3. Given that the switching speed of HZnO RRAM is faster than that of $HfO_2$ RRAM (40 ns vs 50 ns), and the former switches at a lower pulse amplitude as compared to that of the latter (0.9 V vs 1.6 V), this translates to a >1 order reduction in switching energy (3 pJ vs 96 pJ).

The improvement in both operating voltages and switching speed is likely to be attributed to the lower effective Gibbs free energy of the HZnO oxide system in creation of oxygen vacancies for switching. The pristine device is likely to already possess much larger concentration of $V_o$, resulting in a lower forming voltage. This is further investigated via XPS measurements of both types of films, where the O1s spectra in Figure 4 clearly shows a larger amount of $V_o$ in the HZnO film represented by the peak intensity located at binding energy of 531.7 eV as compared to that of $HfO_2$ film.

Figure 5(a) shows the retention of the HZnO RRAM at 85 ℃ exceeding $10^5$ s, indicating its stability at high temperature, which is essential for reliable non-volatile memory operation. In addition, the HZnO RRAM also exhibited synaptic behavior with good reliability. Figure 5(b) shows the LTP and LTD of the device under the identical pulses (LTP: 0.9 V/100 μs, LTD: -1 V/100 μs). Figure 5(c) shows 5 consecutive P/D cycles of the HZnO

RRAM under the same set/reset pulse conditions with good repeatability. Finally, we employed NeuroSim to evaluate online learning/offline classification of the MNIST handwriting recognition database in a two-layer-based multilayer perceptron (MLP)[15]. In this simulation based on our extracted programming characteristic, our simulated accuracy can reach >89% as shown in Figure 5(d).

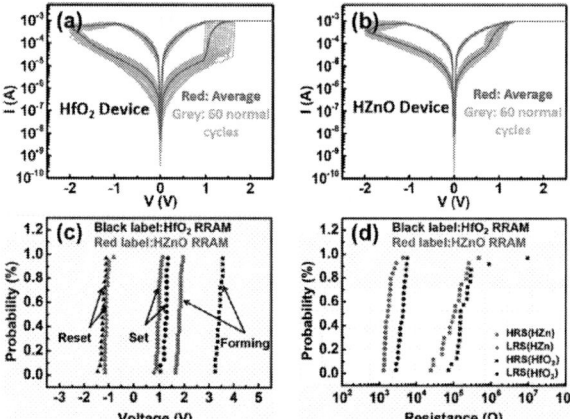

Figure 2: Typical I-V curve of $HfO_2$ RRAM (a) and HZnO RRAM (b). The DC voltage sweep from -2 V to 3 V and compliance current are 100 μA. Cumulative distribution of the (c) Set/reset/forming voltage, and (d) resistance of the set/reset states measured from 20 different devices.

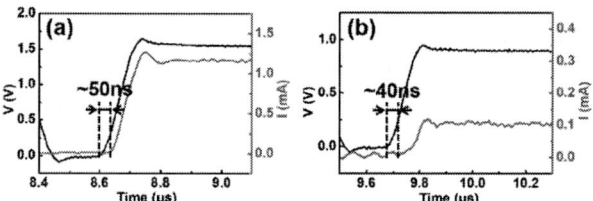

Figure 3: Comparison of switching speed of (a) $HfO_2$ RRAM and (b) HZnO RRAM tested using voltage pulses of 1.2 μs width.

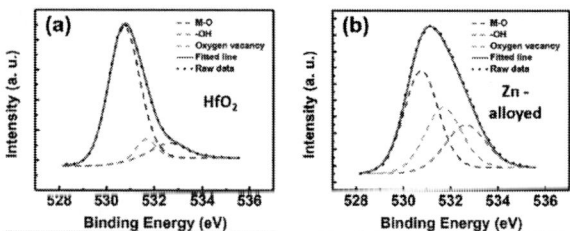

Figure 4: XPS O1s core level spectra of (a) $HfO_2$ and (b) Zn-alloyed film.

Table I compares the proposed device with the various typical RRAM reported in [9]-[13]. The forming voltage (~1.7 V) and set voltage (~0.9 V) demonstrated in this work are lower than the reported RRAMs as tabulated in the table. And benefited from smaller switching voltage and corresponding current, energy of the HZnO RRAM (~3 pJ) is much lower than the other two mentioned in the table (62.5 pJ and 400 pJ respectively).

*TABLE I : BENCHMARK OF THE PERFORMANCE OF DIFFERENT REPORTED RRAM*

| Switching Layer | $V_{form}$ (V) | $V_{operate}$ (DC)(V) | Energy (J) | Reten -tion(s) |
|---|---|---|---|---|
| GZⅢ[9] | 2 | 2/-0.5 | NA | >$10^4$ |
| LSO[10] | 2.8 | 1/-1 | ~62.5p | >$10^3$ @280℃ |
| TaON /WO$_x$[11] | 1.8 | 1.8/ -1.1 | NA | >$10^6$ @150℃ |
| HfO$_x$ (NE)[12] | 3.72 | 1.2/ -0.8 | NA | NA |
| TiO$_x$ /NiSi[13] | 4.37 | 2.34/ -1.61 | ~400p | >$10^4$ @85℃ |
| HZnO (This work) | 1.7 | 0.9/ -1.1 | ~3p | >$10^5$ @85℃ |

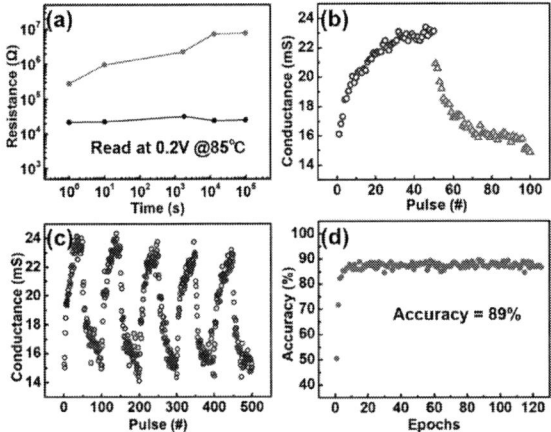

*Figure 5: (a) Retention characteristic read at 0.2 V under the temperature of 85 ℃. LTP and LTD of HZnO RRAM with 50 identical positive pulses and 50 identical negative pulses for (b) a single P/D cycle and (c) 5 consecutive P/D cycles. (d) Handwriting recognition accuracy of two-layer neural network simulation.*

## CONCLUSION

In summary, we report on a HZnO RRAM prepared via an ALD process with enhanced performance. As compared to HfO$_2$ RRAM, the alloyed RRAM boasts a smaller forming and set voltage (1.7 V/1 V) and lower switching energy (3 pJ) due to the presence of higher $V_o$ content as revealed from XPS. In addition, the proposed device also exhibits good stability (retention >$10^5$ s, at 85℃) and synaptic behavior. Based on the synaptic characteristics, the handwriting recognition accuracy was

obtained and evaluated through simulation. Hence, our results provide us with a feasible avenue to engineer RRAM in a controllable manner for various emerging applications such as multi-bit storage and neuromorphic computing.

## ACKNOWLEDGEMENTS

This work was supported by the National Natural Science Foundation of China (Grant No. 62174074), and in part by the Shenzhen Fundamental Research Program (Grant No. JCYJ20190809143419448), NSQKJJ (Grant No. K21799123, K21799131) and Engineering Research Center of Three-Dimensional Integration in Guangdong Province. We would also like to acknowledge the Core Research Facilities (CRF) at SUSTech for the facilities used, and the technical support provided by the staff and engineers at the CRF.

## REFERENCES

[1] H. Y. Chen et al. , *Journal of Electroceramics* (2017): 1-18.
[2] H. Wu et al., *Proceedings of the IEEE*, vol. 105, no. 9, 2017, pp. 1770-1789.
[3] Y. Xi et al., *Proceedings of the IEEE*, vol. 99, 2020, pp.1-29.
[4] Y. Yang, M. Yin, Z. Yu, Z. Wang, T. Zhang, Y. Cai, W. D. Lu, and R. Huang , *Adv. Electr. Mater.*, vol. 3, 2017, pp.1700032.
[5] Q. Duan, L. Xu, J. Zhu, X. Sun, Y. Yang, and R. Huang, *China Semiconductor Technology International Conference (CSTIC)*, 2018, pp. 1-3.
[6] Z. Wang, M. Yin, T. Zhang, Y. Cai, Y. Wang, Y. Yang, and R. Huang, *Nanoscale*, vol. 8, 2016, pp.14015-14022.
[7] Y. Fang, Z. Yu, Z. Wang, T. Zhang, Y. Yang, Y. Cai*, R.Huang, *IEEE Electron Device Letters*, vol. 39, no. 6, 2018, pp. 819-822.
[8] Y. Qin et al., *China Semiconductor Technology International Conference (CSTIC)*, 2021, pp. 1-3.
[9] Xing Li et al., *ACS Applied Materials And Interfaces* 12.27(2020):30538-30547.
[10] X. Zhao et al., *IEEE Electron Device Letters*, vol. 40, no. 4, pp. 554-557, April 2019.
[11] L. Tai et al., *IEEE International Conference on Integrated Circuits, Technologies and Applications (ICTA)*, 2018, pp. 170-171.
[12] B. Zhang et al., *China Semiconductor Technology International Conference (CSTIC)*, 2020, pp. 1-3.
[13] D. K. Lee et al., *IEEE Transactions on Electron Devices*, vol. 68, no. 1, pp. 438-442, Jan. 2021.
[14] J. Yang et al., *China Semiconductor Technology International Conference (CSTIC)*, 2020, pp. 1-4.
[15] P. Chen, X. Peng and S. Yu, *IEEE International Electron Devices Meeting (IEDM)*, 2017, pp. 6.1.1-6.1.4.

# NOVEL NEGATIVE-FEEDBACK METHOD FOR WRITING VARIATION SUPPRESSION IN FEFET-BASED COMPUTING-IN-MEMORY MACRO

*Weikai Xu[1], Jin Luo[1], Yide Du[1], Qianqian Huang[1,2]\*, and Ru Huang[1,2]\**

[1] Key Laboratory of Microelectronic Devices and Circuits (MOE), School of Integrated Circuits,
Peking University, Beijing 100871, China
[2] Chinese Institute for Brain Research (CIBR), Beijing 102206, China
*Corresponding Author's Email: hqq@pku.edu.cn; ruhuang@pku.edu.cn

## ABSTRACT

In this work, a novel one-shot negative-feedback writing method is proposed for suppression of writing variation in ferroelectric FET (FeFET) based analogy computing-in-memory (CIM) macro for high-accuracy artificial neural network (ANN). By utilizing the source-voltage negative-feedback mechanism and the voltage and time dependent multi-domain switching dynamics, a source-follower writing structure with an FeFET and a NMOS is proposed and simulated to reduce variation of target programming $V_{TH}$ for FeFET with even multi-level state, revealing significant decrease of FeFET synaptic conductance variation. Furthermore, based on the proposed FeFET writing variation suppression method, the CIM macro of FeFET array is demonstrated to improved accuracy of image recognition ANN by 40% compared with direct writing, showing great potential for high-accuracy neural network system.

## INTRODUCTION

Emerging non-volatile memory devices (NVM) [1][2] based computing-in-memory (CIM) architectures which perform matrix vector multiplication (MVM) operations of artificial neural network (ANN) have triggered a lot of interests. Ferroelectric FET (FeFET) is considered as a NVM candidate for synaptic weight cell of CIM macro due to low write power consumption and high on/off ratio [3][4]. However, the writing variation of FeFET caused by non-uniformity of domain distribution or stochasticity of domain switching [5], resulting in memory window collapse [6] and network accuracy decline [7], which is one of the main bottlenecks in the realization of high-accuracy large-scale ANN. Recently, a current-limiting circuit with an FeFET and a resistor was proposed to restrain $I_{on}$ variation for 1-bit storage [8], while sacrificing the dynamic range. The write-and-verify scheme was proposed to restrain FeFET $V_{TH}$ variation [9], while suffering from complex sequence control circuit and high program energy consumption.

In this work, a one-shot negative-feedback writing method based on source-follower structure and voltage and time dependent multi-domain switching dynamics, is proposed to suppress writing variation of FeFET based multi-level weight cell. Without influencing the FeFET dynamic range, only one-shot write operation is required

to form the target state. Furthermore, based on the proposed writing variation suppression method, FeFET based CIM macro for image recognition with ANN is demonstrated to achieve accuracy improvement by 40%.

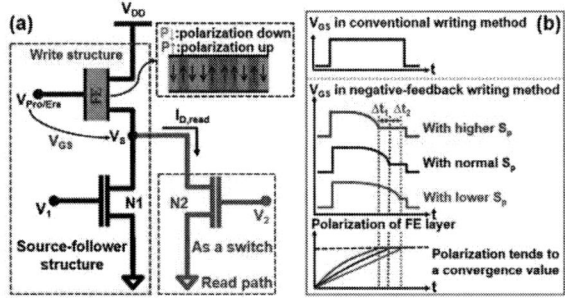

*Figure 1: (a) The novel one-shot negative-feedback writing circuit; (b) The principle of suppressing variation*

## THE NEGATIVE-FEEDBACK WRITING METHOD FOR WRITING VARIATION SUPPRESSION OF FEFET-BASED CELL

### Principle of the negative-feedback writing method

As shown in Fig.1(a), the one-shot negative-feedback writing circuit based on FeFET is proposed, which is composed of one FeFET and two NMOSs. During the writing process, N2 is turned off so that the source of FeFET is connected to the drain of N1 with a fixed gate voltage, which forms a source-follower structure. During the reading process, N2 is turned on with sufficiently high conductivity to short the N1 and the FeFET channel conductance is obtained. Taking the program process as an example to analyze the principle of suppressing writing variation, as shown in Fig.1(b). The amount of polarization switching per unit time during program pulse applied is decided by the polarization program speed ($S_p$). Sp is different due to the variation of coercive field ($E_c$) distribution and remanent polarization ($P_r$) of multi-domain ferroelectric (FE) layer, leading to different polarization switching amount per unit time. For conventional open loop direct writing operation where the source of FeFET is connected to GND and $V_{GS}$ remains constant during pulse programming process, the FeFET programming $V_{TH}$ state shows a severe deviation due to the different $S_p$ of FE layer. For the proposed negative-feedback method of source-follower structure,

978-1-6654-9759-6/22 $31.00 © 2022 IEEE

the source voltage of FeFET will change along the dynamic switching process of FE polarization, which will play a negative-feedback role to adaptively adjust $V_{GS}$ of FeFET. Higher $S_p$ of FeFET results faster $V_{TH}$ decrease and channel conductance increase during positive pulse, and $V_{GS}$ decreases faster to inhibit excess polarization switching as shown in the red line in Fig.1(b). On the contrary, for FeFET with lower $S_p$, $V_{GS}$ decreases slower to allow longer programming duration as shown in the blue line in Fig.1(b). Therefore, the final polarization of FE layer tends to a relatively convergence value due to the dynamic negative-feedback process of $V_{GS}$, which suppresses the variation of programmed FeFET $V_{TH}$. For the erase process, the principle is the same as above.

## Modeling of FeFET variation

As shown in Fig.2(a), the FeFET model composed of the multi-domain dynamic Preisach model for FE layer and BSIM model for MOSFET [10][11] is established to analyze the function of the proposed negative-feedback writing method by adding additional Gaussian distribution to reflect the variation of dispersion degree of $E_c$ and $P_r$ for FeFETs. As the FE layer is integrated within the MOSFET gate stack in FeFET, the Eq.1 governing the law of charge conservation and voltage division are solved on iterative method in HSPICE.

$$\begin{cases} Q_{Fe}(V_{Fe}) = Q_{MOS}(V_{MOS}) \\ V_G = V_{Fe} + V_{MOS} \end{cases} \quad (1)$$

The $E_c$ and Pr of the FE model is calibrated with experimental results in [12]. Fig.2(b) shows the *P-V* saturation loop and corresponding *I-V* curve of metal-ferroelectric-metal (MFM) capacitor. There are various variation sources which mainly change the $E_c$ distribution and $P_r$ in the *P-V* loop of MFM capacitor. The variation of dispersion degree of multi-domain $E_c$ distribution and FE polarization density is analyzed respectively, and the simulation results are shown in

## Variation suppression with negative-feedback method

When applying the program/erase voltage pulses with the conventional open-loop direct writing operation and the proposed negative-feedback writing operation respectively, $I_D$-$V_{GS}$ read-out curves of FeFET with high/low $V_{TH}$ states are obtained with backward/forward gate-voltage scanning, as shown in Fig.3. For the variation of Ec distribution and Pr, the negative-feedback writing method reduces the standard deviations ($\sigma$) of FeFET $V_{TH}$

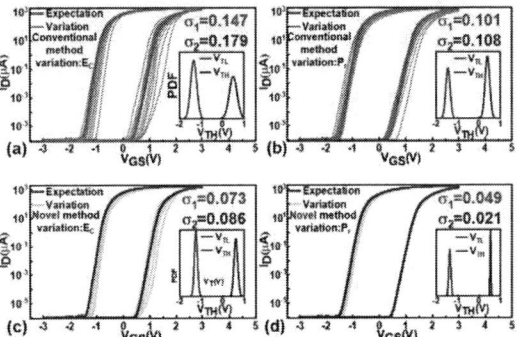

*Figure 3: $I_D$-$V_{GS}$ curves with the two types of variation mentioned above (a)(b) In conventional writing method; (c)(d) In one-shot negative-feedback writing method*

distribution by more than 50% from Fig.3(a)(b) to Fig.3(c)(d), without significantly reducing of the memory window width compared with the conventional method, indicating that the negative-feedback writing method can effectively suppress the writing variation of FeFET.

Furthermore, because the multilevel characteristics of FeFET can improve the performance of FeFET-based CIM macro [3], we further study the influence of negative-feedback writing method on the multilevel states of FeFET with variation. We use the characteristic that FE polarization follows the minor-loop under unsaturated applied voltage to obtain four different $V_{TH}$ states by

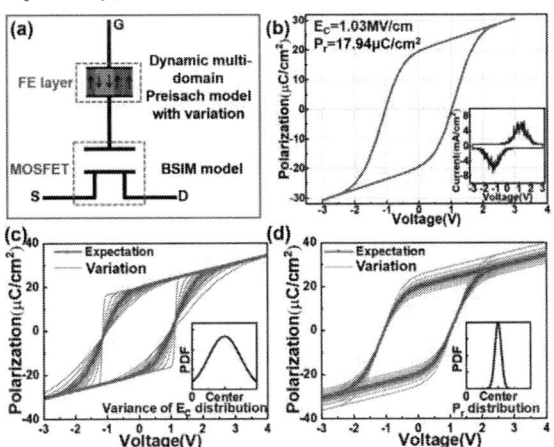

*Figure 2: (a) The FeFET model; (b) P-V saturation loop of MFM capacitor; (c)(d) P-V loops with variation Fig.2(c)(d).*

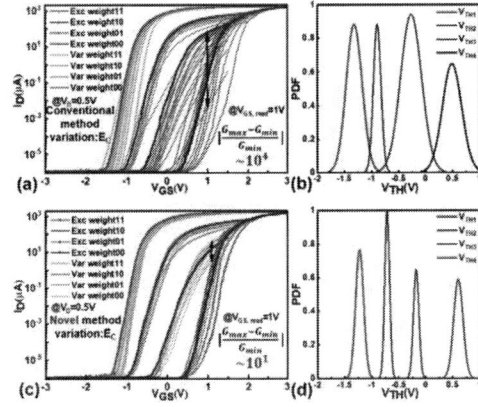

*Figure 4: Simulation results of multilevel $I_D$-$V_{GS}$ curves and threshold voltages distribution of FeFET with variation (a)(b) In conventional writing method; (c)(d) In one-shot negative-feedback writing method*

978-1-6654-9759-6/22 $31.00 © 2022 IEEE

applying four different program voltages, representing and storing four different weight values respectively. The $I_D$-$V_{GS}$ curves with the variation of dispersion degree of $E_c$ distribution of FE layer are shown in Fig.4. The four threshold voltage states in conventional method have obvious overlapped and are difficult to distinguish, while there is still a significant sense margin in negative-feedback method. For different FeFETs with device-to-device variation or different write operations in one FeFET with cycle-to-cycle variation, the maximum shift of conductance state which is defined as $|\frac{G_{max}-G_{min}}{G_{min}}|$ at $V_{GS}=1V$ in this work, is reduced by about 1000 times from nearly $10^4$ in conventional method to 10 in negative-feedback method, The variation of FeFET based multi-weight on MVM operation is significantly reduced, which is of great significance to improve the accuracy of analog synaptic ANN based on FeFET CIM macro.

## FEFET-BASED CIM MACRO FOR IMAGE RECOGNITION

Based on the proposed negative-feedback writing method and multiplexing principle, a negative-feedback FeFET pseudo-crossbar array architecture is proposed as shown in Fig.5(a).The memory cell consists of a NMOS for selection and an FeFET for storage, which can be accurately programmed/erased separately. Therefore, the whole array only needs to share a feedback transistor through a multiplexer for the negative-feedback write operation, which has negligible area and energy efficiency loss compared with the conventional method.

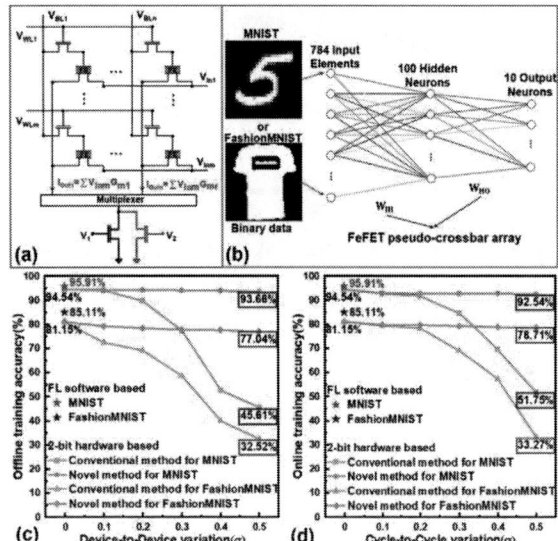

*Figure 5: (a) FeFET pseud-crossbar array based on negative-feedback writing method; (b) Two-layer MLP neural network; (c) Offline training accuracy with device-to-device variation; (d) Online training accuracy with cycle-to-cycle variation*

A two-layer multilayer perceptron (MLP) ANN with quantifiable weight accuracy is established, which can further add device-to-device during inference and cycle-to-cycle variation during training respectively, as shown in Fig.5(b). The classification accuracy of 2-bit FeFET weight-based hardware without variation reaches 94.54% and 95.91% respectively, close to 95.91% and 85.11% of full precision weight based on software for MINST and FashionMNIST databases. However, the classification accuracy in conventional method will decline severely when adding the variation of FeFET. Based on the proposed novel negative-feedback writing method, the image classification accuracy of ANN with variation is demonstrated with significantly improved compared to the conventional method in a large variation range, as shown in Fig.5(c)(d). Furthermore, the improvement of accuracy is more significant with variation increasing. The offline training accuracy improvement can reach 48.05% and 44.52% based on MNIST and FashionMNIST databases respectively with the biggest device-to-device variation considered in this model as shown in Fig.5(c), and the online training accuracy improvement can reach 40.79% and 45.44% with the biggest cycle-to-cycle variation as shown in Fig.5(d).

## CONCLUSION

In this work, a novel one-shot negative-feedback writing method of FeFET is proposed, which can suppress the writing variation effectively utilizing the negative-feedback mechanism and the voltage and time dependent multi-domain switching dynamics. Furthermore, based on the proposed FeFET-based CIM macro design, high-accuracy image recognition ANN with variation is demonstrated, which is of great significance to the implementation of high-accuracy large-scale neural network system.

## ACKNOWLEDGEMENTS

This work was supported by National Key R.D Program of China (2018YFB2202800), NSFC (61927901, 61822401, 61851401, 61421005) and 111 Project (B18001).

## REFERENCES

[1] C. Xue et al., *ISSCC*, pp. 388-390, 2019.
[2] L. Gallo et al., *IEEE TED*, pp. 99.1-99.9, 2018.
[3] S. Yu, *Proceedings of the IEEE*, pp. 260-285, 2018.
[4] J. Luo et al., IEDM, pp.19.5.1-19.5.4, 2021.
[5] K. Ni et al., *IRPS*, pp. 1-5, 2020.
[6] K. Ni et al., *VLSI*, pp. T40-T41, 2019.
[7] P. Chen et al., *IEDM*, pp. 6.1.1-6.1.4, 2017.
[8] T. Soliman et al., *IEDM*, pp. 29.2.1-29.2.4, 2020.
[9] H. Zhou et al., *IEDM*, pp. 18.6.1-18.6.4, 2020.
[10] J. Luo et al., *IEDM*, pp. 6.4.1-6.4.4, 2019.

[11] Z. Fu et al., *CSTIC*, pp. 1-4, 2020.
[12] V. Gaddam et al., *IEEE TED,* pp. 745-750, 2020.

# A Study on the $Hf_{0.5}Zr_{0.5}O_2$ Ferroelectric Capacitors fabricated with Hf and Zr Chlorides

*Yujin Kim, Zhan Liu, Hao Jiang, and T.P. Ma*
Department of Electrical Engineering, Yale University, 15 Prospect St, New Haven, CT 06511
*Jun-Fei Zheng\*, Phil Chen, Eric Condo, Bryan Hendrix, and James A. O'Neill*
Entegris Inc., 7 Commerce Drive, Danbury, CT 06810
Corresponding Author's email: Jun-Fei.Zheng@Entegris.com

## ABSTRACT

Ferroelectric capacitor memory devices with carbon-free $Hf_{0.5}Zr_{0.5}O_2$ (HZO) ferroelectric films are fabricated and characterized. The HZO ferroelectric films are deposited by ALD at temperatures from 225 to 300°C, with $HfCl_4$ and $ZrCl_4$ as the precursors. Residual chlorine from the precursors is measured and studied systematically with various process temperatures. 10nm HZO films with optimal ALD growth temperature at 275°C exhibit remanent polarization of $25\mu C/cm^2$ and cycle endurance of $\sim 5x10^{11}$. Results will be compared with those from HZO films deposited with carbon containing metal-organic precursors.

## INTRODUCTION

Since the discovery of ferroelectricity in doped $HfO_2$ [1], ferroelectric $HfO_2$ films with many different dopants have been extensively studied. Among them, $ZrO_2$ doped $HfO_2$ films with a Hf and Zr ratio close to 1:1 (known as $Hf_{0.5}Zr_{0.5}O_2$, or HZO) are favored for their relatively large doping margin and low processing temperature. HZO is compatible with BEOL CMOS process using MFM (Metal-Ferroelectric-Metal) capacitor structure as ferroelectric memory. It is also possible to incorporate HZO in the FEOL process for making FE-FET or capacitor-less DRAM devices [2], which can also be used to make Ferroelectric 3D NAND memory if we can stack the FE-FET in vertical direction. In these applications, improvement of remanent polarization and cycle endurance are critical. It is well known that $HfO_2$ and $ZrO_2$ used in modern transistors are deposited with $HfCl_4$ and $ZrCl_4$ precursors, instead of carbon containing metal-organic precursors such as TEMA-Hf (tetrakis-ethylmethylamino hafnium) and TEMAZr (tetrakis-ethylmethylamino hafnium Zirconium). This is to avoid carbon contamination and gain better device performance. Recently, a few studies have been reported to grow ferroelectric HZO films using Hf and Zr chloride precursors at 300°C [3-5]. To maximize the formation of Orthorhombic phase and avoid undesirable Monoclinic or Tetragonal phases, the deposition temperature of HZO films is limited to below 300°C. However, it is well known that at 300°C and below, chlorine from the ALD process using $HfCl_4$ and $ZrCl_4$ precursors will be trapped inside the HZO films. Thus, a systematic study of chlorine residual from Hf and Zr Chloride in HZO and its impact to MFM devices is needed.

In this study, we report a systematic study of the key device performance parameters of MFM capacitors with ferroelectric HZO films deposited by Hf an Zr chloride precursors. We first show the change of chlorine impurity as a function of various deposition temperatures. Then we show the performance characteristics of the HZO films in Ferroelectric capacitors in terms of remanent polarization, P-V loops, and cycle endurance. And we will also compare that with the reference HZO films deposited with metal-organic precursors

## EXPERIMENTS

HZO films were deposited using an ASM Pulsar 3000 using $HfCl_4$ and $ZrCl_4$ precursors with $H_2O$ as oxidant co-reactant. 10nm fully mixed HZO films and laminated HZO films (with 1 nm $HfO_2$ and 1 nm $ZrO_2$ alternative layers) are made at deposition temperature of 225°C, 250°C, 275°C and 300°C. Cl content was measured by X-ray Fluorescence Spectrometer (XRF) for samples deposited at various temperature (Fig.1). The atomic concentration of the Cl in the samples is calibrated to $HfO_2$ sample with known chlorine content measured by SIMS (Secondary Ion Mass Spectroscopy). MFM capacitor devices were fabricated by depositing 10nm HZO films on 20nm CVD TiN on Si as bottom electrode followed by 150nm PVD TiN films as top electrodes. The top electrodes were patterned via either shadow masks or lithographic lift-off process. Samples were annealed at 400-500°C for up to 10 minutes in $N_2$ inside a RTA (Rapid Thermal Annealing).

## RESULTS AND DISCUSSIONS

Fig. 1 shows different levels of chorine contents inside HZO films deposited at 225°C, 250°C, 275°C and 300°C. Calibration and calculation show that, at 225°C and 275 °C. Cl is 6.6 atomic % and 2 atomic %, respectively. At 300°C the Cl content drops to 1.4%. In Logic device application, the process temperature for HfCl4 and ZrCl4 can be as high as 350°C, in order to drive down the Cl level. In MFM capacitor application, we prefer to grow the HZO film below 300°C to maximize the formation of Orthorhombic phase for Ferroelectric property.

The MFM capacitor devices with fully mixed HZO and laminated $HfO_2$ and $ZrO_2$ HZO films as illustrated in Fig. 2(a) and 2(b) were fabricated. Both types of MFM capacitor devices were measured in aixACCT TF analyzer 2000. P-V loops of the MFM capacitors with HZO films were measured after wake-up cycles of typically $1x10^4$ with 10KHz triangular waves with +/-3V amplitude. Fig. 3 (a), (b), and (c) are results of the P-V loops of MFM capacitors with HZO deposited at 225°C, 250°C, and 275°C. Lower temperature deposited HZO films show anti-ferroelectric behavior. At 275°C, the MFM device shows typical characteristic ferroelectric memory behaviors, with remanent polarization of ~25$\mu C/cm^2$. MFM capacitor with fully mixed or laminated HZO films deposited at 275°C show nearly identical P-V loops, as seen in Fig.4.

While MFM capacitors with fully mixed and laminate HZO films exhibit identical P-V loops, the fully mixed sample shows better projected endurance of ~$5x10^{11}$ cycles, determined by $5\mu C/cm^2$ demarcation of remaining polarization. The laminated sample has ~ $1x10^{10}$ projected endurance.

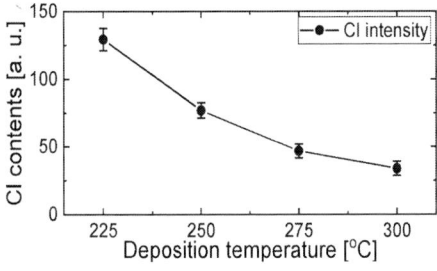

Fig. 1 Cl content in HZO film at various ALD deposition temperatures

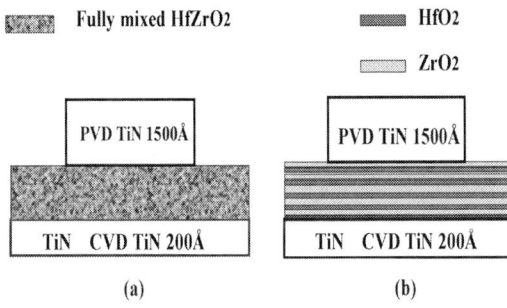

Fig. 2 MFM Capacitor structures. (a) MFM capacitor with 10nm fully mixed HFO and (b) MFM capacitor with laminated ~1nm $HfO_2$ and 1nm $ZrO_2$ alternatively for up to 10nm total thickness

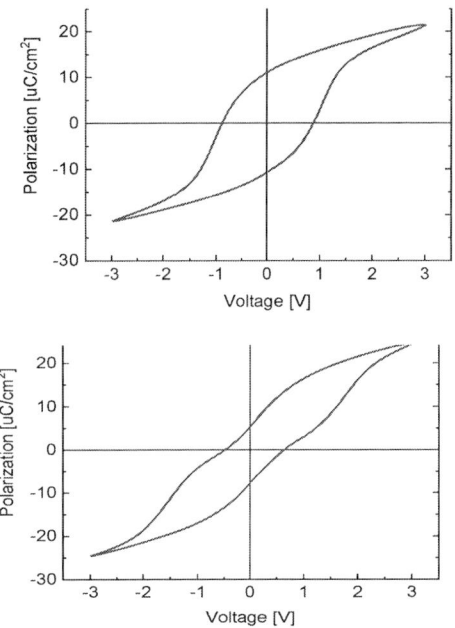

Fig. 3 (b) . P-V loops of the MFM capacitors deposited with HZO films deposit at 250°C

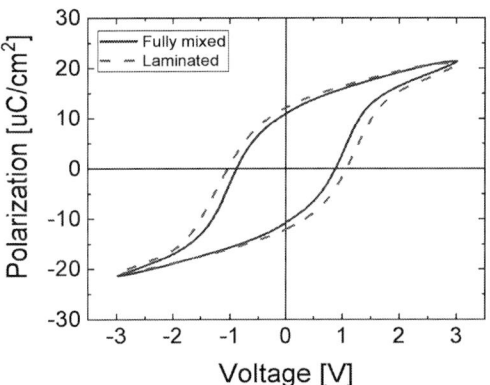

Fig. 4. P-V loops of fully mixed and laminated HZO capacitors after $10^4$ wake-up cycle, samples are annealed at 500°C for 10min

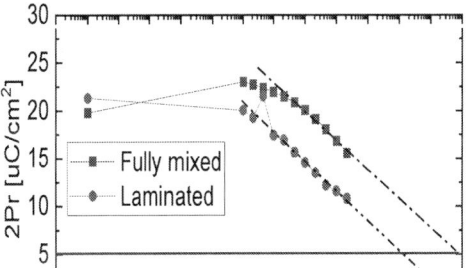

Fig. 5 Endurance characteristics of fully mixed and laminated HZO MFM capacitors. The measurement condition is 2.5MV/cm

The endurance of a reference MFM capacitor with HZO films deposited by TEMAHf and TEMAZr precursors shows best cycle endurance of $\sim 1 \times 10^{10}$. This result suggests that MFM capacitors with HZO films deposited by $HfCl_4$ and $ZrCL_4$ precursors have achieved better cycle endurance performance of $5 \times 10^{11}$ than HZO films deposited by TEMAHf and TEMAZr precursors. The achieved remanent polarizations of the MFM capacitors at starting of cycle endurance test are similar for MFM with HZO films deposited from TEMAHf and TEMAZr precursors and HZO films deposited from $HfCl_4$ and $ZrCL_4$ precursors.

## SUMMARY

MFM capacitors with 10nm HZO films produced by ALD using $HfCl_4$ and $ZrCl_4$ precursors show ferroelectric memory effect with 2Pr of $25\mu C/cm^2$ in P-V loop measurements. MFM capacitor with fully mixed 10nm HZO film shows projected cycle endurance of $\sim 5 \times 10^{11}$ under measurement condition of 2.5MV/cm field.

## REFERENCES

[1] T. S. Böscke, J. Müller, D. Bräuhaus, U. Schröder and U. Böttger, "Ferroelectricity in Hafnium Oxide Thin Films", Appl. Phys. Lett., **99**, p.102903, 2011

[2] T.P. Ma, "FEDRAM: A capacitor-less DRAM based on ferroelectric-gated field-effect transistor", 2008 9th International Conference on Solid-State and Integrated-Circuit Technology, Beijing, China, Oct. 20-23, 2008

[3] Kristjan Kalam et al, "Electric and Magnetic Properties of Atomic Laye\r Deposited $ZrO_2$-$HfO_2$ Thin Films", ECS J. Solid State Sci. Technol. **7** (9), N117-N122, 2018

[4] D. Lehninger et al "Back-End-of-Line Compatible Low-Temperature Furnace Anneal for Ferroelectric Hafnium Zircobium Oxide Formation", Phys. Status Solidi A, **217** (8), 1900840, 2020

[5] S. Abdulazhanov, et al, "Influence of Antiferroelectric-like behavior on tunning properties of ferroelectric HZO-based varactors", MRS Advance **6**, pp. 530-534, 2021

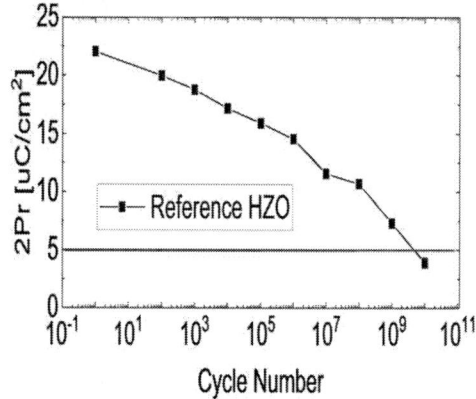

*Fig. 6. Endurance of the reference MFM capacitor with HZO deposited with carbon-containing precursors measured under the same condition as that in Fig. 4.*

# A NOVEL ANTI-FERROELECTRIC NEGATIVE CAPACITANCE TUNNELING FET WITH MITIGATED SUBTHRESHOLD SWING DEGRADATION ISSUE

*Shaodi Xu[1], Chang Su[1], Yimei Li[1], Qianqian Huang[1,2]\*, and Ru Huang[1,2]\**

[1]Key Laboratory of Microelectronic Devices and Circuits (MOE), School of Integrated Circuits, Peking University, Beijing 100871, China

[2]Chinese Institute for Brain Research (CIBR), Beijing 102206, China

*Corresponding Author's Email: hqq@pku.edu.cn, ruhuang@pku.edu.cn

## ABSTRACT

In this work, a novel anti-ferroelectric (AFE) negative capacitance tunneling FET (ANC-TFET) is proposed and simulated based on our developed device model. In ANC-TFET, the amplification coefficient ($A_V$) of AFE layer is designed to be increasing when the underlying TFET in series operates in subthreshold region, and thus the subthreshold swing ($SS$) degradation issue in TFETs can be mitigated. Comparing with conventional TFET, the ANC-TFET shows not only steeper $SS$ due to $A_V>1$ but also larger current range of sub-60mv/dec $SS$ for more than one decade at 0.4V operation voltage, showing its great potential for low power application.

## INTRODUCTION

The power consumption of traditional MOSFETs cannot be reduced by continuously reducing operation voltage ($V_{DD}$) while maintaining high ON current due to its fundamental subthreshold swing ($SS$) limitation of 60 mV/dec at room temperature. Negative capacitance FET (NCFET) and tunneling FET (TFET) can achieve sub-60 mV/dec $SS$ at room temperature through negative capacitance (NC) effect and band-to-band tunneling (BTBT) mechanism respectively [1, 2], showing great potential for low-power applications. However, TFET faces $SS$ degradation issue, that is, the $SS$ increases with the increase of gate voltage [3]. The $SS$ degradation issue may result in a large average subthreshold swing ($SS_{avg}$), which is undesirable for low-power applications. For NCFET, the physical origin of NC effect is still not clear. The "S" shaped curve of P-E loop derived from L-K equation [1] cannot be directly measured. Now it is widely acknowledged that NC effect originates from the dynamic behavior of domain switching. Our previous work has proposed and experimentally verified the prerequisite for NC effect named as "dynamic polarization matching" condition [4]. Moreover, we have found that the voltage amplification coefficient ($A_V$) of NCFET will increase at first and then decease during NC effect, and the ferroelectric based NCFET also faces the $SS$ degradation issue [5]. Therefore, by combining the NC effect and BTBT mechanism, the ferroelectric based negative capacitance TFET (NC-TFET) has been proposed and fabricated which shows steeper $SS$ and higher ON-current because the $A_V$ is larger than 1 [6]. However, the $A_V$ of NC-TFET will also decrease with increased gate voltage in the subthreshold region and exacerbate the $SS$ degradation issue of TFET.

In this work, we propose a novel anti-ferroelectric negative capacitance tunneling FET (ANC-TFET) structure. The designed AFE layer can result in the increase of $A_V$ in the subthreshold region of the underlying TFET thus alleviating the $SS$ degradation issue. We further simulate the proposed device with AFE model based on multi-domain Preisach theory of ferroelectric switching and our previously-developed TFET model. The simulation results show mitigated $SS$ degradation issue with steeper $SS$ than conventional TFET.

## DEVICE STRUCTURE AND MODELING FRAMEWORK

The schematic structure of the proposed ANC-TFET and its equivalent circuit are shown in Figure 1. An equipotential internal metal layer is inserted between the ferroelectric layer and dielectric layer, and thus the device can be considered as an M-AFE-M capacitor in series with a TFET. The voltage drop on the internal gate of TFET is identified as internal gate voltage ($V_{int}$). The gate charge of TFET is $Q_{TFET}$ and the total charge of AFE layer is $P_{AFE}$.

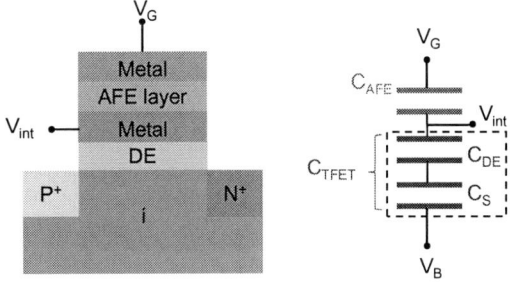

*Figure 1: Device structure of ANC-TEFT and corresponding equivalent circuit.*

According to the device structure, the following two equations should be satisfied.

$$\begin{cases} P_{AFE} = Q_{TFET} \\ V_G = V_{AFE} + V_{int} \end{cases} \tag{1}$$

The experimentally calibrated current and capacitance model of TFET based on surface potential and the impacts of both gate and drain voltages [7. 8] is used to calculate $Q_{TFET}$ and drain current ($I_D$).

In order to solve the equations above, an AFE model that is consistent with experiment data and reflects the dynamic multi-domain switching is needed. Same as FE layer, the total polarization of AFE layer is composed of spontaneous polarization ($P_{re}$) and dielectric polarization ($P_t$) as shown in Figure 2 (a). The relationship between $P_{re}$ and $V_{AFE}$ can also be described by a tanh-based model [9]. The dielectric polarization part is calculated by thickness of AFE layer $t_{AFE}$, dielectric constant $\varepsilon_{AFE}$ and $V_{AFE}$ as shown in Figure 2 (b).

The AFE model in this work also combines a RC-like delay unit in order to describe the dynamic domain switching behavior. Since the polarization switching cannot respond instantly to the change of $V_{AFE}$, an effective voltage $V_{eff}$ is put forward to calculate the spontaneous polarization of FE layer through the tanh function mentioned above. As shown in Figure 2 (c), $\tau$ is the time constant of the delay unit and it reflects the response speed of domain switching to $V_{AFE}$.

The AFE parameters in Figure 2 (b) are calibrated with the P-V loop in [9] which fits experimental results well as illustrated in Figure 3.

Since NC effect is a dynamic process, the simulation method should be able to describe changes in device variables over time by solving Eq.1 through iteration as shown in Figure 4. To make the solution more accurate, the time difference ($\Delta t$) between the two states should be small. The gate voltage ($V_{G,i+1}$) can be determined first with its change rate ($dV_G/dt$) and $V_{eff,i+1}$ can be calculated based on the current state. Then iteratively find the voltage drop on TFET ($V_{int,i+1}$) that satisfies the charge equilibrium condition through following steps. The value of $V_{int,i+1}$ should be assumed first, then $V_{AFE,i+1}$ can be obtained according to $V_{AFE,i+1}=V_{G,i+1}-V_{int,i+1}$. $Q_{TFET}$ and $P_{AFE}$ can be calculated based on TFET model and AFE model respectively. If the difference between $Q_{TFET}$ and $P_{AFE}$ is smaller than the tolerance, the assumed value of $V_{int,i+1}$ can be used to further calculate the drain current of next state $I_{D,i+1}$. If not, the value of $V_{int,i+1}$ needs another guess until the results match. As the gate voltage changes with time, the aforementioned simulation flow calculates $I_D$ corresponding to each $V_G$, thus obtaining the $I_D$-$V_G$ relationship.

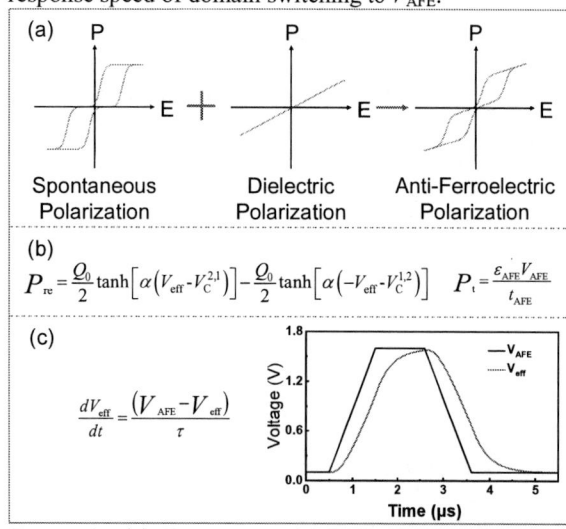

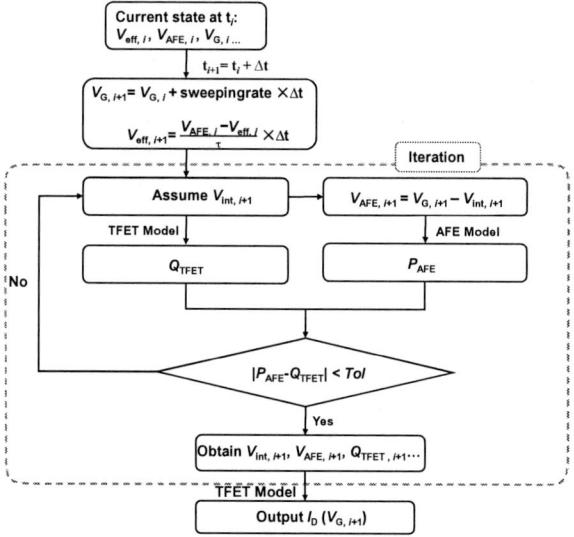

Figure 2: (a) Two parts of AFE polarization; (b) the calculation of $P_{re}$ and $P_t$; (c) a RC type delay between $V_{AFE}$ and $V_{eff}$.

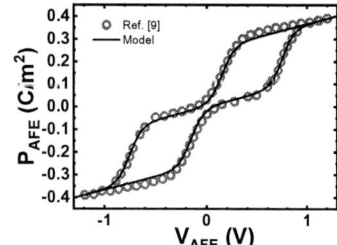

Figure 3: Calibrated P-V loop of AFE layer.

Figure 4: Simulation flow for obtaining the $V_G$-$I_D$ relationship of ANC-TFET.

## RESULS AND DISCUSSION

On one hand, the *SS* degradation issue of TFET is mainly due to that its tunnel width ($\lambda$) decrement tends to be smaller as the gate voltage increases [10]. On the other hand, for NC-TFET, the NC effect can improve the *SS* of TFET through voltage amplification as shown in Eq. 2 and Eq. 3.

$$A_V = \frac{dV_{int}}{dV_G} = 1 - \frac{dV_{FE}/dt}{dV_G/dt} \qquad (2)$$

$$SS_{NCTFET} = \left[\frac{\partial V_{int}}{\partial V_{GS}} \cdot \frac{\partial \lg I_{DS}}{\partial V_{int}}\right]^{-1} = \frac{1}{A_V} \cdot SS_{TFET} \propto \frac{1}{A_V} \cdot \qquad (3)$$

$$\left[\frac{\partial \log T(\lambda)}{\partial V_{int}}\right]^{-1} \approx \left|A_V \cdot \frac{\partial \lambda}{\partial V_{int}}\right|^{-1} \cdot \ln(10) \cdot \left[\frac{4 \cdot \sqrt{2m^*} \cdot E_g^{1/2}}{3q\hbar}\right]^{-1}$$

Both FE layer and AFE layer can induce NC effect when "dynamic polarization matching" condition (Eq. 4) is satisfied [4], which means the change rate of charge of polarization ($\partial P_{FE}/\partial t$ or $\partial P_{AFE}/\partial t$) exceeds that of $C_{TFET}$ in response to the total increasing gate voltage. During the NC effect, $P_{re}$ dominates the change of total charge $\partial P/\partial t$ at the beginning, leading to the increase of $A_V$. The continuous decrease of $V_{FE}$ or $V_{AFE}$ slows down and eventually reverses the change of $\partial P/\partial t$, resulting in the decrease of $A_V$ [5].

$$\frac{\partial P}{\partial t} > \frac{dV_G}{dt} \times C_{TFET} \qquad (4)$$

Based on the above analysis, ferroelectric based NC effect occurs with the rapid increase of $P_{FE}$ which happens when the value of $V_{eff}$ is near $V_C$. Since the NC effect introduced by FE layer begins with negative $P_{FE}$, although the $SS$ is improved due to $A_V>1$, its degradation issue is severer because $A_V$ is decreasing in the subthreshold region. For ANC-TFET, NC effect occurs with positive $P_{AFE}$, and therefore the $A_V$ is increasing in the subthreshold region, during which the decrement of $\lambda$ ($|\partial\lambda/\partial V_{int}|$) is amplified and thus can mitigate the $SS$ degradation issue according to Eq. 3. Hence AFE layer is more suitable for NC-TFET devices.

The proposed n-type ANC-TFET is simulated based on our above developed model. The $I_{off}$ for both devices are set to be $10^{-14}$ A/um. The initial values of $V_{eff}$ and $V_{AFE}$ at zero gate voltage are not the same considering that the sweeping range is started from the negative voltage, and the difference is preset to be 0.4V according to [9].

The simulation results are shown in Figure 5 and Figure 6. $A_V$ increases at first and then decreases as shown in Figure 5.

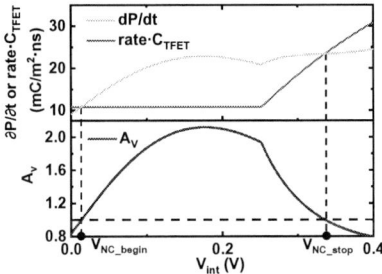

*Figure 5: Polarization matching situations and the corresponding voltage amplification coefficient.*

As illustrated in Figure 6, the $I_{ON}$ of ANC-TFET is larger than conventional TFET at ultra-low voltage of 0.4V, and it achieves better $SS$ within almost 5 decades of current range. Figure 6 (d) shows that the increment of BTBT generation rate (proportional to $\lambda$) of ANC-TFET degrades slower when $A_V$ rises, leading to a smaller $SS_{avg}$.

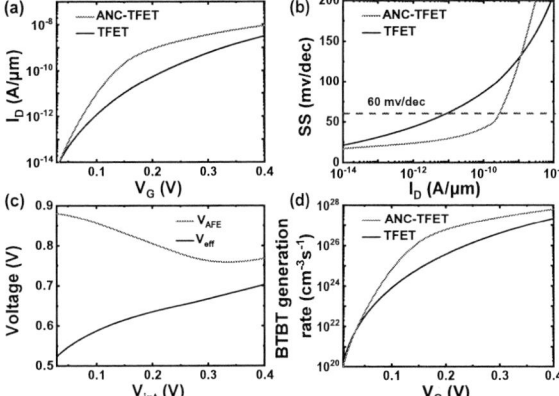

*Figure 6: (a) Transfer characteristics of ANC-TFET and TFET; (b) comparison of SS; (c) $V_{AFE}$ and $V_{eff}$ during the turn-on process; (d) extracted BTBT generation rate of ANC-TFET and TFET.*

## CONCLUSION

We have proposed and simulated a novel ANC-TFET in which the AFE layer can result in the increase of $A_V$ in the subthreshold region of TFET. The increase of $A_V$ enlarges $|\partial\lambda/\partial V_{int}|$ and thus alleviates the $SS$ degradation issue of TFET. Simulation results show the alleviation of $SS$ degradation and a decade more current range of sub-60 mv/dec $SS$, showing the great potential of ANC-TFET for future low power applications.

## ACKNOWLEDGEMENTS

This work was supported by National Key R&D Program of China (2018YFB2202800), NSFC (61822401, 61851401, 61927901, 61421005), Beijing Nova Program of Science and Technology (Z191100001119101) and 111 Project (B18001).

## REFERENCES

[1] S. Salahuddin, et al. *Nano lett*, vol. 8, no. 2, p. 405-410, 2008.
[2] A. M. Ionescu, et al. *Nature*, vol. 479, no. 7373, pp. 329-337, 2011.
[3] R. Jia, et al. *Sci China*, vol. 63, pp. 241-243, 2020.
[4] H. Wang, et al. *IEDM*, pp. 31.1.1-31.1.4, 2018.
[5] M. Yang, et al. *Sci China*, in press.
[6] Y. Zhao, et al. *IEEE EDL*, vol. 40, no. 6, pp. 989-992, 2019.
[7] J. Wang, et al. *J. Appl. Phys.*, vol. 116, no. 9, 2014.
[8] C. Wang, et al. *Sci China*, vol. 58, no. 2, pp. 022402, 2015.
[9] A. K. Saha, et al. *Device Research Conference*, pp. 1-2, 2018.
[10] C. Wu, et al. *IEEE TED*, vol. 63, no. 12, pp. 5072-5076, 2016.

# AIR STABLE HIGH MOBILITY ALD ZNO TFT WITH HFO$_2$ PASSIVATION LAYER SUITABLE FOR CMOS-BEOL INTEGRATION

*Wenhui Wang[1], Jiqing Lu[1,2], Jun Lan[1], Bing Zhou[1], Yiyang Zhang[1], Jinxuan Liang[1], Muhammad Zaheer[1], Mei Shen[1\*], and Yida Li[1\*]*

[1]Southern University of Science and Technology, Shenzhen 518055, China

[2] Harbin Institute of Technology, Harbin 150001, China

*Email: shenm@mail.sustech.edu.cn, liyd3@sustech.edu.cn

## ABSTRACT

An air-stable ALD ZnO TFT with high field-effect mobility of 82 cm$^2$/V·s and high on-off ratio of $5 \times 10^7$ is realized with an 18 nm thick channel. This is the highest reported mobility value using an ALD process with a low temperature of 200℃. With HfO$_2$ as a passivation layer, the TFT exhibits negligible electrical degradation even after 90 days of exposure in ambient air environment, paving the path for BEOL integration.

## INTRODUCTION

With the growing demand for data-driven applications such as the next-generation machine learning accelerators and Internet of Things (IoT), the conventional von-Neumann architecture in which the memory and computing units are separated, suffers from huge memory latency and limited data bandwidth. In order to break through this bottleneck, Three-Dimensional (3-D) integration with combined memory and logic, or in-memory computing has emerged as a promising solution. However, the utility of silicon (Si) technology is difficult to realize 3-D integration due to high thermal budget (> 900℃) of Si. Beyond-Si devices, such as Carbon Nanotubes (CNT) and oxide FET, can overcome the low thermal budget limitations [1]. Oxide semiconductor, in particular, with its process maturity for large scale growth and decent electrical performance (e.g., mobility, leakage) may offer an even better proposition for 3-D integration.

Among all kinds of oxides, ZnO has attracted intensive research attention as a wide bandgap semiconductor material for thin-film transistor (TFT) applications due to its excellent mobility that is comparable to that of silicon [2-4]. Furthermore, its low processing temperature makes it suitable for CMOS Back End of Line (BEOL) integration [5]. However, ZnO suffers from several issues, impeding its commercial adoption. Firstly, the electrical properties of ZnO channel are known to be sensitive to various factors such as morphology and body defects [6-7]. Secondly, the hygroscopic nature of ZnO makes it susceptible to electrical performance degradation over time [3]. Lastly, the sputter approach that has been widely reported possesses disadvantages such as non-conformal step coverage and poor thickness control.

In this work, we report on an air-stable, high mobility,

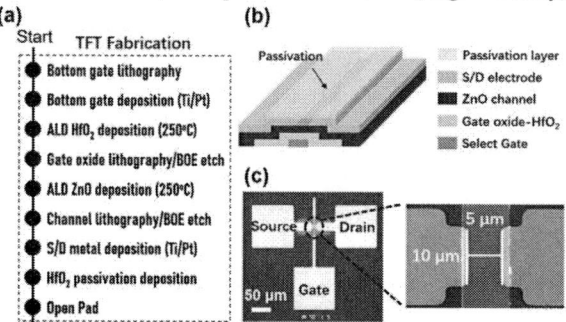

*Figure 1: (a) Process flow of TFT fabrication, (b) schematic of the TFT cell with layers indicated in the legend, and (c) microscope image of a fabricated TFT (left) and the corresponding zoomed-in SEM image of the channel region (right)*

thin-body (18 nm) ALD ZnO TFT with a low process temperature of 200℃. The ZnO TFT exhibits a field-effect mobility ($\mu_{FE}$) of 82 cm$^2$/V·s, the highest reported in literature based on ALD process at one of the lowest processing temperatures. With a 5 nm ALD HfO$_2$ passivation layer, the ZnO TFT exhibits negligible electrical degradation > 90 days in ambient air, thus showing its suitability for BEOL integration for high performance circuits.

## EXPERIMENT

In our TFT, the bottom-gate electrode consisting of 5 nm Ti and 20 nm Pt was firstly deposited using e-beam evaporation onto a 285 nm SiO$_2$ layer on Si substrate. ALD HfO$_2$ with ~11 nm thickness deposited at 250℃ was used as the gate dielectric, while the 18 nm ALD ZnO channel was deposited at an optimized temperature of 200℃. All layers' thicknesses were confirmed via ellipsometer. Both the gate dielectric and ZnO channel regions were defined via standard lithography followed by a buffered oxide etch (BOE). Thereafter, the Ti/Pt Source/Drain electrodes were then deposited using e-beam evaporation followed by a lift-off process. A 5nm ALD HfO$_2$ passivation layer was deposited to enhance the long-term electrical performance stability of the TFT, followed by a final contact pad opening via lithography

978-1-6654-9759-6/22 $31.00 © 2022 IEEE

and BOE etch to complete the fabrication. The fabricated TFTs have channel width ($W_{ch}$) and channel length ($L_{ch}$) of 10 µm and 5 µm respectively. Figure 1 (a) shows the basic process flow. The schematic of the TFT (layers indicated), and the microscopic image with zoomed-in channel region are shown in Figure 1 (b) and (c), correspondingly. All electrical characteristics were measured using a Keithley 4200-SCS semiconductor parameter analyzer.

## RESULTS AND DISCUSSION

### Material Characterization

AFM image of the ZnO channel region on $HfO_2$ layer shown in Figure 2 (a) reveals the root mean square roughness ($R_q$) to be 0.78 nm, indicating the high quality ALD deposition process. Figure 2 (b) shows the TFT's TEM cross-section image with all the different layers and thickness labelled, confirming the device structure. Figure 2 (c) shows the grazing-incidence XRD (GI-XRD) scan of the ALD ZnO film. Despite the underlying amorphous $HfO_2$ film, clear polycrystallinity of the wurtzite ZnO was formed during the deposition process without the need for subsequent annealing. Figure 2 (d) shows the XPS O1s spectra of the ZnO film, with extracted Zn:O stoichiometry of 1:1.08 and oxygen vacancies ($V_o$)/M-O ratio of ~0.6.

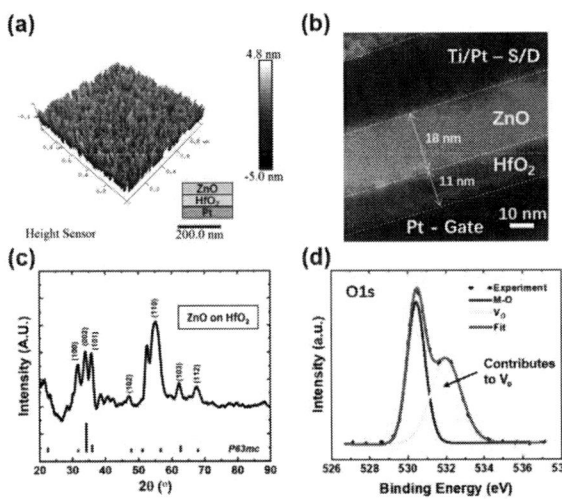

Figure 2: (a) AFM scan of the ZnO channel layer, (b) TEM cross-section image of the TFT stack with all layers and thicknesses labelled, (c) GIXRD of polycrystallinity of the wurtzite ZnO, and (d) XPS O1s spectra of the ZnO film.

### TFT Characterization

Figure 3(a) and 3(b) show the normalized to $W_{ch}$ transfer curves ($I_{DS}$-$V_{GS}$) at $V_{DS}$ from 0.2 to 2 V and output curves ($I_{DS}$-$V_{DS}$) at $V_{GS}$ from 0 to 3.5 V with dual sweep of the ZnO TFT, respectively. Negligible hysteresis in both forward and backward sweep directions was observed,

indicating stable performance due to the excelled quality of the interfaces. The corresponding gate leakage current ($I_{GS}$) is shown in Figure 3(a). A high drive current of ~80

TABLE I. BENCHMARK OF RECENT REPORTED ALD ZNO TFT

| Process | Spin Coating [8] | ALD [4] | ALD [9] | ALD [10] | ALD (This work) |
|---|---|---|---|---|---|
| Temperature (°C) | 350 | 350 | 200 | N/A | 200 |
| ZnO Thickness (nm) | NA | 11 | 27.2 | 33.1 | 18 |
| Equivalent Oxide Thickness (nm) | NA | 90 | 30 | 30 | 3.1 |
| On-Off ratio | $10^8$ | $5 \times 10^9$ | $3 \times 10^9$ | $3 \times 10^9$ | $5 \times 10^7$ |
| SS (mV/dec) | 238 | NA | 250 | 127 | 113 |
| $\mu_{FE}$ (cm²/V·s) | 31.6 | 43.2 | 12 | 7.8 | 82 |

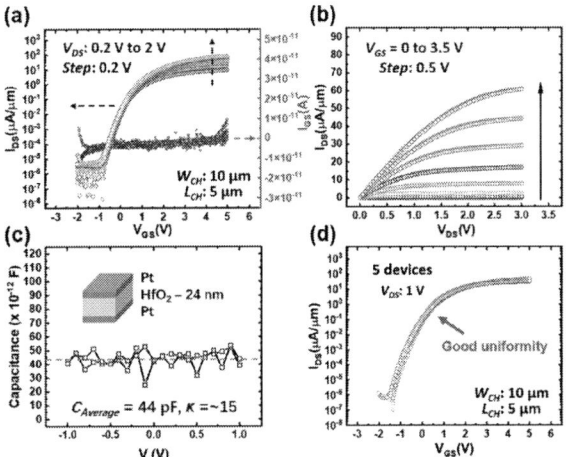

Figure 3: (a) Transfer curves (dual sweep) and (b) output curves (dual sweep) of the fabricated ZnO TFT, with no hysteresis. (c) CV curve of a $HfO_2$ capacitor, and (d) $I_{DS}$-$V_{GS}$ of 5 different devices, showing good uniformity.

µA/µm was measured, accompanied by a large $I_{ON/OFF}$ ratio of $5 \times 10^7$ at $V_{DS} = 2$ V, with negligible $I_{GS}$. Figure 3(c) shows the CV curve of a 24nm thick $HfO_2$ capacitor with average capacitance of ~44pF, from which the relative dielectric constant (k) of $HfO_2$ film of ~15 was extracted. The $\mu_{FE}$ of the TFT was then extracted via (1),

$$\mu_{FE} = \frac{G_m}{C_{ox}\frac{W}{L}V_{DS}} \quad (1)$$

, where $G_m$ is the calculated maximum transconductance from Figure 3(a) and $C_{ox}$ is extracted experimentally. This results in a $\mu_{FE}$ of 82 cm²/V·s, the highest value reported so far in literature utilizing ALD process. Figure 3(d) shows $I_{DS}$-$V_{GS}$ of 5 different devices with minor variations, exhibiting good uniformity. Table 1 benchmarks the ZnO TFT reported here with recently reported works, showing the highest mobility achieved using ALD process.

### Air Stable ZnO TFT with HfO₂ Passivation

Figure 4 (a) and (b) show the visible degradation of the ZnO film after 30 days in ambient air environment, resulting in the TFT being non-functional. On the other hand, the passivated TFT shows none of such visual

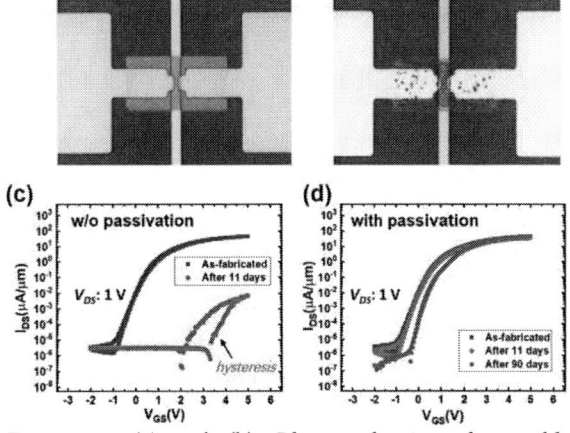

*Figure 4: (a)* and *(b) Photos showing the visible degradation of the non-passivated TFT. (c) and (d) Transfer curves comparison of the non-passivated and passivated TFT for as-fabricated case and after several days, respectively.*

degradation. Figure 4 (c) shows the $I_{DS}$-$V_{GS}$ curves of non-passivated TFT for as-fabricated and after 11 days. A performance degradation of reduced current and the appearance of hysteresis is evident after 11 days, and the TFT failed to function after 30 days. On the contrary, the passivated TFT shows excellent stability and a slight right shift of $I_{DS}$-$V_{GS}$ curve after 90 days (figure 4 (d)), demonstrating ALD HfO$_2$'s superiority as a passivation layer for device reliability optimization.

## CONCLUSION

In summary, a CMOS-BEOL compatible, as-deposited low temperature (200°C) ALD ZnO TFT with a high $\mu_{FE}$ of 82 cm$^2$/V·s and a high on-off ratio of 5 × 10$^7$ has been demonstrated. Equipped with a thin ALD HfO$_2$ passivation layer, the TFT exhibits negligible electrical degradation after 90 days in ambient air exposure as compared to the non-passivated TFT that deteriorated to being non-functional. Our results pave a potential path for ALD ZnO TFT as a high performance, beyond-Si transistor, suitable for CMOS-BEOL integration.

## ACKNOWLEDGEMENTS

This work was supported by the National Natural Science Foundation of China (Grant No. 62174074), and in part by the Shenzhen Fundamental Research Program (Grant No. JCYJ20190809143419448), NSQKJJ (Grant No. K21799123, K21799131, K21799124, K21799128)

and Engineering Research Center of Three-Dimensional Integration in Guangdong Province. We would also like to acknowledge the Core Research Facilities (CRF) at SUSTech for the facilities used, and the technical support provided by the staff and engineers at the CRF.

## REFERENCES

[1] M. M. Shulaker, G. Hills, R. S. Park, R. T. Howe, K. Saraswat, H.-S. P. Wong, and S. Mitra, "Three-dimensional integration of nanotechnologies for computing and data storage on a single chip," *Nature*, vol. 547, no. 7661, pp. 74–78, 2017.

[2] H. Frenzel, A. Lajn, H. von Wenckstern, M. Lorenz, F. Schein, Z. Zhang, and M. Grundmann, "Recent progress on ZnO-based metal-semiconductor field-effect transistors and their application in transparent integrated circuits," *Advanced Materials*, vol. 22, no. 47, pp. 5332–5349, 2010.

[3] C. W. Shih and A. Chin, "Remarkably high mobility thin-film transistor on flexible substrate by novel passivation material," *Scientific Reports*, vol. 7, no. 1, 2017.

[4] M. Wang, D. Zhan, X. Wang, Q. Hu, C. Gu, X. Li, and Y. Wu, "Performance optimization of atomic layer deposited ZnO thin-film transistors by vacuum annealing," *IEEE Electron Device Letters*, vol. 42, no. 5, pp. 716–719, 2021.

[5] J. Wu, F. Mo, T. Saraya, T. Hiramoto, and M. Kobayashi, "A monolithic 3-D integration of RRAM array and Oxide Semiconductor FET for in-memory computing in 3-D Neural Network," *IEEE Transactions on Electron Devices*, vol. 67, no. 12, pp. 5322–5328, 2020.

[6] T. Muneshwar, G. Shoute, D. Barlage, and K. Cadien, "Plasma enhanced atomic layer deposition of ZnO with diethyl zinc and oxygen plasma: Effect of precursor decomposition," *Journal of Vacuum Science & Technology A: Vacuum, Surfaces, and Films*, vol. 34, no. 5, p. 050605, 2016.

[7] F. Gunkel, D. V. Christensen, Y. Z. Chen, and N. Pryds, "Oxygen vacancies: The (in)visible friend of oxide electronics," *Applied Physics Letters*, vol. 116, no. 12, p. 120505, 2020.

[8] J. K. Saha, R. N. Bukke, N. N. Mude, and J. Jang, "Significant improvement of spray pyrolyzed ZnO thin film by precursor optimization for high mobility thin film transistors," *Scientific Reports*, vol. 10, no. 1, 2020.

[9] H. Li, D. Han, J. Dong, W. Yu, Y. Liang, Z. Luo, S. Zhang, X. Zhang, and Y. Wang, "Enhanced electrical properties of dual-layer channel ZnO thin film transistors prepared by atomic layer deposition," *Applied Surface Science*, vol. 439, pp. 632–637, 2018.

[10] H. Li, D. Han, Z. Yi, J. Dong, S. Zhang, X. Zhang, and Y. Wang, "High-performance ZnO thin-film

transistors prepared by atomic layer deposition," *IEEE Transactions on Electron Devices*, vol. 66, no. 7, pp. 2965–2970, 2019.

# THE LOW-FREQUENCY NOISE BEHAVIOR OF ADVANCED LOGIC AND MEMORY DEVICES

*Eddy Simoen[1,2*], Romain Ritzenthaler[1], Hans Mertens[1], Eugenio Dentoni Litta[1], Naoto Horiguchi[1], Adrian Vaisman[1], Nouredine Rassoul[1], Gouri Sankar Kar[1] and Cor Claeys[3]*

[1]Imec, Kapeldreef 75, B-3001 Leuven, Belgium
[2]also at Solid-State Sciences Depart., Ghent University, Gent, Belgium
[3]ESAT-INSYS Depart., KU Leuven, Kasteelpark Arenberg 10, B-3001 Leuven, Belgium
*Corresponding Author's Email: eddy.simoen@imec.be

## ABSTRACT

The paper gives an overview of low-frequency (LF) noise studies in advanced logic and memory devices, focusing on two case studies. First, the LF noise of forksheets (FS) transistors is compared with double nanosheets devices fabricated on the same wafer. As will be shown, no degradation of the $1/f$ noise Power Spectral Density (PSD) is observed. Second, the LF noise of InGaZnO (IGZO) nFETs with different gate dielectric and thin-film materials is investigated. It is demonstrated that the $1/f$ noise in the above-threshold-voltage range studied is dominated by trapping and detrapping through shallow defects in the IGZO film.

## INTRODUCTION

In many cases, flicker or $1/f$ noise dominates the low-frequency (LF) current fluctuations in state-of-the-art devices. Assuming number-fluctuations or, in other words, a $\Delta n$ origin, caused by trapping/detrapping through defects in the gate stack, geometrical scaling predicts an $(EOT)^2/WL$ dependence [1]-[3], with EOT the Equivalent Oxide Thickness and $W \times L$ the transistor area (width times length). This implies that for a given materials' system and device architecture, geometrical scaling automatically results in a reduction of the $1/f$ noise Power Spectral Density (PSD), in proportion with $(EOT)^2$.

However, in the strive for scaling logic devices to the 5 nm node and beyond, reduction of the device feature size is only part of the story and does not suffice to achieve a higher performance. In addition, the device architecture has changed from planar to FinFET and more recently to Gate-All-Around (GAA) nanowire (NW) or nanosheet (NS) devices [4]-[6]. On top of that, the traditional $Si/SiO_2$ combination has been replaced by other materials, like high-mobility semiconductor channels or metal-oxide thin films in combination with high-$\kappa$ gate dielectrics. This may result in a much more pronounced change in the LF noise, as the latter parameter is very sensitive to the material quality and more in particular to processing-induced defects and imperfections. It is, therefore, mandatory to investigate the LF noise behavior of advanced logic and memory devices, which are based on a novel architecture or implementing new process modules.

In this work, the LF noise properties will be investigated of forksheets (FS) transistors and compared with nanosheets (NS) counterparts fabricated on the same wafer. In a second part, the focus will be on IGZO nFETs, developed for 3D memory applications. Devices with different material quality and gate dielectric will be investigated.

## LF NOISE OF FORKSHEETS DEVICES

FS transistors can be regarded as nanosheet devices with a forked gate structure [7],[8]. Due to the physical separation of the n- and p-type devices by a dielectric wall, as illustrated by the cross-sectional Transmission Electron Microscopy (TEM) view in Fig. 1a, n-p space scaling is enabled, resulting in a sheet width optimization, compared with a regular NS architecture. Processing has been performed as outlined elsewhere in more detail [7],[8]. Forksheets and nanosheets with identical channel width (23 nm) and height (7 nm) have been fabricated. The gate stack consists of Interfacial Layer $SiO_2/HfO_2$, followed by dual work function metal gates.

LF noise measurements have been carried out on wafer in linear operation ($|V_{DS}|$=0.05 V), stepping the gate voltage ($V_{GS}$) from weak to strong inversion. Devices with $L$=28 nm have been evaluated.

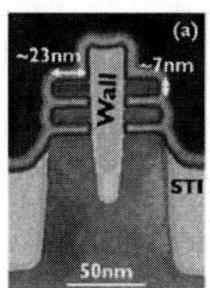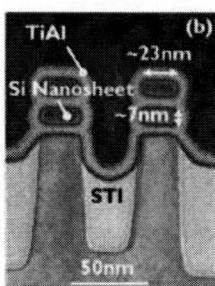

*Figure 1: Cross-sectional TEM view of a FS (a) and (b) NS FET.*

Typical drain current noise Power Spectral Density (PSD) $S_I$ for a FS and NS n- and pFET is compared in Fig. 2a and 2b, respectively. In most cases, the spectra are $1/f$-like, with occasionally Generation-Recombination

978-1-6654-9759-6/22 $31.00 © 2022 IEEE

(GR) Lorentzian humps on top. This is the case for the FS nMOSFET in Fig. 2a, where the GR noise is present above 10 Hz.

Overall, very similar $1/f$ noise PSD can be found for the FS and NS MOSFETs, as shown in more detail in Fig. 3. This applies to both p- and n-type devices. Again, GR humps may occasionally be observed in Fig. 3, most likely originating from Random Telegraph Signals (RTS) in the gate stack. The normalized noise in Fig. 3 exhibits a plateau in weak inversion, followed by a roll-off at higher absolute drain current. It indicates the dominance of $\Delta n$ fluctuations in the $1/f$ noise [1]-[3].

This is also confirmed by the input-referred voltage noise PSD ($S_{VG}$) in Fig. 4. The increase in the normalized noise PSD for the FS pMOSFET in Fig. 3b at higher $|I_D|$ can be explained by the onset of the access resistance effect [9]. In Fig. 4, the $S_{VG}$ is rather constant for both types of devices, in agreement with the $\Delta n$ origin of the $1/f$ noise. In addition, very similar PSD values are found, so that the implementation of FS does not affect the LF noise behavior [10].

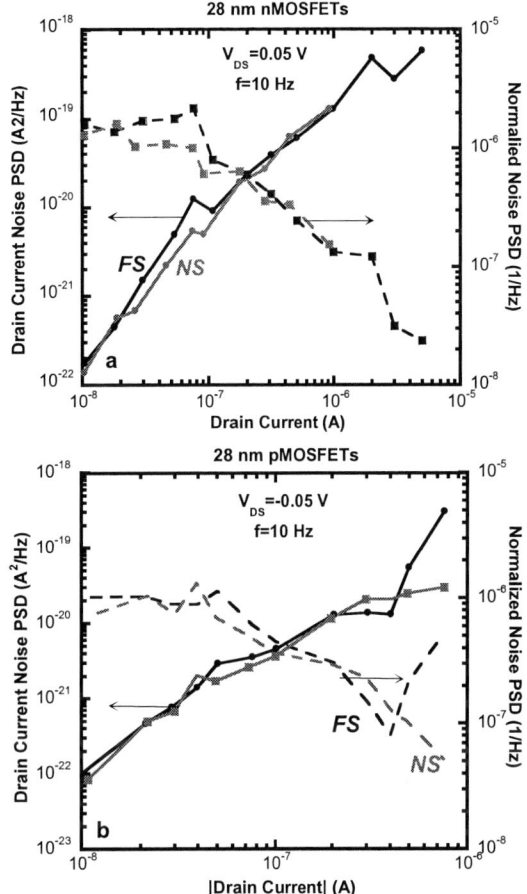

*Figure 3: Drain current noise PSD and normalized noise PSD at 10 Hz versus drain current for a NS and FS nFET (a) and (b) a NS and FS pFET.*

## LF NOISE IN IGZO NMOSFETS

IGZO-based thin-film transistors are currently of interest for 3D integrated nano-electronic systems, including stacked DRAMs, owing to their low off-state leakage and Back-End-Of-Line compatibility [11],[12]. In this work, 12 nm IGZO films have been deposited on a highly-doped p$^+$ silicon wafer, serving as gate electrode, with on top the gate dielectric (GoX). A schematical cross section is shown in Fig. 5, while processing details can be found elsewhere [13].

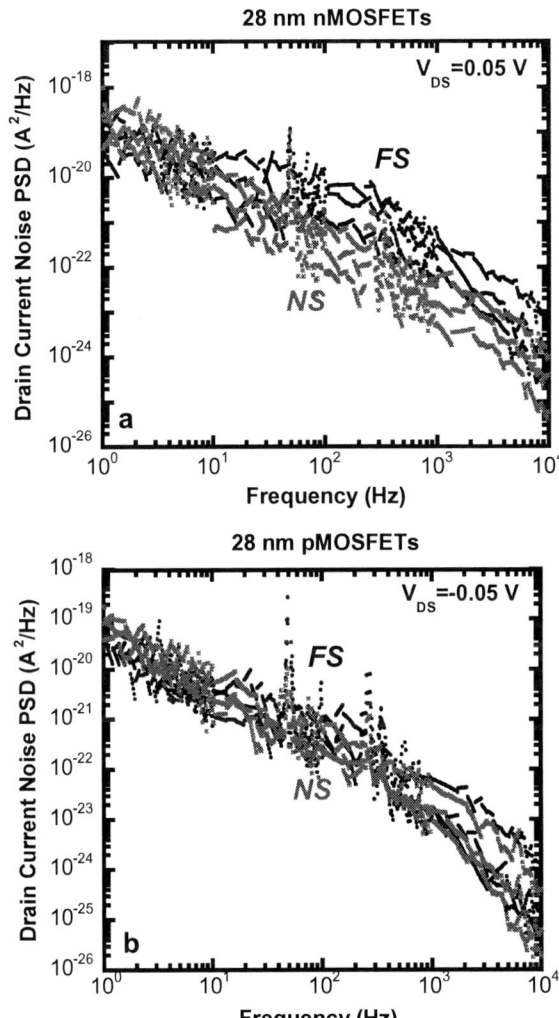

*Figure 2: Noise spectra in linear operation for (a) a NS and FS nFET and (b) pFET, for several gate voltages $V_{GS}$ around threshold voltage $V_T$.*

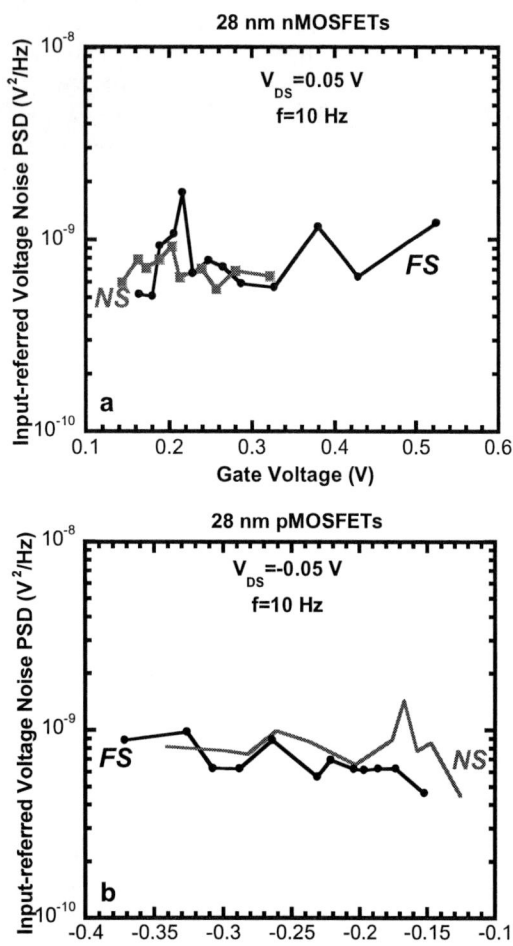

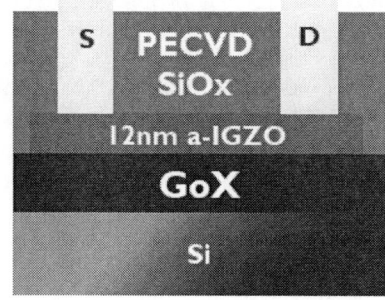

Figure 5: Schematic representation of the IGZO nMOSFETs investigated.

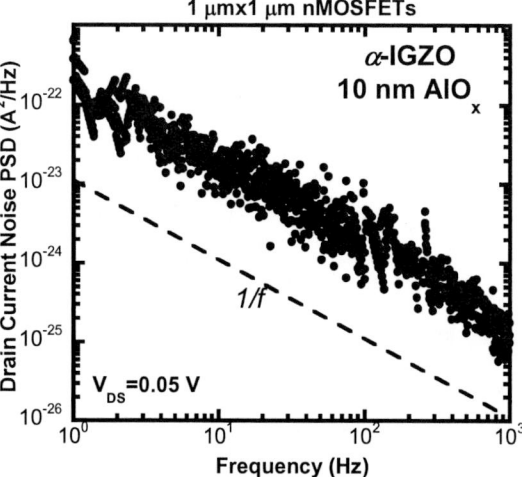

Figure 6: $S_I$ spectra for a 1 $\mu m \times 1$ $\mu m$ nMOSFET with $\alpha$-IGZO and 10 nm AlO$_x$.

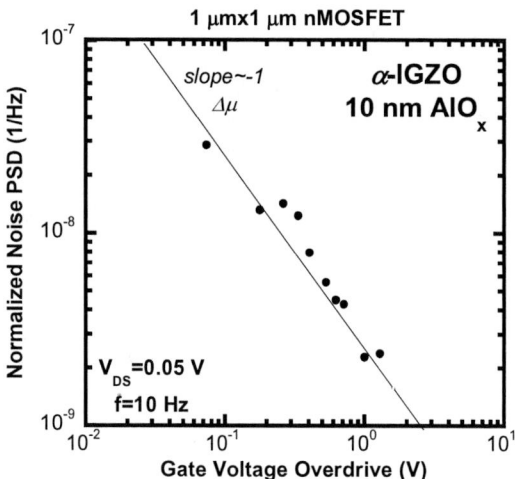

Figure 7: Normalized Noise PSD at 10 Hz versus gate voltage overdrive for a 1 $\mu m \times 1$ $\mu m$ nMOSFET with $\alpha$-IGZO and 10 nm AlO$_x$.

Figure 4: Input-referred voltage noise PSD at f=10 Hz versus gate voltage for a NS and FS nFET (a) and (b) a NS and FS pFET.

Three types of n-channel transistors have been evaluated: with 10 nm AlO$_x$ or 10 nm SiO$_2$ gate dielectric and an amorphous-IGZO film, and, thirdly, 10 nm AlO$_x$ combined with a $c$-Axis-Aligned Crystalline (CAAC) IGZO channel [14]. In the latter case, a higher defectivity in the thin film is expected. At least 6 transistors per wafer with area 1 $\mu m \times 1$ $\mu m$ have been evaluated in linear operation ($V_{DS}$=0.05 V), with $V_{GS} > V_T$.

As represented in Fig. 6, the spectra are 1/$f$-like for the frequency and gate voltage range investigated. This applies for the three types of IGZO nFETs. Representing the normalized drain current PSD versus gate voltage overdrive in Fig. 7 reveals a slope of -1 in a log-log plot. This is representative of so-called mobility fluctuations dominated flicker noise [15],[16].

Combining the results for the three different splits reveals that the gate dielectric has relatively small impact on the $1/f$ noise PSD, while the CAAC devices exhibit a clearly higher input-referred noise PSD, increasing from $S_{VG} \sim 5 \times 10^{-10}$ V$^2$/Hz (Fig. 8) to $\sim 10^{-9}$ V$^2$/Hz, for 10 nm AlO$_x$ gate dielectric. This is related to the expected higher defect density in the films, resulting in more carrier scattering and, hence, two times higher $\Delta\mu$ noise. This originates from a band of shallow defect states close to the conduction band in the IGZO layers [15],[16].

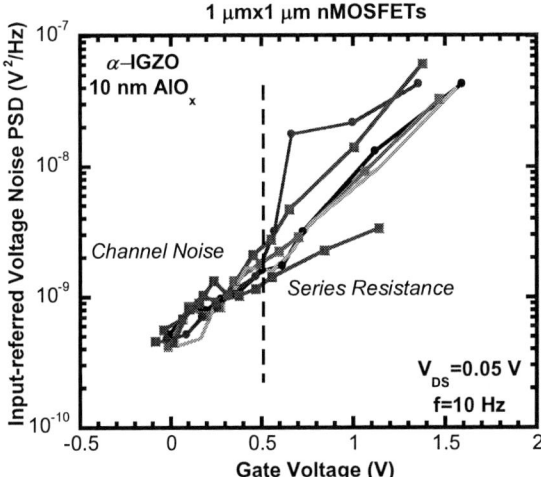

Figure 8: Input-referred voltage noise PSD at 10 Hz versus gate voltage for a 1 μm×1 μm nMOSFET with α-IGZO and 10 nm AlO$_x$.

## CONCLUSIONS

It has been shown that LF noise is a versatile and powerful tool to investigate the quality of novel state-of-the-art semiconductor devices. Defects both in the material and in the gate dielectric may contribute excess $1/f$ noise. It was shown that the implementation of a FS structure does not compromise the flicker noise behavior. In the case of the IGZO nMOSFETs, the quality of the channel material plays a crucial role in the $1/f$ noise magnitude above threshold voltage.

## ACKNOWLEDGEMENTS

This work has been performed in the frame of imec's Core Partner Programs on advanced logic devices and on advanced memory devices.

## REFERENCES

[1] G. Ghibaudo, O. Roux, Ch. Nguyen-Duc, F. Balestra and J. Brini, *Phys. Status Solidi (a)*, vol. 124, 1991, pp. 571-581.

[2] E. Simoen and C. Claeys, *Solid-State Electronics*, vol. 43, 1999, pp. 865-882.

[3] G. Ghibaudo and T. Boutchacha, *Microelectron. Reliab.*, vol. 42, 2002, pp. 573-582.

[4] T. Imamoto, Y. Ma, M. Muraguchi and T. Endoh, *Jpn. J. Appl. Phys.*, vol. 54, 2015, p. 04DC11.

[5] E. Simoen, A. Vinicius de Oliveira, P. Ghedini Der Agopian, R. Ritzenthaler, H. Mertens, N. Horiguchi, J. Antonio Martino, C. Claeys and A. Veloso, *Solid-St. Electron.*, vol. 184, 2021, p. 108087.

[6] E. Simoen, A. Veloso, B. O'Sullivan, K. Takakura and C. Claeys, *2021 China Semicond. Technol. Intern. Conf. (CSTIC)*, IEEE Xplore, 2021, pp. 1-4.

[7] P. Weckx *et al.*, *IEDM Tech. Dig.*, 2019, pp. 871-874.

[8] H. Mertens *et al.*, *2021 Symp. on VLSI Technol. Dig. of Techn. Papers*, 2021, pp. 1-2.

[9] D. Boudier *et al.*, *Solid-St. Electron.*, vol. 128, 2017, pp. 102-108.

[10] R. Ritzenthaler, to be published in *Tech Dig. IEDM2021*.

[11] K. Han *et al.*, *2021 Symp. on VLSI Technol. Dig. of Techn. Papers*, 2021, pp. 1-2.

[12] S. Subhechha *et al.*, *2021 Symp. on VLSI Technol. Dig. of Techn. Papers*, 2021, pp. 1-2.

[13] G. Hublot *et al.*, *2021 IEEE Int. Reliab. Phys. Symp.*, 2021, pp. 1-8.

[14] H. Kunitake *et al.*, *J. Electron Devices Soc.*, vol. 7, 2019, pp. 495-502.

[15] J.C. Park *et al.*, *Appl. Phys. Lett.*, vol. 97, 2010, p. 122104.

[16] T.-C. Fung *et al.*, *J. Appl. Phys.*, vol. 108, 2010, p. 074518.

# EFFECTS OF GATE METAL AND CHANNEL SHAPE ON THE VARIABILITY OF JUNCTIONLESS FIELD-EFFECT TRANSISTOR

*Xinhe Wang[1], Feng Xu[1,2], Zhigang Zhang[1], Xinyi Li[1], Jianshi Tang[1,2], Bin Gao[1,2], Huaqiang Wu[1,2], He Qian[1,2]\**

[1]School of Integrated Circuits, Beijing Innovation Center for Future Chips (ICFC), Beijing National Research Center for Information Science and Technology (BNRist), Tsinghua University, Beijing, 100084, China

[2]Beijing Innovation Center for Future Chips (ICFC), Tsinghua University, Beijing, 100084, China

*Corresponding author: qianh@tsinghua.edu.cn

## ABSTRACT

This work comprehensively investigates the influence of gate metal work function and its grain size variations as well as channel shape on the electrical characteristics of junctionless field-effect transistors (FETs) through 3D TCAD simulations. It is found that, for a certain gate length, the smaller the gate metal grain size, the more compact the distribution of the threshold voltage. Two different channel shapes concerning ellipse and cuboid are further analyzed.

## INTRODUCTION

Conventional metal-oxide-semiconductor field-effect transistor (MOSFET) is based on the formation of junctions between channel and source/drain. As the transistor size continues scaling down driven by Moore's law, it faces critical challenges from short-channel effects. In order to mitigate these deleterious effects, many novel device structures are proposed, such as multi-gate MOSFET[1], FinFET[2, 3], and Gate-All-Around (GAA)[4, 5]. Recently, the proposal of junctionless transistor has attracted much attention[6, 7]. Junctionless FET has a heavily doped channel to increase the current of the channel and the doping type is the same with source and drain. It also requires gate metal with the appropriate work function to fully deplete the channel and turn off the transistor when the gate voltage is at zero. However, the depletion region width of a heavily doped semiconductor is very small, which hence requires an ultrathin channel. Also, the metal gate usually consists of multiple grains, which could have different sizes and crystalline orientations as well as surface densities, resulting in considerable variations in the gate work function[8]. Due to the difficulty in precisely controlling the gate metal grains during deposition, it is critical to investigate the effects of metal grain size (GSV) and work function variations (WFV) on the device characteristics of junctionless FETs. In this work, we use TCAD simulations to study junctionless FETs with different channel shapes, investigate the impact of their work function fluctuation on the device performance, and compare the impacts of grain size and different channel shapes.

## SIMULATIONS OF WFV-GSV IN JUNCTIONLESS FET

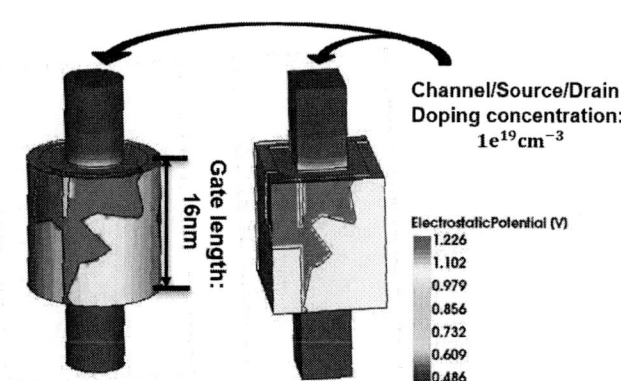

*Figure 1: Junctionless FET structures with different channel section shapes and metal grains*

Figure 1 illustrates junctionless FET structures with different channel shapes. In order to accurately investigate the impact of WFV-GSV on $V_{TH}$ variation in junctionless FET, two different device structures were carefully designed to have identical doping profiles, using the device design parameters as follows: gate length ($L_G$) of 16 nm, equivalent oxide thickness (EOT) of 1 nm, the same N-type doping concentration of $10^{19} cm^{-3}$ for the source/drain and channel. The 3D TCAD simulations were carried out using various physics models: high- field saturation model, band-to-band tunneling model, bandgap narrowing model, and Shockley–Read–Hall model. As shown in Figure 1, the WF of each grain was defined according to the grain orientation (i.e., WF of 4.6 eV for (100) orientation and 4.4 eV for (111) orientation). In order to investigate the impact of grain size on

978-1-6654-9759-6/22 $31.00 © 2022 IEEE

the induced threshold voltage $\sigma V_{TH}$, we studied the grain size in the range of 5-14 nm. For each given grain size, the

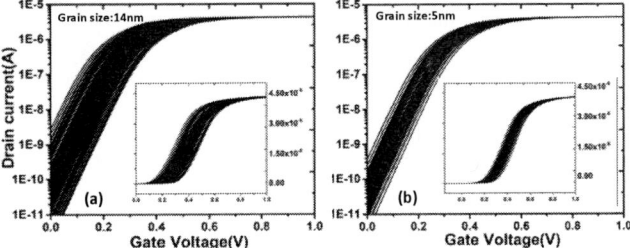

Figure 2: (a) and (b) shows the drain current ($I_d$) versus gate voltage ($V_g$) of 300 simulations for different grain sizes. The simulated junctionless FETs have a dimension of $L_G$ = 16 nm, EOT = 1 nm, N-type doping of $10^{19} cm^{-3}$ for the source/drain and channel

WFV-GSV induced $\sigma V_{TH}$ was extracted from the output characteristics of at least 100 simulations with randomly generated metal grains based on a Gaussian distribution.

**RESULTS AND DISCUSSION**

Figure 2 (a) and (b) shows the drain current ($I_d$) versus gate voltage ($V_g$) of 300 simulations for different grain sizes of 14 nm and 5 nm, respectively. It is obvious that different grain sizes have an obvious influence on the distribution of $V_{TH}$. Figure 3 shows the $V_{TH}$ distributions for four different grain sizes. The $V_{TH}$ variation of 5 nm grain size ($\sigma V_{TH}$ = 18 mV) is smaller than that of 14 nm grain size. ($\sigma V_{TH}$ = 44.47 mV). Also, the range of its $V_{TH}$ distribution increases with the grain size, as the probability of having entirely (100) or (111) gate metal increases. This suggests that the smaller the grain size for the junctionless FETs with a fixed gate length, the less the impact of the metal gate function fluctuation on the threshold voltage, so in practice it is recommended to prepare the metal gate composed of as small grains as possible.

Figure 4 shows the influence of the grain size on the two different channel shapes (ellipse and cuboid), which indicate that the influence of the channel shape on the $V_{TH}$ distribution is much less than that of the grain size. Also, for different channel shapes, it is found that the cuboid channel has a smaller DIBL than ellipse as shown in Figure 5 (a). To compare the effect of the shape on the device performance, we fixed the area of the gate by varying the nanowire diameter. As the grain size decreases, the DIBL value of the cuboid channel is reduced. To analyze the causes of this phenomenon, the potential distribution along the channel is shown in Figure 5 (b). Figure 6 (a) and (b)shows the electron density of the channel at Vg = 0 V, showing higher density for ellipse shape. The increasing DIBL

performance. These results suggest that the cuboid channel is less affected by the drain voltage (red line), and hence the

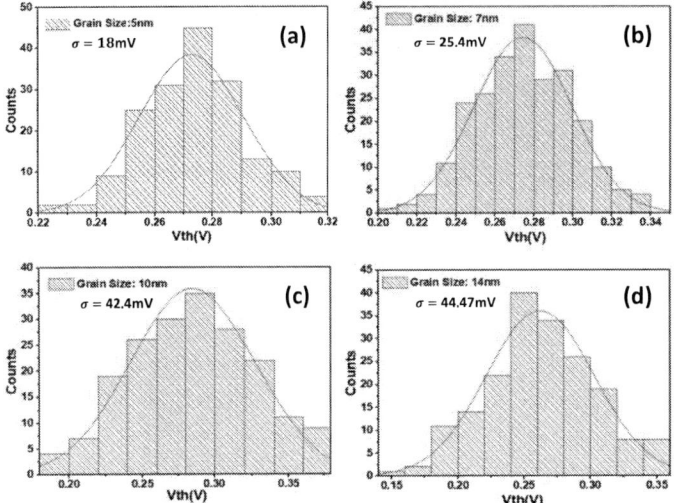

Figure 3: Threshold voltage $V_{TH}$ distribution of the four grain sizes: (a) 5nm; (b) 7nm; (c) 10nm, and (d) 14nm.

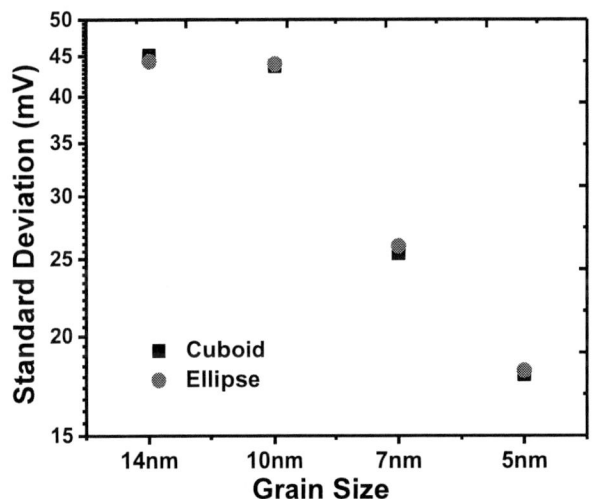

Figure 4: The influence of the grain size on the $V_{TH}$ variation for two different channel shapes.

value of DIBL is smaller.

**CONCLUSION**

In summary, this work used TCAD simulations to study junctionless FET with different channel shapes to investigate the impact of their work function fluctuation on the device performance, and compared the impact of grain size and different channel shapes. It is found that the gate metal grain size has a dominant effect on the transistor threshold voltage variation, which decreases as the grain size reduces. It is also shown that as the grain size decreases, the DIBL value of the cuboid channel is smaller, as it has a better ability to suppress the short-channel effect. The findings in this work could provide a useful guidance for the design and optimization of junctionless FETs.

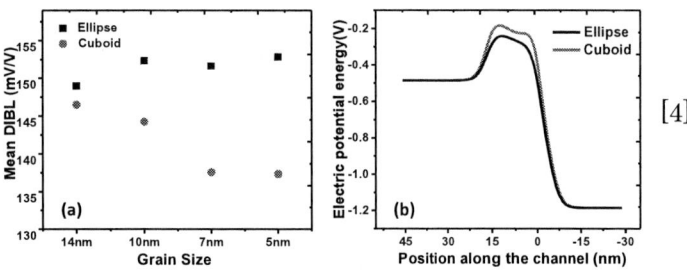

*Figure 5: (a) The influence of different channel shapes for the different grain sizes on the DIBL; (b) The electric potential along the channel at Vg = 0V.*

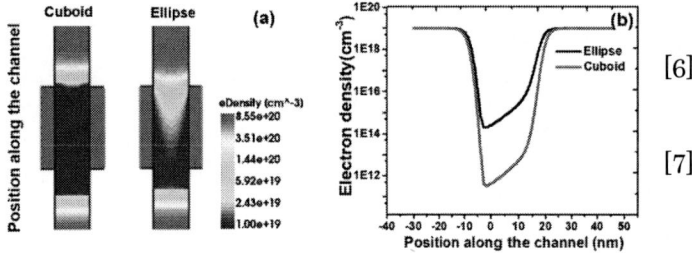

*Figure 6: (a) The influence of different channel shapes for the different shapes on the electron density; (b) The electron density along the channel at Vg = 0 V*

### ACKNOWLEDGMENT

This work was in part supported by the National Science and Technology Major Project of China (2017ZX02315001-005)

### REFERENCES

[1] Y. Lee and C. Shin, "Impact of equivalent oxide thickness on threshold voltage variation induced by work-function variation in multigate devices," *IEEE Transactions on Electron Devices,* vol. 64, no. 5, pp. 2452-2456, 2017.

[2] S. Mathew, K. Bhat, and R. Rao, "Investigations on the effect of Dual Material Gate work function on DIBL and Subthreshold Swing in Junctionless FinFETs," in *2020 IEEE International Conference on Electronics, Computing and Communication Technologies (CONECCT),* 2020: IEEE, pp. 1-5.

[3] T. Matsukawa *et al.,* "Suppressing V t and G m variability of FinFETs using amorphous metal gates for 14 nm and beyond," in *2012 International Electron Devices Meeting,* 2012: IEEE, pp. 8.2. 1-8.2. 4.

[4] K. Nayak, S. Agarwal, M. Bajaj, P. J. Oldiges, K. V. Murali, and V. R. Rao, "Metal-gate granularity-induced threshold voltage variability and mismatch in Si gate-all-around nanowire n-MOSFETs," *IEEE Transactions on Electron Devices,* vol. 61, no. 11, pp. 3892-3895, 2014.

[5] H. Mertens *et al.,* "Vertically stacked gate-all-around Si nanowire CMOS transistors with dual work function metal gates," in *2016 IEEE International Electron Devices Meeting (IEDM),* 2016: IEEE, pp. 19.7. 1-19.7. 4.

[6] J.-P. Colinge *et al.,* "Nanowire transistors without junctions," *Nature nanotechnology,* vol. 5, no. 3, pp. 225-229, 2010.

[7] C.-W. Lee *et al.,* "Performance estimation of junctionless multigate transistors," *Solid-State Electronics,* vol. 54, no. 2, pp. 97-103, 2010.

[8] H. F. Dadgour, K. Endo, V. K. De, and K. Banerjee, "Grain-orientation induced work function variation in nanoscale metal-gate transistors—Part I: Modeling, analysis, and experimental validation," *IEEE Transactions on Electron Devices,* vol. 57, no. 10, pp. 2504-2514, 2010.

# CFET 6T HD SRAM DESIGNS WITH 3NM DESIGN RULE

*Xiaona Zhu[1], RongZheng Ding[1], Yanli Li[1,2\*], Qiang Wu[1,2\*], Shaofeng Yu[1,2\*]*
[1]School of Microelectronics, Fudan University, Shanghai 200433, China
[2]National Integrated Circuit Innovation Center, Shanghai 201203, China

*Corresponding Author's Email: li_yanli@fudan.edu.cn, wu_qiang@fudan.edu.cn,
shaofeng_yu@fudan.edu.cn*

## ABSTRACT

As the critical dimensions continue to shrink, CFET structures which have alternating NMOS and PMOS stacking structures have been widely considered as one of the main device structures after gate all around(GAA) devices. This paper gives two designs of 6T high density(HD) SRAM with (Complementary Field-Effect-Transistor) CFET unit. Compared to 9 Fin pitch (FP) cell height of GAA consisted 6T SRAM, our CFET design achieve 6FP and even a 5FP cell height, thus shrinking the cell area to ~11520 nm$^2$, 44.4% shrinking compared to GAA constructed SRAM. The resistance and capacitance performance of our 2 designs also discussed. The decreased BL and WL resistance and capacitance revealed that our CFET design could achieve better SRAM read and write speed. This work gives a comprehensively analysis including layout, process flow and device performance co-optimization.

## INTRODUCTION

As the area scaling capability of traditional transistor approaches its limit, Complementary Field-Effect-Transistor (CFET) with stacked structure provides a new path for device density increasing[1-4]. The specific flow of CFET is recently widely investigated beyond 3 nm node, and regarded as the next generation after Gate-All-Around (GAA) FET [5-8]. Even though there is work reported ~50% scaling or standard cell and SRAM area[1], the detailed design rules for every patterning layers and their process feasibilities have not been reported.. From our point of view, CFET with 50% area shrink compared to GAA is still too hard to achieve for SRAM. The bottleneck in the area shrink for CFET SRAM is not coming from the found-end of line(FEOL) loop layers, such as the fin or poly-silicon (poly later) gate, but from the space limitation of the interconnect metals for the storage node.

In this work we discussed on the layout, process flow and RC performance of a conventional CFET SRAM Design1(D1 later) and an aggressively area shrunk CFET SRAM Design2 (D2 later). The conventional CFET SRAM could almost achieve a cell height~6FP (33.3% shrink to GAA-SRAM), but with no robust process. The D2 could achieve a ~5FP (44.4% shrink to GAA-SRAM) cell height with 45-degree angled M0G design and power rails in separated layers. The D2 is believed to be a practical design for the CFET structured SRAM aiming for area shrinking.

## Design rules and layout design

Table 1 listed the 3 nm design rules used in this work. The FEOL layers (fin and gate) adopted 193i SAQP patterning solution, while the Fin cut and Gate cut need EUV lithography. The fin pitch is set to 24 nm, whereas the gate pitch is set 48 nm, so that contacted poly pitch (CPP) is 48 nm. The backend of line(BEOL) vias and metal wires are set with width=12 nm, the minimum metal pitch is 24 nm. The metal and via layers all adopted EUV lithography due to the limited width and pitch.

Figure 1 shows the layout designs of GAA and 2 designs of CFET for 6T SRAM based on 3 nm design rules. As it is shown that, GAA structure-based SRAM has 9FP cell height and 2 CPP=96 nm cell width which is comparable to previous report[2,3]. The unit cell area is 20736 nm$^2$. The D1 as shown in Figure 1(b) is a commonly seen design with BPR , which could almost achieve a 6FP cell width, so that it can shrink the unit cell area to ~13824 nm$^2$. The D2, shown in Figure 1(c), with only Vdd buried as BPR, could aggressively continue to shrink the cell width to 5FP, so that the unit cell area can be reduced to ~11520 nm$^2$. This unit cell area is 16.7% shrink compared to the commonly seen CFET design(D1) and 44.4% compared to 9FP GAA designed 6T SRAM.

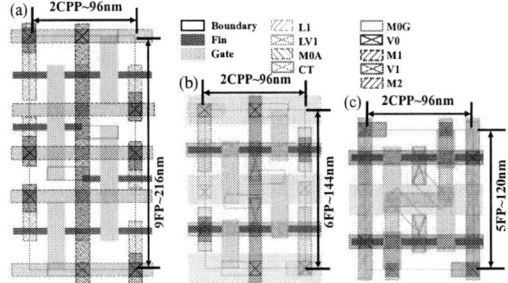

*Figure 1: 6T HD SRAM with (a) GAA, (b) CFET D1, (c) CFET D2 designed with 3 nm logic design rules.*

TABLE I. DESIGN RULES AND LITHOGRAPHY REQUIREMENTS FOR 3 NM NODE.

| Key Parameters | Width | *Pitch* | *Pattern Option* |
|---|---|---|---|

| Key Parameters | Width | Pitch | Pattern Option |
|---|---|---|---|
| **Fin** | 6 | 24 | 193i SAQP EUV LEn Cut |
| **Gate** | 16 | 48 | 193i SAQP EUV LEn Cut |
| **Via** | 12 | 48 | EUV LE |
| **Metal** | 12 | 24 | EUV LE2 |

CFET, compared to GAA, stacked NMOS/PMOS upon each other, is significantly different in process flow. Due to this, interconnect metal wires are designed to link the 1st floor PMOS to 2nd Floor NMOS (this work), or verse visa. Moreover, most CFET designs used BPR so that extra contact vias are designed to connect to the BPR floor. Thus, it is worth to be noticed that the L1, as shown in Figure 1, is the metal layer which has same height to 1st floor of 2-floor CFET, and the CCT layer, as shown in Figure 1, is used as a via that connects the L1 to the BPR layer. Similarly, M0A and M0G are the metal layers which have same height to 2nd floor of the CFET, and the CT is the via connecting to 1st floor CFET.

Figure 2 shows the detailed layout information of the 2 CFET design for the SRAM up to M0A and M0G layers, where the process flow has been completed up to the local contact layer. As shown is Figures 2 (a) and 2(c), The L1 and CCT layers of the two CFET designs are significantly different, and it is one key limitation for area shrinking. The D1 uses BPR for Vdd and Vss, so that we need to connect both NMOS and PMOS source or drain to the buried layers. The distance between Vss CCT to Vdd metal should be above 16 nm (EUV cut limitation), so that it needs 1.5FP to place this Vss CCT(marked as arrow1 in Figure 2a). In comparison, the CFET D2 only uses BPR for Vdd, the 2nd floor Vss does not need to use L1 and CCT layer to be connected to the Buried layers. Thus, D2 can save 0.5FP for single cell per side, so that it can achieve a 5FP cell height compared to the 6FP of the D1. Besides, as shown in Figures 2(b) and 2(d), D2 has a M0G with 45 angled shared contact connected to both M0 and the gate, which can increase the contact area of M0G to gate. In comparison, CFET D1 with normal horizontal M0Gs has narrow space ~10 nm (marked as arrows in Figure 2(b)) in both M0G to gate and space between M0Gs, which is very tight for overlay(OVL) and CD control. If the process control was not so perfect, the D1 will easily result in contact open or M0G to neighboring contact short.

Let's consider the every mask layer: The D1's L1 layer actually needs 2 masks: one for the vertical Vdd metals, and the other for the horizontal Vss metals. The M0 layer can use 1 mask with EUV litho, whereas the M0G should use 2 EUV masks in both D1 and D2, because of nearest M0Gs has space~10 nm. The CT layer also needs 2 EUV masks in D1, because it has narrow space ~16 nm, whereas D2 only needs one due to the angled M0G can provide larger space. In conclusion, the interconnect

layers with CFET D1 need 8 masks in total whereas D2 needs only 6 masks. It indicates complicated patterning solution and expensive cost of D1 CFET consisted SRAM. The D2 with Vdd buried, putting Vss upstairs, could decrease the patterning pressure of the FEOL interconnect wire routing.

*Figure 2: FEOL layouts of CFET D1: (a) L1 and CCT, (b) M0, CT, M0G, and D2:(c)L1 and CCT, (d)M0, CT, M0G to achieve the full function of the HD SRAM bit cell.*

Figure 3 shows the BEOL layers of CFET D1 and D2. D1 has large space for M1 and M2 for WL and BL routing, because it uses BPR for Vdd and Vss. The D2 uses BEOL layers for BL and WL and Vss. The V0 has minimal space of 34 nm (<48 nm of minimum 0.33NA EUV SE space) so that it needs two masks for this layer. In conclusion, the BEOL metal and via layers, the D2 needs totally 5 masks, one more than D1.

Considering patterning layers for CFET designs 1 and 2, D1 needs 19 masks whereas D2 needs 17 masks. Moreover, the D2 releases the routing space pressure of the CFET FEOL interconnect layers.

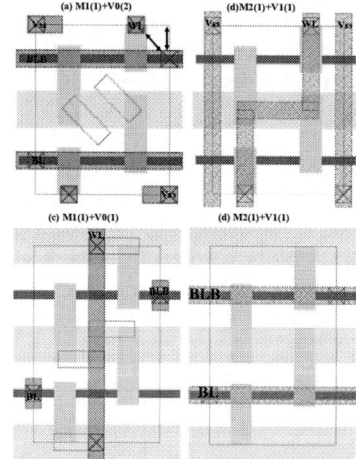

*Figure 3: BEOL layouts of CFET D1: (a)M1 and V0, (b)*

*and D2:(c)M1 and V0, (d)M2 and V1 to achieve the full function of the HD SRAM bit cell.*

## Process and flow design

Figure 4 gives a brief process flow for the CFET structured 6T SRAM. GAA structured SRAM is also shown here for comparison. As it is shown, the D1 and D2 both put PMOS in the 1$^{st}$ floor, and NMOS in the second floor. It is worthy of being noted that we use NMOS for passing gate(PG). It is challenging to disable the PMOS under the NMOS-PG. In this work, we use one additional mask to cover this PMOS for PMOS epitaxial growth.

Table II illustrates the distribution of the signal wires of different CFET designs, and GAA also listed for comparison. We regard the BPR as the -1F of the CFET-SRAM building, the PMOS as the 0F (ground) floor, then the NMOS as the 1F, and the M1 and M2 layers as the 2F, and 3F, respectively. From this table, it is clear the main difference between terminal locations for the 6T SRAM in CFET designs 1 and 2: (1) in CFET-D1, the Vdd is buried in -1F, while the Vss is connected to the top metal, and in the CFET-D2, the Vdd and Vss are both buried; (2) the WL and BL signals are also in the different floor.

TABLE II.    FLOOR INFORMATION OF GAA AND DESIGNED CFET.

| Floor | GAA | CFET-D1 | CFET-D2 |
|---|---|---|---|
| **3F** | BL, Vss | BL | WL,Vss |
| **2F** | WL | WL | BL |
| **1F** | | N-ch | N-ch |
| **0F** | N-ch; P-ch | P-ch | P-ch |
| **-1F** | | Vdd,Vss | Vdd |

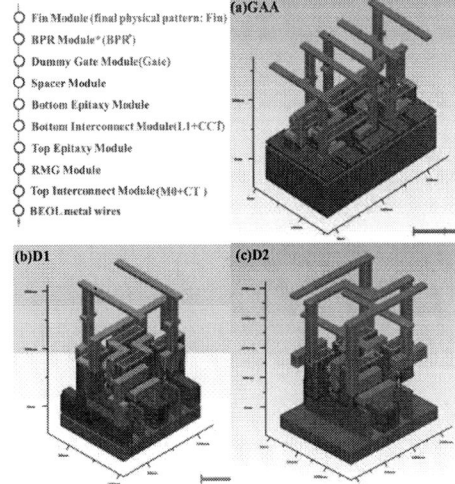

*Figure 4: The CFET basic process flow including some important lithography layers, and (a) Diagram of GAA CFET SRAM, (b) CFET SRAM D1, (c)CFET SRAM D2.*

## Device performance

The 2 structurally different CFET SRAM designs will expectively result in SRAM device performance difference. The read access delay of an SRAM is composed of WL propagation and BL evaluation delay. The WL and BL capacitance and resistance are shown in Figure 5. The CFET-D1 SRAM has comparable BL and WL capacitance compared to GAA SRAM. The WL resistance of CFET-D1 decreased remarkably compared to GAA, mainly because the CFET structure stacked the N/PMOS, and make the cross-coupling node (Q,Qb) closer to be connected. The CFET-D2 has increased unit cell capacitance, and WL resistance also increased. It is related to that the WL has twisted metal wiring and is also placed in M2, the top metal in the design.

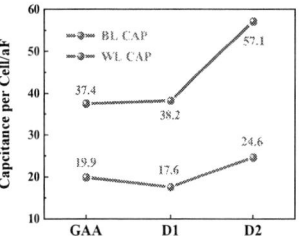

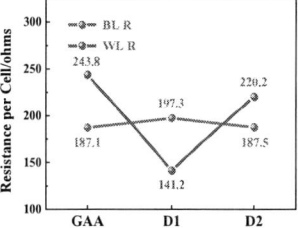

*Figure 5: SRAM bitcell and wordline parasitic components of different designs.*

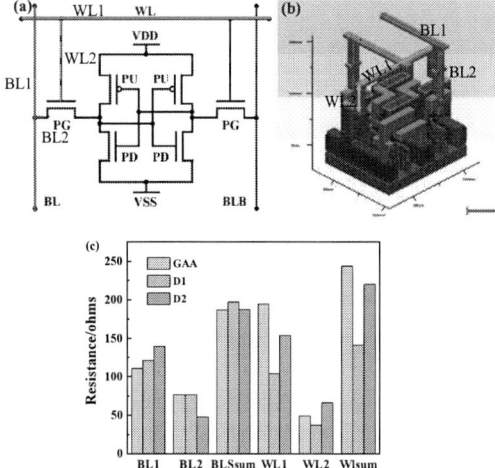

*Figure 6: (a) schematic of 6T SRAM, (b) 3D structure of CFET D1 with separated WL and BL marked corresponding to (a), (c)resistance of WL and BL*

*distribution.*

A comparison in the breakdown of BL and WL resistance into components of the 2 CFET designs are shown in Figure 6. The Figure 6(a) illustrates the BL and WL interconnect part (BL2, WL2) and external part (BL1,WL1) from schematic view. The Figure 6(b) shows the resistance calculation part from 3D structure view, using CFET-D1 as an example. The total BL resistance is comparable to the GAA counterpart. The total WL resistance of D1 decreased by 41.5%, and of D2 decreased by 9% compared to their GAA counterpart. Compared to the above external parts, the internal interconnect resistance (BL2, WL2), is the limiter in advanced technology nodes. The D2 has 37.3% reduced BL2 resistance compared to D1 and GAA design, while WL2 has slightly increased resistance. The reason for the interconnect resistance value change, is due to the floor locations of the WL and BL out wires.

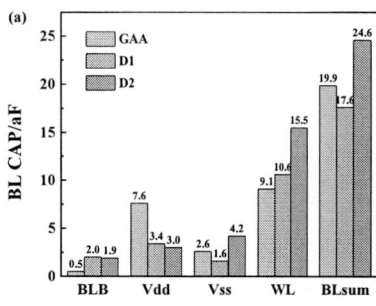

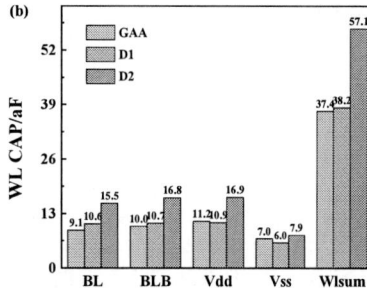

*Figure 7: (a)BL capacitance components and (b)WL capacitance components of three SRAM structure*

Figure 7 shows the BL and WL capacitance components of three kinds of SRAM structure. As it is shown, CFET-D1 constructed SRAM has lowered 10% BL capacitance, while D2 has increased capacitance.

## CONCLUSIONS

Two designs of 6T HD SRAM with CFET unit have been proposed. Compared to 9FP cell height of GAA consisted 6T SRAM, our CFET design achieve 6FP and even a 5FP cell height, thus shrinking the cell area to ~11520 $nm^2$, a 44.4% shrink compared to GAA structured SRAM. The patterning design rules have been discussed in detail, the CFET with 5FP cell height could also save 2

masks for pattern layers. The decreased BL and WL resistance revealed that our CFET design could achieve better SRAM read and write speeds. This work provides a comprehensively solution including layout, process flow, and device performance co-optimization.

## REFERENCES

[1] Ryckaert J, Schuddinck P, Weckx P, et al. "The Complementary FET (CFET) for CMOS scaling beyond N3." 2018 IEEE Symposium on VLSI Technology. IEEE, 2018: 141-142.

[2] Spessot A, Parvais B, Rawat A, et al. "Device Scaling roadmap and its implications for Logic and Analog platform." 2020 IEEE BiCMOS and Compound Semiconductor Integrated Circuits and Technology Symposium (BCICTS). IEEE, 2020: 1-8.

[3] Moore, M. "International Roadmap for Devices and Systems. 2020."

[4] Samavedam S B, Ryckaert J, Beyne E, et al. "Future logic scaling: Towards atomic channels and deconstructed chips." 2020 IEEE International Electron Devices Meeting (IEDM). IEEE, 2020: 1.1.1-1.1. 10.

[5] Subramanian S, Hosseini M, Chiarella T, et al. "First Monolithic Integration of 3D Complementary FET (CFET) on 300mm Wafers." 2020 IEEE Symposium on VLSI Technology. IEEE, 2020: 1-2.

[6] Vincent B, Boemmels J, Ryckaert J, et al. "A benchmark study of complementary-field effect transistor (CFET) process integration options done by virtual fabrication." IEEE Journal of the Electron Devices Society, 2020, 8: 668-673.

[7] Vincent B, Ervin J, Boemmels J, et al. "A Benchmark Study of Complementary-Field Effect Transistor (CFET) Process Integration Options: Comparing Bulk vs. SOI vs. DSOI starting substrates." 2019 IEEE SOI-3D-Subthreshold Microelectronics Technology Unified Conference (S3S). IEEE, 2019: 1-2.

[8] Huang, C. Y., et al. "3-D Self-Aligned Stacked NMOS-on-PMOS Nanoribbon Transistors for Continued Moore's Law Scaling." 2020 IEEE International Electron Devices Meeting (IEDM), 2020.

## ACKNOWLEDGEMENTS

Xiaona Zhu thanks the Shanghai Sailing Program.

# IC TECHNOLOGIES AND SYSTEMS FOR GREEN FUTURE

*Min-hwa Chi[1, 2]*

[1]SiEn (Qindao) Integrated Circuits Cor., Shandong, China 266500
[2]Micro-Nano Technology College, Qindao University, Qindao, Shandong, China 266071
Email: minhwa39@yahoo.com

## ABSTRACT

The "green" IC technologies and systems at AI/IoT era include broad concepts in smart systems (HW/SW, energy efficient, reconfiguration and modularization, data centers and cloud/edge computing, 5G communication, etc.), IC designs and 3DIC (for low power, reusable IPs, reconfigurable and modularized devices/blocks, 3DIC/SIP), and smart manufacturing (for low Carbon generation, low consumption of energy and materials, automation/robotics, waste material treatment /recycling). Similar "green" technologies are also actively leveraged and progressed in clean/renewable energy generation, power grid, automotive, transportation, smart building/city, and agriculture, etc. These "green" technologies are in active progressing for greener future.

*Keywords—Green technologies, smart systems, green manufacturing, green design and 3DIC.*

## INTRODUCTION

The climate change [1] as related to the earth warming effect is clearly threatening human life. Both the earth warming effect as well as the advancement of high-tech especially in Artificial-Intelligence and Internet-of-Things (AI/IoT) are accelerating, but in an opposite direction of impacts. We all intend to pursue a great future life but not the weather change to destroy our future. This paper overviews how green ICs and systems at AI/IoT era can be designed and manufactured with advanced green technologies. The "green" technologies include broad concepts in design, manufacturing, and systems; e.g. energy efficient, low power design, reconfigurable and reusable IPs, flexible 3DIC/SIP, low Carbon generation, low energy/material consumption, efficient waste material treatment /recycling.

The AI/IoT as a driving force for future business needs ICs with massive capabilities in sensing, computing and transmitting with superior performance and quality in wide range, i.e. from ULP to high performance, from wearable/mobile devices to embedded components in systems. These complex ICs (hardware/software) [2-4] need fast upgrade and frequent revision in cycles of development and manufacturing. The green technology for high performance AI/IoT systems (e.g. Data centers, Smart cars and autonomous driving, Robotics, Industry 4.0, etc.) need to achieve low power, high performance and superior quality/reliability as achievable by utilizing AI technologies extensively for smart design and manufacturing toward multi-level co-optimization, con-current design, fast failure analysis, fab automation, and 3D packaging. Modern foundry are also progressed toward green technologies and manufacturing on logic platforms with specialty devices and capability of SOC, Chiplet, 3DIC/SIP for advanced applications (Fig. 1).

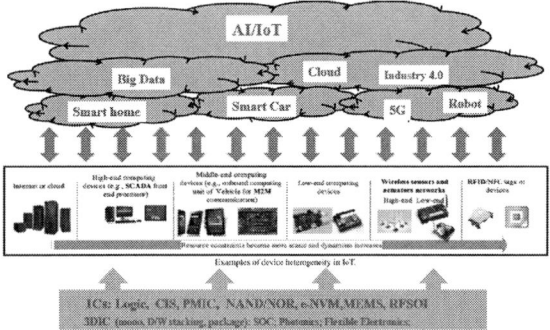

*Fig. 1. Green IC's and systems at AI/IoT era can be based on advanced technologies of design and manufacturing with low power, energy/material efficient, and multi-level optimization.*

## GREEN IC DESIGN AND 3DIC/SIP

The green IC design lies on the ultra-low power device with concurrent design platform in cloud, multi-level optimization, reconfiguration and reuse IPs and modules. The debug and failure analysis of ICs is performed effectively by utilizing AI techniques on big-data analysis.

**Ultra-low power (ULP) device and design**:
Green AI/IoT systems [5-6] need to operate CMOS ICs or 3DIC with ultra-low-power (ULP) with energy-harvesting for data processing/compression and RF communications. The FinFET on bulk and FDSOI technologies [3-8] are more suitable for ultra-low-voltage (ULV) or ULP operations as related to the better gate control on channel and less variability in device parameters (vs planar transistors on bulk).

Near-threshold circuit (NTC) design has been considered as promising to achieve energy efficient for ULP though the performance is seriously degraded [9]. Fortunately, the advancement of 3DIC technology (for fast signal transmission) as well as neural circuit (for non von Neumann circuit) can enhance its performance significantly. Thus, a combination of NTC designs with neuromorphic circuit (for effective computing) and 3DICs (for signal transmission at lower power) appears a best compromise for achieving both high performance and ULP. Various types of neuromorphic computing [10-11] can be implemented by analog circuits, mixed-signal, NVM (E2PROM, Flash, RRAM, PCRAM, MRAM), or elements with memory (mem-resistor, mem-capacitor, and mem-inductor). As each solution has own trade-offs, there is no commonly accepted best one. Neuromorphic computing algorithms on spiking neural networks (SNNs) can realize computations extremely energy efficient (Fig.2).

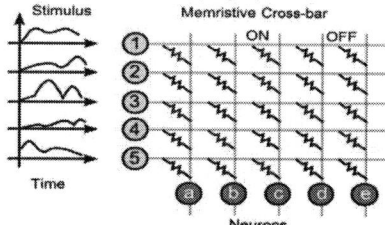

*Fig. 2. Schematic of a cross-bar array based SNN. The analog currents (as input) are integrated by down-stream neurons (as output spike) [10].*

**Parallel design platform in cloud and co-optimization**

The green concepts include advanced design and EDA platform on Cloud including a full portfolio of low power low leakage IPs, complete EDA tools (e.g. PDK,

SPICE, TCAD, Ref. flows, data, etc.) for concurrent design and verification among multiple design groups (at worldwide locations or organizations). Foundry business model is efficient and standardized in design as a green operation. One example of a complex AI chip is illustrated in Fig. 3.

*Fig. 3. A complex AI chip (Nvidia 2017) is enabled by multi-levels of co-optimization to achieve high performance, yield, low-power, and reliability [12].*

**Power devices**

Power devices and PMICs can leverage advanced IC

design methodology, e.g. technology platform (PDK, models, etc.) with EDA tools (e.g. design, simulation, TCAD, layout, reference flows, etc.) in cloud for concurrent design and co-optimization. The proto-typing and manufacturing of power semiconductor in fully automated Fab can achieve minimum defect, stable and consistent processing (especially for thin wafers) [13]. The product debug, FA, yield enhancement and

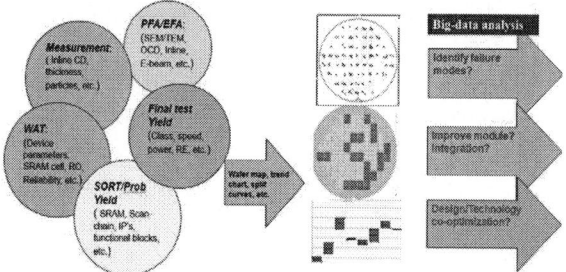

co-optimization can be performed effectively by big-data analysis and deep learning (e.g. in-line monitors, WAT, and yield data, etc.) (Fig. 4). The deep learning and judgment with AI techniques are useful for multi-level co-optimization, simulation, adjustment of process window, failure analysis, yield enhancement, etc.

*Fig. 4. The AI techniques are adopted in all aspects of IC design and manufacturing for product debug, FA, and yield, etc.*

**3DIC and SIP:**

The monolithic 3DIC integration including NVM (e.g. RRAM, PCRAM, MRAM, C- nanotube RAM, ...etc.) as well as logic circuits (e.g. magnetic spin-logic, C-nanotube FET, etc.) at BEOL together with CMOS at FEOL promises digital systems with massive connectivity and data communication with high security, high speed, and low-power. The monolithic 3DIC has advantages of precision alignment of contacts at transistor levels with performance gain (wire length reduction) and independent optimization of stacked devices as in [14]. The monolithic 3DIC with its high contact density is a powerful solution for heterogeneous co-integration with high density of via. The main process flow is illustrated in Fig. 5, with top transistors processed at low temperature and similar structure as FDSOI [15].

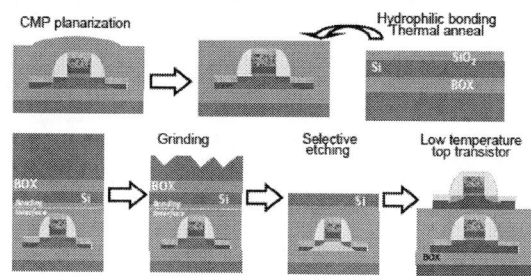

*Fig. 5. Main process flow for monolithic 3DIC [15].*

The 3D packaging or system-in-package (SIP) technologies [16] as based on chiplet, chip-to-chip or wafer-to-wafer stacking by using through-Si-via (TSV) technique can offer significant advantages in low power, high speed, data security, etc. The 3D-DRAM high-bandwidth memory (HBM) [17] is a recent success based on stacking-up DRAM chips through TSV technology. All these are effective to achieve high performance, modularization, and lower power as green schemes.

## GREEN AI/IoT SYSTEMS

An infrastructure of networks [18] connecting users to a cloud and the data center is illustrated in Fig. 6 below.

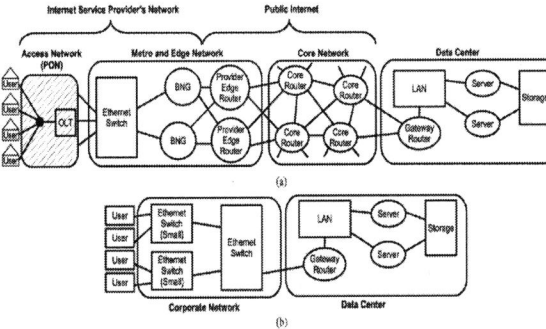

*Fig. 6. Schematic of networks connecting users to a cloud and the data center (a) Public cloud. (b) Private cloud [18].*

### Data centers

A modern state-of-art data center has three main components: data storage, servers, and a local area network (LAN). The data center connects to the rest of the network through a gateway router, as shown on the right-hand side of Fig. 6 (a) and (b).

### Cloud/Edge computing

As cloud computing is more widespread, the energy consumption of the public or private network and computing resources includes those in switching and transmission as well as data processing and data storage. The energy consumption in transport and switching can be a significant percentage of total energy consumption in cloud computing. Cloud computing can enable more energy-efficient use of computing power, especially when the computing tasks are of low intensity or infrequent. Cloud computing may consume more energy than on individual personal computer (PC). A user may use a range of devices to access a cloud computing service, including a mobile phone (cell phone), notebook or desktop computer, or a laptop computer.

### Hardware/software

An environmentally sustainable methodology [19] of green HW/SW includes developing technologies (for both user and power friendly), minimize high tech waste/trash in manufacturing, and also built-in green design for reuse, reconfiguration, and modularization. Firstly, the human computer interaction (HCI) shall be effective with good human interaction with other systems to produce effective results. Secondly, the e-waste and techno-trash (e.g. disposal of outdated computers) is to be minimized by implementing the techniques of green communication and IT. Green communication mainly focuses on energy saving and emission reduction for interaction of computers and servers. Green IT focuses on reducing the usage of hazardous materials in manufacturing and non-renewable energy resources. Thirdly, the HW/SW has built-in green design with capability of reuse, reconfiguration, and modularization.

### 5G Communication

The "green" 5G communication systems need to support higher data rates, broader bandwidths, and massive connectivity (than 4G systems), but also to minimize energy consumption. There are numerous technologies for friendly power allocation and energy efficient including massive multiple input multiple output (MIMO), internet of things (IoT), ambient energy harvester (EH), etc. There are new schemes proposed recently, e.g. a bidirectional absorptive common-mode filter (A-CMF) [20] using balanced-to-balanced structure with inter-digitally coupled lines for absorbing the CM noises from both directions within a broad band, and a heterogeneous cloud radio access network (H-CRAN) [21] by adding extra number of remote radio heads (RRHs) powered by energy harvest (EH) modules within the vicinity of one macro base station.

## GREEN MANUFACTURING

For semiconductor manufacturing Fab's, the green concepts need to apply to all aspects of inputs (energy, materials), outputs (solid/liquid waste and emissions of $CO_2$ and PFC's), facilities (gas, DI), process equipment/tools, wafer fabrication system (HW/SW), testing/monitors, yield debug and failure analysis (FA), etc. Optimization of all these related processes and logistics toward energy efficiency in a sustainable manner are the keys to a greener and successful Fab; furthermore, it needs advanced technology in automation, robotics, AI technique, big-data base analysis, etc. for greater success.

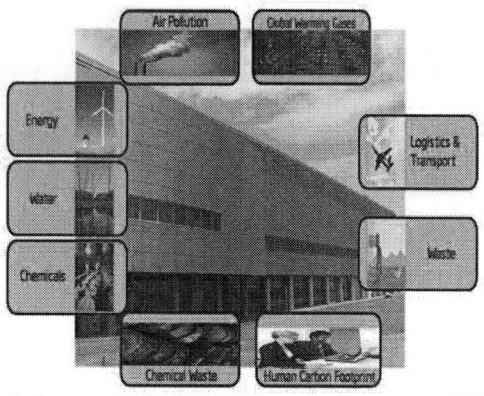

Fig. 7. Factors with highest impact to the environment [22].

**Fab Automation / Robotics**

A fully automated Fab [23] includes the following features: Fully automated workflow and material handling at floor level for all production, experiment, dummy/test wafers, decision making at all metrology steps, and tool down booking and hold management. 100% paperless, lot traceability at all locations, and single wafer tracking. The ability to prevent human error or MO (mis-operation) and achieving high yield/quality of wafer processing already justify the cost for automation. Furthermore, it is a highly green Fab by easily optimizing through advanced technology in AI/Lot and big-data analysis.

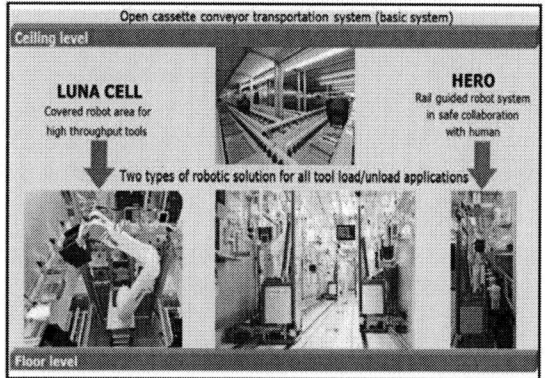

Fig. 8. Automation hardware solutions for automated tool load/unload [23].

**Fab facilities**

The vacuum pumps and abatement systems are critical in reducing the utility consumption (and also running cost). Significant savings can be realized by using green modes, i.e. these supporting equipment in low energy states [24]. By using simulation models for the green modes, it was also noted that additional savings are possible in the facility, e.g. reduced process cooling water at reduced thermal load of pump and abatement.

Chiller system can be optimized by using Machine Learning (ML) of AI technology for reducing power consumptions effectively [25]. The physical models are based on the fundamental thermodynamic cycles of refrigeration in compressors, electro-mechanical characteristics of the water pumps, and thermo-properties of humid air for the cooling towers. With variable frequency devices in motor controls, the absolute theoretical minimum of power consumptions can be predicted and the control adjustments can be tuned accordingly to reduce power consumptions significantly.

# CONCLUSIONS

To achieve green future, the IC technologies and systems are all progressing toward advanced and green technologies with AI/IoT extensively for design and manufacturing in multi-level co-optimization, failure analysis, and 3D packaging. These "green" technologies are in active progressing toward high performance but also greener future with Carbon reduction and neutrality..

# REFERENCES

[1] D.Fork and R.Koningstein, "How Engineers Can Disrupt Climate Chang", IEEE Spectrum, p.25, 2021. [2] M.A. Razzaque, et.al., "Middleware for Internet of Things: A Survey", IEEE Internet of Things J. v.3, no.1, p.70, 2016. [3] J. R. Hu, et.al. "Systematic Co-Optimization From Chip Design, Process Technology To Systems For GPU AI Chip", IEEE, VLSI-TSA, 2018. [4] A. Sharma, et.al., "Circuits and Systems for Energy Efficient Smart Wearables", IEDM, p.147, 2016. [5] D. Bol, G.de Streel, D. Flandre, "Can We Connect Trillions of IoT Sensors in a Sustainable Way? A Technology/Circuit Perspective", S3S, 2015. [6] A.Sharma, et.al., "Circuits and Systems for Energy Efficient Smart Wearables", IEDM, p.147, 2016. [7] O.Weber, "FDSOI vs FinFET: differentiating device features for Ultra Low Power & IoT applications", ICICDT, p.1-3, 2017. [8] T.Dry and T.Letavic; "Semiconductor Platforms for Ultra Low Power IoT Solutions", VLSI-technology, p.T164, 2017. [9] V.M. van Santen , et.al., "Reliability in Super- and Near- Threshold Computing: A Unified Model of RTN, BTI, and PV", IEEE Circuits & Systems I: REGULAR PAPERS, p.1-14, 2017. [10] B.Bajendran, et.al., "Neuromorphic computing based on emerging memory technologies", IEEE J. on Emerging and Selected Topics in circuits and systems, v.6, no. 2,p.198, 2016. [11] G.Volanis, et.al., "Toward Silicon-Based Cognitive Neuromorphic ICs—A Survey", IEEE Design and Test, p.91, 2016. [12] J. R. Hu, et.al. "Systematic Co-Optimization From Chip Design, Process Technology To Systems For GPU AI Chip", IEEE, VLSI-TSA, 2018. [13] G.Schneider, et.al., "Use of Simulation Studies to Overcome Key Challenges in the Fab Automation of a 300 mm Power Semiconductor Pilot Line Comprising Thin-Wafer Processing", ASMC, p.42, 2015. [14] C.Fenouillet-Beranger, et.al., "Recent advances in 3D VLSI integration", ICICDT, 2016. [15] G.Ghibaudo, "Electrical characterization of FDSOI CMOS devices", ESSDERC, p.53, 2016. [16] J.Lau, "Evolution, Challenge, and Outlook of TSV (Through-Silicon Via) and 3D IC/Si Integration", Keynote at IEEE Japan ICEP, 2011. [17] H. Jun, et.al., "High-bandwidth memory (HBM) test challenges and solutions", IEEE Design &

Test, Jan/Feb, p.16-25, 2017. [18] J. Baliga, R. W. A. Ayre, K. Hinton, and R. S. Tucker "Green Cloud Computing: Balancing Energy in Processing, Storage, and Transport", Proceedings of the IEEE, v.99, no.1, p.149, 2011. [19] K Vishrutha, "A Green Infrastructure for HCI and IT- A Review", Intl. Conf. on emerging trends in IT and engineering (ic-ETITE), 2020. [20] Y.Guan, Y.Wu, M.M. Tentzeris, "A Bidirectional Absorptive Common-Mode Filter Based on Interdigitated Microstrip Coupled Lines for 5G ''Green'' Communications", IEEE Access, p.20759, 2020. [21] N.A.Chughta, et.al., "Energy Efficient Resource Allocation for Energy Harvesting Aided H-CRAN", IEEE Access, v.6, p.43990, 2018. [22] S.Basu and S.Viarengo, "Looking Towards a Sustainable and Green Future", ASMC, 2011. [23] H. Heinrich, A. Deutschländer, "The long journey from standardization to full automation of a mature 200 mm fab", ASMC, p.353, 2018. [24] M. Czerniak1 , D. Hacker1 , and H. Schneider, "Reducing the environmental impact of semiconductor manufacture", CSTIC, 2016. [25] K. Hui, et.al., "AI applications for green manufacturing", e-Manufacturing and Design Collaboration sym, eMDC, 2018.

978-1-6654-9759-6/22 $31.00 © 2022 IEEE

# INFLUENCE OF ION IMPLANTATION ON SILICON PITS DEFECT FORMATION IN OXIDE ETCH PROCESS

*Zhiqiang Xiao[1]\*, Long Feng[1]\*, Jiaxing Xiao[1], Haitao Yan[1], Zhenchao Sui[1], Xin Zhang[1]*

[1]Semiconductor Manufacturing North China (Beijing) Corp. Beijing, China

\*Corresponding Author's Email: ZhiQiang_Xiao@smics.com; Frank_Feng@ smics.com

## ABSTRACT

In this paper a systemic investigation was performed to understand the effect of ion implantation on pits defect formation in the semiconductor fabrication. The pits defect is discovered after hydrofluoric acid etch process by dark or bright field optical inspection (DFI or BFI), whereas cannot be detected after ion implantation. Two typical pits defect formation model are proposed based on the defect position located by scanning electron microscope (SEM) and solution to prevent pit defects formation is also discussed and provided.

## INTRODUCTION

With the critical dimension continuous shrinking and moving to larger diameter substrates, device yields become more sensitive to contamination (visible and non-visible) [1]. For ion implantation process, the most familiar contamination type is surface particle defect (visible), which is easily detected by DFI and removed by the subsequent wet process. However, a new type of invisible defect of implantation, which can grow up to visible pits defect in post process, is demonstrated that can impact yield or act as a precursor to physical defects on the semiconductor fabrication [2]. The pits defect is discovered after hydrofluoric acid (HF) etch process by DFI, whereas cannot be detected after ion implantation.

Previous study show that pits defect is formed at HF electrochemical etching process catalyzed by the remain wafer charge on dual gate photoresist [2-4]. However, our recent manufacturing data revealed that ion implantation is another precursor of pits defect and can enhance the electrochemical etch process. In this study, a series of experiments were performed to demonstrate the effect of implantation on the formation of pits defect. Moreover, based on the defect position on chip layout, two model for the effect of implantation on pits defect formation were proposed and the preventing solutions were also provided

## EXPERIMENTS

Figure 1 shows a standard flow from ion implantation to HF etch process to form pits defect. A silicon wafer was first masked with photoresist (PR) and then implanted with phosphorus (2 steps implantation, with high energy and low energy respectively) (Figure 1a). After the PR removed, the silicon sample was annealed to activate the implanted ions (Figure 1b). Then $SiO_2$ was formed by thermal oxidization (Figure 1c). After the dual gate photoresist (DG_PH, Figure 1d) was etched by an HF-based etch solution (DG_ET), the pits defect was formed (Figure 1e). DFI and SEM were used to scan and review the defect respectively. Transmission electron microscope (TEM) was used to collect the depth of the pits.

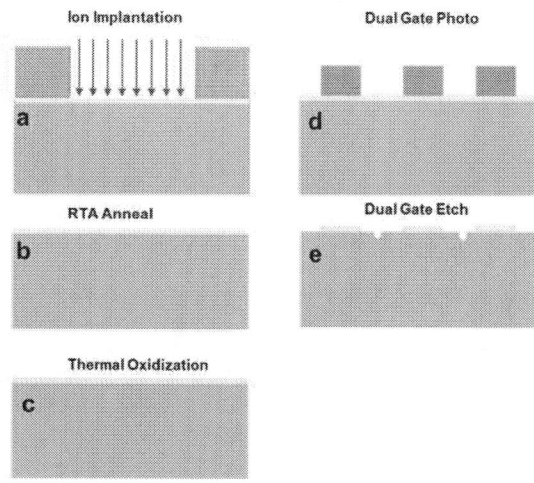

*Figure 1: Standard pits defect formation process flow diagram from ion implantation to HF-based etch*

TABLE I. EXPERIMENT RESULTS SUMMARY

| Experiment No. | 1-Baseline | 2-Skip IMP | 3-Skip DG_PH | 4-Change ESC Setting | 5-Change ESC Setting +vacuum time |
|---|---|---|---|---|---|
| Flow Condition | abcde | bcde | abce | abcde | abcde |
| Defect Map | | | | | |
| Review Image | 1. Wafer center area | NA | NA | 1. Wafer center area | 1. Wafer center area |
| | 2. Line shape area | NA | NA | NA | NA |

## RESULTS AND DISSCUSSION

Five set of experiments were designed based on the flow. The condition and result summary are listed on TABLE I. With the normal baseline condition (1-Baseline condition with standard process flow), there are two type of pits defect formed based on their location. The first type defect is located in the center area within wafer but in edge area of active area (AA) within die. The other type is two line-shaped defect within wafer but located in center area of ion implantation open area within die (Figure 2).

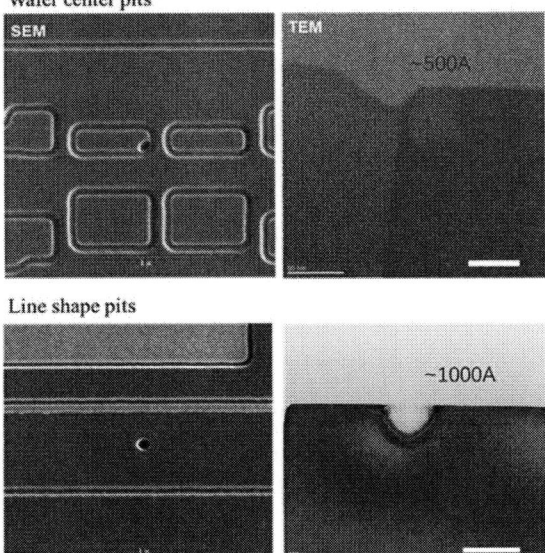

*Figure 2: Two type of pits defect SEM and TEM image*

The second and third experiment skipped the ion implantation process and DG_PH process respectively. The result show the pits defect almost disappeared. It means implantation and DG_PH are the two key factors for the formation of pits defect.

In order to distinguish the two type defect, the fourth experiment was performed with changing implantation E-chuck (ESC) setting. It is found that the line-shape pits defect disappeared, which means the implantation step is the dominant process for the line-shape pits defect formation.

Previous studies [2] show that the presence of a surface charge on the photoresist film is the catalyst where other non-visible "footprint" defect on wafer (such as the non-uniform thickness of the oxide or the roughness or damage etc.) is the "breakthrough" or "weakness" point for pits defect formation. Based on this mechanism and the above experiment result, we proposed two typical model for the influence of implantation on pits defect formation.

The first one is the implantation photo-resist (PR) outgas corroding model (Figure 3a), corresponding to the defect position located on the AA edge area. Ion implantation, especially high energy implantation, can produce large amount of PR outgas. If the outgas cannot vent out timely. its acidic hydronium easily corrodes the

AA edge area which has weak Si-Si bond and forms non-visible micro-pores. Then after the charging process of DG_PH and electrochemical etching of DG_ET, the macro-pits defect formed. As the fifth experiment show in TABLE 1, by adding the duration of wafer in vacuuming load lock or chamber, this type pits defect decrease evidently.

The second one is the implantation charging model, corresponding to the defect position located on the implant open center area (Figure 3b). As the fourth experiment show, changing implantation ESC setting can remove the line-shape pits defect. As the ESC clamp wafer based on coulombic force, the charging release process after implantation has big influence on the wafer surface charge distribution. If the ESC discharge ability is not enough, the positive charge of ion can trap in the center area of AA. Then in the electrochemical etching of DG_ET, the trapped areas are easily etched by HF-based solution and form pits defect.

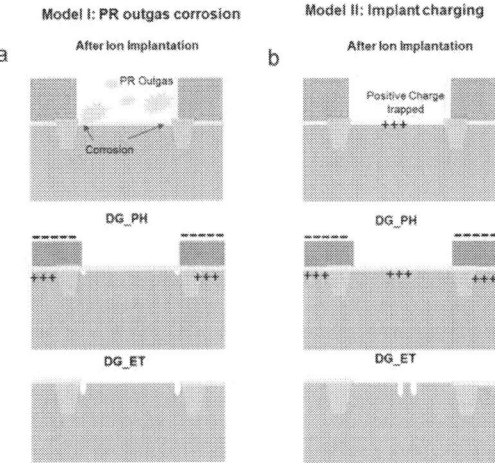

*Figure 3: a) Diagram of wafer cross section and flow for model of PR outgas corrosion model; b) Diagram of wafer cross section and flow for model of implant charging model*

## CONCLUSIONS

In this paper the influence of ion implantation on the pits defect formation is presented. The pits defect is discovered after HF etch process whereas cannot be detected after ion implantation. Two typical pits defect formation model are proposed based on the defect location of the production layout. The first one is implantation PR outgas corroding model, corresponding to the defect position located on the implant AA edge area; the other one is ion implantation charging model, corresponding to the defect position located on the implant AA center area. Both PR outgas and implant charging can form silicon non-visible defect and then are enhanced by the subsequent electrochemical HF oxide etch, resulting to the pits defect. The above two models were verified by a series of experiments. And by prolonging PR outgassing duration and changing ESC setting, the two type of pits defect decrease rapidly even disappear. The prevention methods mentioned in this paper focused on decreasing the effects of implantation on pits defect formation. Besides, there are also other methods to prevent the pits defect by fine tune DG_PH condition, which is under further studying.

## REFERENCES

[1] K. Mori; N. Nguyen; D. Keeton; R. BurnsB. Noble, and I. N. Sneddon, *IEEE '94 Defect and Fault Tolerance., 1995*

[2] J. Park, S. Cho and J. Hawthorne, *Semiconductor Manufacturing, vol. 26, pp. 315-318, 2013.*

[3] J. Park, W. Kang, P. Bang, T. Hwang, S. Cho, J. You, *Proc. Sematech Surf. Prep. Cleans Conf., pp. 174-182, 2012.*

[4] V. Lehmann, *J Electrochem Soc, vol. 140, no. 10, pp. 2836-2843, 1993.*

# Wafer to wafer bonding to increase memory density

Dube Belinda Langelihle Yolanda
*System Plus Consulting*
Nantes, France
bdube@systemplus.fr

*Abstract*— **Fulfilling huge memory chip demand in strong growth applications like Internet of Things, mobile telephones and datacenters requires memory manufacturers to constantly evolve the manufacturing process of 3D NAND flash memories. Wafer to wafer hybrid bonding has been introduced in new generation memories to overcome scaling limit and eliminating several 3D NAND manufacturing challenges. Wafer to wafer bonding in memories involves joining a NAND array wafer to the logic wafer. The wafer-to-wafer hybrid memory design was introduced to continue Moore's law in the third dimension. Copper metals from the two wafers are joined together to form one component. This hybrid bond between two individual wafers permits the manufacturer to produce denser and smaller dies with high-speed data transfer rate. Benefits of wafer-to-wafer bonding could attract manufacturers to adopt this design to continue miniaturization of NAND memory components. Wafer to wafer hybrid bonded NAND chips is compared to conventional 3D NAND memories to reveal density and cost benefits of the novel design. Relentless advancements in 3D NAND designs, together with the emergence of hybrid architectures have enabled continuous NAND memory bit cost reduction while improving the memory characteristics.**

*Keywords—NAND, Flash, Non-Volatile Memories (NVM), Wafer, Bonding, Hybrid bonding, Xtacking*

## I. INTRODUCTION

The digital era has caused a drastic increase in data production and consumption. Consumer applications and high-power computing demand high speed, reliable and low power consuming memories. To cater for this revolutionary demand, memory manufacturers focus on new methods and adapt innovative methods in memory manufacturing. The emergence of 3D NAND memories have enabled continuous advancement in bit-density growth via vertical scaling of memory cells[1]. With each generation of 3D NAND memories high-capacity memories are manufactured. This 3D architecture can address density limitations in planar memory cells. The Memory architecture comprises of a memory array and peripheral devices for commanding signals to and from the memory array[2]. Wafer to wafer bonding is a method of manufacturing 3D NAND memory array cells and CMOS transistors on separate individual wafers. Wafer to wafer hybrid bonding is designed to increase memory capacity. The CMOS periphery circuit wafer is bonded directly to the 3D NAND memory array wafer. This fabrication method delivers several advantages and demonstrates potential to supply the industry with high-density Non-Volatile Memories (NVMs) merged with advanced packaging design. Wafer-to-wafer hybrid integrates a higher number of memory cells on a single chip and provides the semiconductor industry with 3D NAND memories with exceptionally high interconnect density[3]. The process of joining the CMOS wafer and NAND Array wafer

does not use any intermediate layers between wafers. CMOS Image sensors have successfully made use of this technological innovation[4]. This bonding technic needs an extremely high level of accuracy. Complexity of hybrid bonding technic could result in reduced manufacturing yields. In this investigation, 3D NAND Memories with different designs are torn down to determine the advantages of bit density increase including bit cost reduction in hybrid bonded NVMs.

## II. NON VOLATILE MEMORY SCALING

### A. Layer Count Increase

NAND flash memory evolved from 2D to 3D structure as NAND memory faced scaling and lithography limitations. Manufacturers of 3D NAND continue to scale the NAND memories by constantly increasing the layer count with each generation without any change in lithography technology node. Increase in layer count results in increased memory density. Samsung started with 32-layer 3D NAND and increased to 64-layer, 92-layer and currently their 3D NAND commercial products use 128-layers. This technic does not only increase memory density but also caters for a significant bit cost decrease. Fig 1 shows a 3D NAND die cross section revealing the stacked layers. Continual increase of layers could face a few technical challenges in the manufacturing process like asymmetric etching and bow during high aspect ratio etching[5].

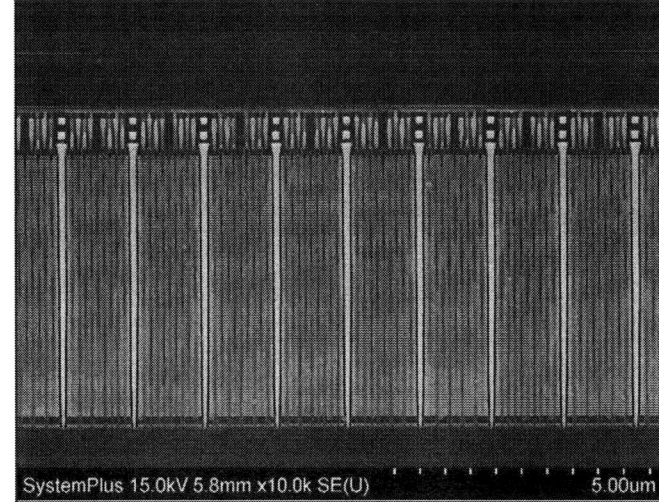

Fig. 1. *Samsung 92 layer 3D NAND Cross Section*

### B. Memory Bit Increase

Data is stored as bits in the memory cells, the bits represent an electrical charge contained within the cell. The first NAND memory produced had only one bit referred to as a Single Level

Cell (SLC). The bit count can be modified to increase the bits that can be stored per memory cell[6]. Increase in bit count increases the memory density without an increase in manufacturing steps and cost. Bit augmentation in a single memory cell increases the number of states a cell can have, thereby exponentially increasing its capacity. Numerous commercial devices use Triple Level Cell (TLC) storage that allows three bits of information per memory cell. The Multi-level Cell (MLC) provides the industry with higher performance and reliable NVMs, whilst the TLC benefits the consumer by providing low-cost 3D NAND memory.

### C. CMOS Under Array/ Periphery Under Circuit

NAND design and architecture differs as different manufacturers modify the CMOS circuit positioning. The CMOS transistors are positioned strategically to reduce the area occupied by the CMOS transistors on the dies. CMOS under array architecture is a technic used by Micron in the 96-layer 3D NAND shown in Fig 2. The CMOS under Array (CuA) design is implemented to increase area that can be occupied by the NAND Array. CMOS Under Array architecture achieves a minimal footprint with logic/CMOS transistors located below the array. This technology design sustains the industry with low-cost advantages. YMTC introduced 3D NAND build on two wafers, one wafer integrates the CMOS wafer, and the second wafer integrates the NAND array. The CMOS wafer is placed under the NAND array wafer and later bonded. YMTC produced smaller 3D NAND dies that can also be fitted into smaller packages advantageous for mobile device purposes.

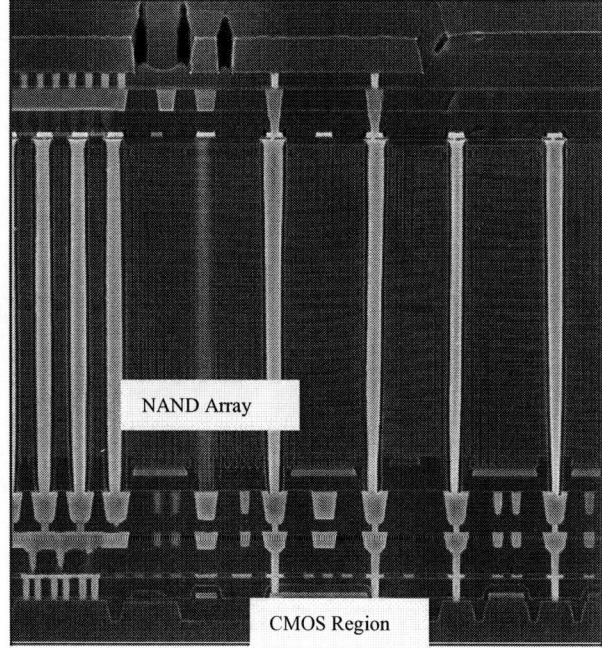

Fig. 2. *Micron CUA 96 layer 3D NAND Cross Section*

### III. WAFER TO WAFER DIRECT BONDING

Direct wafer to wafer hybrid bonding is a heterogenous process which involves stacking and electrically connecting wafers from different production lines. The bonding process makes use of chemical bonds between two wafer surfaces.

Wafer-to-wafer hybrid bonding is an important process step to enable 3D stacked devices. This process has been referred to this as the Xtacking process. Two bonded wafers can be differentiated on YMTC's die cross section. Fig 3. shows a Scanning Electron Microscope (SEM) Image of YMTC's 64-layer 3D NAND die cross section using hybrid bonding design.

This design strategy aims to produce denser and higher capacity 3D NAND dies which results in reduced NAND die area. In this technic One 300mm wafer is purely used for high performance CMOS transistors and the other another 300mm wafer has the total area occupied by the NAND array. This strives to elevate the NAND Cell production per wafer. The two wafers are joined together before wafer thinning and dicing. Producing the two wafers separate could be advantageous as the manufacturer is able to produce the wafers simultaneously and the CMOS wafer can be manufactured in high temperature without risk of degrading the NAND Array or NAND Cells. Therefore, allowing reduced product cycle development.

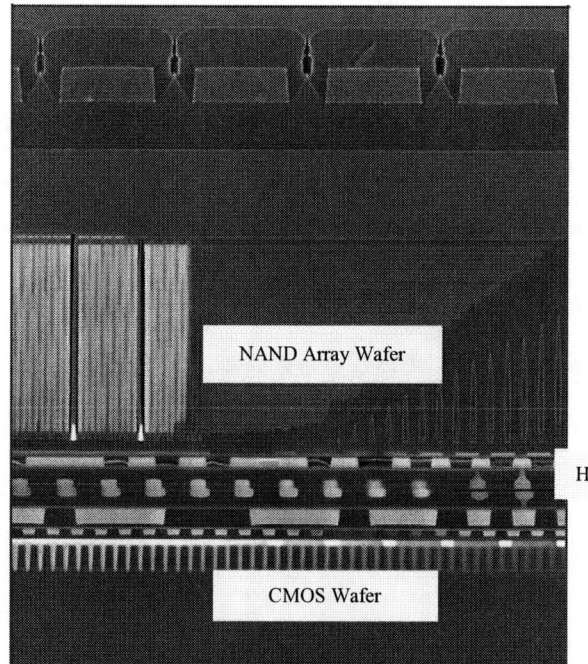

Fig. 3. *YMTC Xstacking 3D NAND Cross Section*

This bonding process used in 3D NAND manufacturing is divided into three major steps. The wafer processing, Prebonding and annealing steps. Direct hybrid bonding allows exceptionally fine pitch 3D interconnect. It also uses low temperature bonding with fine pitch interconnect scalable to 1μm. The technology also enables high bandwidth and high-speed interface. This process is complex as it requires a clean, flat, and smooth wafer surface. Defects are prone to occur when the wafer surface requirements are not fulfilled, these defects could be in a form of voids or internal bubbles resulting from unbonded surfaces. This kind of defect could result in electrical malfunction of the device. However, good electrical contact between the wafers must be achieved by precise alignment and extreme overlay accuracy. The interconnect mismatch can result in lower manufacturing yields in direct

978-1-6654-9759-6/22 $31.00 © 2022 IEEE

hybrid bonding. In Fig 4, the SEM image shows the hybrid bond mismatch that could results from bonding two wafers. The wafers are bonded by connecting the copper metals from the NAND array wafer and the CMOS wafer.

Hybrid bonding process involves metal bond pads deposition, and the wafer surface is polished to remove roughness. Plasma surface activation is attained by use of a Nitrogen gas to facilitate strong direct bonds. Wafers are precisely aligned before joining them together, an instant attraction and formation of strong chemical bonds is created between the two wafer surfaces. An anneal process creates copper to copper metal direct bond, the two copper metals expand and merge into each other forming a strong homogeneous attachment.

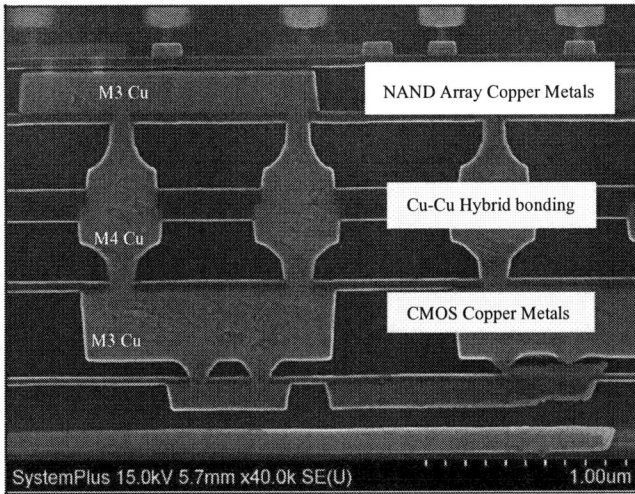

Fig. 4.   *YMTC 3D NAND Cross Section- Cu-Cu hybrid bonding*

Two copper layers are used in the direct bonding interconnection. Metal 3 from NAND array wafer connects to Metal 4 of the CMOS Wafer. There is approximately 30 nanometers misalignment or offset between the copper metals. The alignment accuracy is less than 30 nm as illustrated in Fig 5.   The difference in copper metal size could increase the misalignment.

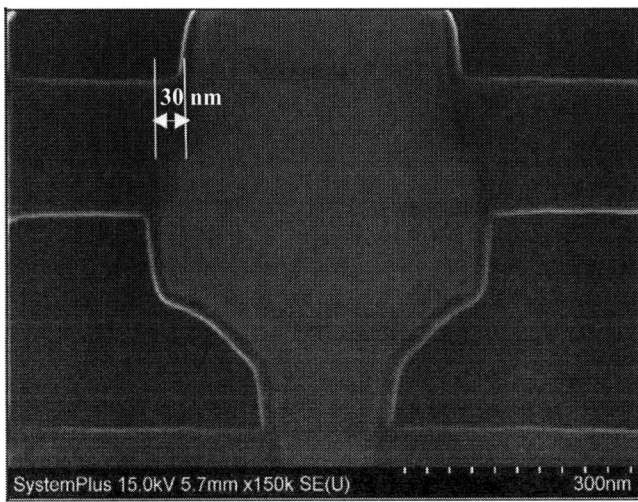

Fig. 5. *YMTC 3D NAND Cross Section- Cu-Cu hybrid bonding misalignement*

## IV. EFFECT OF WAFER TO WAFER BONDING ON DIE SIZE AND DENSITY IN COMMERCIAL PRODUCTS

Revealed in the study is the impact of wafer-to-wafer hybrid bonding on 3D NAND wafer density and bit cost reduction. Hybrid bonded wafers are compared to the CMOS under Periphery design and the conventional Periphery on side wafers. The technology shift facilities production of denser dies that use less silicon area, therefore producing more dies per 300mm wafer. Different designs show different results of potential capacity produced per wafer.

Table 1. details comparison results between 64-layer hybrid bonded NAND wafers, CMOS under array wafers and conventional periphery on side wafers. Die density is significantly increased when the CMOS transistors are positioned on a separate wafer or under the array. The effective memory area is increased because the integral area on the wafer is occupied by the NAND array memory cells. Hybrid bonding gives a successful design in producing more storage cells per wafer compared to the other designs. The study results demonstrate significantly higher density in hybrid bonded 3D NAND dies with approximately 94% effective memory area. 30% more memory capacity is attained from hybrid bonded wafers compared to periphery on side wafers. Hybrid bonded NAND dies have approximately 25% smaller die size compared to the periphery on side NAND dies.

TABLE 1. 3D NAND COMPARISON

| | 64 Layer 3D NAND Comparison | | |
|---|---|---|---|
| | *Wafer to Wafer Hybrid Bonding (YMTC)* | *CMOS Under Array CuA (Micron)* | *Normal Periphery on side* |
| Die Capacity | 32GB | 32GB | 32GB |
| Die Size | 57.3 mm² | 57.7 mm² | 74.7 mm² |
| Potential Dies per wafer | 1108 | 1104 | 852 |
| Potential Capacity/ GB per Wafer | 35 400 | 35 300 | 27 200 |
| Die Density | 4.46 Gb/mm² | 4.43 Gb/mm² | 3.42 Gb/mm² |
| Effective memory Area | 94 % | 83% | 52% |

*Table 1. Comparison of Wafer to wafer hybrid bonding 3D NAND to Pheriphery under Circuit and Periphery on side*

The YMTC 64-layer 3D NAND die is then compared the 96 layers from two manufacturers using the conventional 3D NAND design with the Periphery on the side. Fig 6. shows a graph comparing the three commercial dies from three different manufactures. The three dies compared have the same capacity of 32Gigabyte. YMTC's 64-layer 3D NAND using hybrid bond design is competitive even with less layer count. Hybrid bonded die is 5% bigger and only 10% less density compared to the 96-layer dies. This proves that wafer to wafer bonding is equally advantageous as adding layers to the 3D NAND Wafer design. A scenario where wafers are manufactured in the same fab and having the same layer count, the wafer-to-wafer bonded dies would achieve a higher density and therefore bit cost effective compared to the conventional 3D NAND using periphery on side design.

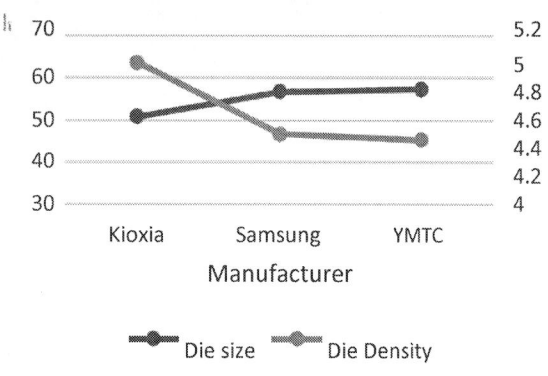

Fig. 6. *Comparison of Wafer to wafer hybrid bonding 3D NAND to Pheriphery under Circuit and Periphery on side*

## V. STRING STACKING SOLUTION COMPARED TO WAFER TO WAFER BOND

NAND Manufacturers overcome the challenge to produce higher memory storage per wafer by constantly increasing layer count. Adding number of layers increases the number of NAND cells hence an increase in Gigabytes produced in a single wafer. This process presents several manufacturing challenges like high aspect ratio etching that could result in low yields and increased manufacturing cost. Manufacturers are using string stacking to reduce the high aspect ratio etching, as more string would be stacked in the future, the challenges could build up. Wafer to wafer hybrid bonding presents a solution to the challenge. Figure 7 illustrates a design of wafer-to-wafer bonding that could be used in future in 3D NAND Memories. Each stack could be integrated in an individual wafer.

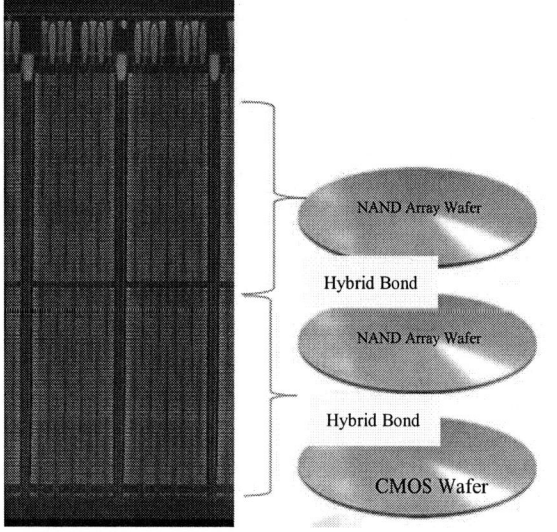

Fig. 7. *Kioxia 96 layer string stacking, illustration of future string stacking usingwafer to wafer hybrid bonding*

## I. CONCLUSION

Wafer to wafer bonding is becoming popular in the semiconductor industry. Since Moore's law is limited, to fulfill the need for high performance and smaller semiconductor devices hybrid bonding technology design presents an effective solution. Future NVM memory devices could integrate two or more wafers in the manufacturing process to continue miniaturization of semiconductor products and increase memory density. The memory market is escalating, and the dense interconnect created by wafer-to-wafer bonding enables faster interconnect speed. The YMTC 64-layer 3D NAND using wafer to wafer bonding indicate excellent die shrinkage compared to other designs. The die shrinkage results in a significant bit cost reduction. As the consumers actively search for low-cost memories, the xtacking design could potentially be adopted by various players to improve high density output per die. The wafer to wafer stacking gain is not only beneficial for improved bit transfer rate but also lower memory cost.

### A. Authors and Affiliations

*1) Belinda Langelihle Yolanda Dube*

### ACKNOWLEDGMENT *(Heading 5)*

I would like to acknowledge System Plus Consulting for trusting me with memory technology analysis and reports. I would like to extend my gratitude to my department supervisor for her constant support. This report would have not been possible without System Plus Consulting laboratory and Veronique Le Troadec who carried out the physical analysis.

### REFERENCES

[1] H. Kim, S. Ahn, Y. G. Shin, K. Lee and E. Jung, "Evolution of NAND Flash Memory: From 2D to 3D as a Storage Market Leader," 2017 IEEE International Memory Workshop (IMW), Monterey, CA, 2017, pp. 1-4, doi: 10.1109/IMW.2017.7939081.

[2] J. Lee, J. Jang, J. Lim, Y. G. Shin, K. Lee and E. Jung, "A new ruler on the storage market: 3D-NAND flash for high-density memory and its technology evolutions and challenges on the future," *2016 IEEE International Electron Devices Meeting (IEDM)*, San Francisco, CA, 2016, pp. 11.2.1-11.2.4, doi: 10.1109/IEDM.2016.7838394.

[3] J. A. Theil, L. Mirkarimi, G. Fountain, G. Gao and R. Katkar, "Recent Developments in Fine Pitch Wafer-To-Wafer Hybrid Bonding with Copper Interconnect," 2019 International Wafer Level Packaging Conference (IWLPC), San Jose, CA, USA, 2019, pp. 1-6, doi: 10.23919/IWLPC.2019.8913862.

[4] Y. Kagawa et al., "Novel stacked CMOS image sensor with advanced Cu2Cu hybrid bonding," 2016 IEEE International Electron Devices Meeting (IEDM), San Francisco, CA, 2016, pp. 8.4.1-8.4.4, doi: 10.1109/IEDM.2016.7838375.

[5] Z. Yang et al., "Asymmetric etching profile control during high aspect ratio Plasma etch," 2018 29th Annual SEMI Advanced Semiconductor Manufacturing Conference (ASMC), Saratoga Springs, NY, 2018, pp. 211-215, doi: 10.1109/ASMC.2018.8373146.

[6] H. Lue et al., "A 128Gb (MLC)/192Gb (TLC) single-gate vertical channel (SGVC) architecture 3D NAND using only 16 layers with robust read disturb, long-retention and excellent scaling capability," 2017 IEEE International Electron Devices Meeting (IEDM), San Francisco, CA, 2017, pp. 19.1.1-19.1.4, doi: 10.1109/IEDM.2017.826841

# PATHFINDING BY PROCESS WINDOW MODELING: ADVANCED DRAM CAPACITOR PATTERNING PROCESS WINDOW EVALUATION USING VIRTUAL FABRICATION

*Qingpeng Wang, Yu De Chen, Jacky Huang, Benjamin Vincent and Joseph Ervin*
Coventor, Inc., A Lam Research Company, Shanghai, China
*Corresponding Author's Email: Qingpeng.Wang@lamresearch.com

## ABSTRACT

Virtual fabrication was used for process window evaluation and optimization during pathfinding in an advanced DRAM capacitor process. Our results show that the in-spec ratio for an SADP approach will be 10% higher than that of using SAQP under optimized conditions and tightened specifications. The best capacitor area pass-rate (in-spec ratio) using an SADP approach can be enhanced to nearly 100%, while it only reaches 91% for the SAQP process. This research offers a quantified process window comparison for different patterning approaches. Using this approach, the optimal process target combinations and the largest process window can be achieved, prior to performing any wafer-based experimentation.

## INTRODUCTION

With continuous device scaling to smaller sizes, integration and patterning schemes become more and more complicated at advanced nodes. At the same time, process windows have become narrower and narrower due to smaller feature sizes and process step variability [1]. Choosing a good integration scheme, with a relatively large process window, is a key task during the R&D stage. Traditionally, in order to identify a process window, trial and error Si wafer experiments across a wide range of DOE inputs have been needed. These series of wafer-based tests can be very costly and produce long cycle times. Given these challenges, virtual fabrication has become an effective, direct, and inexpensive way to specify process windows and perform pathfinding at advanced nodes [2].

In DRAM, the capacitor is the key component for data storage. Thus, capacitance and its variations are key parameters used to determine the DRAM noise margin. The capacitor hole CD and its variation directly determine capacitance levels and their variation. In advanced DRAM, capacitors with closely packed patterning are designed to increase cell density (Fig. 1) [3]. Thus, advanced patterning schemes, such as multiple litho-etch, SADP and SAQP processes may be needed.

In this paper, we systematically evaluate a capacitor hole formation process sequence that includes SADP and SAQP patterning, using virtual fabrication and statistical analysis in SEMulator3D®. The purpose of this analysis is to obtain a quantified process window comparison between the SADP and SAQP patterning schemes then provide clear guidance for a process developer.

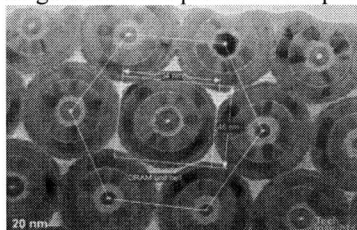

*Figure 1: Top view of 1x DRAM capacitor packed hole pattern (Courtesy: TechInsights).*

## FLOW AND VIRTUAL METROLOGY

The available patterning schemes to print a 40 nm hole array include EUV LE, LE4, double SADP (80 nm mandrel pitch), and double SAQP (160 nm mandrel pitch). We selected immersion double SADP and SAQP as candidate run paths and compared their process sensitivity and process windows. Figure 2 (a) and (b) displays the design for the SADP and SAQP mask. Figure 2 (c) shows a fin marking layer (HMX, HMY) and wafer cut mask (Cut). A virtual flow was built for each process as shown in Figure 3. After the final structure was generated, the wafer cut mask shown in Figure 2(c) was used to crop out a 4x4 array for downstream virtual metrology. The CD in both the x and y directions are important factors in determining capacitance. Instead of checking both CDs, the capacitor hole area is used as a metric for the capacitance and its uniformity analysis. Structure search in SEMulator3D was used to find the minimum and maximum capacitor hole area from the 4x4 hole array, and then calculate the mean area, so that the delta from maximum to minimum area can be calculated. Figure 4 displays the metrology results for one output structure, with the minimum and maximum area holes identified.

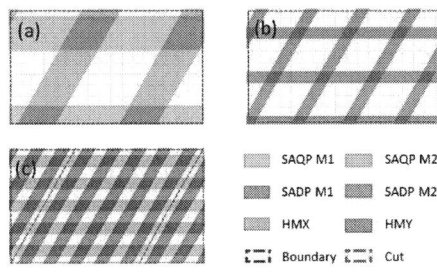

*Figure 2. (a) SAQP Mandrel 1 and 2, (b) SADP Mandrel 1 and 2, (c) HMX and HMY marking layer.*

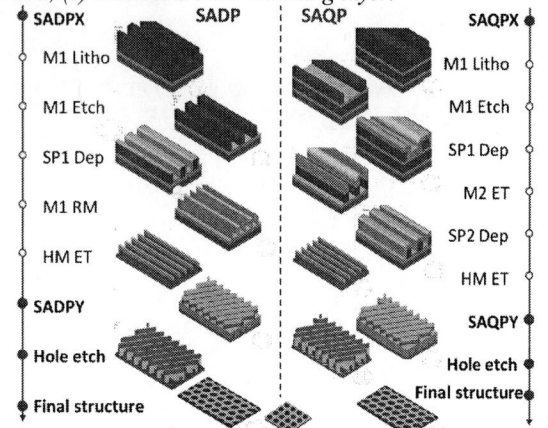

*Figure 3. Major process steps of SADP and SAQP*

(a) Output structure  (b) Minimum area  (c) Maximum area

*Figure 4. Virtual metrology results for minimum and maximum area.*

## DOE DESIGN AND ANALYSIS

Based on this virtual flow and metrology, a Monte Carlo study with 3000 trials was performed using the SEMulator3D's Analytics module. This study was used to check the process sensitivity and process window for both the SADP and SAQP approach, respectively. Mandrel CDs and spacer thicknesses were selected as DOE input parameters, while mean area and delta area were selected as the output parameters. Table 1 lists the ranges of input parameter values for the SADP and SAQP processes. The virtual DOE results guided our investigation regarding the impact of each input on variations in the mean and delta area. In Table 1, MX is the X direction mandrel CD; MY, the Y direction mandrel CD; SPX1, the X direction 1st spacer thickness; SPX2, the X direction 2nd spacer thickness; SPY1, the Y direction 1st spacer thickness; and SPY2, the Y direction 2nd spacer thickness. In the SADP process (Fig. 5 (a-b)), the mean area is impacted by spacer thickness, while the delta area is impacted by both mandrel CDs. In the SAQP process (Fig. 5 (c-f)), the mean area is impacted by both spacer thicknesses, while the delta area is impacted by both mandrel CDs and the 1st spacer thickness. This dependence is depicted Figure 6.

*Table 1: DOE variables and input ranges.*

| | SADP | | SAQP | |
|---|---|---|---|---|
| | Min | Max | Min | Max |
| MX CD (nm) | 26 | 38 | 42 | 50 |
| SPX1 THK (nm) | 5 | 9 | 30.7 | 34.7 |
| SPX2 THK (nm) | - | - | 7 | 9 |
| MY CD (nm) | 26 | 38 | 44 | 52 |
| SPY1 THK (nm) | 5 | 9 | 30.2 | 34.2 |
| SPY2 THK (nm) | - | - | 8 | 10 |

*Figure 5: Contour of mean area and delta area with respect to input process parameters. Plots a-b show SADP results, while plots c-f show SAQP results.*

## PROCESS WINDOW CHECK

A larger mean area and smaller delta are preferred for a higher and more uniform capacitance distribution. A mean area between 900 and 1100 nm$^2$ and delta smaller than 200 nm$^2$ were defined as the success criteria that determined which trials passed or failed. The ratio of passed to failed simulation runs (known as the in-spec ratio) can be calculated for the process window under a specific set of conditions, to generate a mean value and 3-sigma (±3*standard deviation) distribution. This ratio indicates the fraction of input combinations that produce mean and delta areas within the success criteria ranges. Figure 7 shows the calculated in-spec curves for each process input parameter with a 3-sigma specification of 3 nm for the mandrel CDs and 0.9 nm for the spacer thicknesses for both the SADP and SAQP processes.

The in-spec ratio can be optimized by shifting the mean values of the input process parameters in order to maximize the number of successful runs included in the mean±3 sigma window [4]. If the optimized in-spec ratio is still not high enough, specification (3 sigma) tightening can further enhance it. Table 2 shows the calculated In-spec ratio for both SADP and SAQP processes under different conditions. With the same 3 sigma distribution, the in-spec ratio of the SADP process will be about 10%

higher than that of the SAQP process. After the 3-sigma specification for the mandrel CD is tightened from 3 nm to 1.5 nm, the in-spec ratio of the SADP process is close to 100%. The in-spec ratio is only 91% for the SAQP process at this mandrel CD, highlighting that the SAQP process window needs further tightening.

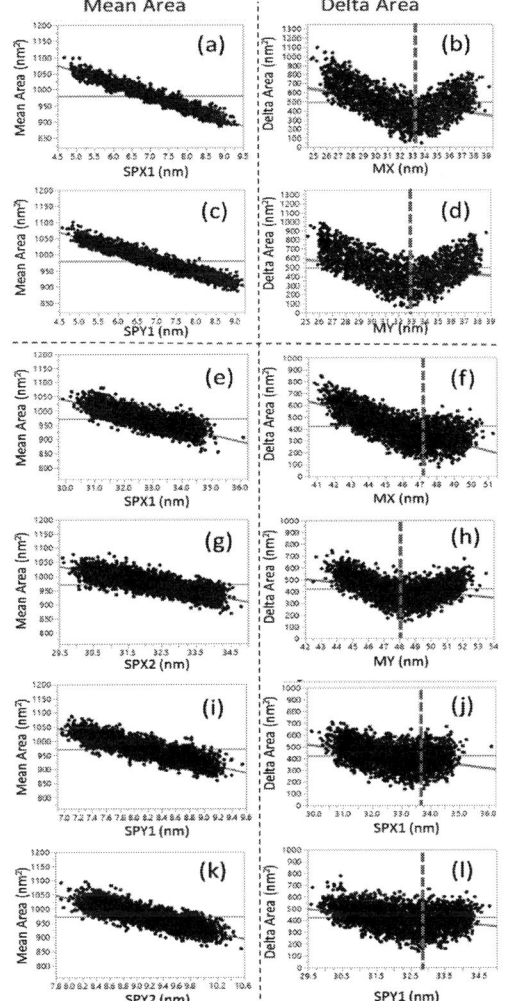

*Figure 6: Leverage plots showing the impact of mandrel CDs and spacer thicknesses on mean and delta area.*

*Table 2: In-spec ratio of the POR, using optimized and specification tightened conditions.*

| Condition \ Variable | SAQP | | | | | | SADP | | | | | |
|---|---|---|---|---|---|---|---|---|---|---|---|---|
| | POR | | Optimization | | Tighten | | POR | | Optimization | | Tighten | |
| | Mean (nm) | 3 S (nm) | Mean (nm) | 3 S (nm) | Mean (nm) | 3 S (nm) | Mean (nm) | 3 S (nm) | Mean (nm) | 3 S (nm) | Mean (nm) | 3 S (nm) |
| MX CD | 45.8 | 3.0 | 46.8 | 3.0 | 46.8 | 1.5 | 32 | 3.0 | 33.8 | 3.0 | 33.8 | 1.5 |
| MY CD | 48 | 3.0 | 48.1 | 3.0 | 48.1 | 1.5 | 32 | 3.0 | 32.4 | 3.0 | 32.4 | 1.5 |
| SPX1 THK | 32.7 | 0.9 | 32.6 | 0.9 | 32.6 | 0.9 | 7 | 0.9 | 6.3 | 0.9 | 6.3 | 0.9 |
| SPY1 THK | 32.2 | 0.9 | 32.6 | 0.9 | 32.6 | 0.9 | 7 | 0.9 | 7.9 | 0.9 | 7.9 | 0.9 |
| SPX2 THK | 8.2 | 0.9 | 8.1 | 0.9 | 8.1 | 0.9 | - | - | - | - | - | - |
| SPY2 THK | 9.2 | 0.9 | 9.3 | 0.9 | 9.3 | 0.9 | - | - | - | - | - | - |
| in-Spec ratio | 40.2% | | 82.7% | | 90.9% | | 65.1% | | 92.4% | | 99.7% | |

## CONCLUSION

In this paper, virtual fabrication was used to perform process window evaluation and optimization during the capacitor formation process of an advanced DRAM structure. The results show that the in-spec ratio of an SADP patterning approach will be 10% higher than that of an SAQP approach under optimized conditions. By tightening the 3-sigma mandrel CD specification, the SADP in-spec ratio can be enhanced to nearly 100% while it is only 91% for the SAQP process, demonstrating the difficulty of optimizing the SAQP process. Unfortunately, an immersion lithography based SADP process will have insufficient physical space when the capacitor hole pitch is scaled to less than 40 nm. Thus, a virtual evaluation of an SAQP patterning approach can provide clear and quantified guidance to help gauge process difficulties in next generation DRAM technologies. Most importantly, virtual fabrication can be used to recommend optimal process target combinations and the largest allowable process window prior to wafer-based experimentation.

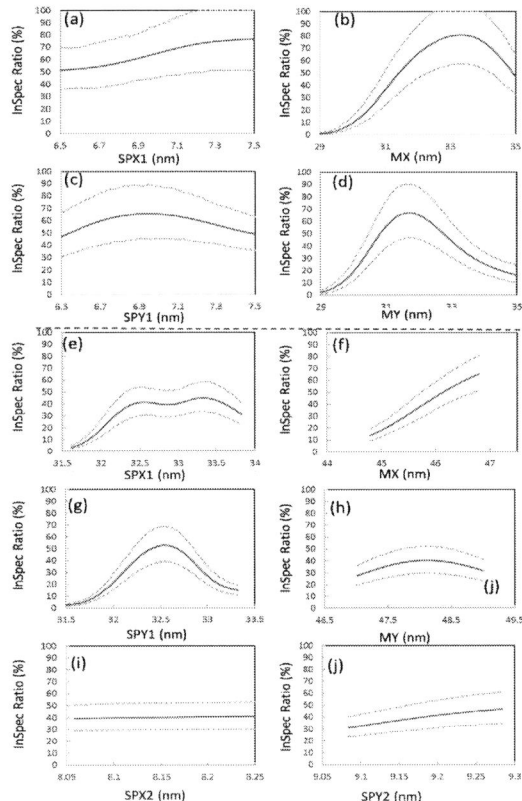

*Figure 7: In-spec ratio curve for each of the different process parameters*

## REFERENCES

[1] A.J., Strojwas, *2006 IEEE International Symposium*

*on Semiconductor Manufacturing* (pp. xxiii-xxxii).

[2] http://www.coventor.com/products/semulator3d

[3] TechInsights Hynix1x DRAM tear down report.

[4] Q. Wang, Y. D. Chen, J. Huang, W. Liu and E. Joseph, *2020 China Semiconductor Technology International Conference (CSTIC)* (pp. 1-3).

# SYSTEMATIC STUDY OF TEMPERATURE AND VDD IMPACT TO READ CURRENT AND STANDBY LEAKAGE OF SRAM IN FINFET TECHNOLOGY

*Yijun Zhang, Yu Li, Yuan Wang*

Technology R&D Center, SMIC

18 Zhangjiang Road, Pudong New Area, Shanghai, P.R. China 201203

E-mail: YiJun_Zhang@smics.com

## ABSTRACT

Static random access memorie (SRAM) are used as cache memory in modern embedded system applications. Read current and standby leakage are used to characterize SRAM speed and power consumption. Read current mainly dominated by N-type transistor device performance, while standby leakage will be an accumulation of all single transistor leakage. SRAM single transistor performance and leakage have strong correlation with $V_{DD}$ and temperature. In this paper, $V_{DD}$ and temperature impact to SRAM read current and standby leakage is systematically studied. When $V_{DD}$ is set from 0.90V to 0.50V, both SRAM read current and standby leakage will become smaller, which can be ascribed to single transistor Idsat and leakage reduce. When temperature change from -40℃ to 125℃, both HD/HC standby leakage increase drastically, while read current show no obvious change. Standby leakage increase is induced by single device leakage trend up.

*Keywords: SRAM, VDD, temperature.*

## INTRODUCTION

SRAM cell stability is one of the key constraints in our ability to continue the constant electric field scaling of semiconductor technology. While providing beneficial concerning area efficiency, it is associated with the challenge of read stability and write ability in low-voltage domains toward process, voltage, and temperature variations [1]. Traditional 6T SRAM cell structure comprises of two cross-coupled CMOS inverters for storing data and two access transistors for data transfer, as shown in Fig.1.

The sizing of access transistors calls for a trade-off between the write ability during write operation and the data retention in read mode. In subthreshold region, a memory circuit faces more challenges because of the high standby power and instability. The main goal at lower power supply is that SRAM cell should perform write, hold and read operations conveniently and achieve power saving as well [2]. However, the mismatch in the strength between transistors of 6T SRAM cell due to process variations can results failure during read operation, especially at low-$V_{DD}$ level. In addition, as CMOS technology scaled down, the total leakage of SRAM chip will be increased since it include large number of transistors. The SRAM leakage has also become a more significant component of total chip leakage in scaled CMOS technology [3]. In this study, different $V_{DD}$ and temperature impact to SRAM read current and standby leakage is systematically researched.

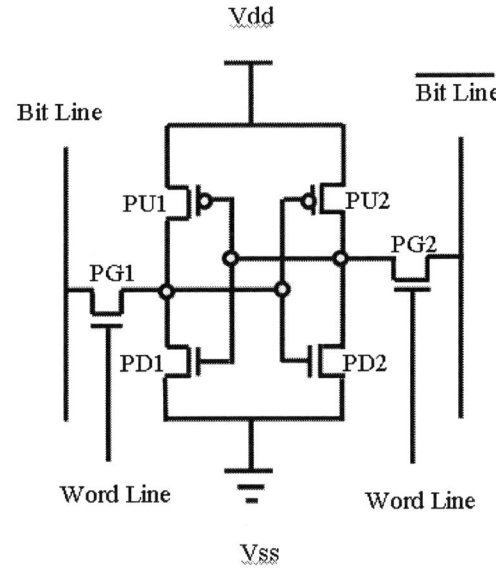

*Fig. 1: Illustration of 6T SRAM circuit.*

## EXPERIMENTAL

The MOSFET samples in this paper were fabricated by FinFET process. Experimental condition of different $V_{DD}$ and temperature is setting as Table. 1 and Table. 2.

Table. 1. Experimental design details of different Vdd

| HD | 0.5V | 0.68V | 0.75V | 0.90V |
|----|------|-------|-------|-------|
| HC | 0.5V | 0.68V | 0.75V | 0.90V |

Table. 2. Experimental design details of different temperature

| HD | -40℃ | 25℃ | 85℃ | 125℃ |
|----|------|-----|-----|------|
| HC | -40℃ | 25℃ | 85℃ | 125℃ |

Both high density (HD) and high current (HC)

978-1-6654-9759-6/22 $31.00 © 2022 IEEE

SRAM cell is studied, layout diagram is shown as Fig. 2. HD SRAM cell possesses of 1Fin PU, 1Fin PD and PG, while HC SRAM cell possesses of 1Fin PU, 2Fin PD and PG.

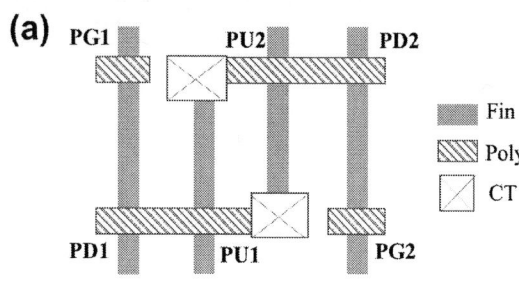

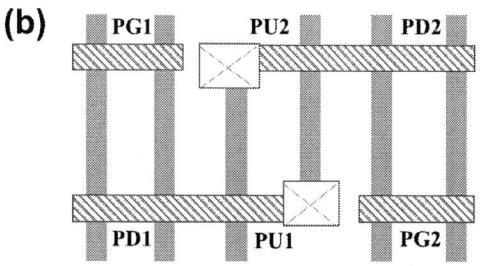

Fig. 2: SRAM 6T cell layout diagram: (a) high density SRAM with 1Fin PU/PD/PG; (b) high current SRAM with 1Fin PU and 2Fin PD/PG.

## RESULT AND DISCUSSION

### I. Different $V_{DD}$ impact to 6T SRAM

In SRAM design, the retention mode has drawn attention because substantial leakage reduction can be achieved by applying aggressive power saving strategies. A common approach is using SRAM array $V_{DD}$ scaling, where a voltage regulator supplies the SRAM either internally or externally based on a scaled reference voltage source [4]. Different $V_{DD}$ impact to SRAM HD read current and single device Idsat are shown in Fig. 4a. When $V_{DD}$ change from 0.9V to 0.5V, both PU/PD/PG Idsat reduce, which can be explained by below equation:

$$\text{Idsat} = \frac{1}{2}\frac{W_{eff}}{L_{eff}} \times C_{ox} \times (V_{GS} - V_T)^2 \qquad (1)$$

$W_{eff}$: effective device width;
$L_{eff}$: effective device length;
$C_{ox}$: device gate oxide capacitance;
$V_{GS}$: gate voltage;
$V_T$: threshold voltage.

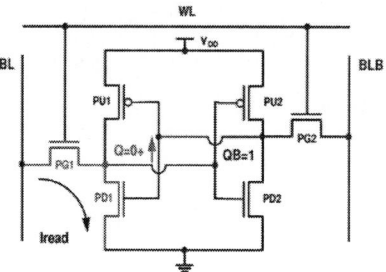

Fig. 3: 6T SRAM work states in read operation

Decreased $V_{DD}$ represents decreased $V_{GS}$, which lead to smaller device Idsat. According to Fig. 3, SRAM read current is dominated by PD and PG Idsat, which can explain the decreased read current of HD SRAM.

Different $V_{DD}$ impact to SRAM HD single device and standby leakage are shown in Fig. 4b. When $V_{DD}$ reduce from 0.9V to 0.75V, HD standby leakage and single device leakage show obvious reduce, while they keep stable if $V_{DD}$ declined further. Main sources of leakage current in CMOS transistor are reverse-biased junction leakage current, sub-threshold leakage current, oxide tunneling current, gate current due to hot carrier injection, gate induced drain leakage and channel punch through current. Out of these, sub-threshold leakage and gate leakage currents are the dominant ones in a SRAM cell [5]. During WAT test, both $V_{GS}$ and $V_{DS}$ are connected to $V_{DD}$. In high voltage, both sub-threshold leakage and gate leakage will reduce when $V_{DD}$ change from 0.90V to 0.75V, which can explain the decline of single device leakage and standby leakage. Beyond that, the experiment results indicate that SRAM leakage is not sensitive to lower voltage (from 0.75V to 0.50V).

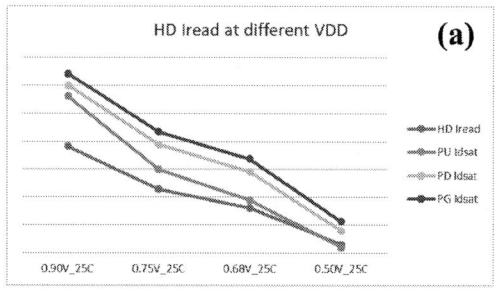

Fig. 4(a): HD read current and PU/PD/PG Idsat at different VDD

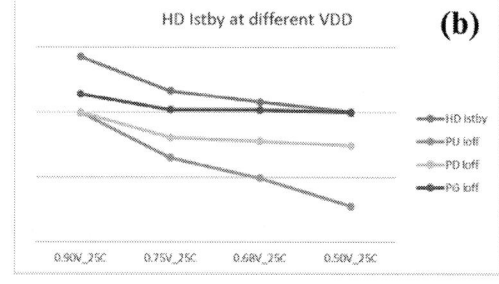

Fig. 4(b): HD standby leakage and PU/PD/PG Idsat at

*different VDD*

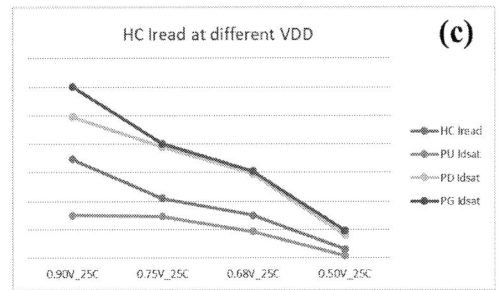

Fig. 4(c): HC read current and PU/PD/PG Idsat at different

*VDD*

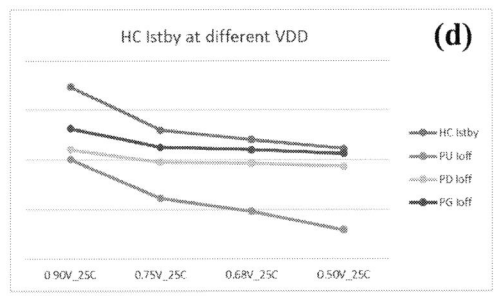

Fig. 4(d): HC standby leakage and PU/PD/PG Idsat at

*different VDD*

Different $V_{DD}$ impact to SRAM HC read current and standby leakage are shown in Fig. 4c and Fig. 4d. And the experimental results are similar to HD SRAM.

## II. Different temperature impact to 6T SRAM

The temperature impact to SRAM HD/HC read current and standby leakage is shown in Fig. 5. According to Fig. 5a and Fig. 5c, HD/HC read current is not sensitive to temperature change, which can be explained by Equation. 1, single device Idsat not directly related to temperature. Temperature impact to HD/HC standby leakage are shown in Fig. 5b and Fig.5d. It is obvious that SRAM single device leakage will significantly increase when temperature increase from -40℃ to 125℃. In high temperature, both sub-threshold leakage and gate leakage currents will trend up as doping ions diffusion become worse. As SRAM standby leakage is the sum of single device leakage, it also will trend up when temperature increase.

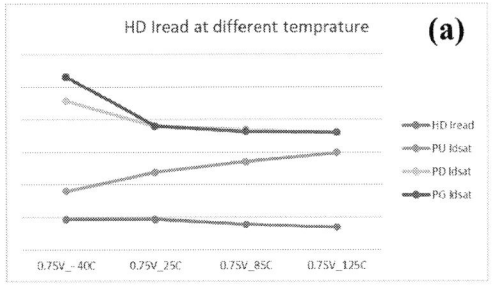

*Fig. 5(a): HD read current and PU/PD/PG Idsat at different*
*temperature*

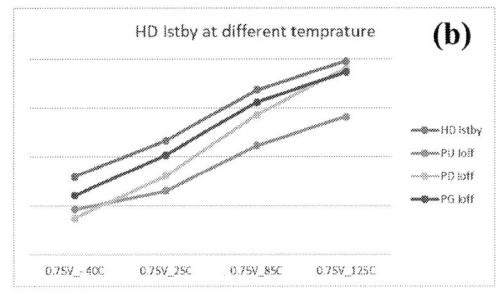

Fig. 5(b): HD standby leakage and PU/PD/PG Idsat at
*different temperature*

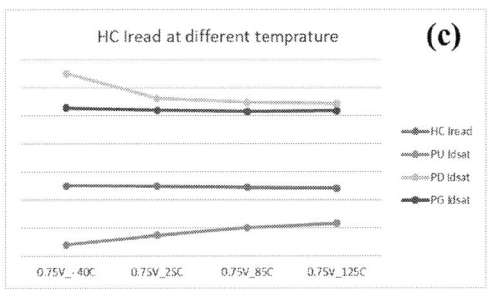

Fig. 5(c): HC read current and PU/PD/PG Idsat at different
*temperature*

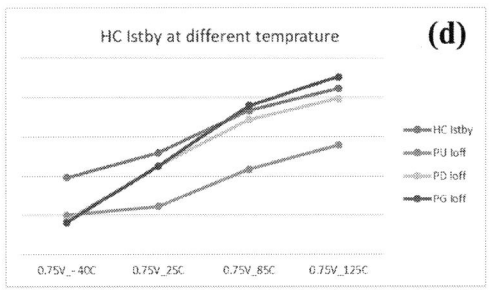

Fig. 5(d): HC standby leakage and PU/PD/PG Idsat at
*different temperature*

## SUMMARY

Different $V_{DD}$ and different temperature impact to read current and standby leakage of FinFET 6T SRAM are systematically studied. As single device Idsat has positive correlation with $V_{DD}$, both HD and HC read current will reduce when $V_{DD}$ turns down. HD/HC standby leakage also will reduce when $V_{DD}$ decrease, while it is obviously at high voltage but not sensitive at low voltage. The temperature impact to SRAM HD/HC read current and standby leakage also studied. According to the experimental results, HD/HC read current show no obvious sensitivity to temperature change. As single device leakage will increase when temperature turns up, both HD and HC standby leakage will significantly enlarge.

## REFERENCE

[1] Liang Wen, Yuejun Zhang, Xiaoyang Zeng, IEEE Transactions on Very Large Scale Integration (VLSI) Systems, pp. 1470-1474, 2019.

[2] Rajani Suthar, Kirti S. Pande, Murty N.S., 2017 IEEE International Symposium on Nanoelectronic and Information Systems (iNIS), pp. 215-219, 2017.

[3] Arash Azizi Mazreah, Mohammad Taghi Manzuri Shalmani, Ali Mehrparvar, 2009 Fifth International Conference on MEMS NANO, and Smart Systems, pp. 33-36, 2009.

[4] Cyrille Dray, Nabil Badereddine, Christophe Chanussot, 2010 IEEE International Memory Workshop, 2010.

[5] Akanksha Pandey, Praveen Kumar Sahu, R. Dwived, et al. 2019 International Conference on Computing, Power and Communication Technologies (GUCON), 2019.

# T2000 ISS FT SOLUTION USING CP LIGHT SOURCE

*Hiroyuki.Kobayashi [1*], Bin.Gong [1], Senter.liu[1], Shenqi.Cai[1],*
[1]Advantest (China) Co., Ltd, Shanghai, China
*Email: hiroyuki.kobayashi@advantest.com , bin.gong@advantest.com

## ABSTRACT

We introduce the T2000 ISS final test (FT) solution using a chip probing (CP) light source for high-end image sensor package devices. The image sensor market is expanding day by day due in part to the influence of smartphones. Although charge-coupled devices (CCD) used to be the mainstream, CMOS Image Sensor ( CIS ) now account for the great majority of the market. While image sensor for smartphones now account for the great majority of shipments, demand for high-performance digital cameras such as single-lens reflex cameras is still alive and well, with higher image quality, faster data rates, and increased output pins. Progress has been made in developing technology that can capture the same or higher level of image quality as such images, and this technology has now matured.

C-PHY 3.5Gsps simultaneous testing of up to 64 has been successfully measured in CP on an experimental trial. However, in the mass production of CIS, few tests are using a light source in FT. Or, even if a light source is used, most of the tests are done using a simple light source. In this paper, we introduce an automated test equipment (ATE) developed in cooperation with handler, socket, and light source vendors to meet the demand in the high-end digital camera market for the same level of testing in FT as CP. By using the same light source as that used in CP, there is no difference in performance. The light source is installed inside the test head, CIS need to orientation is the opposite of normal. In terms of temperature control, if the test items are the same as CP, it will take longer to device test and it will be difficult to keep the temperature constant in high temperature test. This solution solves these problems and achieves the same level of testing in FT as CP.

## INTRODUCTION

The traditional package device test is to directly lead the resources from the test head to the socket board on the performance board (PB). However, the CIS test needs the illumination of the light source. We need to consider the location and specifications of the light source. To ensure uniformity, our solution uses a halogen light source in CP. The light source is installed in TH to save space. So, we need to flow out an area on PB to allow the light source to pass through. To receive the light irradiated by the light source, we need to adjust CIS position so that the light-receiving area faces the light-emitting area. We specially designed a shield and socket, to make the light source better illuminate the packaged devices. To achieve a better effect of the light shining on the devices, we installed a lens under the socket.

T2000 ISS includes 4.8GICAP as an image capture module specially designed for image sensor testing. MIPI D-PHY Ver2.1 is supported as image protocol(up to 4.8Gbps),and MIPI C-PHY Ver1.2 is supported(up to 3.5Gsps).Beside, 4.8GICAP supports testing C\D-PHY Combo without hardware switching. T2000 ISS is specially equipped with IP Engine3 with extremely high processing speed for image processing test items.

## DISCUSSION
### FT & CP Solution Structure Compare

As we know, the test objects of FT and CP are different, so the construction of the test system is different. For the wafer test, in addition to the ATE test system (mainly including the Mainframe cabinet, Test head), a handler and a light source are also needed. The handler is responsible for controlling and moving the wafer and the probe to the next area to be measured in the wafer. The light source provides light excitation for the pixel. The light source and ATE communicate through GPIB.

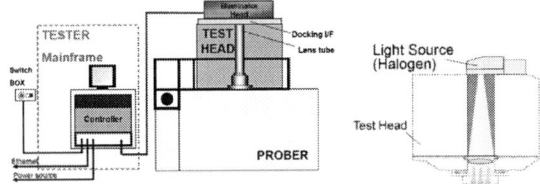

*Figure 1: Structure diagram of CP*

The output of the light source can be controlled by T2000 ISS in the program. To meet the capture image needs during image processing, the selection of light source should also meet a certain specification. Generally speaking, Halogen and LED are used in the selection of light sources, To ensure that the prober is moving, Light covers the measured area on the wafer all the time, so the lens tube of the light source needs to cover the whole wafer. Compared with LED, the halogen wavelength is closer to natural light, which will be more beneficial to the CIS test. Besides, The image quality of the halogen lamp is higher.

978-1-6654-9759-6/22 $31.00 © 2022 IEEE

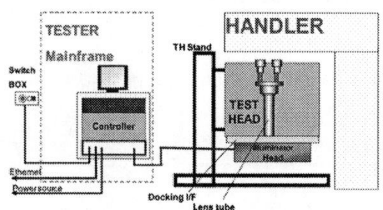

*Figure 2: Structure diagram of FT*

For FT, The optical fiber branch is introduced in the light source part because the packaging makes the devices under test need to be set in different sockets, so the light intensity received by each device is much smaller than that of CP. Which is hard to solve by the general solution. We would like to take the following MIPI D-PHY device as an example to introduce how this solution solves this concern.

### Solution for CP light source

By comparison, it can be found that the light source configuration scheme under the wafer test and FT is greatly different. Another fact is that at present, the domestic CIS test for CMOS is mainly focused on-wafer test. If the light source scheme of the wafer test is modified to meet the demand of the FT light source scheme, the development cost can be greatly reduced.

It can be found that in the above FT, the light received by each socket device is through a branch. However, inevitably, the introduction of branch will lead to a decrease in light intensity. Therefore, in order to ensure that the light intensity is not used by the branch, the device under test (DUT) must be placed within the illumination area of the light source. The following diagram shows the device schematic diagram of the current test project. After size comparison, up to 4 sites can be tested at the same time.

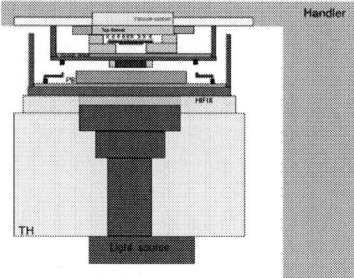

*Figure 3: Structure diagram of using CP light source for FT*

Since the light source is housed in the test head, the orientation of the device is the opposite of normal. For this reason, a Printed circuit board (PCB) is built into the socket to contact the device.

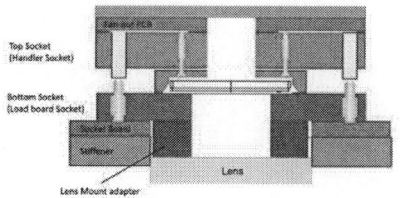

*Figure 4: Structure diagram of socket*

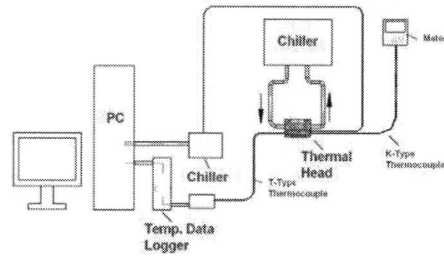

*Figure 5. Temperature control*

The size of the device is also called medium size, which requires a solution to keep the temperature uniform during high-temperature testing. If the device is heated and then transported for testing, the temperature cannot be maintained at a constant level. A certain amount of time is required for image processing. The package type of this FT device is land grid array (LGA), and we adopted a solution that uses warm water to heat the device from the back.

## STATUS AND RESULT

Because of the introduction of this "cooperation-type" structure, the final MIPI signal has been captured need to through a longer path, which is from the socket contact pad, pass through socket board and Cable, to PB. For this long trace, we adopted a thin, flexible, high-speed cable，which can support up to 32Gbps high-speed signal transport. By checking the S parameter result of PB and cable, the internal loss after the long trace can be ignored, besides, we measure the eye diagram of the target tested speed(1.8Gbps) signal, from the result we can know the signal integrity was not compromised. Furthermore, we increase the signal rate and finally confirmed the 2-channel D-PHY 2.5Gbps signal still can be captured by the 4.8GICAP module.

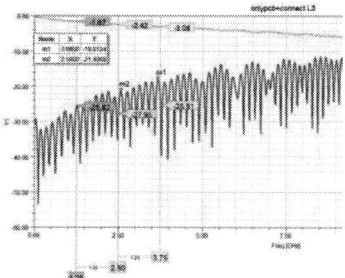

*Figure 6: PB and cable S11&S21*

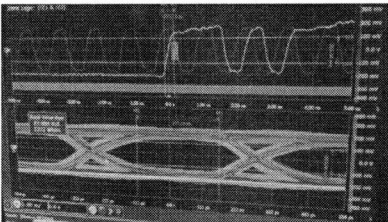
*Figure 7. 1.8Gbps device output eye*

## FUTURE CHALLENGE

For us, it was the first time to use a device that requires high uniformity and luminous intensity for the light source, and we will make sure that the FT solution using CP light source can stably capture images and be mass-produced. CIS requires faster and lower-cost test solutions. In the future, we will take on the challenge of multi-site measurement and develop a solution with a better cost of the test (COT). This time, it is a high-end product with single sight measurement, but it will support multi sites measurement in the future.

## CONCLUSION

In this paper, the T2000 ISS FT solution uses CP light source for high-end image sensor package devices, allowing the FT solution to perform the same testing as the CP solution. This means that T2000 ISS is a solution that can support both CP and FT. We evaluated the ability to capture high-speed signals in the current simulation. We will confirm the accuracy and reproducibility online to be ready for mass production. In the future, we will conduct the multi-sites test to evaluate D-PHY 4.8Gbps C-PHY 3.5Gsps, which is within the specification range of 4.8GICAP. We will also consider replacing such a cable-less method, to support high-speed devices.

## ACKNOWLEDGEMENTS

We would like to acknowledge the support from the company for providing us with the tester to professional advice. We would like to acknowledge every colleague who had helped us but cannot be listed here.

## REFERENCES

[1] S.Soma *"Introduction of 4.8Gbps Image Capture Module"*, Global SE Workshop for T2000 and T6391, 2019.

# IMPACT OF TOPOLOGY OF TRENCH GATE BOTTOM CORNER FOR POWER MOSFET AND IGBT

*Min-hwa Chi[1], Janifer Liu, and Perry Li*
SiEn (Qingdao) Integrated Circuits Cor., Shandong, China 266500
[1]Micro-Nano Technology College, Qingdao University, Qingdao, Shandong, China 266071
Email: 18017378580@qq.com

## ABSTRACT

Power ICs and devices, as key components for high performance systems with Artificial Intelligence and Internet-of-Things (AI/IoT) (e.g. Data centers, Smart cars and autonomous driving, Robotics, Industry 4.0, etc.), need superior quality and reliability but also super low on-state resistance (Ron) with good breakdown voltage (BV). This paper illustrates how to enhance the BV and reduce Ron of power MOSFETs and IGBTs by optimizing the shape of the bottom corner of trench gate.

## INTRODUCTION

Various prior arts for achieving better trade-offs among bottom oxide thickness, Ron (Vsat) and BV of power MOSFETs and IGBTs are complicate in processes and structures. We need a deeper understanding and smarter scheme of the trench-gate structure and process. The conventional concept of rounded bottom (inner) corners does not seem the main factor to enhance the oxide thickness than others, (e.g. plasma damage, implant species, crystal surface orientation, etc.); furthermore, the rounded inner bottom corner is not clearly improving BV and Ron. On the contrary, we found through TCAD study by simulation that an outward notch shape of the trench gate bottom corner can boost BV significantly and may enhance oxide reliability (due to thicker oxide at the inner corner) and reduce Ron (due to enhanced field at the sharper notch tip of poly gate in the bottom corner of trench gate).

## POWER MOSFET AND IGBT OVERVIEW

The Si-based power devices, such as power MOSFETs and IGBTs are established their important role from low to high voltage and power applications. Currently, they are widely used in power modules based on CMOS technology platforms. IGBT was developed in ~1980 as a superior alternative to bipolar power transistors. IGBT [1] is an integrated structure of power MOS and bipolar transistors; when the power MOS structure is turned on for providing the base current and turn-on the integrated PNP bipolar power transistor; then the PNP bipolar transistor can further modulate the conductivity of the drift region of the MOS structure and achieve low forward voltage drop. In recent years, IGBTs are applied to most power electronic applications, especially medium and high power equipment, such as AC drive motion control, UPS, renewables (wind and solar), and others. In the last quarter of centuries, IGBT has been playing a leading role for controlling the power electronics equipment through the improvement of IGBT chips manufacturing and packaging technologies. A state-of-art IGBT (with integrated power MOSFET) structure and process flow with critical process modules are illustrated in Figure 1.

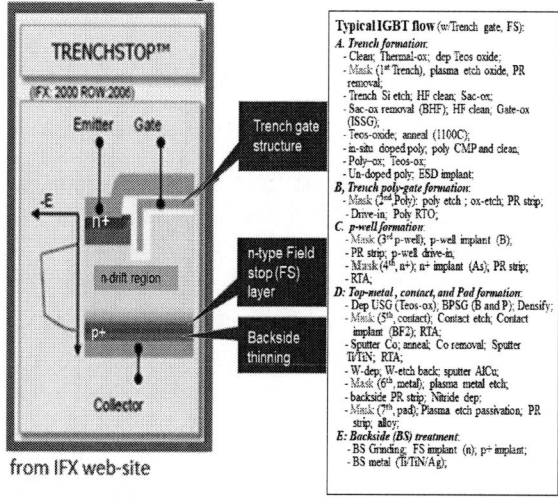

*Figure 1: (a) Modern IGBT (with integrated power MOS) has advanced features of trench-gate (TG), field-stop (FS), and backside thinning and metal sintering for optimized performance [2]. (b) A simplified IGBT process flow.*

A typical modern IGBT cross-section is illustrated in Figure 1 with process flow and critical process modules (e.g. Trench-gate formation, Backside (BS) thinning and metal sintering, and Field-stop (FS) layer formation). The trench sidewall angle (~87 - 88°) and bottom corner profiles are critical for the Vt and stress near the trench bottom corners. The uniformity of trench parameters (e.g. critical dimension (CD), trench depth, trench sidewall slope, the thickness of drift and FS layers, etc.) are all critical for IGBT's reliability and performances.

## MODERN TRENCH GATE PROCESS AND DEVICE SIMULATION

The trench gate is formed by firstly patterning the substrate and plasma etching Si as shown in Figure 2 (left); then, after forming the gate oxide and filling with doped poly-Si and followed by CMP planarization, the final trench structure with sharp bottom corner is shown in Figure 2 (right). The conventional process results in thinner oxide at the bottom of inner corners [3]; this in turn may lead to device break down at the bottom corners as illustrated in Figure 3. In this case, the simulation of this trench MOSFET showed that BV is ~739V (@turn-off), and $V_{CE}$=2.7V (@turn-on). The oxide on the trench sidewall (110 surface) is thicker than bottom (100) surface as well known. For smaller CD of the trench bottom, the weak spots of thinner oxide at trench bottom corners are related to the thermal oxidation process (with the stress at corner as well as limited oxygen diffusion into the inner corner). For larger CD of trench bottom, the thinner oxide (then sidewall) is mainly due to the (100) surface of the trench bottom. The oxide thickness at bottom corner is same as on sidewall.

The Si oxidation rate can be enhanced by ion implantation at the trench bottom (e.g. implanting by Ar, As, P, O, etc.) using angled implantation techniques. A double Sac-ox can also reduce the bottom oxide thinning effect [4].

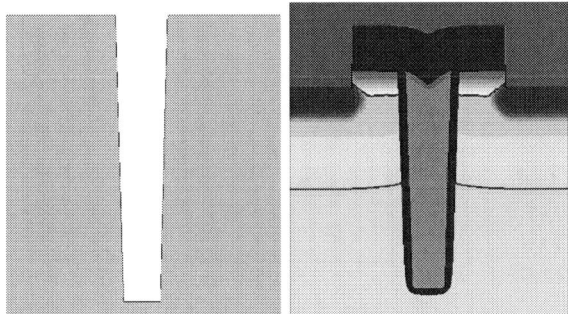

*Figure 2: Conventional trench etching with sharp bottom corner (left) and final structure after doped poly-Si filling and planarization (right).*

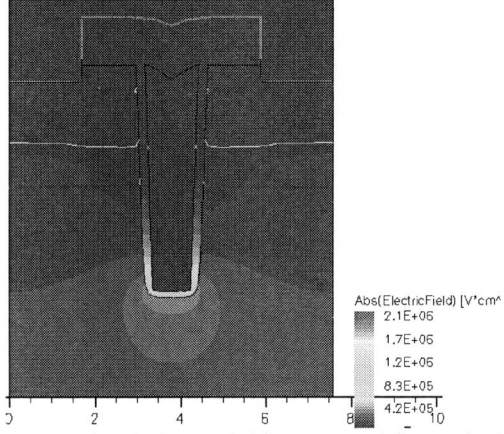

*Figure 3:* Simulated *electric field intensity of the resulted trench structure with sharper bottom corner. Higher field occurred at the inner corner portion.*

## TRENCH GATE BOTTOM CORNER ROUNDING PROCESS AND DEVICE SIMULATION

One typical way to improve the bottom oxide thinning effect is to form the trench gate with enough bottom corner rounding e.g. tuning etching parameters. The improved process results in less thinner oxide at the inner corners but still thin on the bottom flat portion as illustrated in Figure 4. The simulation of the improved trench MOSFET showed that BV is ~729V (@turn-off), and $V_{CE}$=2.65V (@turn-on) with higher field occurred at the bottom flat portion as in Figure 5. The BV and $V_{CE}$ are not improved with the rounded trench bottom corners.

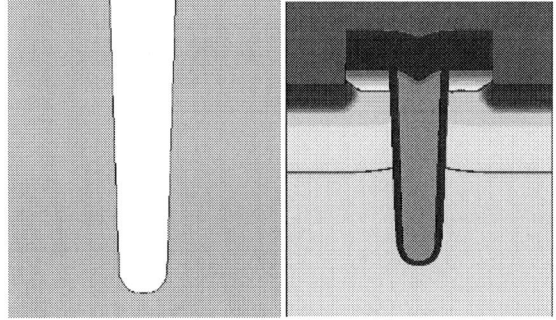

*Figure 4: Improved trench etching with rounded bottom corner (left) and final structure after doped poly-Si filling and planarization (right).*

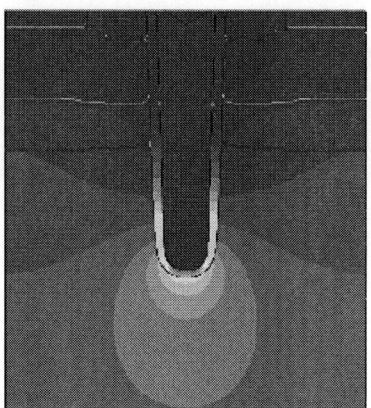

*Figure 5:* Simulated *electric field intensity of the resulted trench structure with rounded bottom corner. Higher field occurred at the bottom flat portion.*

## NOTCH SHAPE AROUND TRENCH BOTTOM AND DEVICE SIMULATION

We found a new method to eliminate the bottom thinning effect by forming an outward notch shape (with characteristic angle $\alpha$) at the trench bottom corner by tuning etching parameters as in Figure 6. (Left). The simulation of the new notch-shape trench MOSFET showed that BV is ~788V (@turn-off), and $V_{CE}$=2.73V (@turn-on) with lower field around the notch of the trench bottom as in Figure 6. More simulation data with $\alpha$ as a parameter are shown in Table 1. The notch angle ($\alpha$) is proposed to be <80° for best BV and process capability.

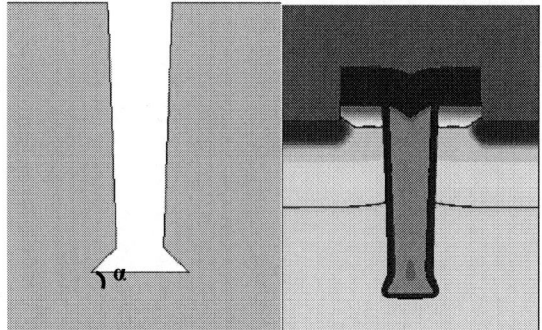

*Figure 6: New trench etching with outward notch shape at the bottom corner (left) and final structure after doped poly-Si filling and planarization (right).*

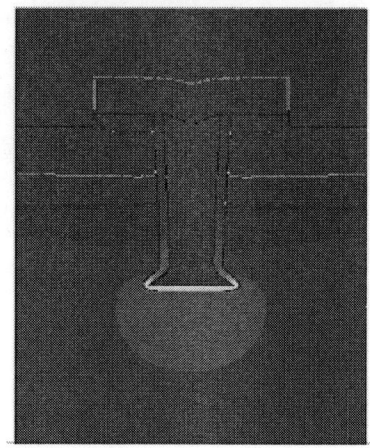

*Figure 7: The simulation shows that lower electric field intensity around the new notch shape ($\alpha=44^o$) at the trench bottom corner.*

Table-1: The electrical performances of notch shape trench by changing the Angle of alpha.

| N. | TYPE | BV/V | $V_{CE}$/V | Max E./V*CM-1 |
|---|---|---|---|---|
| 1 | RIGHT ANGLE | 739 | 2.700 | 2.10E+06 |
| 2 | ROUNDING | 729 | 2.650 | 1.40E+06 |
| 3 | $\alpha$=84 | 765 | 2.500 | 2.20E+06 |
| 4 | $\alpha$=74 | 770 | 2.600 | 1.60E+06 |
| 5 | $\alpha$=64 | 772 | 2.640 | 2.40E+06 |
| 6 | $\alpha$=52 | 782 | 2.690 | 8.80E+06 |
| 7 | $\alpha$=44 | 788 | 2.730 | 8.00E+06 |

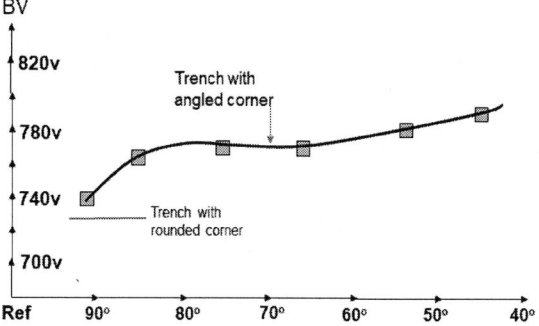

*Figure 8: The simulated BV vs notch angle ($\alpha^o$). Smaller notch angle result in significantly larger BV with only small effect on Ron.*

The mechanism of the BV enhancement by the outward notch shape at bottom corners is discussed further here. After turn on into on state, the higher electric field around the sharper poly-gate can induce more accumulation electrons in the drift region adjacent to the notch corner, thus, the Ron is reduced slightly, though the notch corner occupied a little bit room of the n-drift region. The net effect appears a small effect on Ron. After turn-off, the breakdown voltage increased. Firstly, the electric field in the depletion region is uniformly distributed for improving the BV. Secondly, the Si-oxide inside the notch shape corner appears thicker and more resistant to breakdown than Si. The sharper corner at the notch does not contribute to the electrical field during off as the field around the notch is determined by the ionized dopants in the depletion layer outside the notch.

## CONCLUSIONS

As illustrated in this study, the notch shape at trench bottom corner for power MOSFET and IGBT is effective to enhance BV with negligible effect on Ron. It is simple and easy to implement and there is no need of special trench bottom rounding, double oxidation, or complex epi process for super junction formation, etc.

## ACKNOWLEDGEMENTS

The authors would like to thank SiEn and Qingdao University for their supporting.

## REFERENCES

[1] N. Iwamuro and T. Laska, "IGBT History, State-of-the-Art, and Future Prospects", IEEE Trans. Electron Devices, V.64, No.3, p.741, 2017.
[2] http:\\www.Infineon.com.
[3] Ng Hong Seng , "Review on Methods for Trench MOSFET Gate-ox Reliability and Switching Speed Improvement", ECS, 27 (1), p.21-26, 2020.
[4] Ng Hong Seng , "Review on Methods for Trench MOSFET Gate-ox Reliability and Switching Speed Improvement", ECS, 27 (1), p.21-26, 2020.

# INVESTIGATION OF REMOVING STANDING WAVE EFFECT DURING LITHO PROCESS

*Xueqiang Liu[1]\*, Biqiu Liu[1], Xiaobo Guo[1], Shirui Yu[1]*
[1]Shanghai Huali Integrated Circuit Corporation,Shanghai 201317,China
\*Corresponding Author's Email: liuxueqiang@hlmc.com

## ABSTRACT

Standingwave effect could produce poor PR profile thus influence following etch process. Here, we try to decrease reflectivity of PR bottom to ensure low standingwave effect. In order to controll reflectivity of PR bottom, several factors including source, film scheme and n, k of SOC materials are investigated thoroughly. It is found that reflectivity of PR bottom is greatly increased when polarization is introduced. This is because TE mode light contributes to total reflectivity much larger than TM mode light reflectivity. Compared to bi_layer scheme, tri-layer scheme not only has lower reflectivity, but also is seldom influenced by films under SOC. Thus, tri-layer scheme is more picked up in HuaLi advanced processes. For Huali MEOL litho process, through optimization n, k of SOC, low reflectivity of PR bottom (<0.5%) is got. Finally, it proves both good PR profile in litho process and vertical hardmask profile in etch process.

## INTRODUCTION

AS Mole law predicts, the size of IC chip will be half smaller in dimension every one and half year. LITHO process is the first patterning process. The resolution of litho process determines dimension of full chip, thus lithographer has undergone iline, KrF, ArF and now EUV[1]. For litho process, there are several fundamental requirements to mass_production, including good PR profile, large process window, less defect, good overlay control and so on. The first indispensable requirement is good PR profile in that following etch process will probably fail if PR profile is poor. It is ideal situation that PR profile is vertical so that the following etch process could also produce vertical pattern profile.

LITHO process could be decomposed into two procedures, photon absorption and chemical reaction amplification. Chemical reaction amplification during development is a complicated process. Here, we mainly focused on photon behavior in photon resist. During advanced technology litho process development in Huali, several cases appear that profile is poor induced by standingwave effect. Standing wave effect is caused by strong interference and diffraction between incident light and reflective light on the bottom of PR, as shown in fig1.

The key to remove standing wave effect is to reduce the magnitude of reflective light, which is influenced by several factors including source information, PR scheme, refractive index of film material n, and absorption coefficient of film material k. Two kinds of PR scheme are popularly picked up in FAB nowadays. One is PR and BARC Bi-layer scheme, which is usually picked up before 28nm node. The other one is PR, BARC and SOC tri_layer scheme, which is widely picked up in advanced node. In this paper, reflectivity of PR bottom of both these two schemes are investigated. Influence of n, k of PR, BARC and SOC material are also investigated in detail. Through optimizing these parameters, good PR profile is obtained in advanced technology node MEOL process.

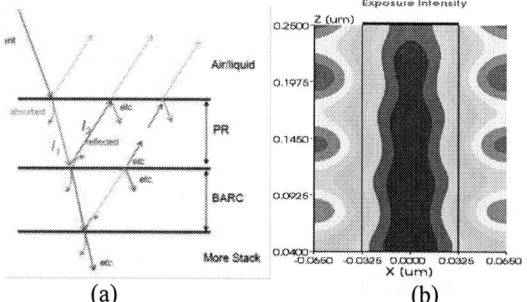

Fig.1:(a) Incident light diffracted with reflective light. (b). Light intensity distribution among PR in vertical dimension.

## BASIC THEORY AND EXPERIMENT METHOD

The reflectivity is simulated by SLITHO software from Synopys company. Litho process is perfomed in scanner ASML 1980i tool. PR image is collected on Hitachi CDSEM 6300.

## RESULTS AND DISCUSSION

### Influence of source polarization on reflectivity

BARC is short for bottom anti-reflective coating. It is widely used in film scheme to reduce reflective light into PR because it has high absorption k[2]. Although composition of BARC material is similar to that of PR, it is not Photon-sensitive. Thus it should be removed by etch process. When BARC is removed during etch process, similar thickness of PR is also consumed due to similar property. Therefore, we should use as thin as possible BARC material to get a low reflectivity of resist bottom.

Fig.2 shows reflectivity of resist bottom varies as thickness of BARC changes. Reflectivity of resist bottom firstly decrease to a valley, then increase to a peak and at last decrease to a second valley, which is called "swing curve". Trend of this reflectivity curve is same with light intensity distribution in PR as shown in fig.1(b). It is caused by diffraction and interference between incident light on top of BARC and reflective light from top of BARC, as shown in fig1. Thus the ideal thickness of

BARC material is usually location of the first valley in fig.2. And it is 0.028μm here. Its corresponding reflectivity is 0.1%.

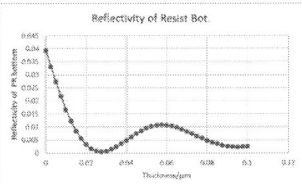

Fig.2: Reflectivity of resist bottom variation with thickness of BARC

Light is composed of TE and TM mode, which is distinguished by oscillation direction. TE mode oscillated perpendicular to the incident light plane while TM mode oscillated parallel to the incident light plane. Fig.2 shows reflectivity of resist bottom, TE mode reflectivity of resist bottom and TM mode reflectivity of resist bottom vary as thickness of BARC changes. It could be seen that reflectivity of resist bottom is sum of reflectivity of TE and TM mode. Furthermore, is is obvious that TE mode light contributes more to overall reflectivity of resist bottom.

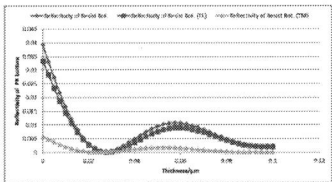

Fig.3: Reflectivity of resist bottom, TE mode reflectivity of resist bottom and TM mode reflectivity of resist bottom variation with thickness of BARC.

Source polarization is widely adopted to improve resolution in advanced lithography technology. There are several kinds of polarization method, including X-direction, Y-direction, circular-direction and Sector-direction. And sector direction polarization method is most widely adapted in that it could improve contrast of patterns oriented in both X and Y directions. However, polarization is not friendly to the reflectivity of resist bottom in that TE mode light intensity will be largely improved thus overall reflectivity will be also improved. Fig.4 are reflectivity comparision between no-polarizaton and sector polarization. It could be seen that after sector polarization is introduced, TE mode reflectivity of resist bottom is largely increased while TM mode reflectivity is decreased substantially. As a result, total reflectivity of resist bottom is largely improved after introduction of sector polarization. Thus it is more challenging to control reflectivity of resist bottom in advanced technology where polarization is widely adapted.

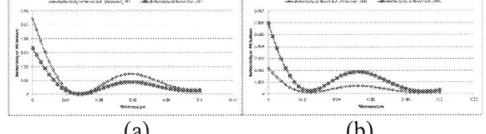

(a)　　　　　　　(b)

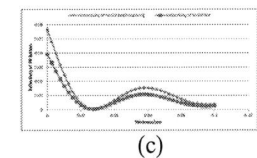

(c)

Fig.4: (a) Comparision of TE mode reflectivity between no_polarization and sector_polarization; (b) Comparision of TM mode reflectivity between no_polarization and sector_polarization; (c) Comparision of overall reflectivity between no_polarization and sector_polarization;

**Comparision of bi_layer and tri-layer scheme**

It has been discussed that BARC is used as the anti-reflective layer to reduce reflectivity of PR bottom. PR and BARC bi_layer scheme is widely picked up before 28nm mode. When it comes to below advanced technology node, only BARC material could not satisfy the requirement of reflectivity. Thus a second anti-reflecitve layer is introduced, SOC(spin-on-carbon). On one side, reflectivity of resist bottom could be controlled below 0.5%. On the other, it is easier to transfer pattern from PR downwards in that its etch rate is 20~30 times larger than that of BARC.

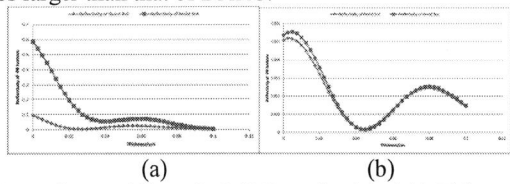

(a)　　　　　　　(b)

Fig.5. Influence of film under BARC on reflectivity in(a) Bi-layer scheme and (b) tri-layer scheme

Here, reflectivity of bi_layer and tri-layer scheme are investigated in detail. In bi_layer scheme, as shown is fig.5(a), reflectivity of resist bottom increased dramatically when film under BARC is removed. It means that reflectivity of resist bottom in bi_layer scheme is greatly influenced by under-BARC films. In bi_layer scheme, in order to get a low reflectivity, high k material film should be considered, thus choices of films under BARC are limited. While in tri-layer scheme, as shown in fig.5(b), reflectivity of resist bottom is easy to keep below 0.5%. Furthermore, reflectivity is little influenced by films under SOC. Thus, there are more choices of films for tri-layer scheme. As a result, tri-layer scheme is more popular below advanced technology node.

**Tri-layer scheme in advanced technology Finfet MEOL process**

Table.1: SEM image of CMD layer pattern with different tri-layer scheme

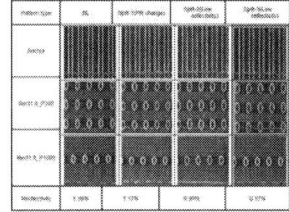

978-1-6654-9759-6/22 $31.00 © 2022 IEEE　　　61

In advanced technology Finfet process, CMD layer is used as block layer to cut lines into holes. Thus profile of CMD layer is very critical. It should be as vertical as possible. When litho process of CMD layer is developed, a very terrible case appeared that profile of patterns are poor as shown in table 1. Obviously, such poor profile could not meet requirement of process. Here, reflectivity of resist bottom is 1.16%. Thus, it is doubted that the bad profile is caused by standing-wave effect owing to high reflectivity of resist bottom. In split-1 experiment the PR material is changed. However, the PR profile is still poor, which demonstrates that it is not related to PR material. This furthermore leads us to believe that it is connected to standing-wave effect.

Here influence of n, k of SOC on reflectivity is firstly investigated, which could pose a direction to choose material. The relationship between reflectivity of resist bottom and N, K of SOC is investigated in fig.6. Here, PR and BARC material are not changed while N, K of SOC is changed. This N,K map is very helpful to find appropriate SOC material. N, K of BL SOC material located in the red color area where the reflectivity is too high(>1.5%). The blue color area shown in fig.6 is ideal choice. It shows that SOC should has relatively larger refractive index, 1.45-1.70, and extinction at 193nm should be small, -0.2~0.3. At last, one kind of SOC material meets the requirement, of which N, K are 1.66, -0.26, respectively.

After the appropriate SOC material is determined. Thickness of BARC and SOC are co-optimized to get an as low as possible reflectivity. As shown in fig.5(b), ARC(1) is SOC while ARC(2) is BARC. It could be seen that when thickness of SOC increases above 0.16μm, reflectivity of resist bottom is below 0.5% and changes little though thickness of SOC increases. As a result, thickness of SOC is determined to be 0.2 μm. The best thickness of BARC is 0.040μm, where reflectivity is below 0.1%. For etch process consideration, thickness of BARC is determined to be 0.038μm. This tri-layer scheme is tried in experiment split-3 and it proves successful. PR profile is greatly improved compared to other choices. The profile in 3D showed the profile is vertical and could meet the requirement of mass-production.

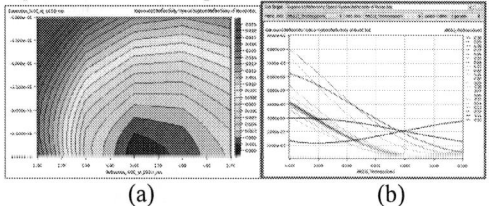

(a)                              (b)

Fig.6. (a).Relationship between N, K of SOC and reflectivity; (b). Reflectivity with BARC and SOC thickness;

For advanced technology MD and MP layer, same N, K map is simulated as shown in fig.6. There are mall differences between MD, MP and CMD because of different PR material. It could be seen from fig.7 that N, K

of MD layer should in the range of 1.45-1.7, -0.2-0.3 while N, K of MP layer should be in the range of 1.4-1.65, -0.2—0.3, respectively.

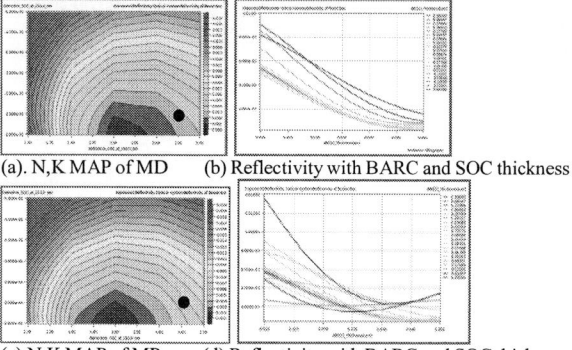

(a). N,K MAP of MD        (b) Reflectivity with BARC and SOC thickness

(c).N,K MAP of MP        (d) Reflectivity with BARC and SOC thickness

Fig.7. N,K Map of MD (a) ,(b)and MP (c),(d)

## Conclusions

In this paper, influence of several factors including source, film scheme and n, k of SOC materials on reflectivity of PR bottom are investigated thoroughly. It is found that TE mode light contributes more to reflectivity than TM mode. Reflectivity of PR bottom is greatly increased when polarization is introduced. This is because TE mode light reflectivity is greatly increased while TM mode light reflectivity is little decreased. Compared to bi_layer scheme, tri-layer scheme not only has lower reflectivity, but is also seldom influenced by films under SOC. Thus, tri-layer scheme is picked up in HuaLi advanced technology processes. For MEOL litho process, through optimizatoin n, k of SOC, appropriate SOC material is found. Furthremore, thickness of BARC and SOC are determined to be 0.038 and 0.200μm, respectively. As a result, reflectivity of PR bottom is controlled to be 0.1%. Finally good PR profile on wafer is obtained.

## REFERENCES

[1] V. Singh, "Lithography at 14nm and beyond: Choices and challenges," 2011 48th ACM/EDAC/IEEE Design Automation Conference (DAC), New York, NY, 2011, pp. 459-459.

[2] H. L. Chen, M. C. Shih, C. F. Hsieh, B. C. Chen and F. H. Ko, "Reduction of substrate alkaline contamination by utilizing multi-layer bottom antireflective coating structures in ArF lithography," Digest of Papers. Microprocesses and Nanotechnology 2001. 2001 International Microprocesses and Nanotechnology Conference (IEEE Cat. No.01EX468), Shimane, Japan, 2001, pp. 230-231

# A NOVEL METHOD USING BARC EB PROCESS TO IMPROVE SPLIT GATE FLASH ERASE CAPABILITY

*Song Zhang [1\*], Yaohui Zhou [1], Qun Liu [1], DeJin Wang [1]*

[1]Technology Development, Central Semiconductor Manufacturing Corporation (CSMC), Wuxi, China

*Corresponding Author's Email: zhouyaohui@csmc.crmicro.com

## ABSTRACT

Typical Erase time in the flash memory is 30ms, shortening the Erase time will lead to a decrease in Erase capacity and bring reliability problems meanwhile. In this paper, we propose a new solution use Barc EB method to decrease the tunnel ox thickness without impact the device. The Flash erase capability get better performance, and the erase time can decrease to 5ms. Also the yield and reliability evaluation is ok. We also study the Barc EB experiment split results, get the best Barc EB process condition..

***Keywords- Erase time; Reliability; Tunnel ox; Barc EB***

## INTRODUCTION

The Erase operation of split gate flash is performed by applying high voltage to the word line, inject the electrons to tunnel from the floating gate(FG) through the tunnel oxide layer via the FG tip to the word line WL[1], leaving holes in the floating gate(FG) with a positive charge and turning on, as shown in Figure 1.

The Erase ability is the integral of the Erase capacity and the Erase time, as the Erase time shortened ,The Erase capacity should increase to achieve the same Erase result[2] , which can be achieved by the sharper FG tip, the thinner tunnel oxide, or the higher voltage applied on word line(WL)[3]. Since the FG structure fix, Sharper FG tip is difficult to realize in process. And, Increasing the WL voltage will downgrade the reliability of Flash devices

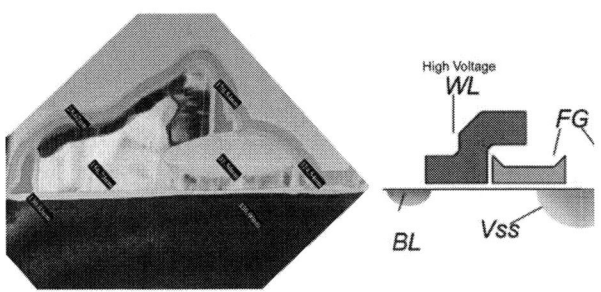

Figure. 1 SEM micrographs and Schematic of Split Gate Flash Cell, the electrons was injected from FG tip to the word line (WL) : (a) SEM micrographs of Flash Cell, (b) Schematic of Flash Cell

The most feasible way is to reduce the tunnel oxide thickness, but as the tunnel oxide is shared by high-voltage devices and Flash (this shared layer is to reduce the photomask layer, reducing costs), high-voltage devices require thick enough oxide to ensure breakdown voltage. So the key factor is how to reduce the tunnel oxide thickness of the Flash device while keeping the gate oxide of the high-voltage device not change without add new masks.

### Barc EB Structure

To reduce the thickness of the tunnel oxide of the Flash while keeping the GOX of the high voltage device unchanged, it is necessary to separate the Flash region from the high voltage region. Consider the difference between the Flash device and the high voltage device is the step height caused by the special FG stack bulging field ox. We propose the BARC Etch Back solution to to do separate treatment for different step height region .

Specifically, as shown in Figure 2, after the tunnel oxide is coated with a layer of BARC, the thickness of BARC on the flat Logic/HV Silicon substrate will be thicker than that of on the bulging Memory cell FG tip. Take advantage this Barc thickness difference feature, Use dry etch to decrease the tunnel oxide thickness of the FG tip, while other HV area ox will not be impact as the thicker barc coating. Thus, the Flash Erase time can be shortened without affecting other devices.

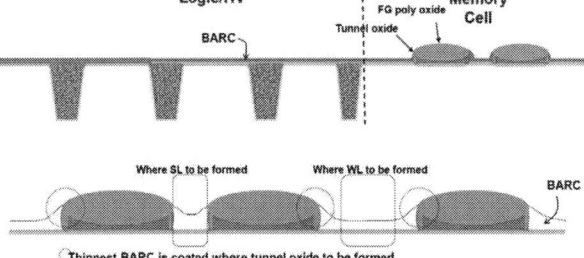

Figure. 2. Barc etch back solution schematic of the high voltage region and the memory cell region

## EXPERIMENT

The Barc EB solution requires a suitable BARC coating process condition so that the BARC coating step coverage can get the obvious difference between on the flat Silicon substrate and on the FG tip, and the resulting thickness difference is the process window for the subsequent etching steps. The Barc thickness on flat substrate is thicker than on FG tip, a high Oxide to BARC

978-1-6654-9759-6/22 $31.00 © 2022 IEEE

etch selection ratio for dry etching is required. Considered the BARC thickness of the sidewall cannot be monitored, sufficient process stability and process window are required.

1) **BARC Coating Process:** As shown in Table I a BARC thickness is inversely proportional to the rotational speed during coating, the BARC thickness can be adjusted by rotational speed.

TABLE I
Barc process condition: rotate speed vs Barc thickness

| Rotate speed/r ps | Barc coating region | Barc Tk(A) | | | ΔTK(A):Barc between FG – Barc on fox |
|---|---|---|---|---|---|
| | | *center* | *edge* | *average* | |
| 1818 | On Fox | 245 | 235 | 240 | 563 |
| | Between FG | 764 | 842 | 803 | |
| 4000 | On Fox | 117 | 127 | 122 | 274 |
| | Between FG | 440 | 352 | 396 | |
| 4500 | On Fox | 117 | 98 | 107.5 | 338 |
| | Between FG | 421 | 470 | 445.5 | |

The BARC thickness on top of the Silicon substrate and the BARC thickness at the FG tip are varied together, when the rotate speed set 1818 rps, the barc ΔTk between the flat area and the flash FG tip region can reach 563Å as the figuer3 data show，which meets the requirement.

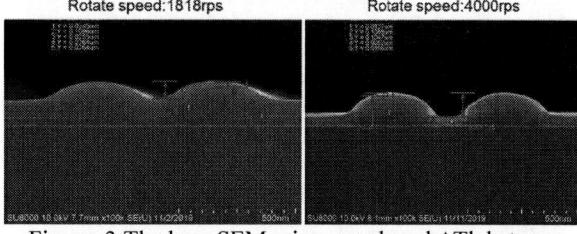

Figure. 3 The barc SEM micro graph and ΔTk between the flat area and the flash FG tip region use different coating condition, : (a) Rotate speed 1818 rps, (b) Rotate speed 4000 rps,

2) **BARC etch process:** The etch process including BARC etch and oxide etch, BARC etch requires high Barc/Oxide selectivity, low oxide etch rate avoid ox loss; While in Oxide etch process requires less barc loss to avoid substrate damage by etch plasma. So the etch time control is important, longer BARC etch time lead to less barc remain , which may suffer over etching risk in oxide etch process and will not protect the substrate. On the other hand, the tunnel ox on FG tip can not be thinned in oxide etch process as the barc residue existed when the BARC etch time is not enough. Barc etch time split cross section SEM show in figure4, base the SEM data we can get the correspondence between Barc remain and etching time shown in Figure 5. The final BARC etch time selection is 28s, which ensures the tunnel oxide on FG tip

be exposed while enough BARC is retained on the flat area

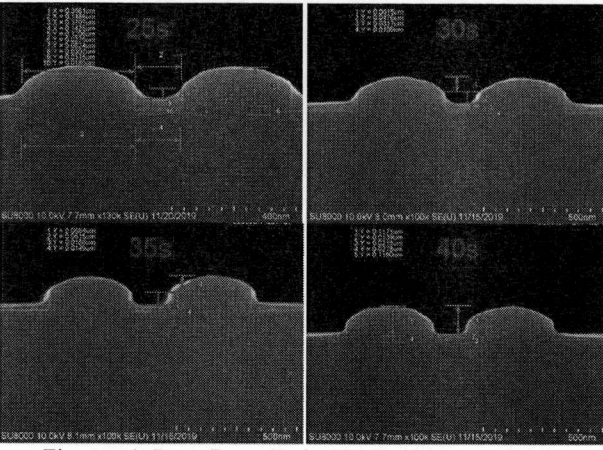

Figure. 4 Post Barc Etch SEM micro graph use different etch time : (a) Barc Etch time 25s,(b) Barc Etch time 30s, (c) Barc Etch time 35s, (d) Barc Etch time 40s

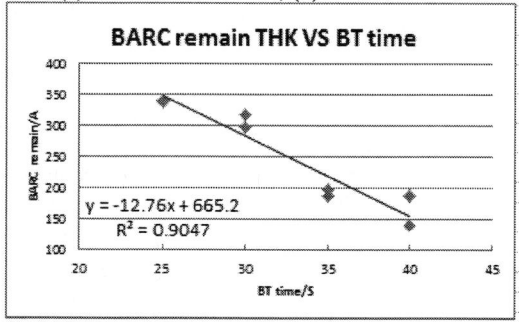

Figure. 5 The correspondence between Barc remain of flat area and Barc etching time

3) **Oxide etch process:** The Oxide etch time is determined by the relationship between the Erase time and the Erase voltage, as shown in Table II, the decrease of the Erase time from 30ms to 5ms corresponds to an increase of ~0.3V Erase voltage, which needs to be achieved by thinning the tunnel oxide thickness. Base on the formula 1, we can calculate the Oxide etch time to make the tunnel ox on FG meet the requirement. The details are shown in Figure 6

TABLE II
The relationship between Erase time and Erase voltage

| Erase Time | Cycle | Vee@0sigma(V) |
|---|---|---|
| 30ms | 0K | 9.10 |
| 30ms | 100K | 10.79 |
| TUR (Trap up ratio, ΔVee) | | 1.69 |
| 5ms | 0K | 9.36 |
| 5ms | 100K | 10.94 |
| TUR (Trap up ratio, ΔVee) | | 1.58 |

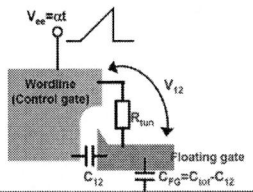

Vee-VFG=(1-CR)·Vee(t)-QFG(t)/Ctot;
The CR(Couple Ratio)=C12/Ctot,
Formula1[4]:

$$Vfg = \frac{Q_{fg} + C_{wl} \times V_{wl} + C_{vss} \times V_{vss}}{C_{tot}}$$

Figure. 6 Erase circuit diagram of Split Gate Flash and the relationship between Tunnel Oxide thickness and erase voltage

## RESULTS AND DISCUSSION

After implementing the BARC etch back scheme, Flash cell erase time decrease to 5ms and the erase capability keep the same level with the old scheme of 30ms erase time , as show in Figure7.

At the same time the yield and reliability are evaluated under the Erase time of 5ms, as shown in Figure 8 and Figure9. The yield histogram chart show that the Barc EB solution yield is comparable with old baseline. And the BARC etch back time is proven to have a certain process window.

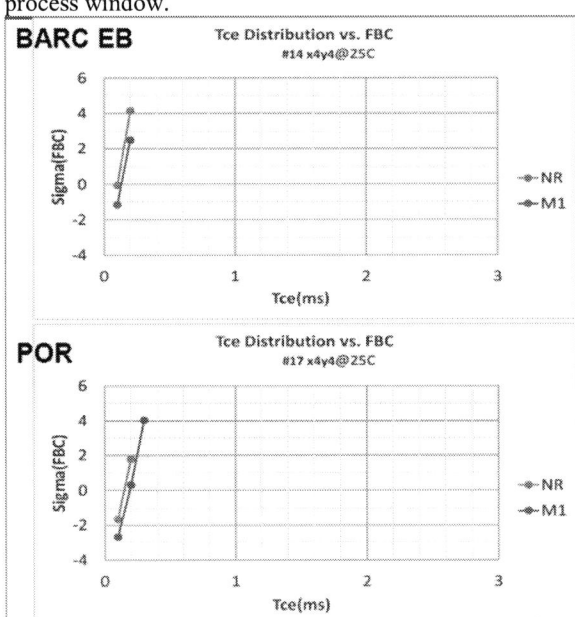

Figure. 7 Flash cell erase speed comparison between Barc EB scheme with POR scheme

Cycling evaluation data in Figure9 compare the endurance of BARC etch back solution with the POR

scheme, The BARC etch back scheme can achieve the same durability with the condition of erase time of 30ms

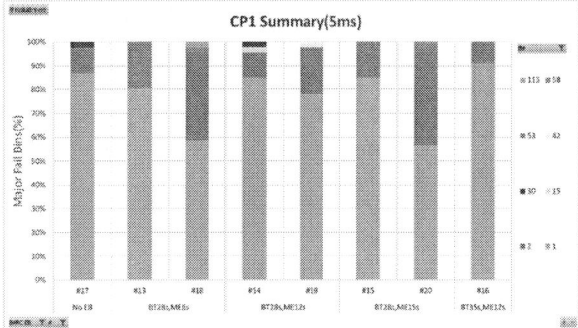

Figure. 8 Flash Yield trend chart of Barc EB scheme

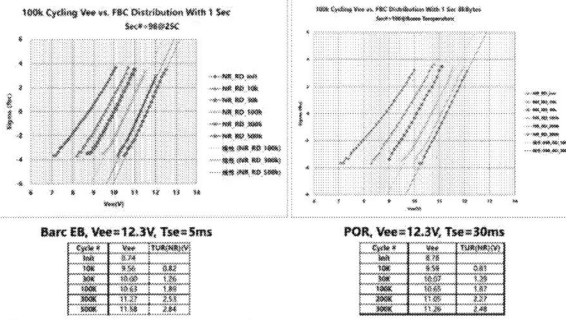

Figure. 9 The endurance data comparison between the Barc EB scheme and old POR scheme:(a) Barc EB scheme cycling data, (b) POR scheme cycling data

## CONCLUSION

Split gate Flash erase result is the integral of the Erase capacity and the Erase time, erase time decrease will bring the worse erase speed and endurance risk. In this paper, we propose a BARC etch back process method to decrease the Erase time base on 0.11um flash memory process. The erase time decrease from 30ms to 5ms , while the erase speed comparable with old scheme. Flash evaluation data show Barc EB solution Cp and endurance all ok. We also study the Barc EB split experiment result and find the best process condition: Barc coating use rotate speed 1818rps, Barc Etch time 28s

## REFERENCES

[1] S. Senturia. *Proceedings of Transducers2003*, Boston, June 8-12, 2003, pp. 10-15.
[2] T. Tsuchiya, O. Tabata, J. Sakata and Y. Taga. *J. Microelectromech. Syst.*, vol. 7, 1998, pp. 106-113.
[3] R. P. Feynman, *Lectures on Physics*, Addison Wesley, 1989.
[4] K. Elissa. unpublished.
[5] R. Nicole. *J. Name Stand. Abbrev.*, in press.

# OPTIMIZATION OF METAL LINE THICKNESS & CD AND EFFECT ON RC DELAY

*Pengfei Lyu[1]\*, Qingpeng Wang[1], Lifei Sun[1]*
[1]Lam Research Service Co., Ltd, Shanghai 201203, China
\*Corresponding Author's Email: PengFei.Lyu@lamresearch.com

## ABSTRACT

In order to improve device performance, smaller resistance (R) and capacitance (C) in back-end-of-line (BEOL) metallization are pursued as two key technological boosters in minimizing time delay in lead-edge CMOS devices. R is inversely proportional to metal line cross sectional area, which is approximately the product of CD and thickness, while C is proportional to metal line thickness and reciprocal to metal line space (pitch - CD). Thus, a trade-off exists with respect to metal line CD and thickness to obtain desirable RC delay performance. Additionally, with continuous metal line scaling, R is further impacted by irregular trench shapes and electron surface scattering, which poses further challenges in CD and line thickness optimization. In this paper, the effects of metal line thickness and CD on RC is investigated through virtual fabrication method and electrical analysis of structures obtained. A systematic investigation is carried out to evaluate the impact of metal line thickness and CD calibrated on wafer structures.

***Keywords—Metal Line Profile; Resistance; Capacitance; RC delay; SEMulator3D***

## INTRODUCTION

Nowadays, with leading edge CMOS device dimension continuously scaling down, metal line resistance-capacitance (RC) control has become increasingly important in back-end-of-line (BEOL) as a main limiter in increasing chip speed and performance [1]. Resistance increases rapidly with narrower line CDs, due to not only shrinking metal line cross-sectional area with node advancing, but also surface scattering effect: electrons scatter off metal lines close to the surface. This effect starts to dominate resistivity when metal line CD decreases to below 50 nm, resulting in an exponential increase in resistance [2]. Capacitance also increases with smaller line-to-line space. At a specific metal layer of a specific process node, metal line pitch is typically fixed, meaning the sum of metal line CD and line-to-line space is constrained. This leads to a trade-off between R and C to obtain the minimum RC value. Similar trade-off also exists with line thickness in minimizing RC. For simplified rectangular or trapezoidal shapes, it seems that the best combination of metal line CD and thickness can

be obtained by calculation. However, metal line trench profiles on real Si wafer are often not simply rectangular or trapezoidal. The rounding corner and other irregular features, together with electron surface scattering effect, make manual calculation complicated and unattainable.

In this paper, we will use virtual fabrication tool to rebuild metal line profile matched to real Si wafer, and systematically investigate the impacts of metal line CD and thickness on resistance and capacitance. The best set of CD and thickness will be also given for a desirable RC delay under assumed line pitch.

## EXPERIMENT

The traditional way to obtain the best metal trench dimensional design is to do CD and thickness design of experiments (DOE) on actual wafer to obtain measurements on R and C. It is an expensive way because a lot of wafers and time will be consumed to accumulate sufficient data for conclusion. On the contrary, virtual fabrication using SEMulator3D® is faster and a more feasible. By setting up virtual decks and metrologies, R and C information can be collected from numerous DOE splits with many possible parameter combinations, which saves both time and cost in the fab.

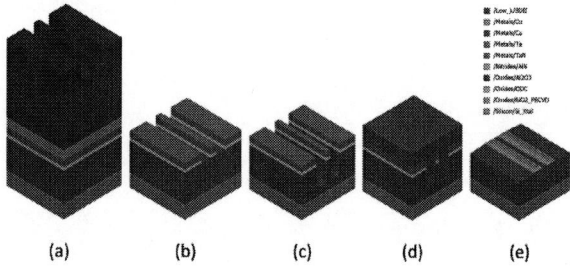

*Figure 1: Fabricated BEOL M1 metal line: (a) Post lithography, (b) Post hard mask etch, (c) Post all-in-one etch, (d) Post metal fill, (e) Post CMP.*

For capacitance measurements, two adjacent 200nm-long metal lines were modelled and studied. The formation of trenches is through litho-etch approach, as shown in Figure 1, with CDs modulated by litho bias in the simulation. The pitch is fixed at 53 nm, and the trench sidewall angle is fixed at 85°. To best reproduce the irregular features on real-Si wafer, the rounded corner at trench bottom and Cu recess at top surface are calibrated

978-1-6654-9759-6/22 $31.00 © 2022 IEEE

as shown in Figure 2. The metal line top CD and thickness are taken as variables, and the range tested in the studied is listed in Table 1. The electron surface scattering effect is also taken into consideration by inputting two size correction factors: surface correction and decay length.

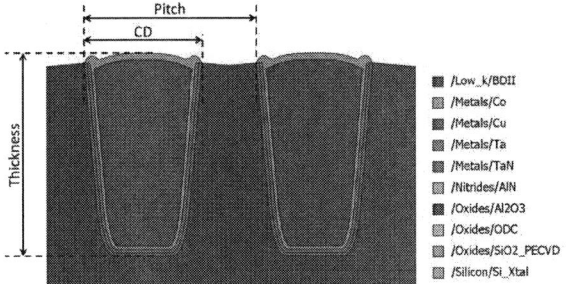

*Figure 2: The calibrated Cu metal lines cross-section with fixed pitch 53 nm and sidewall angle 85°.*

| Variable | Min (nm) | Max (nm) |
|----------|----------|----------|
| CD | 20 | 50 |
| Thickness | 30 | 80 |

*Table 1: The range of variable metal line CD and thickness tested in the virtual DOE.*

## RESULTS

500 runs of virtual RC extraction DOE are carried out with CD and thickness set in the range as shown in Table 1. Figure 3 and Figure 4 are contour maps of R and C separately, as function of metal line CD and thickness. It can be observed that the minimum R values converge at upper right corner, while C value shows the opposite trend, which is fully consistent with theoretical trends with thickness and CD.

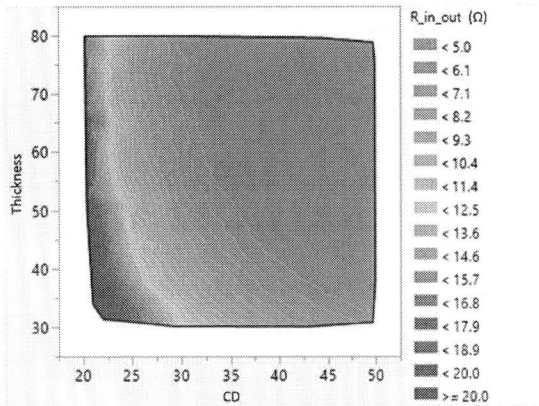

*Figure 3: Contour maps of resistance as a function of metal line thickness and CD.*

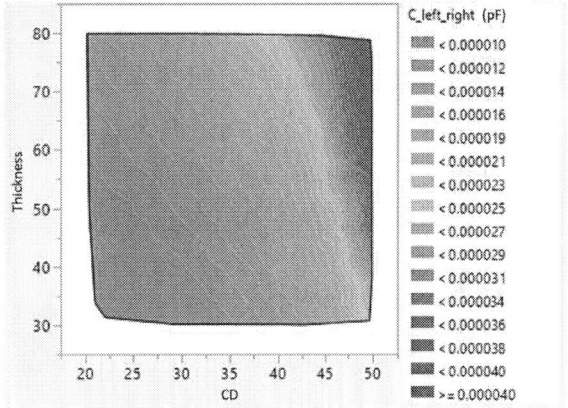

*Figure 4: Contour maps of capacitance as a function of metal line thickness and CD.*

The time delay T is acquired by multiplying R and C, and its contour based on CD and thickness is shown in Figure 5. Different from individual behavior of R and C, there exists a specified region (in darkest green in Figure 5) for a minimum T value, where CD ranges from 35 nm to 40 nm and thickness ranges from 70 nm to 75 nm.

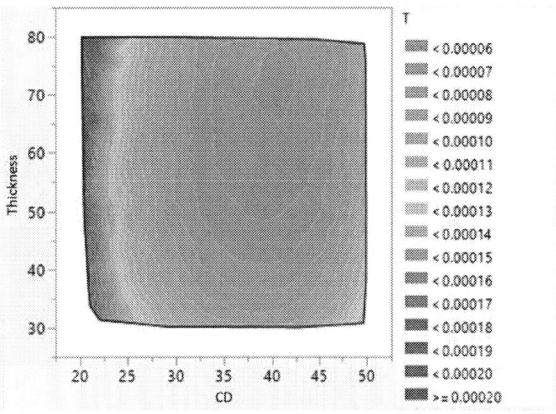

*Figure 5: Contour maps of time delay as a function of metal line thickness and CD. Best point exists in region CD 35~40 and thickness 70~75.*

Neutral network analysis is also utilized to locate the best set of R and C for minimum time delay T in a more qualitative and accurate manner. As shown in Figure 6, the predicted minimum T contour line given by neutral network analysis falls in the same region in the coordinate. The CD and thickness are predicted to be 40.7 nm and 70.7 nm. It can be concluded that with given line pitch and trench profile, there is a corresponding combination of CD and thickness that can make time delay the smallest. This virtual experiment could be a reference in optimization of metal line CD and thickness.

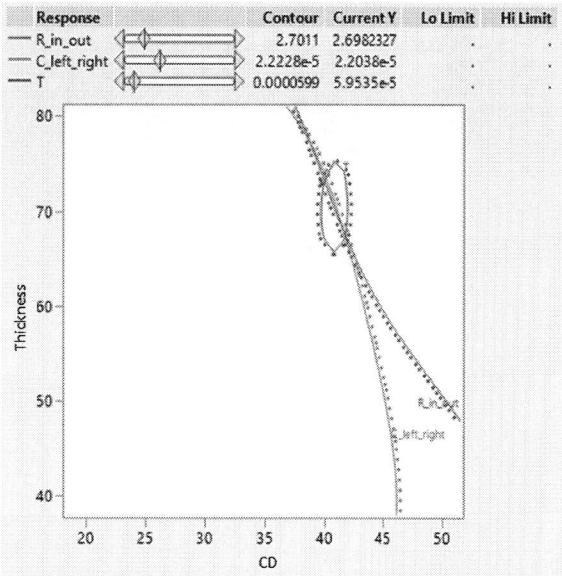

| Response | | Contour | Current Y | Lo Limit | Hi Limit |
|---|---|---|---|---|---|
| R_in_out | | 2.7011 | 2.6982327 | . | . |
| C_left_right | | 2.2228e-5 | 2.2038e-5 | . | . |
| T | | 0.0000599 | 5.9535e-5 | . | . |

*Figure 6: Neutral network analysis gives out the minimum time delay T range (blue cycle) and the corresponding R and C contour line (red line and green line).*

## CONCLUSION

This paper introduces a virtual fabrication method using SEMulator3D to precisely predict optimized metal line R and C to minimize RC delay. Compared with traditional test on real Si wafer, virtual DOE provides a faster and cheaper methodology with a more complete dataset. In our study, SEMulator3D deck is setup with fixed metal line pitch, and the trench profile is calibrated to real case for accuracy. The trend of R and C with metal line CD and depth are acquired through 500 DOE runs and found to be consistent with theory. RC time delay is minimized for a CD range of 35~40 nm and a thickness range of 70~75 nm. The best point is predicted to be 40.7 nm and 70.7 nm by neutral analysis. The virtual DOE method is a more convenient and precise way to perform BEOL dimensional design to optimize IC electrical performance.

## REFERENCES

[1] Chen H. C., Fan S. C., Lin J. H., et al. The impact of scaling on metal thickness for advanced back end of line interconnects. Thin Solid Films, 2004, 469(1):487-490.

[2] Van D., Heyler N., Pedreira O. V., et al. Damascene Benchmark of Ru, Co and Cu in Scaled Dimensions. IEEE International Interconnect Technology Conference (IITC), 2018: 172-174.

[3] http://www.coventor.com/products/semulator3d.

# NOVEL SOI BASED PSEUDO-INVERTER: EXPERIMENTAL AND SIMULATION RESEARCH

*Sherzod Khaydarov[1], Kai Xiao[1], Haihua Wang[1], Bo Li[2], Fanyu Liu[2*], and Jing Wan[1*]*
[1]School of Information Science and Technology, Fudan University, Shanghai 200433, China
[2] Institute of Microelectronics, Chinese Academy of Sciences, Beijing 100029, China
*Corresponding Author's Email: liufanyu@ime.ac.cn, jingwan@fudan.edu.cn

## ABSTRACT

This paper reveals the Silicon-on-Insulator (SOI) substrate based pseudo-MOSFET (Ψ-MOSFET) device as a novel pseudo-inverter structure. Experimental device was fabricated and voltage transfer characteristics (VTC) were measured. The pseudo-inverter (Ψ-inverter) shows similar, but not exactly the same, behaviour as conventional inverter with the gain higher than 1000. TCAD simulation is further used to validate the experimental results and study its physics. The simulation shows that the inverter gain is very sensitive to the top surface trap density ($D_{it}$) which significantly increases the maximum electric field in the channel. The Ψ-inverter demonstrated in our work might be useful for extraction of SOI parameters and sensors application.

## INTRODUCTION

Silicon-on-Insulator (SOI) technology has proven to be one of the promising candidates to substitute the conventional planar bulk Si technology to match the future needs of Moore's Law, as the latter was confronted with power consumption and scaling down challenges. The importance of evaluation of the quality of SOI wafers without complicated CMOS processing gave rise to the non-source/drain doped Ψ-MOSFET structure utilized as in-situ characterization device [1-2].

Previous studies proposed point contact Ψ-MOSFET technique as a tool for rapid extraction of the flat-band ($V_{FB}$) and threshold voltages ($V_{TH}$) [3-4]. Furthermore, the conventional MOSFET parameter extraction Y-function method suggested by Ghibaudo was adapted into the Ψ-MOSFET to exclude the effect of the source/drain series resistance $R_{SD}$ and extract the effective carrier mobility [5]. Moreover, interface traps density $D_{it}$ extraction model was also developed based on the Ψ-MOSFET technique [6-7].

To the best of our knowledge, the investigations of Ψ-MOSFET as a novel pseudo-inverter approach hasn't been discussed before. In this work, we propose and investigate the pseudo-inverter scheme which shows similar behaviour as the conventional CMOS inverter with extremely high gain. We further study its physical mechanism and the impact of interface trap density ($D_{it}$) on its gain.

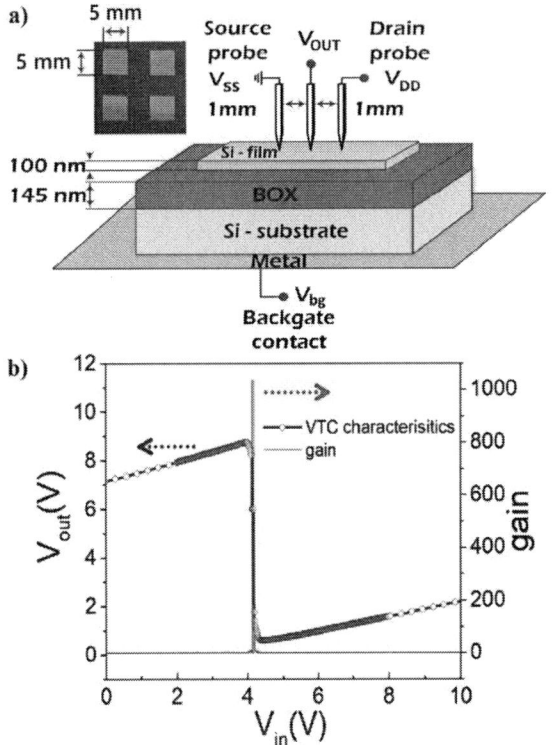

*Figure 1: (a) Experimental device structure and (b) VTC characteristics diagram at $V_{DD}=10$ V*

## EXPERIMENTAL RESULTS

The schematic illustration and the sample of pseudo–inverter (Ψ-inverter) are presented in Figure 1(a). The sample was fabricated on SOI substrate with 100 nm top Si layer and 145 nm buried oxide (BOX). Both top Si layer and substrate are p-type doped ($10^{15}$ cm$^{-3}$). Photolithography and etching were used to define the square pads (5 mm × 5 mm) on the Si device layer. The substrate of the device was used as backgate which is the input of the inverter and denoted by $V_{in}$. Unlike in the Ψ-MOSFET where only two probes are needed as the source and drain, three probes are used in Ψ-inverter with

978-1-6654-9759-6/22 $31.00 © 2022 IEEE

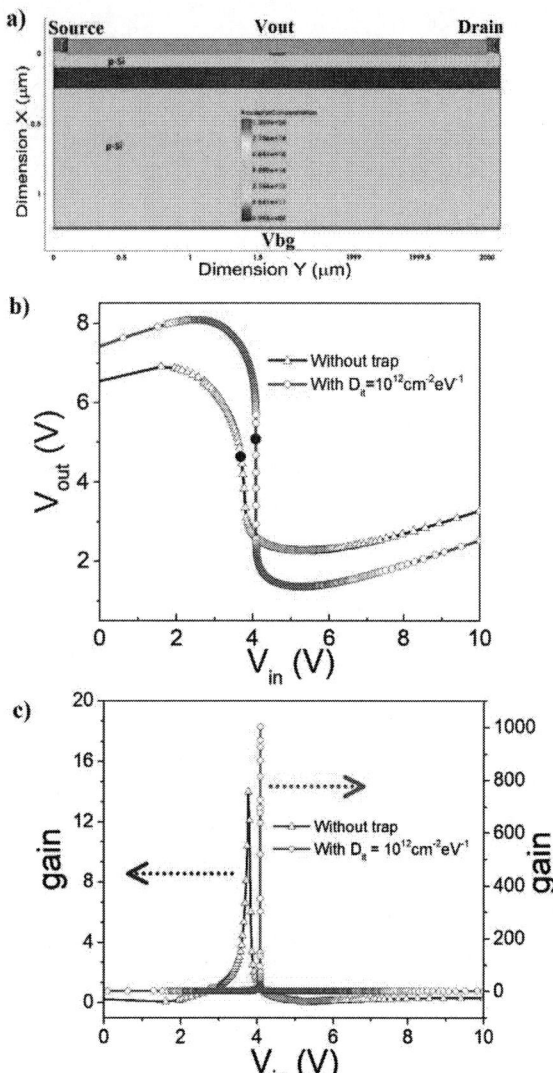

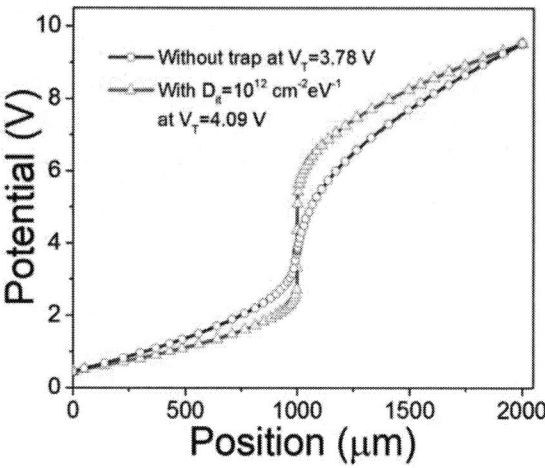

*Figure 3: Electrostatic potential distributions from devices without (blue) and with (red) interface trap density of $D_{it}=10^{12}cm^{-2}eV^{-1}$*

Sentaurus to understand the physical behaviour of the experimental device. Figure 2(a) shows the structure of the Ψ-inverter built on SOI substrate with the same dimensions as the experimental device.

The source/drain contacts of the simulated structure consist of N+ and P+ regions, simulating the bipolar behaviour of the real pseudo-MOS contacts. Moreover, we allocated the output port as the potential point on top of the Si-film interface, in the middle of the device. Si substrate is defined as the back gate contact.

Figure 2(b) shows the simulated voltage transfer characteristics of the device, which looks similar to our experimental data. However, the maximum gain is only around 14 which is much lower than our experimental data. In real pseudo-MOS device, the top surface of the Si device layer is typically not passivated and induces high interface trap density. To simulate this effect, trap density of $D_{it}=10^{12}$ $cm^{-2}eV^{-1}$ is placed on the top Si surface in our simulation device. A much sharper transfer characteristics is observed in the transfer characteristics, as shown in Fig. 2(b). Figure 2(c) shows that the gain with the interface trap reaches as high as 1000, close to our experimental data.

To understand the mechanism of this, further investigations of the potential distribution are conducted. Figure 3 compares the potential distributions extracted from the devices without and with trap density of $D_{it}=10^{12}$ $cm^{-2}eV^{-1}$. The potential distributions are extracted right at the transfer voltages ($V_T$) with maximum gain in both cases, whose values are $V_T=3.78$ V and $V_T=4.09$ V, respectively. As observed, potential in both cases changes abruptly at the middle of the device, corresponding to the sharp transfer in the voltage transfer characteristics. However, with $D_{it}=10^{12}$ $cm^{-2}eV^{-1}$, the change of the

*Figure 2: (a) Simulated device structure. (b) Voltage transfer characteristics and (c) gain comparison without and with top surface trap at $V_{DD}=10$ V*

the middle probe used as the output of the inverter, see Fig. 1(a). The left and right probes are used as source and drain, and are biased at $V_{SS}=0$ V and $V_{DD}=10$ V respectively.

Voltage transfer characteristics with corresponding gain curve is depicted in Figure 1(b). When $V_{in}$ is swept from 0 V to 10 V, initially the output shows high voltage, then at some point it drops abruptly to the minimum $V_{out}$, similar to a conventional inverter. The gain, defined as the $dV_{out}/dV_{in}$, reaches a maximum value of 1038 at $V_{in}$ of 4.174 V. However, the transfer behaviour is not exactly the same as a conventional inverter. To understand the mechanism and unique behaviour of the Ψ-inverter, we proceed to device simulation by TCAD.

TCAD simulation device is modeled in Synopsys

978-1-6654-9759-6/22 $31.00 © 2022 IEEE

potential is much more abrupt, indicating higher electric field, compared to that without interface trap density. This is the root cause of the high gain of the device with interface trap. The interface defects can trap carriers in the channel, which depletes the junction and thus enhance the electric field in the middle of the channel. This is similar to that in a conventional MOSFET, where interface trap increases threshold voltage and leads to high electric field in the drain junction.

## CONCLUSIONS

In this paper, we revealed a $\Psi$-inverter structure whose behaviour is similar to a conventional inverter through both experiments and simulations. The device with three probes setup shows a gain as high as 1038. Simulation validates our experimental results and further shows that the high gain is originated from the interface trap density of the top Si surface which is typically not passivated in a $\Psi$-inverter or $\Psi$-MOS measurement. The $\Psi$-inverter demonstrated in this paper might be interesting for parameters extraction and sensor applications.

## ACKNOWLEDGEMENTS

This work was supported by the National Key R&D Program of China (2021YFA1200500), the National Natural Science Foundation of China (61904032), Shanghai Science and Technology Commission "explorer project"(21TS1401300) and Pioneering Project of Academy for Engineering and Technology, Fudan University (gyy2021-001).

## REFERENCES

[1] S. Cristoloveanu and S. Williams, *IEEE Electron Device Lett.*, vol. 13, no. 2, pp. 102–104, 1992.

[2] S.G. Kang, and D.K. Schroder, *IEEE Trans. on Electron Dev.,* vol. 49 (10), pp.1742-7., 2002.

[3] N. Rodriguez, S. Cristoloveanu, and F. Gamiz, *IEEE Trans. Electron Dev.*, vol. 56(7), pp. 1507-1515, 2009.

[4] S. Cristoloveanu, D. Munteanu, M. Liu, *IEEE Electron Device Lett.,* vol.47(5), pp. 1018-1026, 2000.

[5] G.Ghibaudo, *Electron. Lett.*, vol.24, pp.543-545, 1988.

[6] L. Pirro, I. Ionica, G. Ghibaudo, X. Mescot, L. Faraone, and S. Cristoloveanu, *J. Appl. Phys.* 119, 175702, 2016.

[7] G. Hamaide, F. Allibert, H. Hovel, and S. Cristoloveanu, *J. Appl. Phys.*, 101, 2007, 1144513.

# SIGE EPITAXIAL GROWTH ON SIGE SUBSTRATE FOR ADVANCED FDSOI TECHNOLOGY

*Yongyue Chen\**

Research and Development Dept., Shanghai Huali Integrated Circuit Corporation, Shanghai 201314, China

\*Corresponding Author's Email: chenyongyue@hlmc.cn

## ABSTRACT

FDSOIs with SiGe channel have drawn significant attention due to their advantages in uniaxial strain, higher mobility and better capability to tune threshold voltage over conventional Si channel MOSFETs. In addition, raised SiGe source and drain can boost the electrical performance of PMOS by reducing the access resistance. Fabrication of SiGe material on SiGe channel which is formed by Ge condensation method is challenging due to the different surface condition and lattice property compared to Si channel. The SiGe material are successfully grown on SiGe substrate by tuning the process parameter by selective epitaxial growth.

## INTRODUCTION

Over the past years, CMOS scaling has been continuing to achieve higher performance and lower power consumption with lower cost. As traditional bulk planar device structure reached its physical limitation, fully depleted devices such as FINFETs and planar FDSOIs are being applied into production. Additional process complexity of FDSOIs compared to bulk planar devices is relatively smaller than FinFETs due to its planar structure [1]. The interest of the Dual Channel CMOS integration scheme, which features SiGe channels for p-type and Si channels n-type FDSOI, has conclusively been demonstrated [2-3].

Introducing SiGe as a raised source/drain (RSD) material of the pMOSFETs has also been reported to improve the performance by reducing the access resistance on FDSOI [8]. Fabrication of the raised SiGe source and drain on SiGe channel is challenging due to the different surface condition and lattice property of SiGe channel which is formed by Ge condensation method compared with Si channel [4-6].

In this paper, the SiGe selective epitaxial growth was optimized by tuned process parameters. The surface condition of core area and IO area in the wafer are tuned by wet clean method. The SiGe material are successfully grown on SiGe substrate at the location of pad by tuning the process parameter of SiGe material by selective epitaxial growth. However, some area of PMOS source and drain based on SiGe channel still can't be grown. Further study was needed to epitaxy SiGe material on SiGe channel that cover all the area of PMOS source and drain.

## EXPERIMENT AND DISCUSSION

Figure.1 shows schematics of FDSOI with SOI channel for nMOSFET and c-SiGe channel for pMOSFET. The in-situ boron doped SiGe acts as PMOS raised source and drain, and the SIP act as NMOS source and drain. Figure 2 shows a brief process flow for fabrication of SiGe raised source and drain in FDSOI devices integration. The experiments were carried out on 12 inches SOI wafers with BOX thickness of 20nm based on the advanced planar FDSOI devices integration. The PMOS area was covered up by lithography, the pad oxide and nitride were etched. Remaining SOI as the initial layer for selective SiGe epitaxy process. Following the SiGe epitaxy layer been formed, the oxidation of SiGe layer was utilized to obtain the high Ge fraction SiGe channel layer. Hafnium-oxide-based high-k metal gate is then formed and patterned. The spacer of oxide and nitride is formed. After SiGe hard-mask deposition, the growth area of SiGe is patterned by photo and etch process. Then, the SiGe RSD with a high Ge content and high boron doping concentration is grown.

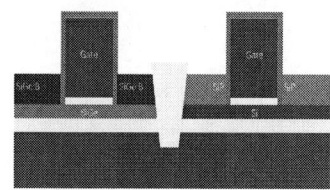

*Figure 1: Schematics of FDSOI with SOI channel for nMOSFET and c-SiGe channel for pMOSFET)*

*tuning*

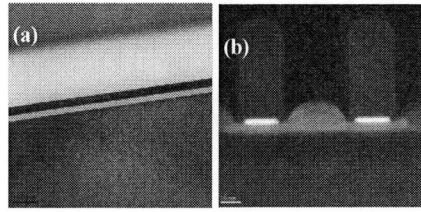

*Figure 4: (a) Epitaxial SiGe layer on pad of IO area after tuned process(b) Epitaxial PMOS SiGe layer on core area after tuned process*

## SUMMARY

Fabrication of SiGe material on SiGe channel which is formed by Ge condensation method is challenging due to the different surface condition and lattice property compared to Si channel. The SiGe material are successfully grown on SiGe substrate by tuning the process parameter by selective epitaxial growth. Further study was needed to epitaxy SiGe material on SiGe channel that located at the area of source and drain.

## ACKNOWLEDGEMENTS

The author wishes to express gratitude to Jiaqi Hong, Li Peng, Jian Lv, Wangxin Nie, Huojin Tu, Qiang Yan, Jun Tan, Qing Sun, Xinhua Cheng, Jingxun Fang and Yu Zhang at HLMC for their expertise, feedback and advice during the whole work. Author also thanks Lan Jiang, Xinhao Wu, Tao Wang, Duo Shan and Yan Li for their kind support. HLMC failure analysis department supplies the supports of TEM microscopy.

## REFERENCES

[1] S. Yamaguchia , L. Wittersb, J. Mitardb, G. Enemanb, G. Hellingsb, A. Hikavyyb, R. Loob, N. Horiguchi, *Microelectronics Reliability*, vol. 83 (2018) 157–161.

[2] A. Khakifirooz, K. Cheng, T. Nagumo, N. Loubet, T. Adam, A. Reznicek, J. Kuss, D. Shahrjerdi, R. Sreenivasan, S. Ponoth, H. He, P. Kulkarni, Q. Liu, P. Hashemi, P. Khare, S. Luning, S. Mehta, J. Gimbert, Y. Zhu, Z. Zhu, J. Li, A. Madan, T. Levin, F. Monsieur, T. Yamamoto, S. Naczas, S. Schmitz, S. Holmes, C. Aulnette, N. Daval, W. Schwarzenbach, B.-Y. Nguyen, V. Paruchuri, M. Khare, G. Shahidi, and B. Doris, in Proc. VLSI Symp. Technol, Jun. 2012, pp. 117–118.

[3] Tsutomu TEZUKA*, Naoharu SUGIYAMA, Tomohisa MIZUNO, Masamichi SUZUKI1 and Shin-ichi TAKAGI, *Jpn. J. Appl. Phys.*, vol. 40 (2001) pp. 2866–2874.

[4] Yongyue Chen*, Qiang Yan, Jiaqi Hong, Qiuming

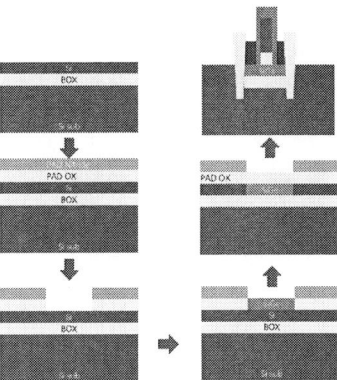

*Figure 2: A brief process flow of PMOS fabrication in FDSOI devices integration*

In our work, as Figure 3(a) and (b) shows, when the SiGe growth condition is applied in the same wafer, the pad area PMOS based on SiGe channel failed to the growth of SiGe for source and drain and the IO area PMOS based on Si channel can be successfully epitaxial growth. A key parameter that is evaluated to measure the quality of pre-epi clean is the interfacial contamination between the substrate and epitaxial layer. It is suspected that the interfacial oxygen concentration between the epitaxial SiGe and Si substrate at IO area is lower than the interfacial oxygen concentration between the epitaxial SiGe and SiGe substrate of the pad area. For the optimized condition of pre-clean, the interfacial oxygen concentration arrived the target interfacial contamination [5]. Then, the growth parameter of SiGe is fine tuned. The nucleation layer growth parameter such as Ge concentration is fine tuned to match with the lattice property of SiGe substrate at the location of pad. As illustrated in Figure 4 (a), the epitaxial SiGe layer are successfully grown on SiGe substrate after tuning the process parameter by selective epitaxial growth. Only a few PMOS source and drain based on SiGe channel can be successfully grown as shown in Figure 4 (b). Further study was needed to epitaxy SiGe material on SiGe channel that cover all the area of PMOS source and drain.

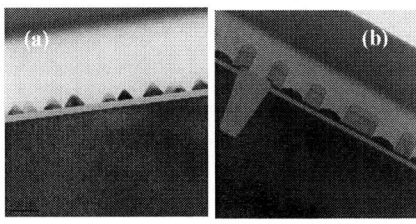

*Figure 3: (a) Epitaxial SiGe layer fail on SiGe pad of core area before process tuning (b) Epitaxial SiGe layer on PMOS of IO area based on Si channel before process tuning*

Huang, Jun Tan, Haifeng Zhou, Jingxun Fang, *CSTIC2020*, Shanghai, March 11-12, 2020, pp. 1-3.

[5] Pouya Hashemi, Takashi Ando, Karthik Balakrishnan, Siyuranga Koswatta, Kam-Leung Lee, John A. Ott,Kevin Chan, John Bruley, Sebastian U. Engelmann, Vijay Narayanan, Effendi Leobandung and Renee T. Mo, *VLSI Tech. Symp.*, p. 38, 2017

[6] Yiqun Liu*, Lan Jin, Kunshan Song, Qiong Wu, Youfeng He, Yonggen He, *CSTIC2018*, Shanghai, March 11-12, 2018, pp. 1-3.

# A PIN PHOTODETECTOR BASED ON A RF-SOI SUBSTRATE

*JJ. Chou and J. Wan\**

State key lab of ASIC and System, School of Information Science and Engineering, Fudan University,
Shanghai, China
*Corresponding Author's Email: jingwan@fudan.edu.cn

## ABSTRACT

In this work, A lateral PIN diode is fabricated based on the RF-SOI substrate. The diode is further used in UV photodetection and high-speed optical communication. It exhibits excellent linearity under various light intensity. The response spectrum of the photodiode shows UV-enhanced characteristics and its quantum efficiency reaches up to 89% at wavelength of 300nm. Thanks to the RF-SOI substrate which reduces the parasitic capacitance dramatically, the capacitance of the fabricated photodiode is as low as 0.4pF at -1V bias. The demonstrated photodiode based on RF-SOI substrate might be attractive for high speed optical communication system using short wavelength light.

## INTRODUCTION

The PIN diode based on the SOI substrate has received widespread attention for higher sensitivity and lower parasitic capacitance than the conventional PIN diode [1-3]. It has a wide range of applications, including high-speed optoelectronic systems, ultraviolet detection and so on [4,5]. In UV sensing, the absorption of UV radiation is very close to the surface of the detector due to the light absorption properties of the silicon material. To achieve a high sensitivity in the wavelength range of UV radiation, the absorption layer, that is the depleted region of the diode, must be as near as possible to the surface of the detector. Thus, a lateral PIN diode based on the SOI substrate with a thin top layer Si is the most suitable for UV detection. Besides, the theoretical and previous studies have shown that the photoelectric properties of the device are optimal by carefully choosing the parameters of SOI substrate and depleted region. The adjustment of the thickness of the Si absorption layer can improve the response of a certain wavelength. Meanwhile, the width of the appropriate depletion region is also important for the absorption of light and the operating speed. With the absorption layer of 210nm, the sensitivity of the device reaches 7.5mA/W at 850nm [6]. However, the high-speed detectors based on RF-SOI in the short wavelength band are rarely studied.

In this work, we propose and study the photodiode fabricated on the RF-SOI substrate which shows excellent linear correlation between the photocurrent and the optical intensity. The response spectrum and the capacitance feature of the device are further investigated, which appear UV enhanced characteristics and very low parasitic capacitance, thanks to the SOI substrate perspective.

## THE STRUCTURE OF RF-SOI PIN

Figure 1(a) schematically shows the structure of the PIN photodiode built on RF-SOI substrate with 75nm top silicon film, 500nm buried oxide (BOX) and the trap rich layer. The N+ type acts as the cathode of the device while the P+ type is defined as the anode of the device, with 3 μm length of the intrinsic layer. The mesa isolation of the device is fabricated by photolithography followed by wet etching of top Si layer. Lithography and ion implantation steps are used to form P+ and N+ regions. After high temperature anneal to activate the dopants, metal layers are deposited with evaporation process to form anode and cathode contacts. As the microscope image in Figure 1(b) shows, the device uses an interdigitated electrode and the effective detection area of the device fabricated is 0.04cm².

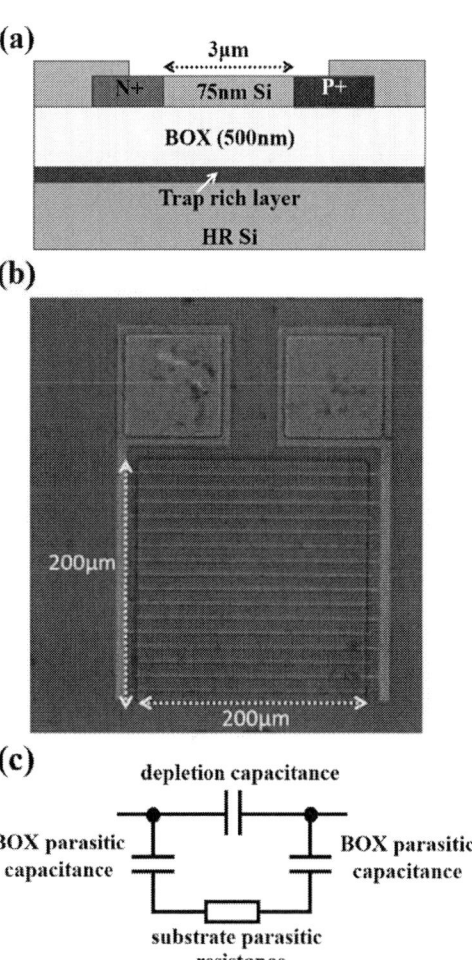

Fig. 1. (a) the basic schematic cross-sectional view of RF-SOI PIN photodiode. (b) microscope image and (c) the equivalent circuit of the device.

In the operating state, the equivalent circuit appears in Figure 1(c). the capacitance of the PIN diode depletion region is mainly determined by the intrinsic region while the parasitic capacitance is greatly affected by the thickness of BOX. We carefully choose the RF-SOI substrate with 500nm BOX and the trap rich layer in the substrate to effectively suppress parasitic capacitance and increase the substrate parasitic resistance.

## PHOTOELECTRIC CHARACTERISTICS

At the light wavelength of 450nm, the I-V characteristics of the device were measured at different light intensity in Figure 2 (a). The dark current of the device is as low as 2.5nA/cm², so the device based on the SOI substrate has a very low noise associated with the dark current. Moreover, the light current under illumination significantly increases with various light intensity. Figure 2(b) is the dependence of the photocurrent versus optical intensity, which shows excellent linearity from 10μW/cm² to 600μW/cm².

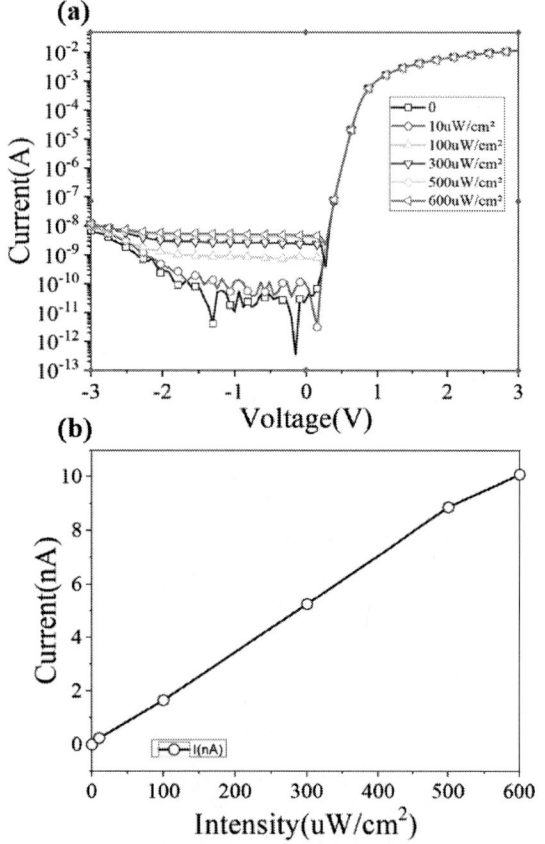

Fig. 2. (a) I-V characteristics of the device under various light intensity. (b) the dependence of the forward photocurrent versus optical intensity.

To understand the response spectrum of the device, further investigations of the RF-SOI PIN are conducted. Figure 3 shows the sensitivity and quantum efficiency (QE) of the photodiode at both V=-1V and at an optical intensity of 100μW/cm². Compared to the conventional SOI PIN detectors, the RF-SOI PIN detector shows a significantly high response in the UV range with the QE reaching up to 89% at wavelength of 300nm. This is because the thin-top Si layer is more efficient to absorb light with short-wavelength, leading to a high sensitivity and QE in the UV band.

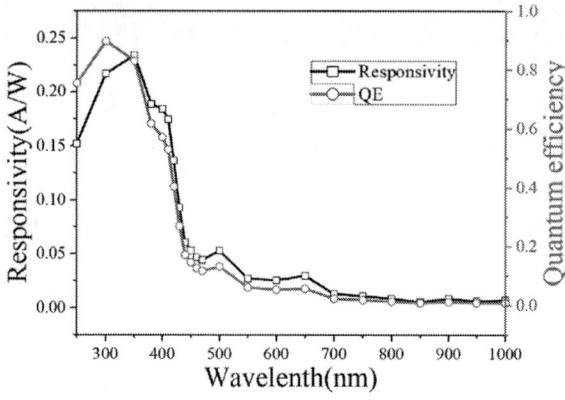

Fig. 3. The responsivity and quantum efficiency of the device vs. wavelength in the λ=250-100nm range, at a fixed reverse bias of V=-1V and light intensity of 100μW/cm².

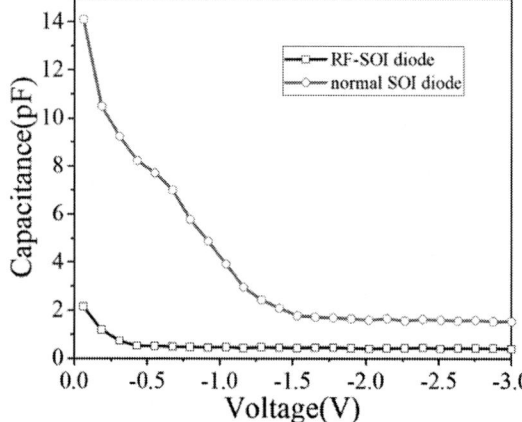

Fig. 4. The capacitance of the device vs. voltage with both RF-SOI diode and normal SOI diode.

The C-V curves are presented in Fig. 4 to understand the advantage of the RF-SOI substrate. Figure 4 compares the capacitances of PIN diodes based on RF-SOI substrate and conventional SOI substrate which has 145nm BOX without trap-rich layer. The C-V measurements are performed at dark condition with frequency of 1MHz. Compared to the normal SOI diode, The RF-SOI PIN diode exhibits distinctly lower parasitic capacitance, which is as low as 0.4pF at fixed -1V bias. Therefore, the RF-SOI based PIN photodetection could have excellent frequency characteristics, which is more suitable for high-speed application.

## CONCLUSION

This work revealed a PIN diode based on the RF-SOI substrate for photodetection application. The device shows remarkable UV-enhanced characteristics with the quantum efficiency up to 89% at 300nm. In addition, we further show that the low parasitic capacitance is due to the RF-SOI substrate which significantly reduces the parasitic capacitance of the fabricated photodiode. The demonstrated photodiode in this paper RF-SOI-based substrate might be useful for high-speed optical communication and imaging system in short wavelength band.

## ACKNOWLEDGMENTS

This work was supported by the Shanghai Science and

978-1-6654-9759-6/22 $31.00 © 2022 IEEE

Technology Commission "explorer project"(21TS1401300) and Pioneering Project of Academy for Engineering and Technology, Fudan University (gyy2021-001).

## REFERENCES

[1] J Liu, S. Cristoloveanu , and J Wan. "A Review on the Recent Progress of SOI‐based Photodetectors." Physica Status Solidi (2021).

[2] Colinge, and J.-P. "p-i-n photodiodes made in laser-recrystallized silicon-on-insulator." IEEE Transactions on Electron Devices 33.2(1986):203-205.

[3] Afzalian, A. , and D. Flandre . "Physical Modelling and Design of thin film SOI lateral PIN Photodiodes for Blue DVD-applications." IEEE International Soi Conference IEEE, 2004.

[4] Mitsuno, H. , T. Maruyama , and K. Iiyama . "Sub-μm electrode spacing SOI-PIN photodiode fabricated by CMOS compatible process." Optoelectronics & Communications Conference IEEE, 2016.

[5] Afzalian, A. , and D. Flandre . "Measurements, Modelling and Electrical Simulations of Lateral PIN Photodiodes in Thin Film-SOI for High Quantum Efficiency and High Selectivity in the UV range. " Conference on European Solid-state Device Research IEEE, 2003.

[6] Li, G. , et al. "Over 10 GHz lateral silicon photodetector fabricated on silicon-on-insulator substrate by CMOS-compatible process." Jpn.j.appl.phys 54.4s(2015):04DG06.

# 55NM ULTRA-LOW-POWER PLATFORM 3.3V I/O DEVICE RELIABILITY IMPROVEMENT

*Zhao Guo*

Shanghai Huali Microelectronics Corporation (HLMC)
Shanghai, China
15221318291, guozhao@hlmc.cn

## ABSTRACT

The demand of Ultra-Low-Power (ULP) chips is increasing gradually with the development of intelligent society. ULP chips are widely used in Bluetooth and WIFI. In order to further expand the application scenarios, we developed 3.3V I/O device based on Huali-55nm ULP platform by adjusting the gate oxide thickness and implant (IMP) conditions (source, angle, energy and dosage). Then, a series of process optimizations were performed to achieve a high reliability level for the 3.3V I/O device. The Hot Carrier Injection (HCI) Direct Current (DC) $t_{0.1\%}$ lifetime of NMOS and PMOS can reach 0.63year and 24.5year respectively (DC criteria is 0.2year), while Negative Bias Temperature Instability (NBTI) DC $t_{0.1\%}$ lifetime can reach 127year (DC criteria is 5year).

*Key words： 3.3V; I/O device; ultra-low power; reliability; HCI; NBTI*

## INTRODUCTION

The common working voltage of I/O device in HLMC 55nm ULP platform is 2.5V, and the corresponding gate oxide thickness is 58.5Å. In order to coexist with CORE device (gate oxide thickness is 26Å), we adopt Dual Gate (DG) production process: Firstly, a layer of 58.5Å thick oxide is grown globally, and then the oxide on the surface of CORE device is removed by dual gate mask and WET process, finally, a layer of 26Å thin oxide is grown.

Because of the higher operating voltage of 3.3VIO devices, the corresponding gate oxide is also thicker. Referring to 3.3V IO device on other platforms, the gate oxide thickness was finally set at 76Å. In order to allow the co-existence of 2.5V I/O device, 3.3V I/O device and 1.2V CORE device, we finally decided to adopt triple Gate to grow gate oxide. The specific technological process is shown in Figure 1.

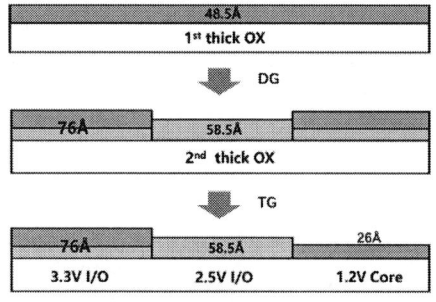

*Fig.1. Triple gate process*

Firstly, a layer of gate oxide with a specific thickness is grown, and then the oxide on the surface of 2.5V I/O device is removed by DG mask and WET process, and then a layer of 58.5Å thick oxide is grown. At this time, the combined thickness of the first and second gate oxide is 76Å. Then the oxide on the surface of CORE device is removed by Triple Gate (TG) mask and WET process, and finally a thin layer of 26Å oxide is grown. At this point, gate oxide has been generated on the surface of the 2.5V I/O, 3.3V I/O and CORE devices.

TABLE 1. GATE OXIDE THICKNESS IN TRIPLE GATE PROCESS

| EXP | Thick OX-1 (Å) | Thick OX-2 (Å) | Total (Å) |
|---|---|---|---|
| First | 58.5 | 58.5 (48%) | 86.5 |
| Second | 58.5 | 44.5 (33%) | 73.3 |
| Third | 42.5 | 58.5 (50%) | 71.3 |
| Fourth | 48.5 | 58.5 (47%) | 76.1 |

However, since the superposition of the first and second gate oxide is not linear, the thickness of the first layer of gate oxide is critical in order to make the superposition thickness reach 76Å. A series of experiments were conducted to determine the first gate oxide thickness. The results are shown in the table 1, the first layer gate oxide thickness is finally determined to be 48.5Å. Due to the change of the growth conditions of gate oxide, the thermal of the device will be affected. By referring to the electrical parameters of other platforms, we adjusted the IMP dosage to make the saturation current ($I_{dsat}$) of NMOS typical size device reach 600µA/µm. Then the reliability test was carried out. It was found that the DC $t_{0.1\%}$ lifetime of PMOS HCI and NBTI is 33.3year and 14.7year respectively, far higher than the industry standard. However, the DC $t_{0.1\%}$ lifetime of NMOS HCI is only 0.15year, which is lower than the industry standard. It is necessary to improve the production process to improve the reliability of NMOS HCI [1-6].

## EXPERIMENT AND CHARACTERIZATION

All experiments are based on HLMC 12 inch 55 nm ULP logic process. We first adjusted the NIO LDD energy to improve the reliability of 3.3VIONMOS. The IMP condition used in the first round of experiments is 25KeV. According to previous experience, NIO HCI is inversely proportional to the $I_{sub}$, which is the substrate current. The increase of LDD energy can reduce the substrate current, but it will increase the leakage.

By splitting the energy of the LDD, we hope to find an optimal condition that can not only reduce the substrate current, but also make the leakage level not increases significantly. The energy of the LDD is divided into 22KeV, 25KeV, 28KeV, 31KeV, 34KeV, and based on these conditions Perform experimental verifications.

The experimental results are shown in Figure 2. The red circle is the substrate current with the increasing energy of the LDD IMP, and the blue circle is the leakage current ($I_{off}$). This result shows that when the energy of LDD IMP is 28KeV, $I_{sub}$ and $I_{off}$ have reached an optimal state. In the end, we set the energy of the LDD IMP as 28KeV, while the HCI of IONMOS is increased to 0.35year, and the leakage level did not increase significantly.

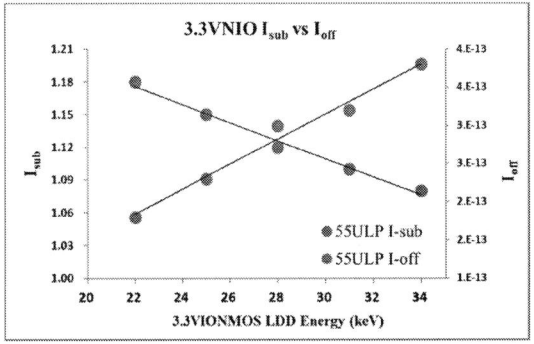

*Fig.2. $I_{sub}$ and $I_{off}$ of 3.3VIONMOS with increasing LDD IMP energy*

After the energy is determined, the device speed needs to be adjusted. Because the $I_{dsat}$ and reliability are also related to a certain degree, we still use different energy in the second round of experiments, and divide the IMP dosage to different degrees. The saturation current ranges from 550μA/μm to 630μA/μm. Then perform reliability tests on different current. The test results are shown in Figure 3. It can be seen from the figure that the reliability reaches the maximum under the condition of 28keV, which verifies our previous conclusions. The HCI reaches the highest at 600μA/μm. We use 28keV and 600μA/μm saturation current as the final condition of LDD IMP.

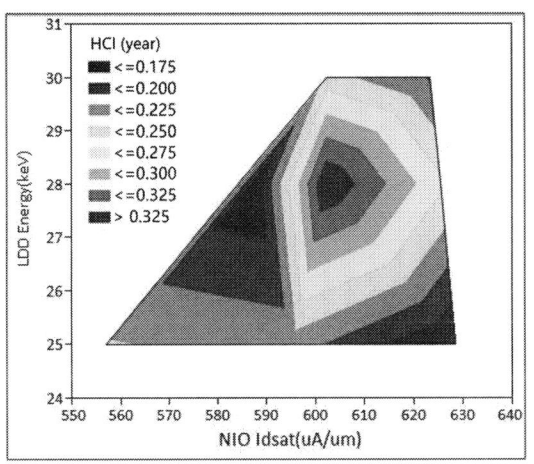

*Fig.3. Correlation between $I_{dsat}$ and HCI*

In addition to the LDD IMP energy and current, adding fluorine (F) IMP can play a role in repairing lattice damage, thereby increasing the reliability of IONMOS. The two parameters of F IMP need to be carefully selected. One is energy and the other is ion concentration (dosage). Refer to 3.3VIO devices on other platforms. In the third round of experiments, we divided the energy and dosage of F IMP to find an optimal condition to improve the reliability. F IMP energy is divided into 14keV, 20keV and 23keV, dosage is divided into 1x, 1.5x and 2x, where 1x is equal to $10^{15}$/cm3 ion concentration. The experimental results are shown in Figure 4. It can be seen from the figure 4 that the result of HCI (0.57year) is the highest under the conditions of 20keV and 1x dosage.

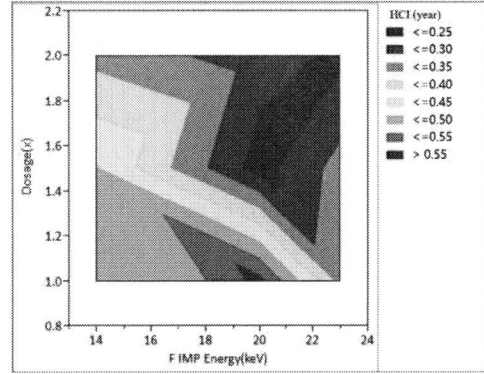

*Fig.4. Correlation between F IMP conditions and HCI*

## RESULTS AND DISCUSSION

After determining the LDD IMP energy and the conditions of the F IMP, we combined all the conditions for the final experiment. The purpose is to adjust the NIO $I_{dsat}$ to 600μA/μm by changing the LDD IMP dosage, which is the best condition we verified before. The final result It is shown that under this condition, the 3.3VNIO HCI reaches 0.63year, which has far exceeded the reliability standard. The overall reliability improvement measures are shown in Figure 5.

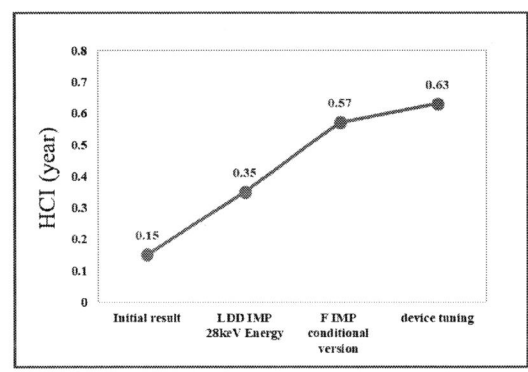

*Fig.5. 3.3VIO NMOS reliability improvement road map*

There are other ways to increase the reliability of 3.3VIO devices, such as increasing the size of Poly, etc. We will conduct experimental verification in the follow-up.

## CONCLUSION

Based on the HLMC 55ULP platform, we developed a 3.3VIO device using a triple gate process. However, under the initial conditions, NIO HCI is only 0.15year, which is seriously lower than the industry standard. In order to improve the reliability of NIO, we have made a series of improvements, including the adjustment of LDD IMP energy, adding F IMP and determining IMP conditions. Finally, the device was adjusted so that the HCI of 3.3VIO NMOS reached 0.63year, and under this condition, the HCI and NBTI of PMOS has also met the demand for mass production.

## REFERENCES

[1] B. Kaczer F. Crupi R. Degraeve Ph. Roussel C. Ciofi and G. Groeseneken "Observation of hot-carrier-induced nFET gate-oxide breakdown in dynamically stressed CMOS circuits" *IEDM Tech*. Dig. pp. 171 2002.

[2] H. C.-H. Wang C. H. Diaz B.-K. Liew J. S.-C. Sun and T. Wang "Hot carrier reliability improvement by utilizing phosphorus transient enhanced diffusion for input/output devices of deep submicron CMOS technology" *IEEE Electron Device Lett.* vol. 21 pp. 598-600 2000.

[3] Oxide reliability of drain engineered I/O NMOS from hot carrier injection [J]. *IEEE Electron Device Letters*, 2003

[4] D. Nayak M. Y. Hao J. Umali and R. Rakkhit "A comprehensive study of performance and reliability of P As and hybrid As/P n LDD junctions for deep-submicron CMOS logic technology" *IEEE Electron Device Lett.* vol. 18 pp. 281-283 June 1997.

[5] I. Chen J. Y. Choi T. Chan and C. Hu "The effect of channel hot-carrier stressing on gate-oxide integrity in MOSFET's" *IEEE Trans. Electron Devices* vol. 35 pp. 2253-2258 Dec. 1988.

[6] T. Hori K. Kurimoto T. Yabu and G. Fuse "A new submicron MOSFET with LATID (large-angle-tilt implanted drain) structure" *VLSI Symp*. Tech. Dig. pp. 15-16 1988.

# MBIST REPAIR MECHANISHM AND IMPLEMENTATION

*Haijing Wu[1]\**

[1] Global Application and Development Center,
Advantest (China), Co., Ltd., Shanghai 201203, China
*Corresponding Author's Email: haijing.wu@advantest.com

## ABSTRACT

Today's deep submicron technologies allow the implementation of multiple memories on a single chip. Due to their high density, memories are more prone to faults. These faults impact the total chip yield [1]. Keeping the memory cores at a reasonable yield level is thus vital for SOC products. For such purpose, memory designers usually employ redundancy repair logic using spare rows and columns to improve the yield. The BISR (Built in Self Repair) technique is a promising and popular solution for enhancing the yield of memories with the redundancy logic. Furthermore, redundancies of memories are not just for defects, redundancies can also be used to recover yield due to process variation in addition to yield recovery for defects.

## INTRODUCTION

### MBIST repair mechanism introduction

One way to solve this yield problem is to enhance the memory by redundant memory locations. The address mapping of the fault free working memory is programmable within certain limits. In order to do so, a memory test is needed to identify the faulty regions. The memory is tested by external test hardware or by which on chip dedicated hardware.

This paper will share learnings for a reconfigurable BISR scheme for repairing RAMs and a test flow on ATE testing. An efficient redundancy analysis algorithm is proposed to allocate redundancies of defective RAMs. In BISR, a reconfigurable built − in redundancy analysis (BIRA) circuit is designed to perform the redundancy algorithm for carious RAMs. The BISR structure has been synthesized and found that the area cost when compared with the dedicated BISR structure is very small.

## THEORY AND METHODOLOGY
### BUILT-IN SELF-TEST

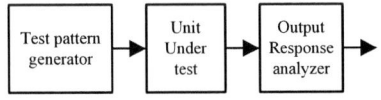

*Figure 1. BIST Block diagram*

A normal BIST block diagram is given in figure 1. It consists of a test pattern generator, unit under test and a response analyzer. The test pattern generates different test patterns which can be applied to the unit under test. The output response analyzer ascertains the faults in the unit under test and provides the faulty information as the output.

### MBIST Introduction

The basic BIST model block diagram is shown below. A memory BIST unit consists of a controller to control the flow of test sequences and other components to generate the necessary test control and data, generates patterns to the memory and reads them to log any defects. Memory BIST also consists of a repair and redundancy capability.

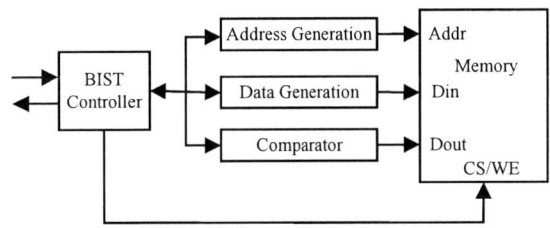

*Figure 2. MBIST Model*

The process of testing the fabricated chip on ATE involves the use of external test patterns applied as stimulus. The device's response is analyzed on the tester, comparing it against the golden response which is stored as part of the test pattern data. MBIST makes this easy by placing all these functions within a test circuitry surrounding the memory on the chip itself. It implements a finite state machine (FSM) to generate stimulus and analyze the response coming out of memories.

This extra self-testing circuitry acts as the interface between the high-level system and the memory. The challenges of testing embedded memories are minimized by this interface as it facilitates controllability and observability. The FSM provides test patterns for memory testing; this greatly reduces the need for an external test pattern set for memory testing [2].

### Memory Built-in Self Repair (MBISR)

The block diagram of a MBISR scheme for a RAM is shown in figure 3, which consists of four major components.
Repairable RAM: A RAM with redundancies and reconfiguration circuit is called as a repairable RAM.
BIST Circuit: It can generate test patterns for RAMs under test. While a fault in a defective RAM is detected by the

978-1-6654-9759-6/22 $31.00 © 2022 IEEE

BIST circuit, the faulty information is sent to the BIRA circuit.

BIRA Circuit: It collects the faulty information sent from the BIST circuit and allocates redundancies according to the collected faulty information using the implemented redundancy analysis algorithm.

Fuse Macro: It stores repair signatures of RAMs under test.

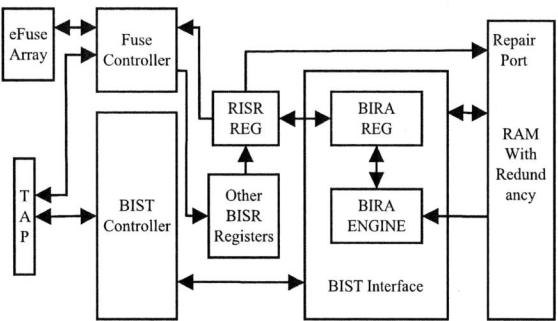

Figure 3. BISR Architecture

Memory repair is implemented in two steps. The first step is to analyze the failures diagnosed by the MBIST Controller during the test for repairable memories, and the second step is to determine the repair signature to repair the memories. All the repairable memories have repair registers which hold the repair signature.

BIRA (Built-In Redundancy Analysis) module helps to calculate the repair signature based on the memory failure data and the implemented memory redundancy scheme. It also determines whether the memory is repairable in the production testing environments. The repair signature will be stored in the BIRA registers for further processing by MBIST Controllers or ATE.

The repair signature is then passed on to the repair register's scan chain for subsequent Fusebox programming, which is located at the chip level. The reading and writing of a Fusebox is controlled through TAP (Test Access Port) and dedicated repair registers scan chains connecting memories to fuses. The repair information is then scanned out of the scan chains, compressed, and is burnt on-the-fly into the e-Fuse array by applying high voltage pulses.

On chip reset, the repair information from the e-Fuse is automatically loaded and decompressed in the repair registers, which are directly connected to the memories. This results in all memories with redundancies being repaired. Finally, BIST is run on the repaired memories which verifies the correctness of memories.

# MBIST REPAIR PROCESS
## Test Flow to Activate the Redundancy

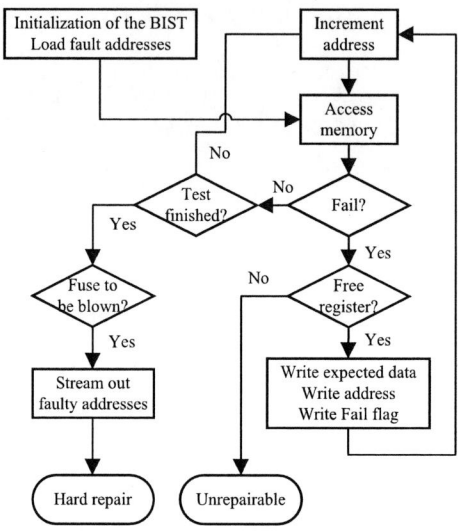

Figure 4. Test flow to activate the redundancy

The test program flow to active the redundancy is shown in figure 4. Two different results are possible (Another method of software repair is not widely used).

- A hardware repair is done including the process to blow the fuses. This is normally done a wafer level test because most of the fuses are activated with laser before packaging.
- A repair overflow when too many failures been encountered. This allows the test to identify failures that cannot be repaired.

## MBIST BISR Process

Based on the previous introduction to MBIST repair, we have generally understood the process of memory repair. The figure 5 and figure 6 show the summarized repair process and memory BISR flow respectively. The whole test process is as follows:

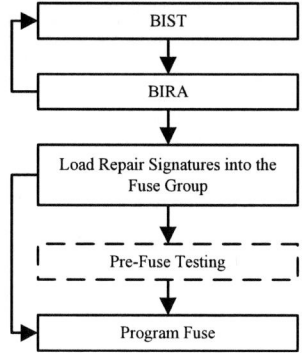

Figure 5. Repair Process

978-1-6654-9759-6/22 $31.00 © 2022 IEEE

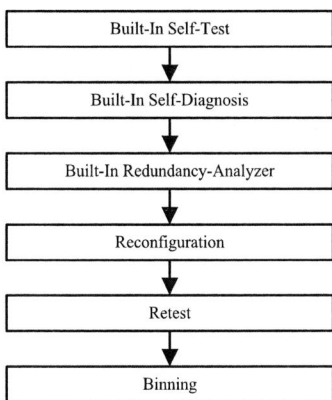

*Figure 6. Memory BISR Flow*

## SUMMARY

There are 6 steps that correspond to the steps on ATE:
1. MBIST test
2. Fault Location
3. Redundancy allocation
4. Swap defective cells
5. Verification
6. Binning

As discussed in the article, using the MBIST model along with the algorithms and memory repair mechanisms, including BIRA and BISR, provides a low-cost but effective solution. However, whether MBIST repair is needed depends on other factors such as demand and budget.

## ACKNOWLEDGEMENTS

The author wishes to thank Advantest for the support on this paper.

## REFERENCES

[1] V. SRIDHAR, M. RAJENDRA PRASAD. Built-in self -repair (BISR) technique widely Used to repair embedded random-access memories (RAMs)

[2] Milind Priyadarshi. Memory Testing – An Insight into Algorithms and Self Repair Mechanism.

# THE RESEARCH OF DECREASING SIGE LOSS IN FULLY DEPLETED SILICON ON INSULATOR (FDSOI) DEVICES

*Siyuan Che[1]\*, Xiangguo Meng[1], Lian Lu[1], Quanbo Li[1], Jun Huang[1], Yu Zhang[1]*

[1]Shanghai Huali Integrated Circuit Corporation, Shanghai, China

*Corresponding Author's Email: chesiyuan@hlmc.cn

## ABSTRACT

SIGE-On-Insulator (SGOI) structure in P-channel MOSFETs of Fully Depleted Silicon on Insulator (FDSOI) devices is widely used for mobility enhancement and junction leakage current reduction. However, it is a greater challenge to obtain relatively complete SIGE layer, due to excessive SIGE loss, which is caused by etching Spacer-1 and EPI layers. This paper introduces an optimization approach which is 2 in 1 step to decrease SIGE loss from 75A to 17A, reducing by 77.3%. In addition, the comprehensive analysis of SIGE loss is also proposed. The result provides an effective strategy to fabricate high-performance FDSOI devices with safety SIGE thickness.

## INTRODUCTION

As Silicon on Insulator (SOI) process technology moves in nano-scale, Partially Depleted Silicon on Insulator (PDSOI) device exposed serious floating-body effects and short channel effects. At the same time, Fully Depleted Silicon on Insulator (FDSOI) devices are introduced with the aim of inhibiting negative effects [1]. Moreover, in the fabrication of devices, the SIGE-On-Insulator (SGOI) structures of P-channel MOSFETs, which are advantageous for enhancing hole mobility and suppressing junction leakage current, have been intensively used [2-4].

However, when Si is replaced by SIGE in P-channel MOSFETs, Spacer-1 and EPI process encounters new challenges. It is difficult to obtain relatively complete SIGE layer. Excessive SIGE loss results in failure of epitaxial growth. Further, Contact process punch through insulator layer, which decrease performance of FDSOI device.

Therefore, it is obvious that the formation of relatively complete SIGE layer would take a crucial role on device performance. In this paper, an optimization approach is proposed to decrease SIGE loss. Also, comprehensive analysis of SIGE loss should be pointed out and studied.

## EXPERIMENT

The schemes of Spacer-1 and EPI process are described respectively, as shown in Figure 1 (A) and (B). The film stack is SICN for Spacer-1 and BARC/SIN for EPI.

The films are etched by ICP plasma (Inductively Coupled Plasma) named Kiyo EX from LAM Corporation.

The morphology characterization is taken using TEM named JEM-2100F from JEOL Corporation.

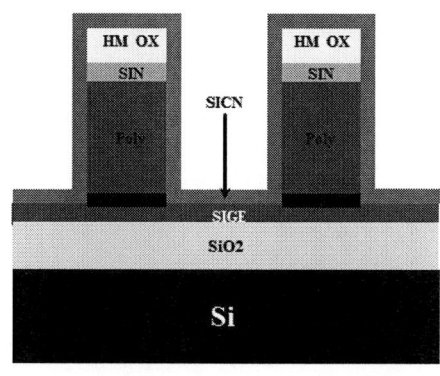

*(A)*

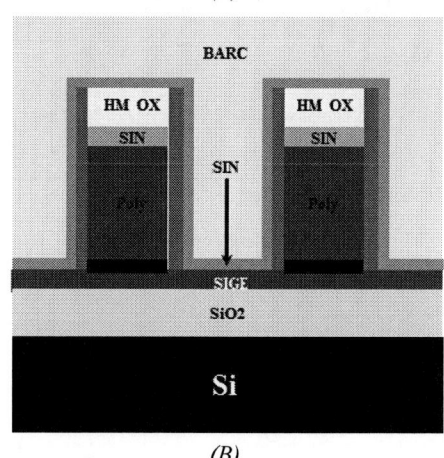

*(B)*

*Figure 1: (A) Spacer-1 process with SICN film stack; (B) EPI process with BARC/SIN film stack*

## RESULT AND DISCUSSION

Fig.2 shows NMOS and PMOS substrate condition after Spacer-1 and PEPI etching. NMOS using Si substrate shows <10A loss (Figure 2 (A)). However, SIGE in PMOS substrate decreases from 73A to 48A after Spacer-1 etch process, showing 34% loss (Figure 2 (B)). Seriously, SIGE film is fully etched after PEPI etch process (Figure 2 (C)).

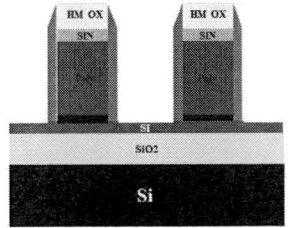

*(A) The diagram (left) and TEM (right) of Si substrate condition in NMOS area after Spacer-1 process.*

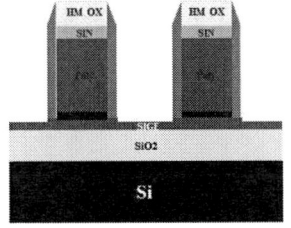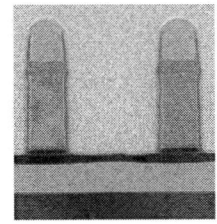

*(B) The diagram (left) and TEM (right) of SIGE substrate condition in PMOS area after Spacer-1 process.*

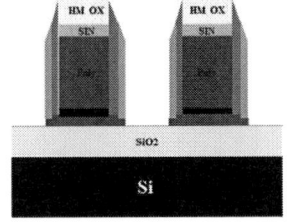

*(C) The diagram (left) and TEM (right) of SIGE substrate condition in PMOS area after PEPI process.*

*Figure 2: NMOS and PMOS substrate condition*

SIGE shows a large number of loss compared to Si. In order to get relatively complete SIGE layer, multi kinds of cross-linked reasons should be considered, such as flow design, dry etch process and wet process.

*A. Flow Design*

Generally, Spacer-1 and PEPI etch process are independent steps. As shown in Figure 3 (A), Spacer-1 SIN is etched at first. Then, etching PEPI BARC/SIN is executed. Separate two steps cause superimposed SIGE loss severely. In that case, the method of combination of Spacer-1 and PEPI process is proposed, which etch BARC/SIN and SICN together. The change of flow design by reducing one etch step shows a huge boost to SIGE loss (Figure 3 (B)).

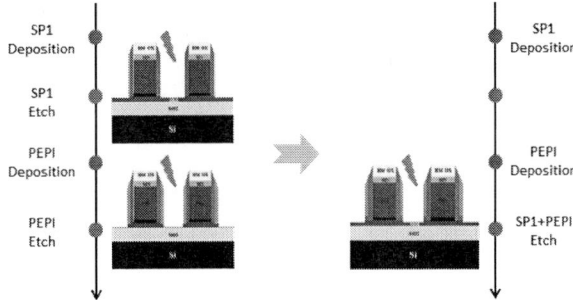

*(A) The change of flow design*

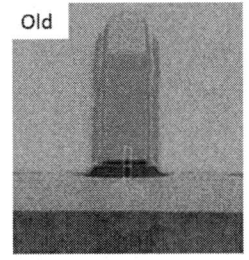

*(B) TEM comparison*

*Figure 3: The flow design change condition*

*B. Dry Etch Process*

As we known, CF4, CH2F2 and CH3F that involved F content gases are usually used for etching nitride. Selectivity of film to film is a very necessary factor to consider while choosing the type and flow rate of gas. Normally, CH3F and O2 are perfect gases as SIN OE step, which exhibits higher selectivity in SIN/SI film stack. However, using the same gas for SIN/SIGE, SIGE is etched simultaneously. Lower SIN/SIGE selectivity is the root cause of it. It is obvious that the increasing of SIN/SIGE selectivity would take a crucial role on SIGE loss.

We attempted to use half of original CH3F to tuning selectivity. Results are shown in TABLE I, the SIN/SIGE selectivity is 2.8 by using new OE step, 1.3 times higher than BSL. The lower flow rate of CH3F shows a higher selectivity of SIN/SIGE, which is beneficial for SIGE loss.

TABLE I
Etch rate and selectivity for different OE step

| OE step | SIN ER | SIGE ER | SIN/SIGE Selectivity |
|---------|--------|---------|----------------------|
| BSL | 82.2 A/min | 38.4 A/min | 2.14 |
| New | 43.8 A/min | 15.6 A/min | 2.81 |

*C. Wet Process*

The whole PEPI process contains film deposition, dry etch and wet process. Obviously, wet process is an indispensable step for polymer strip after etching process. The solution selection of wet process is also one of the most important points for SIGE loss control. It is necessary to check wet effect in SIGE loss.

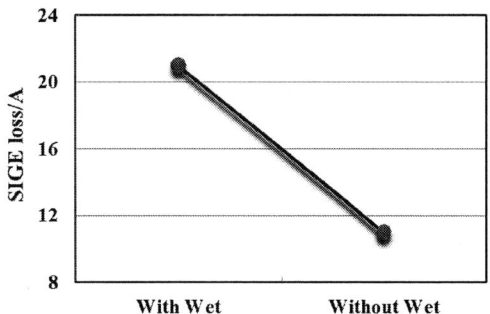

Figure 4: SIGE loss & wet condition correlation

We compared remaining SIGE thickness with wet and without wet, respectively. As shown in Figure 4, SIGE loss with wet is 90% more than that without wet. Wet solution consisted of HF which is widely used to strip polymer and oxide causes SIGE loss severely. For this condition, improved wet recipe is used, instead of BSL.

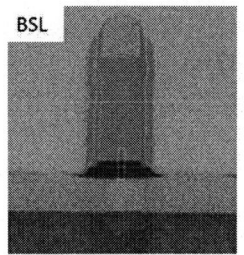

Figure 5: Current SIGE loss compared with BSL

Finally, SIGE loss after PEPI step decreases from 75A to 17A, reducing by 77.3%, under the combined action of improved flow design, etch and wet process, as shown in Figure 5.

## CONCLUSIONS

In this paper, the comprehensive analysis of SIGE loss caused by Spacer-1 and PEPI etch has been investigated in detail. Flow design, etch and wet processes are the main issues which have been considered for get a relatively complete SIGE. We delivered expected SIGE layer with safety SIGE thickness to fabricate high-performance FDSOI devices.

## ACKNOWLEDGMENTS

Thanks for all the authors for great work and collaboration. Thanks for FA (failure analysis) for TEM support.

## REFERENCES

[1] S. Takagi, in Symp. VLSI Tech. Dig., 2003, pp. 115–116.
[2] T. Irisawa, S. Tokumitsu, T. Hattori, N. Nakagawa, S. Koh, and Y. Shiraki, Appl. Phys. Lett., vol. 81, pp. 847–850, Jul. 2002.
[3] M. L. Lee and E. A. Fitzgerald, in IEDM Tech. Dig., 2003, pp. 429–432.
[4] T. Low, Y. T. Hou, M. F. Li, C. Zhu, A. Chin, G. Samudra, L. Chan, and D.-L. Kwong, in IEDM Tech. Dig., 2003, pp. 691–694.

# INVESTIGATION OF EPOXY MOLDING COMPOUND ON SOP DEVICE

*Hongjie Liu, Wei Tan, Xingming Cheng, Lanxia Li, Yangyang Duan, Dandan Fan, Ming Gu,*
*jianglong Han*
JiangSu HHCK Advanced materials Co., Ltd.
Lianyungang, 222047, China
hongjie.liu@hhck-em.com

## ABSTRACT

With the rapid development of the semiconductor market, semiconductor device turns to more and thinner, lighter and smaller. Correspondingly, the performance requirement of epoxy molding compound (EMC) which used as packaging material for semiconductor device has become more and more stringent too. Temperature Cycle Test (TCT), Pressure Cooker Test (PCT), High Temperature Storage Test (HTST) as well as delamination test after precondition Moisture Storage Level (MSL), etc. are the normal reliability performance test. Here, from the material point of view, we investigate different type of epoxy and hardener on the influence of the performance of epoxy molding compound. The result showed that epoxy molding compound with modified multi-functional type epoxy resin1 displayed highest glass transition temperature (Tg), as well as the worst delamination performance, and then is the EMC with modified multi-functional type2 epoxy molding compound. While EMC with MAR type epoxy resin 3 in exhibited the best delamination performance with the lower Tg, lower high temperature modulus as well as lower water absorption. Furthermore, additives including prepolymer with terminal isocyanate groups 1, maleic anhydride modified polybutadiene 2 as well as photopolymer of polybutadiene 3 have been tried, and the results disclosed that maleic anhydride modified polybutadiene showed good delamination performance and then is the photopolymer of polybutadiene, while prepolymer with terminal isocyanate groups exhibited the worst performance.

*Keywords—epoxy molding compound; resin; delamination; water absorption; additive*

## INTRODUCTION

Epoxy resins have relatively low volumetric shrinkage upon polymerization, good thermal and dimensional stability, excellent moisture and chemical resistance, and superior electrical and mechanical properties [1]. Epoxy molding compounds (EMCs) are often employed to encapsulate microelectronic devices. With the rapid development of the semiconductor market, semiconductor device turns to more and thinner, lighter and smaller. Correspondingly, the performance requirement of epoxy molding compound (EMC) which used as packaging material for semiconductor device has

become more and more stringent too. Especially with the application the lead-free solder, the reflow temperature increased to about 260 degree centigrade for surface mounting package [2]. And delaminations often happened during the reflow process for the coefficient of thermal expansion between epoxy molding compound, chip, die-attach and the chip; the higher water absorption of the epoxy molding compound during the pre-condition according to the JEDEC standard, as well as the insufficient adhesion strength between epoxy molding compound and the leadfream[3].

G. H. Oh et al.[4] have tested the change of adhesion strength between EMC/chip interface with respect to plasma treatment number, and it was found that plasma treatment could induce higher adhesion strength than that of without- plasma treatment case. However, excessive plasma treatment results in a decrease in the adhesion strength by reduction of effective area. It was also found that the moisture absorption could be prevented by the plasma treatment due to stronger chemical bonding between EMC and Si chip. C. Chao et al. [5] have investigated the surface oxidized layers of Cu-Fe-Zn-P (C194) and Cu-Ni-Si-Mg (C7025) lead frames under different oxidation conditions and their influence on the adhesion of epoxy molding compounds with lead frames. And they found that in addition to surface roughness, surface oxides and their thickness play more important roles in the interfacial strength between lead frames and epoxy molding compounds. In the case of the surface with one single $Cu_2O$ layer, an increase in oxide layer results in a decrease in bonding strength. Once the topmost surface forms CuO, the bonding strength could be enhanced due to a greater surface energy. Compared with electroplated C194, the thinner surface oxide layer and thus better performance in adhesion strength for electroplated C7025 can be ascribed to the immersion Ag surface treatment.

Here, from the material point of view, we have tried three different type of epoxy resins and three different types of additives on the influence of the performance of epoxy molding compound.

## EXPERIMENTAL
### Materials

Preporties of epoxy resin 1-3 used in this study was showed in the table I, and structures of additives 1-3 was displayed in table Ⅱ. Phenol hardeners was MAR type phenol MEHC-7851SS obtained from MeiWa Plastic Industries, Ltd. Triphenylphosphine as accelerator was got from HOKKO's Fine Chemicals. 2,4-diamino-6-[2'-undecylimidazolyl-(1')-ethyl-s-triazine (C11-ZA) was obtained from SHIKOKU Chemical Corporation. Filler was chosen as the spherical fused silica from Novoray Corporation.with the mean diameter as 23 μm and maximum filler size as 75 μm. Silanes were form shin-Etsu Chemical Co. Ltd., mold releasing agent was carnauba wax obtained from Foncepi Comercial Exportadora Ltd.

TABLE I.    PROPERTIES OF EPOXY RESIN

| Type | Details of raw materials | | |
|---|---|---|---|
| | *EEW g/eq* | *Soft point /℃* | *Viscosity @150℃/P* |
| Epoxy resin 1 | 168 | 66 | 1.6 |
| Epoxy resin 2 | 217 | 57 | 1.5 |
| Epoxy resin 3 | 276 | 56 | 0.6 |

TABLE II.    PROPERTIES OF ADDITIVES

| Type | Details of additives | |
|---|---|---|
| | *Name* | *Viscosity@25℃ /mpa.s* |
| Additive 1 | | 8000 |
| Additive 2 | | 200000 |
| Additive 3 | | 40000 |

**EMC preparation**

All the ingredients were weighted up according to the formulation and mixed with the high speed mixing machine, And then the mixture was feeding to double screw extruder under the heating condition for extrusion, after that, the discharged material go through sheet formation, after cooling, pre-braker, granulation, post blend, and then storaged at 5℃.

**Measurements**

**Dynamic Mechanical Analyzer (DMA) modulus and Tg**

The specimens of 50mm×13mm×3mm were prepared through transfer molding method and post mold cure (PMC) at 175℃ for 6 hours. The Dynamic Mechanical Analysis (DMA) of the cured specimens was performed in the three point bending mode under 1Hz sinusoidal strain loading on the Dynamic Mechanical

Analyzer with the temperature ramping from room temperature to 265℃ at the rate of 5℃/min. The storage modulus at room temperature, 175℃ and 260℃ and tan δ were recorded.

**Moisture absorption**

The specimens with the diameter of 50mm and thickness of 3mms were prepared through transfer molding method and post mold cure (PMC) at 175℃ for 6 hours. And then the samples were weighted and then placed in an 85℃/85%RH chamber for 168 hours. The samples were taken out of the chamber and weighted after the time run out. Weight gain of the EMC during aging at 85℃/85%RH chamber was defined as moisture absorption.

**Delamination**

EMC samples was molded on SOP8 packages with double die, after PMC at 175℃ for 6 hours, trim and form, plating, C-SAM Scanning Acoustic Microscope(SAM) inspection was carried out to check the delamination, after baked at 125℃ for 24 hours, the packages were placed in the 60℃/60%RH chamber for 40 hours for precondition, and then reflowed in the 260 reflow oven for three times according to JEDEC standard. And the then C-SAM inspection was carried out again.

## RESULTS AND DISCUSSION

**Influence of epoxy resin on the performance of EMC**

EMC with the filler content of 88% were prepared. The performance of EMC samples with different epoxy resins were tested, and the results were disclosed in table Ⅲ and Ⅳ.

TABLE III.    INFLUENCE OF EPOXY RESIN ON THE PERFORMANCE OF EMC

| properties | EMC samples | | |
|---|---|---|---|
| | *EMC A* | *EMC B* | *EMC C* |
| Epoxy resin | Resin 1 | Resin 2 | Resin 3 |
| SM @ RT/Mpa | 26658 | 27532 | 24864 |
| SM@175℃/Mpa | 1834 | 650 | 428 |
| SM@175℃/Mpa | 1152 | 396 | 376 |
| Peak of LM/℃ | 137 | 127 | 111 |
| Peak of tanδ/℃ | 149 | 141 | 119 |
| PCT 24h/% | 0.32 | 0.29 | 0.26 |

SM: storage modulus, LM: loss modulus

The results displayed that EMC A with epoxy resin 1 exhibited the highest high temperature modulus and Tg as well as the worst delamination performance, EMC C with epoxy resin 3 showed the lowest Tg, modulus, the lowest water absorption and no delamination was discovered

978-1-6654-9759-6/22 $31.00 © 2022 IEEE

before and after MSL3. While the performance of EMC B with epoxy resin 2 was between the EMC A and EMC C.

TABLE IV.    DELAMINATION PERFORMANCE OF EMC WITH DIFFERENT EPOXY

| Sample | Delamination | |
|---|---|---|
| | *Before MSL3* | *AFTER MSL3* |
| EMC A | | |
| EMC B | | |
| EMC C | | |

**Influence of additives on the performance of EMC**

EMC samples with epoxy resin as MAR type, hardener as MAR type and the filler content of 88% were prepared, and the additives were prepolymer with terminal isocyanate groups 1, maleic anhydride modified polybutadiene 2 and photopolymer of polybutadiene 3 respectively.

TABLE V.    INFLUENCE OF ADDITIVES ON THE PERFORMANCE OF EMC

| Properties | EMC samples | | |
|---|---|---|---|
| | *EMC D* | *EMC E* | *EMC F* |
| Epoxy resin | Additive 1 | Additive 2 | Additive 3 |
| SM @ RT/Mpa | 27884 | 26950 | 26647 |
| SM@175℃/Mpa | 592 | 606 | 673 |
| SM@175℃/Mpa | 395 | 534 | 432 |
| Peak of LM/℃ | 107 | 118 | 110 |
| Peak of tanδ/℃ | 122 | 127 | 117 |
| PCT 24h/% | 0.34 | 0.29 | 0.26 |

SM: storage modulus, LM: loss modulus

Table V and table VI disclosed the performance of the EMC samples with different additives. And the results showed that EMC D with prepolymer with terminal isocyanate groups 1 in possessed the highest water absorption and the worst delamination performance, while EMC E with maleic anhydride modified polybutadiene 2 exhibited the best delamination performance with no delamination before and after MSL3, the delamination performance of EMC E with photopolymer of polybutadiene 3 in displayed partial delamination after MSL3 although it possessed the lowest water absorption.

TABLE VI.    DELAMINATION PERFORMANCE OF EMC WITH DIFFERENT ADDITIVES

| Sample | Delamination | |
|---|---|---|
| | *Before MSL3* | *AFTER MSL3* |
| EMC D | | |
| EMC E | | |
| EMC F | | |

## CONCLUSION

The delamination performance of the epoxy molding compound on the SOP packages are relating properties of the EMC such as the water absorption, the Tg, the modulus, and from the results we can see that low modulus, low water absorption, relatively lower Tg was good for the delamination. And in this study, the result showed that epoxy molding compound with modified multi-functional type epoxy resin1 displayed highest glass transition temperature (Tg), as well as the worst delamination performance, and then is the EMC with modified multi-functional type2 epoxy molding compound. While EMC with MAR type epoxy resin 3 in exhibited the best delamination performance with the lower Tg, lower high temperature modulus as well as lower water absorption. For additives, maleic anhydride modified polybutadiene showed good delamination performance and then is the photopolymer of polybutadiene, while prepolymer with terminal isocyanate groups exhibited the worst performance.

## REFERENCE

[1] H.J. Tai, H. L. Chou. *European Polymer Journal,* vol. 36, 2000, pp. 2213-2219.

[2] H. Y. Lee. *Transactions on Electrical and Electronic Materials,* vol.1, 2000, pp. 23-28.

[3] K. Cho, E. C. Cho, C. E. Park. *J. Adhesion Sci. Technol.,* Vol. 15, 2001, pp. 439-456.

[4] G. H. Oh, S.J. Joo, J. W. Jeong and H. S. Kim. *Microelectronics Reliability,* Vol. 92, 2019, pp.63–72.

[5] S. C. Chao, W. C. Huang, J. H. Liu, J. M. Song, P. Y. Shen, C. L. Huang, L. T. Hung and C.H. Chang. *Microelectronics Reliability,* Vol. 99, 2019, pp.161–167.

978-1-6654-9759-6/22 $31.00 © 2022 IEEE

# IMPROVEMENT OF CMOS DEVICE PERFORMANCE BY A COMBINATION OF SPIKE AND FLASH ANNEALING

*Kecheng Chen[*], Lan Jiang, Yaoting Shen*

Research and Development Dept., Shanghai Huali Integrated Circuit Corporation, Shanghai 201314, China

*Corresponding Author's Email: chenkecheng@hlmc.cn

## ABSTRACT

As the size of semiconductor devices scales down, the fabrication of devices calls for highly activated source/drain with as little diffusion as possible. Continuing a trend that began with rapid thermal annealing (RTA), anneals continue to shrink in time and increase in temperature. This paper describes a fabrication process that uses a combination of spike RTA, laser spike annealing (LSA) and flash lamp annealing (FLA) and the device characteristics of product wafers showed positive results. The superior electrical characteristics of NMOS transistors whose gate length is less than 50nm, which are fabricated by using the FLA process, are demonstrated and leads to better threshold-voltage (Vt) roll-offs. Vt roll-off, Vt-Ion, Ion-Ioff correlation have been reviewed, with FLA combined, core N and PMOS device performance gains 2.5% and 5% respectively. The main benefits of flash annealing for junction formation are low sheet resistance junctions with a high level of activation together with no or limited diffusion.

## INTRODUCTION

The continuing progress of scaling CMOS technology to smaller feature sizes faces a severe challenge from performance loss cause by increasingly significant parasitic resistances in the transistor source/drain regions. Threshold voltage (Vth) variation control with junction scaling is a key limitation factor for Si planar device scaling [1]. A very high concentration of electrically active doping would help, but traditional doping methods face difficulties, including the need to exceed the equilibrium dopant solubility while strongly limiting the thermal budget to prevent excessive dopant diffusion. Continuing a trend that began with rapid thermal annealing (RTA), anneals continue to shrink in time and increase in temperature [2-3]. Millisecond annealing at high temperatures allows efficient dopant activation with well controlled limits on diffusion. During the last few years, rapid thermal annealing technologies like non-melt laser spike annealing (LSA) or flash lamp annealing (FLA) have been widely investigated in order to achieve high performance devices. Both of these anneal technologies use the wafer as a heat sink for the energy pulse provided to achieve rapid cooling. But as a certain amount of

well-controlled diffusion of the extension implantation under the gate is necessary to achieve good transistor performance, conventional spike annealing at lower peak temperatures is a feasible and robust technology. But conventional RTA is not favorable in future device process for its de-activation and dopant diffusion. So in the advanced node, LSA and FLA these new annealing strategies are mainly applied in combination with a conventional RTA [4-6]. In this work, firstly we compare the effects of Spike RTA, LSA and FLA on RS of the control wafer respectively, and then combination LSA and FLA after spike RTA to achieve the desired levels of dopant diffusion and activation for both blanket and device wafers.

## EXPERIMENT

300mm n- and p-type (100) Si wafer were used for both blanket and device studies. On blanket control wafer, the $5 \cdot 10^{15} cm^{-2}$ dose of B (Boron) was implanted into n-type crystalline silicon under 3Kev energy with 7° tilt and 22° twist. The device coming wafers were fabricated with the gate-last process flow for NMOS and PMOS are shown in Figure 1.

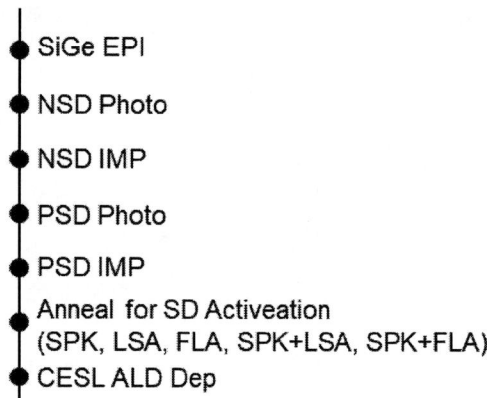

- SiGe EPI
- NSD Photo
- NSD IMP
- PSD Photo
- PSD IMP
- Anneal for SD Activeation (SPK, LSA, FLA, SPK+LSA, SPK+FLA)
- CESL ALD Dep

*Figure 1: The process flow of device coming wafers*

The spike RTA, LSA and FLA process was performed in the AMAT Vulcan, Ultratech 101LP and DNS LA-3000U machines respectively. The peak temperature of the spike RTA anneal is baseline (BSL) condition RTA1 (BSL) and higher RTA2, the LSA is LSA1/LSA2/LSA3 and temperature increases successively(LSA1<LSA2<LSA3), FLA Ta (Temp-assist) is 750℃, Tp (Temp-peak) FLA1/FLA2/FLA3 is the same as the LSA temperature. All wafer was annealed with single thermal (spike-RTA, LSA or FLA only) or a combination of double thermal is also investigated with first spike RTA then LSA or FLA (RTA+LSA, RTA+FLA). Blanket control wafer sheet resistances were measured by a KLA-Tencor RS-100 four-point probe (4PP) using probe type D with a circular 121 sites pattern and 3 mm edge exclusion.

**RESULT AND DISCUSSION**

In our work, as Figure 2 shows, under spike RTA (BSL) sheet resistance was the highest, while thermal was the lowest, and sheet resistance decrease with increasing temperature. At the same peak temperature, sheet resistance of LSA is lower than FLA. As mentioned in the previous paper, the process with slow cooling will deactivate the activated ions, and there is no deactivation of LSA process.

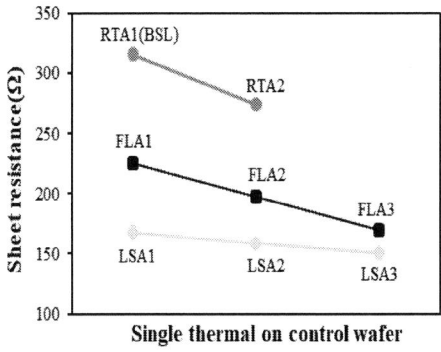

*Figure 2: Sheet resistance on implanted control wafer with single thermal*

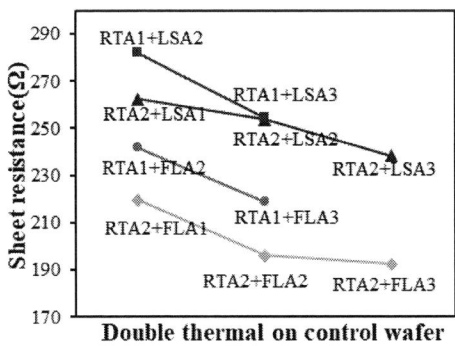

*Figure 3: Sheet resistance on implanted control wafer with double thermal*

As shown in Figure 2 and 3, with the thermal increase,

the more ions are activated and there is the lower sheet resistance. Compared with LSA, FLA can better activate the cluster generated by RTA deactivation at the same peak temperature after RTA. This is because the process time of LSA is too short and the thermal is too small to activate the cluster generated by RTA deactivation.

Figure 4 (a)/(b) shows that device wafer with RTA2 has the best performance; contrary to the result on control wafer, the LSA condition inline performance is the worst, indicating that LSA thermal is small and ion activation is too low. As shown in Figure 5 (a)/(b), LSA and FLA processes are combined with RTA2, respectively. The results show that the performance of core N/PMOS is significantly improved after RTA2 combined with FLA3, which is 2.5% and 5% respectively. However, the device combines LSA, the performance is comparable with baseline (RTA2).

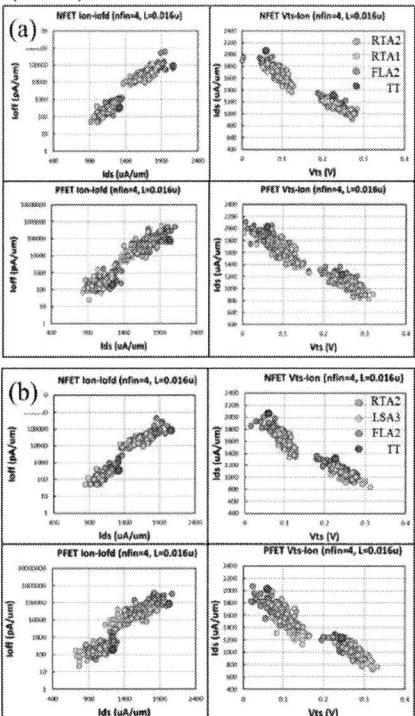

*Figure 4: N/P device WAT result with single thermal*

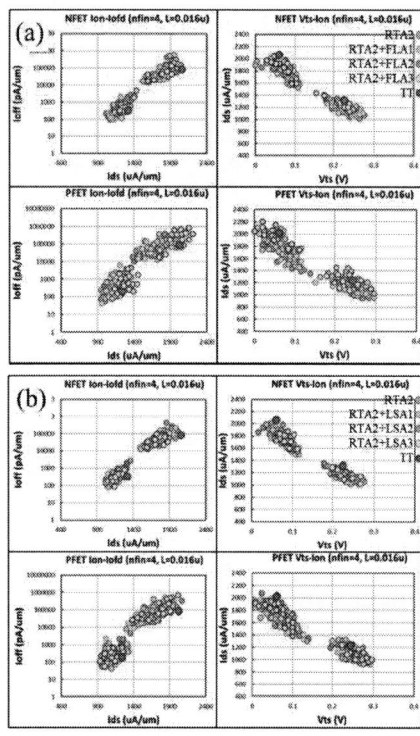

*Figure 5: N/P device WAT result with double thermal*

## CONCLUSION

In this paper, we describe a fabrication process that uses a combination of spike RTA, LSA and FLA, and the device characteristics of product wafers showed positive results. The superior electrical characteristics of NMOS transistors whose gate length is less than 50nm, which are fabricated by using the FLA process, are demonstrated and leads to better threshold-voltage (Vt) roll-offs. Vt roll-off, Vt-Ion, Ion-Ioff correlation have been reviewed, with FLA combined, core N and PMOS device performance gains 2.5% and 5% respectively. The main benefits of flash annealing for junction formation are low sheet resistance junctions with a high level of activation together with limited diffusion.

## REFERENCES

[1] T. Noda et al., "Analysis of pocket profile deactivation and its impact on Vth variation for laser annealed device using an atomistic kinetic Monte Carlo approach,". IEEE IEDM,2010, pp. 383–386.

[2] S. Kato, T. Aoyama, T. Onizawa, K. Ikeda and Y. Ohji, Extended Abstracts of 17th IEEE International Conference on Advanced Thermal Processing of Semiconductors – RTP 2009, pp. 163-168.

[3] T. Ito. K. Suguro, T. Itani, K. Nishinohara, K. Matsuo and T. Saito, *VLSI Tech. Symp.*, 2003, pp. 53-54.

[4] P. J. Timans, G. Xing, J. Cibere, S. Hamm, and S. McCoy, in Subsecond Annealing of Advanced Materials, W. Skorupa and H. Schmidt, *Springer Series in Materials Science,* vol. 192, 2014, pp. 229-270.

[5] P. J. Timans, N. Acharya, and I. Amarilio, in Rapid Thermal and Other Short-Time Processing Technologies, F. Roozeboom, J. C. Gelpey, K. Reid, and D. L. Kong, *The Electrochemical Society*, 2000, pp. 375-382.

[6] H. W. Kennel et al., 14th IEEE International Conference on Advanced Thermal Processing of Semiconductors, Kyoto, 2006, pp. 85-91.

# NEGATIVE CAPACITANCE DOUBLE-GATE VERTICAL TUNNEL FET WITH IMPROVED SUBTHRESHOLD CHARACTERISTICS

*Guoliang Tian[1,2], Gaobo Xu,2[1*], Huaxiang Yin[1,2], Lianlian Li[1,2], Gangping Yan[1,2], Yanna Luo[1,2], Xiaoting Sun[1]*

[1] Institute of Microelectronics, Chinese Academy of Sciences, Beijing 100029, China
[2] University of Chinese Academy of Sciences, Beijing 100049, China
*Corresponding Author's Email: xugaobo@ime.ac.cn

## ABSTRACT

Negative capacitance (NC) double-gate vertical tunnel FET (DG-NCTFET) is proposed and the impacts of ferroelectric critical parameters on the device's performance have been investigated. When the capacitance matching condition is satisfied, NC effects substantially improve the electrical characteristics of TFET to achieve ultra-steep subthreshold swing and high drivability without hysteresis.

***Keywords*** □ SOI double-gate NCTFET, ferroelectric, and negative capacitance

## INTRODUCTION

Tunnel FET (TFET) is envisaged as one of the most potential candidates for next generation low-power devices due to its excellent electrical properties, such as steeper subthreshold swing, ultralow off-state current and larger on/off current ratio [1]. Ferroelectric (FE) gate insulators can boost the channel potential because they exhibit negative capacitance (NC) effects under some conditions [2]. In addition, the application of FE materials and silicon-on-insulator (SOI) on transistors enables them to have excellent radiation hardness for space applications [3].

In this paper, in order to improve the performance of TFET, we have integrated the ferroelectric material into the vertical TFET and proposed a novel silicon-on-insulator double-gate negative capacitance tunnel FET (SOI DG-NCTFET). We also analyzed the influence of the variation ferroelectric critical parameters on its performance.

## MODELING

The structure of the SOI DG-NCTFET is shown in Fig. 1 with detailed parameters for simulation shown in Table 1.

For the DG-NCTFET simulation, first, we extract charge-voltage ($Q_{int}$-$V_{int}$) and current-voltage ($I_{DS}$-$V_{int}$) of the underlying DG-TFET by Synopsys Sentaurus TCAD simulator. Then Landau-Khalatnikov (L-K) ferroelectric equation was employed to obtain the voltage drop across the ferroelectric layer. The quasi-static L-K equation is as follows [4]:

$$V_{FE} = V_G - V_{int} = -\frac{3\sqrt{3}}{2}\left(\frac{E_c}{P_r}Q_{int} - \frac{E_c}{P_r^3}\right)T_{FE} \quad (1)$$

The SS of DG-NCTFET and ferroelectric capacitor ($C_{FE}$) are given by [4]:

$$SS = 60 \text{ mV/decade}\left(1 + \frac{C_{dep}}{C_{ox}} - \frac{C_{dep}}{|C_{FE}|}\right) \quad (2)$$

$$C_{FE} = \frac{1}{2\alpha T_{FE}} = \frac{2}{3\sqrt{3}}\frac{P_r}{E_c T_{FE}} \quad (3)$$

The decrease of SS can be attributed to the amplification of voltage due to negative capacitance effect, the amplification factor $A_V$ is given by [5]:

$$A_V = \frac{|C_{FE}|}{|C_{FE}| - C_{int}} \quad (4)$$

In order to get a higher $A_V$ and smaller SS, $|C_{FE}|$ must be close to $C_{int}$ and greater than $C_{int}$.

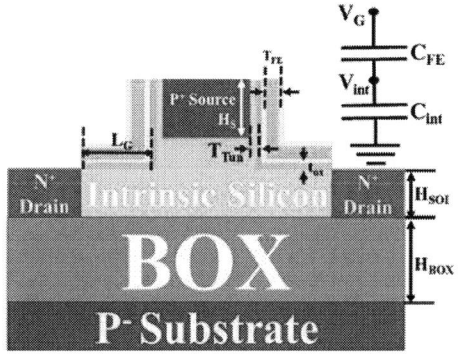

*Fig. 1: Schematic of the simulated DG-NCTFET.*

TABLE I. Main device parameters for the TCAD simulation

| Parameters | DG-NCTFET |
|---|---|
| **Gate oxide thickness ($t_{ox}$)** | 2 nm |
| **Tunneling layer thickness ($T_{Tun}$)** | 3 nm |
| **Gate length ($L_G$)** | 38 nm |
| **Source height ($H_S$)** | 40 nm |
| **Drain height ($H_{SOI}$)** | 20 nm |
| **Source doping ($N_A$)** | $1 \times 10^{20}$ cm$^{-3}$ |
| **Drain doping ($N_D$)** | $1 \times 10^{18}$ cm$^{-3}$ |

## RESULTS AND DISCUSSION

Table 2 summarizes the key parameters and $C_{FE}$ of ferroelectric Si:HfO$_2$ [6]. Fig. 2(a) shows the variation in $I_{DS}$-$V_{GS}$, $SS_{min}$ and $I_{DS}$ of SOI DG-TFETs with respect to

TABLE II. The parameters of Si:HfO₂ layer with different thicknesses and corresponding capacitance.

| Parameters | $T_{FE}$ = 5 nm | $T_{FE}$ = 10 nm | $T_{FE}$ = 15 nm | $T_{FE}$ = 20 nm |
|---|---|---|---|---|
| $P_r$ ($\mu C/cm^2$) | 11.37 | 9.5 | 7.63 | 5.75 |
| $E_c$ (MV/cm) | 1.15 | 1.1 | 1.05 | 1 |
| $C_{FE}$ ($F/cm^2$) | $7.61 \times 10^{-6}$ | $3.32 \times 10^{-6}$ | $1.86 \times 10^{-6}$ | $1.66 \times 10^{-6}$ |

different Si:HfO₂ ferroelectric layer thickness ($T_{Si}$:HfO₂). It is shown that the NC effect can enhance the $I_{DS}$, for example when $V_{GS}$ = 0.75 V and greatly improve the SS, especially when $I_{DS}$ ranges from $10^{-8}$ to $2\times10^{-7}$ A/$\mu$m. As can be seen from Fig. 2(b), decrease in $SS_{min}$ and increase in $I_{DS}$ will be obtained at thicker $T_{Si}$:HfO₂, which can be attributed to the reduction of $|C_{FE}|$, resulting in an increase of $A_V$. The hysteresis starts appearing when $|C_{FE}|$ is less than $C_{int}$, such as when $T_{Si}$:HfO₂ $\geq$ 10 nm, which is unstable for the device's electrical characteristics.

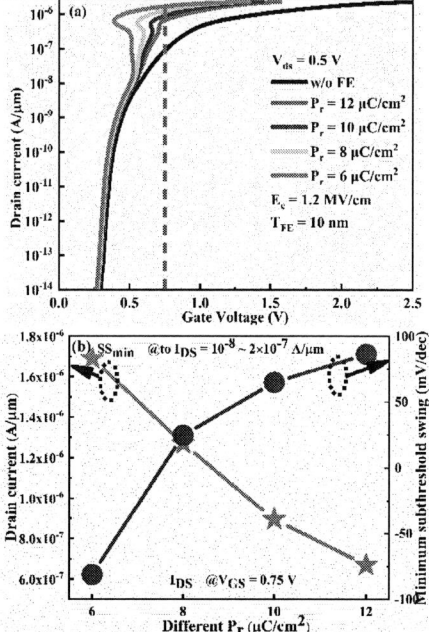

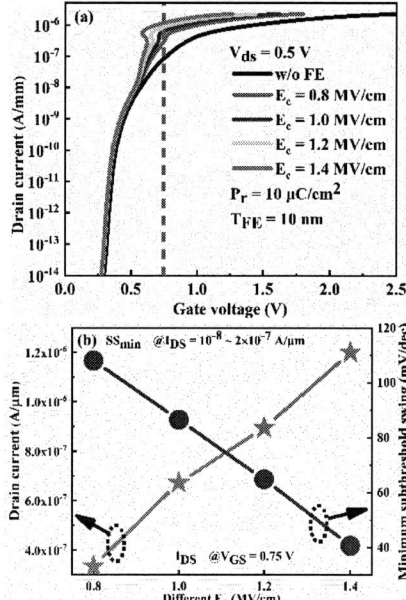

Fig.3: (a) $I_{DS}$-$V_{GS}$, (b) $I_{DS}$ and $SS_{min}$ curves for different $P_r$ values with the constant $T_{FE}$ and $E_c$.

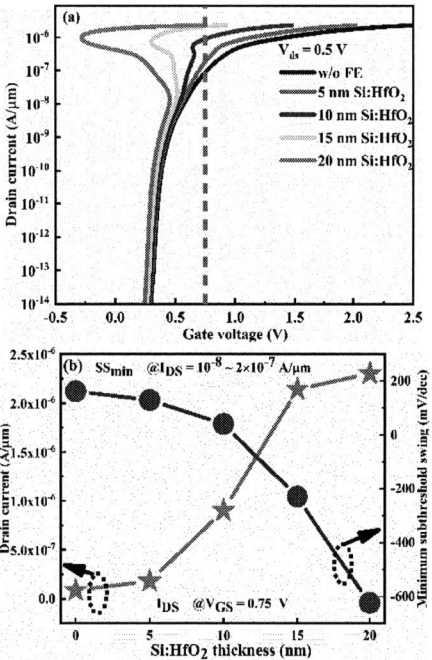

Fig.2: (a) $I_{DS}$-$V_{GS}$, (b) $I_{DS}$ and $SS_{min}$ curves for different ferroelectric layer thickness ($T_{Si}$:HfO₂).

Fig. 3(a) shows the variation in $I_{DS}$-$V_{GS}$, $SS_{min}$ and $I_{DS}$ of SOI DG-NCTFET with different $P_r$ when the thickness and $E_c$ of Si:HfO₂ are constant. $I_{DS}$ will increase and $SS_{min}$ will decrease with the reduction of $P_r$, as shown in Fig. 3(b), which can be attributed to $|C_{FE}|$ decreases with the reduction of $P_r$, resulting in an increase of $A_V$. When $P_r$ is smaller than 9 $\mu$C/cm2 at $E_c$ = 1.2 MV/cm and $T_{FE}$ = 10 nm, the $SS_{min}$ is negative and hysteresis come out obviously.

Fig.4: (a) $I_{DS}$-$V_{GS}$, (b) $I_{DS}$ and $SS_{min}$ curves for different $E_c$ values with the constant $T_{FE}$ and $P_r$.

As shown in Fig. 4, decrease in $I_{DS}$ and increase in

$SS_{min}$ will be achieved at smaller $E_c$ when $T_{FE}$ and $P_r$ are fixed. This is because $|C_{FE}|$ increases with the reduction of $E_c$, resulting in a decrease of $A_V$.

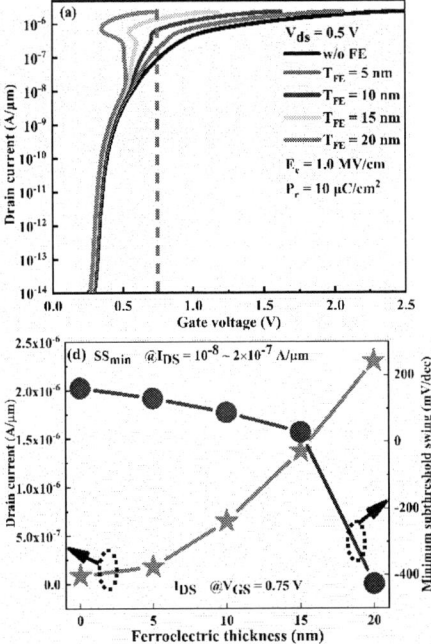

Fig.5: (a) $I_{DS}$-$V_{GS}$, (b) $I_{DS}$ and $SS_{min}$ curves for different $T_{FE}$ values when the $E_c$ and $P_r$ are constant.

When $E_c$ and $P_r$ are constant, $|C_{FE}|$ decreases with the increasing of $T_{FE}$, resulting in an increase of $A_V$. Fig. 5 shows the increase in $I_{DS}$ and decrease in $SS_{min}$ with the increase of $T_{FE}$. If $T_{FE}$ is larger than 15nm, the $|C_{FE}|$ is smaller than $C_{int}$, which results into negative $SS_{min}$ and the appearance of hysteresis.

## CONCLUSION

In summary, a SOI DG-NCTFET was proposed and the effects of ferroelectric parameters on the device's electrical characteristics have been investigated. The SOI DG-NCTFET shows ultra-steep SS and higher $I_{DS}$ compared with DG-TFET because NC effect can increase the probability of BTBT. In addition, the excellent performance without hysteresis can be achieved only if $|C_{FE}|$ is greater than $C_{int}$ and close to $C_{int}$.

## REFERENCES

[1] Vallett, A. L., et al. "Fabrication and Characterization of Axially Doped Silicon Nanowire Tunnel Field-Effect Transistors." Nano Letters, vol. 10, 2010, pp. 4813-4818.

[2] Salahuddin, S., and S. Datta. "Use of negative capacitance to provide voltage amplification for low power nanoscale devices. " Nano Letters, vol. 8, 2008, pp. 405.

[3] Yannan, Xu., et al. "Total ionizing dose effects and annealing behaviors of $HfO_2$-based MOS capacitor." Science China Information Sciences, vol. 60, 2017, pp. 1-3.

[4] Lin, Cheng-I., et al. "Effects of the variation of ferroelectric properties on negative capacitance FET characteristics." IEEE transactions on electron devices, vol. 63, 2016, pp. 2197-2199.

[5] Yeung, C. W., et al. "Low power negative capacitance FETs for future quantum-well body technology." VLSI Technology, Systems, and Applications (VLSI-TSA), 2013, pp. 1-2.

[6] Yurchuk, Ekaterina, et al. "Impact of layer thickness on the ferroelectric behaviour of silicon doped hafnium oxide thin films." Thin Solid Films, vol. 533, 2013, pp. 88-92.

# RESEARCH ON VT WINDOW IMPROVEMENT PROCESS OF 55NM SONOS EFLASH CELL

*Shipu Li \*, Jun Qian, and Yufei Peng*
*Shanghai Huali Microelectronics Corporation*, Shanghai 201203, China
*Corresponding Author's Email: lishipu@hlmc.cn

## ABSTRACT

In this work, two kinds of Vt window improvement directions are confirmed. For one thing, the overall improvement of Vt is confirmed which widens the difference between Vtp and Vtpi. For another, the uniformity of Vt is improved which makes the Vt distribution of all bits in the Cell more convergent. Based on the double liner process, STI corner rounding is adjusted according to STI step height and ONO film profile is optimized so that the effective area of ONO film on AA is greatly expanded. Secondly, well merge scheme is adopted. The well implant of 2T structure in Cell area is performed together and the Vt implant is performed separately, which significantly improves the uniformity of well implant in the whole cell area. At the same time, the thickness of photoresist applied on 2T structure in Vt IMP process is thinned experimentally, which is expected to effectively improve the CD uniformity within wafer and the process stability. After the above process optimization conditions are online, WAT parameters related to cell can be significantly improved, of which Vtp is increased by 15% and standard deviation is decreased by 37%.

## INTRODUCTION

Huali 55nm eFlash platform is developed by embedding flash storage structure in 55LP standard process platform. Fig.1(a) shows the standard 2T storage unit of SONOS cell, consisting of a CG and a SG. As seen from the partial enlargement of CG structure in Fig.1(b), the structure of CG is consisted of poly silicon at the top, middle oxide/nitride/oxide layer and silicon substrate at the bottom. Top oxide acts as a barrier which prevents electrons from entering the poly. Nitride is the actual electron storage layer, which is significantly different from FG storage medium of 65Nor Flash. The bottom oxide acts as tunnel oxide. For the ONO layer is the actual storage layer, the thickness and film quality of ONO are on high requirements and require special stricter control. SONOS eFlash stands for Silicon Oxide Nitride Oxide Silicon embedded flash.

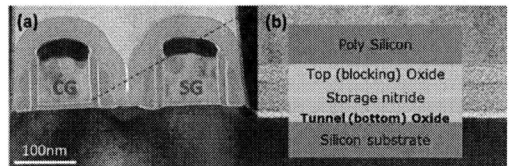

*Figure 1: (a) Storage cell of 2T structure and (b) the side view of local enlarged CG structure.*

## RESULTS AND DISCUSSION

Fig. 2(a) shows the cross section of the Cell area of the BSL process along the Poly direction. It can be seen that due to the process problem, there is a certain depression at the connection between STI and AA, which cannot be avoided in the process. Meanwhile, the corner of AA is relatively straight. As can be seen from the figure, the ONO film covered on the AA corner is thinner than the profile on the AA plane, which causes the reduction of the actual effective area of ONO on AA, even smaller than the size of AA.

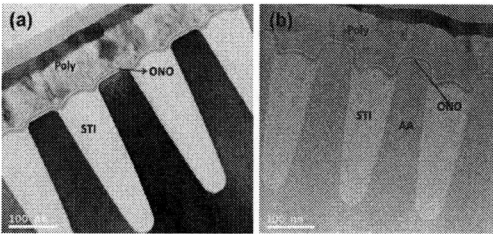

*Figure 2. The cross section of the Cell area of the BSL process along the Poly direction CG (a) before and (b) after the process improvement*

In order to improve the profile of ONO on AA, we try the double liner process in STI loop firstly, which means processing twice STI liner oxidation for AA corner rounding. As depicted in Fig. 2(b), after twice STI liner oxidation process, the AA corner rounding is obviously, from the original straight feature to the sleek corner. This is beneficial to maintain the continuity and uniformity of the entire film at the corner position when the ONO film is generated. Secondly, in order to further improve the profile of ONO film and increase the effective area of ONO film on AA, we adjust the step height of STI. Since the TUN loop only affects the Cell area and has no effect on the peripheral logic areas and the peripheral devices of the product. Fig. 2(b) shows that after adjusting the step height, the step height of STI is significantly lower than the AA area of BSL, which provides more attachment points for ONO film deposition. It can be seen from the results that ONO film wraps the top of entire AA area, and combined with the corner rounding effect caused by the double liner process, the film quality of the entire ONO film wrapping AA is continuous and uniform greatly. Compared with the profile of BSL, the effective area of ONO film on AA is greatly increased, thus effectively improving the electronic storage capacity of product and then improving the overall Cell Vt window.

In Well loop of BSL process, the IMP process of the CG/SG of the 2T structure in the Cell area is performed separately. However, due to the high energy and dose of Well implant, the phenomenon of uneven doping concentration caused by sputtering of implanted ions and Well photoresist is very obvious. In addition, the size of Cell area is small, so the process fluctuation in photolithography have significant impact on the actual implant effect. Figure 3 shows the IMP injection pattern of the 2T structure in the cell area is exactly complementary to the entire Cell area, and merged two layers can form the complete Cell area. Moreover, during the whole process, the mask of CD layer is well-matched to the whole Cell area, needing no more mask. At the same time, the condition of the IMP process of two layers are almost the same, which can fully support to being implanted together. Since the entire Cell area is opened, the impact of process fluctuations on CD will be greatly reduced, leading to the great improvement of process stability. Moreover, the doping concentration of the entire implant area will be uniform greatly basing on Well merge method, which can effectively improve the uniformity of the entire Cell Vt, thereby increasing the Vt window.

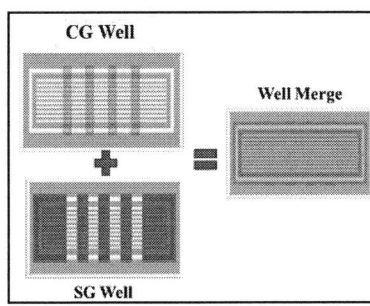

*Figure 3. The schematic map of CG/SG Well merge in cell.*

The obvious improvement can be seen by comparing the relevant WAT parameters of Cell. Take product A as an example. Vtp is the Vt when the Cell is written. After the window improvement process, the mean value of Vtp increases from 0.86 V to 0.98 V, exhibiting an overall increase of 15%. At the same time, the overall standard deviation also decreases significantly with the margin of 37%. Vtpi is the Vt of the inhibit bit when the Cell is written. The overall standard deviation is significantly reduced, which is reduced by 34%. Although the mean value of Vtpi dropped slightly by 5%, the overall Cell window increases from initial 2.11V to 2.17V with consideration of the improvement of Vtp. Idspi and Idsp are another two WAT parameters which can characterize the Cell. Idspi is the current when the Cell is written. After the window improvement process, the mean value of Idspi is increased by 12%, and the standard deviation is decreased by 21%. Idsp is the leakage when the Cell is written. The mean value of Idsp can be effectively reduced

by 14% through the window improvement process, and the standard deviation is significantly reduced by half. It can be seen that after process optimization, the Vt window of the entire Cell area is significantly improved.

## CONCLUSION

Firstly, ONO film profile which is the core storage structure is optimized. Based on the double liner process, STI corner rounding is adjusted according to STI step height and ONO film profile is optimized so that the effective area of ONO film on AA is greatly expanded. Secondly, according to the 2T structure of the Cell area, the two IMP process are optimized. Well merge scheme is adopted. The well implant of 2T structure in Cell area is performed together and the Vt implant is performed separately, which significantly improves the uniformity of well implant in the whole cell area. At the same time, the thickness of photoresist applied on 2T structure in Vt IMP process is thinned experimentally, which effectively improves the CD uniformity within wafer and the process stability. Combined with the above process optimization conditions, it can be found that WAT parameters related to Cell have been significantly improved and the overall window has been significantly improved. Vtp mean value is increased by 15% and standard deviation is decreased by 37%; Overall standard deviation of Vtpi is reduced by 34%; Idspi mean value is increased by 12%, STD is decreased by 21%; Idsp mean value is decreased by 14% and overall STD is decreased by 50%. In addition, improvement experiments are conducted on ONO film and other process, as well as the overall process window confirmation experiments are in progress.

## REFERENCES

[1] Seung-Hwan Song, *IEEE TRANSACTIONS ON ELECTRON DEVICES*, VOL. 61, NO. 11, NOVEMBER 2014

[2] Seung-Hwan Song, *IEEE JOURNAL OF SOLID-STATE CIRCUITS*, VOL. 49, NO. 8, AUGUST 2014

[3] S.-H. Song, *IEEE J. Solid-State Circuits*, vol. 48, no. 5, pp. 1302–1314, May 2013.

[4] R. Strenz, *IEEE Int. Electron Devices Meeting (IEDM)*, Dec. 2011, pp. 211–214

[5] H. Kojima, *IEEE Int. Electron Devices Meeting (IEDM)*, Dec. 2007, pp. 677–680.

[6] S.-T. Kang, *in Proc. 4th IEEE Int. Memory Workshop (IMW)*, May 2012, pp. 1–4.

[7] T. Kono, *in IEEE Int. Solid-State Circuits Conf. Dig. Tech. Papers (ISSCC)*, Feb. 2013, pp. 212–213.

# Suppression of TSV Leakage of Stacking CIS by Optimizing TSV Profile and Uniformity

*Zherui Cao, Weiwei Yu, Chunshan Zhao and Wuzhi Zhang*
Shanghai Huali Integrated Circuit Corporation, Shanghai, China
021-61871360-55840, caozheriui@hlmc.cn

## ABSTRACT

TSV (through-silicon via) technology plays a vital role in the application of stacking CIS. Nevertheless, TSV leakage and breakdown bring challenges in the performance improvement and scale production. The potential contributing factors to the TSV leakage are discussed and analyzed in this study. It was found that the most contributing factor is TSV profile and uniformity. Therefore, we developed a simple method as CIP process and reduced the yield loss.

Keywords—Stacking CIS; TSV leakage; etching uniformity

## INTRODUCTION

As one of the most important supporting technologies for 3D IC package and integration, TSV has been widely used in the stacking CIS to keep electrical connection between the pixel and logic wafer. However, the TSV leakage and breakdown also occur, which destroy the stable connection and cause yield loss. The wafers which suffered from TSV defect present special CP maps in the center regions. The major fail bins are Power short or open, IOUT and IDDS. Although some efforts have been made to avoid this defect by adjusting the stress in the deposition of barrier layer, the improvement is still limited in the wafer center region (Fig. 1). [1-3]

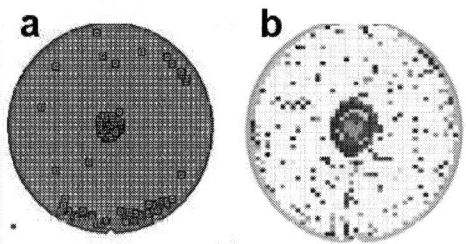

Fig. 1: Defect (a) and CP (b) maps of wafers with TSV leakage.

## EXPERIMENT

The morphologies were characterized through PFA test. As shown in Fig. 2, the abnormal TSV structure suffered from void in Cu. Due to the mismatch of the coefficient of thermal expansion between metal, oxide and Si, the inhomogeneous deformation of different materials led to the destruction of the TSV structure after thermal treatment, which seriously affected the mechanical and electrical performance of the device.

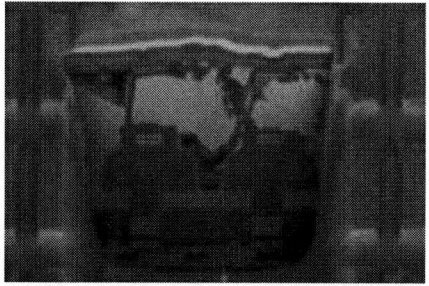

Fig. 2: Abnormal TSV structure in the wafer center region.

## DISCUSSION

Through reviewing the inline results of normal and abnormal wafers, the appearance of TSV leakage in the center region exhibits an obvious relationship with the thickness of post Si etch remaining oxide. Fig. 3a exhibits the map of the offline etching rate in the process, the etching rate of the center region is much smaller than that of the edge region. Therefore, the thickness of post Si etch remaining oxide in the wafer center is higher compared with the thickness in other areas. To find out the impact degree of the issue, the fail ratio was calculated by zone analysis (Fig. 3b).

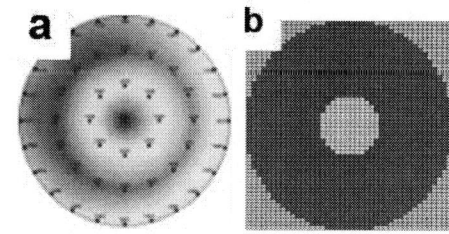

Fig. 3: (a) Offline etching rate map in the Si etch process. (b) Zone map in the analysis of fail ratio in the wafer center.

Fig. 4 presents the variation of the fail ratio caused by TSV leakage versus the highest thickness of the post Si etch remaining oxide in the wafer center. When the maximal thickness was under a certain thickness, the fail ratio was low and irrelevant with the inline thickness. When the maximal thickness of oxide increased over the thickness, the TSV leakage occurred and its fail ratio showed almost linear correlation with the thickness ($R^2$=0.93). It was suspected that the thickness of the post Si etch remaining oxide had possessed a considerable influence on the mechanical and electrical connection between the pixel and logic wafers, which should be

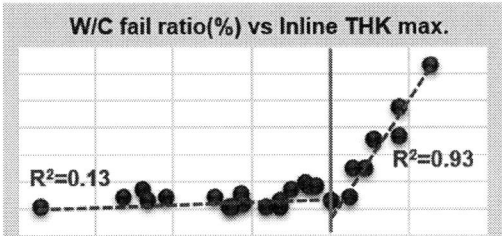

fine-tuned precisely.

*Fig. 4: Variation of fail ratio caused by TSV leakage versus inline maximal thickness.*

As the difference of the thickness between the center and edge regions is caused by the non-uniformity of etching rate, we developed a simple method to optimize TSV profile and uniformity. Fig. 5 presents the schematic diagram of the CIP process. The one-step process was divided into 3 steps with the increased selection ratio between Si and oxide. After the improvement, the maximum and range of thickness decreased with a remarkable reduction of TSV leakage.

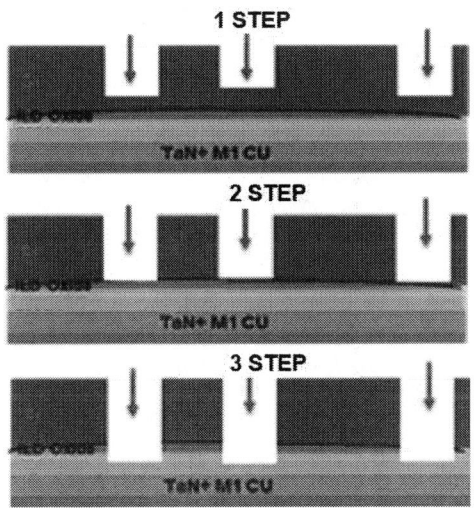

*Fig. 5 Schematic diagram of the CIP Si etch process.*

Meanwhile, the specification and control limit of the inline chart had been modified as TSV leakage occurred when the thickness was high. We retargeted the thickness of the post Si etch remaining oxide. The fail ratio in the wafer center showed a rapid decrease (Fig. 6). The yield loss was further reduced after the use of the CIP condition.

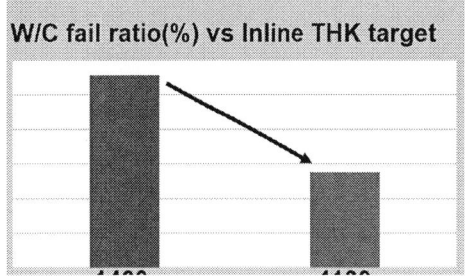

*Fig. 6: Variation of fail ratio caused by TSV leakage versus inline target thickness.*

## CONCLUSION

In this work, we observed the morphologies of the wafers with TSV leakage and found out the correlation with inline performance. The fail ratio was calculated by zone analysis and reduced by the adjustment of inline control. Furthermore, the CIP process of Si etch was developed. This simple method has a potential in optimizing TSV profile and uniformity and reducing the wafer center yield loss caused by TSV leakage.

## REFERENCES
[1] Geert Van der Plas, Paresh Limaye, Igor Loi, et. al., Design Issues and Considerations for Low-Cost 3-D TSV IC Technology [J]. IEEE Journal of Solid-State Circuits, 2011.
[2] Xianghua Hu, Guangzhi He, Xiaofang Gu, et. al., Stress adjusting for hillock size reduction in UTS CIS base on graphics analysis [J]. 2021 5th IEEE Electron Devices Technology & Manufacturing Conference (EDTM), 2021.
[3] Gudi Chen, Zhiqi Wang, Weiyin Wang, Influence of different materials on thermal stress of conical TSV [J]. 2018 19th International Conference on Electronic Packaging Technology (ICEPT), 2018.

# ALD CHARACTERISTIC STUDY OF AL2O3 FILM DEPOSITED BY A DUAL SINGLE-WAFER PROCESS CHAMBER

*Ge Zhang, Qihang Chen, Gi Kim, and Sean Chang*

Piotech Inc., No. 900, Shuijia, Hunnan District, Shenyang, China 110171

*Corresponding Author's Email: +86-24-24188000, zhangg@sypiotech.cn

## ABSTRACT

The technological importance of ALD (atomic layer deposition) process in the semiconductor business is more evident than ever before as the device structures become more and more challenging to deposit with various thin films. Therefore, it is highly beneficial to have more capable ALD equipment in the semiconductor manufacturing.

The performance of a newly designed dual single-chamber ALD system has been evaluated by examination of basic process results obtained with the system. Al2O3 film was selected for the process characterization with the ALD chamber. The ALD growth linearity, and pulse-and-purge saturation behavior were evaluated. The degree of ALD saturation was also confirmed by step coverage test.

*Keywords—ALD, Atomic Layer Deposition, Thin films, $Al_2O_3$, Pulse, Purge, Saturation, Step coverage, Single Wafer Chamber*

## INTRODUCTION

Aluminum oxide is one of the most widely used materials in semiconductor devices. For example, the $Al_2O_3$ film has been used as a dielectric material in metal-insulator-semiconductor (MIS) or metal-oxide-semiconductor (MOS) [1 and 2]. It has also been studied for the application of passivation layer of photoelectrodes [3] and permeation barrier for organic device like organic light emitting diodes [4].

Therefore, the deposition mechanism and characterization of aluminum oxide films with atomic layer deposition using TMA and $H_2O$ has been well established [5 and 6].

The goal of this study is to evaluate the performance of a newly designed ALD system via studying basic ALD process characteristics and the qualities of films deposited with the system. We have selected aluminum oxide process as a case study to evaluate the system initial performance.

## EXPERIMENTAL METHOD

The aluminum oxide deposition test was performed with the ALD system, PF-300T Altair dual chamber system, designed by Piotech Inc. The deposition was performed on 12-inch wafers at 180C of susceptor heater temperature and 0.9 torr of chamber pressure. TMA (trimethylaluminum) and $H_2O$ were used as precursors. The deposited film thickness and thickness uniformity was measured and calculated based upon averaged numbers of 49 points detection using KLA Aleries 8500 system. The step coverage of the deposited film was measured with scanning electron microscopy, Hitachi S4800.

The most of the process data in this report were collected from the newly designed system to evaluate the initial system performance per design concept via initial basic process tests. The collected initial data in this study will be used for future system and process refinement.

## RESULTS AND DISCUSSION

The TMA pulse saturation test results are summarized as shown in Figure 1 (a). TMA pulse time was tested from 0.01 sec to 0.2 sec. Other conditions were fixed at 1 sec of TMA purge, 0.1 sec of $H_2O$ pulse, and 1.0 sec of $H_2O$ purge time respectively. As shown in Figure 1(a), the ALD growth (Growth Per Cycle, GPC) at 0.01 sec of TMA pulse time was about 0.53~0.6 Å/c, which is about 57~65% of the fully saturated GPC, 0.92~0.93A/c at 0.2 sec. At 0.02 sec of TMA pulse time, the GPC was reached to 0.87~0.88 Å/c, which is almost 94~95% of fully saturated GPC. The GPC continued to increase up to 0.09 sec where already reach 98% of GPC, 0.9~0.91 Å/c. There was no GPC change from 0.12 sec to 0.2 sec of TMA pulse time.

Figure 1 (b) is the summary of the film non-uniformity change during the TMA pulse split tests. The non-uniformity percent, NU%, was around 9~11% at 0.01 sec of TMA pulse time. The high NU% numbers should be due to significantly under saturated TMA condition during the process. At this time the GPC was only about 57~65% of the fully saturated film GPC as shown in Figure 1(a). However, the NU% of the film deposited from 0.02 sec of the TMA pulse time and longer was all near or below 1% of NU%. By considering Figure 1 (a) and (b), it is clear the TMA saturation was almost fully reached at 0.02 sec of TMA pulse time on blanket Si wafer.

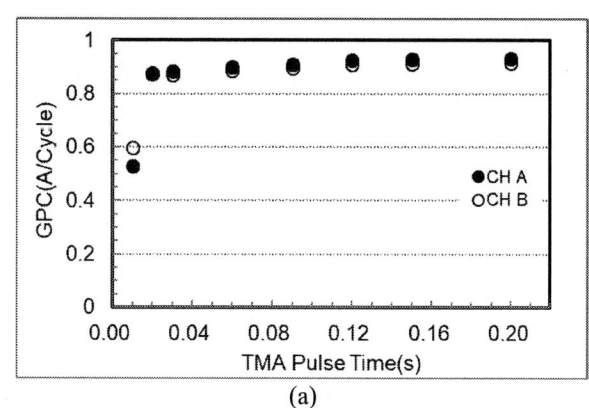

(a)

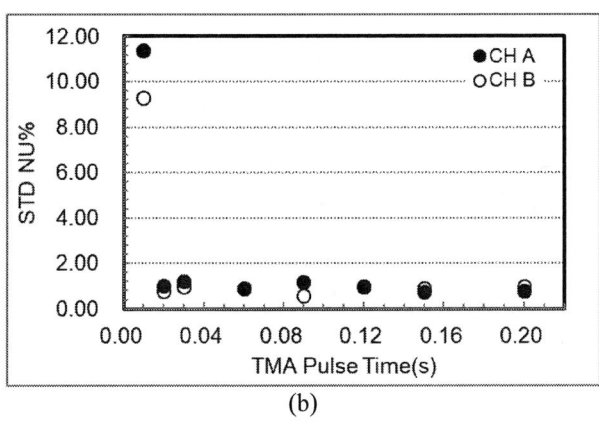

(b)

Figure 1. (a) Growth per cycle (GPC) results at different TMA pulse time with t(x)s//1s//0.1s//1s of pulse and purge conditions. (b) Thickness non-uniformity results at different TMA pulse time.

TMA purge saturation test results are as shown in Figure 2 (a) and (b). The TMA purge time was tested from 0.45 sec to 1.85 sec. Other conditions were fixed at 0.08 sec of TMA pulse, 0.1 sec of $H_2O$ pulse, and 1.0 sec of $H_2O$ purge time respectively. The GPC at 0.45 sec of purge time was 0.93~0.95 Å/c, which was about 3.5~4% thicker than the GPC at fully saturated purge time, 0.90~0.91 Å/c, as shown in Figure 2 (a). On the other hand, there was no clear trend of the NU% change of the film with TMA purge time as shown in Figure 2(b). All the data points showed near or below 1% of film NU% at all purge times tested.

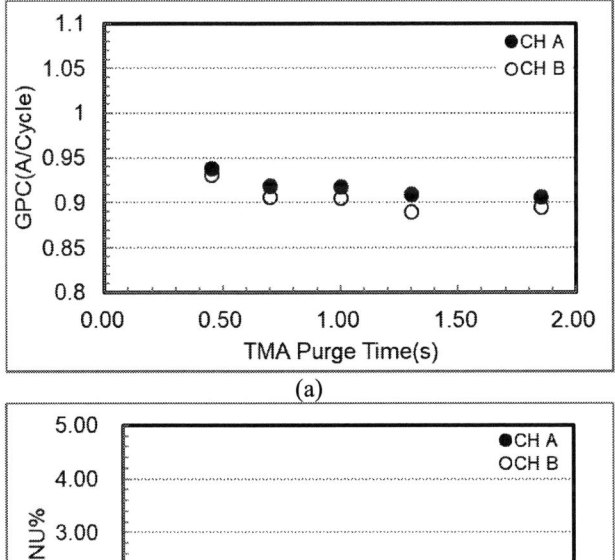

Figure 2. (a) GPC results at different TMA purge time. (b) Thickness non-uniformity results at different TMA purge time.

The $H_2O$ pulse saturation test results are summarized as shown in Figure 3 (a). $H_2O$ pulse time was tested from 0.01 sec to 0.2 sec. Other conditions were fixed at 0.08 sec of TMA pulse, 1.0 sec of TMA purge, and 1.0 sec of $H_2O$ purge time respectively. As shown in Figure 3 (a), the GPC at 0.01 sec of $H_2O$ pulse time was about 0.45~0.47 Å/c, which is about 45~47% of the GPC, 0.99~1.01 Å/c, at 0.2sec. The GPC increment was much slower than that of TMA, and it took almost 0.1 sec to reach the GPC 0.89~0.90 Å/c, about 90% of the GPC at 0.2 sec. This slower saturation of $H_2O$ is typical as reported [5 and 6].

The film NU% change also showed slower change over $H_2O$ pulse time than that of TMA as shown Figure 3 (b). The NU% was about 8~12% at 0.01 sec of $H_2O$ pulse time, and gradually improved until 0.09 sec from where both chambers started to produce the NU% less than 1 %. This gradual NU% improvement in conjunction to the GPC increment clearly showed that the $H_2O$ pulse saturation is much slower than that of TMA.

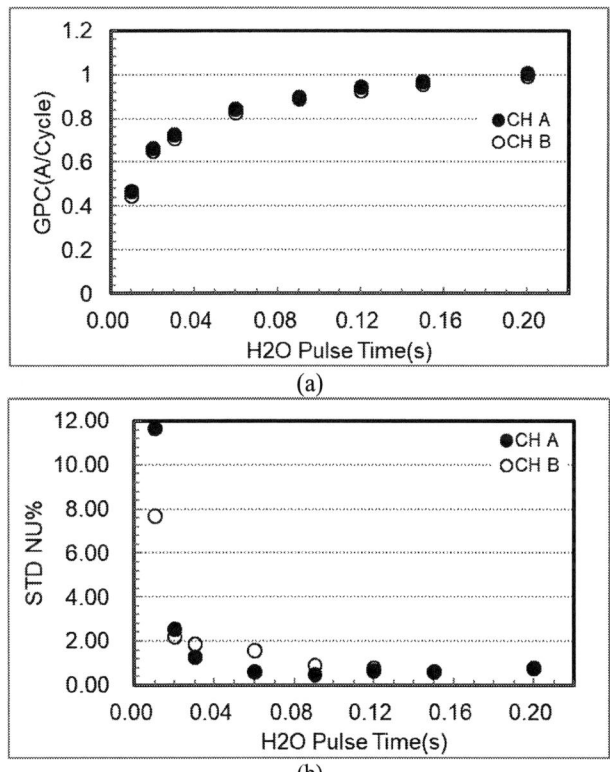

Figure 3. (a) GPC results at different $H_2O$ pulse time with 0.08s//1s//t(x)s//1s of pulse and purge conditions. (b) Thickness non-uniformity results at different $H_2O$ pulse time.

Water purge saturation test results are as shown in Figure 4 (a) and (b). The $H_2O$ purge time was tested from 0.45 sec to 1.85 sec. Other conditions were fixed at 0.08 sec of TMA pulse, 1 sec of TMA purge, and 0.1 sec of $H_2O$ pulse time respectively. The GPC at 0.45 sec of purge time was 1.01~1.03 Å/c, which was about 12.6~12.7% higher than the GPC at fully saturated conditions, 0.90~0.91 Å/c, as shown in Figure 4 (a).

The thickness NU% change over $H_2O$ purge time as shown in Figure 4 (b) was also slow as seen in the GPC change over purge time. The NU% was about 3% at 0.45 sec of purge time, and improved down to 1% or better from the $H_2O$ purge time at 0.7 sec and longer.

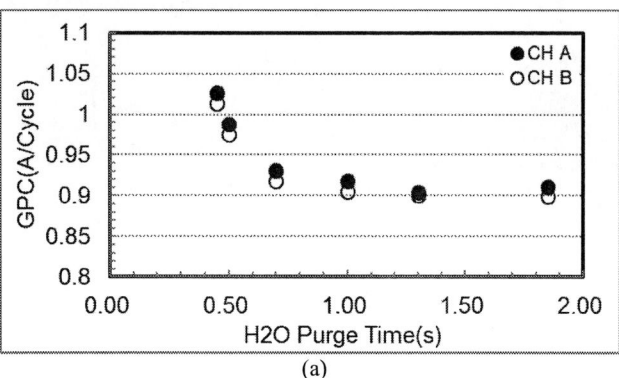

(a)

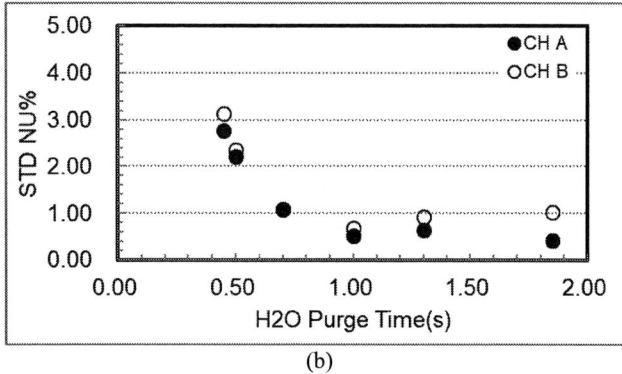

(b)

Figure 4. (a) GPC results at different $H_2O$ purge time. (b) Thickness non-uniformity results at different $H_2O$ purge time.

Figure 5 shows the growth linearity of the ALD process. The measured film thickness includes native thickness before deposition. As shown in Figure 5, the thickness was linearly increased with ALD cycle.

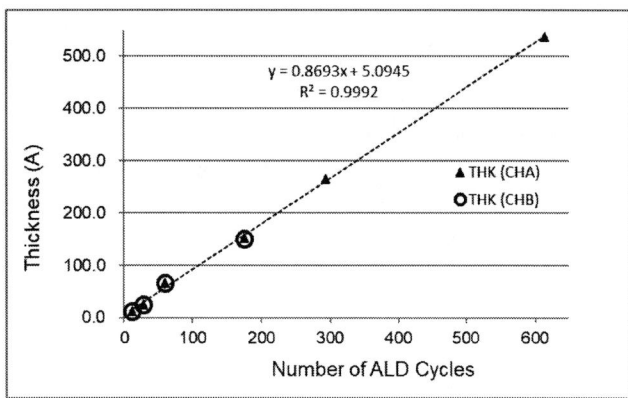

Figure 5. The linearity test results prepared by the conditions of TMA/Purge/$H_2O$/Purge = 0.08s/1.0s/0.1s/1.0s.

Step coverage test of $Al_2O_3$ films was performed on structure wafer coupon as shown in Figure 6. The pulse and purge time was 0.08 sec of TMA pulse, 1 sec of TMA purge, 0.1 sec of $H_2O$, and 1 sec of $H_2O$ purge condition (total 2.18 sec of ALD cycle). As shown in Figure 6 (b)~(d), the step coverage was confirmed around 100% with test conditions, indicating full saturation on the trench structure with aspect ratio of 12

with 230 nm top opening (and 183nm bottom).

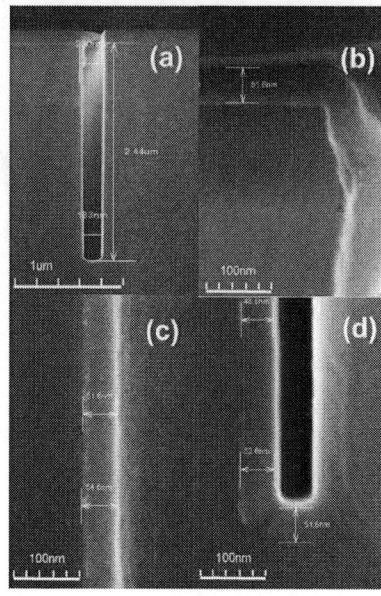

Figure 6. SEM image of $Al_2O_3$ films on the trench, (a) overall image of trench structure before deposition, and (b) $Al_2O_3$ film on top of the trench, (c) $Al_2O_3$ film on the mid portion of the trench, and (d) $Al_2O_3$ film on the bottom of the trench after deposition.

GPC variation between chamber A and chamber B was noticed only at one test condition, 0.01 sec of TMA pulse time, where the TMA dosing amount was significantly under saturated condition as shown in Figure 1 (a). However, this GPC variation between twin chambers quickly disappeared from 0.02 sec and longer TMA pulse time.

NU% variation between dual chambers was also observed at only at 0.01 sec of TMA and $H_2O$ pulse time, but the variation disappeared quickly from 0.02 sec of pulse time as shown in Figures 1(b) and 3(b). This NU% variation between the twin chambers should be the same reason as GPC variation, significant under saturation mode of the precursor and reactant.

## CONCLUSION

The evaluation of a twin-chamber ALD system was performed with $Al_2O_3$ film process analysis. The pattern of pulse and purge saturation observed in this study was as expected, fast TMA saturation and relatively slow $H_2O$ saturation.

The step coverage data indicated that 2.18 sec of ALD cycle time was more than enough to fully saturate the surface of the trench-structure coupons with aspect ratio of 12. It is highly possible to reduce ALD cycle time further by minor process and system refinement.

The process variations between two chambers were minimal except only one condition, when TMA and $H_2O$ were significantly under dosed condition (0.01 sec of pulse time), and this variation started to disappear quickly from 0.02 sec of pulse time. System and process refinement is in progress for further improvement.

# REFERENCE

[1] Wang et al. Nanoscale Research Letters (2016) 11:21

[2] Masahito et al., Phys. Status Solidi C **11**, No. 3–4 (2014)

[3] Kim, et al., ACS Appl. Mater. Interfaces 2015, 7, 8572−8584

[4] Klumbies et al., Organic Electronics 17 (2015) 138–143

[5] Puurunen, J. of Appl. Phys. 97, 121301 (2005)

[6] Miikkulainen et al., J. of Appl. Phys. 113, 021301 (2013)

# A NOVEL SIGE HETEROJUNCTION PHOTOTRANSISTOR APPLICABLE IN WIDE SPECTRAL RANGE

*Yin Sha[1], Hongyun Xie[1*], Yan Wang[1], Yang Xiang[1], Fu Zhu[1], Ruilang Ji[1], Weicong Na[1], and Wanrong Zhang[1]*

[1]Micro-Electronics Department, Beijing University of Technology, Beijing 100124, China

*Corresponding Author's Email: xiehongyun@bjut.edu.cn

## ABSTRACT

This paper reports on a novel SiGe heterojunction phototransistor (HPT) with a wide applicable spectral range covering 405-940 nm wavelength. By introducing the Ge component into the base region and designing its content and distribution, the absorption band and applicable spectral range are greatly broadened. Under the illumination of a light source with 14.3 μW optical power, the responsivity of HPT is 1.72 A/W, 12.38 A/W, 3.69 A/W, and 0.77 A/W at 405 nm, 650 nm, 850 nm, and 940 nm wavelength respectively.

## INTRODUCTION

Recently, with the development of optical communication and optical interconnection, SiGe heterojunction phototransistor (HPT) is getting more and more attention due to its advantage of low cost, low power consumption, easy monolithic integration, and compatibility of complementary metal-oxide-semiconductor (CMOS) technology [1]. Researches have been performed to realize high-performance HPTs under certain wavelength [2-4]. Reference [2] demonstrated an edge illuminated SiGe HPT with 0.4A/W responsivity for radio-over-Fiber (RoF) application under 850 nm wavelength. Reference [3] fabricated an HPT and obtained 0.805 A/W responsivity at 850 nm wavelength. Lately, Nanni et al. manufactured an HPT with 0.4 A/W responsivity for standard single-mode fiber (SSMF) RoF link under 850 nm wavelength in Reference [4]. However, the relatively low responsivity and the narrow spectral band limited their further application and there is an increased demand for HPTs which can be used in a wide spectral range with good performance.

In this paper, we design and manufacture a novel SiGe HPT for application in a wide spectral range. The testing results indicate that the HPT presented in this paper can achieve high responsivity under the illumination of light ranging from 405 nm to 940 nm wavelength.

## DEVICE DESIGN AND MANUFACTURE

Si-based photodetectors are currently used predominantly in the visible wavelength and infrared regime. However, the absorption coefficient α of Si material drops sharply at the near-infrared wavelengths, which are close to silicon bandgap wavelength (1.1 μm),

as shown in Figure 1. The absorption coefficient of Ge material is also presented in Figure 1. As can be seen from Figure 1, α for Si is 2600 cm$^{-1}$ and greater for the wavelength of 650 nm and shorter. However, for longer wavelengths, α for Si is 487 cm$^{-1}$ at 850 nm and drops to 145 cm$^{-1}$ at 850 nm wavelength. With such weak absorption, the Si-based HPTs can hardly absorb and detect longer wavelengths efficiently and then show low responsivity. While Ge material exhibits a great absorption coefficient up to 1.55 μm owing to its small bandgap. The absorption coefficient of Ge is as high as 11200 cm$^{-1}$ at 940 nm wavelength. When introducing some Ge component into Si material, the photon absorption of compounded SiGe increases consequently. However, it should be noted that the lattice constants of Si and Ge are different and there must be strain between Si and SiGe epitaxy layer. Considering the dependence of the critical thickness of SiGe material on the Ge component fraction, the content and distribution of Ge component in the SiGe epitaxy layer are designed. $Si_{0.8}Ge_{0.2}$ material with Ge component in uniform distribution is utilized in the base of SiGe HPT. The absorption coefficient of $Si_{0.8}Ge_{0.2}$ is shown in Figure 1 too. The absorption improvement for longer wavelengths can broaden the absorption spectrum and improve HPTs' performance in near-infrared regime.

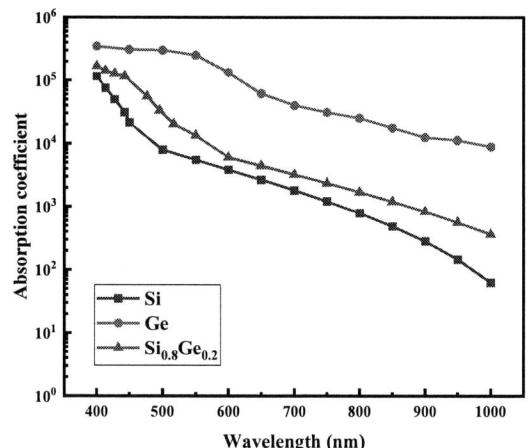

*Figure 1: Absorption coefficient of Si, Ge, and $Si_{0.8}Ge_{0.2}$ material.*

The epitaxial structure of the proposed

surfaced-illuminated SiGe HPT is shown in Figure 2. The resistivity of the $n^+$ doped ($1 \times 10^{19}$ cm$^{-3}$) silicon substrate is 0.002-0.004 $\Omega$/cm and used as the sub-collector. Using low-pressure chemical vapor deposition (LPCVD), a 700 nm $n^-$ doped ($1 \times 10^{17}$ cm$^{-3}$) silicon layer as the collector is grown on the substrate, followed by a 10 nm intrinsic silicon buffer layer. A 40 nm $p^+$ doped ($1 \times 10^{19}$ cm$^{-3}$) Si$_{0.8}$Ge$_{0.2}$ layer and a $p^-$ doped ($1 \times 10^{18}$ cm$^{-3}$) silicon cap layer as the base are sequentially and epitaxially deposited by reduced pressure chemical vapor deposition (RPCVD). Finally, a 300 nm $n^+$ doped ($1 \times 10^{20}$ cm$^{-3}$) polycrystalline silicon layer as the emitter is deposited by LPCVD.

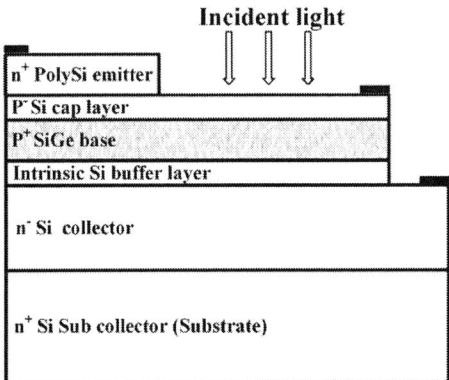

*Figure 2: Vertical structure of the SiGe HPT.*

The scanning electron microscope (SEM) photo of the top view of the manufactured HPT is shown in Figure 3. The area of emitter mesa, base mesa, and collector mesa is $6 \times 50$ $\mu$m$^2$, $64 \times 50$ $\mu$m$^2$, and $70 \times 54$ $\mu$m$^2$ separately. The square optical window is set in the base region with an area of $50 \times 50$ $\mu$m$^2$. The light is vertically illuminated onto the base region.

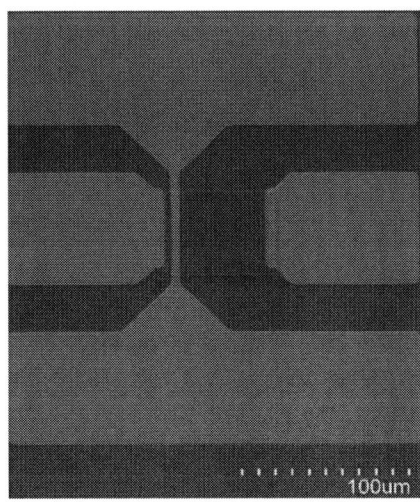

*Figure 3: SEM photo of the top view of the SiGe HPT.*

# RESULTS AND DISCUSSION

We measured the HPT's collector current under the illumination of a semiconductor laser which can provide a light source with a wavelength of 405 nm, 650 nm, 850 nm, and 940 nm. The collector voltage ranges from 0 V to 1 V, the emitter is grounded and the base is floated during the whole testing.

Figure 4 shows the collector current of HPT under 405 nm wavelength. There is the typical output characteristic of a transistor and the collector current increases with input optical power. When the optical power is 14.3 $\mu$W, the collector current is 24.56 $\mu$A. As is shown in Figure 5, for 850 nm wavelength，52.78 $\mu$A collector current is observed in HPT under the illumination of light with 14.3 $\mu$W optical power.

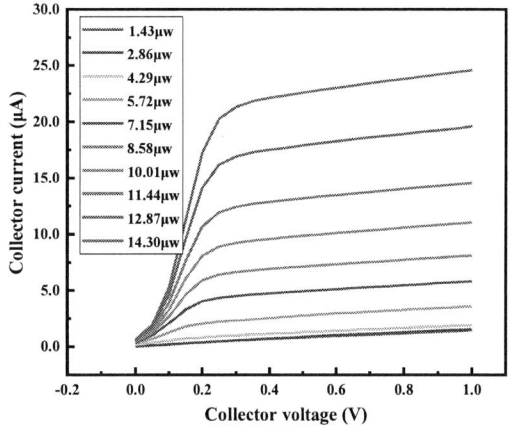

*Figure 4: Collector current of HPT at 405 nm wavelength.*

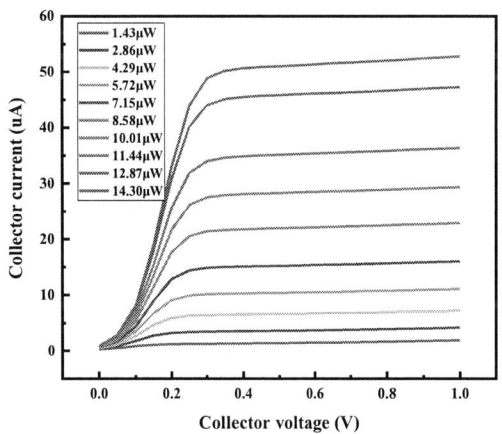

*Figure 5: Collector current of HPT at 850 nm wavelength.*

The curves of the collector current of HPT at 650 nm and 940 nm are similar to the conditions at 405 nm and 850 nm wavelength. But when the wavelength raises to 940 nm where the absorption of material decreases, the collector current under small optical power is very small and is hard to detect. So we only measured the performance of HPT under big optical power as 14.3 μW. Figure 6 presents the collector current when optical illustrated power is 14.3 μW for different wavelengths. The maximum achieves 177 μA for 650nm and the minimum gets 11 μA at 940 nm.

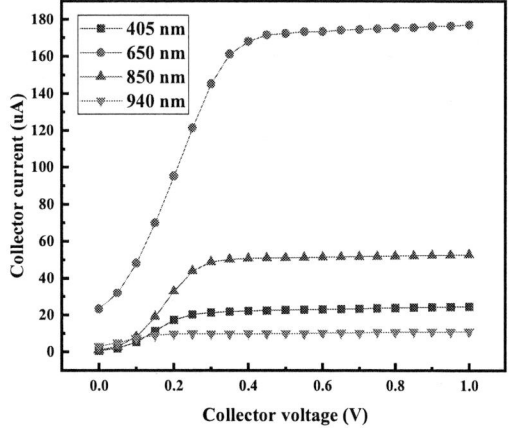

*Figure 6: Collector current of HPT under 14.3 μW optical power at 405 nm, 650 nm, 850 nm, and 940 nm wavelength.*

The responsivity of HPT under different optical powers is calculated from measurement data. The responsivity of HPT in a wide spectrum covering 405 nm to 940 nm is shown in Figure 7. The responsivity improves with the increase of optical power. When the incident optical power is 14.3 μW, the responsivity of HPT at 405 nm, 650 nm, 850 nm, and 940 nm are 1.72 A/W, 12.38 A/W 3.69 A/W, and 0.77 A/W respectively.

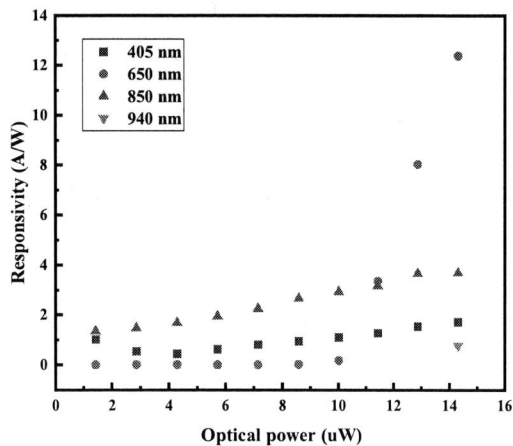

*Figure 7: Responsivity of HPT under different wavelengths and optical powers.*

It is distinct that with the increase of wavelength, the responsivity increases firstly and decreases finally. The reason for it is that the photon absorption is determined not only by absorption coefficient but also by the absorption length. Though the absorption coefficients under short wavelenght are always big, their penetration depth is shallow and only a small part of the absorption region is utilized and contributes to the photocurrent. The maximal responsivity is achieved at 650 nm where both absorption coefficient and penetration depth are relatively high. When the wavelength increases continuously, the absorption coefficient under 850 nm and 940 nm drops sharply and the responsivity also decreases accordingly.

## CONCLUSION

A novel SiGe HPT is designed, manufactured, and measured in this paper. After the $Si_{0.8}Ge_{0.2}$ material with Ge component in uniform distribution is used in the base of SiGe HPT, the absorption band of SiGe HPT is broadened and then HPT's performance in the near-infrared regime is improved. The testing results indicate that the proposed HPT can achieve high collector current and responsivity in the whole band ranging from 405 nm to 940 nm. Under the illumination of light source with 405 nm, 650 nm, 850 nm, and 940 nm wavelength and 14.3 μW optical power, the responsivity of the HPT can reach 1.72 A/W, 12.38 A/W, 3.69 A/W, and 0.77 A/W respectively.

## ACKNOWLEDGEMENTS

Supported in part by the National Natural Science Foundation of China (61604106, 61774012, 61901010), Beijing Municipal Natural Science Foundation (4192014, 4204092), and Shandong Province Natural Science Foundation (ZR2021MF077).

## REFERENCE

[1] C. Li et al. *Sci Rep*, vol. 6, 2016, p. 27743.
[2] Z. G. Tegegne et al. *International Journal of Microwave and Wireless Technologies*, vol. 9, no. 1, 2015, pp. 17-24.
[3] Z. G. Tegegne, C. Viana, J.-L. Polleux, M. Grzeskowiak, and E. Richalot. *IEEE Transactions on Electron Devices*, vol. 65, no. 6, 2018, pp. 2537-2543.
[4] J. Nanni, Z. G. Tegegne, C. Viana, G. Tartarini, C. Algani, and J.-L. Polleux. *IEEE Journal of Quantum Electronics*, vol. 55, no. 4, 2019, pp. 1-9.

# DEFECT REDUCTION WITH ADVANCED LITHOGRAPHIC FILTER

*Robb Fang[1]\*, Alexander Zhu[1], Yoshiaki Yamada[2]*
[1]Hangzhou Cobetter Filtration Equipment Co.Ltd. , Hangzhou, China
[2]Nippon Cobetter Co.Ltd. , Osaka, Japan
\*Corresponding Author's Email: syfj@cobetterfilter.com

## ABSTRACT

It is well known that Nylon membrane can significantly reduce defects on a wafer when it is used in filtration of lithography resist by removing metal ion, particle and gel [1,2]. With the development of advanced technology nodes, improvement of higher cleanliness and higher retention rating of filters for point-of-use (POU) applications is becoming increasingly necessary. In this study, we present 2 different types of 1nm rating Nylon filter for photoresist filtration, namely standard type and thick type. Filter cleanliness confirmed by liquid particle counter (LPC) and ICP-MS and defect reduction performance on wafer are compared. Evaluation of metal reduction and polymer removal behavior are also studied to clarify the mechanism of defect reduction.

*Keywords—Filtration; Filter; Nylon66; ArF; PHS polymer; Gel*

## INTRODUCTION

As lithographic technical node continues to update, the CD of pattern keeps shrinking. Especially after the introduction of EUV lithography application, the process significantly reduces the tolerance for defects and particle contamination, to which, cleanliness of photoresist is one important factor. Far now, Chemically Amplified Resist (CAR) is still largely used in KrF, ArF and EUV processes. Normally, polymer in CAR material has a side protection group and *t*-butyloxycarbonyl is commonly used. However, *t*-butyloxycarbonyl consists of carbonyl structure that is strongly polar. This polar structure may cause polymerization and even cross-linking effect with the presence of metal catalyst [3]. These oversized polymer may further cause some special defects. The filtration of CAR, both at bulk and point-of-use (POU),

has already been demonstrated in KrF and ArF lithography to play a significant role in defect reduction by efficiently removing particle and gels [4]. Hereby, Nylon66 has been chosen as the filtration material for CAR due to the presence of strong polar group in its structure, being able to remove those polar oversized polymer, and thus, reducing pattern defects. Initial cleanliness, defect reuduction performance as well as the mechanism of Nylon66 filter are studied and the results are discussed in this report.

## EXPERIMENTAL

### Solution

*t*-Butyloxycarbonyl(*t*-BOC) protected Polyhydroxystyrene (PHS), MW≈6500, was dissolved in OK73 solvent. And 10% concentration solution was prepared. Then the solution was pre-filtered by modified PE plus 50nm UPE to make sure of low metal and particle concentration.

### Filter and Membrane

Cobetter standard 5nm Nylon66 membrane, standard 1nm Nylon66 membrane and new 1nm Nylon66 membrane were used in SW44 structure capsule with HDPE material. Track etch membrane with PI material was used in 47mm diameter disk (PFA material).

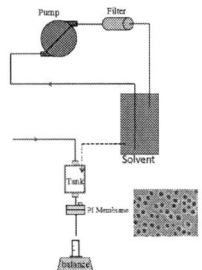

*Figure 1: Sketch of gel/particle removal test*

**Gel/Particle removal test**

2L of raw PHS solution and metal spiked PHS solution were stored in a 4L HDPE bottle. A diaphragm pump was used to provide power. Then the solution was filtered by Nylon66 filter for 20 times circulation. The filtrate was collected and then continued to be filtered by PI track etch membrane, and the filtrate of PI membrane was weighed by electronic Balance. The sketch was showed in Figure 1.

**Molecular weight of solution**

The molecular weight of solution was evaluated by a Gel Permeation Chromatography (GPC, TOSOH, HLC-8420).

**Metal in solution**

The concentration of metal in solution was evaluated by a ICP-MS (Agilent, 7900).

## RESULTS AND DISSUCION

### New 1 nm Nylon66 Filter

Under the help of capillary flow porometer tool, we can learn pore parameters of the membrane. The results are shown in Figure 2.

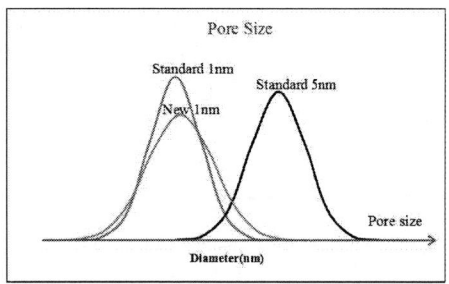

*Figure 2: Normalized pore size of membrane*

The new 1 nm Nylon66 membrane has very close (slightly larger) mean pore size, as compared with standard 1 nm Nylon66, with slightly wider pore size distribution. As the thickness of the membrane is increased, the contact time of liquid with the filtration medium is also extended, leading to higher particle retention and less particle penetration. Therefore it can be expected that thick membrane will lead to better defect reduction. A 60% increase of membrane thickness was

found in new 1nm Nylon66, as is shown in Figure 3(a). Figure 3(b) shows comparison of the differential pressure at 1 mL/s DIW flow rate, surprisingly, the pressure drop of the filter made from new 1 nm Nylon66 membrane is even lower than that of standard 1 nm Nylon66 membrane, despite the increase of thickness. Such effect is benefited from the higher porosity of new thick 1 nm Nylon66.

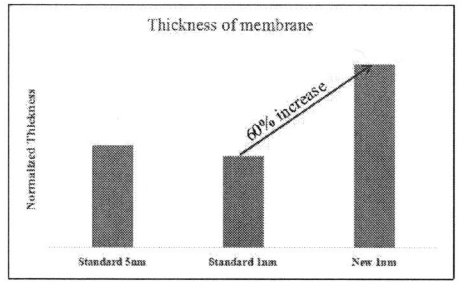

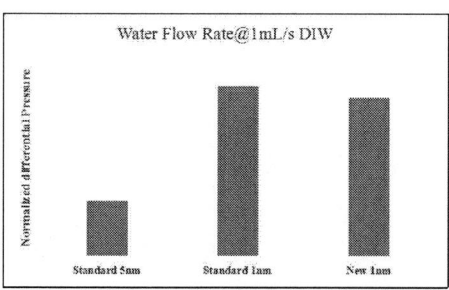

*Figure 3: (a) Normalized thickness of membrane; (b)Normalized differential pressure of filter*

In order to meet the stringent defects requirements of advanced node process, higher cleanliness of filter itself is highly important. Particle shedding and metal extractable of filters with new cleaning technology was evaluated. Metal extractable was evaluated by soaking the filter in 100 mL OK73 for fixed time and then the soaking solvent was analyzed by ICP-MS. The result is shown in Figure 4. Metal extractable of filter made from new 1 nm Nylon66 with new cleaning tehnology has a further reduction of about 12% as compared with that of standard 1 nm.

Figure 5 shows the count of particle above 30nm shedding from filter measured by a liquid particle counter. Filter with new thick 1nm Nylon66 membrane cleaned by new tehnology shows the lowest initial particle shedding level and shortest time consumption to reach baseline.

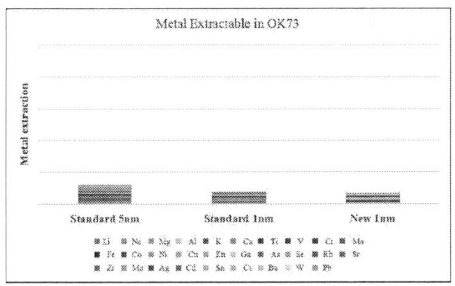

*Figure 4: Metal extractable in OK73 solvent*

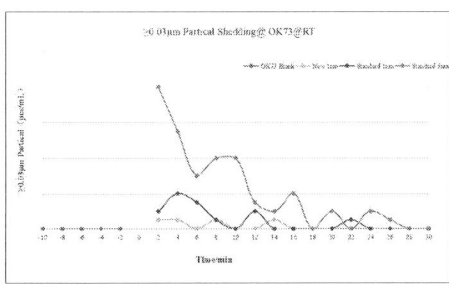

*Figure 5: Particle Shedding in OK73 solvent*

**Metal removal test**

300 ppb Mg and Ca was spiked into pure OK73 and 10% PHS/OK73 solution. Metal removal test was conducted in solvents and the results are compared. Metal removal efficiency on different metal was calculated by equation (1):

$$Removal\ Efficiency = \frac{concentration\ in\ influent}{concentration\ in\ enfluent} \quad (1)$$

Figure 6 shows filters made from standard 5nm, standard 1nm and new thick 1nm Nylon66 can all effectively reduce Mg and Ca in both OK73 solvent and PHS/OK73 solution. Furthermore, all Nylon66 filters has a increased metal removal efficiency in PHS/OK73 solution compared to those in pure OK73 solvent. This phenomenon is more obvious on Mg than Ca. It is speculated that some cross-linking reaction happened in polar PHS polymer molecule with the presence of certain metal (in this case, Mg and Ca), forming the so-called gel,

refer to Figure 7. As a polar material, Nylon66 can adsorb metal and polar organics species, leading to the enhanced metal removal performance.

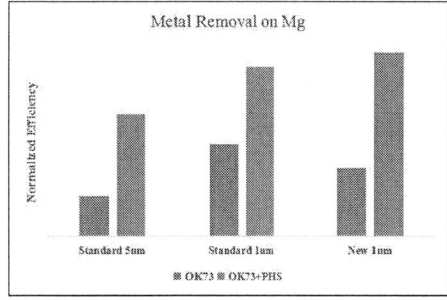

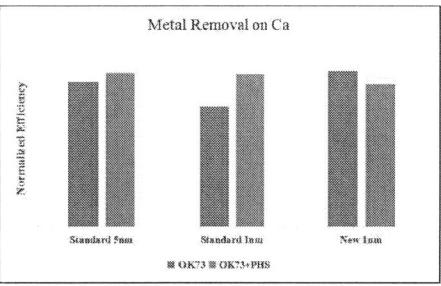

*Figure 6: Metal Removal Efficiency in OK73 solvent and PHS solution*

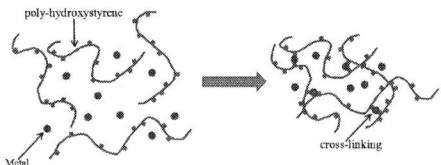

*Figure 7: Sketch of gel's formation*

**Gel/Particle removal test**

Figure 8 shows the molecular weight of PHS polymer after filtered by different Nylon66 filters. The peak value in X-axis shows the mean M.W. of polymer, the starting and ending of peak represent the smallest and highest M.W. in the solution, respectively. Meanwhile, the peak area represents the concentration of polymer in the solution. Larger peak area means higher concentration of polymer in the solution.

978-1-6654-9759-6/22 $31.00 © 2022 IEEE

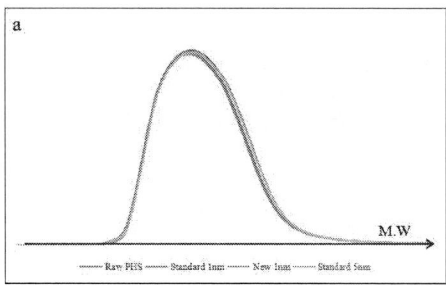

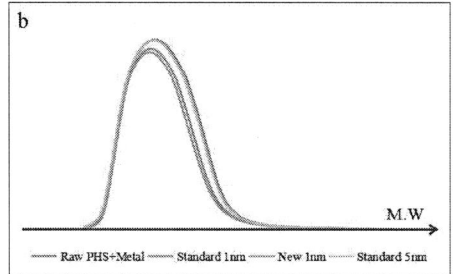

*Figure 8: M.W. of different PHS filtrate by different filters: a) Raw PHS; b) 300ppb metal spiked PHS*

In the raw PHS filtration test, the M.W. of filtered polymer and the area of peak both has no obvious change compared to raw PHS polymer. But in the filtrate of metal spiked PHS solution, the M.W. of filtered polymer as well as the concentration decreased a little, mainly resulting from the decrease of the polymer with higher M.W.. The decreasing trend for different membranes goes like this. new thick 1nm＜standard 1nm＜standard 5nm.

It is speculated that after metal spike in PHS solution, cross-linking occurs among some PHS polymer molecules around the metal site and gel with higher M.W formed. Then both standard 1 nm and new thick 1 nm Nylon66 filters can remove this gel/particle due to their higher rating, while 5 nm Nylon66 filter can only partially remove this gel/particle due to some adsorption effect. Furthermore, because of thickness increase, new thick 1nm Nylon66 membrane works better than standard 1nm in gel/particle removal due to extended contact time. To further verify the removal efficiency of gel/particle of Nylon66 membrane, PHS solutions filtered by different filters were filtered again by track etch membrane (PI membrane) under same pressure. The result is showed in

Figure 9.

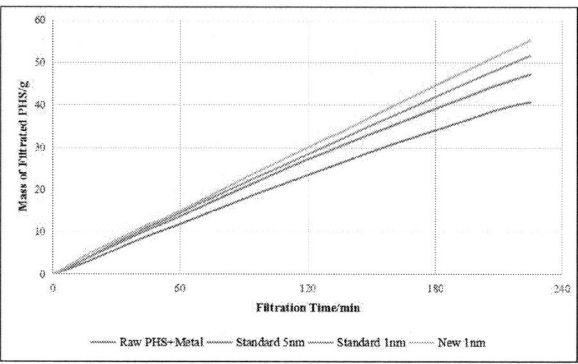

*Figure 9: Filtration yeild in different Nylon66 filter*

Assuming that filtered PHS solution is totally clean, as the filtrate pass through track etch membrane, the mass of the throughput should increase linearly as no impurity can be intercepted. On the contrary, the higher the concentration of gel/particle is in the solution, the sooner the clog and the slower the increase of throughput may occur. So the value of increase of throughput *VS.* filtration time can approximately indicates the cleanliness of solution. Lower value indicates higher concentration of gel/particle in the solution. From the result of Figure 9, we can clearly observe filtered solution is more cleaner than raw metal spiked PHS solution due to the removal of gel/particle by the Nylon66 membrane. 1nm Nylon66 filter showed the best gel/particle removal performance among all the filters due to longest contact time.

**Testing on wafer**

Filtered ArF photoresist was dispensed on testing wafers. Then After Etch Inspection (AEI) was studied. The defect counts were obtained by scanning the test wafers on a KLA tool. The result is showed in Figure 10. Compared to POR 1 nm Nylon66, new thick 1 nm Nylon66 filter has a 50% decrease on AEI defect count, clearly demonstrating the superior performance of new 1nm thick Nylon 66 membrane on defect reduction in KrF and ArF, due to the longer contact time and more tortuous flow channels.

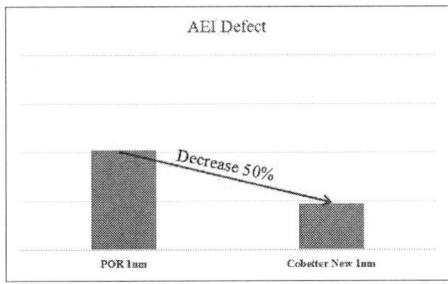

*Figure 10: After Etch Inspection Defect Count*

## CONCLUSIONS

Gel formation mechanism and how polymer gel/particle is removed by Nylon66 membrane was studied in this report. Thicker membrane with larger porosity demonstrates higher flow rate and better gel/particle removal performance. The newly designed filter with 1 nm new thick Nylon66 membrane can significantly reduce AEI defects, offering new solution to ArF and KrF photoresist filtration applications.

## REFERENCES

[1] Toru Umeda, "Defect reduction by using point-of-use filtration in a new coater/developer," *Proc. SPIE 7273*, Advances in Patterning Materials and Processes XXXVII

[2] Andre Xiao, Kalon Ke, Yoshiaki Yamada, "Development of metal purifiers specific to lithography materials," *Proc. SPIE 11612*, Advances in Patterning Materials and Processes XXXVII

[3] Toru Umeda, S Tsuzuki, "Defect reduction using POU filtration in a new coater/developer," Advances in Patterning Materials and Processes XXXVII, 2009.

[4] Lucia D'Urzo, Toru Umeda, Takehito Mizuno, "Defectivity modulation in EUV resists through advanced filtration technologies," *Proc. SPIE 11326*, Advances in Patterning Materials and Processes XXXVII.

# FCVD ANNEAL PROCESS FOR FIN CD LOSS REDUCTION STUDY

*Jun Yin\*, Yanxia Hao, Yinshuai Wang*

Research and Development Dept., Shanghai Huali Integrated Circuit Corporation, Shanghai 201314, China

\*Corresponding Author's Email: yinjun@hlmc.cn

## ABSTRACT

In the Fin Field-Effect Transistor (FinFET) process，the aspect ratio between Fin trench and Fin pitch has reached more than 5:1. STI is formed by flowable CVD(FCVD) deposition with steam ox anneal. The size of Fin CD has a great influence on the process window of the subsequent process.

We can reduce Fin CD loss by adjusting the FCVD STI anneal process. In this paper, we discuss how to reduce Fin CD loss 6% by adjusting the temperature of FCVD STI anneal process and the properties of FCVD films change at different annealing temperatures. Finding the optimal conditions for process integration.

## INTRODUCTION

With the requirement of smaller and smaller channel size, the design of logic devices has entered the 3D era. FCVD deposition has become the mainstream process with its excellent gap filling ability.

Under the action of plasma, NH3 reacts with O2 and TSA. then through O3 cure and UV cure, FCVD deposition film is deposited on wafer. In addition to the Si-O bond, such thin films also have Si-H bonds and Si-N bonds. The Si-H bond and Si-N bond can be effectively reduced by FCVD STI anneal process.

FCVD STI anneal is a steam thermal anneal process. At 400 ℃, steam fully diffuses in FCVD film. Then raise the temperature to 650 ℃ to rearrange the lattice of Si-O bonds. Finally, raise the temperature again and bake with high-temperature nitrogen to remove Si-H bond and Si-N bond and make STI OX denser under the action of high temperature.

The effect of the anneal process at different temperatures on the physical properties of STI FCVD deposition films and the difference in AA region consumption after the anneal process are formally discussed in this paper.

## EXPERIMENT AND ANALYSIS

Based on 28nm technology experience,We chose 12 inches blanket wafers with 2500A FCVD deposition film, anneal thermal temperature is 1000 ℃.900 ℃ and 800 ℃ as two splits plan.

We shorten anneal time from 120min to 30min.We get the wafer stress, the film shrinkage and H concentration SIMS data.

| Condition | Anneal | Temp | Time |
|---|---|---|---|
| BSL | Furnace | 1000 ℃ | 120min |
| 1 | Furnace | 900 ℃ | 30min |
| 2 | Furnace | 800 ℃ | 30min |

*Table 1. Split condition for FCVD STI anneal*

Stress and Shrinkage data shows the wafer stress data increases with the decrease of anneal temperature. And the film shrinkage data decreases with the decrease of the anneal temperature (Fig. 1).

The wafer stress with 900 ℃ anneal shows-77.35 Mpa and with 800℃ show 49.38 Mpa.Why dose stress data reverse?

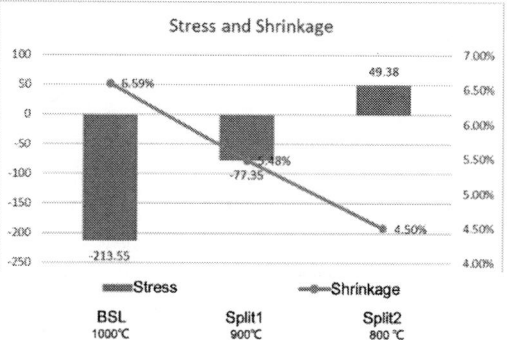

*Fig. 1. Post STI anneal Stress and Shrinkage*

After passing through the furnace steam anneal, the FCVD film is reorganized to remove free H, and Si-O bond and Si-H bond are broken to form Si-O-H, and the film shrinks to form tensile stress;

When the anneal temperature rises above 725℃, the SiO2 viscous flow partially eliminates the tensile stress, but it is still the tensile stress;

In the cooling down stage of furnace tube cooling, due to different coefficient of thermal expansion , the volume shrinkage of SiO2 is small，the shrinkage of Si substrate is large, and SiO2 produces compressive stress on Si substrate. (the higher the anneal temperature, the greater the shrinkage of Si substrate when cooling, and the greater the final compressive stress.)

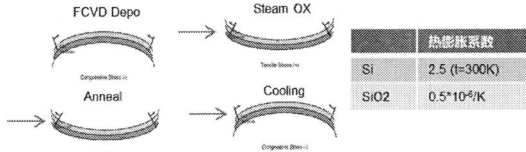

Fig. 2: Stress trend in anneal process

As shown in (Fig.3), The H concentration gradient reflects the adequacy of anneal process. After anneal process, H concentration between 1.99E19 to 2.68E20C/S. H concentration decreases with the rise of anneal temperature. , H concentration is lowest with anneal 1000℃, H concentration gradient is maximum by depth with 900℃, H concentration gradient is most equal by depth with 800℃. The level of H concentration directly affects E/R. After E/R test, all E/R data is acceptable.

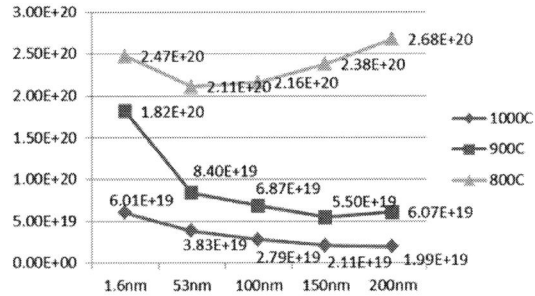

Fig. 3: H concentration SIMS data

Based on the blanket wafer data, we repeat the same split plan on structure wafers.We check FIN pitch TEM profile after FIN recess stage in (Fig. 4).Base on 1000 ℃ FIN profile, anneal with 900 ℃ gain Fin CD Loss 0.2nm , anneal with 800 ℃ gain Fin CD Loss 0.5nm (Fig 4).

Because high temperature will accelerate the diffusion of oxygen in FCVD and ALD OX film.This oxidizes the side wall of FIN,resulting in FIN CD loss.Therefore,the decrease of FCVD anneal temperature can effectively protect FIN profile.

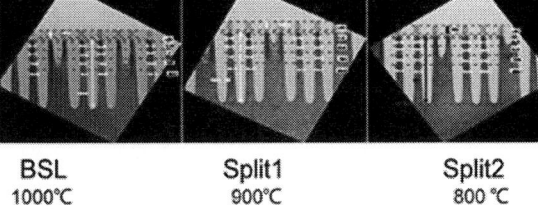

Fig. 4: Structure Wafer TEM

## CONCLUSION

In summary,The premise that FCVD deposition film

after anneal process is fully reacted and the wafer stress, the film shrinkage , H concentration SIMS data are acceptable.FCVD Anneal temperature reduction can well prevent oxygen diffusion and thus reduce FIN CD loss.

## REFERENCES

[1] T. Ghani et al., "A 90-nm high volume manufacturing logic technologyfeaturing novel 45-nm gate length strained silicon CMOS transistors,"in IEDM Tech. Dig., 2003, pp. 978–980.

[2] Xu H, Chengen Y, Dingyu L, et al. In A simple nano-scale patterning technology for FinFET fabrication, 2008 9th International Conference on Solid-State and Integrated-Circuit Technology, 20-23 Oct. 2008; 2008; pp 1340-1342.

[3] Khatkhatay F, Singh S, Wenner S, et al. In Gross substrate defects caused by thermal gradients in high temperature furnace processes, 2018 29th Annual SEMI Advanced Semiconductor Manufacturing Conference (ASMC), 30 April-3 May 2018; 2018; pp 382-385.

[4] M. H. Liao, "Local stress determination in shallow trench insulator struc tures with one side and two sides pad SiN layer by polarized micro-Raman spectroscopy extraction and mechanical modelization," J. Appl. Phys., vol. 105, no. 9, pp. 093511-1–093511-4, May 2009.

# THE INFLUENCE OF THE FILM ADJUSTMENT FOR DEEP TRENCH ISOLATION (DTI) FILLING ON THE THERMAL QUALITY IN CMOS IMAGE SENSOR

*Yaguo Cai, Chunshan Zhao, Wuzhi Zhang*
Huali Microelectronics Corp.Shanghai, China
Email:caiyaguo@hlmc.cn

## ABSTRACT

Backside deep trench isolation (B-DTI) technology for crosstalk reduction has been widely adopted in CIS production. The filling status and film stack are important for the thermal process. No crack or bubble is acceptable. As the DTI depth increase, the filling and thermal stability become difficult to control. In this paper, we adjusted the film stack and thickness for different DTI depth to realize crack-free, thus increasing the yield and improving the performance..

*Keywords—film adjustment; Deep trench isolation (DTI); crack-free; CMOS image Sensor (CIS); Backside illumination (BSI)*

## INTRODUCTION

Backside illumination (BSI) technology for higher sensitivity and backside deep trench isolation (B-DTI) technology for crosstalk reduction have been widely adopted in CIS production. However, the thinning and plasma etching process-induced silicon surface damage, resulting in the dark current (DC) increases. Therefore, after thinning the Si surface has to be passivated. Surface passivation usually is achieved by a shallow boron implant followed by a laser anneal step to activate the boron in the Si lattice near the Si surface[0]. And the field-effect passivation (FEP) method was also developed by Sung-Kun Park et al.[1] to suppress DC by accumulating holes in the damaged Si surface using a negative fixed charge of a high-k material, which also affects the anti-refraction (AR) ratio. Francois Roy et al.[2] designed the positively charged ONO stack (20 nm $SiO_2$/60 nm $Si_3N_4$/170 nm $SiO_2$)by PECVD, the first oxide layer plays a role of chemical surface passivation by forming lower interface state density at the $Si-SiO_2$ interface, which can also buffer the Si/SiN interface stress, the second SiN layer retain charges in the form of SiN_x:H, inducing field-effect passivation and antireflection, the third oxide layer is deposited to enhance the optical anti-reflective effect.

In our case, to decrease the optical and electrical crosstalk, backside deep trench isolation (DTI) structure was fabricated, at the same time, a shallow deep trench in the middle of the pixel was also adopted to increase the optical distance, which is beneficial for the light of long wavelength to be absorbed, thus increasing the quantum efficiency of near-infrared light. For special application, DTI depth varies. The filling material and thereafter film stack may not stable during anneal, crack or bubble defect will appear. In order to meet the mass production requirement, the worse test is needed. In this paper, we studied

the influence of film stack on the thermal stability. By optimizing the film stack, the manufacturable process was developed.

## EXPERIMENT

Fig. 1(a) shows the process flow for DTI filling and anneal. In mass production, $Al_2O_3$-$Ta_2O_5$ dual-layer film was selected as a high k layer to trap electrons. The DTI depth and filler varied with different performance requirements. For OmniVision's production, OV8856 was designed for a normal mobile camera with silicon thickness of 2.32 µm and DTI depth of 0.4µm filled with tungsten, while OS04C10 aimed for near-infrared imaging contains 3.5 µm silicon and 2 µm DTI for better isolation, which was filled with silica film formed by atomic layer deposition (ALD) method. Although the ALD shows an excellent filling capability, there still exists a void for the imperfect DTI profile as shown in Fig. 1(b), in which the middle part is wider than the top critical dimension (CD), inducing the opening closed early, and the overhang of $Ta_2O_5$ also accelerated the closing. As depicted in Fig. 1(a), the flow includes buffer layer anneal, the high temperature (400 °C) anneal was used for repairing the silicon damage caused by plasma etch and drive H in SiN_x:H into the interface of silicon to remove the dangling bond by forming Si-H. During the anneal, the gas in the void expanded and formed crack if the stress is too high, as shown in Fig. 1(c).

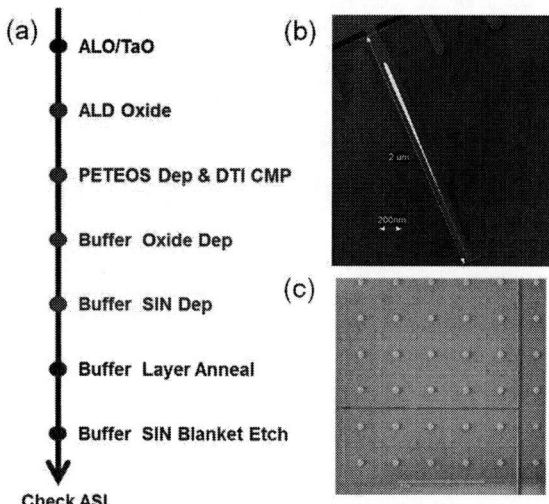

Fig. 1 (a) process flow for DTI filling and anneal, (b) SEM image of DTI, (c) Optical image of DTI crack

## DISCUSSION

For the DTI of 2 μm depth, as the thickness of ALD oxide increased 4%, the white pixel become better, while decreasing the thickness of ALD oxide deteriorate the white pixel obviously, the white pixel fail bin map was shown in Fig. 2, which means that the thickness of ALD oxide plays an important role in compressing the crack formation for that the thicker the oxide, the larger the compress stress is. However, the thickness setting should be a tradeoff between throughput and performance.

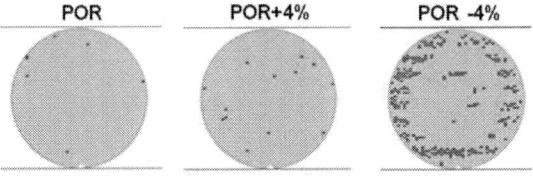

Fig. 2 Fail Bin map of different ALD oxide thicknesses for 2 μm DTI

In order to increase the quantum efficiency of near-infrared light, 6 μm silicon was introduced, of which the depth of DTI is 4 μm. As the DTI became deeper, the damage of silicon sidewall is more serious than normal, and the filling also became difficult along with a bigger void. In Fig.1(a), buffer oxide (PECVD) was deposited on the ALD oxide and followed by chemical mechanical polish (CMP), which import stress damage to the bottom film. In the concern of thermal expansion difference, the PETEOS deposition and CMP as well as buffer oxide deposition were skipped, the result depicted no crack was reviewed and the white pixel box chart was shown in Fig.3, which verified our theory. So the flow was simplified, and the ALD oxide mainly controlled the performance of the device.

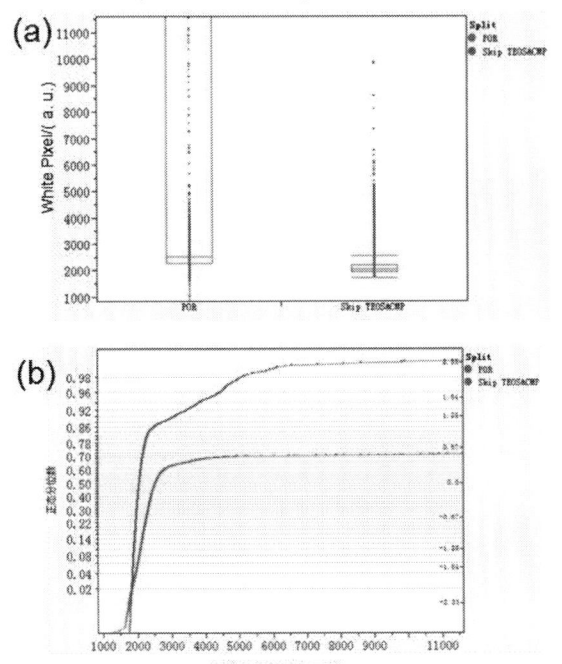

Fig. 3 (a) box chart and (b) P-chart of white pixel for skip TEOS and CMP

The white pixel performance for different ALD oxide thickness was shown in Fig. 4, as the thickness increase, the performance trends better. The dishing shown in Fig. 5 is not visible, which is crucial for the crack for that the crack trend expand vertically the dishing and the DTI bottom.

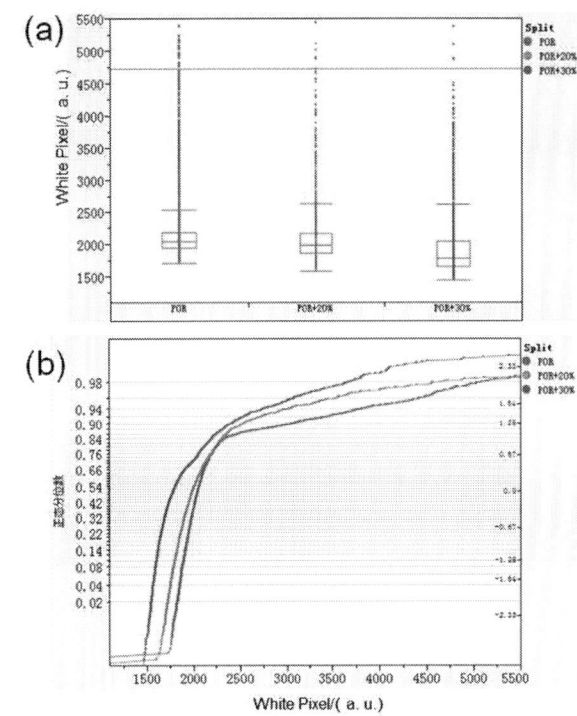

Fig. 4 (a) box chart and (b) P-chart of white pixel for different ALD oxide thickness of 4 μm DTI

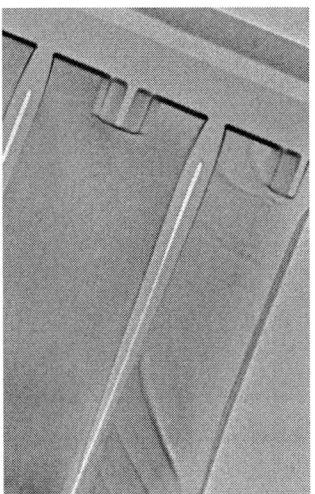

Fig. 5 The profile of DTI with 4 μm depth

## CONCLUSION

By the adjustment of the film stack of DTI filling, we obtained BSI chips with acceptable white pixel performance for different DTI depth. Although crack-free was achieved in the present process condition, the thick ALD oxide throughput is not suitable for the mass production, especially for the deeper DTI structure. Based on Ref.[3], the profile of DTI decided the stress distribution, therefore, the tuning of the DTI profile is ongoing in our subsequent work, which is believed to be better for process optimization.

# REFERENCES

[1] S. K. Park, M. Li, and Y. W. Do, et al, "An investigation of field-effect passivation layer characteristics using second harmonic generation measurement," IEEE Transactions on Semiconductor Manufacturing, vol. 33, no. 4, p109-115, February, 2020.

[2] F. Roy, A. Suler, and T. Dalleau, et al, "Fully depleted, trench-pinned photo gate for CMOS Image Sensor applications," Sensors, vol.20, pp. 727–734, 2020.

[3] D. H. Kim, S. Park, and D. Jung, et al, "Analysis of structural effect on material stress at backside deep trench isolation using finite element method," Microelectronic Engineering, vol. 154, pp. 42-47, 2016.

# FDC HIGH ORDER ANALYSIS

*He Yiwen, GONG Danli,and Shao Xiong*

SCHOOL OF ELECTRONIC INFORMATION, SHANGHAI UNIVERSITY OF ENGINEERING SCIENCE, CHINA

ENGINEERING SECTION 1, HUALI IC MANUFACTURING CO., LTD., SHANGHAI, CHINA

CORRESPONDING AUTHOR'S EMAIL: HEYIWEN@HLMC.CN

## ABSTRACT

At present, the FDC （ Fault Detection and Classification ） system is controlled by the engineers manually setting one-dimensional SPEC according to the process experience. This control mode only controls the vertical data in a rough range, and cannot accurately control the horizontal data. At the same time, due to too many parameters received by the sensor, engineers cannot manually set the SPEC one by one, which may lead to certain risks of omission and error.

## INTRODUCTION

Through the research of FDC high-order algorithm discussed in this paper, it mainly adopts the analysis of data distribution type, then sets reasonable SPEC, and combines linear correlation to detect and analyze anomalies. This study is mainly to discuss whether FDC data analysis technology can help manufacturing factories improve manufacturing capacity, its significance is to improve product yield, reduce scrap risk for reference research and exploration.

## SPEC SETTING ALGORITHM

Study the distribution characteristics and types of data, divide quantitative data and qualitative data to distinguish basic statistics, and find data rules quickly.

### Normal distribution

One of the most intuitive ways to judge a normal distribution is the probability density function,The probability density function is:

$$F(x) = \frac{1}{\sigma\sqrt{2\pi}} e^{-\frac{(x-\mu)^2}{2\sigma^2}} \qquad (1)$$

The mathematical expected value or expected value μ of the normal distribution is equal to the position parameter, which determines the position of the distribution; The square root or standard σ of the variance $\sigma^2$ is equal to the scale parameter, which determines the magnitude of the distribution.

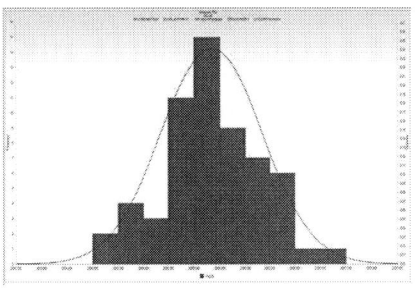

*Fig1:Normal distribution diagram*

### The skewness distribution

Skewness distribution means that the frequency distribution is asymmetric and the concentration is biased to one side. If the concentration is biased to the side with small value, it is positive skewness distribution. Otherwise, it is negative skewness distribution. If the peak of the frequency distribution shifts to the left, the long tail extends to the right, which is a positive skewness distribution (right skewness distribution). Otherwise, it is negative skewness distribution (left skewness distribution).

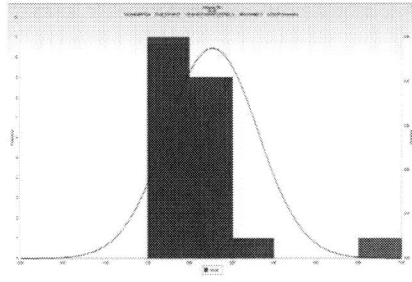

*Fig2:Skewness distribution diagram*

### Uniform distribution

A uniform distribution is a symmetric probability distribution, and distributions of the same length interval are equally likely.

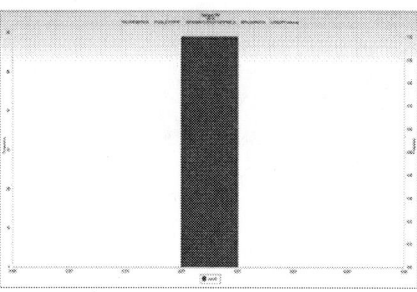

Fig3: Uniform distribution diagram

# SPEC RANGE CARD CONTROL METHOD

Normal distribution, skewness distribution and uniform distribution are set by using the concept of historical data mean, STD, kurtosis, skewness and quantile. Abnormal data is defined as follows: normal distribution, if the data exceeds the historical dispersion by 3 times; Skewness distribution, if the data exceeds 3 times the peak; Uniform distribution. Due to the characteristics of stable data without fluctuation, the upper quartile, median and lower quartile are combined with quantile to set, and a certain range of data fluctuation is given at the same time.

## The SPEC verification

According to the characteristics of the process data, the SPEC is calculated with the history of 6 months. Data with normal distribution, skewness distribution and uniform distribution were selected for verification.

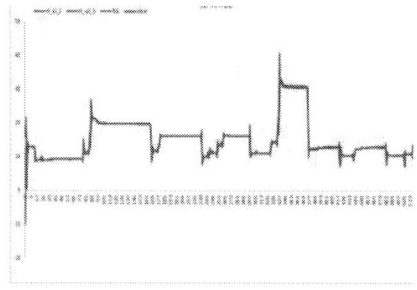

Fig4: It has normal distribution, skewness distribution and uniform distribution

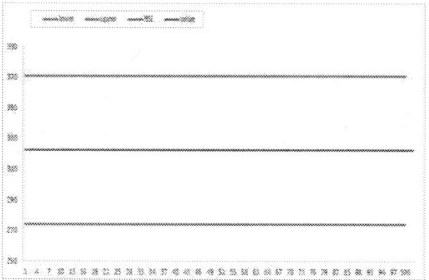

Fig5: Uniformly distributed

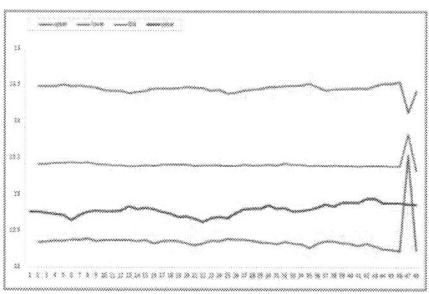

Fig6: It has normal distribution and uniform distribution

From the above three figures, we can see that the range and trend of the SPEC can be predicted well by extracting the historical data distribution type. From Figure 6, we can clearly see that the 47th closing point exceeds the lower limit of our SPEC card control and meets our requirements for detecting outliers.

# ANOMALY DETECTION

By means of correlation analysis, the relationship between the sensor's received value and the predicted BSL is analyzed whether there is a deterministic relationship. If the relationship between the two is strong, the alarm information is valid

## Application and Analysis

We select the data of 6 RUNIDs received by VAR5 sensor and divide them into 3 process stages (Table 1) for testing and verification. These 3 process stages correspond to the 3 stages (marked by dotted boxes) of the sensor's received value in Figure 7 respectively.

TABLE I.  PARAMETERS OF THE PROCESS STAGES

| Step 1 | Step 2 | Step 3 |
|--------|--------|--------|
| Stable | Dep | Discharge |

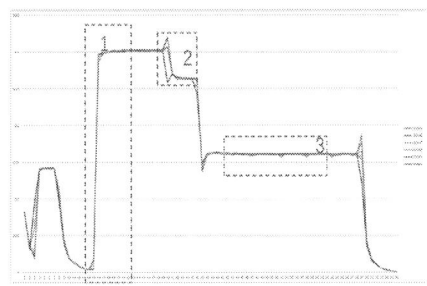

*Fig7:VAR5 sensor received value*

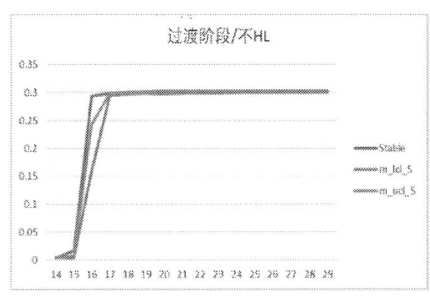

*Fig8:Excessive stage（Stable）*

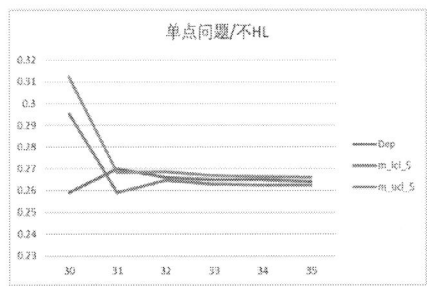

*Fig9:A single point of the problem（Dep）*

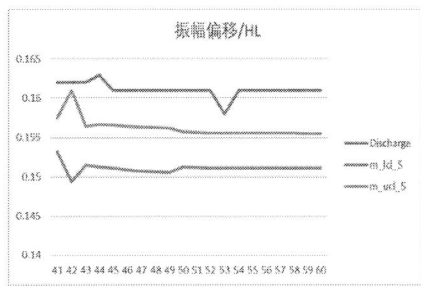

*Fig10:Amplitude deviation（Discharge）*

From the analysis of the above three pictures, it can be seen that the linear correlation algorithm can effectively filter the process stages unrelated to the BSL obtained by us (See FIG. 8 (excessive stage) and FIG. 9 (a single point of the problem)), and can effectively capture the process stages with real anomalies on the actual line (see FIG. 1 (amplitude offset)). Therefore, the correlation analysis algorithm can capture useful alarm information effectively and remove invalid data.

## CONCLUSION

Through the analysis of data, etc, has established the effective correlation historical data and existing data, and then established a two-dimensional SPEC control standard, implements to monitor all the equipment in the FAB, can fully use the FDC effective analysis of original data, when the machine run goods effective monitoring of different period of abnormal.

## ACKNOWLEDGEMENTS

The author would like to thank the authors for their important comments and ideas and the anonymous reviewers for their valuable comments and suggestions.

## REFERENCES:

[1] Normal Distribution, Gale Encyclopedia of Psychology

[2] Shaou-Gang Miaou; Jin-Syan Chou. 《Fundamentals of probability and statistics》

[3] John Aldrich. Earliest Uses of Symbols in Probability and Statistics.

[4] 4.R. J. Herrnstein and Charles Murray (In 1994). The Bell Curve: Intelligence and Class Structure in American Life. Free Press. ISBN 978-0-02-914673-6.

# A STUDY OF LDMOS WITH HIGH BREAKDOWN VOLTAGE AND LOW ON-RESISTANCE IN 22NM TECHNOLOGY

*Zhenchao Sui[12*]; Yongqiang Che[1]; Xiaoxi Liu[1]; Yongjin Tang[1]; Yongsheng Yang[1]*

[1]Semiconductor Manufacturing International Corporation(SMIC), Beijing 100176, China

[2] School of Software and Microelectronic, Peking University, Beijing 102600, China

*Corresponding Author's Email: mark_sui@smics.com

## ABSTRACT

The characteristics of LDMOS (Laterally Diffused Metal Oxide Semiconductor) with various layout structures have been studied based on 22nm HKMG technology in this paper. A lightly drift region Psub is designed for high BVDS performance, which induces the impact ionization drive further away from the STI edge and $Si/SiO_2$ interface. To improve the trade-off parameter Rdson, systematical TCAD simulations and experiments have shown optimization layout structure for small Ron with optimized Lw (Psub region), Lg (effective channel), and Lp (a NW region and gate overlap). The optimized layout devices showed BV>13.4V&Rdson of 2.21 $mohm \cdot mm^2$ for NLDMOS and BV>14.1V &Rdson of 6.51 $mohm \cdot mm^2$ for PLDMOS.

## INTRODUCTION

LDMOS is an attractive device in high voltage, smart power integrated circuits, and display drivers, which are investigated with existing technology and handle a wide range of operating voltages [1]-[4]. To achieve the performance of BVDS and Rdson need to optimize LDMOS transistor layout and characteristics within standard technology process, which decrease not only a manufacturing cost but also responding time to market [5]-[8]. An LDMOS with STI structure or heavy doping in the drift region has been applied to reduce Ron, which was universality applied on planar nodes [9]-[11]. But along with critical size scaled, LDMOS optimization would be challenged. In this paper, to achieve large BVDS and low Ron LDMOS for 22nm planar HKMG technology, a slight body doping Psub region was introduced for better performance, which makes the BVDS higher twice than operation voltage. At the same time, the LDMOS with different layouts is characterized to understand the layout dependence. The various parameter about effective channel length Lg, Psub length Lw, NW region with gate overlap length Lp, and STI size Ls. Based on the experiment and TCAD simulation, small Lp and optimized Lg/Lw would get good BVDS and Low Rdson.

## EXPERIMENTS

The LDMOS with shallow trench isolation on the N-type well and Psub area under Gate was fabricated in 22nm HKMG technology. The TCAD simulation of the LDMOS process is compatible with normal CMOS processes. The process flow is to follow the 22nm HKMG general flow, first, the shallow trench isolation(STI) was fabricated, then the poly dummy gate with a gate oxide and offset spacer were formed, LDD implanted followed by the selective deposition of in situ boron-doped SiGe on PMOS source/drain. Then after the nickel silicide is formed, the MG loop removes the dummy poly and fills the metal gate with a different N/P gate. Finally, the W contact and Cu backend lines are applied to complete the process.

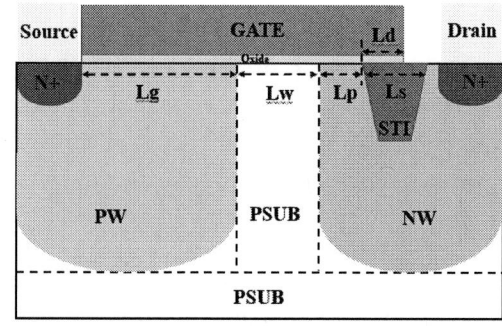

*Figure1. Cross section of NLDMOS with layout parameters with layout parameters Lg, Lp, Lw, and Ls.*

The cross-section of n-channel LDMOS design with 2.5V IO gate oxide was shown in Fig.1. Various parameters of layout including Lg, Lw, Lp, and Ls, could be adjusted to optimize the device characteristics. Lg is the effective channel length. Lw is defined as the Psub region, which is the body doping region between N-drift and P-drift. Lp is the Pwell area below the metal gate. Ls is the STI size, which is in the N-drift region and close to the drain. The range of critical sizes was designed by well DOE masks. To evaluate the LDMOS performance, the BVDS and Idlin (linear drive current) were measured and analyzed.

## RESULTS AND DISCUSSIONS

### Lw Dependence on Device Characteristics

In this part, Lp dependence of LDMOS characteristic is studied with fixed other layout parameters. As presented in Fig2, along with the Lp size increase, BVDS showed a move up first and then move downtrend. At the same time, the Rdson almost indicated a contrary trend. In this parameter study, we can find a high BVDS voltage and low Rdson for one certain Lw size. The BVDS improves due to Lw increasing the channel resistor, which can be understood easily, but when Lw size is big enough, the BVDS begin degrade may be another factor impact. TCAD simulation may make clear this doubt. The TCAD simulation shown with Lw size increase, the impact ionization has been driving further away from the corner between STI and gate oxide, which contributes to a higher BVDS value. But along with Lw large to a certain value, as shown in fig3(c), the impact ionization would close with the STI edge and Si/SiO$_2$ interface, which would impact device BVDS degradation.

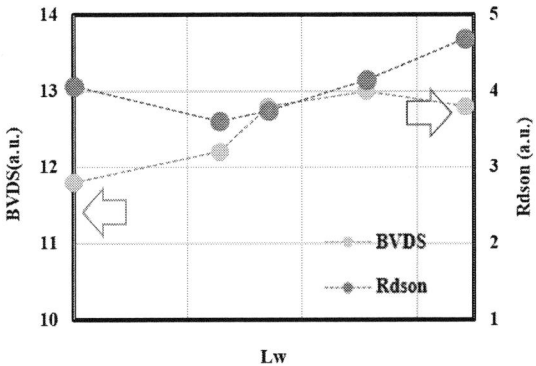

*Figure2. Normalized n-type LDMOS BVDS and Rdson dependence of Lw from 0 to big size, the yellow line is BVDS and blue line is Rdson.*

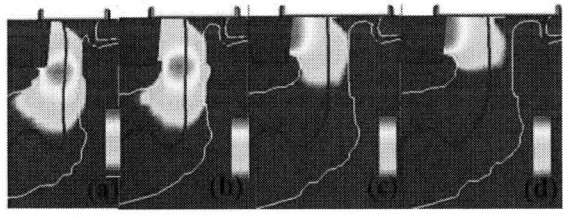

*Figure3. Impact ionization of different NLDMOS simulated by TCAD (a) Lp 1x and Lw 0; (b) Lp 1x and Lw 1x; (c)Lp 1x and Lw 2x; (d) Lp 1x and Lw 3x.*

## Lp Dependence on Device Characteristics

Lp is the P drift region below the metal gate, which is one of the critical parameters for LDMOS characteristics. Other parameters are fixed and the influence of Lp was shown in Fig4. With Lp size up, the BVDS is rapidly degraded, when the Lp size is up from 1x to 2x, the BVDS degrade ~1.8V, and Rdson increase ~5%. TCAD simulation shown on Fig5, the Lp change from 3x to 1x, the impact ionization has been driving further from

Si/SiO$_2$ interface, and further increase Lw size from 1x to 3x, the impact ionization close to STI edge, which would goto BVDS worse condition.

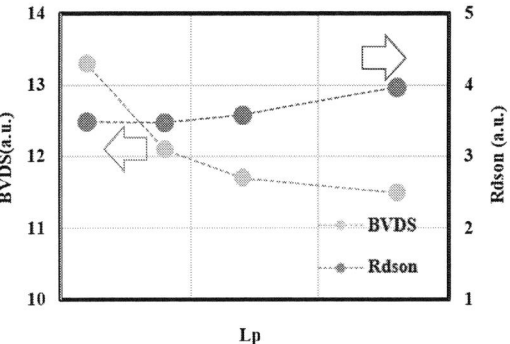

*Figure4. Normalized n-type LDMOS BVDS and Rdson dependence of Lp from small to big size, the yellow line is BVDS and blue line is Rdson.*

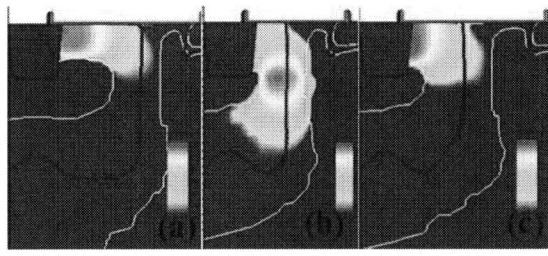

*Figure5. Impact ionization of different NLDMOS simulated by TCAD (a) Lp 3x and Lw 1x; (b)Lp 1x and Lw 1x; (c)Lp 1x and Lw 3x.*

## Lg/Ls Dependence on Device Characteristics

Lg is the effective channel of LDMOS, keeping other parameters with one fixed value, the BVDS and Rdson changes were shown in Fig6. Along with Lg size increase, at Lg small, the BVDS improved obviously from ~4V to ~13V, then continue to increase Lg, the BVDS is no obvious change and Rdson keep increase linearly. On the other hand, the STI size of Ls shows a singleness trend in Fig7. Along with Ls size larger and larger, the BVDS showed no obvious change and the Rdson slight increase linearly. Suspect the root cause is the channel resistor increase along with Ls size larger, but there is no influence on impact ionization location, which arises at the STI edge and Si/SiO$_2$ interface. So to optimize the BVDS and Rdson of LDMOS could design the Ls with small size for small pitch.

978-1-6654-9759-6/22 $31.00 © 2022 IEEE

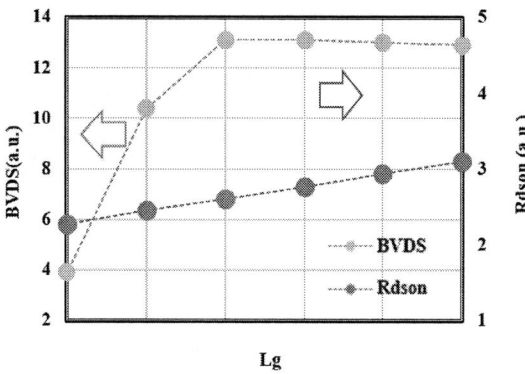

*Figure6. Normalized n-type LDMOS BVDS and Rdson dependence of Lg from small to big size, the yellow line is BVDS and blue line is Rdson.*

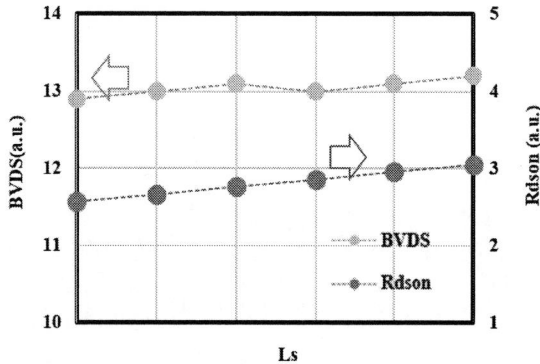

*Figure7. Normalized n-type LDMOS BVDS and Rdson dependence of Ls from small to big size, the yellow line is BVDS and blue line is Rdson.*

## SUMMARY

A novel LDMOS structure with Psub region under the metal gate is proposed and fabricated with the standard foundry 22nm HKMG CMOS process. At the same time, the dependences of layout parameters Lw, Lp, Lg, and Ls is the critical size for LDMOS characters. In this study, we introduce Lw for LDMOS, which would further improve the BVDS accounting for size shrink for 22nm, and through layout optimize the Rdson performance of LDMOS could be maintained. Based on the experiment data, Lw and Lp play an important role in device BVDS performance, the optimal Lw and Lp size could drive the impact ionization far away from the STI edge and Si/SiO$_2$ interface. On the other hand, the Lw, Lg, and Ls would greatly increase the Rdson. The experiment results show the LDMOS in our study can be improved by optimizing the layout parameters.

## REFERENCES

[1] A. N. Tallarico, S. Reggiani, R. Depetro, G. Croce, E. Sangiorgi and C. Fiegna, "Full Understanding of Hot Electrons and Hot/Cold Holes in the Degradation of p-channel Power LDMOS Transistors," *International Reliability Physics Symposium*, USA, 2020, pp. 1-5.

[2] B. Duan, Z. Cao, X. Yuan, S. Yuan and Y. Yang, "New Super junction LDMOS Breaking Silicon Limit by Electric Field Modulation of Buffered Step Doping," *IEEE Electron Device Letters*, Jan. 2015, vol. 36, pp. 47-49.

[3] Z. Dong, B. Duan, C. Fu, H. Guo, Z. Cao and Y. Yang, "Novel LDMOS Optimizing Lateral and Vertical Electric Field to Improve Breakdown Voltage by Multi-Ring Technology," *IEEE Electron Device Letters*, vol. 39, no. 9, Sept. 2018 ,pp. 1358-1361.

[4] A. Saadat, P. B. Vyas, M. L. V. d. Put, M. V. Fischetti, H. Edwards and W. G. Vandenberghe, "Channel Length Scaling Limit for LDMOS Field-Effect Transistors: Semi-classical and Quantum Analysis," *The 32nd International Symposium on Power Semiconductor Devices and ICs (ISPSD)*, 2020, pp. 443-446.

[5] G. Wang, B. Lee, N. Wang, K. Ding and G. Ma, "On-Resistance Improvement Impacted by Trapping Effects in Fin-LDMOS Technology," *2019 China Semiconductor Technology International Conference*, pp. 1-3, 2019.

[6] J. Yuan and L. Jiang, "Evaluation of Hot-Electron Effect on LDMOS Device and Circuit Performances," *IEEE Transactions on Electron Devices*, June 2008, vol. 55, pp. 1519-1523.

[7] S. Yuan, B. Duan, H. Cai, Z. Cao and Y. Yang, "Novel LDMOS with assisted deplete-substrate layer consist of super junction under the drain," *The 29th International Symposium on Power Semiconductor Devices and IC's (ISPSD)*, 2017, pp. 279-282.

[8] Z. Cao et al., "Novel superjunction LDMOS with multi-floating buried layers," *The 29th International Symposium on Power Semiconductor Devices and IC's (ISPSD)*, 2017, pp. 283-286.

[9] G. Ma et al., "A study of 28NM LDMOS linear drain current degradation induced by hot carrier injection," *China Semiconductor Technology International Conference,* 2018, pp. 1-3.

[10] Ruo Yuan Li, Yongsheng Yang and Fang Chen, "A Study of 28nm LDMOS HCI Improvement By Layout Optimization", *China Semiconductor Technology International Conference*, 2017, pp. 1-4.

[11] A.F.M. Alimin, H.H. Hizamul-din, S.F.W.M. Hatta and N. Soin, "The influence of Shallow Trench Isolation Angle on Hot Carrier Effect of STI-Based LDMOS Transistors", *RSM of IEEE*, 2017,pp. 248-251.

# A STUDY OF THE EFFECT OF SIGE ON THE INVERSE NARROW WIDTH EFFECT IN 28NM PROCESS

*Yang Li[1]\*, Weiwei Ma[1], Ran Huang[1], Yamin Cao[1], Wei Zhou[1]*
[1] Shanghai Huali Integrated Circuit Corporation, Shanghai, China
\* Corresponding Author's Email: liyang_e1-1@hlmc.cn

## ABSTRACT

The profile of SiGe plays a dominant role in determining inverse narrow width effect of sub-28 nm PMOS. In this paper, different SiGe profiles, which are illustrated in multiple parameters, are achieved by altering the SiGe etch process and deposition processes in 28 nm HKMG technology. Then the relationship between the inverse narrow width effect and SiGe processes are obtained. It is found that the drive current becomes higher when the sigma-shaped SiGe structure is smaller or the raised SiGe height is larger. Furthermore, the inverse narrow width effect is closely related to the SiGe profile.

## INTRODUCTION

Strain engineering is nowadays widely used as a necessary solution to improve the performance of silicon-based complementary metal-oxide-semiconductor filed-effect transistor (CMOS). By changing lattice parameters to reduce the interaction atomic force of Si substrates, the channel mobility is enhanced and therefore the drive current is improved. Because of the difference between the effects of strain on the carrier mobility, stressor techniques are used extensively for NMOS and PMOS.

As for PMOS, it is well-known that providing compressive stress can effectively improve hole mobility, and such compressive stress is commonly derived from the embedded silicon–germanium in 28nm technology. [1-2]. The introduction of embedded SiGe in adjacent source-drain regions provides paramount compressive strain as the lattice parameter of relaxed SiGe is larger than Si channel.

However, the amount of tensile stress provided by SiGe is determined by various parameters, such as the Ge content in the virtual substrate[3] and the profile of SiGe. Hence the carrier mobility changes significantly as the physical dimension of Si channel or embedded SiGe varies, and the inverse narrow width effect (INWE) becomes more complicated, which leads to the threshold voltage decreasing and drive current increasing with decreasing channel width. To illustrate the INWE of PMOS in 28nm technology, it is necessary to clarify the effect of SiGe profile on the INWE.

In this paper, different sigma-shaped SiGe structures are designed by adjusting the etch process and depositions process of SiGe, then the drive current dependent on the channel width of PMOS transistors are obtained. It can be found that the drive current of large-width PMOS is less sensitive to the change of SiGe process because compressive stress oriented by SiGe becomes more neglected.

## EXPERIMENT

The profile of SiGe is illustrated in figure 2. The SiGe parameters can be classified in 2 groups, which are controlled by the etch process and depositions process separately. The sigma-shaped SiGe structure is described by tip-to-gate (T2G) and recess depth (RCD). The embedded SiGe becomes larger as the T2G decreases and the RCD increases. The other parameters, the thickness of SiGe cap and the overfill of bulk SiGe, determines the ratio of the seed layer, bulk layer and cap layer of SiGe. Different etch process and depositions process is applied to obtain various SiGe profiles and WAT is used to measure the device performances. All wafers analyzed in this study were manufactured based on HLIC 28 nm HKMG technology.

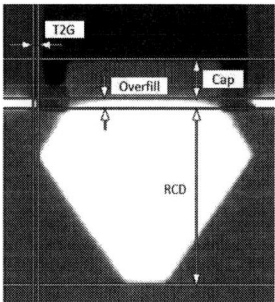

*Figure 1: Parameters of SiGe profile.*

## RESULTS AND DISCUSSION

The relationship between the drive current of PMOS and the parameters of sigma-shaped SiGe structure is shown in Fig. 2(a). It reveals that when T2G increases and RCD decreases, which means the sigma-shaped SiGe structure becoming smaller, the drive current decays. The reason is that the stress generated by SiGe is negatively related to the distance between Si channel and SiGe[5], consequently leading to the reduced hole mobility.

978-1-6654-9759-6/22 $31.00 © 2022 IEEE

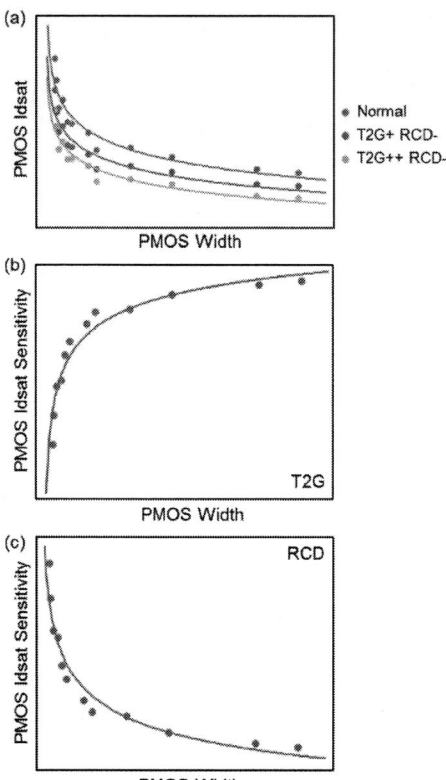

Figure 2: (a) INWE of PMOS with different SiGe profiles, (b) the relation between PMOS Idsat sensitivity of T2G and PMOS width, (c) the relation between PMOS Idsat sensitivity of RCD and PMOS width.

Furthermore, it can be seen from Fig.2(b-c) that the drive current changes more dramatically when channel width becomes smaller. ANSYS simulations by Wang[4] also presented the same trend. Such a phenomenon is derived from the fact that the compressive stress along the channel is found to dominate drive current improvement when the Si channel is narrow, which decreases significantly with the increasing channel width. As the channel width is large enough, the drive current is mainly effected by the compressive stress along the channel and the tensile stress along the width direction, both of which is too small to enhance the hole mobility.

The relation between drive current of PMOS and the cap thickness and overfill of SiGe is shown in Fig. 3(a). It clearly shows that the smaller the total raised SiGe height is, the smaller the drive current improvement is. It is because the compressive stress is mainly derived from the bulk of SiGe. When the total raised SiGe height increases, the bulk of SiGe is away from the channel, leading to the reduced stress and decreased driving current.[6] Moreover, as shown in Fig.3(b), Compared with the drive current of narrow-width PMOS, the drive current of large-width

PMOS changes less because the compressive stress is not the decisive parameter of the drive current.

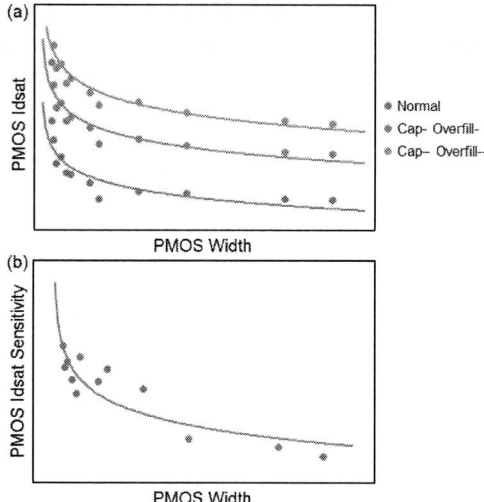

Figure 3: (a) INWE of PMOS with different SiGe profiles, (b) the relation between PMOS Idsat sensitivity of the raised SiGe height and PMOS width.

## CONCLUSION

The effect of SiGe profiles on the inverse narrow width effect is discussed. It is found that the drive current becomes higher when the sigma-shaped SiGe structure is smaller or the raised SiGe height is larger. The drive current also decays when the Si channel turns to be wider. The reason is that the increasing drive current is mainly determined by the compressive stress generated by SiGe, which changes with SiGe profiles and decreases with channel width.

## ACKNOWLEDGEMENTS

I would like to appreciate senior engineer Weiwei Ma, Ran Huang, Yamin Cao, Wei Zhou for their support and advice on this experiment.

## REFERENCES

[1] Arvind Sharma; Naushad Alam; Sudeb Dasgupta and Anand Bulusu, "Multifinger MOSFETs' Optimization Considering Stress and INWE in Static CMOS Circuits," IEEE Trans. vol. 63, no. 6, pp. 2517-2523, Mar. 2016

[2] Tara Prasanna Dash; Suprava Dey; Sanghamitra Das; Eleena Mohapatra; Jhansirani Jena, "Strain-engineering in nanowire field-effect transistors at 3 nm technology node", Phisical E: Low-dimensional Systems and Nanostructures, vol. 18, pp. 113964, Jan. 2020

[3] G. Tsutsui; S. Mochizuki; N. Loubet; S. W. Bedell; D. K. Sadana; "Strain engineering in functional materials", AIP Advances, vol. 9, pp. 030701, Mar. 2019

[4] Xin Wang; M. Huang; C. Bowen; L. Adam; S. Singh and C. Chiu; J. Wu, "Exploring Transistor Width Effect on Stress-induced Performance Improvement in PMOSFET

with SiGe Source/Drain", 2005 International Conference On Simulation of Semiconductor Processes and Devices, December 2005

[5] Chia-Feng Lee; Ren-Yu He; Kuan-Ting Chen; Shu-Ying Cheng; Shu-Tong Chang; "Strain engineering for electron mobility enhancement of strained Ge NMOSFET with SiGe alloy source/drain stressors", Microelectronic Engineering. vol. 138, pp. 12-16, Jan. 2016.

[6] H.C. Lo, C.T. Li, Y.T. Chen, C.T. Yang, W.C. Luo, W.Y. Lu, C.F. Cheng, T.L. Chen, C.H. Lien, H.T. Tsai, M.C. Chen, Samuel K.H. Fung, C.C. Wu; "CMOS on dual SOI thickness for optimal performance", Microelectronic Engineering, vol. 87, pp. 2531-2534, Jun. 2010

# IMPROVING THE DEVICE PERFORMANCE OF LDMOS THROUGH THE OPTIMIZATION OF STRUCTURE

*Shuang Jiao[1]\*, Chenchen Qiu[1], Jun Qian[1], and Chang Sun[1]*
[1] Shanghai Huali Microelectronics Corporation, Shanghai 201203, China
*Corresponding Author's Email: jiaoshuang@hlmc.cn

## ABSTRACT

Considerable effort has been put into the development of LDMOS. The main issue in the development of LDMOS is to obtain the best trade-off between specific on-resistance $R_{sp}$ and BV. In this paper, the influence of key size on device performance is systematically studied based on 55nm platform. By optimizing key dimensions, Rsp and BV are balanced. The BV of LDMOS is greater than 50V, and the performance of Rsp reaches the advanced level in the industry.

*Keywords—Lateral diffused MOS (LDMOS), 55 nm, BV, specific on-resistance $R_{sp}$*

Since the 1980s, CMOS has become the leading technology of very large scale integration (VLSI) due to its outstanding characteristics of simple process, high device integration and low power consumption. However, with the advent of ultra large scale integration (ULSI), a single technology can no longer meet the requirements of increasingly complex integrated systems. Modern microelectronic system hopes to realize the comprehensive integration of information receiving, storage, transmission, processing and execution. However, due to the high power consumption and low integration, the bipolar technology is not competitive in USLI. It still plays an irreplaceable role in the design of ultra-high speed and high current drive chips, such as power management chips. Therefore, it is necessary to develop a process technology compatible with bipolar and device characteristics and design technology of new structure devices, which is Bipolar-CMOS-DMOS (BCD) process technology. BCD technology focuses bipolar transistors with low noise, high precision and high current density, CMOS transistors with high integration, low power consumption, easy logic control, and DMOS devices with high voltage resistance, fast switching speed, high input impedance, strong driving ability, good thermal stability, etc.

Since its birth, BCD technology has been widely concerned by the semiconductor industry, and the characteristic size of BCD keeps shrinking. However, due to the need to balance between the high density of digital circuits and the high voltage and high current characteristics of power devices, the pace of BCD technology feature size reduction is relatively slow. At present, when the MOSFET based on 3nm process node has been developed, the process node of BCD still stays at 90/65nm.

In order to be more compatible with CMOS technology, DMOS usually use a horizontal double diffusion MOS transistor (LDMOS). Specific on resistance and breakdown voltage show an exponential growth relationship, therefore, it is

necessary to systematically study the geometry structure, process flow, impurity distribution of LDMOS, so as to better improve device performance. The work of this paper is mainly based on the 55 nm process platform of HLMC to study LDMOS devices. We systematically studied the influence of key dimensions of LDMOS device on device performance, and obtained a breakdown voltage of more than 50 V and a specific on-resistance of as low as 25 by optimizing the key performance of the device.

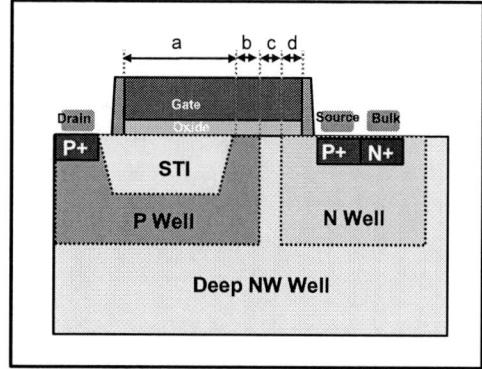

Fig. 1. The device structure of LDMOS.

The schematic diagram of device structure of LDMOS is shown in Figure 1. In the schematic diagram, we mainly name several key dimensions that affect the performance of our device respectively. Where, a is the length of the field board, b is the length of the STI wrapped by the P-well, c is the distance from the P-Well to the N-well, and d is the length of the channel.

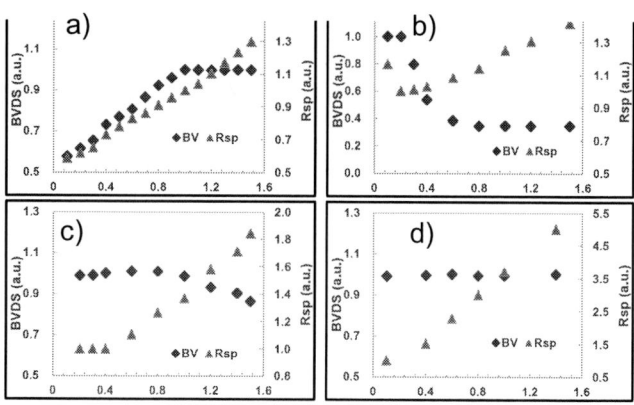

Fig. 2. BVDS and Rsp under different key size. Pictures a, b, c and d correspond to the key dimensions a. b. c. d, respectively.

As shown in Figure 2, Rsp increases significantly with the increase of a size. BV first increases with the lengthening of a size, then stops increasing and tends to be stable. In order to meet the requirements of the breakdown voltage, a size needs to meet a minimum value. On this basis, a size needs to be reduced as much as possible to obtain a smaller Rsp and a smaller device size. The influence of b size on the device is more responsible. With the increase of b size, BV stays stable and then decreases gradually, while Rsp decreases first and then increases correspondingly. Based on the effect of B size on device BV and Rsp, we can optimize the key size to obtain both larger BV and smaller Rsp. The increase of c size will decrease and the increase of Rsp. According to our experimental results, c is not necessary. In order to reduce the device size, we can reduce c size to the minimum. d size has no obvious effect on the breakdown voltage of the device, but with the increase of size, the increase of on-resistance is very obvious. The influence of each key size on the device performance is the basis of our device debugging. Through the optimization of each size, we can obtain the best device performance

## CONCLUSION

Based on the 55nm process platform, we carried out relevant research on LDMOS. This paper systematically studies the influence of key dimensions of LDMOS devices on the on-resistance and breakdown voltage of devices, and summarizes the influence trend of different sizes on devices. By optimizing the device structure, we have obtained over 50V breakdown voltage and on-resistance comparable to that of advanced factories. This research also provides the basis for the subsequent device development and provides the idea of refining the device structure to obtain better performance.

## REFERENCES

[1] R. Pan, B. Todd, P. Hao, R. Higgins, D. Robinson, V. Drobny, W. Tian, J. Wang, J. Mitros, M. Huber, S. Pillai, and S. Pendharkar, "High voltage (up to 20V) devices implementation in 0.13 um BiCMOS process technology for System-On-Chip (SOC) design", ISPSD 2006, pp. 349-352.

[2] R. Pelliconi, D. Iezzi, A. Baroni, M. Pasotti, and P. L. Rolandi, "Power efficient charge pump in deep submicron standard CMOS technology", IEEE Journal of Solid-State Circuits, Vol. 38, NO. 6, June 2003, pp. 1068-1071.

[3] Weifeng Sun, Longxing Shi, Zhilin Sun, Yangbo Yi, Haisong Li, and Shengli Lu, "High-voltage power IC technology with nVDMOS, RESURF pLDMOS, and novel level-shift circuit for PDP scan-driver IC", IEEE Trans. Electron Devices, Vol. 53, No. 4, April 2006, pp. 891-896.

[4] P. Moens, F. Bauwens, M. Nelson, and M. Tack, "Electron trapping and interface trap generation in drain extended pMOS transistors", IRPS 2005, pp. 555-559.

[5] M. M. Iqbal, F. Udrea, and E. Napoli, "On the static performance of the RESURF LDMOSFETS for power ICs," in Proc. IEEE 21st Int. Symp. Power Semiconductor Devices ICs, Jun. 2009, pp. 247–250.

[6] H. Cha, K. Lee, J. Lee, and T. Lee, "0.18μm 100V-rated BCD with large area power LDMOS with ultra-low effective specific resistance," in Proc. 28th Int. Symp. Power Semiconductor Devices ICs (ISPSD), Jun. 2016, pp. 423–426.

978-1-6654-9759-6/22 $31.00 © 2022 IEEE

# BOTTOM DIELECTRIC ISOLATION TO SUPPRESS SUB-FIN PARASITIC CHANNEL OF VERTICALLY-STACKED HORIZONTAL GATE-ALL-AROUND SI NANOSHEETS DEVICES

*Lei Cao[1,2], Yang Liu[1], Zhenhua Wu[1,2], Qingzhu Zhang[1,2*], Jiaxin Yao[1,2], Yanna Luo[1,2], Haoqing Xu[1,2], Peng Zhao[1,2], Kun Luo[1], Yongqin Wu[3], Weihai Bu[3], and Huaxiang Yin[1,2*]*

[1]Key Laboratory of Microelectronics Devices and Integrated Technology, Institute of Microelectronics of Chinese Academy of Sciences, Beijing 100029, China

[2]University of Chinese Academy of Sciences, Beijing 100049, China

[3]Semiconductor Technology Innovation Center Corporation, Beijing, 100176, China

*Corresponding Author's Email: yinhuaxiang@ime.ac.cn, zhangqingzhu@ime.ac.cn

## ABSTRACT

In this paper, to suppress the parasitic channel of sub-fin in vertically stacked gate-all-around (GAA) Si nanosheet FETs (NSFETs), the bottom dielectric isolation (BDI) was designed and simulated using TCAD simulation. The results show that the sub-threshold characteristics of the $n$-type NSFETs are improved by the BDI approach, and the gate capacitance is reduced by 16.98% compared with that of the NSFETs with sub-fin. In addition, the BDI technology can increase the process and electrical stability of the NSFETs.

**Keywords** — Nanosheet FETs (NSFETs); TCAD simulation; sub-fin; bottom dielectric isolation (BDI)

## INTRODUCTION

Vertically-stacked GAA Si NSFETs have been considered as one of the most promising candidates beyond FinFETs technology due to its excellent ability of gate control and short-channel effects immune [1-2]. However, compared with the traditional bulk FinFETs, the fabrication of NSFETs suffers from a lot of challenges, such as the high selective of channel release process, the suppression of parasitic sub-fin channel, the complex module process of inner spacers and low temperature integrated processes [3-5]. In order to reduce the influence of sub-fin, the bottom dielectric isolation (BDI) technology can be used to isolate the devices and Si substrates [6-7]. However, a detailed investigation of BDI properties with ground plane (GP) engineering has not been extensively studied.

In this work, the NSFETs with BDI was studied by the process and electrical simulation. Compared with GP engineering, the BDI technology shows the advantages of significantly suppressing the parasitic sub-fin channel and reducing the gate capacitance. Meantime, the BDI technology can improve the process stability of NSFETs.

## DEVICE STRUCTURE AND TCAD SIMULATION

3-nm-node NSFETs were simulated using Sentaurus TCAD tools. The structure of conventional NSFETs with

Table 1: Geometrical Parameters for 3-nm-node NSFETs

| Symbol | Geometrical Parameters | Values |
|--------|------------------------|--------|
| $L_g$ | Gate length | 16 nm |
| $L_{SD}$ | Source/Drain (S/D) length | 8.5 nm |
| $L_{sp}$ | Spacer length between S/D and Metal gate | 6 nm |
| $W_{NS}$ | Nanosheet width | 20 nm |
| $T_{NS}$ | Nanosheet thickness | 5 nm |
| $N$ | The number of nanosheets | 3 |
| $EOT$ | Effective oxide thickness | 1 nm |
| CPP | Contacted poly pitch | 45 nm |

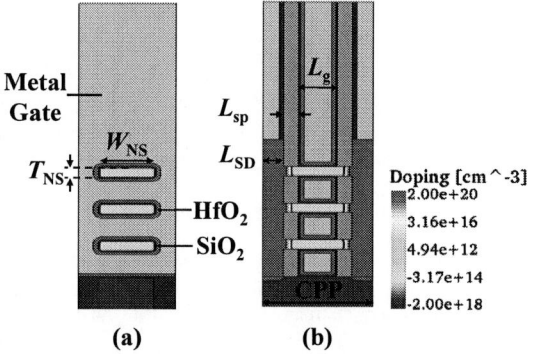

*Figure 1: The cross-sectional image of the conventional NSFETs along the vertical channel direction (a) and the channel direction (b).*

three-staked nanosheets (NSs) was shown in Fig. 1. The geometric parameters of NSFETs were referenced according to the IRDS 2020 [8], as shown in Table Ⅰ. Among them, the $L_g$ was set to 16 nm, and the $W_{NS}$ and $T_{NS}$ were set to 20 nm and 5 nm, respectively. For the $n$-type NSFETs, the boron concentration of channels was $1 \times 10^{15}$ cm$^{-3}$, the well concentration ($N_{well}$) was set to $2 \times 10^{18}$ cm$^{-3}$, and the arsenic concentrations of S/D was $2 \times 10^{20}$ cm$^{-3}$. In addition, the simulation model was calibrated using the experimental data of 10-nm-node FinFET [9]. In the simulation model, the diffusion-drift equation was solved

978-1-6654-9759-6/22 $31.00 © 2022 IEEE

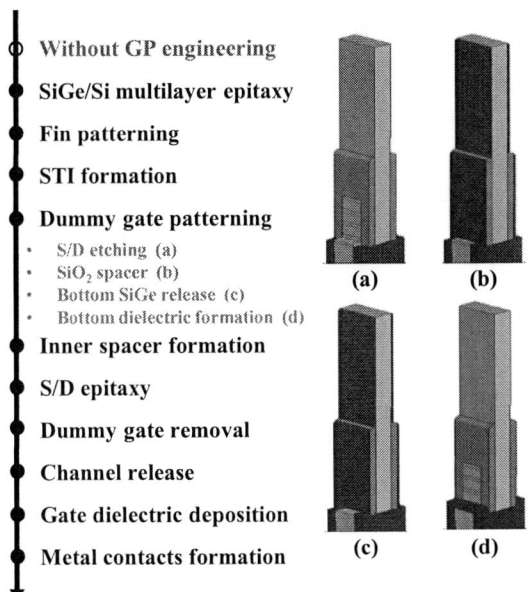

- Without GP engineering
- SiGe/Si multilayer epitaxy
- Fin patterning
- STI formation
- Dummy gate patterning
  - S/D etching (a)
  - SiO₂ spacer (b)
  - Bottom SiGe release (c)
  - Bottom dielectric formation (d)
- Inner spacer formation
- S/D epitaxy
- Dummy gate removal
- Channel release
- Gate dielectric deposition
- Metal contacts formation

*Figure 2: The process flows of GAA Si NSFETs with the BDI structure, including (a) S/D etching, (b) SiO₂ spacer formation, (c) bottom SiGe release and (d) bottom dielectric formation.*

by Poisson and carrier continuity equation considering quantum correction, mobility degradation, stress effecting and the generation and recombination of carriers.

## RESULTS AND DISCUSSION

The process flow of the stacked GAA Si NSFETs is compatible with the main process flows of the FinFETs. In the shallow trench isolation (STI) formation, SiO₂ is first deposited to a certain height above the Si substrate, and then etched to the edge of Si substrate. However, due to the challenge of etching accuracy, SiO₂ is generally etched below the Si substrate, which is conducive to the selective release of NS channels. Similarly, the STI is slightly corroded during the selective release of NS channels. Therefore, there is a fat sub-fin under the gate, resulting in large leakage and performance variation. In order to reduce the leakage of the parasitic sub-fin, a SiGe sacrificial layer with high Ge concentration is first epitaxial, then the stacked SiGe/Si multilayer is formed. After the dummy gate is formed, the BDI structure is realized by the processes, including S/D etching, SiO₂ spacer, bottom SiGe release and bottom dielectric formation, as shown in Fig. 2. The *n*-type NSFETs with sub-fin are shown in Fig. 3(a) and (b). The height of sub-fin is 10 nm, which is caused by the increase in the SiO₂ etching depth during the STI formation. The *n*-type NSFETs with BDI are shown in Fig. 3(c) and (d). The device and Si substrate are completely isolated by the bottom dielectric.

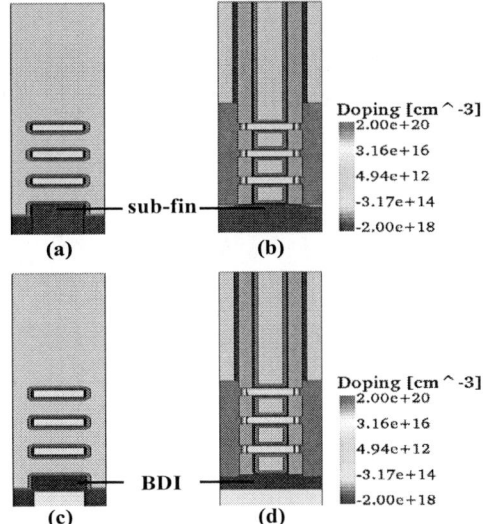

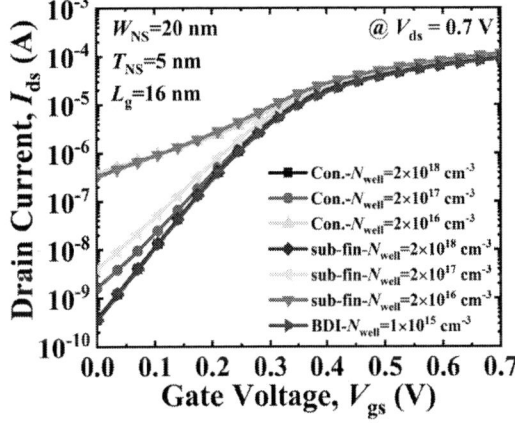

*Figure 3: Cross-sectional images of the n-type NSFETs with sub-fin cut across the stacked NSs channel direction (a) and the top of the NSs direction (b). (c) and (d) are the cross-sectional images of n-type NSFETs with BDI cut across the stacked NSs channel direction and the top of the NSs directions, respectively.*

*Figure 4: Transfer curves of the conventional NSFETs and the NSFETs with sub-fin and BDI.*

Fig. 4 shows the transfer curves of conventional NSFETs and the NSFETs with sub-fin and BDI. The results show that the conventional NSFETs and the NSFETs with sub-fin are obviously affected by the $N_{well}$ and have the better electrical characteristic at $2 \times 10^{18}$ cm⁻³, demonstrating the effect of GP engineering to suppress the leakage. Meanwhile, the *n*-type NSFETs with BDI overcome the degradation of electrical characteristics caused by the sub-fin, which is because the bottom dielectric layer cuts off the leakage ways of the parasitic channel. The parameter comparisons between the conventional NSFETs and NSFETs with sub-fin and BDI

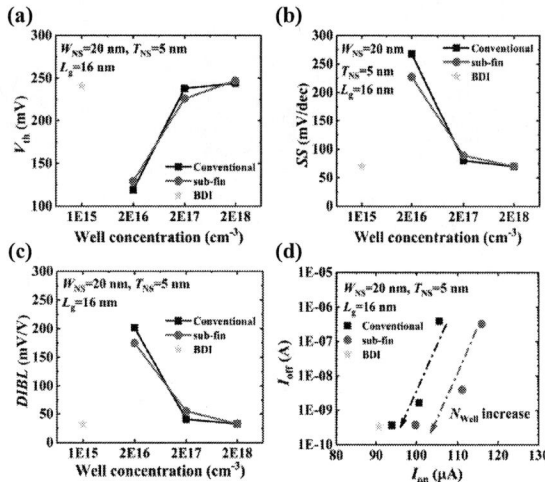

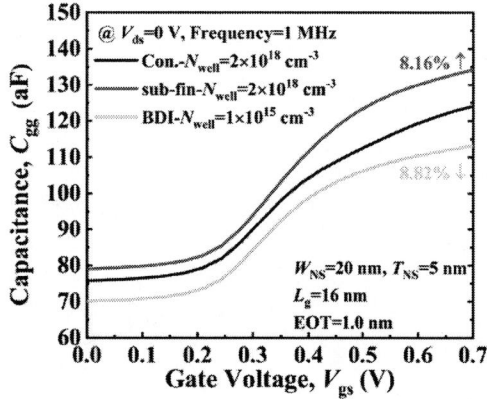

*Figure 5: Comparisons of (a) $V_{th}$, (b) SS, (c) DIBL, and (d) $I_{on}$-$I_{off}$ mapping among the conventional NSFETs and NSFETs with sub-fin and BDI.*

*Figure 6: The comparison of the gate capacitance between the conventional NSFETs and NSFETs with sub-fin and BDI.*

are shown in Fig.5. Fig. 5(a) exhibits that the $V_{th}$ reduces as the $N_{well}$ decrease, and the BDI technology guarantees the $V_{th}$ stability at lower $N_{well}$. Furthermore, the SS and DIBL are obviously improved for the NSFETs with BDI by suppressing the bottom parasitic channel, while the conventional NSFETs and the NSFETs with sub-fin deteriorate the sub-threshold characteristics at the lower $N_{well}$, as shown in Fig. 5(b) and (c). Fig. 5(d) shows that the $I_{off}$ of the conventional NSFETs and the NSFETs with sub-fin decrease significantly with increasing the $N_{well}$, and the NSFETs with BDI have the minimal leakage. Moreover, the results indicate that the $I_{on}$-$I_{off}$ ratio and the GP influence are improved by the BDI technology. For the integrated circuits, the RC delay is affected by the capacitance characteristics [10]. The gate capacitance of NSFETs with BDI is reduced by 8.82% and 16.98% compared with those of the conventional NSFETs and the

NSFETs with sub-fin, respectively, which helps to reduce the parasitic capacitance of NSFETs, as shown in Fig. 6. As the $L_g$ of transistors continues to decrease, the NSFETs with BDI can improve the stability of devices and suppress the bottom parasitic channel at smaller nodes.

## CONCLUSION

The BDI technique of NSFETs is investigated by the TCAD simulation, and the BDI approach provides a good solution for reducing the leakage of the parasitic sub-fin and improving the sub-threshold characteristics. In addition, compared with GP engineering, the NSFETs with BDI have the advantage of reducing the parasitic capacitance and overcoming the influence of GP concentration. The BDI technology is expected to promote the development of NSFETs towards smaller nodes.

## ACKNOWLEDGEMENTS

This work was supported in part by the Pilot Projects of the Chinese Academy of Science under grants E1XDC2X002, in part by the Joint Development Program of Semiconductor Technology Innovation Center (Beijing), Corp, in part by the Youth Innovation Promotion Association, Chinese Academy of Sciences under grant Y9YQ01R004, and in part by the National Natural Science Foundation of China under Grant 91964202 and 61904194.

## REFERENCES

[1] N. Loubet et al. *2017 Symposium on VLSI Technology*, Kyoto, Jun. 5-8, 2017, pp. T230-T231.

[2] Q. Z. Zhang et al. *Nanomaterials*, vol. 11, 2021, pp. 1-16.

[3] R. Liu, X. Li, Y. Sun and Y. Shi. *IEEE Transactions on Electron Devices*, vol. 67, 2020, pp. 2249-2254.

[4] J. J. Li, et al. *Nanomaterials*, vol. 10, 2020, pp. 1-11.

[5] Q. Z. Zhang et al. ECS *Journal of Solid State Science and Technology*, vol. 7, 2018, pp. 671-676.

[6] J. Zhang et al. *2019 IEEE International Electron Devices Meeting (IEDM)*, San Francisco, Dec. 7-11, 2019, pp. 11.6.1-11.6.4.

[7] Dielectric isolation for nanosheet devices, by K. Cheng, R. Xie, T. Yamashita and C. Yeh. (2020 Jul. 16). US 2020/0227305 Al

[8] IEEE, International Roadmap for Devices and Systems (IRDS™) 2020 Edition, 2020.

[9] C. Auth et al. *2017 IEEE International Electron Devices Meeting (IEDM)*, San Francisco, Dec. 2-6, 2017, pp. 29.1.1-29.1.4.

[10] J. Yoon, J. Jeong, S. Lee and R. Baek. *IEEE Journal of the Electron Devices Society*, vol. 6, 2018, pp. 942-947.

# Influence of Parasitic Capacitance and Resistance on Performance of 6T-SRAM for Advanced CMOS Circuits Design

*Yanna Luo[1,2], Gangping Yan[1,2], Lei Cao[1,2], Jiali Huo[1,2], Xuexiang Zhang[1,2], Yanzhao Wei[1,2], Guoliang Tian[1,2], Qingzhu Zhang[1], Zhenhua Wu[1], and Huaxiang Yin[1,2*]*

[1]Key Laboratory of Microelectronics Devices and Integrated Technology, Institute of Microelectronics of Chinese Academy of Sciences, Beijing 100029, China

[2] University of Chinese Academy of Sciences, Beijing 100049, China

*Corresponding Author's Email: yinhuaxiang @ime.ac.cn*

## ABSTRACT

In this paper, the impacts of parasitic capacitance and resistance on the performance of 6T-SRAM are investigated comprehensively by a design-technology co-optimization (DTCO) method and TCAD simulation. It is found that the impact of parasitic capacitance on 6T-SRAM read and write operations is all adverse. However, the influence of parasitic resistance on 6T-SRAM is double-sided and can be used to improve the performance of 6T-SRAM. An optimization strategy of layout and route is proposed for SRAM with high speed, high noise stability or high read and write capabilities. These results provide a useful guidance for the 6T-SRAM layout design toward advanced node CMOS circuits.

*Keywords—CMOS; 6T-SRAM; parasitic resistance; parasitic capacitance; layout;*

## INTRODUCTION

In the current situation where nodes continue to shrink and routing continues to become crowded, parasitic resistance increases with scaling due to S/D and contact are shrinkage, also parasitic capacitance increases with smaller wiring spacing [1-2]. The parasitic capacitance and resistance in SRAM cell tend to rise sharply and limits the transistors performance. However, due to the complexity of layout, the impacts of parasitic parameters among all key nodes on 6T-SRAM remain unexplored extensively.

In this paper, the influences of parasitic capacitance and parasitic resistance on the performance of 6T-SRAM are investigated by simulation. A device to circuit design-technology co-optimization (DTCO) method is used to obtain device and circuit performance [3-4]. After 6T-SRAM's working principle and current transport analysis, it is interesting that part of parasitic parameters can improve SRAM performance. The preferable layout and routing schemes are respectively proposed for 6T-SRAM cells with high speed, high noise stability or high read and write capabilities.

## ENVIRONMENT SETTING

Inspired by design-technology co-optimization (DTCO)

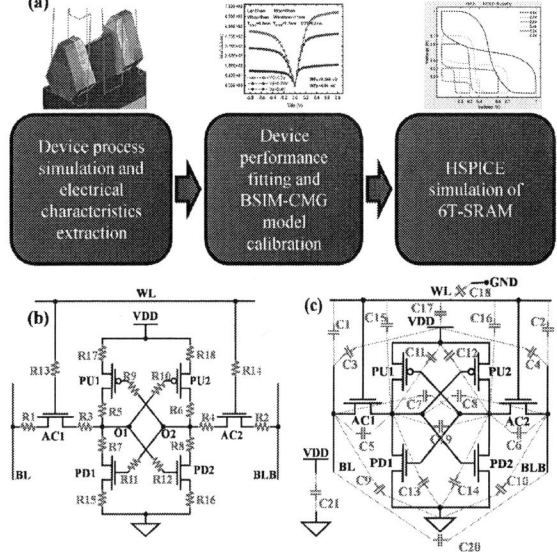

**Fig.1.** *(a) Device and circuit performance evaluation methodology. 6T-SRAM cell network with (b) parasitic resistances and (c) parasitic capacitances.*

flow, we obtained the performance of a single device, constructed a 6T-SRAM cell circuit, and analyzed the influence of the parasitic parameters between each node on the 6T-SRAM cell circuit based on the netlist obtained from the analysis. The simulation process is shown in Fig. 1(a). The characteristics of 14 nm SOI FinFET-based NMOS and PMOS are simulated by Synopsys Sentaurus TCAD software. These models are validated with the experimental data of the nanosheet transistors.

In order to perform the HSPICE simulation of the circuit, devices' TCAD-calibrated compact models are extracted by fitting the I-V and C-V curve of devices with Berkeley short-channel insulated-gate field effect transistor model - common multi-gate (BSIM-CMG). Other voltage sources, capacitances, and resistances are constructed with HSPICE internal components. The circuit framework of the 6T-SRAM cell including pull-up (PU), access (AC), and pull-down (PD) devices. The parasitic capacitance and resistance of 6T-SRAM networks, shown in Fig. 1 (b) and (c), are listed in Table I and Table II. In order to consider the performance of the 6T-SRAM in an ideal parasitic-free state and after being affected by a huge parasitic effect, we have considered a large range of parasitic capacitance and parasitic resistance values. The parasitic capacitance varies from 0 to an extreme value of 1pF. What's more, the parasitic resistance varies from 0 to an extreme value of 10M ohm. The 6T-SRAM circuit is simulated under 1.0V supply voltage. Hold/Read/Write static noise margin (H/R/WSNM) is extracted from the maximum inner square of each butterfly curve. The Write Margin (WM) is defined as the maximum bit-line (BL) voltage at which the write '0' operation is successfully enabled. Write time (WT) is defined as the delay between 50% word-line (WL) signal rising edge and 90% $V_{DD}$ drop of Q node while writing '0'. Similarly, read time (RT) is defined as the delay between 50% WL signal rising edge and 90% $V_{DD}$ drop of BL while reading '0'.

## RESULTS AND DISCUSSIONS

### Influence of Parasitic Capacitances

In order to fully consider the coupling effect between each node, we have added parasitic capacitance between each node. In view of the 6T-SRAM circuit topology and the high rotational symmetry of the three-dimensional physical structure, the parasitic capacitance in the 6T-SRAM circuit can be classified into 13 categories, as shown in Table I. The capacitive coupling effect between the BL/BLB and the WL is represented by $C_{bw}$, and the remaining capacitive coupling effects can be deduced by analogy. Fig. 2 shows that the increase of the parasitic capacitance results in the degradation in read and write time. The main parasitic capacitances that affect the reading and writing speed of 6T-SRAM are $C_{bo1}$, $C_{bo2}$, $C_{ow}$, $C_{bd}$, $C_{os}$, $C_{bs}$, $C_{bw}$, $C_{od}$, $C_{oo}$, and $C_{bb}$. The parasitic capacitance introduces a path to charge the bit-line, which hinders the bit-line from reading data, so that the read time increase. In addition, the excessive parasitic capacitance leads to an increase in the charging and discharging time on the storage node. Therefore, in order to obtain high-speed characteristics in the high-density 6T-SRAM, the values of the above capacitances need to be reduced as much as possible. By separating the connection layer of O1 and O2 from the BLs, WL and power lines, the coupling effect between internal nodes and BL, WL or power lines can be reduced, thus reducing the charging effect on internal nodes and improving write speed. By distinguishing the BLs from the power lines layer and WL layer, the equivalent parasitic capacitance on the BLs can be reduced and the read speed can be improved. Placing the two BLs farther away can also reduce the parasitic capacitance on BL. Based on the above analysis, we propose the following layout suggestions:

1) separating the connection layer of O1 and O2 from the BLs, WL and power lines;

2) distinguishing the BLs from the power lines layer and WL layer;

3) distributing a pair of bit-lines on the two side of the 6T-SRAM cell.

TABLE I
TYPES OF PARASITIC CAPACITANCES STUDIED

| Cap. | Start Net | End Net | Label | Range |
|---|---|---|---|---|
| $C_{bw}$ | BL/BLB | WL | C1, C2 | |
| $C_{bd}$ | BL/BLB | VDD | C3, C4 | |
| $C_{bo1}$ | BL/BLB | O1/O2 | C5, C6 | |
| $C_{bo2}$ | BL/BLB | O2/O1 | C7, C8 | |
| $C_{bs}$ | BL/BLB | VSS | C9, C10 | |
| $C_{od}$ | O1/O2 | VDD | C11, C12 | |
| $C_{os}$ | O1/O2 | VSS | C13, C14 | 0 to 1 |
| $C_{ow}$ | O1/O2 | WL | C15, C16 | (pF) |
| $C_{dw}$ | VDD | WL | C17 | |
| $C_{sw}$ | GND | WL | C18 | |
| $C_{oo}$ | O1 | O2 | C19 | |
| $C_{bb}$ | BL | BLB | C20 | |
| $C_{ds}$ | VDD | GND | C21 | |

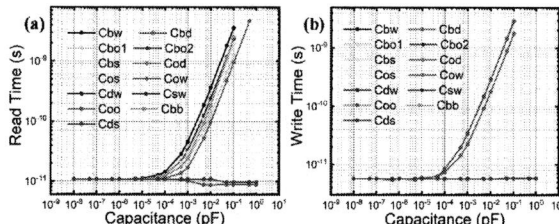

*Fig.2. The influence of parasitic capacitances on the (a) read time and (b) write time.*

### Influence of Parasitic Resistances

For parasitic resistance, we can divide it into 9 categories, as shown in Table II. $R_{bax}$ comprehensively considers the parasitic resistance of the S/D of AC, the contact resistance of the S/D, and the connection resistance of the AC to the BLs. The definition of the remaining parasitic resistance is similar to $R_{bax}$. Different from the influence of parasitic capacitance, the parasitic resistance has a two-sided effect on the performance of SRAM. For the 6T-SRAM with high noise tolerance, the most vulnerable RSNM is considered. $R_{pds}$ and $R_{pud}$ lead to the internal node voltage increase or decrease, which will reduce the ability of the pull-down path to maintain the '0' node, as depicted in Fig. 3(b). This effect can be mitigated by placing the power rail on M1 layer and using wider power lines. The $R_{oax}$ could be increased by separating the source and drain of the AX and the PD, so that first layer of contact can be used to connect the NMOS and PMOS in the inverter to reduce the $R_{opd}$ and $R_{opu}$, which will help the RSNM increases. In order to achieve high write stability in 6T-SRAM, reducing $R_{bax}$ can be achieved by arranging bit-lines with M1 or

TABLE II
TYPES OF PARASITIC RESISTANCES STUDIED

| Res. | Start Net | End Net | Label | Range |
|------|-----------|---------|-------|-------|
| $R_{bax}$ | BL/BLB | AX_D | R1, R2 | |
| $R_{oax}$ | O1/O2 | AX_S | R3, R4 | |
| $R_{opu}$ | O1/O2 | PU_D | R5, R6 | |
| $R_{opd}$ | O1/O2 | PD_S | R7, R8 | |
| $R_{opug}$ | O1/O2 | PU_G | R9, R10 | 0 to 1e4 |
| $R_{opdg}$ | O1/O2 | PD_G | R11, R12 | (ohm) |
| $R_{wax}$ | WL | AX_G | R13, R14 | |
| $R_{pds}$ | GND | PD_S | R15, R16 | |
| $R_{pud}$ | VDD | PU_S | R17, R18 | |

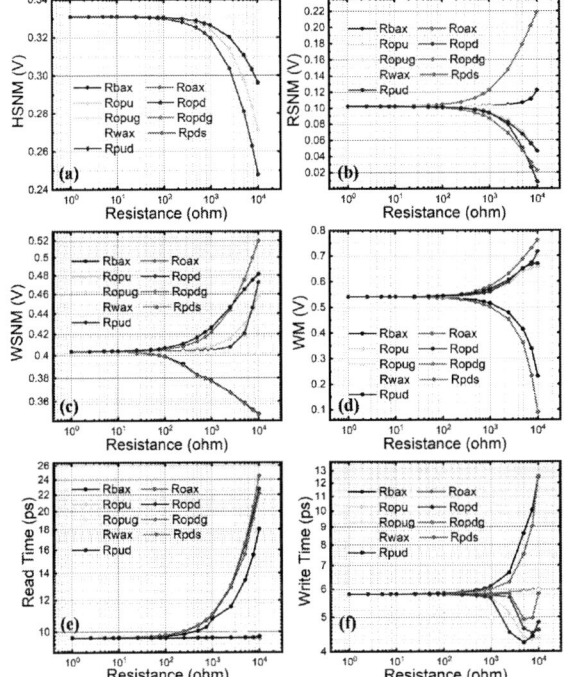

**Fig. 3.** *The influence of parasitic resistances on (a) HSNM, (b) RSNM, (c) WSNM, (d) WM, (e) Read Time, and (f) Write Time.*

using wider bit-lines, and reducing $R_{oax}$ can be achieved by sharing source and drain between AX and PD. Apart from this, adding $R_{opd}$, $R_{opu}$, $R_{pds}$, and $R_{pud}$ will help increase write stability, which can be achieved by arranging NMOS away from PMOS and routing the power rail in higher layer of metal. For high-speed 6T-SRAM, the values of $R_{bax}$, $R_{pds}$, $R_{oax}$, and $R_{opd}$ need to be reduced. In this case, the bit-line and ground line need to be arranged on the M1 layer, and the source and drain of AX and PD need to be shared. Based on the above analysis, we provide the following layout suggestions for different need:

1) High noise tolerance: placing the power lines on M1 layer and using wider power lines; separating the source and drain of the AC and PD; connecting PU and PD in the first layer of contact.

2) High write margin: arranging BL with M1 or use wider BL; sharing the AC source with PD drain; arranging NMOS away from PMOS and routing the power rail in higher layer of metal.

3) High speed: arranging BL and ground line in M1; sharing the AC source with PD drain.

## CONCLUSION

In this paper, the influences of parasitic capacitance and resistance in 6T-SRAM cells on the cell performance were investigated extensively. It is found that the impact of parasitic capacitance on 6T-SRAM read and write operations is all adverse. However, the influence of parasitic resistance on 6T-SRAM is double-sided and can be used to improve the performance of 6T-SRAM. According to our analysis results, the optimization strategy of the layout and routing in the compact 6T-SRAM is proposed for application scenarios with high read and write speed, high noise stability, or high read and write capabilities. This research is very instructive for the 6T-SRAM layout design for the advanced CMOS circuits.

## ACKNOWLEDGEMENTS

This work was supported in part by the Pilot Project of the Chinese Academy of Sciences (E1XDC2X002), in part by the Joint Development Program of Semiconductor Technology Innovation Center (Beijing), Corp. Thanks to all the teachers and students of the research group, as well as the scholars who provided consultation and help.

## REFERENCES

[1] G. Yeric, "Moore's law at 50: Are we planning for retirement?," *2015 IEEE International Electron Devices Meeting (IEDM)*, 2015, pp. 1.1.1-1.1.8, doi: 10.1109/IEDM.2015.7409607.

[2] M. K. Gupta et al., "A Comprehensive Study of Nanosheet and Forksheet SRAM for Beyond N5 Node," *in IEEE Transactions on Electron Devices*, vol. 68, no. 8, pp. 3819-3825, Aug. 2021, doi: 10.1109/TED.2021.3088392.

[3] Q. Huo et al., "Physics-Based Device-Circuit Cooptimization Scheme for 7-nm Technology Node SRAM Design and Beyond," *in IEEE Transactions on Electron Devices*, vol. 67, no. 3, pp. 907-914, March 2020, doi: 10.1109/TED.2020.2964610.

[4] J. Yao et al., "Physical Insights on Quantum Confinement and Carrier Mobility in Si, Si0.45Ge0.55, Ge Gate-All-Around NSFET for 5 nm Technology Node," *in IEEE Journal of the Electron Devices Society*, vol. 6, pp. 841-848, 2018, doi: 10.1109/JEDS.2018.2858225.

# OPTIMIZATION AND DEMONSTRATION OF LOW-VOLTAGE NMOS FOR RF SWITCH APPLICATION IN 65NM SOI TECHNOLOGY

*Zhaozhao Xu[1,*], Tianyu Han[1], Wan Song[1], Fei Meng[2], Wensheng Qian[2]*
[1]Huahong Semiconductor (Wuxi) Limited, Wuxi 214029, China
[2]Shanghai Huahong Grace Semiconductor Manufacturing Corporation, Shanghai 201203, China
*Corresponding Author's E-mail: Zhaozhao.xu@hhgrace.com

## ABSTRACT

A novel low-voltage n-type metal-oxide-semiconductor (NMOS) structure for radio frequency (RF) switch application in 65nm Silicon-on-Insulator (SOI) technology is proposed in this work. The degradation of device induced by impact-ionization in NMOS transistors at the lightly-doped drain (LDD) and halo junction of top-silicon surface is fully investigated and optimized. It revealed that significant improvement of the device performance can be obtained without increasing manufacturing cost. Additionally, the physical mechanism behind the results is also analyzed from the technology computer-aided-design (TCAD) simulation.

*Key Words—Silicon-on-Insulator (SOI); n-type metal-oxide-semiconductor (NMOS); impact-ionization; lightly-doped drain (LDD)*

## INTRODUCTION

CMOS technology has become a perfect choice for radio frequency (RF) due to its high integration density. Because of superior RF performance, the Silicon-on-Insulator (SOI) technology has been frequently used for application of low-power microwave circuits [1, 2]. Lower specific on-resistance ($R_{sp}$) as well as higher breakdown voltage ($BV$) is preferred for RF switch applications. In this paper, a 2.5 V low-voltage n-type metal-oxide-semiconductor (NMOS) for RF switch application have been fabricated at a commercial 65nm SOI technology node. Currently and however, 2.5 V switch NMOS exhibits lower breakdown voltage as well as larger OFF-state leakage current. To improve the device performance, in this paper, we design a novel device structure through process optimization without increasing the manufacturing cost. Furthermore, traditional and novel switch NMOS devices are fully compared through technology computer-aided-design (TCAD) simulation.

This work is organized as below. Firstly, the device structure and simulation details are presented. Secondly, the main mechanism is investigated and discussed. Then, the characteristics of this novel device are fully demonstrated. It shows that significant improvement can be achieved compared with the traditional one. Finally, we present a conclusion about this work.

## DEVICE STRUCTURE AND SIMULATION

Figure 1 shows the schematic cross-sectional views of traditional and novel low-voltage NMOS for switch application with SOI technology. Conventional 300mm SOI wafer has been used, which features a p-type substrate, a buried oxide (BOX) of 2000Å, and a top-silicon layer of 750Å. Shallow trench isolation process was adopted to isolate different CMOS and power switch devices, which features insensitivity of logic

blocks to parasitic disturbances from the power areas [3]. The channel is formed by p-type well (PW) and both self-aligned LDD and source/drain implants to the gate. As shown in the Figure 1(a), traditional LDD and halo (or pocket) are implanted into drain and source regions to enable device with 0.1μm-level channel length.

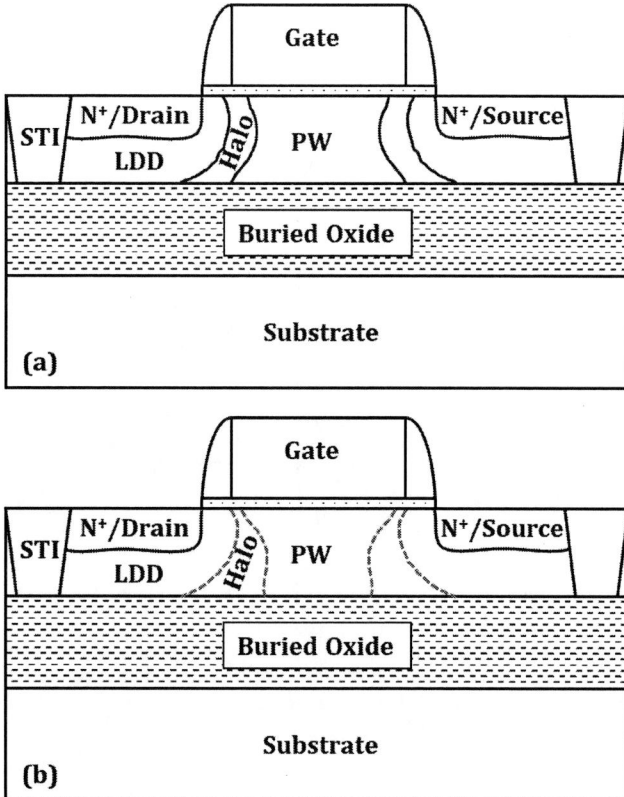

*Figure 1: Schematic cross-sectional drawings of (a) traditional and (b) novel switch NMOS devices with SOI technology.*

However, as illustrated in Figure 1(a), the traditional LDD and halo implants form a doping profile, which features a heavier concentration at both top-silicon/BOX interface and top-silicon surface. That is, heavier halo implant exits at top-silicon that constructs a wider halo region at top-silicon surface. As also shown in Figure 1(b), lightly-doped drain and halo are also employed in our novel low-voltage NMOS for decreasing on-resistance and depressing short-channel effect (SCE), respectively. In contrast to the traditional one, new LDD and halo implants build up a doping profile, that feature a heavier/wider halo region at top-silicon/BOX interface. Meanwhile, Figure 1(b) presents a lighter/narrower halo region at top-silicon surface. Above two devices is implemented with our 65nm RF-SOI technology.

978-1-6654-9759-6/22 $31.00 © 2022 IEEE

TCAD simulations are also performed by Sentaurus TCAD simulation tools from Synopsys technology [4]. For the electrical simulations of SOI device, Shockley–Read–Hall, Auger generation–recombination, OldSlotboom, and Okuto-Crowell avalanche model were turn on with default parameters. In addition, mobility models including PhuMob, HighField-Saturation, Enormal are employed for device simulation with calibrated parameters. The main solving models are Poisson, Electron, and Hole.

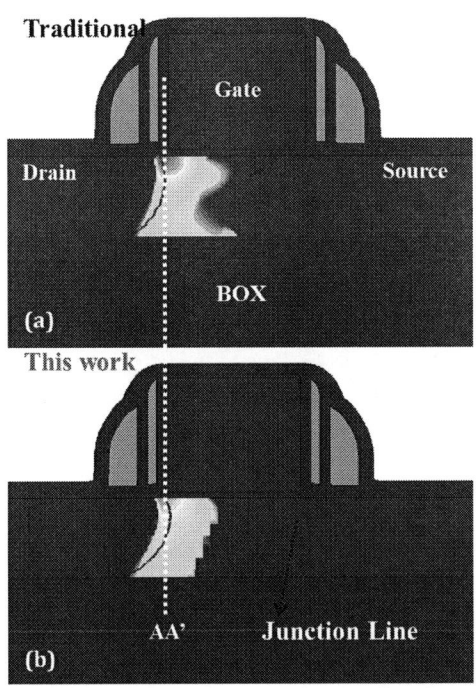

*Figure 2: The simulated impact-ionization distribution of (a) traditional and (b) novel RF switch NMOS devices with $V_{GS}/V_{DS}= 0.0/2.75$ V.*

## RESULTS AND DISCUSSION

For low-voltage RF switch application, the specific $R_{sp}$ and $BV$ are two most important electric parameters. However, the $R_{sp}$ of low-voltage switch NMOS device is mainly dominated by the channel length [5] and threshold voltage. Additionally, the $BV$ is strongly dependent on the LDD/halo junction doping distribution and weakly dependent on the channel length. Therefore, optimization of LDD/halo junction doping distribution is critical for increasing $BV$. As illustrated in Figure 2, the strongest impact-ionization is located at the LDD/halo junction of top-silicon surface in both devices. However, the traditional one demonstrates a wider halo region at the top-silicon surface, as shown in Figure 1(a). This configuration is advantageous for depressing SCE, but it is detrimental to breakdown voltage because it would forms a less graded junction.

For this reason, a novel low-voltage RF switch NMOS studied and demonstrated in this paper. As indicated by the Figure 1(b), the proposed device features wider halo region located at the top-silicon/BOX interface. Both wider halo at top-silicon/BOX and BOX layer can also enable excellent depression of SCE. Additionally, proposed device presents a lighter/narrower halo region at the LDD/halo junction at the top-silicon surface, as shown in Figure 1(b). This brings us a more graded doping distribution at the weak point of breakdown, which located at the LDD/halo junction of top-silicon surface under the gate, as presented in Figure 2. The impact-ionization ratio at the weak point along line AA' under $V_{GS}/V_{DS}= 0.0/2.75$ V are plotted in Figure 3. The strength of impact-ionization is significantly decreased in drain-side LDD/halo junction at the top-silicon surface. Which means that novel device shows released electric field crowding effect at LDD/halo junction of top-silicon surface.

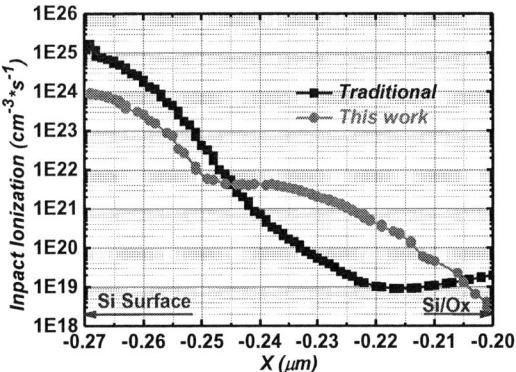

*Figure 3: The impact-ionization ratio at the LDD/halo junction of low-voltage RF switch NMOS device along line AA' (Ref. Figure 2) under $V_{GS}/V_{DS}= 0.0/2.75$ V.*

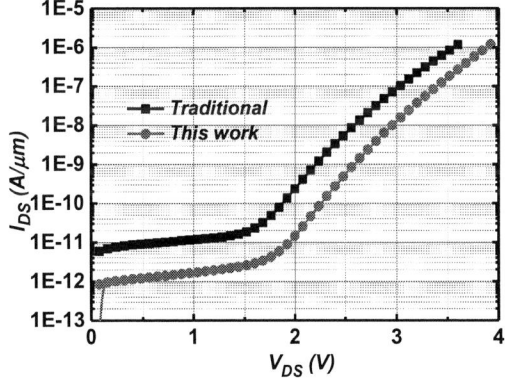

*Figure 4: The OFF-state $I_{DS}$-$V_{DS}$ characteristic of 2.5 V switch NMOS devices under $V_{GS} = 0.0$ V.*

The OFF-state $I_{DS}$-$V_{DS}$ characteristics of 2.5 V switch NMOS devices are depicted in Figure 4. $BV$ ($V_{DS}$ at 0.1 µA/µm with $V_{GS}$ of 0.0 V) of 3.0 V and 3.4 V are obtained for traditional and proposed devices, respectively. It was also found that the OFF-state leakage is decreased by 89%. The OFF-state leakage is defined as the $I_{DS}$ with $V_G/V_D/V_S/V_B = 0.0/2.75/0.0/0.0$ V. Figure 5 shows $I_{DS}$-$V_{GS}$ curves with $V_{DS}= 2.5/0.1$ V for both traditional and novel devices. It was revealed that improvement of BV is obtained without sacrifice of the saturation current. Furthermore, the linear current is increased by 9.5%. That is to say, the $R_{sp}$ of novel device is decreased by 9.5% while the $BV$ is increased by 13%, which indicated that significant achievement of figure of merit (FOM) [6] of power switch device can be obtained. The lower $R_{sp}$ is probably more a result of lighter doped halo at the top-silicon surface at the channel edge, which gives rise to higher mobility of carrier.

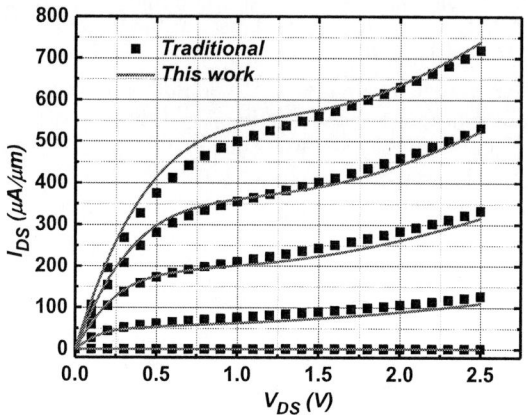

Figure 7: The ON-state $I_{DS}$-$V_{DS}$ characteristics of 2.5 V RF switch NMOS devices under $V_{GS}$ = 0.5/1.0/1.5/2.0/2.5 V.

## CONCLUSION

A novel low-voltage RF switch NMOS was proposed and investigated. This proposed device features a heavier/wider halo region at top-silicon/BOX interface while a lighter/narrower halo region at top-silicon surface. With this configuration, it was found that the *BV* of proposed device is increased by 13% without sacrifice of the saturation current. Meanwhile, the $R_{sp}$ is also decreased by 9.5%. Consequently, the optimized device exhibits much better $BV$-$R_{sp}$ relationship, compared to the traditional one.

## REFERENCES

[1] J. Ren, R. Dai, J. He, J. Xiao, W. Kong, and S. Zou, "A novel stacked class-E-like power amplifier with dual drain output power technique in 0.18 um RFSOI CMOS technology." pp. 1-4.

[2] R. Dai, J. Ren, J. He, J. Xiao, and W. Kong, "RFFE Integration Design for 5G in 0.13um RFSOI Technology." pp. 1-3.

[3] R. Rudolf, C. Wagner, L. O'Riain, K. H. Gebhardt, B. Kuhn-Heinrich, B. V. Ehrenwall, A. V. Ehrenwall, M. Strasser, M. Stecher, and U. Glaser, "Automotive 130 nm smart-power-technology including embedded flash functionality," vol. 19, no. 3, pp. 20-23, 2011.

[4] U. Manual, *Sentaurus Process/Device Version G-2012.06 Synopsys*, Mountain View, CA, USA, Jun. 2012.

[5] A. B. Joshi, S. Lee, Y. Y. Chen, and T. Y. Lee, "Optimized CMOS-SOI process for high performance RF switches." pp. 1-2.

[6] W. Ge, X. Luo, J. Wu, M. Lv, J. Wei, D. Ma, G. Deng, W. Cui, Y. H. Yang, and K. F. Zhu, "Ultra-Low On-Resistance LDMOS With Multi-Plane Electron Accumulation Layers," *IEEE Electron Device Letters*, vol. 38, no. 7, pp. 910-913, 2017.

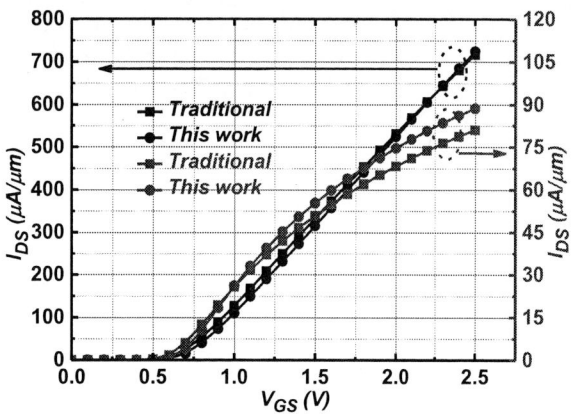

Figure 5: The $I_{DS}$-$V_{GS}$ curves (linear scale) with $V_{DS}$= 2.5/0.1 V (black/blue) for both traditional and novel devices.

The sub-threshold region of $I_{DS}$-$V_{GS}$ curves with $V_{DS}$= 2.5 V are shown in Figure 6. Noted that the proposed device in this work illustrates a larger sub-threshold slop, which demonstrated that novel device exhibits better control capability of gate except for the above mentioned benefits.

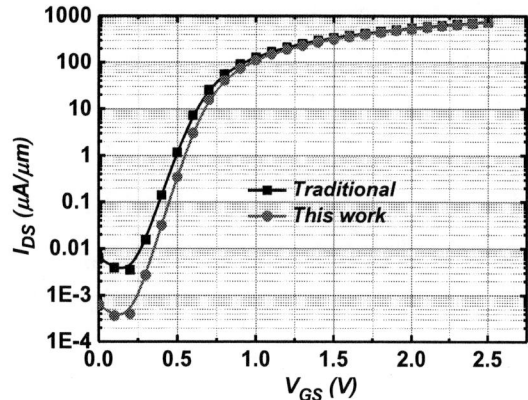

Figure 6: The $I_{DS}$-$V_{GS}$ curves (log scale) with $V_{DS}$= 2.5 V for both traditional and novel devices.

The ON-state $I_{DS}$-$V_{DS}$ characteristics of 2.5 V RF switch NMOS devices under $V_{GS}$ = 0.5/1.0/1.5/2.0/2.5 V are shown in Figure 7. It manifested that the proposed device exhibits less quasi-saturation under high $V_{GS}$. This is also attributed to lighter doped halo at top-silicon surface, which forms a more graded LDD/halo junction. Less impact-ionization and lower resistant of LDD region can be achieved. Both explains the less quasi-saturation in this novel device.

# THE OPTIMIZATION OF BREAKDOWN VOLTAGE AND SPECIFIC ON-RESISTANCE OF 30V N-VDMOS WITH SHORT CHANNEL

*Xiaoqing Cai, Donghua Liu, Wensheng Qian*

HuaHong Grace Semiconductor Manufacturing Corporation, Shanghai 201206, China

*Corresponding Author's Email: Xiaoqing.Cai@hhgrace.com

## ABSTRACT

This article presents a way of optimizing the breakdown voltage (BV) and specific on-resistance (Rsp) of 30V n-type vertical double-diffused metal-oxide-semiconductor field-effect transistor (n-VDMOS) with short channel. Firstly, the depth of deep trench is shortened. Secondly, the BV and Rsp of 30V n-VDMOS are both decreased by matching and regulating the process conditions on the premise of ensuring the BV. Finally, the Rsp is further reduced by thinning the epitaxial layer. The simulation result indicates that the performance of 30V n-VDMOS is greatly improved.

***Key Words: n-VDMOS; low Rsp; high-BV.***

## INTRODUCTION

DMOS (Double-diffused metal-oxide-semiconductor field-effect transistor) is widely used in industry, automobile and consumer electronics due to its high voltage resistance, high current drivability and ultra-low power consumption [1]. Similar to the structure of CMOS, DMOS also has source, drain and gate electrodes. DMOS mainly has two kinds of devices that are VDMOS (vertical double-diffused MOS) and LDMOS (lateral double-diffused MOS) according to the diffusion direction [2, 3]. This paper focuses on the characteristics and optimization of the VDMOS. The typical n-channel VDMOS structure is shown in Figure 1.

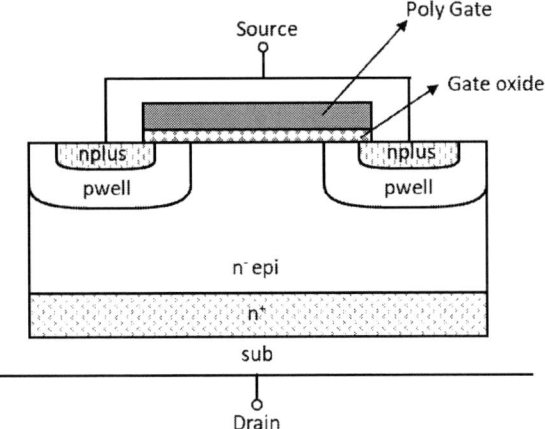

*Figure 1: typical n-channel VDMOS structure*

Different from the traditional working mode of the CMOS, the VDMOS is one type of device in which the current flows vertically under the control of the gate voltage. When the voltage on the gate is set in an appropriate range, the semiconductor material on the surface is inverted and formed a conductive channel where the carrier flows from the source to drain [4].

Besides the threshold voltage and transconductance, the BV and Rsp are the most significant parameters of the VDMOS [5]. When the Rsp is small enough, the VDMOS will provide an excellent switching characteristics and current drive capability because of the large output current. In this paper, the optimization of the BV and Rsp is mainly discussed by regulating the geometrical and technological parameters.

## Device structure and fabrication

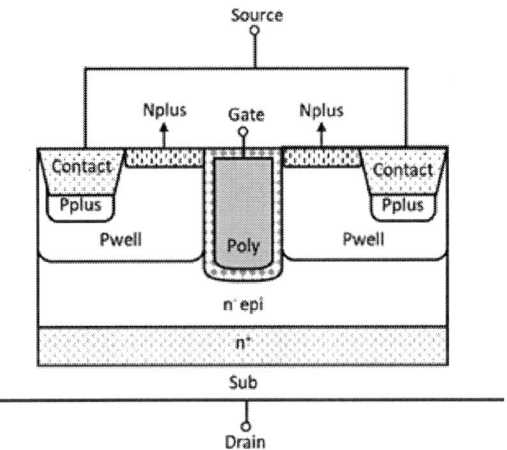

*Figure 2: improved n-channel VDMOS structure*

The structure of the transistor discussed in this paper is altered lightly as shown in Figure 2. The n-type silicon wafer with heavy Arsenic doped is prepared. A layer with lightly phosphorus doped is deposited on the wafer mentioned above at the beginning. After depositing a thin layer of oxide served as the hard-mask on the silicon wafer, the area of the oxide defined by photoresist on the wafer is etched to expose the silicon area that will be etched to form the trench. After trench etching process finishing, the gate oxide is grown along the side wall of the trench by a thermal process. The chemical mechanical polishing (CMP) is implemented to remove redundant oxide on the surface of this wafer to ensure that the next process can be carried out smoothly [6]. In the next step, polysilicon is deposited to fill the trench and form the poly gate. The CMP step is applied once again to remove superfluous polysilicon on the wafer. After the polysilicon annealing process, n-type impurity, boron with appropriate energy and dose is implanted into the wafer by single of multiple times to constitute P-well. A thermal process with high temperature is used for an expected impurity density distribution. The next step is to form N-plus which is achieved by implanting Arsenic and boron with high concentration and low energy. After an annealing process of N-plus, the oxide is deposited on the wafer to protect the area that do not need etching. After exposed contact area for etching oxide and silicon to shape contact structure, $BF_2$ and Boron are implanted to realize ohmic contact. Finally, contact activation, TiN deposition, TiN annealing, top metal alloy and passivation alloy are carried out in order to accomplish n-VDMOS structure.

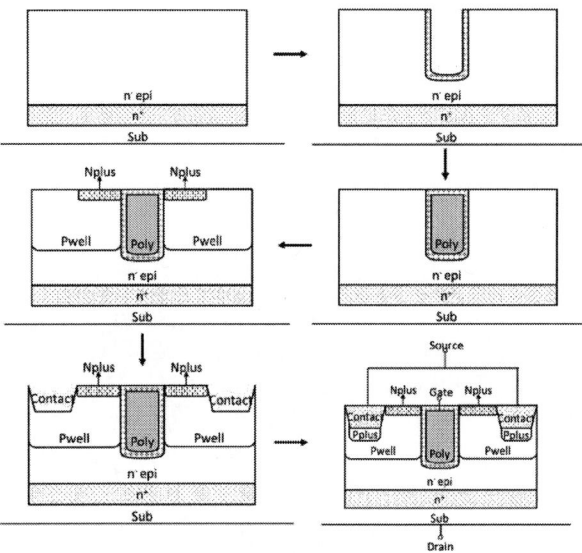

Figure 3: The fabrication sequence for n-VDMOS transistor

## RESULT AND DISCUSSION

The n-VDMOS transistor fabricated by the process mentioned above have high-Rsp and high-BV with thick n-type epitaxial layer and deep trench depth. This structure also has been created and compared by using Sentaurus TCAD simulation tools from Synopsys technology. Compared with actual silicon data, the simulation result indicates that the trend of the simulation is consistent with the actual silicon data.

The simulation structure of the transistor before optimization has been shown in Figure 4 in which the critical process conditions haven't matched well. The BV of the device is high enough for 30V n-VDMOS, so partial of the BV can be sacrificed for decreasing Rsp. In order to reduce Rsp, P-well injection conditions, the contact

etching depth and contact injection are optimized respectively.

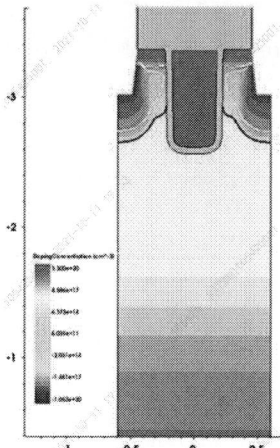

Figure 4: The simulation structure of the n-VDMOS transistor

In order to optimize the structure of 30V n-VDMOS transistors, etching depth of the trench, the injection energy and dose of P-well reduce appropriately for creating short channel. In addition to guarantee reliability of the device, an additional boron implantation with low energy and lightly doped is added in N-plus process. From the simulation result (Table 1), threshold voltage (Vt), Rsp and BV have deteriorated in different degree after regulating the structure and process of the device.

Table 1: Optimization result of Trench, P-well and N-plus

| | Optimization Category | Vt (V) | BV (V) | Rsp (mohm) |
|---|---|---|---|---|
| 1 | No optimization | 1.259 | 37.84 | 11.97 |
| 2 | Trench depth | 1.173 | 28.01 | 7.486 |
| 3 | Trench depth and P-well | 0.891 | 27.98 | 7.71 |
| 4 | Trench depth, P-well and N-plus | 1.537 | 28.804 | 8.06 |

The optimization of contact based on previous simulation result is divided into two parts, etching depth and process conditions as shown Table 2 and Figure 5.

Table 2: Optimization result of Contact

| | Optimization Category | Vt (V) | BV (V) | Rsp (mohm) |
|---|---|---|---|---|
| 1 | No optimization | 1.259 | 37.84 | 11.97 |
| 2 | Contact depth | 1.523 | 31 | 7.762 |
| 3 | Contact depth | 1.493 | 34.3 | 7.566 |

| 4 | Depth and process conditions of contact | 1.486 | 34.31 | 7.539 |

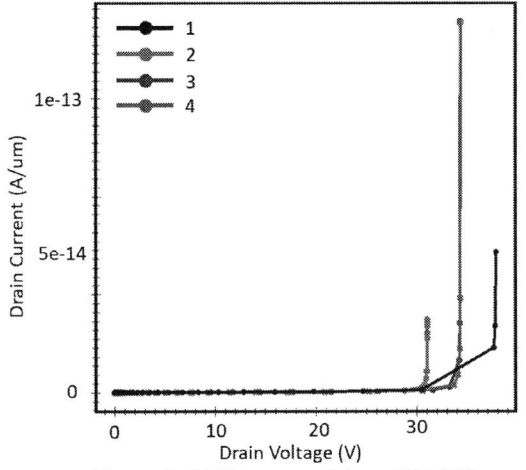

*Figure 5: Id-Vd curve of the n-VDMOS*

By decreasing the etching depth of the contact, the BV of the devices has decreased by 3 V. Then, the energy and dose of Boron implantation in contact process conditions are regulated in order to match the variety of trench, P-well, N-plus and contact. The Rsp has decreased by about 37% as shown in Table 2. For promising the BV and Vt, the thickness and resistivity of the epitaxial layer is optimized finally as shown in Table 3. The Rsp has decreased by about 17% with satisfied BV.

*Table 3: Optimization result of epitaxial layer*

| | Optimization Category | Vt (V) | BV (V) | Rsp (mohm) |
|---|---|---|---|---|
| 1 | No optimization | 1.259 | 37.84 | 11.97 |
| 2 | Epi thickness and resistivity | 1.607 | 38.88 | 9.947 |
| 3 | Epi thickness and resistivity | 1.474 | 37.34 | 9.005 |

## CONCLUSION

In this paper, a way in optimizing Rsp and BV of 30V n-VDMOS transistors has studied. By regulating process conditions of trench and P-well, the BV and Rsp decrease rapidly. The contact process conditions is furtherly optimizing to increase the BV and Vt. To ensuring the BV of the devices, the epitaxial layer is adjusted finally to obtain ideal performances of the n-VDMOS transistors.

## REFERENCES

[1] Bornhorst T, Ritchie J, Sheehan L. Determinants of tourism success for DMOs & destinations: An empirical examination of stakeholders' perspectives [J]. Tourism Management, 2010, 31(5):572-589.

[2] Raman A, Walker D G, Fisher T S. Simulation of nonequilibrium thermal effects in power LDMOS transistors [J]. Solid State Electronics, 2003, 47(8):1265-1273.

[3] Gao B, Liu G, Wang L X, et al. Research on the total dose effects for domestic VDMOS devices used in satellite [J]. Acta Physica Sinica Chinese Edition, 2012, 61(17):21-21.

[4] Qian Q, Sun W, Jing Z, et al. A Novel Charge-Imbalance Termination for Trench Superjunction VDMOS [J]. IEEE Electron Device Letters, 2010, 31(12):1434-1436.

[5] Dyer T, Mcginty J, Strachan A, et al. Monolithic integration of trench vertical DMOS (VDMOS) power transistors into a BCD process [C]. Power Semiconductor Devices and ICs, 2005. Proceedings. ISPSD '05. The 17th International Symposium on. IEEE, 2005.

[6] Ma W, Zhao W. Research on Several Manufacturing Methods of SRC Area of Trench VDMOS [J]. Semiconductor Technology, 2011.

# STUDY OF BREAKDOWN VOLTAGE IMPROVEMENT OF HIGH-VOLTAGE NLDMOS IN WIDTH DIRECTION

*Wenting Duan, Wensheng Qian, Donghua Liu, Haiyang Ling, Feng Jin, Ying Cai*

HuaHong Grace Semiconductor Manufacturing Corporation, Shanghai 201206, China

*Corresponding Author's Email: Wenting.Duan@hhgrace.com

## ABSTRACT

The impact factors of High-voltage NLDMOS Breakdown Voltage in width direction is studied in this paper. The Breakdown Voltage is improved by retracting gate poly into the body area which is high-voltage P-type well (HVPW) and adjusting the size and ratio of HVPW and deep N-type well (DNW). Finally the High-voltage NLDMOS Breakdown Voltage in width direction is improved by 10V in TCAD simulation.

***Key Words—High-voltage NLDMOS; Breakdown Voltage; in width direction***

## INTRODUCTION

For the high voltage LDMOS (Lateral Double-diffused Metal Oxide Semiconductor field effect transistor) in BCD (Bipolar, CMOS and DMOS) platform implicated in 80V and even higher voltage, LDMOS devices need to be able to withstand high voltage when in the off state. RESURF (Reduce Surface Field) technology [1][2] and adjusting of drift region concentration in device structure have been widely adopted for breakdown voltage (BV) improving. [3][4] However, the device peripheral structure is also important. The BV of device peripheral structure is required higher than that of device structure itself. [5] In this paper, the method of high voltage NLDMOS peripheral structure BV improving is described.

## DEVICE STRUCTURE AND LAYOUT

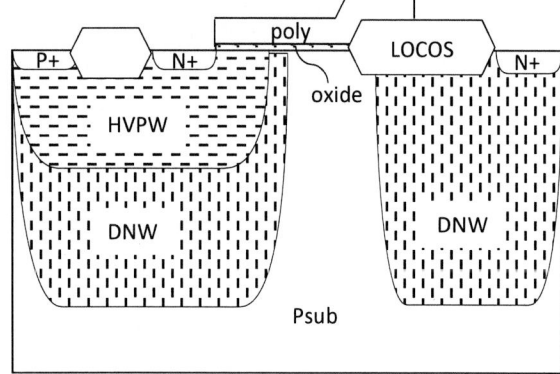

*Figure 1: NLDMOS device structure in this paper*

NLDMOS device structure in this paper is shown in Fig. 1, the NLDMOS drift region is formed by deep N-type well (DNW). The gap in DNW is for lower concentration, which is disappeared after DNW thermal process. The high-voltage P-type well (HVPW) is for more DNW depletion. The both DNW and HVPW are contributed to high BV. The NLDMOS device BV 103V for 80V application meets target in simulation, but DNW/HVPW junction BV 93V of NLDMOS peripheral structure in width direction is not enough. The original NLDMOS main layout is shown in Fig. 2, the poly extends HVPW in width direction, which is restrained HVPW further depletion.

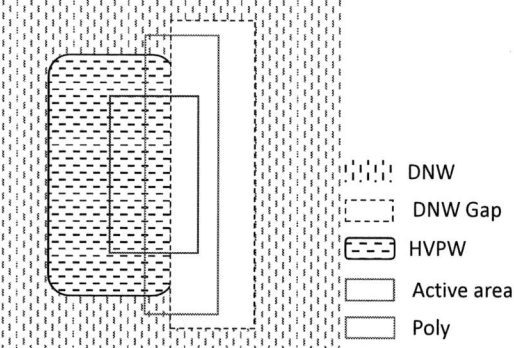

*Figure 2: original NLDMOS main layout*

## RESULTS AND DISCUSSION

The new NLDMOS main layout is shown in Fig. 3, in which poly is retracted into HVPW, it helps HVPW to deplete completely. The simulation comparison of cutline1(the boundary of poly in HVPW) is in Fig. 4, the grey area is depletion region, the depletion region of new poly location is obviously larger than that of original poly location, DNW/HVPW junction BV in width direction is improved to 122V after poly retracting into HVPW. The surface electrical field at bird-beak in HVPW is raised in new poly location, as shown in Fig. 5.

978-1-6654-9759-6/22 $31.00 © 2022 IEEE

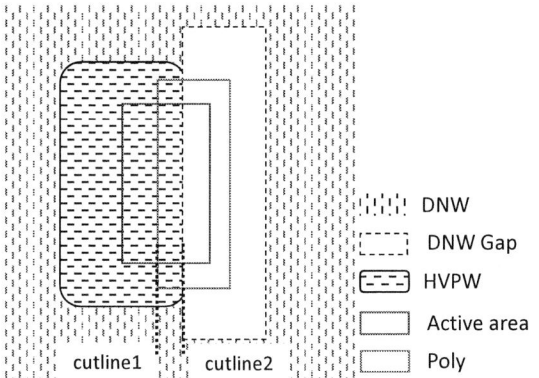

Figure 3: new NLDMOS main layout

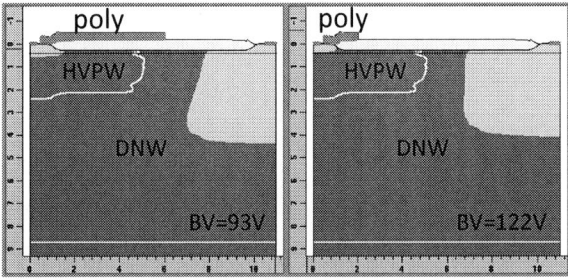

a. original poly location    b. new poly location

*Figure 4: depletion region comparison of DNW/HVPW junction BV in width direction of cutline1*

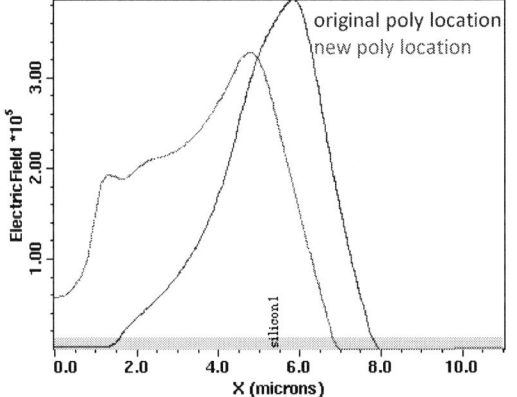

*Figure5: Lateral electric field distribution of different poly location of surface cutline in Fig. 4*

The distance of DNW gap extending HVPW in width direction is another influencing factor of BV in width direction. The HVPW has deep junction depth and low concentration which is even lower than DNW for higher BV, so low DNW concentration is needed for matching that of HVPW. The DNW concentration decrease with the distance increasing, which causes BV in width direction increasing. The DNW/HVPW junction BV in width direction of cutline2 (the boundary of HVPW and DNW gap) in 2D simulation which increases with the distance of DNW extending HVPW is shown in Fig. 6.

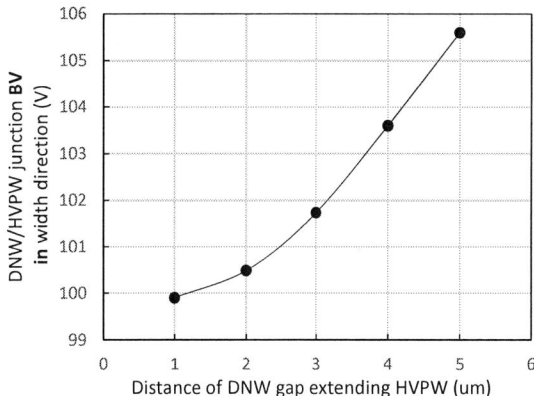

*Figure6: BV in width direction with distance of DNW gap extend HVPW in width direction*

The device width direction BV is the lower one of that of cuteline1 and cuteline2. It is 93V of cutline1 BV originally which is lower than cutline2 BV, then it is cutline2 BV when cutline1 BV is improved to 122V by poly size retracting into HVPW. After the distance of DNW gap extending HVPW increasing to 4um, the device width direction BV is above 103V which is higher than NLDMOS device itself BV(103V).

With the optimization of poly and DNW gap size in width direction, the BV of device peripheral structure is improved obviously. The high voltage NLDMOS device BV is not restricted by that in width direction, and the method of improving device BV like drift region concentration adjusting and field plate optimization is effective.

## CONCLUSION

In this paper, the method of the high voltage NLDMOS BV in width direction is studied. Poly size and DNW gap size optimizing are both beneficial for device BV improving in width direction. Finally, the high voltage NLDMOS BV is improved by 10V in TCAD simulation.

## ACKNOWLEDGEMENTS

The authors would like to thank all members of device group and high voltage group of HHgrace for their great support on this work.

## REFERENCES

[1] J.A.Appels, H.M.J. Vaes, "HV thin layer devices (Resurf devices), " IEDM, 1979, pp. 238-241
[2] A.W. Ludikhuize, "A review of RESURF technology," Proceedings ISPSD, 2000, pp. 11-18
[3] Yang K , Guo Y , Pan D Z , et al. A Novel Variation of Lateral Doping Technique in SOI LDMOS With Circular Layout[J]. IEEE Transactions on Electron Devices, 2018, PP(4):1-6.

[4] He J , Xi X , Chan M , et al. Linearly graded doping drift region: a novel lateral voltage-sustaining layer used for improvement of RESURF LDMOS transistor performances[J]. Semiconductor Science & Technology, 2002, 17(7):721-728.

[5] Shaari S , Hanim A R , Mardiana B , et al. Modeling And Analysis Of Lateral Doping Region Translation Variation On Optical Modulator Performance[J]. AIP Conference Proceedings, 2010, 1325(1):297-300.

# IMPROVE THE ON-RESISTANCE FOR 80V NLDMOS WITH STI TECHNOLOGY
# IN 0.18 UM BCD PROCESS

*Xiaoming Zhang[1,2*], Donghua Liu[2], Haiyang Ling[2], and Wensheng Qian[2]*

[1]School of Information Science and Technology, ShanghaiTech University, Shanghai 201210, China
[2] HuaHong Grace Semiconductor Manufacturing Corporation, Shanghai 201206, China
*Corresponding Author's Email: Xiaoming.Zhang@hhgrace.com

## ABSTRACT

In recent years, devices used in power application are required to operate at high voltage, high frequency conditions and at low current to save energy. As a result, it is important to improve the specific on-resistance (RON) for switch LDMOS. In this paper, we will aim at 80VnLDMOS in 0.18um BCD process to discuss the composition of on-resistance and analysis the dominated part of RON. This paper also proposes a simple method to improve the RON. Based on no additional masks and complex process flows, the combination of adjusting ratio of field plate and multiple ion implantation in drift region can improve BV and reduce Ron. The method can improve the on-resistance, which are proved by TCAD simulation results.

## INTRODUCTION

Typical mixed-signal applications, which are in automotive, power management and etc. require high voltage capability and fast switch velocity. High voltage lateral double-diffused MOS (LDMOS) are widely used in these applications to meet requirements. Because the mechanism of LDMOS is the transport of majority carriers, the fast switching speed is the significant advantages compared with the bipolar power transistor. To get high breakdown voltage (BV), it is usually to increase the drift region length or decrease the doping concentration of the drift region, leading to a significant increase of the specific on-resistance (Ron,sp). Therefore, the main focuses of designing power device such as LDMOS are breakdown voltage and specific on-resistant. But it is no doubt that the electrical performance has the limit for silicon material (Ron,sp$\propto$ BV$^{2.4-2.6}$). In order to get higher BV and lower possible Ron, several new structures and new materials have been developed in recent years, such as RESURF technology [1-2], field plates technology [3-5], high-k dielectric [6-7] and Super Junction (SJ) technology [8-9]. However, the BV is sensitive to the doping concentration of the drift region. In this work, a method is proposed to improve the 80V n-LDMOS with STI. With the help of long drift region and modulating doping concentration, the on-resistance decreases obviously. In the meanwhile, by adjusting the position of the field plate, the breakdown

voltage will not decrease with the increase of doping concentration. To some extent, this method can break the trade-off between breakdown voltage and on resistance when adjusting the drift region. At the same time, no additional masks and complex process flows are required.

This paper is organized as follows. In Section II, the device structure, the NetActive of the device and specific parameters of the device are presented. In Section III, the TCAD simulation and results are presented.

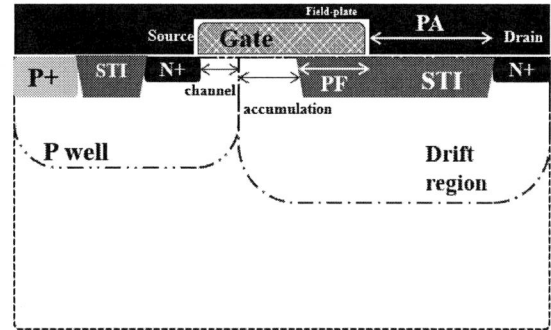

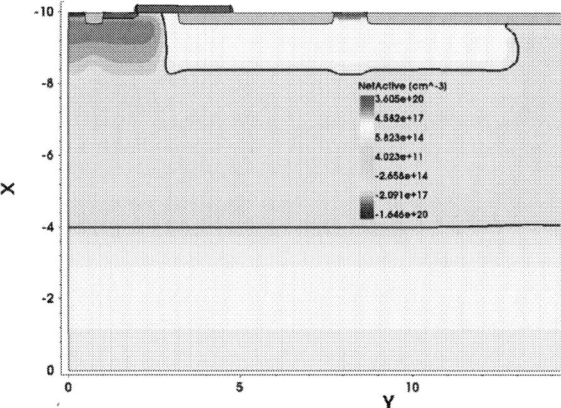

*Figure 1: (Top) Cross-section of the STI based LDMOS device. The main geometrical features are reported; (bottom) NetActive simulation result of the device.*

## DEVICE STRUCTURE

Fig. 1 shows the 2-D cross section of the STI based LDMOS used 0.18 µm process and the NetActive cross-section in simulation. The device is built on a N-type buried layer to connect with other devices. The source region is in the P-well and a STI is used to isolate the contact. To balance the doping concentration in P-well, several implantations with different energy are used. For

TABLE I.    THE KEY PARAMETERS IN DEVICE SIMULATION

| Parameter | Definition | Values |
|---|---|---|
| $L_{ch}$ (μm) | the length of the channel region | 0.5 |
| LA (μm) | the length of the accumulation region | 0.7 |
| PF+PA (μm) | the width of the shallow trench isolation | 4.5 |
| Ndrift Si ($cm^{-3}$) | the concentration of Si in drift region | 1e16 |
| Pwell Si ($cm^{-3}$) | the concentration of Si in P well | 1e18 |
| $N^+$ Si ($cm^{-3}$) | the concentration of Si in source and drain | 1e20 |
| Sub Si ($cm^{-3}$) | the concentration of Si in substrate | 1e15 |

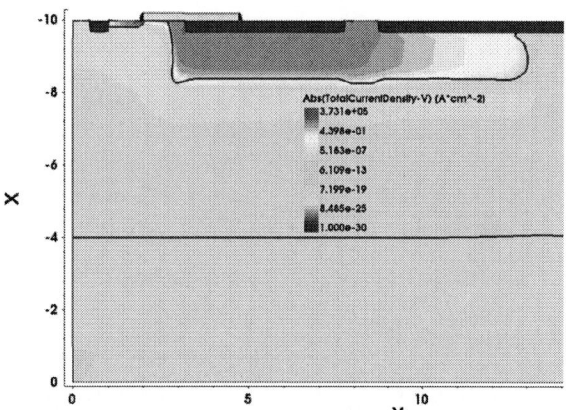

Figure 2: The cross-section of the total current.

LDMOS, the main region to withstand the high voltage is the drift region. Therefore, several implantations with different energy are also used to get better implantation effect. The gate oxide thickness is 20 nm. The width of the STI region covered with field plate is defined as PF and the width of the STI region without field plate is defined as PA, which are indicated in Fig. 1. The specific parameters in the NetActive simulation result are reported in Table. 1.

## RESULTS AND DISCUSSION

According to conventional specific parameters in Table. I, the simulation results are: breakdown voltage is 120 V, the on-resistance is 130 mOhm*mm². To use for

80V n-LDMOS, there is a large margin in BV region. However, it is necessary to optimize the $R_{on}$ to use for the switching application. The initial simulation result of total current is indicated in Fig. 2. The current path in drift region is much larger than the current path in channel and accumulation region. Therefore, the main optimization

TABLE II.    SIMULATION RESULTS OF $R_{on}$ AND BV

| Simulation results | | | | | |
|---|---|---|---|---|---|
| *P En1* | *P Do1* | *P En2* | *P Do2* | $R_{on}$ *(mOhm *mm²)* | *BV (V)* |
| 1000 | 1e12 | 800 | 1.5e13 | | |
| *P En3* | *P Do3* | *P En4* | *P Do4* | 128.9 | 120 |
| 200 | 2e12 | 50 | 3e12 | | |
| *P En1* | *P Do1* | *P En2* | *P Do2* | $R_{on}$ *(mOhm *mm²)* | *BV (V)* |
| 1000 | 1e12 | 400 | 1.5e13 | | |
| *P En3* | *P Do3* | *P En4* | *P Do4* | 119.9 | 120.3 |
| 200 | 2e12 | 50 | 3e12 | | |
| *P En1* | *P Do1* | *P En2* | *P Do2* | $R_{on}$ *(mOhm *mm²)* | BV (V) |
| 1000 | 1e12 | 800 | 1.5e13 | | |
| *P En3* | *P Do3* | *P En4* | *P Do4* | 125.3 | 121 |
| 200 | 3e12 | 50 | 3e12 | | |
| *P En1* | *P Do1* | *P En2* | *P Do2* | $R_{on}$ *(mOhm *mm²)* | BV (V) |
| 1000 | 1e12 | 800 | 1.5e13 | | |
| *P En3* | *P Do3* | *P En4* | *P Do4* | 123.8 | 119.7 |
| 200 | 3.2e12 | 50 | 3e12 | | |
| *P En1* | *P Do1* | *P En2* | *P Do2* | $R_{on}$ *(mOhm *mm²)* | BV (V) |
| 1000 | 1e12 | 800 | 1.5e13 | | |
| *P En3* | *P Do3* | *P En4* | *P Do4* | 132.1 | 121 |
| 200 | 2e12 | 70 | 3e12 | | |
| *P En1* | *P Do1* | *P En2* | *P Do2* | $R_{on}$ *(mOhm *mm²)* | BV (V) |
| 1000 | 1e12 | 400 | 1.5e13 | | |
| *P En3* | *P Do3* | *P En4* | *P Do4* | 112 | 104 |
| 200 | 2e12 | 50 | 3e12 | | |

method is to modulate the doping concentration in drift region based on no additional masks and process flows. There are four implantations to form the drift region. In Table. II, it shows the simulation results of $R_{on}$ and BV with different implantation condition. From the results, $R_{on}$ decreases obviously when the second implantation reduced energy and decreases a bit when the third implantation increased doping concentration. Obviously, the main influence of the $R_{on}$ is the second and third implantation. In the last case of Table. II, the most optimization result is reported. Considering the voltage

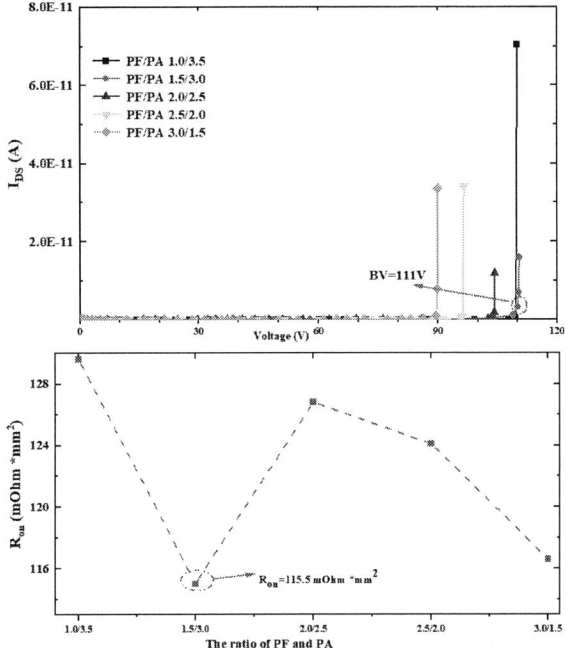

*Figure 3: $I_d$-$V_d$ curve based on different ratio of PF and PA; Dependences of $R_{on}$ on the ratio of PF and PA.*

margin during specific use, the BV is better to above 110V while the $R_{on}$ is no more than 120 mOhm*mm$^2$.

Based on no additional masks and process flows, the main method to improve breakdown voltage is to adjust the field plate. Fig. 3 shows the simulation results of $I_d$-$V_d$ curve based on different ratio of PF and PA and the corresponding $R_{on}$ respectively. Different ratio of PF and PA change the location of electric field peak, leading to a bit change of current path in drift region. The best case of the ratio of PF and PA is 1.5/3.0. At this case, the breakdown voltage is 111 V, which could meet the requirement of the voltage margin. In the meanwhile, the $R_{on}$ is 115.5 mOhm*mm$^2$, basically maintained at the same level compared with the application requirement.

## CONCLUSION

In this paper, a simple method is proposed to improve the device with STI and used 0.18 μm process. Based on no additional masks and process flows, the combination of adjusting ratio of field plate and multiple ion implantation in drift region can give consideration to improve BV and reduce Ron. According to the simulation results, it has

effect to improve the device. At the same time, the improved devices will be put into practical productions and applications.

## REFERENCES

[1] A. W. Ludikhuize, in 12th ISPSD, May 2000, pp.11-18.

[2] T. Tsuchiya, O. Tabata, J. Sakata and Y. Taga. *J. Microelectromech. Syst.*, vol. 7, 1998, pp. 106-113.

[3] N. Fujishima, M. Saito, A. Kitamura, et al, in Proc. ISPSD, 2001, pp. 255-258.

[4] Kenji Hara, Tomoko Kakegawa, Shinichiro Wada, Tomoyuki Utsumi, Tetsuo Oda, in Proc. 29th ISPSD, Sapporo, Japan, 2017, pp. 307–310.

[5] Guangsheng Zhang, Wentong Zhang, 2, Junqing He, Xuhan Zhu, Sen Zhang, Jingchuan Zhao, Zhili Zhang, Ming Qiao, Xin Zhou, Zhaoji Li and Bo Zhang, in Proc. 31th ISPSD, Shanghai, China, 2019, pp. 507–510.

[6] Junhong Li, Ping Li, Weirong Huo, et al, IEEE Electron Device Letters, vol. 32, no. 9, pp. 1266–1268, Sept. 2011.

[7] Zhen Cao, Baoxing Duan, Haitao Song, Fengyun Xie, and Yintang Yang, IEEE Transactions on Electron Devices, vol. 66, no. 5, pp. 2327-2332, May 2019.

[8] Zhen Cao, Baoxing Duan, Song Yuan, et al, in 29th ISPSD, May 2017, pp.283-286..

[9] Jhen-Yu Tsai and Hsin-Hui Hu, IEEE Transactions on Electron Devices, vol. 63, no. 6, pp. 2482-2487, June 2016.

# LOW CROSSTALK AND HIGH Q-FACTOR INDUCTOR BASED ON MULTIPLE GUARD-RING DESIGN

Xiaodong Wang*, Xining Wang, Weihong Qian, Xi Huang, Muyi Liu, and Byung Sup Shim

Semiconductor Manufacturing International Corporation, Shanghai, China

*Corresponding Author's Email: xiaodong_wang2022@163.com

## ABSTRACT

The RF inductor suffers from larger area and low Q-factor in CMOS process. With technology scaling down, the guard-ring design of inductor is limited by BEOL double pattern process, which induces higher crosstalk coupling noise in RF circuits. A novel closed multiple guard-ring inductor is proposed based on SMIC FinFET process. The crosstalk of the novel design is about 60% lower, equivalent to about 10% area saving. Benefiting from higher shielding effect, the Q-factor has about 2-5% improvement.

*Keywords—RF inductor; Low crosstalk; Q-factor; Guard-ring; BEOL; Double pattern*

## INTRODUCTION

The inductor is widely spread in Voltage Controlled Oscillator (VCO), Low Noise Amplifier (LNA) circuit as magnetic stored component [1-2]. The low Q-factor and oversize make the inductor as bottleneck device in these circuits. As one of the basic components of inductor, the guard-ring (GR) is designed beneath the top metal coil to ground the pattern ground shielding (PGS), and suppress the signal leakage and noise crosstalk between devices [3-4]. In mature node, the half-continuous metal parallel substrate is used in GR. Whereas, as technology scaling down, the continuous metal design is forbidden in double patterning process. Stacked metal lines by different layers are used for GR. Compared with continuous metal GR, signal leakage shielding capability becomes worse, leading to higher loss and lower Q-factor. The crosstalk between adjacent inductors in circuit increases at the same time. The interval of inductors needs to be increased to avoid signal interact with each other.

In this paper, a novel inductor with approximate closed multiple GR structure is proposed based on SMIC FinFET process. The multiple GRs form approximate closed structure by one metal layer. The crosstalk between inductors is greatly mitigated by using this novel GR that effectively equivalent to saving die area for multi-inductors application. This paper is arranged as follows. We firstly introduce the inductor device design and fabrication. Then discuss the experiment results. Finally, we draw the conclusion according.

## DEVICE DESIGN AND FABRICATION

Fig.1 shows the schematic of inductor. The GR is usually designed by active area (AA), polysilicon, and metal layers. The PGS works by grounding to GR to prevent electric filed flux from penetrating conductive semiconductor silicon substrate, which is helpful to improve the Q-factor. Meanwhile, the GR can suppress the leakage of stored magnetic energy. The lower signal energy leakage, the lower crosstalk between adjacent inductors.

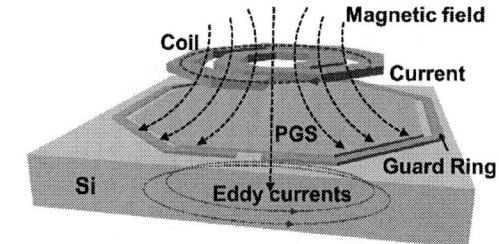

*Figure 1: The schematic of inductor.*

The schematic of mature node and advanced node GRs structure are shown in Fig.2 (a) and (b), respectively.

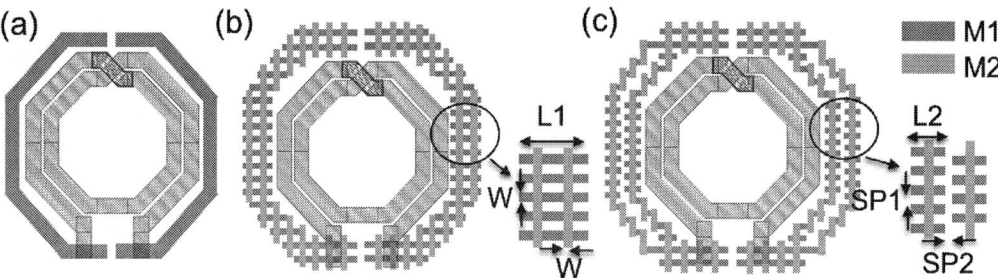

*Figure 2: The layout of GRs in (a) mature node, (b) advanced node, and (c) the novel design for advanced node.*

$$Crosstalk = 20*lg\ (S.14)$$

In mature node, the first metal layer (M1) is designed in half-octagonal shape connected to below AA structure. The continuous M1 and AA structures show better shielding effect. In advanced node, the GR is fabricated by perpendicular M1 and M2 due to the double pattern process limitation. Compared with continuous semi-closed GR in mature node, the shielding effect of GR in advance node for signal leakage becomes worse, resulting stronger crosstalk and lower Q-factor.

The basic cell of proposed multiple staggered GR for advanced node is shown in Fig.2 (c). This novel GR consists of four semi-circles and two of the adjacent semi-circles on the same side are connected. The M1 in different semi-circles is complementary to form approximate closed structure. The width of novel GR is the same with typical GR. The different width and space can be used in different circles as long as complementary is guaranteed.

## EXPERIMENT RESULTS

The RF performance of the inductors based on different GRs are evaluated by a commercial electro-magnetic solver named EMX from Cadence Corp. We use SMIC 14 nm process Techfile for the electro-magnetic simulation. Fig.3 shows the simulation structure for crosstalk. Two inductors are placed adjacently with different intervals split.

In our study, we simulate 4-ports s-parameters with port definition as P1, P2, P3 and P4 in the range of 0.1 GHz to 40 GHz. The P1 and P2 are input/output ports of the first inductor; P3 and P4 are ports of the adjacent inductor. The GRs of the two inductors are grounded. The crosstalk is calculated by the equation as below:

The S.14 represents the signal intensity coupled from P1 to P4, which can be extracted from 4-ports s-parameters matrix.

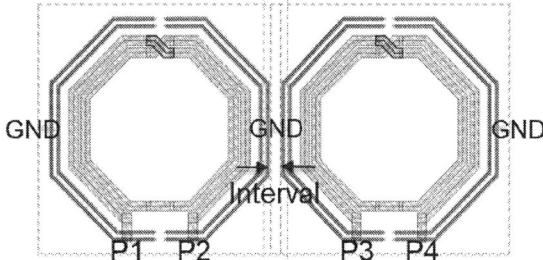

*Figure 3: The layout for inductor crosstalk in EMX simulation.*

The Q-factor of the inductors are calculated from s-parameters simulated with single inductor. The L1, L2, W, SP1 and SP2 are fixed at 1.5 μm, 0.6 μm, 0.6 μm, 0.5 μm and 0.3 μm, respectively, the definition of these parameters are shown in Fig.2. The simulation results of the inductors with different GRs are shown in Fig. 4, the two inductors have the same dimension, Radius = 60 μm, N = 2, SP = 2 μm and W1 = 8 μm. The interval is 10 μm. It can be seen that the inductor with novel GR achieves 2.2 dB to 4.1 dB crosstalk lower in the range of 0.1 to 38 GHz due to the better shielding effect. The shielding effect increases with frequency. At 16 GHz, the Q-factor of the inductor with novel GR also has 2.8% improvement. There is no obviously difference for the resonant frequency ($F_R$). The Q-factor improvement can reach 5% with inductor geometry parameters variation.

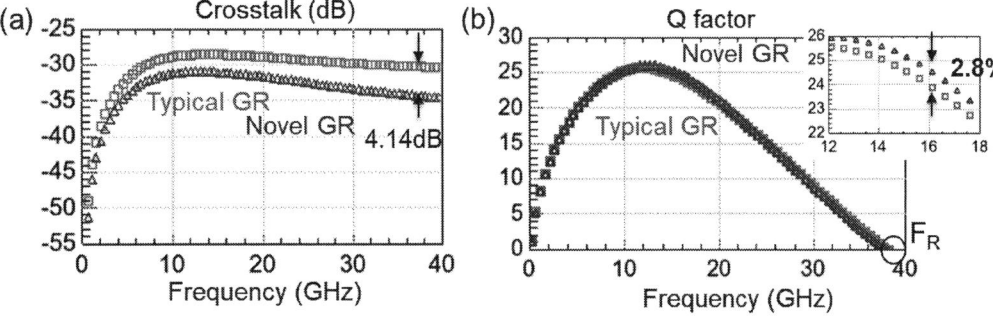

*Figure 4: (a) The crosstalk for inductors based on different GRs with 10 μm intervals, and (b) the Q-factor for the inductors with different GRs.*

The crosstalk performance at 38 GHz for different intervals are shown in table1. As we can see from this table, the crosstalk performance of novel GR is always lower than the typical one when the interval changes from 5 to 30 μm. At the same interval, the improvement is about 60%. By comparison, we find that the crosstalk performance of novel GR design with 5 μm interval is even lower than typical with 30 μm interval. That means the RF circuit layout design using our novel GR can achieve higher integration with closer inductor placement. Compared with the typical design, the total area for two adjacent inductors based on novel multiple GR with parameters in Fig .4 has 12.5% reduction.

| Table1 The crosstalk (dB) @ 38 GHz for different intervals | | | |
|----------|---------|--------|------------------|
| Interval | Typical | Novel | Gap |
| 5μm | -29.58 | -33.73 | 61.3% (4.15dB) |
| 10μm | -30.29 | -34.39 | 61.1% (4.14dB) |
| 30μm | -32.52 | -36.6 | 60.7% (4.12dB) |

The proposed novel inductor with new GR design in advance node exhibits excellent crosstalk and Q-factor isolation effect.

## CONCLUSION

In conclusion, we proposed and demonstrated a low crosstalk and high Q-factor inductor based on staggered multiple guard-ring. Benefiting from the lower signal leakage, the crosstalk for adjacent inductors reduce more than 60%, equivalent to about 10% area saving. In addition, the Q-factor has 2 - 5% improvement.

## REFERENCES

[1] K. Kawabe, H. Koyama, and K. Shirae. "Planar inductor." IEEE Transactions on Magnetics 20.5 (1984): 1804-1806.

[2] M. Tiebout, "Low-power low-phase-noise differentially tuned quadrature VCO design in standard CMOS." IEEE Journal of Solid-State Circuits 36.7 (2001): 1018-1024.

[3] Y. Seong-Mo, and T. Chen. "The effects of a ground shield on the characteristics and performance of spiral inductors." IEEE Journal of Solid-State Circuits 37.2 (2002): 237-244.

[4] Kim, C. S., Park, P., Park, J. W., Hwang, N., & Yu, H. K.,"Deep trench guard technology to suppress coupling between inductors in silicon RF ICs." 2001 IEEE MTT-S International Microwave Sympsoium Digest (Cat. No. 01CH37157). Vol. 3. IEEE, 2001.

# FLICKER NOISE CHARACTERIZATION OF LDMOS IN FINFET TECHNOLOGY

*Junwei Guo[1*], Yuning Guo[1], Yali Liu[1], Jing Tong[1]*

[1]Semiconductor Manufacturing International Corporation (SMIC), Shanghai, China

18 Zhangjiang Road, Pudong New Area, Shanghai, P.R. China 201203

*Corresponding Author's Email: JunWei_Guo2@smics.com

## ABSTRACT

The flicker noise (1/f noise) behavior of FinFET LDMOS (Lateral Double-diffused MOS) transistor varies significantly from that of a conventional MOS transistor because of the extended drain region in LDMOS. Therefore, through an investigation of a 14 nm FinFET LDMOS, two distinct phenomena were discovered, discussed and first reported in this paper. Firstly, the trend of the Sid vs. Vd (drain voltage) curves was studied. The Sid vs. Vd curves exhibited an increase with drain voltage at relatively low drain voltage, then followed by a significant drop once the drain voltage reached a certain value. Space Charge Modulation (SCM) and hot spot formation theories were employed to interpret the mechanism behind this trend and this trend was linked to Quasi-saturation (QS) effect in LDMOS. Secondly, the unparalleled Sid vs. Frequency curves due to the existence of the extended drain in the LDMOS were studied. On the basis of this finding, a new BSIM-CMG LDMOS subcircuit model was proposed, which consisted a MOS transistor and a terminal voltage dependent resistor. SPICE simulation results revealed that this subcircuit model was able to accurately describe the noise behavior of the FinFET LDMOS and could be applied to tremendous fields where LDMOS plays a role.

## INTRODUCTION

LDMOS, by virtue of its advantages including its compatibility with standard CMOS processing and its capability to withstand high drain voltage, has been intensively used in a vast of fields. It could be served as a crucial part in smart power devices, I/O devices, power amplifiers and HV drivers, etc. [1-2] Moreover, with the rise of electric vehicles in recent years, the prosperous applications of LDMOS in power management chips in the near future is promising.

Among all properties of interest, flicker noise property is vital for LDMOS because flicker noise constitutes a large portion from low to moderate frequencies, thus determining the performance of LDMOS in actual circuits and putting limits to the dynamic range of a device, especially in digital circuits. Some published literatures on the flicker noise behavior of LDMOS focused mainly on LDMOS with large channel length [3-4] and the flicker noise behavior of FinFET LDMOS in

advanced technology node has never been reported. Besides, FinFET LDMOS has dramatic changes both in channel length and structure. Therefore, it is worthwhile to investigate the flicker noise behavior of FinFET LDMOS. In the present work, a comprehensive analysis of the Sid vs drain voltage and Sid vs. Frequency curves was carried out. In addition, a BSIM-CMG subcircuit model was proposed to simulate the flicker noise phenomena found in FinFET LDMOS, to provide an accurate model support in IC industry.

## DEVICE AND EXPERIMENT

The device used in this study was an 14 nm n-type FinFET LDMOS, manufactured in SMIC through a series of delicate processes: 1) fin process to form fin structure; 2) poly process to form poly gate; 3) EPI process for channel stress engineering; 4) high-k metal gate process to remove poly gate and deposit metal layers; 5) middle end of line (MEOL) and back end of line (BEOL) processes. The cross-section schematic of the LDMOS is shown in Fig.1. Similar to conventional MOS transistor, four terminals - bulk, source, gate and drain, could be easily identified, while the channel region features gradient dopant concentration and is separated into three parts – channel, drift region and drift extension region. Shallow trench isolation (STI) sat in NWell was carried out during fin loop.

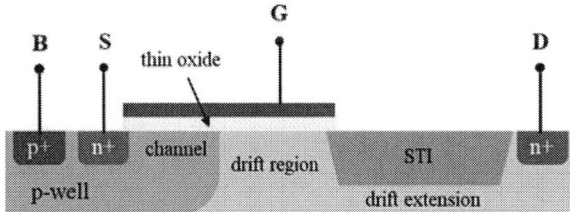

*Figure 1: Schematic showing the cross-section of an 14 nm n-type FinFET LDMOS manufactured in SMIC*

For flicker noise detection, measurements were performed on ProPlus 9812DX low frequency noise characterization system. Sid spectral density vs. Frequency curves were measured with frequency ranging from 5 to 100k Hz. Subsequently, flicker noise values at certain frequency were extracted from these measured noise spectrum curves for further Sid vs. Vd trend analysis with drain voltage varying from 0.1 to 5 V.

## RESULT AND DISCUSSION

The relationship between flicker noise and drain voltage was investigated as can be seen from the Sid vs. Vd curves shown in Fig. 2. Sid increases with drain voltage and this trend is typical in MOSFET transistor. However, once drain voltage reaches around 2 - 3 V, a sharp drop in Sid can be detected and this is valid over a large frequency range from 100 to 100k Hz, as shown in Fig. 2(a) – (d). Despite the physical mechanism of flicker noise is not yet fully understood, it is generally believed that the capture and emission of electrons by interface traps at the Si-SiO$_2$ interface plays an important role. [5] Therefore, the structural difference between LDMOS (extended drain) and MOS transistor is speculated to cause changes in trapping/de-trapping and accounts for the Sid vs. Vd trend found in LDMOS. The following discussions will focus on this issue in more detail.

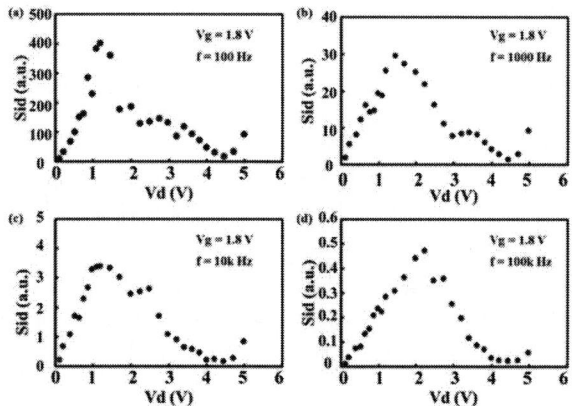

Figure 2: Sid vs. Vd curves measured at 1.8 V gate voltage over a frequency range from 100 to 100k Hz.

Initially, LDMOS works in the linear region when drain voltage is low as shown in the Id-Vd curve (Fig. 3). In this stage, noise behavior is quite similar to MOS transistor. Subsequently, with the increasing of drain voltage, the transistor gradually enters into quasi-saturation state and the carriers in the channel and extended drift region have undergo Space Charge Modulation (SCM, also called Kirk Effect). [6-7] After SCM, the electrical field in LDMOS at channel end and drain end has experienced opposite changes. At drain end, the electrical field soars with drain voltage and forms a hot spot, while slightly decreases at the channel end. Moreover, the electrical field at the channel region shifts deeper into the LDMOS from its surface. [8] As a consequence, current path in the FinFET LDMOS deviates from the surface, hindering the trapping/de-trapping process happened close to the Si-SiO$_2$ surface. Hence, Sid vs. Vd curve drops after SCM.

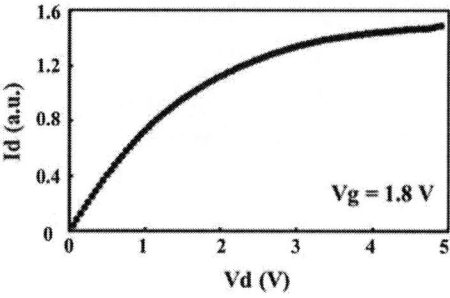

Figure 3: Id-Vd curve measured at a gate voltge of 1.8 V

The influence of SCM could be further verified by the phenomenon found in a FinFET LDMOS reliability study. In another ongoing research on the Hot Carrier Instability (HCI) effect of FinFET LDMOS, it is shown that the threshold voltage degrades less under the stress condition with SCM (~2.2 mV) compared to vt degradation under the stress condition without SCM (~9.6 mV) (data not included in this paper). Since HCI effect depicts the process that carriers gaining high kinetic energies in the lateral electrical field inject into gate oxide and cause charge trapping and interface states and this effect could be linked with SCM in LDMOS [9]. It is reasonable to deduce that the alleviation of HCI effect due to SCM will lead to the dropping trend in Sid vs. Vd curves.

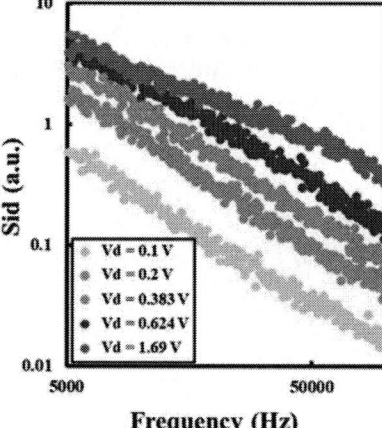

Figure 4: Sid vs. Frequency curves measured at a series of drain voltages with Vg = 1.8 V

The Sid vs. Frequency curves shown in Fig. 4 were measured at a gate voltage of 1.8 V with drain voltage ranging from 0.1 to 1.7 V. In general, the Sid vs. Frequency curves (in log-log scale) follow $(1/f)^n$ behavior as the name 1/f noise indicated. In conventional MOS transistors, the frequency exponent n is typically in the range of 0.8 to 1.2 [10], and terminal bias barely has any impact on the value of n, which results in a group of

978-1-6654-9759-6/22 $31.00 © 2022 IEEE      150

paralleled curves. However, in FinFET LDMOS, the Sid vs. Frequency curves exhibited unparalleled trend with the increase of drain voltage as can be seen from Fig. 4, which explicitly indicates that n is dependent on drain voltage. n value drops from 1.14 down to around 0.77 with increasing Vd (from 0.1 to ~1.7 V). Considering the structural difference between LDMOS and MOS, the resistance of the extend drain in the LDMOS due to non-homogeneous doping inevitably has an impact on the 1/f noise.

On this basis, the flicker noise behavior of FinFET LDMOS could be expressed as a linear superposition of a MOS transistor and a terminal voltage controlled resistor as shown in Fig. 5(a). With BSIM-CMG model, which is the most widely used in FinFET technology, using a subcircuit model including MOS and voltage-dependent resistor, we have successfully reproduced the drops in the Sid vs. Vd curves. As in Fig. 5 (b)-(d), SPICE simulation data has shown that the newly proposed subcircuit model is able to accurately describe the Sid vs. Vd curves over a wide frequency range from 1000 to 100k Hz.

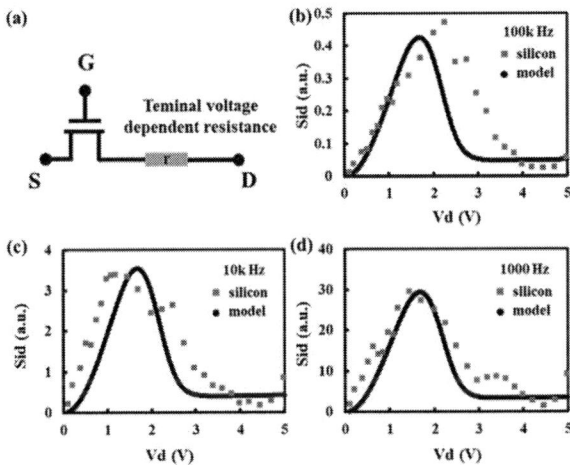

Figure 5: (a) Schematic of FinFET LDMOS consists a MOS transistor and a terminal voltage dependent resistor for FinFET modeling; (b)-(d) Sid vs. Vd curves simulated with the proposed subcircuit model at 100k, 10k and 1000 Hz, respectively.

## CONCLUSION

In this paper, the flicker noise behavior of a 14 nm FinFET LDMOS was investigated through a profound discussion of the Sid vs. Vd and Sid vs. Frequency curves. A decrease trend was found in the Sid vs. Vd curves, and this trend was explained with space charge modulation and hot spot formation theories. With the increasing of drain voltage, LDMOS gradually enters quasi-saturation state and SCM happens. In this state, the corresponding electrical field in LDMOS shifts deeper in to the device, blocking the trapping/de-trapping process near the surface,

thus leading to Sid drop. In the Sid vs. Frequency study, the unparalleled Sid vs. Frequency curves led us to realize the unneglectable function of the extended drain. Accordingly, a BSIM-CMG subcircuit model consisting a MOS transistor and a terminal voltage dependent resistor was proposed in order to simulate the unique flicker noise phenomena in FinFET LDMOS. The subcircuit model has proven to be accurate compared to primary silicon data from low to moderate frequencies, which could potentially be used in LDMOS related IC designs.

## REFERENCE

[1] J. A. van der Pol et al., *in Proc. ISPSD*, 2000, pp. 327–330.
[2] Aarts, Annemarie CT, and Willy J. Kloosterman. *IEEE Transactions on Electron Devices* 53.4 (2006): 897-902.
[3] Mahmud, M. Iqbal, et al., *IEEE Transactions on Electron Devices* 60.2 (2012): 677-683.
[4] Mavredakis, Nikolaos, et al., *2015 International Conference on Noise and Fluctuations (ICNF)*. IEEE, 2015.
[5] Ytterdal, Trond, Yuhua Cheng, and Tor A. *John Wiley & Sons*, 2003.
[6] M. Knaipp, G. Rohrer, R. Minixhofer and E. Seebacher, *IEEE Transactions on Electron Devices*, vol. 51, no. 10, pp. 1711-1720, Oct. 2004.
[7] A. W. Ludikhuize, *Proceedings of the 6th International Symposium on Power Semiconductor Devices and Ics*, 1994, pp. 249-252,.
[8] B. S. Kumar and M. Shrivastava, in *IEEE Transactions on Electron Devices*, vol. 65, no. 1, pp. 191-198, Jan. 2018.
[9] Kumar, B. Sampath, and Mayank Shrivastava. *IEEE Transactions on Electron Devices* 65.1 (2017): 199-206.
[10] Dikshit, A. A., et al., *2012 24th International Symposium on Power Semiconductor Devices and ICs*. IEEE, 2012.

## ACKNOWLEDGEMENTS

The authors would like to thank the financial support provided by Shanghai Pujiang Program.

# LOW ON-RESISTANCE LDMOS WITH STEPPED FIELD PLATES FROM 12V TO 40V IN 300-MM 90-NM BCD TECHNOLOGY

*Hualun Chen[1], Zhaozhao Xu[1,*], Yu Chen[1], Mingxu Fang[1], Li Wang[1], Li Xiao[1], Yuanyuan Qian[1], Wan Song[1], Tian Tian[1], Ziquan Fang[2], Donghua Liu[2], Wensheng Qian[2], Weiran Kong[1,2]*
[1]Huahong Semiconductor (Wuxi) Limited, Wuxi 214029, China
[2]Shanghai Huahong Grace Semiconductor Manufacturing Corporation, Shanghai 201203, China
*Corresponding Author's E-mail: Zhaozhao.xu@hhgrace.com

## ABSTRACT

A laterally double-diffused metal–oxide–semiconductor (LDMOS) field-effect transistor with stepped filed plates (SFP) from 12V to 40V was proposed in this work. A scheme of multi stepped field plates were proposed to obtain better comprehensive performance. Furthermore, two stepped oxides (SO) has been achieved and fabricated with only one additional mask for 2nd SO. With SFP, the trenched current flow path can be removed. Therefore, lower specific on-resistance ($R_{sp}$) can be achieved in low-voltage LDMOS devices. Experimentally, the SFP-LDMOS transistors in this process have very competitive specific on-resistance.

*Key Words—Stepped filed plate (SFP); stepped oxide (SO); laterally double-diffused metal-oxide-semiconductor (LDMOS); specific on-resistance ($R_{sp}$)*

## INTRODUCTION

Highly compatible with advanced CMOS, laterally double-diffused MOSFET (LDMOS) devices has been widely used in a variety of smart-power applications. Recently, LDMOS has been studied for sub-14 nm system on chip applications [1], which exhibited compatible with 3D-FinFET CMOS device. LDMOS was always preferred because of low specific on-resistance ($R_{sp}$). Especially, LDMOS transistors for low-to-median voltage (5 to 40V) are very competitive due to its planar structure. Shallow trench isolation (STI) -based LDMOS [3] is also frequently used because the ease of co-integration with the conventional CMOS technologies [4] without additional mask request for field oxide formation. However, the depth of STI was mainly determined by the isolation rule of CMOS and was about 3000 Å at 90-nm node. A field oxide of 3000 Å was much thicker to low voltage LDMOS. Consequently, an LDMOS with ultra-shallow trench isolation (USTI) concept [2, 5] has been proposed and studied for achieving better breakdown voltage ($BV$)-$R_{sp}$ relationship. And, LDMOS with stepped-STI was also presented to reduce $R_{sp}$ and improve HCI effect [6, 7]. Another kind of LDMOS with stepped oxide (SO) [8-10] was also introduced by many authors to reach the same destination of better $BV$-$R_{sp}$ trade-off.

In this article, a LDMOS with two stepped filed plates (SFP-LDMOS) was presented. Based on this device structure, we have developed a new Bipolar-CMOS-DMOS (BCD) process platform in 300-mm 90-nm CMOS technology. This SFP-LDMOS was obtained by using a novel process integration to reduce the cost. It enables only one additional mask for this two stepped oxides. The device size and drift implant for the key switch were optimized to maximize device performance. To save the cost and time, process and device simulations were carried out by using Sentaurus TCAD tools

from Synopsys technology [11]. The actual fabricating process was exactly followed for simulations. These process and device simulations can be matched with silicon data [8] with calibrated parameters.

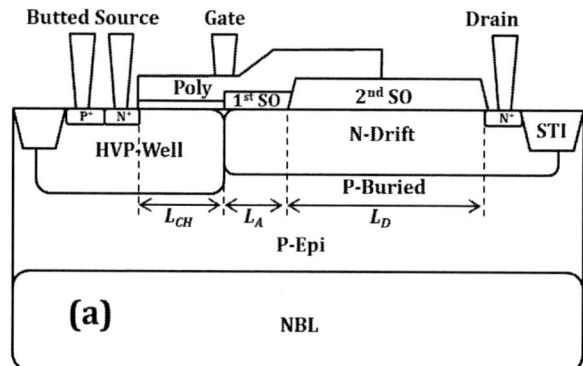

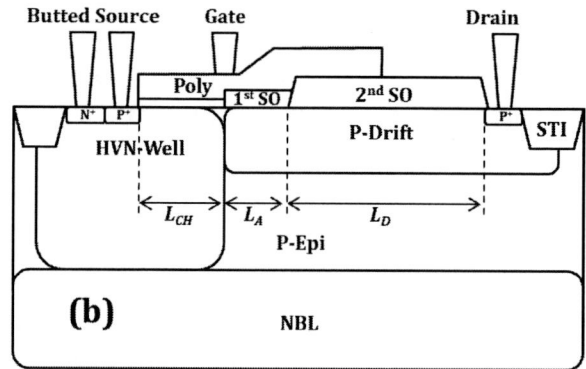

*Figure 1. Schematic cross-sectional drawings of (a) NLDMOS and (b) PLDMOS with two stepped filed plates.*

This work is organized as below. Firstly, the device structure and process details were introduced. Secondly, the overview of this platform technology and the outline of the SFP-LDMOS were presented. Finally, we present a conclusion about this work.

## DEVICE AND PROCESS

Figure 1 shows the schematic cross-sectional views of the N-type and P-type SFP-LDMOS device in this work. Two SOs (1st SO and 2nd SO) were employed in this work. Thus, oxide layers with three different thickness were integrated into the LDMOS. The other oxide layer was the thick gate oxide (TGO) except for the two stepped oxides. As illustrated in Figure 1, high-voltage (HV) N-type and P-type wells implants were employed for NLDMOS and PLDMOS, respectively. The HV wells are self-aligned with the 1st SO because of cost-effective process. After the implantation of HV-wells, the pad oxide for implant were separately etched with the un-striped photoresist of HV-wells. Thus, the pad oxide was remained at the masked regions. Then, the TGO for HV device was grown at the HV-wells region after the removal of photoresist. Consequently, the 1st SO was formed by pad oxide formation followed by the treatment from TGO thermal. However, the thickness of 1st SO was mainly determined by the thickness of pad oxide of implant. Therefore, the thickness of 1st SO can be changed according to the request.

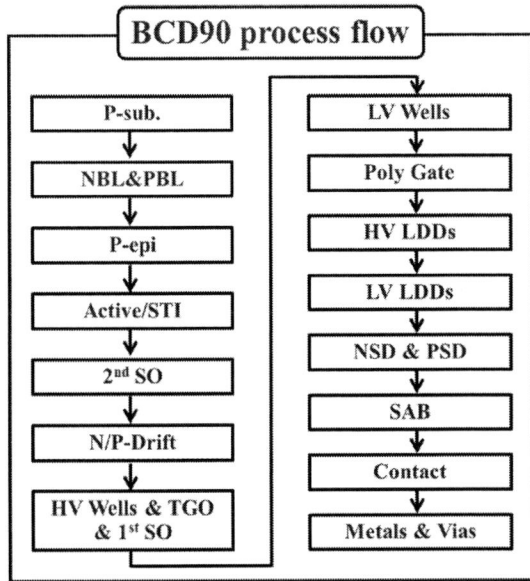

Figure 2. Key process flow of BCD90.

The key process flow was shown in Figure 2. As shown in the Figure 1(b), high energy implant was also included in the HVN-Well for peak-up of NBL. Therefore, the deep N-Well was removed for lower cost. Traditionally, N-type buried layer (NBL) was implanted on P-type substrate followed by the enhanced thermal drive-in. Then, a blanket P-type buried layer (PBL) was implanted followed by epitaxial growth. After conventional STI process, 2nd SO module was performed. Then and as shown in Figure 1, phosphorus and boron were implanted into drain-side to form N-Drift and P-Drift regions, respectively. During N-Drift implant, boron with higher energy was also co-implanted to form a P-Buried layer under N-Drift at the same masking step, which helps the depletion of the drift region. After drift implant, HV-Wells, TGO, and 1st SO are processed. Figure 3 shows the scanning electron microscope (SEM) cross sections of 24V SFP-NLDMOS device in 300-mm 90-nm BCD Technology. The angle of trapezoid transition in 2nd SO was optimized and 60° was obtained for our SFP-LDMOS. Currently and as shown in Figure 3, the difference of thickness between the 1st SO and TGO was only 30 Å. As mentioned above, the thickness of 1st SO can be changed according to the request with only slight modification of HV-wells. In this work, 5V to 9V devices have the same structure, which was defined as drain-extended MOS [12-14]. The 12V to 40V devices have the same SFP-structure as illustrated in Figure 1 and Figure 3.

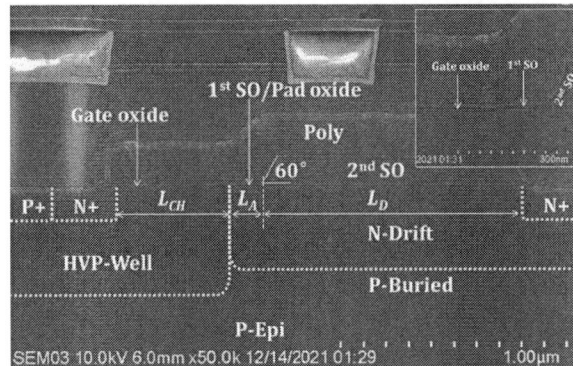

Figure 3. SEM cross-sectional view of the 24V SFP-NLDMOS device.

Table I. CHARACTERISTICS OF BCD90 DEVICES

| Device | Type | | Rated(V) | Characteristics |
|---|---|---|---|---|
| SFP-LDMOS | Nch | | 5V | Vth=0.7V, Rsp =1.6mΩ·mm2, BVdss=13.5V |
| | | | 7V | Vth=0.7V, Rsp =3.07mΩ·mm2, BVdss=18V |
| | | | 9V | Vth=0.7V, Rsp =3.66mΩ·mm2, BVdss=20.6V |
| | | | 12V | Vth=0.7V, Rsp =4.6mΩ·mm2, BVdss=28.7V |
| | | | 16V | Vth=0.7V, Rsp =6.22mΩ·mm2, BVdss=32V |
| | | | 20V | Vth=0.7V, Rsp =8.04mΩ·mm2, BVdss=37V |
| | | | 24V | Vth=0.7V, Rsp =10.8mΩ·mm2, BVdss=42V |
| | | | 30V | Vth=1.2V, Rsp =15.0mΩ·mm2, BVdss=48V |
| | | | 36V | Vth=1.2V, Rsp =19.5mΩ·mm2, BVdss=54V |
| | | | 40V | Vth=1.2V, Rsp =24.4mΩ·mm2, BVdss=58V |
| | Pch | | 5V | Vth=-0.95V, Rsp =3.96mΩ·mm2, BVdss=-10.0V |
| | | | 7V | Vth=-0.95V, Rsp =8.56mΩ·mm2, BVdss=-14V |
| | | | 9V | Vth=-0.95V, Rsp =11.3mΩ·mm2, BVdss=-22V |
| | | | 12V | Vth=-0.95V, Rsp =14.4mΩ·mm2, BVdss=-27V |
| | | | 16V | Vth=-0.95V, Rsp =18.5mΩ·mm2, BVdss=-32V |
| | | | 20V | Vth=-0.95V, Rsp =22.4mΩ·mm2, BVdss=-36V |
| | | | 24V | Vth=-0.95V, Rsp =26.3mΩ·mm2, BVdss=-42V |
| | | | 30V | Vth=-0.95V, Rsp =54.5mΩ·mm2, BVdss=-55V |
| | | | 36V | Vth=-0.95V, Rsp =69.1mΩ·mm2, BVdss=-58V |
| | | | 40V | Vth=-0.95V, Rsp =79.1mΩ·mm2, BVdss=-60V |
| BJT | NPN | | 10V | Beta(Ic=100uA)=75, BVceo=32V, BVcbo=35V |
| | PNP | | 10V | Beta(Ic=100uA)=13, BVceo=-32V, BVcbo=-35V |
| CMOS | Nch | | 1.5V | Vth=0.63V, IdSat=410uA/um |
| | | | 5V | Vth=0.615V, IdSat=605uA/um |
| | Pch | | 1.5V | Vth=-0.61V, IdSat=-180uA/um |
| | | | 5V | Vth=-0.73V, IdSat=-290uA/um |
| Capacitor | MOS | | 1.5V | Cap./Area=9.7fF/um2 |
| | | | 5V | Cap./Area=2.85fF/um2 |
| | MIM | | 15V | Cap./Area=1.0fF/um2 |
| | | | | Cap./Area=1.5fF/um2 |
| | | | | Cap./Area=2.0fF/um2 |
| Diode | Schottky | | 9V | Vf=0.35V, Vr=-25V |
| | | | 24V | Vf=0.35V, Vr=-50V |
| | Zener | Lateral | | BV(@I=-1uA)=-6.25V |
| | | Vertical | | BV(@I=-1uA)=-5.75V |
| Poly Resistor | | | | Sheet Resistance=10Ω/□, 295Ω/□, 1KΩ/□, 2KΩ/□ |

## RESULTS AND DISCUSSIONS

The electrical parameters for those main devices are listed and summarized in the Table I. Experimentally, $R_{sp}$ of 4.6 mΩ·mm$^2$ with a $BV$ of 28.7 V was obtained for 12 V SFP-NLDMOS. And $R_{sp}$ of 11.4 mΩ·mm$^2$ with a $BV$ of 43.3 V was obtained for 24 V SFP-NLDMOS. Therefore, an ultra-low $R_{sp}$ has been achieved in our low-voltage SFP-NLDMOS. For 40 V device, $R_{sp}$ of 24.4 mΩ·mm$^2$ with a $BV$ of 58 V was also obtained. This performance was also competitive. The measured OFF-state and ON-state $I_{DS}$-$V_{DS}$ characteristics for 12 V and 24 V N-channel and P-channel SFP-LDMOS are shown in Figure 4 to Figure 7. Both LDMOS devices exhibited slight quasi-saturation under high $V_{GS}$. As illustrated in Figure 6, mild self-heating effect can be observed in 24 V device. The current of $I_{DS}$ at $V_{GS}$ of 5 V finally increases with $V_{DS}$ due to kirk effect.

978-1-6654-9759-6/22 $31.00 © 2022 IEEE      153

However, the *BV* at ON-state was larger than 26.4 V, as illustrated in Figure 6.

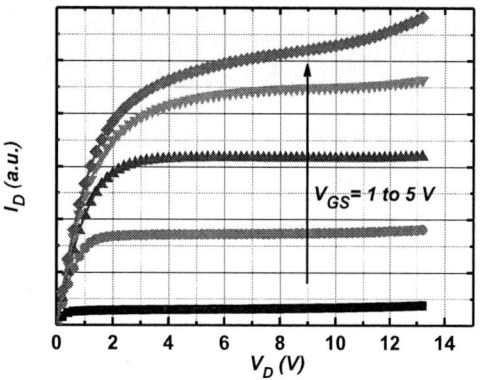

Figure 4. The measured $I_{DS}$-$V_{DS}$ characteristics for 12 V SFP-NLDMOS.

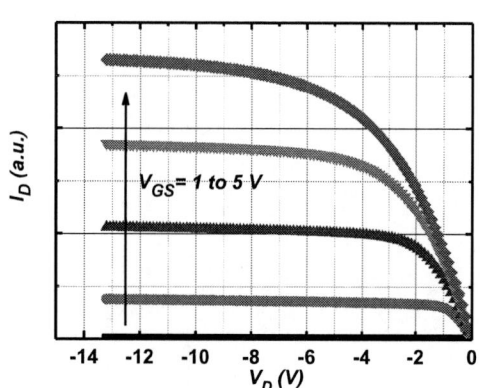

Figure 5. The measured $I_{DS}$-$V_{DS}$ characteristics of 12 V SFP-PLDMOS.

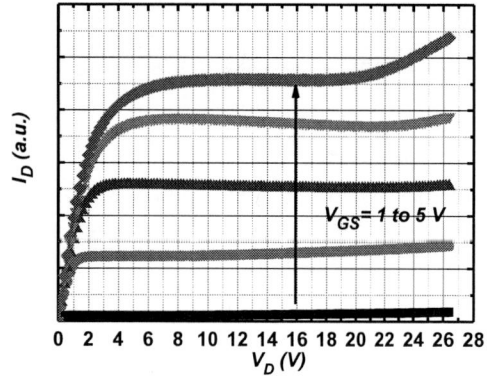

Figure 6. The measured $I_{DS}$-$V_{DS}$ characteristics of 24 V SFP-NLDMOS.

The electrical safe operating area (SOA) was also evaluated by using transmission line pulse (TLP) measurement. Figure 8 shows the measured electrical SOA of the single device of 24V SFP-NLDMOS with a width of 20 μm. The electrical SOA boundary can be obtained, as illustrated in the

Figure 8. The TLP characteristics of 12V/24V SFP-NLDMOS with power array of 200*100 μm was also shown in Figure 9 and Figure 10, respectively. It was indicated that the device can switch without suffering damage.

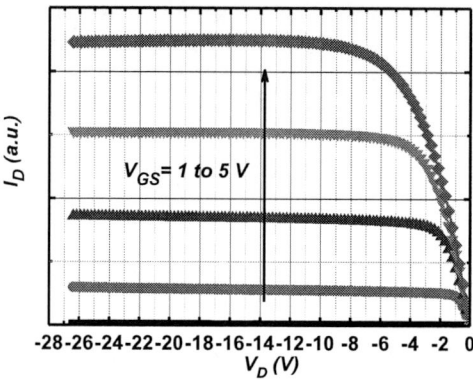

Figure 7. The measured $I_{DS}$-$V_{DS}$ characteristics of 24 V SFP-PLDMOS.

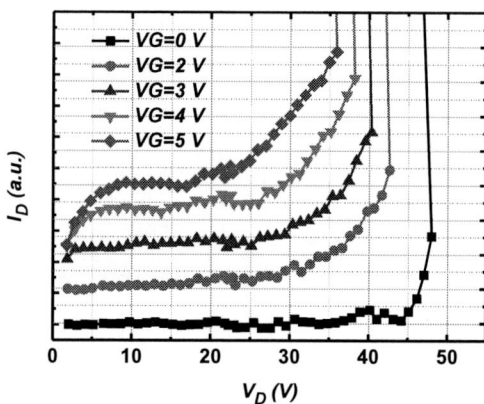

Figure 8. The measured TLP characteristics of 24 V SFP-NLDMOS with a width of 20 μm. (@ TLP pulse width of 100ns, rising time of 10ns)

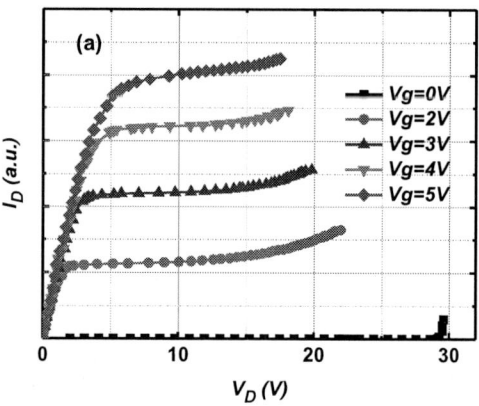

Figure 9. The measured TLP characteristics of 12V SFP-NLDMOS with array of 200*100 μm.

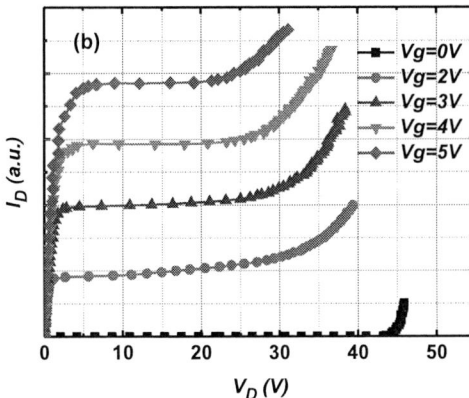

*Figure 10. The measured TLP characteristics of 24V SFP-NLDMOS with array of 200\*100 µm.*

The performance plot of $R_{sp}$ versus $BV$ of the SFP-NLDMOS was illustrated in Figure 11. The measured $R_{sp}$ was lower than those published results [2, 15, 16]. The $R_{sp}$ shows a competitive performance in the low-to-median voltage application.

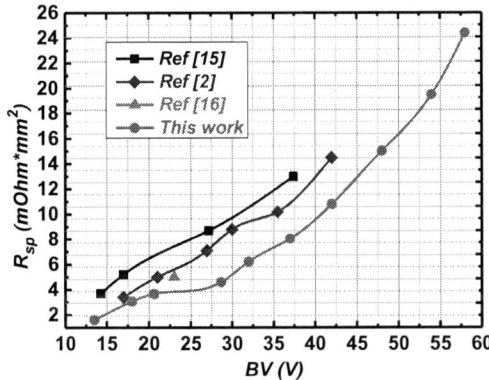

*Figure 11. The benchmark of SFP-NLDMOS and other existing LDMOS technologies.*

## CONCLUSION

A new BCD process was developed in a 300-mm 90-nm technology. It integrated standard 90 nm CMOS, 5 V CMOS, and 5 V to 40 V LDMOS device. Both drain-extended and SFP-based structures were employed for the LDMOS devices. Those LDMOS devices exhibited a very competitive on-resistance and BV. Experimentally, it was found that the 12 V SFP-NLDMOS showed $BV = 28.7$ V and $R_{sp} = 4.6 \text{m}\Omega \cdot \text{mm}^2$, which is lower than earlier published values. And $R_{sp}$ of 24.4 $\text{m}\Omega \cdot \text{mm}^2$ with a $BV$ of 58 V was obtained for 40 V SFP-NLDMOS. This BCD90 process platform can also provide analog components such as MIM capacitor, Zener diode, and high resistor for smart power IC.

## REFERENCES

[1] B. S. Kumar, Ajay, M. Paul, J. Somayaji, and M. Shrivastava, "Device, Circuit, and Reliability Assessment of Drain-Extended FinFETs for Sub-14 nm System on Chip Applications," *IEEE*

*Transactions on Electron Devices,* vol. 67, no. 11, pp. 4728-4735, 2020.

[2] F. Jin, D. Liu, J. Xing, X. Yang, J. Yang, W. Qian, W. Yue, P. Wang, M. Qiao, and B. Zhang, "Best-in-class LDMOS with ultra-shallow trench isolation and p-buried layer from 18V to 40V in 0.18µm BCD technology." pp. 295-298.

[3] T. Mori, S. Kubo, and T. Ipposhi, "A novel divided STI-based nLDMOSFET for suppressing HCI degradation under high gate bias stress." pp. 299-302.

[4] A. F. M. Alimin, H. H. Hizamul-din, S. F. W. M. Hatta, and N. Soin, "The influence of shallow trench isolation angle on hot carrier effect of STI-based LDMOS transistors." pp. 248-251.

[5] Z. Xu, J. Hu, Z. Fang, W. Duan, and W. Qian, "Investigation and Demonstration of Hot Carrier Effect in LDMOS Transistors with Ultra-Shallow Trench Isolation."

[6] S. Yanagi, H. Kimura, T. Nitta, T. Kuroi, K. Hatasako, S. Maegawa, K. Onishi, and Y. Otsu, "0.15µm BiC-DMOS technology with novel stepped-STI N-channel LDMOS," *Proc Ispsd*, pp. 80-83, 2009.

[7] T. Y. Huang, W. Y. Liao, C. Y. Yang, C. H. Huang, W. C. V. Yeh, C. F. Huang, K. H. Lo, C. W. Chiu, T. C. Kao, and H. D. Su, "0.18um BCD technology with best-in-class LDMOS from 6 V to 45 V," pp. 179-181, 2014.

[8] Z. Xu, D. Liu, J. Hu, F. Jin, X. Yang, W. Duan, W. Yue, Z. Fang, W. Qian, and W. Kong, "Demonstration of improvement of specific on-resistance versus breakdown voltage tradeoff for low-voltage power LDMOS," *Microelectronics journal,* vol. 88, no. JUN., pp. 29-36, 2019.

[9] D. G. Lin, S. L. Tu, Y. C. See, and P. Tam, "A novel LDMOS structure with a step gate oxide."

[10] S. Sharma, T. Letavic, Y. Shi, A. Loiseau, J. E. Monaghan, N. Feilchenfeld, R. Phelps, C. Lamothe, D. Cook, and J. Dunn, "Planar dual gate oxide LDMOS structures in 180nm power management technology." pp. 405-408.

[11] U. Manual, *Sentaurus Process/Device Version G-2012.06 Synopsys*, Mountain View, CA, USA, Jun. 2012.

[12] D. Varghese, P. Moens, and M. A. Alam, "on-State Hot Carrier Degradation in Drain-Extended NMOS Transistors," *IEEE Transactions on Electron Devices,* vol. 57, no. 10, pp. 2704-2710, 2010.

[13] S. Liu, W. Sun, Q. Qian, J. Wei, J. Fang, T. Li, C. Zhang, and L. Shi, "A Review on Hot-carrier-induced Degradation of Lateral DMOS Transistor," *IEEE Transactions on Device and Materials Reliability,* vol. PP, no. 2, pp. 1-1, 2018.

[14] I. Y. Park, Y. K. Choi, K. Y. Ko, C. J. Yoon, B. K. Jun, M. Y. Kim, H. C. Lim, N. J. Kim, and K. D. Yoo, "BD180 - a new 0.18µm BCD (Bipolar-CMOS-DMOS) Technology from 7V to 60V." pp. 64-67.

[15] T. Huang, W. Liao, C. Yang, C. Huang, W. V. Yeh, C. Huang, K. Lo, C. Chiu, T. Kao, H. Su, and K. Chang, "0.18um BCD technology with best-in-class LDMOS from 6 V to 45 V." pp. 179-181.

[16] J. Jang, K. Cho, D. Jang, M. Kim, C. Yoon, J. Park, H. Oh, C. Kim, H. Ko, K. Lee, and S. Yi, "Interdigitated LDMOS." pp. 245-248.

# Suppression of Poly Depletion Effect by Optimizing Ion Implantation and Rapid Thermal Annealing Process in 55nm Nor Flash Technology

*Hualun Chen[1], Jiahui Jiao[1*], Hu Wang[1], Haochun Xiao[1], Zhaozhao Xu[1], Lin Gu[1]*

[1]Huahong Semiconductor (WUXI) Limited, Wuxi 214028, China

[*]Corresponding author. Tel.: +86-15261805755. E-mail addresses: Jiahui Jiao@hhgrace.com

## ABSTRACT

Poly depletion effect (PDE) results in unstable PMOS device performance in 55nm nor flash technology. Optimized boron ion implantation and rapid thermal annealing process are proposed to suppress poly depletion effect by increasing the concentration of dopants at poly-gate oxide interface. It is experimentally revealed that the variation of PMOS devices can be significantly improved by 60%~95% through these optimizations. Among them, returning spike RTA, advanced process typically adopted at 55nm node, to soak RTA has great potential for completely suppressing PDE and improving device performance.

***Keywords— PMOS; poly depletion effect; device performance; optimized ion implantation; soak RTA***

## INTRODUCTION

Nor Flash memories with low power consumption have become particularly interesting due to their potential applications for mass storage electronic devices, such as mobile devices, Internet of things and wearable devices [1], [2]. For threshold voltage (Vth) reduction and process simplicity, the polysilicon in a PMOS device are typically doped after patterning by the same ion implantation processes used to form the source-drain regions [3], [4]. Large doses of P-type Boron impurities and short time scales of anneal process, such as spike rapid thermal annealing, are commonly employed on the source-drain structures in a 55nm nor flash technology [5]. However, this process does not yield sufficiently high active dopant concentration at the poly-gate oxide interface, leading to poly depletion effect [6], [7]. The PDE results in a strong correlation between the thickness of polysilicon and the Vth of PMOS devices. This problem is exacerbated as the gate oxide becomes thinner [8], [9]. When the gate oxide thickness is thinned to about 30 Å, the Vth drift is close to 50 mv, which results in unstable device performance. Reducing the thickness of poly can improve the PDE to a certain extent. Nonetheless, high-voltage devices and low-voltage devices share the same polysilicon in 55nm nor flash technology, which results in the necessary requirements for poly thickness. In this work, optimized ion implantation conditions and rapid thermal annealing (RTA) process are investigated in a 55nm nor flash technology. Experimental results show that these methods are authentically effective to suppress PDE, which is of great significance for the improvement of device performance.

## EXPERIMENTA

PMOS devices were fabricated on (110) p-type silicon wafers. A multi-step phosphorous ion implantation was performed through ~175 Å sacrificial oxide for n-well formation. After removing the sacrificial oxide, thin gate oxide was grown in a rapid thermal oxidation process. Subsequently, the gate poly films were deposited in a conventional LPCVD system with a final thickness of 0.2 μm. Following this, arsenic and fluorine were performed for halo and LDD formation. Finally, Boron was then implanted and annealed to form source-drain regions. The samples were separated into three groups according to different conditions. The first group of samples were implanted at different doses. The second group of samples were implanted at different energies. Both groups of samples were then thermally annealed using spike RTA. The third group of samples were annealed by two different processes of spike RTA and soak RTA.

## RESULTS AND DISCUSSIONS

Fig. 1 shows the correlation between poly thickness and Vth of the PMOS devices. There is basically a liner relationship between them due to the PDE, which results in a really bad sigma of wafer to wafer. Therefore, the PDE is also preliminarily characterized by the drift of the Vth caused by the variation of poly thickness in this experiment.

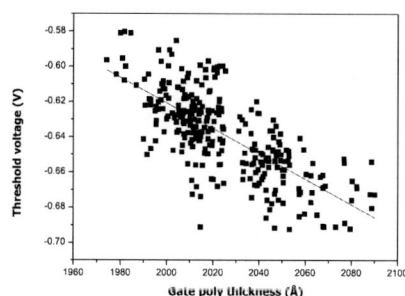

*Fig. 1. The correlation between the thickness of polysilicon and the Vth of the PMOS devices*

Concentration distributions of boron ions along the direction of poly-gate oxide of PMOS devices implanted at different conditions obtained by TCAD simulation are displayed in Fig. 2. As dose increases, the concentration of at the poly-gate oxide interface increases which may result in a reduction in PDE. The concentration at the poly-gate oxide interface also increases with increasing energy. Compared with the increase of boron ion dose, the change in ion concentration caused by the increase of energy is smaller. Increasing the annealing thermal of boron ions can also make the distribution of boron ions more uniform, as exhibited in Fig. 2(c). The concentration distribution of boron ions in poly basically keeps a straight line annealed by soak RTA.

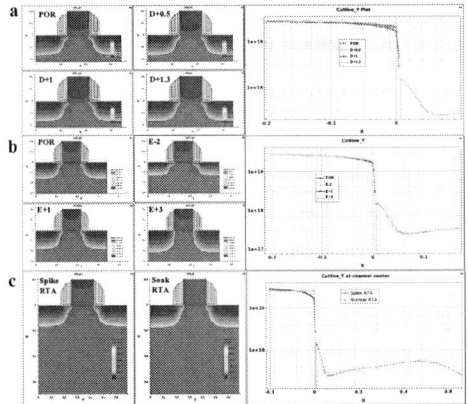

Fig. 2. Concentration distribution of boron ions along the direction of poly-gate oxide of samples with different (a) doses, (b) energies (c) RTA processes

To verify the TCAD simulations, Vths of the PMOS devices are tested. Fig. 3 demonstrates the Vth of the samples with different conditions and the Vth drifts are calculated. When boron ions are performed in process record (POR) condition, Vth drift up to 110 mv due to 200 Å reduction in poly thickness. From Fig. 3(a), it can be seen that the Vth drift becomes smaller and smaller as the boron ions dose increases, which means the PDE is reduced. Particularly, the variation of the Vth also becomes smaller with the dose increase, and tends to be stable finally. This result suggest that more incremental doses do not absolutely eliminate the Vth drift. The trend of Vths of samples with different energies is consistent with the trend of samples with different doses. Moreover, soak RTA shows excellent effect on Vth drift and the effect gets better with the increase in temperature. The Vth drift is reduced to only 5 mv. The improvement effect is greater than 90%.

The full mapping of Vth for PMOS devices with different implantation and annealing conditions are shown in Fig. 4. For POR condition, the Vth of PMOS located at the wafer far edge becomes abnormal high and the sigma within wafer is pretty bad. Although increasing the dose or energy can improve the sigma of wafer to wafer, the sigma within wafer cannot be improved sufficiently. There is still a high Vth of devices at wafer far edge. After replacing spike RTA with soak RTA, Vth of wafer far edge becomes normal and the sigma within wafer reduced to

5mv, as displayed in Table 1. The poor uniformity of the PMOS devices within the wafer is suspected to be caused by insufficient thermal. By the way, the optimization of ion implantation and annealing processes has no effect on NMOS and cell devices. After annealed by soak RTA, the p-type poly resistance shows acceptable variation.

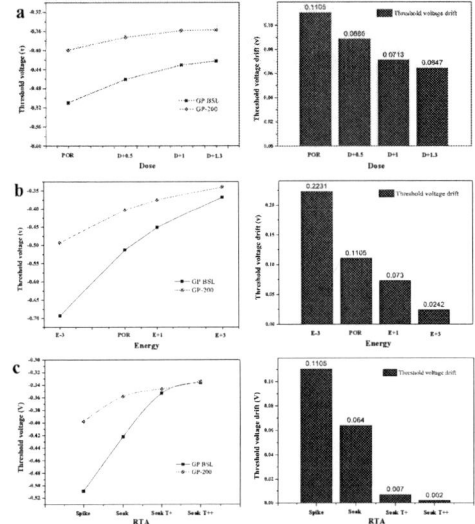

Fig. 3. Vth and Vth drift of the samples with different (a) doses, (b) energies (c) RTA processes

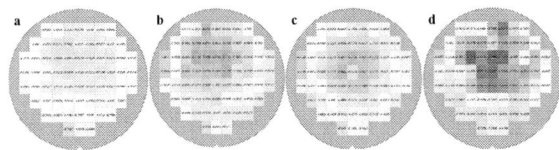

Fig. 4. Full mapping of Vth for PMOS devices with (a) POR, (b) boron dose+, (c) boron energy+ and (d) soak RTA

Table 1. Sigma of LVPMOS Vth between wafer to wafer and within wafer

| Condition | Vth (V) | Sigma (wafer to wafer) | Sigma (within wafer) |
|---|---|---|---|
| POR | -0.509 | 0.019 | 0.045 |
| Dose+1 | -0.431 | 0.011 | 0.017 |
| Energy+1 | -0.450 | 0.013 | 0.016 |
| Soak RTA | -0.331 | 0.004 | 0.005 |

A further convincing argument for the decrease in PDE after optimizing boron ion implantation conditions and annealing process is provided by capacitance-voltage curves with low frequency shown in Fig. 5. For PMOS, when the gate bias sweeps to a negative value with low frequency, the effect of poly depletion is observed obviously. Therefore, the inversion capacitance can be used to measure PDE. From Fig. 5, it can be seen that PMOS device with POR condition has the minimal inversion capacitance, which means the device suffers severe PDE. The optimization of boron ion implantation conditions can effectively suppress the PDE. Moreover, replacing spike RTA with soak RTA shows a better effect on suppressing PDE. These conclusions are consistent with the Vth drift results above.

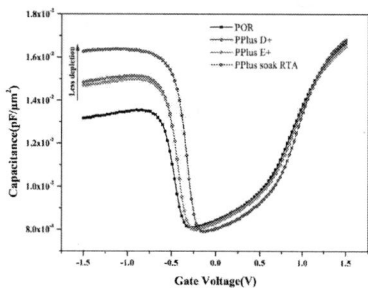

*Fig.5. C-V curves of the samples with different implantation and annealing conditions*

The PED can be explained by the model demonstrated in Fig. 6. $C_{dep}$ is the capacitance due to the poly depletion, $R_{dep}$ is the resistance in the depletion region, $R_{gate}$ is the resistance of poly, and $C_{ox}$ is the gate oxide capacitance. Such a structure is equivalent to connecting a capacitor in series due to poly depletion. When the concentration of boron ions is nonuniformity, the width of the depleted poly is larger. A higher Vth is required to turn on the channel resulted from higher $R_{dep}$. The measured capacitance will also be smaller because a capacitor is connected in series. Either increasing the dose and energy of boron, or using soak RTA, can reduce the width of the depleted poly, resulting in lower Vth and larger capacitance in inversion.

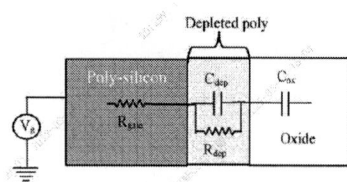

*Fig. 6. Poly-silicon depletion region model*

As the result of the Vth above, while suppressing the PDE in various ways, the device of PMOS will experience a large shift. In order to meet the device requirements, it is necessary to adjust the device to target through n-well, halo and LDD implantation. Such an alteration may cause reliability problems for the device, such as hot-carrier injection (HCI) and negative bias temperature instability (NBTI). The wafer level HCI and NBTI results of the PMOS devices adjusted to target are shown in Fig. 7.

Compared with POR condition, the HCI results of the samples with different boron dose or energy and annealing process are slightly worse. The current HCI has a very long lifetime and the margin is sufficient. The NBTI results of wafers with different splits are better than POR condition, as exhibited in Fig. 7(b).

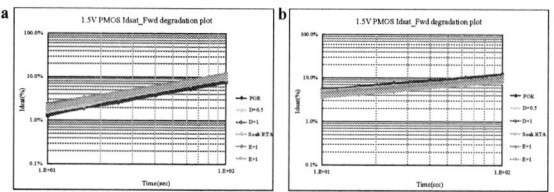

*Fig. 7 Wafer level HCI and NBTI Idsat degradation plot of samples with different conditions*

## CONCLUSION

The poor PMOS device performance due to PDE is undesirable in mass-produced 55nm nor flash technology. TCAD simulation results show that increasing the dose and energy of ion implantation or employing soak RTA instead of spike RTA can suppress PDE by enhancing the concentration of boron ions at the poly-gate oxide interface. Wafer acceptance test results indicate that the Vth drift of PMOS results from poly thickness variation is reduced with these methods. Meanwhile, the uniformity of device within wafer has been greatly improved. The C-V curve intuitively notarizes the improvement for PDE by optimizing ion implantation conditions and annealing process. Among these methods, soak RTA has the most significant improvement in PDE and the expected sigma is less than 5 mv. Considering the improvement effect and process cost, soak RTA is probably the best way to suppress PDE and improve device performance.

## REFERENCE

[1] Lai, K. Stefan. Journal of Nanoscience & Nanotechnology 12.10(2012): 7597.

[2] A. B. Vavrenyuk, V. V. Makarov, V. A. Shurygin. Advanced Technologies in Robotics and Intelligent Systems (2020).

[3] A. Blosse, M. Wei, V. Bhachawat. International Symposium on Vlsi Technology IEEE, 2007.

[4] E. Napolitani, G. Impellizzeri. Semiconductors and Semimetals 91(2015): 93-122.

[5] C. L. Huang, N. D. Arora. Electronics Letters 29.13(1993): 1208-1209.

[6] T. Aoyama, K. Suzuki, H. Tashiro. International Electron Devices Meeting IEEE, 1997.

[7] T. J. King, Electron Devices Meeting, 1990. IEDM '90. Technical Digest. International IEEE, 1990.

[8] T. J. King. Cheminform 25.48 (2010).

[9] S. D. Kim, C. M. Park, J. Woo. International Electron Device Meeting 2000.

# DATA RETENTION ENHANCEMENT OF MODERN 55NM NOR FLASH MEMORY

*Hualun Chen, Botong Liu\*, Lin Gu*

Huahong Semiconductor (Wuxi) Limited, Wuxi 214029, China
*Corresponding Author's E-mail: Botong.Liu@hhgrace.com

## ABSTRACT

Charge loss of NOR flash cell originates from H species diffused to interface of tunnel oxide and silicon substrate from CESL (contact etch stop layer) was reported. On the contrary to data loss from intrinsic defects of growth tunnel oxide, the observed data loss is directly related to CESL layer thickness. Hydrogen species released from PECVD (plasma enhanced chemical vapor deposition) of silicon oxy-nitride and silicon nitride diffuse into $Si/SiO_2$ interface, which may significantly affect interface states and bulk traps of tunnel oxide, which are responsible for NOR flash memory data loss. Stress exhibit negligible influence on data retention for presented study. By choosing proper CESL layer thickness, greatly enhanced data retention performance is achieved.

## INTRODUCTION

NOR type Flash, one of the most important non-volatile memories, which is vital for many applications that taking advantages of its fast read speed and random access properties [1]. The greatest demand for NOR flash memory is high density, high performance and reliability with the technology node keeps scaling [2]. Shrinking of cell device dimensions means small floating gate area, which induces decreasing numbers of electrons could be stored in FG and exhibits very narrow margins of data retention window [3].

Charge loss generally is regarded to electrons get out of FG through surrounding dielectric block layers. Since program and erase operation for NOR flash memory are electrons striking tunnel oxide, the charge loss mostly happened through bulk traps and interface of tunnel oxide [4]. The intrinsic traps generated during tunnel oxidation is well studied, which are dangling Si and Si-H bonds.

In addition to intrinsic defects of tunnel oxide growth, there are several processes that could generate bulk traps and interface states. Lee, et al [5] reported that sodium ions from bit line contact movement in sidewall spacers are responsible for charge loss of NOR flash cell. PECVD is used for deposition of CESL, which release H species during silicon oxy-nitride/silicon nitride (SiON/SiN) deposition process, since $SiH_4$ and $NH_3$ were used. As a result, some of released $H^+$ can diffuse to the oxide/Si interface or into bulk oxide [6]. There are some papers have been reported that the released H species from PECVD will lead to large device shift and reliability problems (TDDB and NBTI) of MOSFET devices [7]. H species diffuse to tunnel oxide through SiN liner significantly affect P/E endurance on split-gate type flash has also been reported [8]. CESL deposition released H species might have similar disadvantages on NOR type flash memory cells. However, there are barely reported about the influence of released H species on data retention of NOR flash, which is vital important for NOR flash memory of advanced technology node (below 55 nm). In this study, we reported the investigation regarding to H species from SiON/SiN deposition influence on charge loss and insights into mechanism of modern 55nm NOR flash memory.

## EXPERIMENTS AND METHODS

The experiments were done on 55nm-node 12 inch wafer. The main process flow for presented 55nm NOR flash device is shown in Figure1. The cell exhibits a ~10nm thick tunnel oxidation by ISSG (in-situ steam generation process), followed by FG poly deposition and control gate (CG) poly deposition, which are stacked poly layers and separated by inter-poly dielectric ONO. TEM image in Figure1b shows the structure after CG deposition. CESL was deposited by PECVD, including silicon oxy-nitride and silicon nitride films shows in Figure1c. The detailed experiments splits were shown in table1. HTDRB performance was evaluated by baking at 250 °C for 24h.

*Table1: Experimental splits combinations.*

| SIN \ SION | BSL | +30% | +60% | -10% | -20% |
|---|---|---|---|---|---|
| BSL | V | | | V | V |
| -20% | | V | | V | |
| -40% | | | V | V | |
| Stress+ | V | | | | |
| Stress ++ | V | | | | |

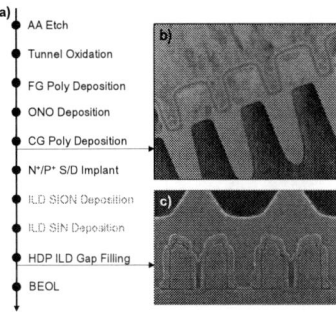

a)
- AA Etch
- Tunnel Oxidation
- FG Poly Deposition
- ONO Deposition
- CG Poly Deposition
- N⁺/P⁺ S/D Implant
- ILD SION Deposition
- ILD SIN Deposition
- HDP ILD Gap Filling
- BEOL

*Figure1: 55nm NOR Flash main process flow and corresponding SEM image after certain process.*

## RESULTS AND DISCUSSION

Figure 2a shows the programed and erased cell threshold voltage Vt at ultraviolet (UV) exposed initial state. HTDRB (high temperature data retention bake) was measured by checking of programmed cell Vt distribution after baking at 250 °C for 24h. As shown in Figure 2b, Vt distribution peak shifted about 0.7V after baking for the baseline condition, and when SiON thickness increased 30% and 60%, post bake Vt distribution peak shifted about 0.8 and 0.9V respectively. On the contrary, when the SiON film become thinner, Vt distribution peak shifted about 0.6V, indicating enhanced HTDRB performance. To illustrate influence of stress on HTDRB, post-bake Vt distribution of splits with stressed CESL layers were measured. From Figure 2c, the stressed wafers exhibit almost the same distribution and no obvious trend along with tress, indicating stress has negligible influence on HTDRB in presented experiments.

For HTDRB charge loss mechanisms, there are some well accepted models were reported, showing in Figure 3a. Trap assisted tunneling is regarded as main charge loss of post P/E cycling cells, which generally happened when there are lots of bulk traps exist. The interface trap are mainly caused by Si-X bonds (X might be H, OH, B), and electrons trapped and released during Si-X dangling bonds break and recovery [9]. Based on above models, we might propose that post-bake Vt shift of presented fresh cells mainly arise from interface states.

To further investigate the interface states induced charge loss and simplifying, wafer level time dependent dielectric break down and negative bias temperature instability measurements of test MOSFET were conducted. As shown in Figure 3b, $H^+$ ions can be released during PECVD SiN and SiON deposition. Considering the two reactions:

$$SiH_4 + N_2O + N_2 \rightarrow SiON \quad (1)$$
$$SiH_4 + NH_3 \rightarrow SiN \quad (2)$$

SiN deposition might release more hydrogen containing species than SiON, because of $SiH_4$ and $NH_3$ were used. As a result, some of $H^+$ can diffuse to the silicon/oxide interface, reacting with silicon dangling bonds form

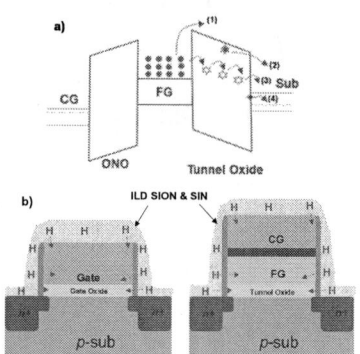

*Figure3: a) Schematic of charge loss mechanisms. (1) Thermionic emission, (2) De-trapping, (3) Trap assisted tunneling, (4) Interface states. b) Schematic of H species diffusion generated from PECVD SiON and SiN layers.*

interface states or bulk oxide trap states [10].

Figure4 a) and b) shows the TDDB Weibull distribution and NBTI lognormal distributions of test MOSFET with CESL SiON thickened 30% and 60% respectively. SiON+60% has no early fail and the distribution is more convergent than the other two conditions, suggesting low bulk traps. In addition, NBTI distribution for SiON+30% and SiON+60% are all worse than baseline condition, demonstrating increased interface states. Based on the above results, thicker SiON film lowers the bulk traps due to prevention of H species generated by SiN deposition, while more interface states were introduced because of increasing amount of $H^+$ released during thicker SiON deposition. Therefore, larger post-bake Vt shift was observed due to more interface states were introduced.

For the cases of SiON-10% and SiON-20%, TDDB become much worse as SiON thickness gets down, showing in Figure4c. Indicating numerous bulk defects formed from SiN deposition. NBTI is expressed as Id degradation, Id degradation for SiON-10% and SiON-20% are consistent with baseline condition (Figure4d). It is

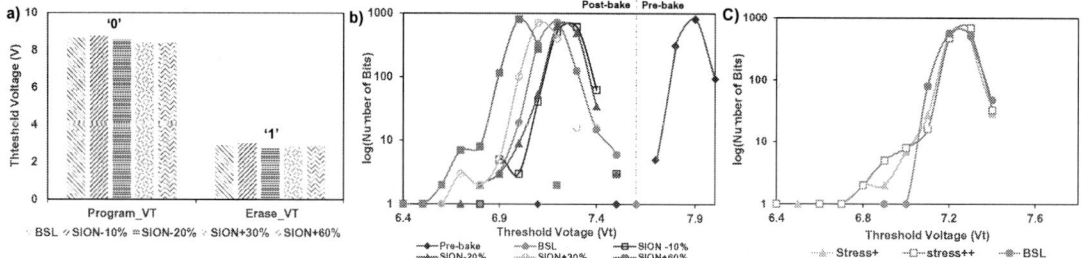

*Figure2: a) P/E window of fresh Cell. b) Threshold voltage distribution after high temperature bake 250 °C for 24h. Pre-bake Vt distributions are very similar for each other, so only one pre-bake curve was shown here. c) Post-bake threshold voltage distribution of stressed CESL splits.*

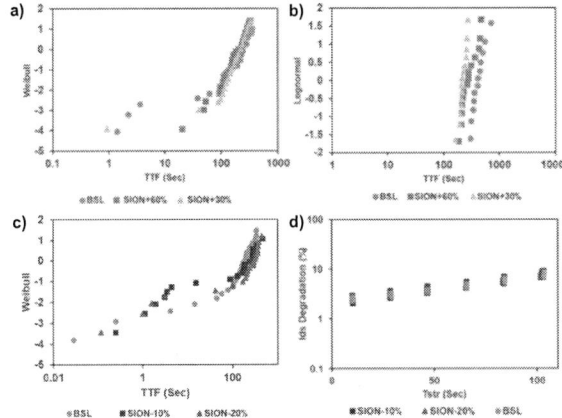

*Figure4: Wafer level reliability measurement. Weibull distribution a, c) and TTF lognormal distribution b) and Id degradation d) for test MOSFET.*

deduced that H species released from PECVD SiN diffuse to SiO₂ generates bulk oxide traps with thin layer of SiON block. Thin layer of SiON is not enough to prevent H species from SiN diffuse to Si/SiO₂ interface. H species from SiN contributes more on bulk traps generation since more H species released.

Thinner SiON layer is good for fresh cell HTDRB because of reduced interface states, while is bad for prevention of H species from SiN. In order to minimize H species diffused to Si/SiO₂ interface, SiN layer thickness need to be reduced simultaneously. Therefore, experiments with combination of thinner SiON and SiN layers were carried out. Figure5a shows the post-bake Vt distribution of splits with decreased SiN thickness. Post-bake Vt shift for SiN-20% is little bit smaller compared with baseline condition. Vt shifted more when further reducing SiN thickness, which may affect by plasma. Figure5 b) and c) are the TDDB and NBTI results, suggesting lower interface states and bulk traps formation. In summary, since lower interface states and bulk traps generated from PECVD released H species, decreasing SiON and SiN thickness greatly enhanced HTDRB performance without sacrifice endurance in the meantime.

## CONCLUSION

In conclusion, H species released from PECVD SiON and SiN deposition could diffuse into Si/SiO₂ interface, which may significantly affect interface states and bulk traps of oxide, which further affecting data retention performance. Stress exhibits negligible improvement on data retention under presented experimental conditions. It is revealed that SiN deposition generated H species contribute more to bulk traps formation since silane and ammonia were used, while from SiON deposition may introduce more interface states. By optimization of SiON and SiN thickness, greatly enhanced HTDRB performance and better reliability were obtained. All the above findings in this paper may have great potentials of applications in IC and memory device manufacture.

## REFERENCES

[1] Chih-Yuan Lu, Kuang-Yeu Hsieh, Rich Liu. *Microelectronic Engineering,* Vol 86, 2009, 283–286.
[2] Jang-Sik Lee. *Electronic Materials Letters,* Vol 7(3), 2011, 175-183.
[3] Helena Handschuh, Elena Trichina. *Workshop on Fault Diagnosis and Tolerance in Cryptography (FGTC),* 2007.
[4] Jae-Duk Lee, Jeong-Hyuk Choi, Donggun Park and Kinam Kim. *IEEE Transactions on Device and Materials Reliability,* Vol 4(1), 2004, 110-117.
[5] Wook H. Lee, Dong-Kyu Lee, Keon-Soo Kim, et al. *IEEE Transactions on Device and Materials Reliability,* Vol 1(2), 2001, 128-132.
[6] Yung-Yu Chen. Materials, Vol 7, 2014, 2370-2381.
[7] N.S. Saks, D.B.Brown. *IEEE Transactions on Nuclear Science,* Vol 37(6), 1990, 1624-1631.
[8] Ziyuan Liu, Shinji Fujieda, Fumihiko Hayashi, et al. *45th Annual International Reliability Physics Symposium,* 2007, 190-196.
[9] Matchima Buddhanoy, Sadman Sakib and Biswajit Ray. *IEEE International Reliability Physics Symposium,* 2021.
[10] M J. Chen, J.R. Shih, K.F. Yu, et al. *6th International Symposium on Plasma Process-Induced Damage,* 2001, 12-15.

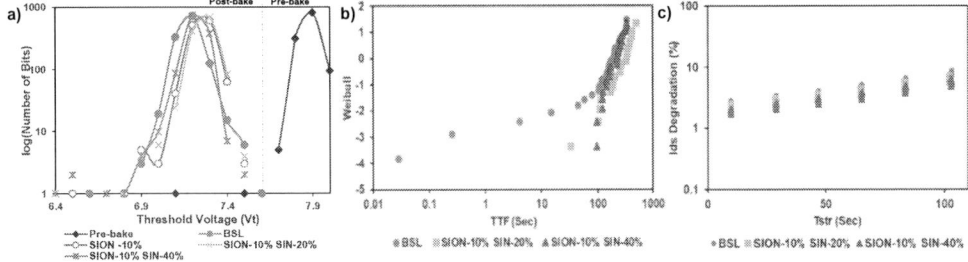

*Figure5: a) Threshold voltage distribution after high temperature bake 250 °C for 24h. Weibull distribution b) and ID degradation c) for test MOSFET.*

# Optimization of Integrated Process by reducing Dual Gate and Cell Lightly Doped Drain Photo Layer in ETOX NOR Flash Technology

*Hualun Chen[1], Yihang Du[1]\*, Lin Gu[1], Zhuangzhuang Wang[1], Hongjie Shen[1], Hu Wang[1], Jiuli Hu[1], Jiaming Pu[1]*

Wuxi Huahong Grace Semiconductor Manufacturing Corporation, Wuxi 214028, China

\*Corresponding Author's E-mail Addresses: *Yihang Du@hhgrace.com*

## ABSTRACT

The optimization of integrated process by reducing dual gate (DG) and cell lightly doped drain (CLDD) photo layer in ETOX-type NOR flash is proposed, which not only improves the reliability of periphery device, but also reduces two photo layer in the manufacture process. Further the production cycle and manufacturing cost are saved, which greatly elevated the competitiveness of ETOX-type flash as it scaling down to below the 45nm generation.

## INTRODUCTION

NOR-flash memory plays a dominant role in the code storage usage for the embedded non-volatile memory (eNVM), true wireless stereo (TWS), augmented reality (AR), virtual reality (VR) personal computer (PC), active matrix/organic light emitting diode (AMOLED) and the Internet of Things (IoT), thanks to its high-speed random access, specific byte write capability and fast read speed [1, 2]. Compared with other type of eNVM, the advantage of process is not obvious because of the fabrication procedures of erase through oxide (ETOX) NOR flash is too complicated and costly, which is predominantly in the multiple mask layer and photo layer. In order to promote the competitiveness of the ETOX-type flash in flash market, the reduction of the mask layer or photo layer is necessary. Simultaneously, the characteristics of flash cell and peripheral device after optimized process is needed to match process record (POR) condition.

In this paper, the optimization of integrated process by reducing DG and CLDD photo layer was proposed, and then the related characteristics of flash cell and peripheral device was investigated. The electrical performance and yield of samples with optimized process are not less than the POR samples, which is beneficial to reduce the production cycle and save manufacturing cost. Simultaneously, the competitive advantage of flash production is greatly elevated as the ETOX-type NOR flash memory scaling down to below 45nm generation.

## EXPERIMENTA

The schematic cross-section of POR and optimization process were depicted in Fig.1. The low voltage (LV) area was common defined by the LV p-well and LV n-well photo layer. After forming the implantation of LV area, the cell and high voltage (HV) area were sheltered by the DG photo resistance (PR), and then high voltage gate oxide (HV-GOX) which was used as the screen oxide on the LV area was removed by the DG wet etch, as depicted in Fig.1(a). As compared as POR, the optimized process flow was shown in Fig.1(b), after forming the LV p-well and before the PR was removed, the LV p-well oxide wet etch process was executed, which aim to remove the HV GOX on the LV p-well section. Similarly, the HV-GOX on the LV n-well section was removed by the LV n-well oxide wet etch process. The aforementioned process was used to skip the DG photo layer, and the process of DG loop of peripheral section were optimized. Subsequently, the LV gate oxide (LV-GOX) was deposited and the control gate (CG) was deposited.

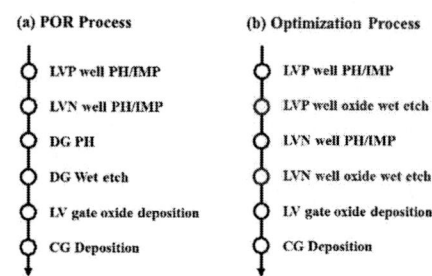

***Fig.1*** *Schematic cross-section of **(a)** POR and **(b)** optimization process*

Subsequent processing step were CG deposition, CG etch and CLDD implantation, as depicted in Fig.2 (a) and 2(b). Significantly, the cell area LDD and peripheral LDD process were defined by two photo mask and two photo layer in the ETOX structure. The CLDD PH was skipped in the process of CLDD implantation, as presented in Fig.2(c), and then the gate poly (GPL) was implanted the arsenic ions in peripheral section, taking into consideration the sheet resistance of poly (Rs p-poly) in the practical application, and the deviative degree of $R_S$ p-poly severely shifted target, the adjusted p-plus process was adopted, as depicted in Fig.2(d).

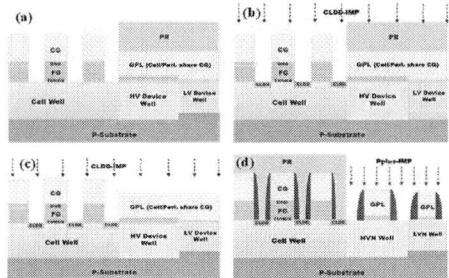

***Fig.2 (a)** CLDD PH and **(b)** CLDD p-type IMP of POR CLDD implanting process; **(c)** Skipped CLDD PH and **(d)** adjusted pplus-IMP of optimized process*

Subsequent processing step defined disconnected control gate at Y direction, gate poly, spacer, source/drain, contact, and all metal layer. Finally, the electrical characteristics of sample with optimization process were tested and estimated.

## RESULTS AND DISCUSSIONS

In order to explore the best condition of removing HV gate oxide on the LV area, the diluted hydrofluoric acid (DHF) and buffered oxide etch (BOE) were used to etch screen oxide on the LV p-well and n-well area after the corresponding well implantation. In the process of LV p-well oxide wet etch, the defect result of the sample with DHF treatment was poor, as depicted in Fig.3 (a). Judging from the defect image, the defect type was PR lifting, as shown in Fig.3 (b). By comparison, the defect map of the sample with BOE treatment was shown clear, as presented in Fig.3(c). According to the results of LV p-well screen oxide wet etch, the BOE was also used to etch the screen oxide on the LV n-well area, the results of cleared defect map were exhibited in Fig.3 (d). These results indicated that the samples was treated by BOE can effectively remove the screen oxide on the LV area.

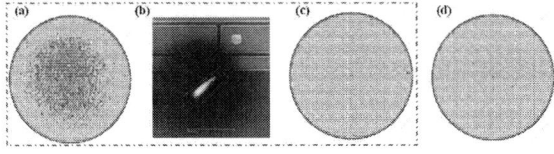

***Fig.3 (a)** DHF wet etch in LV p-well, **(b)** Defect review image, **(c)** BOE wet etch in LV p-well and **(d)** BOE wet etch in LV n-well*

The DG wet oxide etch was removed the HV gate oxide on the LV section, thus the electrical characteristics of LV device was checked, as presented in Fig.4. As compared with the POR samples, the $I_{DSC}$ degradation can be 6.4% and the $V_T$ promotion can be 7.6% of LV NMOS of samples with DG skip, as shown in Fig.4 (a) and 4(b). For the LV PMOS

of samples with DG skip, the $I_{DSC}$ degradation can be 7.2% and $V_T$ promotion can be 9.3% as compared with the POR samples, as exhibited in Fig.4(c) and 4(d).

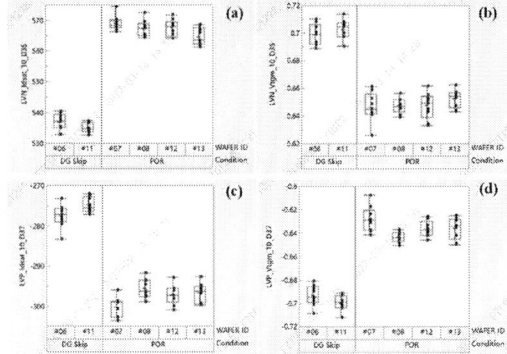

***Fig.4** The $V_T$ and $I_{DSC}$ of samples of DG Loop skip and POR. **(a)** $I_{DSC}$ and **(b)** $V_T$ of LVNMOS **(c)** $I_{DSC}$ and **(d)** $V_T$ of LVPMOS*

The equivalent oxide thickness (EOT) and inline thickness of LV-GOX are checked, as shown in Fig.5 (a) and 5(b). The EOT and inline thickness of samples with DG skip is thicker than POR samples. The dry strip process is executed after removing the HV-GOX on the LV area, which is the PR ashing process, triggering the growth of native oxide, which remain as part of LV-GOX, thus the LV-GOX thickness of samples DG skip is thicker than the POR sample.

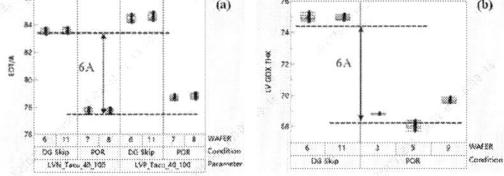

***Fig.5 (a)** EOT and **(b)** inline thickness of samples*

For the sake of adjusting the LV device characteristics to match the POR samples, the design of experiment (DOE) is implemented by adjusting the dosage of implanted ions. For the samples with DG skip, the $I_{DSC}$ and $V_T$ of the samples with LVP well $V_T$ dose-0.4 and LVN well $V_T$ dose-0.8, as shown in Fig. 6(a) and 6(b), is match the proposed target.

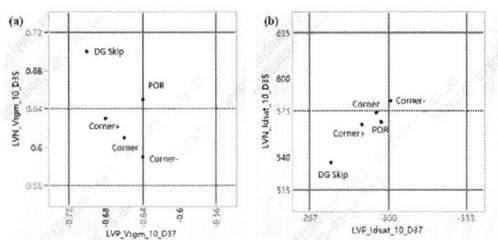

***Fig.6** LV device adjustment corner of **(a)** $V_T$ and **(b)** $I_{DSC}$*

The wafer level reliability (WLR) of samples with DG skip and POR are displayed in Fig.7, the LVN and LVP TDDB of samples with DG PH skip is better than the POR, as shown in Fig.7 (a) and 7(b). The LVN HCI of samples with DG photo skip is better the POR, as exhibited in Fig.7 (c). The LV PMOS NBTI of samples with DG PH skip and POR is matched, as depicted in Fig. 7(d).

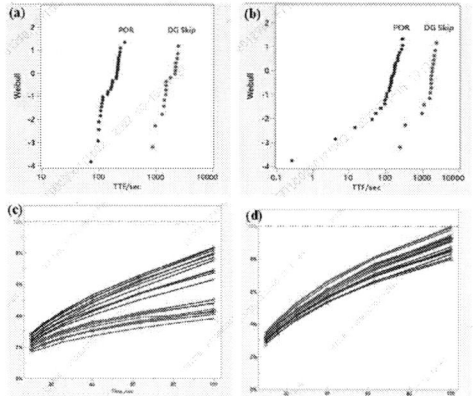

**Fig.7 (a)** *LVN TDDB,* **(b)** *LVP TDDB,* **(c)** *LV NMOS HCI and* **(d)** *LV PMOS NBTI of wafer level reliability*

The CP yield results of samples with DG skip and POR are depicted in Fig.8. The yield of samples with DG skip is not less than the POR samples, indicates that it is possible with the DG loop skip in the fabrication of ETOX flash.

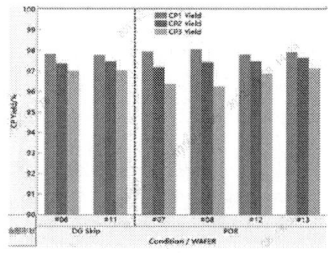

**Fig.8** *The CP yield of the DG Loop skip and POR samples*

The CG was deposited after growing the LV-GOX, which was shared by CG on the cell array and the gate poly (GPL) on the peripheral section. After CG dry etch, the CLDD was implanted. Traditionally, LDD process is utilized in MOSFET to reduce E-filed by forming a gradient drain junction near channel [3] and consequently decrease the HCI effect at drain side [4], nevertheless, the HCI was used for programming memory cell. Consequently, the LDD of cell and periphery section is necessary to process separately in order to meet the different applications.

Owing to the CLDD PH was skipped, the periphery section was exposed to CLDD blanket implantation, thus the GPL of periphery section was doped by CLDD implantation, and the device characteristics of periphery section would be effected. It's for that reason, as well as for periphery device adjustment, that the split of p-plus implanted ions combined with CLDD PH skip was arranged at several dosages. As presented in Fig.9(a) and 9(b), the characteristics of NMOS was no significant difference with the increasing dosage of p-plus implantation. Nevertheless, the increasing $V_T$ of

PMOS was presented with the increasing dosage of p-plus implantation, as exhibited in Fig.9 (c) and 9(d). The reason for that the poly depletion effect (PDE) of PMOS was dominated, the flat GPL implanted by arsenic species when the CLDD PH was skipped, the implanted arsenic ions of GPL were compensated by the subsequent p+ dopant plus implantation, leading to the increased PDE.

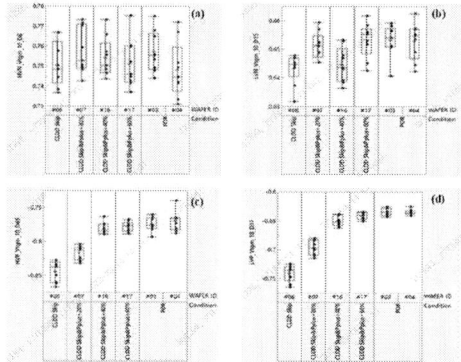

**Fig.9** *The $V_T$ of samples of CLDD PH skip and POR.* **(a)** *HV NMOS,* **(b)** *LV NMOS,* **(c)** *HV PMOS and* **(d)** *LV PMOS*

Compared with POR condition, the BIN49 fail ratio of the samples with CLDD PH skip is 66.3%, as depicted in Fig.10(a). The BIN49 is monitored the variation of p-type poly resistance, the results $R_S$ p-poly of samples with CLDD PH skip and POR are presented in Fig.10(c), the $R_S$ p-poly of samples with CLDD PH skip is larger than POR.

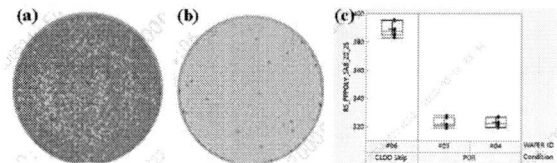

**Fig.10** *The yield distribution of samples* **(a)** *CLDD PH skip and* **(b)** *POR condition,* **(c)** *the $R_S$ p-poly with SAB*

The voltage needed for array operations have to be regulated within rather narrow ranges to ensure reliable sensing and narrow programmed and erased cell $V_T$ distributions, hence the bandgap reference (BGR) circuits is adopted [5], as shown in Fig.11. The basic idea of BGR is to add a proportional to absolute temperature (PTAT) voltage to the emitter-base voltage ($V_{BE}$) of a parasitic PNP transistor, so that the first-order temperature dependency in $V_{BE}$ is compensated by the PTAT voltage, resulting in a nearly temperature-independent output voltage [6]. The bandgap reference voltage can be expressed as [7]:

$$V_{ref} = V_{BE1} + \frac{R2}{R3}\Delta V_{BE} = V_{BE1} + \frac{R2}{R3}\ln(n \cdot \frac{R2}{R3}) \cdot V_T \quad (1)$$

where $V_{BE}$ is the voltage of emitter-base voltage of transistor, $\Delta V_{BE}$ is $V_{BE}$ difference between Q1 transistor and Q2 transistor, R1, R2 and R3 is fixed resistor, *n* is the emitter area ratio of the BJT transistor Q1 and Q2, $V_T = kT/q$, in which $k$, $T$ and $q$ are respectively represent the boltzmann's constant, absolute temperature and charge of the electron.

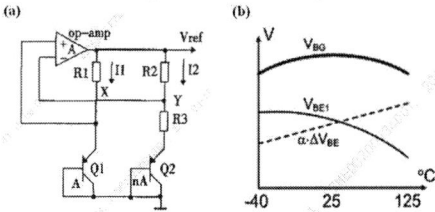

**Fig.11 (a)** *The structure and* **11 (b)** *the principle of normal circuit bandgap voltage reference*

According to the above conducted results, the $R_S$ p-poly is played an important role in fixed resistor of the BGR circuit. The implanted arsenic ions of GPL in the samples with CLDD PH skip are compensated the p+ dopant plus implantation, consequently, the $R_S$ p-poly is larger than POR. When the $R_S$ p-poly is shifted severely, which will decrease the BGR output voltage variation against temperature, leading to the supply voltage of periphery circuit is become unstable. Therefore, the samples with CLDD PH skip is presented the BIN49 fail in CP1 test.

In order to adjust the $R_S$ p-poly to match POR, and then the condition of increasing dosage of p-plus implanted ions was executed. The $R_S$ p-poly is significantly decreased with the increasing dosage of p-plus implanted ions, increasing the implanted dosage to p-plus+40% or 60% can effectively match the resistance of POR samples, and then the samples have no suffered BIN49 fail ratio, as exhibited in Fig.12(a). The CP yield of samples with DOE is shown in Fig.12 (b), the yield of sample with CLDD PH skip combine with p-plus+40% dosage is match the samples with POR when the $R_S$ p-poly are matched.

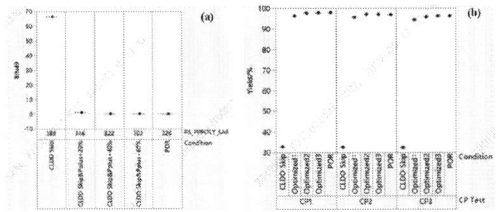

**Fig.12 (a)**. *The correlation of BIN49 fail ratio and dose of p-plus implantation* **(b)** *The CP yield of the CLDD PH skip and POR samples*

In the light of the results of periphery device and the yield, the characteristics of PMOS is more sensitivity than NMOS in the condition of CLDD PH skip. Taking these factor into consideration, and WLR of PMOS samples with CLDD PH skip and POR are displayed in Fig.13. For the samples with CLDD PH skip combine with p-plus+40% and +60% dosage, the TDDB characteristics of Core PMOS is better than POR,

as presented in Fig.13 (a). As for the TDDB characteristics of HV PMOS is match the POR, as depicted in Fig.13 (b). The NBTI characteristics of LV and HV PMOS is match POR, as shown in Fig.13(c) and 13(d).

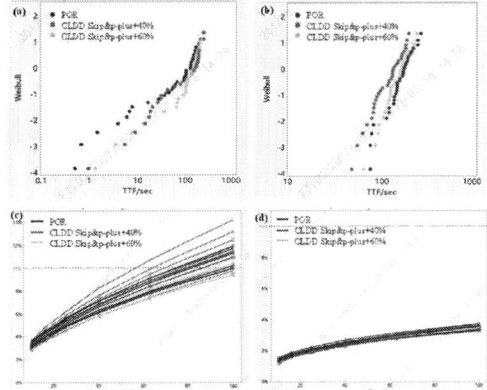

**Fig.13 (a)** *LVP TDDB,* **(b)** *HVP TDDB,* **(c)** *LV PMOS and* **(d)** *HV PMOS NBTI of wafer level reliability*

## CONCLUSION

The integrated process by reducing the DG and CLDD photo layer and related process optimization is proposed and analyzed in this paper, as a technology candidate to improve the competiveness of ETOX-type NOR flash by shortening production time and saving manufacturing cost. Experimental results of DG PH skip and related optimize process demonstrated the better performance and reliability than the POR, and the results of CLDD PH skip and related optimize process testified the matching characteristics and WLR of POR. Optimally integrated process by reducing the two photo layer and related ameliorative process can be a technology option of the next generation of ETOX-type flash.

## REFERENCE

[1] J. Meena, S. Sze, U. Chand, and T. Y. Tseng, Nanoscale Res. Lett., vol. 9, 2014, pp. 526.

[2] J.M. Yu, J.Y. Park, J. Hu, D.H. Yun and Y.K. Choi, IEEE TRANSACTIONS ON NANOTECHNOLOGY., vol. 18, 2019, pp. 1110-1113.

[3] Y.Cai, Z. Xing, H. Ru, IEEE., vol. 11, 2011, pp. 207-210.

[4] D. R. Nair, S. Mahapatra and J. D. Bude, IEEE Trans. Electron Devices., vol. 51, 2004, pp. 701–707.

[5] X.K. Guan, X. Wang, A. Wang and B. Zhao, Analog Integr Circ Sig Process., vol. 62, 2010, pp. 113-119.

[6] H. Banba, H. Shiga, A. Umezawa and S. Atsumi, IEEE Journal of Solid-State Circuits., vol. 34,1999, pp. 670–674.

[7] P. Malcovati, F. Maloberti and M. Pruzzi, IEEE Journal of Solid-State Circuits., vol. 36, 2001, pp. 1076–1081.

# IMPROVEMENT OF DISTURB AND ENDURANCE IN NOR FLASH MEMORY

Hualun Chen[1], Ran Xu[1*], Hui Wang[1], Xiaobing Ren[1], Xiaojun Xu[1] and Jian Zhang[1]

[1]Wuxi Huahong Grace Semiconductor Manufacturing Corporation, Wuxi 214028, China

*Corresponding Author's Email: Xu.Ran@hhgrace.com

## ABSTRACT

The optimization methods of embedded NOR flash memory disturb and endurance characteristics are discussed in this paper. By optimizing the germanium implant (Ge IMP) process, the dislocations under flash cell can be reduced to improve the flash memory disturb. Additionally, the flash memory endurance is enhanced by optimizations of world Line (WL) erase efficiency and control gate (CG) and floating gate (FG) coupling efficiency, which can expand the window of read currents.

## INTRODUCTION

NOR Flash memory, which is suited for code and word addressable data storage, is one of dominate forms of nonvolatile memory (NVM) [1]. It is a benefit to both automotive-level applications and consumer-level applications, such as various integrated circuit (IC) smart-cards and system-on-chip applications. Reliability characteristics, especially disturb [3-5] and endurance [6-8], are the key parameters to evaluate the performance of flash memory. In order to meet more complex electronic application, improvement of reliability of flash memory is an important topic for us to discuss.

In this paper, the optimization methods of embedded NOR flash memory disturb and endurance characteristics are proposed. By optimizing the Ge IMP process, the dislocations under flash cell can be reduced to improve the flash memory disturb. Additionally, the flash memory endurance is enhanced by optimizations of WL erase efficiency and coupling efficiency between CG and floating gate FG, which can expand the window of read currents depending on the readout-current ($IR_{10}$) and leakage-current ($IR_{01}$).

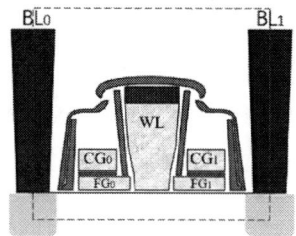

Figure 1: Schematic diagram of select-gate shared split-gate flash memory cell

We discuss WL shared split-gate flash memory cell in this paper showed in Figure 1, which realizes the optimization of bit size by sharing a WL and a group of source and drain for two bits in a flash cell unit. Additionally, over-erase [9] problem is also avoided with split-gate structure compared with traditional electron tunnel oxide (ETOX) stacked flash memory.

## IMPROVEMENT OF DISTURB

The bin2 failure is found caused by program disturb in chip probe (CP) test. The test condition is showed in Table1. By checking flash bits state under mass punch through (MPT), the bit is disturbed if the state is changed from erased status to programed status.

Table1 MPT stress condition

| Item | BL₀ (V) | BL₁ (V) | WL (V) | CG₀ (V) | CG₁ (V) | Stress time(ms) | Background |
|------|------|------|------|------|------|------|------|
| MPT | 5.9 | 0.2 | 0 | 0 | 0 | 500 | Erased |

In order to investigate the fail mode of bin2 failure, the electronic fail analysis (EFA) is used, and the results as showed in Figure 2. The contact (CT) via of failure bit is obviously brighter than others, which means that the leakage current of failure bit is bigger than other bits. Meanwhile, there is a dislocation under the flash cell observed in Figure 2 (b). What's more, we find there are several similar dislocations under other failed bits. Therefore, we tend to suspect that the dislocations is the root cause of the bin2 failure.

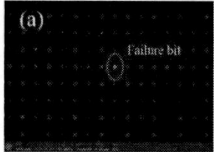

Figure 2: (a) The top view scanning electron microscope (SEM) of CT via arrays and (b) the cross section transmission electron microscope (TEM) of flash memory cell

By using physical failure analysis (PFA) in relevant process steps, we observe that the process of dislocations is between Spacer-1 Etch (ET) and Spacer-2 ET, as showed in Figure 3.

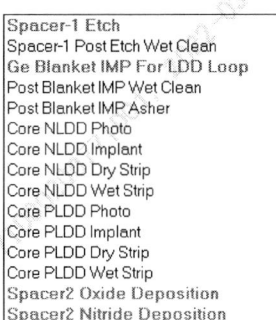

Spacer-1 Etch
Spacer-1 Post Etch Wet Clean
Ge Blanket IMP For LDD Loop
Post Blanket IMP Wet Clean
Post Blanket IMP Asher
Core NLDD Photo
Core NLDD Implant
Core NLDD Dry Strip
Core NLDD Wet Strip
Core PLDD Photo
Core PLDD Implant
Core PLDD Dry Strip
Core PLDD Wet Strip
Spacer2 Oxide Deposition
Spacer2 Nitride Deposition

*Figure 3: The Flow of NOR flash*

In order to improve the dislocations under failed flash cells, we designed three groups of experiments as showed in the Table2.

*Table2 Three groups of experiment splits*

| No. | Process | Experiment Splits | Result |
|-----|---------|-------------------|--------|
| 1 | Spacer-1 ET | (a)Main ET RF Bias Power - | Not work |
|   |   | (b)Skip over ET | Not work |
| 2 | Ge IMP | (a)Ge IMP Dose - | Good |
|   |   | (b)Skip Ge IMP | Good |
| 3 | Spacer-2 Ox Dep | (a)Ox film thickness + | Not work |
|   |   | (b)Ox Dep by HTO | Good |

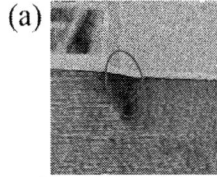 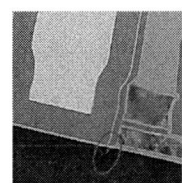

*Figure 4: The TEM of Spacer-1 ET splits: (a) Main ET RF- and (b) Skip over ET*

Considering that the dislocations may cause by dry etch damage, we design Spacer-1 Etch splits. The dislocation are not significantly improved by reducing main etch RF bias power, which can reduce the energy of etch ions, showed in Figure 4(a), or by skipping the over etch step showed in Figure 4(b). Therefore, there is no strong correlation between dislocation formation and Spacer-1 Etch process.

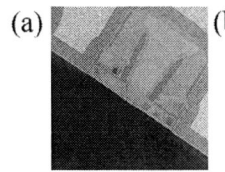 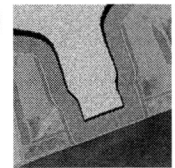

*Figure 5: The TEM of Ge blanket IMP splits: (a) Ge IMP Dose - and (b) Skip Ge IMP*

The dislocations are obviously optimized by reducing the dose of Ge blanket IMP showed in the Figure 5. By

forming pre-amorphization lattices, the Ge blanket IMP can prevent core lightly doped drain (LDD) IMP channeling in peripheral logic circuit regions. While in flash cell regions, Ge blanket IMP induces forming of dislocations.

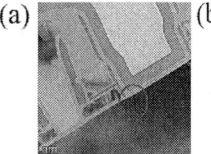 

*Figure 6: The TEM of Spacer-2 deposition splits: (a) Ox film thickness+ and (b) Ox Dep by HTO*

As showed in the Figure 6(a), the dislocation are not optimized by increasing Spacer-2 Oxide (Ox) film thickness. However, by improving the quality of Spacer-2 Ox films with high temperature oxide (HTO) process, the dislocations can be optimized obviously showed in Figure 6(b). The film stress between Spacer-2 SIN and substrate can be released by extra thermal budget in HTO process, which optimized dislocations under flash cell. However, it is not a good way to improve dislocations, for device characteristics may shift due to extra thermal budget.

To sum up, the dislocations can be improved by optimization of Ge blanket IMP and Spacer-2 Ox deposition. To balance the side effect on logic devices, we add a photo mask before Ge IMP to optimize dislocations, which ensures that Ge IMP only occurs in logic circuit regions instead of flash cell regions.

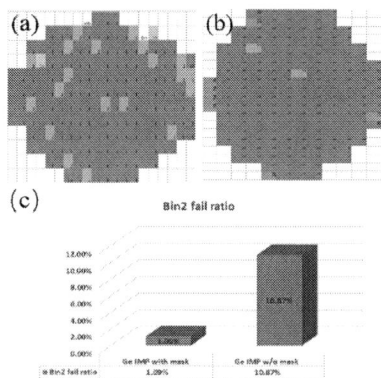

*Figure 7: CP1 results of samples (a) without Ge IMP mask and (b) with Ge IMP mask, and (c) histogram of bin2 fail ratio.*

The CP1 results are showed in Figure 7. Compared with the samples without Ge IMP mask, bin2 failure ratio of sample with Ge IMP mask decreased from 10.87% to 1.09%. By adding a photo mask before Ge IMP, which ensures that Ge IMP only occurs in logic circuit regions, the dislocation under flash cells can be obviously reduced to improve the flash program disturb.

## IMPROVEMENT OF ENDURANCE

The flash memory endurance is enhanced by optimizations of coupling efficiency between CG and FG, which expand the window of read currents depended on the $IR_{10}$ and $IR_{01}$.

NOR flash devices typically use channel hot electron (CHE) injection to program FG and Fowler – Nordheim (FN) tunneling to erase FG. The NOR flash memory transistor can be viewed as a capacitively coupled network as illustrated in Figure 10. The voltage on the CG will be capacitively coupled to FG, which cause the movement of electrons to FG. Coupling efficiency between CG and FG depends on coupling ratio $\lambda_g$, would be $C_g/C_t$, where $C_t = C_g + C_w + C_b + C_s$.

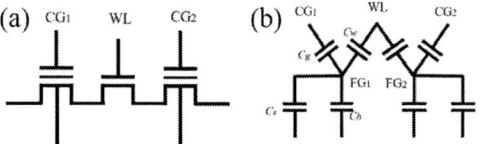

*Figure 10: Nor flash: Schematic diagram of (a) circuit symbol and (b) capacitor model*

To improve coupling efficiency, we optimize flash cell structure as showed in the Table 3. By shrinking offset spacer (OFFSP) length and optimize flash cell profile, we improve ratio of $\lambda_g$ from 69.57% to 73.33%, which dominates coupling efficiency between CG and FG.

*Table 3 Optimizations of flash cell structure*

| Condition | CG (nm) | FG (nm) | OFFSP (nm) | WL (nm) | $\lambda_g$ ratio |
|-----------|---------|---------|------------|---------|-------------------|
| Baseline  | 80      | 115     | 20         | 65      | 69.57%            |
| Improved  | 77      | 105     | 16         | 65      | 73.33%            |

The WAT results of optimized flash cell are showed in the Figure 11. Compared with the baseline samples, $IR_{01}$ of optimized flash cell decreases one order of magnitude, while $IR_{10}$ stay at the same level. It means that the window of read currents has expanded by optimizing flash cell, which improve NOR flash endurance.

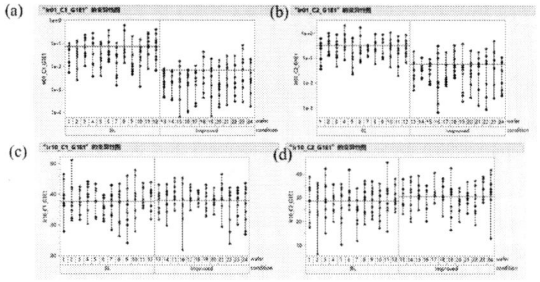

*Figure 11: WAT results of optimized and baseline samples: $Ir_{01}$ of (a) bit1 and (b) bit2 and $Ir10$ of (c) bit1 and (d) bit2*

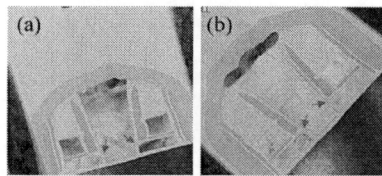

*Figure 12: The TEM of flash memory cell: (a) baseline sample and (b) FG tips improved sample.*

Another important optimization of flash cell is improving FG tips between FG and WL showed in the Figure 12, which can increase WL erase efficiency by enhancing WL localized electric field around FG tips. For FN tunneling, in fact, is a field – assisted electron - tunneling mechanism. In flash reliability evaluation, all optimized FG tips samples with sharper point could pass 300K endurance evaluation, while a few samples with baseline FG tips fail 100K endurance evaluation.

## CONCLUSION

In this paper, the optimization methods of embedded NOR flash memory disturb and endurance characteristics are proposed. By optimizing the Ge IMP condition, the dislocation under flash cells can be reduced to improve the flash memory disturb. Additionally, the flash memory endurance is enhanced by optimizations of WL erase efficiency and coupling efficiency between CG and FG. Finally, those optimization methods provide a solution for more complex electronic applications, such as automotive-level applications, which require high reliability.

## REFERENCES

[1] R. Bez, E. Camerlenghi, *Proc IEEE*, Apr. 2003, pp. 489-502.

[2] T. Endoh, H. Iizuka, *IEDM Tech. Dig.*, Dec. 1992, pp. 603-606.

[3] T. Endoh, K. Shimizu, *IEEE Trans.Electron Devices*, Jan. 1998, pp. 98-104.

[4] J. W. McPherson, H. C. Mogul, *Proc. IRPS*, 1998, pp. 47-56.

[5] M. Kato, N. Migamota, H. Kume, A. Sato, T. Adachi, M. Ushigama, *IEDM Tech. Dig.*, 1994. pp. 45-48.

[6] K. Robinson, *Electronic Component News*, Nov. 1989, pp. 167 – 169.

[7] R. Bez, D. Cantarelli, L. Moioli, *IEEE Electron Device Lett.*, 1998, Vol. 19, No. 2, p. 37.

[8] N. Tsuji, N. Ajika, K. Yuzuriha, Y. Kunori, M. Hatanaka, *IEDM Tech. Dig.*, 1994, p. 53.

[9] Y. Ma, C. S. Pang, K. T. Chang, S. C. Tsao, J. E. Frayer, K. Taehyoung, *IEEE International Electron Devices Meeting*, Dec.1994,pp. 57-60, 11-14.

# PROCESS OPTIMIZATION AND YIELD IMPROVEMENT FOR ERASE FAILURE IN EMBEDDED FLASH MEMORY

*Hualun Chen[1], Yang Dang[1]\*, Hui Wang[1], Xiaojun Xu[1], Jian Zhang[1] and Yunlong Huang[1]*
[1]Wuxi Huahong Grace Semiconductor Manufacturing Corporation, Wuxi 214028, China
\*Corresponding Author's Email: Yang.Dang@hhgrace.com

## ABSTRACT

This work is aimed at the yield improvement with root cause definition on erase failure inside flash cell by novel process optimization in embedded flash memory device. It is analysis that the root cause the very gross erase failure was mainly due to the abnormal short circuit between bit lines (BL) and floating gate (FG) ploy inside flash matrix cell. Furthermore, it turned out that coupling ratio of FG and control gate (CG) which impacts programing and erasing of flash cell is one of big modulators to flash performance. To tackle these problems, a new approach to eliminate the short circuit by simply process optimization without increasing cost. More importantly, the coupling ratio of FG/CG inside flash cell was carried out so as to increase read current window, leading to the successful performance improvement. Our efforts demonstrate that the erase defect has been eliminated and the flash yield has reached 95.6% through process and structure optimizations.

## INTRODUCTION

Recently, it has been paid significant attention with strong efforts to develop the flash memory devices. The explosively growing demand of the electronics forces device manufacturers to integrate together on the same chip embedded flash memory with high performance logic functions for low voltage and low power operations. Therefore, a lot of studies have been done to explore new structures and high performance for industrial application [1-3]. Among these researches, dual-bits split-gate flash memory has received extensive attention due to its excellent application prospects. It shows not only the high density due to the 2-bits cell storage capability by sharing the select-gate with two separated bits, but also its competitive bit size by using advanced 55nm node [4, 5].

However, the process difficulty sharply increases with the improvement of process node [6]. Some disadvantages give the roadblocks to mass production. Therefore, high yield is expected and yield up has become an important task to cope with the increasingly complex process and strict requirements.

This paper will mainly focus on the yield improvement with successful root cause definition on the one of major failures, erase failure, inside the flash matrix cell by key process optimization in embedded flash memory device. By discussing the mechanism behind on the failure, effective process and structure optimization is performed to perfectly solve the erase failure and mass production has been realized.

## EXPERIMENTAL

The devices in this work were fabricated using embedded flash memory. A brief cross-section drawing of the embedded flash memory was illustrated in Figure 1a. There are two FGs (FG1 and FG2), two CGs (CG1 and CG2) and a common word line (WL) in one flash cell. The first step of flash cell fabrication is coupling oxide deposition. The FG poly-Si is deposited on the oxide, and subsequently the oxide-nitride-oxide (ONO) film is formed which act as insulator between CG and FG. Then the CG and a thick nitride film are deposited. The region of flash cell is defined by a lithographic step (FG photo). A tetraethylorthosilicate (TEOS) spacer (FGSP) is formed by etching back as a hard mask to etch CG poly-Si. After forming offset nitride and offset oxide spacers, the FG is etched and the length of FG is determined by the two spacers. With tunneling oxide and WL poly-Si deposition, the flash cell structure has been completely formed. As shown in Figure 1b, a hard mask for logic area etching is deposited before CG contact (CGCT) formation. After logic area is formed, the flash area is exposed by lithography. Then the CGCT etching is executed to remove the CG and FG in flash area.

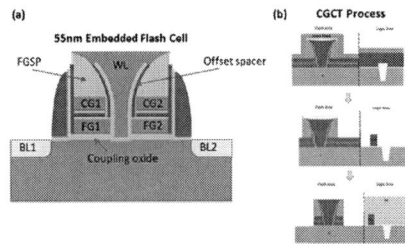

*Figure 1 (a) Cross-section drawing of embedded flash memory cell (b) CGCT process diagram*

## RESULTS AND DISCUSSIONS

From the production LOTs, the very massive erase failures with circular distribution were observed (Figure 2). Such a failure mode causes a yield loss of 95%. The CP result showed that the read current of twin bits with a common word line (WL) is low after erase (Figure 2b). Based on electrical analysis on a failure die, we could observe that erase failure is mainly manifested in low read current after erase. In order to reveal the root cause of these problems, product failure analysis (PFA) was executed. The PFA results indicated that a residue

978-1-6654-9759-6/22 $31.00 © 2022 IEEE

between BL and FG can be seen (Figure 2c), which can be attributed to the erase failure. EDS was using to analysis the residual components and the result (Figure 2d) revealed that the residue is FG poly-Si. The mechanism of the FG poly-Si residue formation was presented in Figure 3. With the flash area nitride was removed by using phosphoric acid, the CG poly-Si was corroded and voids were generated. These voids were filled with the hard mask oxide and the oxide residue was formed in the voids after logic region poly-Si etching. Due to the high selection of silicon oxide for CG poly-Si contact (CGCT) etching, the FG poly-Si below the voids cannot be completely removed and finally form the poly-Si residue between BL and FG. As a result, we could conclude that the root cause of erase failures can be FG poly-Si residue after CGCT etching, resulting in the low read current.

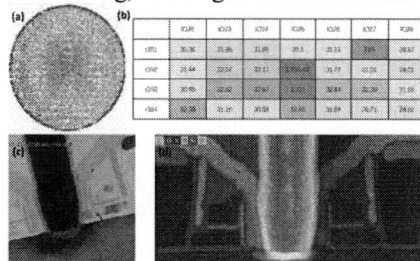

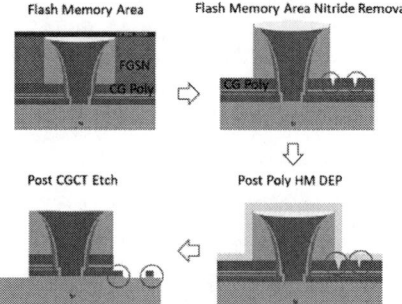

*Figure 2 (a) CP erase failure distribution (b) Read current of erase failure (c) PFA result of erase failure (d) EDS analysis of residue*

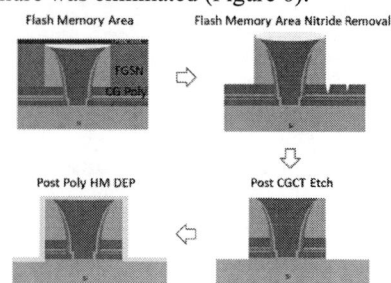

*Figure 3 Mechanism of FG poly-Si residue formation*

To demonstrate the inference, scanning electron microscope (SEM) and transmission electron microscope (TEM) were used to characterize the flash cell area. As shown in Figure 4b, the voids can be observed on the CG poly-Si after the flash memory area nitride was removed. The cross-sectional view of the voids was shown in Figure 4c. The size of the hole is 10nm and the depth is 25nm. The oxide in the voids acted as a barrier layer for CGCT etching. The FG poly-Si below the voids was not completely etched, and therefore the FG residue which bridge FG poly-Si and BL was formed. Figure 4d presents the top view of FG ploy residue in flash memory area. In order to accurately confirm the exact process step of the voids formed, we performed SEM characterization for the flash memory area after FG nitride dry etching. As can be

seen in the SEM image (Figure 4a), there is no voids on the top of CG poly-Si, indicating the voids were generated during flash memory area nitride removal.

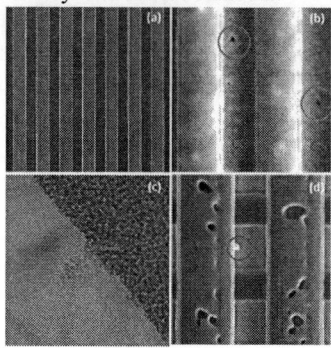

*Figure 4 (a) Top view SEM after FG nitride dry etching. (b) Top view image after flash memory area nitride removal. (c) Cross-sectional view of the voids. (d) FG poly-Si residue*

In order to fundamentally solve the erase failure, a simple and cost-effective solution was proposed. The critical reason for the FG poly-Si residue is that the gate poly-Si hard mask filled the void and formed a block layer in flash memory area. Therefore an effective process solution is to perform the CGCT etching before the gate poly-Si etching (Named as CGCT-pre). Figure 5 shows the CGCT-pre process flow diagram. The CGCT etching was executed to remove CG and FG poly-Si after flash memory area nitride removal. As shown in Figure 5, CG voids were formed after the flash memory area nitride removal, CG voids were formed. Firstly, the CGCT etching was performed to etch off the CG and FG poly-Si. Then the hard mask oxide was deposition on the flash memory and logic area. This optimized process avoids the formation of CG block layer in the voids, thus successfully eliminate the erase failure. Through the introduction of this novel process, significant improvements were observed. The yield of embedded flash increases from 0.1% to 95.6% after the novel process introduction and the erase failure was eliminated (Figure 6).

*Figure 5 CGCT-pre process flow diagram*

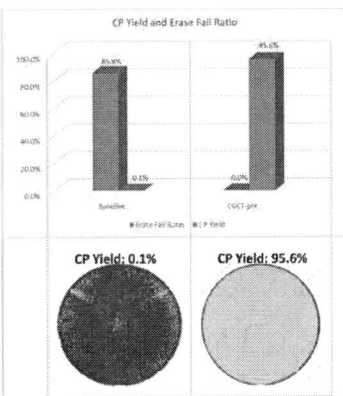

*Figure 6 CP yield of CGCT-pre process*

However, CGCT-pre process could reduce the length of the flash memory, leading to the program and erase performance of flash. The length of CG and FG is 68nm and 109nm in CGCT-pre flash memory, respectively, smaller than that of 82nm and 115nm in baseline cell (Table 1). This is because of the hard mask oxide formed a spacer which can block CG and FG poly-Si from partially etched by the CGCT process. The shortened CG and FG lengths lead to the reduction of coupling ratio, which is defined as the ratio of CG length to FG length. The coupling ratios of baseline and CGCT-pre process are presented in Table 1. The CGCT-pre process shows a coupling ratio of 62%, 9% lower than the baseline process, which results in the increase of 0 current (Ir01) and reduce of 1 current (Ir10). As shown in Figure 7a, the Ir01 increased 2~3 orders by using CGCT-pre process. The narrowing of Ir01 and Ir10 current window indicates a significant decrease in the programming capability of the flash memory, which can be demonstrated by the increased programming time (Tpgm) (Figure 7c).

*Table 1 Coupling ratio and length of CG and FG*

| Condition | CG (nm) | FG (nm) | CG/FG Coupling Ratio |
|---|---|---|---|
| Baseline | 82 | 115 | 71% |
| CGCT-pre | 68 | 109 | 62% |

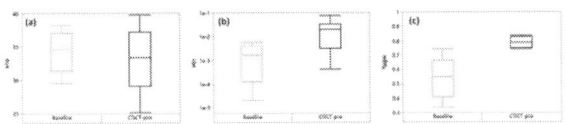

*Figure 7 Performance parameters of baseline and CGCT-pre flash memory cell*

After thickening the FG spacer (FGSP) by 100A, the length of CG and FG can be effectively increased, which successfully improves the coupling ratio. Table 2 shows the coupling ratio after process optimization. The coupling ratio after process improvement has increased to be comparable to the baseline. The flash cell shows a better performance than the baseline (Figure 8). The Ir01 current

is $6\times10^{-3}$uA, matched the baseline cell. Moreover, the optimized flash memory shows a yield up to 95.6%. Finally, by optimizing key processes with employing process and structure adjustment, the dramatic yield improvement (~95%) was achieved by resolving erase failure.

*Table 2 Coupling ratio and length of CG and FG after structure optimization*

| Condition | CG (nm) | FG (nm) | CG/FG Coupling Ratio |
|---|---|---|---|
| Baseline | 82 | 115 | 71% |
| CGCT-pre | 72 | 97 | 74% |

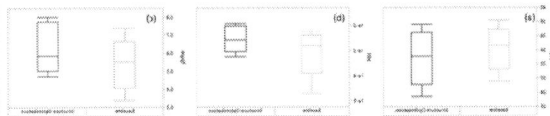

*Figure 8 Performance parameters of baseline and structure optimized flash memory cell*

## CONCLUSION

In conclusion, we could confirm that the root cause of the very gross erase failure was due to the FG poly-Si residue, so that the failed flash cell could not be erased, resulting in an incorrect read current in flash cell. Our optimization results indicated that the FG poly-Si residue was successfully eliminated. Moreover, the coupling ratio reduction was improved by structure optimization. Finally, an impressive yield of 95.6% was achieved and the mass production wafers has reached 3000 pieces.

## REFERENCE

[1] Roberto Bez. Proc. of the IEEE, Vol. 91, No. 4, April 2003, pp. 489-502.

[2] Khouri, O, et al. Industrial Electronics, 2002 Vol. 4, pp. 1322 - 1326.

[3] Maex, K. IEEE TED, Vol. 46, No.7, July 1999.

[4] P. Nicosia. IEEE NVSMW 2007, pp. 11-18.

[5] R. Chandramouli. IEEE Asian Test Symposium 2005, pp. 452-452.

[6] Kiyonori Ohyu. Electron Devices, 1995 Vol 42, Issue 8, pp. 1404 – 1412.

# A WAFER-ON-WAFER NON-UNIFORM HIGH POWER THERMAL MODEL FOR 3D CHIP PACKAGE

*Song Wang[1,3] , Xiping Jiang\*[1,2,4], Fengguo Zuo[1], Huimei Wang[1], Xiaofeng Zhou[1], Jingrui Chai[1], Yubing Wang[1], Bin Hou[1], Liang Zhong[1], Wenxin Li[1], Yi Kang[3]*

[1]Xi'an UniIC Semiconductors, Xi'an, China, [2]Institute of Microelectronics of the Chinese Academy of Sciences, Beijing, China,[3]University of Science and Technology of China, Hefei, China,[4] University of Chinese Academy of Sciences, Beijing, China
*Corresponding Author's Email: xiping.jiang@unisemicon.com

## ABSTRACT

Three-dimensional (3D) stacked embedded DRAM (SeDRAM) relies on Wafer-level logic and DRAM hybrid bonding. While this advanced technology provides the high performance computing, it also has significant challenges in thermal management. An equivalent model for non-uniform temperature and power distribution in SeDRAM with multiple power sources is proposed. Based on HFCBGA package, the junction temperatures and thermal resistances of SeDRAM are investigated. Furthermore, the model is validated by experimental results, and the temperature distribution simulate by equivalent model compares well with the test results. A 2.2℃ temperature difference between the proposed model and realistic test demonstrate its availability.

## INTRODUCTION

With the growth of electronics products, an advanced packaging technology is needed, A variety of Wafer-level three-dimension advanced packaging for manufacturing like direct bonding, TSV, hybrid bonding, or using micro-bumps have become the emphasis and frontier for industrial research. Hybrid bonding technology (HBT) of W2W, C2W, C2C used mainly in realize I/O interconnection [1]. The great advantage of HBT is that it can decreases parasitic effect, reduces thermal resistance and improve chip performance.

While 3D packaging technology provides some useful benefits in the chip performance, it still faces the challenges of power consumption. The power distribution of the chip is showing the great differences as more and more functional modules are integrated in one chip. The thermal investigation and non-uniform power consumption of the whole chip are not studied in detail in previous. In order to accurately evaluate the thermal characteristic of the HFCBGA package with 3D wafer-on-wafer (WoW) technology, we develop an equivalent model for analysis the junction temperature of different model. And the simulation results indicate that the extracting accuracy as high as 98% compared to the actual observations. Based on our method, it can easily increase simulation speed and improve simulation quality.

## THERMAL MODEL

### WoW Structure with Non-uniform Power

Figure1 shows the structure of WoW with hybrid bonding technology. Two wafers achieve metal interconnection by Cu hybrid bonding, The DRAM is the top wafer with uniform power distribution and the non-uniformly distributed logic wafer is below in the SeDRAM. The DRAM wafer and the logic wafer are of the same size, but the thickness is very different. In the stacked chip, thickness that measured 750um from the DRAM surface to the end of the bonding interface. The measured logic thickness is just 8um.

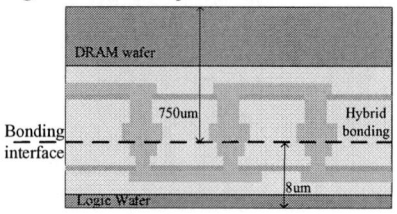

*Figure1: The structure of WoW*

The logic chip is divided into 9 partitions. Each partition has 4.5W power dissipation, with the exception of 45W onto Partition9. The power dissipation of DRAM is 3.6W, thus the total power dissipation of SeDRAM is 74.6W. Figure2 shows the power distribution of the logic chip.

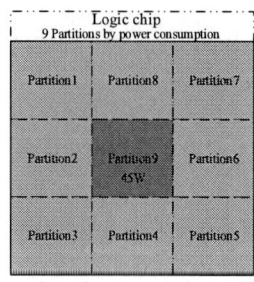

*Figure2: The power distribution of the logic chip*

### HFCBGA Model Based on WoW Non-uniform Power

Figure3 shows that the WoW structure is equivalent to two independent chip without consideration of hybrid bonding and SiO2. The HFCBGA package model that are adaptive for WoW. In order to reduce the heat dissipation, the lid is mounted directly on the top of package.

978-1-6654-9759-6/22 $31.00 © 2022 IEEE

This package with this structure can enhance the heat dissipation of HFCBGA [2]. In this work, thermal interface material (TIM2) between the lid and the memory chip, the thickness and thermal conductivity of TIM2 are 100um and 1.92W/m-K.

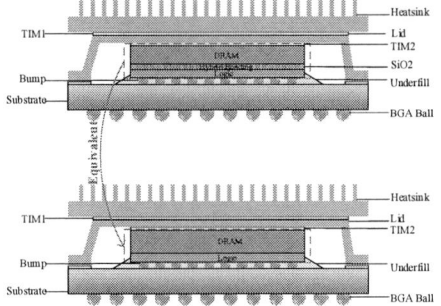

*Figure3: The FCBGA Model of WoW and equivalent*

Heat sink is a conventional technique to dissipate high flux chips. Thermal interface material (TIM1) is used between lid and heat sink that can further enhance heat dissipation. Because of the high power logic chip, special attention should be paid to the thickness of thermal interface material (TIM1). Even if there is little difference in one of the mail components of the thermal resistance, it will cause a big error in maximum junction Rja, Rja represents the thermal resistance between the chip and lid. Both logic and DRAM chips are modeled as homogeneous blocks with equivalent thermal conductivities methods [3]. The WoW stack with Cu interconnects enclosed by SiO2. The equation is used to transform the true structure and the equivalent structure is as follows:

$$K^m \delta_{ij} \left( \frac{\partial T^c}{\partial x_j} + \frac{\partial T^0}{\partial x_j} - \frac{\partial T^f}{\partial x_j} \right) = K_{ij}^f \left( \frac{\partial T^c}{\partial x_j} + \frac{\partial T^0}{\partial x_j} \right) \quad (1)$$

Here $x_j$ is the $j$th element of the vector. $\delta_{ij}$ is the Kronecker symbol and the subscripts j=1, 2, 3 indicate three coordinate axes. $K^f$ and $K^m$ indicates the heat conductivity of the true structure and the equivalent structure. The left of the equation represent the thermal flow of the equivalent structure, and the right side of the equation shows the true structure. We obtain the lateral equivalent heat conductivity K11=K22 and the longitudinal equivalent heat conductivity K33.

$$K_{11} = K_{22} = K^m \left[ 1 - \frac{1 - FQ_1^f(\theta)}{1 + FQ_1^c(\theta)} \right] \quad (2)$$

$$K_{33} = K^m \left[ 1 - \frac{1 - FQ_3^f(\theta)}{1 + FQ_3^c(\theta)} \right] \quad (3)$$

The logic chip and substrate are connected together through solder bumps. Solder bump and underfill are simplified as compact blocks based on the equivalent thermal conductivity model. The establishment of equivalent thermal conductivity for bump and underfill

follows equation (3).

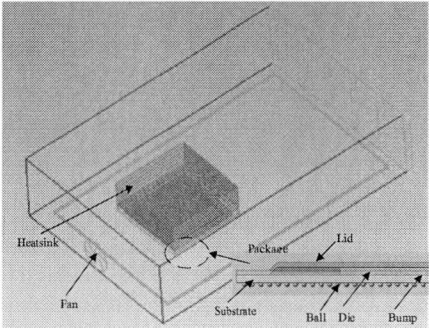

*Figure4: The thermal simulation model*

The thermal model of the package is shown in Figure 4. Because of the high power of logic chip, a fan is mounted to provide forced convection (20CFM) during operation. Strong convection can greatly improve the heat dissipation of package, gas flow along the fin direction and keep the walls perpendicular to y axis opened. The detail parameters of the model are shown in Table1.

TABLE I.  THE DETAIL PARAMETERS OF MODEL

| Items | Size(mm) |
|---|---|
| PCB(length ×width× height) | 165×102×2.1 |
| Package | 45×45×1.8 |
| Die | 30.8×25.4×0.79 |
| Lid thickness | 1.0 |
| Substrate thickness | 1.2 |
| Bump diameter/Height | 0.1/0.08 |
| Ball diameter/Height | 0.6/0.5 |

To analyze the heat characteristic of the HFCBGA models, the thermal conductivity of package material are listed in Table 2.

TABLE II.  THERMAL CONDUCTIVITY OF MATERIALS

| Items | Thermal Conductivity(w/m. k) |
|---|---|
| Die | 148 |
| Bump | 387 |
| Under fill | 0.4 |
| Solder Mask | 0.21 |
| Substrate | K_X:46.23 K_Y:46.23 K_Z:0.408 |
| Solder Ball | 60 |
| Lid | 387 |
| TIM1 | 1.92 |

# THE EQUIVALENT MODEL OF WOW
## The Simulation Result of Different Model

The equivalent model is mentioned previously.

978-1-6654-9759-6/22 $31.00 © 2022 IEEE

Simulation results indicated that the influence of natural convection and radiation heat transfer can be neglected and the forced air dominated thermal dissipation. The heat sink is divided into discontinuous mesh, the grid type and size of model do not affect the background area. The grid size in the background area is larger than heat sink, which greatly reduces the number of grid.

**The WoW Accurate Model Based on HFCBGA with Uniform Chip Power Distribution**

The power of logic chip is uniformly distribution. And the model is established and imported into ANSYS for simulation. The maximum junction temperature of the logic chip and DRAM chip are 60.1052°C and 59.9454°C respectively in Figure5. According to the junction temperature and power of the chip the thermal resistance $R$ja of the package can be calculated, which represents the thermal resistance between the junction of the chip and the air, $R$ja=($T$j-$T$a)/P, $T$j is the junction temperature of the chip, $T$a is the ambient temperature. The thermal resistance generally includes natural convection and forced air convection. In this paper, forced air convection is used by actual environment.

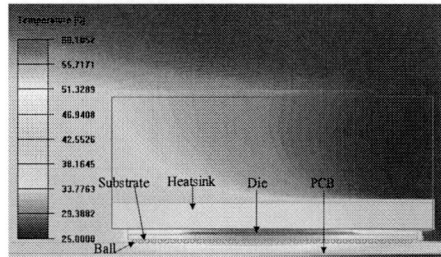

*Figure5: Cross section of temperature distribution*

**The WoW Accurate Model Based on HFCBGA with Non-uniform Chip Power Distribution**

WoW Structure still use the accurate model, but the chip's power is non-uniformly distributed situation depending on the actual environment.

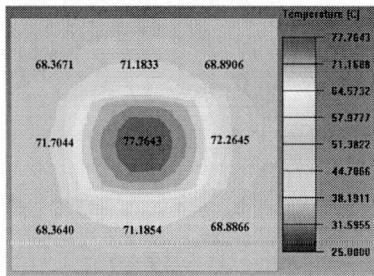

*Figure6: temperature distribution of logic chip*

The DRAM chip's temperature is almost the same as logic chip due to the 80%-90% thermal dissipation via the upper surface of the package. The temperature of partition9 in logic chip is higher than other partitions. The temperature distribution of each partition in logic chip is displayed separately in Figure6. The junction temperature at the center of logic chip is 77.7643℃, while the junction temperature of other partitions is from 68.364 to 72.2645℃. The logic chip has the maximum temperature difference 9℃ at different area.

**The WoW Equivalent Model Based on HFCBGA with Uniform Chip Power Distribution**

The structure of WoW is equivalent to fabricated wafer, without consideration of gap between wafers. The power of equivalent model is the sum of power on logic and DRAM with uniform distribution. As can be seen in Figure7, the maximum junction temperature of the equivalent chip is 59.3870℃，and the error between the equivalent values and the accurate values is less than 1℃. The main reason is that the difference of chip structure cannot cause a significant change of the junction temperature under the effect of enhanced heat sink and forced air convection.

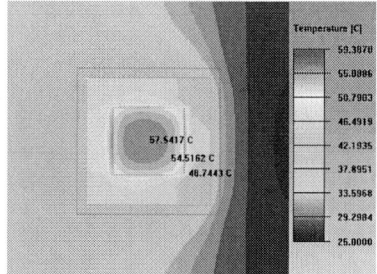

*Figure7: Temperature distribution of uniform power*

**The WoW Equivalent Model Based on HFCBGA with Non-uniform Chip Power Distribution**

The structure of WoW adopts the equivalent model with non-uniform power distribution in the logic chip. The temperature cloud diagram is shown in Figure8. The maximum junction temperature of the equivalent model is 76.4755℃, which is less than 1.3℃ the accurate model.

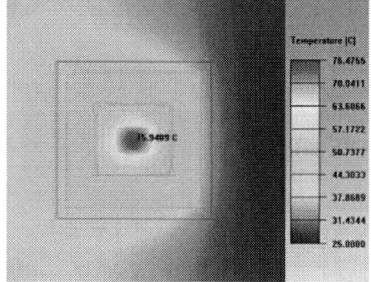

*Figure8· Temperature distribution of non-uniform power*

It can be concluded from Table3:

(1)Whether it is accurate model or equivalent model, there is a great difference between the uniform and non-uniform distribution of chip power

(2)When the chip power distribution is the same, the chip junction temperature difference between the accurate model and the equivalent model is only 1℃.

TABLE III.  SIMULATION RESULTS OF FOUR MODELS

| Model | Model description | | $Tj(℃)$ | R ($℃/W$) |
| --- | --- | --- | --- | --- |
| | WoW structure | Power distribution | | |
| A | Accurate | Uniform | 60.1052 | 0.47 |
| B | Accurate | Non-uniform | 77.7643 | 0.71 |
| C | Equivalent | Uniform | 59.3870 | 0.46 |
| D | Equivalent | Non-uniform | 76.4755 | 0.69 |

## COMPARISON OF SIMULAITON RESULT AND TEST VALUE

The packaging structure and parameters in the actual application are the same as in the Model D, as shown in Table3. In real application, the power consumption of each partition in logic chip is different due to different computing functions of logic. However, the total power consumption of logic chip remains unchanged. The power distribution of logic chip is shown in Table4.

TABLE IV.  THE POWER OF EACH MODEL IN LOGIC CHIP

| Item | Model1-8(W) | Model9(W) |
| --- | --- | --- |
| 1 | 4.5 | 35 |
| 2 | 4.75 | 33 |
| 3 | 5 | 31 |
| 4 | 5.25 | 29 |
| 5 | 5.5 | 27 |
| 6 | 5.75 | 25 |

In this work, the thermocouple is placed in the center of the heat sink to test, which is compared with the simulation results in Figure9. Although the total power consumption of the chip remain unchanged, the junction temperature of chip changes greatly due to the different power distribution. The maximum difference between the actual test data and the simulation values is 2.2℃. However, the junction temperature of the chip is about 7℃ higher than the heat sink temperature.

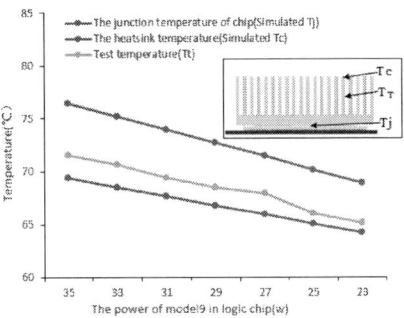

*Figure9: simulation and test results at different power*

In addition, in real application, different wind speeds will be used in different environments, the simulation values and test result are also compared at different wind speeds.

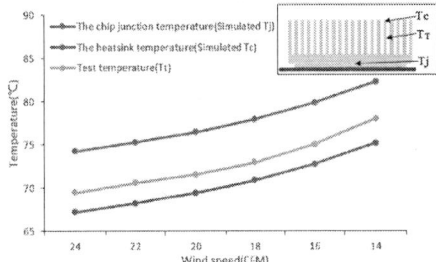

*Figure10: simulation and test results at different wind speeds*

As can be seen in Figure10, the wind speed of the fan has great influence on the temperature of product. The difference between simulation temperature and test result is about 3℃.

## CONCLUSION

An equivalent model of advanced packaging is proposed according to the WoW structure. Because of the non-uniform power distribution of the chip, the proposed model has two features including uniform distribution and non-uniform distribution. The accurate model and equivalent model are established by real chip structure. The junction temperature of chip are investigated. By comparison, the results of different power distribution and different wind speed is obtained.

(1) The non-uniform power distribution is used in the model according to actual conditions. With the uniform power distribution, the junction temperature of the chip is much lower than the true scenes and the reliability problem of the device is introduced.

(2) With enhanced heat sink and forced air convection, the junction temperature of the chip has a slight difference, even using an equivalent model.

(3) The simulation results of the chip are in good accordance with the measured values, which means that the proposed model is accurate.

[1] Bai Fujun, Jiang Xiping, Wang Song. *A Stacked Embedded DRAM Array for LPDDR4/4X using Hybrid Bonding 3D Integration with 34GB/s/1Gb 0.88pJ/b Logic-to-Memory Interface*, 2020 IEEE International Electron Devices Meeting (IEDM).

[2] Hengyun Zhang, Zhifeng Zhou. *Thermal Modeling and Design of 3D Memory Stack on Processor with thermal Bridge structure*, ICEPT, 2020.

[3] Jingrui Chai, Gang Dong, and Yintang Yang. *An Effective Approach for Thermal Performance Analysis of 3-D Integrated Circuits With Through-Silicon Vias.* IEEE Transactions on Components, Packaging and Manufacturing Technology (TCPMT), vol. 9, pp: 877-887, May 2019.

# EVALUATION OF DIFFERENT CUT APPROACHES ON ADVANCED BEOL SELF-ALIGNED DOUBLE PATTERNING SCHEME

*Chia Lin Lu[1*], Peng Fei Lyu[1], Qing Peng Wang[1] and Yu Shan Chi[1]*
[1]Lam Research Service Co., Ltd, Shanghai 201203, China
*Corresponding Author's Email: Tank.Lu@lamresearch.com

## ABSTRACT

Self-aligned double patterning (SADP) has been evaluated and employed as a back-end-of-line (BEOL) patterning solution for 10/7-nm logic technology and beyond. In this paper, we study metal line end shapes created by both cut-first and cut-last approaches in a SADP BEOL flow, and investigate their impacts on capacitance and dielectric leakage. Very different line end shapes were obtained from the two approaches investigated, due to post 1st patterning silicon surface oxidation differences. The cut-first approach results in a horn-shaped line end shape, while the cut-last approach yields a bowl-shaped line end shape. By using virtual fabrication on Coventor Inc.'s SEMulator3D®, a systemic investigation was performed to understand the impact of metal line end shapes on electrical performance. Our results indicate that capacitance and dielectric leakage performance of metal line to metal line and metal tip to metal tip, cut-first approach is reverse to cut-last approach. This study demonstrates that capacitance and dielectric leakage performance must be considered during selection of a BEOL SADP approach.

## INTRODUCTION

With increasing pressure for pitch down scaling, the conventional litho-etch-litho-etch (LELE) patterning scheme faces significant challenges for back-end-of-line (BEOL) integration. Instead, self-aligned double patterning (SADP) is expected to be the primary solution [1].

BEOL metal line end shape is very important for capacitance and dielectric leakage performance. Capacitance is important for time delay control, and dielectric leakage for reliability and yield performance. In this work, we explore two types of SADP approaches and discuss device behavior's dependence on metal line end shapes by constructing a virtual design of experiment (DOE) and virtual wafer fabrication using Coventor Inc.'s SEMulator3D® [2-4]. The impacts of both cut-first and cut-last schemes are investigated, to provide guidance for integration scheme selection.

## EXPERIMENT DESCRIPTION

In our experiment, the SADP patterning was simulated by using Coventor Inc.'s SEMulator3D® virtual fabrication software. A process model was built and calibrated in the software.

The purpose of cut pattern is to define the space of metal line end to metal line end. The cut pattern is defined during 1st patterning, called cut-first approach. Cut-last approach is during 2nd patterning. Figure 1 shows 3D schematic of two SADP approaches on the 2nd patterning. Due to post 1st patterning surface oxidation differences on Si between these two approaches, different line end shapes were obtained (Figure 2). The cut-first approach results in horn-shaped line end shape, and the cut-last approach yields bowl-shaped line end shape.

To explore the impacts of the line end shapes on the electric performance such as capacitance and dielectric leakage, a layout was designed, as shown in Figure 3 to perform further DOEs based on different horn- and bowl-shapes (Figure 4). Capacitance and dielectric leakage performance of the structures were investigated with electrical analytics capabilities on Coventor Inc.'s SEMulator3D®.

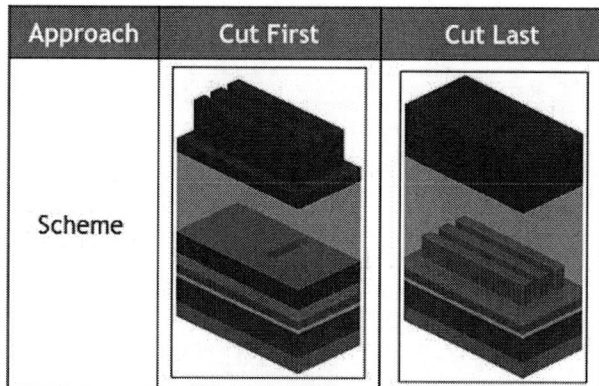

Figure 1: 3D schematic of 2nd pattering Cut-First and Cut-Last structure.

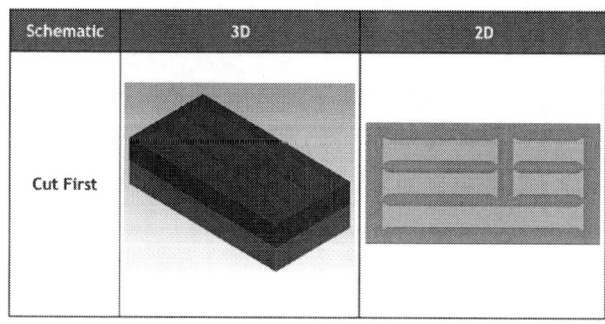

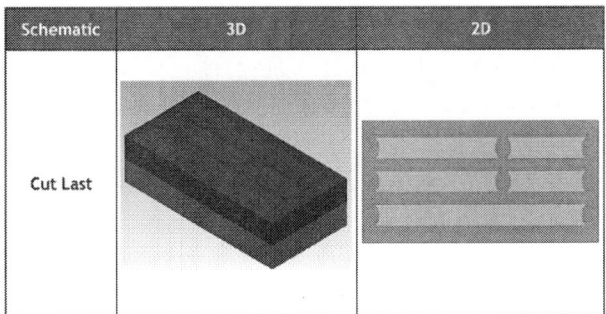

Figure 2: 3D and 2D schematic of Cut-First and Cut-Last structure.

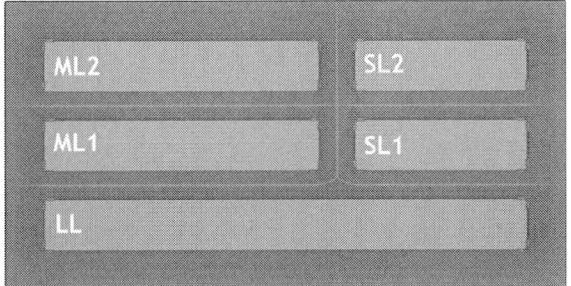

Figure 3: Layout of BEOL SADP

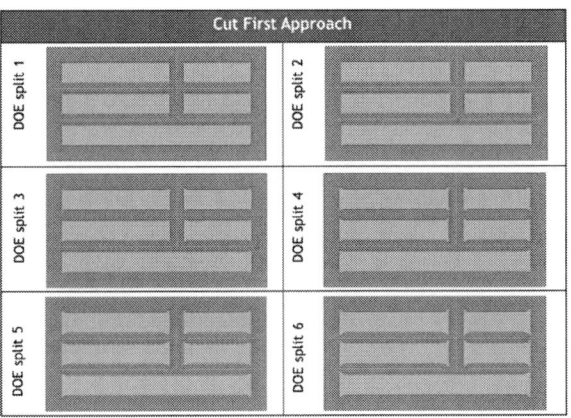

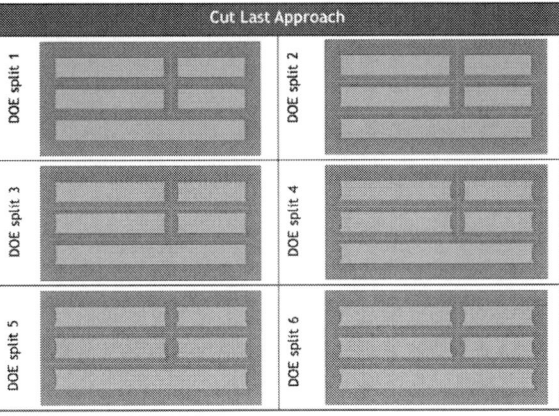

Figure 4: DOE split of Cut-First and Cut-Last approach

# RESULT AND ANALYSIS

The capacitance and dielectric leakage are extracted individually by using virtual fabrication on Coventor Inc.'s SEMulator3D® (Figure 5). The electric performance between metal line to metal line (ex: ML1-ML2) and between metal tip to metal tip (ex: ML1-SL1) is demonstrated straightforwardly.

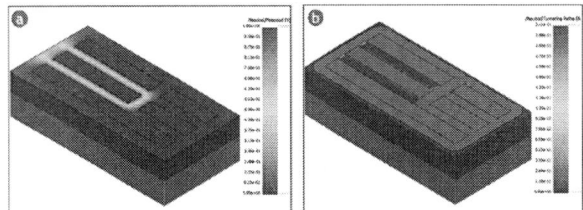

Figure 5: Capacitance and dielectric leakage calculation

The extracted capacitance is obtained by electric charge stored on a conductor dividing the applied electric potential, and it is inversely proportional to the distance to the guide plate. Therefore, both the capacitance between metal line to metal line and between metal tip to metal tip increase with CD increasing regardless of horn- or bowl-shape CD resulted by cut-first or cut-last approaches (Figure 6). However, the capacitance between metal line to metal line using the cut-first approach is higher than that using the cut-last approach, contracting to the trend of the metal tip to metal tip (Figure 7). This is due to the difference of pattern stick section, i.e., the cut-first approach is located at metal line to metal line while cut-last approach is at the metal tip to metal tip.

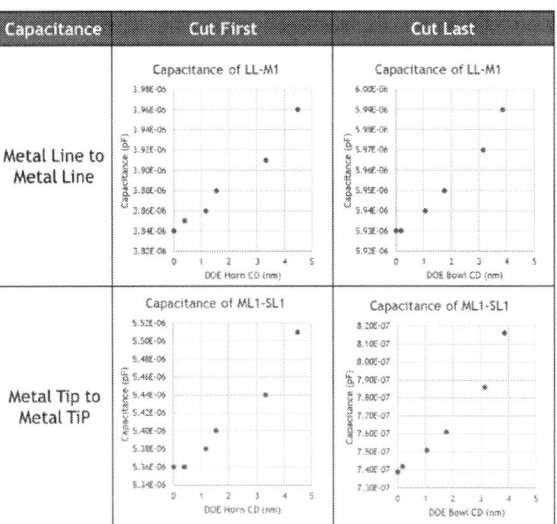

Figure 6: Capacitance DOE results of Cut-First and Cut-Last approach

| Capacitance DOE split6 | Cut first | Cut last |
|---|---|---|
| Metal Line to Metal Line | Higher | Smaller |
| Metal Tip to Metal Tip | Smaller | Higher |

| Capacitance (pF) DOE split6 | LL-SL1 | LL-ML1 | SL2-SL1 | ML2-SL2 | ML2-SL1 | ML1-SL1 |
|---|---|---|---|---|---|---|
| | Metal Line to Metal Line | | | Metal Tip to Metal Tip | | |
| Cut first | 3.96E-06 | 6.05E-06 | 3.42E-06 | 1.19E-06 | 4.55E-07 | 7.76E-07 |
| Cut last | 3.90E-06 | 5.99E-06 | 3.34E-06 | 1.23E-06 | 4.57E-07 | 8.16E-07 |

Figure 7: Capacitance DOE split6 result of Cut-First and Cut-Last approach

As known, the dielectric leakage is the gradual transfer of electron across boundary. Weak point, especially like sticks section on pattern, can increase the chance for the leakage. Hence, the dielectric leakage for both metal line to metal line and metal tip to metal tip increases with CD increasing regardless of the approaches (Figure 8). Moreover, similar to the trend of the extracted capacitance, the dielectric leakage is also higher for cut-first approach for the metal line to metal line whereas metal tip to metal tip shows the opposite trends (Figure 9).

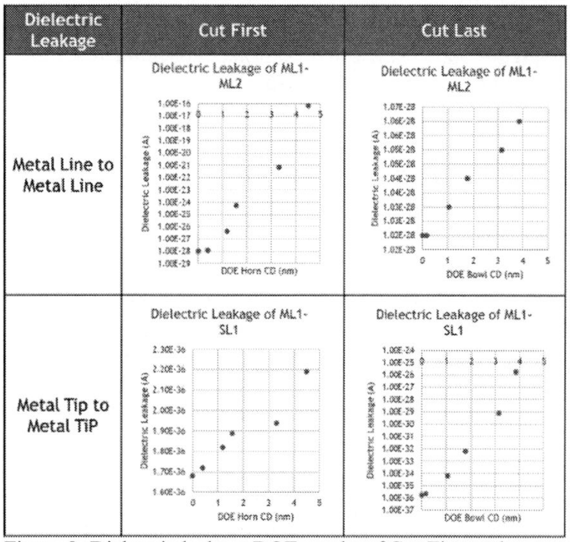

Figure 8: Dielectric leakage DOE results of Cut-First and Cut-Last approach

| Dielectric Leakage DOE split6 | Cut first | Cut last |
|---|---|---|
| Metal Line to Metal Line | Higher | Smaller |
| Metal Tip to Metal Tip | Smaller | Higher |

| Dielectric Leakage (A) DOE split6 | ML1-ML2 | ML1-LL | ML1-SL2 | ML1-SL1 |
|---|---|---|---|---|
| | Metal Line to Metal Line | | Metal Tip to Metal Tip | |
| Cut first | 6.57E-17 | 4.34E-17 | 9.01E-44 | 2.19E-36 |
| Cut last | 1.06E-28 | 1.11E-28 | 1.79E-41 | 1.77E-26 |

Figure 9: Dielectric leakage DOE split6 result of Cut-First and Cut-Last approach

## CONCLUSION

In this paper, the BEOL SADP process is simulated and the virtual DOE is conducted using the Coventor Inc.'s SEMulator3D® virtual fabrication platform. A virtual DOE was completed to understand the capacitance and dielectric leakage performance of different approaches. Our study reveals that cut-first and cut-last approaches yields very different line end shapes, resulting in opposite device performance. Both capacitance and dielectric leakage performance is influenced by pattern stick section that cut-first approach is located at metal line to metal line and cut-last approach at metal tip to tip. Hence, by comparing of these two approaches, the performance shows the opposite trends. This study is useful to select suitable BEOL SADP approach for capacitance and dielectric leakage concern.

## REFERENCE

[1] Ma, Y., Sweis, J., Yoshida, H., Wang, Y., Kye, J., Levinson, H. (2012). Self-Aligned Double Patterning (SADP) Compliant Design Flow. SPIE.

[2] http://www.coventor.com/products/semulator3d

[3] Chen, Y. D., Huang, J., Zhao, D., Yim, D., Ervin, J. (2019). Advanced 3D Design Technology Co-Optimization for Manufacturability. CSTIC.

[4] Wang, Q., Chen, Y. D., Huang, J., Liu, W., & Joseph, E. (2020). Yield Enhancement by Virtual Fabrication: Using Failure Bin Classification, Yield Prediction and Process Window Optimization to Identify and Prevent Process Failures. In 2020 China Semiconductor Technology International Conference (CSTIC) (pp. 1-3). IEEE.

# ADDRESSABLE WAT TEST OF DOMESTIC SEMITRONIX TESTER

*Yangyang Xing[1]\*, Yanping Zhu[1], Xiong Shao[1]*
[1]Shanghai Huali Integrated Circuit Manufacturing Co., LTD
\*Corresponding Author's Email: xingyangyang@hlmc.cn

## ABSTRACT

Traditional WAT test is executed on a series of testkeys which is located in the scribe line of wafers. A transparent disadvantage of this kind of test is large amount of test time and limited number of DUT(device under test). So a different kind of WAT test is urgently needed, which combines the advantages of testing efficiency and accuracy. This paper will introduce a novel WAT test, it adopts parallel test based on specially designed testkey structure. What's more, domestic Semitronix tester is able to conduct this test, huge amount of time is saved, and testing accuracy is guaranteed.

## INTRODUCTION

Wafer acceptance test (WAT) is responsible for shipping qualified wafer to customers. The purpose of WAT is to reduce unnecessary cost by testing the collected electrical parameters. It measures the electrical characteristics of the process control monitoring (PCM) test keys such as transistors, resistors, and capacitors and compares the results with the fabrication specification to make the pass/fail decision [1]. PCM test keys are usually placed in the scribe lines on the wafer whose widths are no wider than 60 um. Traditionally, evaluation of local process variations requires statistically measuring numerous devices-under-test (DUTs). Placed in scribe lines of a wafer, each DUT needs four I/O pads to measure its characteristics. Due to the limitation of I/O pad count and layout area [2], it is thus not feasible for this kind of traditional method to place a large number of DUTs.

Thus it is urgent to come up with a new method: addressable test. Addressable test means multiple DUTs share the same set of prober pads by using decoder and circuits [3-4]. Decoder circuits provide an address at every selection, and then circuits select the addressed DUT to be tested, and isolate the other DUTs at the same time.

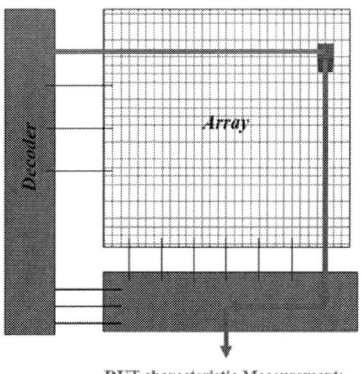

*Fig.1: DUT characteristic measurements*

Addressable test has 4 main key features:

1. High area efficiency

Pad areas will reduce by using addressable, so it is possible to fulfill complete DOE implementation within a small wafer area, even a scribe line module.

2. High area efficiency

Semitronix's IP is able to eliminate the effects of parasitic resistance, leakage or capacitance introduced by addressable circuits. Such as MOSFET: Idsat/Ioff/Gate Leakage, etc; Parametric/Yield: resistance/leakage; Ring Oscillator: IDDQ/IDDA/frequency.

3. Fully automated test chip layout design

Customized addressable test chip design platform – ATCompilerto reduce design complexity and improve design efficiency.

4. Fast testing cycle

　　Multiple Metal testing options: M1/M2/M3/M4/M5.

5. Flexible chip size

　　Large scale chip to be placed in MPW, or small chip to be placed in scribe line.

　　A few real cases are exhibited as follows:

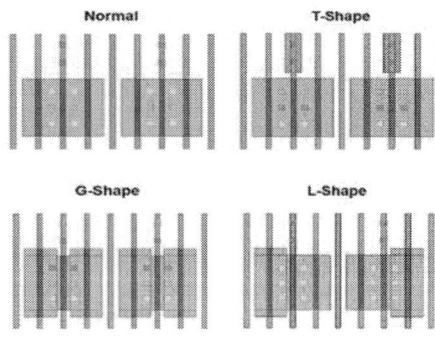

*Fig.2: 22 pad test structure designed by semitronix*

　　All the DUTs can be placed in scribe line and only between pads. Flexible array size to fit different applications. In Fig2, it has 10 blocks. 12 DUTs are placed in each block. So it is up to 240DUTs which can be placed on one scribe line module; N/P transistors can be placed in different modules or the same module; Usually M3 testable, M1/M2 testable is available if needed; As shown in Fig3, We can measure different types of basic transistor simultaneously, which can only be tested in order through traditional WAT test.

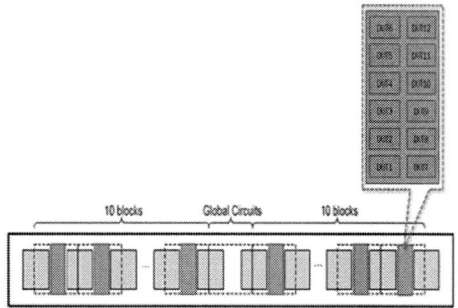

*Fig3: Different types of basic transistors*

　　For accuracy, it is exhibited in Fig4, Idsat is less than 1% bias; While Vt is less than 1 mv bias.

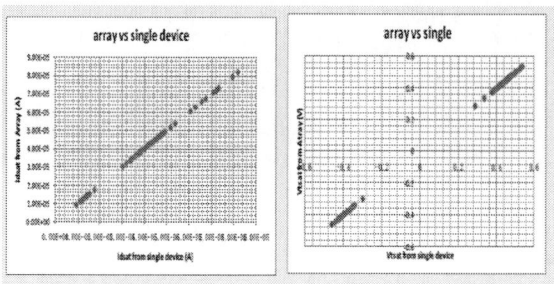

*Fig4: Idsat and Vtsat correalation results between addressable and traditional test*

　　So it is a trend that addressable test will play a more and more important role in WAT test. With an outstanding tester and powerful design, a huge rise of DUTs will be tested in the future.

## REFERENCES

[1] T. C. Luo, M. C. T. Chao, M. S. Y. Wu, K. T. Li, C. C. Hsia, H.C. Tseng, P. A. Fisher, C.U. Huang, Y. Y. Chang, S. Pan, and K. K. L. Young, "A novel array-based test methodology for local process variation monitoring," IEEE Trans. on Semiconductor Manuf., vol. 24, no. 2, 2011, pp. 280–293.

[2] C. Ferri, S. Reda and R. I. Bahar. Parametric yield management for 3D ICs, ACM J. Emerg. Technol. Comput. Syst., vol. 4, no. 4,2008, pp. 1-22.

[3] H. Jafari, L. Soleymani, and R. Genov, "16-Channel CMOS Impedance Spectroscopy DNA Analyzer With Dual-Slope Multiplying ADCs," IEEE Trans. on . Biomedical Circuits and Systems, vol. 6, 2012, pp. 468–478.

[4] T. Sato, H. Ueyama, N. Nakayama, and K. Masu, "Accurate array-based measurement for subthreshold-current of MOS transistors," IEEE Journal of Solid-State Circuits, vol. 44, no. 11,2009, pp. 2977–2986.

# 40nm BACKSIDE OPTIMIZATION AND IMPROVEMENT

Lu Xu, Wei Lu, Li Ding

Shanghai Huali Integrated Circuit Manufacturing Co., Ltd, Shanghai, China.

*Corresponding Author's Email: xulu@hlmc.cn

## ABSTRACT

Wafer backside condition is a major and everlasting problem during IC manufacturing process in recent years, which may induce yield loss or increase energy consumption due to tool alarm. In this paper we design a simple way to maintain backside nitride film to protect wafer backside from high etch rate acid erosion during cleaning. The leveling and backside flatness are efficiently improved, which is extremely promising for wafer backside uniformity improvement and wafer yield enhancement.

*Keywords: backside surface, SMT nitride remove, leveling abnormal*

## INTRODUCTION

In IC manufacturing process, the quality and integrity of wafer front is highly concerned due to the direct impact to yield and chip performance; However, with the rapid development of semiconductor technology, the condition of wafer backside can also induce yield loss according to massive lesson learn. For example, in photo process, poor wafer backside condition will directly impact leveling and result in defocus; Or in etching and CMP process, it may influence the vacuum degree and leads to equipment tool alarm or shut down;

Below 40nm technology, SMT (Stress Memorization Technique) is introduced for NMOS improvement. Nitride is deposited on NMOS area and removed after S/D Anneal. Traditional method of SMT nitride remove is to put the wafer in hot H$_3$PO$_4$ batch, SiN film on wafer front and backside will be removed indistinguishably due to the high etch rate acid; After nitride remove, backside poly is exposed without any protection; In the BEOL process, backside clean(BSC) before photo has a high etch rate to poly silicon, the center surface of backside become rough during repeated BSC process and leads to leveling abnormal. Engineers have tried to dilute the cleaning solution and reduce cleaning time to alleviate backside erosion. But this method may cause etching tool alarm and the metal ion out of limits. Meanwhile, wafer backside is frequently in contact with various equipment such as Chuck on photoetching machine. The contact points are easily contaminated and will reduce etch rate compare with other parts and will also impact leveling[1-3]. The worse flatness causes litho leveling abnormal and defocus, finally resulted in yield loss with special ring map(Fig. 1).

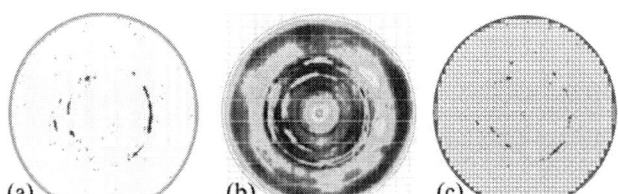

*Figure 1: (a)CP image. (b)Leveling image. (c)Defect image*

In this paper, one-step batch SMT Nitride remove is replaced with two steps(one single step+ one batch step) to retain the nitride film on the wafer backside surface, which could protect the surface from the erosion of BSC. The result shows nitride film is reserved and can effectively improve leveling and yield performance.

## EXPERIMENT

We set up a demo lot with single clean step and batch clean to verify the feasibility of two step clean method. Then we design deterioration experiments by repeating Mx BSC times and Mx CMP rework times in short time to simulate the worse condition in practical manufacturing process. In this paper we do BSC 2 times and CMP rework at Metal 2 layer. Backside defect image, leveling, PFA image, inline and WAT data of baseline(BSL) and CIP wafer are collected in the experiments.

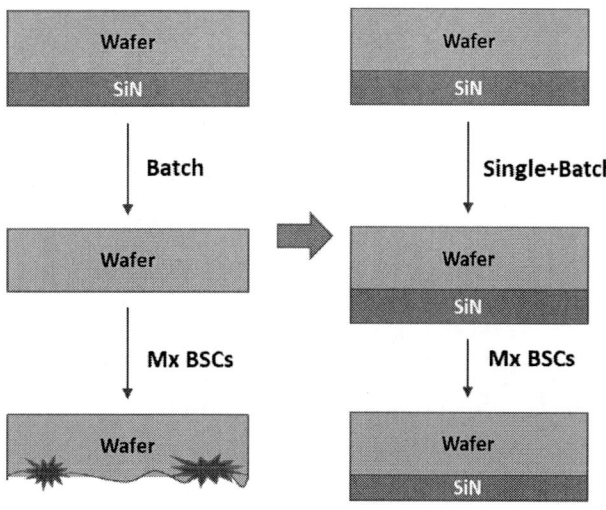

*Figure 2: Schematic diagram of the process step change.*

## RESULT AND DISCUSSION
### DEMO EXPERIMENT

Table 1 shows the result of BSL and split condition in demo experiment. The SMT Oxide Post SiN-RM thickness of CIP condition is thicker than BSL condition. This is due to the change of surface clean condition and result in difference of oxide loss. After SMT Oxide Remove process, SMT Oxide Post OX RM thickness are comparable between BSL and CIP wafer.

The darkfield backside inspection image of BSL wafer shows there are ring maps at center area, which confirms that with the increase of backside cleaning times, the backside flatness especially the meddle area will because worse. And the backside image of CIP wafer keeps clean and integrated after multi BSCs. Leveling data of Mx layers are also shown in Table1. CIP Wafer's leveling range is 0.079 at M5-PH, while BSL wafer is 0.128. The integrated backside inspection image and smaller leveling range shows that the backside condition of CIP wafer is better than BSL wafer.

*Table 1 Result of CIP Condition*

| Condition | BSL | CIP |
|---|---|---|
| SMT Oxide Post SiN-RM | 31.8 | 61.8 |
| SMT Oxide Post OX-RM | 10.2 | 9.8 |
| Backside Defect | | |
| Leveling Image | | |
| Leveling Range | 0.128 | 0.09 |

Table 2 displays the TEM image of backside film remove of CIP wafer. After Mx BSCs process, the center area still maintains 79.5nm poly silicon and 8.8nm SiN film, and the edge area remains 85.4nm poly silicon and 12.6nm SiN film. This proves the SiN film on the wafer backside can be efficiently retained after two steps nitride remove method. The compact and strong nitride film can effectively protect poly silicon film from acid erosion.

*Table 2: The TEM figure of the backside film stack of CIP condition.*

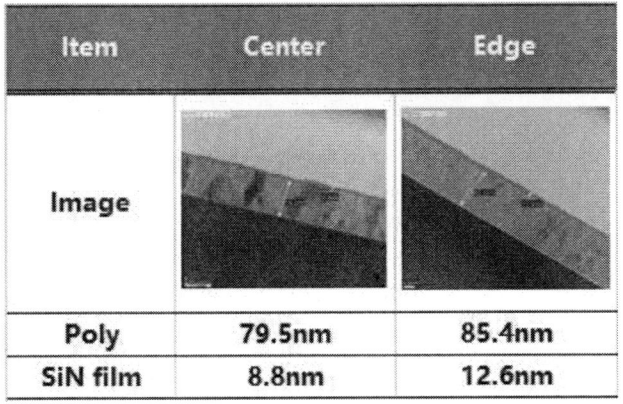

| Item | Center | Edge |
|------|--------|------|
| Image | | |
| Poly | 79.5nm | 85.4nm |
| SiN film | 8.8nm | 12.6nm |

DETERIORATION EXPERIMENT

The results of deterioration experiment are shown in Table 3. The darkfield backside inspection image show no difference before M1 BSC. But after multi BSCs, BSL wafer shows ring map, while CIP and deterioration wafer stay flat and clean. Leveling of deterioration wafer is 0.174, which is comparable with CIP wafer. No tool alarm occurred during litho or etching process.

*Table 3: Result of Deterioration Condition*

| Condition | CIP | Deterioration |
|-----------|-----|---------------|
| Backside Defect | | |
| Leveling Image | | |
| Leveling Range | 0.183 | 0.174 |

Table 4 is the TEM of CIP and deterioration wafer. After Mx BSCs process, deterioration wafer 's center area maintains 80.3nm poly silicon and 10.9nm SiN film, and the edge area remains 81.04nm poly silicon and 14nm SiN film. This result proves that the reserved SiN film is capable to cover the actual manufacturing process and ensure the flatness of backside condition.

*Table 4: The TEM figure of the backside film stack of deterioration condition.*

| Item | Center | Edge |
|------|--------|------|
| Image | | |
| Poly | 80.3nm | 81.04nm |
| SiN film | 10.9nm | 14nm |

**CONCLUSION**

We develop a simple method to maintain backside SiN film by replacing one step batch SMT Nitride remove with two step single + batch method. TEM show after multi BSCs, the SiN film can also be retained to protect poly silicon from high etch rate acid erosion. The leveling and flatness are efficiently improved, which is extremely promising for wafer backside uniformity improvement and wafer yield enhancement.

**REFERENCES**

[1] N. Wang, H. Chang, C. Chang and T. Wang, "Backside Wafer Damage Induced Wafer Front Side Defect and Yield Impact," 2007 IEEE/SEMI Advanced Semiconductor Manufacturing Conference, 2007, pp. 58-60.

[2] J. G. Zhou, H. Chen, Y. Long, K. Wang, H. Guo and F. Liu, "Backside Defect Monitoring Strategy and Improvement in the Advanced Semiconductor Manufacturing," 2021 China Semiconductor Technology International Conference (CSTIC), 2021, pp. 1-5.

[3] E. Balu, W. -T. Tseng, D. Jayez, J. Mody and K. Donegan, "Wafer backside cleaning for defect reduction and litho hot spots mitigation: DI: Defect inspection and reduction," 2018 29th Annual SEMI Advanced Semiconductor Manufacturing Conference (ASMC), 2018, pp. 216-221.

# METHODS OF REDUCING ETCHING DEFECTS IN BACK END OF LINE FOR SEMICONDUCTOR IN 28NM TECHNOLOGY

*Shanshan Chen[1], Hunglin Chen[1]\*, Yin Long[1], and Kai Wang[1]*

[1]Shanghai Huali Integrated Circuit Corporation, Shanghai 201314, China

*Corresponding Author's Email: chenhunglin@hlmc.cn

## ABSTRACT

As technology keeps shrinking to 28nm and below, Plasma etching is a key method to define high-resolution patterns in integrated chip manufacturing. To create structures in a chip, a pattern is formed in a photoresist by lithography and then the pattern is transferred to the device materials by plasma etching. In this paper, the defect adders in hard mask post dry etching is identified, and the removal characteristics of these defects in the back-end etching process are studied. Here we describe three ways to reduce the etching defects. Firstly, Argon purge and chamber pump down are adopted after process etching to remove wafer edge particle. Secondly, Performing reverse bias voltage discharge on the chuck of the wafer, the purse is to inhibit charge adsorption and prevent etching defects from adhering to the surface. Thirdly, the addition of scrubber after hard mask process also helps to remove large-size defects, with a removal rate of about 60%, which is conducive to reducing the impact on yield loss.

## KEYWORDS

Etching defects; Argon purge; Pump down; Reverse bias voltage discharge; Scrubber.

## INTRODUCTION

As the CD of advanced integrated circuits scaling down, 28nm metal line are remarkably thinner in cross-sectional area than previous generation devices, which lead to more increasing challenging to minimize RC delays and crosstalk in interconnects. Various parameters that would affect final device yield need to be optimized while new process technologies successful implementation and development. From the perspective of defect team, one of a key challenges is the detection of systematic yield-limiting defects and effectively solving potential process issues [1].

Plasma dry etching is the most difficult and comprehensive step in the manufacturing of semiconductor wafers, and the requirements for materials have a very high standard. A pattern can be taken shape in a photoresist by exposure and then redirected to the device materials by plasma etching, which helps create structures in a chip. With the ability to approach microcircuit structures on the order of 28nm, this technology faces grave challenges. Therefore, we should be better comprehending the surface reaction mechanisms that happened on the wafer and chamber walls during plasma etching to select the appropriate recipe for etching application [2, 3, 4].

A major challenge of dielectric etching in advanced chip design is the phenomenon of plasma damage. Defects usually occur at the edge of the wafer. Two phenomena are suspected to be the root cause of etching particle. The first is defects caused by sensitive structures at the wafer edge. The second maybe the instability of the plasma. Here we illustrate three methods for reducing etching defects. Firstly, adding Argon purge and pump down after etching process for removing surface defects. Secondly, a reverse bias voltage discharge is performed on the wafer's chuck and extending the period of time to suppress charge adsorption and prevent particles from sticking to the surface. Finally, by adding scrubber after Metal hard mask process also help removing large-size defects.

## RESULTS AND DISCUSSION

The architecture of metal loop integration is presented. An Etch stop layer and hard mask are deposited, which locate above low-k films, inter-metal structure is formed after hard mask (TiN) opening and dielectric etching.

The distribution of failed chips is wafer edge signature map. Typical etching defects wafer map signature post hard mask is shown in Figure 1(a), defects distribute at the wafer edge. Defects falling on the surface is judged that it induced after etching process, which would lead to copper loss at CMP process. TEM elemental cross-section shows good profile after line etching, but etching-induced particle on the surface leading to the copper difficult to fill in, as shown in Fig1(b),(c).

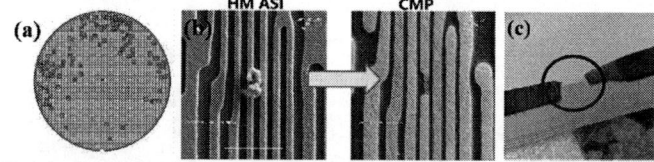

*Fig.1. Failure mechanism of hard mask standard physical structure, (a) Etching defect wafer map signature; (b)Top view defect Image; (c) Cross-sectional TEM image.*

Metal hard mask open is one of the metal etch applications that draws hard mask patterns for low-k trench etch applications. Meanwhile insert a cap oxide as an extra breakthrough step. For hard mask TIN main etching step, in order to consume more PR accompanied by lower selectivity, chlorine-based chemistry is selected as the source. In this situation, after the two processes, PR and BARC will be largely consumed. Since etching defect particles are usually observed falling on the oxide hard mask patterning, two possibilities are suspected to be the cause of etching particles. One is defects caused by sensitive structures at the wafer edge. The other maybe the instability of the plasma. The DC bias generated by RF coupling to wafer far edge maybe different from the wafer center. By impeding any possible wafer edge dielectric/metal exposure or conducting on the surface. Figure2. Shows the fail mode of this mechanism [4, 5].

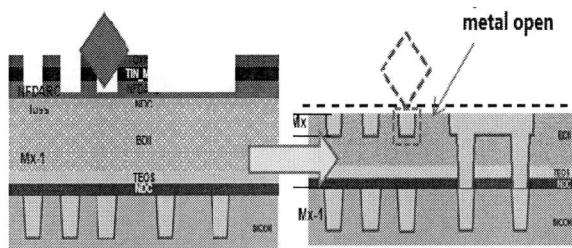

*Fig.2. Etching defects' fail mode diagram.*

A series of actions are conducted to solve these problems. Fig.3 shows the experimental conditions of the study. From the split results, for the purpose of removing particles, we first extend EBR to wafer edge. However, after the action, its defect signature is the same worse as baseline condition, which is not a valid solution. Base on this, process optimization is adopted. While we try adding Argon purge and chamber pump down after process etching, particles can be significantly reduced. Then by performing reverse bias voltage discharge on the chuck of the wafer with extending its time, the purpose is to inhibit charge adsorption and prevent particle from sticking to the surface. Adopting this method is also conducive to the great improvement of defects, as shown in Fig.3. Finally, because defects are falling on the surface, adding scrubber after hard mask process also help removing large-size defects, with a removal rate of about 60%. According to Fig.4, with these improvements, the etching defects have been greatly improved [5, 6].

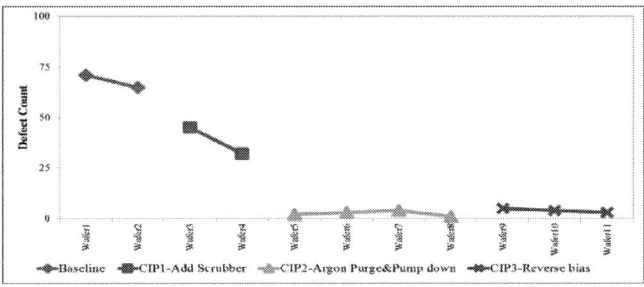

*Fig.3 Results of experimental conditions for etching defects improving.*

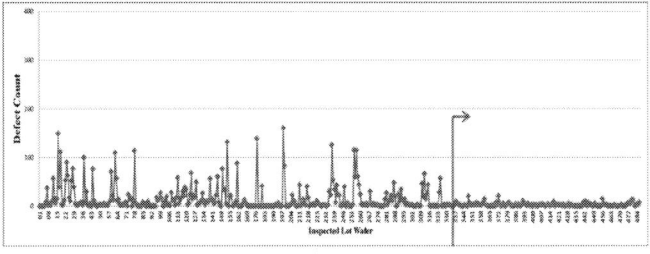

*Fig.4 Trend Chart after actions for etching defects.*

## CONCLUSION

This paper describes the removal of etching defects. The defects are commonly observed after the TIN hard mask patterning followed by resist ashing. Particles can be initiated at wafer edge and propagate inward through the conductive current path on the surface while wafer surface structure extends to the wafer edge. A series of actions have been conducted to improve defects, and help greatly reducing defect density and better enhancing yield.

## ACKNOWLEDGEMENTS

The authors would like to thank our YE teams for their support and HLMC FA department for the supports of TEM/EDS analysis, also Etch team for their technical discussion

## REFERENCE

[1] S.S Chen, H.L. Chen, "Methods of Reducing Metal Damager Defect in Back End of Line for Semiconductor in 28nm Technology," CSTIC, 2021.

[2] K. Nojiri, "Dry Etching Technology for Semiconductors," Springer International Publishing,

[3] Lieberman, Michael A., "Principles of Plasma discharge and material processing," Wiley Interscience.

[4] S. Ma, N. Hanabusa, B. Mays, S. Shoji, M. Kutney, T. Detrick, et al., "Backend Dielectric Etch Induced Wafer Arcing Mechanism and Solutuon," IEEE 2003 8th International Symposium Plasma and Process Induced Damage, pp. 178-181, 2003.

[5] Thorsten, Lill, Olivier, Joubert, "The cutting edge of plasma etching," Science 319, 1050 (2008).

[6] L Hong-Ji, H Che-Lun, CH Leng, NT Lian, CY Lu, "Etch Defect Characterization and Reduction in Hard-Mask-Based Al Interconnect Etching," International Journal of Plasma Science and Engineering, 2008, 1-5.

# HARD MASK OPEN TILTING IMPROVEMENT WITH ADVANCED SOURCE COIL

*Xiantao Luo\*, Qiang Ge, Ying Huang*
Applied Materials (China), Inc. China
E-mail: Xiantao_luo@amat.com; Qiang_ge@amat.com

## ABSTRACT

When 3D NAND technology developing from 64pair to 96pair and 128pair. Increasing ON/OP pairs will create higher aspect-ratio structures requiring vertical profiles and no tilting and resulting in more challenges for Etch. The increasing gate stack pairs and thicker film thicknesses will also require an optimized hard mask etch. Hard masks etch also plays a role with the profile and tilting which will influence the ON and Slit etch.

For Carbon hard mask etch, if the mask open profile is not vertical, it can impact the ON etch profile and worse case impact device properties. So, for vertical bow-free profiles and low/no tilting performance, Applied Materials® innovated Advanced Source Coil to improve the plasma uniformity and distribution to improve the profile and tilting performance.

This paper details the structure, the principle, and the etch profile and tilting improvement as a result of the Advanced Source Coil.

Keywords—Carbon mask, tilting, Advanced Source Coil, higher Aspect ratio

## INTRODUCTION

In recent years, flash storage is becoming more widely used and important, so in microelectronic fabrication, to have better yield, less defects, especially better electrical properties are more important at the 3D NAND flash storage array area. At customer sites for 3D NAND, the main challenges come from the high aspect etching such as silt trench ON etching, so the profile directly influences the properties, and further determines the device electrical properties, such as if tilting leads to ON damage, which also leads to electrical leakage. So, having a better profile is very important, especially for the high aspect etching which is also determined by hard mask profile and performance.

1. When performing the hard mask profile etch, the big challenge mainly comes from tilting, especially edge tilting performance, which mainly influence the mask profile and further directly influence the slit profile and performance as showed Figure 1. For the traditional methods with a Conventional ICP Coil to etch slit hard mask, the tilting is worse even with tuning by source current ratio (Ra), process gas distribution(FRC) and pressure the tilting still remained for the high-aspect ratio hard mask etch, such as thicker more than 3um thickness film.

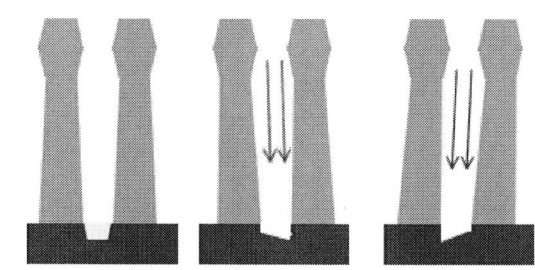

*Figure 1: Hard mask open tilting profile carton*

2. Advanced Source Coil was developed to improve the tilting performance from the structure and plasma distribution, especially the wafer extreme edge improvement for the plasma distribution and tilting direction, ON etch tilting and performance more matched with mask tiling, and further to improve the final performance combined with ON slit etch.

With the film thicknesses getting more and more pairs, such as the >3um or thicker film, the tilting performance getting more and more important and much enough influence on slit etch. So, the traditional etch chamber plasma distribution cannot meet the request even its best tilting performance. That was because the traditional plasma distribution decided by the source coil structure, which inner and outer coil vertical structure. That means the electricity and magnetic field distribution worse uniformity, the detail structure showed as Figure2. Combined the Figure 1, Figure 2 and the on-site tuning data, found that the tilting was worse, the hard mask profile is worse, especially since the top CD is larger while the bottom CD is smaller, which leads the mask worsening the CDU, so when performing ON slit etch, the sidewall damage worsens, top CD larger and bottom CD are smaller for the mask tilting is worse and the profile is worse, the plasma damages the top, the sidewall and cannot reach to the bottom as Figure1 showed.

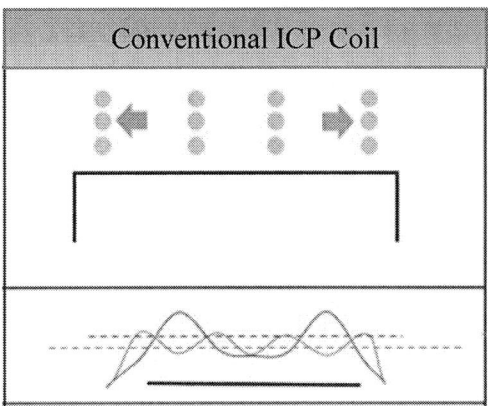

*Figure 2: Conventional source coil structure*

To improve the condition, the new designed source coil-Advanced Source Coil, especially optimize the structure to improve the plasma distribution more uniformity and tunable to make the plasma get vertical and tunable. The coil structure changed to inner vertical, but the outer coil changed to horizontal, which made inner and outer coil to "L" structure as showed Figure3, and this structure made plasma distribution more stable and better uniformity as showed Figure4. And at the same time, for the old source coil, the current ratio tunable range wider, but less sensitivity. But on the contrary, new designed Advanced Source Coil, the current ratio more sensitivity and the tilting improvement too much as Figure5 showed.

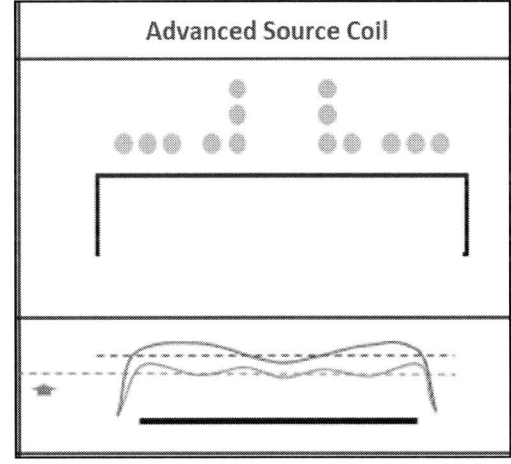

*Figure 3: Advanced Source Coil structure*

Some reference as below detail showed the new designed Advanced Source Coil on the mask tilting improvement:

1. The chamber more sensitive to the chamber condition and hardware, if the chamber condition or the hardware issue, the Advanced Source Coil result will show difference, which also help do chamber matching and process tuning.

2. The tilting improvement can help widen the ON etch window, and improve the ON etch profile and performance to improve the electrical properties, such as leakage caused by the sidewall damage, bottom vigor.

3. The tilting improvement versus the Conventional ICP Coil. Compared tilting performance result showed as Figure6.

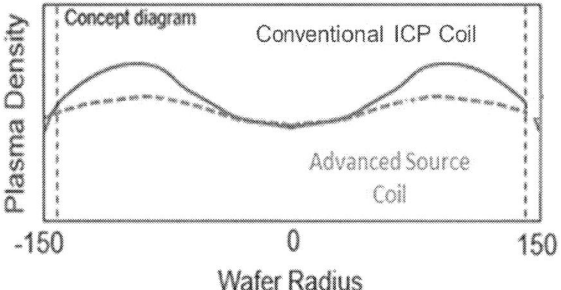

*Figure 4:  Plasma distribution comparison*

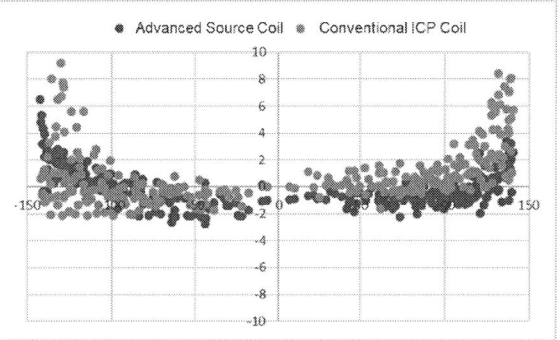

*Figure 5: Tilting performance of Conventional ICP Coil vs Advanced Source Coil tilting*

## ACKNOWLEDGEMENTS

In summary, as for the slit hard mask tilting with film getting thicker and thicker, Advanced Source Coil showed performance improved too much. And the result showed the totally tilting performance improved. We will further develop the performance with Advanced Source Coil.

Last, great appreciation and thanks for Applied Materials and customer provide the chance and source to do process development. And great thanks ETCH BU Xiaosong Ji, Liming Yang and Zhiqi Hu, they provide great coupon and process help to help us get the POR and great advice for the paper.

**REFERENCE**

[1] Xiaosong Ji, Qiang Ge. 3D flash process challenge and Insolution report, unpublished

[2] T. Tsuchiya, O. Tabata, J. Sakata and Y. Taga. *J. Microelectromech. Syst.*, vol. 7, 1998, pp. 106-113.

[3] Xiao. Hong. *3D IC Device and technologies*, 2016 published

[4] Andrew J. Walker. 3D Flush.

# CONSIDERATIONS IN THE SETTING UP OF INDUSTRY STANDARDS FOR PHOTOLITHOGRAPHY PROCESS, HISTORICAL PERSPECTIVES, METHODOLOGIES, AND OUTLOOK

*Qiang Wu*, Yanli Li*, Xianhe Liu, Xiaona Zhu, Shaofeng Yu, Wei Zhang*

School of Microelectronics, Fudan University, Shanghai 200433, China
*Corresponding Author's Email: wu_qiang@fudan.edu.cn

## ABSTRACT

Photolithography has been the driving force for the continuous design rule shrink where the manufacturing linewidth has been evolving from the early a few micrometers in the 1970's to the current several tens of nanometers and is due to develop into the future single-digit-nanometer technologies. Throughout the past years, both the Front-End-Of-the-Line (FEOL) device performance and the Back-End-Of-the-Line (BEOL) metal routing performance have been related to the photolithography process quality. There are several key parameters that we use to characterize the quality of a lithography process, such as, Exposure Latitude (EL), Depth of Focus (DoF), and relatively later added Mask Error Factor (MEF). All of these parameters are related to the linewidth uniformity, including, the within exposure field uniformity, the across wafer uniformity, and the across pattern uniformity (Optical Proximity Effect, OPE). For photoresists, the important parameters are the photo-sensitivity, the photoacid diffusion length, the PhotoAcid Generator (PAG) loading, the dissolution contrast, etc. Among all the parameters, there are trade-offs and possible areas for mutual compensation. The industry standard is a set of rules or practices that can best reflect the needs for performance, reliability, cost, and short time-to-market. In this paper, we will present our studies on the setting up of industry standards for photolithography with the analysis of the photolithographic process at typical past technology nodes and Extreme Ultra-Violet (EUV) technology nodes with future outlook and proposal.

## INTRODUCTION

### About Industry Standards

Any industry standard in semiconductor integrated circuit consists of a set of equipment, material, computation and automation algorithms, methodology in manufacture, and product/product platform that represents a good compromise in Performance (P), Power (P), Area(A) and Cost(C). It is a result of many years of focused research and optimization on equipment, material, computation and automation algorithms, and methodology in manufacture. Examples are x86 CPU platforms, 193 nm immersion process, Chemically Amplified photoResist (CAR) with low/high activation energy platforms, specific material and processes, such as Copper, Cobalt,

Ruthenium and Back-End-Of-the-Line (BEOL) processes, simulation algorithms, such as Transmission Cross Coefficient (TCC), Abbe, Source-Mask co-Optimization (SMO), Optical Proximity Correction (OPC) algorithms, etc., processes, process development and mask tapeout flows, process qualification methodologies, process development roadmaps, i.e., ITRS and IRDS roadmaps, "China's Integrated Circuit Development Roadmap"[1].

### Advantages in Setting Up Industry Standards

The advantages in setting up industry standards include, to have the quickest path to product development and production (returns); can maximally utilize all contributors in given eco-systems (efficiency); through the set up of reasonable specifications in equipment, material, and processes, can enable parallel efforts in technology development (time to market); through the set up of publicized standards, can encourage broadest participation including universities and research institutes (social and resource support); through the set up of publicized standards, can flag errors in time and keep the development on track and can avoid redundant or meaningless developing efforts.

### Who Needs Industry Standards

Anyone who needs to develop new technologies, to participate in technology development, to develop and to market any product, and to train students, engineers, and to teach classes can benefit from industry standards.

## THINGS INCLUDED IN PHOTOLITHOGRAPHY STANDARDS

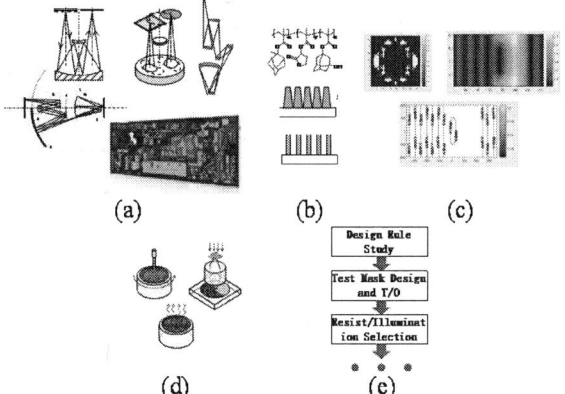

*Figure 1: Schematic diagrams and images showing (a) equipment, (b) material, (c) computation, (d) process, and (e) methodologies in lithographic process development.*

As we have pointed out, industry standards include equipment, material, computation, process, process window, process evaluation, methodologies in process development, etc., as depicted in Figures 1(a) through 1(e).

# HISTORICAL EXAMPLES IN PHOTOLITHOGRAPHY PROCESS DEFINITION WITH FUTURE OUTLOOK
## Historical Examples

0.13μm gate: Pitch: 310 nm, Exposure Tool: 248 nm Krypton Fluoride (KrF), Photoresist：CAR. The 0.13 mm technology uses KrF exposure tool and CAR photoresist extended from 0.18 μm technology. Due the pitch shrink from 430 nm to 310 nm, the Exposure Latitude (EL) dropped to about 10% from about 18% in 0.18 μm. As it has been tried, the improvement in the exposure tool, i.e., Numerical Aperture (NA) can not restore the EL to the 0.18 μm level, one Fab historically even used 193 nm (dry) exposure (3× more expensive!) for the first production of the 0.13 μm logic technology. Finally, the improvement of CAR photoresist has savaged the situation, achieving: EL=20.8%, Mask Error Factor (MEF)=1.53, as shown in Figure 2, simulated by CF Litho®.

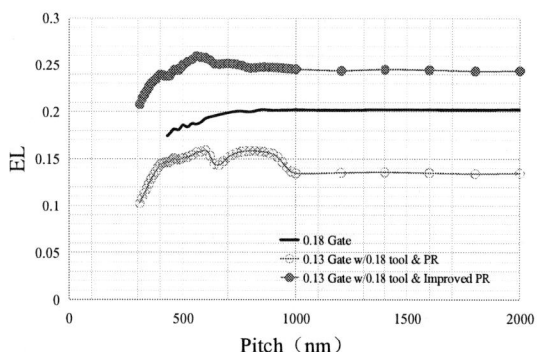

*Figure 2: Simulated 0.13 μm node line CD through pitch under 2 conditions vs.0.18 μm node reference.*

65 nm gate: Pitch: 180 nm~210 nm, Exposure Tool: 193 nm Argon Fluoride (ArF) (Dry), Photoresist：CAR. The 65 nm technology uses ArF exposure tool and CAR photoresist extended from 90 nm technology. Due the pitch shrink from 240 nm to an initially 180 nm, the EL dropped to about 10.6% from about 18.6%. Even the NA was later improved from 0.75 to 0.82, the EL improved to only 12.6%, With photoresist (PR) improvement, the EL climbed to 16.1%. Finally, with minimum pitch relaxed to

210 nm, the EL reached 19.9% (and MEF to 1.51), as shown in Figure 3.

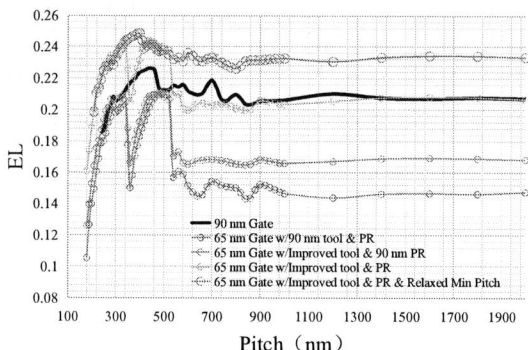

*Figure 3: Simulated 65 nm node line CD through pitch under 4 conditions vs. 90 nm node reference.*

28 nm gate: Pitch: 117~118 nm, Exposure Tool: 193 nm ArF immersion, Photoresist：CAR. The 28 nm technology uses ArF immersion exposure tool and CAR photoresist extended from 40 nm technology. Due the pitch shrink from 240 nm to ⩽118 nm, the EL dropped to about 14.8% (Annular illumination) and 13.1% (C-Quad illumination) even with PR improved to extreme. With XY polarization, the EL improved to 15.8% (Annular) and 14.4% (C-Quad). Finally, with Dipole illumination and XY polarization, the EL reached 23.2% (and MEF to 1.39), which requires uni-directional design for the gate lines, as shown in Figure 4.

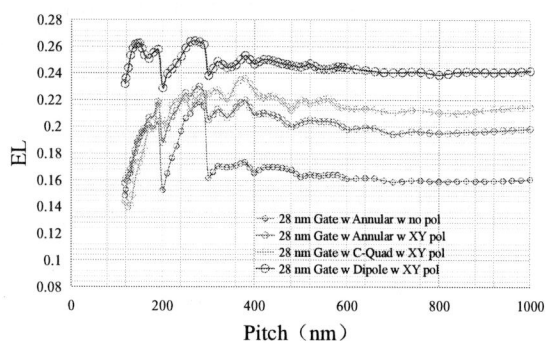

*Figure 4: Simulated 28 nm node line CD through pitch under 4 conditions.*

14 nm node: Gate Pitch: 84~90 nm, Metal Pitch: 64 nm, Exposure Tool: 193 nm ArF immersion, Photoresist: CAR. At 16/14 nm technology node and beyond, the Physical Gate Length (PGL) reaches below 20 nm with a gate Critical Dimension Uniformity (CDU) requirement of <2 nm, as shown in Table I. It also has a Line Edge Roughness (LE) requirement of <1.6 nm (difficult). Also, a gate pitch of 84~90 nm also pushes 193 nm immersion lithography with good contrast to the limit~ 18% EL, ~1.5

978-1-6654-9759-6/22 $31.00 © 2022 IEEE

MEF. The 14 nm is the last technology node that uses single exposure for gate. For metal layers, the 64 nm pitch (Double patterning with 90 nm single patterning pitch) with 32 nm trench (exposure trench width=45 nm) requires an etch bias of 45-32=17 nm, making it difficult to avoid residual. Finally, Negative Toned Developing (NTD), which can print 37/50 nm at 90 nm/isolated pitch, reduced the etch bias from 17 nm to 5 nm.

TABLE I. GATE PITCH AND CDU REQUIREMENT FOR VARIOUS TECHNOLOGY NODES

| Tech Node (nm) | Gate Pitch (nm) | Gate CDU (nm) | Ratio: Gate CDU/Gate HP | Physical Gate Length (nm) | Ratio: Gate CDU/Physical Gate Length |
|---|---|---|---|---|---|
| 130 | 310 | 5.3 | 3.4% | 65 | 8.2% |
| 90 | 240 | 3.7 | 3.1% | 45 | 8.2% |
| 65 | 210 | 3.3 | 3.1% | 32 | 10.3% |
| 45 | 180 | 2.6 | 2.9% | 25 | 10.4% |
| 40 | 162 | 2.3 | 2.8% | 23 | 10.0% |
| 32 | 130 | 2.4 | 3.7% | 29 | 8.3% |
| 28 | 118 | 1.8 | 3.1% | 24 | 7.5% |
| 22 | 90 | 2 | 4.4% | 20 | 10.0% |
| 20 | 90 | 1.8 | 4.0% | 18 | 10.0% |
| 16/14 | 87 | 2 | 4.6% | 17 | 11.8% |
| 10 | 66 | 2 | 6.1% | 20 | 10.0% |
| 7 | 54 | 1.8 | 6.7% | 18 | 10.0% |
| 5 | 50 | 1.6 | 6.4% | 16 | 10.0% |
| 3 | 42 | 1.4 | 6.7% | 14 | 10.0% |
| 2.1 | 32 | 1.2 | 7.5% | 12 | 10.0% |
| 1.5 | 32 | 1.2 | 7.5% | 12 | 10.0% |
| 1 | 32 | 1.2 | 7.5% | 12 | 10.0% |

10~7 nm node: Gate Pitch: 66/54 nm, Metal Pitch: 44/40 nm, Exposure Tool: 193 nm ArF immersion, Photoresist：CAR. To meet the stringent LER<1.4/1.3 nm (very difficult), and to make 66/54 nm gate pitch, Self-Aligned Multiple Patterning (SAMP) is used for the Front-End-Of-the-Line (FEOL) layers, i.e., the Fin and Gate. To meet the tight On-Product-Overlay (OPO) of 3.5~2.5 nm, the BEOL metal also adopts self aligned process: Self-Aligned Litho-Etch Litho-Etch (SALE2) process.

**Future Outlook**

For Extreme Ultra-Violet (EUV) and future logic processes, both front and back ends will use self-aligned methods to meet shrinking LER, LWR, and overlay specs. EUV will be used for contacts and metals. There has been many studies on EUV process [2-7], We have done a summary of exposure methods for various technology nodes based on many studies, as shown in Table II. Shown in Table III are pitches and lithography methods that extended to fin and via layers for the advanced technology nodes starting at 14 nm to 1 nm [1].

With the above photolithographic methods, the process window parameters, e.g., EL and MEF for FEOL gate and BEOL metals are studied through calibrated simulation with physical modeling from 0.25 µm through 1 nm logic technology nodes, as shown in Figures 5(a) and 5(b) [3]. The data show that for Deep-UV (DUV) process, the EL for the Gate is ≥18%~20% for the gates and is ≥13% for the metals and the MEF is ≤1.5 and ≤3.5, respectively, for the gate and metal layers. For EUV process, all layers are recommended to have EL ≥18%~20% and MEF≤1.5.

TABLE II. GATE AND METAL LITHOGRAPHY METHODS FOR VARIOUS TECHNOLOGY NODES

| Tech Node (nm) | Gate Pitch (nm) | Gate Lithography Method | Metal Pitch (nm) | Metal Lithography Method |
|---|---|---|---|---|
| 250 | 500 | 248 nm | 640 | 248 nm |
| 180 | 430 | 248 nm | 460 | 248 nm |
| 130 | 310 | 248 nm | 340 | 248 nm |
| 90 | 240 | 193 nm | 240 | 193 nm |
| 65 | 210 | 193 nm | 180 | 193 nm |
| 45 | 180 | 193 nm immersion | 160 | 193 nm immersion |
| 40 | 162 | 193 nm immersion | 100 | 193 nm immersion |
| 32 | 130 | 193 nm immersion | 90 | 193 nm immersion |
| 28 | 118 | 193 nm immersion | 90 | 193 nm immersion |
| 22 | 90 | 193 nm immersion | 80 | 193 nm immersion |
| 20 | 90 | 193 nm immersion | 64 | 193 nm immersion LELE |
| 16/14 | 87 | 193 nm immersion SADP | 64 | 193 nm immersion LELE |
| 10 | 66 | 193 nm immersion SADP | 44 | 193 nm immersion SALELE |
| 7 | 54 | 193 nm immersion SADP | 40 | 193 nm immersion SALELE |
| 5 | 50 | 193 nm immersion SADP | 30~32 | 0.33 NA EUV SALELE |
| 3 | 36~42 | 193 nm immersion SADP | 20~24 | 0.33 NA EUV SALELE |
| 2.1 | 32 | 193 nm immersion SAQP | 14~18 | 0.55 NA EUV SALELE |
| 1.5 | 32 | 193 nm immersion SAQP | 14 | 0.55 NA EUV SALELE |
| 1 | 32 | 193 nm immersion SAQP | 14 | 0.55 NA EUV SALELE |

TABLE III. FIN, GATE, METAL, AND VIA LITHOGRAPHY METHODS FOR VARIOUS TECHNOLOGY NODES

| Logic Tech Node | 14 nm 2015 | 10 nm 2017 | 7 nm 2019 | 5 nm 2021 | 3 nm 2024 | 2.1 nm 2027 | 1.5 nm 2030 | 1.0 nm 2033 |
|---|---|---|---|---|---|---|---|---|
| Fin Pitch (nm) | 48 193i SADP | 33 193i SAQP | 27~30 193i SAQP | 22.5~25 193i SAQP | 20 (?) 193i SAQP | 18 0.33NA EUV SADP | 14 0.33NA EUV SAQP | 14 0.33NA EUV SAQP |
| for Nano-plate/Nano-wires (nm) | | | | | 21 193i SAQP | 18 0.33NA EUV SADP | 14 0.33NA EUV SAQP | 14 0.33NA EUV SAQP |
| Gate Pitch (nm) | 78/84~90 193i SP | 66 193i SADP | 54~58 193i SADP | 48~50 193i SADP | 36~42 193i SADP/SAQP | 32 193i SAQP, 0.33NA EUV SADP | 32 193i SAQP, 0.33NA EUV SADP | 32 193i SAQP, 0.33NA EUV SADP |
| Metal Pitch (nm) | 64 193i LE2 | 44 193i SALE2 | 40 193i SALE2/ 0.33NA EUVSP | 30~32 0.33NA EUV SALE2 | 30~32 0.33NA EUV SALE2 | 14~18 0.55NA EUV SALE2 | 14 0.55NA EUV SALE2 | 14 0.55NA EUV SALE2 |
| Via Pitch (nm) | 64~80 193i LE2 | 44~62 193i LE2 | 40~56 193i LE3/ 0.33NA EUV SP | 36~50 0.33NA EUV SP/LE2 | 30~36 0.33NA EUV LE2 | 25 0.55NA EUV LE3/ EUV+DSA | 20 0.55NA EUV LE4/ EUV+DSA | 20 0.55NA EUV LE4/ EUV+DSA |

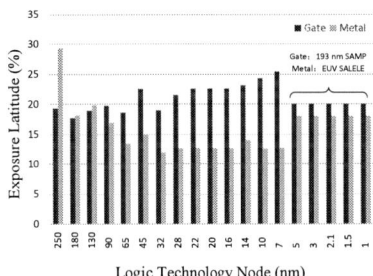

(a)

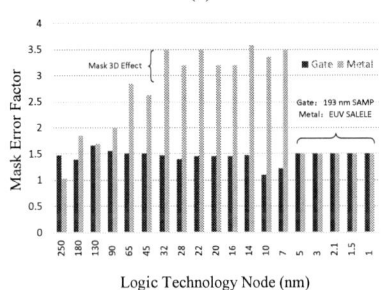

(b)

*Figure 5: Simulated gate and metal (a) EL and (b) MEF for various technology nodes.*

Shown in Figures 6(a) and 6(b) [3] are Gate LER, Metal LWR and On Product Overlay (OPO) requirements, respectively, for logic technology nodes 14, 10, 7, 5, 3, 2, 1.5, and 1 nm. All these can be well controlled through the use of self-aligned processes.

Shown in Figure 7 is the most important photoresist parameter, the Effective Photoacid Diffusion Length (EPDL) for various logic technology nodes. Numbers involving EUV process in Figures 5(a)-7 are numbers with the consideration to keep photon absorption stochastic induced defectivity $<10^{-12}$. An example of such study is shown in Figure 8 [4].

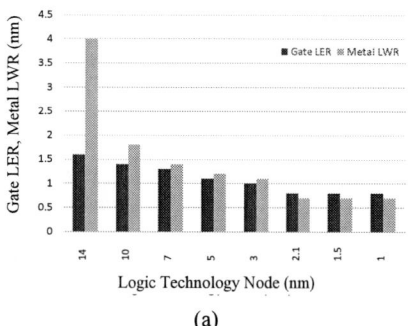

(a)

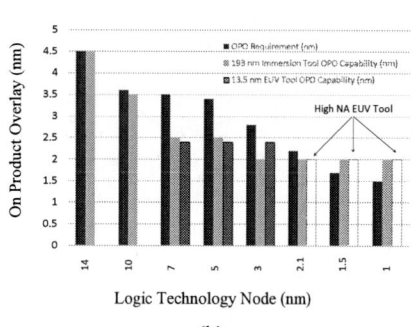

(b)

*Figure 6: Simulated gate and metal (a) Gate LER, Metal LWR and (b) On Product Overlay for various advanced technology nodes.*

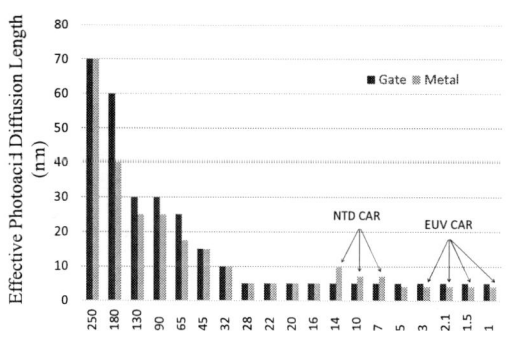

*Figure 7: Effective Photoacid Diffusion Length for various logic technology nodes.*

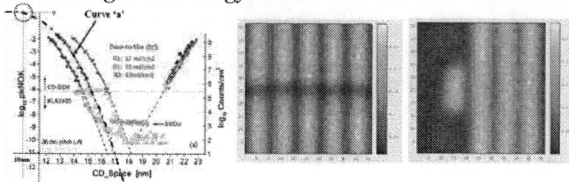

*Figure 8: Simulation with EUV Photon absorption stochastics from previous study.*

## Effect of Source Mask Optimization (SMO)

Here we discuss the effect of the SMO. Shown in Figure 9 are Source Optimization (SO) result for 8 typical patterns (with mask pre-optimized by experience). For the 1st one, the 90 nm pitch dense trench, the SO'ed source is a dipole in the X direction, the 2nd (pitch 150 nm trench), 3rd (Pitch 450 nm trench), and 4th (short line at dense array edge) are also similar: the SO has <u>significantly</u> improved EL, Depth of Focus (DoF) , and MEF. Also this is true for the last 2 patterns: the SO'ed sources are closer to dipoles with <u>significant</u> improvement in EL, DoF, and MEF.

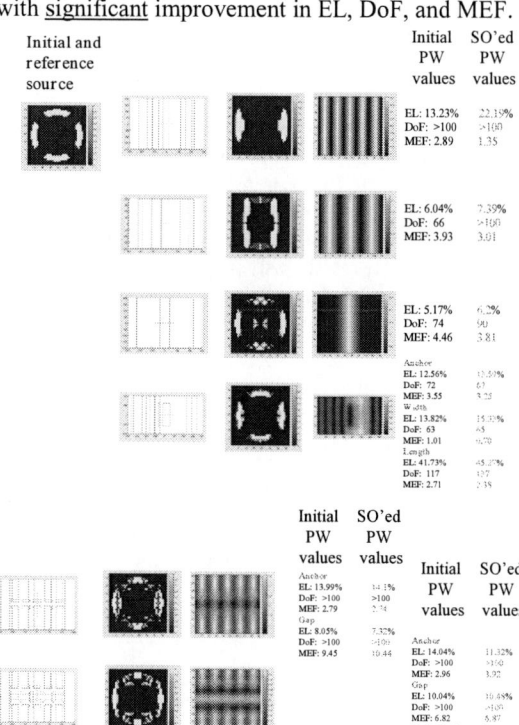

*Figure 9: Source optimization result for 8 typical patterns from top top to bottom, respectively, 90 nm pitch dense trenches, 150 nm pitch trenches, 450 nm pitch trenches, array edge 45×200 nm short line, 90 nm pitch dense trench ends, 90 nm dense trench ends to horizontal trench, dense 45×90 nm short trenches, and staggered dense 45×90 nm short trenches.*

But for the 5th and 6th (counted from the top) patterns, where significant cost function weightings are applied to the Y dimensions, the gain in process window through SO is <u>NOT</u> significant where the SO'ed source is not too different from the initial C-Quad. Experience says that on the average, an overall 8~10% process window gain can be expected from SMO at a given design. For example, if the pre-optimized EL is 12%, the optimized EL can be 12% (1.08~1.1) ≈12.96~13.2%, or about 1 percentage point gain. For process standardization consideration, the SMO is an important technique, but it will not affect major process strategy.

## PHOTOLITHOGRAPHY STANDARDS
### Process: EL

According to past and current practice, for DUV process, for the device making layers, such as Gate, the EL is >18%; for the interconnect layers, for lines/trenches, the EL is >13%, for line ends/trench ends, the EL is >10% with worst case 8%. According to current studies and practice, for EUV process, due to the existence of photon absorption stochastics, for all layers, for lines/trenches, the EL is >18%; for line ends/trench ends, the EL is >13% with worst case 10%.

### Process: DoF

DoF is determined by imaging condition, mechanical stage control, and leveling. A summary of DoF for various technology nodes are shown in Figure 10.

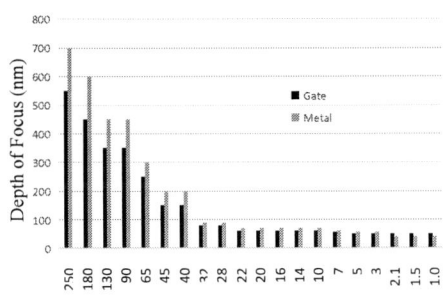

*Figure 10: DoF for various technology nodes for gate and metal layers.*

### Process: MEF

According to past and current practice, for DUV process, for the device making layers, such as Gate, the MEF is <1.5; for the interconnect layers, for lines/trenches, the MEF is <3.5, for line ends/trench ends, the MEF is <7 with worst case 10, for contact/vias, the MEF is <7.

According to updated studies and practice, for EUV process, due to the existence of photon absorption stochastics, for all layers, for lines/trenches, the MEF is <1.5~2, for line ends/trench ends, the MEF is <3~4% with worst case 5, for contacts/vias, the MEF is <3~4. Since the stochastics is highly related to CD, EL, MEF, and exposure dose. An overall consideration needs to be done.

### Exposure: Tooling, RETS

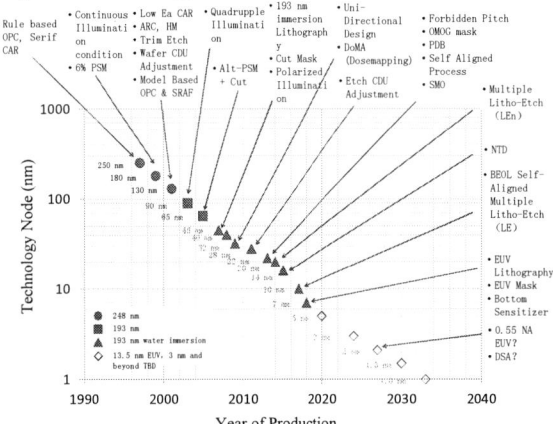

*Figure 11: RETs added to the photolithography process over the years. Acronyms: PSM: Phase Shifting Mask ARC: Anti-Reflection Coating, HM: Hard Mask, SRAF: Sub-Resolution Assist Features, OMOG: Opaque Molybdenum silicon On Glass, PDB: Photo Decomposable Base, DSA: Directed Self-Assembly.*

Over the years, quite many Resolution Enhancement Technologies (RETs) has been invented and implemented in photolithography process, as depicted in Figure 11, they become industry standards once the respective processes and integrated circuit products become mature.

EUV has been chosen as the primary exposure method for future logic generations starting at the 5 nm with following characteristics.

- as a high k1 process, not much RETs is needed
- with self-aligned patterning eco-systems: uni-directional design, pattern splits, and cut masks
- with acceptable photon-chemical deprotection reaction mechanisms in photoresist
- with scalable exposure energy-stochastics caused defectivity mitigation mechanism
- with acceptable exposure tool complexity and cost
- with acceptable pre- and post-tapeout mask processing complexity and cost, i.e. OPC, fracturing, mask making, mask inspection, etc.

DSA is competitive in cost. But the primary issue is to build a defectivity evaluation vehicle, such as a mature chip manufacturing flow with layer(s) processed with DSA, an infrastructure that can support from chip design to wafer manufacturing flow.

## Material: CAR, UL

There are 4 basic types of photo-chemical reaction for photoresists, as depicted in Figure 12. Within each type, the reactions, such as photochemical deprotection, catalyzed deprotection, ionization, photo-chemical cross-link reaction, etc. can be modeled.

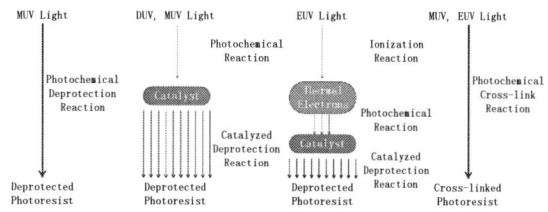

*Figure 12: 4 types of photo-chemical reaction for photoresists.*

Shown in Table IV is a sample physical photo-chemical reaction model for CAR developed based on Figure 12 for various logic technology nodes. The detailed numbers are proprietary and masked on purpose.

TABLE IV.  A PHYSICAL PHOTORESIST MODEL (EXAMPLE) THAT TABULATES REACTION DETAILS FOR VARIOUS LOGIC TECHNOLOGY NODES

| | Technology Node (nm) | 250 | 180 | 130 | 90 | 65 | 45 | 32/28 |
|---|---|---|---|---|---|---|---|---|
| Basic Process Conditions | Wavelength (nm) | 248 | 248 | 248 | 193 | 193 | 193 | 193 |
| | Refractive Index | 1 | 1 | 1 | 1 | 1 | 1.436 | 1.436 |
| | NA | | | | | | | |
| | BEOL CD (nm) | | | | | | | |
| | 焦深空间像 (nm) | | | | | | | |
| | 光刻胶膜厚 (nm) | | | | | | | |
| | 焦深 (nm) | | | | | | | |
| Photoacid Diffusion | PEB Temperature (℃) | | | | | | | |
| | EPDL (nm) | | | | | | | |
| | Molecules in Diffusion Sphere | | | | | | | |
| | Molecular Weight | | | | | | | |
| | Total Deprotection Volume/Unit Area (cm3) | | | | | | | |
| | Number of Molecules/Unit Area/Deprotection Volume | | | | | | | |
| Number of Photons-Photoacid Density | Exposure Energy (mj/cm2) | | | | | | | |
| | Total Absorption | | | | | | | |
| | Absorption QE | | | | | | | |
| | Number of Photons-Photoacid/Deprotection Volume | | | | | | | |
| Single Polymer Molecule | Maximum Number of Deprotection Reaction Needed for each Polymer Molecules | | | | | | | |
| | Deprotection Ratio (PTD) Number of Deprotection Reaction Needed for each Polymer Molecules | | | | | | | |
| | Number of Deprotection Needed/Unit Area | | | | | | | |
| | Number of Photons-Photoacid/Deprotection Volume/Unit Area | | | | | | | |
| | Number of Deprotection Reactions/Photoacid | | | | | | | |

Within the physical model, we can build industry standards for the following parameters: *absorption/ Quantum Efficiency (QE), number of catalyzed reactions/ catalyst, Effective Photoacid Diffusion Length (EPDL), PAG loading/blending uniformity, quencher loading/*

*blending uniformity, dissolution contrast in developing solution, BARC residual reflectivity, immersion topcoat water permeability, contact angles, resist model deviation from optical model, especially for NTD, EUV high energy photoelectron thermalized distance, EUV bottom sensitizer, or underlayer (EL) contribution,* etc.

## SUMMARY

We have presented a study of the photolithography technology over the past 20 years. We have analyzed key technology nodes, process requirement, and provided a review of the key process technologies that are instrumental to meet the process requirement. Our conclusion is that the photolithographic processes have always made efforts to meet the requirements that propels the continued success of the integrated circuit industry following the Moore's law tightly. We propose that we build industry standards for the photolithography technology based on the sets of requirement, relevant process technologies, process development/optimization methodologies, process evaluation parameters, materials, etc. We propose continued research on the standards and extend the standards and the ways of working in the standard research to the entire photolithographic technology eco-system development efforts in China.

## ACKNOWLEDGEMENTS

The authors of this paper would like to thank School of Microelectronics Fudan University for the support of this work.

## REFERENCES

[1] National Integrated Circuit Innovation Center, "2019 China's Integrated Circuit Development Roadmap".

[2] Qiang Wu, Yanli Li, et al., "A Study of Image Contrast, Stochastic Defectivity, and Optical Proximity Effect in EUV Photolithographic Process under Typical 5 nm Logic Design Rules", *Proc. CSTIC2020, IEEE Xplore.*

[3] Qiang Wu, Yanli Li, et al., "The Law that Guides the Development of Photolithography Technology and the Methodology in the Design of Photolithographic Process", *Proc. CSTIC2020, IEEE Xplore.*

[4] Yanli Li, Qiang Wu, et al., "A Simulation Study for Typical Design Rule Patterns and Stochastic Printing Failures in a 5 nm Logic Process with EUV Lithography", *Proc. CSTIC 2020, IEEE Xplore.*

[5] D. De Simone, L. Kljucar, P. Das, et al., "28nm pitch single exposure patterning readiness by metal oxide resist on 0.33NA EUV Lithography", *Proc. SPIE 2021, 11609, 116090Q.*

[6] J. Guo, J. Church, L. Meli et al., "EUV single exposure via patterning at aggressive pitch", *Proc. SPIE 2021, 11609, 116090U.*

[7] Q. Lin, T. Hisamura, N. Chong et al., "Optimization

of the EUV Contact Layer Process for 7nm FPGA Production", *Proc. SPIE 2021, 11609, 116090V.*

# THE STATUS OF STOCHASTIC ISSUES OF EUV LITHOGRAPHY
## -PHOTON STOCHASTIC AND CHEMICAL STOCHASTIC-

*Toru Fujimori*
Electronic Materials Research Laboratories
FUJIFILM Corporation
4000 Kawashiri, Yoshida-cho, Haibara-Gun, Shizuoka, 421-0396, Japan
toru.fujimori@fujifilm.com

## ABSTRACT

Extreme ultraviolet (EUV) lithography is ready for realize 7 nm generation node manufacturing and beyond. With recent rapid progress on the source power improvement, process and material explorations are more and more accelerated to achieve HVM requirements. Therefore, a key factor for the realization of EUV lithography is the choice of EUV resist materials that are capable of resolving below 15 nm half pitch with high sensitivity. However, the performance of EUV resist is still not enough for the true HVM requirements, even by using the latest qualifying EUV resist materials. One critical issue is the stochastic issues, which will be become 'defectivity', like nano-bridge or nano-pinching.

We report herein the status of stochastic issues of EUV lithography, 'Photon Stochastic' and 'Chemical Stochastic'.

## INTRODUCTION

EUV lithography is most promising process for alternative to 193 nm immersion multiple lithography, realizing 7 nm generation node manufacturing and beyond. With recent rapid progress on source power improvement [1], process and material explorations are more and more accelerated to achieve HVM requirements. Therefore, a key factor for the realization of EUV lithography is the choice of EUV resist material that is capable of resolving below 15-nm half pitch with high sensitivity [2]. However, the performance of EUV resist is still not enough for the true HVM requirements, even by using the qualified EUV resist materials. One critical issue is 'stochastic issues', which will be become 'defectivity' [3]. Therefore, the status of stochastic issues of EUV lithography, photon stochastic and chemical stochastic, have to be understood [4].

## RESULTS AND DISCUSSION

### Concept of the reducing stochastic effect

In the past, speaking of the stochastic issue of EUV lithography was basically considered from low photon number from EUV light source, which means 'photon shot noise'. It was still critical concerning point of the stochastic issue, even with recent progress on source power improvement. However, the stochastic issue is not only from them but also from EUV materials and processes, called 'Chemical Stochastic' [5]. The 'Chemical stochastic' means caused from resist materials and processes for lithography, materials uniformity in the film, catch the photon efficiency, reactive uniformity in the film, and dissolving behavior with the developer. Each step must be reduced each stochastic issue to improve the quality of lithography performance (Figure 1, 2).

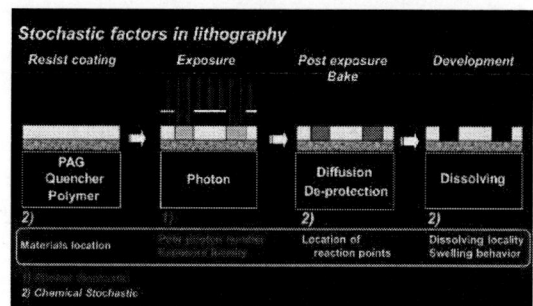

*Figure 1: The stochastic factors in lithography.*

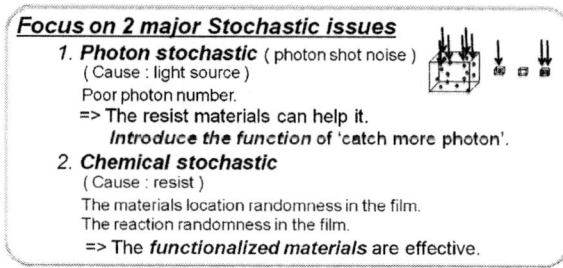

*Figure 2: 2 major stochastic issues and their points.*

**The results of photon shot noise reducing by using 'Organic high EUV absorption materials'.**

Photon stochastic, photon shot noise effect, on EUV lithography was well-known issue and various studies have emphasized its impact on LWR performance [6]. As well as LWR, the photon shot noise effect on defectivity must be considered to obtain pattern quality satisfying HVM requirements. By using the high absorption material observed higher image contrast by simulation study. So, the high absorption materials should be effective for improving the lithographic performance. Therefore, 'Organic high EUV absorption materials' was designed and synthesized, which expected to be able to catch photon more efficiency from the source (Fig. 3) [7].

As a result of the lithographic performance, 'Organic high EUV absorption materials', showed 20% dose reduction with keeping its LWR value. Besides that, nano-bridge is clearly decrease on the high absorption resist (Fig. 4) [7].

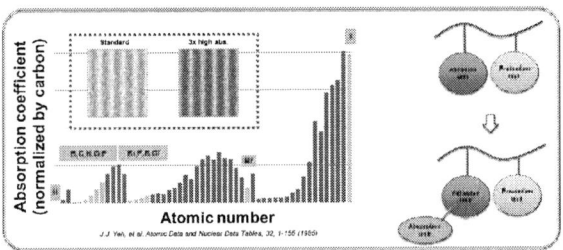

*Figure 3: The simulation study of high EUV absorption materials and design concept of 'Organic high EUV absorption materials'.*

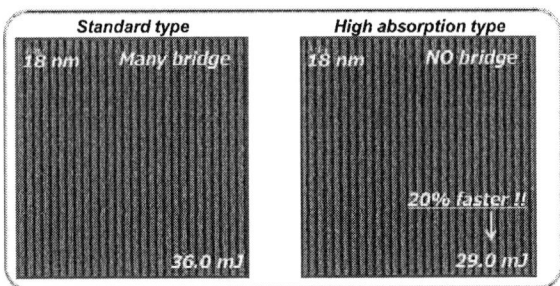

Excellent bridging performance with high sensitivity !!

*Figure 4: The lithographic performance of '*
*'Organic high EUV absorption materials'.*

**The results of 'Chemical Stochastic' improvement by using novel functionalized materials.**

The other hand, 'Chemical stochastic', the materials location randomness, could be observed in the film. A traditional CAR material is mainly composed from polymers, PAGs (Photo Acid Generator) and quenchers, which locate variously in the film. The location locality of the materials in the film induced solubility randomness during the development process, which observed worse LWR and defectivity, like nano-bridge and nano-pinching. 'The novel functionalized materials' was designed and synthesized to introduce connection unit, interaction unit and bounded to approach the uniform film preparation. 'The novel functionalized materials' observed better lithographic performance (Fig. 5) [8] [9].

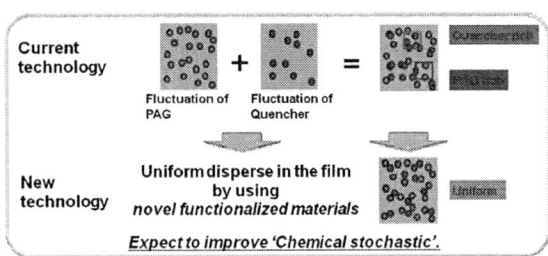

*Figure 5: The concept of novel functionalized materials.*

As a result, new resist, which introduced new designed materials, observed better lithographic performance [8] [9].

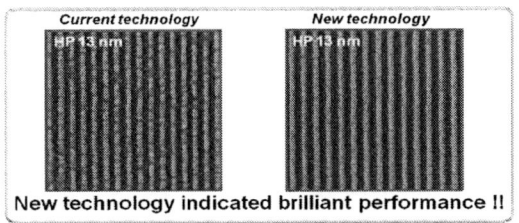

*Figure 6: The lithographic performance of new materials.*

**The results of 'Chemical Stochastic' improvement by using negative-tone imaging (NTI) processes.**

Also, the dissolving stochastic with developer, a kind of 'Chemical Stochastic', could be observed during the development process due to their swelling behavior. One of the key items to improve them was Negative-tone imaging (NTI, using organic solvent-based developer) process.

Negative-tone imaging (NTI) process has been developed to expand ArF immersion lithography by FUJIFILM [10]. Previously, negative-tone system means crosslink system with conventional TMAH developer. However, the crosslink system with TMAH developer observed huge swelling, so it does not suit for preparing the small CD patterns. NTI process is a different idea, which is developed with organic solvent to prepare the negative images (Fig. 7).

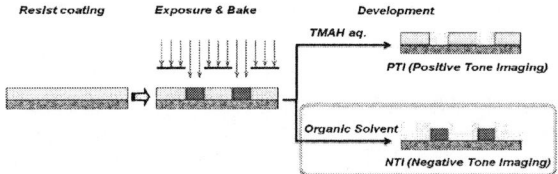

*Figure 7: The process flow of lithography, PTI and NTI.*

NTI process provided lower swelling and smoother dissolving behavior than Positive-tone imaging (PTI) process, which reduced the dissolving stochastic. Therefore, negative-tone imaging with EUV exposure (EUV-NTI) has huge advantages for the performance, especially for improving LWR, which will be expected to resolve RLS trade off. Also, NTI system has already introduced to manufacturing with ArF exposure. So, it seems not so difficult to use EUV-NTI for manufacturing.

Comparison of dissolving behavior between positive-tone and negative-ton by using in-situ high speed AFM, NTI process observed less swelling than PTI process (Fig. 8). The lithographic performance of NTI process observed 60% better LWR performance than PTI process under the same exposure dose owing to their smooth-dissolving behavior and non-swelling properties (Fig. 9). [11][12][13][14].

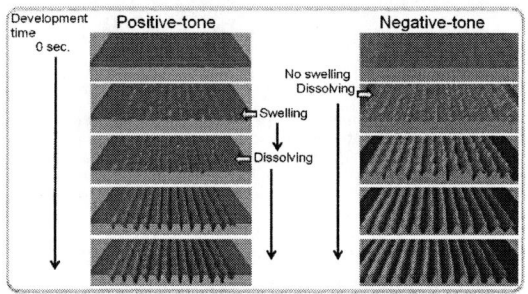

*Figure 8: The dissolution behavior comparison between PTI and NTI by using in-situ high speed AFM.*

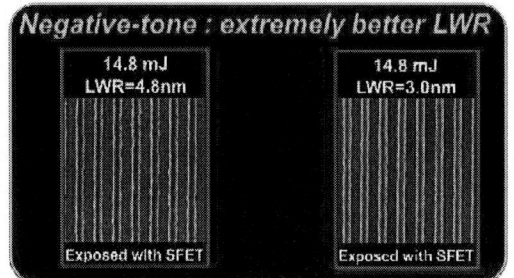

*Figure 9: The lithographic performance comparison between PTI and NTI.*

Finally, the latest lithographic performances of both PTI and NTI were introduced, which observed brilliant performance for each application.

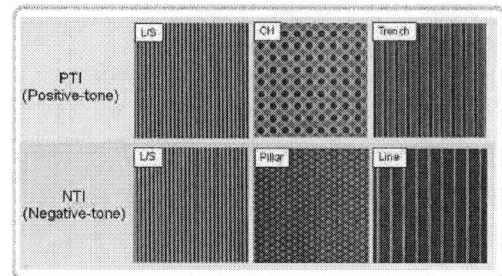

*Figure 10: The latest lithographic performances of both PTI and NTI for each application, example.*

## CONCLUSION

This study hereby showed excellent improving the lithographic performance due to reduce the stochastic issue, 'Photon stochastic' and 'Chemical stochastic'. These technologies were expected to apply the real EUV lithography HVM.

## REFERENCES

[1] A. A. Schafgans, D. J. Brown, I. V. Fomenkov, Y. Tao, M. Purvis, S. I. Rokitski, G. O. Vaschenko, R. J. Rafac, D. C. Brandt, *Proc. SPIE 10143, 1014311 (2017).*

[2] H. Furutani, M. Shirakawa, W. Nihashi, K. Sakita, H. Oka, M. Fujita, T. Omatsu, T. Tsuchihashi, N. Fujimaki, and T. Fujimori, *J. Photopolym. Sci. Technol. (2018).*

[3] P. De Bisschop, J. Van de Kerkhove, J. Mailfert, A. Vaglio Pret, J. Biafore, *Proc. SPIE 9048, 904809 (2014).*

[4] T. Fujimori, *34th International Microprocesses and Nanotechnology Conference (2021).*

[5] T. Fujimori, *International symposium on extreme ultraviolet lithography*, **11147** (2019).

[6] Suchit Bhattarai, Weilun Chao, Shaul Aloni, Andrew R. Neureuther, Patrick P. Naulleau, "Analysis of shot noise limitations due to absorption count in EUV resists" *Proc. SPIE 9422, 942209 (2015).*

[7] T. Fujimori, *International symposium on extreme ultraviolet lithography* (2018).

[8] T. Fujimori, *CSTIC 2020.*

[9] T. Fujimori, *IWAPS 2020.*

[10] S. Tarutani, S. Kanna, H. Tsubaki, *Proc. SPIE 6923, 69230G (2008).*

[11] T. Fujimori, **2-7**, *Advanced Technologies for Functional Materials and Process Optimization, CMC books* (2017).

[12] T. Fujimori et al., *Proc. SPIE* **9425**, 942505 (2015).

[13] T. Fujimori et al., *J. Photopoly., Sci. Technol.*, **Vol. 28**, Issue 4, 485 (2015).

[14] T. Fujimori, *CSTIC 2021.*

# PROCESS WINDOW, AND PROCESS OPTIMIZATION IN BOTH LOW AND HIGH NA EUV PROCESSES FOR ADVANCED LOGIC TECHNOLOGIES NODES

*Yanli Li\*, Qiang Wu\*, Xianhe Liu, Xiaona Zhu, Shaofeng Yu*

School of Microelectronics, Fudan University, Shanghai 200433, China

\*Corresponding Author's Email: li_yanli@fudan.edu.cn

## ABSTRACT

As we know, if we consider Extreme Ultra-Violet (EUV) stochastics induced defectivity, the minimum pitch for the line/space is around $36 \sim 40$ nm and $48 \sim 50$ nm for the contact holes and vias in 0.33 NA EUV lithography. From the current roadmap, the ultimate minimum pitch of the metal layer in EUV lithography is 14 nm, which will be realized by double patterning with high NA EUV tool. After some studies, we believe that the optimum pitch for the line/space is believed to be around 28 nm and $38 \sim 40$ nm for the contact holes and vias with 0.55 NA EUV lithography. We have done a simulation study for some typical patterns with the anchoring metal pitch of 28 nm in high NA EUV lithography with self-developed program. Generally speaking, critical structures in advanced logic technologies nodes are Tip-to-Tip structures, the minimum area structures, etc. For these patterns, we have briefly studied the influence of Source Optimization (SO) on and the impact from coma Y to the process window and we have found that the contribution of SO is not significant (<5%) and we have determined the specification on the optical aberration based on our understanding in the requirement from process window. Due to the high k1 nature of EUV lithography, we have also done a study on a 45° local interconnect pattern in a 3 nm CFET SRAM with both low NA and high NA EUV processes.

## INTRODUCTION

Recent literature indicates that the optimum pitch for the line/space is around $36 \sim 40$ nm and $48 \sim 50$ nm for the contact holes under 0.33 Numerical Aperture (NA) Extreme Ultra-Violet (EUV) single exposure processes [1]. With the further advances of logic process nodes, the high NA EUV will be introduced at more advanced technology nodes, such as 2 nm, 1 nm nodes. According the IRDS roadmap, the minimum pitches of the metal and via layers, respectively, will ultimately be 14 nm and 20 nm [2]. Therefore, the anchor pitch for the line/space and via, respectively, is around 28 nm and $38 \sim 40$ nm in 0.55 NA EUV lithography in this paper.

In our previous papers [3], we have described the algorithm and multi-layer mask film stack used by our self-developed simulation program. Compared with the 6° of 0.33 NA, the Chief Ray Angle at Object space (CRAO) of incidence in 0.55 NA EUV lithography is 5.355° [4].

High NA EUV lithography uses the anamorphic 8:1 and 4:1 reduction ratios along the scanning direction (Y-axis) and perpendicular direction (X-axis), respectively. While low NA EUV lithography uses the same 4:1 reduction ratios along both Y- and X-axis. In this paper, we focus on the role of Source Optimization (SO) and coma Y (Z8) on the process windows of Tip-to-Tip (TtT) and isolated trench patterns in both low and high NA EUV processes. And we will give the specification of aberration for high NA EUV. Starting from the 3 nm logic technology node, in order to realize a smaller SRAM unit cell, Complementary Field Effect Transistor (CFET) structure has been introduced [5]. As an application of our lithography study, we have built a tentative layout and a flow of a 3 nm CFET SRAM, within which a key 45° local interconnect pattern is used and studied with both low and high NA EUV lithography simulations.

## RESULTS AND DISCUSSION

### H-V Bias and M3D in both Low and High NA EUV

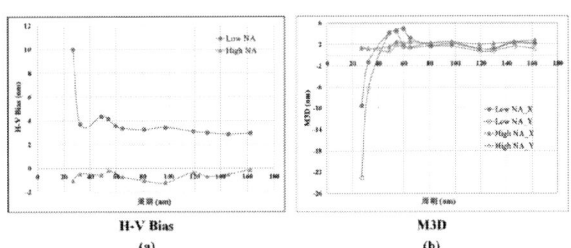

*Figure 1: The comparison of the H-V Bias and M3D between low and high NA EUV.*

Figure 1 shows the Horizontal and Vertical (H-V) Bias and Mask 3D (M3D) scattering effect of 1D through pitch from 27 to 162 nm in both low and high NA EUV conditions. Owing to the CRAO, there is a Critical Dimension (CD) bias for horizontal and vertical patterns, or so-called H-V Bias, as shown in Figure 1 (a). For low NA EUV lithography, the CD along the Y direction is larger than that along the X direction with about $3 \sim 4$ nm. However, for high NA EUV, the H-V Bias is about within ± 1 nm, which is not as significant. This is due to the 8:1 reduction ratios along the Y direction. Therefore, EUV light is more easily to be reflected from the mask stack in high NA. Similarly, from Figure 1 (b) we can find that the M3D effect (focus deviation) for the 1D through pitch in low NA EUV is greater than 15 nm, while in high NA EUV, this value is less than 2 nm.

## SO simulation results of typical metal layer patterns in both Low and High NA EUV processes

In this section, we study the influence of SO on process window of typical metal layer patterns in both low and high NA conditions.

### The simulation results of TtT pattern in both Low and High NA EUV processes

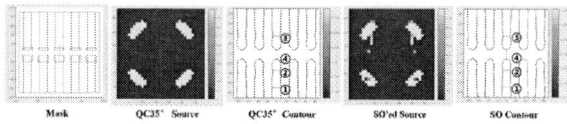

*Figure 2: The mask, initial Source, SO'ed Source and simulation contours of TtT pattern in low NA EUV lithography.*

TABLE I.    THE SIMULATION RESULTS OF FIGURE 2

| SO'ed Source | | | | |
|---|---|---|---|---|
| Line cut | ① | ② | ③ | ④ |
| CD (nm) | 18 | 16.24 | 15.6 | 19.7 |
| EL | 20.31% | 16.77% | 15.49% | 15.96% |
| DOF(nm) | >100 | >100 | >100 | >100 |
| QC35° Source | | | | |
| Line cut | ① | ② | ③ | ④ |
| CD (nm) | 17.54 | 15.24 | 14.55 | 19.73 |
| EL | 18.07% | 14.20% | 12.90% | 16.08% |
| DOF(nm) | >100 | >100 | >100 | >100 |
| ΔCD (nm) | 0.46 | 1.00 | 1.05 | -0.03 |
| Δ CD (%) | 2.62 | 6.56 | 7.22 | -0.15 |
| Δ EL | 2.24% | 2.57% | 2.59% | -0.12% |
| Δ EL (%) | 12.40 | 18.10 | 20.08 | -0.75 |

A typical mask of TtT pattern, initial Source, SO'ed Source and simulation contours are shown in Figure 2. The initial Source is a Quasar 35° with 0.9-0.7 partial coherence factor and 0.33 NA. The PhotoResist (PR) is a typical 30 nm thick Chemically Amplified photoResist (CAR). The main pattern pitch is 36 nm and mask trench CD is 18 nm. The anchor point is also set to 36 nm pitch and 18 nm After-Developing-Inspection (ADI) CD. As documented previously [1], the line cut ① is the dense pattern, taking EUV stochastics impact into account, the Exposure Latitude (EL) must be ≥ 18%, ④ is the TtT line cut whose EL must be ≥ 13%. Line cuts ② and ③ are the trenches width near the tip, which are used to assure that manual Optical Proximity Correction (OPC) is adequate for the study of 2D process limit.

The process window data of these two sources are shown in Table I, from which, we have found that the EL at all line cut positions satisfy our target requirement with TtT CD close to 20 nm in 0.33 NA EUV lithography. In the simulation, the simple OPC of the mask does not make the CD of the dense pattern to 18 nm, but it (line cut ①) can be achieved through source optimization, and the EL

can be improved by 2 percentage points. At the same time, the CDs of line cuts ② and ③ can be improved by 1 nm. Whether before or after SO, the Depth of Focus (DoF) of these four line cuts is greater than 100 nm. This shows that it is not easy to quantify the effect of SO, because the EUV process is a high k1 process. Therefore, when the mask optimization is adequate, source optimization becomes less important compared to 193 nm immersion process. This conclusion applies to the following part of this paper.

Figure 3 shows a typical mask of TtT pattern, initial Source, SO'ed Source and simulation contours in high NA EUV lithography. The initial Source is a Quasar 35° with 0.7-0.5 partial coherence factor and 0.55 NA. The PR is a typical 30 nm thick CAR. The main pattern pitch is 28 nm and mask trench CD is 15 nm. The line cut ① is the dense pattern, taking EUV stochastics impact into account, the EL must be ≥ 18%, ④ is the TtT line cut whose EL must be ≥ 13%.

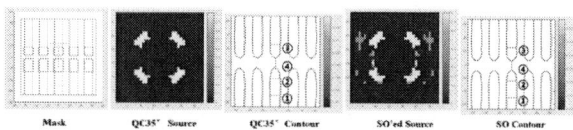

*Figure 3: The mask, initial Source, SO'ed Source and simulation contours of TtT pattern in high NA EUV lithography.*

TABLE II.    THE SIMULATION RESULTS OF FIGURE 3

| SO'ed Source | | | | |
|---|---|---|---|---|
| Line cut | ① | ② | ③ | ④ |
| CD (nm) | 15 | 14.75 | 14.56 | 15.76 |
| EL | 18.30% | 17.59% | 17.19% | 13.07% |
| DOF(nm) | 100 | 90~100 | 90~100 | 90~100 |
| QC35° Source | | | | |
| Line cut | ① | ② | ③ | ④ |
| CD (nm) | 15 | 14.32 | 14.35 | 15.66 |
| EL | 18.62% | 16.97% | 17.04% | 13.18% |
| DOF(nm) | 100 | 60 | 60 | 60 |
| ΔCD (nm) | 0.00 | 0.43 | 0.21 | 0.10 |
| Δ CD (%) | 0.00 | 3.00 | 1.46 | 0.64 |
| Δ EL | -0.32% | 0.62% | 0.15% | -0.11% |
| Δ EL (%) | -1.72 | 3.65 | 0.88 | -0.83 |

The process window of these two sources is shown in Table II, from which, we have found that the CD and EL at all line cut positions satisfy our target requirement with TtT CD close to 16 nm in 0.55 NA EUV lithography. After SO, the CD and EL of line cut ② can be improved by 0.5 nm and 0.6 percentage points, respectively. However, the CDs and ELs of other line cuts remain basically the same with that of the initial source. With the NA increasing from 0.33 to 0.55 NA, the DoF decreased significantly. Line cut ① still has a larger DoF, which is about 100 nm, while the DoFs of other line cuts are only ~ 60 nm. After SO, the DoF of all line cuts can be increased to larger than 90 nm. From our simulation, when the mask optimization is adequate, the improvement of CDs and

ELs is not significant, while the improvement of DoF is notable and the lines are more straight along Y direction.

From above studies, we conclude that the significant advantage of SO under high NA process is to improve DoF. This conclusion also applies to the following part of this paper.

**The simulation results of isolated trench pattern in both Low and High NA EUV processes**

*Figure 4: The mask, initial Source, SO'ed Source and simulation contours of isolated trench pattern in low NA EUV lithography.*

TABLE III. THE SIMULATION RESULTS OF FIGURE 4

| SO'ed Source | | | | |
|---|---|---|---|---|
| Line cut | ① | ② | ③ | ④ |
| CD (nm) | 18 | 18.05 | 19.95 | 18.7 |
| EL | 19.41% | 14.97% | 18.48% | 15.97% |
| QC35° Source | | | | |
| Line cut | ① | ② | ③ | ④ |
| CD (nm) | 18 | 18.01 | 20.19 | 18.7 |
| EL | 19.30% | 14.78% | 18.87% | 15.89% |
| ΔCD (nm) | 0.00 | 0.04 | -0.24 | 0.00 |
| Δ CD (%) | 0.00 | 0.22 | -1.19 | 0.00 |
| Δ  EL | 0.11% | 0.19% | -0.39% | 0.08% |
| Δ EL (%) | 0.57 | 1.29 | -2.07 | 0.50 |

Shown in Figure 4 is another typical metal layer design situation, which is a minimum isolated area on one side of dense line/space. The anchor point is set to 36 nm pitch and 18 nm ADI CD. In our previous papers [3], we have learnt that in order to achieve an EL more than 13% for the isolated trench at line cut ② position, the minimum area in low NA EUV must be at least 5 times that of minimum pixel squared, i.e. $5 \times 18 \times 18$ nm$^2$, with about 18 nm contour CD. Line cut ① is the dense pattern, and the EL must be $\geq$ 18%. It can be seen that SO'ed source has a little change from the initial source and the improvement of CDs and ELs is not significant. The reason for the above phenomenon is that k1 is large and it's easy to achieve our target by OPC.

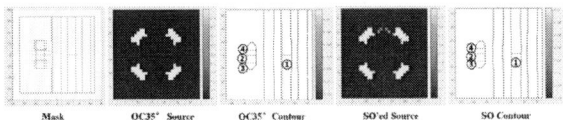

*Figure 5: The mask, initial Source, SO'ed Source and simulation contours of isolated trench pattern in high NA EUV lithography.*

TABLE IV. T THE SIMULATION RESULTS OF FIGURE 5

| SO'ed Source | | | | |
|---|---|---|---|---|
| Line cut | ① | ② | ③ | ④ |
| CD (nm) | 15 | 15.03 | 15.85 | 15.22 |
| EL | 18.05% | 18.49% | 18.98% | 18.57% |
| QC35° Source | | | | |
| Line cut | ① | ② | ③ | ④ |
| CD (nm) | 15.16 | 15.16 | 16 | 15.36 |
| EL | 18.19% | 18.81% | 19.31% | 18.89% |
| ΔCD (nm) | -0.16 | -0.13 | -0.15 | -0.14 |
| Δ CD (%) | -1.06 | -0.86 | -0.94 | -0.91 |
| Δ  EL | -0.14% | -0.32% | -0.33% | -0.32% |
| Δ EL (%) | -0.77 | -1.70 | -1.71 | -1.69 |

Figure 5 and Table IV show the simulation results of the isolated trench pattern in high NA EUV lithography. The anchor point is set to 28 nm pitch and 15 nm ADI CD. In our previous papers [6], we have known that in order to achieve an EL more than 13% of the isolated trench at line cut ② position, the minimum area in high NA EUV must be at least 3 times that of minimum pixel squared, i.e. $3 \times 15 \times 15$ nm$^2$, with about 15 nm contour CD. Line cut ① is the dense pattern, and the EL must be $\geq$ 18%。 The minimum area of high NA EUV is about half that of 0.33 NA EUV lithography. Similar to the low NA case, the target requirement of each line cut can be easily achieved by OPC due to the large k1 value, and the need for source optimization is not necessary.

**The aberration effect of typical metal layer patterns in both Low and High NA EUV processes**

In this section, we study the influence of Z8 on process window of typical metal layer patterns in both low and high NA EUV processes.

**The simulation results of TtT pattern in both Low and High NA EUV processes**

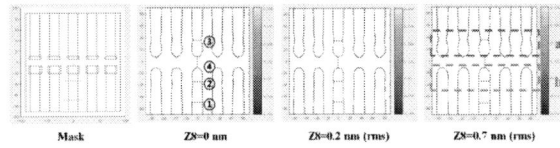

*Figure 6: The mask, the simulation results in different Z8 under SO'ed source as shown in Figure 2 of TtT pattern in low NA EUV lithography.*

The masks in Figure 6, 7, 8, 9 are the same to the masks in Figure 2, 3 ,4, 5, respectively. Taking the SO'ed source shown in Figure 2 as the illumination condition, we have made a split for Z8, which are 0, 0.2 and 0.7 nm root-mean-square (rms) respectively, and carried out a set of simulations in 0.33 NA EUV lithography. The simulation contour images show in Figure 6 and the corresponding simulation results show in Table V. It can be seen from the simulated contours that when Z8 is 0.2

nm rms, there is no obvious change in contour image. As can be seen from Table V, the CD and EL at line cut ② increase by 0.6 nm and 1 percentage point, respectively, and the CD and EL at line cut ③ decrease by ~ 0.6 nm and 1 percentage point, respectively. When Z8 is 0.7 nm rms, obvious "necking" occurs in contour image at line cut ③ as shown in red dotted box with CD and EL decreasing of 2.5 nm and ~ 6 percentage points. The CD and EL at line cut ② increase by ~ 2 nm and ~ 3 percentage points, respectively.

TABLE V.     THE SIMULATION RESULTS OF FIGURE 6

| | | SO'ed Source | | | |
|---|---|---|---|---|---|
| | Line cut | ① | ② | ③ | ④ |
| Z8=0 nm | CD (nm) | 18 | 16.24 | 15.6 | 19.7 |
| | EL | 20.31% | 16.77% | 15.49% | 15.96% |
| | CD (nm) | 18.11 | 16.83 | 14.93 | 19.67 |
| | EL | 20.45% | 17.83% | 14.31% | 15.87% |
| Z8=0.2 nm(rms) | ΔCD (nm) | 0.11 | 0.59 | -0.67 | -0.03 |
| | Δ CD (%) | 0.61 | 3.63 | -4.29 | -0.15 |
| | Δ EL | 0.14% | 1.06% | -1.18% | -0.09% |
| | Δ EL (%) | 0.69 | 6.32 | -7.62 | -0.56 |
| | CD (nm) | 18.2 | 18.06 | 13.14 | 19.69 |
| | EL | 19.77% | 19.53% | 9.77% | 15.14% |
| Z8=0.7 nm(rms) | ΔCD (nm) | 0.20 | 1.82 | -2.46 | -0.01 |
| | Δ CD (%) | 1.11 | 11.21 | -15.77 | -0.05 |
| | Δ  EL | -0.54% | 2.76% | -5.72% | -0.82% |
| | Δ EL (%) | -2.66 | 16.46 | -36.93 | -5.14 |

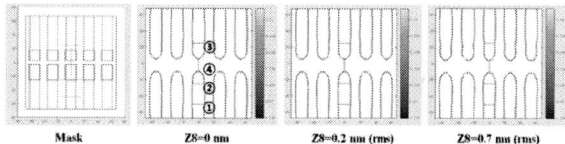

Figure 7:  The mask, the simulation results in different Z8 under SO'ed source as shown in Figure 3 of TtT pattern in high NA EUV lithography.

TABLE VI.     THE SIMULATION RESULTS OF FIGURE 7

| | | SO'ed Source | | | |
|---|---|---|---|---|---|
| | Line cut | ① | ② | ③ | ④ |
| Z8=0 nm | CD (nm) | 15 | 14.75 | 14.56 | 15.76 |
| | EL | 18.30% | 17.59% | 17.19% | 13.07% |
| | CD (nm) | 14.85 | 15.05 | 14.17 | 15.68 |
| | EL | 18.02% | 18.26% | 16.38% | 12.64% |
| Z8=0.2 nm(rms) | ΔCD (nm) | -0.15 | 0.3 | -0.39 | -0.08 |
| | Δ CD (%) | -1.00 | 2.03 | -2.68 | -0.51 |
| | Δ  EL | -0.28% | 0.67% | -0.81% | -0.43% |
| | Δ EL (%) | -1.53 | 3.81 | -4.71 | -3.29 |
| | CD (nm) | 14.58 | 15.46 | 13.14 | 15.31 |
| | EL | 16.81% | 18.38% | 13.92% | 11.62% |
| Z8=0.7 nm(rms) | ΔCD (nm) | -0.42 | 0.71 | -1.42 | -0.45 |
| | Δ CD (%) | -2.80 | 4.81 | -9.75 | -2.86 |
| | Δ  EL | -1.49% | 0.79% | -3.27% | -1.45% |
| | Δ EL (%) | -8.14 | 4.49 | -19.02 | -11.09 |

Taking the SO'ed source shown in Figure 3 as the illumination condition, we have made a split for Z8, which are 0, 0.2 and 0.7 nm rms respectively, and carried out a set of simulations in 0.55 NA EUV lithography. The simulation contour images show in Figure 7 and the corresponding simulation results show in Table VI. It can be seen from the simulated contour that when Z8 is 0.2 nm rms, there is no obvious change in contour image. As can be seen from Table VI, the CD and EL at line cut ② increase by 0.3 nm and~ 0.7 percentage point, respectively, and the CD and EL at line cut ③ decrease by ~0.4 nm and 0.8 percentage point, respectively. When Z8 is 0.7 nm rms, the CD and EL at line cut ③ have a significant reduce of ~ 1.5 nm and ~ 3 percentage points. The CD and EL at line cut ② increase by 0.7 nm and 0.8 percentage points, respectively.

From the above results, the influence of aberration of Z8 on near tip part of the trenches is particularly significant, and the influence of aberration in low NA EUV is basically twice or even more than that of in high NA EUV.

**The simulation results of isolated trench pattern in both Low and High NA EUV processes**

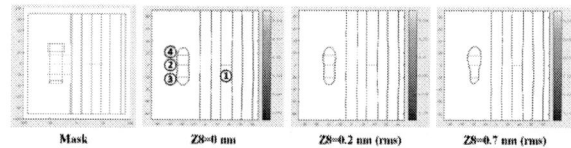

Figure 8: The mask, the simulation results in different Z8 under SO'ed source as shown in Figure 4 of isolated trench pattern in low NA EUV lithography.

TABLE VII.     THE SIMULATION RESULTS OF FIGURE 8

| | | SO'ed Source | | | |
|---|---|---|---|---|---|
| | Line cut | ① | ② | ③ | ④ |
| Z8=0 nm | CD (nm) | 18 | 18.05 | 19.95 | 18.7 |
| | EL | 19.41% | 14.97% | 18.48% | 15.97% |
| | CD (nm) | 17.99 | 18.43 | 18.45 | 19.89 |
| | EL | 19.34% | 15.56% | 15.51% | 18.27% |
| Z8=0.2 nm(rms) | ΔCD (nm) | -0.01 | 0.38 | -1.5 | 1.19 |
| | Δ CD (%) | -0.06 | 2.11 | -7.52 | 6.36 |
| | Δ  EL | -0.07% | 0.59% | -2.97% | 2.30% |
| | Δ EL (%) | -0.36 | 3.94 | -16.07 | 14.40 |
| | CD (nm) | 17.44 | 19.34 | 13.74 | 22 |
| | EL | 17.81% | 16.43% | 1.80% | 21.38% |
| Z8=0.7 nm(rms) | ΔCD (nm) | -0.56 | 1.29 | -6.21 | 3.30 |
| | Δ CD (%) | -3.11 | 7.15 | -31.13 | 17.65 |
| | Δ  EL | -1.60% | 1.46% | -16.68% | 5.41% |
| | Δ EL (%) | -8.24 | 9.75 | -90.26 | 33.88 |

Figure 8 and Table VII show the simulation results of the isolated trench pattern with different Z8 in low NA EUV lithography. It can be seen from the simulated contour that when Z8 is 0.2 nm rms, there is a little change in contour image. As can be seen from Table VII, the CD and EL at line cut ④ increase by 1 nm and ~ 2 percentage points, respectively, and the CD and EL at line cut ③ decrease by ~ 1.5 nm and 3 percentage points, respectively. When Z8 is 0.7 nm rms, there is a drastic change in CD and EL that distorts the contour image. Especially at line cut ③, its CD and EL are reduced by ~ 6 nm and ~ 17 percentage points, respectively.

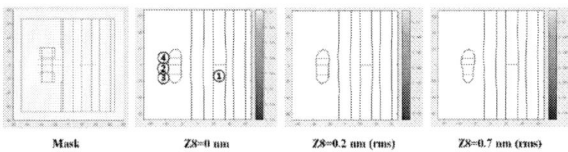

Figure 9: *The mask, the simulation results in different Z8 under SO'ed source as shown in Figure 5 of isolated trench pattern in high NA EUV lithography.*

TABLE VIII.　THE SIMULATION RESULTS OF FIGURE 9

| | Line cut | SO'ed Source | | | |
|---|---|---|---|---|---|
| | | ① | ② | ③ | ④ |
| Z8=0 nm | CD (um) | 15 | 15.03 | 15.85 | 15.22 |
| | EL | 18.05% | 18.49% | 18.98% | 18.57% |
| Z8=0.2 nm(rms) | CD (nm) | 15.01 | 15.22 | 14.98 | 15.88 |
| | EL | 18.07% | 18.86% | 17.08% | 20.01% |
| | ΔCD (nm) | 0.01 | 0.19 | -0.87 | 0.66 |
| | Δ CD (%) | 0.07 | 1.26 | -5.49 | 4.34 |
| | Δ EL | 0.02% | 0.37% | -1.90% | 1.44% |
| | Δ EL (%) | 0.11 | 2.00 | -10.01 | 7.75 |
| Z8=0.7 nm(rms) | CD (nm) | 15.2 | 15.62 | 12.15 | 16.83 |
| | EL | 18.05% | 19.08% | 11.09% | 21.71% |
| | ΔCD (nm) | 0.20 | 0.59 | -3.70 | 1.61 |
| | Δ CD (%) | 1.33 | 3.93 | -23.34 | 10.58 |
| | Δ EL | 0.00% | 0.59% | -7.89% | 3.14% |
| | Δ EL (%) | 0.00 | 3.19 | -41.57 | 16.91 |

Figure 9 and Table VIII show the simulation results of the isolated trench pattern with different Z8 in high NA EUV lithography. Similar to the results of low NA, when Z8 is 0.2 nm rms, the contour image changes a little, and when Z8 is 0.7 nm rms, the contour changes so significantly resulting in the distortion of isolated trench. The specific values are as follows, when Z8 is 0.2 nm rms, the CD and EL at line cut ④ increase by 0.6 nm and ~ 1.5 percentage points, respectively, and the CD and EL at line cut ③ decrease by ~ 0.9 nm and 2 percentage points, respectively. When Z8 is 0.7 nm rms, the CD and EL at line cut ③ are reduced by ~ 4 nm and ~ 8 percentage points, respectively.

Similar to the results for TtT in 2.3.1, the influence of aberration in low NA EUV is basically twice or even more than that of in high NA EUV.

In summary, with the NA increasing from 0.33 to 0.55 NA, keeping the same aberration specification of 0.2 nm rms [3], the lithography process is still acceptable and there is no need to further improve the lens aberration.

## A 3 nm SRAM CFET Flow and a 45° local interconnect Design

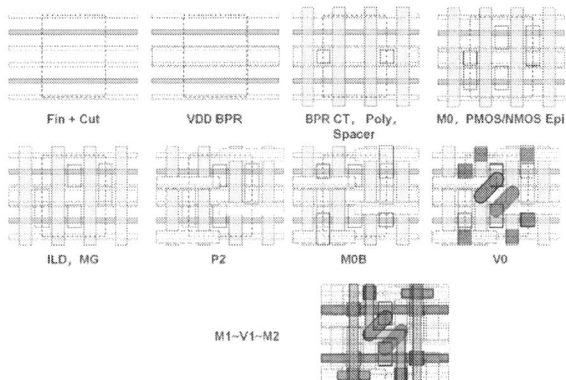

Figure 10: *The flow of a 3 nm CFET SRAM [7].*

In this section, we introduce a tentative 3 nm CFET SRAM layout which contains a 45° local interconnect design and the corresponding flow [7] as shown in Figure 10. From the figure, the Fin Pitch (FP) is 24 nm, a contact-to-poly pitch (CPP) is 48 nm, and the minimum contact pitch that can be produced by 0.33 NA EUV is 48 nm. There is a challenge to area utilization if we keep the original square local interconnect design. However, if 45° local interconnect is adopted, the SRAM area will be decreased from > 0.015 to ~ 0.011 $\mu m^2$.

Therefore, we will study the process window of the 45° local interconnect design in both low and high NA EUV processes.

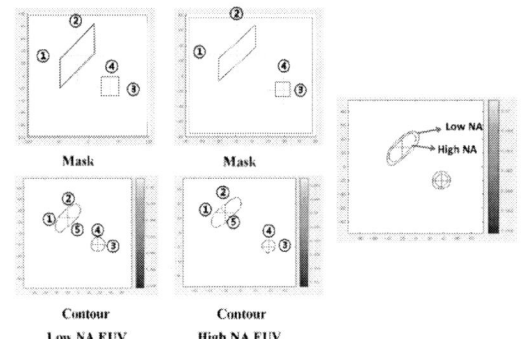

Figure 11: *The mask, simulation contours of a 45° local interconnect pattern in both low and high NA EUV processes.*

TABLE IX.　THE SIMULATION RESULTS OF FIGURE 11

| Low NA EUV | | | | | |
|---|---|---|---|---|---|
| Line cut | ① | ② | ③ | ④ | ⑤ |
| CD (nm) | 33.3 | 32.5 | 24.5 | 24.4 | 23.5 |
| EL | 19.59% | 18% | 19.04% | 18.35% | 19.20% |
| High NA EUV | | | | | |
| Line cut | ① | ② | ③ | ④ | ⑤ |
| CD (nm) | 25.2 | 24.5 | 18 | 17.6 | 17.6 |
| EL | 30.00% | 29.00% | 24.00% | 23.00% | 29.50% |

Figure 11 and Table IX show the simulation results of the 45° local interconnect design in both low and high NA EUV processes. From line cuts ① and ②, we can calculate the CD of line cut ⑤, which is along the 135° direction.

For low NA, the anchor pitch is 48 nm with ADI CD of 24 nm. The illumination condition is an annular with 0.9-0.7 partial coherence factor and 0.33 NA. In order to achieve an EL more than 18% at line cut ⑤ position, the ADI CD must be at least 24 nm. And this area for low NA EUV satisfy the minimum area rule for 0.33 NA lithography with ~ 5 squares (~ 5 squares) and need large etch bias ~ 5 nm on one side for a final etch CD of 14 nm.

For high NA, the anchor pitch is 38 nm with ADI CD of 18 nm. The illumination condition is an annular with 0.7-0.5 partial coherence factor and 0.55 NA. The ADI CD of line cut ⑤ is 17.6 nm with EL of 29%, which is much greater than 18%. And this area for high NA EUV satisfy the minimum area rule for 0.55 NA lithography with ~ 3.6 squares (>3 squares) and need little etch bias ~ 2 nm on one side for a final etch CD of 14 nm.

As can be seen from the simulation results, in addition to the larger etch bias, the EL of each line cut in low NA EUV is also smaller than that of each line cut in high NA EUV.

In summary, low NA EUV can be used in the 45° local interconnect design for the 3 nm CFET. After mass production of 3 nm logic node, high NA EUV machines can first be introduced into the 3 nm process to get "warmed up" before applied to more advanced logic technology nodes.

## CONCLUSION

In this paper, we have performed a simulation study on some typical metal layer patterns with both low and high NA EUV processes. We have briefly studied the influence of source optimization and the impact from coma Y (Z8) to the process window. We find that due to the high k1 nature of EUV lithography, it's easy to achieve ADI target by mask optimization. In low NA EUV, the effect of source optimization is not significant, while in high NA EUV, source optimization can significantly improve the DoF of near tip part of the trenches and TtT patterns. With the NA increasing from 0.33 to 0.55 NA, keeping the same aberration specification of 0.2 nm rms, the lithography process is still acceptable and there is no need to further improve the lens aberration.

We have displayed a 3 nm CFET SRAM Layout and the corresponding flow which involve 45° local interconnect patterns. These patterns can be realized by both low and high NA EUV processes. If we use the low NA EUV, there is a larger etch bias with ~ 5 nm on one side for a final etch CD of 14 nm compared with the 2 nm etch bias in high NA EUV process. Actually, after mass production of 3 nm logic node, high NA EUV machines can first be introduced into the 3 nm process to get "warmed up" before applied to more advanced logic technology nodes.

## ACKNOWLEDGEMENTS

We thank the higher management team from FUDAN University for the support of this work.

## REFERENCES

[1] Qiang Wu*, Yanli Li, Yushu Yang, Yuhang Zhao. "A Photolithography Process Design for 5 nm Logic Process Flow", J. Microelectron. Manuf. 2019, 2(4):11.

[2] Shaofeng Yu, China's integrated circuit technology development roadmap, 2019.

[3] Yanli Li, Qiang Wu, Yuhang Zhao. "A Simulation Study for Typical Design Rule Patterns and Stochastic Printing Failures in a 5 nm Logic Process with EUV Lithography", IEEE Xplore, 2020.

[4] Jan van Schoot, Kars Troost, el al, "High-NA EUV lithography enabling Moore's law in the next decade", Proc SPIE. 10450, Extreme Ultraviolet Lithography, 2017, 104500U.

[5] S. Subramanian et al., "First Monolithic Integration of 3D Complementary FET (CFET) on 300mm Wafers," 2020 IEEE Symposium on VLSI Technology, 2020, pp. 1-2, doi: 10.1109/VLSITechnology18217.2020.9265073.

[6] Yanli Li, Xiaona Zhu, Shaofeng Yu, Qiang Wu*. "A Study of the Advantages to the Photolithography Process brought by the High NA EUV Exposure Tool in Advanced Logic Design Rules", IEEE Xplore, 2021.

[7] Qiang Wu*, Yanli Li*, Xiaona Zhu, and, Shaofeng Yu, "The Discussion of the Typical BEOL Design Rules from 3 nm to 2 nm Logic Process with EUV and High NA EUV Lithography", IEEE Xplore, 2021.

# A STUDY OF IMPROVED DESIGN RULES THROUGH ALLOWING 45-DEGREE METAL LINES

*Xianhe Liu\*, Yanli Li, and Qiang Wu\**

School of Microelectronics, Fudan University, Shanghai 200433, China

\*Corresponding Author's Email: xianheliu@fudan.edu.cn

## ABSTRACT

In advanced integrated circuit manufacturing technology, the continuous drive for chip economics has been pushing the chip design toward high area utilization. For metal layers under logic design rules, there is a need for interconnection between metal tracks. In advanced technology nodes, most design rules are square-like with restriction on the oblique line placement. And the mask fracturing algorithm favors designs with only horizontal and vertical lines. In the designs that push the limit of optical lithography, due to the limitation in optical imaging, corner rounding becomes significant compared to the metal track, e.g., in the 7 nm metal design, where the Self-Aligned Litho-Etch Litho-Etch (SALELE) is used, the single pattering metal pitch is around 80 nm, the best corner rounding radius under 193 nm immersion lithography is 30~35 nm, if a "Z"-like turn is designed under neighboring metal tracks, it is difficult to keep square turns. The current fix is to make the turn through adding two vias and a perpendicular line at the next metal layer. In the 7 nm Back-End-Of-the-Line (BEOL), this method will need many via and metal layers since both metals and vias need multiple patterning steps. We propose a study of adding one type of 45-degree patterns to the 193 nm immersion design rules since we believe it has cost-effective merits. The study was conducted with a self-developed aerial image simulator based on the Finite-Difference Time-Domain (FDTD) and the Abbe imaging algorithm. We first establish a typical metal lithography condition, then we will approximate the 45-degree line with a series of rectangles, which can overlap and treated by e-beam algorithm developed by D2S; Next, we will evaluate the performance of different densities of the rectangles under the lithography condition as well as mask manufacturing cost; finally, we will report the result of our study.

## INTRODUCTION

7 nm logic technology node is believed to be the last one with critical layers that can be made entirely with 193 nm immersion process. However, the current design rules are mostly uni-directional with cuts. An example under 7 nm BEOL design rule with a metal pitch of 40 nm is shown in Figure 1(a).

In Back-End-Of-the-Line (BEOL), a Z-shape square turn is commonly used to connect two neighboring metal tracks, however this would easily produce congestion

because the connecting pattern would have to be enlarged enough to meet the process window requirement, i.e. Exposure Latitude (EL) > 9% [1]. As a potential problematic weak point, or so-called hotspot, several additional approaches in mask synthesis and correction flows may be needed [2, 3]. An alternative is to make the turn by adding two vias and a perpendicular line at the next metal layer, but it requires multiple patterning steps, as well as additional mask cost. A 45-degree improved design is therefore taken into consideration, with the aim of reaching greater design density with enough process window.

Under traditional design rules, only axis-parallel lines oriented at 0° or 90° are allowed. However, various shapes or angles lines can be formed on the reticle, with the current mask making techniques, such as the Variable Shaped Beam (VSB) writer. Some attempts have been made in critical layers to create a 45-degree line in Z-like hotpots utilizing irregular shapes or curvilinear forms [4-7].

In this study, we propose a type of design in the logic 7 nm BEOL employing a 45-degree connecting line as shown in Figure 1(b), consisting of overlapping rectangles placed along the 45-degree direction. We will present later in this report, that when compared to the traditional rectilinear connection, this design yields a positive outcome, reducing pattern area occupation, and as a result effectively saving mask manufacturing cost.

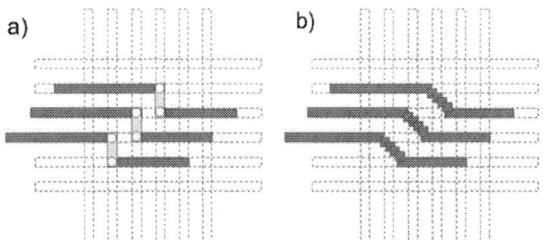

*Figure.1: Introduction of 90-degree design (a) and 45-degree design (b) in Z-like turns.*

## RESULT AND DISCUSSION

An extreme case is explored under the 193 nm immersion lithography condition, with a pitch of 80 nm and targeting Critical Dimension (CD) of 40 nm. Three Z-like connections are established in the mask design. A self-developed aerial image simulator is used in the

978-1-6654-9759-6/22 $31.00 © 2022 IEEE

simulation, with the setting of Negative Tone Development (NTD) and dipole illumination mode.

**90-degree connection**

As illustrated in the Figures 2(a)-2(d), when the connecting square pattern between two closest metal tracks is 67-70 nm wide (line cuts 2&3), the generated EL is too low, ~5-6%, compared to our understanding of the industry standard, ~13% [8]. The only way to solve the problem is to increase the connecting pattern dimension. Meanwhile, the tip-to-edge spacing between two adjacent tracks (line cuts 4&5) must be kept to ~90 nm in order for the EL to be appropriate, ~10%. The difference between the 10% and the above-mentioned industry standard of 13% is due to the fact that we have used NTD with no Photo Decomposable Base (PDB), which will have slightly lower EL. And the addition of PDB, according to our experience, can contribute about 2~3 percentage points to the EL.

The improved result with several iterations, is shown in Figures 3(a)-3(d). In this case, the Z-like connecting turn is widened to 92-94 nm, and the tip-to-edge spacing 95-100 nm. The linewidth that correlates to the electrical current flow is the line cut that is perpendicular to the contour line in the aerial image, as represented as red lines in Figure 3b, their accurate CD and EL, presented as $CD_\perp$ and $EL_\perp$ in Figure 3d, are ~76-80 nm and ~9-10% respectively.

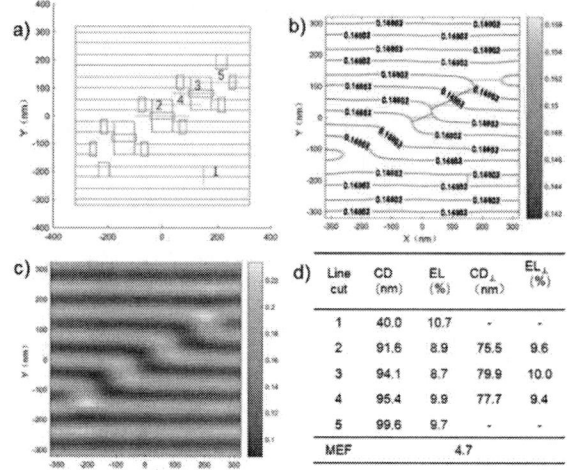

*Figure.3: a) Simulated 2D tested mask patterns, b) contour and c) aerial image, Z-like square connecting turn is set to 75 nm wide in the mask. d) the simulated CD and EL values in different line cuts. Red lines refer to the perpendicular line cuts in these positions, where the CD and EL are remeasured, noted as $CD_\perp$ and $EL_\perp$.*

**45-degree connection**

In this case, a 45-degree connecting line between two tracks is created via a succession of aligning rectangles. It is worth mentioning that these rectangles serve as not only the fractures of main features by D2S [7], but also assistant patterns for overcoming the optical proximity effect. As a result, fewer assistant features are required here than in case of 90-degree connection design. At least 3 rectangles in connection lines are needed for obtaining appropriated EL values in all line cuts. It will be tough to counterbalance the EL (line cuts 4&5 in Figure 4(a)) in both lines and spaces if the number of rectangles is reduced to 2, as the generated EL values are not acceptable for fabrication process as shown in Figure 4(d).

With 3 rectangles-composing connecting lines, as seen in Figures 5(a)-5(d), EL values are >9% for all cut lines. The perpendicular critical dimensions, or $CD_\perp$ of 45-degree lines, collected in the Figure 5(d), are ~67-69 nm (line cuts 2&3), tip-to edge spacing ~74 nm (line cut 4). A simple analysis of this outcome suggests that the minimum area occupied for the same Z-like pattern in 45-degree design is lower than in a standard 90-degree design, which will be carefully investigated in the next section.

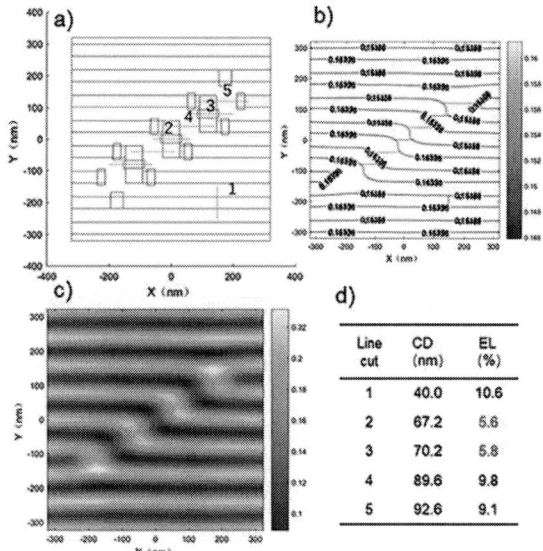

*Figure.2: a) Simulated 2D tested mask patterns, b) contour and c) aerial image, Z-like square connecting turn is set to 55 nm wide in the mask. d) the simulated CD and EL values in different line cuts.*

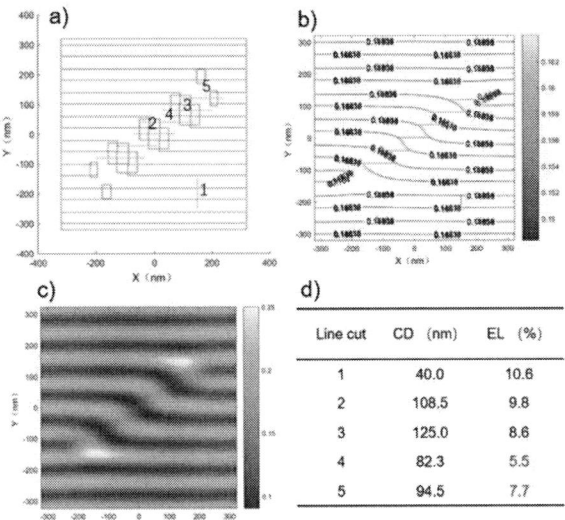

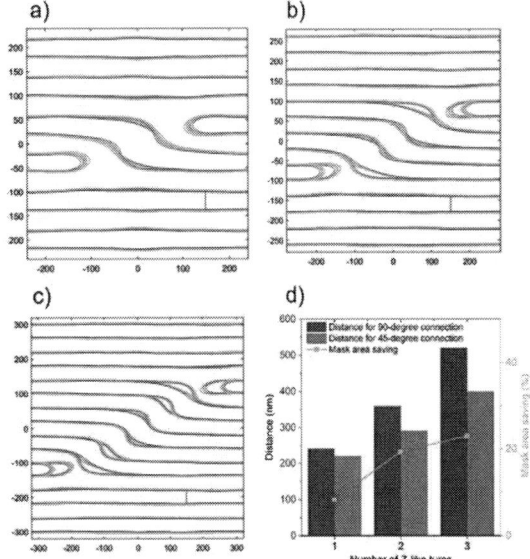

number of Z-like turns, ranging from 1-3, is considered. The horizontal distance between the two trench ends in Figure 6(a) is used for area calculation, in order to quantify the minimum area occupation of Z-like patterns. The result collected in Figure 6(d) shows that in a 45-degree design, the minimum mask area is always less than in a 90-degree configuration. More importantly, when there is only one Z-like turn, the mask area reduction is 8%, but it significantly increases to ~20% when there are 2 or 3 turns. The 45-degree design can therefore potentially offer a substantial advantage in terms of mask area savings for a series of dense Z-like turns.

Figure.4: a) Simulated 2D tested mask patterns, b) contour and c) aerial image, 2 overlapping rectangles consisting the 45-degree connecting line, with size of rectangles of $70 \times 70$ nm$^2$ in the mask. d) the simulated CD and EL values in different line cuts.

| Line cut | CD (nm) | EL (%) |
|---|---|---|
| 1 | 40.0 | 10.6 |
| 2 | 108.5 | 9.8 |
| 3 | 125.0 | 8.6 |
| 4 | 82.3 | 5.5 |
| 5 | 94.5 | 7.7 |

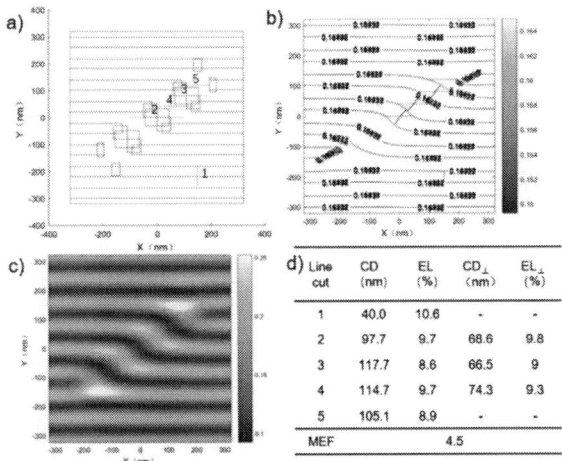

| Line cut | CD (nm) | EL (%) | CD$_\perp$ (nm) | EL$_\perp$ (%) |
|---|---|---|---|---|
| 1 | 40.0 | 10.6 | - | - |
| 2 | 97.7 | 9.7 | 68.6 | 9.8 |
| 3 | 117.7 | 8.6 | 66.5 | 9 |
| 4 | 114.7 | 9.7 | 74.3 | 9.3 |
| 5 | 105.1 | 8.9 | - | - |
| MEF | 4.5 | | | |

Figure.5: a) Simulated 2D tested mask patterns, b) contour and c) aerial image, 3 overlapping rectangles consisting the 45-degree connecting line, size of center rectangle is set to $62 \times 88$ nm$^2$, and others $50 \times 50$ nm$^2$. d) the simulated CD and EL values in different line cuts. Red lines refer to the perpendicular line cuts in these positions, where the CD and EL are remeasured, noted as CD$_\perp$ and EL$_\perp$.

**Mask manufacturing cost**

The pattern contours for these two designs are overlapped in Figures 6(a)-6(c), where the effect of the

Figure.6: a-c) 1-3 Z-like connecting turns with 90-degree and 45-degree designs are simulated and compared by overlapping their contour images in each case, orange line corresponding to the 90-degree design, and blue to the 45-degree. d) The horizontal distance between the two trench ends in 1-3 Z-like turns are presented in the bar chart, and the mask area saving in percentage in 45-degree design is shown as the green line.

## CONCLUSIONS

A 45-degree line connection in Z-like turn made up of a series of overlapping rectangles, in contrast to the usual 90-degree square connection, is investigated in this paper. The line/space dimensions and EL values in these two connection designs under 193 nm immersion resolution limit condition (80 nm pitch, 40 nm CD) are determined using the self-developed aerial image simulator. In these two situations, the minimum occupied areas are compared with enough process window (e.g., EL~ 9-10% in all line cuts). The result shows that when compared to typical 90-degree design, 45-degree design saves at least 8% mask area, and this value jumps to over 20% in dense

Z-like patterns, implying that 45-degree design is a recommended solution to reduce mask costs.

## ACKNOWLEDGEMENTS

The authors of this paper thank School of Microelectronics Fudan University for the support of this work.

## REFERENCES

[1] Y. Hou and Q. Wu, "Optical Proximity Correction, Methodology and Limitations," in *2021 China Semiconductor Technology International Conference (CSTIC)*, 14-15 March 2021, pp. 1-5.

[2] J. F. Bradley *et al.*, "Integrating enhanced hotspot library into manufacturing OPC correction flow," in *Proc.SPIE*, 2020, vol. 11328.

[3] J. F. Bradley *et al.*, "Rule-based hotspot correction using a pattern matching flow," in *Proc.SPIE*, 2021, vol. 11614.

[4] K. Rehab, H.-F. Ahmed, and W. James, "Exploring patterning limit and enhancement techniques to improve printability of 2D shapes at 3nm node," in *Proc.SPIE*, 2020, vol. 11328.

[5] P. Ryan, O. R. Don, U. Jeff, N. Mariusz, P. Leo, and F. Aki, "How utilizing curvilinear design enables better manufacturing process window," in *Proc.SPIE*, 2020, vol. 11328.

[6] D. Amit, S. Syed Muhammad Yasser, R. K. Ryoung-Han, T. Darko, G. Werner, and D. Youssef, "A study of curvilinear routing in IN5 standard cells: challenges and opportunities (Poster Presentation)," in *Proc.SPIE*, 2019, vol. 11148.

[7] C. Yohan, F. Aki, and S. Abhishek, "Curvilinear masks: an overview," in *Proc.SPIE*, 2021, vol. 11855.

[8] Q. Wu, Y. Li, Y. Yang, S. Chen, and Y. Zhao, "The Law that Guides the Development of Photolithography Technology and the Methodology in the Design of Photolithographic Process," in *2020 China Semiconductor Technology International Conference (CSTIC)*, 26 June-17 July 2020, pp. 1-6.

# GAN-BASED FAST MASK NEAR-FIELD CALCULATION

*Yijiang Shen[1], Zhijie Jiao and JiaxiangZhuo*

School of Automation, Guangdong University of Technology,
Mega Education Center South, Guangzhou, 510006, P. R. China
*Corresponding Author's Email: yjshen@gdut.edu.cn

## ABSTRACT

Near-field calculation of thick masks in extreme ultra-violet (EUV) lithography is one of the fundamental tasks for process modeling and physical verification. Rigorous simulations are not applicable because of the enormous computation load, whereas the perturbation models (M3D) are susceptible to inaccurate computation for mask patterns with random edge corner and feature size. In this paper, we propose a generative adversarial network (GAN) based deep-learning approach to calculate the mask near-field to address the regarding efficiency and accuracy issues. We describe the GAN structure including the encoder, the discriminator network and the training strategy explicitly with sufficient detail, so others can follow, analyze and improve. The network is trained based on a set of mask samples where the corresponding mask near-field data are obtained by the EMF simulator. Simulation results show the merits of the proposed GAN-based approach.

*Keywords—mask near-field; deep learning; generative adversarial network (GAN); encoder; discriminator; training strategy*

## INTRODUCTION

Lithography is the core technology of very large scale integration (VLSI) [1] manufacturing, whose continuous improvement extends the lifecycle of the Moore's Law [2]. However, with the ever diminishing features size of semi-conductor devices, the diffraction effect reaches the point where, especially in the extreme ultra-violet (EUV), the thin mask model (TMM) defined by the Hopkins assumption, is simply not competent. Therefore，mask near-field computation is becoming extremely important in computational lithography.

Rigorous Maxwell equations solvers including finite difference time domain (FDTD) [3] method and rigorous coupled wave analysis (RCWA) [4] are hardly applicable to full-chip simulation because of the computation complexity. Recently, the rapid development of machine learning has encouraged practitioners to engage in design-for-manufacturing (DFM) applications. For example, generative adversarial networks (GAN) [5] based on deep learning has been used prevalently and brought algorithmic and practical insights to improving the performance and efficiency of hotspot detection, compact Lithographic process models [6] and optical proximity correction (OPC) [7].

Motivated by the recent development of cycle consistent adversarial network (CycleGAN) [8], we elect in this manuscript to address the mask near-field computation from a new perspective. Inspired by machine translation dual learning [9], CycleGAN uses reciprocal translators $G: X \rightarrow Y$ and $F: Y \rightarrow X$, where G and F are inverses of each other and are both bijections; G and F are further trained simultaneously by adding a cycle consistent loss to encourage $F(G(X)) \approx X$ and $G(F(Y)) \approx Y$. Upon this understanding, the authors of Ref. [8] proposes a CycleGAN based method to learn the transformation of images from the source domain to the target domain and vice versa.

Marrying the idea to lithographic modeling, we apply the CycleGAN[8] network paired with ResNet [10] to predicting the diffraction matrices with respect to the mask context. Experimental results show the merit of the proposed Gan-based method with lower verification loss function value and around 30 times faster near-field generation as compared to fully convolution network (FCN) [11] with dilated convolution [12].

In the remainder of this manuscript, we first introduce the detailed structure and training details of the GAN based architecture for mask near-field computation. After that, experiment results are presented and compared with the results in Ref. [11]. Finally, some conclusion remarks are drawn.

## NETWORK FRAMEWORK AND DETAILS
### Data Preparation

361 periodic contact arrays and 696 rectangle arrays are included in the mask pattern data and their corresponding near-field are calculated rigorously with waveguide-based electromagnetic field simulator in Sentaurus[TM] Lithography software. CD = 14nm are set for the contact and rectangle arrays in the Y coordinate, while CD = 14nm and 28nm are defined for the contact and rectangle arrays in the X coordinate. The pitches are in the range [40nm, 100nm]. Examples of rectangle and contact arrays are presented in Fig. 1. The numerical aperture (NA) of the EUV lithography system is 0.33 and coherent illumination is used for simplicity.

Since the GAN network has very large requirements for the amount of data, we enhance the obtained data set through splicing and cropping operations, making the number of data sets 5 times larger. Because there are less

978-1-6654-9759-6/22 $31.00 © 2022 IEEE

contact arrays in the data set, the contact arrays are mixed with the rectangular arrays, therefore, the trained network will not be biased.

The output data of the network can be represented the diffraction matrices

$$E(UV) = E_{real}(UV) + iE_{imag}(UV), \qquad (1)$$

in which $E(UV)$ represents the complex amplitude of the near-field of the mask polarized in the U direction generated by a unit incident electric field polarized in V direction, where $U = X$ or $Y$, and $V = X$ or $Y$. $E_{real}(UV)$ and $E_{imag}(UV)$ are the real and imaginary part of the complex diffraction matrices $E(UV)$. Under permutation and combination, we have total four diffraction near-field diffraction matrices, namely $E(XX)$, $E(YY)$, $E(XY)$ and $E(YX)$. However, it is duly noted that the $E(XY)$ or $E(YX)$ components are negligible comparing to $E(XX)$ and $E(YY)$, therefore, we will be focusing on the $E(XX)$ and $E(YY)$ diffraction matrices in this work.

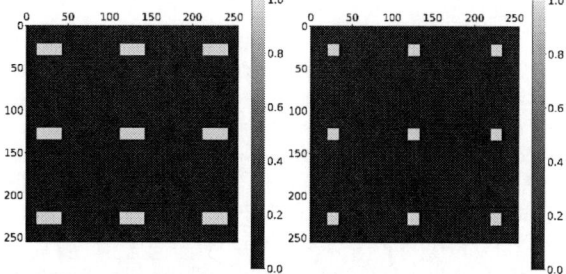

*Figure 1: Examples of rectangular array and contact array mask patterns.*

Finally, we divide the enhanced data set into training, validation and testing sets, in which 70% (3699) of them are the training set, 20% (1057) are the validation set, and the remaining 10% (529) are the test set to verify the quality of the network model.

**Network Framework**

CycleGAN [8] contains a pair of generators, one of which maps the mask to the near-field and the other vice versa, and a discriminator. The loss function formula of the generator and the discriminator is given as

$$L_{GAN}(G, D_Y, X, Y) = E_{y \sim P_{data}(y)}[\log(D_Y(y)] + E_{x \sim P_{data}(x)}[\log(1 - D_Y(G(x))], \quad (2)$$

and

$$L_{GAN}(F, D_X, Y, X) = E_{x \sim P_{data}(x)}[\log(D_X(x)] + E_{y \sim P_{data}(y)}[\log(1 - D_X(F(y))], \quad (3)$$

where E is the mathematical expectation, $\sim$ is the obedience relation, and $P_{data}(y)$ is the mask near-field datadistribution, $P_{data}(x)$ is the mask data distribution, the generation network G is the mapping from the mask to the near field which is forward mapping; the generation

network F is the mapping from the mask near field to the mask, which is the reverse mapping. The discriminator $D_Y$ is to discriminate the true and false of the real mask near field and the near field generated by the mask, and the discriminator $D_X$ is to discriminate the authenticity of the mask and the generated mask.

Since the adversarial loss alone cannot guarantee the one-to-one mapping of the mask and the near-field, we include a cycle consistency loss

$$L_{Cycle}(G, F) = E_{x \sim P_{data}(x)}[\|F(G(x)) - x\|_1] + E_{y \sim P_{data}(y)}[\|G(F(y)) - y\|_1] \quad (4)$$

to give the CycleGAN loss function

$$L_{GAN}(G, F) = \lambda L_{Cycle}(G, F) + L_{GAN}(G, D_Y, X, Y) + L_{GAN}(F, D_X, Y, X). \quad (5)$$

The architecture of the generator is depicted in Fig. 2 including 2 down-sampling blocks, 2 up-sampling blocks, and 9 residual blocks. The architecture of the discriminator is presented in Fig. 3 to include 4 down-sampling blocks.

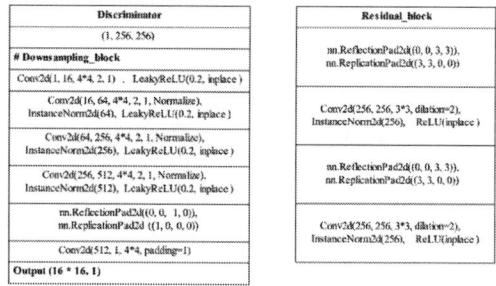

*Figure 2: The architecture of the generator of the proposed CycleGAN*

*Figure 3: The Resblock and discriminator architecture*

**Network Improvement**

Different from the direct zero-padding strategy in [8], because the mask pattern in the enhanced data set is symmetric in the Y direction and asymmetrical in the X direction, we use the mirror reflection padding method and duplicating padding strategy in the according direction. This improved padding strategy avoids the result of the direct padding value being zero at the beginning, which might lead to the disappearance of the generated pattern in the later training stage, and the failure of the training stage. This padding strategy is also applied in the discriminator..

To improve the effect of convolution kernels, we do not directly convert the grayscale input image from 1 channel to 64 channels as suggested in Ref. [8]. Alternatively, we succeed a $7 \times 7$ convolution into three $2 \times 2$ convolutions. Progressively zoomed in three times to 64 channels and applying dilation convolution [11] to expand the receptive field compensate for the reduction in the feature capture range caused by the reduction in the size of the convolution kernel. This improvement also avoids the vanishing of the generated patterns and makes the quality of the generated pattern more stable.

**Training Details**

The networks are implemented in Python with PyTorch on a Windows PC with an 8-core 3.60-GHz CPU and 16-GB memory. The batch size is set to 1, which is deliberately selected for the stability of the model training stage. The optimizer uses Adam [15], and the learning rate is set to 0.0002 .When doing standardization, after carefully observing the training data, for the imaginary part, we normalize it to the range of $-1$ to 1; for the real part, we normalize it to the range of $-1$ to 0. The number of epoch is set to be 20.

**EXPERIMENTAL RESULT**

The metric to measure the accuracy of near-field prediction is the mean square error (MSE) loss. Due to limited manuscript pages, we only present the real part $E_{real}(XX)$ and imaginary part $E_{imag}(XX)$ near-field patterns of polarization to show the efficacy of the proposed GAN-based approach. $E_{real}(YY)$ and $E_{imag}(XX)$ can be readily calculated with similar training strategies.

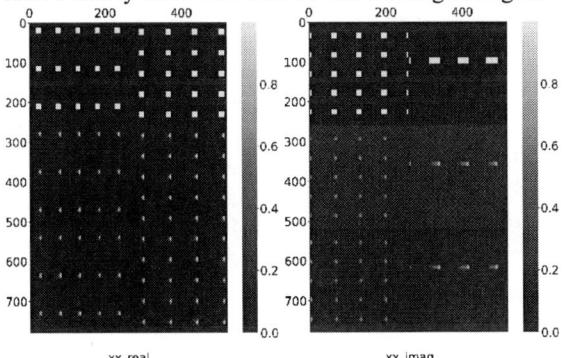

*Figure 4: Generated $E_{real}(XX)$ and $E_{imag}(XX)$ near fields.*

Figure 4 visualizes the performance of the trained generator. In Fig. 4, the first row and the second row are the mask pattern and the near-field pattern in the dataset, and the third row are the generated near-field real and imaginary part respectively learned by the proposed network. While naked eye can hardly tell the true

near-field from the generated one, some quantitative metrics are presented to show the accuracy and stability of the proposed CycleGAN based network.

It is observed from Fig. 5 that convergence is fast, and the loss values of the generator and the discriminator also converge after confrontation. The test loss curve in Fig. 6 shows the stability of the network. From the comparison of the data in Table 1 and Table 2, it is obvious that the proposed CycleGAN based network outperforms the FCN model [11] with reduced validation and test values. The runtime comparison in Table 3 shows that the computation of mask near-field is about 30 times faster than the FCN model.

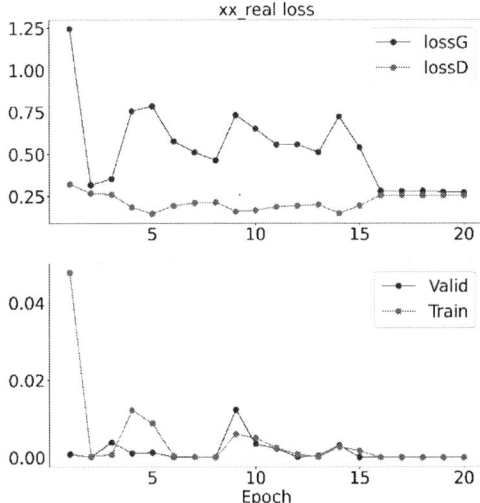

*Figure 5: Training curve*

**CONCLUSION**

In this paper, we propose a method for the fast calculation of the mask near-field based on the CycleGAN model. In the training process, we employ dilation convolution and revised padding strategies together with ResNet and other network structures enabling stable generations of mask near-fields, while improving the accuracy of the generated results, and greatly improving the calculation speed.

**ACKNOWLEDGEMENTS**

This paper is partially supported by Natural Science Foundation of China (62174037), Natural Science Foundation of Guangdong Province, China (2020A1515010633, 2021A1515012000). The authors would also like to thank the support from the Institute of Microelectronics, the Chinese Academy of Sciences for providing rigorously computed mask near-fields.

| Method | Test Loss Value | |
|---|---|---|
| | $E_{real}(XX)$ | $E_{imag}(XX)$ |
| CycleGAN | 0.107E-3 | 0.095E-3 |
| FCN | 0.24E-3 | 0.69E-3 |

TABLE III.     RUNTIME COMPARISON

| Method | Run Time |
|---|---|
| CycleGAN | 0.012 |
| FCN | 0.33 |

## REFERENCES

[1] W. Oldham, S. Nandgaonkar, A. Neureuther and O. Michael. *IEEE. T. Electron Dev.*, vol. 26, 1979, pp. 717-722.

[2] R. Schaller. *IEEE Spectrum*, vol. 34, 1997, pp. 52-59.

[3] S. Burger, R. Köhle, L. Zschiedrich, W. Gao, F. Schimdt, R. März and C. Nölscher. *25th annual BACUS Symposium on photomask technology,* Monterey, 2005, pp. 599216.

[4] P. Schiavone, G. Granet and J. Robic. *Microelectronic Eng.*, vol. 57, 2001, pp. 497-503.

[5] I. Goodfellow, J. Pouget-Abadie, M. Mirza, B. Xu, D. Warde-Farley, S. Ozari, A. Courville and Y. Bengio. *Proceedings of NIPS*, 2014, pp. 2672-2080.

[6] I. Elfadel, D. Boning and X. Li. *Machine learning in VLSI computer-aided design.* Cham: Springer, 2019.

[7] H. Yang, S. Li, Y. Ma, B. Yu and E. Young. *IEEE T. Aid. D.*, vol. 39, 2019, pp. 2822-2834.

[8] JY. Zhu, T. Park, P. Isola and A. Efros. *Proceedings of the IEEE I. Conf. Comp. Vis.*, 2017, pp. 2223-2232.

[9] D. He, Y. Xia, T. Qin, L. Wang, N. Yu, T. Liu and W. Ma. *Proceedings of NIPS*, 2016, pp. 820-828.

[10] K. He, X. Zhang, S. Ren, and J. Sun. *Proc. CVPR. IEEE*, 2016, pp. 770-778.

[11] J. Lin, L. Dong, T. Fan, X. Ma, R. Chen and Y. Wei. *IEEE International Workshop on Advanced Patterning Solutions*, 2020, pp. 1-4.

[12] J. Long, E. Shelhamer and T. Darrell. *Proc. CVPR. IEEE*, 2015, pp. 3431-3440.

[13] J. Johnson, A. Alahi and L. Fei-Fei. *Lecture Notes in Computer Science*, Springer, Cham, 2016.

[14] P. Isola, JY. Zhu, T. Zhou and A. Efros. *Proc. CVPR. IEEE,* 2017, pp. 1125-1134.

[15] D. Kingma and J. Ba. *arXiv e-prints*, 2014, pp. arXiv:1412.6980.

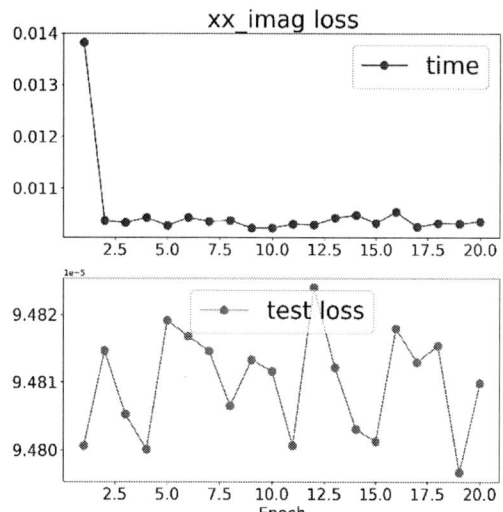

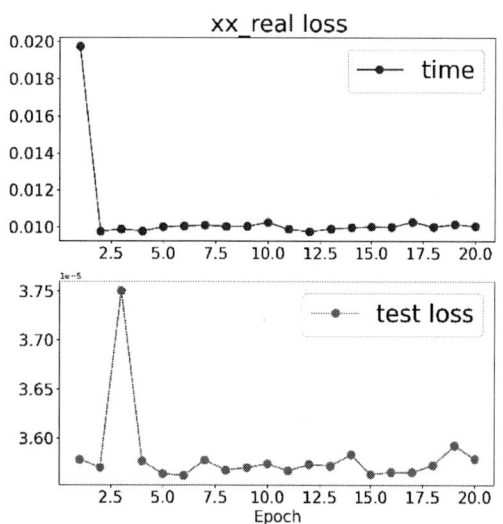

*Figure 6: Test curve*

TABLE I.     VALIDATION LOSS VALUE COMPARISON

| Method | Validation Loss Value | |
|---|---|---|
| | $E_{real}(XX)$ | $E_{imag}(XX)$ |
| CycleGAN | 0.108E-3 | 0.094E-3 |
| FCN | 1.00E-3 | 1.46E-3 |

TABLE II.     TEST LOSS VALUE COMPARISON

| Method | Test Loss Value | |
|---|---|---|
| | $E_{real}(XX)$ | $E_{imag}(XX)$ |

# DOSE CONTROL STRATEGY USING RANDOM LOGIC DEVICE PATTERNS AND MASSIVE METROLOGY IN A FOUNDRY HIGH VOLUME MANUFACTURING ENVIRONMENT

*Kan Zhou\*, Xin Guo\*, Yu Yang Bian\*, Wen Zhan Zhou\*, Yu Zhang \**

Advanced Module Technology Development, Shanghai Huali Integrated Circuit Corporation, Shanghai 201314, China

*Corresponding Author's Email: zhoukan@hlmc.cn

## ABSTRACT

One of the most critical challenges for lithography process is to effectively control all critical patterns over one field, across wafer, and from lot to lot consistently. With design rules shrink, the lithography process has become more vulnerable, and the critical patterns control becomes more and more challenging. ASML Imaging optimizer (IMO) dose control has been widely adapted to control CD uniformity (CDU) of critical patterns, with designed marks or patterns in most of the situations [1-3]. However, this traditional method has a few weaknesses at logic foundry environment. First, the designed marks are often not representative of random logic critical device patterns, thus leads to a situation that marks are controlled well, while device patterns are not. The designed marks normally put more constrains on device design layout, which is not wanted in some cases. In large die lay out, where there is limited space to place marks, the traditional dose control method can become even more limited. Moreover, the traditional method using limited data point is due to limited CD-SEM tool throughput. Weak point (WP) patterns are typically 2D patterns in critical BEOL layers. To measure just a single WP, metrology noise is normally high. Thus, to reduce the metrology noise, to measure more 2D weak point patterns and make an average are preferred for dose control. This requires massive amount of measurement. A high throughput metrology SEM tool to make an averaging to reduce noise is thus required.

To address the problems, we have developed a method to use yield limiting device patterns to directly control dose thus improve CDU. To reduce the metrology noise of 2D patterns, a lot weak points per die have been selected which can well cover the full die. A high speed eBeam metrology tool, EP5, is used to measure all these identified weak points. To make this method to work effectively, a CDU budget breakdown (BB) has analyzed to identify and quantify CDU contributor [4-5], such as Local CDU (LCDU), metrology noise, reticle fingerprint, etc.. This can lead to optimized number of WP measurements to achieve optimal CDU correction balancing metrology tool time at HVM application.

Keywords — Weak points; E-beam metrology; ASML IMO; CDU; Budget breakdown (BB).

## I. INTRODUCTION

### A. Dose control technologies and pattern selection introduction

A full wafer CDU fingerprint consists of interfiled and intra-field and residues, where global shape of a full wafer is called inter-field fingerprint; and repeating shape for each field is called intra-field fingerprint, as indicated in Fig.1. The residues are the final achievable CDU after correcting both interfiled and intra-field fingerprint [1-3]. The intra-field CDU contributors are normally from scanner and reticle including OPC errors; inter-field CDU contributors from photoresist processing and etch processing if after etch CDU is considered.

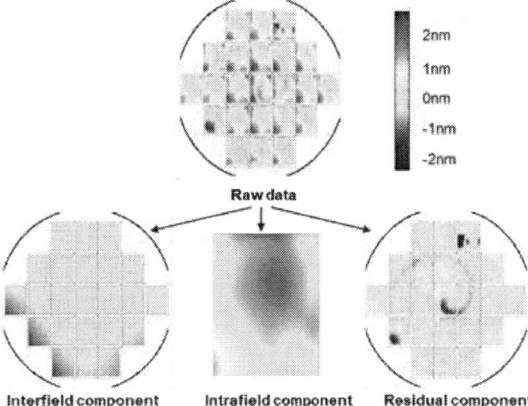

Fig. 1. CDU fingerprint breakdown to inter-field, intra-field and residue components

In advanced technology nodes, ASML imaging optimizer (IMO) and dose mapper (DOMA) have been adopted to correct both intra-field and interfiled fingerprint thus to improve full CDU, and for both after lithography and after etch process steps [6,7], depending on technology and layers.

978-1-6654-9759-6/22 $31.00 © 2022 IEEE

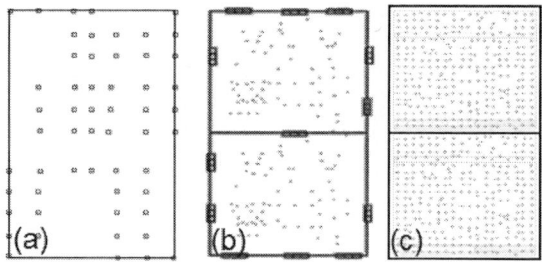

Fig. 2. Traditional dose control measure designed marks in scribe line (a) or in-die (b). Using device WP for dose control, measure point distribution as shown in (c)

Traditional dose control is to apply designed marks representing critical pattern in a layer. Marks normally are placed at scribe lines or in-die, distributed dispersive across die/field for full die/field coverage [8-10], as Fig.2 shown. There are a couple limitations with this method. First, the designed marks are often not representative of random logic critical patterns. The second, the designed marks normally put more constrains on device design layout, which is not wanted in some cases. In some production which die size is large and has limited space to place marks, the traditional dose control method can become more inapplicable.

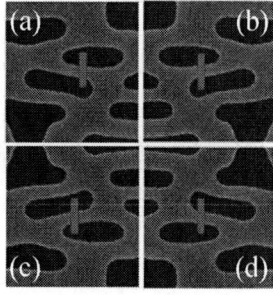

Fig. 3. Yield limiting original, and rotated and mirrored WPs. Red line highlights the metrology locations

In this study, on the contrary to the traditional designed marks, yield limiting device weak points are selected and used, as Fig.2(c). The selected WP can evenly distribute across the die depending on device layout. Two requirements are needed for the WP selection. Firstly, there should be identical WP with enough quantity. Secondly, the WP distribution should cover the device areas. In order to do so, Pattern Search through processing full GDS is utilized. For example, as Fig.3 indicated, the yield limiting WP is identified as top-left SEM image. In pattern search function through the full die, user can search patterns the same as shown in Fig. 3(a), and can also extend the pattern search to its rotated and/or mirrored images, as the other three images in Fig.3 shown. These four patterns are identical from imaging point of view except that they are rotated or mirrored as first order

estimation. The pattern search functions then search these four defined patterns through the full chip GDS. Data filtering is necessary when there are more than 1 WP in 1mm-by-1mm field grid, which is the maximum correctable grid by IMO or DOMA.

## II. SMALL DIE GATE LAYER DOSE CONTROL

### A. Experimental setup

In this work, one BEOL metal layer is chosen for WP based dose control. One focus and exposure matrix (FEM) wafer is used to define best energy (BE) and best focus (BF) per WP.

As shown in Fig.4, this is an 8x14 dies/field layout, 59 full fields over the wafer. In this case, there are 112 dies in one field. This is called small die case, in contrast to large die case where there are a few dies per field. 4 dense dies WPs were measured for intra field dose control and total 59 dies WPs were measure for inter dose control. The total measurement locations are 9184 per die.

Pattern search is performed, and the WP patterns are identified, and they distribute over device area in the die. As a starting point, 82 WP are selected per die. As shown in Fig. 4, WP layout shows the distribution by die and by field. The metrology points are not evenly distributed as shown in Fig. 4(c). Because in these gray gap, no such little patterns was found by pattern search. EP5, ASML latest generation high speed e-Beam metrology tool is employed. Process window metrology tool analyzes all WP data and calculate the dose and focus contribution simultaneously.

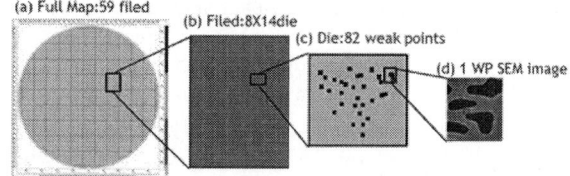

Fig. 4. WP measured map. (b-c) WP layout by field and by die. (d) one WP SEM image.

### B. Experimental data analysis

All selected 82 WP process window is analyzed to study the dose and focus contribution. This data can assess if dose or focus contributes more to the CDU. Process window metrology is a software tool to analyze pattern process window. The EP5 CD measurement output is automatically read by PWM to analyze process window of each WP, per die and across field. Per WP, the best energy (BE), best focus (BF), depth of focus (DOF) and energy latitude (EL) results are analyzed, as shown in Fig.5. From the BE, BF, EL and DOF field maps, the data show the dose variation in the field is about 23.6%, while focus variation about 4.5%. Relative to EL and DOF range, this

978-1-6654-9759-6/22 $31.00 © 2022 IEEE     214

indicates the dose contribution to the CDU is much larger and impact more significant than focus contributions. To reduce the complexity, we choose to skip the focus correction, and focus on dose corrections through IMO or DOMA to improve CDU, though these two functions can correct both dose and focus.

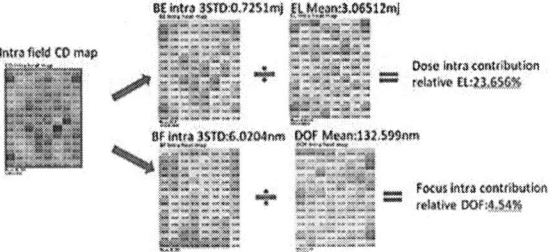

Fig. 5.   Intra field CD map, BE/EL map, BE/DOF field map

IMO is a well-established tool to correct the interfiled and intra-field wafer CD profile with a dose sub recipe. In this analysis we focus on the complexity of correcting a logic in-device measured pattern. The complexity in in-device dose optimization is:
1. The measured weak point patterns are complex in shape (2D patterns);
2. The stochastic and metrology CD variance is large on a single weak point position;
3. There is no guaranteed pattern repeatability within the FOV / field part (e.g., 1x1mm).

This paper investigates the possibilities to create an accurate intra-field dose recipe with pre-processing CD data before use them in IMO. Special attention is given to investigate different number of feature set per die area. Several dose control scenarios are described in Table. 1.

Table. 1. Dose control scenarios description

| # | naming | Method | Approach | Same set of features per point | Comment |
|---|--------|--------|----------|-------------------------------|---------|
| 1 | IMO-4 | IMO | Standard IMO of value of 4x feature's | UC depended Yes with dense die layout | Con: input is limited to 4x features |
| 2 | IMO-die-avg | limitation 4x feature | Average per die of value of multiple feature's | Yes | Con: Requires a dense die coverage Con: Higher order field information is lost |
| 3 | IMO-1x1mm-avg | | Average per die part 1x1mm of value of multiple feature's | No | Con: loses the exact metrology position Con: different features per field point |
| 4 | IMO-FoV-avg | | Average per FoV of value of multiple feature's | No | Con: Requires measure multiple PID's per FoV Con: Different features per field point |

Simulations and experiments are performed to predict the CDU performance gains, as Table.1 shown. IMO-4 means using 4 features for IMO sub-recipe calculate, as method 1. And the following method means average CD value per die (IMO-die-avg., method 2), per 1x1mm grid (IMO-1x1mm-avg.) and per FOV (IMO-FOV-avg.), method 3 and 4. For this use-case with 112 dies per field, method 1&2 are selected for the investigation. The data generated can be used for learning on possibilities to measure and control on a random feature (method 3&4).

The main target for optimization is AEI intra-field CDU. In case of additional inter-field CDU correction, the applied methodology is less trivial, variations beyond the scanner e.g., etcher should be considered. Including this in the optimization could put constraints on the control process flow as scanner & etch plate dedication or optimization on average etch fingerprint.

## C.  Experimental results of IMO optimizations

Predictive simulations of method 1&2 IMO dose recipe are created. As input data, a CDU wafer exposure with 4 shots dense field measurements is used as input. The predictive simulations indicate a clear improvement in intra-field CDU compared to POR performance, with applying normal IMO-4 (method 1) 6% and an additional +2% improvement with IMO-die-avg. (method 2).

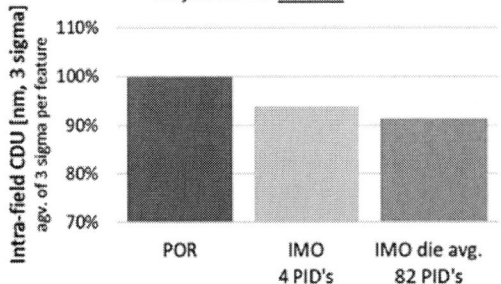

Fig. 6.   Simulated intra-field CDU results at POR, IMO-4-WPs, and IMO-die-avg with 82 weak points patterns

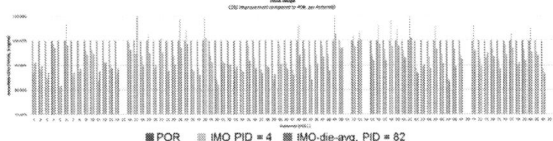

Fig. 7.   Individual WP, CDU improvement at POR, IMO-4-WP, and IMO-die-avg-82-WPs

For IMO-4 the weak points #1-4 are used as input. Improvement on selected features is achieved, ~80% of the additionally measured features benefits from the same dose recipe. However, for a minor 20% degradation in CDU performance is observed, compared to a 4% degrading with IMO-die-avg. using all weak points per WP. The results imply that the measured intra-field CDU data contains more sources of CD variance than the intra-field profile. To better judge the improvement results a deep dive in the various CDU contribution is analyzed and discussed in the following sessions.

## D.  CDU budget breakdown and analysis

A breakdown of CDU data in the functional buckets "LCDU (local CDU)", "OPC error", "global"/"local" reticle fingerprint is performed to better understand how the intra-field CDU variance is build-up [10,11]. This budget breakdown is determined by massive metrology on FEM wafer, all weak points in all dies over the wafer. The data is processed in an intra-field dose profile, and the dose is selected to mitigate the delta in dose sensitivity between features. To achieve this goal, the CD data is fitted with a simple Bossung model and filtered with an +/- 2sigma residual data illustrated in filter Fig.8.

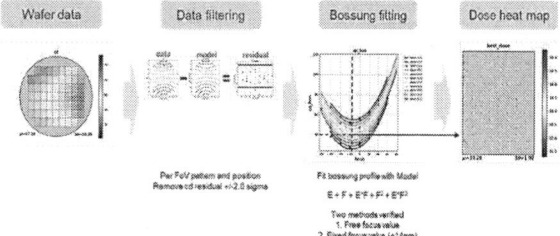

Fig. 8.  Approach to create intra-field dose heat maps

The dose heat map data can be broken down in dose variance and from dose a CDU contribution is determined, for simplified multiplication with an average dose sensitivity is considered. Table 2 shows budgeting approach from dose heat map. Fig.9 and Fig.10 illustrate the breakdown on dose heat map data in different formats.

Table. 2.  CDU budget breakdown approach from dose heat map

| # | Contributor | Approach to determine |
|---|---|---|
| 1 | Global OPC | Average dose per feature |
| 2 | Global reticle profile | Average dose per die |
| 3 | Local reticle profile (+noise) | Variance in dose in per die (averaged over all die's) Noise contributor is removed from budget |
| 4 | Noise | LCDU/$\sqrt{n}$ residual in best dose/focus |

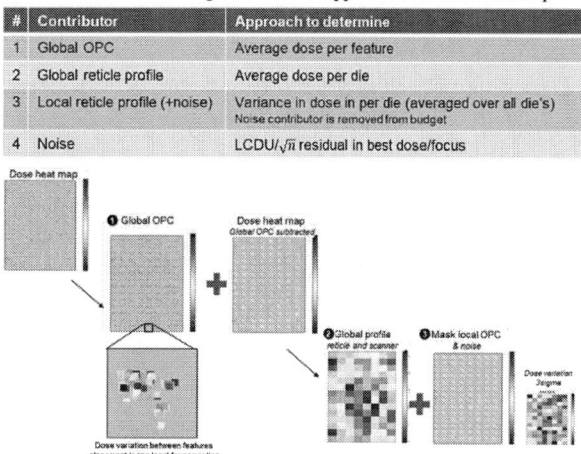

Fig. 9.  CDU breakdown dose heat map in CDU variance budget

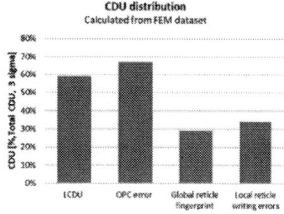

Fig. 10. Budgeting approach from dose heat map

The CDU budget breakdown indicates that the primary target for dose optimization "global reticle fingerprint" is a small budget item relative to the rest. The budget posts causing inaccuracy in the optimization are relatively high "LCDU" and "OPC -error". The LCDU component

consists variance of resists stochastically and metrology noise. The OPC-error is the largest and an issue in case a varying feature set per measurement (as Method 3&4) is used.

*a) OPC errors on various field layout*

In logic in device metrology the CD is extracted from various features. A separation can be made in device critical features WP and non-critical features. The OPC-error is the delta between the measured CD value and the target GDS. The imaging system will create the discrepancy even with an aggressive OPC [11-12].
To avoid optimization towards an OPC profile select a repeating feature or a mitigate action should be taken e.g., averaging many features. A secondary risk is that non-critical feature selection for optimization could improve the selected feature at a cost of the weak point feature.

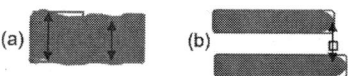

Fig. 11. (a) Mesh grid jumps in target GDS. (b) line end retraction compared to target GDS

*b) Noise reduction*

As a target for a good dose optimization recipe, we aim to have the CDU noise contributors limited to max 20% of the to-be corrected profile. The noise reduction depends on the optimization approach in case of a random feature set (method 3&4) the OPC error should be considered as a noise contribution, for repeating features this is not (method 1&2). The scaling factor depends on the repeats per pattern, LCDU scales with the number of fields and WPs, OPC-error (local/global) with the amount of PID's. For scaling we assume random Gaussian distributed variance $/\sqrt{n}$.
The analysis indicates that the method-1 using only 4x WP's and 4x fields the noise is ~80% equivalent to the measured profile, for an accurate dose recipe further scaling is required up to at least 20 features. OPC error reduction per field point is critical in case there are no repeating die's (and features), averaging over ~50 WP required to suppress the OPC error per field point. The optimization of the number of measured WPs will improve dose control and save metrology tool time for HVM production needs.

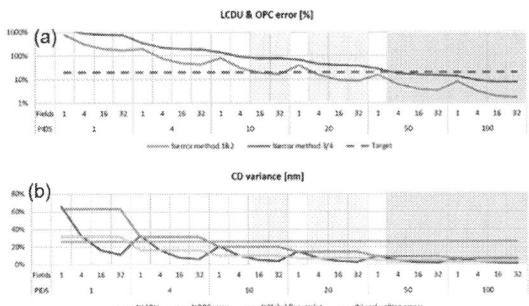

Fig. 12. (a) LCDU and OPC error vs #field and number of WP. (b) CD variance vs # fields and number of WP

### c) Dose field distribution at two scenarios

The dose sub recipes are simulated and verified on wafer, to determine the dose recipe the same 4x field dataset is used as input, as shown in Fig.13. From the simulation results the IMO-4 already captures the global profile, between the IMO-4 and IMO-die-avg. a slight difference in higher order profile variation is seen, as shown in Fig.14.

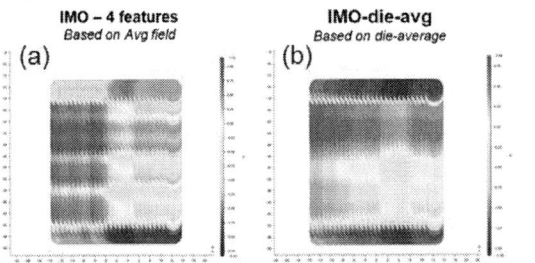

Fig. 13. (a) IMO with 4 WP selected, and (b) IMO-die-avg with 82 WP

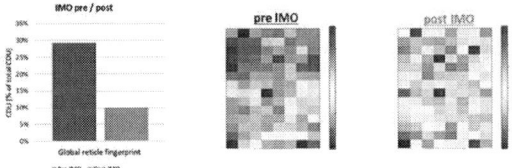

Fig. 14. Pre and post intra-field CDU with IMO-die-avg dose control to reduce ~60% global CDU profile

Verification on CD data indicates that the IMO-die-avg. recipe correct ~60% of the global CDU profile. The illustrated CDU field plot is averaged cd over 82 weak points. The global low order profile is significantly corrected. But in the CD data it is clear that a higher order profile is significant and maintained between the pre and post IMO. This is in line with the expectation in earlier budget exercise that all the noise contribution to the CDU is limited.

## III. LARGE DIE GATE LAYER DOSE CONTROL

### A. Experimental setup

As discussed before, using critical device WP to control dose which gain better CD uniformity, has a special advantage in large die production. So as an initial study, a 2Xnm gate layer in large die production is chosen to test WP dose control. There are only 2 dies per field, with die size of 14mm in X direction, and 16mm in Y direction.

At the large die dose control use case with a gate layer, the lines of the most critical pitch are selected, as seen in Fig.15. The lines are nested and the middle ten lines measured to avoid semi-dense lines in the edge. The intra-field sampling for two dose control methods, Imaging optimizer and Dose mapper, is shown in Fig.16. For the IMO method (left of Fig.16), there are 600 dense devices patterned identified for the dose control. It is obvious that this sampling is very dense. Next step, down sampling is to achieve similar performance while with much shorter metrology time. At the other dose control method, DOMA mark dose control (right of Fig.16), the marks are placed and measured. Because of the device constrain, there are 41 marks placed in the device, with no reasonable space for more marks.

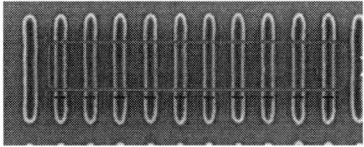

Fig. 15. Lines in most critical pitch are measured in gate layer

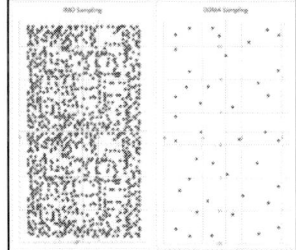

Fig. 16. Intra-field sampling for IMO method (left), and dose mapper method (right). Down-sampling optimization to be made for IMO method

At the full wafer sampling with Imaging optimization method, 4 fields selected, dense sampling with 600 locations for intra-field corrections. All fields with 5 locations are selected for interfiled corrections. This is shown in Fig.17. At the full wafer sampling with dose mapper method, all the fields are measured with 41 marks. There is no difference in sampling for intra-field vs. inter-field.

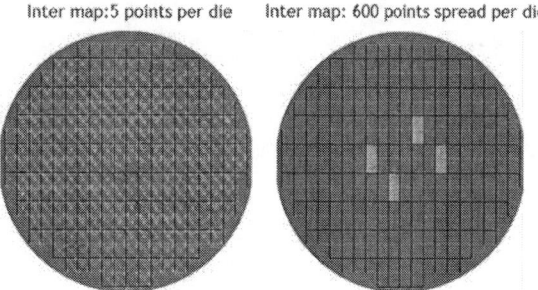

Fig. 17. Gate layer metrology map for inter and intra field.

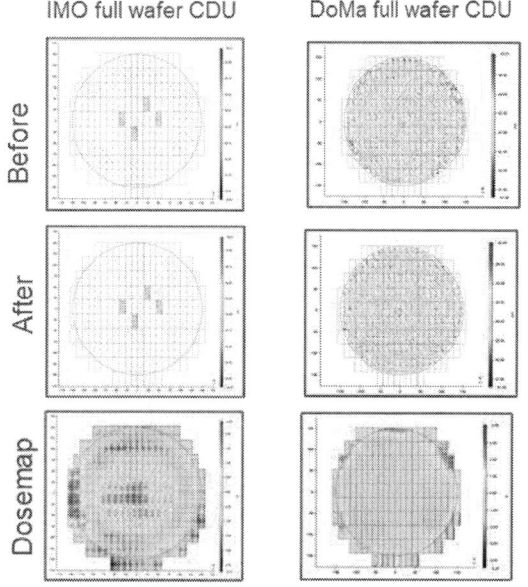

Fig. 18. Full wafer, inter-field and intra-field CDU results at both IMO method (IMO) and dose mapper (DOMA) methods

### B. Dose control results by 2 methods –IMO with device pattern & DOMA with mark

Fig.18 shows the dose control results, full wafer CDU, interfiled CDU and intra-field CDU, at both IMO with device pattern and DOMA with design mark methods, before and after dose corrections. It clearly shows the IMO method brings ~30% improvement on three CDU after dose correction. The interfiled and intra-field both contribute to the full wafer CDU in similar manner. The DOMA delivers less than 10% improvements at all three CDU matrix, with more improvement in inter-field correction than the intra-field correction. One could observe that the intra-field contribution is close to 30% with IMO method, while DOMA method only correct 2% of that. For the intra-field CDU improvement difference at IMO and DOMA method, the dramatic different in sampling, 600 vs. 41, has caused this. The sparse sampling cannot fully capture field profile, thus cannot make the corrections.

For the interfiled corrections, one can observe from Fig.18 that there is a donut shape at the wafer. The CD at wafer center is bigger than CD at other regions. The maximum correction is 8th order at DOMA; thus, it cannot correct the donut shape, which is still visible after correction. While IMO method can correct higher order. This can also be explained by the dose map applied to the wafer for correction. In the IMO method, the dose applied in wafer center is higher, thus it can correct the donut shape, while the dose at wafer center is like other areas except wafer edge which leads to non-corrected wafer donut profile. This is shown in the two figures at the last row in Fig.19 The above analysis shows that the IMO method used in this gate layer combined with dense sampling in intra-field, can correct significantly both intra-field and interfiled fingerprints. The sparse sampling cannot capture the intra-field fingerprint thus cannot correct it. DOMA with sparse sampling cannot correct donut shape wafer CDU fingerprint.

### C. Simulation results by other 2 methods

Based on previous 2 methods dose control results, single bar as a pattern search source as new method 1 and IMO-FOV-avg. as new method 2 were also used on large die gate layer dose control. Because the two new methods are all using IMO not DOMA, so the IMO with device pattern, IMO with single pattern, and IMO-FOV-avg. three methods are compared. The patterns used in three methods are shown in Fig. 19.

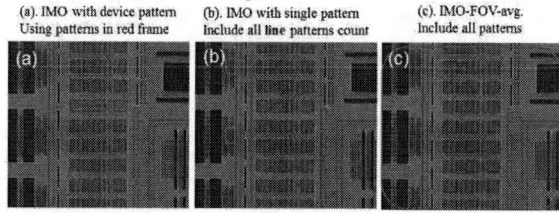

Fig. 19. The pattern used in 3 IMO methods

The IMO with device pattern method for dose control result was already shown in the previous part. For the purpose of using all line patterns for dose control is due to the blank area for IMO with device pattern measurement map as shown in Fig. 16. Using all line patterns was increased the distribution of measurement points. However, the blank area was still exit, as shown in Fig. 20. Sampling point cannot cover full field area. Even more, check the CD distribution align X direction indicate a through scan behavior, which may induced by scanner possibly. And especial local high CD points were also observed. Based on above facts, using all line patterns for dose control induce much noises than using device pattern. It is not suitable for our gate layer case.

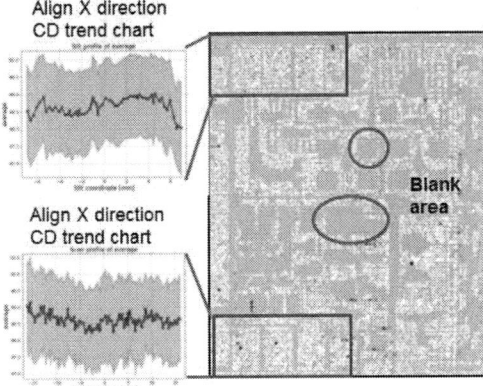

Fig. 20. Line patterns distribution and CD trend chart align X direction.

As IMO-FOV-avg. method, because a mass of patterns type in different locations, metrology CD minus GDS CD was used as the dose finger print. The CD average map and minus GDS CD map are shown in Fig. 21. Average CD map is not match with the To-Target map, which may induce by different OPC modification. Also, the massive CD value distribution shows multi peaks, which may also cause by OPC.

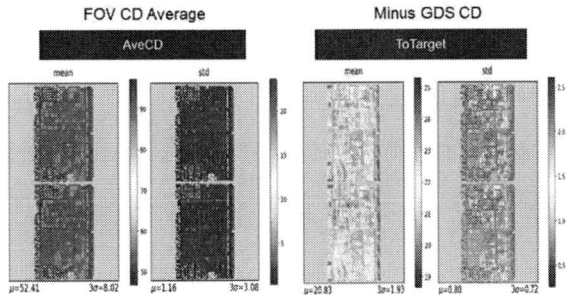

Fig. 21. FOV CD average value heat map (left) and To-Target CD value heat map (right).

**IMO sub-recipe improvement**

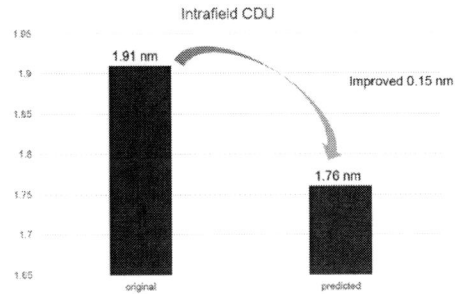

Fig. 22. Intra CDU improve result by IMO-FOV-avg.

Therefore, the simulation result by IMO-FOV-avg. method is shown in Fig. 22. Compared with the previous method using IMO with device pattern, only 7.8% intra

CDU was improved.

## IV. SUMMARY

Innovative dose control method using device yield limiting patterns has been studied in both small die and large die cases.

In small die case, a BEOL metal layer is selected. We've found that this requires sufficient suppression of noise. A CDU performance breakdown indicates that the noise is induced by LCDU resist stochastic / metrology noise, OPC errors and global/local reticle writing errors. A separation in performance can be made for a use-case with and without repeating features over the field, for use-cases without repeating features an additional suppression of the OPC error is required. In the selected metal layer, the study indicates that the standard IMO method with 4 weak points is not sufficient to reach the performance target of noise contribution <20%. Extension to ~20 features is required; and in case of no repeating features this needs to be extended further to ~50 features. In line with our observation in a novel method where we average the cd of 82 weak points measured in the die to a single data point before IMO optimization we improve the CDU performance up to 60% of the global reticle profile, leading to a total 6% CDU gain for individual patterns.

In the large die case with the selected gate layer, 600 critical weak points per die, dense sampling, are measured for dose control with IMO method. This is compared with sparsely 41 placed marks per die using dose mapper method. It clearly shows that the dense sampling with IMO method gains ~30% CDU improvement for both intra-field and interfiled; while sparse sampling with dose mapper method only gains <10%. This is because sparse sampling cannot capture intra-field fingerprint thus cannot correct it. And the dose mapper can only correct lower order inter-field fingerprint while the IMO function can correct higher order inter-field fingerprint.

We also try 2 other method for large die case. Using all line patterns for dose control will induce too much noises to competent this job. As an outlook in small die case, a better approach would be the IMO-FOV-avg. by using all the information in the EP5 large field of view SEM image to reduce metrology and stochastic noise, as method4 in Table.1. However, the performance is worse than expected. More detailed works are going to do for clarification.

## REFERENCES

[1] Marinus Jochemsen, Roy Anunciado, Vadim Timoshkov, et al., "Process window limiting hot spot monitoring for high-volume manufacturing," Proc. SPIE 9778, (2016).

[2] ZY Chen, TY Chen et al., "Effective Epi Process Window Monitoring by High Resolution Massive CDU Metrology" Proc. of IEEE 2018

[3] Mohamed Saib, Gian Francesco Lorusso, et al., "Multivariate analysis methodology for the study of massive multidimensional SEM data," Proc. SPIE 11611,(2021).

[4] Le-Gratiet , Mermet, Gardin,Desmoulins , et al., "Investigating process variability at ppm level using advanced massive eBeam CD metrology and contour analysis," Proc. SPIE 10959, (2019).

[5] Fanton,La Greca, Jain, Prentice, Simiz, et al., "Process window optimizer for pattern based defect prediction on 28nm metal layer," Proc. SPIE 9778, (2016).

[6] Honggoo Lee , Sangjun Han, Minhyung Hong, et al.,"Clean focus, dose and CD metrology for CD uniformity improvement," Proc. SPIE 10585, (2018).

[7] Shmoolik Mangan, Erik Byers, Dan Rost, et al., "Novel lithography approach using feed-forward mask-based wafer CDU correction increase fab productivity and yield," Proc. SPIE 7272, (2009).

[8] Lei Sun, Tsunehito Kohyama, et al., "High throughput and dense sampling metrology for process control," Proc. SPIE (2017).

[9] Bernd Geh, Marija Djordjevic Kaufmann, Rolf Seltmann, et al., "Hot spot variability and lithography process window investigation by CDU improvement using CDC technique," Proc. SPIE 10587, (2018)

[10]Boo-Hyun Ham, Il-Hwan Kim, Sung-Sik Park, et al., "The use of computational inspection to identify process window limiting hotspots and predict sub-15nm defects with high capture rate," Proc. SPIE 10145, (2017).

[11]Simiz, Hasan, Staals, Le-Gratiet, et al., "Predictability and impact of product layout induced topology on across-field focus control," Proc. SPIE9424,(19 March 2015).

[12]Junji Miyazaki, Orion Mouraille, Jo Finders, etal., "Impact of mask CDU and local CD variation on intra-field CDU," Proc. SPIE 8522, (2012).

# PATTERNING CAPABILITY OF SURFACE PLASMON IMAGING

Lihong Liu[1], Huwen Ding[1,2], Libin Zhang[1], Lisong Dong[1], Le Ma[1,2], Yayi Wei[1,2*]

[1]Institute of Microelectronics, Chinese Academy of Sciences, Beijing, China
[2]University of Chinese Academy of Sciences, Beijing, China

*Email: weiyayi@ime.ac.cn

## ABSTRACT

This article concentrates on the patterning capability of surface plasma based nanolithography, giving concerns of the several prestigious structure. We focused on the one-dimensional periodic line space patterning. With parallel sweeping of several chosen parameters at the same time, the light intensity contrast is discussed, which gives a vital reference of the patterning capability evaluation for future industrial application of surface plasma based nanolithography.

## INTRODUCTION

The advantage of surface plasma based nanolithography is the high resolution imaging capability with low fabrication cost of the optical elements [1,2]. Although with high resolution advantage, the near field imaging capability or low depth of focus has limited its application in semiconductor industry [3]. The demand of enlarging the process window to adapt its application to semiconductor industry has becoming more urged recently.

We concentrate on the imaging capability of the surface plasma based nanolithography. For surface plasma nanolithography, the so-called 'superlens' functions in the high resolution imaging. The physical model includes the source, the mask, superlens, photoresist and the substrate. With one-periodic line space pattern on the object plan, the superlens acts as the imaging part, and the center of the photoresist functions as the image plane. The aerial image in the photoresist is studied with rigorous electromagnetic modeling based on the Finite Element Method.

For the system of the superlens based photolithography, the different parameters act very important role in the imaging and lithographic quality. The different parameters impact are studied in this article, based on the light intensity contrast in the center of the photoresist. The parameters include the off-axis illumination angle, the air distance, the period of mask, etc. Based on the imaging quality evaluation, this article gives an basic evaluation and reference for the surface plasma based nanolithgraphy system.

## BASIC PHYSICAL MODELING AND NUMERICAL COMPUTATION

The basic 'superlens' is modelled physically, followed with the rigorous electromagnetic numerical simulation. Based on the numerical simulation, the impact of different parameters on the imaging quality are evaluated.

### Physical model

The basic physical model is given in figure 1. Off-axis illumination source is assumed to be monochromatic with wavelength of 365nm. The material of the mask substrate is quartz, with dielectric constant of 2.25. Mask is assumed to be one-dimensional metallic grating with chromium, and dielectric constant of -8.55+8.96i; air spacer is between grating and the photoresist. Due to the exponential decay of the evanescent wave in air space, the thickness of the air should be lower than the decay length. The interface of metal and dielectric functions for evanescent wave stimulating and coupling. The functionality of dielectric depends on the matching of the dielectric constant between metal and dielectric. In accordance with the matching of the dielectric constant, the dielectric of Ag is -2.4+0.45i, the dielectric of photoresist is 2.59. The substrate of photoresist uses silica, the dielectric constant is 2.17.

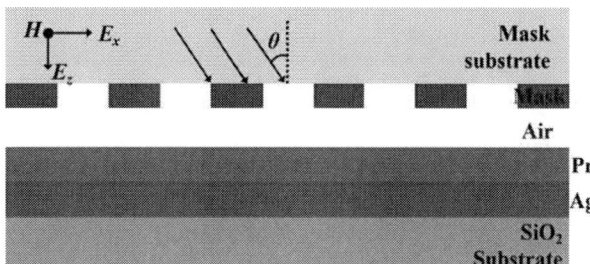

Figure 1: Basic physical model for patterning capability evaluation.

**Numerical computation**

The numerical computation is based on the setting up of the physical model. The rigorous electromagnetic modeling is performed based on FEM with commercial software Comsol Multiphysics 5.6. In Comsol Multiphysics, the periodic boundary condition is applied, with the occupation of the periodic input and output port. TM polarized light incidents on the front face of the system. In our model, the two dimensions of $x$-$z$ plan is used to evaluate the imaging property. Propagation constant of $\beta$ is defined with $2\pi/\lambda$. Poynting vector $S=E\times H$ is evaluated, which indicates the electromagnetic energy density in average time.

In figure 2, the poynting vector distribution presents the negative transportation direction of the evanescent wave in the interface of metal and dielectric. The period of the mask is 60nm. The unit of the poynting vector is W/m$^2$. The normalized poynting vector distributed in the range of (-0.2, 0.8) W/m$^2$. The 'superlens' imaging property of 1:1 imaging ratio is clearly given in figure 2. From the poynting vector distribution, we can clearly see the energy flow of the electromagntic field in the unit time, which is vital for lithographic quality evaluation.

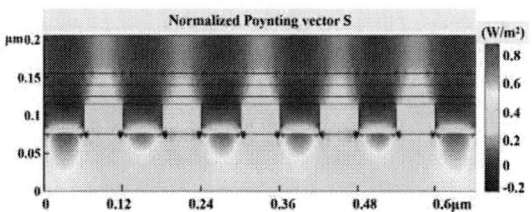

*Figure 2: Poynting vector distribution in the x-z plane.*

In figure 2, the two-dimensional normalized poynting vector distribution is given in the $x$-$z$ plan. The thickness of the substrate of the mask is 75 nm. The incident face is the $z$=0 plan. Parallel line space mask is used with 6 bars of chromium. The thickness of chromium bars is 40 nm. The material between the mask and the photoresist is air, the air distance is 10 nm, the thickness of the photoresist is 30 nm, one transverse line is positioned in the center of the photoresist to evaluate the imaging quality. The thickness of Ag film is 50nm. In the $x$-$z$ plan of the propagation direction, the energy distribution is periodic, and 1:1 imaging ratio between the mask and the imaging plan is obvious. The reflection of the Ag film plays an import role in the periodic energy distribution and the light intensity contrast in the photoresist. Meanwhile, in the photoresist, the energy distribution is a bit decreased in the edge of each periodic line.

Figure 3 gives the one dimensional curve of the normalized poynting vector in the center of the photoresist. The period of the normalized poynting vector distribution is 120nm, which indicates the patterning capability of the surface plasma imaging of grating object in equal proportion.

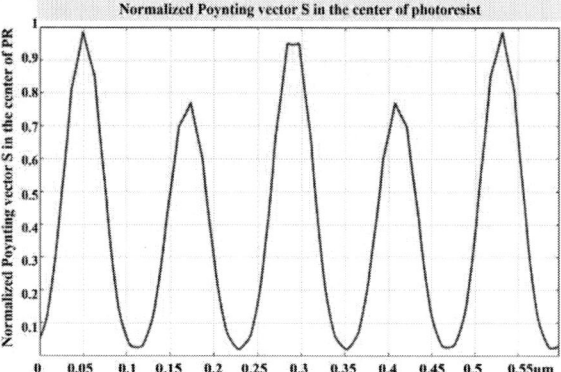

*Figure 3: One-dimensional curve of the normalized poynting vector in the center of photoresist.*

Based on the quantitative evaluation of the energy distribution of the poynting vector and the light intensity contrast, parameters impact are evaluated, the imaging quality and lithographic quality are quantified, which are given in the following section.

## PARAMETERS EVALUATION ON IMAGING AND LITHOGRAPHIC CAPABILITY

In surface plasma imaging, several parameters play an important role in the imaging quality. Here, we present the influence of the incidence angle, the air distance and the period of the mask to the imaging quality. The incidence angle is one important parameter to be responsible for the coupling of the evanescent wave to transmit through the interface of the metal and dielectric. Theoretically, the imaging resolution depends on the incident angle. The parallel parameters sweeping include the incidence angle and the air distance. With the sweep of the incidence angle in the range of (0, 56) degrees, the impact of the incidence angle on the imaging quality is evaluated. The air distance is also an important parameter, after leaving the mask, the intensity of the evanescent wave decreased exponentially. With the increase of the air distance, the imaging quality decreased significantly. In the final section, the imaging quality for different period of the mask is given.

Figure 4 gives the impact of the incidence angle and the air distance to the imaging quality. Light intensity contrast is applied to demonstrate the imaging quality, as in

$$\text{Contrast}=(I_{max}-I_{min})/(I_{max}+I_{min}) \qquad (1)$$

As given in figure 4, the air distance impact is evaluated in the range of (10, 40) nm. When the air distance increased to 40 nm, the light intensity contrast decreased to about

0.82 for normal incidence. When the air distance is 10 nm, the light intensity contrast is bigger than 0.98 for different incidence angle. When the air distance is 40nm, the value of the image contrast is in the range of (0.82, 0.96) for incidence angle ranging from 0° to 56°. For air distance of 40nm, the image contrast reaches the maximum when the incidence angle is 56°, the image contrast reaches the minimum with the normal incidence. Thus, the incidence angle has a non-negligible impact on the surface plasma based nanolithography. When the air distance increased from 20 nm to 40 nm, the impact of the incidence angle on the imaging and lithogrpahic quality becomes bigger.

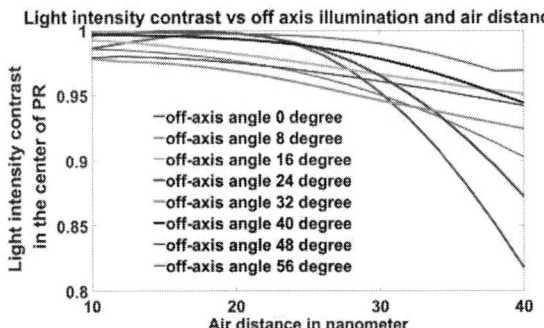

Figure 4: Light intensity contrast for air distance in the range of (0, 40)nm, and off axis angle in the range of (0, 56)°.

In the below figure 5, the one-dimensional curve of the light intensity contrast to the period of the mask in the range of (0.09, 0.24) μm is given. Normally, with the increasing of the period of the mask, the imaging quality becomes better. We can see in figure 5, with the increasing of the period of the mask, several local minimum exist. But, the light intensity contrast is still higher than 0.985. When the period of the mask is 92nm, the image contrast is 0.94, which demonstrates the 28nm lithographic node could be achieved. We can see that with the increasing of the period of the mask, the light intensity contrast becomes higher. When the period of the mask is 240nm, the light intensity contrast is about 0.992. We can conclude that surface plasma based nanolithography could fulfill the lighographic imging quality requirement for 28 nm node.

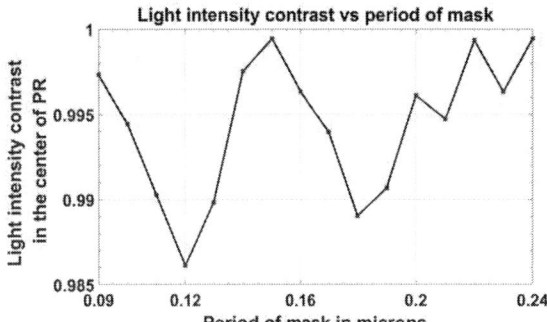

Figure 5: Light intensity contrast for period of mask in the range of (0.09, 0.24) μm.

## CONCLUSION

The basic configuration is studied for surface plasma patterning, with poynting vector distribution in 2D and 1D direction. Parameters impact of off-axis illumination and air distance on the pattering quality is evaluated. The imaging quality on 28nm lithographic node is evaluated with different period of mask. In conclusion, this article gives a reference of the future patterning quality optimization for surface plasma based nanolithography.

## ACKNOWLEDGEMENTS

This article is supported by the National Key Research and Development Program of China (No. 2017YFA0206002). This article acknowledges the thorough advice from professor Yayi Wei.

## REFERENCES

[1] X Luo, T Ishihara, *Applied Physics Letters*, 2004, 84: 4780-4782.

[2] Wei Zhang, Hao Wang, Changtao Wang, Na Yao, Zeyu Zhao, Yanqin Wang, Ping Gao, Yunfei Luo, Wenjuan Du, Bo Jiang, Xiangang Luo, *Plasmonics*, 2015, 10: 51–56.

[3] Zhen Guo, Qizhao Huang, ChangtaoWang, Ping Gao, Wei Zhang, Zeyu Zhao, Lianshan Yan, Xiangang Luo, *Plasmonics*, 2014, 9: 103–110.

# UTILIZING BOSSUNG PLOT TO CALIBRATE OPC OPTICAL MODEL

*Jian Wang\**

Semiconductor Manufacturing International Corporation, Shanghai, China

*Corresponding Author's Email: ken_wang2@smics.com

## ABSTRACT

Compact OPC model calibration consists of three parts: the mask model, the optical model and the resist model. Resist model is "empirical" but optical model is "physical" which strictly simulates scanner's illumination and projection system. As the "physical" property, optical model tuning is a very important step in the three parts calibration. If optical part is not well calibrated, resist part probably force to minimize the merit function RMS beyond physical range, that means model over-fitting. This paper presents a method of calibrating optical model utilizing Bossung plot. Different merit functions were studied: RMS, GRADIENT and FOCUS CENTER of Bossung plot fitted with quadratic function. Model candidates selected by these functions were analyzed and results showed that this method is a good way to search the optical model.

## INTRODUCTION

Typically, compact OPC model calibration consists of three parts: the mask model, the optical model and the resist model. Mask tuning involves bias and corner chop optimization, with additional 3D effect involved in advanced nodes. Resist tuning is to minimize the cost function of a linear sum of multiple terms that represent different resist chemical effects, such as diffusion and neutralization of acid and base. In general, optical model is "physical" and resist model is "empirical", resist part has high risk of inducing model over-fitting than optical part. If the physical optical model is not well calibrated, the probability of model over-fitting will increase a lot as the resist part may try to repair the optical offset by hand of their powerful empirical nature, the consequences are: the resist terms become weird and the model becomes unpredictable.

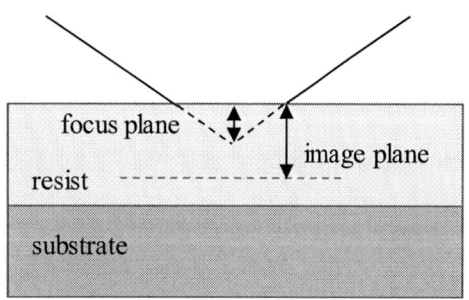

*Figure 1: Illustration of Focus plane and image plane.*

Optical model considers optical conditions such as wavelength, NA, sigma, etc. It simulates the image distribution of light source propagating through illumination and projection system. As shown in Fig 1, there are two main parameters to be calibrated generally: one is focus plane and the other is image plane. Focus plane is the position of film stack with respect to the scanner while image plane is where the image is been thresholded in the resist. The aim of optical model tuning is to find the best pair of focus and image plane value.

RMS is the common merit function of model tuning whether for optical or resist search, slight difference is that resist search uses nominal Litho condition while optical search usually uses FEM Litho condition. It has been proven that FEM measurements can help model explore other defocus regions and improve model prediction ability[1,2], such defocus-aware model is more "physical". In practice, as FEM measurements has big image noise, the common model tuning method is to import FEM measurements at step of optical search, while resist tuning still uses nominal measurements. Figure 2 shows a typical example of the output of a focus-exposure matrix using CD as the response in what is called a Bossung plot[3,4]. Bossung plot is usually used to define the process window, but here is to evaluate optical model performance by analyzing matching degree of curve simulated and measured.

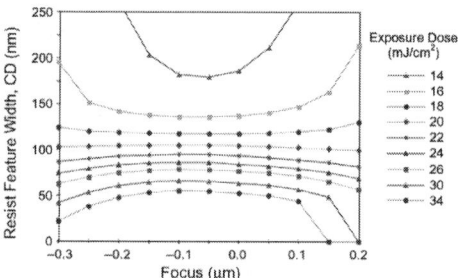

*Figure 2: Example of Bossung plot. Ref[4]: Figure 8.17*

## METHODS

The common way to calibrate optical model is using defocus measurements. Bossung plot of points measured and simulated are analyzed to find the best optical model. The essential of making model defocus aware is trying to match the simulated curve with measured curve. Figure 3 shows examples of good matching and bad matching. A good matching curve indicates model has good prediction ability for defocus conditions. Noticed that the simulated value is not close to measured value, the reason is that the simulated is aerial image that does not contain resist terms at optical search step, but it does not affect optical model tuning as the importance here is to catch the curve trend.

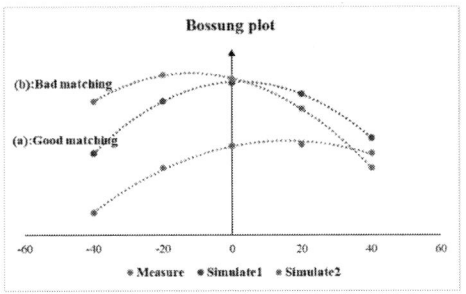

*Figure 3: (a) a good matching Bossung curve; (b) a bad matching Bossung curve.*

Firstly, the Bossung plot is updated by using quadratic function to fit the values of different defocus conditions:

$$f = aX^2 + bX + c \#(1)$$

Then as illustrated in Figure 4, we can extract 3 types of parameters from the updated Bossung plot: (1) the absolute value of each point: $X$. (2) the gradient value of each point derived from quadratic function: $G$. (3) the extreme point (minimum or maximum) of the quadratic function: $F$. Here, $X$, $G$, $F$ are used to calculate RMS, GRADIENT and FOCUS CENTER respectively, details are described in below sections.

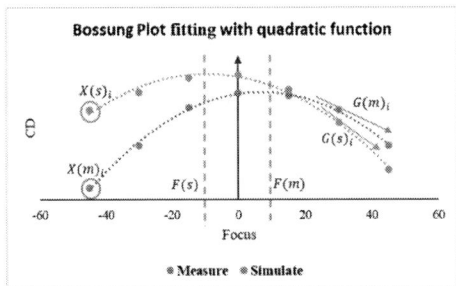

*Figure 4: illustration of updated Bossung plot with 3 parameters: (1) absolute value X, (2) gradient value G, (3) focus center F.*

### RMS

Root mean square (RMS) is the golden merit function for model evaluation especially in resist model tuning task. It is defined as below:

$$RMS = \sqrt{\frac{\sum_{i=1}^{N}(X(m)_i - X(s)_i)^2}{N}} \#(2)$$

In which $m$ represents measured value, $s$ represents simulated value.

It is still suitable for optical model tuning even with defocus measurements. This function will consider each measurement as an independent individual regardless of focus conditions. As shown in Figure 4, actual absolute measurements $X(m)$ and simulated values $X(s)$ participate in RMS calculation.

### GRADIENT

$G$ is the derivative of quadratic function at each point:

$$G = 2aX + b \#(3)$$

In order to distinguish with traditional RMS, GRADIENT is used to represent the difference of $G$ between simulated and measured values with a similar formula:

$$GRADIENT = \sqrt{\frac{\sum_{i=1}^{N}(G(m)_i - G(s)_i)^2}{N}} \#(4)$$

### FOCUS CENTER

$F$ is the extreme point (minimum or maximum) of the quadratic function:

$$F = \frac{-b}{2a} \#(5)$$

As the function is with respect to focus, $F$ is explained as focus center of Bossung plot. Similarly, FOCUS CENTER (FC) is used to represent the difference of $F$ between simulated and measured values:

$$FC = \sqrt{\frac{\sum_{g=1}^{N}\left(F(m)_g - F(s)_g\right)^2}{N}} \#(6)$$

Different with RMS and GRADIENT, each Bossung plot only has one $F$ value, so here subscript $g$ represents a group of defocus points plotting one Bossung plot.

### AVE

Besides the 3 merit functions mentioned above, the weighted average value AVE of the three functions was also calculated:

$$AVE = \frac{w_r * RMS + w_g * GRADIENT + w_f * FC}{w_r + w_g + w_f} \#(7)$$

It is a balance of RMS, GRADIENT and FC, and determined by users demand through adjusting weight $w_r$, $w_g$, $w_f$.

## EXPERIMENTS AND RESULTS

A line/space layer was tested in this paper. The total 4

merit functions described above were calculated at step of optical model tuning and further used to select candidates for resist model tuning. The method of tuning optical model is ranging focus plane and image plane from resist top (0nm) to bottom (max) with a step $s$ (e.g. 5nm) separately. At step of optical model tuning, besides traditional RMS, the distribution of focus center difference was analyzed to help judge the quality of optical model, which is defined as below:

$$errF = F(s) - F(m) \#(8)$$

Table 1 shows all pairs of focus plane and image plane (here marked as P(F, I)). For convenience, below table only shows two RMS values for comparison and omits the GRADIENT, FC and AVE values. P(F45, I60) is the minimum RMS location with RMS value of $a$, while all minimum GRADIENT, FC and AVE location point to P(F40, I55) with RMS value of $b$, while $a < b$. Optical model with P(F45, I60) is called OM1, and the other model with P(F40, I55) is called OM2.

*Table 1. Pairs of focus plane and image plane.*

| Focus / Image | 0 | 5 | ... | 40 | 45 | ... | max |
|---|---|---|---|---|---|---|---|
| 0 | x | x | x | x | x | x | x |
| 5 | x | x | x | x | x | x | x |
| ... | x | x | x | x | x | x | x |
| 55 | x | x | x | *b* | x | x | x |
| 60 | x | x | x | x | *a* | x | x |
| ... | x | x | x | x | x | x | x |
| max | x | x | x | x | x | x | x |

Figure 5 shows the histogram of errF of OM1 and OM2. We can observe that though RMS of OM2 is bigger than that of OM1, more points of OM2 have converged to ZERO regarding errF. Such improvement indicates more simulated Bossung plots match measured Bossung plot. In fact, Figure 3 is an actual example of same gauge extracted from the two models (Figure 3(a) (good matching) is from OM2 while Figure 3(b) (bad matching) is from OM1). Comparing of Bossung plot of Figure 3(a) and 3(b), *errF* is improved from OM1 with ~25nm to OM2 with ~10nm.

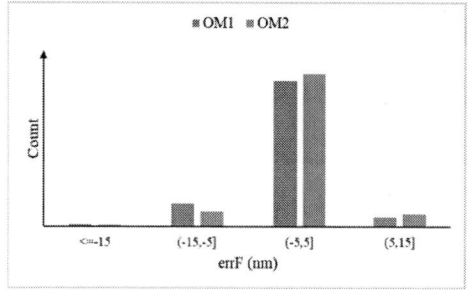

*Figure 5: errF histogram of OM1 and OM2.*

Lastly, the two optical candidates were used to tune resist model, corresponding resist model were called RM1 and RM2. Figure 6 shows the final RMS of different pattern types for RM1 and RM2. These types consist of 1D and 2D type, but still almost all types RMS are improved with RM2 except for Type2 with negligible difference. RM1 failed though the corresponding OM1 won.

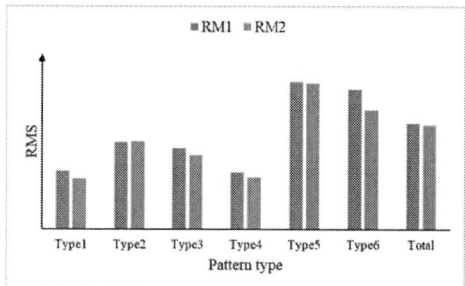

*Figure 6: resist model RMS of RM1 and RM2.*

## CONCLUSIONS

By exploring Bossung plot, GRADIENT and FOCUS CENTER are proposed in this paper. It is demonstrated that trying to match Bossung plot similarity is a better way to search the best optical model comparing to traditional RMS. The better matching Bossung plot, the better prediction of defocus conditions. What's more, the more physical optical model is probably to bring a more robust resist model with smaller final RMS.

## REFERENCES

[1] J. Schacht, K. Herold, R. Zimmermann, J. A. Torres, W. Maurer and Y. Granik. "Calibration of OPC models for multiple focus conditions", Proc. SPIE 5377, Optical Microlithography XVII, 2004.

[2] J. A. Torres, T. Roessler, and Y. Granik. "Process window modeling using compact models", Proc. SPIE 5567, 24th Annual BACUS Symposium on Photomask Technology, 2004.

[3] J. W. Bossung. "Projection Printing Characterization", Proc. SPIE 0100, Developments in Semiconductor Microlithography II, 1977.

[4] C. Mack. "Fundamental Principles of Optical Lithography", The Science of Microfabrication, 2007.

# CD-SEM CONTOUR EXTRACTION FOR COMPLEX FEATURES MEASUREMENT

*Ting He, Yingchun Zhang, Jian Wang*
Semiconductor Manufacturing International Corporation
18 Zhangjiang Road, Shanghai, China
Ting_He2@smics.com

## ABSTRACT

Scanning Electron Microscope (SEM) metrology is the key and useful method for the critical dimension (CD) after developing inspection (ADI) and after etching inspection (AEI) measurement of regular and periodic features, but it is challenged when it comes to multi-point measurement and pattern shift inspection of complex features. In this work, a SEM contour based measurement strategy for complex features is proposed. A robust and reliable multi-step contour extraction and alignment method is studied to extract ADI and AEI SEM image contours and contours are converted into layout format. The solution is verified on complex patterns measurement with high accuracy and precision. The method also demonstrates the capability of lithography to etching bias measurement and contour to target layer edge placement error (EPE) measurement, which is effective for asymmetric phenomenon inspection. With GDS environment, more contour and design information that is useful is extracted to analysis lithography and etching steps patterns performance and the transfer behavior of complex structures.

## INTRODUCTION

Scanning Electron Microscope (SEM) are widely used in semiconductor manufacturing companies for critical dimension metrology, process control inspection and defects analysis[1]. For Optical Proximity Correction (OPC) engineering CDSEM is an important and effective method for wafer ADI and AEI CD metrology and performance evaluation. OPC model development, mask verify, weak point analysis and etch bias compensation rule establishment are based on data measured by CDSEM.

As the feature size node gets smaller, design rule becomes more aggressive and features are designed more complex, conventional CDSEM metrology cannot meet the requirement. CDSEM is typically used to measure regular and periodic 1D features with high accuracy and precision. It is challenged when it comes to complex features and multi point measurement. For complex structures, it is not easy to determine what to measure and design appropriate measurement settings to obtain the desired data. In addition, how to secure stable measurement at correct position and capture asymmetric phenomenon are other difficulties for CDSEM metrology. In recent years, SEM image contouring has attracted the attention of relevant enterprises to develop a robust metrology for complex features. 2D structures have been a difficulty for OPC modeling. To produce more accurate OPC model for complex patterns, Hitachi High-Technologies had continued to develop SEM contouring technology for improving OPC model calibration and verification[2~6]. CDSEM image contouring, in addition, has been studied by more and more researchers with various advanced approaches and applied to unique requirements requirements. TASMIT, Inc has been investigated on a machine learning trained contour extraction algorithm. The method has been utilized to solve e-beam scanning direction effect and improve the accuracy and precision of EPE measurement[7-8]. GLOBALFOUNDRIES has been worked on SEM-contour and standard SEM-CD matching[9]. A SEM contour metrology has been developed by STMicroelectronics after a lot of research and analysis on SEM contouring[10-12]. The SEM contour metrology has integrated a model-based contour extraction solution, a contour manipulation environment and a comprehensive metrology toolbox., which was expected to be used by metrology, process, OPC, Integration and design rule engineers.

The objective of this work is to present a solution for ADI and AEI measurement of complex feature based on SEM image contouring. For this purpose, firstly, we propose a multi-step SEM image contouring method, which includes two steps of contour extraction, two steps of alignment, and a step of matching between SEM contour measurement and CD-SEM measurement. Then, a measurement rule is exemplified for complex feature measurement. Finally, the approach is evaluated on the ADI and AEI measurements of complex features and applied to the analysis of asymmetrical structure.

## METHOD

### Preprocessing

Firstly, we set up a CD SEM recipe for complex structures, measuring the CD of a polygon in each complex structure. The CD SEM image is prepared for contour extraction and the CD is used for SEM contour measurement and CDSEM measurement matching. The recipe controls SEM metrology conditions such as e-beam intensity, voltage, scanning mode, the field of view (FOV), pixel size, magnification, measure point, sum lines, smoothing grade, and threshold. Secondly, a CDSEM equipment is used for capturing ADI and AEI images of complex structures. Thirdly, we summarize the files for contour extraction, alignment and matching, including the original GDS layout of measured patterns, ADI and AEI measurement result documents, and ADI and AEI SEM images. The measure result document contains coordinates of complex features and CDSEM measurement results. SEM images of complex structures contain raw SEM images without measure information and measured SEM images. Before contour extraction, the images with poor quality are filtered.

### Contour extraction and alignment

SEM images are greyscale images, on which pattern edges are highlighting. In the first step of contour extraction, the

978-1-6654-9759-6/22 $31.00 © 2022 IEEE

SEM contour is extracted by the gray level of the SEM image. Then, the extracted contours are aligned to corresponding GDS clips which are clipped from GDS layout according to the coordinate of each pattern with an appropriate size. The extracted ADI and AEI contours are overlapped with ADI target layer and design layer on GDS clips, respectively (see Fig. 1). The first step alignment is based on the overlap ratio between the extracted contours and GDS polygons in the same FOV of SEM image. The best alignment is achieved when the overlap ratio reaches the maximum. The purpose of the first step contour extraction and alignment is to identify the location of patterns and align contours to the target polygons.

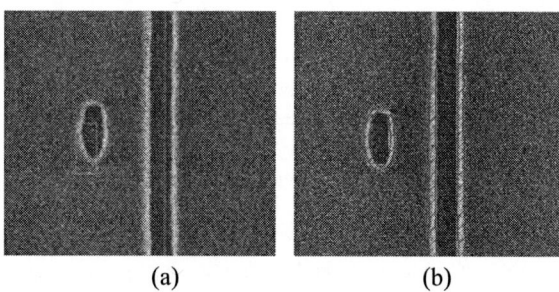

(a)            (b)

Fig. 1. (a) SEM image overlap with GDS layout (green line) before $1^{st}$ contour extraction and $1^{st}$ alignment. (b) After $1^{st}$ contour extraction (red slash) and $1^{st}$ alignment.

In the second step of contour extraction, the SEM contours are extracted by processing the intensity signal of the SEM images. For each polygon on the SEM image, the amplitude and gradient of the signal of cutline is used to detect edge (see Fig. 2). A slope point is define as the edge position where the signal is to be extracted. Smoothing and averaging are carried out to improve the signal-to-noise ratio before extraction. A threshold is used to find slope points, which calculated from the amplitude and the gradient. A design tracking method is used to fit the "black" edges where the signal is lost, so that the extract contour of a polygon of a complete closed loop.

A second step alignment is applied to optimize the accuracy of the metrology for EPE measurement and asymmetry analysis. During this step, the second extracted contours are overlapped with the GDS layout based on $1^{st}$ alignment. Then, the EPE between contours and target of all polygons in an SEM image are measured horizontally (left and right edges) and vertically (top and bottom edges). An average EPE of each direction is calculated (see equation 1, 2). The second step alignment of a SEM image finished by moving extracted contours $X_{shift}$ horizontally and $Y_{shift}$ vertically (see Fig. 3).

$$Y_{shift} = \left( \frac{\sum_i^n EPE_{top}}{n} - \frac{\sum_i^n EPE_{top}}{n} \right) / 2 \qquad 1$$

$$X_{shift} = \left( \frac{\sum_i^n EPE_{left}}{n} - \frac{\sum_i^n EPE_{right}}{n} \right) / 2 \qquad 2$$

**Contour CD and SEM CD matching**

To further ensure the reliability of the metrology, a SEM CD and contour CD matching step is taken. For each image, the contour CD is measured at the same location of CDSEM measured and kept the same measurement settings of CDSEM (see Fig. 4). Considering the possible measurement deviation of CDSEM measurement, contour measurement location is corrected to the actual measurement location of CDSEM. The CD difference between contour measured and CD-SEM measured is defined (see equation 3). If the average CD difference of a recipe is larger than a preset value, the extracted contour will be resized by $\Delta CD_{diff}$ (see equation 4). After resizing the CD difference still unacceptable, it returns to the first step of contour extraction.

$$CD_{diff} = CD_{contour} - CD_{SEM} \qquad 3$$

$$\Delta CD_{diff} = \sum_i^n CD_{diff} / 2n \qquad 4$$

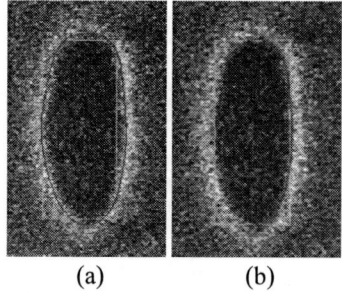

(a)            (b)

(c)            (d)

Fig. 2. (a) A polygon example before $2^{nd}$ contour extraction, (b) an amplitude of the signal of a cutline, (c) slop points of a polygon and (d) contour of the polygon after $2^{nd}$ contour extraction (red line).

(a)            (b)

Fig. 3. A polygon example (a) before $2^{nd}$ alignment and (b) after $2^{nd}$ alignment.

Finally, after completing all steps of contour extraction, alignment and matching, the extracted contours of each SEM image are merged into corresponding layer and converted into layout formant. To ensure the accuracy and reliability of measurement, generally, for one pattern it is often repeatedly measured on different dies under the same process condition. For this case, an average contour is generated for the pattern and dumped into the layout.

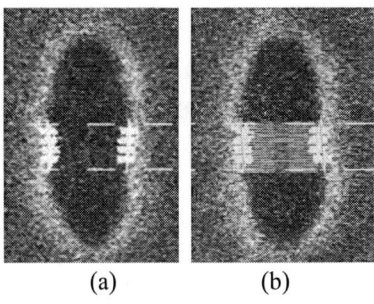

(a)           (b)

Fig. 4. A polygon examples (a) CDSEM CD measurement and (b) Contour CD measurement.

### Measurement for complex feature

Once the extracted contours are available in layout format, the measurement for complex feature becomes easier and flexible. It is possible for edge-to-edge measurements including CD, space, pitch at expected location. As extracted contours are precisely aligned to the original layout, the target layer can be worked as a reference for EPE measurement. In addition, contours of ADI and AEI can be merged into difference layers of a GDS layout, so that it can be applied to contour bias analyzing, such as etch bias (ADI-AEI). More information can be measured in terms of requirements.

We take a complex feature (see Fig. 5) as an example. The short bar A and long bar B are a designed asymmetric structure. The asymmetric effect of the short bar on long bar is difficult to be quantitatively analyzed for conventional CDSEM metrology. For dense pattern, the asymmetric displacement may be possibly be calculated indirectly by measuring the CD and space of more adjacent polygons. However, setting up an appropriate multiple measurement recipe and measuring at the right locations are other difficulties of conventional CDSEM measurement. For SEM contour metrology, in addition to measuring CD and space of short bar and long bar, with reference layers in GDS layout environment it is possible to measure the EPE of ADI contour to ADI target, EPE of AEI contour to design, and the bias between ADI contour to AEI contour at specific locations automatically, as shown in yellow mark in Fig. 5. With these measurement data, the influence of short bar on long bar can be clearly analyzed.

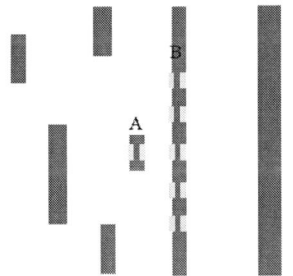

Fig. 5 An example of measurement rule of a complex feature.

## Results

### Measurement accuracy and precision

The SEM contour metrology is applied on a CDSEM recipe that contains 263 different types of complex structures. All patterns are on the same mask and processed under same condition. For each pattern, the ADI and AEI are repeat measured on five different dies respectively.

The CD difference between CDSEM measurement and SEM contour measurement is compared die by die of every pattern. The 3Sigma of CDSEM CD measurement and SEM contour CD measurement for every pattern is calculated and compared. Recognized the accuracy and precision of CDSEM measurement, the accuracy and precision of SEM contour measurement are evaluated through the CD difference and 3Sigma difference between two approaches. It showed the SEM contour measurement ADI CD is 0.13nm smaller than CDSEM CD on average. All ADI CD difference is within -2nm to 2nm and 90% pattern CD difference is within -1nm to 1nm (see Fig. 6(a)). The SEM contour measured AEI CD is 0.29nm smaller and the AEI CD difference of 98% pattern is within -1nm to 1nm (see Fig. 6(b)). Both ADI and AEI CD 3Sigma of SEM contour measurement are comparable to CDSEM measurement (see Fig. 6(c-d)).

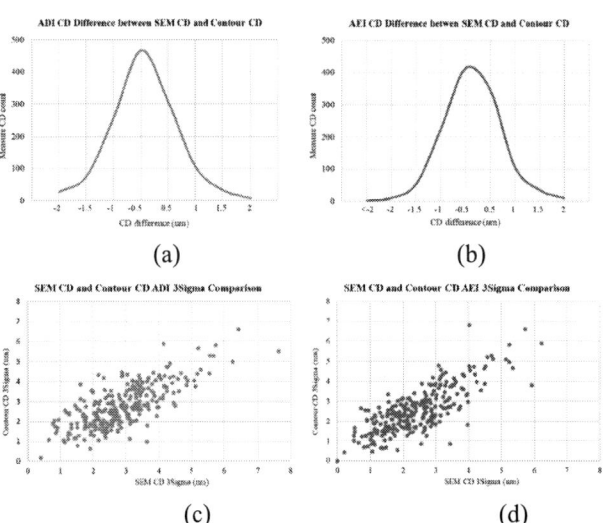

(a)          (b)

(c)          (d)

Fig. 6 (a) The ADI CD difference of SEM CD and contour, (b) the AEI CD difference of SEM CD and contour CD, (c) the ADI CD 3Sigma of SEM CD and contour CD, and (d) the AEI CD 3Sigma of SEM CD and contour CD.

### Complex feature measurement

ADI and AEI SEM contours of the measured recipe are converted into layout. We use extracted contours to analysis the aforementioned asymmetrical pattern. With the aid of ADI and AEI contours, long bar shifting to short bar can be clearly observed (see Fig. 7(a) and Fig. 7(b)). Analyzed a large number of automatic measured data, the influence range and intensity of short bar on long bar are determined. Long bar projecting short bar edge ADI and AEI EPE is positive value and the opposite edge of long bar EPE is negative value. However, the ADI and AEI CD of long bar projecting short bar part and non-projecting part are comparable. It reflects long short shifting to short bar. Adjacent interval measurement illustrates that the shifting is weaker further away from the short bar. The short bar effect exists both on lithography and etching process. It disappears when the space of short bar to long bar is larger than a certain value (see Fig. 7(d)). The shifting can be compensated by adding asymmetrical etch bias value of the long bar when projecting to a short bar (see Fig. 7(c)).

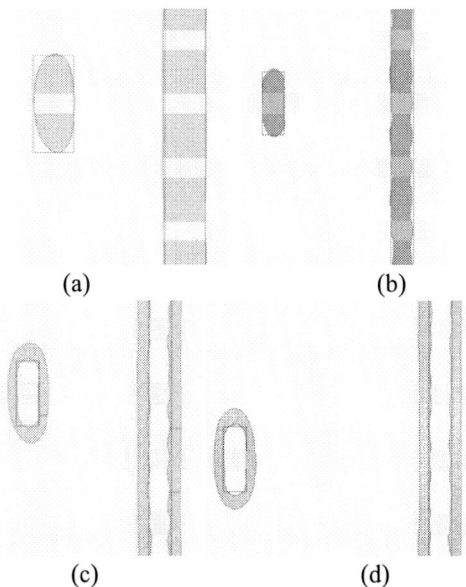

Fig. 7 Extracted contours of short bar effect pattern and EPE and CD are measured at yellow mark position. (a) ADI target (green line) and ADI SEM contour (red fill), (b) design (blue line) and AEI SEM contour (purple fill), and (c) ADI to AEI SEM contour bias (brown fill) of a short bar projecting long bar pattern1. (d) ADI to AEI SEM contour bias of a larger short bar to long bar space pattern.

## CONCLUSIONS

In this work, a SEM contour metrology was investigated for complex features. To ensure the measurement accuracy and precision, a robust and reliable multi-step contour extraction solution was developed. The method was qualified on real wafer ADI and AEI measurement, which showed that the measured CD and 3Signma are comparable with CDSEM metrology.

The extracted contours were fine aligned to original layout and converted into GDS layout, which made the approach a powerful automatic measurement tool for complex features. In the GDS environment, in addition to CD and space measurement, it was also applied to EPE measurement at specific locations. It was successfully used to analysis asymmetrical structures and solve the measurement difficulties of conventional CDSEM metrology.

## REFERENCES

1. Le Gratiet B, Bouyssou R, Ducoté J, et al. "Contour based metrology: getting more from a SEM image," *Metrology, Inspection, and Process Control for Microlithography XXXIII.* International Society for Optics and Photonics, 2019, Vol. 10959, pp. 109591M.

2. Granik Y, Kusnadi I. "Challenges of OPC model calibration from SEM contours," *Metrology, Inspection, and Process Control for Microlithography XXII.* International Society for Optics and Photonics, 2008, Vol. 6922, pp. 69221H.

3. Hibino D, Shindo H, Abe Y, et al. "High-accuracy OPC-modeling by using advanced CD-SEM based contours in the next-generation lithography," *Metrology, Inspection, and Process Control for Microlithography XXIV.* International Society for Optics and Photonics, 2010, Vol. 7638, pp. 76381X.

4. Hibino D, Shindo H, Hojyo Y, et al. "The assessment of the impact of mask pattern shape variation on the OPC-modeling by using SEM-Contours from wafer and mask," *Photomask Technology 2011.* International Society for Optics and Photonics, 2011, Vol. 8166, pp. 81661S.

5. Fuchimoto D, Hibino D, Shindo H, et al. "Weighting evaluation for improving OPC model quality by using advanced SEM-contours from wafer and mask," *Optical Microlithography XXV.* International Society for Optics and Photonics, 2012, Vol. 8326, pp. 83262Q.

6. Miller M, Hitomi K, Halle S, et al. "Application of SEM-based contours for OPC model weighting and sample plan reduction," *Optical Microlithography XXVIII.* International Society for Optics and Photonics, 2015, Vol. 9426, pp. 94260Y.

7. Okamoto Y, Nakazawa S, Kawamura A, et al. "Contour extraction algorithm for edge placement error measurement using machine learning," *Metrology, Inspection, and Process Control for Microlithography XXXIV.* International Society for Optics and Photonics, 2020, Vol.11325, pp. 113252A.

8. Okamoto Y, Nakazawa S, Kawamura A, et al. "Improvement of EPE measurement accuracy on ADI wafer, the method of using machine learning trained with CAD," *Metrology, Inspection, and Process Control for Semiconductor Manufacturing XXXV.* International Society for Optics and Photonics, 2021, Vol. 11611, pp. 116111W.

9. Weisbuch F, Schatz J, Mattick S, et al. "Investigating SEM-contour to CD-SEM matching, Metrology," *Inspection, and Process Control for Semiconductor Manufacturing XXXV.* International Society for Optics and Photonics, 2021, Vol. 11611, pp. 116110Y.

10. Lakcher A, Le-Gratiet B, Ducoté J, et al. "Robust 2D patterns process variability assessment using CD-SEM contour extraction offline metrology," *Metrology, Inspection, and Process Control for Microlithography XXXI.* International Society for Optics and Photonics, 2017, Vol. 10145, pp. 1014514.

11. Lakcher A, Schneider L, Le-Gratiet B, et al. "Advanced metrology by offline SEM data processing," *33rd European Mask and Lithography Conference.* International Society for Optics and Photonics, 2017, Vol. 10446, pp. 104460L.

12. Le Gratiet B, Bouyssou R, Ducoté J, *et al.* "Contour based metrology: getting more from a SEM image," *Metrology, Inspection, and Process Control for Microlithography XXXIII.* International Society for Optics and Photonics, 2019, Vol. 10959, pp. 109591M.

# NEW MODEL-RULES-HYBRID SBARS PLACEMENT STRATEGY FOR 2D PATTERN

*Ge Zhang[1], Yibin Huang[1], Zhongwen Yana*
[1]Semiconductor Manufacturing International Corporation
18 Zhangjiang Road, Shanghai, China
*Corresponding Author: 13371931202, zhangge_94@163.com

## ABSTRACT

With the development of semiconductor technology and the reduction of critical dimensions, the sub-resolution assist features (SRAFs) such as scattering bars (SBARs) has become an effective solution to improve the lithography process window of isolated pattern, and the insertion strategy of scattering bars is the key to achieving optimization effects. The current mainstream sbar insertion methods are divided into two types: model-based sbar and rule-based sbar. In this paper, we introduce a new model-rule-hybrid sbar insertion method for 2D patterns, which inherits the both advantages of the rule-based and model-based sbar: short running time, reasonable sbar placement and effectively printing avoidance. This new method can cover the weak point with insufficient process window more comprehensively at the full chip level and is applicable to all tech nodes. Firstly, we use the OPC model to determine the best placement parameters and MRC limit parameters for the sbar generation of the 2D pattern, then, prioritize these rules according to the process window of the main pattern types, this prioritized sbar insertion approximately turned the model sbar result into a rule based placement strategy. This sbar insertion strategy can fully consider any possible weak point and ensure that at least one side sbar can be generated around the weak point pattern in the optimal position to increase lithography process window. The simulation result of the full chip shows that this new sbar has less potential weak points (the number of PVband weak points is significantly reduced) than traditional rule based sbar.

*Keywords—Optical proximity correction; sub-resolution assist features; Scattering bar; placement strategy*

## INTRODUCTION

With the development of semiconductor technology, the critical size of transistors is constantly decreasing, which makes greater distortion between mask pattern and final wafer image, that is, the pattern transferred on the wafer is more different from the pattern defined on the mask. At present, Optical Proximity Correction (OPC) has been used as a standard optimization technique to reduce the pattern distortion on the wafer mainly caused by proximity effects and non-linear effects [1]. With the continuous decrease of the k-factor in lithography technology, the process window of the full chip has become an important limiting factor for achieving high yields. In this case, a kind of resolution enhancement techniques (RET) that insert sub-resolution assist features (SRAF) such as scattering bars (SBARs) [1,2] around the main pattern will play an important role in high nodes, because they can improve the image contrast on the wafer and enhance the depth of focus (DOF) range of isolated features. Therefore, designing an optimal SRAF

placement strategy for different types of main patterns is the key to realizing the improvement of the process window.

For the contact layer, the main patterns are all 2D patterns, which makes the structure environment very complex with thousands of different pitches in full chip layout. It is a huge challenge to get best SRAF placement strategy in this very random environment. At present, the commonly used SRAF placement strategies in semiconductor industry are model-based sbar [3] and rule-based sbar. Based on model, the EDA tool places sbar at the best position according to the simulation results and effectively prevent sbar printing. However, this method has two limits to applicate in real case: firstly, model based calculation will cost significantly computer resource and unacceptable runtime. Secondly, the shape of model-based sbar is intricate, meanwhile, the consistency of the sbar generated in the period area is also very poor. Rule-based sbar can effectively improve the running time, but in order to optimize the parameters of sbar under different pitch, it takes a lot of time and machine resources to measure enough wafer data on test mask in the early stage, including change the main pattern configuration and assist features configuration, dose condition and defocus condition. Furthermore, the coverage of the final sbars generation table is poor for the 2D pattern with very random environments. Therefore, it is not the optimal choice for the contact layer of higher tech nodes.

In this paper, we describe a novel SRAF placement strategy for 2D patterns. This method combines the advantages of model-based sbar and rule-based sbar, and can reduce the weak point (structures with insufficient process window) as much as possible at the full chip level. At the same time, this strategy will not be affected by technology nodes change, and can be applied across the entire range of technology nodes.

### The new model-rule-hybrid sbar placement strategy

For the contact layer, the environment of each edge of the 2D Main pattern is very different, and the corresponding sbar insertion is also very different, but in general, the sbar side edge has a better enhancement effect on the DOF and contrast of main pattern, while the sbar end edge is harmful to the process window of main pattern. In full chip layout, most of weak points with insufficient process windows are caused by the insufficient number of side sbars generated around. Some inappropriate sbar placement strategies will cause conflicts between the sbars generated by multiple patterns. After conflict resolve, some final sbars are difficult to cover main patterns in multiple pitch environments at the same time, or even worse situations may occur, some main patterns are surrounded by all line-end sbars (see Fig.1), which will become a new weak point.

978-1-6654-9759-6/22 $31.00 © 2022 IEEE

In order to reduce the number of weak points as much as possible, it is necessary to perform multiple special handles for different types of weak points, which will undoubtedly increase the OPC complexity and running time.

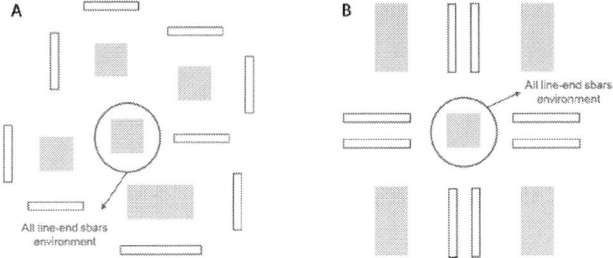

Fig.1. The schematic diagram of all line-end sbars environment

In order to avoid the above problems, we try to add the best sbar at one time to reduce the OPC complexity and effectively optimize the potential weak point environment within the full chip range. We propose a new model-rule-hybrid sbar placement strategy to meet the above requirements. Fig.2 is the brief flow chart of new model-rule-hybrid sbar placement strategy.

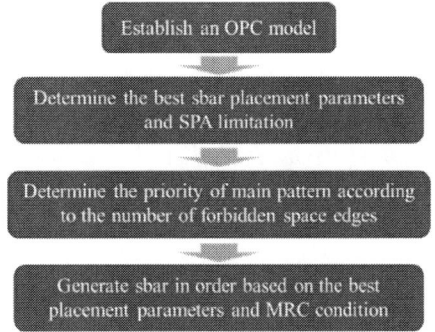

Fig.2. Brief flow chart of new model-rule-hybrid sbar placement strategy

First, similar to model-based sbar operation, an OPC model will be prepared based on the existing light source, mask stack, wafer film stack, and test mask pattern to simulate various behaviors of the actual lithography process. After that, the contour obtained by the OPC model simulation will replace the wafer data collecting operations in the rule-based sbar strategy. Then, plan a DOE (design of experiment) to generate a series of sbars with different offsets, widths, and extensions around isolated 2D pattern. After that, we use the OPC model to simulate the inverse intersection of the minimum and maximum contours of each isolated 2D pattern under FEM (focus energy matrix) condition , that is, the process variation band (PVband), which has been widely used to identify image quality based on process variation [4,5] . The smaller PVband means better imaging quality and larger depth of focus (DOF), therefore, the set of sbar addition parameters with the smallest PVband will be selected as golden parameters, including the best offset, best width and best extension of the side sbar and corner sbar from the first cycle to the third cycle, as shown in Fig.3. The sbar printing risk is also considered during the DOE process, and the final golden parameters can ensure that sbar will not be transferred to wafer as much as possible. Meanwhile, according to the mask write capability, the mask rule checking (MRC) will be formulated to add a series of restrictions to the sbar placement process to facilitate the production of the mask, including the minimum width and the maximum length of the sbar, and the minimum space between the sbar and other surrounding patterns, as shown in Fig.4.

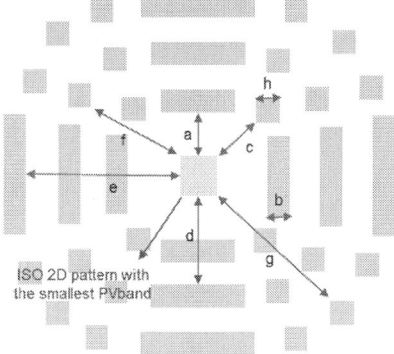

Fig.3. The schematic diagram of sbars addition golden parameters. a=1st side sbar best offset, b=side sbar best width, c=1st corner sbar best offset, d=2nd side sbar best offset, e=3rd side sbar best offset, f=2nd corner sbar best offset, g=3rd corner sbar best offset, h=corner sbar best width

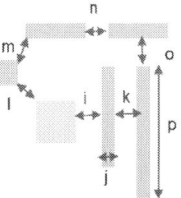

Fig.4. The schematic diagram of the mask rule checking. i=sbar to main min offset, j=sbar min width, k=sbar to sbar min space, l=sbar corner to main corner min offset, m=sbar corner to corner min space, n=sbar end to end min space, o=sbar to sbar end min space, p=sbar max length

The structure of 2D pattern layout is flexible and complex. When two main patterns are too close, that is, they are in the forbidden space (FS), the side sbar of the corresponding side will not be generated, as shown in Fig.5, the forbidden space is defined as the space less than (2i+j). If there are multiple patterns that are too close around a certain main pattern, the number of side bars that can be generated around the main pattern will be reduced. Compared with other patterns, the pattern in this structure is easier to form a weak point. In order to eliminate this potential weak point risk caused by the layout structure, we first prioritize the patterns in the full chip according to the number of FS edges, the greater the number of FS edges, the higher the risk of the pattern forming a weak point, and set the higher priority, as shown in Fig.6. Furthermore, for structures with the same priority, we customize different first-circle side sbar placement strategies according to the space of the environment where the free edge (the edge that can generate side sbar) is located, to ensure that at least one side sbar will be generated around each pattern, as shown in Fig.7. When the space between two opposite free edges is (2i+j), a side sbar with the smallest width can be generated in the middle. As the space increases, the width of the center sbar can gradually increase, the offset between sbar and main patterns can also gradually become farther, as shown in Fig.7A. When the space between two opposite free edges reaches (2i+2j+k), two side sbars with the smallest width can be generated in the middle, as the space continues to increase , the width of the sbar and the offset to the main pattern can also be increased, as shown in Fig.7B. When the distance between two opposite free edges reaches (2a+2b+k), a side sbar with the best width can be generated at the best offset, in this case, the side sbar parameter will no longer change, as shown in Fig.7C. The above strategy can ensure that no matter how the space changes, the first-circle side sbar that is optimal for the existing environment can be generated in the middle, which can maximize the enhancement effect on two main patterns at the same time, in order to minimize the possibility of weak point generation in complex random environment.

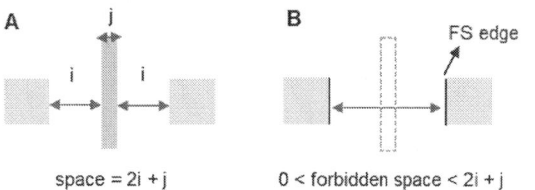

Fig.5. The schematic diagram of the forbidden space edge. (A) when the space between two edges is 2i+j, the smallest side sbar can just be generated; (B) when the space between two edges is less than 2i+j, side sbar cannot be generated, the edge in this environment is called FS edge.

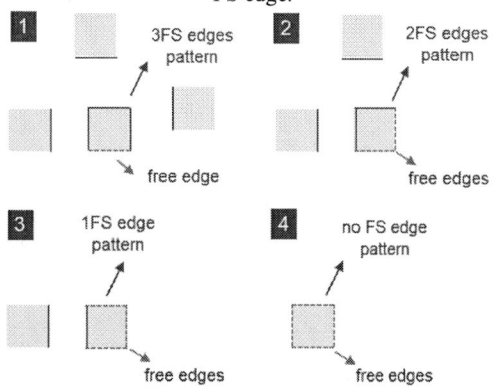

Fig.6. The schematic diagram of the priority of different structures in the full chip: 3FP edges pattern > 2FP edges pattern > 1FP edge pattern > no FP edge pattern

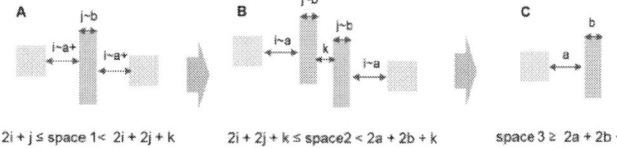

Fig.7. The different first-circle side sbar placement strategies for different space when the main patterns priority are the same.

In general, the side sbar in first circle has the most obvious enhancement effect on main pattern, and the enhancement effect of the corner sbar and the peripheral side sbar will be much weaker. Therefore, after generating the optimal first circle side sbar according to the above strategy, the subsequent corner sbar and peripheral sbar will be generated in the remaining space in sequence, in the order of 1st circle corner sbar, 2nd circle side sbar, 2nd circle corner sbar, 3rd circle side sbar and 3rd circle corner sbar. The placement strategy is to start from the optimal offset and optimal width, if the remaining space is insufficient and the optimal sbar cannot be generated, then the offset and width will be reduced according to a certain step (within the MRC constrain), and then try several times until the sbar can just generate.

## EXPERIMENTS ON REAL CHIP

In order to verify the optimization effect of the new model-rule-hybrid sbar placement strategy on main patterns in different environments, and to further study the applicability of the strategy in different technology nodes, we selected different full chip of 20nm+ technology node and clip chip of technology node of 14nm for testing. Meanwhile, the differences between the maximum PV band range, the number of PV band weak points and the running time of different sbar placement strategies compared.

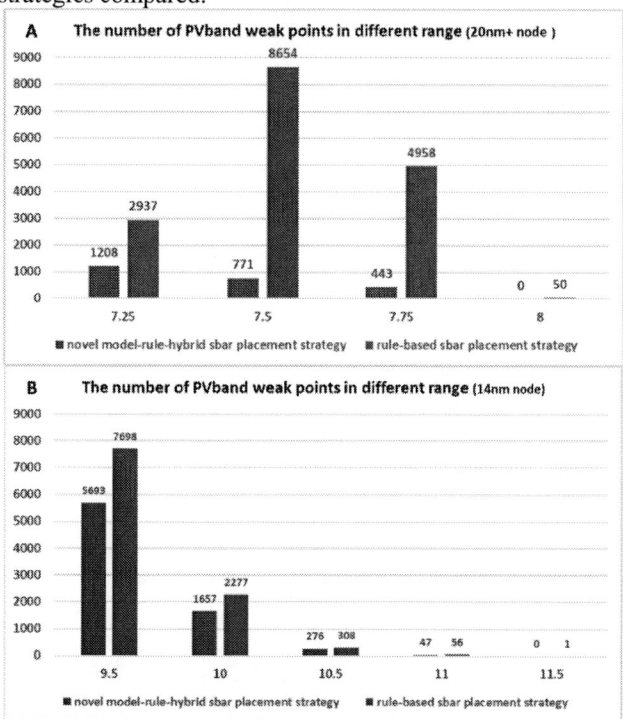

Fig.8. The PVband result of model-rule-hybrid and rule-based sbar placement strategy under different technology nodes

Figure 8 shows the difference in PVband performance between the traditional rule-based sbar method and the new model-rule-hybrid sbar placement strategy. At the 20nm+ technology node (Fig.8A), the model-rule-hybrid sbar placement strategy reduces the maximum value of the main pattern PVband in the full chip from 8nm to 7.75nm, and the number of weak points in different PVband ranges is greatly reduced compared to the traditional rule-based sbar method. The overall number of PVband weak points significantly reduced from 16.6k to 2.4k and the potential weak point structure in the full chip is reduced by 85%. At technology nodes of 14nm (Fig.8B), the maximum value of PVband is reduced from 11.5nm to 11nm, and the overall number of PVband weak points reduced from 10.3k to 7.6k, that is, the potential weak point structure in the clip chip is reduced by 25.8%. At the same time, we also compared the difference in running time between the model-based sbar method and the new model-rule-hybrid sbar placement strategy under different nodes. As shown in Figure 9, the new sbar placement strategy has significantly optimized the running time, reducing the running time by 66.6% from 24 hours to 8 hours at the 20nm+ technology node (Fig.9A), and reducing the running time by 67.8% from 14 hours to 4.5 hours at technology nodes of 14nm (Fig.9B).

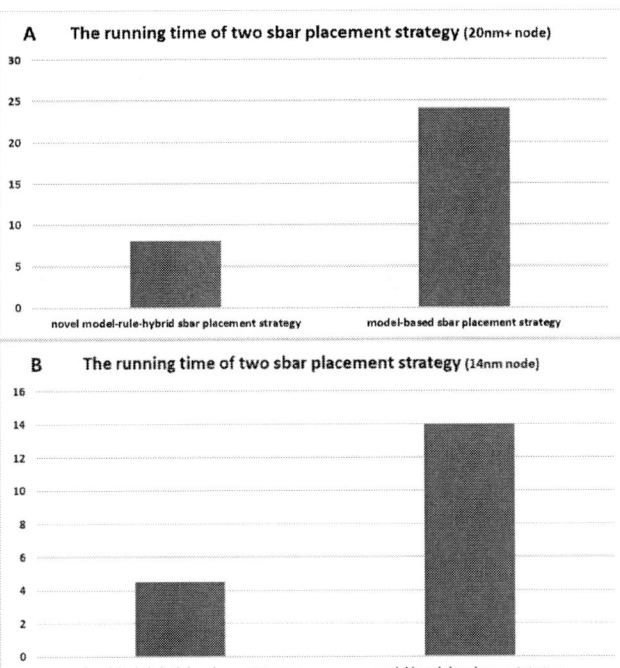

Fig.9. The running time of model-rule-hybrid and model-based sbar placement strategy under different technology nodes

All of the above shows that compared with the traditional sbar insertion flow, the new sbar placement strategy introduced in this paper has a better optimization effect on the complex structure of the contact layer, which can greatly reduce the number of potential weak points and reduce the maximum value of PVband. It also shows that this new sbar insertion strategy has good applicability in a wide range of technical nodes. In addition, this new strategy has a simple structure and clear logic, and is no need for the special handle to perform multiple weak point compensation, which can make the OPC recipe as clear as possible.

## SUMMARY

In this paper, a novel model-rule-hybrid sbar placement strategy is presented. First, the best sbar placement parameters will be determined from a series of DOE results. Then determine the forbidden space through MRC constrain, that is, the distance between the two edges that cannot generate sbar in the middle, and classify the priority according to the number of forbidden space edges of main pattern from more to less. For main patterns of the same priority, based on the distance from the free edge to the opposite edge, three strategies will be formulated to generate one center sbar, two center sbars, and the optimal sbar in the middle of the edge pair to ensure that at least one first-circle side sbar can be generated around each main pattern. Finally, three circles of side sbar and corner sbar will be generated in order according to the priority. For To compare three different sbar placement methods, the full chip layout of different technology nodes was selected for testing. The results showed that compared to the rule-based sbar method, the new model-rule-hybrid sbar placement strategy reduces the maximum value of PVband in the full chip, and the number of PVband weak points is significantly reduced 85% at 20nm+ node and 42.5% at 14nm node. Compared with the model-based sbar method, the new model-rule-hybrid sbar placement strategy has significantly reduced the running time

of the full chip by 66.6% at 20nm+ node and 67.8% at 14nm node. All of the above indicate that the new model-rule-hybrid sbar placement strategy can better optimize complex 2D patterns in a short running time, and can be widely used at different technical nodes. At the same time, it can also make the recipe more logical and clear.

## REFERENCES

[1] A.K. Wong. "Resolution Enhancement Techniques in Optical Lithography", SPIE Press, 2001.
[2] US patent 5,821,014, 1998.
[3] L.D. Barnes, B.D. Painter, L.S. Melvin III, "Model-based placement and optimization of subresolution assist features," Proc. SPIE 6154, Optical Microlithography XIX, 61542C, March, 2006.
[4] M. L. Kempsell, E. Hendrickx, A. Tritchkov, K. Sakajiri, K. Yasui, S. Yoshitake, et al, "Inverse lithography for 45-nm-node contact holes at 1.35 numerical aperture," J. Micro/Nanolith. MEMS MOEMS, vol. 8(4), pp. 043001, October, 2009.
[5] S. Jayaram, A. Yehia, M. Bahnas, H. A. M. Omar, Z. Bozkus, and J. L. Sturtevant, "Automatic assist feature placement optimization based on process-variability reduction," Proc. SPIE 6730, Photomask Technology 2007, pp. 67302E, October, 2007.

# Simulation-based source and mask optimization for advanced nodes

## Lianbo Luo[1], Yu Yan[2]

Semiconductor Manufacturing International Corporation (SMIC), Shanghai, 201203, China,
Email: luolianbo2008@sina.com
2 Mentor Graphics, Wilsonville, OR 97070, USA

### Biography

Lianbo Luo, OPC engineer of SMIC, received the PhD. degree in material science from Shanghai Jiaotong University, Shanghai, China, in 2019.

### Abstract

Lithography is a key and fundamental technology to drive the development of the integrated circuit. For advanced nodes, source mask optimization (SMO) has become a more and more promising technique to push the limits of the lithography. Thus, it is important to achieve the optimal settings in SMO to achieve the best source for lithography. In this paper, based on two target patterns, min pupil illumination efficiency of the source is optimized. Besides, better balance for dose latitude through focus is obtained. The effect of the target patterns weight on the result is demonstrated. Finally, we get a tuned source with the optimal parameters to cover the target patterns.

*Keywords — lithography, SMO, NILS, pupil illumination efficiency*

### Introduction

As the technology node in semiconductor manufacturing continuously shrinks its feature size and boosts the transistor density, lithography technique becomes cornerstone of the semiconductor industry. With advances in microlithography now pushing towards 22 nm and beyond, the engineering of how to print circuit layouts on wafers has become more intricate and complicated. The lithography community has relied on the lithography tool manufacturers to provide the next generation toolsets to move to smaller geometries. As an effective resolution enhancement techniques (RET) widely used in the 22nm node and below, source mask optimization is first proposed in 2001 by Rosenbluth et al[1].Since then, a variety of SMO methods have been proposed in literature[2-4]. The primary goal of SMO is to provide the optimum source and mask shapes for a given ideal wafer image or images. SMO aims at maximizing the process window, simultaneously optimizing the pattern on the mask and the angular distribution of the illumination settings[5]. SMO provides intensively optimized wave distributions that illuminate both the mask and the wafer, exploiting all possible benefit from the available degrees of freedom in the band limited exposure process. The illumination patterns that SMO generates will typically have a far more detailed structure than previous methods could consider, and the mask shapes that are designed by SMO to diffract the optimal imaging waves to the wafer are likewise highly nonintuitive, differing radically from conventional mask

solutions. The work covers optimization of an illuminator and masks to print two typical test patterns in advanced nodes. The effect of the target patterns on the result is demonstrated.

### Methodology

Lithography can be modeled as a partially coherent imaging system. The intensity of the aerial image can be calculated by Abbe model[6].

$$I(x,y) = \int\int_{-\infty}^{\infty} J(f,g) \left[ \left| \int\int_{-\infty}^{\infty} H(f+f',g+g')M(f',g')e^{-i2\pi[f'x+g'y]}df'dg' \right|^2 \right] dfdg$$

where $(x, y)$ represents the normalized spatial coordinates on the image plane, $(f, g)$ and $(f, g')$ represents the normalized spatial frequency coordinates of the pupil plane and the diffraction lights, respectively. $I$ is the intensity of the aerial image, $S$ is the illumination source, $H$ is the pupil function of the projector and can be regarded as a low pass filter, $M$ is the diffraction spectrum of the mask.

$$ICC(x,y,f,g) = \left| \int\int_{-\infty}^{\infty} H(f+f',g+g')M(f',g')e^{-i2\pi[f'x+g'y]}df'dg' \right|^2 \tag{1}$$

For SO, Eq. (1) can be transformed into:

$$I(x,y) = \sum_f \sum_g ICC(f,g,x,y)S(f,g) \tag{2}$$

where $ICC$ represents the illumination cross coefficient. $ICC$ stores the aerial image intensity of each source point and is pre-calculated before SO. For MO, Eq. (1) can be rewritten as:

$$ICC(x,y) = \sum_k^K \mu_k |F^{-1}\{\Phi_k(f',g')B(f',g')\}|^2 \tag{3}$$

where $\mu_k$ and $\Phi_k$ represent the $k$-th singular value and corresponding kernel function of the transmission cross coefficient (TCC) after singular value decomposition (SVD), respectively. $K$ is the truncate order of SVD, $F^{-1}\{\cdot\}$[7] represents the inverse Fourier transform. The calculation of the mask diffraction spectrum $B$ via a fast and accurate mask model is essential for the imaging model.

The optimization and evaluation functions of SMO mainly include normalized image log slope (NILs), process window (PW), mask error enhancement factor (MEEF) and so on[8].

978-1-6654-9759-6/22 $31.00 © 2022 IEEE

## Results and Discussions

Calibre WorkBench[TM] from Mentor Graphics is used throughout this study. In this paper, the wavelength of the source is 13.5 nm. The polarization mode is TE. The incident angle of the source is $\theta = 6°$ and the azimuth angle is $\varphi = 0°$ The material of the absorber is TaN, whose complex index is 0.954-0.027j and thickness is 60nm. The multilayer is composed of 40 bilayers of Mo/Si whose thicknesses are 3/4nm. The complex indexes of Mo and Si are 0.9238-0.0064j and 0.999-0.00183j, respectively. The material of mask substrate is SiO$_2$, whose complex index is 0.976-0.00906j. Two typical target patterns are used in the simulations, as shown in Fig.1. The pattern1 is 1D dense line space feature with 16nm width and 32nm pitch. The pattern2 is line end feature with 16nm width, 32nm pitch and 20nm gap (head to head, HTH). The initial source is shown in the Fig.1(c).

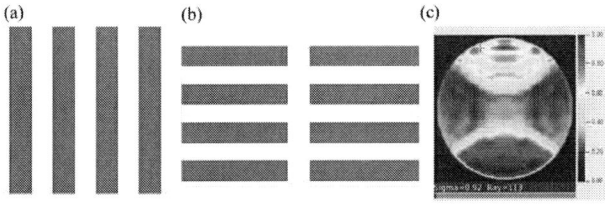

Fig.1. Initial state. (a) Pattern1 (b) Pattern2 (c) Source

The SMO flow includes four steps: start, decorate, improve and polish. The final SMO results are demonstrated in the Fig2. Since SRAF is not considered, the optimized mask obtained by the proposed mask is still composed of line space with only the edges changed. However, the shift of the pattern position in the line end region is obvious (as seen in Fig. 2(b)). It is indicated that 2D constraint should be taken into consideration in the line end and the optimized target becomes complicated. In addition, the nominal condition contours of the two target patterns are both on target under this optimized source. The pupil fill of the optimized source is 0.35.

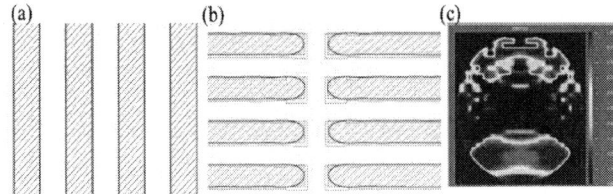

Fig.2. SMO results. (a) Mask of pattern1 (b) Mask of pattern2 (c) optimized Source. The blue line is initial target. The green line is optimized mask target. The red line is the simulation contour.

The optimization and evaluation of SMO is taken. The critical dimension (CD) tolerance is 10%. For example, if the target CD is 50nm, the CDs between 45nm and 55nm can be treated as on target. Dose latitude is 5%. Under this constraint, the dose and focus are changed. Fig.3 provides the process window evaluation of the two target patterns. For the pattern1 and pattern2, the CD measurements are the line and gap, respectively. As shown in Fig.3(a), the dose latitude of patten1 is smaller than that of pattern2. Common process window is calculated to better balance for dose latitude through focus. Fig.3(b) shows the exposure latitude (EL) changes with the depth of focus (DOF). The common DOF is 0.15μm (Focus +/-70nm) and EL is 10%(0.9%~1.1%).

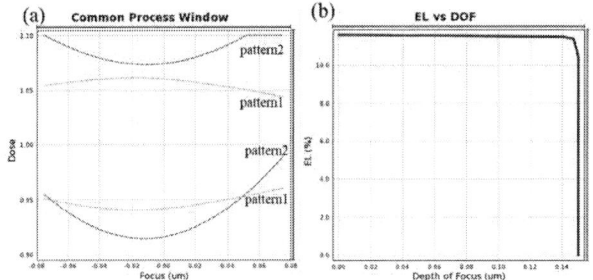

Fig.3 the process window evaluation of the two target patterns.
(a) Common process window, (b) a gain in both exposure latitude and depth of focus

The optimization results of these two target patterns are presented in Table 1. Although the two target patterns has the same DOF, the performance is different. The pattern2 with high ILS and NILS could have higher image contrast and image quality than pattern1. However, the pattern2 has larger MEEF and EPE, which means that the line end feature has poor CD uniformity (CDU) on the wafer. To improve CDU of the pattern2, the weight of patten2 should be increased to drive the optimization to maximize ILS in the process window conditions. In this way, the quality of the patten1 may be sacrificed slightly. At the same, more attention should be paid to optical proximity correction (OPC) model and OPC recipe setup to consider the line end patterns.

Table 1 Performance of the SMO

| Type | MEFF | ILS | NILS | PVBAND (%) | TOTAL EPE(nm) | DOF (μm) |
|------|------|-----|------|------------|---------------|----------|
| Pattern1 | 2.95 | 72.36 | 1.16 | 20.63 | 0.62 | 0.15 |
| Pattern2 | 3.35 | 79.66 | 1.59 | 16.50 | 1.57 | 0.15 |

ILS: image log slope
NILS: Normalized image log slope
EPE: edge placement error

## Conclusions

In this paper, we propose a source mask optimization for the two typical target patterns in the advanced nodes. We get a tuned source with the optimal parameters to cover the target patterns. A series of simulation results show that the dense line space feature and line end feature have 0.15μm common DOF and 10% EL. However, the line end feature has larger MEEF and EPE. Therefore, the line end feature will have worse CDU on the wafer. We demonstrate that the line end feature may benefit from increasing the weight in the SMO flow.

## References

[1] A.E. Rosenbluth, S.J. Bukofsky, M.S. Hibbs, K. Lai, A.F. Molless, R.N. Singh, A.K. Wong, Optimum mask and source patterns to print a given shape, Optical Microlithography XIV, International Society for Optics and Photonics, pp. 486-502, 2001.

[2] A. Erdmann, T. Fuehner, T. Schnattinger, B. Tollkuehn, Toward automatic mask and source optimization for optical lithography, Optical Microlithography XVII, SPIE, pp. 646-657, 2004.

[3] Y. Peng, J. Zhang, Y. Wang, Z. Yu, Gradient-based source and mask optimization in optical lithography, IEEE Transactions on image processing 20(10)2856-2864 (2011).

[4] X. Ma, Y. Pan, S. Zhang, J. Garcia-Frias, G.R. Arce, Informational lithography approach based on source and mask optimization, IEEE Transactions on Computational Imaging 732-42 (2020).

[5] R. Socha, X. Shi, D. LeHoty, Simultaneous source mask optimization (SMO), Photomask and Next-Generation Lithography Mask Technology XII, International Society for Optics and Photonics, pp. 180-193, 2005.

[6] A.K.-K. Wong, Optical imaging in projection microlithography, SPIE press2005.

[7] A. Rosenbluth, S. Bukofsky, M. Hibbs, K. Lai, A. Molless, R. Singh, A. Wong, Optimum mask and source patterns to print a given shape, SPIE2001.

[8] A.E. Rosenbluth, N. Seong, Global optimization of the illumination distribution to maximize integrated process window, Optical Microlithography XIX, International Society for Optics and Photonics, p. 61540H, 2006.

# PROCESS WINDOW INTEGRATION RULE CHECK FOR BEOL IN ADVANCED TECH NODE

Xiaoyan Wang[1]*, Qing Yang[1], Honghong Ren[1], Xuanbo Chen[1], Qingwei Liu[1]

[1]Semiconductor Manufacturing International Corporation, Shanghai, CHINA

*Corresponding Author's Email: Rebecca_Wang@smics.com

## ABSTRACT

As the technology node continuously shrinks in semiconductor manufacturing, the multi-patterning method is used extensively, especially for BEOL layers. In a process of LELE (litho-etch-litho-etch) with cut layers, the relationship between each sub-patterning layer becomes critical for the fidelity of the original design transition, as well as the patterning process window.

This paper introduces a defect detection method based on resist model simulation. The method detects the exact location according to the OPC simulation result to find the potential defects. Compared with the traditional method of image comparison, the new method could improve the detection speed and accuracy greatly. Besides, traditional defect detection methods lack quantitative defect evaluation criteria, but the new method introduced in this article can provide a numerical basis for defect detection and the continuous optimization for defect evaluation algorithm. In addition, the actual wafer defect data, not only could be used to optimize OPC target and reduce defect generation, but also could provide wafer data support for searching potential weak points based on OPC simulation, and optimizing OPC weak points spec continuously, and further improving defect detection efficiency.

This paper takes a cut-missing defect as an example, and designs splits of cut enclosure length on mask to find out the minimum safe enclosure of cut pattern when cut layers and metal lines overlaid, especially in consideration of BEOL production process window. The experiment of PWIRC (Process Window Integration Rule Check) is of great significance to the good transition and graphic function of the original design, as well as the improvement of mass production process and yield.

## INTRODUCTION

Due to the resolution limit of the lithography tools, multiple patterning technologies are introduced to the back-end of the line (BEOL). For example, LELELE (or LE3, triple litho-etch) or SADP (self-aligned double patterning) are already implemented in 10nm node technology based on a polygon's geometry, its orientation and pitch requirement [1]. Moreover, the line cut is recommended due to overwhelming advantage in mask reduction [2]. Therefore , for the LELE (litho-etch-litho-etch) process that introduces the cut layer, not only the graphics of the metal line and cut layer should be ensured to transfer smoothly after multiple lithography, but also the final process window will not be reduced due to the gap in the process and the interaction of them, especially when metal and cut overlapping.

In the actual process, we found that the OPC (Optical Proximity Correction) model simulation result could effectively predict the potential risks and defects and feed the prediction results back to the lithography process, and propose optimization suggestions on the ADI (After Development Inspection) target to avoid defects. Therefore, this paper takes the cut partial missing defect that appeared in BEOL process of the advanced technology node as an example, and proposes a method of defect calculation that combines OPC model simulation prediction with defect detection, which not only improves the accuracy of defect scanning, but also applies in the subsequent inspection of mass production.

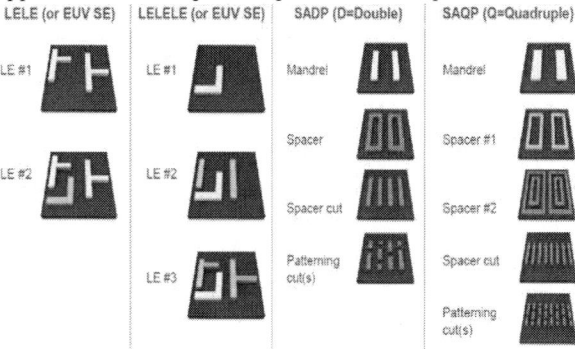

*Figure 1: Multiple patterning technologies of BEOL*

## RESULT AND DISSCUSSION

### Cut partial missing issue

In a process of BEOL with cut layers (as shown in Figure 2(a), 2(b)), when the metal layer overlapping with its cut layer [3], not only the width of cut pattern, but also the length of the enclosure (the length of cut extending across the metal line, as shown in Figure 2(c)) has a great impact on the cut efficiency. In other words, the length of enclosure determines whether the cut pattern could completely interrupt the metal line or not. When the enclosure is too short, the effective cutting length is less than the length of ADI target due to its rounding line-end, so some edge parts of cut may be missing, resulting in a metal bridge defect (as shown in Figure 2(d)). When the enclosure is too long, the cut pattern maybe extend the metal line too long and cut the next metal line accidentally, causing the overcut defect. Therefore, the length setting of the cut pattern enclosure is very important in the OPC re-target operation, and this step requires a large amount of real and effective wafer data to feed back to the OPC and continuously optimize the OPC simulation result.

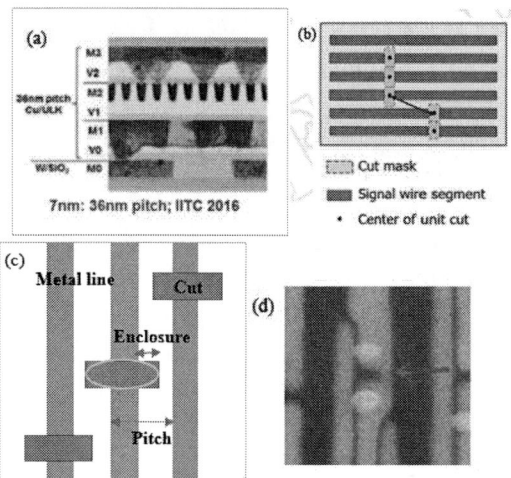

loss of yield.

Figure 2: (a) BEOL copper metal interconnection; (b) The LELE process with cut layers; (c) Overlapping relationship between metal line and cut layer; (d) Metal bridge issue caused by cut partial missing

However, when collecting wafer defect data, we found that the current defect scanning detection method could not accurately capture the potential weak points highlighted by OPC simulation results, and even problems of omissions or misreporting are prone to occur. These problems make it difficult to catch defects in the try run stage, and ultimately lead to yield loss in mass production. In this regard, this paper proposes an improved defect algorithm based on traditional defect scanning detection, the new algorithm could greatly improve the matching and accuracy of defect detecting judgments.

**Improvement of defect detecting algorithm**

In the current process, the principle of defect detection is based on the image matching of wafer data and layout design. There are two kinds of criteria for judging. One is to treat the highlighted wafer location of the mismatching (missing or residue) image as a hard defect; the other is that, for the wafer location where the center CD (critical dimension) of the pattern could not reach the target spec, it is regard as a soft defect. Based on these judgment criteria, some pattern whose wafer CD are too small to reach the target spec, but there is no risk in the actual process, it is easy to misreport (as shown in Figure 3(b)) and causes the loss of humanity and material resources. However, for some irregular defects with a large enough center CD and a small line-end CD (as shown in Figure 3(c)), the current detecting algorithm usually fail to catch and result in omission. Finally, in the product tape out process, due to the defect omission in the try run stage, the potential risk was not caught in time, resulting in a serious missing defect of cut pattern and a

Figure 3: (a) The critical dimension of the center is used as the criterion of defect judgment; (b) Defect types that can be caught by the traditional algorithm; (c) Defect types that cannot be caught by the traditional algorithm

In order to solve problems of misreporting and omission, a new defect algorithm of measurement and judgment is proposed based on the traditional defect scanning. One is the optimization of the measurement location. The measuring point is changed from the center of cut pattern to the line-end of both sides (as shown in Figure 4(a) b/c point) to take the uneven size of pattern into consideration. The other is the optimization and quantification of the principle of the defect judgment. The traditional method is to compare the CD of center location with a fixed empirical value. In contrast, the new algorithm is to calculate the ratio of the minimum CD of b/c location to target and then use the ratio as the criterion for judging the defect, and successfully capture the defect of cut partial missing issue (as shown in Figure 4(b)). The new algorithm takes the size of the original design into account when calculating the CD of wafer data, making the judgment result more reasonable and effective. According to our actual case and empirical values, when $min(b,c)/target<0.6$, the point would be defined as a hard defect; when $0.6< min(b,c)/target<0.72$, the point would be defined as a soft defect; when $min(b,c)/target>0.72$, this location would no longer defined as a defect. Compared with the original defect judgment method, the new algorithm is more accurate and reasonable, reducing the occurrence of defect omissions and misreporting problems. At the same time, the new algorithm also quantifies the severity of the defect and makes detecting results clearer, it is of great significance to the research and mass production of the chips.

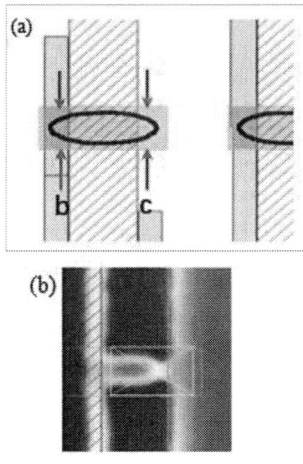

Figure 4: (a) The critical dimension of the b/c location is used as the criterion of defect judgment; (b) Defect types that can be caught by the improved algorithm

### DOE split for PWIRC

Based on the improved algorithm of defect detection, we designed the DOE (Design on experiment) split of cut enclosure length under various conditions on the mask. The purpose is to find the minimum allowable enclosure under the premise of satisfying the process window of metal line and cut pattern at the same time, which is regarded as the critical size of cut missing or metal bridge.

First of all, in view of the current issue in the process, we divided the environment in which the cut is located into two types based on the characterizes of metal line. One is when the cut is located in a dense metal line environment (as shown in Figure 5(a)), the etch bias on both sides of the metal line is uniform at this time, so the enclosures at both sides of the cut could not be too short or too long. If it is too short, there will be the risk of cut missing, and if it is too long, it is likely to cause overcutting to the next metal line. The other is when the cut is located at the metal line with at least one side isolate (as shown in Figure 5(b)), the etch bias on both sides of the metal line is different widely at this time, so the safe length of the enclosure at both ends of the cut will also change accordingly. Secondly, in order to ensure the variable unity of each split, we set the enclosure at one end of the cut pattern to a fixed length to keep it is no risk of missing, and set 2nm as a step at the other end. According to the current empirical length of enclosure, a total of 10 length splits are set for experiment to determine the final enclosure spec. Finally, we can select a proper enclosure spec through two methods: defect scanning and electrical signal testing. Defect scanning can use the double-sided calculation algorithm mentioned above to judge the safe enclosure

spec, and the electrical signal testing method can judge the proper enclosure spec based on the process window of the full-loop, so the accuracy and feasibility of the experiment results could be guaranteed.

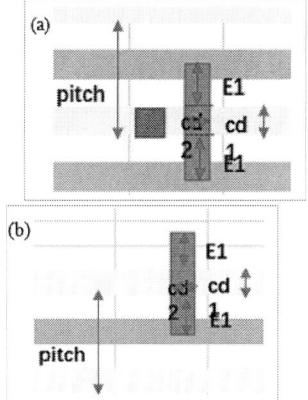

Figure 5: (a) The cut is located in the environment of dense metal line; (b) The cut is located at the metal line with at least one side isolate

At present, the graphic design, OPC correction, and mask writing of this experiment have been completed, and the next working plan is to collect and analyze wafer data and draw conclusions about the setting of appropriate enclosure spec.

## CONCLUSION

This paper introduces an actual case of cut partial missing issue in LELE process of the advanced technology node, and points out the reason of such defects, that is, the potential risk of insufficient enclosure length when the cut layer and metal line layer overlapping. Therefore, we can revise and optimize the re-target process of OPC to avoid this type of risk. In response to such risks, the paper also proposes an improved defect algorithm based on the traditional defect scanning detection, which can accurately capture potential weak points highlighted by OPC simulation in the try run stage, and reduce the occurrence of defect omission or misreporting problems. Furthermore, the paper designs the DOE split of cut enclosure length under various conditions on the mask, aiming to find out the minimum allowable enclosure of cut line-end to metal line, especially under the premise of satisfying the both process window of metal layer and cut layer at the same time. This work is of great significance to the risk control of research and mass production and we will continue to follow up.

## REFERENCE

[1] Y. Woo, M. Ichihashi, S. Parihar, L. Yuan, S. Banna and J. Kye, "Design and process technology co-optimization with

SADP BEOL in sub-10nm SRAM bitcell," 2015 IEEE International Electron Devices Meeting (IEDM), 2015, pp. 11.2.1-11.2.4.

[2] W. Gillijns, S. M. Y. Sherazi, D. Trivkovic, B. Chava, B. Vandewalle, V. Gerousis, P. Raghavan, J. Ryckaert, K. Mercha, D. Verkest, G. McIntyre, K. Ronse, "Impact of a SADP flow on the design and process for N10/N7 metal layers," Proc. SPIE 9427, Design-Process-Technology Co-optimization for Manufacturability IX, 942709 (18 March 2015).

[3] P. Debacker, K. Han, A. B. Kahng, H. Lee, P. Raghavan and L. Wang, "MILP-Based Optimization of 2-D Block Masks for Timing-Aware Dummy Segment Removal in Self-Aligned Multiple Patterning Layouts," in IEEE Transactions on Computer-Aided Design of Integrated Circuits and Systems, vol. 36, no. 7, pp. 1075-1088, July 2017.

# EFFECT OF HIGH ENERGY IMPLANTATION ON THE PHOTORESIST FOR SMALLER SIZE CMOS IMAGE SENSOR

*Hui Chen\*, Xiaoyu Li, Zhengying Wei, Chang Sun, Jiong Xu, and Yufei Peng*

Shanghai Huali Microelectronics Corporation, Shanghai 200433, China

\*Corresponding Author's Email: chenhui@hlmc.cn

## ABSTRACT

High energy implantation has become a key process to form n- and p-type well structures in small size CMOS image sensors (CIS) for higher full well capacity (FWC) purpose. However, high energy application needs relatively thicker photo-resist (PR) above 3μm in thickness than conventional ion implantation process depending on the ion energy and ion dosages. This thick PR will shrink during the ion implantation, which contribute additional side effect on the PR critical dimension and PR effective thickness to against next implantation step. Such effect will has an impact on the CIS performance, especially on smaller pixel products. To study the effect of high energy implantation on the PR for small size CIS, the shrinkage behavior of 3μm PR under different implantation energy and dosages was investigated in this work.

## INTRODUCTION

In recent years, the demand for the high resolution CMOS image sensors (CIS) is rapidly increasing due to the expansion of smartphones [1-2]. In the high resolution CMOS image sensors, the pixel size continued to decrease to lower than 0.7 μm [3]. As the pixel size decreases, the light receiving area decreases, resulting in the full well capacity (FWC) reduced. Therefore, it is very important to improve FWC performances of the image sensor while reducing the pixel size.

High energy implantation has become a key process in high resolution image sensor for higher FWC purpose. In order to improve the FWC, the required implantation energy and dosage should be higher for smaller pixels, and high energy implantation needs relatively thicker photo-resist (PR) above 3μm in thickness than conventional ion implantation process depending on the ion energy and dosages. However, the thick PR will shrink rapidly after implantation because of the bombardment of high energy ion beam [4-5], which contribute additional side effect on the PR critical dimension (CD) and PR effective thickness to against next implantation step. Such effect will has an impact on the CIS performance, especially on smaller pixel products. So it is very important to investigate the shrinkage behavior of the PR under high energy and dosages exposure for small size CIS. In this paper, the shrinkage behavior of 3μm PR under different implantation energy and dosages was studied and the PR shrink model was also proposed.

## EXPERIMENTALS

To investigate the shrinkage behavior of photoresist (PR) after high energy implantation, a 3μm thickness PR was performed on 12-inch Si wafer and PR critical dimension (CD) after development is designed to 250 nm in this work. The wafer was exposed with different implantation energy and dosages. The implant energy range of $1.0 \times 10^3$- $2.1 \times 10^3$ keV and dose range of $4.5 \times 10^{12}$-$1.0 \times 10^{13}$ cm$^{-2}$ were used. Before and after implantation, the PR CD was measured on a metrology tool.

## RESULTS AND DISCUSSION

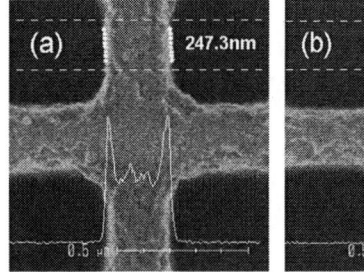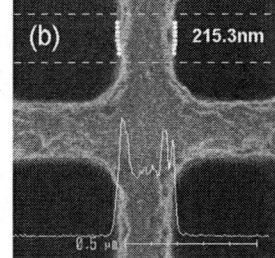

*Figure 1: (a) SEM image before implantation, (b) SEM image after implantation*

Figure 1 shows the SEM image of a 3μm PR before and after implantation. It can found that the PR CD after implantation is about 32nm smaller than before implantation. The PR CD will continually shrink during the next implantation step. H. Shi et al. proposed a radical mechanism of PR shrinkage, which suggest that the PR shrink is caused by high energy ion beam bombard the resist film to induce polymer crosslinking and chain scission [6]. However, the correlation between PR shrink and ion implantation has not clear until now [7].

**TABLE 1. THE ENERGY AND DOSAGE FOR IMPLANTATION USED IN THIS WORK**

| IMP | Energy (keV) | Dose (cm$^{-2}$) | Energy×Dose (Work) | Normalized work | Total Work |
|---|---|---|---|---|---|
| IMP1 | 2.10E3 | 1.00E13 | 2.10E+16 | 7.00 | 7.00 |
| IMP2 | 1.50E3 | 2.00E12 | 3.00E+15 | 1.00 | 8.00 |
| IMP3 | 1.00E3 | 4.50E12 | 4.50E+15 | 1.50 | 9.50 |

To study the implantation effect on the PR shrinkage, the implantation was made in three steps and energy range of $1.0 \times 10^3$-$2.1 \times 10^3$ keV and dosage range of $4.5 \times 10^{12}$-$1.0 \times 10^{13}$ cm$^{-2}$ were used, as list in Table 1. The PR CD before implantation is 247.3nm. After IMP1, IMP2 and IMP3 step, the PR CD is 221.1nm, 218.4nm and 215.3nm as shown in Table 2. They were reduced by 26.2nm, 28.9nm and 32.0nm compared with those before ion implantation respectively.

**TABLE2. THE PR CD AND CD BIAS BEFORE AND AFTER IMPLANTATION**

| IMP1-2-3 | After IMP CD (nm) | CD bias (nm) |
|---|---|---|
| IMP1 | 221.1 | 26.2 |
| IMP2 | 218.4 | 28.9 |
| IMP3 | 215.3 | 32.0 |

A new parameter 'Work' defined as energy × dosage was proposed to study the correlation between PR shrink and ion implantation. The work was normalized for convenience. It is worth mentioning that since PR CD bias is the result of multiple IMPs, the work of all completed IMPs should be superimposed which defined as total work. The correction between PR CD bias and total work is shown in Figure 2. As can be seen, the PR CD bias has strong correction with total work. The factor ion implantation affecting PR CD shrink was found.

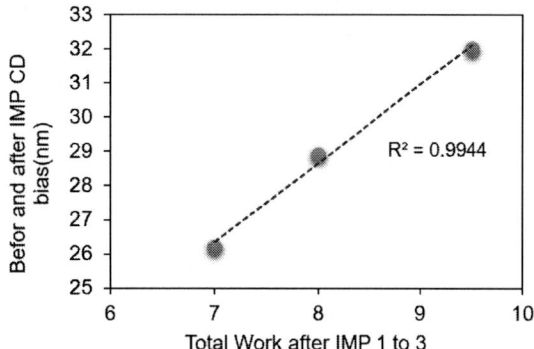

*Figure 2: The correction between PR CD bias and total work*

To further verify the correctness of this factor, the ion implantation order was changed from IMP1-2-3 to IMP2-1-3 and IMP3-1-2, respectively. The PR CD bias before and after implantation for IMP2-1-3 and IMP3-1-2 is shown in Table 3. Figure 3 illustrates the correction between PR CD bias and total work for different implant conditions. It was also observed that different work has different effect on PR bias reduction. Although the implantation order is different, the PR bias caused by the same implantation total work is consistent. The findings of this work can provide guidance to adjust FWC performance in a way that implantation injection affects PR shrink, especially for smaller pixel CIS.

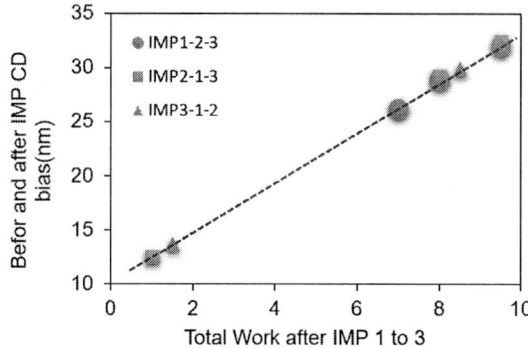

*Figure 3: The correction between PR CD bias and total work for different implant conditions*

**TABLE3. THE PR CD BIAS BEFORE AND AFTER IMPLANTATION FOR DIFFERENT IMPLANT CONDITIONS**

| IMP 2-1-3 | After IMP CD (nm) | CD bias (nm) | IMP 3-1-2 | After IMP CD (nm) | CD bias (nm) |
|---|---|---|---|---|---|
| IMP2 | 234.8 | 12.5 | IMP3 | 233.6 | 13.7 |
| IMP1 | 218.1 | 29.2 | IMP1 | 217.4 | 29.9 |
| IMP3 | 214.9 | 32.4 | IMP2 | 215.1 | 32.2 |

## CONCLUSIONS

In this paper, the shrinkage behavior of 3μm PR under different implantation energy and dosages was investigated and the correlation between PR shrink and ion implantation has been established. The 'Work' defined as energy×dosage was proposed for the first time to study the correlation between PR shrink and ion implantation. It was found that the PR CD bias has strong correction with total work. The results of this work can provide guidance for adjusting FWC performance by influencing PR contraction through implantation injection, especially for smaller pixel CIS.

## ACKNOWLEDGEMENTS

The authors would like to expresses their deep gratitude to the Huali Microelectronics Corporation for proving an excellent research environment to complete this work.

## REFERENCES

[1] J. Xu, Q. Chen and Z.Y. Gao, *IEEE Electron Devices Society*, vol. 123, 2021, pp. 27-35.
[2] S.Y. Chai and S.H. Cho, *Crystals*, 2021, pp. 1106.
[3] S.S Park, B.J. Shin and H.S. Uh, *J. Microelectron. Packag. Soc.*, vol. 27, 2020, pp. 1-10.

[4] T.M. Parrill, M. Jones and A. Jain, *IEEE Proceedings of 11th International Conference on Ion Implantation Technology*, 1996, pp. 178-181.

[5] H. Kamiyanagi and S. Shibata, *IEEE Extended Abstracts of the Fifth International Workshop on Junction Technology*, 2005, pp. 61-64.

[6] H.G. Shi, Z. Fang and D.X Yang, *Proc. of SPIE*, vol. 10586, 2018, pp.105861H1-H9.

[7] T.N. Horsky, *IEEE 1998 International Conference on Ion Implantation Technology*, 1999, pp. 654-657.

# STITCHING PROCESS DEVELOPMENT ON 300MM WAFER CMOS BEOL FOR HIGH PERFORMANCE CHIP APPLICATION

*Ming Li[1], Xiaoxu Kang[1*], Kaiyan Zang[1]*

[1]Process Technology Department, Shanghai IC R&D Center, Shanghai, 201210, China

*Corresponding Author's Email: kangxiaoxv@icrd.com.cn

## ABSTRACT

In this work, stitching process was developed on 300mm wafer CMOS BEOL for high performance chip application. Different stitching pattern and process was designed and implemented to verify the process performance, including different pattern overlap length, hammerhead dimensions, etc. Based on process data, optimized process condition was obtained to achieve better process performance.

## INTRODUCTION

With CMOS technology developing and consumer market growing, more and more high-end applications for large chip size are emerging, such as large chip for X-ray image sensor, wafer-size CMOS-Image-Sensor or AI chip, etc. [1-3] As known to all, chip size should be smaller than shot size of the litho tool, which is commonly 33mm X 26mm. To develop a chip beyond the size limitation, stitching is a promising choice [4-6]. By establishing electrical interconnections between neighboring shots, chips in different shot can be merged into a bigger one.

In this work, stitching process was developed on 300mm wafer CMOS BEOL. Different stitching patterns and process were designed and implemented to evaluate the process performance. The profile and process window of the stitching patterns was compared. Based on the process data, optimized process condition, especially the design rule of stitching line ends, was obtained to achieve better process performance.

## EXPERIMENTS

To reduce the cost and achieve higher yield, the chip stitching was realized on 4X Cu BEOL interconnect layer of 55nm technology node, whose line/space design rule is 0.36um/0.36um. The film stack is as follows: substrate \ $SiO_2$-3500A \ SiN-1000A \ $SiO_2$-9500A \ SiON-700A \ photoresist-12000A, in which the SiON film was used as an anti-reflection layer.

The stitching patterns were designed on mask with different critical dimension (CD). As shown in Fig.1, "a" is min non-stitching line CD, "b" is the half of the hammerhead enlarged CD, "c" is the hammerhead length, "d" is the stitching pattern overlapped CD, "e" is the space between two adjacent hammerhead, "f" is the space between two adjacent non-stitching lines with hammerhead on the line end. And "c" is fixed at 0.36um, and "d" is fixed at 0.12um and can be adjusted by tuning the overlap area of litho process condition. In the experiment, different stitching patterns were put on the same mask to compare the performance. Neighboring dies were exposed with separate and precise position control by the scanner. And they were exposed using the same energy dose at which 360nm dense lines outside the stitching area were on the target, so that lines without stitching could be used as reference.

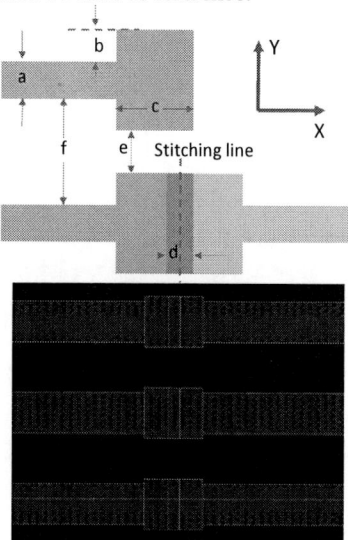

Figure. 1: *Schematic stitching pattern with critical dimension (upper) and designed stitching pattern layout (lower)*

All the process was developed on 300mm wafer BEOL LAB of Shanghai ICR&D Center. The litho tool used is ASML XT860M scanner. Following the litho process, dry etch and wet clean was conducted on the wafers. Both after-development inspection (ADI) and after-etch inspection (AEI) were applied to check the stitching process performance with CD-SEM Hitachi CG4000.

## RESULTS AND DISCUSSION

To achieve higher metal wire density, smaller hammerhead Line/Space is required. Thus the stitching of dense lines without hammerhead was checked first, in which line/space = 0.36um/0.36um, same as the original design rule. The patterns with different overlap length d along the X direction were exposed and checked with a rotation of 90°. As shown in Fig.2a, when d was set to zero,

978-1-6654-9759-6/22 $31.00 © 2022 IEEE

the trench CDs around the stitching line were greatly narrowed, accompanied with rising risk of bridging problem. And as can be seen in Fig.2b-2d, the stitched trench CDs were enlarged with d increasing, which in the opposite will add to the possibility of breaking resist lines. Both of the risks should be avoided in the design and process. As a result, it is necessary to have a control SPEC for the overlap length d to achieve a smooth and stable profile.

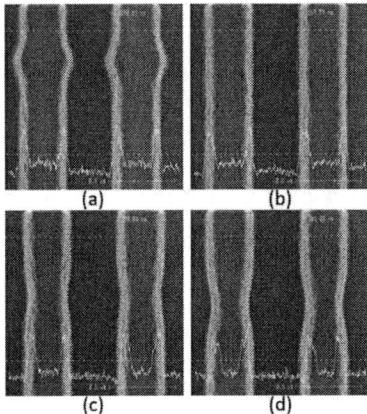

Figure. 2: ADI photos of stitched patterns without hammerhead and with different overlap d: (a) d=0; (b) d=120nm; (c) d=240nm; (d) d=360nm

In order to check the process window, above processes were repeated with an overlay of 40nm in the Y direction. As can be seen in Fig.3, the lower part of the patterns were slightly shifted rightward and the CDs around the stitching line were affected.

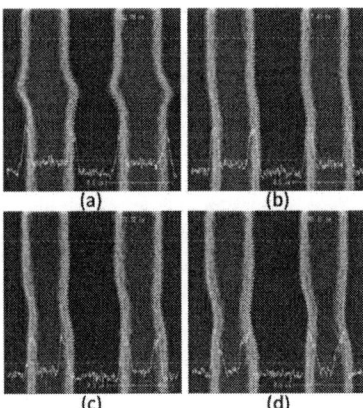

Figure.3: ADI photos of stitching patterns without hammerhead, but with an extra Y-direction overlay of 40nm and different overlap d: (a) d=0; (b) d=120nm; (c) d=240nm; (d) d=360nm

The minimum CDs of the photoresist lines and trenches were measured and plotted in Fig.4, and a 330nm

red line was plotted as CD SPEC for both line and space, below which risk of bridging problem will raise. As can be seen in the figure, the allowed overlap d was limited from about 90 to 180nm, and only about 40nm process window can be obtained when setting 135nm as the optimized overlap d value.

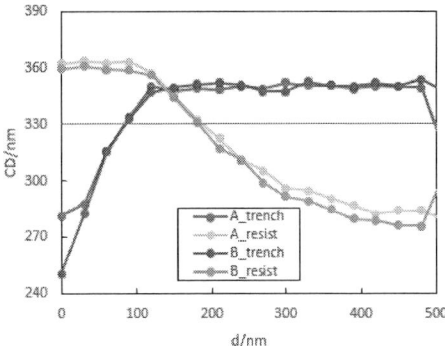

Figure.4: Minimum resist line and trench ADI CDs with different d: "A" was for patterns without overlay in Y-direction and "B" was for patterns with a 40nm overlay in Y-direction

However, 40nm overlay tolerance might not be sufficient because of increasing BEOL layer number and related process variations. Thus stitching patterns with different hammerhead dimension @ fixed a =0.36um was checked, which was shown in Fig.5. Compared with the patterns without hammerhead, stitched lines with widened hammerheads and space had robust CD performance. For the dense lines with hammerhead line/space = 0.4um/0.4um, process window of overlap d was ranging from 60nm to 500nm, and was much larger than that of the patterns without hammerhead. Further increasing hammerhead dimension to 0.45um/0.45um, the process window and performance will be even better, but at the cost of reducing metal wire density.

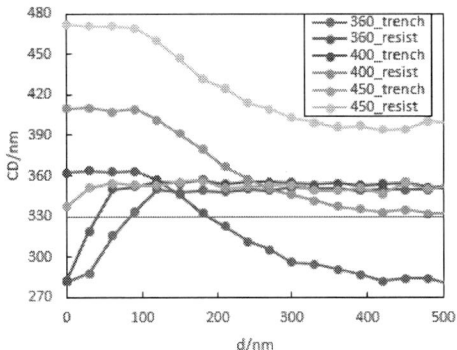

Figure.5: Minimum resist line and trench ADI CDs of different hammerhead width b and overlap d

Based on above ADI data, patterns with hammerhead of line/space = 0.4um/0.4um will be a practical choice for stitching process of 55nm 4X BEOL interconnect layer. The stitched patterns are expected to endure certain CD variance and positioning mismatch either radially or tangentially, as long as there is a pre-defined overlap of more than 100nm.

Subsequently, the wafers proceeded to the etch process. The AEI was checked and part of the results were shown in Fig.6. & Fig.7 in comparison to the ADI photos. As can be seen in these figures, the resist patterns were successfully transferred downward and no extra defects were found. The AEI CDs were close to the ADI CDs except for a small etch bias, which could be covered by the enlarged process window using a hammerhead of 0.4um/0.4um.

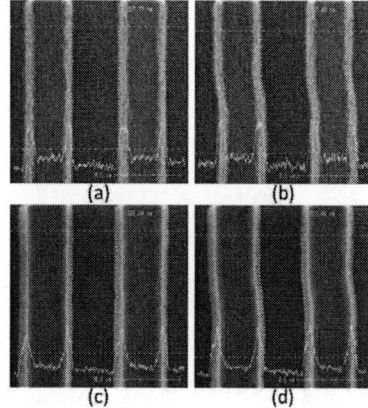

*Figure.6: photos of 0.36um/0.36um stitched patterns without hammerheads: (a) ADI without Y-direction overlay; (b) ADI with an extra Y-direction overlay of 40nm; (c) AEI without Y-direction overlay; (d) AEI with an extra Y-direction overlay of 40nm*

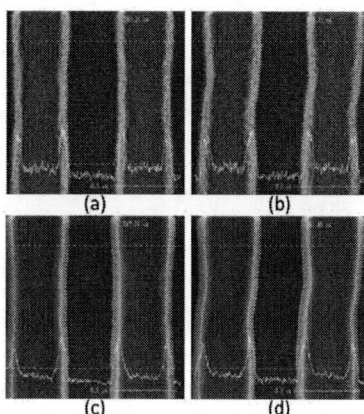

*Figure.7: photos of stitched patterns with hammerheads of 0.4um/0.4um: (a) ADI without Y-direction overlay; (b) ADI with an extra Y-direction overlay of 40nm; (c) AEI without Y-direction overlay; (d) AEI with an extra Y-direction overlay of 40nm*

Based on the above ADI and AEI data, stitching with hammerhead of line/space = 0.4um/0.4um and an overlap of larger than 100nm was tested to be robust in the 55nm BEOL process.

## CONCLUSION

In this work, stitching process was developed on 4X BEOL interconnect layer of 55nm technology node. Patterns with different hammerhead size and overlap dimension was designed and evaluated. Based on the measured data, for dense patterns of line/space = 0.36um/0.36um, the stitching process will require a hammerhead line/space = 0.4um/0.4um.

## ACKNOWLEDGMENTS

The author would like to thank R&D department of Shanghai IC R&D Center for the support in this work. Special thanks to Xiaolan Zhong, Zhenghui Chu, Ran Nie, Yifei Lu and Bin Jiang for their strong support in this work.

## REFERENCES

[1] Y. Yamashita, et al. *2011 IEEE International Solid-State Circuits Conference*, 2011, pp. 408-410.

[2] P. Morey-Chaisemartin, E. Beisser, F. Brault, F. Benzakour. *SPIE Photomask Technology + EUV Lithography*, 2019.

[3] A. T. Clark, N. Guerrini, N. Allinson, S. E. Bohndiek, R. Turchetta. *2008 IEEE Nuclear Science Symposium Conference Record*, 2018, pp. 4540–4543.

[4] S. Zihir, O. D. Gurbu, A. Karroy, S. Raman, G. M. Rebeiz. *2015 IEEE Radio Frequency Integrated Circuits Symposium (RFIC)*. IEEE, 2015.

[5] J. Zhu, D. Liu, W. Zhang, et al. *IEICE Electronics Express*, 2016, 13(15):20160441-20160441.

[6] R. N. Das, V. Bolkhovsky, C. Galbraith, D. Oates, & L. Johnson, et al. *2019 IEEE 69th Electronic Components and Technology Conference (ECTC)*. IEEE, 2019.

# PITCH WALKING SIMULATION AND PROCESS WINDOW EVALUATION OF SELF ALIGNED QUADRUPLE PATTERNING FOR ADVANCED NODES

*Xing Ke[1], Changcheng Jiang[1], Shiliang Ji[1], Zhenyang Zhao[1], Bo Su[1], Fengmei Li[1], Chongchong Zheng[1], Haiyang Zhang[1]\**

[1] R&D Infrastructure, Semiconductor Manufacturing International Corp. Pudong New Area, Shanghai 201203, P.R.China

*Corresponding Author's Email: Steven_Z@smics.com

## ABSTRACT

Self-Aligned Quadruple Patterning (SAQP) scheme has been widely used in FinFET manufacturing process. As the technology node reaches 14nm and beyond, Pitch Walking (PW) control is becoming more and more critical for the correct transferring of Critical Dimension (CD) down to the desired patterns. In this work, we study the process window for patterning Fins on both logic and memory regions using SEMulator3D® software suits. A quantitative evaluation of the process window and insight into the impact of the added processes are discussed.

## INTRODUCTION

### Self-Aligned Quadruple Patterning

A virtual fabrication platform SEMulator3D® from Coventor Inc. has been developed for semiconductor process modeling [1]. The virtual lab built in SEMulator3D® has been widely applied to predict the results before real wafers start to run in a fab.

In this paper, SEMulator3D® is applied to simulate the Fin forming process with a Self-Aligned Quadruple Patterning (SAQP) scheme, as in Figure 1.

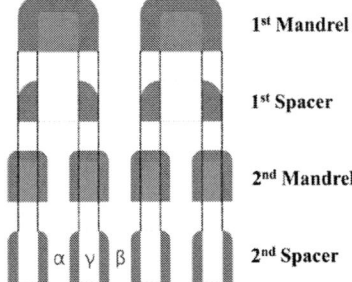

Figure. 1: CD Transferring Process of SAQP Scheme

SAQP patterning has been an important approach both in the Front End of Line (FEOL) [2] and Back End of Line (BEOL) [3]. With one photolithography step, fin pattern can be formed with a combination of deposition and etching steps. With the advance of technology nodes, the SAQP is becoming more and more critical due to the CD limitation of scanners. At the same time, the process requirements of deposition and etching will be continuously increasing. The geometrical data are transferred down by thin film deposition and etching

process. And the fidelity of CD transferring is strongly dependent on process variation controlling. For example, a local temperature variation will affect the uniformity of deposited film thickness or etch rate, resulting in a CD deviation from the designed value. The Pitch Walking (PW) is a key representation of process capabilities.

In our work, the parameters of PW are defined as shown in Figure 1. Position α corresponds to the Fin spacing below 1st Mandrel (MD1), the β position corresponds to Fin spacing between two SAQP patterns generated from neighboring MD1 blocks, and the γ position corresponds to the Fin spacing below 2nd Mandrel (MD2). PW is defined as in

$$PW = |\alpha - \beta| + |\gamma - (\alpha + \beta) / 2| \quad (1)$$

A previous study by Baudot et al. has applied SEMulator3D® to calculate the PW of Fin on 7nm FinFET node [4]. The MD1 Spacer and MD2 Spacer and core line CD are identified as key process parameters of PW. The pattern-dependent Fin etching is applied to reproduce the Fin height variability. And the PW is confirmed to impact the Fin height non-uniformity by 0.5%/nm.

In real manufacturing, different patterns have to be taken into consideration at the same time. We proposed an advanced SAQP scheme to pattern Fins on three regions: an SRAM region and two logic regions (Logic 1 and Logic 2) of different CDs. The logic 1 region has a smaller Fin CD than the Logic 2 region. Two trimming processes are introduced into the SAQP flow to satisfy the CD requirements of different patterns.

## SIMULATION

### Self-Aligned Quadruple Patterning Flow

In this work, the process flow deck is built in SEMulator3D® with 3 different patterns: SRAM, Logic 1, and Logic 2, as shown in Figure 2. To simplify the CD transferring during SAQP patterning, certain process-related details are omitted, for example, corner rounding and non-conformal deposition are not taken into consideration. To satisfy target CD of different patterns, we introduce two trimming processes named 1st Mandrel Trimming (MT) as shown in Figure 2(b) and 1st Mandrel Spacer Trimming (MST) Figure 2(d) into the advanced SAQP flow.

978-1-6654-9759-6/22 $31.00 © 2022 IEEE

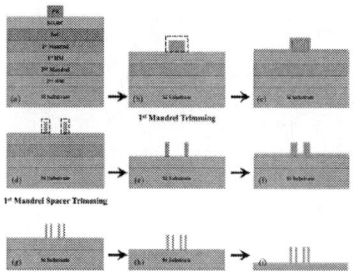

*Figure. 2: Self-Aligned Quadruple Patterning (SAQP) flow with advanced trimming scheme. (a) The film stack of SAQP flow. (b) 1st Mandrel etching with trimming. (c) 1st Mandrel hard mask deposition. (d) 1st Mandrel spacer etching with trimming. (e) 2nd Mandrel etching. (f) 2nd Mandrel spacer deposition. (g) 2nd Mandrel removal. (h) Fin hard mask etching. (i) Fin etching*

**Trimming Processes for CD Controlling**

The trimming process in the flow is designed to trim specific regions to target CDs as shown in Figure 3. MT is intended to cover SRAM region with mask and trim 1st Mandrel CD of the Logic region to the target CD, as shown in Figure 3(b). MST is designed to cover the Logic 2 region and trim the 1st Mandrel Spacer of SRAM and Logic 1 to target CDs, as shown in Figure 3(c).

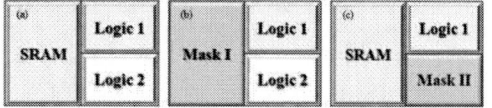

*Figure 3: Illustration of masks for different regions. (a) Regions of SRAM, Logic 1, and Logic 2. (b) CD control Mask I for 1st Mandrel Trimming of Logic 1 and Logic 2 Regions to target CD. (3) CD control Mask II for 1st Mandrel Spacer Trimming of Logic 1 Region to target CD*

# RESULTS AND DISCUSSION
## DoE of MT and MTS

The Design of Experiment (DoE) is applied using the built-in function inside SEMulator3D®. The process parameters are designated as Monte Carlo variables within a certain range. And the PW of each region can be calculated and analyzed. The two variables MT and MST are distributed randomly around target values respectively, with the range from 0 nm to 8 nm. 1000 Monte Carlo runs are completed in SEMulator3D® with PW of the 3 regions calculated as the final output. Large PW values are expected when MT and MST deviate from their targets.

In Figure 4, the relationship of MT and MST with PW of 3 regions are shown, the first row of Figure 4(a)-(c) shows the effect of MT on the PW, and the second row of Figure 4(d)-(f) shows the effect of MST on the PW. In Figure 4(a), MT does not influence the SRAM region since the region is covered with Mask I as shown in Figure

3(b). Thus, the PW distribution of MT in the SRAM region is completely random. A similar conclusion can be drawn for MST in the Logic 2 region as shown in Figure 4(f). In Figure 4(b)-(e), PW can be close to 0 nm when the MT/MST value is located around the targets. When MT and MST deviate from the ideal value, two linear relationships with PW are observed around the designed values with a V-shaped profile observed. For SRAM and Logic 2 region, only 1 trimming process is applied, either MT or MST, the V-shaped profiles exhibit narrow deviation ranges. For the Logic 1 region, where both MT and MST processes have been applied, the data have a much larger deviation range. Based on Figure 4, an estimation of the process window can be obtained. We choose the upper limit of the PW to be 2 nm and define the data range inside the 2 nm limit as the process window. The results are summarized in Table I. For the Logic 2 region, the process window of MT is 0.8 nm; for the SRAM region, the process window of MST is 1.0 nm; the Logic 1 region is estimated to have the most stringent process window of 0.62 nm for MT and 0.48 nm for MST.

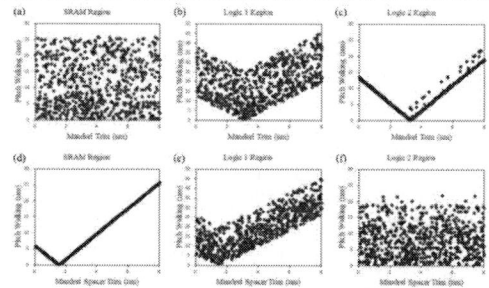

*Figure 4: PW distribution of 3 different patterns (SRAM, Logic 1, and Logic 2) for the MT and MST. (a)-(c) PW vs. MT. (d)-(f) PW vs. MST*

TABLE I. PROCESS WINDOW SUMMARY OF 1000 MC RUNS.

| PW Window (nm) | SRAM | Logic 1 | Logic 2 |
|---|---|---|---|
| MT | N/A | 0.62 | 0.80 |
| MST | 0.99 | 0.48 | N/A |

From the result of the simulation, it is found that, for patterns with only 1 trimming process, SRAM and Logic 2 regions have a larger process window compared to Logic 1, which goes through both MT and MST processes. For the MT step, the process window of Logic 1 is 22.5% smaller than that of Logic 2. For the follow-up MST step, the process window is even smaller, reaching 40.0% than that of Logic 2. It is a clear indication that introducing the extra MST step will result in a window shrinkage up to 40.0%

## DoE of 4 Process Variables with ±0.5 nm Window

For the above simulation, we applied a rather large window of MT and MST. In the next DoE simulation, we introduced 4 variables: MT, MST, 1st and 2nd Mandrel

CDs as the DoE parameters by setting their variations to ±0.5 nm around their designed value. The 2nd DoE is intended to discover the distribution of the PW if the process capability is 1 nm during manufacturing.

Introducing extra parameters, the CD variations of 1st Mandrel and 2nd Mandrel could result in the deviation of ideal values from design values. To determine a good starting point of targets, we apply a medium variation range around ideal design values, where MT, MST, MD1 and MT have ranges of ±2nm, ±1.5nm, ±2nm and ±1nm, respectively. With 200 Monte Carlo runs, one set of four variables (MT, MST, MD1, MD2) can satisfy PW requirements (≤ 2nm) for all 3 regions. Starting with this data set, we introduce a variation of ±0.5 nm for the above four parameters of interest.

With 100 MC runs, it is found that only 42 runs can satisfy the PW requirement (≤ 2 nm) of all four regions. Following the same analysis as in 1st DoE, it is found that variation range of 1st Mandrel CD remains the same; MT decreases by 10%, from ±0.5 nm to ±0.45 nm; 2nd Mandrel CD variation decreases by 26%, from ±0.5 nm to ±0.37nm; MST decreases by 24% from ±0.5 nm to ±0.38 nm, as in Figure 5. With a process window control capability of 1 nm for all 4 parameters, if the requirement of PW is imposed, the allowed process window can decrease as much as 26%. It is an indication that to satisfy the requirement of different regions with extra process stages, our process control capabilities will also need to be increased. Otherwise, the yield rate would drop dramatically down to 42%.

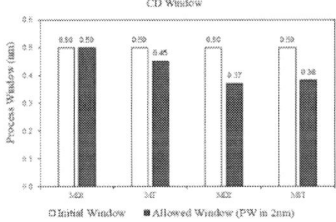

*Figure 5: CD window shrinkage*

The contour plot of PW of the Logic 1 region is plotted against MS and MST as shown in Figure 6. The red region has a higher PW while the green region has a lower PW. The allowed process window is located around the lower-left corner of Figure 6, centered around the designed values where MT is 3 nm and MST is 1.5 nm, as shown in the blue marker. It is worth noting that, even if inside the blue marker, there are a smaller number of areas that make PW larger than 2 nm; and a smaller number of cases satisfying PW requirement outside of the marker. The main reason is that four different processes are considered in the simulation: MT, MST, MD1, and MD2. For certain rare cases, the cooperation of the 4 processes will make the outlier cases possible.

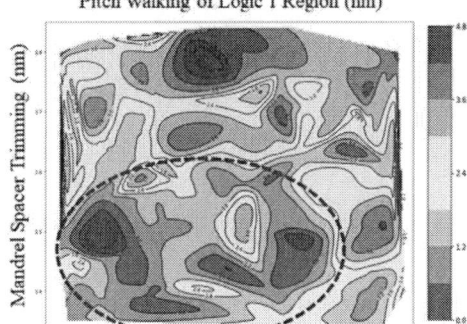

*Figure 6: Pitch Walking of Logic 1 region vs. MT and MST. The process window is identified inside an ellipse-shaped marker with blue color*

## CONCLUSION

In our work, we propose an advanced SAQP scheme to form 3 different Fin patterns and apply Monte Carlo based simulation in SEMulator3D® to evaluate the feasibility of the scheme by calculating the Pitch Walking when considering the variation of 1st Mandrel Trimming, 1st Mandrel Spacer Trimming, 1st Mandrel CD and 2nd Mandrel CD. It is found that 1) the MST stage introduced to trim Logic 1 region results in the process window shrinkage up to 40.0%; 2) in case of considering all four parameters, with the process window capability of 1 nm, the actual window satisfying PW requirement can be decreased as much as 26% due to extra MST process. The process window controlling capability needs to be improved, otherwise, a decrease in the yield rate down to 42% will be expected.

## ACKNOWLEDGMENTS

Sponsored by Shanghai Pujiang Program. The authors would like to show their appreciation to the colleagues in Etch Department of Technology Development & Research Center in SMIC for their valuable discussions.

## REFERENCES

[1] https://www.coventor.com/

[2] R. Kim et al. Design-Process-Technology Co-optimization for Manufacturability XI. International Society for Optics and Photonics, vol. 10148, 2017, pp. 10148:101480V.

[3] J. Bekaert et al. Extreme Ultraviolet (EUV) Lithography VIII. International Society for Optics and Photonics, vol. 10143, 2017, pp. 10143:101430H.

[4] Baudot S, Guissi S, Milenin A P, et al. 2018 International Conference on Simulation of Semiconductor Processes and Devices (SISPAD), 2018, pp. 344-347.

# Bitline etch process development for Advanced process DRAM manufacturing

Xing-Jun Yao[1]*, Li-Tian Xu[1], Jan-Kun Zhang[1], Zun-Hua Zhao[1], Guang Yang[1], Chen Chen[1], Jing-Lun Ma[2], Zheng-Qing Sun[2], You-Quan Yu[2], Ming Cheng[2], Fei Li[2], Bing-Hui Lin[2], Xin-Wen Huang[2], Ying-Yi Chen[2], Xian-Wen Su[2]

[1] Beijing NAURA Microelectronics Equipment Co. Ltd
No. 8 Wenchang Avenue, Beijing Economic-Technological Development Area. Beijing City, China
[2] ChangXin Memory Technologies, Inc. Ltd
No. 388, Xingye Avenue, Economic and Technological Development Area, Hefei City, China

*Corresponding Author's Email: yaoxingjun@naura.com

## Abstract

DRAM continues to play a major role in the semiconductor industry. As the design rule scaled down, one of the key technologies is to obtain the sufficient sensing margin for high performance by decreasing the leakage current from bitline (BL) to storage node contact (SNC). In the bitline manufacturing, we overcome ACL hard mask collapse, SiN hard mask taper and W metal bowing issues to create 10nm~14nm CD capability of bitline for the advanced process node.

*Keywords —Advanced process DRAM, bitline, plasma etch, dry etch, patterning, collapse, W metal, SiN hard mask, poly-Si, TiN, NMC612E.*

## Introduction

Key design features for DRAM (Dynamic Random Access Memory) cells are a high storage capacitor and low leakage current at the storage node connected to the capacitor [1]. As the design rule scaled down, DRAM performance can be improved by decreasing the bit line parasitic capacitance [2]. The bitlines are formed of a W/WN/TiN/WSi/polySi stack for 32-nm Mobile DRAM [3]. Since the bitline film structure is complex, the pattern needs to be accurately transferred down layer by layer. In this paper, we demonstrate the bitline etch process for advanced process node DRAM Manufacturing.

## Experiment

Fig.1 illustrates the schematic film structure in the bitline etch process. There are eight kinds of film layers as below list：
1) Oxide Hard Mask,
2) HM1 act as under layer ARC
3) HM2 such as APF，ACL
4) HM3 most formed with SiN
5) Tungsten Bit line
6) Glue layer
7) Poly Si
8) Ox stop layer
We use NMC612E dual frequency chamber for the etching.

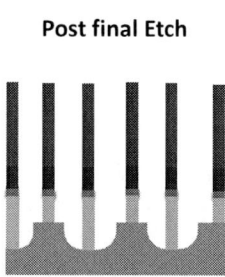

Fig.1 The film structure of Bitline.

Fig.2 shows the step-by-step morphology of bitline etch process. However, the morphology of the remaining SiN hard mask requires not only straight but also reaching narrow CD regime 10~14nm. Then, the graphics can be well transferred to W and poly-Si layer.

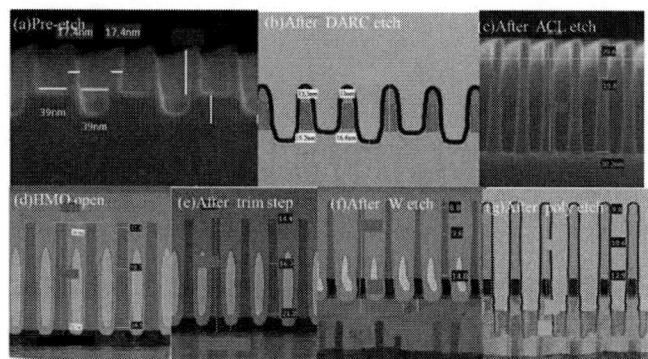

Fig.2   Bitline profile step-by-step check for etch process.

## Result and discussion

### A.   Adding novel passivation gas to solve ACL collapse issue

The main component of ACL film is carbon, which is relatively soft and easy to fall and collapse. We propose adding a novel passivation gas to form SiOx polymer on the sidewall in ACL etch step. It can protect the ACL sidewall from the plasma damage and solve the profile collapse issue as showed in Fig.3.

978-1-6654-9759-6/22 $31.00 © 2022 IEEE

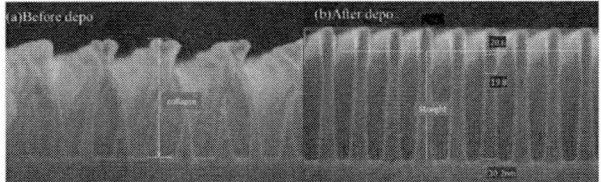

Fig.3 ACL Etch profiles comparison between (a) without the novel gas addition and (b) with 5sccm novel gas addition in ACL step.

### B. SiN hard mask CD trimming

In order to meet the design rule of the bitline CD, a trimming step with CHF3-based plasma is leveraged to shrink the SiN hard mask CD as Fig.2(e) after SiN hard mask opening. In calculation, the trim rate is 0.3nm per sec for the SiN hard mask lateral pull back as shown in Fig.4. It is critical to form a vertical and uniform SiN profile for the device current leakage reduction between the bitline to its neighbor storage node contacts (SNC).

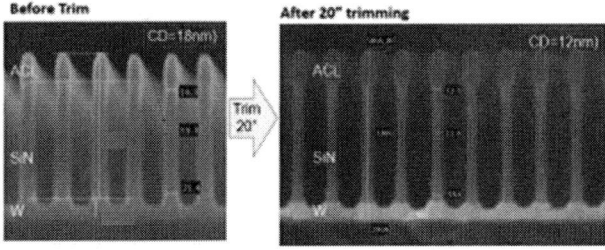

Fig. 4 SiN hard mask CD trimming by CHF3-based plasma

### C. SiN remain thickness tuning

For the process integration rule, the thickness of SiN hard mask must maintain HM less erosion after bitline etch. We suggest to reduce the plasma bias power in W metal etch step to improve the etch selectivity of W to SiN. Fig.5 shows SiN HM thickness erosion improvement by reducing bias power from 250Wb to 50Wb.

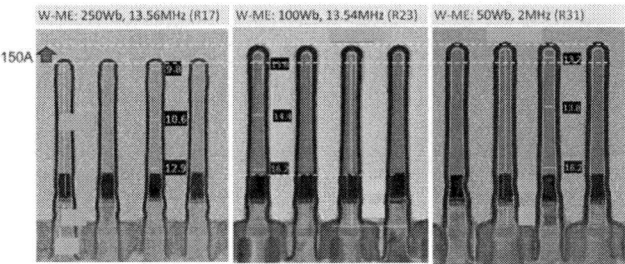

Fig.5 The remaining SiN hard mask thickness depending on the plasma bias power in W etch step.

### D. W metal profile slope control by NF3 flow rate

Fig.6 shows that W profile can be controlled by NF3 flow rate. When NF3 flow rate is 12sccm, W profile looks tapered. When NF3 is increased to 15sccm, W profile is close to vertical. We found that the morphology of W profile may become bowing when NF3 is further increased to 18sccm.

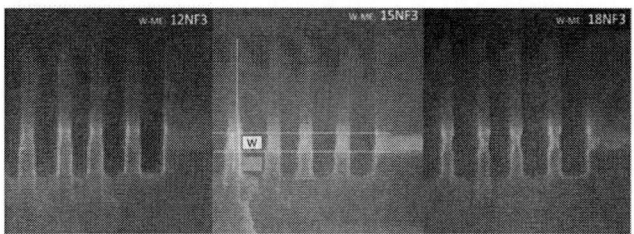

Fig.6 Bitline W profile tuning by NF3 gas flow rate.

### Conclusion

We have developed the bitline pattering technology. It is possible to demonstrate a vertical bitline morphology as well as reach 10nm~14nm CD control capability for the requirement of advanced process DRAM manufacturing.

### References

[1] Kinam Kim et al., "Memory technologies in Nano-era:Challenges and Opportunities,"(invited) ISSCC Dig. Tech. Papers, pp. 576-577 ,2005.
[2] B.J. Park et al., Symp. on VLSI Tech., p.182-183, 2002.
[3] Dick James, "Recent Advances in Memory Technology", ASMC2013

# SADP etch process development using PR core for sub 17nm DRAM

Li Tian Xu[1]*, Shuai Zhang[2], Ling Feng Li[1], Hao Liu[1], Xin Wen Huang[2], Ying Yi Chen[2], Xian Wen Su[2],
Zhong Ning Guo[2], Chun Yu Xiu[2], Tian Lei Mu[2], Bing Hui Lin[2], Zhong Yi He[1] and Qing Jun Zhou[1]

[1] Beijing NAURA Microelectronics Equipment Co. Ltd
No. 8 Wenchang Avenue, Beijing Economic-Technological Development Area. Beijing City, China
[2] ChangXin Memory Technologies, Inc. Ltd
No. 388, Xingye Avenue, Economic and Technological Development Area, Hefei City, China

*Corresponding Author's Email: xulitian@naura.com

## Abstract

Self-aligned double patterning (SADP) process has become the standard patterning technology for extending the half-pitch resolution beyond current ArF lithography tool's limit. In this paper, we mitigate the compressive stress of ALD SiO2 spacer in SADP etching and solve the SADP spacer pattern collapse problem to minimize the pitch walking deviation. The SADP technology is applied for 17nm of buried word line (bWL) process to meet DRAM scaling requirements.

*Keywords — spacer-defined double patterning, self-aligned double patterning, resist mandrel, resist core, oxide spacer, ALD SiO2, compressive stress , wiggling, leaning , carbon hard mask, buried word line, pitch walking, DRAM, and NMC612E.*

## Introduction

Spacer-defined double patterning (SDDP) which is also called self-aligned double patterning (SADP) involves using sidewall spacers to create hard masks as a means of doubling the printed line density [1]. It has big advantage over other pitch splitting techniques such as litho-etch-litho-etch (LELE) in terms of overlay control and pitch walking [2,3]. Aimed at reducing cost, a scheme which integrates photo resist (PR) pattern as core mandrel [4] with a low temperature ALD SiO2 as spacer [5] has been proposed. The key challenge of resist-mandrel scheme is heavily impacted by the irregularity of mandrel shape [6,7].

In this study, photo resist (PR) and ALD SiO2 were selected for core and spacer materials, respectively. We developed a plasma etch process to suppress the ALD SiO2 spacer collapse during SADP process. Furthermore, we apply this SADP technology to create 17nm of critical dimension (CD) for buried line (bWL) process to meet DRAM manufacturing requirements [8].

## Experiment

Fig.1 illustrates the schematic SADP process flow. The film scheme started with a bottom-SiON hard mask (HM) as an etch stop layer, followed by 500~600A carbon mask and 150~250A dielectric anti-reflection coatings (DARC). PR pattern of 450~550A thickness was selected as a core mandrel from

lithography process. ALD SiO2 spacer was deposited to target a designed sidewall width (20~25nm) as shown in Fig.1 (a). The etch process was carried out by NAURA NMC612E tool for ALD SiO2 spacer etch back (Fig.1 (b)), selective PR core removal (Fig.1 (c)), DARC open and carbon mask patterning (Fig.1 (d)) in the sequence of each plasma step.

As for DRAM manufacturing, the negative tone of bWL pattern was created by reversing the SADP defined carbon line to SiO2 HM trench. Then, the bWL trench profile in memory array area was formed after etching through Si substrate and STI dielectric layers simultaneously. We measured the pattern CD uniformity of the final bWL trenches as well as the SADP defined carbon lines. In addition, pitch walking performance was also the key factor of success (KFS), which was related to the asymmetric spacer shape of the ALD SiO2 film.

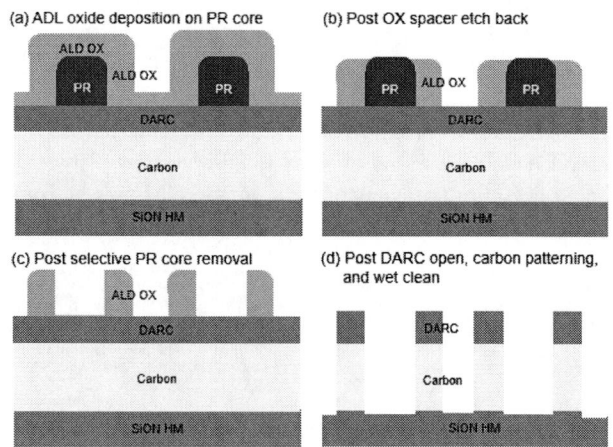

Fig.1 SADP process flow: (a) pre-etch, (b) post ALD SiO2 spacer etch back, (c) post selective PR core removal, and (d) post DARC open, carbon mask etch and wet clean.

## Result and discussion

### A.  OX spacer etch process development

In order to understand the interaction of RP mandrel with SiO2 spacer before PR core removal, DOE-L4 method was leveraged to compare the spacer etching profiles depending on the spacer etch back process parameters, such as gas pressure,

978-1-6654-9759-6/22 $31.00 © 2022 IEEE

RF bias power and CF4/O2 gas ratio. An inward collapse issue intended to happen for the defined SiO2 spacer profile as shown in Fig.2 (a~c). It will cancel out the line CD reduction when the leaning and bending SiO2 spacer profile further transfers to the under-layer of carbon mask film. As a result of DOE-L4, Fig.2 (d) revealed that the profile collapse issue was suppressed in the case of using lower pressure, lower bias power and lower O2 flow rate. It indicates that reducing the plasma damage on the soft PR core during spacer etch back process can help to prevent PR deformation. We believe that the compressive stress always exists on the PR core from the neighbor spacer SiO2 film. The stress might be enhanced through the strong plasma ion bombardment.

Fig.3(a) which implemented a pulsed bias power in SiO2 spacer etch back step showed severe spacer leaning issue, compared to Fig.3(b) using continuous wave (CW) plasma. Since pulsed plasma may reduce the surface charging on wafer, it proves that the plasma charging effect is not the root-cause for PR core deformation. Therefore, the pulsed plasma might enhance the compressive stress because of the slower etch rate and longer process time during spacer etch back.

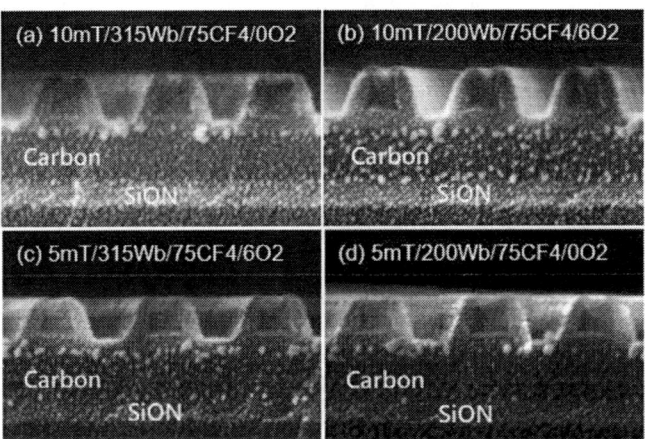

Fig.2 SADP spacer profiles comparison depending on SiO2 spacer etch back parameters, such as gas pressure, bias power and CF4/O2 ratio, by DOE-L4 method before PR core removal.

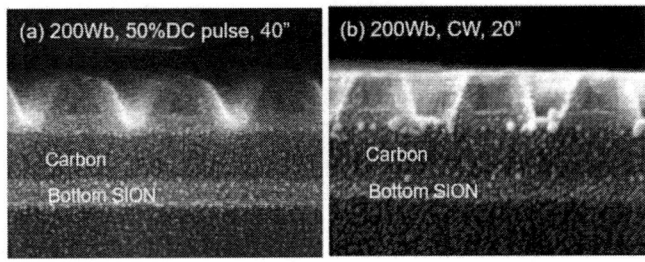

Fig3. SADP spacer profiles comparison between (a) 1000Hz pulsing and (b) continuous wave (CW) bias power during SiO2 spacer etch before PR core removal.

Eventually, we got rid of the spacer line collapse problem after minimizing the plasma induced compressive stress in the spacer etch back process. Fig.4 (a) and (b) reveal the integral SiO2 spacer profiles, respectively, for post spacer etch back and post PR core removal.

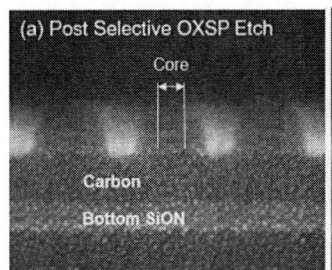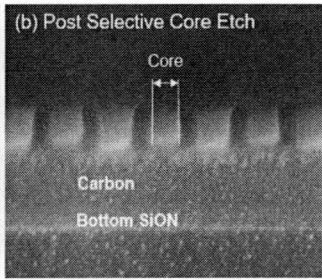

Fig.4 Integral SADP spacer profiles etched by the optimized spacer etch back process: (a) before PR core removal, (b) after PR core removal.

*B. SADP line CD control technology*

For the device mass production, a tuning knob for the SADP line CD is required to compensate the etch chamber deviations. We studied the CD control factors for each SADP etch steps and proposed a novel polymer deposition step inserted after core removal as shown in Table 1. This sidewall deposition technology can help to not only gain the line pattern CD but also improve the surface roughness which is transferred from the non-uniformed shape of soft PR core. In calculation, the final CD tuning rate for the downstream bWL trench pattern is 0.4 nm per 6 sec of deposition time. The additional deposition process may contract much more space CD in gap area than that in core area because of different polymer build up on the asymmetric shape of spacer sidewall. However, the pitch walking offset can be compensated simply by enlarging the incoming PR CD in lithography process.

After SADP etch process for carbon mask patterning, AEI CD range deviation can meet the requirement of <0.6nm according to top view CDSEM images at the different locations within a whole wafer as Fig.5. Fig.6 revealed the post-etched cross-section SEM profile with 500~600A carbon lines and >250A dielectric hard mask remaining before wet clean process. And its pitch walking was controlled well within less 0.3nm offset for memory array area. Besides, a slight notch was obvious at the bottom of SiO2 spacer in the core mandrel site. These systematic notch profiles came from the bottom footing of the upstream PR shape.

| Etch step | Chemistry | CD control factor | CD trend |
|---|---|---|---|
| OX spacer | CF4, CH2F2, O2 | CH2F2 (sccm) | Up |
| PR Core | O2, N2 | (None) | (n/a) |
| Depo | CHF3, CH2F2 | Process time (sec) | Up |
| DARC | CF4, Cl2, HBr | HBr (sccm) | Up |
| Carbon | O2, SO2 | Over etch (%) | Down |

Table 1. SADP process control factors for the in-line CD of carbon hard mask line.

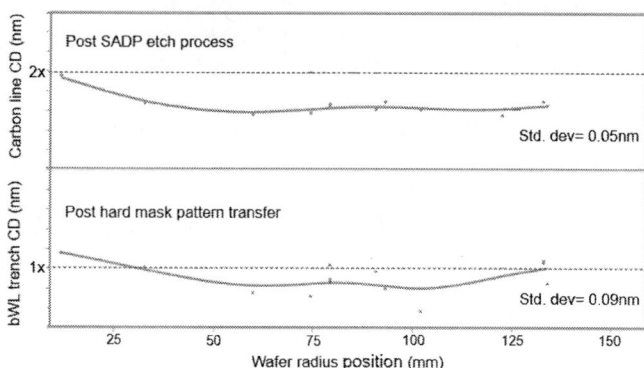

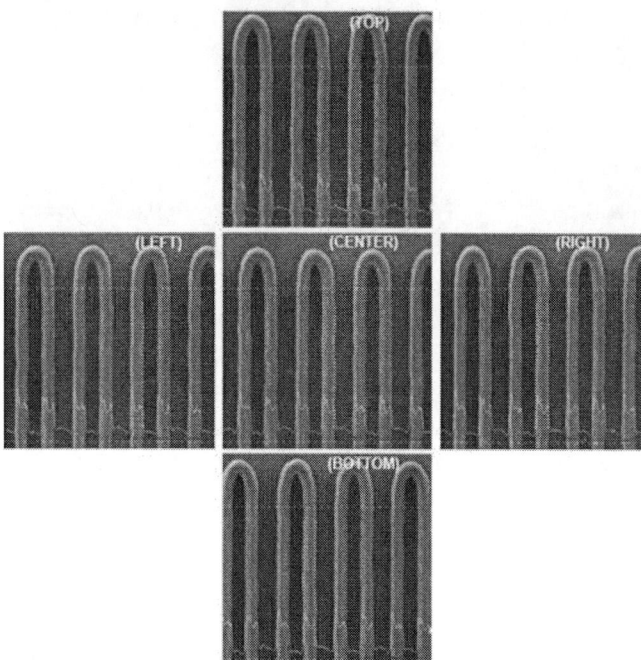

Fig.5 AEI SEM CD inspections on a whole wafer for the optimized SADP etch process before wet clean process.

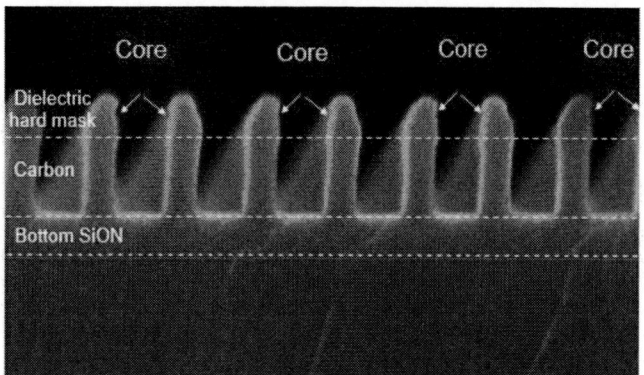

Fig.6 Cross-section SEM profile of the defined carbon lines by the optimized SADP etch process before wet clean process. (The arrows represent a slight notch which appears at the bottom of SiO2 spacer in PR core mandrel site.)

*C. Final bWL trench CD performance*

In term of DRAM bWL manufacturing, the mask CD of carbon lines was shrunk to a target by post etch treatment (PET). Then, we leveraged an available negative tone technology for the mask pattern transfer from carbon lines to SiO2 trenches. Next, bWL trenches were formed after etching through Si substrate and STI dielectric with the hard mask of SiO2 trench pattern. The final bWL trench CD can meet the requirement for sub 17nm in memory array area. And the CD uniformity was as good as less 0.1nm sigma deviation as shown in Fig.7.

Fig.7 In-line AEI CD distribution charts for the lines of carbon hard mask and the spaces of bWL trench with the optimized SADP process on one whole wafer.

## Conclusion

We proposed the low-cost SADP process that implements soft litho PR instead of hard mask for the core mandrel and developed a low damage etch process to prevent the PR deformation from the compressive stress of the asymmetric oxide spacer shapes during spacer etch back process. This technology can meet the requirements of the advanced DRAM manufacturing for 17nm dimension and beyond.

## References

[1] Christopher Bencher, et al., "22nm half-pitch patterning by CVD spacer self alignment double pattering (SADP)", Proc. of SPIE, vol.6924, 69244E-1~7, (2008)

[2] Huayong Hu, et al., Self-aligned double patterning (SADP) process even-odd uniformity improvement", 2016 China Semiconductor Technology International Conference (CSTIC), pp.1-4 (2016)

[3] T. Yang and D. Yim, "SAQP Pitch Walking Improvement Path Finding by Simulation", 2019 International Symposium on Dry Process (DPS), P-22, pp.99-100 (2019)

[4] Nihar Mohanty, et al.,"Challenges and mitigation strategies for resist trim etch in resist-mandrel based SAQP integration scheme", Proc. of SPIE, Vol.9428, 94280G-1~14 (2015)

[5] Qingqing Wu, et al., "Optimization of the CD uniformity (CDU) in silicon oxide spacer process for 5nm FIN SAQP process flow", 2020 China Semiconductor Technology International Conference (CSTIC), pp.1-4 (2020)

[6] Angélique Raley, et al., Self-aligned Quadruple Patterning Integration using spacer on spacer pitch splitting at the resist level for sub 32nm pitch applications", Proc. of SPIE, vol. 9782, 97820F-1~14 (2016)

[7] Efrain Altamirano-Sanchez, et al, "Self-aligned quadruple patterning to meet requirements for fins with high density", SPIE, 14 May (2016)

[8] Min Nee Cho, et al., "An innovative indicator to evaluate DRAM cell transistor leakage current distribution", IEEE J. Electron Devices Soc., vol. 6, pp. 494-499 (2018)

# Beyond 20nm DRAM Capacitor Etch Challenge and Process Solution

*Jianqiu Hou[1], Jun Xia[2], Ya Zhou[1], Kangshu Zhan[\*], Zengwen Hu[\*]*

[1]Advanced Micro-Fabrication Equipment Inc. China, Shanghai 201201, China
[2] Chang Xin Memory Technologies Corporation, Hefei 230093, China
\* Correspondence to: Zengwen Hu (AMEC), ZengwenHu@amecnsh.com
Kangshu Zhan (CXMT), Kangshu.Zhan@cxmt.com

## Abstract

With the smaller size of the dynamic random access memory, the capacitor etch becomes more and more difficult due to the smaller critical dimension (CD) and higher aspect ratio. We found that mask selectivity, CD control, missing holes and bottom distortion will become very margin when capacitor aspect ratio is higher than 30. To address the challenges, we propose some process solutions in three directions, including lower pressure, higher bias power with low duty cycle and new chemistry are all effective to etch very small and deep capacitor holes. We also try to explain the mechanism of observed performance caused by optimal process conditions. It will be a good reference of capacitor etch for the next generation of DRAM.

**Key words**: Dynamic random access memory; Capacitor etch; Capacitive coupled plasma etch; High-aspect-ratio etch; Dielectric etch.

## Introduction

When the dynamic random access memory (DRAM) device has been scaled down beyond 20nm i.e. 1x nm, the requirement of plasma etch, especially the high aspect ratio (HAR) capacitor etch, becomes more and more strict. Capacitor is used to store charges in DRAM device [1]. Figure 1 shows the structure of a capacitor, including two electrodes and one dielectric layer. They are designed to be like cylinder or pillar holes to achieve the highest superficial area. The volume and density of capacitor hole decide the capacitance storage value. The distance between adjacent holes decides the risk of leakage current [2]. Therefore, to get over the trade-off effect between capacitance and leakage, we should make capacitor holes very deep and small by HAR plasma etching.

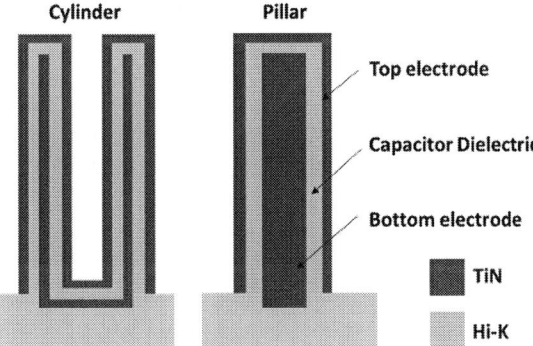

*Figure 1. The cylinder and pillar like capacitor*

We here list the common size of capacitor in different DRAM generations in Table 1. When the size of DRAM device is reduced to 1α nm, the appropriate aspect ratio of capacitor holes should be higher than 50 meanwhile the critical dimension i.e. diameter of capacitor holes should be around 20nm. It is close to the technical limitation of dielectric HAR etch [3].

We use the advanced capacitive-coupled-plasma (CCP) etch tool to achieve dielectric HAR etch due to the high collimation and energy of adjacent ions. The plasma will be formed in two plate electrodes with additional radio frequency (RF) power supplies and specific gases. We usually use two kinds of RF power to de-coupled control ion density and ion energy. The high frequency i.e. source power can increase the collision frequency between electron and molecule thus increase the density of reactive ions and radicals. The low frequency i.e. bias power can accelerate ions to bombard the dielectric surface. By combining physical bombardment and chemistry reaction, we can successfully break the Si-O and Si-N bonds then form volatiles, thereby remove the target materials with anisotropy and high etch rate (ER).

*Table 1. The size of capacitor in various DRAM generations*

| DRAM | Capacitor Pitch | Capacitor CD | Capacitor aspect-ratio |
|---|---|---|---|
| 1x nm | ~70 | ~50 | ~30 |
| 1y nm | ~55 | ~40 | >30 |
| 1z nm | ~45 | ~30 | >40 |
| 1α nm | ~35 | ~20 | >50 |

With the decreased CD of holes, the transportation of etchant and the pumping out of byproduct or polymer will become more and more difficult. When etch to deep location, many CF polymers will condense at the etch front. In addition, the ions will positively charge the bottom. It is difficult to be neutralized due to the lack of electrons. Therefore, the ion energy will be much reduced by charge repulsing force. To avoid under etch in very small but deep holes, we can decrease the chamber pressure meanwhile increase bias power to intensify the capability of pumping out as well as ion bombardment. However, constantly increasing power is not an economically strategy because most power are wasted on removing polymer and thermal loss but not promote the reaction with target layers. Besides, when we raise power, the mask ER will be also much increased, leading to the lower mask selectivity, which is defined by the ratio of in-hole ER to mask ER. Considering the fact that the deposition and opening of mask will also be much more difficult with increased thickness, we cannot further increase the mask thickness to compensate the drop of mask selectivity.

To get over the side effect of increasing power, we can explore new hard mask materials, optimize the mode of power input and change gas ratio or spices. The traditional mask material for capacitor etch is Poly-Si. By doping other elements, such as boron, we can obviously increase the selectivity to OX. However, to date we still have not found a good solution to remove the boron byproduct. It will contaminate chamber thus reduce the process stability. Therefore, we prefer other two process solutions. RF pulsing mode can much increase the efficiency of power [4]. It can switch the power between high level and low level at a very high rate. The frequency is usually higher than 1 kHz. In some extreme conditions, the power can be switched off at low level. Ideally, when power is at high level, the dielectric layer is etched by physical bombardment and chemical reaction. Meanwhile, the side-wall and bottom of holes are charged by positive ions. When power comes to low level, polymer deposition is dominated, resulting in the good protection of mask. According to some researches, the positive charges can even be neutralized by electrons and negative ions when power is switched off. Consequently, at RF pulsing mode, we can optimize the distribution of power to separately control charge/discharge and etch/deposition, resulting in a much larger process window. The percentage of high-level power in one cycle is called duty cycle (DC). By tuning DC and two-level power, we can find the best point that shows both acceptable ER and high mask selectivity.

However, when the DRAM generation comes to 1z nm, the process window is not enough again even using RF pulsing mode. Except not enough mask and some missing holes i.e. under etch, bottom circularity also becomes much worse. It can be ascribed that the charges and polymer residuals are not uniform at sharp etch front. To address the challenges, we should change C4F6/C4F8 ratio or even add some untraditional gases. For example, a gas that shows the higher mask selectivity than traditional C4F6 is feasible. On the other hand, a gas that can release the charges close to bottom is also attractive. It can increase in-hole ER and etch front circularity.

In addition the challenges caused by smaller size and higher aspect ratio, we should also consider the challenges caused by layer thickness and material. Figure 2 shows the structure of capacitor holes. Two silicon oxide (OX) layers are sandwiched by three silicon nitride (SiN) layers. From the top to the bottom, we usually call the five dielectric layers as T-SiN, OX1, M-SiN, OX2 and B-SiN, respectively. Among them, T-SiN, OX1 and OX2 is usually much thicker than others. We should etch them step by step to form a capacitor hole.

The first step is always critical. The profile and CD of T-SiN decide the following layers, especially when T-SiN becomes much thicker in more advanced DRAM generations. Commonly, SiN etch is dominated by chemical reaction and OX etch is dominated by physical bombardment. It is decided by the variation of bond energy and surface defect. Therefore, etching SiN always requires some H-contained gases but not very high power. However, for very thick T-SiN layer, lower power cannot well control Top CD (TCD) and straight profile. We should try our best to make T-SiN etch more smooth i.e. increase in-hole ER and anisotropy in T-SiN etch.

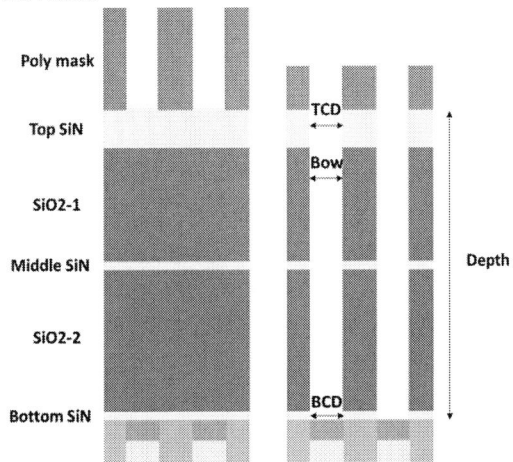

*Figure 2. The pre and post etch structure of capacitor holes*

According to the previous introduction, we here provide four process optimizations to address challenges, caused by the smaller CD, higher aspect ratio and thicker T-SiN, beyond 20nm DRAM device. It includes the better capacity of pumping out byproduct and polymer, the better control of RF-pulsing power, the optimally gas component and the strategy change for T-SiN etch. Our target is to find the useful trend and best point of capacitor etch process to follow the continuously update DRAM generation. Through our chip-etch experiments, we found that low pressure, from 15mT to 20mT, is appropriate for etching the capacitor whose aspect ratio is higher than 30. Higher pressure will increase the risk of blocking and too low pressure will much decrease mask selectivity and in-hole ER. To balance the trade-off effect between mask remain and straight profile, we suggest decreasing DC meanwhile increasing bias power. It can much better protect mask while sacrifice some in-hole ER. The appropriate bias power should be higher than 15kW in 1x nm and 1y nm generation. For 1z nm and 1α nm, >30kW bias power will be better. For chemistry, when capacitor CD is smaller than 30nm, the traditional view that more C4F6/C4F8 flow can increase mask selectivity is invalid. Excessive C4F6 flow will more decrease in-hole ER rather than decrease mask ER in very small and deep holes. Based on our experiments, the appropriate C4F6/C4F8 ratio should be lower than 1. We also find a specific gas that is effective to increase in-hole ER without excessive consuming mask, improve bottom circularity and avoid missing holes, when the capacitor aspect ratio is higher than 40. For thicker T-SiN layer, relative high power, low C4F6 flow and low temperature are necessary. They can make SiN etch smoother even though slightly decrease mask selectivity and etch front circularity. All the process trends are listed in Table 2. By analyze them, we can predict the direction of tuning capacitor etch process when we comes to the next generation of DRAM. We have to admit that some performances still lack the solid demonstration of mechanism. By sharing the results of our experiments on advanced and critical capacitor etch, we hope the future cooperation on mechanism analysis approaches with other institutes or colleges.

*Table 2. The process trend of capacitor etch in different DRAM generations*

| DRAM | Pressure (mT) | Bias power (kW) | Duty cycle (%) | C4F6/C4F8 | New gas |
|---|---|---|---|---|---|
| 1x nm | 15-25 | 7-15 | 35-50 | >2 | N |

| | | | | | |
|---|---|---|---|---|---|
| 1y nm | 15-20 | 15-25 | 25-35 | 1-2 | N |
| 1z nm | 15-17 | 25-35 | 20-25 | ~1 | Y |
| 1α nm | 13-17 | 35-50 | 15-20 | <1 | Y |

## Experiment

All etching experiments were conducted in a capacitive coupled plasma (CCP) etch tool with two radio frequency (RF) generator (60 MHz source and 400 kHz bias power). The etch tool was AMEC HD plus RIE. We etched DRAM capacitor in five steps, including T-SiN, OX1, M-SiN, OX2 and B-SiN, due to five dielectric layers in capacitor film stack. $C_4F_6/C_4F_8/CH_2F_2/O_2$ chemistry were used to etch SiN film. $C_4F_6/C_4F_8/O_2$ chemistry were used to etch OX film.

To avoid wasting resources, we used "coupon", a small chip sample cut from whole wafer, for experiment. The 2cm x 3cm coupon has already been pre-coated with designed film stack and exposed for following etching process. We then pasted the chip on a carrier Si wafer as a sample in AMEC lab. When we obtained a satisfied result, we would transfer the condition on a 300mm-structure whole wafer in CXMT fab to confirm the capacity of quantity production.

Scanning Electron Microscopy (SEM), Hitachi 4800, in AMEC lab, was applied to observe the cross-section and top view of samples. To observe the bottom circularity, we immersed post-etch coupon in 1% HF solution for 5-10 min then applied ultrasonic to clean it before SEM analysis.

Transmission Electron Microscope (TEM) and Energy Dispersive spectroscopy (EDS) analysis were carried out with the help of MSSCORPS CO., LTD.

## Result and discussion

### Chamber Pressure

The byproduct and polymer are difficult to be pumping out with decreased CD and increased depth of capacitor holes, resulting in the lower in-hole ER, the high risk of blocking and under etch. We detected the in-hole OX ER of four different capacitor CDs with the same etching depth, as shown in Figure 3a. When CD is decreased from 50nm to 20nm, the OX in-hole ER is dropped from 5.18nm/s to 2.33nm/s. We also compared the in-hole OX ER of various depth with ~30nm initial CD, as shown in Figure 3b. We observe a decrease of ER from 4.00nm/s to 2.67nm/s when etching depth is increased from ~300nm to ~1000nm. Besides, as CD is below 30nm and depth is more than 800nm, we observe a few missing holes covered by polymer in the top view (Figure 3c) and some under-etch holes in the cross-section image (Figure 3d). We have analyzed the element composition of blocking holes in Figure 3c by EDS. The result shows that the blocked polymer is composed of 53 wt. % carbon, 32 wt. % oxygen, 6 wt. % fluorine and 9 wt. % silicon. All the evidences prove that the byproduct and polymer will resist etching in deep hole with smaller CD.

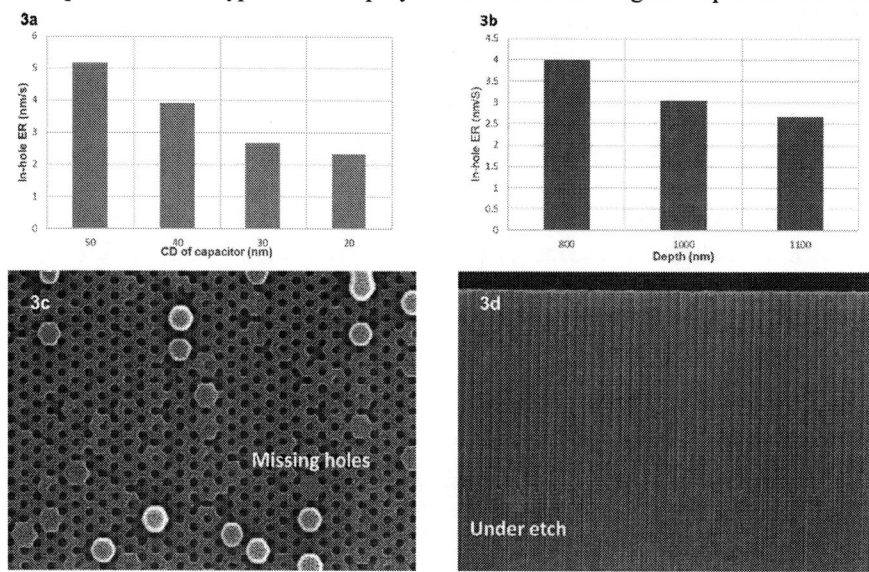

***Figure 3.*** *The trend of in-hole ER with (a) decreased CD and (b) increased depth; (c) Missing holes in the top view of capacitor; (d) Under-etch holes observed by the SEM cross-section.*

To overcome it, we first try to decrease the chamber pressure. Theoretically, lower pressure can accelerate the pumping out of polymer and byproduct. It can promote the forward reaction of dielectric etch, especially when the ER is dominated by polymer-deposition rate. Figure 4 shows the in-hole ER and the mask selectivity of ~30nm CD capacitor with decreased pressure. When we decrease pressure from 25mT to 20mT, ER is increased and mask selectivity is comparable. However, when we further decrease pressure from 17mT to 12mT, both ER and mask selectivity are dropped. It can be ascribed that ion, radical and mask-protection polymer density are all much decreased below 15mT due to the less residence time. Therefore, we suggest the appropriate chamber pressure for capacitor etch should be 15mT-20mT.

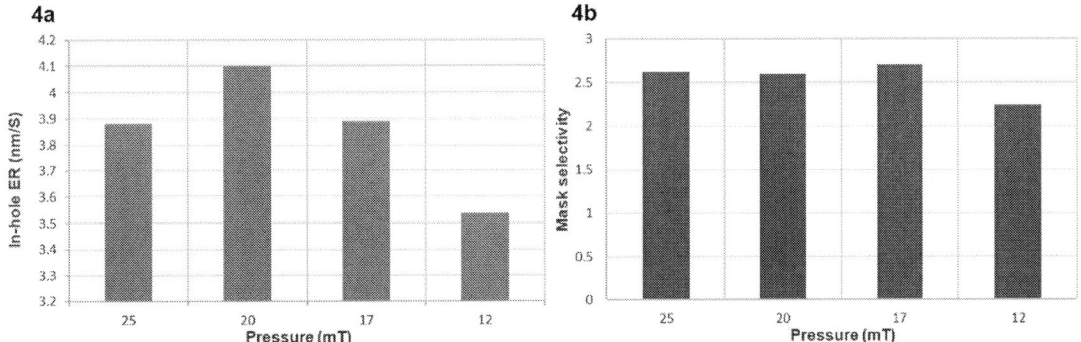

***Figure 4.*** *The trend of (a) in-hole ER and (b) mask selectivity with decreased pressure.*

**RF pulsing power**

Only controlling pressure still cannot meet the target of very HAR capacitor etch. When we etch to the bottom in 1z nm generation, the mask remain is not enough, leading to poly-Si damage and TCD blowing up, as shown in Figure 5a. Besides, the process window of enlarging bottom CD (BCD) by over etch (OE) will be very shallow with few mask remain. Large BCD i.e. straight profile of capacitor hole is crucial to increase the capacitance storage value. Therefore, we plan to increase mask selectivity by tuning RF pulsing power.

As previously introduced, RF pulsing mode can separate the etching and deposition by fast switching high/low level. To demonstrate that polymer can be deposited on the mask in low level power, we designed a deposition experiment to observe the location of polymer deposition by TEM. We first etch a capacitor hole with ~30nm CD and strip the remaining CF polymer by O2. Then we use the low-level power, 600W source power and 100W bias power, to deposit CF polymer with C4F6 and C4F8 in 10s. By TEM analysis, we can observe C polymer layer (white one) on the mask, as shown in Figure 5b. On the contrary, the polymer on the side-wall and bottom of dielectric layer are not obvious. It indicates that the polymer deposition under low-level power prefers to protect mask rather than resist etching target layers.

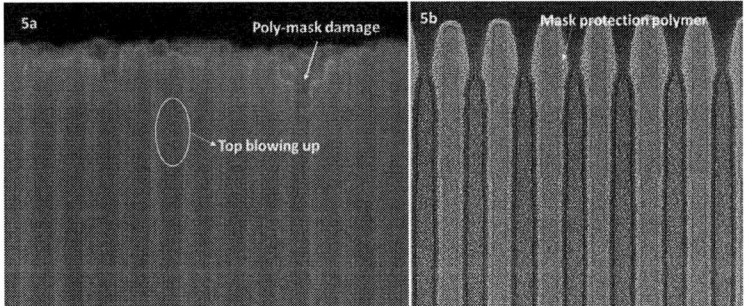

**Figure 5**. (a) Poly-mask damage and top blowing up of ~30nm capacitor hole in SEM cross-section; (b) Polymer deposition on the mask of capacitor in TEM cross-section.

Therefore, to intensify the protection of mask, we can decrease the high-level power ratio i.e. DC in RF pulsing mode. However, only decrease DC will much decrease the in-hole ER. To compensate it, we should increase the high-level bias power at the same time. We here defined a concept called total power. It is the result that high-level source power add high-level bias power then multiple DC. When we decrease DC and increase bias power, we usually keep the similar total power and the same ratio of bias power to source power. It can mitigate the negative effect on ER and polymer deposition. For example, to etch the capacitor hole in 1z nm DRAM, we decrease the DC from 40% to 20%, the in-hole ER is dropped by ~ 9.1% and the mask ER is dropped by~12.2%. Then we raise the bias power and source power from 17kw to 34kW and 4kW to 8kW. The in-hole ER is comparable and the mask selectivity is increased by~ 14.4%, in comparison with the initial condition. The strategy, decreasing DC and increasing power, is very useful when the aspect ratio of capacitor hole is raised up. We predict that the appropriate DC and bias power will reach to 15% and 40kW-50kW, respectively, in 1α nm DRAM capacitor etch.

**Gas component and spices**

We usually use $C_4F_6$ and $C_4F_8$ to produce polymer in HAR etch [5]. Based on the functions of polymer, it can be roughly divided into three kinds, including mask protection polymer, side-wall polymer and bottom residual polymer. Generally, $C_4F_6$ is more likely to form high-molecular-weight polymer due to the double bonds. The mechanism is close to radical chain polymerization. Oppositely, $C_4F_8$ prefers to form low-molecular-weight polymer because the four membered ring is readily broken and form two radical sites. The mechanism is radical coupling polymerization. The reaction rate of radical coupling polymerization is much slower than the other. Consequently, more $C_4F_6$ will form more sticky and stubborn polymer. It can effectively protect mask while make bottom residual difficult to be removed. In very small holes, the side effects of heavy polymer, such as lower ER and even blocking or under etch, become more severe, as discussed before. Therefore, as the aspect-ratio of capacitor hole is increased, we should tune the distribution of polymer molecular weight and structure by reducing the ratio of $C_4F_6$ to $C_4F_8$.

Table 3 shows the result of tuning $C_4F_6/C_4F_8$ ratio in ~30nm capacitor etch. When we decrease the $C_4F_6/C_4F_8$ ratio from 0.7 to 0.5, the in-hole OX ER is increased by 18.5% and the mask ER is increased by 12.1%. Consequently, the mask selectivity is instead increased by 5.8%. However, when $C_4F_6/C_4F_8$ ratio is reduced to 0.25, the in-hole OX ER is increased by 2.9% and the mask ER is increased by 34.8%. The mask selectivity starts to be decreased due to the lack of mask protection. The result indicates that $C_4F_6/C_4F_8$ ~0.5 is appropriate for ~30nm capacitor. The ratio will be trend to below 0.5 when the CD of capacitor hole is decreased to ~20nm.

*Table 3. In-hole ER, mask ER and mask selectivity depend on C4F6/C4F8 ratio*

| C4F6/C4F8 ratio | 0.7 | 0.5 | 0.25 |
|---|---|---|---|
| In-hole ER (nm/s) | 3.19 | 3.78 | 3.89 |
| Mask ER (nm/s) | 1.41 | 1.58 | 2.13 |
| Mask selectivity | 2.26 | 2.39 | 1.83 |

Only decreasing $C_4F_6$ ratio is still not enough to solve other problems, such as under-etch holes, as shown in Figure 3d. It can be ascribed to mask blocking or bottom charging. Fortunately, we found a new metal halide gas that can work. Figure 6 shows the cross-section by SEM with and without the specific gas. After adding the new gas, all holes are opened and hole to hole uniformity is also good. Besides, in-hole ER thus mask selectivity are both increased >10%, implying a larger process window to make profile more straight. We suspect that the gas can discharge by surface deposition. The investigation of detail mechanism is ongoing. We will systematically illustrate it in the future.

*Figure 6. The TEM cross-section of capacitor holes with and without the specific gas*

**Strategy change for thick T-SiN etch**

The aforementioned discussions focus on OX etch because OX layer is generally much thicker than SiN layer. The performance of SiN makes less effect on final profile and CD. However, as T-SiN becomes thicker in 1y nm DRAM capacitor, the influence cannot be regardless. Compared with OX Etch, we usually etch T-SiN layer by low power and H-contained gases due to the reaction is chemically dominated. Besides, the polymer in SiN hole is always heavier than that in OX hole because OX surface can desorb O radical, which can remove residual polymer. Consequently, taper profile and severe depth variation readily appears in a deep SiN hole, as shown in Figure 7a. According to the discussion on the effect of power and chemistry, by moderately increasing power and decreasing C4F6/C4F8 ratio, we can straighten T-SiN profile, as shown in Figure 7b. To mitigate depth variation, we found that low temperature (-20C or even below) is helpful (Figure 7c and 7d). In that case, SiN ER is much increased due to the higher surface adsorption coefficient and the more flexible bottom residual polymer. We suspect the low-temperature etch of SiN layer will be more and more important in capacitor or other higher aspect ratio etch.

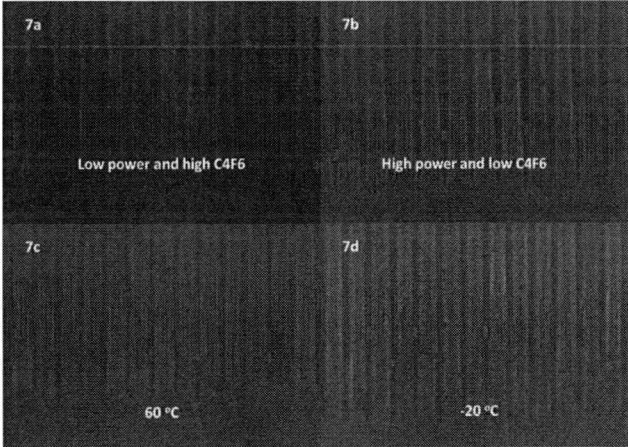

*Figure 7. (a) The SEM cross-section of T-SiN profile with (a) low power and high C4F6/C4F8 ratio, (b) high power and low C4F6/C4F8 ratio, (c) 60 °C and (d) -20 °C*

**Conclusions**

To meet the more and more strict requirement of capacitor etch in advanced DRAM, we propose four directions for optimizing process. Firstly, chamber pressure should be decreased to ~15mT when CD of capacitor hole is decreased below 30nm. It can help to pump out the residual byproduct and polymer while make less effect on mask selectivity and etchant density. Secondly, we should decrease DC and increase power to enlarge the process window of mask selectivity. The appropriate DC and bias power for >40 aspect-ratio capacitor etch are 15%~25% and 25kW~50kW, respectively. Thirdly, decreasing C4F6/C4F8 ratio is helpful to increase in-hole ER as well as mask selectivity, especially in <30nm capacitor etch. We also found a specific metal halide gas is very effective to remove missing holes and under etch. Finally, to straighten T-SiN profile and mitigate depth variation, which will be more important as aspect ratio is much increased, we suggest using higher power, lower C4F6 and lower temperature. All the strategies, provided here, can guide engineers to efficiently find process solutions when update DRAM generation. We will focus on mechanism analysis as we obtain the help from other institutes or colleges in the future.

**Acknowledgements**

The authors thank the cooperation between Chang Xin Memory Technologies Corporation (CXMT) and Advanced Micro-Equipment Corporation (AMEC). The authors acknowledge the helpful collaboration with the colleagues in the companies.

**Reference**

[1] Kotecki and E. David *Integr. Ferroelectrics,* vol. 16, 1997, pp. 1-19.

[2] J. M. Park, Y. S. Hwang, S.-W. Kim, S. Y. Han, J. S. Park, J. Kim, J. W. Seo, B. S. Kim, S. H. Shin, C. H. Cho, S. W. Nam, H. S. Hong, K. P. Lee, G. Y. Jin and E. S Jung *2015 IEEE International Electron Devices Meeting (IEDM)*, Washington, December 7-9, 2015, pp. 26.5.1-26.5.4.

[3] Y. Kim, S. Lee, T. Jung, B. Lee, N. Kwak and S. Park *Proceedings of SPIE 9428, Advanced Etch Technology for Nanopatterning IV*, March 17, 2015, 942806.

[4] D. J. Economou *J. Phys. D: Appl. Phys.*, vol. 47, 2014, 303001.

[5] W. W. Stoffels, E. Stoffels and K. Tachibana *J. Vac. Sci. Technol. A*, vol. 16, 1998, pp. 87-95.

# LOCALIZED PATTERN LOADING IMPROVEMENT OF SOC RECESS PROCESS FOR AIRGAP SPACER

*Bo Su[1], Abraham Yoo[1], Hansu Oh[1]*

[1] Semiconductor Manufacturing International Corporation, Shanghai 201203, China

*Corresponding Author's Email: Bo_Su2@smics.com

## ABSTRACT

Airgap spacer is one of the promising transistor performance knobs for next-generation logic semiconductor applications. Existence of airgap spacer between metal gate and source/drain contact leads to a further reduction of parasitic capacitances. Nevertheless, a well-controlled air gap shape with low-k material surrounded, which is mainly defined by integration scheme and dummy spacer removal step, is not trivial and represents the main realization challenge in the field. Here, a novel airgap spacer scheme featuring with SOC recess step is introduced. However, SOC recess process is still a challenge for manufacturing, due to SOC thickness variation between dense and isolate regions. In this investigation, local uniformity improvement of SOC remaining thickness is achieved via combined advances in two-step plasma recess and DEDE method (deposition-etch-deposition-etch method). It provides new insights into SOC recess process involved in airgap spacer formation scheme and offers ideal guideline and framework for improvement of local uniformity in SOC recess process.

## INTRODUCTION

Effect of parasitic capacitance is continuously increasing with scaling down in FinFETs, but also inevitable for Gate-All-Round (GAA) transistor. FinFETs integrated with airgap spacer are under the spotlight in recent years, which shows COAG integration friendly and 15% of effective capacitance reduction. [1,2] However, realization of such structure for IC volume production is still very challenging due to metal gate CD expansion and metal exposure during airgap spacer formation, consistent airgap spacer shape control, and EPI loss or missing defect. A promising approach for airgap spacer with fully surrounding low-k material is introduced to well-controlled airgap spacer shape and ensure sufficient equivalent k value reduction. SOC, which shows a good gap-filling property has been firstly studied in 1980s, and widely used in semiconductor manufacturing as planarization-less process. Furthermore, recent designed SOC can be used in the process under 400°C, which makes it possible to deposit dielectric material sequentially after SOC recess process. Accordingly, SOC recess is applied in our patented airgap spacer realization scheme as shown in Figure 1, in which high thermal resistance SOC material is chosen. Three concerns are addressed for SOC recess process: (1) local uniformity of remaining SOC thickness when reaching around 5 nm (SOC remaining thickness should be consistent in active area); (2) residual polymer on sidewall, which means sidewall polymer should be removed completely during SOC recess step; (3) process time of SOC recess should be acceptable (< 5 min), meanwhile, considering SOC recess uniformity.

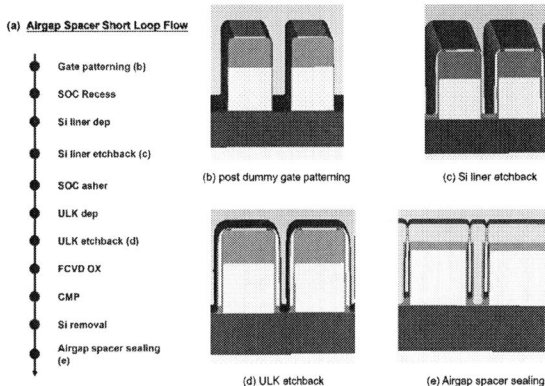

*Figure 1: Illustration of key steps for airgap spacer formation with L shape ULK surrounding via SOC recess. (a) airgap spacer short loop flow; (b)-(e) represent the illustration for step of post dummy gate patterning, Si liner etch back, ULK etch back and airgap spacer sealing, respectively.*

In this study, the effectiveness of optimized two-step plasma recess and DEDE (deposition-etch-deposition-etch) method is demonstrated by tailoring of plasma gases, pressure and bias on both blanket and patterned wafers. Local uniformity of SOC remaining thickness is improved from 10 nm to nearly zero loading using this approach. These results expand the knowledge on the improvement of SOC local uniformity allowing to further apply this process technique into FinFETs and GAA with integrated low-k surrounded airgap spacer scheme.

## RESULTS AND DISCUSSION

SOC plasma recess conditions are studied on 300 mm blanket Si wafer covered with 230 nm SOC, which is prepared using one-step spin-coating approach followed by 350°C baking condition. Gas combination of $H_2/N_2$, $H_2/Ar$, $O_2/Ar$, $H_2/He$ are chosen for SOC recess step, moreover, varied pressure and bias power are also studied.

The etch rates for different combinations are shown in Figure 2. Fastest etch rate (ER=32 A/s) is achieved in the condition of $O_2/Ar$ under 5 mTorr with bias. However, the uniformity shows the moderate performance (3sigma = 0.58 A/s). In contrast, $H_2/N_2$ without bias power shows the smallest etch rate (ER= 1 A/s) with 3sigma uniformities are 0.11 A/s and 0.08 A/s for 10 mTorr and 5 mTorr, respectively. In case of $H_2/N_2$, the pressure shows a different uniformity tuning trend between with bias and without bias power, in which relative lower pressure is beneficial for uniformity improvement.

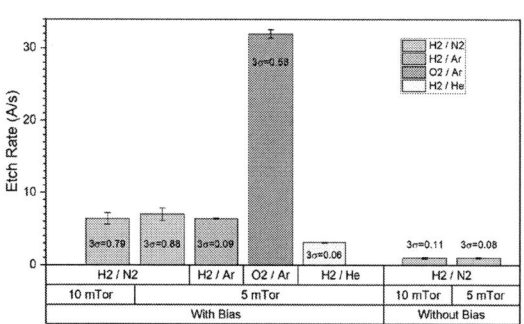

*Figure 2: Experimental conditions for SOC plasma dry etch on blanket wafer. Colum represent etch rate in each condition, and error bar represent 3sima value of etch rate within wafer.*

One-step SOC recess approach is investigated on a pattern wafer using $O_2/Ar$ with bias power under 5 mTorr as main etch step (ME). SEM result of edge (right) to center (left) patterns are shown in Figure 3a, which indicated clearly that the SOC localized loading is seriously exist by this one step etch. In particular, no SOC remaining on patterns on the edge area, while 9.8 nm and 6.6 nm SOC thickness are observed on patterns in center area and in-between, respectively. To improve local uniformity of SOC remaining thickness, we modulate SOC process protocols with multiple steps, which consists of ME step, $N_2/H_2$ at 10 mTorr as OE (w. or w.o. bias), and optional pull-back step. The pull-back step, using non-bias OE, is added to clean residual polymer byproduct on adjacent sidewalls. Nearly a similar range of SOC thickness is observed in pattern center area for both conditions. However, although we optimized SOC recess process we cannot obtain constant SOC thickness in different pattern area and no SOC left on pattern edge area is indicated in Figure 4c. Therefore, only dry etch recipe tuning is insufficient to balance the initial SOC thickness loading between isolated and dense regions after SOC spin-coating. SOC planarization is strongly depending on the pattern density difference across the designed structure based on experimental data and simulations.[3] This

loading can be minimized by choosing an appropriate etch gas source, in which more byproducts can be deposited on isolated area than dense area, however, it is still not sufficient. Other methods, such as changes of SOC material filling property, optimization of spin-coating process to increase planarization length, define a process sequence, need to be developed further.

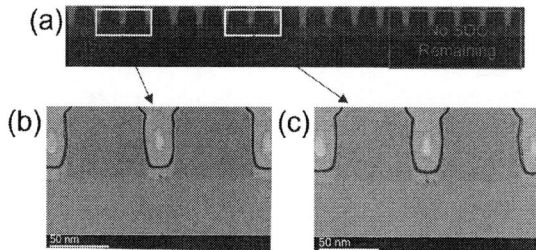

*Figure 3: Cross-sectional view of SOC recess on patterned wafer. (a) SEM result from pattern edge (right) to pattern center (left); (b) TEM result in pattern center area; (c) TEM result between pattern edge and pattern center.*

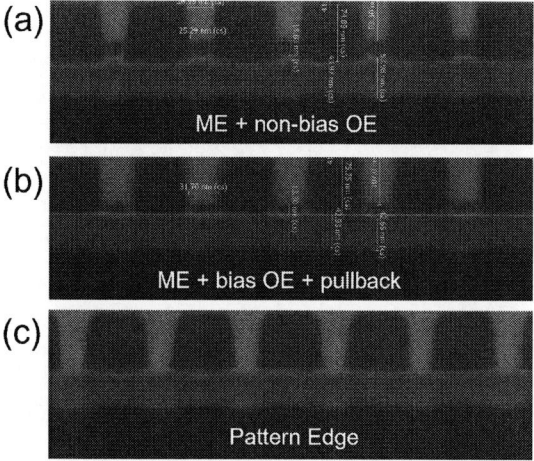

*Figure 4: Cross-sectional SEM view of SOC recess on patterned wafer using multiple steps for SOC recess. (a) SEM result in pattern center area using ME step followed by non-bias OE for SOC recess; (b) SEM result in pattern center area using ME step followed by bias OE and non bias pullback for SOC recess; (c) SEM result in pattern edge area.*

To demonstrate the optimized SOC dense-iso loading process, a new process scheme (see in Figure 5), which named DEDE method, has been proposed. SOC is firstly spin coated on patterned wafer, followed by ME step with fast etch rate recipe to stop on certain thickness above pattern top. The second SOC spin coating is applied followed by two-step SOC recess because the pattern area is not exposed before the second SOC coating step, hence the difference between dense and isolated area can be further minimized.

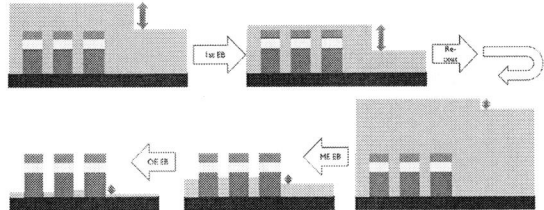

*Figure 5: Process flow of DEDE method for SOC recess on patterned wafer.*

Combining DEDE and two-step SOC etch back shows most potential method to reduce the remaining SOC thickness differnece between pattern center and edge as shown in Figure6. Constant SOC remaining thickness is confirmed by TEM image (see in Figure 6b).

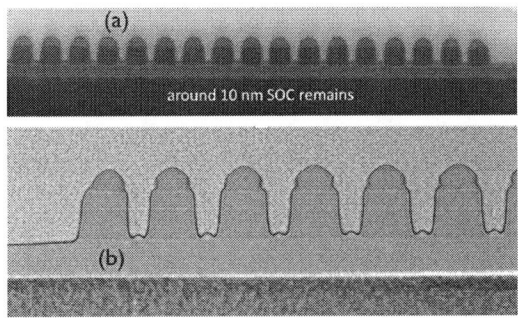

*Figure 6: Cross-sectional view of SOC recess on patterned wafer using DEDE method. (a) SEM results from pattern edge (right) to pattern center (left); (b) TEM result in pattern edge area.*

## CONCLUSION

In summary, the experimental study of remaining SOC thickness improvement via a DEDE method and two-step SOC recess process, have been demonstrated successfully. The initial SOC thickness loading after SOC spin-coating is benefited by DEDE method, furthermore, the optimized $H_2/N_2$ for ME and OE step can further improve the SOC thickness loading from pattern edge to center. This result shows a great potential for the realization of fully low-k surrounded airgap spacer structure using SOC recess scheme in FinFET and GAA architectures.

## ACKNOWLEDGEMENTS

This work was partially supported by "Shanghai Pujiang Program".

## REFERENCES

[1] K. Cheng et al., "Improved Air Spacer Co-Integrated with Self-Aligned Contact (SAC) and Contact over Active Gate (COAG) for Highly Scaled CMOS Technology," in 2020 IEEE Symposium on VLSI Technology, 16-19 June 2020, pp. 1-2.

[2] K. Cheng et al., "Air spacer for 10nm FinFET CMOS and beyond," in 2016 IEEE International Electron Devices Meeting (IEDM), 3-7 Dec. 2016, pp. 17.1.1-17.1.4.

[3] S. Decoster, X. Piao, W. Gillijns, and F. Lazzarino, "Modeling the topography of uneven substrates post spin-coating," Journal of Vacuum Science & Technology B, vol. 36, no. 3, p. 03E102, 2018.

# A STUDY OF WAFER CENTER VIA ETCH ARCING MECHANISM AND SOLUTION

*Jun Wang[1], Yuan Li[1], Xinruo Su[1]\**

[1] Semiconductor manufacturing North China Corporation

*Corresponding Author's Email: Jacob_su@smnchina.com

## ABSTRACT

In 40nm and sub-40nm manufacturing process, inter metal dielectric thickness shrink for process requirement, as a result, arcing defect on wafer during plasma etch start attract interests of researchers. Arcing defect, result from violent charge release process, occur on wafer edge usually. However, wafer center arcing is not a common phenomenon. In this paper, we report an arcing defect occur on wafer center and proposed a possible formation mechanism. Furthermore, a solution is also proposed and proven in both yield and reliability.

## INTRODUCTION

Arcing defect can be wildly detected in BEOL process with the application of CCP (Capacitance coupled plasma) chamber. Arcing defect can destroy an entire die by causing and open circuit in metal wiring. Moreover, other nearby die may also be impacted by sputtered copper residue or leakage path in a junction of a device.

The factors which will affect arcing defect can be summarized as below. Firstly, most severe arcing defect is triggered by unstable plasma. Unstable plasma can be detected with plasma strike step or unexpected metal exposure during process. These phenomena will induce on-wafer horizontal DC voltage and will trigger arcing on wafer. This type of arcing often results in a glitch of reflect power or Vdc on data log, which indicate the plasma stability.

Therefore, arcing defect of this kind mainly occurs on the edge of wafer, which result from wafer edge metal exposure issue. However, wafer center arcing in dielectric etch happened rarely. Moreover, the solution to this kind of arcing remains a challenge.

Besides, design of circuit also plays an important role in arcing defect endurance. In our previous experiences, different arcing defect performance is revealed in different design house while other conditions are same.

In this paper, a possible mechanism of wafer center arcing with solution in BEOL process is proposed..

## EXPERIMENT

All the wafers in this case were fabricated in a logic technology, the test and invention are studied on AMEC Capacitive Coupled Plasma (CCP) etcher for 12-inch dielectric etching. Arcing defect is characterized by a transmission electron micro scope, and defect scan is taken via dark field inspection (DFI) method.

## RESULTS AND DISCUSSION

During the BEOL via etch, high power steps are employed to achieve high aspect ratio profile, which may induce arcing easily. In all kind of arcing defect, metal ball defect with tail and arcing source is direct evidence of arcing process. Arcing is a violating process during plasma etch with sudden energy release in form of heat and micro-lighting. Arcing is micro explosion on wafer, which result in an arcing hole on wafer, as shown in Fig. 1c. Metal, however, will be melt in high temperature and have a tail head on arcing source. This phenomenon can help researchers judge the existence of arcing and locate arcing source.

In this case, an extraordinary arcing defect occurred during process. As shown in Fig. 1a, unlike traditional arcing defect, this kind of arcing defect is spotted on wafer middle instead of wafer edge. Another special point is that arcing source usually located on dense via boundary.

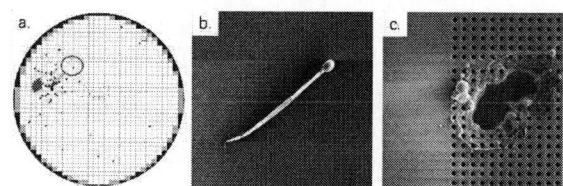

*Fig. 1 a) Wafer arcing map of wafer center/middle in via etch; b) typical defect morphology of arcing defect; c) arcing source of wafer center arcing;*

Further check show wafer center/middle arcing defect show no signal on process data log, which indicated that this kind of defect is micro-arcing.

An investigation of transmission electron morphology (TEM) is taken. As shown in Fig. 2, TEM image show that photo resist and bottom anti reflection coating remain is beneath metal ball detect and its thickness match with previous partial etch PR remain thickness. This result show that arcing defect occurred before photo resist strip step. Moreover, arcing defect mainly happened in primary metal layer such as V1 or V2 layer, after several metal layer stack, arcing ratio trend down rapidly and vanished finally.

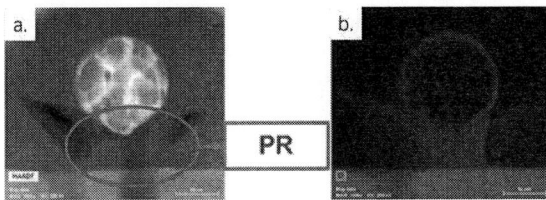

*Fig. 2 a) Transmission electron microscopy image of arcing defect and b) related energy mapping.*

In this paper, a possible mechanism is proposed. CCP chamber is considered as a combination of capacitor. Wafer system, wafer sheath formed by plasma and chamber capacitor is shown on simple schematic diagram. Plasma is formed by capacitive coupled plasma; thus it is supposed to be a RC delay circuit. The admittance of plasm is shown as below:

$$Y_p = j\omega C_0 + \frac{1}{j\omega L_p + R_p} \qquad (1)$$

In this equation, $C_0$ is capacitance in vacuum, $L_p$ is plasma inductance and $R_p$ is plasma resistance.

Here we proposed a hypothesis that the motion of electron is affected by instantaneous potential, thus the displacement current in plasma is much smaller than conduction current. This result mean that during plasma etch, the variance ratio of electric field, which result from displacement current is relatively small.

In sheath area, the situation become quite different. In plasma sheath area, electric field shift periodically, and the main current carrier, electron, can rarely spotted in sheath, which indicate a relatively low conduction current.

In our experiment, CCP chamber upper electrode is grounding, the electric field in wafer sheath ***a*** can be defined by Poisson equation:

$$dE/dx = en/\epsilon 0 \qquad (2)$$

this equation can be explain in integration:

$$E(x,t) = \frac{en}{\epsilon_0}[x - s_a(t)] \qquad (3)$$

Base on plasma theory, electric field is continual between plasma and sheath. In other words, no surface charge is spotted on interface. A boundary condition can be determined on this:

$$E \approx 0 \qquad (x = s_a) \qquad (4)$$

A displacement current in sheath a can also be calculated:

$$I_{ap}(t) = -enA\frac{ds_a}{dt} \qquad (5)$$

Where e is electron charge, n is ion density, A is area of plate condenser. In CCP chamber, a periodic radio frequency power is employed and sheath edge of sheath a will move periodically follow RF frequency.

In this situation, as conductance current is relatively low in sheath, $I_{ap}$ can be considered same with $I_{rf}$ and can be described as below:

$$I_{rf} = I\cos\omega t \qquad (6)$$

Where I represent periodic current peak value. Thus, equation (5) can be integrate as below:

$$s_a = \bar{s} - s_0\sin\omega t \qquad (7)$$

Here, $s_0$ is a constant which represent the vibration range of sheath edge. This result match with previous research that RF generate current can be divided into a direct current part and an alternative current part.

Obviously, the voltage drop between sheath can be calculated by simple integration:

$$V_{rf}(t) = \int_0^{sa} E\,dx = -\frac{en}{\epsilon_0}\frac{s_a^2}{2} \qquad (8)$$

This result show that sheath voltage drop is a nonlinear function if sheath current, and as a result, a quadratic harmonic waves exist in sheath voltage.

However, this model exists only in a symmetric and uniform plasma. In our process, plasma is symmetric but non-uniform. Because of this, voltage drop in plasma sheath become more complicated, especially in transition step. Base on this difference, unstable plasma in transition step will induce extremely high voltage drop on radio frequency loop both on vertical and horizontal direction of wafer surface.

In order to verify this model, an experiment is designed to change transition step to verify arcing performance. As plasma is unstable in transition step, voltage tend to change significantly with different transition step.

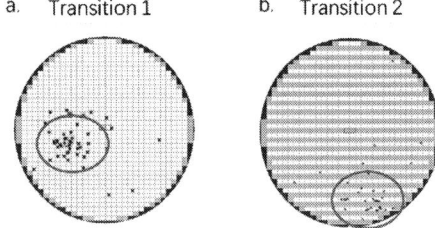

*Fig. 3 VxOX-ET arcing map of a) transition type1 and b) transition type 2.*

The inspection information supports this mechanism. After change transition step, arcing map change from wafer middle (Fig. 3a) to wafer edge (Fig. 3b), result from different voltage drop range in different transition step.

To solve wafer center and wafer middle arcing defect, a wider chamber gap is applied to improve plasma uniformity. To figure out the wafer center and wafer edge difference, rhetorical simulation and calculation is taken. Here we use electron density Ne to verify the difference between wafer center and wafer edge, as shown in Fig. 4. These data are simulated in same gas flow, temperature, pressure and power. According to Fig. 4, electron density range gap is $1.0e+17/m^3$ in base line gap chamber while wider chamber gap one is $0.5e+17/m^3$.

After apply wider gap chamber condition, arcing ratio in via etch process drop from ~1.0% to 0. This result indicated that wide chamber gap between upper electrode and wafer will lead to a more uniform plasma distribution and a possible mechanism of arcing defect on wafer center and wafer middle is proposed.

If wafer center and wafer edge plasma uniformity worse, sheath voltage drop gap between wafer center and wafer edge, which is quadratic to local sheath thickness, will become larger during transition step. This voltage gap will induce horizontal voltage difference on wafer surface and induce arcing in dense via edge area.

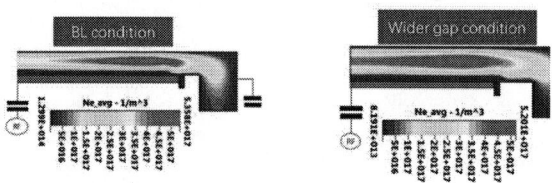

Fig. 4 chamber electron density distribution simulation in different chamber gap condition.

## CONCLUSIONS

In this paper, an extraordinary arcing defect in via etch process is described. Unlike traditional arcing, this kind of arcing is not related to metal exposure, and no signal can be detected according to data log. To investigate this arcing defect, a model of voltage drop in plasma sheath is setup and a correlation between voltage drop and sheath thickness is studied. Base on this theory, we proposed that this kind of arcing defect result from ununiformed plasma horizontal distribution on wafer surface. Experiment is taken and result show that change transition step will induce arcing position switch from wafer middle/center to wafer edge.

Base on this data, we proposed a solution of applied wider chamber gap to solve arcing defect. Simulation and calculation show that wider chamber gap leads to a plasma with better uniformity. Special test production is also taken on this condition and arcing ratio drop from ~1.0% to 0%.

## REFERENCES

[1] PoLi, Yung-ChengWang, Jing-WeiPeng, David WeiZhang et al, Impact of substrate resistance and layout on passivation etch-induced wafer arcing and reliability, *Microelectronics Reliability*, Volume 55, Issue6

[2] C. Mitterer, O. Heuzè, V. H. Derflinger et al, Substrate and coating damage by arcing during sputtering, Volume 89, Issue3

# TRENCH ETCH FOR SIC POWER DEVICES

*Xiaoyu Tan[1]\*, Qiushi Xie[1]\**

[1]Department of Semiconductor Etching, NAURA Technology Group Co., Ltd., Beijing 100176, China

*Corresponding Author's Email: xieqiushi@naura.com; tanxiaoyu@naura.com;*

## ABSTRACT

Nowadays, SiC MOSFET is widely believed to compete with Si-IGBT for its low energy loss during fast switching, in which a trench structure could play an important role to reduce the channel resistance. In this paper, we focus on ICP etch of SiC trench. Gas ratio, radio power and substrate temperature were found to have influence on SiC etch rate, selectivity to SiO2 mask and sidewall profile. Through delicate optimizing, top corner rounding and bottom corner rounding can be achieved in ICP process.

## INTRODUCTION

The climate goal of emission peak and carbon neutrality is boosting the rapid development of the renewable energy industry. SiC power devices, as an excellent means to improve energy efficiency ratio compared with devices based on Si, are being enthusiastically sought after by enterprises and governments. In order to achieve higher current in a limited chip area, SiC must be etched to form complex structures. Given the chemical inertness of SiC, wet etch must be conducted under high temperature over 600 ℃[1], or by photoelectrochemical etching at room temperature[2-4], and either can realize a good control of line-width.

Plasma etch has been widely used in SiC trench process. Given the comparatively low ionization degree in reactive ion etch (RIE), it hard to increase the etch rate over 500 nm/min[5], which is usually realized in an ICP chamber. Besides the high etch rate, a SiC trench profile including bottom rounding and top rounding is also important, which can reduce the electronic field squeezing induced device break-down. In certain cases, a SiC trench with high aspect ratio is desirable to improve the conductance, but it is hard to avoid the sub-trench at the bottom.

In this paper, we present different results from a series of ICP etch experiments, and finally achieve a SiC trench with a high etch rate, high aspect ratio and bottom rounding.

## EXPERIMENT METHODS

Samples to be etched were single crystal of N-type conductive 4H-SiC purchased from CENGOL, with a thickness of 350 μm. $SiO_2$ deposited by PECVD and opened with PR as mask in different pitch was used as

*Figure 1 SEM of the prepared $SiO_2$ mask for SiC etch*

hard mask. Figure 1 shows the SEM of one part of the prepared $SiO_2$ mask.

The etch process was conducted on GDE C200 from NAURA. The coil frequency is 13.56 MHz, and the sample is placed on electrostatic chuck with a backside helium system to control the surface temperature. Etch results were characterized by HITACHI UHR SU8030.

## RESULTS AND DISCUSSIONS

Figure 2 exhibits the SEM images of etch using $Ar/SF_6/O_2$, with a high $O_2/SF_6$ ratio under coil power of 1000W and bias power of 100W. As shown in Figure 2a, a sidewall bowing occurred on upper side of the trench. The aspect ratio is 1.9 if calculates from the widest trench width, and 2.8 for the top trench width. Figure 2b shows the trench with a lower aspect ratio of 0.9, and the sidewall bowing alleviates. A rounded bottom is obtained as shown in Figure 2a.

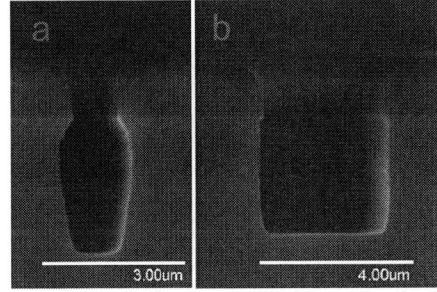

*Figure 2 SiC etched using $Ar/SF_6/O_2$, with a high $O_2/SF_6$ ratio.*

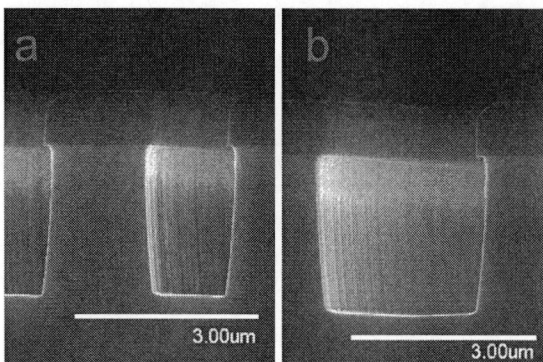

*Figure 3 SiC trench etched using Ar/SF6/O2, with a low O2/SF6 ratio condition.*

There is a change if we increase the $SF_6$ ratio, As can be seen in Figure 3. The sidewall bowing for both higher AR and lower AR is improved compared with those in high $O_2/SF_6$ ratio condition. The sidewall tapered for the trench with lower AR, and the bottom rounding disappeared for the trench with higher AR.

However, CD loss larger than 0.3μm is unacceptable in most SiC power devices, so passivation should be enhanced to minimize it. Under the high $O_2/SF_6$ ratio condition, the CD loss is quite small at the topmost side of the trench, and becomes larger as the depth goes deeper, as shown in Figure 2a. During the etch process, $O_2$ is believed to be possible to form $SiO_x$ by-product, which can redeposit on sidewall and depress the CD loss. The transportation of $O_2$ may more likely to be to hindered by the small opening of trench than $SF_6$, which lead to the different sidewall profiles in higher $O_2/SF_6$ ratio condition, and almost identical sidewall profiles in lower ones.

Based on this assumption, bias power was enlarged to increase the transportation ability for different gas under high $O_2/SF_6$ ratio condition, and Figure 4 shows the SEM images. The top-side of the trench is less bowed compared with the one under lower bias power condition, both for higher and lower AR trench.

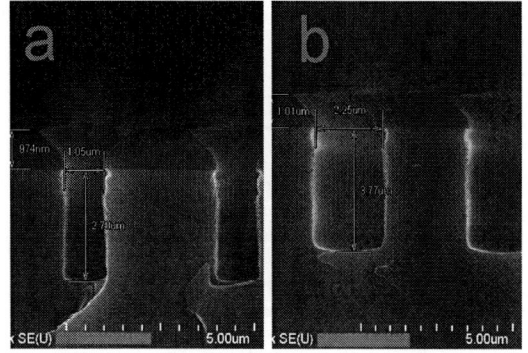

*Figure 4 SiC trench etched using Ar/SF$_6$/O$_2$, with a high O$_2$/SF$_6$ ratio condition and high bias power.*

The nearly identical profiles indicate insufficient

sidewall protection ability from $O_2$. Chlorinated plasma and brominated plasma has been reported for SiC etch[6]. The boiling point of the etch byproducts for chloride and bromide are higher than those of fluoride. Under this consideration, $BCl_3$ is introduced into the $Ar/SF_6/O_2$ system. $BCl_3$ has been used as sidewall protection gas in many dry etch processes, and B is a potential P type acceptor in SiC.

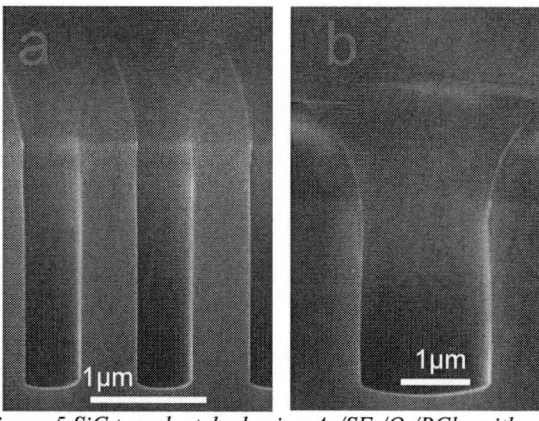

*Figure 5 SiC trench etched using Ar/SF$_6$/O$_2$/BCl$_3$, with a high bias power*

As shown in Figure 5, the sidewall has been protected well using $BCl_3$. Smooth sidewall without striation can be seen in both high AR (4.5) and low AR(1.5), as compared with those in Figure 3. A bottom rounding is inherited from condition in Figure 4. The $SiO_2$ mask on the top has been tapered under strong bombardment from high bias power, which can lead to a top-corner rounding with continued etching. The etch rate of SiC under this condition is 803 nm/min for trench CD of 0.5 μm, and 1065 nm/min for 2.0um. Selectivity of SiC to SiO2 is approximately 5 for trench in Figure 5b.

## CONCLUSIONS

In this paper, a series of experiments for SiC trench etch optimization have been presented. A SiC trench with AR of 4.5 and etch rate of 803 nm/min has been achieved, together with a rounded bottom corner. The etch method and structure may be very helpful in designing and manufacturing high-performance SiC power devices.

## REFERENCES

[1] J. W. Faust, Jr., in: The Etching of SiC, Eds. J. R. O'Connor and J. Smiltens, PergamonPress, London/Oxford 1960.

[2] J. S. Shor and R. M. Osgood, and A. D. Kutz, Appl. Phys. Lett. 60, 1001 (1992).

[3] J. S. Shor and A. D. Kutz, J. Electrochem. Soc. 141, 778 (1994).

[4] Y. Hibi, Y. Enomoto, K. Kikuchi, and N. Shikata, Appl. Phys. Lett. 66, 817 (1995).

[5] P. H. Yih, V. Saxena, and A. J. Steckl, phys. stat. sol. (b) 202, 605 (1997)

[6] N O V Plank，M. A. Blauw, E. W. J. M. van der Drift and R. Cheung, J. Phys. D: Appl. Phys. 36 (2003) 482–487

# REMOVING (SUB)SURFACE DEFECTS INDUCED BY SI WAFER THINNING PROCESSES ENABLES HIGH-PERFORMANCE BACKSCATTERED ELECTRON DETECTOR

*Zhu Chen[1], Lilei Hu[1,2] *, Li Zhang[2], Jingxuan Shen[1], and Chang Chen[1,2,3,4,5]*

[1]School of Microelectronics, Shanghai University, China
[2]Shanghai Industrial µTechnology Research Institute, China
[3]State Key Laboratory of Transducer Technology, Shanghai Institute of Microsystem and Information Technology Chinese Academy of Sciences, Shanghai, China
[4]Shanghai Academy of Experimental Medicine, Shanghai, China
[5]Shanghai Si-Gene Biotech Co., Ltd, Shanghai, China
*Corresponding Authors' Email: hulilei@shu.edu.cn, chang.chen@mail.sim.ac.cn

## ABSTRACT

High-performance backscattered electron detectors (BSDs) with high sensitivity and low dark current require a high-quality wafer substrate surface. Currently, chemical-mechanical polishing (CMP) has become an essential method to obtain a polished flat wafer surface. However, during a CMP process, wafers are subjected to various defect sources attributed to slurries, polishing pads, and polishing machines. Therefore, post-CMP cleaning of a wafer is crucial in fabricating high-performance BSD detectors. In this work, four post-CMP cleaning experiments were taken to identify different (sub)surface defects and to develop proper approaches for such defects' removal. Minority carrier lifetimes of Si wafer before and after various surface-defect removal treatments were measured and found to increase significantly, especially after the surface passivation using amorphous silicon, with an increase from 26 µs to 202 µs, indicating the successful recovery of Si wafer (sub)surface that was previously damaged by conventional wafer thinning processes.

## INTRODUCTION

Backscattered electron detectors (BSDs) are widely applied in electron microscopy including scanning electron microscopy and transmission electron microscopy. The structure of a BSD is typically P-I-N based, while various (sub)surface defects of Si substrates have significant influences on the BSD dark current. A high detector dark current, corresponding to a low signal-to-noise ratio, can deteriorate the detector image quality. The design of high-performance BSDs requires optimizing multiple detector chip parameters including substrate thickness, dopant concentration, surface defect, and detector sensitive area. To achieve the desired substrate thickness, CMP is generally used to realize a thin and surface-polished Si wafer. However, massive or slight (sub)surface defects like hillocks and grooves have been found after a CMP process [1].

Figure 1 shows the surface morphology of a Si wafer (N-type, resistance >10000 Ω•cm, double-polished), characterized by atomic force microscopy (AFM), before and after mechanical grinding and a CMP process. The wafer was mechanically ground from 725 µm to 450 µm, then, a standard CMP process was taken to flatten the surface. It can be seen from Fig. 1(a) that the wafer before CMP is relatively flat, homogeneous, and free of debris. In comparison, after CMP, even after an ultrasound-based cleaning, many debris and scratches from the CMP process attach onto the wafer, Fig. 1(b), resulting in a high-roughness surface. This rough surface morphology is always associated with plenty of (sub)surface defects, acting as recombination centers for charge carriers. Therefore, such wafers exhibit a lower surface carrier lifetime and lead to significant surface leakage currents if used in BSDs [2].

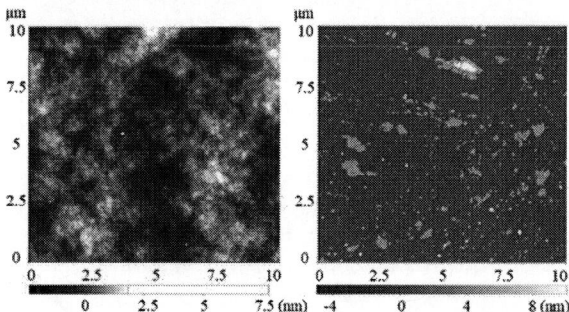

*Figure 1: AFM images of a Si wafer before (a) and after (b) mechanical grinding and a CMP process. The minority carrier lifetime measured for (a) is 40 µs and 26 µs for (b).*

## METHODS for DEFECT REMOVAL

Dark current in a BSD consists of diffusion current, recombination current, tunneling current, and surface leakage. Surface leakage current, especially induced during chip fabrication processes, is one of the main dark current sources. Addressing this, four approaches have been tried to recover the Si wafer surface and improve the minority carrier lifetime in this study: chemical wet etching, deep reactive ion etching (RIE), thermal annealing, and surface passivation with amorphous silicon (a-Si).

For a comprehensive study of the effects of defect-removal approaches on different Si wafers, here, two types of Si wafers were used with notations: Si-wafer A and Si-wafer B. Si-wafer A, purchased from ShinEtsu, is P-type, single-polished, with a resistance of 1~100 $\Omega \cdot$cm. Si-wafer B, purchased from OKMETIC, is N-type, double-polished, with a resistance > 10000 $\Omega \cdot$cm. Besides, this study also compared CMP processes carried out in a laboratory and a micro-fabrication line, and respectively, labeled as CMP_Lab and CMP_Fab through this paper. The CMP_Fab is processed at the 8-inch "More than Moore" R&D and Pilot Line in Shanghai Industrial µTechnology Research Institute (Sitri).

The chemical wet etching was carried out at 0 °C using HNO$_3$: HF: HAC = 50:1:20, and a thickness of 1-2 µm Si was etched away. Figure 2(a) shows that there are massive hillocks formed on the Si wafer surface after CMP_Lab, while hillock seems smoothened and the amount decreases after the chemical etching for 5 s, Fig. 2(b). The etching mechanism can be described by: $4HNO_3 + 2Si + 12HF \rightarrow 4NO \uparrow + 6H_2O + O_2 + 2H_2SiF_6$. Hydrofluoric nitric acetic (HNA) acid can dissolve damaged parts on the silicon wafer surface in an isotropic way, therefore repairing the silicon wafer. This process introduces hydrogen ions for the formation of H-terminated Si-H (a-Si: H/c-Si) surfaces, saturating the surface dangling bonds, therefore, reduce the surface recombination rate, leading to an increase of the minority carrier lifetime [3].

(Sub)Surface damages generally occur inevitably during mechanical grinding and CMP processes [4], for example, as shown in Fig. 2(c), the damaged Si wafer surface with dense speckle-like defects. After etching ~30 µm thickness of Si from the wafer surface through deep RIE, as shown in Fig. 2(d), fewer such speckle-like defects are observed when compared with Fig. 2(c), resulting in a relatively homogeneous surface.

Thermal annealing is generally helpful for wafer stress release and defect recovery, therefore, ultimately increasing the minority carrier lifetime [5]. The wafer was annealed for an hour at 1000 °C in a furnace filled with nitrogen. Fig. 2(e) shows the surface morphology of Si-wafer B before thermal annealing, where there are many sharp hillocks on the surface. However, these hillocks were smoothened significantly with reduced defects as shown in Fig. 2(f), indicating an increased minority carrier lifetime as discussed later.

As for a-Si based surface passivation, 1 nm Si was first removed from the silicon wafer surface through a wet etching process. Then, using chemical vapor deposition, an intrinsic Si layer of 10 nm was deposited and followed by a highly doped 20 nm P-type (or N-type) a-Si. The

whole process was completed with the above-discussed thermal annealing process. Through a-Si passivation, dangling bonds on the surface of the silicon wafer are saturated [6]. Figures 2 (g) and (h) exhibit the surface morphologies of N-type and P-type a-Si, respectively. Both Si wafers present a homogenous and flat surface without obvious debris or cracks.

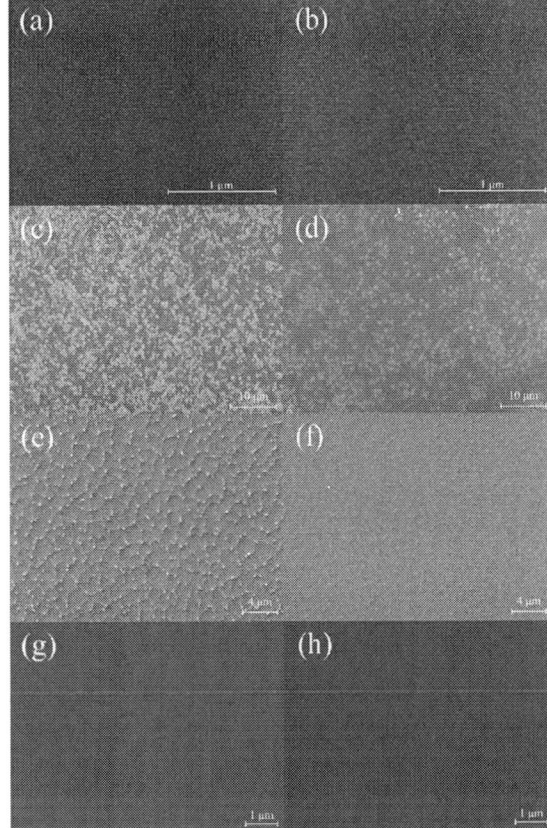

Figure 2: SEM images of Si wafers treated with different defect-removal methods: (a) Si-wafer A after mechanical thinning and CMP_Lab; (b) after 5-s chemical wet etching of (a); (c) Si-wafer A after mechanical thinning and CMP_Fab; (d) after deep RIE of (c); (e) Si-wafer A before thermal annealing; (f) after thermal annealing of (e); and surface morphologies of N-type (g) and P-type (h) a-Si deposited on Si-wafer B.

## CARRIER LIFETIME MEASUREMENTS

The minority carrier lifetimes were measured by Sinton WCT-120 Suns-Voc as shown in Table 1. For each sample, 5 points were measured, and the mean and standard deviation were taken for analysis. The minority carrier lifetime in Si-wafer A decreases from 32 ± 2 µs to <0.1 µs after mechanical thinning and CMP_Lab. In compassion, it improved to 110 ± 13 µs after the wet chemical etching of 5 s. But it should be noticed that when

978-1-6654-9759-6/22 $31.00 © 2022 IEEE        275

the wafer was etched for a longer time of 1 min, a much rough surface was obtained, leading to a significant reduction of minority carrier lifetime to <0.1 µs. The lifetime measured for Si-wafer A decreases to 29 ± 3 µs after CMP_Fab. After etching the Si surface of 30 µm, the carrier lifetime increases to 31 ± 2 µs. Additionally, it further increases 2-fold to 62 ± 4 µs after thermal annealing. Si-wafer B shows a minority carrier lifetime of 40 ± 1 µs, while after CMP_Fab, the lifetime decreases to 26 ± 2 µs. Remarkably, it increases to 202 ± 59 µs (166 ± 61 µs) after the N-type (P-type) α-Si passivation and thermal annealing.

TABLE I. MINORITY CARRIER LIFETIMES FOR SAMPLES AT DIFFERENT STAGES.

| Process | | MINORITY CARRIER LIFETIME | |
|---|---|---|---|
| | | *Wafer A* | *Wafer B* |
| Initial Sample | | 32 ± 2 µs | 40 ± 1 µs |
| CMP_Lab | | <0.1 µs / | |
| CMP_Fab | | 29 ± 3 µs | 26 ± 2 µs |
| Chemical Wet Etching | 5 s | 110 ± 13 µs | / | / |
| | 1 min | <0.1 µs | |
| Deep RIE of 30 µm | | 31 ± 2 µs | / |
| Thermal Annealing | | 62 ± 4 µs | / |
| α-Si Passivation with annealing | N-type | / / | 202 ± 59 µs |
| | P-type | | 166 ± 61 µs |

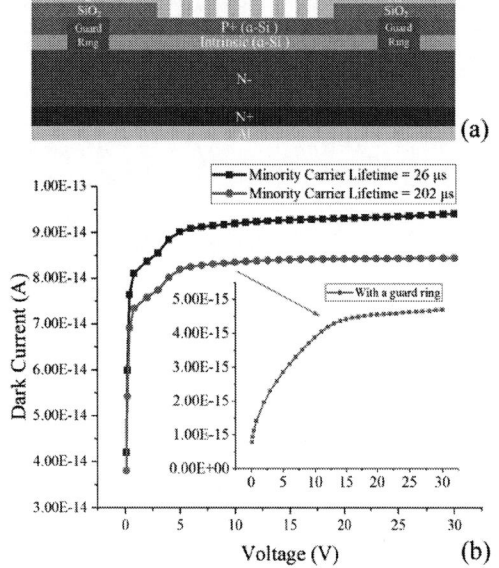

*simulated dark currents for a BSD (without a guard ring structure) with different minority carrier lifetimes; inset: with a guard ring.*

Based on the typical P-I-N structure, this study integrates the α-Si passivation method for (sub)surface defect-removal as well as a guard ring structure to extract the surface leakage current, therefore, a novel BSD structure was designed as shown in Fig. 3(a). Due to the high sheet resistance of 20 kΩ/□ measured for the P+ (α-Si) layer, a grid-shaped Aluminum (Al) top electrode is also designed to reduce the BSD series resistance that contributes to a short device response time. The simulated 2-dimensional detector is 870 µm in width and 300 µm in height. As shown in Fig. 3(b), when the surface minority carrier lifetime increases from 26 µs to 202 µs, the dark current decreases to as low as $8.4 \times 10^{-14}$ A. Using a guard ring, it can be further reduced to $4.69 \times 10^{-15}$ A, according to 0.53 nA/cm$^2$, Fig. 3(b) inset. The simulated BSD cut-off frequency is 6.8 MHz and the capacitance is 0.28 fF.

## CONCLUSIONS

Four (sub)surface defect-removal methods were studied. Chemical wet etching can significantly improve the wafer surface quality for a 5-s short time etching; however, further etching will damage the Si surface. Deep RIE shows negligible help to remove these post-CMP (sub)surface defects, while further thermal annealing can increase the carrier lifetime approx. 2-fold. In comparison, α-Si based passivation demonstrates a successful recovery of Si wafer surface with a prominent lifetime increase to more than 200 µs. Using the as-developed α-Si passivation, a BSD structure with a guard ring was designed and simulated to demonstrate a low detector dark current with high device performance.

## ACKNOWLEDGEMENTS

This work was sponsored by the Shanghai Rising-Star Program (21QB1405700), the Research Instrument and Equipment Development Project of the CAS (GJJSTD20210006), the National Key R&D Program of China (2020YFC0847600), and the Program of Shanghai Technology Innovation Center of IVD Chip (20DZ2220500).

## REFERENCES

[1] Chengqiang Feng et al., Materials Research Express, vol. 8(4), 2021-03-29.

[2] Ivin Varghese, Çetin Cetinkaya. *IEEE Transactions on Semiconductor Manufacturing*, 2012, vol.25, pp. 630-637.

[3] Kihyung Ko et al. *Microelectronic Engineering*, vol. 149, 2016, pp. 85-91,

[4] Weijia Guo et al. *Nanotechnology and Precision Engineering*, vol. 3, 2020, pp. 244-249.

*Figure 3: (a) A BSD structure designed with the integration of the as-developed α-Si passivation. (b) The*

[5] Rohana Perera et al. *Thin Solid Films*, vol. 423, 2003, pp. 212-217

[6] Zhuopeng Wu et al. *Applied Surface Science*, vol. 51, 2019, pp. 504-509

# ADVANCED RUTHENIUM SELECTIVE ETCH FOR MEMS AND SUB-3NM APPLICATIONS

*Chien-Pin Sherman Hsu*

Avantor

No. 38-1, Tai-Yuan St., Chu-Bei, Hsinchu, Taiwan

sherman.hsu@avantorsciences.com

## ABSTRACT

Ruthenium (Ru) etch mechanisms, challenges, and progress are reviewed. Conventional Ru etch chemistries are limited and have substrate compatibility or toxic-material-release concerns. Recently developed Ruthenium Selective Etch (RSE) etchants are oxidative alkaline-based chemistries. They provide tunable Ru selective etch capability for Ru film removal and Ru-containing residue cleans with broad substrate and metal compatibilities in advanced 2nm node applications. Efficient bulk Ru film stripping at micron-level thickness in Micro-Electro-Mechanical Systems (MEMS) device fabrication has also been demonstrated.

## INTRODUCTION

Ruthenium has emerged as an interconnect material to replace Cu in advanced technology nodes [1]. Ruthenium bulk film stripping may be needed in MEMS or other device fabrication. Selective Ru etch chemistries also find uses in Ru liner and Ru-containing residue removal. For fabrication of 3nm and below devices, many selective etch applications require tunable Ru etch vs. various metallization and substrates, including Cu, Al, AlOx, Co, TiN, SiGe, TEOS and SiNx.

This study describes the challenges and development of highly selective Ru etch chemistries with broad substrate and metal compatibilities.

## EXPERIMENTAL

Wet clean experiments were conducted in beakers and the SEZ (LAM) SP304 Single Spin Tool. Film thicknesses were measured with X-ray Fluorescence (XRF), ellipsometry or 4-point probe. Removal of PR films or residues was examined with optical microscopy or High-Resolution Scanning Electron Microscopy (HR-SEM).

## RESULTS AND DISCUSSION

### Ru Etch Challenges and 2nm Technology Solutions

Ru is a hard metal to be etched. There are only a few conventional Ru etchants, including cerium ammonium nitrate-nitric acid and periodic acid [2]. However, both have high metal content or limited compatibility with metals and substrates in advanced 5nm or below technology nodes. Recently, a new series of products based on a novel RSE chemistry has been reported [3,4,5].

RSE etch chemistry uses an alkaline oxidative chemistry to avoid toxic $RuO_4$ generation at acidic oxidative conditions and provides tunable Ru etch or clean capabilities. The proposed chemical reaction is:

$$Ru + OH^- + Oxidant \rightarrow HRuO_4^- \ (major) + HRuO_5^- + RuO_2 + By\text{-}product\ X\ (reduced\ oxidant) \qquad (1)$$

RSE chemistry is aqueous-based and can be made completely metal-free. Ru etch rates vary significantly depending on film deposition processes and annealing conditions. In general, RSE chemistries, such as RSE-1, 2 and 3, etch Ru, W, and Al efficiently, and are compatible with a wide range of metals and substrates, including TiN and AlOx (Table I).

For 2nm and below applications, new chemistries have been developed with both enhanced selectivity on thermally annealed Ru and better electromigration control (Table II). One particular RSE chemistry (RSE-8) demonstrated tunability for Ru-containing residue cleaning with the desired minimal Ru etch rates at lower temperatures (Figure 1).

### TABLE I
### ETCH RATES (Å/MIN) OF SELECTIVE Ru ETCHANTS – RSE-1, -2, and -3.

| Chemistry Description | T | Ru | W | Cu | Al | TiN |
|---|---|---|---|---|---|---|
| RSE-1 | 45°C | 105 | >50 | 0.5 | >900 | <1 |
| RSE-2 | 60°C | 70 | >70 | 0.1 | 2 | <1 |
| RSE-3 | 60°C | >5,000 | >70 | 3.4 | - | <1 |

### TABLE II. ETCH RATES (Å/MIN) OF SELECTIVE Ru ETCHANTS – RSE-7, -8, -9, and -10.

| Chemistry Description | Ru | W | Cu | Co | TiN |
|---|---|---|---|---|---|
| RSE-7, 45°C | 35 | >50 | 0.1 | 0.1 | <1 |
| RSE-8, 60°C | 63 | >70 | 0.4 | 0.6 | <1 |
| RSE-9, 60°C | >3,800 | >3,200 | 3.8 | 1.5 | 1 |
| RSE-10, 30°C | 330 | >70 | 1.9 | 0.7 | <1 |

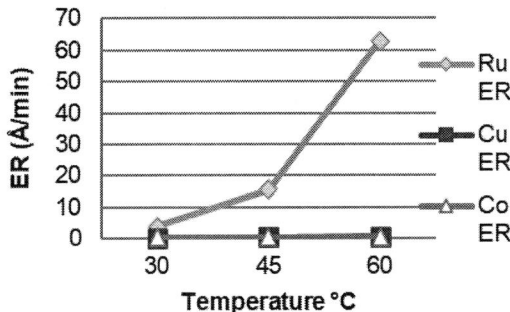

Figure 1: Effect of temperature on etch rates (ER) of RSE-8.

### MEMS Applications with Bulk Ru Removal

Selective Ru etch chemistries may also be used for bulk Ru film removal in MEMS device fabrication. The RSE-9 series chemistry with various pH values (RSE-9A to RSE-9E) at mild 30°C processing temperature shows dramatically increased Ru etch rates at lower pH (Figure 2).

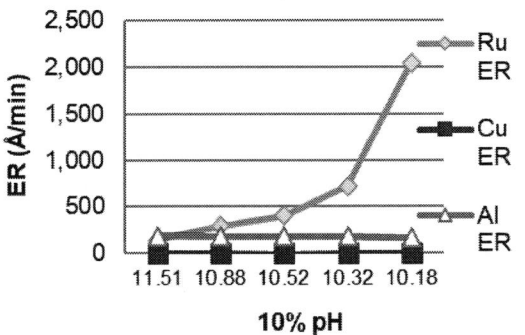

Figure 2: Effect of pH on etch rates (ER) of RSE-9 at 30°C.

One of the proposed mechanisms suggests such Ru etch rate increases come from an equilibrium between Ru etching anions at strong alkaline pH and increasing content of its conjugated acids at lower pH. This conjugated acid is a stronger oxidizer and exhibits greatly increased Ru etch rates even at catalytic concentrations or low amounts. Ru film of micron-level thickness may be removed within a few minutes with RSE-9 or RSE-10 types of chemistries (Figure 3).

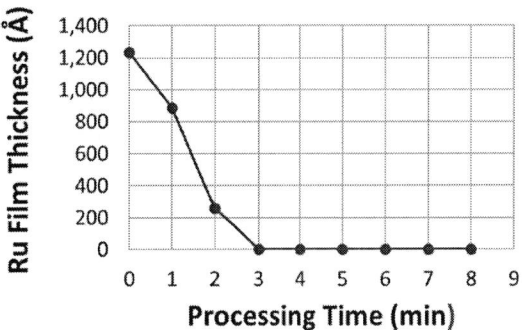

Figure 3: Bulk Ru film removal by RSE-10 at 30°C.

## ACKNOWLEDGEMENTS

The author would like to thank Yi-Ting Polly Chen for carrying out experiments and Dave Cresci for assistance in paper preparation.

## REFERENCES

[1] L. G. Wen et al. *ACS Appl. Mater. Interfaces,* 8(39), pp 26119-26125, 2016.
[2] H. Aoki, K. Watanabe, T. Iizuka, N. Ishikwa and K. Mori, *Extended Abstracts of the 2001 International Conferences on Solid State Devices and Materials,* Tokyo, 2001, pp. 20-21.
[3] C.S. Hsu and P.Y. Chen, *Solid State Phenomena,* Vol.282 (2018) pp.288-292.
[4] C.S. Hsu and P.Y. Chen, *Solid State Phenomena,* Vol.314 (2021) pp.307-311.
[5] Q.T. Le, E. Kesters, H. Philipsen and F. Holsteyns, *21st. Surface Preparation and Cleaning Conference (SPCC)*, Paper 01-06, April 2-3, 2019; Portland, OR, USA.

# TRI-LAYER MASK DRY ETCH PROCESS OPTIMIZING AND WET EFFECT FOR STRAIGHT PROFILE

*Xiaobing Liu\*, Shaojian Hu, Haihua Chen, Haojun Huang, YuShu Yang*

Shanghai IC R&D Center, Shanghai 201210, China
\*liuxiaobing@icrd.com.cn

## ABSTRACT

Photoresist/SiARC/Spin-on-carbon (SOC) trilayer mask is widely used in high density micro-fabrication with high NA ArF lithography. For most applications, trilayer mask is etched with polymer rich chemistry for CD shrinkage, in which taper profile is preferred. And also some special schemes use trilayer mask need straight SOC profile without necking near SiARC/SOC interface.

In this paper, we report a study of dry etch process optimization for straight profile and wet etch process was also introduced for necking improved near SiARC/SOC interface on the top of the via top. ICP process chamber with RF pulsing function is used for decoupled plasma density and ion bombardment energy control. Commonly used sidewall protection gas SO2, N2 and CH4 are added in O2 plasma for several split experiment conditions aim to sidewall passivation and bottom CD controlling capability study.

*Keywords—Trilayer mask; SiARC; Spin-on-carbon; straight profile;*

## INTRODUCTION

As IC densities increasing, the capability of directly pattern transfer by lithography is becoming less likely. A trilayer mask had been investigated as an alternative to single photoresist for CD and pitch shrinkage. In trilayer film stacks, photoresist used as etching mask for under layer pattern transfer, amount of pattern CD and pitch shrinkage controlled by etching process. Therefore, etch process are getting more and more important in IC fabrication, especially for advanced process below 14nm. Silicon anti-reflective coating (SiARC) and Spin-on-carbon (SOC) are widely used in trilayer patterning 193nm ArF immersion lithography and etching process also be studied in many articles [1][2]. The effect of trilayer mask profile was reported by Masatoshi Miyake et al. [3], in which the amorphous carbon layer (ACL) mask tapper angle was found having measurable relationship with bowing/necking ratio and bowing depth of SiO2 underlayer etching profile. Therefore, the controlling of trilayer mask SiARC/SOC profile is highly important for the following pattern transferring. In this article, SO2/O2/N2/CH4 plasma was studied for SOC profile controlling by gas species/ratio split and recipe structure design. WET clean process after dry etching was found helpful for necking improvement, which is beneficial for the following pattern transfer.

## EXPERMENT PROCEDURE

In this study HM918 (SiARC) and MH8102 (SOC) was used as trilayer mask with ArF photoresist patterning. Anchor point pattern is diameter 67nm dense via with 110nm pitch. We used an Inductive coupling plasma (ICP) chamber with 13.56MHz (Source RF) and Bias voltage (Bias RF) power in top and bottom of chamber. Inline critical dimension (CD) was measured in Hitachi CG6300 and SOC profile performance checked by TEM.

## RESULTS AND DISCUSSION

For trilayer mask etching, high SOC/SiARC etching selectivity process was required due to high SOC/SiARC thickness ratio. Polymer rich fluorine carbon etching gas was used for SiARC open for high SiARC/photoresist selectivity and SiARC profile controlling. In this study, 350A SiARC layer was etched with CF4/CHF3/N2 gas and RF power setting at 500W Source RF/50W Bias RF, which was for better photoresist striation. And, SO2 and O2 plasma with Helium dilute gas was used for SOC etching. Sulfur dioxide (SO2) was commonly used side wall passive gas in SOC etch due to sulfur contained polymer formation in plasma. J.K. Kim et al. reported sulfur transition broadband molecular peak and carbon sulfide peak at 257nm in carbonyl sulfide (COS)/O2 plasma optical emission spectroscopy curve. [4] Michel Pons et al. had given the formation mechanism of carbon sulfide in SO2/O2 plasma. [5]

Fig.1 (a) and (b) show cross-sectional TEM images of SOC profile with etching pressure of 8mT and 5mT. It is clear from these images that both conditions show necking and bowing profile, and bowing/necking ratio (BN ratio) is defined as max bowing CD divided by necking CD.

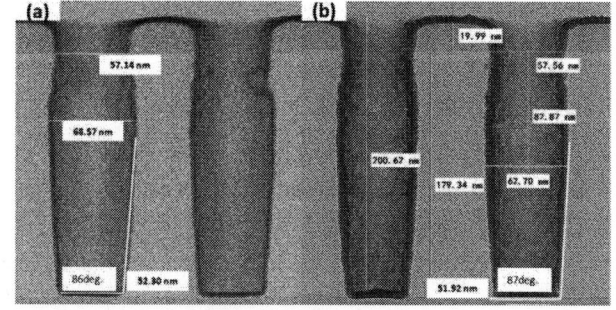

*Fig.1: Pressure effect for bowing/necking ratio in SO2/O2 plasma SOC etching*

The BN ratio is 1.2 for etching pressure of 8mT and improved to 1.1 when process pressure up to 5mT. Pressure always plays as main tuning knob for enchant density control and ions/free radicals residual time adjustment. Michel Pons et al. reported that sulfur had more efficient sidewall passivation in low-pressure [5]. For via etching application, ions mean free run path should be taken consideration. This is because that the incident ion flux had a Gaussian distribution [4], and that ions were reflected by sidewall, which induced the sidewall etch rate high.

It was also found that the BN ratio can be improved by wet process of DHF, which removed the damaged SiARC and polymer near the SiARC/SOC interface. After WET process, the BN ratio decreased from 1.2 to 1.0 showed in Fig. 2(a) and (b). It was induced by different CD changing in top/middle/bottom of the SOC profile. The top CD enlarged over 6nm in comparison with about 3nm and 1nm in middle and bottom, it was also proved that the byproduct polymer mainly deposited near the SiARC/SOC interface.

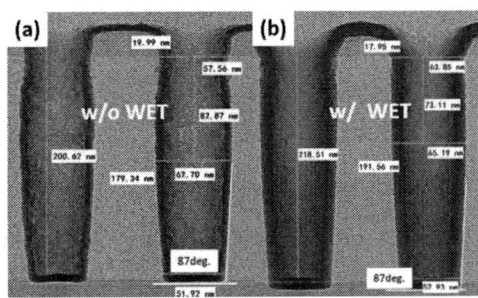

*Fig.2 :DHF wet effect for bowing/necking ratio, which was improved from 1.1 to 1.0 after wet process*

For sidewall protection enhancement, 10sccm and 40sccm CH4 was added into SO2/O2 (180sccm/60sccm) for more carbon-based polymer deposition. It's interesting to find that the condition with 10sccm CH4 had larger CD (Fig.3 (b)) than baseline (Fig.3 (a)) and BN ratio decreased slightly. When CH4 flow increased to 40sccm (Fig.3(c))，the middle CD keep at about 71nm, but top CD decreased to about 64nm,which means that the BN ratio had been got worse.

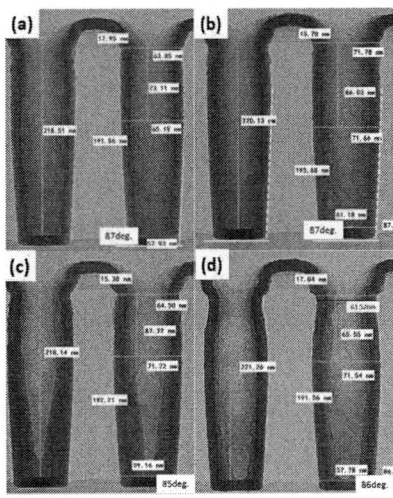

*Fig.3: Different SOC profile changing behavior in SO2/O2 plasma with additional CH4 and N2 gas. (All TEM with DHF wet process)*

In here, polymer gas CH4 showed obviously gas flow effect, this phenomenon may be induced by different dissociated product and recombination rate variation with the gas flow in CH4 plasma. [6] Its well-known that there will be C-N polymer deposition on sidewall when adding N2 into CH4 plasma. [7] When 20sccm O2 had been replaced by N2 with same flow in CH4/O2/SO2 plasma, obviously CD shrinkage was observed in top and bottom of the SOC via (Fig.3(d)).

Due to the polymer gas dissociated product type and charge shading effect, the polymer deposition was mainly on top of the profile. Therefore, it can be found that the top CD shrinkage and BN ratio increased when more polymer gas added into the plasma. Fig.4. shows the SOC profile (TEM) variation with the etching gas, and top CD (CDSEM) having the same trend.

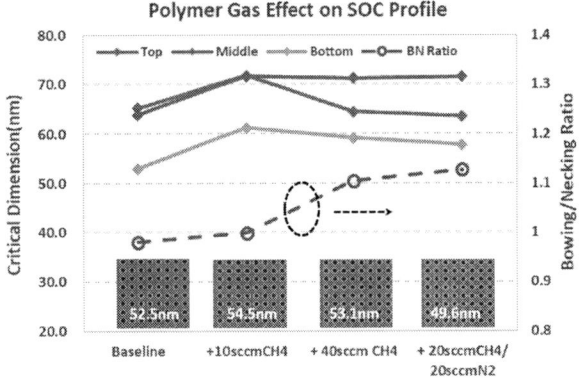

*Fig.4: SOC profile and top CD variation with etching gas*

RF pulsing is helpful changing polymer deposition position on sidewall due to the reduction of charge shading effect and reducing the etching byproduct residual time in structures. Fig.5 shows the comparison of the SOC profile etched with in SO2/N2 plasma with continue wave (CW) RF power (Fig. 5 (a)) and synchronic RF pulsing power (Fig.5(b)). After replaced all the O2 with N2, the SOC profile got bowing with top/bottom CD decreasing obviously. In contrast, when the sync. RF power

978-1-6654-9759-6/22 $31.00 © 2022 IEEE

(Source RF/Bias RF-1kHz, Duty cycle 50%) applied, the profile had been got straighter with top/bottom CD increased about 10nm.

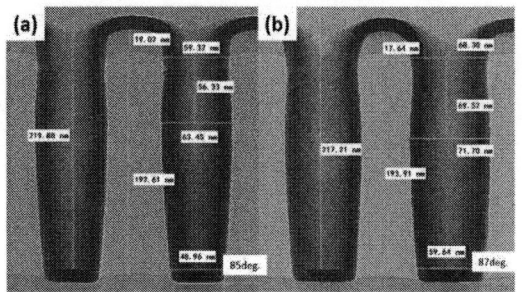

*Fig.5 :Sync. RF pulsing effect on SOC profile, (a) CW RF plasma/(b)sync. RF pulsing (TEM w/o DHF process)*

For some special application (ex. ion implantation mask etching) straight profile is required, in which the gap between middle and bottom CD should be well controlled. Besides the gas flow and RF power, suitable recipe structure design is also very important in etching process tuning. It had been found that the CH4 had gas flow effect for CD tuning in this study, and the profile etched with additional 10sccm CH4 showed CD enlargement obviously in Fig.3(b). Therefore, the SOC etching step was separated into two steps with same etching time. In Fig.6 (a) 10sccm CH4 was added in the first step and same flow additional CH4 only in the second etching process step of Fig.6(b), which showed about 4nm larger of via bottom CD. Two steps SOC etching recipe with CH4 in second step, which mainly etched the SOC near via bottom, has obviously bottom CD enlargement and straighter profile with sidewall angle of 87degree.

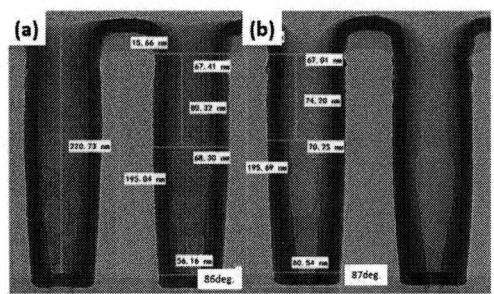

*Fig.6: Two steps SOC etching recipe with 10sccm CH4 in first step(a) and second step(b) (All TEM with DHF wet process)*

## CONCLUSIONS

This study has shown the interest of sidewall passivation by sulfur in the plasma etching of SiARC/SOC trilayer mask. By adding several polymer gases ex. CH4 and N2, it was found the conditions with additional CH4 had obviously flow effect and CH4/N2/SO2/O2 plasma had CN polymer mainly deposition on via top, which made bowing/necking worse. Synchronic RF pulsing power showed much uniform polymer deposition on sidewall, while CW RF power plasma deposited the polymer mainly on top and bottom of the profile. By separated SOC

etching into two steps with additional CH4 in second step, straight profile was observed. And it's very interesting to find that the bowing/necking ratio can be improved by DHF wet process, which removed the damaged SiARC and polymer near the SiARC/SOC interface, then made the profile straighter.

## REFERENCES

[1] Vladimir Bliznetsov, Kelvin Loh Wei Loong, Electron Devices Technology and Manufacturing Conference, pp.127-129, 2019.

[2] David J. Abdallah, Shinji Miyazaki and Aritaka Hishida, et al., Advances in Resist Materials and Processing Technology XXV, edited by Clifford L. Henderson, Proc. of SPIE Vol. 6923, 69230U, 2008.

[3] Masatoshi Miyake et al., *Jpn. J. Appl. Phys.* 48 08HE01,2009.

[4] Jong Kyu Kim, Sung Il Cho, Nam Gun Kim, Myung S. Jhon, Kyung Suk Min, Chan Kyu Kim, and Geun Young Yeom , J. Vac. Sci. Technol. A, Vol. 31, No. 2, 2013

[5] Michel Pons, Jacques Pelletier, and Olivier Joubert, Journal of Applied Physics 75, 4709 ,1994

[6] G Drabner, A Poppe, H Budzikiewicz, International Journal of Mass Spectrometry and Ion Processes, 97, 1990

[7] H Nakagawa, Y Morikawa, M Takano, E Tamaoka, T Hayashi, , Jpn. J. Appl. Phys. Vol. 41, pp. 5775–5781,2002

# EFFECTS OF ION INCIDENT ANGLES ON ETCHING MORPHOLOGY OF BLAZED GRATING BY IBE

*Jie Yuan[1], Xingyu Li[1], Zhongyuan Jiang[2], Yuxin Yang[2], Jiahe Li[2], Kaidong Xu[1,2*], Shiwei Zhuang[1*]*

[1] School of Physics and Electronic Engineering, Jiangsu Normal University, Xuzhou 221116, China
[2] Jiangsu Leuven Instruments Co. Ltd, Xuzhou 221300, China
*Corresponding Authors' Email: zhuangshiwei@jsnu.edu.cn & kaidong.xu@leuven-instruments.com

## ABSTRACT

The blazed grating plays an important role in advanced fields such as spectroscopy, AR, VR, etc. Therefore, a good morphology is critical for its applications. This paper mainly studies the effects of different ion incident angles on the morphology of the etched blazed gratings. The blazed gratings were fabricated by ion beam etching (IBE) and their morphologies were characterized by scanning electron microscopy (SEM). The results show that with the increase of the incident angle of the ion beam, the blazed angle first increased and then decreased, this is caused by different etching selectivities. Our study is of great significance for fabricating blazed gratings with high diffraction efficiency.

*Keywords—Ion beam etching; blazed grating; scanning electron microscopy; incident angle*

## INTRODUCTION

A blazed grating is a high-performance optical element with periodic relief structure. It has been widely used in measurement, display, communication, laser beam shaping and other fields. Among them, augmented reality (AR)[1] and virtual reality (VR)[2] are the fields of science and technology that have attracted much attention in recent years. Their near eye display systems project the pixels on the display to the human eye through a series of optical imaging elements to form a distant virtual image. In order to transmit the virtual image generated by the optical device to the human eye by the optical waveguide, there should be a process of optical coupling in and out of the waveguide[3]. In this process, blazed gratings play a critical role.

At present, the main methods for the fabrication of blazed gratings are mechanical etching, wet etching, electron beam exposure, X-ray lithography and ion beam etching[4]. However, the surface of blazed gratings fabricated by mechanical engraving is not smooth. The roughness is high and is time consuming. Due to the etching characteristics, wet etching is easy to form a platform on the top, which affects its diffraction efficiency. Electron beam exposure depends on the substrate material, and the surface of the blazed grating is rough and the etching morphology is not accurate. X-ray lithography is a high-quality method to fabricate blazed gratings, but there are few X-ray sources that can be used as the exposure light source, and the fabrication cost is high. Ion beam etching removes materials by physical sputtering[5,6]. Its working principle is to inject inert gas argon into the chamber, then energy is applied to ionize the argon into an ionic state. Next, voltage applied in the chamber makes the ionized ions move. Finally, the moving ionized ions are neutralized into neutral particles to etch the material. Because of its high resolution and excellent anisotropy, IBE is widely used in the field of microelectronics, which is a good choice to fabricate the blazed gratings[7-9].

In this paper, IBE was used to fabricate blazed gratings. The samples were etched by etched by argon gas at different incident angles. The morphology of the etched blazed grating was characterized by SEM. The blazed angles, etching rates (ER) and etching selectivities at different incident angles were analyzed.

## EXPERIMENTS

In this work, the samples used were 8-inch silicon substrates with PR masks on them. The period of the PR mask was ~500 nm with the linewidth of 180 nm and mask thickness of 320 nm. According to the calculation, the duty cycle was 36%. As shown in Figure 1, the yellow part represents the diagram of PR mask and silicon substrate. The dotted line indicates the incoming ion beam. The incident angle of the ion beam is the angle between the dotted line and the vertical direction. The ion incident angles were set at 30°, 35°, 40° and 45° respectively. The ion energy, ion beam current, acceleration voltage and the working pressure were set at 400 V, 510 mA, 80 V, 2.5× $10^{-4}$ Torr, respectively.

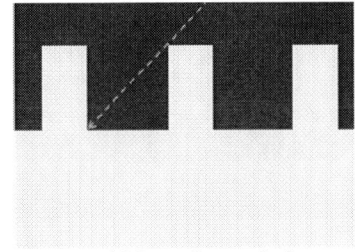

*Figure 1. Diagram of PR mask morphology.*

TABLE 1. EXPERIMENTAL PARAMETERS

| Parameter set | Incident angle(°) | Time(s) |
|---|---|---|
| A1 | 30 | 600 |
| A2 | 30 | 900 |
| B1 | 35 | 600 |
| B2 | 35 | 900 |
| C1 | 40 | 600 |
| C2 | 40 | 900 |
| D1 | 45 | 600 |
| D2 | 45 | 900 |

In order to study the morphology evolution of PR mask and silicon substrate during etching process, etching times were set at different times at different angles. Relevant experimental parameters are shown in the table (1) above. After different time etching, the samples were characterized by SEM. The following figure shows the SEM images when the etching time was 600 s and the ion beam incident angles were 30°, 35°, 40° and 45°.

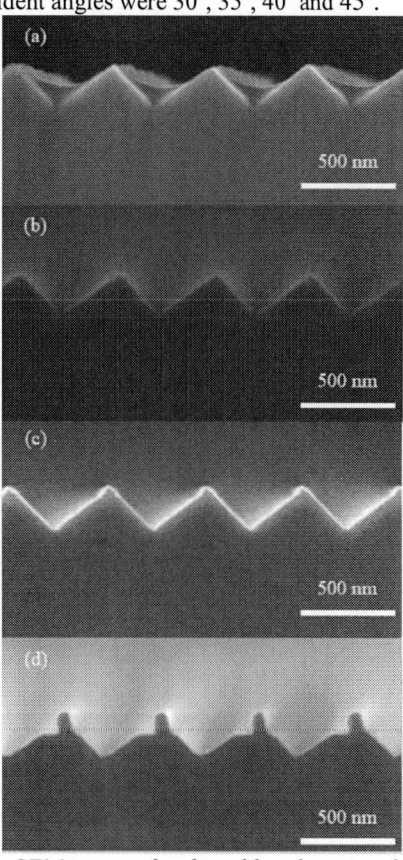

Figure 2. SEM image of etching blazed grating for 600 s and ion beam incident angle of (a)30°, (b)35°, (c)40°, (d)45°

As shown in Figure 2 (a) and (b) the PR mask on the silicon substrate has been removed completely and a triangular morphology similar to the blazed grating was formed. When the incident angle continues to increase, it is obvious that the PR mask was not removed completely, resulting in the morphology far away from the expectation, as shown in Figure 2 (c) and (d) above. This was caused by insufficient etching time. When the etching time was increased to 900 s, some of the situation above disappeared and the silicon substrate can be engraved into an excellent morphology. As shown in the Figure 3 below, when the etching time was increased to 900 s on the basis of Figure 2(d), the photoresist mask was removed completely, and the morphology of a blazed grating appeared.

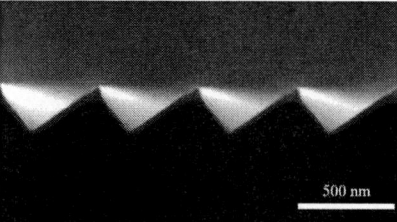

Figure 3. SEM image of etching blazed grating for 900 s and ion beam incident angle of 45°

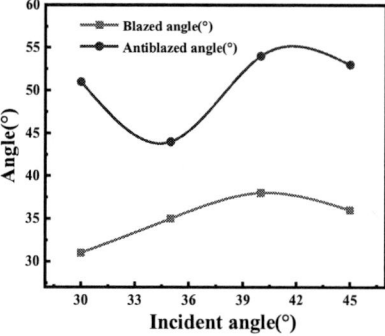

Figure 4. Blazed angles and anti-blazed angles at different incident angles.

Figure 4 shows blazed angles and anti-blazed angles at different incident angles in the case that etching time was 900 s. When the incident angle of ion beam increased from 30° to 45°, the blazed angle increases first and then decreases, while the anti-blazed angle decreases first and then increases. The fundamental reason for this phenomenon is that when different ion beams incident at different angles, the ER of silicon substrate and PR are different, resulting in different selectivities. As shown in Figure 5, the range of ion beam incidence angles were from 30° to 45°, the ER of silicon and photoresist decreased with the increase of ion beam incidence angles. In general, the ER of photoresist was much higher than silicon. This phenomenon shows that in the etching process, the etching amount of photoresist part is more than that of silicon part. As a result, the inclination of the etching slope of the photoresist part is larger and the silicon part is smaller. In the next etching, the two angles will be neutralized to form the final blazed angle.

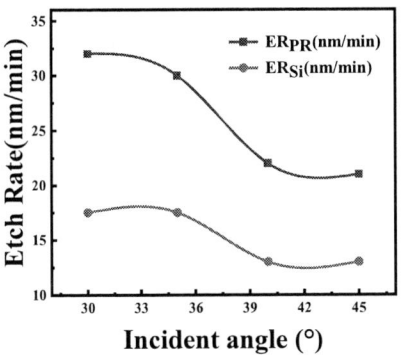

*Figure 5. $ER_{PR}$ and $ER_{Si}$ at different incident angles.*

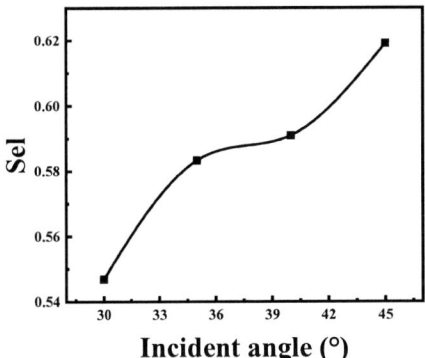

*Figure 6. Selectivities at different incident angles.*

The etching selectivity represents the ratio of the ER of silicon to photoresist, namely $ER_{Si} / ER_{PR}$. It can be seen from Figure 6 that with the increase of ion beam incidence angle, the etching selectivity was also increasing. In the etching process, the formation of blazed angle is related to the incident angle, and the incident angle and grazing incident angle are complementary to each other (The grazing incidence angle represents the angle between the ion beam and the horizontal plane). Theoretically, when a particle enters at a certain angle, the corresponding blazed angle should be consistent with its grazing incidence angle. Because of the increase of ion beam incidence angle, the grazing incidence angle decreases, but the etching selectivity increases. Therefore, as the grazing incidence angle decreases, the difference between the blazed angle and the grazing incidence angle decreases. Therefore, in a certain range, the blazed angle increases with the increase of ion beam incident angle, but the increasing speed decreases gradually. If the grazing incidence angle decreases beyond a certain range, the blazed angle may even decrease, as shown in Figure 4.

The formation of anti-blazed angle is related not only to the ER, but also to the direction of anti-blazed angle. In etching, the anti-blazed surface is opposite to the incident direction of ion beam, and the anti-blazed surface will be etched and deposited at the same time. Therefore, under the dual action of etching and deposition, the formation process of anti-blaze angle is more complex, which needs more experiments to explore.

In general, etching selectivities are significant in the etching process of blazed grating, which can affect the blazed angle to a certain extent.

## CONCLUSION

In this paper, the effects of ion incident angles on etching morphology of blazed gratings by IBE were studied. It found that the variation trend of blazed angles were related to the incident angles of ion beam, and the changes of ER were related to selectivities. Besides, anti-blazed angles also play a key role in the application of blazed gratings. The results provide reference suggestions for the fabrication of blazed gratings.

## ACKNOWLEDGEMENTS

This work was supported by the Key Projects of Ministry of Science and Technology of the People's Republic of China (Grant No. SQ2020YFF0407077), , the Industry-University-Research Cooperation Project of Jiangsu Province (Grant No. BY2020462), Postgraduate Research & Practice Innovation Program of Jiangsu Province (Grant No. 2021XKT1248).

## REFERENCES

[1] M. Zyda, *J. Computer.*, vol. 38, 2005, pp. 25-32.
[2] AS. Merians, D. Jack, R. Boiau, M. Tremaine, CG. Burdea, SV. Adamovich. *J.Physical Therapy.*, vol. 82, 2002, pp. 898-915.
[3] L. Minghuan, F. Xiuhua, W. Fei, *J. Chinese Journal of Liquid Crystals and Displays.*, vol.3, 2021, pp.389-397.
[4] G Jian, C Peng, W Lei, *J. Journal of Physics D: Applied Physics.*, vol.54, 2021.pp. 31.
[5] S. Bin, X. Xiangdong. J. Optics and Precision Engineering., vol.18, 2010, pp.94-99.
[6] T. Lei, L. Wenhao, Q. Xiangdong. *J. Optics and Precision Engineering.*, vol.18, 2010, pp.1536-1542.
[7] H. Xiang, C. Chao, Z Heng. *J. Optical Materials.*, vol.116, 2021. pp.111096.
[8] Z. Faguo, Y. Honglin, *J. Optics Express.*, vol.1, 2008, pp.48-51.
[9] S Reyntjens, R Puers. J. *Journal of Micromechanics and Microengineering.*, vol.11, 2001, pp.287-300.

# INVESTIGATION OF FIN BOWING FORMATION MECHANISM DURING STI ETCHING BY VIRTUAL FABRICATION

Lifei Sun*, Pengfei Lyu, Qingpeng Wang, Qinghua Zhong, Kui Wang, Yushan Chi

Lam Research Service Co., Ltd., Shanghai 201210, China
*Corresponding Author's Email: Lifei.Sun@lamresearch.com

## ABSTRACT

Shallow trench isolation (STI) etch is the process used to form fin channel structures in logic FinFET devices. Fin bowing is a common and key challenge during the STI etch process, especially for advanced nodes with higher aspect ratio. In this study, a virtual STI etch process was established and calibrated using the SEMulator3D® platform by Coventor Inc. A virtual design of experiments (DOE) revealed that ion angular distribution function (IADF) and hard mask (HM) profile taper are two key factors that induce a bowed fin profile. Wide ion angle distribution causes direct ion impingement at the sidewall, and a tapered HM profile will create ion scattering that bombards the adjacent fin sidewall. Both mechanisms result in fin bowing. In this study, we explored process trends and sensitivities of IADF, HM angle, HM height and space CD on fin bowing. Our results provide guidance on fin bowing profile reduction optimization, including enhanced sidewall passivation and implications on etch tool and hardware selection that impact ion distribution angle control and HM profile optimization.

*Keywords—Fin bowing; SEMulator3D; IADF; Hard mask profile*

## INTRODUCTION

Fin and shallow trench isolation (STI) etch are key processes used to fabricate Fin channels and prevent electrical current leakage between devices (Fig.1). As FinFET device scaling continues in more advanced nodes, fin pitch and fin width are further reduced and fin channels must be taller and more vertical. Fin profile control has become an increasingly challenging task, with smaller fin CDs and higher aspect ratio etch requirements. For example, at the 7 nm node the fin pitch is 30-33nm and the fin height is approximately 50nm[1]. The key challenge in fin profile control is in the vertical direction, where there is a tradeoff between taper and bowing profile (Fig. 2a). During the fabrication of an extremely vertical fin (>89°), it is very easy to create a bowed fin profile (Fig. 2b) [1,2], which affects both yield and device performance.

In this work, we will explore the mechanisms that contribute to fin bowing profile formation by executing a virtual design of experiment (DOE) and virtual wafer fabrication using SEMulator3D®. The impacts of ion angular distribution, hard mask (HM) profile and space CD are investigated, to provide guidance for integration optimization and process control.

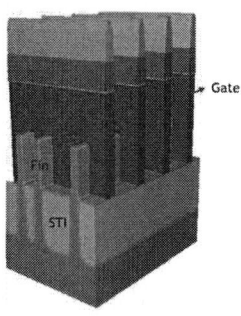

Fig.1. 3D schematic of FinFET structure.

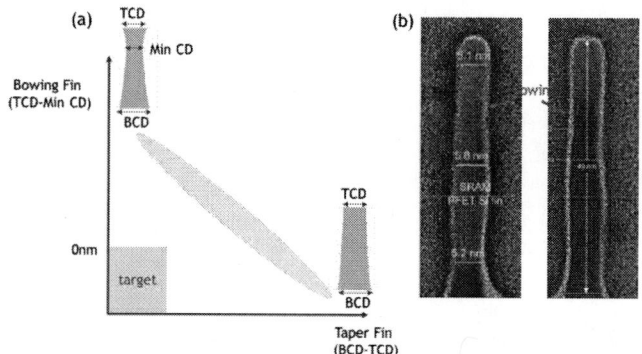

Fig. 2. Key challenges of fin channel etch. (a) Trade-off between fin bowing and taper profile control; (b) A typical TEM image of bowing seen on a 7nm fin profile (Courtesy from Techinsights) [1,2].

## Experiment description

In our experiment, the fin etch process was simulated using SEMulator3D® virtual fabrication software. We established our model in the software, and subsequently calibrated the model using actual silicon wafer data. To explore the impact of ion angular distribution function (IADF) and HM profile, we varied IADF, HM profile angle, HM height and fin-to-fin space CD in our experiment. Only one parameter was modified at a time, with other parameters kept constant. Looking at actual silicon wafer data, we noted that fin bowing became apparent when a fin was partially etched to half-depth. In this paper, to simplify the DOE process and to avoid the complications of second half STI etch processes on the fin bowing profile, the fins in our virtual experiment were only etched to half-depth.

The fin profile curves were extracted and plotted for comparison purposes, to show the process trend (Fig. 4 and 5). The minimum fin CD and position can be identified and extracted, with fin bowing CD defined as the difference between the fin top CD and bowing CD.

## Result and Analysis

IADF is generated due to ion transport from plasma to wafer surface through the plasma sheath. IADF typically depends on the pressure, bias voltage and RF frequency of the etch tool. An IADF with a finite width means that there are ions bombarding on to the substrate at an angle that generates direct ion impingement on the sidewall, resulting in a sidewall bowing profile. An ideal $90^{\circ}$ ion angle is not possible, but certain RF system functions on the etch tool can lead to tighter ion angle and energy control. In more advanced etch hardware, IADF can be tightly controlled to reduce sidewall impingement. The IADF induced bowing profile is more likely to be exhibited at the middle of the FinFET channel in areas with a relatively larger radius of curvature (Fig.3a).

HM profile is another key factor in fin profile control. To achieve more vertical fin profile, high energy ions with tighter IADF are typically used during the fin etch process. These ions may be reflected on the tapered hard mask, bounce to the fin on the other side and etch away material on the sidewall, resulting in a "necking" profile (Fig.3b).

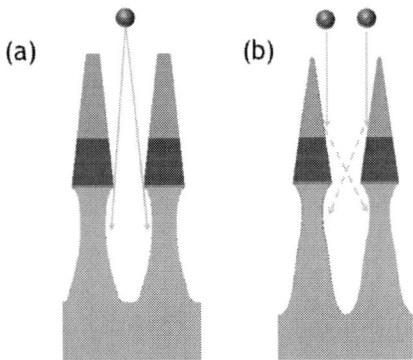

Fig. 3. Ion induced fin bowing profile: (a) Schematic of IADF induced direct ion impingement on Fin sidewall; (b) Schematic of ion scattered off tapered hard mask, resulting in fin sidewall pullback.

To further understand the fin etch process window, a DOE was executed to explore the impact of IADF on fin bowing profile. As shown in Fig. 4, a wider IADF will lead to a larger bowing CD and shallower trench depth. In addition to modifying the bowing profile, a smaller IADF may also induce micro-trenching (Fig. 4a) with the chosen set of simulation parameters. Finally, the bowing position will shift down as the IADF becomes wider.

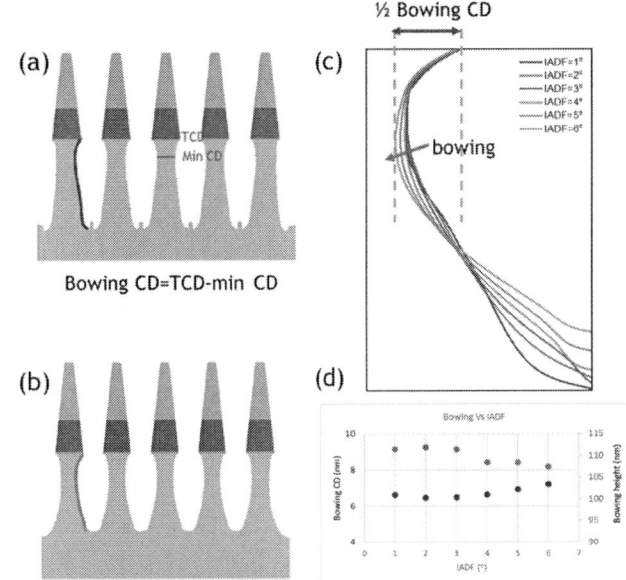

Fig. 4. IADF induced fin bowing profile. (a)-(b) Schematic of fin etch profile with (a) IADF= 1° and (b) IADF= 6°. (c) Extracted fin profile to show impact of IADF on fin bowing profile. (d) Plot of fin bowing CD and position height with different IADF values.

Next, we evaluated the impact of HM profile and space CD on the etch process window. HM profile angle and HM height impact are treated as separate HM profile variables in the virtual experiment. As shown in Fig. 5, the fin bowing profile degrades with a larger taper HM profile, higher HM and smaller fin-to-fin space CD. The fin bowing profile shows highest sensitivity to HM taper angle. When the HM taper angle is decreased from 89° to 82°, the bowing CD increased from 2.5nm to 7.8nm and the bowing position is shifted up (except at 89°). For HM height, the process trend was also clear and linear. A higher HM will induce a larger bowing profile with the bowing position shifting up. In contrast, the fin bowing profile displays smaller sensitivity to fin-to-fin space CD. The bowing CD becomes flat when the space CD is less than 10nm, and the bowing position remains stable for different space CDs. In summary, a vertical HM is needed to reduce the ion scattering and mitigate fin bowing. A higher HM and smaller space CD will enlarge the pattern aspect ratio during etch, increase the ion scattering rate and enlarge the fin bowing.

In addition to optimizing the HM profile and IADF, enhanced sidewall passivation was introduced to mitigate the fin bowing profile. As shown in Fig. 6, when the polymerized gas was added, the fin bowing profile turned to be less bowing and with an optimized sidewall passivation, we can finally deliver a vertical fin profile.

978-1-6654-9759-6/22 $31.00 © 2022 IEEE

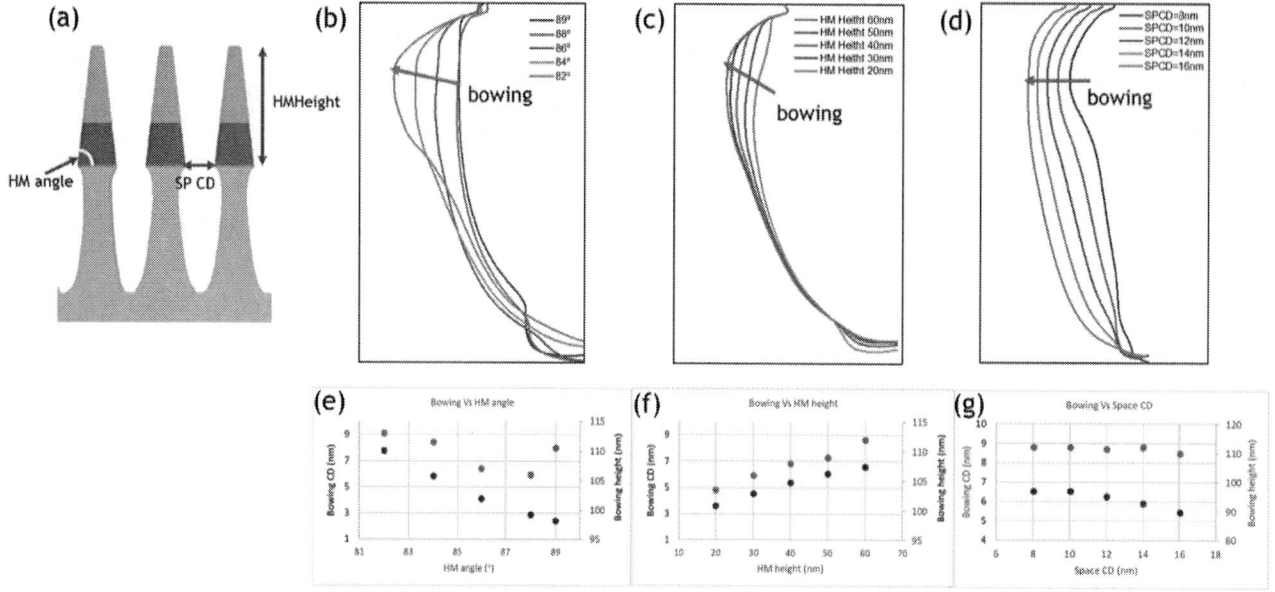

Fig.5. Impacts of HM profile on fin bowing profile. (a) Schematic of a typical HM profile and parameters definition. (b)-(d) Extracted fin profile to show impact of (b) HM angle, (c) HM height and (d) space CD on fin bowing profile. (e)-(g) Plots of fin bowing CD and position height with different (e) HM angle, (f) HM height and (g) space CD.

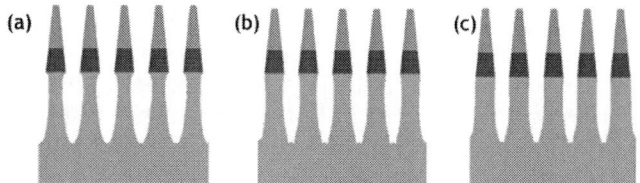

Fig. 6. Enhanced sidewall passivation helps to improve fin profile. Schematics of (a) BSL fin bowing profile; (b) improved fin profile with slight sidewall passivation and (c) vertical fin profile with enhanced sidewall passivation.

## REFERENCES

[1] TechInsights TSMC 7FF FinFET teardown report.
[2] TechInsights TSMC N7P FinFET teardown report.

## CONCLUSION

In conclusion, a fin etch process was simulated and a virtual DOE was conducted using the SEMulator3D® platform. This process modeling exercise strengthened our understanding of the causes of fin bowing profile formation, along with process trends. Our study revealed that a wide ion angle and ion scattering at the taper HM are two of the main factors that create an undesired bowing profile. A DOE based on IADF, HM angle, HM height and space CD was explored in this study, and revealed that creating a narrow ion angle distribution, using a vertical HM (with a smaller aspect ratio) and introducing enhanced sidewall protection are techniques that can mitigate the fin bowing profile. This study highlights a potential direction for etch tool and process tuning improvements to minimize fin bowing. The study was useful not only in understanding bowing profile formation mechanism, but demonstrates the time and cost-saving value of SEMulator3D in process trend exploration studies.

# PSR SILICON TRENCH PROFILE OPTIMIZATION IN FINFET FABRICATION

*Zhengning Li[1], Xing Ke[1], Changcheng Jiang[1], Fengmei Li[1], Yangkui Lin[1], Haiyang Zhang[1]\**
[1]Semiconductor Manufacturing International Corporation, Shanghai 201203, China
*Corresponding Author's Email: Steven_Z@smics.com

## ABSTRACT

In this work a method of PMOS Silicon Recess (PSR) trench profile optimization using silicon oxidation is proposed. TCAD simulations shows that rectangular trench profile is better than two types of U-shaped profile on device performance, such as S/D channel stress 6.2% and 9.2% gaining respectively and Vt 3.6% and 7.1% gaining respectively. Based on different oxidation rates on crystalline faces between 700℃ to 1000℃, a silicon oxidation recess (SOR) flow is designed in order to get more rectangular trench profile. Transmission electron microscope (TEM) images shows the SOR method is effective for trench profile optimization, in which anchor pattern 1 and 2 are improved by 69.4% and 46.7% respectively.

## INTRODUCTION

FinFET architecture is applied in sub-28nm semiconductor manufacturing for superior device performance, compared with planar MOSFET, in which FinFET structure has excellent gate control ability and S/D channel carrier density [1]. SiGe epitaxy is the main source for PMOS S/D channel stress. And both PMOS silicon recess (PSR) trench profile and SiGe epitaxy density contributes to PMOS S/D channel compressive stress. Ge density variation control in SiGe epitaxy for trench stress enhancement mainly relies on S/D trench profile, in which it is the space for SiGe epitaxy [2-3]. In that case, PSR trench profile control can be a prior choice for S/D channel stress improvement. According to current publications' report, PSR trench profile is mostly U-shaped [4-5].

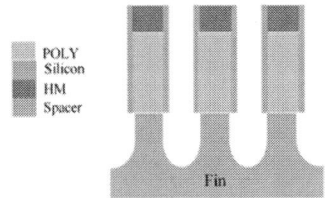

*Figure                                                                    1:*
*FinFET PSR trench profile (cut on fin)*

## METHOD AND CHACTERIZATION

U-shaped trench profile of PSR is determined by directionality of plasma during etching process. Because it is hard to reach the silicon at trench bottom corner for plasma dry etching; however, for wet etch, a round trench bottom corner is easily formed due to surface tension of liquid chemicals, especially if etched materials are not uniform and trench AR (aspect ratio) is small. As shown in Figure 1, PSR trench aspect ratio (AR) calculation includes the dummy gate height, and recessed fin trench is wider than polysilicon dummy gate space due to fin silicon isotropic etching beneath the dummy gate. PSR trench profile can be tuned during etching for device performance optimization. On the one hand, the width of S/D channel is reduced for FinFET on/off capability concern. On the other hand, PSR trench stress is improved for Vt reduction, because work function metal gate needs lower threshold value for chip energy efficiency [6].

### TCAD simulation analysis

Technology computer aided design (TCAD) Synopsys® is a commonly used platform for chip device characteristics simulation. Three types of trench profile U-shaped trench, optimized U-shaped trench and rectangular trench, are set up for TCAD simulation on device-related performance, including trench stress and Vt.

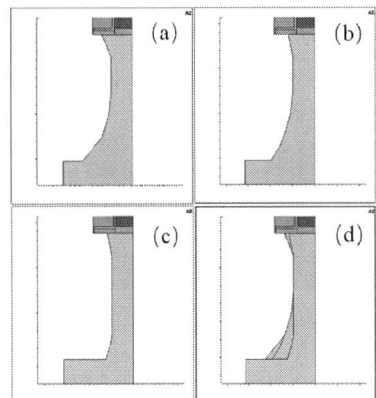

*Figure 2: PSR trench profile (a) U-shaped (b) optimized U-shaped (c) rectangular (d) trench profiles overlap*

Compared to U-shaped trench and optimized U-shaped trench, rectangular trench with a vertical sidewall and a flat bottom, is the final pursuit of ideal trench profile. In terms of TCAD simulation results of U-shaped trench, optimized U-shaped trench and rectangular trench, as shown in Figure 2 and TABLE I, the rectangular trench stress increases by 9.2% and 6.2% respectively than those of U-shaped and optimized U-shape trenches, with a Vt gain of 7.1% and 3.6% respectively, which means rectangular trench profile has better $V_{tlin}$ than U-shaped and optimized U-shaped trenches. The ideal rectangular PSR trench will most effectively improve the $V_t$ drawback of work function metal gate fabrication [7-8].

TABLE I. TCAD SIMULATION OF DIFFERENT TRENCHES

| Items | U-1 | U-2 | Rectangular |
|---|---|---|---|
| $V_{tlin}$ | 0.196 | 0.189 | 0.182 |
| $I_{dlin}$ | 10.8 | 11.4 | 12.1 |
| $V_{tsat}$ | 0.166 | 0.157 | 0.146 |
| $I_{dsat}$ | 60.6 | 64.4 | 69.8 |
| $I_{doff}$ | 131 | 155 | 230 |
| DIBL | 30.4 | 32.2 | 35.9 |
| Channel Stress | 4.22 | 4.34 | 4.61 |

**Trench silicon oxidation**

After PSR process, the exposed silicon facets of the trench should be crystalline face {110} at trench sidewall, crystalline face {100} at trench bottom. Although trench bottom is not an ideal {111} facets, the higher atomic density at the corner can be approximated and treated as equivalent to crystalline face {111}, as shown in Figure 3. Different silicon crystalline faces, like {111}, {110} and {100}, have different thermal oxidation rates [9]. When oxidation temperature is below 700℃ or above 1100℃ under dry $O_2$ atmosphere, silicon crystalline face {110} will be oxidized faster, compared to {111} facet, and the slower one is crystalline face {100}. However, the oxidation rates are also temperature-dependent. When temperature is between 700℃ and 1100℃, crystalline face {111} would be oxidized faster than crystalline {110}. Within the temperature range, the trench bottom corner, crystalline face {111}, will form a thicker oxide layer than trench sidewall (crystalline {110}) and trench bottom (crystalline {100}). And then, all thermal $SiO_2$ in the trench are removed by mean of dry or wet etch methods with high $SiO_2$-to-Si selectivity (>30), a more rectangular PSR trench profile would be obtained with higher $V_t$ performance. Rapid thermal oxidation is applied to PSR trench silicon oxidation for higher throughput [10].

Silicon oxidation thickness can be expressed as

$$x=(kC/N)*(t+\tau) \qquad (1)$$

Where $x$ is the oxide thickness, $\tau$ is initial oxidation period, k is the constant of silicon/$SiO_2$ interface reaction, C and N are the equilibrium concentration of oxygen in $SiO_2$ and diffusion constant of oxygen within $SiO_2$ [11]. Moreover, a straight trench sidewall and trench bottom are beneficial for silicon-germanium (SiGe) epitaxy, in which epitaxial precursor $GeH_4$ and $SiH_4$ tend to react on crystalline face {100}, and precursor $GeH_2Cl_2$ and $SiH_4$ tend to react on crystalline face {110}. SiGe grows on crystalline faces with selectivity to fill up the PSR trench. SiGe epitaxy temperature is about 600℃, a silicon oxidation before is fit for process requirement [12].

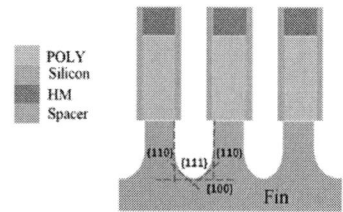

*Figure 3: Crystalline faces of PSR trench*

**Proposed process flow**

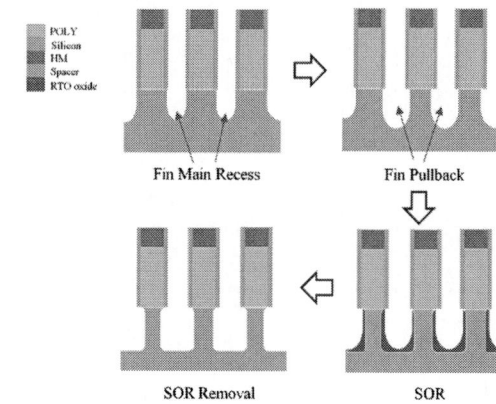

*Figure 4: PSR rectangular trench profile flow (SOR)*

The proposed rectangular PSR trench formation flow is given in Figure 4, in which anisotropic HBr-based plasma is utilized to recess fin vertically until the target trench depth is reached, and isotropic $NF_3$-based plasma is applied to isotropic recess fin to meet S/D channel target, an U-shaped trench profile can be obtained. Silicon oxidation recess (SOR) is applied afterwards to oxidize trench silicon with different crystalline faces. After oxide removal process, a close-to-rectangular profile can be obtained. Considering different oxidation rates, the oxide layer thickness should be over 4nm to reach a difference among crystalline faces of PSR trench. To verify the validity of SOR methods, Two SOR split conditions, in-situ steam generation (ISSG) and in-situ generation (ISG), are utilized in experimental investigation.

Assuming $\tau=0$ for both ISSG and ISG, for a short oxidation time equation (1) is rearranged as

$$x=(k\ C/N)\ {}^*t \qquad (2)$$

The final oxidation thickness of trench sidewall is about 6 nm.

## RESULT AND DISCUSSIONS

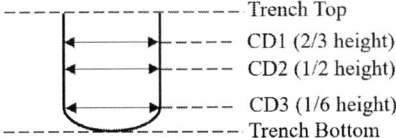

Figure 5: Trench profile definition

TABLE II. PSR TRENCH PROFILE ANCHOR POINTS

| Item | Anchor 1 | Anchor 2 | AR |
|------|----------|----------|-----|
| Baseline | 3.6 | 13.1 | 1.3 |
| SOR1 | 2.6 | 9.5 | 1.4 |
| SOR2 | 1.1 | 7.0 | 1.5 |

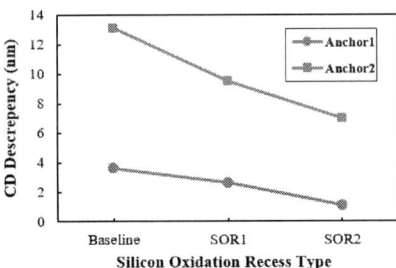

Figure 6: Trench profile comparison

With TEM images, the trench profile can be described with a series of space CDs as in Figure 5. Three lines along the trench vertical direction are defined to characterize the trench profile. The height of the CDs are chosen to be 1/6, 1/2 and 2/3 of total trench bottom. A smaller discrepancies of trench CD at CD 1, CD 2 and CD 3 mean a more rectangular trench profile. The discrepancy between MP1 and MP2 is define as anchor point 1, and discrepancy of MP1 and MP3 is defined as anchor point 2. As shown in

TABLE II and Figure 6, anchor 2 of SOR2 is improved by 47% to 7.0, compared to baseline (U-shaped condition) 13.1. Meanwhile, anchor 2 of SOR1 is 9.5, SOR1 and SOR2 stand for ISSG and ISG oxidation methods respectively. TEM images of SOR scheme proves an effective profile improvement on PSR trench, and the trench bottom corner is more rectangular than PSR process without SOR application.

## CONCLUSION

In this work, we propose a rectangular PSR trench profile for channel stress and epitaxy concern, and also a process flow with SOR for trench profile fabrication is offered. TCAD shows rectangular trench gains 6.2% and 9.2% channel stress than those of U-shaped and optimized U-shaped trench respectively, meanwhile, gains 3.6% and 7.1% Vt respectively. Rectangular trench TEM cut shows the anchor pattern 1 and 2 improves 69.4% and 46.7% than those of U-shaped and optimized U-shaped trench respectively.

## ACKNOWLEDGEMENTS

The work in this paper is sponsored by Shanghai Pujiang Program.

## REFERENCES

[1] D. Wang, T. Liu, X. Sun and et al. *2020 IEEE 14th International Conference on ASIC (ASICON)*, Kunming, October 26-29, 2021

[2] S. Mochizuki, B. Colombeau, L. Yu and et al. *2018 IEEE International Electron Devices Meeting (IEDM)*, San Francisco, December 1-5, 2018

[3] D. Hisamoto, W.C. LEE, J. Kedzierski and et al. *IEEE Trans. on Electron Devices*, vol.47, 2000, pp. 2320-2325.

[4] M. Choi, V. Moroz, L. Smith and et al. *2012 International Silicon-Germanium Technology and Device Meeting (ISTDM)*, Berkley, June 4-6, 2012.

[5] G. Fiorenza, J. Park, L. Anthony and et al, *IEEE Trans. on Electron Devices*, vol.55, 2008, pp. 640-648.

[7] B. Wu, A. Kumar, S. Pamarthy and et al. *J. Appl. Phys.*, 2010, vol.108, 051101.

[8] E. Martinez, P. Ronsheim, J. Barnes and et al. *Microelectron. Eng.*, 2012, vol.101, p182901.

[9] D. Kim, H. Wada, J. Woo. *IEEE Trans. on Semiconductor Manufacturing*, 2004, vol. 17, pp. 192-200.

[10] E. Lewis and E. Irene. *J. Electrochem. Soc.*, 1987, vol.134, pp. 2332-2339.

[11] R. Marcus and T. Sheng. *J. Electrochem. Soc.*, 1982, vol. 129, pp. 1278-1282

[12] A. Lam, Y. Zheng and J. Engstrom. *Surf. Sci.*, 1997, vol. 393, pp. 205-221.

# SIMULATION STUDY ON DIFFERENT INTEGRATION SCHEMES TO FORM SINGLE DIFFUSION BREAK

*Pengfei Lyu[1*], Jian Huang[1], Minxiang Wang[1], Tianhao Zhang[1], Qingpeng Wang[1], Kui Wang[1], YuShan Chi[1]*

[1]Lam Research Service Co., Ltd, Shanghai 201203, China
*Corresponding Author's Email: PengFei.Lyu@lamresearch.com

## ABSTRACT

Single diffusion break (SDB), which provides electrical insulation between two finFET devices, is one of the critical steps in finFET semiconductor fabrication. With device dimension scaling down, SDB has shown a growing importance on chip yield performance. In recent years, various SDB approaches have been developed along with integration evolutions to enable smaller process nodes. In this paper, the 3 main SDB approaches, including fin/STI approach, ILD approach and HKMG approach, are investigated systemically by virtual fabrication with Coventor SEMulator3D®. We find unique advantages and disadvantages present for each approach. The fin/STI approach suffers least from aspect ratio and material complexity but has the most demanding requirements on overlay and CD control. Being self-aligned approaches, ILD and HKMG approaches behave well in overlay and CD control. Nevertheless, both methods are more demanding on etch processes due to higher aspect ratio and presence of multiple materials on the stack.

***Keywords—SDB; fin approach; ILD approach; HKMG approach***

## INTRODUCTION

Double diffusion break (DDB) and single diffusion break (SDB) are two ways to isolate neighboring devices in finFET technology [1]. DDB will cut off the fin between two adjacent gates, resulting in shallow trench isolation (STI) width of gate pitch. In contrast, SDB will cut off the fin under single gate, which narrows down STI width to gate length. SDB integration schemes are scaling boosters and are preferred at advanced process nodes. However, with continuous CMOS device size scaling, critical dimension (CD) and overlay control of SDB becomes more challenging: A widened or misaligned STI will negatively impact the source/drain (S/D) epitaxial growth in proximity and significantly lower device performance and yield. It has become increasingly critical to choose an appropriate SDB approach for advanced node. Among many approaches, the most direct method is to perform SDB post fin formation, which is so-called fin/STI approach [2]. Despite being the most straightforward integration scheme, it is difficult in fin/STI SDB scheme to control CD and overlay, which may negatively impact epi volume and drive current. To avoid these issues, alternative approaches such as post ILD cut and post HKMG cut have been developed [3], which are both self-aligned process and conducted after epi growth. In this paper, we compare the different SDB approaches by virtual fabrication using Coventor's SEMulator3D® [4], and the advantages and disadvantages of each approach are compared and contrasted.

## EXPERIMENT

The model studied is based on one standard AND-OR-Inverter logic cell with 2 fins each in PMOS and NMOS regions with 7 gates. As shown in Figure 1(a), fins are labeled in red, and gates in blue. The SDB masks (yellow bars) are overlapped with the corresponding gates and each cross 2 fins. Physical models are reconstructed in SEMulator3D® from fin loop to high-k metal gate (HKMG) loop. Figure 1(b) demonstrates the insertion points of each approach: fin approach is post fin etch in fin loop, while ILD and HKMG approaches are separately post ILD filling/CMP and post the final metal gate formation.

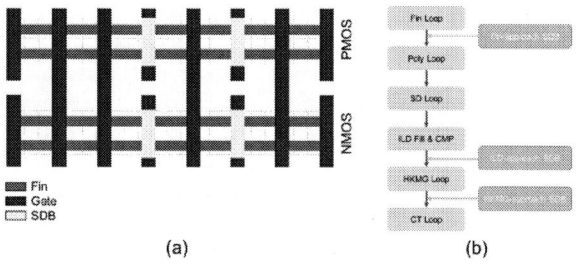

(a)                    (b)

*Figure 1 (a) Layout used in the study, (b) schematic diagram of the 3 approaches SDB (fin/ILD/HKMG) insertion point through the process flow.*

## RESULTS

Models of different SDB approaches are built up in SEMulator3D® and main processes are illustrated for comparison, as shown in Figure 2.

For fin/STI SDB approach, as shown in Figure 2(a-g), a layer of spin-on carbon (SOC) as soft mask is deposited

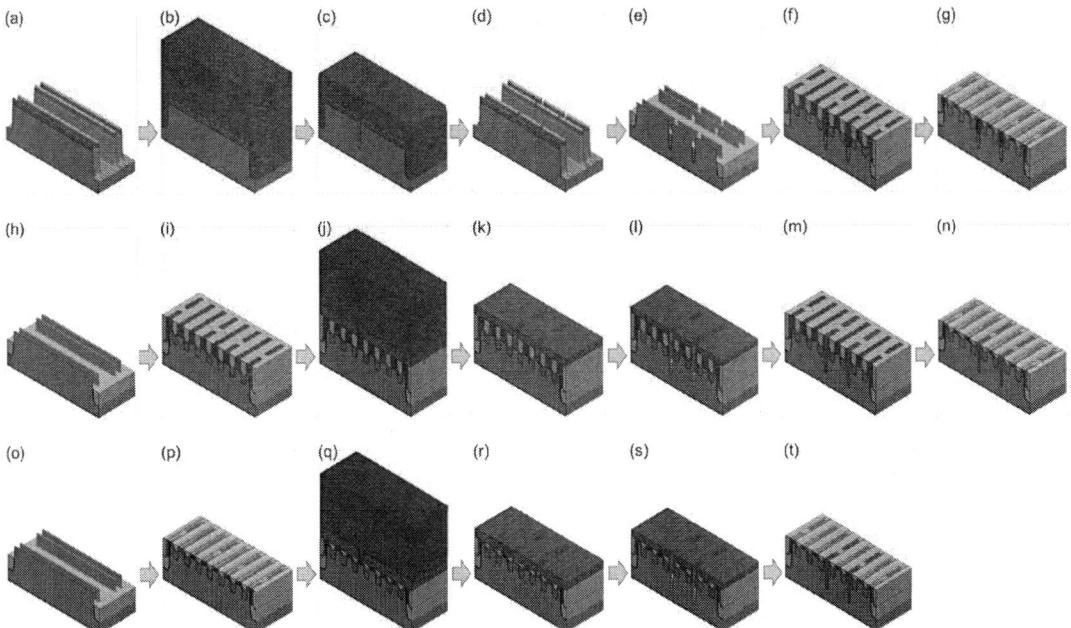

*Figure 2: Different SDB approaches illustrated by SEMulator3D. (a-g) fin/STI approach: (a) post fin STI & fin cut etch, (b) SDB litho, (c) SDB etch, (d) strip, (e) post STI fill & recess, (f) post ILD CMP, (g) post HKMG CMP. (h-n) ILD approach: (h) post STI fill & recess, (i) post ILD CMP, (j) SDB litho, (k) SDB HMO & DPR, (l) SDB etch, (m) SDB CMP, (n) post HKMG CMP. (o-t) HKMG approach: (o) post STI fill & recess, (p) post HKMG CMP, (q) SDB litho, (r) SDB HMO & gate RM, (s) SDB etch, (t) SDB CMP.*

*Figure 3: The final SDB cross-section in X direction of: (a) fin/STI approach, (b) ILD approach, (c) HKMG approach.*

and covers the underlying fin structure after fin formation. SiARC (antireflective coating) and photoresist are subsequently deposited, and lithography is performed. The cut patterns are transferred from layout to tri-layer, and finally into fin silicon. After SDB formation, STI dielectric is backfilled, dummy poly gates are formed above and later removed for metal fill. Figure 3(a) shows the final SDB cross-section post HKMG CMP. The metal gate filled in is dummy, which is not electrically connected. Performing SDB before gate formation provides lower aspect ratio for etch process with simpler integration. However, this approach puts a high requirement on cut CD and overlay control in order not to cause nearby S/D epi missing. The lithography needs to be precise both in dimension and alignment.

To prevent epi damage issues, approaches that shift SDB post epi growth are proposed. The ILD SDB approach, illustrated in Figure 2(h-n), inserts the cut between ILD and HKMG loop. A SiN hardmask and tri-layer masks are applied to define the SDB cut after ILD CMP. After that, a dummy poly removal (DPR) step is used to remove the dummy poly beneath the hardmask. The remained SiN spacer on poly sidewall will help to constrain the following fin recess taking place in the right location, which is so-called "self-aligned" scheme. Similarly, the HKMG SDB approach, which takes place after HKMG formation (Figure 2(o-t)), is also self-aligned scheme. The only difference is that the gate to be removed consists of metal containing materials such as $HfO_2/TaN/TiN/TiAl/W$, and a wet etch process is typically needed in this scheme. The benefit brought by these 2 self-aligned approaches is obvious: CD and overlay are controlled by SiN spacers and are less dependent by lithography. The disadvantages lie in a more complicated integration schemes with additional steps including gate removal, final trench SiN filling and CMP (Figure 3(b, c)). Compared with fin/STI approach, the aspect ratio is also higher by adding in the gate height

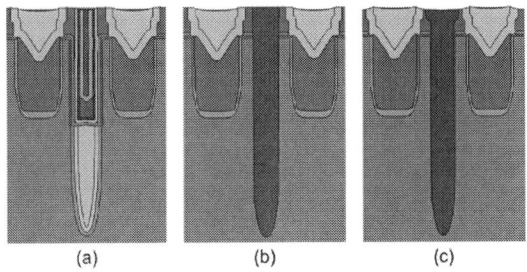

above fin, which makes fin recess more challenging.

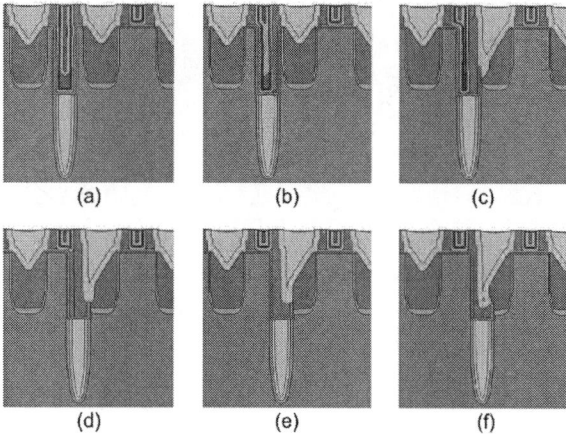

*Figure 4: fin/STI approach SDB Litho overlay window check. SDB cross-section with overlay shifted (a) 0%, (b) 20%, (c) 40%, (d) 60%, (e) 80%, (f) 100% of SDB CD.*

A virtual DOE based on SEMulator3D® is conducted to further survey the margin of lithography overlay for fin/STI approach. The SDB overlay is set to be 0% to 100% of the cut CD and tested until HKMG CMP. As shown in Figure 4, 20% is the maximum overlay shift allowed. Beyond that, epi volume shrinks significantly. The root cause for epi volume shrink can be traced back through the defect evolution function of SEMulator3D®. As shown in Figure 5, in NMOS/PMOS silicon recess step, the epi region is overlapped with shifted SDB cut, leading to epi growth failure on the cut side. Insufficient epi volume will cause drive current variation, resulting in lower yield.

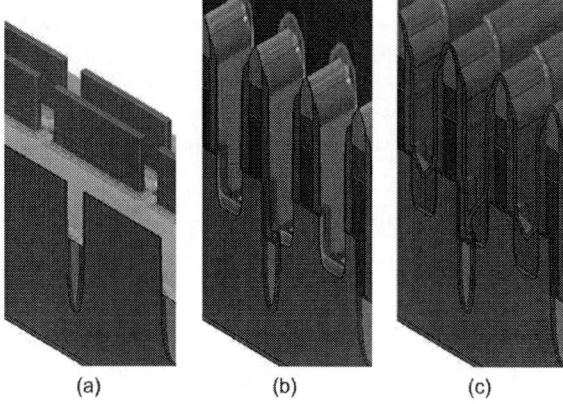

*Figure 5: Epi volume evolution resulted by SDB litho overlay shifting. (a) Post SDB etch, (b) NMOS silicon recess, (c) NMOS epi growth.*

## CONCLUSION

This paper introduces a virtual fabrication method to simulate 3 different integration schemes to form SDB. The benefits and limitations are demonstrated and compared through the virtual results. The fin/STI approach has the simplest integration scheme and the lowest requirement on fin recess process. ILD and HKMG approach have a better control on CD and overlay benefiting from self-aligned scheme. A virtual DOE is also performed to reveal the evolution from SDB overlay shift to epi failure defect formation. The virtual fabrication study method could provide invaluable reference when making selections from different approaches to balance cost and yield.

## REFERENCES

[1] Miyaguchi K., Bufler F., Chiarella T., et al. Single and Double Diffusion Breaks in 14nm FinFET and Beyond. International Conference on Solid State Devices and Materials, 2017.

[2] Wu X., Xiao C., He W., et al. Product comprised of FinFET devices with single diffusion break isolation structures, and methods of making such a product. US 9171752, 2015.

[3] Yu H., Shen H., Hu Z., et al. Method for forming single diffusion breaks between finfet devices and the resulting devices. US9406676, 2016.

[4] http://www.coventor.com/products/semulator3d.

# STUDY ON PROCESS IMPROVEMENT AND YIELD ENHANCEMENT OF 40NM E-FLASH AIO WET STRIP

*Zhiyuan Xu[1*], Youfeng Xu[1], Jin Chen[2], and Zhenghong Liu[3]*
Shanghai Huali Integrated Circuit Manufacturing Corporation,
No. 6 Liang Teng Rd., Pudong New Area, Shanghai, 201315, P.R.China
*Corresponding Author's Email: xuzhiyuan_TD2@hlmc.cn

## BIOGRAPHY

Zhiyuan Xu,1994~, received the Master Degree from WuHan University of Technology in 2019, now is working at Advance Module Development Division of Shanghai Huali Integrated Circuit Manufacturing Corporation as a WET process R&D engineer.

## ABSTRACT

40nm e-flash product CP1 test results show that the wafer has a special map, and the yield is lower. The WAT data shows that Rc is very large, indicating that the chip circuit is open. Through step by step process check and PFA analysis, the root cause of yield loss is copper loss at the bottom of the Via in AIO loop. After AIO Wet Strip, the residual acid in Via will continue to etch copper at the bottom and form serious under cut, resulting in poor connection at the bottom, thus resulting in loss of yield. In this paper, by optimizing the wet cleaning process, a thin passivation layer is formed on the copper surface to avoid further erosion of copper by residual acid and prevent the formation of defects. The results show that this method greatly improves product yield and process reliability.

*Keywords—Copper Loss, AIO Wet Strip, Low Yield, Special Map*

## INTRODUCTION

AIO loop is the important process during integrated circuit manufacturing. By filling the via with copper to form a circuit connection between the different devices. The AIO wet strip process also plays a crucial role in AIO Loop. If the AIO wet strip does not remove dry etch residues and minimize copper loss at the bottom of the via, it will not be well filled in the future, resulting in circuit breaks in extreme cases. With the development of semiconductor manufacturing technology, chip manufacturing technology nodes also follow Moore's law to shrink rapidly. Due to the reduction of chip device size, the geometric size of BEOL (Back End of Line) metal interconnection is also reduced. In the Trench/Via, there is an increasing aspect ratio, which makes the BEOL wet cleaning process a challenge. In order to improve the cleaning ability of BEOL wet process, the researchers of wet process have done a lot of research work[1]-[3].

In this paper, the low yield issue of HLMC 40nm e-flash memory is studied. we study the mechanism of the copper loss at the bottom of via. Combined with the AIO loop process characteristics and wet process conditions are given to improve product yield and process reliability.

## EXPERIMENT

In AIO Wet Strip process for 40nm e-flash products, a cleaning process for post-treatment is proposed. By tuning process parameters and process step, low yield issue was solved. In this paper, chemical A is an organic acid, which is mainly used to remove polymer, chemical B is an inorganic acid with slight polymer removal ability and can etch metal oxides, Scanning Electron Microscope (SEM) was used to detect defect condition after cleaning processes, the pattern CD was measured using HITACHI, the film thickness was measured using SFX200 (KLA-Tencor, USA) after process, and copper loss of Via bottom is judged by cross section TEM (Transmission Electron Microscope).

In addition, the WAT, Yield and RE were also judged by professional test.

## RESULTS AND DISCUSSION

### The formation mechanism of copper loss

40nm e-flash product CP1 test results show that the wafer has a special map. As shown in Figure1(a), a crescent-shaped yield loss appears on the right side of the CP map. Through the analysis of CP failure bin and WAT data, it was found that the CP loss map same as WAT V2 RCKV map, as shown in Figure1(b). Therefore, PFA analysis was performed on the failed wafer, the PFA image shown serious copper loss at the bottom of via of the bad die, but slight copper loss at the bottom of via of the good die, as shown in Figure1(c). Based on the above data, we believe that the failure originated from copper loss at the bottom of via, resulting in connection issues.

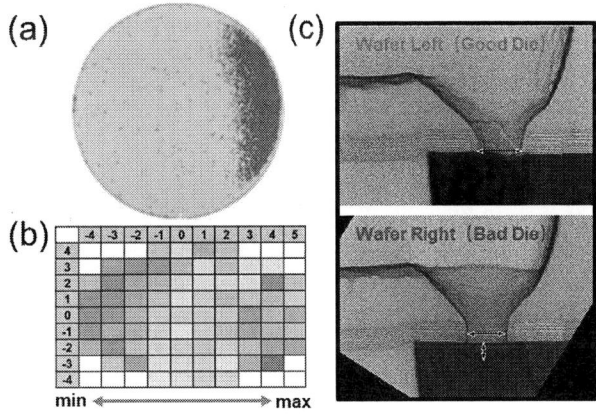

*Figure1. (a)The CP Loss Map; (b) WAT V2 RCKV map; (c) V2 TEM image*

To find out the root cause of the copper loss, we need to understand the process. All-in-One Etch mainly consists of the following steps: First step, via etch and stop on the metal layer below; Second step, wet strip process removed residue and polymer. Therefore, dry etch or wet strip process may cause copper loss at the bottom of via. Based on experimental data and experience[4]-[5], we propose the following model, as shown in Figure2. Because the AIO etching gas is $C_XF_Y$, the polymer after etching contains F element. And there are more polymer on the right side of wafer, resulting in that the residue polymer on the right side cannot be effectively removed after wet cleaning. Under the action of moisture, the residual F-containing polymer reacts with copper to oxidize copper, and then the copper oxide is removed by the residual chemical B to form undercut. Finally, the special failure map is shown on the right side of wafer.

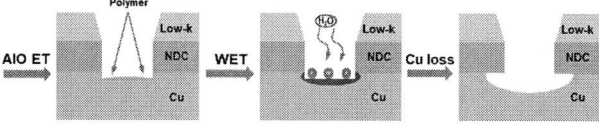

*Figure2. Copper loss model*

The formation mechanism and solution of via open, and the access to resolve via open mainly include the following 2 points: (1) Enhance polymer removal capacity. (2) Reduce further etching of residual chemical B after wet cleaning.

**Research on wet clean to improve copper loss**

Wet clean is a method of removing defects by cleaning the surface of wafers with a variety of chemicals. In this case, for improving copper loss issue, chemical A and chemical B have been used to clean wafer surface after AIO etch, and copper loss is described by CP Map. In order to verify the mechanism of copper loss, four groups of experiments were designed: (1) baseline A+B, (2) increase A process time on the baseline, (3) reduce A process time on the baseline. As shown in TABLE I, the improvement of yield by increasing chemical A process time is limited, and there is still a special failure map on the right of wafer. And reducing the chemical A process time, the yield trends to worse. Based on the above experimental results, we believe that the effect of increasing the chemical A process time is mainly to enhance the removal capacity of polymer residues, and the corrosion of residual acid to copper can not be avoided. This also corresponds to the copper loss model proposed by us, which proves the correctness of our direction.

TABLE I. EFFECT OF CHEMICAL A AND B ON YIELD.

| Wafer ID | 1 | 2 | 3 |
|---|---|---|---|
| Split | Split1 | Split2 | Split3 |
| Condition | Baseline A+B | A(time+)+B | A(time-)+B |
| CP Map | | | |

Since chemical B is placed in the last step of cleaning, the corrosion of copper cannot be avoided, but the cleaning window will be smaller if chemical B is removed. In order to avoid the corrosion of residual acid on copper, it is necessary to form a dense protective layer on copper after cleaning. Therefore, a new recipe structure of A+B+A is proposed and studied. The first step of chemical A cleaning is to remove polymer residue. The second step of chemical B is to increase the cleaning capacity. The third step of chemical A is to form a

thin protective film on the copper surface to avoid the corrosion of residual chemical B on copper. As shown in TABLE II, the split 5 added a step of chemical A cleaning on the baseline, and split 4 and 6 were the process window check. The yield of above three groups of experiments is good, indicating that there is no via open issue. Therefore, we take the condition of split 5 as the new baseline. In order to evaluate the cleaning effect and process reliability of the new process, the wafer cleaned by new process were PFA and tested reliability, the results are shown in Figure3 and Figure4. Figure3 (a) and (b) show the TEM images on the right and left of the wafer respectively. The results show that there is no copper loss, indicating that the failure on the right of wafer has been effectively solved. the new condition better than the baseline. Figure 4 shows that the EM test is passed by new condition but failed by baseline.

TABLE II. WINDOW CHECK.

| Wafer ID | 1 | 2 | 3 |
|---|---|---|---|
| Split | Split4 | Split5 | Split6 |
| Condition | A (time-)+B+A | A+B+A | A (time+)+B+A |
| CP Map | | | |

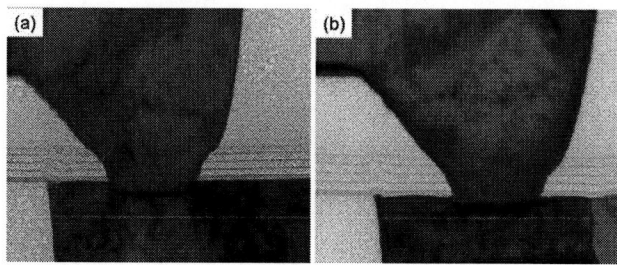

*Figure3. (a) Wafer right; (b) Wafer left.*

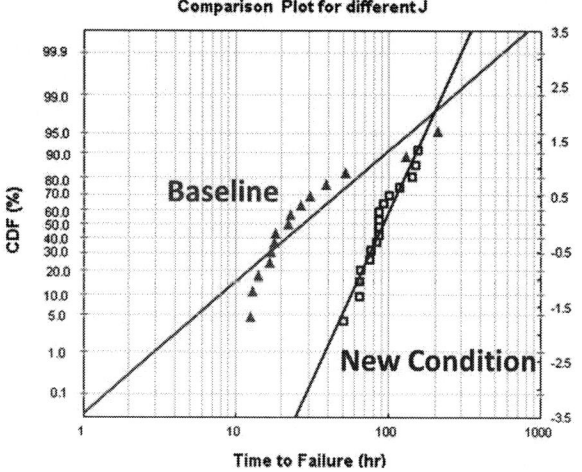

*Figure4. EM results of baseline and new condition.*

## CONCLUSION

In summary, an effective wet cleaning process was used to improve the low yield issue caused by AIO loop in the 40nm e-flash product. By adding process time and post treatment, the process window of AIO wet strip is expanded. Meanwhile the low yield issue caused by this loop is effectively solved. At the end, CP yield shows about 15% improvement and no special CP map. Reliability result shows >10 years, better than the baseline.

## ACKNOWLEDGEMENTS

The authors would like to thank HLMC AMTD Wet and TD1 PIE group for the technical discussions.

## REFERENCES

[1] Mikhail R. Baklanov, Christoph Adelmann, Larry Zhao, and Stefan De Gendt, Advanced Interconnects: Materials, Processing, and Reliability, *ECS Journal of Solid State Science and Technology*, 4 (1) Y1-Y4 (2015).

[2] T. Hattori, R. E. Noval, J. Ruzyllo, P. Besson, P. Mertens, Cleaning and Surface conditioning Technology in Semiconductor Device Manufacturing 10, *ECS Transactions*, vol. 11, No 2.

[3] A. Sharma, Jacob Bulaga, Srishti Agrawal, Ravi Srivastava, Optimization of wet clean and its cost effectiveness in dual Damascene 14 nm BEOL, *J. ASMC*, 2018, pp. 128-130.

[4] Y. Kobayashi, Effects of fluoride residue on Cu agglomeration in Cu/low-k interconnects, *J. Microelectronic Engineering*, vol. 88, 2011, pp. 620-622.

[5] H. Yong, J. Liu, Z. Yang, Optimization of Wet Strip after Metal Hard Mask All-in-One Etch for Metal Void Reduction and Yield Improvement, *2017 China Semiconductor Technology International Conference (CSTIC)*, 2017, pp. 1-3.

# STUDY AND OPTIMIZATION OF PHOTO RESISTOR ETCH BACK LOOP IN HK METAL GATE

*Yajie Li\*, Baichun Zhang, Jianguo Yang, Lei Sun, Quanbo Li, Yu Zhang*

Shanghai Huali Integrated Circuit Corporation

\*Corresponding Author's Email: liyajie@hlmc.cn

## ABSTRACT

Photo resistor etch back (PREB) is an essential process in High-K Metal Gate, which can eliminate the height difference between nFET and pFET, and mitigate the pattern loading occurred in front end of line (FEOL). During PREB process, it is necessary to avoid SiGe damage, especially when photo resistor is being etched in etch back 1 (EB1) step. Moreover, the appearance of oxide residual and higher horn after PREB will ultimately lead to poly residue in dummy poly remove (DPR) process. In this study, those defects were studied systemically to illustrate their generation mechanism and adverse consequences. By suitable adjustment of photo resistor remain thickness, etch rate selectivity and etch amount, most of defects could be reduced or avoid to optimize PREB process.

## INTRODUCTION

In 28nm High-K metal gate, the core nFET and pFET are necessary to give chip the function of calculation. For the generation of SiGe alloys, dry etch or wet etch process will be used to format trenches beside only pFET, which will consume some oxide hard mask on pFET. This difference between nFET and pFET in structure will result in unfavorable pattern loading, which will make it difficult to remove dummy poly in subsequent processes.

photo resistor etch back (PREB) is a process to mainly remove oxide hard mask above the poly in HK flow, which will contribute to flatten the pattern at SiN hard mask layer before dummy poly remove (DPR). Usually, PREB is divided into four steps: photo, etch back 1 (EB1), etch back 2 (EB2) and ash. The purposes of Photo and EB1 are to open the windows and expose the surfaces of n/pFET. EB2 aims to break through the PREB SiN layer and then remove oxide hard mask. Finally, the residual photo resistor (PR) is removed by dry strip in ash step. However, during PREB process, the defects of SiGe damage and poly residue are easily appeared to affect work functions of device. To avoid those defects, it is necessary to explore the influence factors on the defects and optimize the process of PREB.

In this work, the influence of PREB on SiGe damage and poly residue were systemically studied. By adjusting the the PR remain thickness and the time of EB1 and EB2, the defects of higher horn and oxide remain was observed, and their influence on the poly residue was investigated.

## EXPERIMENT

The scheme of PREB process is shown in Fig.1. At first, photo step was performed by lithograghy to expose wider n/pFET. Subsequently, the dry etching experiments of EB1 (PR etch) and EB2 (oxide etch) were taken by inductively coupled plasma (ICP). Finally, residual PR was dry striped and cleaned by diluted hydrofluoric. The test wafer diameter size is 300 mm and film stack is patterned PR/PREB SiN/Oxide/SiN/Poly on substrate as shown in Fig.1a. The vertical view was observed by scanning electron microscope (SEM) and the cross section was obtained by transmission electron microscope (TEM).

The baseline sample was defined as S1. Compared with S1, the EB2 break through (BT) time was reduced for S2 and the EB1 time was reduced for S3. For S4, the EB1 time was reduced, while the EB2 main etch (ME) time was increased.

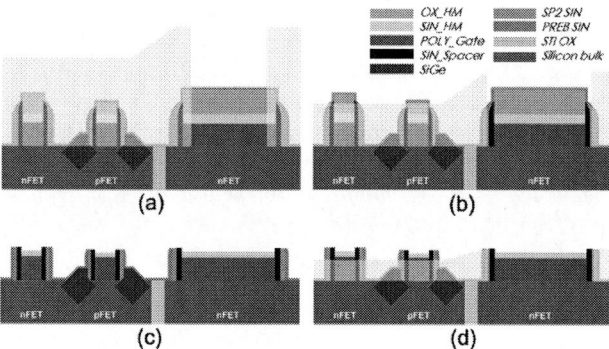

*Figure 1: (a) PH step; (b) EB1 step; (c) EB2 step; (d) Ash*

## RESULT AND DISCUSSION

Before PREB, SiGe has been deposited beside pFET, which makes it easy to be impaired by the subsequence processes. As the drain stressor for pFET, the embedded SiGe source introduce larger atoms during epitaxial deposition, which can squeeze atoms together and create compressive stress in channel for mobility enhancement. However, the thickness of oxide on pFET would be reduced after SiGe trench etch, leading to the different height between nFET and pFET, as shown in Fig.2.

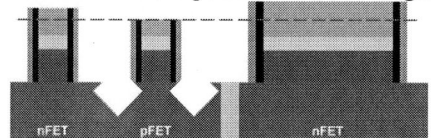

*Figure 2: SiGe trench etch and the appearance of pattern*

*loading*

In 28 nm High-K metal gate, PREB needs to remove the oxide layer to eliminate the pattern loading. To protect the naked SiGe, PREB SiN as etching stop layer and PR are covered on the wafer. During photo step, only the top surfaces of n/pFET are exposed, opening windows for etching oxide. Limited to the precision of mask aligner, the PR on n/pFET with smaller sizes are not able to be accurately opened by lithography, which gives EB1 an assignment to etch PR and reveal the n/pFET. However, the over-etching of PR also reveals the SiGe, and then the PREB SIN and the exposed SiGe will be partially removed by EB2, as shown in Fig. 3.

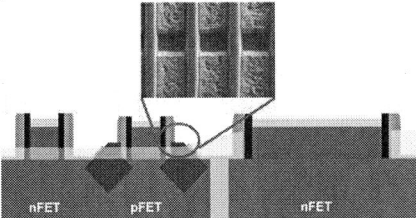

*Figure 3: Damages of exposed SiGe after EB2 step*

In Fig. 4a, the PR thickness between FET was 520A and residual SiN hard mask thickness was 80A. With this condition, there was no PR remain as well as PREB SiN layer on bare Si substrate in Fig. 4b, suggesting the possibility of SiGe damage in low-density pattern.

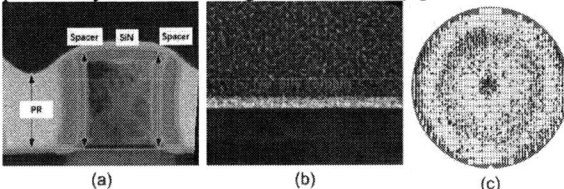

*Figure 4: TEM profiles of (a) nFET and (b) bare Si substrate, and (c) defect scan map after EB2 in S1*

To avoid the SiGe damage, there needs enough PR remain to guarantee the remainder of PREB SiN on SiGe. By reducing the EB2 BT time, the thickness of remaining PR and SiN hard mask after EB2 were increased in S2. As shown in Fig. 5a and 5b, the SiN hard mask thickness was 201A, the PR thickness between FET increased to 561A, and the thickness of PREB SiN on bare Si substrate kept 80A (Fig. 5b). Compared with S1 (Fig. 4c), the number of defects in S2 defect scan map (Fig. 5c) was obviously reduced, which means the good protection of SiGe by increasing PR remain thickness. However, as the Fig. 5a shows, the height of spacer was up to 769A, creating higher horns to embarrass the CMP loop.

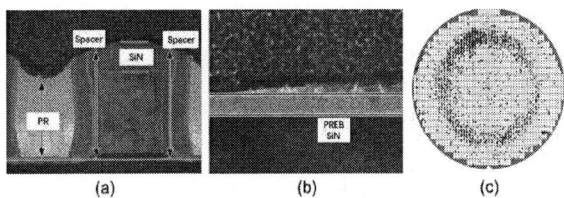

*Figure 5: TEM profiles of (a) nFET and (b) bare Si substrate, and (c) defect scan map after EB2 in S2*

The horn of n/pFET was formed when their spacers were higher than the poly. After PREB loop, the height difference between spacer and poly drew a pit on n/pFET, which would store inter-layer dielectric (ILD) and embarrass the CMP process, as shown in Fig. 6. Eventually, higher horn shelter poly from DPR process and leading to the poly residue defect.

*Figure 6: Poly residue caused by higher horns*

To guarantee the appropriate height of horn, the BT time of EB2 was rose back and the PR remain was increased by only reducing EB1 time. In Fig. 7a, the SiN hard mask thickness was back to 90A. The PR remain thickness between FET in S3 was 553A (33A higher than that in S1), and both PREB SiN and PR were left on bare Si substrate. With the increase of PR remain thickness, the residual SiN and PR was increased, suggesting the good protection for the SiGe under PREB SiN. By scanning the defect, the SiGe damage defect count was rapidly reduced when remaining enough PR. However, if the thickness of PR remain increased without limit, the poly residue issue would also occur after DPR.

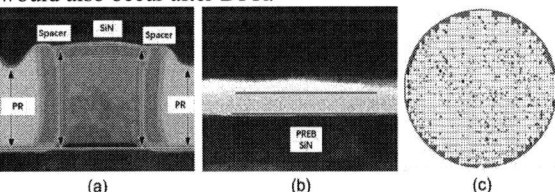

*Figure 7: TEM profiles of (a) nFET and (b) bare Si substrate, and (c) defect scan map after EB2 in S3*

Due to the small window opened by PREB photo step, excessive remaining of PR would lead to oxide residue, which could result in SiN hard mask residue and then also poly residue (Fig. 8). Considering the etch selectivity, the oxide etch rate of the gas used in EB2 ME step was higher than its SiN etch rate, which made it possible to etch a large number of oxide and little SiN at the same time. As shown in Fig. 9a, the increase of EB2 ME time reduce both SiN and PR remain, but not removed them. On the bare Si substrate, the PREB SiN (43A) was thick enough to protect the SiGe. Compared with S3, the amount of oxide residue obviously declined as shown in the defect scanning map (Fig. 10).

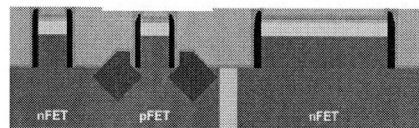

*Figure 8: Poly residue caused by oxide residue*

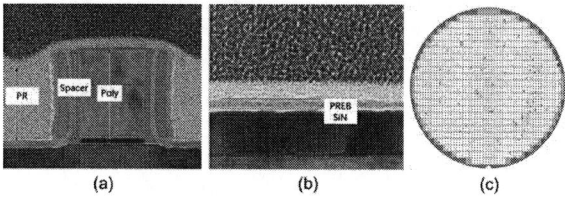

*Figure 9: TEM profiles of (a) nFET and (b) bare Si substrate, and (c) defect scan map after EB2 in S4*

## CONCLUSION

In this work, the PR remain thickness, etch rate selectivity and etch amount were adjusted to achieve the reduction of defect count caused by PREB process. When the BT time of EB2 was shortened, the SiGe under PREB SiN was protected better than that in the base line, but higher horn increased the risk of poly residue. By thickening PR remain, SiGe damage could be entirely avoided. However, the exorbitant PR prevented the oxide and SiN from plasma and eventually led to poly residue after DPR. Due to the high etch rate selectivity in EB2 ME steps, residual oxide hard mask could be thoroughly removed without SiGe damage by increasing the ME time of EB2. By striking a balance among EB1, EB2 BT and EB2 ME, the PREB process was optimized to bring little SiGe damage and poly residue.

## ACKNOWLEDGEMENTS

Thanks for all the authors for great work and collaboration. Thanks for FA (failure analysis) department for TEM image support.

## REFERENCES

[1] W. Lee, M. Wang, S. Wang, W. Lan, C. Li, B. Yang. IEEE Trans. Plasma Sci., vol 42, 2014, pp. 3747-3750.

[2] N. Burham, G. Sugandi, M. M. Nor and B. Y. Majlis. 2016 International Conference on Advances in Electrical, Electronic and Systems Engineering, Putrajaya, Nov. 14-16, 2016, pp. 516-519.

[3] H. Ekinci, N. M. S. Jahed, M. Soltani and B. Cui. IEEE Trans. Nanotechnol., vol 20, 2021, pp. 33-38.

[4] J. M. Regis, A. M. Joshi, T. Lill and M. Yu. 1997 IEEE/SEMI Advanced Semiconductor Manufacturing Conference and Workshop ASMC 97 Proceedings, Cambridge, Sept. 10-12, 1997, pp. 252-256.

[5] Y. Song, Z. Yang, X. Tang, Y. Ma, F. Niu and C. Wang. *2020 China Semiconductor Technology International Conference*, Cambridge, June 26-July 17, 2020, pp. 1-2.

[6] G. Garozzo, S. Colombo, S. F. Lombardo and A. La Magna. *IEEE Trans. Semicond. Manuf.*, vol 28, 2015, pp. 337-344.

# STUDY ON THE OPTIMIZATION OF FINFET ULTRA-SHALLOW JUNCTION ION IMPLANTATION PROCESS

*Li Wenqiang\*, Yang Jianguo, Sun Lei, Quanbo Li, Jun Huang, Zhang Yu*

Shanghai Huali Integrated Circuit Corporation
Shanghai City, China
\*Corresponding Author's Email: liwenqiang@hlmc.cn

## BIOGRAPHY

Li Wenqiang is an etch process engineer. He works in Shanghai Huali Integrated Circuit Corporation, AMTD department. He got master degree of Materials Processing Engineering from Harbin Institute of Technology. He already worked in etch for more than 1years. He focused on middle end of line processes, such as PREB etch, ILD0 etch, BARC etch and so on.

## ABSTRACT

Source/drain ion implantation is the basis of CMOS devices in semiconductor manufacture. In order to ensure good device performance, ultra-shallow junction technology is considered increasingly important at nodes below 16nm. However, ultra-shallow junction is sensitive to thermal budget and temperature as device sizes continue to shrink. In this work, merits and shortcomings of ion implantation in the different stages are analyzed and compared, aiming at SiGe/SiP implant of source/drain epitaxial growth layer in FinFET. It is benefit for Rc value that avoiding implantation region suffer the impact from multi-pass anneal process in the middle of flow. Meanwhile, it can also increase the implant process window and reduce unnecessary loss for implanted epitaxial layer.

**Keywords**—ultra-shallow junction, implant, FinFET, Rc

## INTRODUCTION

In the evolution of integrated circuit manufacturing technology, metal-oxide semiconductor field effect tube (MOSFET) characteristics to follow the "Moore's law" lasts a miniature approaches the physical limits. The gate control of channel current conduction becomes worse which prone to subthreshold leakage, hot carrier effect and other phenomena, because the distance between the source and drain continuously shortening. In order to overcome the technical challenge caused by the short channel effect, the replacement of traditional planar transistor with new device structure is one of the key directions of technological innovation and development. It can break through the disadvantage of traditional planar transistor which can only be controlled in a single dimension which by means of a metal gate (MG) across the top and sides of fin-type silicon (Fin) through FinFET. By making full use of structural advantages to control the on-off of the device in three-dimensional space, it can increase the effective control area of the gate to the channel, suppress the short channel effect, reduce the channel leakage current, and improve the control ability of the gate. Since FinFET has many similarities with traditional MOSFET in manufacturing process and has been mass-produced in advanced process nodes, the research on FinFET process has practical significance in engineering.

At the process node below 90nm, the epitaxial film is grown by local stress technique to improve the carrier migration rate. Embedded germanium-silicon process can optimize device performance for the carrier mobility of PMOS is improved by using the compressive stress generated by the difference of lattice constants. It is a process method of embedded silicon germanium (SiGe) which makes use of dry and wet etching process on the PMOS source drain silicon substrate to form a Sigma groove, and embedded silicon germanium epitaxial film. On the contrary, tensile stress can increase the drain electron migration rate of NMOS source, and the embedded phospho-silicon process (SiP) comes into being. The key step to realize active device PN junction is to inject ion into epitaxial film of source and drain region. In this work, the merits and shortcomings of ion implantation process in different steps are explored in view of the problem of unexpected diffusion in the ultra-shallow junction of source and drain in advanced technology.

## RESULT AND DISCUSSION

It is widely used as ultra-shallow junction ion implantation technology with low injection energy and high dose concentration at advanced nodes below 16nm, in order to achieve the balance between shallow junction, low resistance and activation of implanted ions. The doping activation of source and drain was realized by ultra-shallow junction ion implantation combined with millisecond high temperature annealing. Ultra-shallow junction technology can effectively overcome the short channel effect caused by device size reduction and obtain lower intrinsic working resistance. However, it is important to control the thermal budget of the ultra-shallow junction area. Because high temperature processes (such as CVD, furnace tube, etc.) will accelerate the diffusion of doped impurity atoms, such unexpected diffusion will change the junction depth and concentration of doping region, especially the transverse diffusion of doped atoms will lead to the reduction of channel length, which will seriously affect the device performance.

In the usual process, the source drain region obtained by etching Tri-layer（PR/Si-ARC/SOC）is doped by ion implantation technique, after filling the SiGe/SiP epitaxial layer. This is known as front source drain ion implantation. But in the subsequent process, when the thermal accumulation which came from multichannel annealing and thin film deposition exceeds the thermal budget of the ultra-shallow junction ion implantation zone, it will be intensified that the doped element undergo unexpected diffusion, especially for the diffusion behavior of light element boron. As a result, the concentration of impurity ions in the doped region decreases and the contact resistance Rc increases significantly, as shown in Fig.1. At the same time, the impurity elements may be diffused horizontally into the channel, affecting the doping state of the channel, and even appear short channel effect and hot carrier effect, which ultimately seriously affect the device performance.

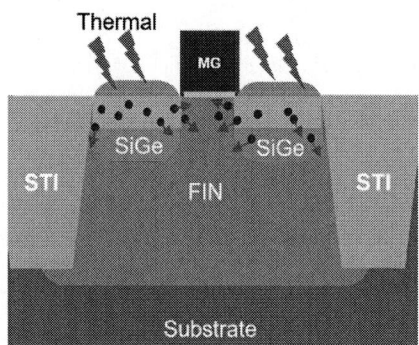

Fig.1 Unexpected diffusion of doped elements

When the doping atoms in the ultra-shallow junction region are subjected to the unexpected diffusion caused by the heat accumulation effect exceeding the thermal budget, the contact resistance Rc will increase as the element concentration in the source-drain region decreases. The reduction of doping concentration caused by unexpected diffusion can be overcome by adding an ion implantation process, that is, by introducing a supplementary ion implantation step (such as BF2/P) before the subsequent metal-filled contact hole process, the impurity concentration in the source drain zone can be increased and the contact resistance Rc can be reduced.

In addition, if the ion implantation process of source and drain is placed in the middle of the whole process (such as contact hole stage), it can effectively avoid the heat accumulation caused by multi-channel high temperature process exceeding the heat budget of ultra-shallow junction, and improve and stabilize the device performance. In the advanced technology node, the contact hole is divided into two steps (M0/V0) to realize the interconnection between the front device and the back copper wire, due to the limitation of device size. For the ion implantation process after the growth of epitaxial film, the M0 step will inevitably consume the epitaxial film in the ion implantation region during the etching process, which will affect the electrical performance of the device.

In order to eliminate the above effects, the ion implantation process is required to accurately increase the source and drain doping depth in advance according to the film thickness of the injection region consumed by the subsequent etching process, so as to ensure the final ultra-shallow junction depth. However, due to the increase of machine RF hours or the change of cavity environment caused by different processes, the etching rate of the etch process will change. Finally, the process window of the whole process is reduced and the difficulty of stabilizing the knot depth is increased.

To optimize the above technological difficulties, if the ion implantation process is placed after M0 etching, the epitaxial layer consumption in the injected region can be avoided at etch process, and the ultra-shallow junction depth is only defined by the ion implantation process which better ensure the performance parameters of the device. At the same time, the trench obtained after M0 etching will grow a layer of silicon oxide film in the diffusion furnace, as shown in Fig.2. Adding a layer of silicon oxide can strengthen the protection of the undoped region, and can increase the process window of ion implantation compared with the previous ion implantation process.

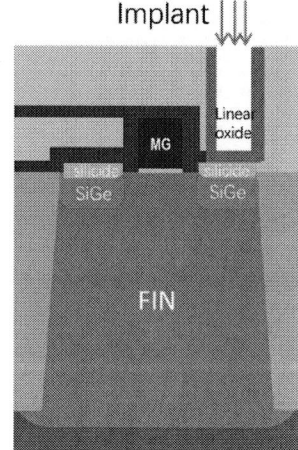

Fig.2 Implant at M0 trench with liner oxide

## CONCLUSIONS

In this work, the ion implantation process of FinFET source drain epitaxial film was optimized. It is benefit for Rc value that avoiding implantation region suffer the impact from multi-pass anneal process in the middle of flow. Meanwhile, it can also increase the implant process window and reduce unnecessary loss for implanted epitaxial layer.

## ACKNOWLEDGEMENTS

Thanks for all the authors for great work and collaboration.

## REFERENCES

[1] J. Kedzierski, International Electron Devices Meeting. Technical Digest (Cat. No.01CH37224), Washington, DC, USA, 2001, pp. 19.5.1-19.5.4.

[2] Yang-Kyu Choi, International Electron Devices Meeting. Technical Digest (Cat. No.01CH37224), Washington, DC, USA, 2001, pp. 19.1.1-19.1.4.

[3] R. Kaneko , 2009 International Workshop on Junction Technology, Kyoto, Japan, 2009, pp. 116-118.

[4] Wen-Hsi Lee, Ming-Han Tsai and Wei-Hsiang Liao, 2014 20th International Conference on Ion Implantation Technology (IIT), Portland, OR, USA, 2014, pp. 1-4.

[5] S. Abo, H. Osae, F. Wakaya, M. Takai and H. Oda, 2014 20th International Conference on Ion Implantation Technology (IIT), Portland, OR, USA, 2014, pp. 1-4.

[6] K. Uejima , 2010 International Workshop on Junction Technology Extended Abstracts, Shanghai, China, 2010, pp. 1-6.

# N/P SPLIT BOUNDARY PROFILE IMPROVEMENT IN HIGH K METAL GATE DUMMY POLY REMOVE PROCESS

*Huang Shan\*; Qu Xiaofeng; Sun Lei; Li Quanbo; Zhang Yu*
Shanghai Huali Integrated Circuit Corporation
\*Corresponding Author's Email: huangshan@hlmc.cn

## BIOGRAPHY

Huang Shan received the B.S. degree in Chemical University of Shen Yang in China in 2011. In 2016, he joined the Technology Development Department of Shanghai Huali Microelectronics Corporation in China, where he has been engaging in etch process development of advanced process High-k Metal Gate for ultra large scale integration.

## ABSTRACT

With the development of VLSI(Very Large Scale Integration) manufacturing, multi-function gate highly integrates in small area, which requests our process with the smaller CD(Critical Dimension) and the more precise profile control. NMOS and PMOS neighboring design can minimize CD, but it's great challenge to get vertical profile on N/P border. We can use TEM(Transmission Electron Microscope) to observe its profile and WAT(wafer acceptance test) to test its performance, mainly about Vt(Threshold Voltage) related parameter. This paper expounds a kind of advanced metal gate control mode. Through dry etch process further studied to improve N/P split metal gate boundary profile, include gas, power adjustment, we can get the vertical profile.

*Keywords—Metal Gate ; N/P split ; boundary ; profile*

## INTRODUCTION

As line width of Integrated Circuit(IC) becomes more and more smaller, the leakage can play an increasingly important role in lowering yield of processor, hindering the device performance improvement and reducing the power consumption. But after entering the 28 nm process, the core area reduces, leading to a dramatic rise in the energy density per unit area. The leakage problem becomes increasingly prominent. The traditional silicon oxide gate has encountered the bottleneck and can not meet the process requirements of 28nm processor. In order to solve this problem, the introduction of high-K metal gate process can reduce leakage as much as 10 times and the power consumption can be well controlled, as well. Nevertheless, the process requirements are also substantially improved. Simultaneously, in order to avoid interaction between NFET and PFET and control the Vt (Threshold Voltage) more accurately, we developed the N/P FET split process of forming metal gate in two times based on the traditional process of forming metal gate in one time. In this paper, the bowl profile encountered in the 28nm Metal Gate DPR(Dummy Poly Remove) etching process was analyzed in detail and the solution was given.

## EXPERIMENT

This work was performed on Kiyo etcher tool from LAM Corporation. The TEM profile image was observed through a Themis TEM from Thermo Fisher Corporation.

## RESULT AND DISCUSSION

### 28nm Metal Gate DPR etching process flow and problems

The 28nm DPR etching process flow is depicted in Figure 1. By the mean of N/PFET split, the gate voltage can be controlled more accurately and mutual influence can be prevented, as well.

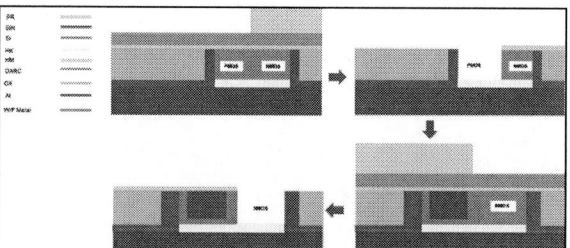

*Fig.1 The process flow scheme of 28nm Metal Gate DPR etch*

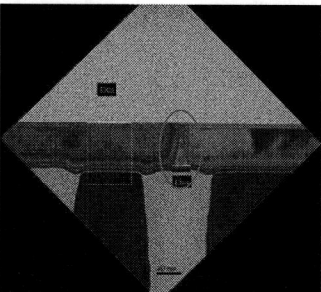

*Fig.2 The poly residue in NP boundary region*

After the DPR process was completed, poly residue was found in the NP boundary region, as shown in Figure 2. It can have a great impact on the metal filling, resulting in higher leakage current and higher power consumption.

**Causes reasoning and verification of Bowl profile**

In consideration of the bowl profile and the process conditions of DPR etching, we think that the profile formation is mainly due to the Cl2/HBr gas used in the PFET DPR etching process. HBr reacts with poly chemically and isotropically, while O2 has insufficient protection on the sidewall. During the NFET DPR etching process, the poly in the bottom corner of the bowl profile could not be removed cleanly, as shown in Figure 3.

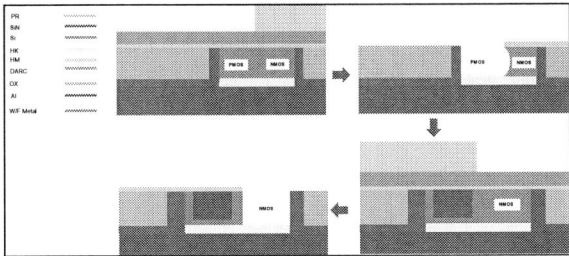

*Fig.3 The causes reasoning of poly residue*

**Bowl profile improvement plans and results**

*Table 1. The experiment plan and mechanism*

| Experiment | Plan |
|---|---|
| 1 | Increase $O_2$ in poly etching |
| 2 | Increase $N_2$ as the sidewall-protected step |
| 3 | Increase bias voltage in poly etching |
| Experiment | Purpose |
| 1 | $O_2$ oxidizes the sidewall to decrease HBr to etch the sidewall |
| 2 | $N_2$ Enhances by-products deposition on sidewall to protect the sidewall |
| 3 | Enhance physical etching and bombardment capability and reduce the chemical etching on the sidewall |

*Table 2. The experiment result and TEM data*

| Experiment | Bowl profile degree | Conclusion |
|---|---|---|
| Baseline | 5.6nm | / |
| 1 | 2.9nm | Improved |
| 2 | 1.0nm | Obviously Improved |
| 3 | 4.04nm | Improved |

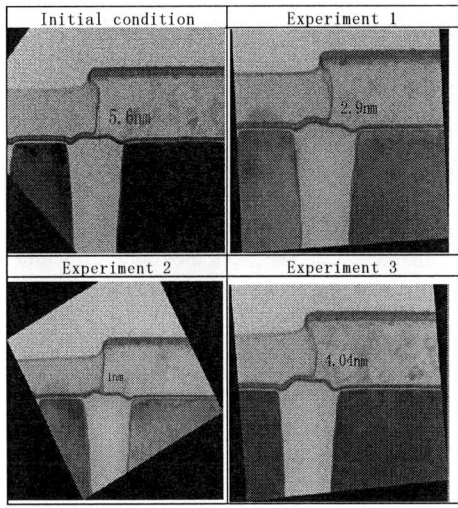

*Fig.4 Comparison data between baseline and new experiments*

Improvement measures were taken for the causes of bowl profile defects, mainly through the optimization of process conditions. The experimental plan and purpose is listed in Table 1. The experiment results and TEM data are in Table 2.

According to the experimental data in Table 2 and Figure 4, the bowl profile can be improved based on all three new experiments, which is consistent with the experimental goal. Furthermore, most improvement can be observed obviously in experiment 2.

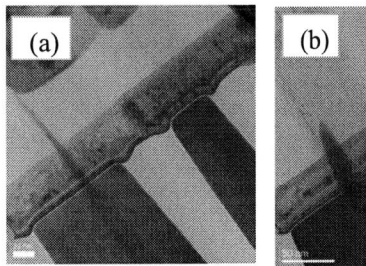

*Fig.5 TEM profile of repeated experiments of best result (a) First repeated experiment of best result (b) Second repeated experiment of best result*

TEM profile of repeated experiments of experiment 2 was observed, as shown in Figure 5. The poly residue disappeared in the boundary area of N/P FET, and the bowl profile was basically eliminated.

**CONCLUSION**

In this work, we found key points to solve bowl profile defects through theoretical deduction analysis and sampling verification. Targeted improvement schemes were then made. The TEM result was used to verify that the improved profile met the requirements. This work provides a mean of N/PFET split to overcome a major difficulty of DPR etching bowl profile defect in metal gate characteristic process.

## ACKNOWLEDGEMENTS

The author sincerely thanks the supports and instructive suggestions from the colleagues in the dry etching module of Advanced Module Technology Development department and in the process integration of Technology Development II department.

## REFERENCES

[1] N. Yamagishi et al, 2003 Dry Process International Symposium, (2003) 105.

[2] T. Yunogami and T. Kumihashi, Jpn. J. Appl. Phys. 37 (1988) 6934

[3] S. Samukawa, J. Vac. Sci. Technol., A17 (1999) 774

[4] K. Nojiri et al, Proc. Symp. Dry. Process (1999) 93.

[5] Mike Barnes et al., Inductively coupled plasma source with controllable power deposition, US Patent 6, 507, 155.

[6] John Holland et. al, 48th AVS Symposium, San Francisco, CA, 2001. Also US Patents 5, 980, 686 and 6, 090, 210.

# EXPLORATION AND OPTIMIZATION OF METAL GATE ETCH BACK PROCESS IN ADVANCED TECHNOLOGY NODE

*Shaoxiong Liu[1] \*, Linpeng Jiang, Junjie Pan, Lian Lu[1], Quanbo Li[1], Jun Huang[1], Yu Zhang[1]*

[1]Shanghai Huali Integrated Circuit Corporation, Shanghai City, China

\*Corresponding Author's Email: liushaoxiong@hlmc.cn

## ABSTRACT

Metal gate etch back process is involved in advanced logic technical node along with FinFET MOS structures in 16/14 logic technology node. Its principal function is etching moderate metal gate composing of TiN/TaN work function metal and W/HfO2 to constitute SAC (Self-Align-Contact) structure. The complex film stack with three-dimensional structure and four kinds of films makes this etch process more challenging than normal two-dimensional etch processes. The etch process is well-designed with etch/flush cycles to obtain ideal W/TiN/TaN/HfO2 profile. Creative design of etch process can achieve relatively independently etching of target film, aiming to accuracy control. Not only the difficult etch profile control of Short-Channel N/P MOS loading, but also the whole wafer uniformity and Short-Channel/Long-Channel pattern loading are very critical for process health. Due to different temperature sensitivity of complex metal gate film stacks, the temperature parameters of different etch step should be control specially to obtain optimized uniformity. Chemical reaction mechanism and physical bombardment mechanism are both playing important roles in SC/LC pattern loading. The systematic thinking of two mechanisms must be involved to decrease these loading as small as possible. Meanwhile, the polymer deposition mechanism also should be play considerable attention for its obvious influence in different patterns. In brief, metal gate etch back is complexity process for its various film stacks and three-dimensional structure. And systematic etch process control is challenging and achieved.

## INTRODUCTION

With the distinguished advantages of higher transistor performance, larger density and lower leakage, FinFET structure device becomes the main stream technology along with the semiconductor device technology advances to 16/14nm node and beyond. As introduced early in 45nm node, the replacement metal gate is still playing important role in FinFET technology node due to its abundant advantages, such as lower gate leakage, thinner Tox and faster electron mobility. [1][2] In replacement metal gate structure, multi-layers work function metal layer consisting of TiN/TaN is used to control different device with different Vt values. [3,4,5,6] Other than 28nm metal gate technology, metal gate etch back process is involved in 16/14nm node for the first time to constitute SAC (Self-Align-Contact) structure, for the sake of contact flow process window.

In this work, metal gate etch back process is investigated in ICP (Inductively Coupled Plasma) etcher equipment. The key requirements for metal gate etch back process are three indexes as Short-Channel N/P MOS loading, the whole wafer uniformity and Short-Channel (SC)/Long-Channel (LC) pattern loading. Taking into consideration of 3D structure metal gate in FinFET device, the challenge of metal gate is much higher than planar structure. The etch steps design should be specially designed with cyclic mode to obtain loading in diverse pattern with different film stacks of SC N/P MOS devices. Chemical reaction mechanism and physical bombardment mechanism should be taken good care to fine-tune etch recess depth between SC and LC, which is sensitive to LC device leakage performance. The accuracy control of patterning loading and uniformity are main subjects of metal gate etch back process.

## EXPERIMENT

All the dry etching experiments were performed in the same types of 300mm commercial etcher of ICP system with tunable ESC (Electrostatic Chuck) and TCCT (Transformer Coupled Capacitive Tuning). TiN, TaN, HfO2 and W were deposited in the trench of dummy gates in which Poly is already removed. Cross section pictures of work function metal and W are performed using SEM (Scanning Electron Microscope), TEM (Transmission Electron Microscope) and EDX (Energy Dispersive X-Ray Spectroscopy). Test wafer condition is as blow: Wafer diameter size is 300mm. Film stack is patterned with HK(HfO2)/WFM(TiN/TaN)/W as Figure 1 (a).

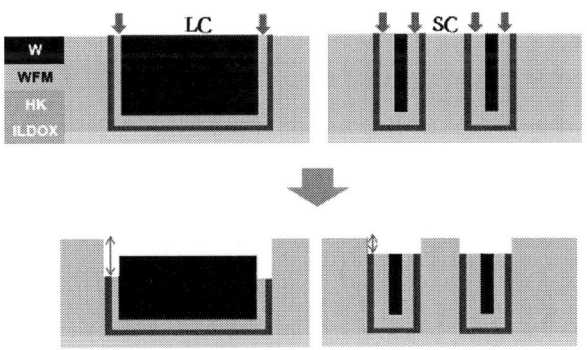

*Figure 1: Metal Gate Etch Back Process in Long Channel and Short Channel*

## RESULT AND DISCUSSION

*A. Metal Gate Etch Back Process Introduction*

In this process, HfO2/TiN/TaN/W are successively deposited into trench after dummy poly gate remove process. Consequently, other than planar structure etch that etching one kind of film and stopping on another kind of film, metal gate etch back process must etch four kinds of film simultaneously, which make this process more complex. In order to obtain smooth TiN/TaN cross profile with acceptable HfO2/W loading, the etch process should be design skillfully.

BCl3/Cl2-compositional gas flow is designed principally to etch TiN/TaN with selectivity to HfO2 and W. As the by-product polymer is heavy, an accompanying etch step is involved to remove by-product. These two etch steps can be cyclic to obtain uniform TiN/TaN profile. While W is etched by NF3-compositional gas flow, a polymer deposition combined with Ar bombardment etch steps aim to etch HfO2 closed to sidewall and smooth W cross profile. The group of above five etch steps is experimented in metal gate etch back process and the optimization of etch steps need systemically consideration of SC N/P MOS loading and SC/LC NMOS depth loading.

### B. SC N/P MOS loading

In FinFET technology node, there are several kinds of metal gate film stacks with different TiN/TaN/W thickness for purpose of modulation of different Vt device. This complex enhance the difficulty of metal gate etch back process and arise the issue of TiN/TaN SC N/P MOS depth loading. Paying attention to the incoming film stacks of SC N/P MOS, it is found that PMOS TiN/TaN is much thicker than that of NMOS. In extreme condition there is no W deposition in PMOS, which lead to different W etch amount in W etch step. In this situation, the placement of W etch step is critical to solve SC N/P MOS loading. If W etch step is placed before TiN/TaN step, then W in NMOS will be etch obviously in NMOS while PMOS TiN/TaN is almost un-damaged due to high selectivity of W to TiN/TaN in this step. Then TiN/TaN in NMOS will be attacked by above/sidewall direction while TiN/TaN in PMOS is only etched by above plasma, which lead to much deeper etch of TiN/TaN in NMOS. The solution of this problem is adjustment the etch sequence of W-TiN/TaN etch steps from W first to TiN/TaN first. Then the TiN/TaN etch of N/P MOS are only from above direction and lower N/P MOS loading can be obtain, as shown in Figure (2).

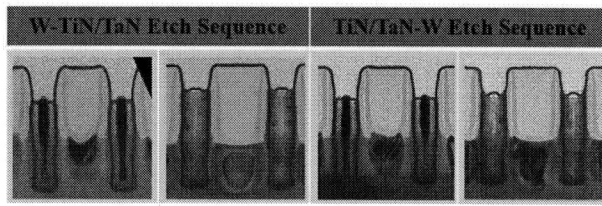

*Figure 2: Short-Channel N/P MOS loading in different etch sequence.*

### C. NMOS SC/LC loading

Except the SC N/P MOS loading due to different film stacks, there is also TiN/TaN depth loading between the NMOS Short-channel and Long-channel. By contrast of the local pattern density, the SC can be considered to be denes area while the LC is isolate area. At this moment, TiN/TaN etch depth of SC in dense area is much lower than that of LC in isolate area. If this level of TiN/TaN depth loading is not decreased, the etch depth of LC will probably too much to touch Fin top and cause Fin damage for the sake of enough etch depth of SC. In order to improve this typical clean mode of etch depth loading, the TiN/TaN etch recipe should be analyzed carefully. In clean mode etch recipe, the main etch mechanism is considered to be chemical reaction, resulting in abundant etchant supply in isolate area with deeper etch depth and minor etchant supply in dense area with shallow etch depth. Therefore, fine-tune chemical ratio of BCl3/Cl2 gas flow in TiN/TaN etch step should be effective. Lower Cl2 or higher BCl3 gas flow will enhance the polymer deposition in TiN/TaN etch step. While polymer deposition mode is relative heavier in isolate area of LC, the deeper LC etch depth will obviously decrease and SC/LC etch loading will be reduced. At the same time, decrease etch amount of chemical reaction and increase physical bombardment etch amount also can be taken into consideration, for the less pattern loading of physical bombardment etch step. In this case of metal gate etch back process, decrease TiN/TaN etch time to decrease the component of chemical reaction and increase Ar step etch time to increase physical bombardment etch amount will greatly lower NMOS SC/LC loading, as shown in Figure (3).

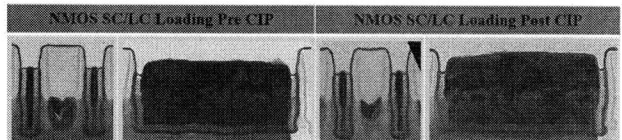

*Figure 3: NMOS SC/LC loading comparison pre and post CIP.*

### D. whole wafer uniformity

After fine-tune the complex pattern loading, the uniformity of single pattern become critical issue, only that TiN/TaN, W and HfO2 uniformity should be consider at the same time. In general, TCCT (Transformer Coupled Capacitive Tuning), gas flow injection ratio and ESC temperature of etch recipe is most effective method. But in metal gate etch back process, it is demonstrated that TCCT and gas flow inject ratio is not efficient for uniformity tuning. The ESC temperature is proved to affect uniformity directly. By ESC temperature tuning, the uniformity of TiN/TaN etch depth of NMOS can be obviously decrease to ~50%, as shown in Figure (4). Due to high selectivity of every single etch step, the accurate tuning of different film can be executed in corresponding etch steps.

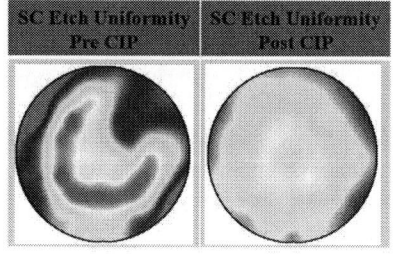

*Figure 4: Wafer uniformity comparison pre and post CIP.*

## CONCLUSIONS

In this work, Metal gate BARC etch have been investigated in detail. BARC etch loop is described by schematic diagram for easy understanding. For a health BARC etch process, pattern loading of dense/isolated area, reasonable over etch amount to balance TiN damage and BARC residue clearance, vertical BARC profile are the main issue. Depending on yield and WAT test, metal gate BARC etch process can be further improved after overall consideration.

## ACKNOWLEDGMENTS

Thanks for all the authors for great work and collaboration. Thanks for FA (failure analysis) for TEM and EDX image support.

## REFERENCES

[1] S. Senturia. Proceedings of Transducers2003, Boston, June 8-12, 2003, pp. 10-15.

[1] W. Wu, M. Chan. *IEEE Electron Device Letters*, Vol. 27, 2006, pp. 68-70.

[2] R. A. Wachnik, S. Lee, L. H. Pan, N. Lu, H. Li, R. Bingert, M. Randall, S. Springer, C. Putnam. *Proceedings of the IEEE 2013 Custom Integrated Circuits Conference*, DOI: 10.1109/CICC.2013.6658494.

[3] J. Xu, A. Wang, J. He, X. Jing, Z. Zhang, B. Zhang. *2017 China Semiconductor Technology International Conference (CSTIC)*, DOI: 10.1109/CSTIC.2017.7919796.

[4] Y. Karzhavin, W. Wu. *1998 3rd International Symposium on Plasma Process-Induced Damage (Cat. No.98EX100)*, DOI: 10.1109/PPID.1998.725579.

[5] A.T. Krishnan, V. Reddy, S. Krishnan. *International Electron Devices Meeting. Technical Digest (Cat. No.01CH37224)*, DOI: 10.1109/IEDM.2001.979650.

# Statistic big data analysis method used for Tilting mismatch problem solving

Qingpeng Wang, Rui Bao, Shaw Zhang, Lisa Wu, Cheng Li

Lam Research Inc.
Shanghai, China
Rui.bao@lamresearch.com

## Abstract

In this paper, we studied the cause of tilting performance in channel hole process. A systemic investigation was performed to understand which component or element leads to tilting mismatch with a real production case. We demonstrated a new statistical methodology, combining big data analysis and semiconductor domain knowledge, to solve this problem. Our results indicate that big data analytics can easily determine the abnormality at tool sensor level and pinpoint the possible root cause, and furthermore to derive corrective actions to solve the problem. It can save additional measurement and testing time. With a closed loop of "Root Cause Corrective Action" (RCCA), we can speed up the troubleshooting process and master a comprehensive view of wafer data, which transcends the traditional data analysis method.

*Keywords—tilting mismatch; 3D NAND; statistic big data analysis; break down to sensor level;*

## Introduction

With advancing semiconductor process development and new technology, structure profile control becomes increasingly important due to its significant impact to yield and device performance. In 3D NAND, channel hole etch is a critical part of structure profiles control. Tilting is a key characteristic in this application, and tilting performance can be directly impacted by lots of factors, such as temperature, pressure, RF power, etc.

Tilting mismatch thus becomes a very common issue in semiconductor production, tool performance can directly impact and cause this phenomenon. Traditional data analysis and detection method have limits, as it requires additional inline measurements and testing, which can reduce the tool availability and throughput. A new methodology is required to speed up the problem solving process.

In this paper, we used a new statistic system called "Equipment Intelligent Data Analyzer" to utilize the comprehensive amounts of tool detected data. Combining with inline data and domain knowledge, it expands big data analysis into the semiconductor industry, and achieves the problem solving objectives without additional inline measurement and testing. It demonstrates a new direction of semiconductor data analysis and troubleshooting.

## Titling mismatch investigation, big data analysis on leading parameters

### A. Titling performance mismatch investigation

Channel hole titling mismatch usually can be found at 3D NAND product (See Figure 1) It is a common phenomenon from channel hole etch process.

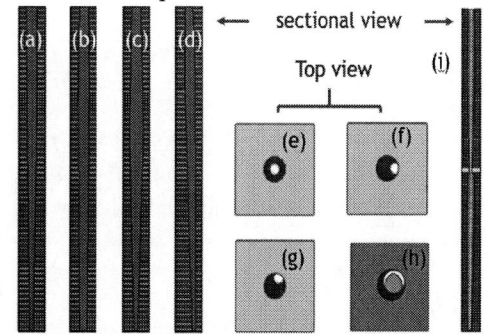

Fig.1. channel hole tilting performance mismatch in 3D NAND

To understand factors leading to channel hole tilting in real production scenarios, vast and valid wafer data must be collected, consolidated, and analyzed. "Equipment Intelligent Data Analyzer" system provides a novel effective data combination and dimensionality reduction method (DRM) to analyze huge amount of data from production tools. The proximity functions collects a mass of different types of data and merge them into one standard format, then Dimensionality Reduction is executed to reduce to 2D or 3D, allowing us to compute and visualize the data patterns.

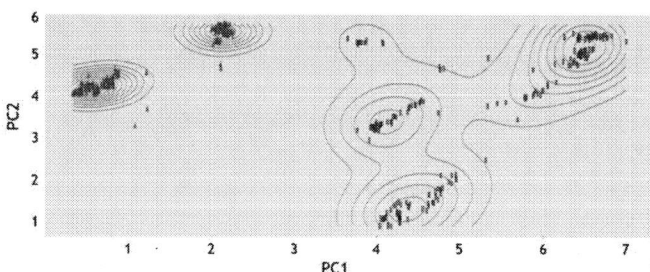

Fig 2. Separation among chambers is seen based on-wafer tilting data and tool sensor data reduced to 2 principal components (PC)

Dimensionality reduction analysis was performed to visualize data patterns. In this study, data from over 100,000 wafers were included. Each point represents one wafer, and contains over 100 vectors in data vector space. Each vector is reflected on independent space axes, and the axis which contains each vector maximum projection is picked as Principle component axis 1 (PC1). Principle component axis 2 (PC2) is selected if it contains second largest vector projection. In such way, the complete wafer data achieve dimensionality reduction through reflection on PC1 and PC2. Data can be visualized in 2D space, and retain its key features in maximum extent.

$$E^T C E = \Lambda = \begin{pmatrix} \lambda_1 & & & \\ & \lambda_2 & & \\ & & \ddots & \\ & & & \lambda_n \end{pmatrix}$$

(eq. (1)

*B. Tilting mismatch analysis against different groups from tool data*

In the analysis system, ungrouping chamber is identified as Class A which represents chamber performance need to be improved. Other grouping chamber is identified as Class B which represent the golden data set targeted, System performance analysis is based on this two data groups, and identifies the major disparate component between them. Top 20 components or parameters are listed, following the descending order. Data processing is based on the algorithm related with decision tree. This is an efficient method to break down entire analysis to parameter or sensor level when combining dimensionality reduction. It finally leads data computing to focus on individual component on tool.

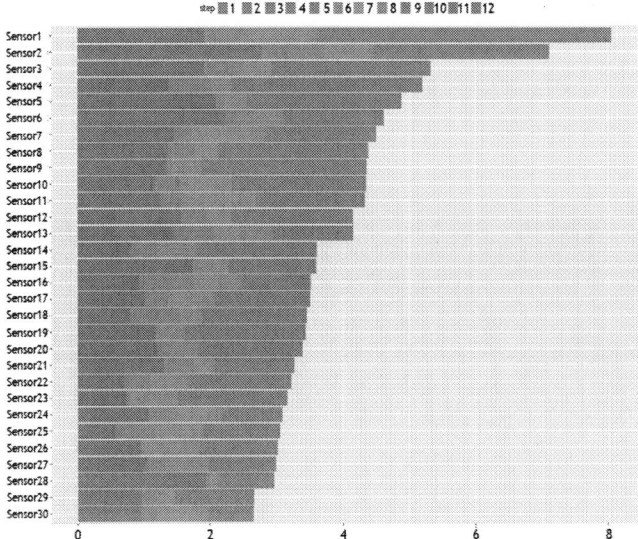

Fig.3. Data analysis break down to individual component level.

To achieve the objectives of this mismatch analysis, the key question is: which component or parameter actually leads to tilting? We investigated this by drilling down to each component's behavior with domain knowledge, and referring the sensor data of other chambers to cross verify. Suspected sensors are reviewed one by one, ruling out the normal

trending noise within one PM cycle. Once a few sensors related to key parameters are identified, extra consideration is needed to carefully check and make sure this selection is valid and trustworthy, as sometimes more evidence is needed to prove the theory behind whole data analysis.

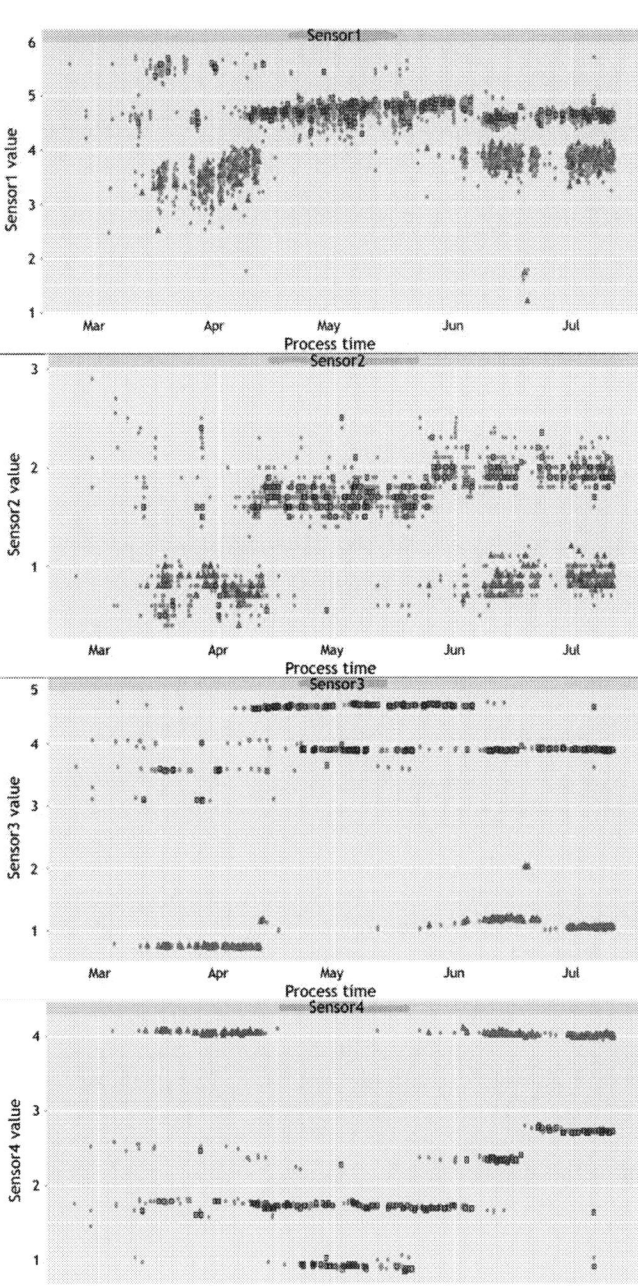

Fig.4. Drilling down to each component's behavior and detail trend of the suspected sensor list.

Using this methodology, we drilled down and checked all Top 20 components' behaviors and detailed trends. Based on domain knowledge to eliminate several noisy signals, we finally narrowed down to highly suspected components as below :1. Sensor1; 2. Sensor2; 3. Sensor3; 4. Sensor4. Considering sensor characteristics and effects, Sensor 3 and Sensor 4 are the major components which may affect etching environment, and thus tilting performance.

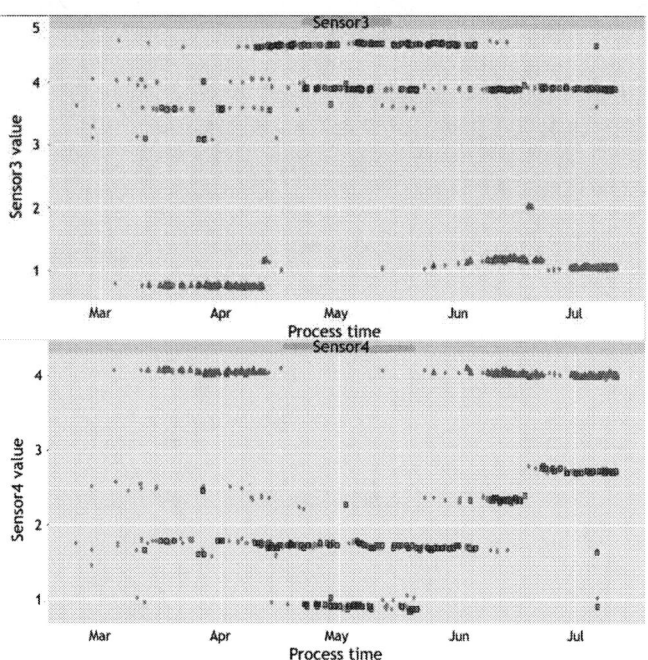

Fig.5. Narrowing down to two major components affecting the etching environment

### RCCA analysis based on statistic data feedback

Using the statistical data analysis, we narrowed down and focused on two major components affecting etching environment to cause the tilting mismatch issue. However, this is far from adequate, the analysis will be not be meaningful if tuning suspected components does not lead to acceptable and effective corrective actions to resolve the tilting mismatch issue. That's why root cause and corrective action must be a closed loop. For Sensor3, it may relate to RF loop performance, other related RF parameters need to be checked to verify the inference. For Sensor4, it may affect by pumping efficiency. In this case, we need to verify if chamber has heavy polymer or residual on chamber wall. When we broke down possible root cause of each component, then the possible corrective actions become obvious. In our case, one preventive maintenance is required. In addition, polymer and parts status check need to be performed at same time, and RF loop optimization is needed during this maintenance.

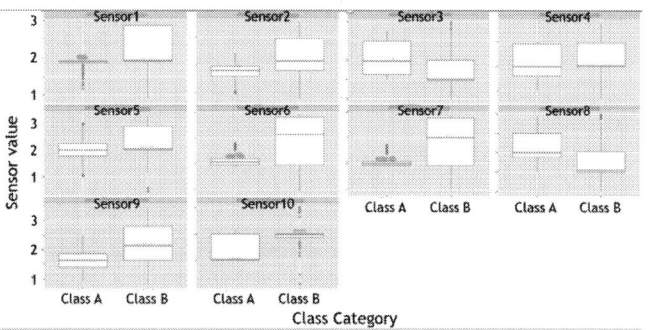

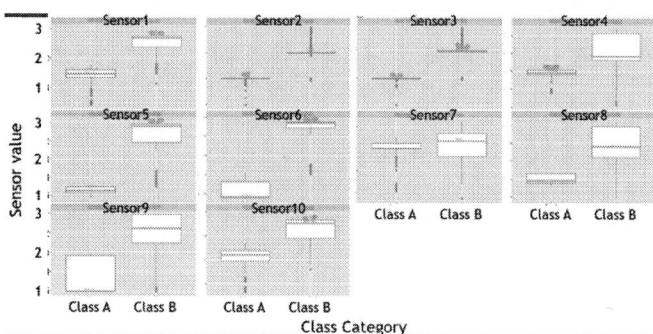

Fig.6. Checking related sensor or components in system to ensure the inference

During the next tool preventive maintenance, RF loop optimization was performed as planned, and the chamber polymer and parts status check were implemented. Certain parts did achieve its limit and need to be replaced. RF loop was found drifting during optimization. After the corrective actions, both tool data and tilting performance were verified. Tool system data were re-loaded and re-grouped, and improvement was shown on the analysis diagram indicating tool performance is getting closer. Tilting performance also showed positive feedback.

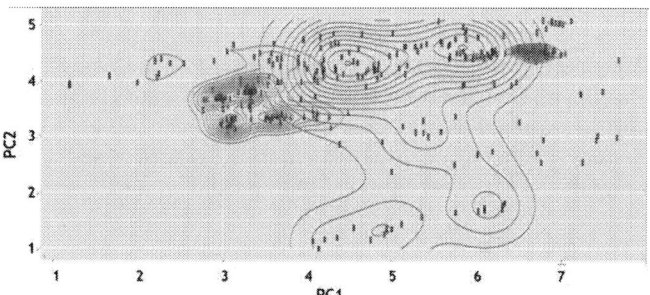

Fig.8. Data set showing grouping trend.

### Conclusion

In this paper, tilting mismatch issue was investigated from a new angle—statistical big data analysis by Lam "Equipment Intelligent Data Analyzer" system, Inline data is closely tied to specific tool data which is successfully narrowed down to sensor level to understand tool performance. It expands and enhances the ability of data analysis in semiconductor industry. Furthermore, with RCCA cycle, big data analysis benefits problem solving and directly pinpoints corrective actions. This case study shows this methodology is effective to verify the implementation and is of greater use in semiconductor manufacturing.

### References

[1] Leszczynski, M., Czernecki, R., Krukowski, S., Krysko, M., Targowski, G., Prystawko, P., Plesiewicz, J., Perlin, P. and Suski, T., 2011. Indium incorporation into InGaN and InAlN layers grown by metalorganic vapor phase epitaxy. Journal of crystal growth, 318(1), pp.496-499.

[2] Sugiyama, M., 2007. Dimensionality reduction of multimodal labeled data by local fisher discriminant analysis. Journal of machine learning research, 8(5).

[3] Wold, S., Esbensen, K. and Geladi, P., 1987. Principal component analysis. Chemometrics and intelligent laboratory systems, 2(1-3), pp.37-52.

[4] Fan, J., Han, F. and Liu, H., 2014. Challenges of big data analysis. National science review, 1(2), pp.293-314.

[5] Okes, D., 2019. Root cause analysis: The core of problem solving and corrective action. Quality Press.

# SELECTIVE WET-ETCHING OF GESI IN MULTI-LAYER GESI/SI STACKS

*Jiajia Tian [1,2], Zhijun Cao [2], Qingzhu Zhang [2]\*, Cinan Wu [1], Zhaohao Zhang [2], Huaxiang Yin [2]*

1. College of Big Data and Information Engineering，Guizhou University Guiyang 550025，China;
2. Key Laboratory of Microelectronic Devices and Integration Technology，Institute of Microelectronics，Chinese Academy of Sciences，Beijing 100029，China;

\*Corresponding Author's Email: zhangqingzhu@ime.ac.cn;  cnwu@gzu.edu.cn

## ABSTRACT

In this paper, the channel release process of stacked gate-all-around (GAA) nanosheet (NS) devices fabricated based on the epitaxial scheme is investigated extensively. The effects of annealing temperatures, thicknesses of the GeSi layers and liquid nitrogen processing on the wet selective etching of GeSi in multi-layer GeSi/Si samples are mainly studied. It is found that the corrosion rate of the GeSi in GeSi/Si stacks samples first decreases and then increases with the increasement of rapid thermal annealing (RTA) temperatures, and the corrosion rate is the slowest at 600℃. In addition, the thicker the epitaxial GeSi, the faster the corrosion rate. Furthermore, we found that the corrosion rate of the samples after liquid nitrogen processing treatment was faster than that without treatment. The research provides a good reference significance for the design and manufacture of advanced stacked GAA Si NS devices.

## INTRODUCTION

With continuous scaling integrated circuits (ICs) along the Moore's law, Conventional fin field effect transistors (FinFET) devices faces more complex fabrication process, non-uniform threshold voltage and serious mobility degradation[1-4]. stacked gate-all-around (GAA) Si nanosheet (NS) FET has been considered as one of the most promising candidates for 3nm node and beyond[5-7]. However, the stacked GAA Si NS FET also faces many challenges, such as selective Si NS channel release, new stack parasitic channel suppression approach and the risk of Ge contaminations[8].

In this paper, the effect of the rapid thermal annealing (RTA)  temperature and thickness of GeSi layer on selective wet-etching of GeSi in GeSi/Si layer was studied in detail. It is found that the corrosion rate of the GeSi layer in laminated samples first decreases and then increases with the increasement of RTA temperatures, and the thicker the epitaxial GeSi, the faster the corrosion rate. In addition, we found that the corrosion rate of the samples after liquid nitrogen processing treatment was faster than that without treatment.

## EXPERIMENTAL

8 inches *p*-type Si <100> wafers were used with 8-12 Ω•cm resistivity. The wafers was load into a ASME2000 plus pressure reduced pursue chemical vapor deposition (RPCVD) tool for epitaxial growth of multi-layer GeSi/Si  stacks after removing the native oxide (see fig 2 (e)). In order to observe and analyze the selective etching results of multilayer GeSi/Si layes, a rectangular arrays with spacing of 4μm were formed by lithography and etching processes (see fig.1 (g)). After epitaxial of multi-layer GeSi/Si stacks, the samples was cut into small species and  annealed at different RTA temperatures (500℃，600℃，700℃，800℃,900℃ and 1000℃) for 30s. In the  flowing step, a mixed solution of HF(6%)：$H_2O_2$(30%)：$CH_3COOH$(99.8%)=1:2:3 holding for 24 hours was used to etch the samples with different annealing temperatures and GeSi thickness [9]. The etched samples were characterized by scanning electron microscope (SEM) to observe the etching  morphology and measure the etch  depths.

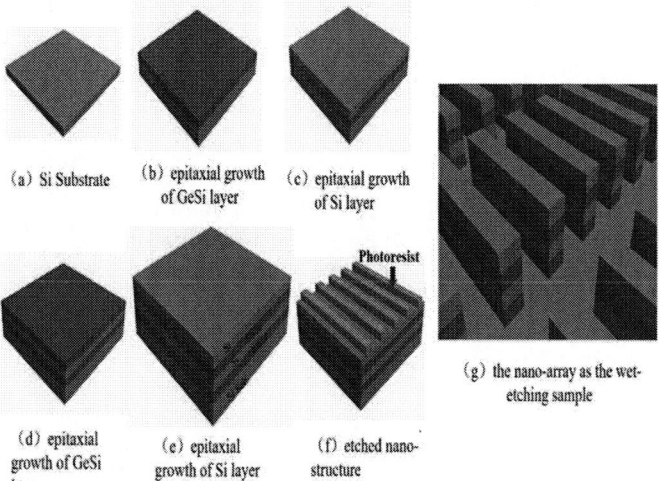

*Fig.1  Fabrication flow of the stacked GeSi/Si samples*

## RESULTS AND DISCUSSION

Fig. 2 is SEM images of the corroded GeSi samples after different RTA temperatures and corresponding etching lengths of the GeSi layers is show in Fig. 3. As can be seen from the images, the etching lengths exhibits firstly decreases, and then increases with the increase of RTA temperatures. The corrosion rate is the slowest at 600 ℃ [10]. As the epitaxial growth temperature of GeSi/Si stacks is 650℃, it deviates from the growth temperature of epitaxial lamination regardless of high or low temperature annealing. The RTA annealing  leads to part of the stress is released, which generates more defects at the GeSi/Si interface and increases the corrosion rate of GeSi layer. When the annealing temperature is 1000℃, the thickness of Si layer becomes thinner and exhibits a little "warping" (see fig. 2 (f)).

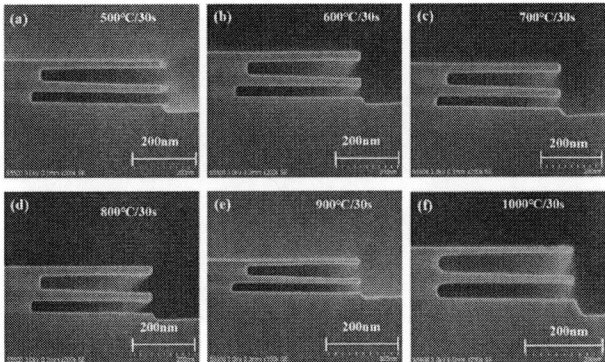

Fig.2 SEM images of the corroded GeSi samples after different RTA temperatures

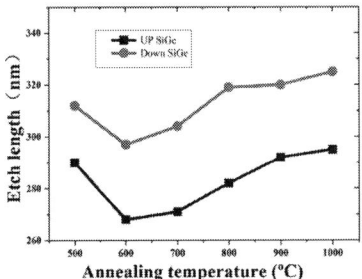

Fig.3 Variation of the corrosion depths of the GeSi layer after different RTA temperatures for 8 min.

In addition, the thickness of *GeSi* layer in the *GeSi*/Si stacks determines the spacing of the final stacked GAA Si NS channel, which would affect channel morphology and high-k/metal gate (HK/MG) filling. So the thicknesses of the GeSi layer (5nm, 10nm and 20nm) on the etching rate is also investigated, and the 20 nm Si layer acts as the interlayer between the GeSi layers. Fig.4 shows the SEM images of different corrosion times (1min, 3min, 5min, 7min, 9min, 11min and 13min, respectively) and the corresponding the etching lengths of different GeSi thicknesses as an function of the etching times is shown in Fig.5 As can been seen from the images, the thicker the epitaxial GeSi, the faster the corrosion rate. In addition, the longer Si layers after more corrosion of GeSi layer are prone to "adhesion", and the thiner GeSi layer is more prone to "adhesion" because a greater surface tension of the liquid in a narrow space [11-12].

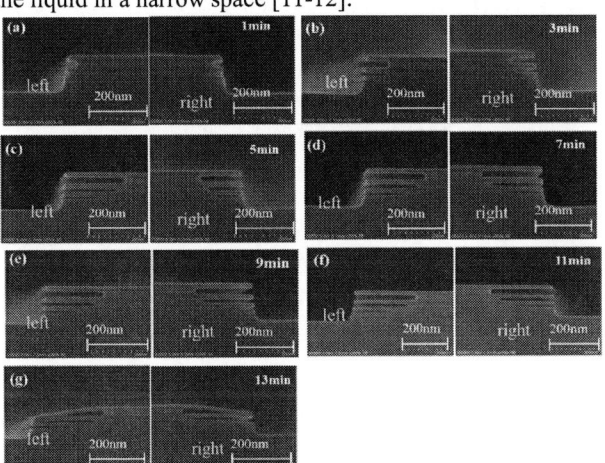

Fig. 4 SEM images of the corroded GeSi samples with different GeSi thicknesses.

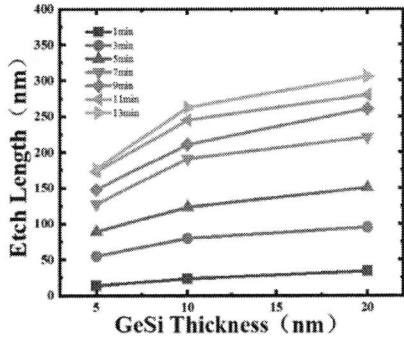

Fig. 5 Variation of the corrosion depth of the GeSi layer with different GeSi thicknesses.

The effect of liquid nitrogen processing on the selective corrosion of GeSi/Si layers was also studied. Then GeSi/Si samples after heated to 125℃, were put into liquid nitrogen processing rapidly. Finally, the samples treated with and without liquid nitrogen were corroded for 1min, 3min, 5min, 7min, 9min, 11min and 13min, and the SEM scanning figure as shown in fig.6 and fig.7, respectively. As can be seen from the images, both samples treated with and without liquid nitrogen achieve high selective ratio GeSi corrosion, but the corrosion footing effect after liquid nitrogen treatment is improved. Fig.8 shows the comparison of corrosion lengths with and without liquid nitrogen treatment. It can be seen that the etching depth with and without liquid nitrogen treatment is almost the same when the corrosion time is less than 2 min. However, the corrosion rate of samples treated with liquid nitrogen is faster than that without treatment when the etching time over 2 min. When the corrosion time exceed 8 min, the depth is greater than 300nm. The corrosion rates of both cavities begin to slow down because the reaction products in the deep cavity cannot exchange quickly, which limit the progress of the reaction. The above results shows that the thermal expansion coefficient of GeSi and Si materials treated with liquid nitrogen is different. The processing of liquid nitrogen caused the rupture of the GeSi/Si interfacial bond, resulting in a faster corrosion rate and improved the "footing" effect.

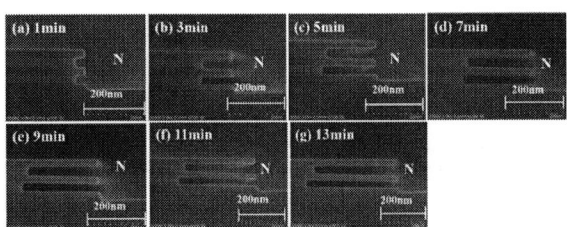

Fig.6 SEM images of stacked GeSi samples with different etching times after liquid nitrogen processing

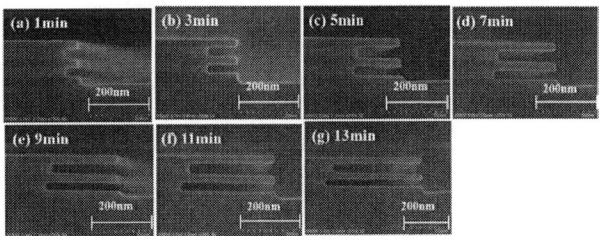

Fig.7 SEM images of stacked GeSi samples with different

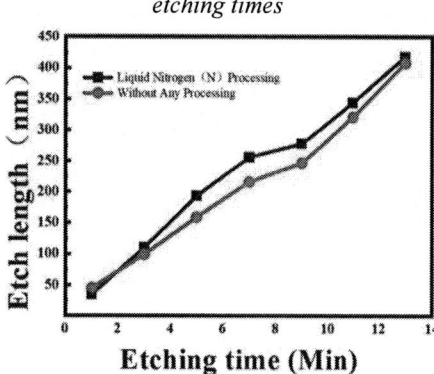

*etching times*

Fig.8 Corrosion of the depths of the GeSi layer with or without liquid nitrogen processing.

## CONCLUSION

In this paper, the effects of annealing temperature and GeSi thickness on the etching of stacked GeSi/Si layers were investigated extensively. It is found that the corrosion rate of laminated samples first decreases and then increases with the increasement of rapid thermal annealing (RTA) temperatures, and the corrosion rate is the slowest at 600 ℃. It is also found that the thicker the epitaxial GeSi layer, the faster the corrosion rate. The effect of liquid nitrogen processing on the selective corrosion of GeSi/Si layers was also studied. The processing of liquid nitrogen caused the rupture of the GeSi/Si interfacial bond, resulting in a faster corrosion rate and improved L "footing". The research provides a good significance reference for the design and manufacture of advanced stacked GAA Si NS devices.

## ACKNOWLEDGEMENTS

This work was supported in part by the Pilot Project of the Chinese Academy of Sciences under grants E1XDC2X002, in part by the Joint Development Program of Semiconductor Technology Innovation Center (Beijing), Corp, in part by the Science and technology program of Beijing Municipal Science and Technology Commission under grants Z201100006820084, in part by the Youth Innovation Promotion Association, Chinese Academy of Sciences under grant Y9YQ01R004, in part by the National Natural Science Foundation of China under grants 61904194 and 6187032253. We thank the institute of microelectronics, Chinese Academy of Sciences (IMECAS) IC advanced technology center (ICAC) on its advanced 200 mm CMOS platform has completed the study.

## REFERENCES

[1] Zhang, Q. , Yin, H. , Luo, J. , Hong, Y. , & Ye, T. . (2017). FOI FinFET with ultra-low parasitic resistance enabled by fully metallic source and drain formation on isolated bulk-fin. 2016 IEEE International Electron Devices Meeting (IEDM). IEEE.

[2] Zhang, Q. , Tu, H. , Yin, H. , Feng, W. , & Wang, W. . (2018). Si Nanowire Biosensors Using a FinFET Fabrication Process for Real Time Monitoring Cellular Ion Actitivies. 2018 IEEE International Electron Devices Meeting (IEDM). IEEE.

[3] Zhang, Z. , Xu, G. , Zhang, Q. , Hou, Z. , Li, J. , & Kong, Z. , et al. (2019). FinFET with improved subthreshold swing and drain current using 3nm ferroelectric Hf0.5Zr0.5O2. IEEE Electron Device Letters, 1-1.

[4] Zhang, Q. , Li, J. , Tu, H. , Yi, H. , & Wang, W. . (2018). Self-aligned metallic source and drain fin-on-insulator FinFETs with excellent short channel effects immunity down to 20 nm gate length. 2018 China Semiconductor Technology International Conference (CSTIC).

[5] Mertens, H. , Ritzenthaler, R. , Hikavyy, A. , Kim, M. S. , Tao, Z. , & Wostyn, K. , et al. (2016). Gate-all-around MOSFETs based on vertically stacked horizontal Si nanowires in a replacement metal gate process on bulk Si substrates. IEEE Symposium on Vlsi Technology. IEEE.

[6] Loubet, N. , Hook, T. , Montanini , P. , Yeung, C. W. , & Khare, M. . (2017). Stacked nanosheet gate-all-around transistor to enable scaling beyond FinFET. 2017 Symposium on VLSI Technology. IEEE.

[7] Tiwari, S. , Dutt, A. , Joshi, M. , Nigam, P. , Beohar, A. , & Mathew, R.(2021).In-silico investigation of cyl. Gate-all-around (GAA) tunnel field effect transistor (t-FET) biosensor. IOP Conference Series: Materials Science and Engineering, 1166(1), 012045.

[8] Zhang, Q., Gu, J., Xu, R., Cao, L., & Luo, J. . (2021). Optimization of structure and electrical characteristics for four-layer vertically-stacked horizontal gate-all-around si nanosheets devices. Nanomaterials, 11(2), 646.

[9] Carns T K , Tanner M O , Wang K L . CHEMICAL ETCHING OF SI1-XGEX IN HF-H2O2-CH3COOH. Journal of the Electrochemical Society, 1995, 142(4):1260-1266.

[10] Zhang Q, Tu H, Gu S, et al. Influence of Rapid Thermal Annealing on Ge-Si Interdiffusion in Epitaxial Multilayer Ge0.3Si0.7/Si Superlattices with Various GeSi Thicknesses. ECS Journal of Solid State Science and Technology, 2018, 7(11): P671-P676.

[11] Alberti G , Narducci R, Sganappa M. Effects of hydrothermal/thermal treatments on the water-uptake of Nafion membranes and relations with changes of conformation, counter-elastic force and tensile modulus of the matrix. Journal of Power Sources, 2008, 178(2): 575-583.

[12] Xia G., Hoyt J L. Si–Ge interdiffusion under oxidizing conditions in epitaxial GeSi heterostructures with high compressive stress. Applied Physics Letters, 2010, 96(12): 122107.

# Optimization of approach for metal contamination reduction

Meng-Yu Xie[1]*, Jian-Kun Zhang[1], Zun-Hua Zhao[1], Zhong-Hai Tang[1]
Jian-Zhu Zhuang[2], Wei Wan[2], Bing-Hui Lin[2], Da-Wei Tao[2], Mei-Hua Liu[2], Pan Wang[2]

[1] Beijing NAURA Microelectronics Equipment Co. Ltd
No. 8 Wenchang Avenue, Beijing Economic-Technological Development Area. Beijing City, China
[2] ChangXin Memory Technologies, Inc. Ltd
No. 388, Xingye Avenue, Economic and Technological Development Area, Hefei City, China

*Corresponding Author's Email: xiemengyu@naura.com

## Abstract

The importance of metal contamination in semiconductor processing and the ultimate yield effects has long been discussed in literature, the ITRS, and most importantly in the fab. Analysis methods including TXRF, VPD ICP-MS, and TOF-SIMS have been used to verify that wafers are properly "clean", In this paper, In order to further reduce the content of metal elements in the plasma etching tools, we have further optimized its detection procedure and explored the results and mechanism of different test preparation procedures on the metal contamination detection. It is worth noting that elements Analysis choice has often been chosen via an instrument that is most convenient and quickest for the fab. Thus, in this work we use an automated vapor phase decomposition (VPD) process to collect and detect the metal contamination. At the same time, in this study, the metal contamination in the plasma etching machine has been significantly reduce after optimization, the actual production process, the problem of excessive Cr element in metal contamination detection was effectively solved, and the success rate of this type of detection increased from about 50% to 98% ~100%.

*Key words: plasma etching; metal contamination detection; vapor phase decomposition (VPD); test preparation procedure optimization;*

## Introduction

Routine and accurate monitoring of trace metal contamination is an important step in semi-conductor processing. As device features shrink and dielectrics scale toward the atomic level, metal concentrations of 5E09 atoms/cm$^2$ can affect semiconductor performance and product yields [1, 2]. Contamination in processing can affect yield in multiple ways. For example, during high temperature processing steps, metals can diffuse into the silicon substrate and act as recombination centers adding electronic states into the band gap of silicon. As a result minority carrier lifetimes are degraded. Metal contamination can also adversely affect silicon oxidation rates and become incorporated in gate oxides where they increase leakage currents by degrading the gate oxide integrity. VPD (for most elements on a 300 mm silicon wafer, sensitivity can reach E07 atoms/cm$^2$) is a proven method of wafer preparation for subsequent measurement of trace

metal contamination [3,4]. The problem of metal contamination has always been an issue that has been paid attention to by various tools vendor and fab. As shown in Fig.1, during the installation of the tools, due to the high content of elements such as Cr and Al, the success rate of this test is not high, and there is still a certain gap compared with well-known international manufacturers. Also due to bad results, we will spend more time re-doing a lot of cleaning on the equipment. This process greatly increases the work cost.

|  |  | 1$^{st}$ test | 2$^{st}$ test |
|---|---|---|---|
| 1$^{st}$ tool | A | √ | - |
|  | B | √ | - |
|  | C | × |  |
|  | D | √ | - |
| 2$^{st}$ tool | A | × | √ |
|  | B | × | × |
|  | C | × | √ |
|  | D | × | √ |
| 3$^{st}$ tool | A | × | √ |
|  | B | √ | - |
|  | C | × | × |
|  | D | √ | - |
| 4$^{st}$ tool | A | √ | - |
|  | B | √ | - |
|  | C | √ | - |
|  | D | √ | - |
| Pass rate: 1$^{st}$ test: 50%; 2$^{st}$ test: 71.2% | | | |

Fig.1 The pass rate of metal contamination detection with the old test preparation procedure

Before the detecting of trace metal contamination, we always perform simple cleaning of the chamber of the tools, such as full prevention maintenance (PM), cycle purge and introducing the corresponding gas to react with pollutant elements. The current practice is to pass chlorine into the chamber and apply electrical energy to the chamber to ionize it. Since chlorine element has a strong electronegativity, it will react with metal elements in the chamber as Eq.1:

$$M + Cl\cdot \rightarrow MCl_x\uparrow \quad (1)$$

The main problem of this method is that the metal contamination components in the chamber cannot be completely cleaned by this kind oxidation-reduction reaction, and the surface of the wafer will still be contaminated with metal elements, making the metal contamination test fail.

Meanwhile, we can learn from related documents and patents that in actual semiconductor production, most chip foundries test the metal contamination of the inductively coupled plasma etching machine as follows:

The first part is contact test between the backside of the wafer and the parts of the tools: the procedure of this test requires the front side of the wafer to be reversed in advance, so that the front side of the wafer contacts all the parts in the transmission path that the machine passes through, and finally transmits it Back, detect the metal content in the wafer. When the metal content on the back side exceeds $10E+10A/cm^2$, the chamber needs to be cleaned again and then tested. When the testing process on the back side of the wafer is completed, we will test the metal elements content on the other side immediately.

## Experimental

In order to understand the effect of different preparation actions before the VPD test on the content of residual metal pollution elements in chamber, we designed two other pre-test preparation procedures based on the common specific test preparation procedures, as shown in Fig.2:

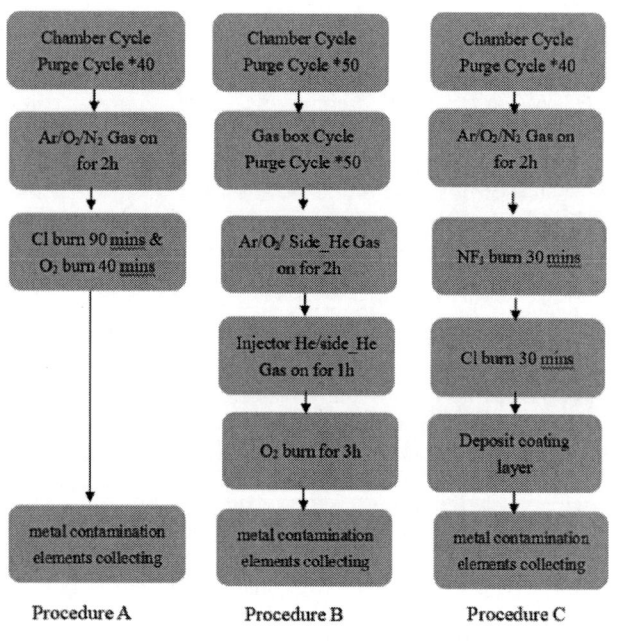

Fig.2 Three kinds of test preparation procedure

1. Procedure A is our commonly used VPD test preparation process. First, we cleaned the chamber by gas circulation, set the chamber swing valve position to the maximum, and let the chamber be filled with nitrogen gas and then pumped out (the flow rate is 500sccm of nitrogen, clean the entire chamber for the first time, lasting an hour) The main purpose of this step is to purge the sloughed off foreign particles out of the chamber through a large flow of gas. Then we introduced oxygen and chlorine to the chamber successively, and ionized it by applying energy (these two steps called $O_2$ burn and $Cl_2$ burn respectively, $O_2$ burn and $Cl_2$ burn recipe are respectively

shown in Table1 and Table2). The high-energy oxygen particles and chlorine particles will react with carbon, hydrogen and metal elements respectively, and the by-products of the reaction will be pumped out of the chamber with turbo pump.

2. Procedure B: In order to further enhance the effect of gas purging and cleaning, we also added a cycle purge step to the gas box, and introduced helium side gas into the chamber. At the same time, we canceled the $Cl_2$ burn step.

3. Procedure C : Based on procedure A, we mainly add the Dep step before collecting the metal elements. In this step, we coat the chamber wall with a Silicon oxide ($SiO_2$) coating layer. The Dep recipe is shown in Table 4.

| Step | Pressure (mT) | Power (SRF/BRF) | Gas (sccm) | VAT position | Time (s) |
|---|---|---|---|---|---|
| trans | 15 | 0W/0W | $300O_2$ | NA | 5 |
| $O_2$ | 15 | 1000W/0W | $300O_2$ | NA | 600 |
| Purge | PV Open | 0W/0W | 300Ar | 1000 | 10 |

Table.1 $O_2$ burn step recipe.

| Step | Pressure (mT) | Power (SRF/BRF) | Gas (sccm) | VAT position | Time (s) |
|---|---|---|---|---|---|
| DC1 | 15 | 0W/0W | $200O_2$ | NA | 10 |
| DC2 | 15 | 1000W/0W | $200Cl_2$ | NA | 1800 |
| Purge | PV Open | 0W/0W | 200He | 1000 | 10 |

Table.2 $Cl_2$ burn step recipe.

| Step | Pressure (mT) | Power (W) | Gas (sccm) | Time (s) |
|---|---|---|---|---|
| Clean1 | 400 | 1800W | $500NF_3/200Ar$ | 600 |
| Clean2 | 65 | 1800W | $200NF_3/100Ar$ | 600 |
| Clean3 | 30 | 1800W | $50NF_3/50Ar$ | 600 |
| Clean4 | 15 | 1500W | $400O_2$ | 600 |

Table.3 $NF_3$ burn step recipe.

| Step | Pressure (mT) | Src Power(W) | Gas (sccm) | Time (s) |
|---|---|---|---|---|
| Dep | 15 | 1200 | $100SiCl_4/200O_2/200Ar$ | 20s |

Table.4 Dep step recipe.

## Result and discussion

Table 5 shows that VPD test results by using procedure A, it can be seen clearly that the residual chromium (Cr) elements content on the surface of the wafer exceeds the maximum value specified by the spec. As we known that the Cr element is one of the main constituent elements in stainless steel, but the gas line in the tools is made of stainless steel containing Cr. At the same time, a large number of documents pointed out that when the Cr in the stainless steel will undergo a certain degree of corrosion reaction under the combined action of hydrogen ions and chloride ions [5]. When the gas line is ventilated, the residual Cr element particles will enter the chamber and fall on the surface of the wafer.

| CHM | Na | Mg | Al | K | Ca | Cr | Fe | Ni | Cu | Zn |
|-----|-----|-----|-----|-----|-----|-----|-----|-----|-----|-----|
| PM1 | 0.003 | 0.011 | 0.029 | 0.002 | 0.026 | 3.979 | 0.049 | 0.007 | 0.035 | 0.029 |
| PM2 | 0.004 | 0.007 | 0.02 | 0.002 | 0.019 | 5.688 | 0.057 | 0.002 | 0.014 | 0.024 |
| PM3 | 0.003 | 0.007 | 0.021 | 0.001 | 0.018 | 5.416 | 0.128 | 0.004 | 0.012 | 0.016 |
| PM4 | 0.006 | 0.009 | 0.033 | 0.03 | 0.017 | 3.633 | 0.041 | 0.003 | 0.013 | 0.029 |

Unit: 1E+10A/cm²; SPEC: 1E+10A/cm²

Table.5. VPD test results of procedure A

In order to explore the role played by $Cl_2$ burn step in the entire preparation procedure, we designed the experimental process of the test B based on the test A. As shown in Fig.1, in procedure B, we enhance the strength of gas purge, cancel the $Cl_2$ burn step directly. The final experimental results are shown in Table 6. It can be seen that the overall content of all detected metal elements has been significantly reduced. For the content of Cr, there is even a change of more than two orders of magnitude, and the reduction of the content of Fe is up to about 1000%. This result also confirms our above analysis from the side. By eliminating the $Cl_2$ burn step, the metal pollution source brought by the gas line can be effectively eliminated, and the residual metal elements in the chamber can be reduced at the same time.

| CHM | Na | Mg | Al | K | Ca | Cr | Fe | Ni | Cu | Zn |
|-----|-----|-----|-----|-----|-----|-----|-----|-----|-----|-----|
| PM1 | 0.005 | 0.016 | 0.130 | 0.004 | 0.068 | 0.407 | 0.027 | 0.008 | 0.011 | 0.004 |
| PM2 | 0.025 | 0.006 | 0.165 | 0.011 | 0.018 | 0.092 | 0.257 | 0.042 | 0.009 | 0.005 |
| PM3 | 0.002 | 0.006 | 0.037 | 0.001 | 0.019 | 0.042 | 0.003 | 0.002 | 0.010 | 0.001 |
| PM4 | 0.002 | 0.004 | 0.024 | 0.001 | 0.022 | 0.041 | 0.004 | 0.002 | 0.011 | 0.001 |

Unit: 1E+10A/cm²; SPEC: 1E+10A/cm²

Table.6. VPD test results of procedure B

The above two sets of comparative experiments have basically proved that the elimination of $Cl_2$ burn in the procedure can effectively reduce the overall level of metal element content in the reaction chamber and meet the Spec. However, due to the continuous shrinking of the size of semiconductor devices in the future, the impact of metal particles on the yield of chip manufacturing will become more and more obvious. In order to further reduce the metal impurity content, we also designed another method based on procedure A. The test results are shown in table 7, which is very gratifying is when we deposit a dense oxide layer on the chamber wall before the metal elements collection, all the detected values have a very obvious drop, and almost all are at the same level. This result shows that the coating layer can effectively prevent impurity particles from falling off during the test. At the same time, the Procedure C is also the most effective way to reduce the concentration of metal elements

| CHM | Na | Mg | Al | K | Ca | Cr | Fe | Ni | Cu | Zn |
|-----|-----|-----|-----|-----|-----|-----|-----|-----|-----|-----|
| PM1 | 0.001 | 0.005 | 0.01 | 0.001 | 0.013 | 0.001 | 0.003 | 0.002 | 0.009 | 0.002 |
| PM2 | 0.001 | 0.008 | 0.019 | 0.001 | 0.018 | 0.001 | 0.003 | 0.002 | 0.001 | 0.003 |
| PM3 | 0.001 | 0.005 | 0.013 | 0.002 | 0.014 | 0.03 | 0.003 | 0.002 | 0.002 | 0.003 |
| PM4 | 0.002 | 0.005 | 0.013 | 0.001 | 0.015 | 0.018 | 0.007 | 0.002 | 0.015 | 0.013 |

Unit: 1E+10A/cm²; SPEC: 1E+10A/cm²

Table.7. VPD test results of procedure C

## Conclusion

In this study, three kinds of test preparation procedure of metal contamination detection are be sued to improve the test results. These results are compared and analyzed separately, it also preliminarily explored the relevant mechanism of result optimization. The conclusion are as below:

(1) The Chlorine has strong electronegativity and is easy to undergo oxidation-reduction reaction with metal elements. This reaction introduces more metal particles into the chamber.

(2) By canceling the Cl2 burn step, the metal element content in the chamber can be effectively reduced.

(3) Depositing a dense SiO2 layer on the chamber wall can effectively protect the chamber and is an effective way to reduce the metal contamination.

## References

[1] Beebe M, Anderson S. Monitoring wafer cleanliness and metal contamination via VPD ICP-MS: Case studies for next generation requirements [J]. Microelectronic Engineering, 2010, 87(9):1701-1705.

[2] Shimazaki A, Sakurai H, Iwase M, et al. Metallic Contamination Control in Leading-Edge ULSI Manufacturing [J]. Solid State Phenomena, 2009, 145-146:115-121.

[3] Maeda A, Kageyama M, Yoshii S, et al. Impurity measuring method [J]. 1991.

[4] Xu Z, Srivatsa A, Samsavar A, et al. Method and system for detecting metal contamination on a semiconductor wafer [J]. US, 2004.

[5] Gong S, Chu M, Zhou Y, et al. Reaction Mechanism for Cr in Stainless Steel Corroded by Hydrogen and Chloride Ion [J]. Corrosion & Protection, 2017.

978-1-6654-9759-6/22 $31.00 © 2022 IEEE

# Optimization of Shallow Trench Isolation CD micro loading in advance CMOS

Guang Yang[1], Zhong-Wei Jiang[1], Jian-Kun Zhang[1], Xiayu-Shi[2] , Su-Sheng Chang[2] , Yang Liu[2] , Ying-Yi Chen[2] ,

Zhong-Ning Guo[2] , Xian-Wen Su[2] , Xin-Wen Huang[2] , Bing-Hui Lin[2]

[1]Beijing NAURA Microelectronics Equipment Co. Ltd
No. 8 Wenchang Avenue, Beijing Economic-Technological Development Area. Beijing City, China
[2]ChangXin Memory Technologies, Inc. Ltd
No. 388, Xingye Avenue, Economic and Technological Development Area, Hefei City, China

*Corresponding Author's Email: xulitian@naura.com

## Abstract

This paper investigates the interest of CD microloading to etch silicon in shallow trench isolation (STI). The influence of different processing parameters like, e.g., pressure, gas flow rate, and flow direction on the CD microloading effect have been investigated. It is also shown that CD microloading are improved by, e.g., decreasing the pressure or RF bias power.

*Keywords—STI, pressure, CD micro loading, bias power*

.

## Introduction

Shallow Trenches Isolation (STI) etching is an important step of integrated circuits fabrication. In this process, trenches are etched in silicon to isolate active areas of different transistors, preventing leakage and parasitic effects. As VLSI technologies were scaled below 28nm, shallow trench isolation (STI) etching becomes more difficult with reductions in device size [1]. STI accommodates much smaller design dimensions, but presents new process integration and control challenges. For STI etching, device's effective electrical width is directly related to STI profile, any variation on critical dimension (CD), sidewall angle (SWA) and height (H) etc could strongly impact on electrostatic performance. Since little difference may result in much different on device performance, more rigorous requirement on STI physical performance and loading control becomes more critical [2-3]. One parameters is the microloading effect dependent variation of critical dimensions (CD) resulting from dry etch. This effect refers to the dependence of the overall etch rate on the density between dense and isolated area. In this paper, the influence of different processing parameters like, e.g., the electrode bias power, and pressure on the microloading effect is discussed between dense and isolated patterns for shallow trench isolation (STI) etching applications. It has been found that the microloading effect can be decrcased by, e.g., decreasing the pressure or bias power of HM and AC film etching.

## Experiment

The STI etch in this paper were completed on one commercial inductively coupled plasma (ICP) reactor (NAURA 612D) from China. The plasma is excited with two RF coils (13.56 MHz) to improve the ions flux uniformity.

Cross section profiles were imaged via Transmission Electron Microscope (TEM).

## Result and discussion

Nitride dioxide was chosen as a mask material in order to minimize the influence of the mask material on the etching process.Advanced pattern film (APF) was chosen as transfer layer between nitride oxide layer and photoresist (PR) layer. The APF and SiN were etched in a standard reactive ion etching (RIE) system. Figure 1 illustrates the process flows of STI formation etching and definition of microloading (isolation and dense area).

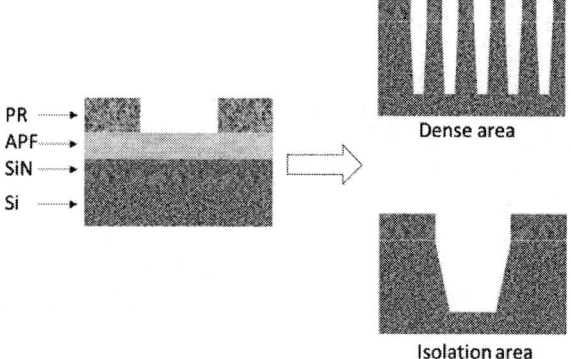

Fig.1 The process of STI formation and definition of microloading area (ISO/Dense)

### A) APF etching

For APF etching, O is necessary, and the etching gas is based on oxygen and sulfur dioxide containing O. Fig.2 illustrates the effects of the etching pressure for partial etch. As the pressure decrease, the mean free path will be longer because fewer molecules exist. In other words, the lower the pressure, the longer the particle will be able to travel without a collision. In comparison with the dimensions of the etched structure, a small microloading effect is occur (from 32nm to 24.8nm). Meanwhile, the generated organic polymer are not stripped off and remain on the pattern sidewall. When APF is etched with O-based gas chemistry, the entire obtained profile is more vertical. The full etch CD variation is plotted in Fig.3. It can be seen that the microloading effect becomes more pronounced as the etch pressure is increased.

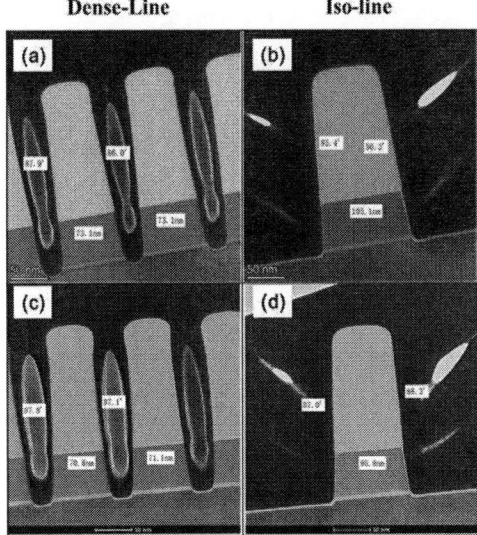

Fig2. TEM image of CD loading in high pressure (a-b) and low pressure (c-d) for partial etch.

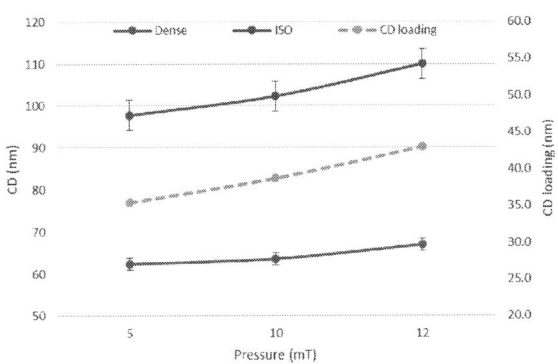

Fig3. CD microloading trend of APF etch pressure

## B) SiN etching

The CD microloading variation is also can be improved by SiN etching. In STI etch process, we chose SiN as hard mask (HM), and the etching gas is based on fluorocarbon containing C, F and H. In our study, we have tried to adjust pressure and RF power to improve microloading phenomenon. The pressure was varied between 5 and 20 mT, andd The power was always kept constant at 200 W,Fig.4 compared microloading effect in high pressure and low pressure after HM step etch. TEM image show microloading between dense and isolated area are improved by low pressure. As shown in table1, in low pressure etched, CD loading is reduced to 32.9 nm, and STI CD loading is reduced to 29.3nm. The final CD loading performance is totally improved about 4nm.

The power was always kept constant at 500 W, but the bias power was varied between 100-200W. It can be seen that the high bias power exhibits a somewhat stronger microloading

effect. The reason may due to a large amount of reaction byproducts on the isolated pattern.

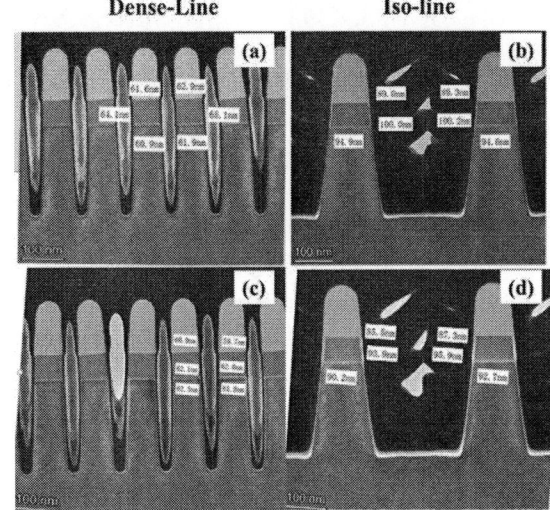

Fig4. TEM image of CD loading in high pressure (a-b) and low pressure (c-d) for full etch

Table 1. Loading comparison in high pressure and low pressure plasma etch process

| Items | High pressure | | Low pressure | |
|---|---|---|---|---|
| | Dense | Iso | Dense | Iso |
| HM TCD loading | 26.9 nm | | 25.9 nm | |
| HM BCD loading | 35.5 nm | | 32.9 nm | |
| STI TCD loading | 33.3 nm | | 29.3 nm | |

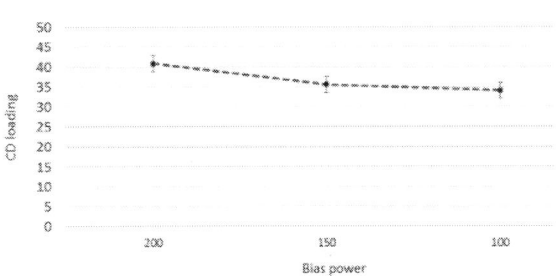

Fig5. CD microloading trend of HM etch bias power

## Conclusion

We have investigated the possibility of different processing parameters like, pressure, bias power on the CD microloading effect for APF and HM etching. It is shown that the CD microloading effect is very dependent on the pressure, and bias power enables a better control of passivation layers deposition. The microloading effect can be decreased by decreasing the pressure or decreasing the bias power. The pressure dependence seems to favor etching with processes that operate at low pressures.

978-1-6654-9759-6/22 $31.00 © 2022 IEEE

## Acknowledgment

The Author would like to thank Dr. Zhong-Wei Jiang in NAURA colleges and Ying-Yi Chen of CXMT, in Hefei, China, for their paten wafer and tool support.

## References

[1] S. Jensen, O. Hansen, "Characterization of the Microloading Effect in Deep Reactive Ion Etching of Silicon" doi: 10.1117/12.524461

[2] C. Hedlund, H.-C. 810m, and S. Berg, 'Microloading effect in reactive ion etching", Sweden J. Vac. ScI. Techno A 12(4), JullAug (1994)

[3] S. Jensen, O. Hansen, "Characterization of the Microloading Effect in Deep Reactive Ion Etching of Silicon" doi: 10.1117/12.524461

# Research of ultra high aspect ratio silicon etching in the advanced process

Zheng Ji[1]*, Jian-kun Zhang[1], Guang Yang[1], Zun-hua Zhao[1], Jing Wang[1], Xue-Sheng Wang[2], Xia-Yu Shi[2], Song-Yu Li[2], Bing-Hui Lin[2], Xin-Wen Huang[2], Ying-Yi Chen[2], Zhong-Ning Guo[2], Xian-Wen Su[2]

[1] Beijing NAURA Microelectronics Equipment Co. Ltd
No. 8 Wenchang Avenue, Beijing Economic-Technological Development Area. Beijing City, China
[2] ChangXin Memory Technologies, Inc. Ltd
No. 388, Xingye Avenue, Economic and Technological Development Area, Hefei City, China

*Corresponding Author's Email: jizheng@naura.com

## Abstract

Integrated circuit equipment is the foundation of the integrated circuit industry, and its technological level represents a country's high-tech research and development capabilities. Since the "Tenth Five-Year Plan", domestic semi vendor has included integrated circuit manufacturing equipment into major national special research and development projects. Ultra high aspect ratio silicon etch process has been demonstrated to build in DRAM process with inductively coupled plasma (ICP) on 612E tool. Key specifications are high aspect ratio, high etch rate, good trench sidewall profile with smooth surface, high aspect ratio dependent etch (ARDE), and low etch loading effects[1]. The results of the development work are implemented in loop process and hardware configuration of 612E.

*Keywords— Ultra-high aspect ratio, STI silicon etching process, ARDE, Loop process, dual-frequency bias cavity, NMC612E*

## 1 Introduction

As the size of integrated circuit components continues to shrink, the limit of Moore's Law is getting closer and closer. The application of high aspect ratio etching processes can prepare finer micro-nano structures, and create more space in the vertical direction of the silicon wafer. High aspect ratio silicon etching is a technical difficulty in the plasma etching process. At present, etching and passivation time-sharing multiplexing processes are often used in the industry, but this causes the problem of uneven sidewall top ography. The through silicon via etching process developed in recent years is increasingly used in the stacking technology of three-dimensional integrated circuits (3D-IC) [2]. This poses a greater challenge for the silicon-based deep etching process. The application meets the requirements of high depth, high precision, high stability, controllable morphology and surface characteristics, and low cost [3].

At present, the research and application of ultra-high aspect ratio silicon etching has achieved a major technological breakthrough in our company's latest generation 612E dual-frequency cavity etching machine, and the whole wafer has been completed on the client, also achieved the process criteria.

The morphology of deep silicon etching (STI) requires that the morphology be straight and strictly control the target size (space CD) to be about 15nm, the etching depth to reach 250nm, and the aspect ratio to reach about 17, which is very technically difficult. In addition to the large aspect ratio, smooth and vertical sidewalls, this process also needs to effectively control or eliminate some negative effects such as load effects, sidewall tilt, and inequality at the bottom of the groove[5, 6]. Therefore, the key point of this process is to effectively control the lateral etching rate. At the same time, it is necessary to increase the vertical etching rate, which brings a huge challenge to the performance of our machine.

The latest generation of our company's 612E etching machine has high frequency and low frequency dual radio frequency bias power supplies. In the ultra-high aspect ratio silicon etching process, the high radio frequency with low Vdc meets the precise requirements of morphology and uniformity. Meanwhile, the newly developed low radio frequency with high Vdc can significantly increase the plasma energy, enhance its directionality, and obtain a higher aspect ratio morphology. In the pulse mode, the electron energy can be reduced to weaken the damage, thereby further improving the shape control ability. The research team developed a shallow trench isolation (STI) etching process using dual radio frequency power on the 612E machine, and finally achieved the process goal of ultra-high aspect ratio silicon etching.

## 3 Process scheme and experiment

The ultra-high aspect ratio silicon etching process has two mask layers, namely amorphous silicon and silicon dioxide. After looping process scheme etch, the remaining mask layer is washed away by wet process. The film structure pre and post etch is shown in fig.1a which enroll ARDE effect between narrow spacer and wide spacer trench. The looping process scheme is shown in fig.1b which is similar as Bosch etch. We use cycling three steps of the main trench etch step, break through de-clogging step and oxidation sidewall to achieve the high aspect ratio profile[5,6].

The HAR STI etching process criteria is shown in Table 1.

## 2 ARDE loading solving by hardware

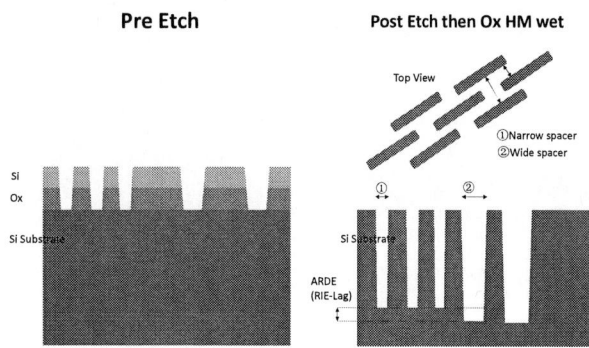

Fig.1a Film structure

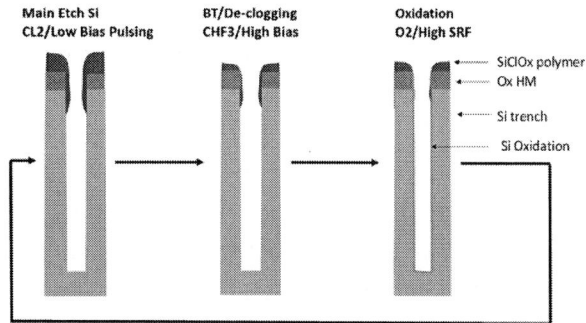

Fig.1b looping process scheme

Table 1. Process Criteria goals

| Check item | Unit | Spec |
|---|---|---|
| STI space CD(narrow) | A | 175~195 |
| STI depth @ narrow | A | 2300~2500 |
| STI depth @ wide | A | 2700~2950 |
| Depth loading | A | < 600 |

## 4 Experiment results

In order to complete the process goal of ultra-high aspect ratio silicon etching, the 612E etching machine has developed a low radio frequency technology on the basis of the original radio frequency.

Fig.2 compared morphology effect in the high radio frequency and the low radio frequency after silicon etching. As shown in table 2 , the STI morphology obtained by etching with low BRF shows the better etch depth loading performance than using the high radio frequency. This is because the DC bias of low frequency is much larger than that of high frequency under the same conditions, and the high Vdc of low frequency can significantly increase the plasma energy and enhance its directivity, thereby obtaining a better STI morphology.

However, the high Vdc in the low radio frequency will increase the risk of the E-chuck arcing, so the 612E machine has designed a "CIP Pin anti-ignition solution". Eventually, the performance of the machine is further optimized and the stability of the 612E machine is enhanced.

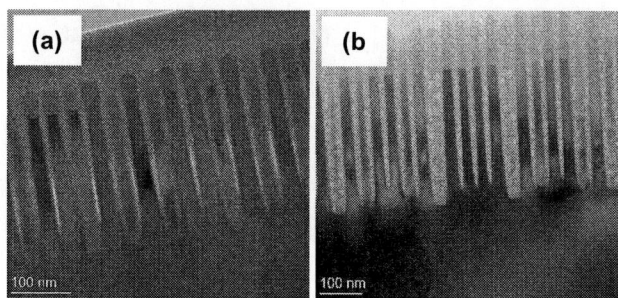

Fig.2 TEM image of morphology in the high radio frequency (a) and the low radio frequency (b) after silicon etching

Table2. Comparison of ARDE depth loading in the high and low BRF etching

| Check item | Spec | High Bias | Low Bias |
|---|---|---|---|
| STI depth @ narrow (A) | 2300~2500 | 2256 | 2361 |
| STI depth @ wide (A) | 2700~2950 | 2794 | 2750 |
| Depth loading (A) @ narrow and wide | 400~600 Less is better | 538 | 389 |

At present, the coupon and whole wafer experiments have been completed on the 612E dual-frequency etching machine, and the result of the Coupon experiment has reached the process goal. The final performance of the ultra-high aspect ratio silicon after etching is shown in table 3.

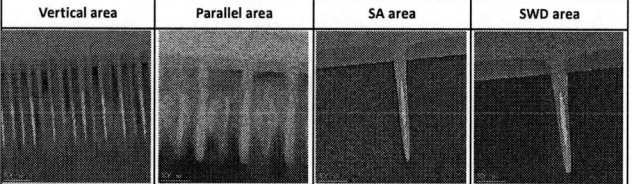

Fig.3 Film morphology and parameters of at different area.

Table 3. STI CD and depth performance for 612E

| Check item (Spec) | Criteria | 612E Result | Comment |
|---|---|---|---|
| STI CD @ narrow (A) | 175~195 | 191 | Meet |
| STI depth@ narrow (A) | 2300~2500 | 2423 | Meet |
| STI depth @ wide (A) | 2700~2950 | 2945 | Meet |

## 5. Conclusion

Deep silicon etch STI production in the advanced process released over the last 3 years, but challenges remain. In Ultra HAR STI etch, 612E HW configuration with dual bias power has demonstrated the straight profile, good depth loading, CD and depth uniformity.

In next phase, 612E more fundamental and productivity work will be enhanced to meet the increasing demands on HAR etch. The main etch, break through and passivation loop mechanisms, includes advanced pulsing technology are deeply developed to fully support the STI in the advanced process.

## Reference

[1] S. Panda, R. Ranade, and G.S. Mathad, "Etching high aspect ratio silicon trenches," J. Electrochem. Soc., 150, G612-G616 (2003).

[2] A. Khan, A. Kumar, and J. Dillard, "A novel process for high aspect ratio (HAR) trench etch on silicon on insulator (SOI) structures," Proceedings of the International Symposium on Plasma Process, ed. G.S. Mathad, Electrochemical Society, pp.230-238, 2000.

[3] K.T. Sung, "Etching of Si with Cl2 using an electron cyclotron resonance source," J. Vac. Sci. Technol. A, 11, 1206-1210 (1993).

[4] H. Kawata, M. Yasuda, and Y. Hirai, "Si etching with high aspect ratio and smooth side profile for mold fabrication," Digest of papers - Microprocesses and nanotechnology: 2005 International Microporcesses and nanotechnology conference, IEEE Computer society, Tokyo, Japan, Oct. 25-26, 2005, pp.198-199 (2005).

[5] Y. Morikawa, T. Murayama, and K. Suu, "Very uniform and high aspect ratio anisotropy thru Si via etching process in magnetic neutral loop discharge plasma," Proc SPIE, 6798, 679812 (2008).

[6] W.C. Tian, J.W. Weigold, and S.W. Pang, "Comparison of Cl2 and F-based dry etching for high aspect ratio Si microstructures etched with an inductively coupled plasma," J. Vac. Sci. Technol. B, 18, 1890-1896 (2000).

# SNC SADP Spacer etch process development using carbon hard mask mandrel for sub advanced process DRAM

Hao Liu[1, *], Jian-kun Zhang[1], Zun-hua Zhao[1], Li-Tian Xu[1], Chen Chen[1], Yang Chen[2], Xin-Ru Han[2], Shi-Ran Zhang[2], Jing-Lun Ma[2], Bing-Hui Lin[2], Xin-Wen Huang[2], Ying-Yi Chen[2], Xian-Wen Su[2]

[1] Beijing NAURA Microelectronics Equipment Co. Ltd
No. 8 Wenchang Avenue, Beijing Economic-Technological Development Area. Beijing City, China
[2] ChangXin Memory Technologies, Inc. Ltd
No. 388, Xingye Avenue, Economic and Technological Development Area, Hefei City, China

*Corresponding Author's Email: liuhao02@naura.com

## Abstract

Self-aligned double patterning (SADP) process has become the standard patterning technology for extending the half-pitch resolution beyond current ArF lithography tool's limit. In this paper, we overcome the oxide spacer shape deformation issue during SADP etch process to minimize the pitch walking deviation. The SADP technology is applied for advanced process of storage node contact (SNC) process to meet DRAM scaling requirements.

*Keywords — spacer-defined double patterning, self-aligned double patterning, carbon HM core mandrel, oxide spacer, wiggling, leaning , carbon hard mask, storage node contact, pitch walking, DRAM, and NMC612E.*

## Introduction

Self-aligned double patterning (SADP) uses sidewall spacers to create hard masks as a means of doubling the printed line density [1]. It has big advantage over other pitch splitting techniques such as litho-etch-litho-etch (LELE) in terms of overlay control and pitch walking [2,3]. Aimed at reducing cost, a scheme which integrates carbon HM pattern as core mandrel [4] with low temperature spacer SiO2 on spacer [5] has been proposed. The key challenge of resist-mandrel scheme is heavily impacted by the irregularity of mandrel shape [6,7].

In this study, carbon HM and spacer SiO2 were selected for core and spacer materials of SADP scheme, respectively. We developed a plasma etch process to suppress the oxide spacer shape deformation issue. Furthermore, we apply this SADP technology to create advanced process of critical dimension (CD) for storage node contact (SNC) process to meet DRAM manufacturing requirements.

## Experiment

Fig.1 illustrates the schematic SADP process flow. The film scheme started with hard mask (HM) about 550A thickness. Carbon HM pattern of 800A thickness was selected as a core mandrel from lithography process. Spacer SiO2 was deposited to target a designed sidewall width (38nm) as shown in Fig.1 (a). The etch process was carried out by NAURA NMC612E

tool for spacer SiO2 etch (Fig.1 (b)), selective carbon HM core removal (Fig.1 (c)), Post HM layer recess (Fig.1 (d)) in the sequence of each plasma step. The HM transfer etch also can be carried out by NAURA NMC612E tool (Fig.1 (e)), the final SNC SADP etch was final SiO2 etch as shown in (Fig.1 (f)).

As for DRAM manufacturing, the positive tone of SNC pattern was created to defined HM layer line to SiO2 HM trench. We measured the CD for both of spacer-defined HM layer masks and final SNC trenches. In addition, the final IMB performance was also the key factor of success (KFS), which was related to the asymmetric spacer shape of the Spacer SiO2 film.

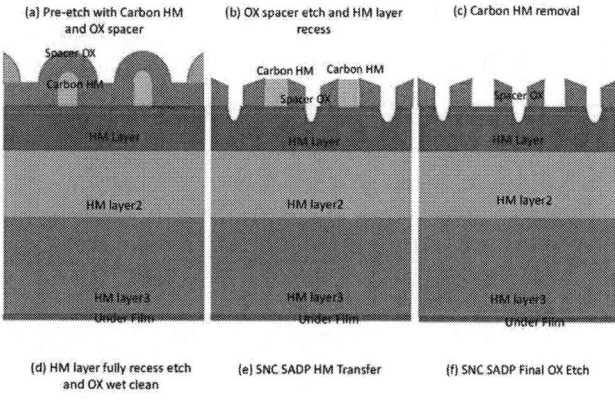

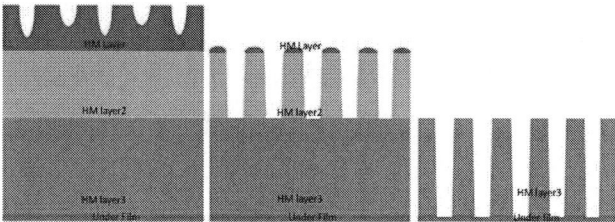

Fig.1 SADP process flow: (a) pre-etch, (b) post SiO2 spacer etch back, (c) post carbon HM core removal, (d) post HM layer fully recess and SiO2 wet clean, (e) SNC SADP HM transfer, and (f) SNC SADP final SiO2 etch.

## Result and discussion

### A. SNC spacer etch process development

In order to understand the impact of HM layer etch on CD and profile, DOE method was leveraged to compare the spacer CD depending on the HM layer etch process parameters, such as RF bias power, process gas. The impact of HM layer etch on profile and CD was shown in the Fig.2 and Table.1. As shown in Fig.2 (b), increasing Bias power makes the profile straighter and smaller spacer CD. Adding passivation gas HBr with the same HM layer Recess will make the HM layer Spacer CD smaller as shown in Fig.2(c). However, HM layer etch become difficult due to CH2F2 polymer too high, so the CD cannot truly reflected (Fig.2 (d)).

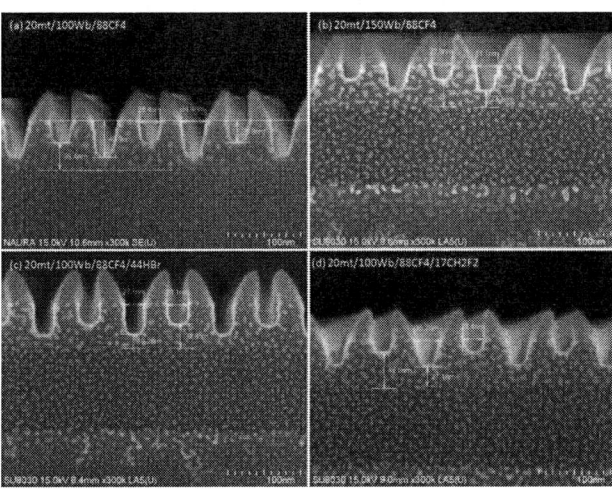

Fig.2 SNC SADP spacer etch DOE method was leveraged to compare the spacer CD depending on the HM layer etch process parameters, such as RF bias power, process gas.

| condition | Gap TCD | Core TCD | Gap Remain | Core Remain |
|---|---|---|---|---|
| a | 34.4 | 28.4 | 17.3 | 36.4 |
| b | 31.1 | 27.8 | 23.1 | 37.7 |
| c | 27.1 | 29.1 | 19.2 | 32.4 |
| d | 35.7 | 28.4 | 27.1 | 48.9 |

Table.1 SNC SADP spacer etch DOE method was leveraged to compare the spacer CD depending on the HM layer etch process parameters, such as RF bias power, process gas

The impact of SiO2 spacer etch on SiO2 spacer profile was shown in Fig.3 (a~c). The higher the SiO2 spacer etch back step Bias power, the smaller the angle of SiO2 sidewall, and the larger the opening angle of the gap area. With the decrease of SiO2 spacer etch bias power and the increase of time, the vertical etch rate will decrease, this will resulting in not enough height difference between core and gap HM layer remain. Meanwhile, the increase of time will also lead to more etch on SiO2 side, and that will cause too large gap spacer top CD.

Fig.3 SNC SADP spacer oxsp open Etch (Bias power)

In future study the impact of SiO2 spacer etch back step on CD, the result of time split as shown in Fig.4 . With the increase of SiO2 spacer etch process time, gap spacer CD increases and gap CD grows become slower, while core CD and depth remain basically unchanged because of carbon HM has not been removed.

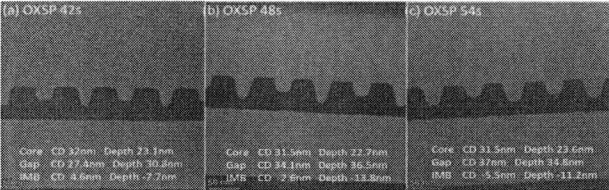

Fig.4 SNC SADP spacer profiles comparison depending on SiO2 spacer etch back parameters (OXSP time)

We also studied the gas ratio influence of SiO2 spacer on CD by partial etch. Under the condition of the same OE amount, it can be seen that spacer CD gradually decreases with the decrease of CF4/CH2F2 gas ratio as shown in Fig.5 top image and Table.2 data. The decrease of spacer CD is caused by the enhancement of Passivation.

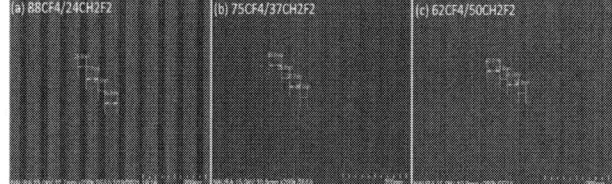

Fig.5 SNC SADP spacer CD comparison depending on SiO2 spacer etch back parameters (OXSP gas ratio)

| Gas Ratio | 88CF4/24CH2F2 | 75CF4/37CH2F2 | 62CF4/50CH2F2 |
|---|---|---|---|
| Line CD | 37.7 | 39.6 | 40.7 |
| Spacer CD | 25.8 | 25.8 | 23.8 |

Table.2 SNC SADP spacer CD comparison depending on SiO2 spacer etch back parameters (SiO2 spacer gas ratio)

Eventually, we also studied the influence of HM layer etch step process time on core CD and Gap CD, and analyzed the HM layer recess trend as shown in Fig.6. With the increase of HM layer etch step process time, Core CD gradually nonlinear increase, but the HM layer recess showed a linear increase of about 1.5nm/s. Similarly, with the increase of HM layer etch step process time, Gap CD also increased nonlinearly, and the HM layer recess showed a linear increase with the ratio of 1.9nm/s.

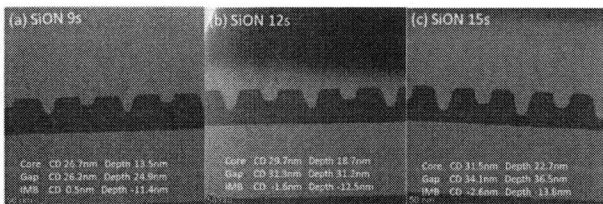

Fig.6 SADP spacer profiles comparison depending on SiO2 spacer etch back parameters (HM layer etch time)

978-1-6654-9759-6/22 $31.00 © 2022 IEEE

## B. SNC SADP spacer SADP power ratio impact for CD performance

In order to understand the interaction of SiO2 spacer etch power ratio with SiO2 spacer before carbon HM core removal, Validation experiment was leveraged to compare the spacer etching profiles depending on the SiO2 spacer etch process power ratio. A clear trend was shown in Fig.7. As we can see from Fig.7 with the Ws/Wb ratio increases, the line CD also increase. This is due to the higher Ws/Wb ratio cause the SiO2 spacer profile more taper that cause the bottom line CD increased.

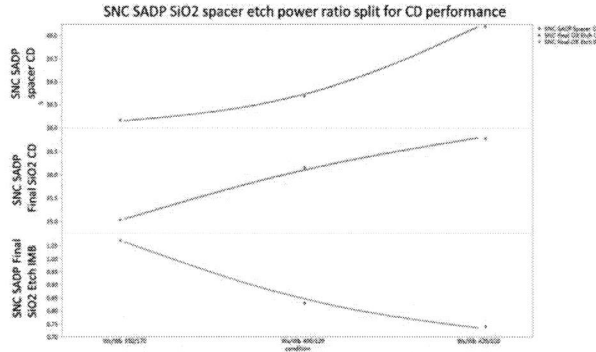

Fig.7 SNC SADP SiO2 spacer etch power ratio for CD performance.

## C. SNC SADP spacer Final SiO2 etch IMB control technology

For the device mass production, a tuning knob for the SNC SADP final IMB is required to compensate the etch chamber deviations. We studied the IMB control factors of SiO2 spacer etch over etch percent as shown in Fig.8. This SiO2 spacer etch over etch percent can not only gain worse IMB but also can improve the IMB. Since Final SiO2 etch IMB is obtained by the difference between the core and gap HM layer remain, so the value of IMB presents a parabolic state. So we need to find best IMB position.

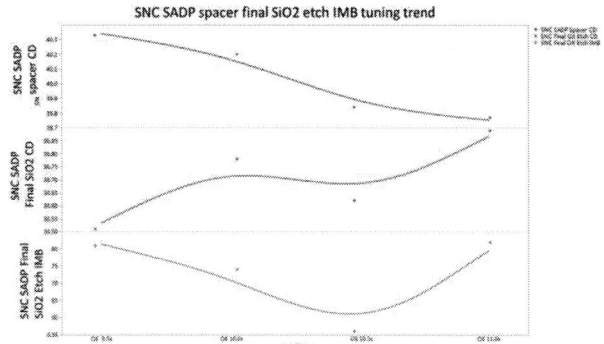

Fig.8. SNC SADP Spacer Final OX Etch IMB tuning trend.

## D. SNC SADP Final SiO2 etch CD performance

In term of DRAM SNC manufacturing, the mask CD of HM layer line CD was shrink to a target by CHF3 flow of HM layer etch step. The final SiO2 etch CD can meet the requirement for sub advanced process in memory array area. It can be seen from the Fig.9 that spacer line CD and Final SiO2

etch line CD both increase with the increase of CHF3 flow. And the Final CD tuning trend about 0.8nm/10CHF3 sccm as shown in Fig.9.

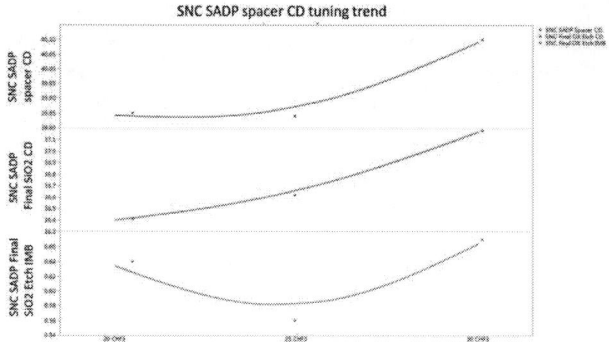

Fig.9 SNC SADP spacer CD Tuning Trend.

## Conclusion

We proposed the SNC SADP spacer process that implements carbon HM for the core mandrel and got a clear method for IMB and CD tuning. This technology can meet the requirements of the advanced DRAM manufacturing for advanced process dimension.

## References

[1] Christopher Bencher, et al., "22nm half-pitch patterning by CVD spacer self alignment double pattering (SADP)", Proc. of SPIE, vol.6924, 6924E (2008)

[2] Huayong Hu, et al., Self-aligned double patterning (SADP) process even-odd uniformity improvement", 2016 China Semiconductor Technology International Conference (CSTIC) , pp.1-4 (2016)

[3] T. Yang and D. Yim, "SAQP Pitch Walking Improvement Path Finding by Simulation", 2019 International Symposium on Dry Process (DPS), P-22, pp.99-100 (2019)

[4] Nihar Mohanty, et al.,"Challenges and mitigation strategies for resist trim etch in resist-mandrel based SAQP integration scheme", Proc. of SPIE, Vol.9428, 94280G (2015)

[5] Qingqing Wu, et al., "Optimization of the CD uniformity (CDU) in silicon oxide spacer process for 5nm FIN SAQP process flow", 2020 China Semiconductor Technology International Conference (CSTIC), pp.1-4 (2020)

[6] Angélique Raley, et al., Self-aligned Quadruple Patterning Integration using spacer on spacer pitch splitting at the resist level for sub 32nm pitch applications", Proc. of SPIE, vol. 9782, 97820F (2016)

[7] Efrain Altamirano-Sanchez, et al, "Self-aligned quadruple patterning to meet requirements for fins with high density", SPIE, 14 May (2016)

# SILICON PARTIAL ETCH DEFECT RESEARCHES IN BSI CMOS IMAGE SENSOR PROCESS PRODUCT

Hebao Liu[1], Xiaoyu Li[2], Qixin Wu[1], Ji Feng[2], Zhe Wang[2], Dongmei Zhai[2], Fanyan Meng[1]*

[1]School of Mathematics and Physics, University of Science and Technology Beijing, China
[2]Semiconductor Manufacturing International Corporation, Beijing 100176, China
*Corresponding Author's Email: 15901019368@163.com, Meng7707@sas.ustb.edu.cn

## ABSTRACT

To reduce ink die loss in BSI CMOS image sensor (CIS), silicon partial etch defect after etching during Backside Silicon Etch (BSSE) process is studied. In this research, the root cause of silicon partial etch defect is clarified and a defect improvement approach is proposed. By verifying the evolution of defects in the Backside Metal Grid (BMG) process step by step, it is shown that the queue time (Q-time) window plays an indispensable role in the formation of silicon partial etch defects. Here provides an effective solution for reducing silicon partial etch defects in BSSE, and gives a method to enlarge the BMG Q-time window.

## INTRODUCTION

In recent years, imaging-dedicated applications have been developed rapidly. Backside illuminated (BSI) sensors, in which the incident light reaches the pinned photodiode (PPD) without passing through metallic routing required by the pixel's electronic components, have a higher quantum efficiency (QE), fill factor, and sensitivity. Benefiting from the back-thinned structure and converse illuminated direction, BSI gradually take place of front-side illuminated (FSI) and slowly becoming a mainstream technology. [1, 2]

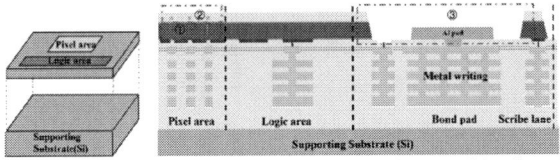

*Fig. 1: simplified BSI CIS structure and partial back-side processes simple flow*

As depicted in Fig. 1, the BSI CIS studied in this paper is to first complete the logic circuit and pixel circuit on the front-side, then turn the wafer over, bond it with the carrier wafer, and make the optical part on the back-side of the wafer. After the thickness of back-side silicon is thinned, the deep trench isolation (DTI) process is carried out, combined with high-k film filling.[3] Then the BMG structure is formed and can get better optical performance when combined with DTI. Finally, pad is started by the BSSE process which completely etches the silicon above the Cu line on the wafer front-side, so that the subsequent Al pad can be connected with the front-side Cu line.

Fig. 2(a) shows a large number of conical silicon partial etch defects observed in the etched area near the edge of the wafer after BSSE etching. As shown in Fig. 2, silicon partial etch defects will affect the surface morphology of the Al pad, and then affect the success of processes such as wafer acceptance testing, reliability testing, and packaging.[4] The ratio of die as inked caused by the defects is about 5% to 20% during outgoing quality inspection (OQI). In order to future reduce failing die loss, it is of great commercial significance to improve this defect.

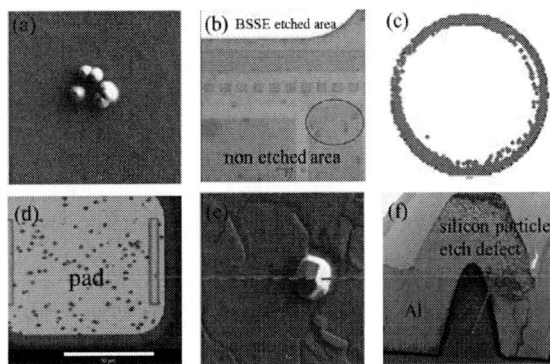

*Fig. 2: After BSSE etching, (a) SEM image; (b) OM image; In OQI stage, (c) map; (d) image; (e) SEM image; (f) TEM cross section.*

Fig. 2(b) shows the black foreign matter observed by Optical Microscope (OM) in the non etched area subsequently. It was speculated that the silicon partial etch defect was caused by the residue of BMG in the previous process.

As shown in Fig. 3, it is found that the defect is related to the Q-time from photoresist (PR) ashing to PR wet strip during BMG hard mask etch (HM ET) process by analyzing the existing data. When the Q-time of lots is greater than 247 min, the silicon partial etch defect will appear after BSSE.

Due to many practical factors of each process machine or production line, the Q-time to the next process is inevitable. In this paper, the cause of silicon partial etch defects is related to the chemical species outgassing of polymer in the closed foup, and the optimal process conditions for enlarging the Q-time window are found through increasing the time of $O_2$ flow in PR ashing and the time of DSP flow in wet strip during BMG HM ET process.

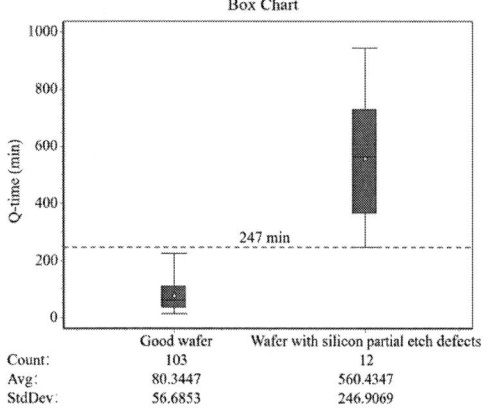

Fig. 3: Box chart of Q-time

## EXPERIMENTAL

The BMG structure is produced by metal film deposition, etching, oxide film deposition, and chemical mechanical polishing (CMP). Fig. 4 shows the simple process flow which is adopted in the BMG etching experiment. The BMG etching process is mainly divided into two steps. The first step is HM ET, which is to etch the hard mask and stop on the top glue layer with PR and bottom anti-reflective coating (BARC) as the mask. In the BMG HM ET process, after dry etching, the PR is removed by oxygen plasma ashing, and then PR wet stripping is carried out by DSP. The second step is aluminum etch (Al ET), which is to etch aluminum film and bottom glue layer with hard mask as the mask and stop on the oxide film. After etching, another layer of oxide film will be deposited to protect the metal grid, and finally CMP.

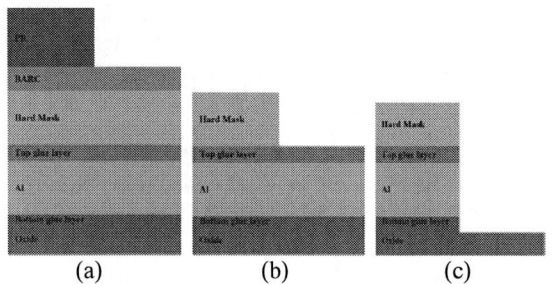

Fig. 4: Schematic diagram of the BMG etching process in BSI CIS. (a) after PR development; (b) after HM ET; (c) after Al ET.

As shown in Fig. 5, the wafer maps scanned after different Q-time are compared. It was found that the Q-time is greater than 4h, the longer the Q-time, the more

defects in the pad area of wafer edge, while the pixel area is normal. In addition, when the Q-time is 46h, the condense defect also appears at wafer center area.

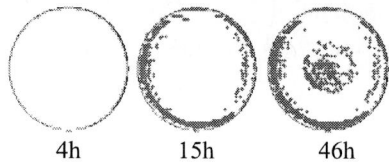

Fig. 5: the wafer map of 4h, 15h, and 46h after BMG HM ET PR ashing.

In order to determine the source of silicon partial etch defects, the wafer was reviewed 8h after BMG HM ET PR ashing. According to the evolution of defects in Fig. 6, the reason for the formation of silicon partial etch defect is that under the original process conditions, Q-time more than 4h after BMG HM ET PR ashing, the polymer volatilizes to form condense defects at the wafer edge, resulting in Al partial etch defects in BMG Al ET and silicon partial etch defects in BSSE etching.

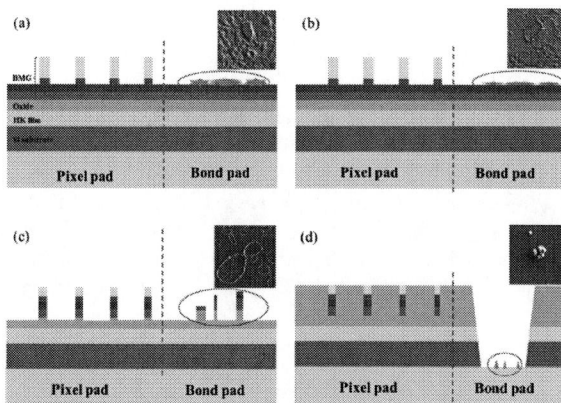

Fig. 6: The evolution of defects after different processes. (a) 8 hours after BMG HM ET PR ashing, SEM image of condense defect in the etching area of the wafer edge. (b) After BMG HM ET wet stripping, the condense residue observed by SEM. (c) After BMG Al ET, SEM image of Al partial etch defect. (d) After BSSE etching, SEM image of Si particle etch defect.

In the BMG HM ET process, the original process condition is that the time of $O_2$ in PR ashing is 30s, and the cleaning time of DSP in the next wet strip is 50s. Since the condense defect formed during BMG HM ET is related to the outgassing of reaction residue, this paper proposes two solutions to this defect. The first is to increase the time of $O_2$ used for PR removal so that the materials can fully react and reduce the polymer. In addition to increasing the post-etch solvent etch polymer stripping effect, the cleaning time of DSP is increased during BMG HM ET wet strip to remove the polymer. Combined with the above two schemes, as shown in Table 1, this paper gives the map obtained after BSSE etching inspection under four groups of different conditions of the BMG HM ET process and the best conditions for increasing the Q-time window while solving the silicon partial etch defect.

## RESULT AND DISCUSSION

TABLE I.        THE BMG HM ET PROCESS SPLIT TABLE

| Split No. | PR ashing $O_2$ time | Q-time | wet strip DSP time | The map scanned after the BSSE etching inspection |
|---|---|---|---|---|
| 1 | 30s | 8h | 50s | |
| 2 | 60s | 8h | 50s | |
| 3 | 60s | 8h | 80s | |
| 4 | 60s | 10h | 80s | |

Compared with Experiment 1 which used the original process conditions of the BMG HM ET process, and the Q-time was 8h, Experiment 2 increased the time of $O_2$ to 60s. After BSSE etching inspection, the silicon partial etch defects at the wafer edge in Experiment 1 were more serious than that in Experiment 2. It shows that increasing the duration of $O_2$ reduces PR residue and the source of outgassing in the foup, so as to reduce the number of condense defects. However, due to the slight defect map observed in Experiment 2, the Q-time window could not be extended to 8h under this experimental condition.

Compared with Experiment 2, Experiment 3 increased the duration of DSP during BMG HM ET wet strip to 80s, which can significantly improve the silicon partial etch defect, indicating that DSP cleaning can reduce the condense defect. Under Experiment 3 condition,

defect free result was achieved after BMG HM ET PR ashing.

Compared with Experiment 3, Experiment 4 increased Q-time to 10h, and the map result was worse. It shows that properly increasing the duration of $O_2$ can reduce the polymer. However, with the increase of Q-time, the higher the concentration of volatile gas in the foup, the more serious the condense defect is. On this condition, increasing the time of DSP can not remove the condense defect completely. Therefore, it is impossible to expand the Q-time window to 10h.

## CONCLUSION

In this paper, the formation mechanism of the silicon partial etch defect and the solution were investigated. The root cause of the silicon partial etch defect was that under the original process conditions, Q-time more than 4h after BMG HM ET PR ashing, the outgassing of polymer to form condense defect near the edge of the wafer, resulting in Al partial etch defect in BMG Al ET process and silicon partial etch defect in BSSE etching step.

The optimum Q-time window was determined to be 8h by optimizing the BMG HM ET process parameters like increasing the duration of $O_2$ flow in PR ashing to 60s and increasing the duration of DSP flow in the wet strip process to 50s.

## REFERENCES

[1] Liu B , Li Y , Wen L , et al. Study of dark current random telegraph signal in proton-irradiated backside illuminated CMOS image sensors[J]. Results in Physics, 2020, 19:103443.

[2] Zhang X , Li Y , Wen L , et al. Displacement damage effects induced by fast neutron in backside-illuminated CMOS image sensors[J]. Journal of Nuclear Science and Technology, 2020(6):1-7.

[3] Yin Z , Li J , Li X , et al. High-K Bubble Defect Researches in Stack-BSI Process Product[C]// 2021 China Semiconductor Technology International Conference (CSTIC). 2021.

[4] Santosh, Kumar, Pani, et al. Crystalline defect formation on aluminum bond pads during CMOS wafer storage and process strategies for defect elimination[J]. Jvst B, 2018.

# Hard mask etch process development for patterning 60nm magnetic tunnel junction

Xiaohui Li*, Hailong Liu, Ruiping Zhu, Jipeng Liu, Zhongyi He and Qingjun Zhou

Beijing NAURA Microelectronics Equipment Co. Ltd
No. 8 Wenchang Avenue, Beijing Economic-Technological Development Area. Beijing City, China

*Corresponding Author's Email: lixiaohui01@naura.com

## Abstract

Magnetic random access memory (MRAM) is one of the most promising solution for the non-volatile memory in the post-Moore era. Its core feature is the magnetic tunnel junction (MTJ), which determines the performance of the device. The main constituency materials of the MTJ are metallic elements such as Ta, Fe, Co, Mg, etc. They generally form a multilayer stack which contains many single-layer material in nanometer scale. The challenges for the plasma etching, which is typically done in an ICP etcher in one pass, mainly include accurate transfer of the etching topography and control of non-volatile metal etching by-products. This paper proposes following solutions to address above challenges: 1. Introduce HBr curing to harden the incoming photoresist surface to reduce the mask shape deformation; 2. Use CF4 gas in the plasma strip step to clean the metal oxide polymer.

***Keywords: MRAM, ICP etching, precise topography transfer, by-product processing, carbon hard mask, TaN, and NMC612M.***

## Introduction

Magnetic Random Access Memory (MRAM) is a new type of memory device with magnetic tunnel junction as its core, which realizes data storage by regulating the direction of the electron spin [1, 2]. Comparing to traditional SRAM and DRAM, MRAM has the advantages of more erasing and writing times; faster reading and writing speed; lower energy consumption; and non-volatile memory [3]. It is widely regarded as one of the most promising memories in the post-Moore era. Since 2016, major semiconductor manufacturers, such as Samsung, Toshiba, TSMC, and Global foundries [4], all have invested heavily in MRAM research and development. Some have announced mass production process of Spin Transfer Torque (STT) MRAM in the past two years. Thus research and develop MRAM process on the domestic etching system carries a great significance for helping Chinese chip maker to take a foothold in the new memory market.

STT-MRAM is fabricated by embedding magnetic tunnel junctions in the middle of the back-end metal connection layers of standard CMOS logic circuits. The magnetic tunnel junction is a sandwich structure composed of a ferromagnetic layer/ barrier layer/ ferromagnetic layer. The ferromagnetic layer is usually a metal or metal alloy such as Co, Fe, Ir, Mg, and Ru. The barrier layer is typically a metal oxide such as MgO. These materials are stacked in monolayers on the order of nanometer. Since metallic materials are very sensitive to corrosive gases, the halogen gases [5] used in etching can easily attack the barrier layer and cause electrical isolation

failure. To prevent the problem, ICP etching process is required to stop on the isolation layer, and subsequent etching is mainly carried out by Ion Beam Etching (IBE) [6] as shown in Figure 1. The treatment of non-volatile metallic byproducts post etching is still a challenge to be solved.

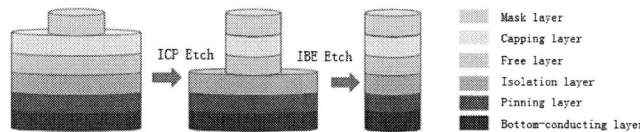

Figure 1.ICP/IBE etching process of magnetic tunnel junction.

The morphology of the magnetic tunnel junction is also an important factor affecting etching performance. The MTJ realizes its function through the inversion of the magnetic anisotropy in the ferromagnetic layer to achieve data recording, in which the magnitude of the inversion magnetic field directly determines the energy consumption of the device. Studies have indicated that, compared with the angular device morphology, the flipping magnetic field of the circular device is smaller and the energy consumption is lower, so the precise transfer of the morphology during etching is also a problem to be considered.

In this paper, we develop an ICP etch process to etch through SiN and TaN films with tri-layer photoresist (PR) approach [7]. This hard mask patterning technology can support the next generation MTJ etch in MRAM device production

## Experiment

In order to inquire into the etching process for MTJ, this paper conducts research and development based on a common MRAM process requirements. The etch system used in this study is NAURA 612M metal etch system. The structure of the MRAM etching process is as Figure 2.

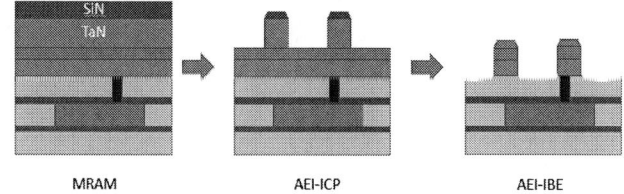

Figure 2. MRAM etching structure and process

The process requirements are shown in Table 1.

| Item | Size |
|---|---|
| HM Sidewall angle | >85deg, no bowing |
| HM AEI CD | 1 sigma<2nm |

Table 1   MRAM hard mask etching process requirements

MTJ etch process using tri-layer etch to achieve the pattern transfer. Then it etches TaN to form MTJ junction. The main bodies of each etching steps are shown in Table 2.

| Step | Main parameters |
|------|-----------------|
| | CHF3/CF4 |
| Pattern | O2/SO2 |
| transfer | CHF3/CF4 |
| | BCl3/Cl2 |
| TaN | BCl3/Cl2 |
| Strip | O2/N2/CF4 based chemistry |

Table 2   Etch steps for the MTJ pattern transfer with PR mask

## Result and discussion

### A.   Hard mask etch process for circular MTJ pattern transfer

The MTJ's cylindrical structure has strict CD uniformity and surface morphology requirements, which poses some etch challenges. The nitride hard mask open is a CHF3 and CF4 based etch process with high bias power, leading to heavy ion bombardment and polymer deposition on the resist mask , creating irregular morphology issue as shown in figure 3. The irregular cylindrical shape will stir up the distortion of the magnetic anisotropy distribution and orientation of the MRAM device, which will enhance the magnetic field required for flipping and energy consumption ultimately. To solve the problem, we introduce a HBr-resist curing step before the nitride etching to solidify the mask. The combination between hydrogen element in HBr and the unsaturated bond in PR can significantly alleviate the pattern deformation in subsequent metal etching.

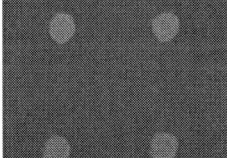

Figure 3 Irregular morphology after SiN hard mask open

As illustrated in Figure 4, the overall morphology is obviously modified compared with Figure 3 after curing step, which builds the foundation for subsequent TaN metal etch.

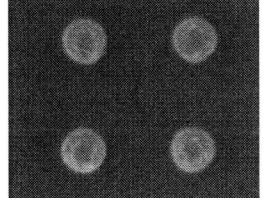

Figure 4   Morphology of post SiN etching with curing step

### B.   TaN etch by-products removal during strip process

Process gases adopted for TaN layer etching are Cl2 and BCl3. The removal of the non-volatile metal by-product $TaCl_X$ generated during period is essential to be considered. Heavy

polymer forms on and around the mask surface after the whole etching (Figure 5), which has an unfavorable influence on the subsequent IBE etching. To modify the phenomenon, the strip scheme needs to be optimized.

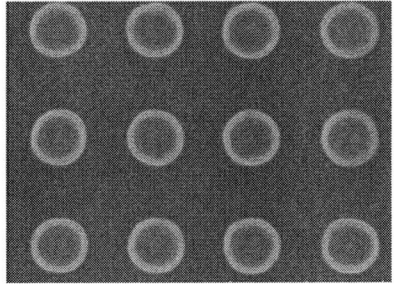

Figure 5 Impurities formed on sidewall after TaN etching

A series of strip DOE experiment (Table 4) was designed to examine its impact on polymer deposition. We compared the function of the O2 strip with N2, CF4 doped addition, as well as RF bias power effect.

| Conditions | BRF | Main parameters |
|------------|-----|-----------------|
| Strip 1 | OFF | O2/N2 |
| Strip 2 | OFF | O2/N2/CF4 |
| Strip 3 | OFF | O2/CF4 |
| Strip 4 | ON | O2/CF4 |

Table 4 DOE test design of the plasma strip conditions

Figure 5 exhibits the post-etch morphologies obtained after applying different strip conditions illustrated in Table 4. The image reveals obvious by-products remaining on the mask surface by the conventional Strip#1. With the replacement of N2 with CF4, the polymer deposition is significantly improved (Strip#3). The effect of adding CF4 to N2 was in between (Strip#2). It is worth noting that increasing bias RF power reduces polymer buildup, but it causes recess on the under layer.

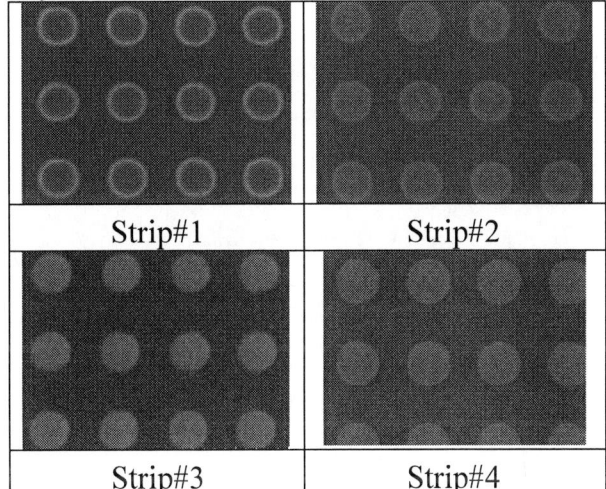

Figure 5 Comparison of the etched morphologies obtained after applying different strip conditions.

We suggest that the TaClx generated from TaN etch tends to

form a non-volatile TaOClx with pure O2 plasma strip. Adding CF4 can convert TaOClx into more volatile TaFx products than TaClx. As a result, TaN sidewall is much cleaner. As seen in Figure 6, the final SiN and TaN profile etched with combination of HBr curing and CF4 added strip are smooth and straight. It meet subsequent MTJ etch requirements.

Figure 6   TEM of SiN hard mask and TaN electrode after ICP etching.

## Conclusion

Irregular morphology and non-volatile metal etching by-products generated during TaN hard mask etching process are issues that must be considered .This study develops an ICP hard mask open process for MTJ front layer of MRAM device. Our data demonstrate that employing a HBr curing step before nitride open can efficiently relieve the mask deformation. Meanwhile, CF4 addition into O2 plasma strip step after TaN etch can remove etching residues on the sidewall effectively.

Through optimization of the etch process, the hard mask profile meets the MTJ integration requirement. We have achieved the metal hard mask etch process development on NAURA NMC612M etcher to support MRAM manufacturing.

## References

[1]   K. Nishioka, et al., "First Demonstration of 25-nm Quad Interface p-MTJ Device With Low Resistance-Area Product MgO and Ten Years Retention for High Reliable STT-MRAM", IEEE TRANSACTIONS ON ELECTRON DEVICES, VOL. 68, NO. 6, pp.2680-2685, JUNE 2021

[2]   K. Nishioka et al., "Novel quad interface MTJ technology and its first demonstration with high thermal stability and switching efficiency for STT-MRAM beyond 2X nm", Proc. Symp. VLSI Technol., pp. C278-C279, Jun. 2019

[3]   Magnetic Memory Technology: Spin-Transfer-Torque MRAM and Beyond, First Edition. Denny D. Tang and Chi-Feng Pai. © 2021 The Institute of Electrical and Electronics Engineers, Inc., Published 2021 by John Wiley & Sons, Inc.

[4]   T. Y. Lee, et al., "Advanced MTJ stack engineering of STT MRAM to realize high speed applications", 2020 IEEE International Electron Devices Meeting (IEDM), pp. 11.6.1-11.6.4, Dec. 2020

[5]   Jang Woo Lee, et al., "High-density plasma etching of CoFeSiB magnetic films with hard mask", J. Magn. Magn. Mater. 304, e282~284 (2006)

[6]   Min Hwan Jeon, et al, "Etch residue removal of CoFeB using CO/NH3 reactive ion beam for spin transfer torque-magnetic random access memory device", J. Vac. Sci. Technol. B 33(6), 061212-1~8 (2015)

[7]   Leonard Hsu, et al., "Tri-layer metal hard mask etch defect control by a novel molecule radical during in-situ chamber clean", 2016 International Symposium on Dry Process (DPS), pp.53-54 (2016)

# Increasing the Post Halo Implantation Anneal Temperature for the Effective Improvement of Threshold Voltage Roll-off Induced by the Unique Non-uniform Boron Diffusion from the Embedded Source/Drain

Run-Ling Li and Yu-Long Jiang*

State Key Laboratory of ASIC and System, School of Microelectronics, Fudan University, Shanghai 200433, China
(*e-mail: yljiang@fudan.edu.cn).

*Abstract* –The severe threshold voltage (Vt) roll-off is revealed for PMOSFETs with the embedded SiGe source/drain (eSiGe S/D) when the gate length decreases. The unique non-uniform boron diffusion into the channel region from eSiGe is found to be the main reason, which is intrinsically caused by the sigma-shaped trench. Lowering the growth temperature of eSiGe is widely adopted to solve the Vt roll-off issue but with a very complicated adjustment of eSiGe growth process. Without the change of eSiGe growth process, a post halo implantation anneal at a higher temperature before the S/D trench etching is proposed and demonstrated to be able to effectively improve the Vt roll-off too, providing a new way to suppress short channel effect.

*Index Terms—threshold voltage roll-off, embedded source/drain, sigma-shaped trench, post halo implantation anneal.*

## I. INTRODUCTION

The embedded SiGe source/drain (eSiGe S/D) process was successfully introduced to boost PMOSFETs' performance [1]. As shown in Fig. 1 (a), the architecture of eSiGe with a high Ge content generally includes a seed layer, a bulk layer and a cap layer. The seed layer is a buffer layer with a low Ge content and no intended boron (B) doping for defect suppression. While the Ge content in bulk layer is high enough to generate the required uniaxial compressive stress in channel region. Also the bulk layer is heavily in-situ doped by B. The sigma-shaped trench is then used in PMOSFETs to further enhance hole mobility [2]. However, such a special shape also results in the possible non-uniform B diffusion into the channel region from eSiGe bulk layer during SiGe growth and the followed S/D activation anneal since the seed layer thickness is non-uniform too. Actually, along the gate length direction it is not easy to obtain a very uniform seed layer even for the eSiGe of P-channel FinFET with an O or U-shaped S/D trench [3].

The diffusion of B into the channel region will result in the degradation of short channel effect (SCE), especially the severe roll-off of threshold voltage (Vt). Although many techniques and processes have been adopted to improve SCE such as lightly doped drain (LDD) [4], halo implantation (IMP) [5], super steep retrograde channel [6], co-IMP [7], pre-amorphization IMP, cryogenic IMP [8], millisecond spike annealing and microsecond laser annealing [9-11], the non-uniform B diffusion makes the SCE suppression more complicated. Lowering the growth temperature of eSiGe is widely adopted to relieve the degradation, but the final S/D activation anneal at a high temperature will still unavoidably enhance the possible non-uniform B out-diffusion into channel region. Besides, the temperature lowering will result in a very complicated systematic adjustment of eSiGe process. In this work, considering the phosphorus (P) loss induced by the sigma-shaped trench etching a simple post halo IMP anneal (PHIA) at a higher temperature is proposed and demonstrated to be able to address the Vt roll-off issue without any change of the existed eSiGe growth process.

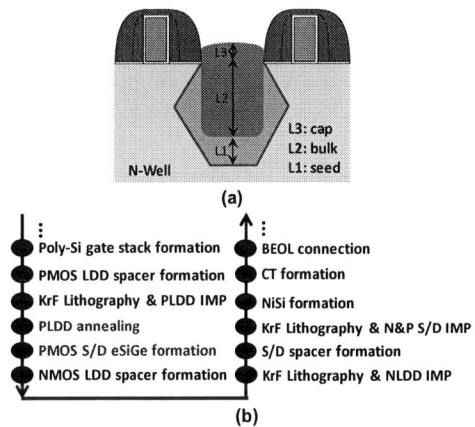

Fig. 1 (a) The typical architecture for sigma-shaped eSiGe with a high Ge content. (b) A typical 28-nm poly-Si gate logic process flow with eSiGe module for PMOSFET [12].

## II. EXPERIMENTAL DETAILS

A typical 28-nm poly-Si gate logic process platform was employed in this work. The corresponding process flow is shown in Fig. 1 (b) [12]. As part of the PMOSFET LDD (PLDD) module, the P halo IMP was performed with a dose of 7E13 $cm^{-2}$, a tilt angle of $27^o$ and an energy of 20 keV. A PHIA at 900 $^oC$ was applied to repair the IMP damage of PLDD process and activate the heavily doped B atoms to avoid the possible B loss in the following process. Then the sigma-shaped trench was formed. The eSiGe deposition was realized by an AMAT Centura RP EPI tool. The eSiGe deposition is the growth combination of the B free SiGe seed layer, SiGe bulk layer with heavily in-situ doped B and Si cap layer, as shown in Fig. 1 (a). Following the process flow in Fig. 1 (b), the corresponding PMOSFETs for electrical tests were prepared as the SiGe control samples.

The secondary ion mass spectroscopy (SIMS) was applied to disclose Ge and B concentration profiles in eSiGe. The cross-sectional transmission electron microscopy (XTEM) was employed to reveal eSiGe structure information. The Sentaurus TCAD tool was used to simulate the process flow and device performance. The Vt and on-state current/off-state current ($I_{on}/I_{off}$) was respectively obtained by a TEL Precio XL prober.

## III. RESULTS AND DISCUSSION

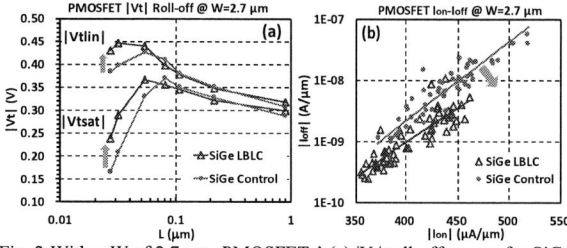

Fig. 2 With a W of 2.7 μm, PMOSFETs' (a) |Vt| roll-off curves for SiGe control and SiGe LBLC samples; (b) $I_{on}/I_{off}$ data corresponding to (a).

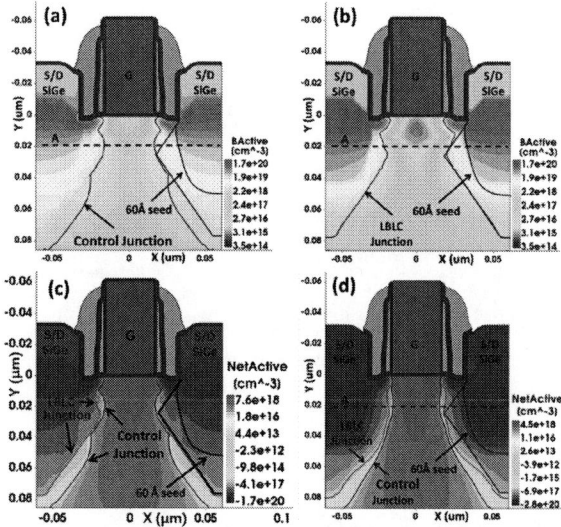

Fig. 3 TCAD simulation results for PMOSFET with a W/L of 2.7 μm/27 nm. The active B distribution just after eSiGe deposition with S/D junction profile for (a) SiGe control sample and (b) SiGe LBLC sample. The net doping distribution with S/D junction profile (c) just after eSiGe deposition and (d) after final S/D activation anneal for both SiGe control and LBLC samples.

Fig. 2 (a) demonstrates the severe |Vt| roll-off of SiGe control samples with a gate width (W) of 2.7 μm. Such a Vt roll-off makes the PMOSFET with a gate length (L) of 27 nm actually fail to work. Since the SiGe bulk layer is heavily doped by B, the B atoms then may diffuse into the gate controlled channel region after eSiGe deposition and S/D activation anneal. The TCAD simulation results shown in Fig. 3 (a) confirms the out-diffusion of B from eSiGe S/D even just after the eSiGe deposition process. Fig. 3 (a) also clearly indicates the non-uniform B distribution in channel region, especially the B atoms lying from 10 to 25 nm in channel depth will significantly affect Vt.

To suppress the B out-diffusion, lowering the eSiGe epitaxy temperature is widely adopted by industry. Since only SiGe bulk layer and cap layer are doped by B, the growth temperatures for

these two layers were respectively reduced by 30 °C. Such a process is called as SiGe LBLC process. However, due to the reduction of epitaxy temperature for cap layer deposition, the addition of a small amount of Ge into the Si cap layer is strongly required as shown in Fig. 4 (b). This further reveals the difficulty of lowering the eSiGe growth temperature. With a complicated systematic adjustment of SiGe epitaxy recipes, Fig. 4 finally confirms the comparable B and Ge distributions of SiGe LBLC process with SiGe control process. The TCAD simulation results shown in Fig. 3 (b) to (d) clearly demonstrate the effective suppression of B out-diffusion even after the final S/D activation anneal. Correspondingly, the Vt roll-off characteristics are greatly improved as shown in Fig. 2 (a). Actually, the |Vt| at linear region (|Vtlin|) and |Vt| at saturation region (|Vtsat|) respectively increases by 12.0% and 43.5% for PMOSFET with a W/L of 2.7 μm/27 nm. Also Fig. 2 (b) reveals an increase by 8.0% for $I_{on}/I_{off}$ performance when LBLC process is applied.

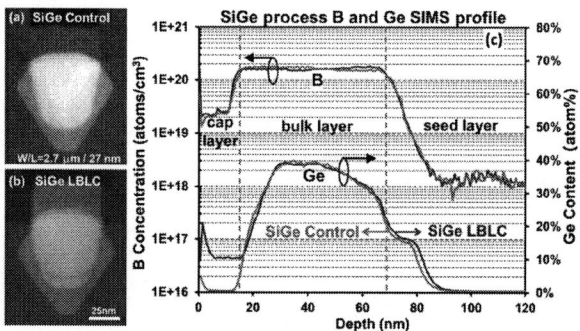

Fig. 4 The XTEM images of PMOSFET eSiGe S/D with a W/L of 2.7 μm/27 nm for (a) SiGe control and (b) SiGe LBLC samples. (c) SIMS depth profiles of B and Ge in SiGe on a large pad for SiGe control and LBLC processes.

However, it is very challenging anyway to modify the eSiGe epitaxy process since the epitaxy process will significantly affect the selectivity, micro-loading and global loading effect [13]. Moreover, the B out-diffusion from eSiGe is actually unavoidable since the final S/D activation anneal temperature cannot be lowered too much at all. Besides the suppression of B out-diffusion by process temperature reduction, doping compensation can be an alternative. As shown in Fig. 5 (a), P atoms are intentionally introduced in halo IMP with a largest angle of 27° limited by the height of adjacent gate lines. However, most P atoms will be etched away after the sigma-shaped trench formation, as shown in Fig. 5 (b). Fig. 5 (a) indicates that the P halo IMP peak is actually beyond the gate-controlled channel region, so increasing halo IMP dose is not an economic way to increase P doping level for B compensation. A higher dose will also cause more severe random dopant fluctuation which will degrade the local variation performance. The PHIA at a higher temperature (HT PHIA) may easily drive more P atoms to diffuse into the critical channel regions for a better B compensation effect. Compared to Fig. 5 (c) for the SiGe control sample, Fig. 5 (d) demonstrates that the HT PHIA does effectively increase the P doping level in the gate-controlled channel regions, especially in the region within 20 nm in channel depth as marked by line A. The active net doping distribution shown in Fig. 5 (e) confirms the effectiveness of HT PHIA, resulting in a reduction of charge sharing. Correspondingly, Fig. 5 (f) demonstrates that the

978-1-6654-9759-6/22 $31.00 © 2022 IEEE                336

increase of PHIA temperature by 70°C can then effectively improve the Vt roll-off characteristics.

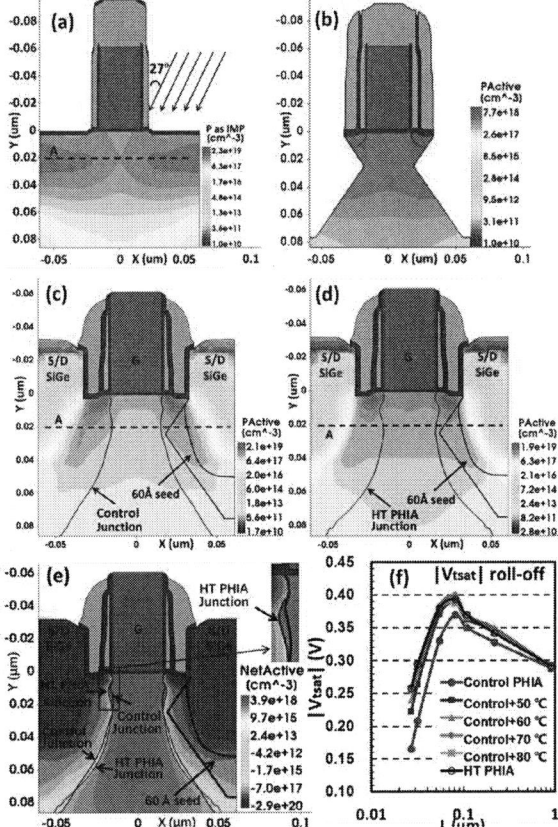

Fig. 5 TCAD simulation results for PMOSFET with a W/L of 2.7 μm/27 nm. (a) The as-implanted P distribution after halo IMP. The active P distribution (b) after control PHIA and S/D trench etching, (c) after final S/D activation anneal for SiGe control sample, (d) after final S/D activation anneal for sample with a HT PHIA. (e) The final pn junction profile after S/D activation anneal for the SiGe control sample and the sample with a HT PHIA. (f) |Vtsat| roll-off curves with PHIA at different anneal temperatures.

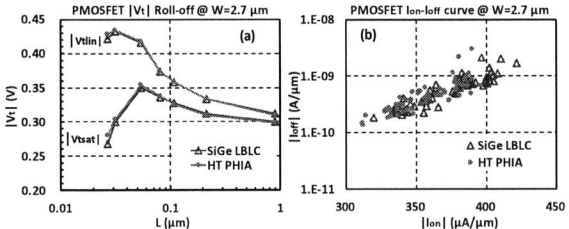

Fig. 6 With a W of 2.7 μm, PMOSFETs' (a) |Vt| roll-off curves for SiGe samples with a HT PHIA and SiGe LBLC samples; (b) $I_{on}/I_{off}$ data corresponding to (a).

According to the TCAD simulation results, following the control process in Fig. 1 (b) but with a HT PHIA at a temperature 70 °C higher than the original control process and a P halo IMP dose decrease by 33% the new PMOSFETs were fabricated. Fig. 6 clearly demonstrates that the HT PHIA can effectively improve

the Vt roll-off characteristics, showing a comparable improvement with SiGe LBLC process. The PHIA was usually employed for B activation in PLDDs and IMP damage recovery. While it can also be applied to drive more P atoms to compensate B atoms from eSiGe. It is simple but effective, providing a new way to suppress SCE as long as the embedded S/D may cause the non-uniform B out-diffusion into the channel region.

## IV. CONCLUSIONS

In summary, it is revealed that the non-uniform B out-diffusion from eSiGe S/D induced by the unique sigma-shaped trench results in a severe Vt roll-off for PMOSFETs. Compared to the widely adopted solution of eSiGe epitaxy temperature reduction with a complicated process adjustment, a simple PHIA at a higher temperature is proposed and successfully demonstrated to be able to obtain the same effective improvement of Vt roll-off characteristics.

## V. ACKNOWLEDGEMENT

The work was supported by the Natural Science Foundation of China (NSFC-61874030).

## REFERENCES

[1] S. E. Thompson, M. Armstrong, and C. Auth. "A 90-nm logic technology featuring strained-silicon." *IEEE Trans. Electron Devices*, vol. 51, no. 11, pp. 1790-1797, Oct. 2004. doi: 10.1109/TED.2004.836648.

[2] N. Tamura, Y. Shimamune, and H. Maekawa. "Embedded silicon germanium (eSiGe) technologies for 45 nm nodes and beyond." *Proc. IEEE IWJT*. Shanghai, China, May 2008, pp. 73-77. doi: 10.1109/IWJT.2008.4540021.

[3] D. James. "Moore's law continues into the 1x-nm era." *27th Annual SEMI Advanced Semiconductor Manufacturing Conference (ASMC)*. Saratoga Springs, NY, 2016, pp. 324-329. doi: 10.1109/ASMC.2016.7491159..

[4] S. Ogura, P. J. Tsang, W. W. Walker, D. L. Critchlow and J. F. Shepard. "Design and Characteristics of the Lightly Doped Drain-Source (LDD) Insulated Gate Field-Effect Transistor." *IEEE Journal of Solid-State Circuits*, vol. 15, no. 4, pp. 424-432, Aug. 1980. doi: 10.1109/JSSC.1980.1051416.

[5] C. F. Codella and S. Ogura. "Halo doping effects in submicron DI-LDD device design." *International Electron Devices Meeting*. Washington, DC, USA, 1985, pp. 230-233. doi: 10.1109/IEDM.1985.190938.

[6] M. Aoki, T. Ishii, and T. Yoshimura. "0.1 mu m CMOS devices using low-impurity-channel transistors (LICT)." *International Technical Digest on Electron Devices*. San Francisco, CA, USA, 1990, pp. 939-941. doi: 10.1109/IEDM.1990.237087.

[7] J. C. Hu, A. Chatterjee, M. Mehrotra, J. Xu, W. -. Shiau and M. Rodder. "Sub-0.1um CMOS with source/drain extension spacer formed using nitrogen implantation prior to thick gate re-oxidation." *Symposium on VLSI Technology. Digest of Technical Papers*. Honolulu, HI, USA, 2000, pp. 188-189. doi: 10.1109/VLSIT.2000.852820.

[8] C. L. Yang, C. H. Tsai, and C. I. Li. "Suppressing Device Variability by Cryogenic Implant for 28-nm Low-Power SoC Applications." *IEEE Electron Device Letters*, vol. 33, no. 10, pp. 1444-1446, Oct. 2012. doi: 10.1109/LED.2012.2209395.

[9] A. Pouydebasque, B. Dumont, and S. Denorme. "High density and high speed SRAM bit-cells and ring oscillators due to laser annealing for 45nm bulk CMOS." *IEEE International Electron Devices Meeting*. Washington, DC, 2005, pp. 663-666. doi: 10.1109/IEDM.2005.1609438..

[10] C. Ortolland, T. Noda, and T. Chiarella. "Laser-annealed junctions with advanced CMOS gate stacks for 32nm Node: Perspectives on device performance and manufacturability." *Symposium on VLSI Technology*, Honolulu, HI, 2008, pp. 186-187. doi: 10.1109/VLSIT.2008.4588612.

[11] T. Noda, W. Vandervorst, and C. Vrancken. "Analysis of pocket profile deactivation and its impact on Vth variation for Laser annealed device using an atomistic kinetic Monte Carlo approach." *International Electron Devices Meeting*. San Francisco, CA, 2010, pp. 15.7.1-15.7.4. doi: 10.1109/IEDM.2010.5703371.

[12] C. W. Liang, M. T. Chen, J. S. Jeng, W. Y. Lien, C. C. Huang, Y. S. Lin, B. J. Tzau, W. J. Wu. "A 28nm poly/SiON CMOS technology for low-power SoC applications." *Symposium on VLSI Technology*, Honolulu, HI, 2011, pp. 38-39. doi: not available.

[13] J. M. Hartmann. "Low temperature growth kinetics of high Ge content SiGe in reduced pressure-chemical vapor deposition." *Journal of Crystal Growth*, vol. 305, no. 1, pp. 113-121, 2007. doi: 10.1016/j.jcrysgro.2007.03.051.

# INVESTIGATION OF NANOSHEET DEFORMATION DURING CHANNEL-RELEASE IN GATE-ALL-AROUND NANOSHEET TRANSISTORS

*Jingwen Yang[1], Kun Chen[1,2], Xin Sun[1], Dawei Wang[1], Qiang Wang[1], Tao Liu[1], Ziqiang Huang[1], Zhecheng Pan[1], Saisheng Xu[1], Chunlei Wu[1,2,\*], Min Xu[1,2,\*], David Wei Zhang[1,2,\*]*

[1]School of Microelectronics, Fudan University, Shanghai 200433, China

[2] Shanghai Integrated Circuit Manufacturing Innovation Center Co., Ltd, Shagnhai 201203, China

*Corresponding Author's Email: wuchunlei@fudan.edu.cn, xu_min@fudan.edu.cn, dwzhang@fudan.edu.cn

## ABSTRACT

In this paper, the nanosheets channel deformation issue in Gate-All-Around (GAA) transistors have been investigated and discussed. Based on simulation study using COMSOL Multiphysics and Sentaurus TCAD tools, it is highlighted that the stress applied to the nanosheets during channel release process plays an important role in Si nanosheets deformation. Three-layer stacked GAA nanosheets devices of different strain conditions have been designed and fabricated to validate the stress impacts on nanosheets deformation. Three-layer uniformly stacked nanosheets channel structure has been successfully fabricated through good control of nanosheets strain during the channel release.

*Keywords—Gate-All-Around (GAA); Nanosheet (NS); Si/SiGe superlattice; Channel release process; Channel stress;*

## INTRODUCTION

Vertically stacked Gate-All-Around Nanosheet transistor has been identified as the most promising candidate as an alternative for 3nm node and beyond applications, owing to its superior electrostatics and short –channel control compared to FinFET [1-5]. In a stacked nanosheet transistor, one of the key challenges in device fabrication is the channel release process - highly selective etch of the SiGe sacrificial layers in Si/SiGe superlattice structure [6]. During channel release process, the stacked nanosheets can be easily deformed or bended, which cannot be adapted in the later-on fabrication process.

In this paper, the root-cause leading to nanosheets deformation in GAA transistor fabrication have been investigated and discussed. It is highlighted that, apart from gravity and the surface tension impacts, the stress applied to the nanosheets during channel release plays an important role in Si nanosheets deformation. Simulation study of the impact of stress on nanosheets channel has be carried out based on Sentaurus Technology-Computer-Aided-Design (TCAD) tool and COMSOL Multiphysics tool, exhibiting obvious nanosheets deformation in strained nanosheets cases.

Moreover, the three-layer stacked GAA nanosheets devices for different strain conditions have been experimentally demonstrated. Highly uniform nanosheets channel as well as inter-channel spaces have been successfully achieved through good control of nanosheets strain during the channel release process.

## SIMULATIONS AND ANALYSIS

### A. Nanosheets deformation simulation

Numerical simulation using COMSOL Multiphysics was conducted first to investigate the factors causing nanosheets deformation. The sheet width $W_{ch} \sim 30nm$ and length $L_{ch} \sim 100nm$ were adopted in this simulation.

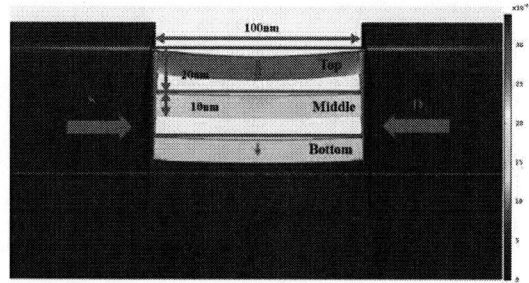

*Figure.1: (a) Simulated results of Si nanosheet channels bending under compressive stress 1.5GPa condition.*

Ideally the surface tension force applied to up and down sides of the nanosheets would be negligible by using IPA and Critical Point Drying (CPD). However, when a compressive stress of 1.5GPa was applied to the Si nanosheets, significant nanosheets deformation has been observed in Fig.1, suggesting that channel stress would be a key factor causing nanosheets bending or even collapse.

### B. Proposed S/D Design for strained nanosheets

In order to verify the impacts of stress on nanosheets deformation, GAA transistors with different source/drain (S/D) sizes have been proposed to provide varied channel stress considering equipment limitations of applying stress by epitaxy. It is known that compressive stress would be delivered from S/D region to the nanosheets channel during the patterning and channel release process, due to great volume reduction in the channel region. For simplicity, in this work, GAA nanosheets devices with different source/drain sizes have been designed and simulated to provide varied compressive stress to the nanosheets during S/D patterning and channel release processes.

3D process simulation of S/D patterning and channel release processes was carried out using Sentaurus TCAD tools. The structure parameters of simulated three-layer stacked nanosheets device are shown in Fig.2 (a) and (b). The S/D regions with varied length of 0.2um, 0.3um, 0.4um, 0.5um, 0.7um, 0.9um, 1.3um, 1.5um, 1.7um, 1.9um and 2.1um and fixed width of 1um were used. In which the same nanosheets width $W_{ch}$ of 30nm and length $L_{ch}$ of 100nm were adopted.

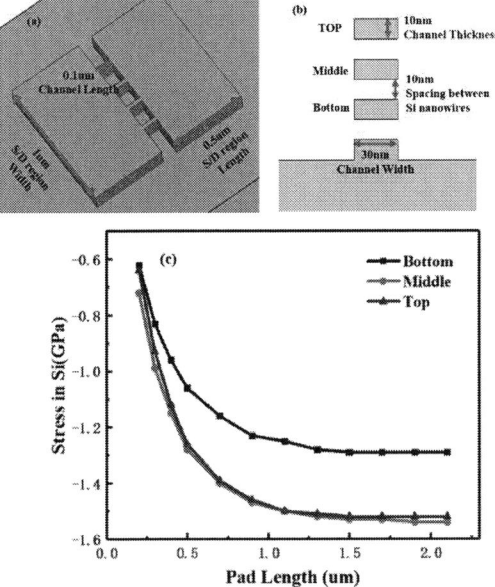

Figure.2: (a) The top view; And (b) The cross-section view of the simulated three-layer stacked nanosheets transistor. (c) The nanosheet channel stress values with varied S/D region sizes.

Fig.2(c) shows the average channel stress values of the top, middle, bottom sheets extracted along the channel direction. The channel stress increases with increasing S/D pad length, and tends to saturate beyond pad length of 1um. As can also be seen in the Fig.2(c), the top and middle sheets exhibit larger compressive stress than the bottom one. This consists with the experimental observations that the bending and stiction tends to occurs in top and middle sheets.

## STACKED NANOSHEETS FABRICATION

Based on above simulations and analysis, fabrication of the designed GAA transistors with different S/D sizes

have been performed.

Fig.3 shows the cross-sectional diagrams of the key steps of GAA device fabrication. Firstly, multilayer Si/SiGe (10nm Si/10nm $Si_{0.7}Ge_{0.3}$) superlattice were epitaxial grown on bulk Si (100) substrate (Fig.3 (a)). The superlattice structure were patterned and etched down to Si substrate by Inductive Coupled Plasma (ICP) to form channel region as well as S/D region (Fig.3 (b)). The fabricated nanosheet width $W_{ch}$ and length $L_{ch}$ are 30nm and 100nm; while two different S/D sizes with fixed width (1um) and different lengths of 2um, 0.5um were used. Then the SiGe sacrificial layers were selectively removed through wet etch (HF: $H_2O_2$: $CH_3COOH$) [7] to release Si nanosheets channel (Fig.3 (c)), following with gate stack deposition of 3nm $HfO_2$ and 10nm TiN (Fig.3 (d)) by Atomic Layer Deposition (ALD).

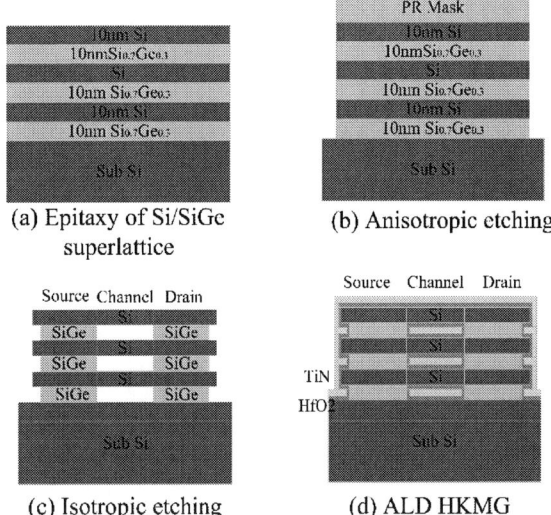

Figure.3: Key fabrication steps of GAA Si nanosheet fabrication: (a) Multilayer epitaxy of Si/SiGe superlattice. (b) S/D and channel patterning. (c) Channel release of Si nanosheets. (d) HKMG deposition.

## RESULTS AND DISCUSSION

The top view and cross-section view of the fabricated GAA transistors with the designed two different S/D sizes are demonstrated in Fig.4. For GAA device with larger S/D size, as seen in Fig.4 (b), obvious sheets stiction occurred between top and middle sheets. While for device with smaller S/D size no visible nanosheets deformation occurred, as observed in Fig.4 (d). The experimental results agree well with the aforementioned simulation analysis. The experimental results strongly verified that strain is one of the key origins causing nanosheets deformation. Therefore, deliberate nanosheets strain engineering during the device fabrication flow is vital in order to achieve uniform nanosheets structure. The results provide useful guidelines for stacked GAA transistors design and fabrication.

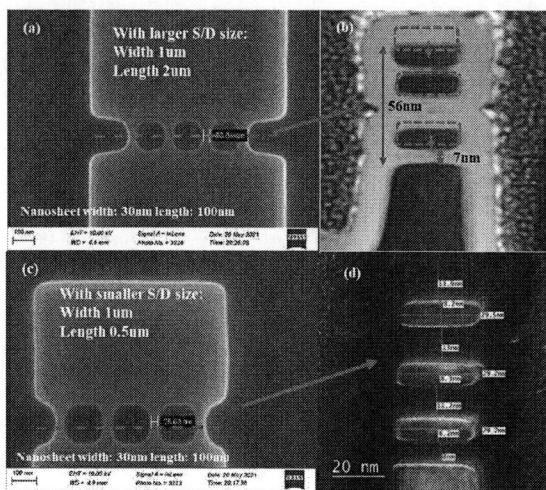

Figure.4: (a) Top view SEM image (b) cross-section view TEM image of 3-layer GAA Si nanosheets with larger S/D size. (c) Top view SEM image (d) cross-section view TEM image of 3-layer GAA Si nanosheets with smaller S/D size.

TABLE I Achieved stacked nanosheets structural dimensions with optimized smaller S/D size

| Nanosheet position | $L_{ch}$ (nm) | $W_{ch}$ (nm) | Thickness of Si nanosheets(nm) |
|---|---|---|---|
| Top | 99.8 | 29.5 | 9.2 |
| Middle | 100.0 | 29.2 | 8.3 |
| Bottom | 99.5 | 29.2 | 8.2 |

TABLE I summarizes the obtained structure dimensions of stacked Si nanosheets with smaller S/D size in Fig.4 (d). The difference of channel width and length among top, middle and bottom Si nanosheets is less than 1nm, attributing to perfect fin etch and wet release. Highly uniform nanosheets thickness has also been achieved through good control of selective SiGe sacrificial layer etching.

## CONCLUSION

In this work, the nanosheets deformation issue in GAA transistors fabrication have been investigated, and uniformly stacked nanosheets channel structure has been successfully demonstrated in experiment. Combining the COMSOL and Sentaurus TCAD simulation, the presented experiment results demonstrated the strain applied to the nanosheets could be an influential factor of the nanosheets deformation. This provides a useful guideline for GAA transistor channel design and fabrication.

## ACKNOWLEDGEMENTS

This work is sponsored by Platform for the development of next generation integrated circuit technology and Shanghai Sailing Program 21YF1402700.

## REFERENCES

[1] C. -H. Jan et al., "A 14 nm SoC platform technology featuring 2nd generation Tri-Gate transistors, 70 nm gate pitch, 52 nm metal pitch, and 0.0499 um2 SRAM cells, optimized for low power, high performance and high density SoC products," 2015 Symposium on VLSI Circuits (VLSI Circuits), 2015, pp. T12-T13.

[2] S. Natarajan et al., "A 14nm logic technology featuring 2nd-generation FinFET, air-gapped interconnects, self-aligned double patterning and a 0.0588 µm2 SRAM cell size," 2014 IEEE International Electron Devices Meeting, 2014, pp. 3.7.1-3.7.3.

[3] C. -H. Jan et al., "A 22nm SoC platform technology featuring 3-D tri-gate and high-k/metal gate, optimized for ultra low power, high performance and high density SoC applications," 2012 International Electron Devices Meeting, 2012, pp. 3.1.1-3.1.4.

[4] H. -J. Cho et al., "Si FinFET based 10nm technology with multi Vt gate stack for low power and high performance applications," 2016 IEEE Symposium on VLSI Technology, 2016, pp. 1-2.

[5] A. Kinoshita, Y. Tsuchiya, A. Yagishita, K. Uchida and J. Koga, "Solution for high-performance Schottky-source/drain MOSFETs: Schottky barrier height engineering with dopant segregation technique," Digest of Technical Papers. 2004 Symposium on VLSI Technology, 2004., 2004, pp. 168-169.

[6] T. Ernst et al., "Novel 3D integration process for highly scalable Nano-Beam stacked-channels GAA (NBG) FinFETs with HfO2/TiN gate stack," 2006 International Electron Devices Meeting, 2006, pp. 1-4.

[7] B. Holländer, D. Buca, S. Mantl, and J. M. Hartmann, "Wet Chemical Etching of Si, Si1−xGex, and Ge in HF:H2O2:CH3COOH," Journal of The Electrochemical Society, vol. 157, no. 6, 2010.

978-1-6654-9759-6/22 $31.00 © 2022 IEEE

# THE APPLICATION OF LAU'S SCHOTTKY-POOLE-FRENKEL THEORY TO DISTINGUISH LEAKAGE CURRENT MECHANISMS IN HIGH-K MIM CAPACITORS BY PATTERN RECOGNITION

*W.S. Lau*

Nanyang Technological University (Retired), School of EEE, Singapore 639798

E-mail: lauwaishing@yahoo.com.sg

## ABSTRACT

Lau's unified Schottky-Poole-Frenkel theory, first proposed in 2012, can be applied to distinguish leakage current mechanisms in high-k MIM capacitors much more effectively than conventional theory by "pattern recognition". The region of Ohmic conduction at low voltage is also discussed.

*Keywords—high-k dielectric; MIM capacitor; leakage current; I-V characteristics; Schottky emission; Poole-Frenkel effect; Ohmic conduction; image force dielectric constant; pattern recognition*

## INTRODUCTION

The physics of the leakage current vs. voltage (I-V) characteristics of high-k MIM capacitors is still not well understood. For example, the leakage current is sometimes due to the Poole-Frenkel (P-F) effect and sometimes due to Schottky emission (SE). The two effects look similar. For more than 50 years, it has been difficult to distinguish them. In this paper, the author would like to apply his unified Schottky-Poole-Frenkel theory, first proposed in 2012, as an efficient method to distinguish the two effects, resulting in much better understanding of the physics underlying the leakage current observed in high-k MIM capacitors.

### Theory

For the P-F mechanism, the leakage current through an insulator is given by

$$J_{PF} = BE\exp\{[\phi_B - ((qE)/(\pi\varepsilon_o K_{PF}))^{1/2}]/(kT/q)\} \quad (1)$$

In equation (1), B is a constant while $E$, $\phi_B$, k, T, q, $\varepsilon_o$ and $K_{PF}$ are the electric field, barrier height of defect state, Boltzmann constant, absolute temperature, electronic charge, vacuum permittivity and the dielectric constant for the P-F effect.

For the SE mechanism, the leakage current through an insulator is given by

$$J_{SK\_TE} = A**T^2\exp\{[\phi_{B\_TE} - ((qE)/(4\pi\varepsilon_o K_{SK\_TE}))^{1/2}]/(kT/q)\}(2)$$

$$J_{SK\_BE} = A**T^2\exp\{[\phi_{B\_BE} - ((qE)/(4\pi\varepsilon_o K_{SK\_BE}))^{1/2}]/(kT/q)\}(3)$$

In equations (2) and (3), A** is Richardson constant while $E$, k, T, q and $\varepsilon_o$ are the electric field, Boltzmann constant, absolute temperature, electronic charge and vacuum permittivity. In addition, $\phi_{B\_TE}$ and $\phi_{B\_BE}$ are the barrier height at metal-insulator interface for

the top electrode and bottom electrode respectively while $K_{SK\_TE}$ and $K_{SK\_BE}$ are the dielectric constant for the Schottky effect for the top electrode and bottom electrode respectively. In general, it is not easy to distinguish between the P-F mechanism and the Schottky emission mechanism because for both cases the logarithm of leakage current plotted against the square root of voltage is a straight line. In this paper, the author points out the two model can be combined into a unified Schottky-Poole-Frenkel (SPF) model. In this model, the SE effect at the top electrode, the PF effect in the bulk of the high-k dielectric and the SE effect at the bottom electrode are 3 effects in series; the author can predict that the I-V characteristics for both polarities will be similar to that shown in Fig. 1. Inspection of equation (1) shows that the current density $J_{PF}$ is proportional to the electric field $E$ at small electric field $E$. Thus the P-F equation can be considered an extension of Ohm's law. The region of Ohmic conduction for both polarities at low voltage can be predicted by equation (1). The author would like to name the I-V characteristics shown in Fig. 1 as O-SPF-1 (Ohmic-SPF-1).

In the equations for SE and P-F, there is a parameter known as the "image force dielectric constant" $\varepsilon_{if}$. Many scientists used $\varepsilon_{if}$ as the basis to distinguish SE and P-F effects, The current dominant theory (Theory X1) regarding the image force dielectric constant is that $\varepsilon_{if} = n^2$, where n is the refractive index of the insulator measured in the visible light range [1]. For a high-k dielectric, n is usually about 2 and so $\varepsilon_{if}$ is about 4. Previously, the author pointed out that there exists a theory (Theory X0) older than Theory X1 with $\varepsilon_{if} = 1$ [1]. In 1991, Li and Lu proposed to use the low frequency dielectric constant as $\varepsilon_{if}$ [2]. In fact, Hota et al. used the low frequency dielectric constant as $\varepsilon_{if}$ and this value is high, for example, about 18 [3]. The author would name this approach as Theory X2. After examining Theory X2, Theory X1 and Theory X0, the author felt that the application of $\varepsilon_{if}$ as the basis to distinguish SE and P-F effects is quite "doubtful". The author proposed his Theory L that the "image force dielectric constant" $\varepsilon_{if}$ can have a range of values from $> n^2$ to slightly $<1$. Previously in 2012, the author pointed out that the SE or P-F mechanism by itself is not sufficient to explain the I-V characteristics; however, a combination of the SE and P-F mechanisms resulting in a unified Schottky-Poole-Frenkel (S-P-F) theory [4] is quite frequently a much better model of a practical MIM capacitor, as shown in Fig. 1. In this paper, the author would like to point out that his S-P-F theory together with his Theory L may be a much better theory to distinguish SE and P-F effects than something like Theory X1.

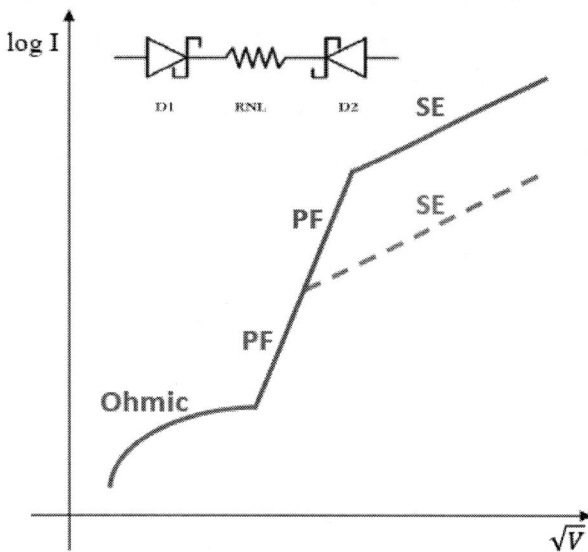

*Figure 1: A plot of the logarithm of the leakage current versus the square root of the applied voltage V for both polarities of V showing the basic Lau's Schottky-Poole-Frenkel theory with two different Schottky barrier heights for the top and bottom electrodes. The region of Ohmic conduction for both polarities is quite frequently observed but the region of PF for both polarities may or may not be seen in experimental results. The region of PF for one polarity and SE for the other polarity is much more commonly seen. The region of SE for both polarities is quite frequently seen if the sample has high enough breakdown voltage. The author would like to name the I-V characteristics shown in Fig. 1 as O-SPF-1 (Ohmic-SPF-1).*

After adopting the new Theory L, the method to distinguish SE and P-F has to be modified. The author noticed that this can be done for MIM capacitors by examining the 2 curves of logI vs $V^{1/2}$ plotted together for both polarities of bias voltage in the same figure instead of just examining 1 curve of logI vs $V^{1/2}$ for only one polarity of bias voltage. The only key assumption is that $K_{SK\_TE} = K_{SK\_BE} = K_{PF} = \varepsilon_{if}$. Theory X1 will not be used. For example, two different but parallel lines imply SE for both polarities, as shown in Fig. 2(c). According to the author's experience, P-F for both polarities, as shown in Fig. 2(a) is a rare case for MIM capacitors with an ultrathin insulator. The majority of experimental observations are similar to Fig. 2(b) or Fig. 2(c). Fig. 2(b) is just the lower portion of Fig. 1 while Fig. 2(c) is just the upper portion of Fig. 1.

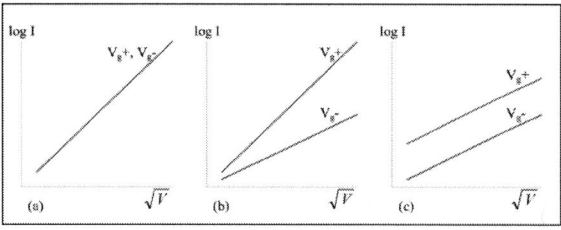

*Figure 2: 3 diagrams of 2 curves of log I vs V $^{1/2}$ plotted together for both polarities of the bias voltage V. (a) P-F for both polarities, (b) P-F for positive bias and SE for negative bias and (c) SE for both polarities.*

### Experimental Support

As shown in Fig. 3, the logarithm of current is plotted against the square root of applied voltage at room temperature for the author's own W/Ta$_2$O$_5$/W capacitor samples [5]. The I-V characteristics are basically symmetrical Ohmic at low voltage. At higher voltage, the I-V characteristics tend to be asymmetric and look similar to Fig. 1; however, the region of PF for both polarities cannot be seen at room temperature. As shown in Fig. 3, the logarithm of current is plotted against the square root of applied voltage at 380 K for the same sample. The I-V characteristics are much more similar to Fig. 1.

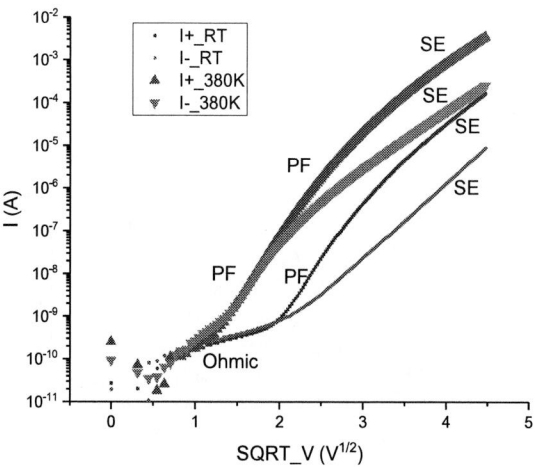

*Figure 3: The logarithm of leakage current versus the square root of voltage characteristics of a W/Ta$_2$O$_5$/W capacitor at room temperature (thin line) and 380 K (thick line). The sample was discussed in reference [5]. The bias is the voltage applied to the top W gate. The interested readers should note the similarity between this figure and Fig. 1. The top electrode has higher Schottky barrier height compared to the bottom electrode because of the Smoluchowski effect.*

The region of Ohmic conduction for both polarities is observed at low voltage. The region of PF for both polarities is observed at slightly higher voltage. At higher voltage, the region of PF for one polarity and SE for the other polarity and then the region of SE for both polarities are observed. Therefore all the 3 types of I-V characteristics shown in Fig. 2 can be observed in the same sample at the higher temperature of 380 K.

Besides the O-SPF-1 type of I-V characteristics, the author has also experimentally observed a slightly different type of I-V characteristics as shown in Fig. 4. The author would like to name the I-V characteristics shown in Fig. 4 as O-SPF-2 (Ohmic-SPF-2).

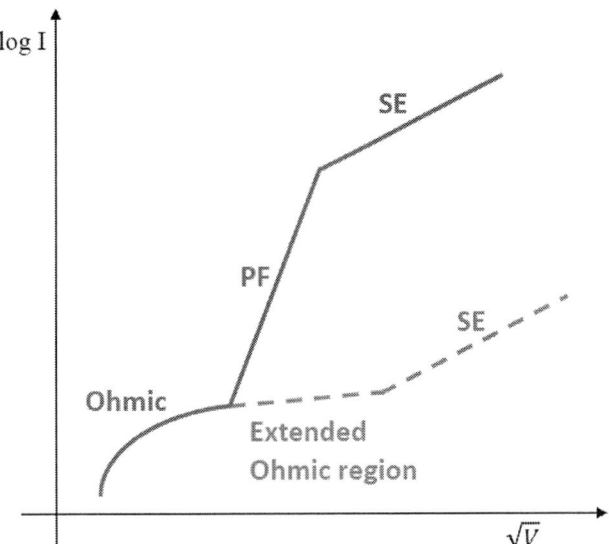

*Figure 4: A plot of the logarithm of the leakage current versus the square root of the applied voltage V for both polarities of V showing the O-SPF-2 (Ohmic-SPF-2) type of I-V characteristics. The region of Ohmic conduction for both polarities is observed but the region of PF for both polarities is absent. The region of PF for one polarity and SE for the other polarity may be observed. The region of SE for both polarities is quite frequently seen if the sample has high enough breakdown voltage. The O-SPF-2 I-V characteristics have an extended Ohmic region compared to O-SPF-1.*

It is easier to predict by theory the existence of O-SPF-1 than O-SPF-2.

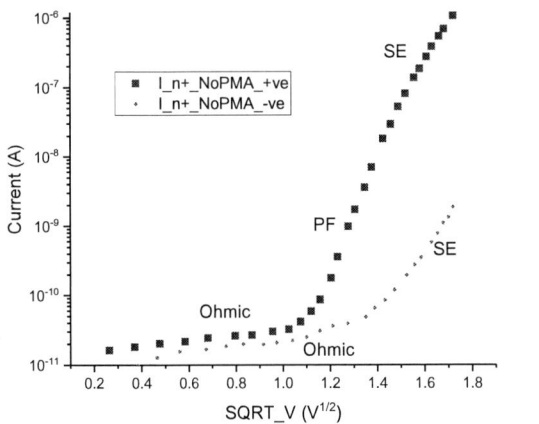

*Figure 5: The logarithm of leakage current versus the square root of voltage characteristics of an $Al/Ta_2O_5/n^+$-Si capacitor at room temperature for positive bias (big blue square) and negative bias (small red dot). The sample was discussed in reference [7]. The bias is the voltage applied to the top Al gate. No post-metallization anneal (PMA) was done.*

The interested readers should note the similarity between this figure and Fig. 4. Previously, the author mentioned the existence of Mechanism A and Mechanism B I-V symmetry in 2021 [8]. The interested readers should note that the region of

P-F conduction for both polarities in Fig. 1 and Fig. 3 is Mechanism A I-V symmetry.

## CONCLUSION

The key point is to examine of the logarithm of the leakage current versus the square root of the applied voltage for both polarities. The author noticed that the I-V characteristics of high-k MIM capacitors are basically symmetrical Ohmic at low voltage. At higher voltage, the I-V characteristics of high-k MIM capacitors tend to be asymmetric and can be explained by a simple series combination of three effects: (1) Schottky emission at the interface of the top electrode and high-k dielectric, (2) Poole-Frenkel effect in the bulk of the high-k dielectric and (3) Schottky emission at the interface of the bottom electrode and high-k dielectric. The author experimentally noticed the existence of two slightly different kinds of I-V characteristics and named them as O-SPF-1 and O-SPF-2. It is easier to predict the existence of O-SPF-1 than O-SPF-2. The Schottky emission and Poole-Frenkel effects can be distinguished by "pattern recognition" using Fig. 1, Fig. 2 and Fig. 4.

## REFERENCES

[1]  W.S. Lau, "Some key modifications of theory required to understand the leakage current mechanisms for MIM capacitors used in DRAM technology," CSTIC 2020 (China Semiconductor Technology International Conference, Shanghai, 2020, IEEE), pp. 1-3, 2020.

[2]  P. Li and T.-M. Lu, "Conduction mechanisms for in $BaTiO_3$", Phys. Rev. B, vol. 43, no. 17, pp. 14261-14264, 1991.

[3]  M.K. Hota, C. Mahata, S. Mallik, C.K. Sarkar and C.K. Maiti. "Impact of top electrodes on HfAlOx-based MIM capacitors", J. Electrochem. Soc., vol. 158, no. 1, pp. H44-H49, 2011.

[4]  W.S. Lau, "An extended unified Schottky-Poole-Frenkel theory to explain the current-voltage characteristics of thin-film metal-insulator-metal capacitors with examples for various high-k dielectric materials," ECS J. Solid State Sci. Technol., vol. 1, no. 6, pp. N139-148, 2012.

[5]  D.Q. Yu, W.S. Lau, H. Wong, X. Feng, S. Dong and K.L. Pey, "The variation of the leakage current characteristics of W/Ta2O 5/W MIM capacitors with the thickness of the bottom W thickness," Microelectronics Reliability, vol. 61, pp. 95-98, 2016.

[6]  W.S. Lau, "The application of the Smoluchowski effect to explain the current-voltage characteristics of high-k MIM capacitors," CSTIC 2017 (China Semiconductor Technology International Conference, Shanghai, 2017, IEEE), pp. 1-3, 2017.

[7]  W.S. Lau, T.S. Tan, N.P. Sandler and B.S. Page "Characterization of defect states responsible for leakage current in tantalum pentoxide films for very-high-density dynamic random access memory (DRAM) applications," Jpn. J. Appl. Phys., vol. 34, Part 1, no. 2B, pp. 757-761, 1995.

[8]  W.S. Lau, "Mechanism B I-V symmetry for MIM capacitors used in microelectronics," CSTIC 2021 (China Semiconductor Technology International Conference, Shanghai, 2021, IEEE), pp. 1-3, 2021.

# Minimizing residual stress of aluminum nitride (AlN) thin films using multi-step deposition of DC pulsed sputtering

*Wei-Lun Chen[1], Shang Shian Yang[1], Ning Hsiu Yuan[1], Wei Yu Zhou[1], Yu-Pu Yang[1], Hsiao-Han Lo[2], Peter J. Wang[2], Walter Lai[2], Yiin-kuen Fuh[1,*], Tomi T. Li[1]*

[1] Department of Mechanical Engineering, National Central University
[2] Delta Electronics, Inc.
Taoyuan City, Taiwan
Center Phone: +886-3-4227151#37313, and *Corresponding Author's Email: michaelfuh@gmail.com

## Abstract

In this study, the minimum residual stress of aluminum nitride (AlN) thin films was obtained controllably using a multi-step deposition technique in a pulsed DC sputtering system. The link between film crystal orientation and residual stress received from X-ray diffraction (XRD) data analyses on this growing AlN experiment was investigated by the comparison of film stress characteristics between one-step deposition and multi-step deposition.

The structure, thickness, crystalline status and residual stress of AlN thin films were measured using scanning electron microscope (SEM), atomic force microscope (AFM) and X-ray diffraction (XRD) respectively. The results reveal that, under the same processing duration, with various intervals of deposition, the AlN film has a distinct structure with various stress properties. It can be concluded from this experiment that the AlN film residual stress can be effectively reduced by multi-step deposition.

In addition, we are aiming to examine the spectrum data acquired by optical emission spectroscopy (OES) in order to see whether there is any association with OES data, the crystalline state, and thin-film quality obtained by the measured result in this study. In our findings, OES data in conjunction with XRD analyses, it is convinced that this multi-step deposition approach can be used to determine the minimum residual stress of thin-film characteristics in AlN thin-film deposition.

## Introduction

Aluminum nitride (AlN) is a III-V compound semiconductor that has a broad bandgap (6.2 eV) with good piezoelectric capabilities, and good thermal conductivity (up to 320 W/mK) as well as outstanding anticorrosive and anti-wear qualities. [1-4] Physical vapor deposition (PVD), sintering procedures [5], and other techniques such as molecular beam epitaxy growth [6] and pulsed radical laser deposition [7] were all commonly used to fabricate aluminum nitride. Various parameters such as distance from target to substrate, reactive gas flow rate, process pressures, and substrate temperature during sputtering have been used for studying AlN deposition [8-9]. However, during DC reactive sputtering, the fallback dielectric compound would deposit on the Al target, electrically insulating the target and producing an arc. Radio frequency (RF) and pulsed DC sputtering techniques with intermittent sputtering have often been used in industry. Compared with RF sputtering, pulsed DC sputtering resulted in a higher deposition rate and higher quality thin films is more used [10-12].

In a recent review paper, the genesis and evolution of residual stress in thin films were deeply studied, and the correlations among stress distributions, microstructures, and process parameters were investigated and studied, thus can obtain the complex gradients of residual stress depth during the PVD thin film growth [13]. Multi-step deposition, on the other hand, has been shown to effectively reduce the residual stress of PVD AlN thin films by varying the applied RF substrate bias and gradients growth [14]. In this research, we used multi-step processes and compared the findings with one-step deposition results. The deposition is carried out at intervals to fulfill the purpose of reducing residual stress by dividing the duration of processing. The microstructure of AlN films formed with multi-deposition stages differs significantly from those deposited with a single deposition step based on the XRD data analyzed in this article.

## Experimental

In this experiment, we deposited AlN thin films atop 4-inch Si (100) p-type substrates in this experiment. Before starting the procedure, clean the substrate with ultrasonic waves for 3 minutes using acetone, isopropanol, and de-ionized (DI) water respectively, and then mix DI water 240 ml with hydrogen fluoride 10 ml for one-minute pickling. For deposition processing, a Delta Electronics pulsed DC power generator (HPP-1K5A01KAT, Delta Incorporated, Taiwan) was employed. The distance between the Al target (10.16 cm in diameter, 0.60 cm in thickness, 99.999 percent purity) and the substrate is 8 cm as shown in Fig. 1.

In these experiments, the spectrum was acquired using OES (SE2020-025-FUV, OtO Photonics Inc., Taiwan), and the cross-section of the sample was observed using SEM (SU8200, Hitachi, Japan). For crystalline observation, an XRD (D2 Phaser, Bruker, USA) is utilized.

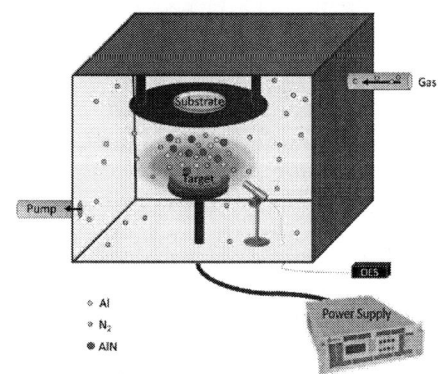

Fig.1. Schematic diagram of the sputter chamber.

978-1-6654-9759-6/22 $31.00 © 2022 IEEE

We utilized a cryopump to raise the chamber vacuum to 8E-6 Torr before starting the PVD process. After that, we cleaned the aluminum target with Ar, which took around 30 minutes. The Al target was cleaned with $Ar^+$ ions to improve coating adherence [15], and injected $N_2$ with 60 sccm for aluminum nitride deposition. Table 1 summarizes the other deposition parameters.

Table. I. The parameters of reactive sputtering for thin films of aluminum nitride.

| Parameter | Value |
|---|---|
| Power | 1200 W |
| Pressure | 2 mTorr |
| Pulsed Frequency | 175 kHz |
| Duty cycle | 85% |
| Distance | 8 cm |
| Substrate temp. | Room temp. |
| Gas flow rate($N_2$:Ar) | 60:0 (sccm) |
| **Deposition steps** | **Deposition time** |
| 1-step | 60 min |
| 2-step | 30/30 min |
| 4-step | 15/15/15/15 min |

Table 1 illustrates the process parameters in 1200W and 2 mTorr using $N_2$ gas. In order to compare with the 60-minute procedure of a single deposition, we performed two-step and four-step processes with separate processing times, and turn off the power for 10 minutes during each interruption in step and step.

## Results and Discussions

The microstructure of the cross-section of AlN thin films generated on a silicon substrate was examined using scanning electron microscopy. Fig.2 is the general view of cross-section SEM in different steps (1, 2 and 4 steps) deposition and processing times. It is observed from the SEM cross-section that every deposition step has a clear gap.

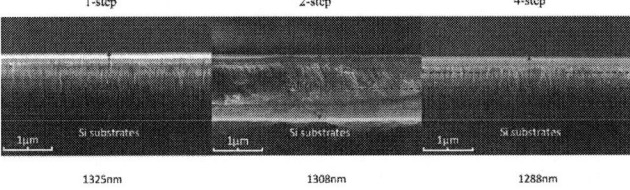

Fig.2. Experimental results are illustrated: The cross-section AlN views of microstructure and thickness of the AlN film deposited via 1-step, 2-step, and 4-step processes respectively.

When AlN film is employed as a piezoelectric material, notably in surface acoustic wave devices, the surface roughness and microstructure of the material have a significant influence on the device's performance. Because the surface acoustic wave travels exclusively on the surface, the range of wavelength from the surface to the interior constrains all the energy [16]. The surface acoustic wave appears to be difficult to pass through when the surface roughness exceeds the wavelength. A non-homogeneous microstructure, on the other hand, such as dislocation fractures and voids, not only does it alter the velocity of the surface acoustic wave and the electromechanical coefficient, but it also causes transmission loss. As a result, the optimized AlN film's surface shape and cross-sectional microstructure were examined using SEM and AFM, respectively [17]. Figure 3 shows AFM pictures of the AlN films. In these figures, each film's root-mean-square (RMS) surface roughness is also displayed, and the value decreases with the steps [18].

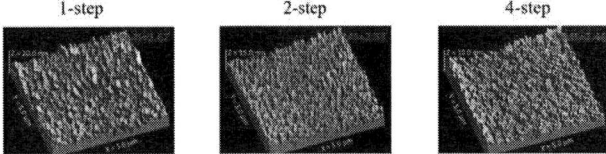

Fig.3. 3D AFM images of the AlN films sputtered at various deposition steps.

X-ray diffraction spectroscopy patterns were utilized to quantify the deposited films' corresponding residual stress, and JADE 6 software was used to analyze the residual stress as formed in the crystal lattice, as shown in Figs. 4 and 5.

The following is a description of the stress estimated using the side-inclination method:

$$\sigma = K\frac{\Delta 2\theta}{\Delta sin^2\varphi} \qquad (1)$$

where $\sigma$ is the residual stress, $\varphi$ is the inclination angle, $\theta$ is the scattering angle, and $K$ is the stress constant. The tensile stress and compressive stress are respectively expressed by the positive and negative values of the slope of $2\theta sin2\varphi$.

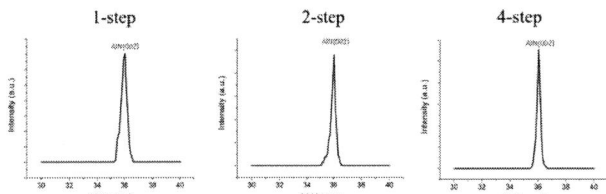

Fig.4. X-ray diffraction spectroscopy patterns of each sample.

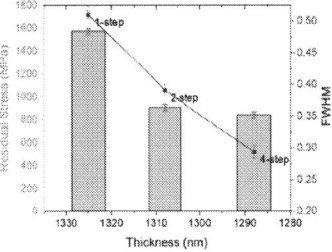

Fig.5. Residual stress and full width at half maximum (FWHM) of XRD with respect to the AlN films vary with the steps of the process.

The full width at half maximum (FWHM) of the $2\theta$ peak narrows when the film stress gradient is lowered. The penetration depth limits of the x-ray source and the crystal disorder of the sputtered films, which result in a large AlN (0 0 2) $2\theta$ peak, are likely to disguise any shortening of the $2\theta$ peaks due to a reduction in the stress gradient. And as the sputtering step increases, the residual stress also changes in a trend. The residual stress declines as the step increases as shown in Figure 6 [14].

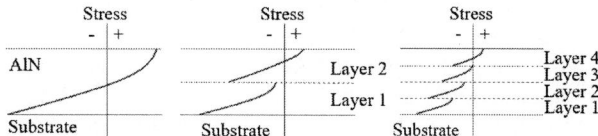

Fig.6. Stress profile of an AlN film formed in a single deposition step, as well as the stress profiles of AlN films produced in two and four deposition stages under the same process circumstances.

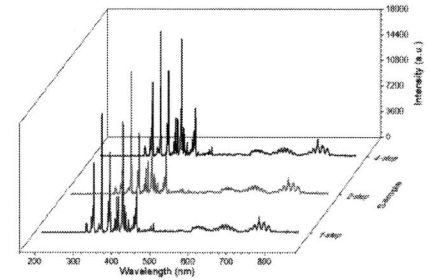

Fig.7. Continuous spectra data of OES are integrated in the processing times of the AlN plasma.

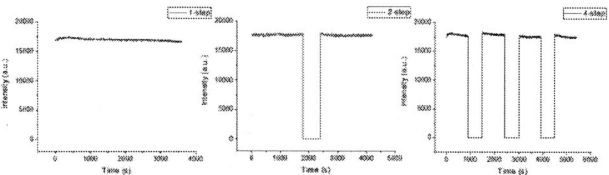

Fig.8. Optical emission spectroscopy patterns showing the intensity of continuous spectra of $N_2$ (336.83 nm) with process time in each process.

Fig.7 shows the continuous optical emission spectra of AlN plasma during the deposition duration of 60 minutes for OES in each process. Fig.8 shows the intensity of $N_2$ (336.83 nm) wavelength which is the maximum intensity of the continuous spectra in each process and the intensity gradually and slightly reduces with process time. This correlates with the trend reduction of film residual stress as shown in Fig. 5.

## Conclusions

In this study, the inherent stress that emerges in PVD AlN thin films was compensated and passivated using a multi-step AlN deposition technique.

The deposition can be separated into several phases in this fashion, with the overall processing time being the same but each step has a varied deposition duration and thickness that is tailored to passivate the average film stress of the deposited film. AlN films with the lowest residual stress and intrinsic stress gradient may be deposited using this approach. The amount of stress gradient decrease is proportional to the number of deposition stages.

We tried to find the correlation between residual stress and OES data during the deposition of AlN thin film. In conventional ways, we can determine the residual stress but after we complete the process. The research in semiconductor

areas is time-consuming, particularly when it comes to determining the appropriate experimental conditions. In the future, we intend to use big data analysis and machine learning to develop in-situ research methods which can predict the residual stress for enhancing the efficiency of processes and experiments.

## Acknowledgements

The authors would like to acknowledge Delta Electronics, Inc., Taiwan, for funding and offering technical support in this project.

## References

[1] P. Vissutipitukul, T. Aizawa " Wear of plasma-nitrided aluminum alloys, " Wear 259 (1-6):482-489, 2005 https://doi.org/10.1016/j.wear.2005.02.119

[2] H. Altun, S. Sen "The effect of DC magnetron sputtering AlN coatings on the corrosion behaviour of magnesium alloys, " Surf. Coat. Technol. 197 (2-3):193-200, 2005 https://doi.org/10.1016/j.surfcoat.2004.06.001

[3] H. Morkoç, S. Strite, G. B. Gao, M. E. Lin, B Sverdlov, M Burns " Large‐band‐gap SiC, III‐V nitride, and II‐VI ZnSe‐based semiconductor device technologies, " J. Appl. Phys. 76 (3):1363-1398, 1994 https://doi.org/10.1063/1.358463

[4] G. Qixin, Y. Akira "Temperature Dependence of Band Gap Change in InN and AlN, " Jpn. J. Appl. Phys. 33 (Part 1, No. 5A):2453-2456, 1994 https://doi.org/10.1143/JJAP.33.2453

[5] J. Y. Qiu, Y Hotta, K. Watari, K. Mitsuishi, M. Yamazaki "Low-temperature sintering behavior of the nano-sized AlN powder achieved by super-fine grinding mill with Y2O3 and CaO additives, " J. Eur. Ceram. Soc. 26 (4-5):385-390, 2006 https://doi.org/10.1016/j.jeurceramsoc.2005.06.016

[6] X. Wang, A. Yoshikawa "Molecular beam epitaxy growth of GaN, AlN and InN, " Prog. Cryst. Growth Charact. Mater. 48:42-103, 2004 https://doi.org/10.1016/j.pcrysgrow.2005.03.002

[7] M. Ishihara, K. Yamamoto, F. Kokai, Y. Koga "Effect of laser wavelength for surface morphology of aluminum nitride thin films by nitrogen radical-assisted pulsed laser deposition, " Jpn. J. Appl. Phys. 40 (Part 1, No. 4A):2413-2416, 2001 https://doi.org/10.1143/jjap.40.2413

[8] T. Y. Lu, Y. P. Yang, H. H. Lo, P. J. Wang, W Lai, Y. K. Fuh, T. T. Li "Minimizing film residual stress with in situ OES big data using principal component analysis of deposited AlN films by pulsed DC reactive sputtering, " J. Adv. Manuf. Technol. 114 (7-8): 1975-1990, 2021 https://doi.org/10.1007/s00170-021-07003-8

[9] M. A. Auger, L. Vázquez, M. Jergel, O. Sánchez, J. M. Albella "Structure and morphology evolution of ALN films grown by DC sputtering, " Surf. Coat. Technol. 180:140-144, 2004 https://doi.org/10.1016/j.surfcoat.2003.10.054

[10] W. D. Sproul "High-rate reactive DC magnetron sputtering of oxide and nitride superlattice coatings, " Vacuum 51 (4):641-646, 1998 https://doi.org/10.1016/s0042-207x(98)00265-6

[11] I. C. Oliveira, K. G. Grigorov, H. S. Maciel, M Massi, C Otani "High textured AlN thin films grown by RF magnetron sputtering; composition, structure, morphology and hardness, " Vacuum 75 (4):331-338, 2004 https://doi.org/10.1016/j.vacuum.2004.04.001

[12] M. H. Park, S. H. Kim "Thermal conductivity of AlN thin films deposited by RF magnetron sputtering, " Mater. Sci. Semicond. Proc. 15 (1):6-10, 2012 https://doi.org/10.1016/j.mssp.2011.04.007

[13] G. Abadias, E. Chason, J. Keckes, M. Sebastiani, G. B. Thompson, E. Barthel, G. L. Doll, C. E. Murray, C. H. Stoessel, L. Martinu "Review Article: Stress in thin films and coatings: Current status, challenges, and prospects, " J. Vac. Sci. Technol. A 36 (2):020801, 2018 https://doi.org/10.1116/1.5011790

[14] K. E. Knisely, B. Hunt, B. Troelsen, E. Douglas, B. A. Griffin, J. E. Stevens "Method for controlling stress gradients in PVD aluminum nitride, " J. Micromech. Microeng. 28 (11):115009, 2018 https://doi.org/10.1088/1361-6439/aad91a

[15] R. Chodun, K. Nowakowska-Langier, K. Zdunek "Methods of optimization of reactive sputtering conditions of Al target during AlN

films deposition, " Mater. Sci.-Poland 33 (4):894-901, 2015 https://doi.org/10.1515/msp-2015-0116

[16] X. H. Xu, H. S. Wu, C. J. Zhang, Z. H. Jin "Morphological properties of AlN piezoelectric thin films deposited by DC reactive magnetron sputtering," Thin Solid Films 388 (1-2):62–67, 2001 https://doi.org/10.1016/S0040-6090(00)01914-3

[17] H. Y. Liu, G. S. Tang, F. Zeng, F. Pan, "Influence of sputtering parameters on structures and residual stress of AlN films deposited by DC reactive magnetron sputtering at room temperature, " J. Cryst. Growth, 363:80-85, 2013 https://doi.org/10.1016/j.jcrysgro.2012.10.008

[18] H. Takeuchi, M. Ohtsuka, H. Fukuyama, "Effect of sputtering power on surface characteristics and crystal quality of AlN films deposited by pulsed DC reactive sputtering, " Phys. Status Solidi B 252 (5):1163-1171, 2015 https://doi.org/10.1002/pssb.201451599

# SILICON PHOSPHORUS PROCESS UNIFORMITY IMPROVEMENT STUDY IN ADVANCED NODE

*Huojin Tu\*[1], Hui Wang[12], Li Peng[1], Jiaqi Hong[1], Wangxin Nie[1], Jun Tan[1], Qin Sun[1], Xinhua Cheng[1]*

[1]Research and Development Dept., Shanghai Huali Integrated Circuit Corporation, Shanghai 201314, China

[2]School of Information Science and Technology, ShanghaiTech University, Shanghai 201210, China

*Corresponding Author's Email: Robert@hlmc.cn

## ABSTRACT

With CMOS devices continuously scaling down, especially from 14nm and below advanced node, embedded silicon phosphorus (SiP) in source and drain regions has been widely used to improve n-MOSFET device performance through electron mobility enhancement. Normally SiP process use N2 as carrier gas, global wafer uniformity almost can meet device requirement, but this SiP process is easy to form tiny particle, this tiny particle will grows larger with SiGe growth, and finally impact device yield; If we use H2 as carrier gas, SiP film defect quality is better relatively, but SiP film global uniformity is worse than SiP film which use N2 as carrier gas formed.

In this paper, we investigated different carrier gas to SiP process effect, including SiP process defect performance and global wafer uniformity performance. According structure lot test result, we found if use N2 as SiP process carrier gas, SiP film uniformity show wafer donut area is thinner than center & edge, but if use H2 as SiP process carrier gas, SiP film uniformity shows the opposite result, that is wafer donut area is thicker than wafer center/edge. Finally, we demonstrated one novel SiP process which combined N2 and H2 carrier gas in one recipe.

**Key words**：SiP, carrier gas, defect, global uniformity

## INTRODUCTION

With advanced CMOS devices continuously scaling down (gL<=28nm), stress engineering plays an important way to improve device performance both in n-MOS and in p-MOS. For n-MOS performance gain, the general way is to use Stress Memorization Technique (SMT), Dual Stress Liner (DSL) and Stress Proximity Technique (SPT). For p-MOS performance gain, the important way for stress enhancement is to use selective epitaxy silicon germanium to replace silicon in PMOS source/drain area. With CMOS device continue develop to FinFET node (gL<=14nm), Silicon phosphorus (SiP) technology is used to boost n-MOS device performance. Because phosphorus atom radius is smaller than silicon, NMOS source/drain area will have compressive stress when phosphorus diffuses into silicon. As a result, it will generate tensile stress in the channel and its mobility will enhancement when electron move in the channel, so device NMOS performance will be improvement by electron moves fast.

Selective epitaxy silicon phosphorus process is a very critical process in semiconductor manufacturing, because it follows "growth" theory more than "deposition" theory. SiP film selectivity and uniformity dominate SiP film quality. For SiP film selectivity, normally we use HCL gas to adjust it, HCL amount is very critical, on the one hand HCL amount should not too small, otherwise SiP particle will exist in dielectric layer (like $SiO_2$ or $Si_3N_4$); on the other hand HCL amount should not too large, because HCL too large will cause SiP film growth very slow even not growth, which shows in figure 1. For SiP film uniformity tuning, tuning method major includes power ratio(top inner, bottom inner), x-flow(x-HCL, x-DCS，x-Dopant), even process temp, pressure, carrier gas etc… also will impact film global uniformity(Show as figure 2 & figure 3). SiP film uniformity direct impact device NMOS performance, so it is an important work to keep SiP uniformity in a low level. [1]

a)            b)

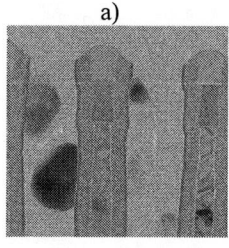 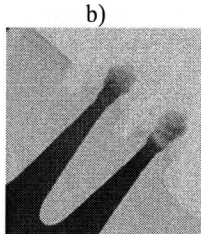

Figure 1: HCL amount to EPI growth impact
a) HCL too small cause selectivity fail
b) HCL too large cause SiP size too small

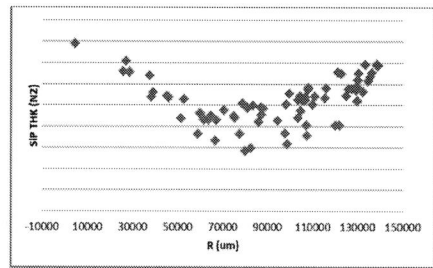

Figure 2: N2 base SiP process profile

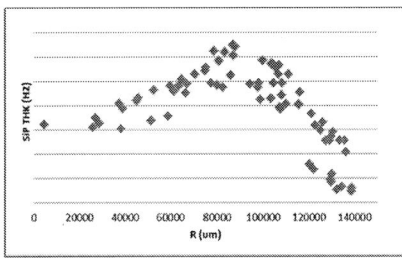

Figure 3: H2 base SiP process profile

In this work, we focus on SiP film uniformity improvement study, including power ratio tuning, x-flow tuning and different carrier gas tuning etc…

## EXPERIMENT

300mm pattern wafers were prepared for process optimization and characterization study according to14nm logic product standard process flow which is shown in figure 4. It includes fin & well formation, gate & LDD formation. SiP hard mask nitride deposition, NMOS source/drain lithograph, NMOS source/drain dry etch, dry strip, wet strip, SiP pre clean & SiP deposition several steps. After SiP process ok we use Nova or KLA OCD tool to measure SiP width. After SiGe process ok we use KLA defect scan tool to measure defect. [2]

Figure 4: 14FF standard process flow

## RESULTS AND DISCUSSION
### 1) N2 & H2 base SiP process comparation
Prepare two pattern wafers which have been done NMOS trench etch and SiP pre clean. One wafer run N2 base SiP process, the other wafer run H2 base SiP process. And we use KT OCD tool to measure SiP size, the data shows in figure 5. From the map we can see, N2 base SiP film donut thickness is smaller, center thickness will large a little than edge. H2 base SiP film donut is the largest, edge is least. We suspect different carrier gas heat conduction is different, it cause SiP profile different.

Above two wafers the following will run SiGe process and scan defect, the defect data shows in figure6. From the data can see N2 base wafer defect is worse than H2 base wafer. The result can be explained that N2 based SiP process growth rate is higher than H2 based SiP process(In this experiment N2 base SiP growth rate is about double than H2 base SiP growth rate) .

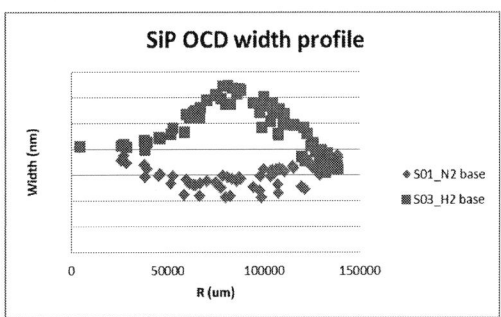

Figure 5: Different base SiP process profile

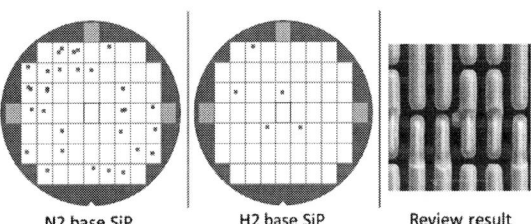

N2 base SiP        H2 base SiP        Review result

Figure 6: Different base SiP process defect result

### 2) H2 base SiP global wafer uniformity tuning result
Six wafers were prepared for H2 base SiP film global wafer uniformity tuning, including power ratio/x-flow etc. Detail splits shows as figure 7. From data we can see tuning bottom power, uniformity almost no change (S02). Only tune TI power (TI-) uniformity has much improvement (S03), it is because of wafer edge thickness larger than before; S04 condition uniformity also better because of tuning TI/BI power; S05 condition(x-DCS tuning) uniformity became worse because of wafer donut thickness large much; S06 condition(x-H2/N2) has minor effect to uniformity improvement. Almost all these splits shows wafer donut area thicker and wafer edge lower.

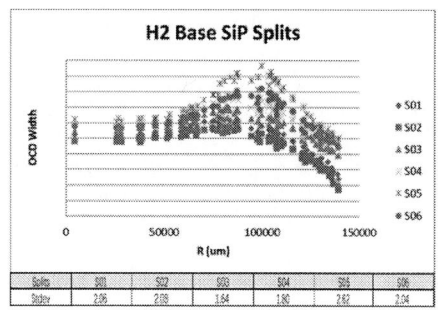

Figure 7: H2 base SiP process profile compare

| Splits | S01 | S02 | S03 | S04 | S05 | S06 |
|---|---|---|---|---|---|---|
| Stdev | 2.06 | 2.09 | 1.64 | 1.80 | 2.82 | 2.04 |

### 3) N2H2 base base SiP global wafer uniformity tuning result

Four wafers were prepared for N2H2 based SiP process uniformity tuning, main focus on power ratio tuning. Detail results shows as figure 8. V02 is based on V01, N2 step & H2 step TI both reduce a little, wafer uniformity no improvement, maybe TI reduce amount is not enough; V03 is N2 step TI reduce much, wafer global uniformity improved much; V04 is H2 step TI reduce much, wafer global uniformity also improved much. From the result can see N2H2 two steps SiP process uniformity is better than single H2 carrier gas SiP process.

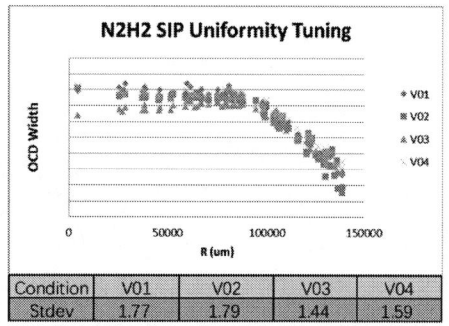

| Condition | V01 | V02 | V03 | V04 |
|---|---|---|---|---|
| Stdev | 1.77 | 1.79 | 1.44 | 1.59 |

Figure 8: N2H2 base SiP process profile compare

N2H2 base carrier gas growth SiP and pure H2 carrier gas growth SiP WAT result shoes in figure 9, the Ion/Ioff data is comparable for different SiP process.

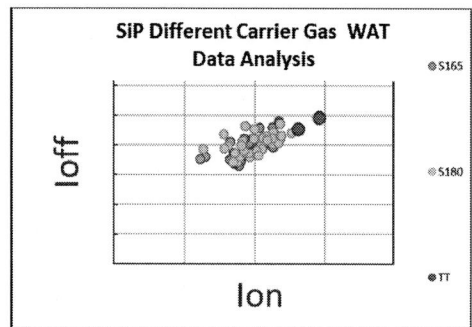

Figure 9: Different base SiP process WAT compare

## CONCLUSION

In this paper, we study the different carrier gas to SiP process film uniformity impact. From our experiment, N2 as carrier gas can get good SiP uniformity but after SiGe DEP defect is not good. H2 as carrier gas can get good defect performance but global uniformity is worse. After combine N2 & H2 carrier gas in one recipe different step, we can get both uniformity & defect better result. [3]

## ACKNOWLEDGMENTS

The author would like to thank Huali Integrated Circuit Corporation Tech R&D EPI group members and AMAT China EPI group members!

## REFERENCES

[1] Huojin Tu, Yonggen He，Youfeng He，Jialei Liu，Lan Jin，Guohui Cai，Yu Liu，Yujian Huang, *CSTIC 2016*，March 13-14, 2016, pp. 1-3

[2] Yiqun Liu*, Lan Jin, Kunshan Song, Qiong Wu, Youfeng He, Yonggen He, *CSTIC2018*, Shanghai, March 11-12, 2018, pp. 1-3

[3] Lan Jin etc. Junction Technology (IWJT), 2012 International Workshop on, Year: 2012, Pages: 234 – 237.

# THE STUDY OF SIGE CHANNEL FORMATION FOR FDSOI

*Lan Jiang*[*]
Shanghai Huali Integrated Circuit Corporation
6 Liangteng Rd., Pudong district, Shanghai, China
*Corresponding Author's Email: jianglan@hlmc.cn

## ABSTRACT

Si1-xGex channel is becoming one of the most promising materials to improve PMOS device performance. In this work, thickness relation between oxide and pre-dep Si1-xGex is proposed. Two step methods, which rapid thermal oxidation (RTO) combined with anneal (RTA), were used to form Si1-xGex channel for PMOS at FDSOI structure. The concentration and thickness of Si1-xGex were measured by SIMS and TEM. It was found that the thickness of Si1-xGex channel was decreased with increasing soak time and temperature. The profile of Ge gets better when bake time become longer. The uniformity of Si1-xGex was affected by pre-layers.

## INTRODUCTION

SiGe-on-insulator (SGOI) MOSFETs have received much attention because of its high hole mobility and superior short-channel control. A critical challenge for SGOI device is to form thin and uniform SiGe channel to achieve good short-channel control and low variability in devices with various widths [1-2]. Tezuka T et al. [3] reported that SGOI-MOSFET structure can be fabricated by using the Ge condensation by oxidation technique. According to Ellingham diagram, the formation energy of SiO2 was lower than GeO2. This difference of the thermal stability between SiO2 and GeO2 suggests a pathway to selectively oxidize Si in SiGe [4]. In this study, two step methods, which rapid thermal oxidation (RTO) combined with anneal (RTA), were used to form Si1-xGex channel for PMOS at FDSOI structure.

## EXPERIMENT

The 300nm FDSOI wafers are used as the starting material. The thicknesses of buried oxide and silicon are about 200A and 120A respectively. Firstly, selective SiGe epitaxial is applied for FDSOI PMOSFET. Then, the Si1-xGex channel is formed by rapid thermal process (RTP). Schematic of fabrication method of Si1-xGex channel is shown in Fig. 1.

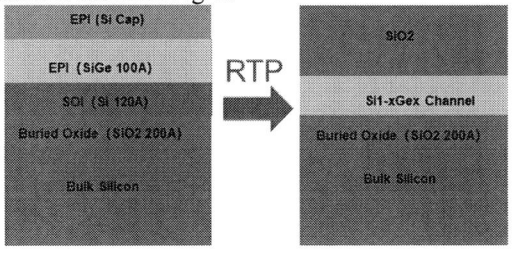

*Figure1: Schematic of fabrication method of Si1-xGex channel*

## RESULT DISCUSSION

*A. Acquirement thermal conditions*

The thermal conditions are mainly depended on pre-dep SiGe layers, including the concentration of Ge, thickness of Si cap and SiGe. It is necessary to get a correlation between them.

At this circumstance, the reaction is proposed as follows:

$$Si_{1-x}Ge_x + O_2 \rightarrow （1-x） SiO_2 + xGe \qquad (1)$$

According to formulas (2) - (5), thickness of SiO2 can be calculated by pre-dep Si1-xGex condition. It means that a suitable thermal condition can be found from oxidation kinetics and oxide properties.

$$\frac{Thickness((1-x)SiO2)*Area}{Thickness(Si(1-x)Ge(x))*Area} = \frac{M_{(1-x)SiO2}/\rho_{SiO2}}{M_{Si(1-x)Ge(x)}/\rho_{Si(1-x)Ge(x)}} \qquad (2)$$

$$\rho_{Si(1-x)Ge(x)} = \frac{Atom\ amount*M_{Si(1-x)Ge(x)}}{V_{Si(1-x)Ge(x)}*NA} = \frac{8*M_{Si(1-x)Ge(x)}}{(aSi(1-x)Ge(x)*10^{-9})^3*(6.02*10^{23})} \qquad (3)$$

$$aSi(1-x)Ge(x)=aGe*x+aSi*(1-x)-0.0272*x*(1-x)$$
$$（aGe=5.646A, aSi=5.431A） \qquad (4)$$

$$\frac{Thickness （1-x） SiO2*Area}{Thickness\ Si(1-x)Ge(x)*Area} = \frac{M_{(1-x)SiO2}*6}{(aSi(1-x)Ge(x))^3} \qquad (5)$$

The oxidation rate of Si on blanket wafer is collected at 1100°C (Fig.2). With increasing soak time, the thickness of oxide increase and the oxidation rate decrease. F. Rozé et al. [5] reported that the oxidation rate of SiGe is just a little higher than the one of Si. Therefore, the data can be used as a reference for SiGe.

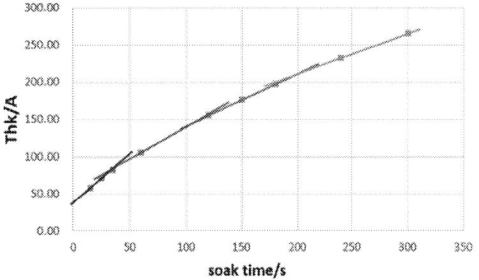

*Figure2: The oxidation rate of Si on blanket wafer*

B. Formation of SiGe channel by different thermal condition

Two step methods, which rapid thermal oxidation (RTO) combined with baking (RTA), were used to form Si1-xGex channel for PMOS at FDSOI structure. The thermal conditions are shown in Table1.

| Condition | RTP1（Oxidation） | RTP2（bake） |
|---|---|---|
| Split1 | 180s | 60s |
| Split2 | 180s | 120s |
| Split3 | 240s | 60s |

*Table1. RTP thermal condition split*

The TEM and SIMS results of SiGe channel at different thermal condition are shown in Fig. 3 and Fig. 4. It can be seen that thickness of SiGe channel reduce and the concentration of Ge increase with rising oxidation time (Fig. 3&4 (b), (d)). It means that thickness SiGe channel is mainly affected by RTO. The profile of Ge gets better when bake time become longer.

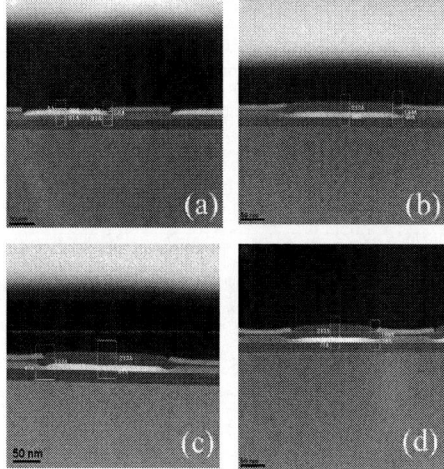

*Figure3: TEM of SiGe channel (a) As SiGe dep (b) split1(c) split2 (d) split3*

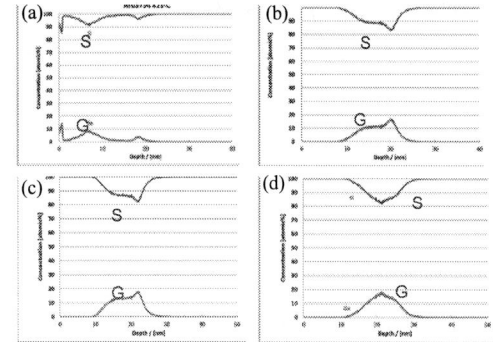

*Figure4: SIMS of SiGe channel (a) As SiGe dep (b) split1(c) split2 (d) split3*

The profile of Ge gets better when bake time become longer. However, although thermal condition changes, the range of thickness between center and edge area still reach nearly 40A (Table2). Hence, uniformity of Si1-xGex channel is affected by pre-layers (SiGe dep).

| Condition | | SiGe dep | Split1 | Split2 | Split3 |
|---|---|---|---|---|---|
| **Center THK/A** | **Channel** | 91(SOI) | 86 (SiGe) | 84 (SiGe) | 74 (SiGe) |
| | **On the Channel** | 5(Si)+99 (SiGe) | 212 (SiO2) | 213 (SiO2) | 243 (SiO2) |
| **Egde THK/A** | **Channel** | 91 (SOI) | 48 (SiGe) | 47 (SiGe) | 38 (SiGe) |
| | **On the channel** | 6(Si)+56 (SiGe) | 134 (SiO2) | 137 (SiO2) | 169 (SiO2) |

*Table2.The thickness of SiGe channel at different thermal condition*

## CONCLUSION

Thermal condition is determined by pre-dep Si1-xGex and the target thickness of SiGe channel. At different thermal conditions, TEM and SIMS results of SiGe channel indicates thickness is mainly affected by RTO and uniformity is attributed to pre-layer.

## REFERENCES

[1] K. Cheng et al., "High performance extremely thin SOI (ETSOI) hybrid CMOS with Si channel NFET and strained SiGe channel PFET," 2012 International Electron Devices Meeting, 2012, pp. 18.1.1-18.1.4.

[2] K.W. Jo et al.,"Hole mobility enhancement in extremely-thin-body strained GOI and SGOI pMOSFETs by improved Ge condensation method," 2018 IEEE Symposium on VLSI Technology, 2018, pp. 195-196.

[3] Tezuka T *et al.*, "Ultrathin Body SiGe-on-Insulator pMOSFETs With High-Mobility SiGe Surface Channels," *IEEE Trans. Electron Devices*, vol. 50, 2003, pp.1328-1333.

[4] C. -T. Chang and A. Toriumi, "Preferential oxidation of Si in SiGe for shaping Ge-rich SiGe gate stacks," 2015 IEEE International Electron Devices Meeting (IEDM), 2015, pp. 21.5.1-21.5.4

[5] F. Rozé et al., "SiGe oxidation kinetics and oxide density measured by resonant soft X-ray reflectivity," 2017 IEEE 12th Nanotechnology Materials and Devices Conference (NMDC), 2017, pp. 177-178

# LOW-TEMPERATURE DIE TO GLASS WAFER BONDING BASED ON AU-AU ATOMIC DIFFUSION

*Shuchao Bao[1,2], Yi Zhong[1,2*], Yimin He[2,3], Ke Li[1,2] and Daquan Yu[1,2*]*

[1] School of Electronic Science and Engineering, Xiamen University, Xiamen 361005, China

[2] Xiamen Sky Semiconductor Technology Co., Ltd., Xiamen 361005, China

[3] School of Materials Science and Engineering, Xiamen University of Technology, Xiamen 361005, China

*Corresponding Author's Email: yudaquan@xmu.edu.cn, zhongyi@xmu.edu.cn

## ABSTRACT

In this paper, the atomic diffusion bonding (ADB) is adopted for silicon dies and glass wafer using. In order to improve the bonding quality, the bonding of glass wafer and silicon dies has been achieved through Au-Au atomic-diffusion bonding. To reduce the surface roughness, copper is deposited on the glass surface to meet the bonding requirements. At the same time, in order to achieve mass production and reduce bonding porosity, an innovative method: die to wafer bonding is proposed, which is also of great help to improve bonding quality.

**Keywords—heterogeneous integration; die to wafer; interposer; ADB; bonding**

## INTRODUCTION

With the development of electronic information technology, devices with single function cannot meet the needs of miniaturization and functional diversification. Therefore, the heterogeneous integration of materials or devices with different functions and the development of multifunctional integrated electronic modules have become the development direction of electronic technology [1]. At present, there are two solutions to realize heterogeneous integration of different functional materials: heteroepitaxy and heterogeneous bonding. However, heteroepitaxy has the problems of crystal form mismatch, lattice mismatch and thermal expansion coefficient mismatch, the epitaxial materials have large dislocation density or polycrystalline properties, and cannot be used to prepare high-performance devices. Heterogeneous bonding can directly combine two kinds of wafers with different materials, which has attracted extensive attention.

Au-Au bonding has no protruding bonding structure, so it has better operation performance. Glass, as an ideal material for high-frequency transmission, has a resistivity of $10^{13}$-$10^{16}\Omega$.cm and strong mechanical stability. Even when the thickness of the glass wafer is less than 100μm, the warpage is still small. Due to its unique properties, glass has been widely used in packaging of MEMS, sensor and RF devices [2]. Based on ADB technology, this study takes the gold layer as the contact surface to complete the bonding between glass wafer and silicon dies [3]. The results show that this way of bonding can achieve good bonding rate, improve bonding quality and reduce the proportion of void area. What is more, the bonding by die to wafer is also ready for large-scale production in the future.

## EXPERIMENT

Before all processes, the silicon dies and glass wafer shall be cleaned by acetone ultrasonic cleaning and alcohol ultrasonic cleaning in turn, and then the silicon dies also should be cleaned by plasma. Different from the standard silicon cleaning process, this cleaning method is simpler, and the surface can be cleaned without acid and alkali. Figure 1 shows the surface morphology of the cleaned silicon dies observed under a microscope with a 5x focal length. As can be seen from the figure, the surface of the silicon die is very clean and smooth without any dirt which shows that this way of cleaning can clean the surface of silicon die effectively.

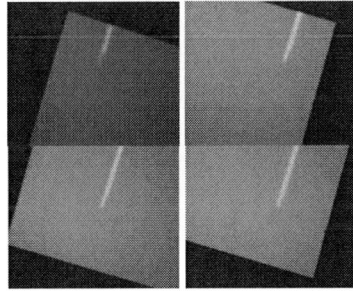

*Figure 1: the surface morphology of the cleaned silicon dies*

Figure 2 is Schematic diagram of silicon dies bonded to a glass wafer. At first, Ti/Cu/Ti/Au was sputtered on the glass by sputtering process, so as to reduce the surface roughness in the deposition process, and Ti/Au was sputtered on silicon dies. Then, the 4 silicon dies are pasted on the foaming film by normal die attach and pre-bonded with the glass wafer to fix the sample to prevent deviation during bonding pressure. Finally, the whole structure is put into the bonding machine and bonded for 30 minutes at 200 ℃ and 6MPa to complete the bonding.

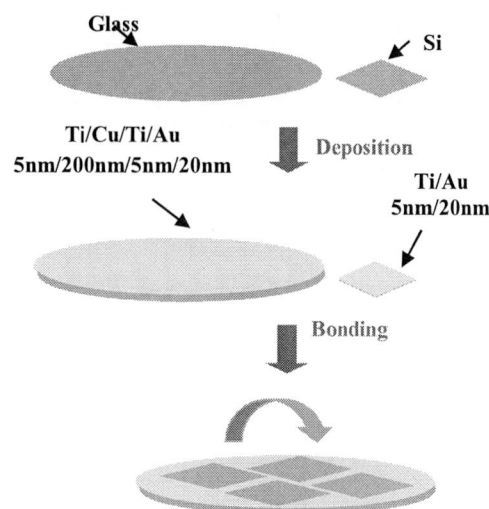

*Figure 2: the schematic diagram of bonding*

## RESULTS AND DISCUSSION

The surface roughness of the bonding surface has a great impact on the bonding quality. In general, physical vapor deposition (PVD) may increase the surface roughness of the sample. However, according to previous studies, the metal layer with specific thickness will not increase considerably and even reduce the surface roughness [4]. Before bonding, a 5/200/5/20-nm-thick Ti/Cu/Ti/Au (from bottom to top) layer was deposited onto glass at room temperature in our study, as shown in Figure 3. Especially, Au films have a significant advantage [5]. An oxide layer does not develop on the free surface of Au, and Au has a high self-diffusion coefficient ($3.3 \times 10^{-29}$ m$^2$/s at 300 K) [6].

Then, the surface roughness of the glass before and after physical vapor deposition was analyzed by a probe profilemeter, as shown in Table 1. Four points are selected on the glass wafer respectively. From the chart, it can be seen that the average Ra of the glass before physical vapor

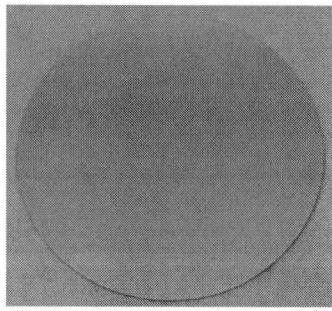

*Figure 3: the surface of the glass wafer*

TABLE 1: The Roughness of Glass

| The Roughness of Glass | | | | | |
|---|---|---|---|---|---|
| **Roughness** | **Ra 1** | **Ra 2** | **Ra 3** | **Ra 4** | **Average Ra** |
| Before PVD | 0.957nm | 1.266nm | 1.348nm | 0.93nm | 1.125nm |
| After PVD | 1.759nm | 2.908nm | 2.408nm | 2.212nm | 2.321nm |

deposition is 1.1252nm, and the average roughness of the glass after physical vapor deposition is 2.3218nm. Although the surface roughness increases during the deposition process, whose range is acceptable. At the same time, 200-nm-thick Cu can be plastic deformed during the bonding compression, so as to bridge the bonding void and reduce the bonding void fraction.

Figure 4 shows the final bonding structure. Different from die to die bonding or wafer to wafer bonding, this bonding method can not only be suitable for mass production, but also realize heterogeneous integration of chips of different sizes, which will be a popular direction in the future.

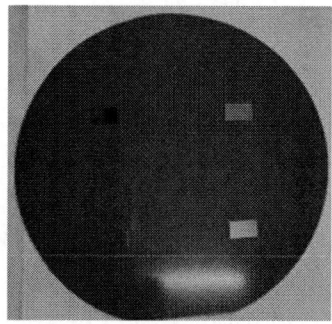

*Figure 4: the bonding structure*

The Scanning Acoustic Tomography (SAT) analysis has been performed to verify the quality of the silicon dies to glass wafer bonding results, as shown in Figure 5 and Figure 6, where the brighter area in the SAT image represents the better bonding quality. Although some black bubbles are observed due to particles inevitably generated during fabrication in the laboratory, most of the bonding area is bright in the image, meaning that successful bonding can uniformly distribute over the wafer. In order to judge the bonding quality more intuitively, each sample is numbered, and then the scanned image is quantified by algorithm. It is worth mentioning that there are no bubbles in the central area of the bonded samples. And the bubbles are mainly distributed in the edge part with a small quantity. In order to judge the bonding quality more intuitively, each sample is numbered, and then the scanned image is quantified by algorithm. The void fraction of the four bonded samples are 1.374%, 9.1676%, 1.4671% and 1.5071% respectively. It can be seen from the results that this bonding is very successful. The maximum porosity of the four samples is no more than 10%, and the minimum porosity is 1.374%. The results show that this way of bonding has great potential.

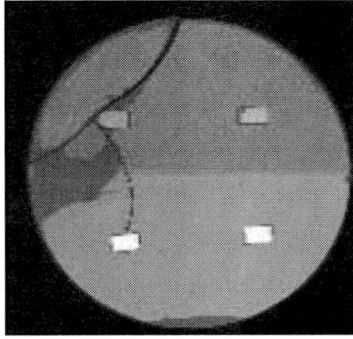

*Figure 5: the overall scanning diagram of SAT*

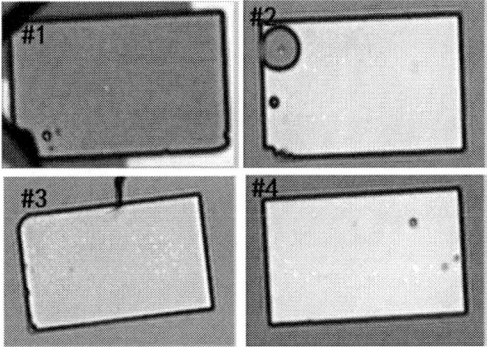

*Figure 6: the specific scanning diagrams of SAT*

## CONCLUSIONS

In this study, four silicon dies are well bonded to a 4-inch glass wafer. The improvement of cleaning process provides a solution for the cleaning of chips containing solder balls in the future, so as to avoid the corrosion of acid and alkali to solder balls. Meanwhile, modest surface roughness was achieved by optimizing the thickness of PVD material, and the maximum roughness Ra is no more than 2nm. The SAT scan shows that the void fraction of three samples was less than 2%, and the yield reached 75%. Samples with voids less than 10% accounts for 100%. This demonstrates the excellent bonding quality and the potential for large-scale industrial production.

Future work needs to use this method to bond more silicon dies to a glass wafer at once, further reducing the number and area of voids in the bonded samples.

## ACKNOWLEDGEMENTS

The research of the project was conducted at Xiamen University and Xiamen Sky Semiconductor Technology Co., Ltd. Thanks very much to colleagues of the research group for their support and help. This research is supported by the National Natural Science Foundation of China (Grant No. 61974121) and Science and Technology Major Project of Xiamen City (Grant No. 3502Z20201004).

## REFERENCES

[1] T. Uhrmann, J. Burggraf and M. Eibelhuber, *Heterogeneous Integration by Collective Die-To-Wafer Bonding*, 2018 International Wafer Level Packaging Conference (IWLPC), 2018, pp. 1-7, doi: 10.23919/IWLPC.2018.8573296.

[2] T. N. Jianye, M. Min, L. Zhensong and L. Qinghai, *Hollow TSV VS solid TSV and the effect of medium filling in the hollow TSV*, 2013 14th International Conference on Electronic Packaging Technology, Dalian, China, 2013, pp. 1308-1311, doi: 10.1109/ICEPT.2013.6756698.

[3] T. Shimatsu, M. Uomoto, *Atomic diffusion bonding of wafers with thin nanocrys-talline metal films*, J. Vac. Sci. Technol. B Microelectron. Nanom. Struct. 28 (2010)706, h

[4] D. Liu, P. -C. Chen, T. -C. Chou, H. -W. Hu and K. -N. Chen, *Demonstration of Low-Temperature Fine-Pitch Cu/SiO₂ Hybrid Bonding by Au Passivation*, in IEEE Journal of the Electron Devices Society, vol. 9, pp. 868-875, 2021, doi: 10.1109/JEDS.2021.3114648.

[5] T. Shimatsu, M. Uomoto, H. Kon, *(Invited) room temperature bonding using thin metal films (bonding energy and technical potential)*, ECS Trans. 64 (2014)317–328.

[6] Takashi Matsumae, Yuichi Kurashima, Hitoshi Umezawa, Yoshiaki Mokuno, Hideki Takagi, *Room-temperature bonding of single-crystal diamond and Si using Au/Au atomic diffusion bonding in atmospheric air*, Microelectronic Engineering, Volume 195, 2018, Pages 68-73, ISSN 0167-9317.

# THE EFFECT OF INTERFACIAL AND BULK FREE ENERGIES ON THE LEAKAGE CURRENT VS VOLTAGE CHARCTERISTICS OF HIGH-K MIM CAPACITORS PREPARED BY ATOMIC LAYER DEPOSITION

*W.S. Lau*

Nanyang Technological University (Retired), School of EEE, Singapore 639798

E-mail: lauwaishing@yahoo.com.sg

## ABSTRACT

Basic thermodynamics principle is used to explain the asymmetrical I-V characteristics of high-k MIM capacitors prepared by atomic layer deposition. The interfacial and bulk free energy differences between the amorphous and polycrystalline forms of high-k dielectric are important.

*Keywords—high-k dielectric; MIM capacitor; leakage current; I-V characteristics; amorphous; polycrystalline; thermodynamics; free energy; surface free energy; grain boundary free energy*

## INTRODUCTION

The physics of the leakage current vs. voltage (I-V) characteristics of high-k MIM capacitors is still not well understood. The majority of experimental results on MIM capacitors showed up asymmetrical I-V characteristics. The author noticed that the I-V characteristics tend to be asymmetrical even though the MIM capacitor has an apparently symmetrical structure with the same metal used for both the top electrode and the bottom electrode. In this paper, the author would like to apply basic thermodynamics principle to explain the asymmetrical I-V characteristics of high-k MIM capacitors prepared by atomic layer deposition.

### Theory

A polycrystalline metal film tends to be rough at the microscopic level while an amorphous metal film tends to be much smoother. Up to now, all metal films used in practical microelectronics are polycrystalline. Previously, the author showed that the deposition of an amorphous $Al_2O_3$ film by ALD on a metallic TiN film has a surface smoothing effect [1]-[2].

ALD of aluminum oxide ($Al_2O_3$) has a large process window such that the as deposited film is amorphous; in addition, the film can tolerate some heat treatment without crystallization. ALD of titanium oxide ($TiO_2$), zirconium oxide ($ZrO_2$) and hafnium oxide ($HfO_2$) is different. If ALD is done in a high temperature (HT) range, the film deposited is polycrystalline independent of thickness. If ALD is done in a low temperature (LT) range, the film deposited is amorphous independent of thickness. However, there exists an intermediate temperature (IT) range such that there is a threshold thickness $t_{threshold}$. When the thickness of the film

$t_{high-k} < t_{threshold}$ the film is amorphous, as shown in Fig. 1. The author's theory is that the "surface smoothing effect" exists for the ALD of an amorphous film. When the thickness of the film $t_{high-k} > t_{threshold}$ the film is polycrystalline, as shown in Fig. 1. The author's theory is that the "surface roughening effect" exists for the ALD of a polycrystalline film. The existence of thickness-dependent crystallization can be explained by thermodynamics according to Nie et al. [3]. It is "disordered crystallization" which can cause roughness. ALD epitaxial growth on a single crystalline substrate is ordered crystallization, resulting in surface smoothing. In general, crystallization of a thin film on a polycrystalline substrate is "disordered crystallization", resulting in a polycrystalline thin film and a surface roughening effect observed.

According to basic thermodynamics principle, a process is spontaneous when the change in free energy G due to the process is negative. The free energy per unit volume is denoted as G/V while the free energy per unit area is denoted as $\gamma = G/A$. Then $\gamma = G/A = G/V \times$ thickness.

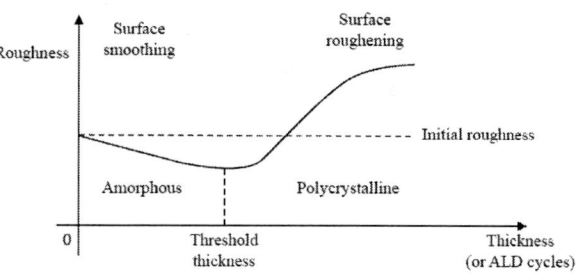

*Figure 1: Surface roughness as a function of film thickness for a film (e.g. $TiO_2$, $ZrO_2$ or $HfO_2$) deposited by ALD on a slightly rough surface according to the author's theory.*

The free energy difference per unit area $\gamma$ for the amorphous form compared to the crystalline form is given by

$$\gamma^{a\text{-}c} = \gamma_{bulk}^{a\text{-}c} + \gamma_{interface}^{a\text{-}c}$$

$$\gamma^{a\text{-}c} = \gamma_{bulk}^{a\text{-}c} - (-\gamma_{interface}^{a\text{-}c})$$

$\gamma^{a\text{-}c}$ is the total free energy difference per unit area for the amorphous form compared to the crystalline form. $\gamma_{bulk}^{a\text{-}c}$ is the free energy difference per unit area for the amorphous form compared to the crystalline form in the bulk of the high-k dielectric film while $\gamma_{interface}^{a-c}$ is the free energy difference per unit area for the amorphous form compared to the crystalline form at the interface between the high-k dielectric film and the substrate. It should be noted that $\gamma_{bulk}^{a-c}$ should be positive while $\gamma_{interface}^{a-c}$ should be negative. Hu et al. gave an explanation why $\gamma_{interface}^{a-c}$ should be negative [4]. They pointed out that the Young's modulus of the amorphous form tends to be smaller than that of the crystalline form such that quite frequently the amorphous form is preferred at the interface between the thin film and the substrate.

When $\gamma_{bulk}^{a-c} > (-\gamma_{interface}^{a-c})$, $\gamma^{a-c} > 0$, i.e. $\gamma^c < \gamma^a$, the crystalline form is preferred over the amorphous form. When $\gamma_{bulk}^{a-c} < (-\gamma_{interface}^{a-c})$, $\gamma^{a-c} < 0$, i.e. $\gamma^c > \gamma^a$, the amorphous form is preferred over the crystalline form. As shown in Fig. 2, $\gamma_{bulk}^{a-c}$ and $(-\gamma_{interface}^{a-c})$ at the deposition temperature are plotted as a function of film thickness. As discussed above, $(-\gamma_{interface}^{a-c})$ is a constant while $\gamma_{bulk}^{a-c}$ is proportional to the film thickness. The intersection of the two lines gives the threshold thickness for the transition from the amorphous form to the crystalline form.

(metal-insulator-metal) capacitors. The combination of the above thermodynamics theory, Smoluchowski effect and the author's S-P-F theory can give a satisfactory explanation of the asymmetrical I-V characteristics of high-k MIM capacitors prepared by atomic layer deposition.

**Experimental Support**

In 2005, Kim et al. published their experimental results on Ru/HfO$_2$/Ru MIM capacitors for various thicknesses of HfO$_2$ deposited by ALD [8]. The I-V characteristics were asymmetrical even though the same metal was used as the top and bottom electrodes. In addition, the author noticed that there as a reversal of the polarity asymmetry when the HfO$_2$ film thickness was increased from 5 nm to 9 nm. The author extracted data from their paper and re-plotted them as shown in Fig. 3. Kim et al. did not mention the reversal of the polarity asymmetry due to the increase of thickness probably because they could not give an explanation of this phenomenon. In this paper, the author attempted to explain this experimental phenomenon by thermodynamics. In addition, Kim et al. did not define the applied voltage V properly; the author believes that the voltage V in Kim et al. is the voltage applied to the bottom electrode with the top electrode grounded according to his theory.

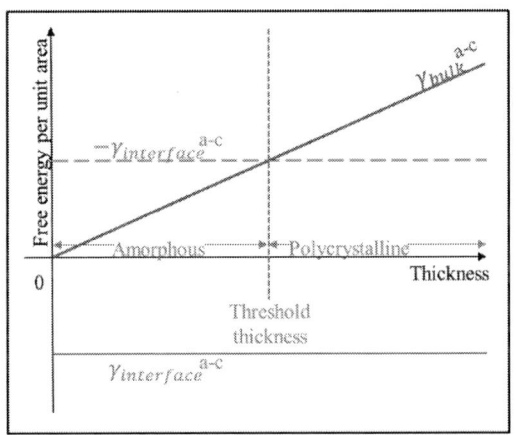

Figure 2: A plot of free energy per unit area at the deposition temperature as a function of thickness for an ultra-thin high-k dielectric film deposited by atomic layer deposition on a metal film.

Therefore basic thermodynamics principle can be used to explain the existence of a threshold thickness for the transition from the amorphous form to the crystalline form. When the film is amorphous, the film can have a surface smoothing effect because a smooth film will minimize the surface free energy. This is not true when the film is polycrystalline. As discussed above, "disordered crystallization" will lead to surface roughening.

Previously the author pointed out that roughness can lead to a change in the effective work function of a metal because of the Smoluchowski effect [5]-[6]. In 2012, the author proposed his Schottky-Poole-Frenkel (S-P-F) theory [7] to explain the I-V (current-voltage) characteristics of thin-film MIM

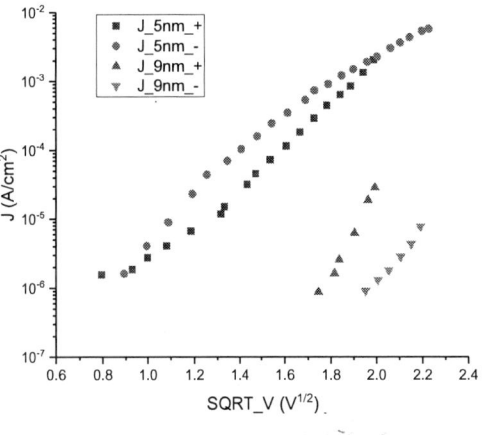

Figure 3: The current density J is plotted against the square root of the voltage SQRT_V for Ru/HfO$_2$/Ru MIM capacitor for 5 nm and 9 nm HfO$_2$ film thickness. The author extracted data from Fig. 3 in Kim et al. [8] and re-analyzed the data. Kim et al. [8] did not give a proper definition of their applied voltage. Based on his own theory, the author believes that the applied voltage in Kim et al. [8] should be the voltage applied to the bottom electrode with the top electrode grounded.

In 2009, Weinreich et al. published their experimental results on TiN/Zr$_x$Al$_{1-x}$O$_2$/TiN MIM capacitors for various thicknesses of Zr$_x$Al$_{1-x}$O$_2$ deposited by ALD [9]. The I-V characteristics were asymmetrical even though the same metal was used as the top and bottom electrodes. In addition, the author noticed that there as a reversal of the polarity asymmetry when the Zr$_x$Al$_{1-x}$O$_2$ film thickness was increased from 7 nm to 9 nm. The author extracted data from their paper and re-plotted them as shown in Fig. 4. Weinreich et al. did not mention the reversal of the polarity asymmetry due to the increase of thickness probably because they could not give an explanation of this phenomenon. In this paper, the author attempted to explain this experimental phenomenon by thermodynamics. Unlike Kim et al., Weinreich et al. defined the applied voltage V properly; the author believes that the voltage V in Weinreich et al. is the voltage applied to the top electrode with the bottom electrode grounded according to his theory.

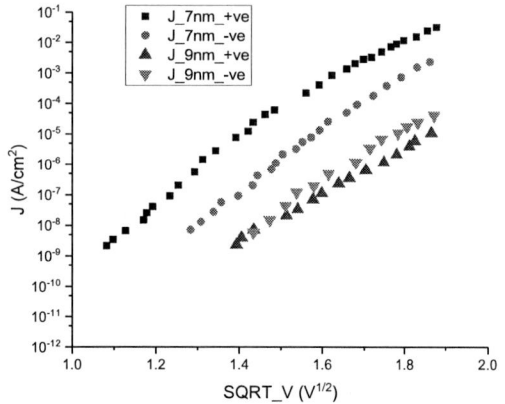

*Figure 4: The current density J is plotted against the square root of the voltage SQRT_V for $TiN/Zr_xAl_{1-x}O_2/TiN$ MIM capacitor for 7 nm and 9 nm $Zr_xAl_{1-x}O_2$ film thickness. The author extracted data from Fig. 1 in Weinreich et al. [9] and re-analyzed the data. Weinreich et al. [9] gave a proper definition of their applied voltage $V_G$, which is the the voltage applied to the top electrode with the bottom electrode grounded.*

The I-V data of Kim et al. [8] and Weinreich et al. [9] are consistent with the author's theory. However, the author can predict that some other scientists may encounter experimental difficulties as follows. As explained by the author before, it may be difficult to detect a small difference in surface roughness by AFM [2]. Another method to study surface roughness is by cross-sectional transmission electron microscopy (XTEM). It may also be difficult to detect a small difference in surface roughness by XTEM. The author noticed that the surface smoothing effect can be more easily observed by XTEM if the XTEM sample thickness is smaller and the thickness of the bottom electrode is increased to increase the grain size of the metal film. For some other authors, the thickness of the bottom electrode is only about 10 nm and the grain size is small; for example, the author used 200 nm thick TiN as bottom electrode and so the grain size is much larger [1]. Actually, the effect of a small difference in surface roughness is more easily noticed by measuring the I-V curve for both polarities of applied voltage. However, it is possible to encounter Mechanism B I-V symmetry for MIM capacitors in experiment [10]; the I-V curves may show up particularly high leakage current and symmetrical I-V characteristics for both polarities independent of the choice of metal or interfacial roughness. This is more likely if the top electrode is deposited by sputtering because of plasma charging during sputtering. Top metal deposited by non-plasma techniques like CVD or ALD is much better. If top metal sputtering cannot be avoided, magnetron sputtering is better because the plasma can be confined to the target far away from the substrate. An additional step of mild post metallization annealing in nitrogen may be helpful to remove Mechanism B I-V symmetry.

## CONCLUSION

A combination of the basic thermodynamics theory, Smoluchowski effect and the author's S-P-F theory can give a satisfactory explanation of the asymmetrical I-V characteristics of high-k MIM capacitors prepared by atomic layer deposition. This may be the first time that basic thermodynamics principle based on the concept of free energy can be used to explain the I-V characteristics of high-k MIM capacitors. Experimental observation of the effect explained above can sometimes be difficult and so some experimental precautions are discussed by the author.

## REFERENCES

[1] W.S. Lau, L. Du, D.Q. Yu, X. Wang, H. Wong and Y. Xu, "The application of a selective etch to conclusively show the surface smoothing effect of an amorphous thin film deposited by atomic layer deposition," ECS J. Solid State Sci. Technol., vol. 6, pp. N111-N116, 2017.

[2] W.S. Lau, L. Du and H. Wong, "Difficulty involved to observe the surface smoothing effect of an amorphous high-k dielectric thin film deposited by atomic layer deposition on a metastable metal film," CSTIC 2019 (China Semiconductor Technology International Conference, Shanghai, 2019, IEEE), pp. 1-3, 2019.

[3] X. Nie, F. Ma, D. Ma and K. Xu, "Thermodynamics and kinetic behaviors of thickness-dependent crystallization in high-k thin films deposited by atomic layer deposition", J. Vac. Sci. Technol. A, vol. 33 (2015), article number 01A140.

[4] M. Hu, S. Noda and H. Komiyama, "Amorphous to crystalline transition during the early stages of thin film growth of Cr on $SiO_2$," J. Appl. Phys., vol. 93, no. 11, pp. 9336-9344, 2003.

[5] R. Smoluchowski, "Anisotropy of the electronic work function of metals," Phys. Rev., vol. 60, pp. 661-674, 1941.

[6] W.S. Lau, "The application of the Smoluchowski effect to explain the current-voltage characteristics of high-k MIM capacitors," CSTIC 2017 (China Semiconductor Technology International Conference, Shanghai, 2017, IEEE), pp. 1-3, 2017.

[7] W.S. Lau, "An extended unified Schottky-Poole-Frenkel theory to explain the current-voltage characteristics of thin-film metal-insulator-metal capacitors with examples for various high-k dielectric materials," ECS J. Solid State Sci. Technol., vol. 1, no. 6, pp. N139-148, 2012.

[8] J.-H. Kim, S.-G. Yoon, S.-J. Seom, H.-K. Woo, D.-S. Kil, J.-S. Roh and H.-C. Sohn, "Electrical properties in high-k $HfO_2$ capacitors with an equivalent oxide thickness of 9 Å on Ru metal electrode," Electrochemical and Solid-State Letters, vol. 8, no. 6, pp. F17-F19, 2005.

[9] W. Weinreich, R. Reiche, M. Lemberger, G. Jegert, J. Muller, L. Wilde, S. Teichert, J. Heitmann, E. Erben, L. Oberbeck, U. Schroder, A.J. Bauer and H. Ryssel, "Impact of interface variations on J-V and C-V polarity asymmetry of MIM capacitors with amorphous and crystalline $Zr_xAl_{1-x}O_2$ films," Microelectronic Engineering, vol. 86, no. 6, pp. 1826-1829, 2009.

[10] W.S. Lau, "Mechanism B I-V symmetry for MIM capacitors used in microelectronics," CSTIC 2021 (China Semiconductor Technology International Conference, Shanghai, 2021, IEEE), pp. 1-3, 2021.

# A HYBRID MODELLING APPROACH FOR THE DIGITAL TWIN OF DEVICE FABRICATION

*Ze Zheng[1], Dong Ni[1]\**

[1]College of Control Science And Engineering, Zhejiang University, Hangzhou 310012, China
*Corresponding Author's Email: dni@zju.edu.cn

## ABSTRACT

With the rapid development of semiconductor technology, it is increasingly difficult to adjust process parameters solely based on engineering experience. To solve the problem, we propose a method for constructing a hybrid mechanism data-driven transistor model. Using appropriate machine learning methods, engineers can build a Digital Twin of device fabrication by combining the first-principle model and actual data. The experiment results also prove that Digital Twin assists engineers in adjusting process parameters promptly.

*Keywords—Transistor model; Data-driven model; Digital Twin*

## INTRODUCTION

With the miniaturization of process technology and the shrinking of device dimensions, the complexity and sensitivity of semiconductor processes have risen sharply. Small changes in process parameters will lead to considerable changes in the key electrical parameters of the transistor[1]. It becomes harder for engineers to adjust the process parameters solely relying on engineering experience to control device performance in technology research and development. Besides, the current research and development method based on physical experiments has a long research and development cycle and brings high costs, which is not conducive to product development and iteration. Therefore, the industry urgently needs a more efficient and systematic method to assist engineers in adjusting process parameters.

As a virtual entity linked to a real-world entity, Digital Twin has received significant attention over the years[2]. Using intelligent algorithms, Digital Twin can provide a means of simulating, optimizing physical manufacturing processes[3]. It has the potential to offer guidance for solving the problem of difficult adjustment of process parameters.

Digital Twin technology relies on a reasonable model that can reflect the actual characteristics of the object. However, due to the entire integrated circuit production process being so complicated, it is challenging to construct a model that is entirely consistent with the actual production process based on first principles.

In order to solve this problem, a method of combining first-principles models and actual data is proposed, which obtains data-driven transistor models that can be calculated in real-time. This method is also capable of assisting process integration engineers in optimizing the process targets to achieve desired device characteristics.

## METHODOLOGY

Constructing a mathematical model between transistor output characteristics and process parameters is the key to realizing digital twin technology. This work uses weighted retraining and machine learning algorithms to construct a data-driven model for transistors by combining simulation data derived from first-principles models and actual experimental data.

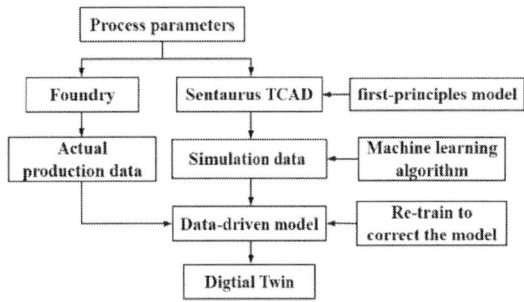

*Figure 1: This is the model building architecture figure of this article.*

Existing methods have some shortcomings. In the field of process simulation, traditional transistor models that utilize physical formulas are widely used[4]. To obtain device output characteristics, people have to solve a large number of partial differential equations. Because the calculation is much too complicated and requires an amount of time, traditional models cannot satisfy the needs of real-time calculation. Moreover, in practical applications, numerical values of many physical quantities are difficult to obtain, contributing to the deviation between the calculated and actual values. Although the experimental data in the real world can fully reflect the situation of the device, it cannot be obtained on a large scale due to the high cost.

Our work combines simulation data with experimental data to ensure the model's accuracy and reduce the cost of calculation. First, to obtain simulation data that fits the actual production, we make use of Sentaurus Topography as the simulator. Using advanced physical models based on first principles and numeric algorithms, Sentaurus can simulate topography-modifying

978-1-6654-9759-6/22 $31.00 © 2022 IEEE

process steps such as deposition, etching, spin-on-glass, reflow, etc.[5]. After that, we used the stepwise regression method to screen key process parameters such as the channel length of the transistor, the thickness of the gate oxide layer, and the ion implantation dose. And then, machine learning algorithms are intended to realize the preliminary construction of a data-driven model.

However, TCAD simulation results often differ from experimental results. This is because engineers cannot guarantee that the device parameters in the simulation match exactly with the actual situation. Therefore, the model constructed from simulation data alone is incomplete. Therefore, the model constructed from simulation data alone is incomplete. We will introduce actual data to retrain the model and give higher weight to the actual data set so that the hybrid model is close to reality. This approach allows for the injection of physical information from the actual process into the data-driven model. Finally, such a hybrid model is applied for calculating in real-time and meets the Digital Twin definition.

## EXPERIMENT

### Data-driven model

In summary, the key is to construct a Digital Twin of device fabrication by establishing a data-driven model between device process parameters and electrical characteristics.

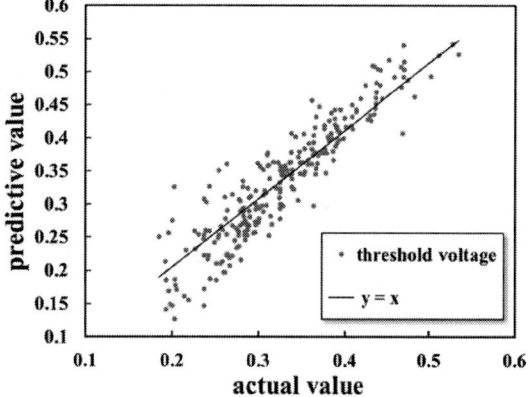

Figure 2: This is a figure that compares the actual value and the predicted value of the threshold voltage. Each dot represents the actual value and predicted value corresponding to a set of process parameters in the test set. The closer the dot is to the diagonal, the more accurate the prediction.

Take 180nm process technology devices as an example. We selected 47 representative process parameters to predict the main electrical parameters of the transistor. Based on the standard CMOS process, we adopt the DOE method to make the process parameters fluctuate.

TABLE I
COMPARISON OF DIFFERENT MACHINE LEARNING ALGORITHMS

| Algorithm | $R^2$ | RMSE | MAPE |
|---|---|---|---|
| PLS | 0.786 | 0.043 | 0.102 |
| SVR | 0.747 | 0.055 | 0.165 |
| MLP | 0.753 | 0.045 | 0.112 |
| GBRT | 0.727 | 0.040 | 0.075 |

Without affecting the regular operation of the device, the fluctuating process parameters are input into Sentaurus software for simulation. Then a data set of electrical parameters of transistors with 2000 sets of different process parameter combinations is obtained. Ultimately, we use machine learning methods to build a data-driven model and use 200 sets of additional weighted actual data for retraining.

In this work, we utilized various machine learning algorithms to create a data-driven model. The comparison results are shown in Table I.

Compared with other algorithms, the partial least squares algorithm has higher accuracy and better performance. More importantly, the PLS method is easy to calculate and has good interpretability, guiding the adjusting transistor process parameters[6]. Take the threshold voltage prediction of the transistor as an example, and the result is shown in Figure 2.

It can be seen from the figure that the data points are better distributed on both sides of the diagonal, indicating that the method we proposed can be applied to establish a data-driven model. With the mapping relationship between device process parameters and electrical properties in the model, device process engineers can also adjust other process parameters.

### Hybrid model

In practical applications, the prediction results obtained from simulation models are often biased. For example, ion implantation is a key semiconductor manufacturing process. However, the practical carrier profile in the device after processing does not exactly follow the theoretical distribution. This can cause the prediction of the initial model not to meet our expectations.

The hybrid model proposed in this paper corrects the data-driven model by absorbing a small amount of actual data and retraining it. The actual physical information is injected into the initial data-driven model to achieve a more accurate prediction.

To validate this approach, In this work, we selected two sets of experiments with different source-drain phosphorus ion implantation concentrations and the same remaining conditions. Simulation results using Sentaurus are used to represent the difference between the simulated experiments and the actual processing. Furthermore, as in the data-driven model section, we also used the prediction of the device threshold voltage to show the model prediction capability.

The two sets of experiments are divided into two groups of high and low levels according to the difference in implantation concentration. We assumed that the source-drain implantation had a high concentration during the simulation and built a data-driven model using the high-level data. However, the actual ion concentration implanted into the device did not reach the theoretical value, but only the low-level concentration due to objective factors. Therefore, the predicted values of the initial data-driven model using high-level data will differ significantly from the true values at low ion concentration levels.

The hybrid model achieves improved prediction accuracy by reintroducing a small number of actual values of low-level concentrations into the calculation. In linear machine learning models, a weighted regression approach can be used. For deep learning models, a meta-learning approach can minimize the loss on a clean unbiased validation set[7]. We introduce a weighting coefficient based on the linear machine learning model and assign a higher weight to a few true values during training. This approach allows the distribution of the model prediction values to be closer to reality.

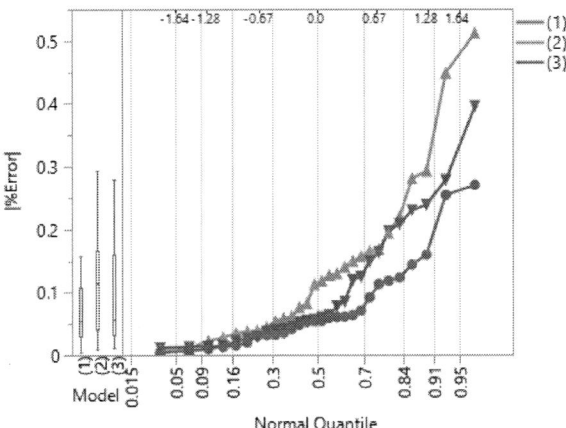

Figure 3: This is a comparison figure of threshold voltage prediction errors (absolute percentage error) under different conditions. (1) The error of the initial data-driven model using the high-level data on the original simulation test set. (2) The error of the initial data-driven model on the actual test set. (3) The error of the hybrid model on the actual test set.

The comparison of the prediction errors of the initial data-driven model on different test sets and the prediction errors of the retrained hybrid model on the actual test set is shown in Figure 3. It can be seen from the figure that after the process parameters change, the predictive performance of the initial model has a significant decline, and the prediction effect of the hybrid model has been more significantly improved than the initial data-driven model. The experimental results indicate that the model combined with the feedback information can better respond to the actual situation.

## CONCLUSION

In this paper, we proposed a hybrid mechanistic data-driven model approach for the Digital Twin of device fabrication. This model has the potential to assist engineers in adjusting process parameters promptly. Experimental results show that appropriate machine learning algorithms can improve model accuracy. After adding actual data for retraining, mechanism knowledge is injected into the data-driven model. The hybrid model can be used for calculating in real-time and meet the definition of the Digital Twin.

## REFERENCES

[1] B. Hoefflinger, "ITRS: The International Technology Roadmap for Semiconductors," in *Chips 2020: A Guide to the Future of Nanoelectronics*, B. Hoefflinger, Ed. Berlin, Heidelberg: Springer, 2012, pp. 161–174.

[2] W. Kritzinger, M. Karner, G. Traar, J. Henjes, and W. Sihn, "Digital Twin in manufacturing: A categorical literature review and classification," *IFAC-PapersOnLine*, vol. 51, no. 11, pp. 1016–1022, Jan. 2018

[3] Y. Lu, C. Liu, K. I.-K. Wang, H. Huang, and X. Xu, "Digital Twin-driven smart manufacturing: Connotation, reference model, applications and research issues," *Robotics and Computer-Integrated Manufacturing*, vol. 61, p. 101837, Feb. 2020

[4] B. Cheng *et al.*, "Statistical-Variability Compact-Modeling Strategies for BSIM4 and PSP," *IEEE Design Test of Computers*, vol. 27, no. 2, pp. 26–35, Mar. 2010

[5] "Synopsys | EDA Tools, Semiconductor IP and Application Security Solutions." https://www.synopsys.com/

[6] P. Geladi and B. Kowalski, "Partial least-squares regression: a tutorial," 1986

[7] M. Ren, W. Zeng, B. Yang, and R. Urtasun, "Learning to Reweight Examples for Robust Deep Learning," in *Proceedings of the 35th International Conference on Machine Learning*, Jul. 2018, pp. 4334–4343.

# EFFECTS OF COIL SURFACE CURRENT DENSITY ON PLASMA CHARACTERISTICS IN A PECVD CHAMBER

*Jiangjie Zeng, Qinrui Zhang, Jiahong Yu, Yongjie Hu, Xingyu Li, Shiwei Zhuang\**

School of Physics and Electronic Engineering, Jiangsu Normal University, Xuzhou 221116, China
\*Corresponding Author's Email: zhuangshiwei@jsnu.edu.cn

## ABSTRACT

Plasma enhanced chemical vapor deposition (PECVD) technology is playing an important role in industrial production, but the costs of its design and development can be very high. In this paper, a PECVD chamber was established and multiple physical fields such as electromagnetic field, temperature field and plasma field were coupled to study the influence of coil current density on plasma characteristics. The simulation results showed that with the increase of coil current density, electron density of plasma increased, electron temperature and electric potential decreased. Moreover, Langmuir probe was used to obtain plasma parameters for experiment validation, and the experimental data accorded with the simulation results, which reflected the rationality of simulation. This paper is of reference value for improving film deposition rate and uniformity.

## INTRODUCTION

PECVD is a technology to produce thin film materials by using RF power supply to generate plasma through glow discharge in the reaction chamber. Gaseous substances containing thin film components decompose under the action of plasma in the chamber and deposit on the substrate. This technology can be carried out at low temperature with a fast deposition rate [1]. The deposited film will have good adhesion to the silicon wafer, which is convenient to prepare the film with large area and high uniformity.

However, the process system of PECVD is very complex. During the film deposition, there are multiple physical fields and various chemical reactions in the chamber, such as plasma field, electromagnetic field, temperature field and so on [2]. Nowadays, the industry still mainly adopts the method of Trial and Error in the design and development of PECVD devices, which can be a waste of time, material and financial resources.

With the advance of computer technology, simulation technology has made great progress and is playing an important role in industrial production and manufacturing. Through the combination of simulation and experiment, the industrial cost will be greatly reduced. In this paper, a PECVD simulation chamber was established to couple multiple physical fields, and the influence of coil current density on plasma electron density, electron temperature and potential was studied [3]. The Langmuir probe was used for experiment validation, and the experimental data

were consistent with the simulation results [4]. This paper can provide theoretical support for the development of PECVD technology.

## CHEMICAL REACTIONS AND THEORETICAL CALCULATION

We established a PECVD simulation chamber with a size of 400×400 mm.

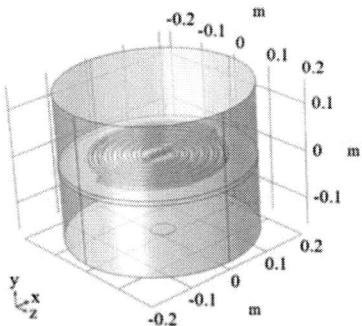

*Figure 1: PECVD chamber model*

A coil was placed in the dielectric layer, and a high frequency current of 13.56 MHz was applied in the center [5]. The plasma was formed in the chamber below the dielectric layer, which contained argon gas with a pressure of 0.02 Torr and a temperature of 300 K. Gas flowed in from the dielectric layer and out of the bottom port, and plasma was generated and maintained by high-frequency alternating electromagnetic fields [6]. The plasma reactions in the chamber were showed in table 1.

*TABLE.1 Plasma reactions in the chamber*

| Reaction | Formula | Type | $\Delta\varepsilon$ (eV) |
|----------|---------|------|--------|
| 1 | e+Ar=>e+Ar | Elastic | 0 |
| 2 | e+Ar=>e+Ars | Excitation | 11.50 |
| 3 | e+Ars=>e+Ar | Excitation | -11.50 |
| 4 | e+Ar=>2e+Ar+ | Ionization | 15.80 |
| 5 | e+Ars=>2e+Ar+ | Ionization | 4.427 |

*TABLE.2 Surface reactions*

| Reaction | Formula | Adhesion coefficient |
|----------|---------|----------------------|
| 1 | Ar+=>Ar | 1 |
| 2 | Ars=>Ar | 1 |

The surface reactions were also analyzed. When the metastable argon atom hit the wall, it would return to the ground state with a certain probability (adhesion coefficient). The reactions were showed in table 2.

Ignoring the electron convection effect caused by fluid motion, the electron density and average electron energy would be obtained by solving a set of drift diffusion equations of electron density and average electron energy. The equations were as follows:

$$\frac{\partial}{\partial t}(n_e) + \nabla \cdot [-n_e(\mu_e \cdot \boldsymbol{E}) - \boldsymbol{D}_e \cdot \nabla n_e] = R_e \quad (1)$$

$$\frac{\partial}{\partial t}(n_\varepsilon) + \nabla \cdot [-n_\varepsilon(\mu_\varepsilon \cdot \boldsymbol{E}) - \boldsymbol{D}_\varepsilon \cdot \nabla n_\varepsilon] + \boldsymbol{E} \cdot \Gamma_e = R_\varepsilon \quad (2)$$

The equation of electrostatic field was as follows :

$$-\nabla \cdot \varepsilon_0 \varepsilon_r \nabla V = \rho \quad (3)$$

The space charge density was calculated according to the chemical composition of the plasma in the chamber. The formula was as follows:

$$\rho = q(\sum_{k=1}^{N} z_k n_k - n_e) \quad (4)$$

# DISTRIBUTION OF PLASMA CHARACTERISTICS AT 800 A/M

**(a)**

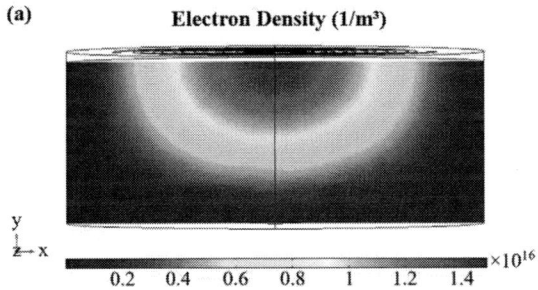

**Electron Density (1/m³)**

**(b)**

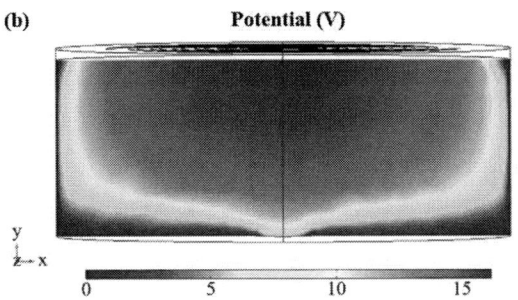

**Potential (V)**

**(c)**

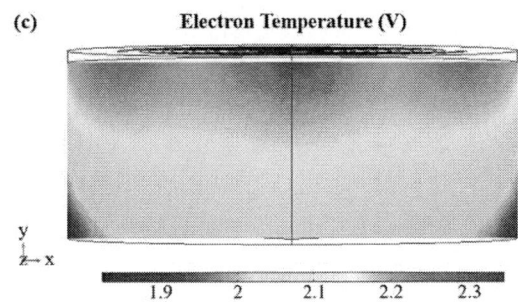

**Electron Temperature (V)**

*Figure 2: (a) Plasma electron density distribution (b) Plasma potential distribution (c) Plasma electron temperature distribution at 800 A/m*

Argon gas was pumped into the chamber. The electron density, electron temperature, and electron potential distribution of the model section were explored when the coil current densities were set from 600 to 1400 A/m.

In the chamber, the electron temperature was much higher than the ion temperature and the electron mass was much less than the ion mass, so the electron migration rate was much higher than the ion migration rate, resulting in the formation of a negative potential on the surface of the cavity wall, negative potential repelled electrons and attracts positive ions, so that the edge electron density was almost zero, forming the plasma sheath [7]. Taking 800 A/m as an example, from figure 2, we can see that the electron density and potential of the section accorded with a Gaussian distribution. The closer it was to the coil center, the greater the value of the electron density and potential was. In the region far from the coil center, the electron density and potential gradually decreased, and the distribution of electron temperature was relatively uniform.

# DISTRIBUTION OF PLASMA CHARACTERISTICS FROM 600 TO 1400 A/M

In order to explore the influence of current density change on RF plasma characteristics in the chamber, we varied the coil current density from 600 to 1400 A/m by controlling other conditions unchanged, and plotted the electron density, electron temperature and electric potential distribution at 50 mm below the dielectric layer.

Figure 3 showed that with the increase of coil current density, electron density increased, while electron temperature and potential decreased. The increase of electron density was due to the enhancement of current density, which enhanced the electric field intensity and RF power in the chamber, and the particles in the chamber obtained more energy and power, leading to the increase of collision frequency and the ionization of more ions and electrons from the plasma, and the increase of electron density. The increase of electric field intensity should have increased the electron temperature, but due to the increase of particle collision frequency, the electron lost more energy, and the energy lost was greater than the energy gained, resulting in the decrease of electron temperature.

**(a)**

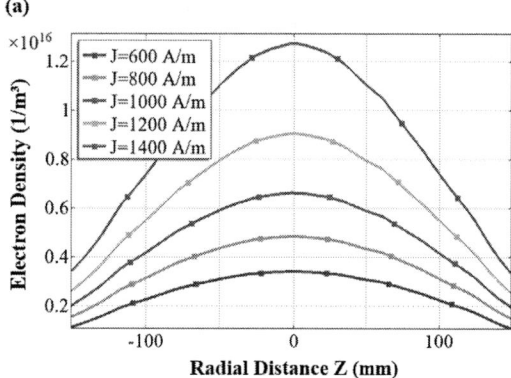

**(b)**

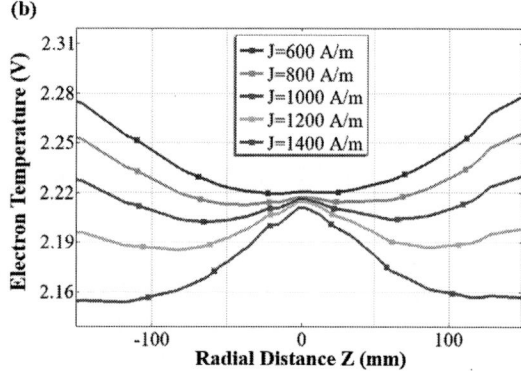

**(c)**

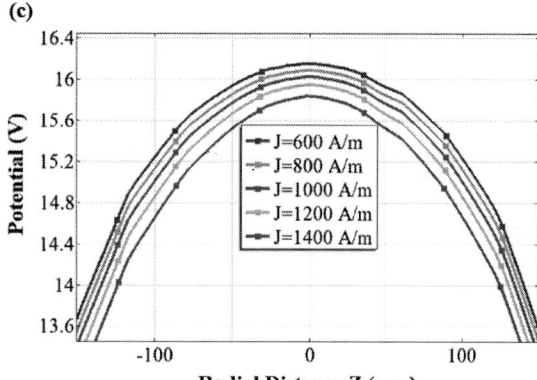

*Figure 3: (a) Plasma electron density distribution (b) Plasma electron temperature distribution (c) Plasma potential distribution from 600 to 1400 A/m*

## CONCLUSIONS

The influence of coil surface current density on plasma characteristics was studied. The electron density, electron temperature, and electron potential distribution of

the plasma were explored when the coil current densities were set from 600 to 1400 A/m. We found that the electron density and potential of the section accorded with a Gaussian distribution. The closer it was to the coil center, the greater the value of the electron density and potential was. In the region far from the coil center, the electron density and potential gradually decreased, and the distribution of electron temperature was relatively uniform. When the coil surface current density varied in a certain range, with the increase of coil current density, electron density of plasma increased while electron temperature and electron potential decreased. The process system of PECVD is very complex with coupling multiple physical fields. This paper can provide theoretical support for the development of PECVD technology.

## ACKNOWLEDGEMENTS

Our work was supported by the Key Projects of Ministry of Science and Technology of the People's Republic of China (Grant No. SQ2020YFF0407077), the Industry-University-Research Cooperation Project of Jiangsu Province (Grant No. BY2020462), Postgraduate Research & Practice Innovation Program of Jiangsu Province (Grant No. KYCX20_2337).

## REFERENCES

[1] L. Guo, Z. Zhang and H. Sun, "Direct formation of wafer-scale single-layer graphene films on the rough surface substrate by PECVD," *J. CARBON*, vol. 129, 2018, pp. 456-461.

[2] Y. Guo, T. M. B. Ong, I. Levchenko and S. Xu, "Inductively and capacitively coupled plasmas at interface: A comparative study towards highly efficient amorphous-crystalline Si solar cells," *J. APPLIED SURFACE SCIENCE*, vol. 427, 2018, pp. 486-493.

[3] Y. Cao, J. Zhou, Y. Ren and W. Xu, "Study on effect of process and structure parameters on SiN x H y growth by in-line PECVD," *J. SOLAR ENERGY*, vol. 198, 2020, pp. 469-478.

[4] W. Qing, B. Dechun and F. Jian, "Diagnosis of the Argon Plasma in a PECVD Coating Machine," *J. PLASMA SCIENCE & TECHNOLOGY*, vol. 10, 2008, pp. 727-730.

[5] Z. Wang, W. Gao, P. Zhang, H. J. Yan and C. S. Ren, "Study of the Characteristics of DC and ICP Hybrid Discharge Plasmas," *J. PLASMA SCIENCE & TECHNOLOGY*, vol. 17, 2015, pp. 191-195.

[6] K. V. Vavilin, M. A. Gomorev, E. A. Kralkina, P. A. Nekludova and V. B. Pavlov, "An experimental study of the plasma parameters of a hybrid low-pressure RF discharge," *J. MOSCOW UNIVERSITY PHYSICS BULLETIN*, vol. 67, 2012, pp. 97-101.

[7] R. Moulick, S. Adhikari and K. S. Goswami, "Sheath formation in collisional, low pressure, and

magnetized plasma," *J. PHYSICS OF PLASMAS*, vol. 26, 2019. pp. 043512.

# Aluminum Gap Fill Improvement For 28 HKMG Process

*Shihao Wang\*, Zhaoqin Zeng, YuBao, Shasha Wang, Haifeng Zhou, Jingxun Fang,Yu Zhang*
Shanghai Huali Integrated Circuit Corporation, Shanghai, China
*Corresponding Author's Email: wangshihao@hlmc.cn

## ABSTRACT

Aluminum (Al) is used as the top fill-in metal layer for forming replacement metal gate (RMG) device at 28nm node. Gap fill and Al diffusion are the main concern of this process. A serials experiments, including top opening CD, Al reflow effect, TBM (Top Barrier Metal) profile, Al process was well studied in this paper to improve gap fill capacity. Results show 50% TiN thickness thin down combine with AC bias power /E-magnet current optimization, Al process temperature 7.5% higher and reflow time 25% longer will significantly improve gap filling. And at the same time, Al diffusion block was well considered.

Keywords :*Aluminum ,Gap fill , Al diffusion ,Top Barrier Metal , metal gate , 28nm*

## INTRODUCTION

When the logic device scaling to 28nm, it evolves from the traditional poly gate to the high-k metal gate. High-k metal gate makes is possible device scaling down and smaller EOT with physical thicker gate-dielectrics. Thicker physical gate dielectrics will reduce the leakage naturally.

Due to the feature of metal deposition, the capacity of Al gap filling depends on the top opening width of the gate hole, which is mainly up to the profile of TBM (Top Barrier Metal).

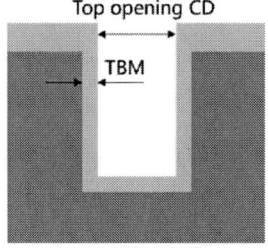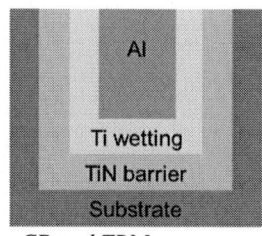

*Fig.1  Top opening CD and TBM*

TBM (Top Barrier Metal) has already taken the method of TiN+Ti bilayer instead of original single-layer TiN. Because Ti has an outstanding wetting capability, which enable the hot Al reflow into the gate bottom smoothly.

The thickness of TBM affect the Al gap filling capacity directly, due to its impact on gate top opening CD, if gate top opening CD getting smaller, it easily causes Al missing defect as Fig.2, serious Al missing defect was found in top CD narrow area.

We plan improve Al gap filling by TBM thin down and Al process parameter (temperature/reflow time) optimization. In addition, in order to keep well Al diffusion performance, TBM AC bias power, E-magnet current was also optimized at the same time.

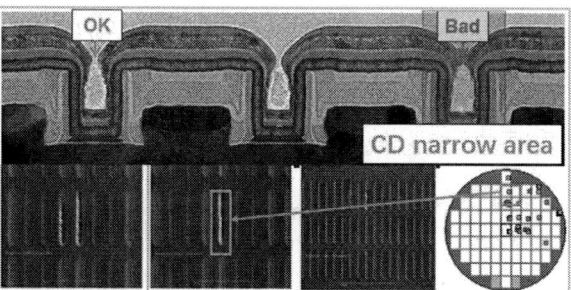

*Fig.2 Al missing defect map and image post CMP Step*

Please make sure your submitted paper is in PDF format and there are no garbled characters during the paper converting from a DOC file into a PDF one. Your PDF will be reproduced exactly as it is submitted, so please make sure your text has been proofread with care.

## EXPERIMENT

Al missing defect is evaluated post CMP process by a KLA-Tencor inspection tool. SEM (Scanning Electron Microscope) review is used to obtain defect type basing on the scanned map and location. TBM (barrier layer) before Al gate, is deposited by PVD (physical vapor deposition), its profile was evaluated by TEM (Transmission Electron Microscopy), and the asymmetry ratio/flat ratio/U% of TBM are also used to judge the Al diffusion and gap filling capacity.:

## RESULTS AND DISCUSSION

(1) The study of TBM thickness spilt for Al gap filling capacity

As mentioned above, barrier metal TiN is used to block the diffusion of aluminum, but inappropriate thickness may lead to narrow top open CD, to investigate the defect performance response on, the TiN thickness, three conditions are designed: baseline/ decrease 30%/ decrease 50%.

Result shown in Fig.3, keep the baseline process temperature setting, when the thickness of TiN decreases to half, Al missing defect almost free.

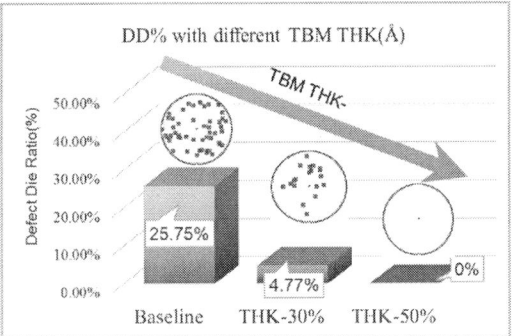

Fig.3 Al missing defect with different thickness spilt

(2) The study of Al process temperature/reflow time spilt for gap filling capacity.

In the process of Al gate, not only including the deposition step, but also has an additional high-temperature reflow step.

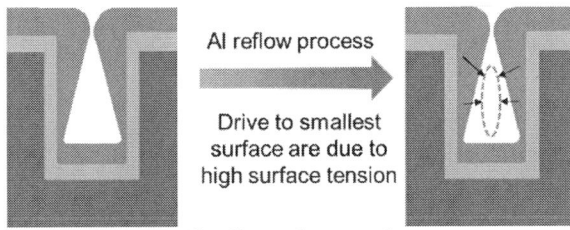

Fig.4 Al reflow schematic diagram

Due to Al high surface tension under proper temperature, deposited Al film in the sidewall are driven to smallest surface area, which lead to the Al reflow.

In view of this feature, we designed about two experiments:

(a) Setting Al deposition temperature, including: decrease 12.5%/baseline /increase 7.5%, to monitor Al missing defect performance difference.

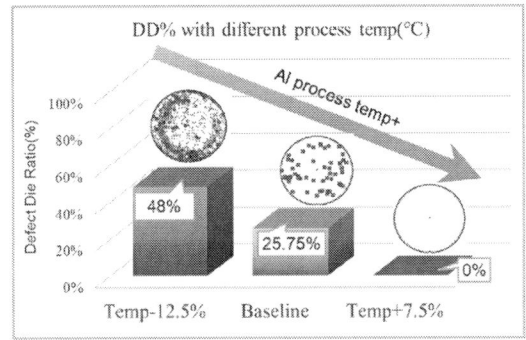

Fig.5 Al missing defect with different temperature spilt

As shown in Fig.5 , when barrier(TiN+Ti) thickness keep constant, Al missing defects decrease with increasing temperature and even eliminate when it reached to 7.5%. Reason for this benefit is the enhanced aluminum reflow effect at higher temperatures.

(b) We also tried to prolong the reflow time to achieve the similar effect. Test condition is set to baseline temperature but with different reflow time.

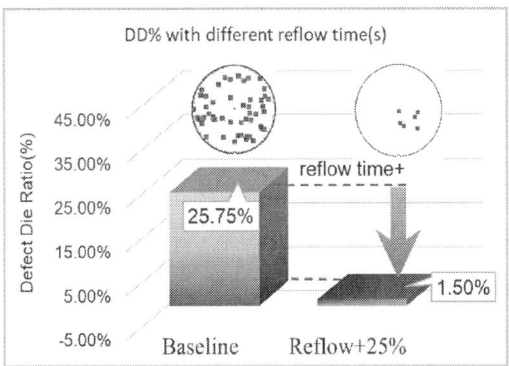

Fig.6 Al missing defect with different reflow time

As expected, reflow time also shows impact on the gap filling. When the reflow time was set to 1.25 times to the baseline, the defect counts decreased approximately 94%.

Although both optimizing TBM thickness and Al reflow time/temperature can improve the capability of Al gap filling, the latter has a serious side effect: increasing R-gate (Metal gate resistivity). From the Fig.7&Fig.8, new Al process condition leads to 5%~7% R-gate degrade on both NMOS and PMOS area, because Al will react with Ti under the enhanced reflow process , which may form higher resistivity product.

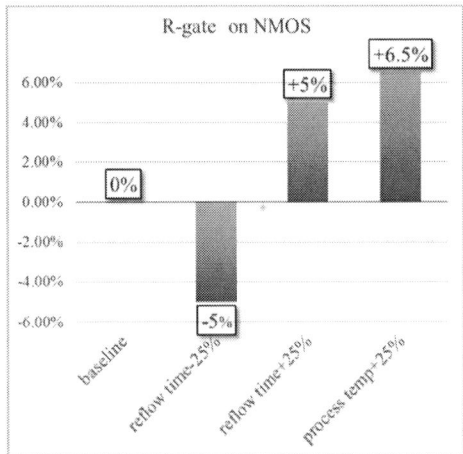

Fig.7 R-gate on NMOS

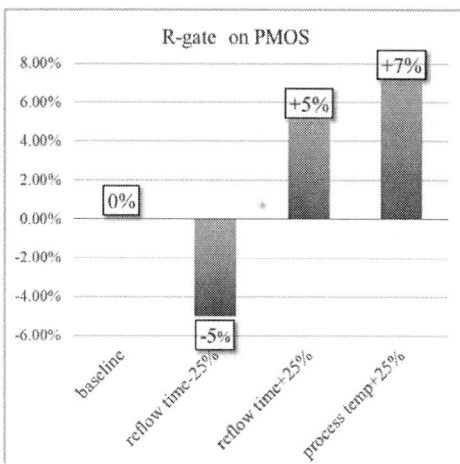

Fig.8 R-gate on PMOS

Thus push us to explore multiple directions to balance the gap filling capacity /Al diffusion and metal gate resistivity.

As we known, the thinner barrier is, the weaker for Al diffusion block but without R-gate degrade. In that case, we explored more experiments (such as the improvement of TBM profile) to prevent Al diffusion even under the thinner TBM thickness condition. Furthermore, overhang also been improved.

(3) TBM Profile Optimization: Prevent Al missing defect and Al diffusion.

The W/E (Wafer Edge) shadow effect was mainly caused by the characteristics of PVD chamber body. In the place of W/E, the angle of plasma treatment is not as vertical as W/C (wafer center), which will easily cause asymmetry at bottom coverage. Therefore, leading to the "weak spot" and overhang, as shown in Fig.9:

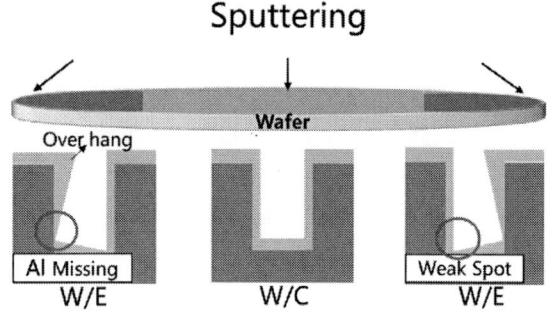

Fig.9 shadow effect and Al missing

These "weak spots" are unfriendly to blocking Al diffusion. Furthermore, the asymmetry effect also easily cause Al missing.

It is necessary to optimize the profile of TBM to avoid the "weak spots" and overhang.

(a)The first method we design to experiment was to adjust AC bias power, in order to improve the re-sputter effect on TiN deposition.

Re-sputter means partial atoms of TiN were re-sputtered from the bottom to the sidewall, thus can get better step coverage (side wall). The power that pulls charged ions of TiN towards bottom was called AC bias power.

Our experiments were divided into several conditions. including no bias/ baseline/ double bias/triple bias. The risk of Al diffusion would be detected by the AR (Asymmetry Ratio) of both corners, which were displayed by TEM results.

$$Asymmetry\ Ratio = \frac{ABS(A-B)}{Average(A\&B)} \quad (1)$$

We define both corner thickness as A,B, asymmetry ratio formula show in（1）,and the less asymmetry ratio was, the slighter asymmetry appearance were.

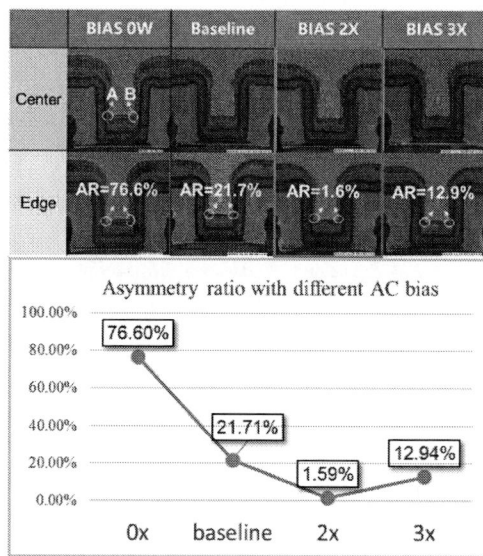

Fig.10 Asymmetry ratio with different AC bias

As Fig.10, we found that when processing without AC bias power, the result was worst, and the AR of the corners is about 76.6% (baseline 21.71%) .On the contrary, when the AC bias was set to double, the asymmetry ratio only show 1.59%. but when continue increasing AC bias power, its AR go back to 12.94% (triple time AC bias), Above all, the asymmetry can be improved significantly by adjusting AC bias.

(b) AC bias also works for tuning Ti wetting layer profile. Ti deposition easily damage the sub-layer TIN, because AC bias would drag the Ti ion to the TIN, which would cause the damage and lead to the TBM bottom non-uniform, like"smile curve" see Fig.11.

Therefore, different AC bias experiment also designed, AC bias power ranging from 1x to 0.4x.

$$Flat\ Ratio = Min\ (\ \frac{L}{C} \& \frac{R}{C})\ (2)$$

We define both sidewall length as "L"&"R", and the center gate high as "C", the flat ratio formula show in (2),and the larger flat ratio was, the slighter bias damage were.

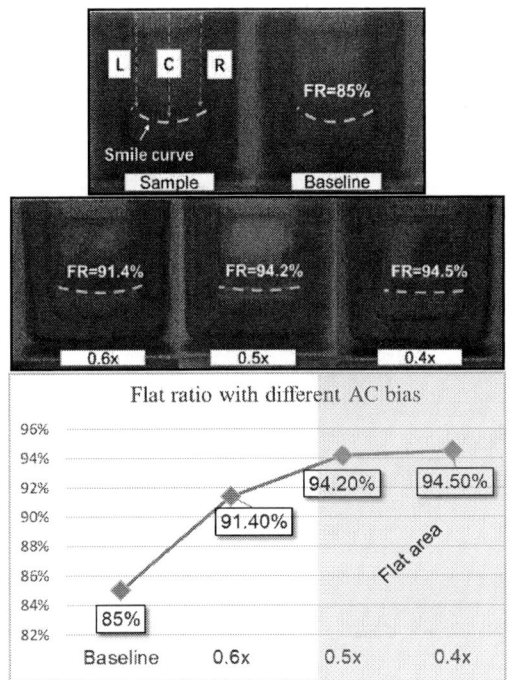

Fig.11 Ti wetting layer flat ratio with different AC bias

From the experimental results we can see that when the AC bias decreasing, the bottom profile become flat: when decreasing to 0.5x, the "smile curve" almost disappear; the flat ratio can reach to 94.2%; but continue decreasing to 0.4x, no obvious improvement found, only degrade the step coverage.

(4) The study of E-magnet effect on improving the uniformity of TBM deposition

Other effort have been tried to reduce the asymmetry, E-magnet current modification is considered as a way to improve the uniformity of TBM deposition.

E-Magnet is a type of magnet in which the magnetic field produced by an electric current, it can be controlled by Ampere setting. E-Mag can be used to tune the direction of metal ions, so as to improve the uniformity of film deposition on wafer.

U% can be used to represent the gap of thickness between W/C and W/E. The smaller U% was, the more uniform TBM were. Different combinations of EM current can obtain different U%.

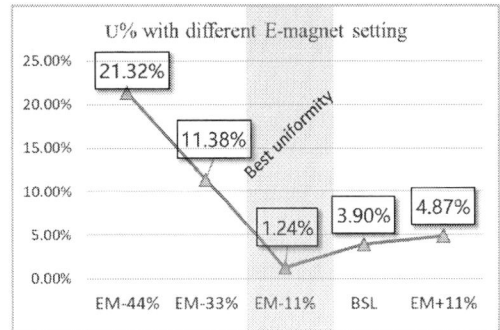

Fig.12 TBM U% with different E-magnet setting

According to the results in Fig12, when the current decrease by 11%, U% can be reduced to 1.24%, which is only one third of baseline.

## CONCLUSION

Increasing Al reflow temperature or extending reflow time can get better gap filling capacity but induce R-gate higher. 50% TBM thin down combine with AC bias power (2x AC bias power for TIN; 0.5x AC bias power for Ti)/ E-magnet current (-11%) tuning during the TBM deposition can achieve less over hang and good bottom profile, which benefit for gap filling capacity, Al diffusion and low R-gate. After all actions taken above, defect and yield is significantly improved.

## ACKNOWLEDGEMENTS

The authors would like to acknowledge HLMC FA department for the supports of TEM analysis and PIE &YE team for data collection and technical discussion.

## REFERENCES

[1] David Williams, Clint Bordelon, Sergei Drizlikh, Paul D. Kirsch, Kin-Sang Lam "Improved 20nm Device Yield and Gate Dielectric Integrity with Optimized Aluminum Metal Fill Process", IEEE.2016
[2] R. P. Huang, "Evaluation of Aluminum Film Properties and Microstructure for Replacement Metal Gate Application at 28nm Technology Node", IEEE2011
[3] Mistry, K., 45nm Logic Tech with High-k + Metal Gate IEDM p. 247-250 (2007).
[4] Hyun, S. Aggressively scaled High-k Last Metal Gate VLSI Symp. p. 33 (2011).

# Research on the oxidation behavior of Titanium nitride thin films and resistance simulation

*Xiaotong Zhang\*, Zhaoqin Zeng, Yanpeng Cao, Yanyan Zhang, Yu Bao, Haifeng Zhou, Jingxun Fang, Yu Zhang*

Shanghai Huali Integrated Circuit Corporation, Shanghai, China

\*Corresponding Author's Email: zhangxiaotong@hlmc.cn

## ABSTRACT

In the preparation of advanced integrated circuits, titanium nitride (TiN) film can be used as a metal hard mask layer, barrier layer and metal gate work function layer. The resistance (RS) varied as queuing time (Q-time) changed, which caused the difficulty to evaluate the true value especially under extremely thin film. The oxidation behavior of this film in clean room environment was studied in the paper. Film component bond and its ratio were explored with XPS. Experiment shows that RS increases logarithmically with the increase of Q-time. What is more, the simulated standard RS ($RS_s$) also correlate to initially measured RS ($RS_m$). After the simulation formula applied in data collection system, a precise monitor has been achieved and the monitor time significantly improved from 5 hours to 1 hours, 80% efficiency benefit is achieved.

**Keywords:** Titanium nitride, oxidation, XPS analysis, sheet resistance

## INTRODUCTION

Titanium nitride (TiN) is an interesting substance. Due to its unique bonding properties, TiN film is endowed with outstanding mechanical, electrical and thermal performances, which plays an important role in semiconductor devices during advanced integrated circuit manufacturing process. So far, TiN film has been widely used as hard coating, diffusion barrier and work function regulating layer in semiconductor devices.

TiN film can be deposited by chemical vapor deposition (CVD), physical vapor deposition (PVD) and atomic layer deposition (ALD) in semiconductor manufacturing. It is easily oxidized for TiN film in air at ambient temperature, leading to its mechanical, electrical and thermal properties altered, which has already inspired many researchers to explore the oxidation behavior of TiN film. The oxidation kinetics of TiN films in air and pure oxygen through different characterization techniques were studied by Harland G. Tompkins[1], who summarized the relationship between film thickness and oxidation time, temperature as well as oxygen concentration. Hong-Ying Chen et al.[2] devoted themselves to making the relationship between the oxidation behavior of TiN film and temperature clearly by X-ray diffraction spectroscopy (XRD), revealing the thickness and porosity of the oxide film ($TiO_2$) on the surface of TiN film increasing with the increase of temperature. A research was that oxidation temperature and oxygen injection amount have obvious effects on oxidation composition of TiN films was carried out through in situ X-ray photoelectron spectroscopy (XPS) by A. Glaser et al[3]. The oxidation behavior of TiN films has been explained well by these researches, while the connection between the oxidation behavior of TiN films and the queuing time (Q-time) in clean room environment has not been studied, besides the effect of the oxidation of TiN films on its resistance (RS).

In this paper, TiN thin films were deposited on silica substrate by means of physical vapor deposition (PVD) in *Fig.1*. By exploring the oxidation behavior of TiN film in clean room environment and variation trend of its RS, we obtained simulated standard RS ($RS_s$) as a function of queuing time (Q-time) and initially measured RS ($RS_m$) which was initially measured. After the simulation formula applied to the data collection system, RS at the any stage of film oxidation can be corrected, meanwhile, an accurate machine monitor can be realized (*Fig.2*). Therefore, the monitor time is significantly improved from 5 hours to 1 hour, 80% efficiency benefit achieved.

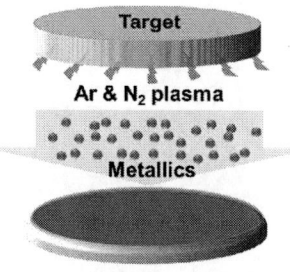

*Fig.1 Diagram of TiN film deposition*

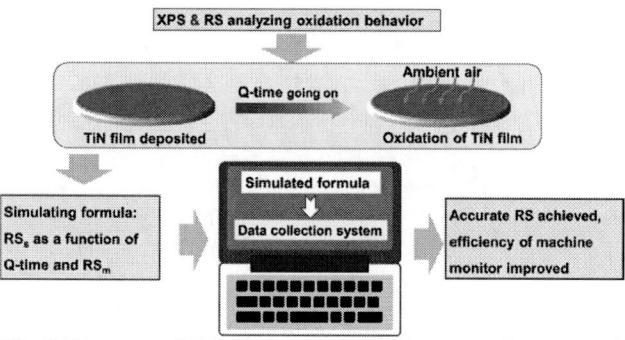

*Fig.2 Diagram of TiN film oxidation behavior and process of achieving accurate RS*

## EXPERIMENT

Preparation and testing of thin films

Firstly, silicon oxide thin films were deposited on a 300mm diameter wafer with a particle number of less than 200 by chemical vapor deposition. Then, TiN thin films were deposited on wafer with silicon oxide substrate by vacuum sputtering (physical vapor deposition, PVD).

Standard four-probe method was used to measure the sheet resistance (RS) of TiN film. The measurement site was 49 sites and distributed in the form of concentric circles (*Formula 1*). The thickness of TiN film was detected by optical measurement method, whose measurement point was

49 dots. The results of the bonding energy of elements in the thin film was given by the application of X-ray photoelectron spectroscopy (XPS), analyzing the oxidation behavior of the thin film.

$$RS=V/I=\rho L/S=\rho L/WT=\rho/T \qquad \textbf{\textit{Formula 1}}$$

$\rho$ is the resistivity of the film, ohm·Å. T is the film thickness, Å. L is film length, Å. W is the film width, Å. RS is the value of the resistance of the film whose length is equal to the thickness, ohm/sq.

# RESULTS AND DISCUSSION

TiN films are easily oxidized in the clean room environment with passage of time, whose oxidation phenomenon can be exactly analyzed by XPS. Here, the C1s XPS spectra of carbon is used as the reference peak for correction. As is shown in *Fig.3* that the corrected XPS spectra for C1s, Ti2p, O1s and N1s of TiN film has been present at different Q-time.

In the spectra of C1s in *Fig.3a*, the bonding energy (BE) of the peak at around 284.8eV could be attributed to C-C bond [4]. When the Q-time increases from 12960s, the density of the peak is gradually increased, indicating the surface of TiN film polluted by carbon, which is a typical phenomenon of the sample exposed to air.

As shown in *Fig.3b*, when the Q-time changes from 4680s to 84240s, the BE of Ti2p$_{1/2}$ at 464.2ev does not show significant changes, while the BE of Ti2p$_{3/2}$ at 458.6ev (m peak) and 457.8ev (n peak) exhibit obvious change [3,5]. We attribute the m-peak at 458.6eV to the Ti-O bond in TiO$_2$. The peak strength of Ti-O bond gradually increases, as well as the ratio of O/Ti (*Table 1*) also keeps increasing, which can be ascribed to gradual oxidation of the surface of TiN film, where the formed titanium oxide compound converted into TiO$_2$ bit by bit, even to titanium oxide with higher oxygen content. The n-peak at 457.8eV is attributed to Ti-N bond, where the BE of Ti-N shift to the lower energy by 0.3 eV and peak strength decrease gradually with the increase of Q-time, which further confirms the oxidation of TiN film.

*Fig.3c* shows the peaks at 530.8eV and 533eV are assigned to the oxygen bond in titanium oxide and in Ti(OH)$_x$ compound, respectively [6]. As Q-time goes on, the peak strength of oxygen bond in Ti(OH)$_x$ compound gradually decreases, whereas the peak strength of oxygen bond in titanium oxide gradually increases. It is clearly indicated that oxygen is capable of doping into TiN film surface under the condition of increasing Q-time, which makes TiN film oxidized and hydroxyl compound decrease.

In *Fig.3d*, the peak at 396.5eV is attributed to the Ti-N bond in TiN film [3], presenting a phenomenon of BE and peak strength decreasing from 4680s to 84240s. It is deeply possible that part of Ti-N bond is broken and the oxide TiN$_x$O$_y$ composed of Ti-O-N bond is formed, due to the introduction of O element. The BE of the other peak in the N1s spectrum is 405eV [5], which can be assigned to the N-O bond in TiN$_x$O$_y$, whose strength drops down slightly, indicating that some nitrogen elements escape into the environment in certain forms, resulting in the increasing Ti/N ratio (*Table 1*), which further proves the oxidation of TiN film.

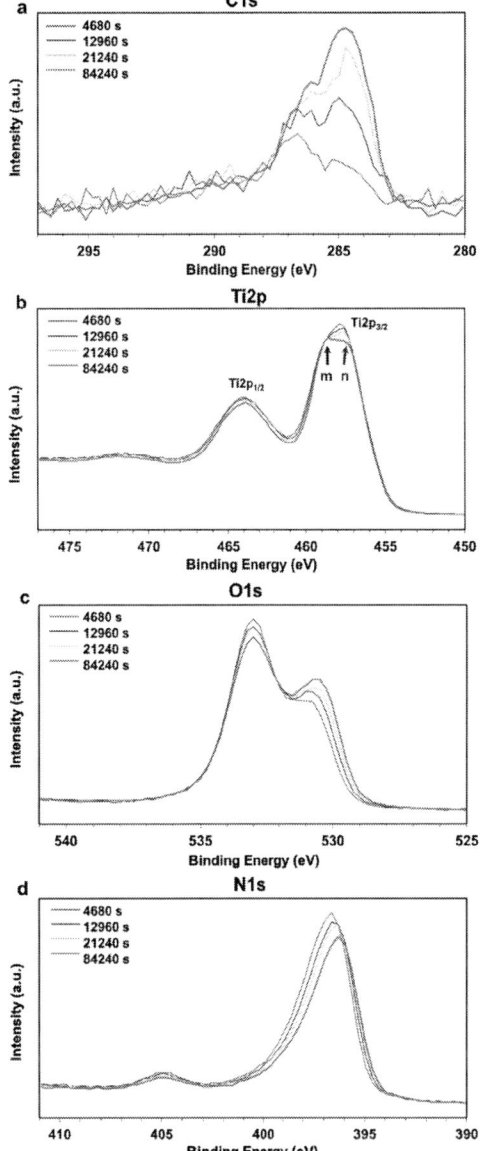

*Fig.3 XPS spectra for C1s, Ti2p, O1s and N1s of TiN film at different Q-time*

*Table1. Atomic ratio of TiN film at different Q-time*

| Q-time (s) | 4680 | 12960 | 21240 | 84240 |
|---|---|---|---|---|
| O/Ti | 1.9132 | 1.9849 | 2.0280 | 2.1137 |
| Ti/N | 1.0783 | 1.2210 | 1.2697 | 1.2796 |

We discuss the relationship between RS and Q-time based on the study of oxidation of TiN films. RS curves about Q-time for different points (Site1, 20 and 40, Fig.4) are drown by measuring the RS of the film with fixed thickness at different Q-time, in which RS increases with the increase of Q-time. It is obviously shown from the *Fig.4* that RS is very sensitive to Q-time, where RS grows rapidly in the first 18000 s (5 h), and then grows slowly after 18000 s, leading to RS at Q-time 18000s taken as the reference value to judge the machine condition for a long time. Nevertheless, 18000s is still a relatively long time, highly affecting the efficiency of daily machine monitoring. In order to reduce the daily monitoring time of the machine and improve the efficiency, it is necessary to eliminate the influence of Q-time on RS as much as possible by using the simulation formula. The scattered points in *Fig.4* are fitted as several functions, which conforms to the model of natural logarithm. However, the RS fitting curves of different sites have different coefficients, so RS measurement of the whole wafer are essential to analyze.

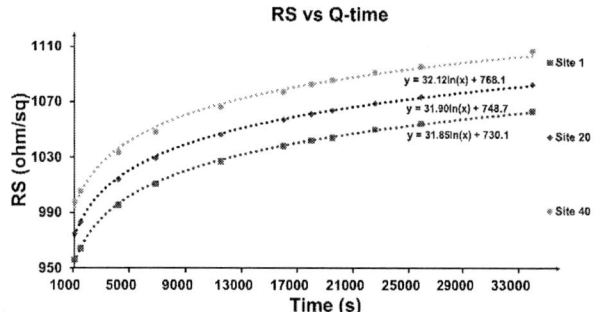

*Fig.4 The fitting curve of RS with respect to Q-time of TiN film at different measuring sites*

The LN(t) and RS' factors are used for statistical analysis of the data of all measuring sites in the whole wafer, and the data is obtained as shown in *Table2*. The P value in the *Table2* is <0.0001, present the factors with extremely significant statistical difference, indicating the reliability of the derived formula.

*Table 2. The coefficients of curve between $RS_m$ and LN(t), RS′*

| Item | Estimate | Std Error | Prob>|t| |
|---|---|---|---|
| Intercept | 747.3599 | 1.4314 | <0.0001 |
| LN(t) | -76.3389 | 0.8568 | <0.0001 |
| RS'*LN(t) | 0.1021 | 0.0008 | <0.0001 |

According to *Table 2*, the following formula can be obtained

$$RS_m = -76.3389*LN(t)+0.1021*RS'*LN(t)+747.3599$$

***Formula 2***

$RS_m$ is experimental measured RS of the film, ohm/sq, where t is the Q-time of the film, s. RS' is the measurement RS value of the film while Q-time is 18000s, ohm/sq.

RS' and $RS_m$ values with Q-time of 18000s and t (t≠18000s) were measured respectively. The corresponding fitting value could be obtained by substituting the RS' value into *Formula 2*, thus, the fitting value and experimental measured value are plotted together to verify *Formula 2*. It can be seen from *Fig.5* and *Table 3* that the fitted values are linearly correlated with the $RS_m$ measured in the experiment, indicating the rationality of *Formula 2*.

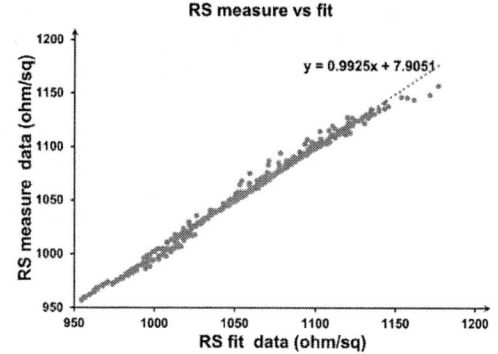

*Fig.5 Correlation between experimental measured value $RS_m$ and fitted values*

*Table 3. Slope and intercept of correlation curves between experimental measured value $RS_m$ and fitted values*

| NO. | 1 | 2 | 3 |
|---|---|---|---|
| slope | 0.9925 | 1.0087 | 0.9905 |
| Intercept | 7.9051 | -9.9590 | 11.0810 |

Since the RS' value at 18000s has been used as the reference value to judge the machine condition for a long time, the simulated value with Q-time equal to 18000s as the standard RS value can be obtained through the conversion of *Formula 2*, its functional relationship as bellow (*Formula 3*):

$$RS_s = \frac{RS_m+76.3389*LN(t)-747.3599}{0.1021*LN(t)}$$

***Formula 3***

$RS_s$ means the standard RS simulation value when RS is tested at different Q-time, ohm/sq.

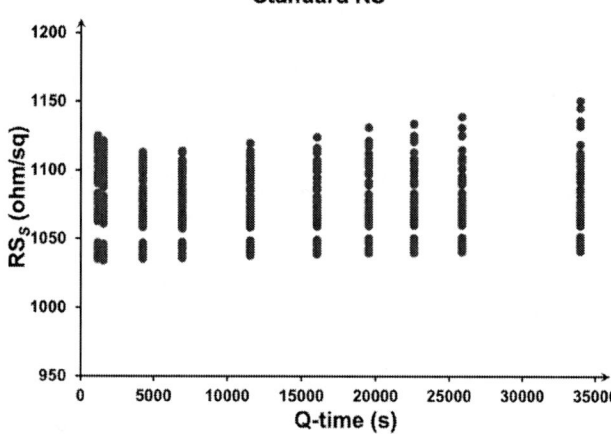

*Fig.6 The standard simulated value $RS_s$ fitted at different Q-time for the same film*

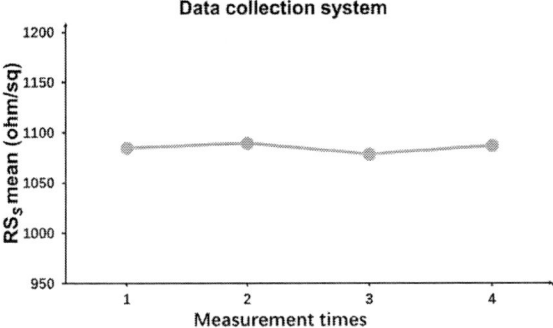

*Fig.7 Average $RS_s$ value obtained after the simulation formula applied in the data collection system*

According to *Formula 3*, we can obtain the $RS_s$ values corresponding to different Q-time. As is shown from *Fig.6*, $RS_s$ values are all within a reasonable fluctuation range (1000-1150 ohm/sq). When the Q-time is close to 3600s (1h), the obtained simulation *Formula 3* is applied to the data collection system to achieve an accurate and reasonable $RS_s$ average value (*Fig.7*). Therefore, we achieved accurate monitoring of the machine, where monitoring time is reduced from previous 5h to 1h.

## CONCLUSION

With the growth of Q-time, the TiN film is easy to be oxidized in clean room environment. During the whole oxidation process, the RS of TiN film increase logarithmically with Q-time going on, and the relation between simulated $RS_s$ and Q-time, measured value $RS_m$ is further achieved. The application of this relation to the data collection system can significantly reduce the testing time of daily monitoring and greatly improve the production efficiency.

## ACKNOWLEDGEMENTS

The authors would like to acknowledge HLMC vendors for the supports of XPS analysis.

## REFERENCES

[1] Harland G. Tompkins., et al. "Oxidation of titanium nitride in room air and in dry $O_2$." *Journal of Applied Physics* 70.7 (1991): 3876-3880.

[2] Hong-Ying Chen., et al. "Oxidation behavior of titanium nitride films." *Journal of Vacuum Science & Technology A: Vacuum, Surfaces, and Films* 23.4 (2005): 1006-1009.

[3] A. Glaser., et al. "Oxidation of vanadium nitride and titanium nitride coatings." *Surface Science* 601.4 (2007): 1153-1159

[4] S. Beldar, et al. "FTIR and X-Ray photoelectron spectral (XPS) evidence for interaction between natural ester and cellulose paper." *2019 IEEE 20th International Conference on Dielectric Liquids* (2019): 1-4, doi: 10.1109/ICDL.2019.8796589.

[5] Xiaosong Zhou., et al. "Preparation of nitrogen doped $TiO_2$ photocatalyst by oxidation of titanium nitride with $H_2O_2$." *Materials Research Bulletin* 46.6 (2011): 840-844

[6] P Verardi., et al. "Characterization of pulsed-laser deposited $Pb(Zr,Ti)O_3$ for piezoelectric thin films devices." *Proceedings of the Eleventh IEEE International Symposium on Application of Ferroelectrics.* (1998): 109-112.

# IMPROVEMENT OF ALUMINUM DIFFUSION IN HKMG PROCESS

*Xiaoyang Xi, Zhaoqin Zeng, Yu Bao, Haifeng Zhou, Jingxun Fang,Yu Zhang*
Shanghai Huali Integrated Circuit Corporation, Shanghai, China

*Corresponding Author's Email: xixiaoyang@hlmc.cn

## ABSTRACT

Excess aluminum diffusion is always the main concern in the high-k metal gate (HKMG) process at 28/22nm node. In this work, we try to add extra ALD TaN film before TiAl and Al eletrode and  the thickness of bilayer TaN, SIMS analysis, Vt variation, fail bin and the mechanism  was well studied in this paper. Experiment result show excess Al was well suppressed with extra TaN added before TiAl. CP bin Trans_AVS and Vmin_HPM performance was significantly improved. A good performance was achieved finally.

## INTRODUCTION

High-k  metal  gate  (HKMG)  integration  using  a replacement metal gate (RMG) approach. TiN was selected as p-MOSFET work function Metal and TiAl as n-MOSFET work function Metal. ALD TaN was used as a wet etching stop layer, which was also well know as an excellent barrier layer material in IC field due to its high thermal stability, high melting point. TiN+Ti multi film was named as TBM(Top barrier Metal), which used  as a barrier layer for aluminum diffusion. [1] In the early stage of HK Metal gate process development, the low yield is suspected caused by Al diffuse  too much from TiAl film or Al electrode to substrate. The schematic diagram of NMOS and PMOS is shown in Figure1.

Aiming at Al diffusion problem, this paper try to add extra TaN with different thickness before NWF layer to verify the blocking effect on Al diffusion by threshold voltage and fail bin analysis. The sensitivity of TaN thickness to Vt was 81mV/nm for  NFET  and  114mV/nm  for  PFET  respectively. [2],[3] Theoretically, higher TaN thickness cause higher Vt shift and lower Al gap filling capacity. however, lower TaN thickness degrade the effect of blocking Al diffusion. So it is crucial to balance the thickness relationship of bilayer TaN.

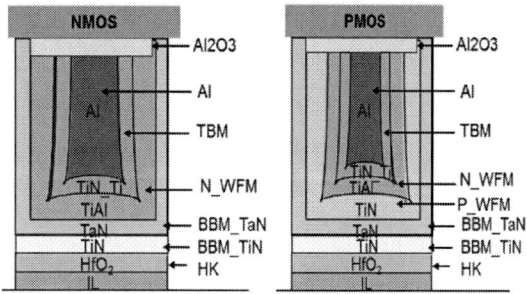

*Figure 1: Schematic View of NMOS and PMOS*

## EXPERIMENT

The  preparation  method  of  TaN  is  ALD  atomic  layer deposition.By comparing single-layer TaN and double-layer TaN, and adjusting the thickness of double-layer TaN, we clarified the diffusion blocking effect of double-layer TaN on Al through TEM and EDS analysis. Secondly, SIMS analysis was performed to identify the material classification of surface

materials  by  measuring  the  mass  of  secondary  ions. Furthermore, the ability of TaN film to block AL diffusion was determined according to the strength of Al signal in each layer of the film. In addition, at the post of CMP (chemical mechanical grinding) station, the N/PMOS region was screened with defect detection to compare single-layer and bilayer TaN.  Finally, the electrical conductivity of the device was verified in the end.

## RESULTS AND DISCUSSION

（1）    The SIMS analysis of Al signal diffusion by extra TaN was explored on MOSCAP

| Wafer | #16 | #17 |
|---|---|---|
| TaN split | BBM TaN + Post P TaN | BBM TaN +skip |

*Table1:The Split of Single-layer and Bilayer of TaN*

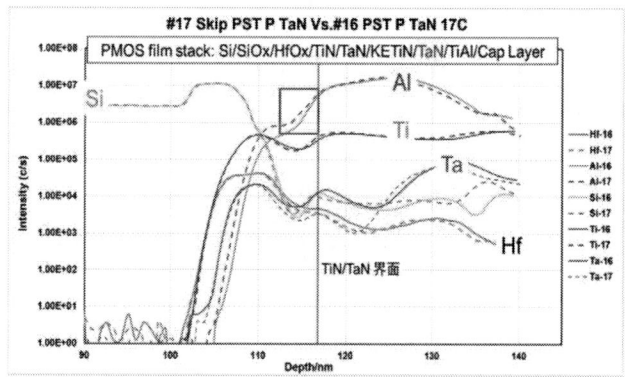

*Figure 2 The Analysis of SIMS Element*

Figure 2 shows the SIMS analysis results. As we know the basic principle is to identify the material classification of surface materials by measuring the mass of secondary ions. The products film stack simulated by MOSCAP experiment is Si/SiO$_x$/HK/TiN/TaN/TiN/TaN/TiAl/Cap Layer. As the result of SIMS icon indicates different elements with the scanning depth change of signal strength, the X  axis for upward from the bottom of the thin film laminated in detecting depth, Solid line (#16) is on behalf of the bilayer TaN which added post PWF TiN, and dotted line (#17) is on behalf of single layer TaN. According to the arc changes of Al element in HfO$_2$ and the slope of signal decline, we could see that the signal of Al element closer to high-K, which meaning that Al is less likely to diffuse downward in the condition of bilayer TaN. Furthermore, it also could illustrate the TaN film has stronger ability in blocking Al than TiN.[5] In conclusion, the bilayer TaN is more effective in preventing downward diffusion of Al than the single-layer TaN.

（2）    Bilayer TaN inline TEM and EDS map

 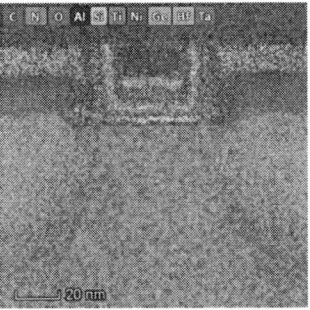

*Figure 3 (a): HKMG TEM Map*     *Figure 3 (b) :EDS of HKMG*

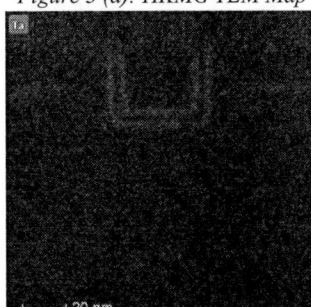 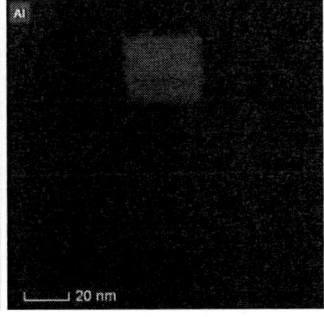

*Figure 3 (c): EDS-Ta Map   Figure 3 (d): EDS of Al Map*

Figure 3(a&b) is the bilayer TaN TEM and EDS map, and Figure 3(c&d) are the EDS map of Ta and Al element respectively in Metal Gate. From the figure 3(d) showing that Al element has no obvious downward diffusion under the condition of bilayer TaN. Furthermore, rhe results show that TaN has a certain blocking effect on Al diffusion under the condition of bilayer, which also layed an effective foundation for the subsequent experiment of bilayer TaN thickness regulation.

**（3）    Bilayer TaN inline split defect comparison**

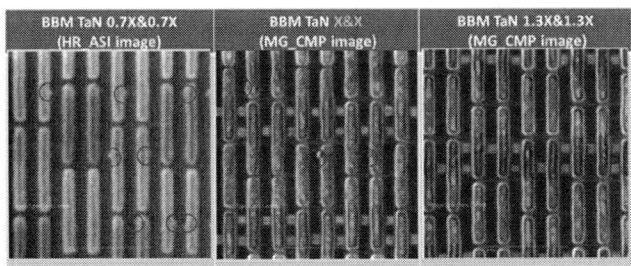

*Table2: MG_CMP Defect Comparison*

### MG_CMP Defect Count Comparison

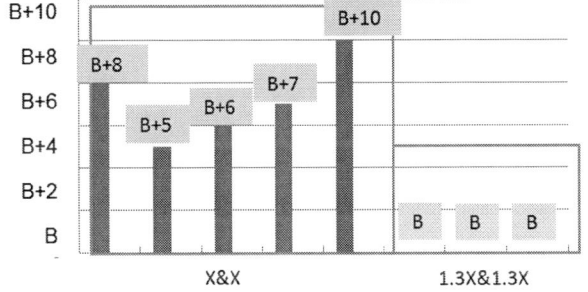

*Figure 4: MG_CMP New Type Defect check*

In the table 2 three conditions were selected in the figure for experimental investigation. Images show the split result of MG_CMP Defect check. We could see that as the decrease of

the thickness, there are varying degrees of defect damage defect under the condition of 0.7X & 0.7X. However, in the condition of 1.3X&1.3X, there is no obvious damage. Suspect that the thinner film was etched excessively more easily. Similarly, Figure4, which the results of the two experimental conditions summarized in FIG.3, shows that the experiment result has repeatability. To sum up, which the condition of 1.3X&1.3X is the best, and the subsequent test excluded the condition of thinner TaN, and continue to verify the Vt comparison of bilayer and single-layer TaN.

**（4）    SC/LC Vt Comparison Between Single-layer and Bilayer TaN**

As shown in Table3, the conditions of X&X and 1.3X&1.3X were chosen to compare with monolayer TaN onVt.

| Vt of TaN split | EOL-N | | | EOL-P | | |
|---|---|---|---|---|---|---|
| | 1.3X&0 | X&X | 1.3X&1.3X | 1.3X&0 | X&X | 1.3X&1.3X |
| SC VTN(30nm) | 0.039 | 0.156 | 0.146 | 0.152 | 0.214 | 0.204 |
| SC VTN(35nm) | 0.078 | 0.186 | 0.183 | 0.181 | 0.249 | 0.239 |
| SC VTN(40nm) | 0.107 | 0.207 | 0.198 | 0.202 | 0.271 | 0.264 |
| LC VTN(1um) | 0.266 | 0.298 | 0.298 | 0.264 | 0.352 | 0.373 |

*Table 3 :SC and LC Vt summary of NMOS&PMOS*

Table3 shows the difference of NMOS and PMOS Vt of different sizes in short channel by combining BBM TaN and extra TaN Layer. Figure5 (a&b) show the comparison of N/PMOS Vt in short channel and (c&d) show the comparison of N/PMOS Vt in long channel. As for the Vt both of NMOS and PMOS, bilayer is higher than that of single-layer. On a theoretical basic, TiN is chosen as the work function of p-MOSFET and TiAl is chosen as the work function of n-MOSFET in HKMG. However, the EWF of TaN is between TiN and TiAl, without considering Al diffusion, when adding extra TaN, for NMOS, combining with TiN will increase the work function of the metal gate, which resulted in the increase of Vt. For PMOS, combined with TiAl, the work function of metal gate will be lowered, and then resulted in the increase of Vt. In the comparison of single layer TaN and bilayer TaN, from the Figure5 result show that with the condition of bilayer TaN, the Vt of both NMOS and PMOS will increase, which is consistent with the conclusion of theoretical calculation. Analysis of the reasons may be that BBM TaN will be affected by Al diffusion under the condition of thinner BBM TaN, resulting in Vt increase. However, due to the influence of SC SCE, IMP and CD variation, Vt in HKMG mainly depends on LC, and the experiment results were basically consistent with the expected guess. In summary, subsequent tests continue to verify the influence of thickness difference of BBM TaN and extra TaN on the yield.

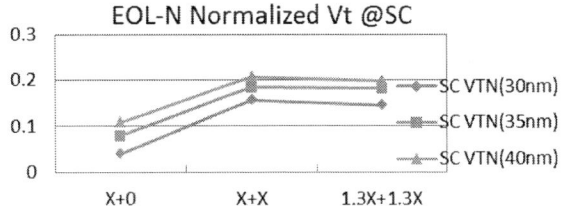

*Figure 5(a): SC Vt Comparison of NMOS*

978-1-6654-9759-6/22 $31.00 © 2022 IEEE          375

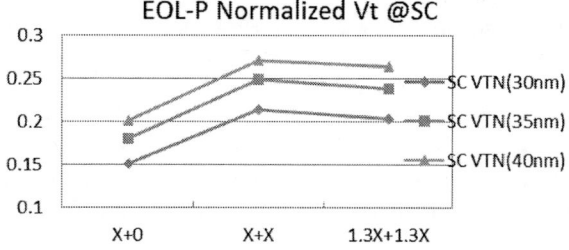

*Figure 5(b):SC Vt Comparison of PMOS*

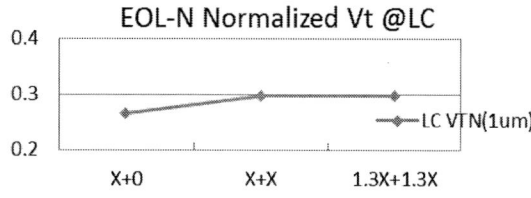

*Figure 5(c) : LC Vt Comparison of NMOS*

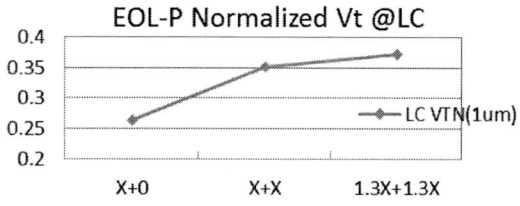

*Figure 5(d): LC Vt Comparison of PMOS*

（5） Extra TaN inline split CP_bin Failure Rate Comparison

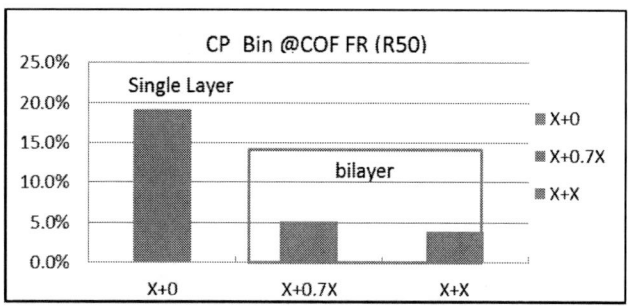

| TaN split | X&0 | X&(X-) | X&X |
|---|---|---|---|
| Trans_AVS (R50) COF Map | | | |
| COF FR | 19.2% | 5.2% | 3.9% |
| Vmin ~ HPM | | | |

*Table 4: The Comparison of Vmin~HPM Between Single-layer TaN and Bilayer TaN*

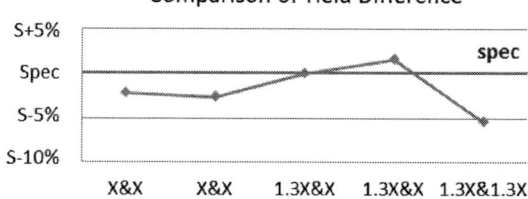

*Figure 6: The comparison of COF Failure Rate Between Single-layer TaN And Bilayer TaN*

As shown in Table 4, according to the distribution

HPM~Vmin, the bilayer TaN is more concentrated than that of single-layer TaN, The parameter HPM~Vmin reflects the running speed of the device. In general, within a certain specification, the larger value indicates better device performance . What's more, Figure 6 shows the comparison of COF (Continue of Fail) failure rate of Trans_AVS between single-layer and bilayer TaN. As the Figure 6 shown from Trans_AVS COF map diagram and histogram analysis, the failure rate of monolayer TaN is much higher than that of bilayer TaN, which illustrated that the performance is optimized obviously by Vmin_HPM parameter in the condition of bilayer TaN.

（6） Bilayer TaN inline split fail bin comparison

| Bilayer TaN Split | | X&X | X&X | 1.3X&X | 1.3X&X | 1.3X&1.3X |
|---|---|---|---|---|---|---|
| Yield | Total | S-2% | S-3% | S+0 | S+2% | S-5% |
| WAT | Idsat_RVT N/P Ratio | S+15% | S+12% | S+18% | S+11% | S+1% |
| CP | SOF FR (R50) | S+167% | S+233% | S-67% | S-33% | S+100% |
| | COF FR (R50) | S+32% | S+12% | S-8% | S-12% | S-12% |
| | COF FR (R40) | S+49% | S-260% | S-7% | S-37% | S-73% |
| | HPM | S+3% | S+6% | S+1% | S+2% | S+0 |

*Table 5:DOE of Bilayer TaN of WAT And CP_bin Comparison*

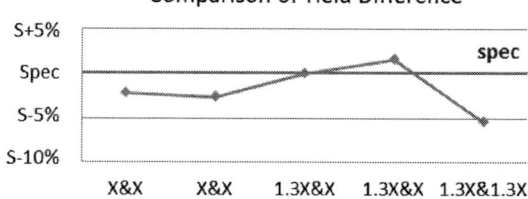

*Figure 7(a): Yield comparison with different split of bilayer TaN*

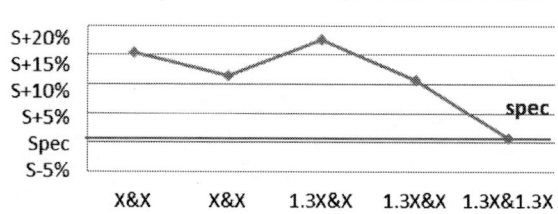

*Figure 7(b): Idsat N/P ratio comparison with different split of bilayer TaN*

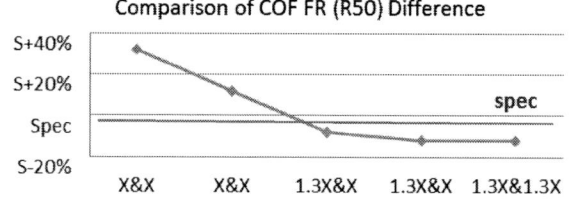

*Figure 7(c): COF Failure Rate(R50) Comparison With Different Split of Bilayer TaN*

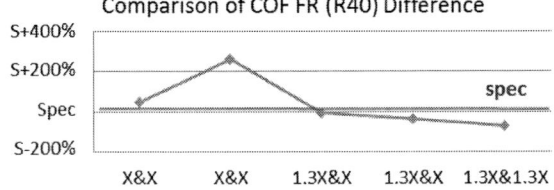

*Figure 7(d): COF Failure Rate(R40) Comparison With Different Split of Bilayer TaN*

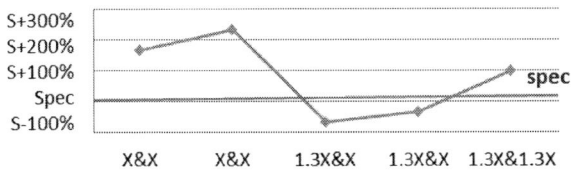

*Figure 7(e): SOF Failure Rate(R50) Comparison With Different Split of Bilayer TaN*

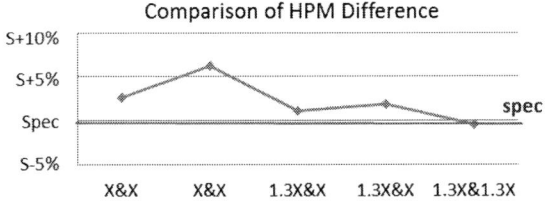

*Figure 7(f): HPM Comparison With Different Split of Bilayer TaN*

| Bilayer TaN split | | X+X | 1.3X+X | 1.3X+1.3X |
|---|---|---|---|---|
| Yield | | S-2% | S+0 | S-5% |
| CP | Trans_NO_CPU Vmin ~ HPM | | | |
| | HPM | S+3% | S+3% | S+0 |
| | Trans_AVS COF Map (R50) | | | |

*Table 6: DOE of Bilayer TaN of Vmin~HPM and Trans_AVS Comparison*

Table 5 is the data summary of the bilayer TaN thickness adjustment comparative test .Figure 7(a)shows the yield chart, the yield of bilayer TaN under the condition of 1.3X+X is better than other split. As is shown in Figure 7(b), According to the electrical parameter Idsat, the N/P ratio of specification was setting to be higher than specification, and the data showed that the three conditions all met the requirements. Figure 7(c&d) show the failure rate of COF(Continue of Fail) in CP bin. R50 and R40 refer to different versions of the scanning specification. According to the requirement, Combining with the two specifications, the X+X condition is worse than the other two conditions. Furthermore, it can be shown the map of Trans_AVS COF (R50) map in Table 6, which the performance of 1.3X+1.3X worse than that of 1.3X+X. Secondly, as shown in Figure 7(e), when the thickness of TaN is too high or too low, the failure rate of SOF (Stop of Fail) will increase. Theoretically, too thick TaN will lead to Vt shift, and even

affect the ability to gap filling, however too thin TaN will lose the effect of blocking aluminum, so it is crucial to balance the thickness relationship of bilayer TaN. As shown in Figure 7(f) and Table6, the current bilayer TaN experiment can meet the requirements according to the specifications set by HPM. Thus, combined with the overall analysis, 1.3X+X combined experimental conditions show the best performance from electrical and yield data summary. Finally, because the final yield was influenced by multiple experimental conditions, the electrical performance and yield failure analysis of the actual product need to be evaluated comprehensively

## CONCLUSION

Base on SIMS data, less Al diffuse to downward with bilayer TaN film when compare to single TaN. And TaN film has better block effect than TiN film. And bilayer TaN thickness split show Al new type defect easily occurred when thickness is low, which indicate TaN film help to prevent Al protrusion happen. However, bilayer TaN cause Vt higher both in NMOS and PMOS. Finally, 1.3X+X bilayer TaN show best WAT(Idsat N/P ratio) and less fail bin ratio. In summary, the results of the above experiments demonstrate block effect on aluminum diffusion play a very key role in the performance improvement of 28HK MG process devices.

## ACKNOWLEDGEMENTS

The authors would like to acknowledge HLMC FA department for the supports of TEM/EDS analysis and PIE &YE team for data collection and technical discussion.

## REFERENCES

[1]Y. Taur and T. H. Ning, *Fundamentals of Modern VLSI Devices*. NewYork, NY, USA: Cambridge Univ. Press, 1998, ch. 3, p. 130.
[2] K. Mistry et aI., IEDM Tech. Dig., 2007, pp.247-25.
[3] S. Pae et al., "Advanced high-k and metal-gate CMOS logic transistor technology", IEEE International Reliability Physics Symposium (IRPS),pp. 5C.1.1-1.5, (2012).
[4] H. Cui, 1. Xu, J. Gao, 1. Xiang, Y. Lu, Z. Tang, X. He, T. Li, J. Luo, X.Wang, B. Tang, J. Yu, T. Yang, 1. Yan,1. Li and C.Zhao, "Evaluationof TaN as the wet etch stop layer during the 22nm HKMG gate last CMOS integrations." ECS Trans. 5S (6), pp. III-lIS. 2013.
[5]H. Shin, W. Zhu,L. Liu,S. Sridhar,V. M. Donnelly, D. 1. Economou, C.Lenox, T. Lii, "Selective etching of TiN over TaN and vice versa inchlorine-containing plasmas" in J. Vac. Sci. Techno!. A 31 (3), 2013.

# INVESTIGATION OF 14 NM CONTACT TUNGSTEN GAP-FILLING PERFORMANCE

*Xiaofang Wang[*], Jiazhang Xu, Zhiqi Yuan, Yingbo Cheng, Zhaoqin Zeng, Yu Bao, Haifeng Zhou, Jingxun Fang, Yu Zhang*

Shanghai Huali Integrated Circuit Corporation, Shanghai 201210, China

*Corresponding Author's Email: wangxiaofang@hlmc.cn

## ABSTRACT

The contact glue layer (CTGL) and tungsten deposition process have been optimized systematically to improve the contact tungsten gap filling performance at 14nm FinFET technology node. The different pre-clean methods (PCXT and SICONI) also have been studied to verify the impact on contact profile. The W seam defect counts were reduced by 99% through optimizing pre-clean, Ti/TiN and tungsten deposition condition. Furthermore, the relationship between process thickness and W seam defect was surveyed by Taguchi method. The results show that the W seam defect is strongly correlated with Ti thickness and weakly correlated with PCXT when its etch amount below 10 Å. The W seam defect counts decrease with lower PCXT EA/thinner Ti thickness and increase with lower SICONI etching amount.

**Keywords:** *Contact, Pre-clean, Ti, TiN, W seam, 14nm Technology*

## INTRODUCTION

Tungsten has been widely used as high aspect ratio contact plug material due to its good gap filling ability. In advanced semiconductor device, the critical dimension (CD) decreases sharply and complicate integration problem impacting tungsten gap filling. In addition, the change of 14nm FinFET contact design makes it more challenge to get better contact gap filling. Comparing to 28nm technology, the contact critical dimension of 14nm technology shrinks by 25% and the design changed from typical cylindrical structure (Fig.1a) to a long strip or even more complex groove structure (Fig.1b). This change makes the contact gap filling more sensitive to the inhomogeneity and the residue of pre-layer. As shown in Fig.1c, 14nm FinFET contact gap filling shows worse W seam defect when adopted the contact gap filling technology of 28nm. The W seam defect will be corroded to big W void during W CMP process, which seriously affects the electrical performance of the device and the yield of the chip.

While the actual CD and morphology of the trench are mainly determined by the etching process. Not only contact CD and profile but also CTGL and tungsten film deposition condition affect total contact gap filling performance. The residual polymer and water vapor can be removed by pre-treatment, including degas and Ar pre-clean process before Ti/TiN deposition. Nucleation and grain growth are the main processes affecting tungsten gap filling during chemical vapor deposition (CVD) process.

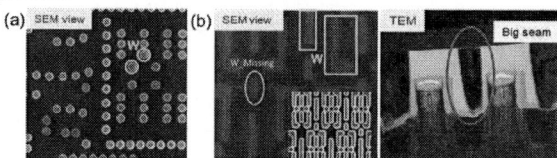

*Figure 1: (a) Contact plane-view SEM of 28 nm technology node; (b) 14nm FinFET W Seam SEM image and (c) W seam cross section TEM image.*

In this work, the CTGL and CVD tungsten processes were investigated step-by-step to improve the 14nm FinFET contact tungsten gap filling performance.

## EXPERIMENT

There are four steps during CTGL process. Degas and pre-clean are the pre-treatment steps which used to remove residual water vapor, oxide and residual polymer. Titanium (Ti) glue layer was deposited by RF PVD (physical vapor deposition) method and TiN was deposited by metal organic precursor and treated by ex-situ $N_2$ plasma to get purify TiN barrier films.

TABLE I: CTGL deposition split table

| Step | Condition |
|---|---|
| Degas | Time+ |
| PCXT | Power+/etching |
| PCXT | EA+- |
| SICONI | EA+- |
| Ti dep | Thickness+- |
| TiN dep | Thickness+- |

The split table of CTGL is shown in Table1. SICONI process is applied to enhance trench bottom oxide removal and ease bottle-neck profile by Ar pre-clean process. In 14nm technology, titanium (Ti) layer is used not only for the contact adhesion layer but also play a role as titanium silicide (TiSi) reactant. In addition, the different pressure, nucleation cycle times and cool fill deposition time of CVD tungsten process also have been investigated to improve tungsten gap filling performance.

## RESULT DISCUSSION

The difference of pre-clean methods was studied for contact profile. As shown in Fig.2, the incoming profile from contact trench open etching process is taper profile (Fig.2a). After PCXT process, the profile changed to vase shape (Fig.2b) due to the higher Ar plasma power. Besides, the cross profile of contact trench had no obvious change after SICONI process, except the enlarged half-width CD of the trench.

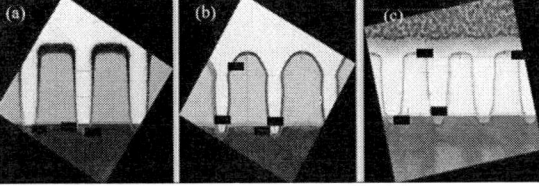

*Figure 2: TEM image of contact profile (a) incoming profile before Ti/TiN deposition; (b) after PCXT 60 Å; (c) after SICONI 60 Å.*

Although the Ar pre-clean method can reduce the contact resistance by effectively removing oxide at the trench bottom. It will lead to the vase shaped profile and the challenge to tungsten gap filling due to necking opening. SICONI method can solve the vase profile problem, but the isotropic etching characteristic will lead to trench sidewall dielectric loss, which increases the short risk between contact and metal gate. Therefore, the combining of PCXT and SICONI process before Ti/TiN deposition has been used for 14nm FinFET contact pre-clean. Therefore, the optimization of PCXT and SICONI etching amount should be studied depending on tungsten gap filling and contact to metal gate short process windows.

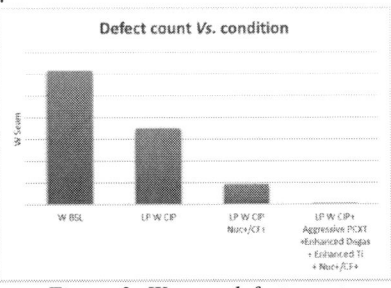

*Figure 3: W seam defect count*

The summary of tungsten seam defect chart is shown in Figure3. Comparing with baseline condition, the lower tungsten deposition pressure (LP W CIP) can significantly improve W seam defect. Based on the low pressure W deposition, the process was optimized by following condition, including degas time, PCXT power and time, Ti thickness, W nucleation cycle and cool fill time. The total W seam can be further reduced by 99%.

The W seam SEM images with baseline and low pressure tungsten deposition conditions are shown in Fig. 6(a) and (b) respectively. It can be seen that the baseline tungsten deposition condition will produce obvious void after WCMP process. Lower tungsten pressure with lower deposition rate is beneficial to the tungsten gap filling based on the conformal growth characteristic. From the plane view SEM, it was significantly improved. Furthermore, there is almost no obvious void defect after W CMP (Fig. 4c) based on the combination of ILB and W optimization conditions including enhanced degas, aggressive PCXT, enhanced Ti and low pressure W deposition, enhanced W nucleation and cool fill, *etc.*

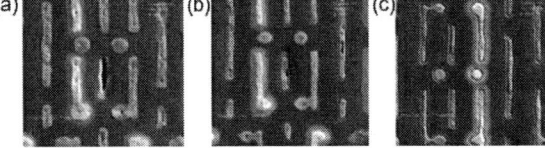

*Figure 4: W seam SEM images with (a) baseline W recipe, (b) low pressure W dep recipe and (c) low pressure W dep with Aggressive PCXT +Enhanced Degas +Enhanced Ti + Nuc+/CF+*

Furthermore, the process window of PCXT, SICONI and Ti based on the optimized conditions were surveyed by Taguchi method. The predicted best condition for W seam defect has been shown in Fig.5. The defect count was predicted to be zero with condition of baseline PCXT, more SICONI etching amount and less Ti thickness. Moreover, the result also shows that the W seam defect count has strong correlation with Ti thickness and weak correlation with PCXT when its thickness below 10 Å (BSL). The W seam defect count decreases with PCXT/Ti thickness reduction, but increases with less SICONI etching amount. On the other hand, the PCXT reduction shows better W gap filling performance owing to the weak bottleneck profile. As the PCXT and Ti process not only influence the contact trench profile, but also affect the contact resistance, the suitable thickness of PCXT and Ti should be chosen combining to the performance of W seam defect and electrical device.

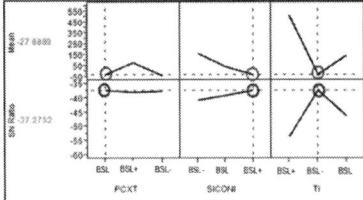

*Figure 5: The predicted best condition for W seam defect*

## CONCLUSION

The different pre-clean methods (PCXT and SICONI) have been studied for contact profile. The combination of PCXT/SICONI is discussed for tungsten gap filling and contact to metal gate short process windows. Besides, the contact glue layer and tungsten deposition process have been investigated systematically. The W seam defect count after W CMP was reduced by 99% with aggressive pre-clean, optimized Ti/TiN and low pressure tungsten deposition condition. The process window survey shows that the W seam defect counts decrease with PCXT/Ti thickness reduction, but increase with SICONI thickness reduction. The defect count was predicted to be zero with suitable condition of PCXT, SICONI and Ti.

## REFERENCES

[1] Kim,S.H., et. al. Characterizations of Pulsed Chemical Vapor Deposited Tungsten Thin Films for Ultrahigh Aspect Ratio W-Plug Process. *Journal of The Electrochemical Society*. 2005.

[2] Jianhua Xu, et. al. An Optimized W Process for Metal Gate Electrode Gap Filling Application. *China Semiconductor Technology International Conference*. 2015.

[3] Kai W, et.al. Improving Tungsten Gap-fill for Advanced Contact Metallization. *Advanced Metallization Conference*. 2016.

[4] Kuang-Wei Chen, et. al. Tungsten plug corrosion on $B_2H_6$-based nucleation layer induced by WCMP process. *IEEE*. 2012.

# REVIEW AND THERMO-FLUIDS NUMERICAL MODELING ON ELECTROSTATIC CHUCK

*Peng Feng[1,2], Taide Tan[2*], Yujie Ji[1], Yaxin Zhang[2], Toshihisa Nozawa[2]*

[1] Department of Mechanical Engineering, Shenyang Ligong University, Shenyang 110159, China
[2] Piotech, Inc, Shenyang 110159, China
*Corresponding Author's Email: edwardw@sypiotech.cn

## ABSTRACT

Electrostatic chucks (ESC) are used in advanced semiconductor manufacturing to clamp wafer combining with backside gas cooling and coolant channel to control wafer temperature during processing. To analyze the complexity and difficulties of ESC, we reviewed published studies on ESC, including chucking principle and de-chuck models. Based on this in-depth analysis, we studied the heat transfer of the ESC system and built numerical models to investigate the thermal performance of coolant channel design. The numerical results have been utilized as guide for high performance ESC design.

Keywords— Review, ESC, Heat transfer, Numerical modeling

## INTRODUCTION

ESC is a key component used to carry and cooling wafers in HDP-CVD device and has a great impact on film uniformity[1]. Previous research has focused more on ESC in the etching process, but the research on ESC in HDP-CVD has never been reported, which is the key factor in improving the quality of the film. In order to further deepen the research on ESC, this work sorts out and summarizes the research results of ESC on chuck and heat transfer. Through the establishment of the domain and simulation analysis of ESC in HDP-CVD, the influence of the heat transfer form and coolant channel of ESC on temperature uniformity is explained. Analysis of other domain will be carried out in the future.

## THE CHUCK AND DE-CHUCK PRINCIPLE OF DEFFERENT TYPE ESC

The traditional mechanical chuck fixes the wafer on the lower electrode through a mechanical structure, which results in that the part of the wafer near the center cannot be closely attached to the electrode, and the part covered by the chuck cannot be deposited or etched. Therefore, electrostatic chucks are widely used today[2].

According to the different ESC system circuits, ESC can be divided into monopolar type and bipolar type. Their chunk mechanism can be equivalent to the chuck between the plates of the parallel plate capacitor[3]. The Fig. 1(a)[2] shows the consists of monopolar type of ESC. The monopolar ESC consists by electrodes dielectric layers, wafer, plasma, and chamber to form a circuit. With

the monopolar structure, the wafer receives electrical charges from the plasma for chuck. Fig. 1(b) shows the consists of bipolar type of ESC. Different from the monopolar ESC, the bipolar ESC has two electrodes and does not require plasma to conduct the circuit. It is equivalent to two capacitors connected in series.

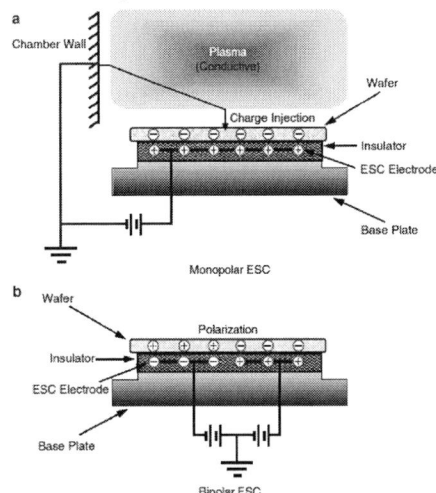

*Figure 1: (a)Schematic of monopolar and (b)Schematic of bipolar ESC*

Since the monopolar-type ESC only has one electrode, the plasma is necessary for generating the electric circuit for chuck the wafer. The advantages of the monopolar type are that the configuration is simple structure, and the clamping force is higher than that of the bipolar type because the relatively large projected area. The disadvantage is that the wafer clamping takes place after the plasma is generated, which makes it impossible to fill with helium when the plasma is generated. This means a low throughput and wafer-temperature controllability. Compared with monopolar ESC, the advantage of bipolar ESC is that no plasma is needed to conduct the circuit. But because the manufacturing precision of the two electrodes is different, it may cause adsorption problems.

Furthermore, the ESC may be categorized into either the Coulomb ESC or the Johnsen-Rahbek ESC, based on the differences in the conductivity of the generating mechanism of chuck force. Fig. 2 show their respective cross-sectional diagrams[4]. With the Coulomb electrostatic chuck, the wafer and the ESC electrode are isolated by an insulator (resistivity $> 10^{14}$ $\Omega$cm), and no charges travel through. The insulating material may be an alumina ceramic or a polyimide. When a high voltage of around 3,000 V is applied on the ESC electrode, charges with opposite polarity are induced at the wafer backside surface, and the wafer is clamped because of the Coulomb force between these charges.

With the Johnsen–Rahbek ESC, a material of some level of electrical conductivity (a resistivity greater than $10^9 - 10^{13}\Omega$cm) is inserted between the wafer and the electrostatic chuck electrode. When a voltage is applied on the electrostatic chuck electrode, the charges move through the ceramic and collect at the surface. For this reason, the distance between the positive and negative charges is smaller, and the clamping force is larger. Both types combine the Coulomb force and J-R effect to some extent, the only difference is which one dominates other[5].

As early as 1993, Davie conducted a detailed theoretical analysis of the adsorption principle of ESC and found through experiments that There is a threshold in electrostatic pressure beyond which no improvement of thermal conduction through the wafer/susceptor interface is observed[3]. Allen McTeer's research obtained the relationship between the thickness of the ESC dielectric layer and the clamp force and de-clamping time[4]. Further, Jae Seok Choi carried out the simple geometrical modeling of the contact surface and simulates the contact resistance, the attractive force and the response time according to the variation of contact surface shape.

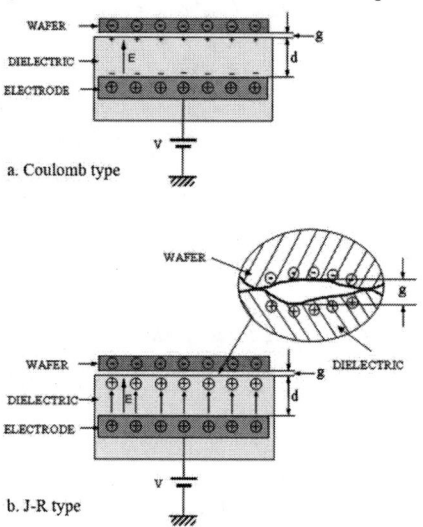

a. Coulomb type

b. J-R type

*Fig 2: The structure of (a) Coulomb and (b) J-R type ESC*

After supply voltage is cut off, residual chuck force usually remains and becomes a serious issue for production efficiency and process reliability. Kesheng

Wang propose a general prediction model and reveal changing laws of residual force with time for both types of bipolar ESC and the error with experiment is within 16%[6]. Gyu Il SHIM by introduced the microscopic bi-layer model introducing three distinct regions to describe the complex de-chuck behavior of the J-R chuck[7]. Xingkuo Wang present a FEA method utilizing COMSOL[8] software to simulate and analyze clamping force of an ESC, compared with experiment, it has high reliability.

## THERMO-FLUIDS NUMERICAL MODELING OF COOLANT CHANNEL

In HDPCVD, because ion bombardment generates very high heat on the wafer, which is undoubtedly fatal to the STI process and Film quality, effective temperature control is essential. Therefore, the research on ESC coolant channels has always been a hot spot. Dae-Hyeon Kim conducted a simulation study on the influence of the coolant channel on the temperature uniformity[9], and further explored its correlation with the viscosity of the coolant. Kuo-Chan Hsu and others have also performed an overall heat transfer simulation analysis on the coolant channel.

However, no one has done the analysis of the flow state of the coolant in the flow channel. In this paper, a spiral block is added to the ESC water inlet to change the flow state and speed of the coolant, and further increase the heat transfer efficiency.

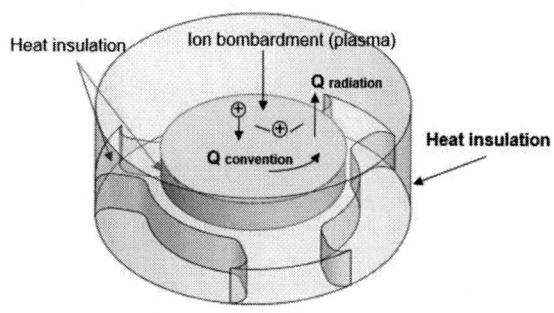

*Fig 3: Domain of the whole ESC system*

In order to further simplify the model and highlight the influence of coolant flow rate and flow state on temperature, several assumptions were made on the operating of the simulation model : Under steady-state conditions of ESC, the wafer surface is evenly heated, ignoring reaction heat and thermal radiation.

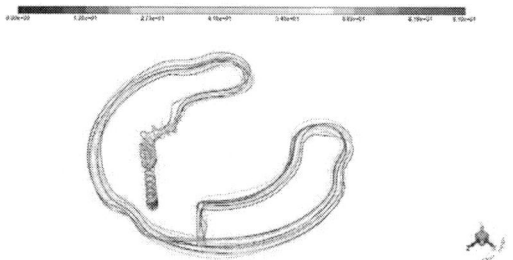

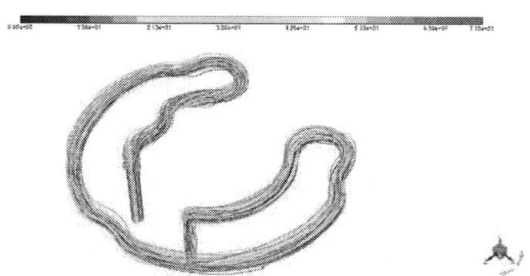

*Fig 4: Path line of coolant without spiral block(down) and with spiral block(up)*

Through Figure 4, it is found that the maximum flow rate of the water channel with spiral block is increased by 15% compared with the normal coolant flow rate. Moreover, a relatively large disturbance occurred at the entrance, which increased the heat dissipation efficiency. In addition, we obtained the temperature distribution of the wafer surface under the two runners, as shown in Figure 4.

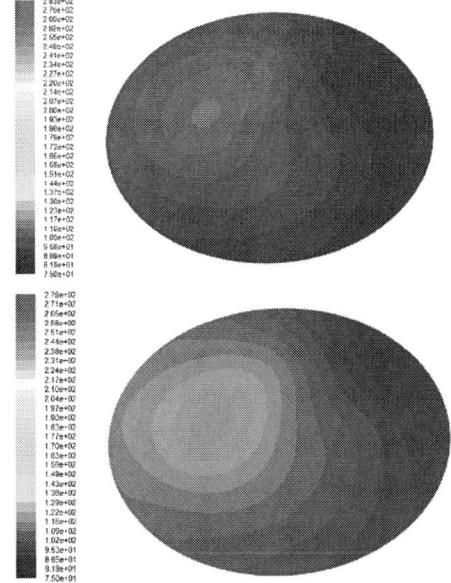

*Fig 5: Wafer surface temperature distribution for without spiral block(up) and with spiral block(down) coolant channel*

## CONCLUSIONS

We summarized the structure, working principle, research progress, advantages, and disadvantages of different types of ESC. Provide assistance to those who have initial contact and understanding of ESC. Lay a solid foundation for further research on the heat transfer of ESC. In addition, we simulated the coolant flow channel with the spiral block. The results show that by adding spiral blocks, the heat transfer efficiency can be increased without changing the shape of the coolant flow channel.

## ACKNOWLEDGFMENTS

This work has been supported by many teachers, colleagues and leaders and appreciated the promptly assistance from everyone.

## REFERENCES

[1] Lee, S.-W., et al. *Thin oxide degradation from HDP-CVD oxide deposition in 300mm process.* in *2004 International Conference on Integrated Circuit Design and Technology (IEEE Cat. No. 04EX866).* 2004. IEEE.

[2] Nojiri, K., *Dry etching technology for semiconductors.* 2015: Springer.

[3] Daviet, J.F., L. Peccoud, and F.J.J.o.t.E.S. Mondon, *Electrostatic clamping applied to semiconductor plasma processing: I. Theoretical modeling.* 1993. **140**(11): p. 3245.

[4] Qin, S. and A.J.J.o.A.P. McTeer, *Wafer dependence of Johnsen–Rahbek type electrostatic chuck for semiconductor processes.* 2007. **102**(6): p. 064901.

[5] Yuchun, S., et al., *Design space of electrostatic chuck in etching chamber.* 2015. **36**(8): p. 084004.

[6] Wang, K., et al., *Prediction of residual clamping force for Coulomb type and Johnsen–Rahbek type of bipolar electrostatic chucks.* 2019. **233**(1): p. 302-312.

[7] Shim, G.I., H.J.P. Sugai, and F. Research, *Dechuck operation of Coulomb type and Johnsen-Rahbek type of electrostatic chuck used in plasma processing.* 2008. **3**: p. 051-051.

[8] Wang, X.K., et al. *Modeling of electrostatic chuck and simulation of electrostatic force.* in *Applied Mechanics and Materials.* 2014. Trans Tech Publ.

[9] Kim, D.-H., K.-s.J.J.o.t.s. Kim, and d. technology, *Study on Coolant Passage for Improving Temperature Uniformity of the Electrostatic Chuck Surface.* 2016. **15**(3): p. 72-77.

# FILM PROPERTY ANALYSIS BY FTIR ON ULK FILM DEPOSITION AND UV CURING PROCESS

*Xinchen Cai1\*; Tianjiao Teng1 ; Xuanyu Lin1; Peipei Li1; Xinyi Chen1; Zhuo Wang1; Huaqiang Tan1; Qing Mi2; Bo Zhang2; Jintao Liu2*

*\*1Piotech Technology Co., LTD., Shuijia 900, Hunnan District, Shenyang city, Liaoning Province, 110171*
*2SMNC Co., LTD., 18 Wenchang Avenue, Beijing Economic and Technological Development Zone, 100176, China*
*\*Corresponding Author's Email: caixc@sypiotech.cn*

## ABSTRACT

Ultra low-k (ULK) film was deposited by plasma enhanced chemical vapor deposition (PECVD) from Diethoxymethylsilane (DEMS, $C_5H_{14}O_2Si$) and α-Terpinene (ATRP, $C_{10}H_{16}$), and cured by UV process to produce porous ULK film. Shrinkage and dielectric constant (k) of ULK film were investigated using ellipsometer, C-V measurement techniques. The film composition was investigated by Fourier Transform Infrared Spectroscopy (FTIR), and results showed lower Si-CH$_3$ and CH$_x$ peaks and conversion from Si-O cage to Si-O network, with longer UV cure time, achieving lower k and higher shrinkage.

*Keywords: ultra low-k; UV cure; FTIR; shrinkage; dielectric constant*

## INTRODUCTION

The semiconductor industry has been improving the ultra-large-scale integrated (ULSI) circuits by shrinking the transistor size according as predicted by Moore's law, and RC delay (the signal delay of BEOL) is increasing as the shrinking of ULSI chips. A significant improvement in the performance of BEOL was achieved by replacing SiO films with low-k films, reducing capacitance and RC delay [1]. There are two major methods of lowering k value: one way is to decrease dipole strength, using lower polarizability than Si-O bonding, where Si-O has been replaced by Si-F, Si-C, and other nonpolar bonds such as C-C, and C-H in organic materials. Another way is to decrease the number of dipoles, meaning lower film density, which increases the free volume through rearranging the material structure or introducing porosity [2]. Porous ULK film was introduced beyond 45 nm technology node, to achieve k =< 2.5 and lower [3].

In this study, porous low-k films were produced from Diethoxymethylsilane (DEMS, $C_5H_{14}O_2Si$) and α-Terpinene (ATRP, $C_{10}H_{16}$) by PECVD and UV curing process, and film structure was investigated by Fourier

Transform Infrared Spectroscopy (FTIR), and other properties, such as dielectric constant and shrinkage were also studied.

## EXPERIMENTAL

The ultra low-k films were deposited using DEMS as framework precursor and ATRP as porogen by PECVD technique (molecular formula shown as *Figure 1*). The precursors DEMS and ATRP are carried to react chamber by He flow, to form composite films, the ratio of ATRP/DEMS is 0.0, 0.2, 0.4, 0.6, 0.8, 1.0, 1.2. O2 is usually used as network structure optimizer. RF power was kept at a level sufficiently to produce SiOCH films and not break organic porogen molecules to obtain pore. The films were deposited at 260℃ on Si (100) substrates on Piotech PF-300T eX LoK II system.

When the composite ULK films were exposed to high temperature, UV light or other types of energy, DEMS formed crosslinked framework structure in the film, and ATRP as porogen molecules escaped from the bulk film, leaving holes with a diameter about 1nm in the cross-linked structure, thus a porous film produced. At the same time, as porosity formed, ULK film was trimmed at high temperature to produce denser films, with Si-O cage transferring to Si-O network gradually. The curing time is from 0 to 320s. The ULK films were UV cured at 385 ℃ on

*(a)*

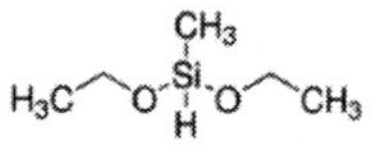

*(b)*

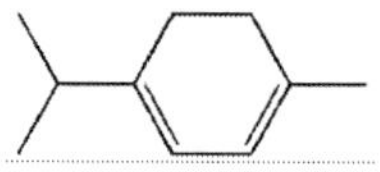

*Figure 1: (a) DEMS and (b) ATRP molecular formula*
Piotech system.

Ultra low-k films thickness and refractive index of

633nm are measured by ellipsometer with KLA-Tencor Aleris 8500. Shrinkage is calculated from the difference in film thickness before and after UV-curing. Films composition is investigated by FTIR, with Nicolet 380 spectrometer using DTGS detector for mid-IR (4000 - 400 cm$^{-1}$) data collection, dielectric constant k is measured by C-V measurement technique with SEMILAB 530CV.

## RESULTS AND DISCUSSION

Structural characterization of amorphous films was extremely difficult and FTIR spectrum was used to analyze the conversion of ULK film before and after UV curing in this study. As shown in *Figure 2 (a) and (b)*, after UV treatment, the absorption peaks Si-CH$_3$ (~1275 cm$^{-1}$) of ULK films were significantly lowered, indicating that a large amount of organic components has been removed from ULK during UV cure process [4]. However, Si-CH$_3$ (~1275 cm$^{-1}$) absorption peaks still exist, indicating that some organic structures are retained in ULK films. The peak of Si-O cage (~1125 cm$^{-1}$) decreased, and Si-O network (~1042cm$^{-1}$) increased significantly. This result showed terminal groups of ULK film decreased, and Si-O Cage gradually changed to Si-O network, and then framework structure was repaired and its mechanical strength could be enhanced.

*Figure 2 (c)* shows the UV curing process reduced C-H$_x$ (2800-3000cm$^{-1}$) peak intensity. That indicates that pore-forming organic molecules escaped from the ULK film partially, and there were still organic terminal groups, such as Si-CH$_3$ in ULK film after UV curing.

DEMS and ATRP flow rate is the key parameter on ULK films properties. *Figure 3* shows the ratio of ATRP / DEMS on ULK films shrinkage and post UV k. k decreased and shrinkage increased with higher ratio of ATRP / DEMS, because porosity and pore size increased with higher ATRP / DEMS.

Films structure was analyzed by FTIR, and peak area ratio of SiCH$_3$ / Si-O and C-H$_x$ / Si-O were summarized in *Figure 4. Figure 4 (a)* shows peak area ratio of Si-CH$_3$ / Si-O pre and post UV curing. Higher ratio indicates more terminal group Si-CH3 exists in ULK film both pre and post UV, using higher ATRP / DEMS ratio. *Figure 4 (b)* shows ratio of C-H$_x$ / Si-O pre and post UV curing. The results indicate that more porogen was composited in ULK as-dep film with higher ATRP ratio, and escaped after UV curing. The remaining C-H$_x$ peak was mainly generated by DEMS, not ATRP.

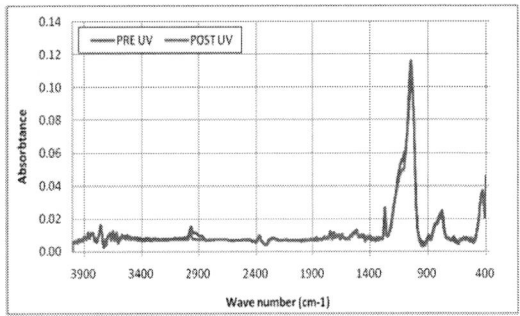

*(a)*

*(b)*

*(c)*

Figure 2: FTIR spectra of ULK film before and after UV curing: (a) Whole FTIR spectrum; (b) FTIR of 1400cm$^{-1}$ ~ 600 cm$^{-1}$; (c) FTIR of 3200cm$^{-1}$ ~ 2600 cm$^{-1}$

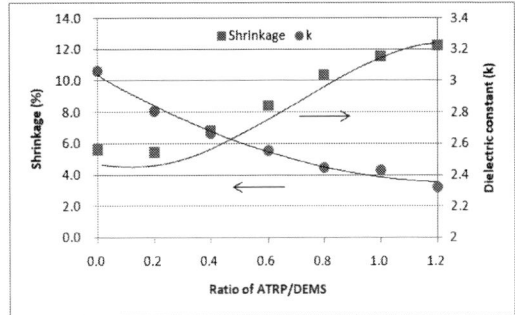

Figure 3: Film shrinkage and k with different ratio of ATRP/DEMS

Shrinkage of ULK films increased with UV cure time, and increased rapidly at the beginning of UV cure process. Dielectric constant (k) decreased with UV curing time, and decreased significantly at beginning of UV cure process. The results were shown in *Figure 5*. Shrinkage and k curves showed pore-forming organic was reacted under UV radiation and escaped from the composited film. And then pore is formed gradually, film structure is re-organized and shrank, with k decreasing and porosity increasing [5].

FTIR spectra were collected to investigate to analyze the composition conversion during UV cure process (shown as *Figure 6*). The FTIR spectra showed C-H$_x$ (2800-3000cm$^{-1}$) and Si-CH$_3$

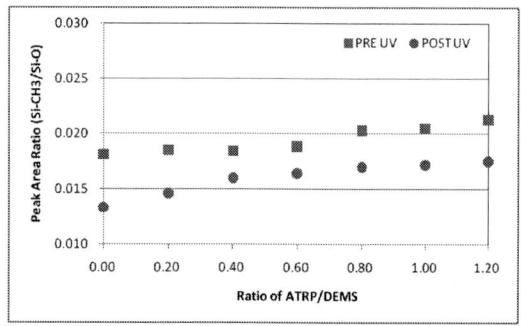

*(a)*

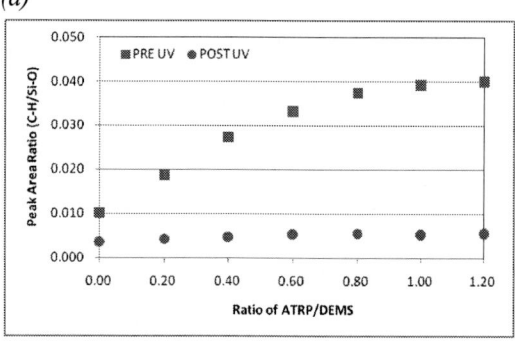

*(b)*

*Figure 4: (a) area ratio of Si-CH$_3$/Si-O on FTIR; (b) area ratio of CH$_x$ / Si-O*

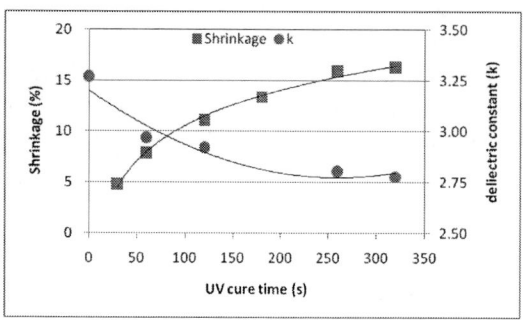

*Figure 5: Film shrinkage and k with different UV curing time*

(~1275 cm$^{-1}$) peak intensities decreased gradually with longer UV cure time. The results also showed that Si-O cage has been converted to Si-O network (~1042cm$^{-1}$) from Si-O cage (~1125 cm$^{-1}$) gradually with longer UV cure time [6]. These results indicated that as the UV cure processing, the organic components has been removed partially and creating pores, and film shrank due to conversion from Si-O cage to Si-O network, to repair framework structure and achieve denser ULK film.

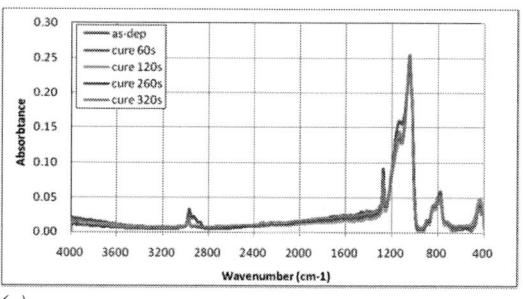

*(a)*

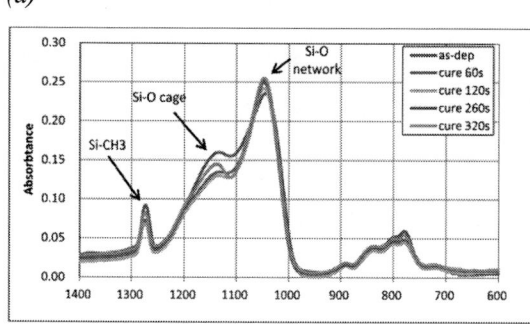

*(b)*

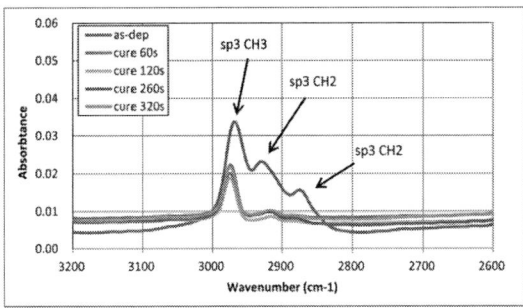

*(c)*

*Figure 6: FTIR spectra of ULK films with different UV curing time: (a) Whole FTIR spectrum; (b) FTIR of $1400cm^{-1} \sim 600 cm^{-1}$; (c) FTIR of $3200cm^{-1} \sim 2600 cm^{-1}$*

## CONCLUSION

In this study, we investigated the films properties such as composition, k and shrinkage on ULK film deposition and UV cure process. The FTIR spectra indicated that organic components have important role in film properties, especially on dielectric constant and shrinkage. The porogen has been removed gradually, and pores formed during UV curing process, and film achieved lower k. ULK films became denser and shrinkage got higher due to conversion from Si-O cage to Si-O network, with skeleton structure repairing.

## REFERENCES

[1] Grill A. Low and ultralow dielectric constant films prepared by plasma-enhanced chemical vapor deposition[J]. Dielectric films for advanced microelectronics, 2007: 1.

[2] Shamiryan D, Abell T, Iacopi F, et al. Low-k dielectric materials[J]. Materials today, 2004, 7(1): 34-39.

[3] Wong T K S, Liu B, Narayanan B, et al. Investigation of deposition temperature effect on properties of PECVD SiOCH low-k films[J]. Thin Solid Films, 2004, 462: 156-160.

[4] Grill A, Neumayer D A. Structure of low dielectric constant to extreme low dielectric constant SiCOH films: Fourier transform infrared spectroscopy characterization[J]. Journal of applied physics, 2003, 94(10): 6697-6707.

[5] Hrubesh L W, Keene L E, Latorre V R. Dielectric properties of aerogels[J]. Journal of materials research, 1993, 8(7): 1736-1741.

[6] E. Todd Ryan1,etal., Journal of Applied Physics, 2014, 115: 144107.

# ANALYSIS OF INFLUENCE TO GATE STACK OF EEPROM USING CL DOPED RE-OXIDATION

*Hong Zhang [1*], and Po-Yu Huang [2]*

[1]School of Microelectronics, Fudan University, Shanghai 200433, China
[2]Technical Marketing, Zing Semiconductor Corporation, Shanghai 201306, China
*Corresponding Author's Email: hzhang18@fudan.edu.cn

## ABSTRACT

Chlorine doped oxidation by adding chloride from HCl or C2HCl2 (DCE) in oxidation process can improve SiO2 quality. The process will lead to chloride-rich zone near the Si-SiO2 interface and bond with dangling Si, which reduces the density of interface charge. In this paper, the different ratio of chloride doped of oxidation by test wafer with pre-Oxide to simulate the process flow of 18um EEPROM is applied to observe and compare the re-oxidation rate and charge change of the final Oxide. The device performance of threshold voltage and breakdown voltage are also observed by 18um EEPROM test vehicle. According to the experimental results of test wafer, as the increase of DCE ratio, the re-oxidation rate and charge first decrease but tends to be flat. When applied to 18um EEPROM, the phenomenon of threshold voltage and breakdown voltage increase at first and slow down, which is match with the result in control wafer.

## INTRODUCTION

With the development of integrated circuits, especially VLSI, the dimensions continue scaling down and the requirement of lower processing temperature to further improving the quality of thermally silicon dioxide [1]. As the purpose, some approach to improve thermal process of the silicon dioxide to reduce the defects, stress and impurity redistribution effect, including, Chlorine doped oxidation, High-pressure oxidation (>1 ATM) to reduce the process temperature (650-950 ℃), N2/Ar2 diluting, and plasma oxidation [1]. Chlorine doping should be the most widely used in the industry to applied to thermal oxidation to improve the film quality compared with HCl for the concerned of corrosion of the gas pipeline. The oxidation rate can be accelerated by adding CL 1-5% in dry oxygen oxidation.〔2,3〕

16K EEPROM (electrically erasable & programmable read only memory) has been developed based on the floating-gate concept and MNOS (metal-nitride-oxide semiconductor) structure. The MNOS memory cells can be electrically erased, while floating-gate cells are usually erased and programmed by F-N electron tunneling or channel hot electron injection（CHEI）. The structure of typical EEPROM is as shows in Figure 1.

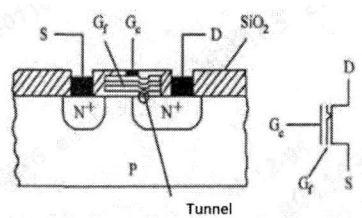

*Figure 1: EEPROM structure*

In this paper, the chlorine-doped oxidation process is applied to the tunnel oxide process of 0.18um EEPROM. The effects of different DCE flow ratio on the re-oxidation rate and charge change are compared. The performance difference for the threshold voltage and breakdown voltage in 0.18um EEPROM also are observed.

## EXPERIEMNT DESIGN

### Flow of Test device

Two transistors contained in EEPROM unit, floating-gate transistor and select-gate transistor, as shown in Figure 2.

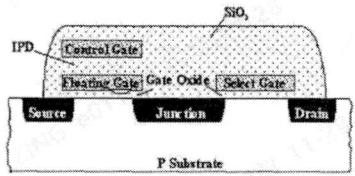

*Figure 2: EEPROM cross section, Wang Fang, 2006*

This study discusses the effect of chlorine doping ratio on re-oxidation and devices for 0.18um EEPROM, and the process flow is introduced as Figure 3

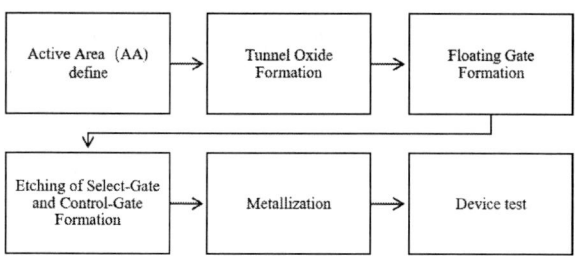

*Figure 3: Process flow of experiment*

Active area (AA) definition: The Active Area (AA) definition is the first step of 0.18um EEPROM process. The depth of AA will affect the final programming voltage and the performance of the device.

Tunnel oxide formation: After the thick gate oxide formed, a small area is opened on the gate oxide by wet etching to expose the silicon surface to re-oxidize to form the tunnel oxide.

Floating gate formation: In EEPROM memory cell, the floating gate realizes electron transfer between the tunnel oxide layer and the silicon substrate. Thus, the floating gate must cover the tunnel oxide layer, and the tunnel oxygen is a part of the floating gate oxide layer.

Etching of select gate and control gate: The EEPROM unit consists of two transistors, floating-gate transistor and select-Gate transistor. The select-Gate transistor is used to select the corresponding floating-gate transistor during programming and erasing.

### Experiment design

In the process of 18um EEPROM, double oxidation process is adopted to complete gate oxygen growth. This paper discusses the result of different micro reactions during the re-oxidation of the front layer of tunnel oxide in a small area, which eventually leads to the threshold voltage drift [4]. The experiment is divided into two parts. In the first part, the control-wafer is apply to simulate the first oxide layer in the device, as shown in Fig. 4. After measuring the thickness, different ratio of DCE-oxygen mixture is set in the tunnel oxide layer recipe to re-oxidation. After re-oxidation, the results are measured by ellipsometer (S200, Rudolph) for thickness and SPV (Quantox，KLA）for charge characteristics of the oxide film. The result of thickness adding and charge characteristic are summarized and compared. The test flow is as Figure 4. The second step is to directly apply it to the 180nm EEPROM device process, and measure the electrical characteristics of the device, including the influence of opening voltage and breakdown voltage.

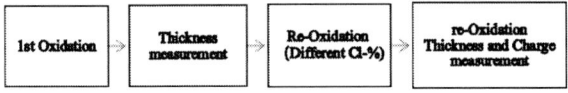

*Figure 4: test flow, prepared by experiment deign*

### Methodology of Oxide charge test

The charge test of oxide layer is carried out by the non-contact dielectric characteristic tester (Quantox) produced by KLA company [5, 6, 7]. It obtains the density of states at the Si-SiO2 interface, the total charge of oxide layer, the mobile charge of oxide layer and the lifetime of carriers in the substrate (related to carrier generation and recombination) by applying thermal, charge electron and optical excitation. At the same time, it can also test the C-V characteristic curve of the oxide layer. Through the charge test, we can comprehensively understand the charge characteristics of the oxide layer [8-11]. It measures the surface voltage of the oxide layer through a reciprocating metal electrode plate.

## RESULT AND DISCUSSION

The correlation experiments of oxidation reactions with different DCE-oxygen flow rates were carried out. The results were observed and compared by the re-oxidation rate and charge changes, and then applied to 18 EEPROM devices to study the effects on the Threshold voltage and breakdown voltage as table 1.

*Table 1 DOE Result at test wafer and device test*

| Condition | Control wafer | | Device Test | |
|---|---|---|---|---|
| DCE Ratio | Re-Ox THK | Charge | Vbd | Vth |
| 0.5% | 69.07 | 1.800 | 27.458 | -0.747 |
| 1.5% | 65.91 | 1.400 | 27.464 | -0.708 |
| 2.0% | 63.62 | 1.327 | 27.462 | -0.696 |
| 3.8% | 64.09 | 1.500 | 27.463 | -0.668 |

The DCE ratio is set as 0.5%, 1.5%, 2.0% and 3.8%, the result of re-oxidation thickness are 69.07A, 65.91A, 63.62A and 64.09A based on the test on control wafer. The trend of the re-oxidation is decreasing as the DCE ratio increase from 0.5% to 2.0% but increase as DCE ratio increase to 3.8%, as shows in Figure 5. The similar phenome shows in result of charge monitor by Quantox (KLA), the result is decreasing as the DCE ratio increasing but increasing when DCE ratio continue increasing from 2.0% to 3.8%, as shows in Figure 6.

Linear Regression and Polynomial Regression are applied to verify the relationship between the factors with re-oxidation vs DCE ratio and charge status vs DCE ratio. The result shows that, the $R^2$ of Linear regression analysis is low relationship (0.54 and 0.21) but high R2 (0.96 and 0.99) by the Polynomial Regression analysis (for the Re-oxidation-DCE Ratio and Charge result- DCE).

When applied to the preparation of Tunnel oxide layer of 18EEPROM, it is found that the threshold voltage and breakdown voltage first rise and then slowdown, which is match to the trend of re-oxidation rate, as shows in Figure 7 &8. The similar regression analysis had been applied to the result of device test. For the Threshold Voltage, the R2 is 0.99 by Polynomial Regression analysis and 0.93 by Linear Regression analysis. For the test of breakdown voltage, the R2 is 0.75 by Polynomial Regression analysis and 0.38 by Linear Regression analysis.

The result above indicate, a chlorine enrichment region is occurred near the Si/SiO$_2$ interface, where a large number of bonding between chlorine and $SiO_2$, which reduces the unsaturated silicon dangling bonds at the interface and reduces the state density at the interface, and reduces the total charge. The doped Chlorine in SiO2 may lead to negative center to capture positive charge of oxide layer and reduces the PMOS threshold voltage （$V_{TP}$）, increases NMOS threshold voltage $V_{TN}$. When the Chlorine doped increasing to more than limit, the dangling bond increasing instead and result to higher re-oxidation rate and charge monitoring [12, 13].

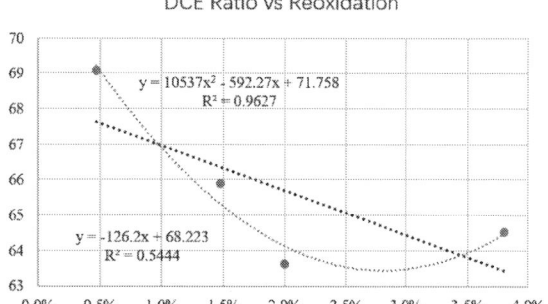

Figure 5: DCE Ratio Vs Re-oxidation

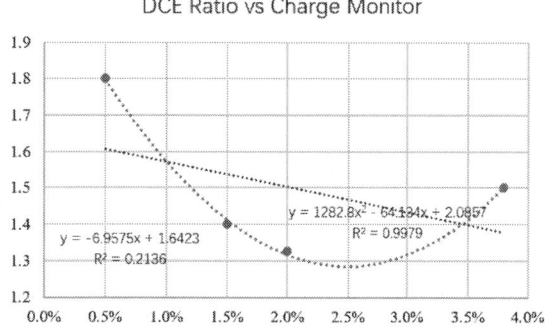

Figure 6: DCE vs charge monitor

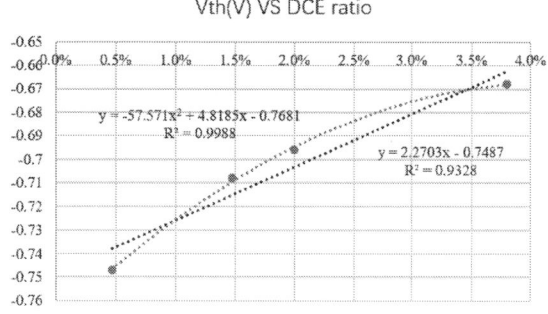

Figure 7: Threshold voltage Vth vs DCE Ratio

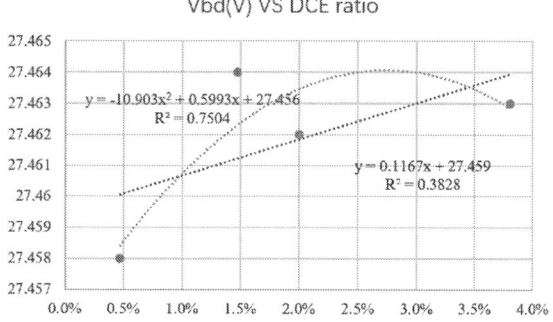

Figure 8: Vbd vs DCE Ratio

## CONCLUSION

Integrated circuits have high requirements for the quality of thermal Silicon dioxide（SiO2）as dielectrics. The most important factors are the control of the pinholes of SiO2, the mobile charge, fixed charges and traps at the Si-SiO2 interface, which affect the state density. The Chlorine doped oxidation will reduce the dangling bond in the interface of SiO2 and result in the better device performance. This paper focus on the chlorine Dropped oxidation process and research for the different DCE ratio doped oxidation for the re-oxidation process, which well applied in the 18EEPROM device manufacturing.

According to the experimental results, it is found that with the increase of DCE flow rate, the re-oxidation rate and charge decrease first and tends to be increase while the threshold voltage and breakdown voltage first increase and then slowdown, which is the match the mechanism. Further research is suggested to verify the result we claimed with more device characteristics of 18 EEPROM and other device.

## REFERENCE

（1）Wang Fang, " Research on size reduction and process integration improvement of flash memory devices", *Shanghai FuDan University 2006*

（2）Zhuang Tongzeng, Zhang Ankang, Huang Lanfang, " principle and practice of integrated circuit manufacturing technology", *Electronic Industry Press*, 1987:83,95-96, 103-105, 110

（3）Ma Z , Chapman G , Ho V., " HCI characterization of Trans-LC and HCl gate oxidation process", *Integrated Reliability Workshop Final Report*, IEEE International. IEEE, 1999.

（4）Zhaoyong Lu, PoYu Huang, Binzong Li, "Impact of

Doped Silicon Oxynitride Gate Dielectric on the 0.13um CMOS Device 1/f Noise Characteristic", *CSTIC 2010*, pp97.

〔5〕Florence Cubaynes, Sophie Passfort, etc. "In-Line Characterization of ultra thin gate dielectric films", *2002 IEEEE/SEM Advanced semiconductor Manufacturing Conference.*

〔6〕L. Kronik, Y. Shapira, "Surfacephotovoltage spectroscopy of semiconductor structures: at the crossroads of physics", *chemistry and electrical engineering. Surf. Interface Anal.*, 2001, pp. 954-965.

〔7〕Siti, Kudnie, Sahari. "Quantox Analysis of SC-1 and SC-2 Variables in wafer Surface Preparation", *IEEE 2008* 1-42442561-7/08

〔8〕Tom G. Miller, etc. "Countless method for measuring total charge of an oxide layer on a semiconductor wafer using charge", *United states patent US006335630 B2* Jan 1,2002

〔9〕LI Qinghua, LIU Jian, LUO Junyi, LI Qing, ZHENG Guoxiang, "The Application of Surface Photovoltage Technology in Monitoring Gate Oxide and Oxidation Furnace System", R*esearch & Progress of SSE*，2003，pp.120-125.

〔10〕Xiong Hai, Kong Xuedong, Zhang Xiaowen, "Application of High Frequency Capacitance-Voltage Characterization of MOS Structure", *Semiconductor Technology* 2010, pp.94-98.

〔11〕2. L. Kronik, Y. Shapira, "Surfacephotovoltage spectroscopy of semiconductor structures: at the crossroads of physics, chemistry and electrical engineering", *Surf. Interface Anal.*, 2001, pp.954-965.

〔12〕Ling Xiaoyu, "study on the effect of gate reoxidation on EEPROM threshold voltage", *Shanghai Fudan University*, 2011

〔13〕Jingxiu Tin, Zhaoyong Lu, Poyu Huang, Bingzong Li, "Study of Nitride-Gate Dielectric Reliability and process stability monitoring of Nitrogen", *CSTIC 2010*, PP86.

# THE EFFECT OF STI DIVOT ON PLANNER LOGIC DEVICE PERFORMANCE STUDY

*Zhenchao Sui[12]\*, Yongqiang Che[1], Wenrong Hou[1], Weichi Cheng[3], Jingang Wang[1]*

[1]Semiconductor Manufacturing North China (Beijing) Corporation, Beijing 100176, China
[2]School of Software and Microelectronic, Peking University, Beijing 102600, China
[3]Semiconductor Technology Innovation Center (Beijing) Corporation, Beijing 100176, China
\*Corresponding Author's Email: mark_sui@smics.com

## ABSTRACT

In this paper, different etching methods and adjustments of the etching amount have been tried and the effect of increased depth of the STI divot on the performance of the planar logic device in the STI process was studied. The results showed that under the same etching amount on a blank wafer, wet etching would obtain larger divot features than dry etching. This paper also studied the relation is about 0.78 between blank wafer etching amount and structure wafer divot depth on the 28nm logic process. There are 3~5% performance improvement by STI divot increased around 30A of Core device nominal P and SRAM PU was been shown on 28nm HKMG platform. And the GOI TDDB of PMOS devices got significant improvement.

## INTRODUCTION

The size of semiconductor chips continues to shrink, but still keeps its high performance and low power consumption. It makes the semiconductor process continue to challenge the limits, so the process parameters of each stage also need to be more finely optimized and adjusted for extreme performance improvement. The focus of this article is the optimization of 28nm HKMG STI divot features for device performance improvement.

As well-known, STI divot will induce some side effects, such as junction leakage, the poly residue, inverse-narrow-width effect, and double-humped I-V curve…etc. [1]-[5].

There are many discussions to reduce STI divot [6][7], but the STI divot is also increasing the short-channel area between poly gate and Si surface, and the advantages of improving channel performance are rarely mentioned [8]. This paper will discuss the effect of etching methods for the STI divot feature and the impact of divot depth on the performance of 28m HKMG logic devices.

## EXPERIMENTS

To study the effect of gate1 oxide(OX) pre-clean process and oxide loss on STI divot, four split conditions were designed as shown in TABLE I. Split 0 represented the DRY etching and WET etching processes. Split1 changed DRY etching to WET etching process while keeping the oxide loss amount the same. Split2 and Split3

increased the WET etching oxide loss amount separately by about 20A and 50A compared to Split1. The experiments were processed on a 28nm HKMG pattern wafer.

TABLE I. STI DIVOT SPLIT CONDITIONS

| Condition | Dry etching OX loss (A) | Wet etching OX loss (A) | Divot (A) | Divot-D0 (A) |
|---|---|---|---|---|
| Split0 | L1 | L2 | D0 | 0 |
| Split1 | 0 | L1+L2 | D1 | 24 |
| Split2 | 0 | L1+L2+20 | D2 | 30 |
| Split3 | 0 | L1+L2+50 | D3 | 63 |

## RESULTS

### Physic Comparing by Etching Splits

Figure1 shows the STI divot TEM results of the different gate1 oxide pre-clean process split. The comparison of Figure1 (a)~(b) shows that under the same etching amount on a blank wafer, wet etching obtained 24A larger divot than dry etching. Figure1 (b)~(d) shows that the STI divot increased by the WET etching oxide loss amount increased, and the ratio of the divot depth increase by the wet etching amount of blank wafer was about 0.78.

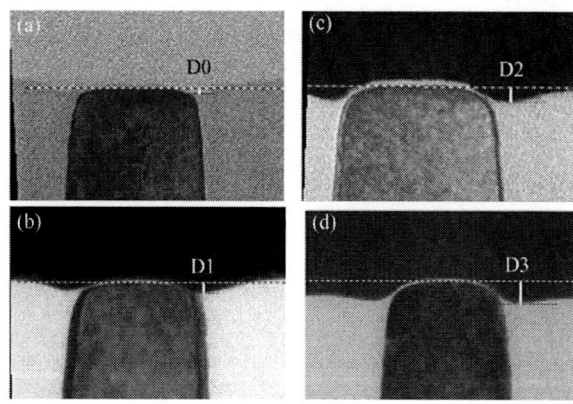

*Figure 1: STI divot TEM results: (a) Split0 (b) Split1 (c) Split2 (d) Split3*

**Core Device by Etching Splits**

Figure2 shows the nominal core device (poly length 0.03um) performance of different STI divot split on 28HKMG pattern wafer. When STI divot increased around 30A, then PMOS performance increased 3~5%, while NMOS device performance kept the same.

**SRAM Device by Etching Splits**

SRAM performance shows the same trend as a core device. The results were shown in Figure3, SRAM PU Vt-Id and Id-Ioff performance gained 3~5%, while PD/PG performance kept the same.

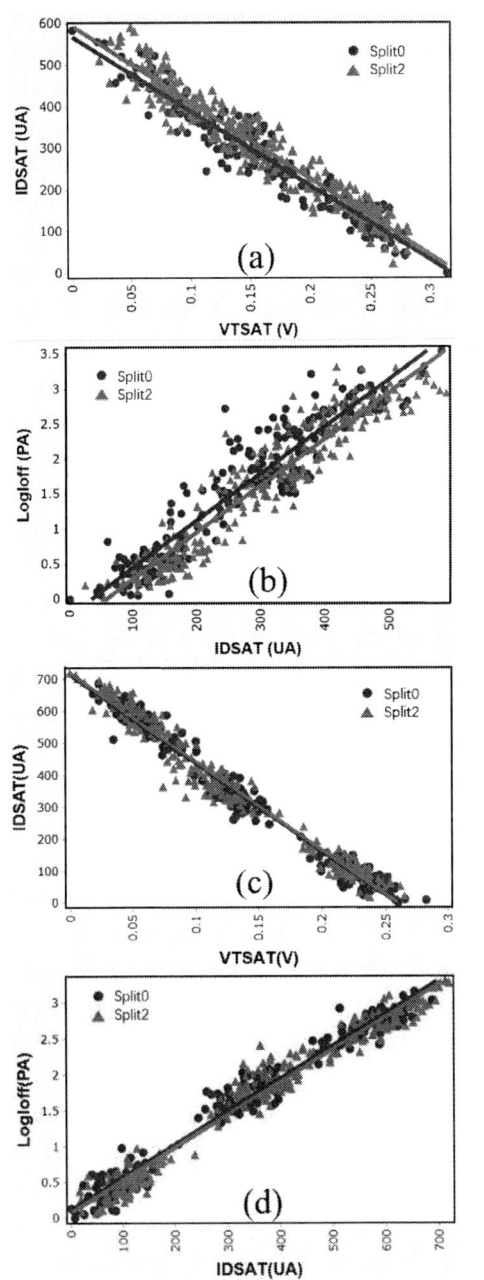

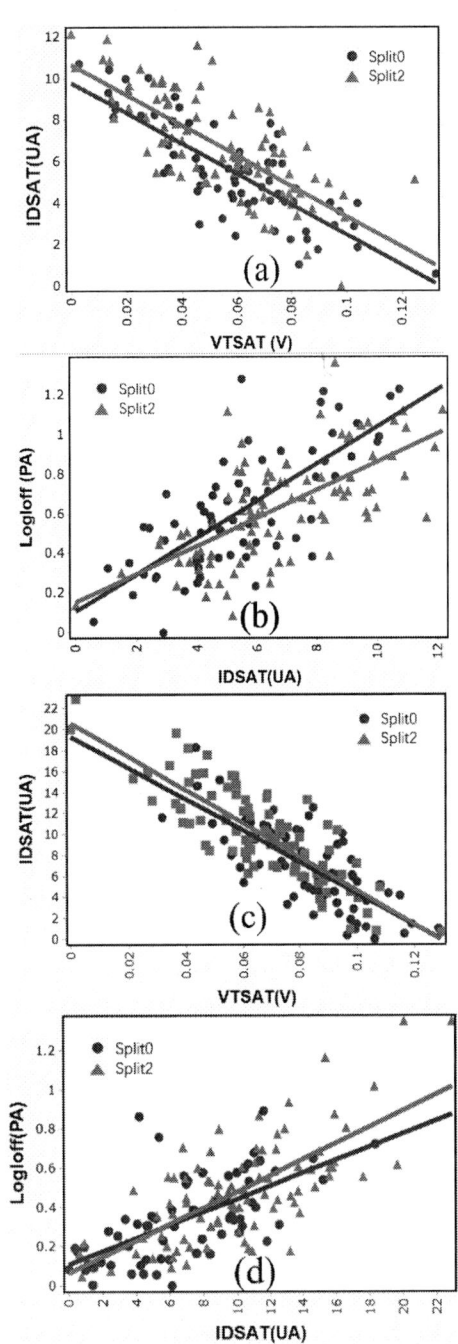

*Figure2: Core device performance of different STI divot split: (a) nominal P Vt vs. Id; (b) nominal P Id vs. Ioff; (c) nominal N Vt vs. Id; (d) nominal N Id vs. Ioff.*

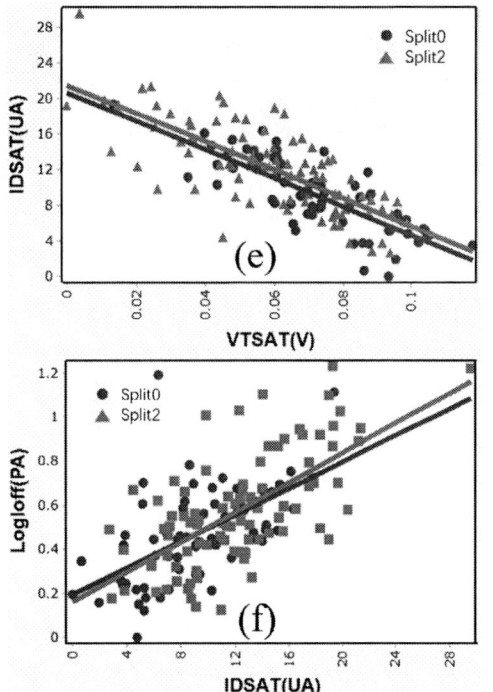

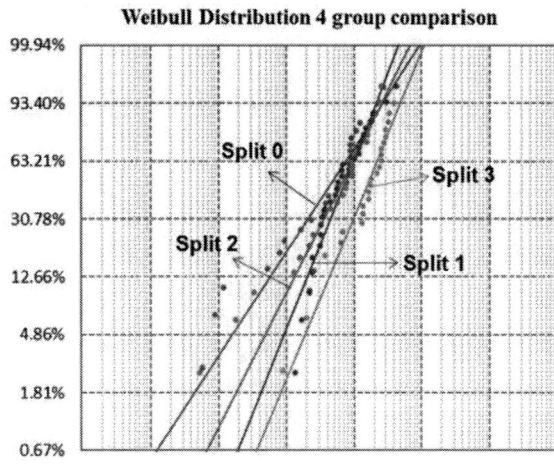

*Figure3: SRAM performance of different STI divot split: (a) PU Vt vs. Id; (b) PU Id vs. Ioff; (c) PD Vt vs. Id; (d) PD Id vs. Ioff; (e) PG Vt vs. Id; (f) PG Id vs. Ioff;*

### Reliability by Etching Splits

Figure4 shows the PMOS GOI TDDB of different STI divot split on 28HKMG. It shows that the T63% of Split1/Split2/Split3 were significantly improved than split0. The improvement of PMOS GOI TDDB was getting obvious when the STI divot depth was highly increased.

*Figure4: PMOS GOI TDDB of different STI divot split*

## CONCLUSIONS

The effect of STI divot on the planner logic device performance was studied in this paper. The results showed that under the same etching amount of blank wafer, wet etching would obtain larger divot depth than dry etching. For a planar 28nm logic process, the ratio of the divot depth increase by the wet etching amount on a blank wafer was about 0.78. In the 28HKMG logic process devices, when STI divot increased around 30A, PMOS SRAM PU performance also increases 3-5% higher, NMOS device performance kept the same, and the GOI TDDB of PMOS devices was significantly improved. The model was STI divot increased the short-channel area between poly gate and Si surface. It is slightly like a FinFET structure.

## REFERENCES

[1] T. Hamamoto, S. Sugiura and S. Sawada, "Well concentration: A novel scaling limitation factor derived from DRAM retention time and its modeling", *IEDM Tech. Dig.,* pp. 915, 1995.

[2] L. A. Akers, "The inverse-narrow-width effect", *IEEE Electron Device Lett.,* vol. 7, pp. 419, 1986.

[3] N. Higyo and T. Hiraoka, "A review of narrow-channel effects for STI MOSFET's: A difference between surface- and buried-channel cases", *Solid State Electron.,* vol. 43, pp. 2061, 1999.

[4] T. Oishi, K. Shiozawa, A. Furukawa, Y. Abe and Y. Tokuda, "Isolation edge effect depending on gate length of MOSFET's with various isolation structure", *IEEE Trans. Electron Devices,* vol. 47, no. 4, pp. 822, 2000.

[5] K. Horita, T. Kuroi, Y. Itoh, K. Shiozawa, K. Eikyu, K. Goto, et al., "Advanced shallow trench isolation to suppress the inverse narrow channel effects for 0.24 um pitch isolation and beyond", *VLSI Tech. Dig.,* pp. 178, 2000.

[6] C.H. Li et al., "A robust shallow trench isolation (STI) with SiN pull-back process for advanced DRAM technology," *The 13th Annual IEEE/SEMI ASMC 2002 Conference,* Boston, MA USA, p.21-26, 2002.

[7] Chienfan Yu et al., "Formation and reduction of embedded contamination defects detected after FEOL poly patterning," *The 17th Annual IEEE/SEMI ASMC 2006 Conference,* Boston, MA USA, p.206-210, 2006.

[8] N. Shigyo et al., "Three dimensional analysis of sub-threshold swing and trans-conductance for fully recessed oxide (trench) isolation 1/4-um-width MOSFET's", *IEEE Trans. Electron Devices,* vol. 35, pp. 945, 1988.

# 300 MM WAFER LASER ANNEAL PROCESS DEVELOPMENT FOR APPLICATIONS OF MULTIPLE PROCESS CONDITION IN DIFFERENT ZONES ON SINGLE WAFER

*Xiaoxu Kang[1]\*, Yuanhao Huang[2], Wen Luo[2], Jianrui Liu[2], Zhenghui Chu[1], Xiaolan Zhong[1], Min Zhang[1], Xiaoqiang Zhou[1], Kaiyan Zang[1], Ming Li[1], Limin Zhu[3], Hanwei Lu[3], Bo Zhang[3]*

[1] Process Technology Department, Shanghai IC R&D Center, Shanghai, 201210, China

[2] Shanghai Micro Electronics Equipment (Group) Co., Ltd., Shanghai, 201203, China

[3] PIE Department III, Shanghai Huahong Grace Semiconductor Manufacturing Corporation, Shanghai 201206, China

*Corresponding Author's Email: kangxiaoxv@icrd.com.cn

## ABSTRACT

With technology developing, laser annealing is attracting more and more interests because of its property of lower thermal budget, near surface activation for shallow junction application, rapidly cooling rate, etc. In this work, the wafer after implantation process was divided into different zones, and process was developed to implement different laser annealing condition in different zones. The effects of different laser annealing parameters on the sheet resistance and RS uniformity of the different zones on implanted wafer were studied and evaluated, and Secondary Ion Mass Spectrometry (SIMS) was used to check its doping profile after annealing. From the measured data, this method can have potential value for tool matching, device splits and related applications.

## INTRODUCTION

With developing of CMOS technology[1], thin film transistor display technology[2] and large area OLED display technology[3], Laser thermal annealing[4] is beginning to replace the traditional rapid thermal annealing tool and become mainstream annealing technology, which is because of its properties of high energy density, very short time and little influence on ion doping profile[5].

Because of these properties of the laser annealing, new requirement is raising to realize different annealing process on different wafer zone. This can be used in tool matching application or device splits application which can avoid wafer-to-wafer process variation influence on split results.

In this work, single wafer after ion implantation process was divided into multiple zones, and different process was implemented in different zones. Process parameters on the sheet resistance of different zones were studied and evaluated. After that, RS matching and SIMS matching were done to check the tool matching performance.

## EXPERIMENTS

300 mm implanted wafers were prepared with Boron element, and the implantation energy and dose was 2 KeV and $1 \times 10^{15}$ ions/cm$^2$ separately. One of the wafers was defined as baseline wafer which was treated with 1200 C/200 μs with mass production tool, and there was no zone division for baseline wafer. The other wafers were defined as tested wafer, and were used to do the experiments on laser annealing tool of Shanghai Micro Electronics Equipment at similar process condition. 4 and 8 zones on single wafer were used in this work, which was schematically shown in Fig.1. Laser annealing process was performed in these zones with different laser power. Four Probe Method was used to measure the sheet resistance of the wafer after annealing by 4 Dimensions 333A. SIMS was used to measure the matching performance between tested wafer and baseline wafer by EAG Laboratories in Shanghai.

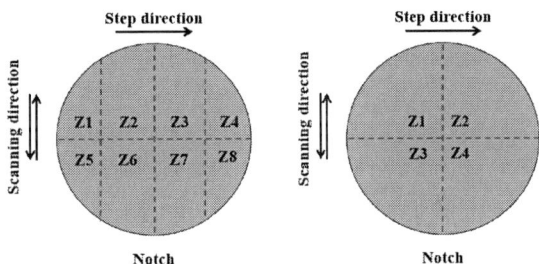

*Figure 1: Schematic diagram of wafer zones*

During the annealing process, each zone will be scanned with a serpent path by focused laser spot, and laser power was adjusted for different zone.

## RESULTS AND DISCUSSION

### 1. Process data for baseline wafer

After laser annealing with mass production tool, RS measurement was performed firstly for baseline wafer for 121 points, which was shown in Fig.2. As shown in the figure, sheet resistance and 1-sigma-uniformity were 292.83 Ohm/sq and 1.20% respectively.

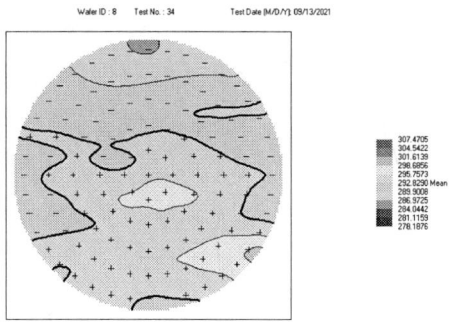

*Figure 2: RS mapping of the baseline wafer*

### 2. Process evaluation for 8 zones on single wafer

The sheet resistance and uniformity of the corresponding zones with different laser power were listed in Table 1. When the power percent was 45.50%, the sheet resistance of the zone 5 was 297.08, which can match the baseline wafer data. However, the uniformity of zone 5 was 3.25%, which cannot meet the process requirement. As shown in Fig.3, the sheet resistance was gradually decreasing with the increase of power percent. But there were great variations for Rs uniformity data of different zones, which may due to the process interaction of adjacent zones because of too many zones division.

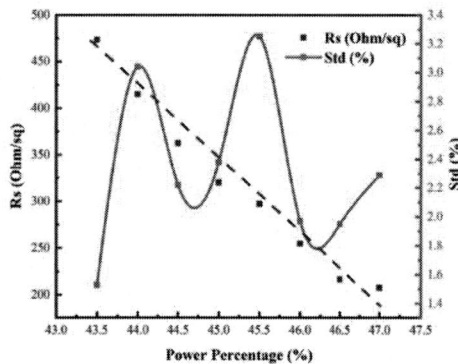

*Figure 3: Sheet resistance variation curve with power percentage in 8 zones.*

*Table 1: Sheet resistance under different power percentage in 8 zones*

| Zone | Power percentage | Rs mean (Ohm/sq) | Stdev (%) |
|------|------------------|------------------|-----------|
| Z1 | 43.50% | 473.15 | 1.53% |
| Z2 | 44.00% | 414.90 | 3.04% |
| Z3 | 44.50% | 362.60 | 2.22% |
| Z4 | 45.00% | 320.01 | 2.38% |
| Z5 | 45.50% | 297.08 | 3.25% |
| Z6 | 46.00% | 254.60 | 1.97% |
| Z7 | 46.50% | 216.31 | 1.95% |
| Z8 | 47.00% | 207.49 | 2.29% |

### 3. Process evaluation for 4 zone on single wafer

*Table 2: Sheet resistance under different power percentage in 4 zones*

| Zone | Power percentage | Rs mean (Ohm/sq) | Stdev (%) |
|------|------------------|------------------|-----------|
| Z1 | 47.50% | 186.12 | 2.05% |
| Z2 | 48.00% | 174.35 | 1.33% |
| Z3 | 48.50% | 168.57 | 1.00% |
| Z4 | 49.00% | 163.04 | 0.58% |

After laser annealing of 4-zone wafer, sheet resistance and Rs uniformity data was shown in Fig.4 and Table 2. As shown in the figure, sheet resistance and uniformity decrease with the increment of power percentage.

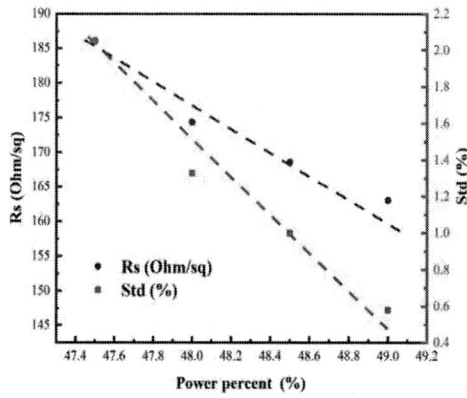

*Figure 4: Sheet resistance variation curve with power percentage in 4 zones.*

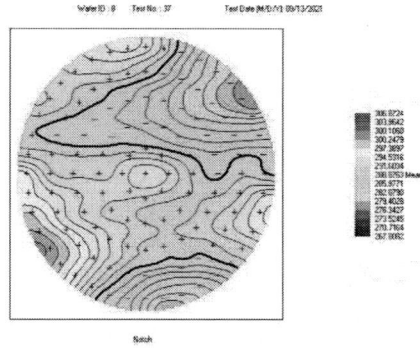

*Figure 5: RS mapping of the single wafer*

According to the data of zone divided wafers in Table 1 and Table 2, the best matched process condition was 45.50% laser power. Then another wafer was treated with this process condition without any zone division. As shown in Fig.5, the sheet resistance and uniformity of this test wafer were 288.88 Ohm/sq and 1.98% respectively, which can well match data of the baseline wafer.

*Table 3: Sheet resistance of baseline wafer and tested wafer*

| Sample | Slot No. | Rs mean(Ohm/sq) | Stdev (%) |
|---|---|---|---|
| Baseline wafer | 20 | 292.83 | 1.20% |
| | 21 | 293.01 | 1.20% |
| Tested wafer | 1 | 288.88 | 1.98% |
| | 2 | 289.85 | 1.80% |
| | 3 | 290.02 | 1.90% |
| | 4 | 291.00 | 1.95% |

After that, repeatability experiments was done for this process condition. Table 3 listed the data of repeated experiments. As shown in Table 3, the mean values of sheet resistance and uniformity of experimental wafers can well match the data of baseline wafers.

After Rs matching, SIMS matching test was done for tested wafer and baseline data, which was shown in Fig.6. As shown in the figure, Cp/Rp were the peak concentration and projected range respectively, and Dose/Cp/Rp variations of tested wafer compared with baseline data in SIMS were 0.10%, 2.84%, 1.17E-6% respectively, which showed good matching performance.

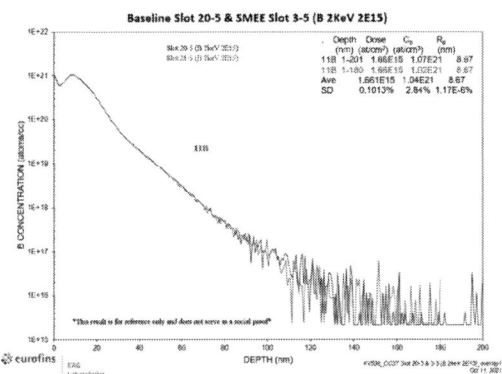

*Figure 6: SIMS matching data between baseline wafer and tested wafer*

Based on above data, 4 zones on single wafer can achieve better performance compared with 8 zones for the different laser power, which may be induced by interaction between adjacent zones during laser annealing. Considering 1200 C is a much higher temperature, 4 zones will be the optimized condition for related application.

## CONCLUSION

In this work, single wafer was divided to multiple zones, and different process was implemented in these zones for the tool matching application. Based on the measured data, sheet resistance and uniformity decreased with the increment of laser power percentage, and the 4-zones scheme will be the optimized condition for related application.

## ACKNOWLEDGMENTS

The author would like to thank R&D department of Shanghai IC R&D Center for the support in this work. Special thanks to Dazheng Xu, Zhengkai Dao, Yu Zhang, Quanchao Gu, Xiao Liu for their strong support in this work.

## REFERENCES

[1] Tolpygo S K, Bolkhovsky V, Weir T J, et al. Advanced fabrication processes for superconducting very large-scale integrated circuits[J]. IEEE Transactions on Applied Superconductivity, 2016, 26(3): 1-10.

[2] Weimer P K. The TFT a new thin-film transistor[J]. Proceedings of the IRE, 1962, 50(6): 1462-1469.

[3] Im J S. A new excimer-laser-annealing method for manufacturing large OLED displays[J]. MRS Online Proceedings Library (OPL), 2012, 1426: 239-249.

[4] Huet K, Toqué-Tresonne I, Mazzamuto F, et al. Laser thermal annealing: A low thermal budget solution for advanced structures and new materials[C]//2014 International Workshop on Junction Technology (IWJT). IEEE, 2014: 1-6.

[5] Young R T, White C W, Clark G J, et al. Laser annealing of boron‐implanted silicon[J]. Applied Physics Letters, 1978, 32(3): 139-141.

# A STRATEGY OF ELIMINATING SALICIDE BLOCK FILM BUBBLES

*Hualun Chen [1], Junwen Liu[1], Chao Bao[1]\**

[1]Huahong Semiconductor (WUXI) Limited, Wuxi 214028, China
*Corresponding Author's Email: Chao Bao@hhgrace.com

## ABSTRACT

Fluorine (F) is a co-implant species, which have numbers of beneficial effects to the semiconductor device. F co-implant has been shown to reduce boron (B) transient enhanced diffusion and deactivation when coupled with conventional spike rapid thermal anneals (RTA). However, salicide block (SAB) film easily suffers bubble defects due to the implant of F. In this paper, SAB bubbles have been observed clearly when the doses of $BF_2$ exceed $3.0\times10^{15}$ cm$^{-2}$ after RTA. Moreover, a defect-free SAB layer has been obtained by adjusting the implant conditions.

*Keywords— rapid thermal anneal; bubble; SAB film*

## INTRODUCTION

F is one of the dopant that is widely been studied by many researchers. The influence of F as a co-implant species also has been studied recently.[1, 2] The effect of F has been studied in the form of $BF_2$ in some works.[3-5] However, with the development of technology, especially the limitations of ultra-shallow junction technology, which makes it impossible to use $BF_2$ implant.[6] Instead, researches improve the performance of positive channel metal oxide semiconductor (PMOS) by implanting pure F because of the F implant prior to the $BF_2$ in suppressing the Negative Bias Threshold Instability (NBTI) effect of this device. However, salicide block (SAB) film easily suffers bubble defects due to the excessive implant of F or $BF_2$.

In this paper, we clearly observe the SAB bubbles by scanning electron microscope (SEM) after RTA process. Then, a defect-free SAB layer has been obtained by adjusting the implant conditions.

## EXPERIMENTAL PROCEDURE

Five Si (110) wafers, named $W_1$, $W_2$, $W_3$, $W_4$ and $W_5$ were used in this paper in order to study the relationship between F implant and defects (SAB bubbles). The implant conditions of the five wafers were shown in table 1. After RTA process, SAB films were observed by SEM.

## RESULT AND DISCUSSION

**Morphologies of SAB films with different F implant conditions.**

The relationship between F implant and SAB bubbles has been studied by adjusting the dose of F implant. The implant conditions of PMOS source/drain area have been presented in table. Figure 1 shows the SEM images of corresponding wafers. All the SAB films show flat surfaces and no defects could be observed, revealing these F implant conditions are all suitable.

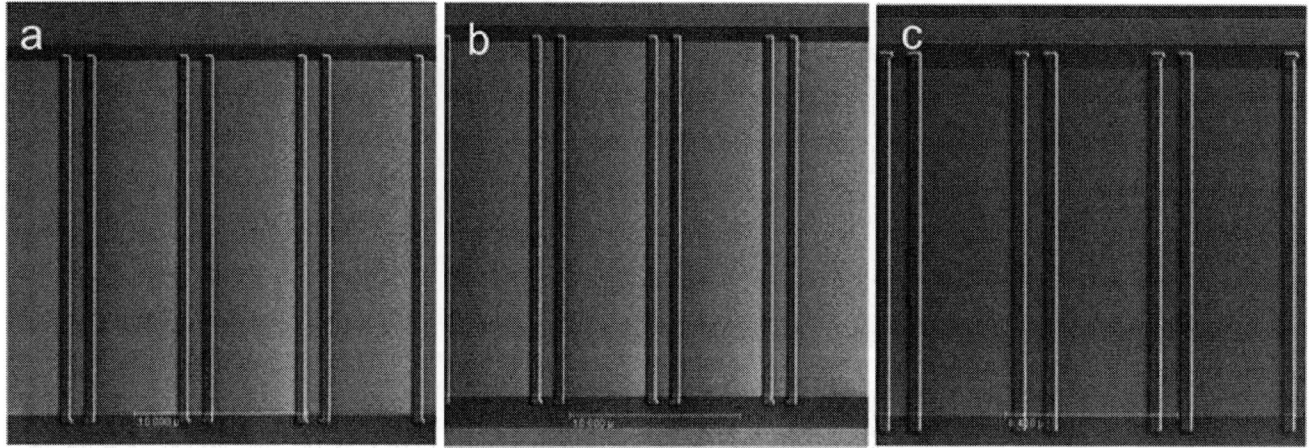

*Figure 1: SEM images of SAB films with different F implant dose. (a) SAB film of $W_1$; (b) SAB film of $W_2$; (c) SAB film of $W_3$.*

When the dose of $BF_2$ was changed to $3.0*10^{15}$ cm$^{-2}$ and $3.5*10^{15}$ cm$^{-2}$, small circular bumps appeared in the implant area after RTA process. Figure $2a_1$ shows the low-magnification SEM image of SAB film in $W_4$. Many circular bumps, which have different sizes, are evenly distributed on the surface. High-magnification image (Figure $2a_2$) shows these circular bumps have a wide distribution of size

and the largest size is about 1.2 μm. When adjusting the dose of $BF_2$ to $3.5*10^{15}$ cm$^{-2}$, many larger circular bumps could be observed after RTA process. Figure $2b_1$ and Figure $2b_2$ show the low-magnification and high-magnification images of implant area, respectively. Comparing with the bumps in $W_4$, the size of circular bumps significantly increased and the maximum size is about 2.5 μm. Moreover, the bumps are more densely    distributed.

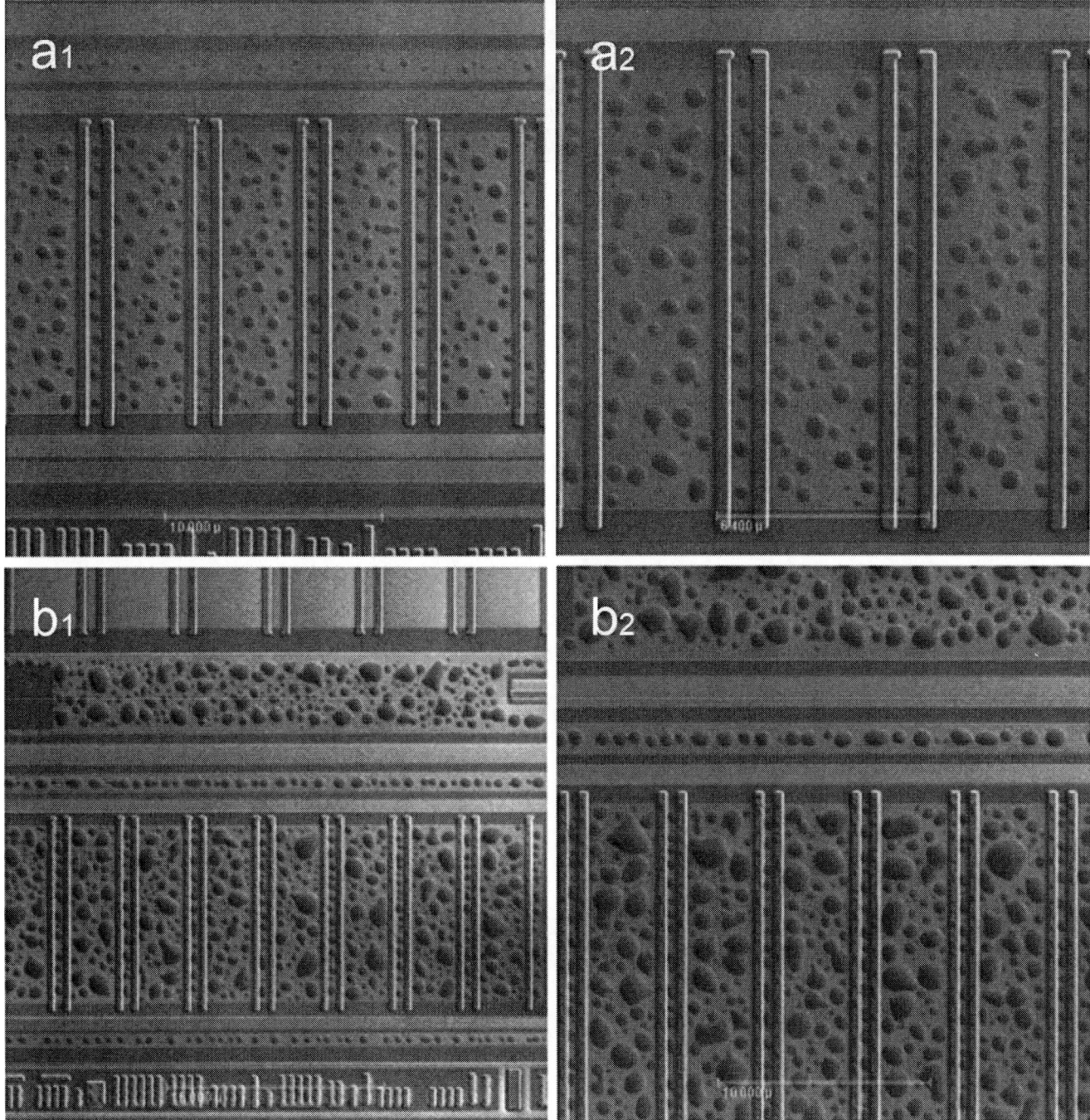

*Figure 2: SEM images of SAB films with different F implant dose. (a₁, a₂) SAB film of $W_4$; (b₁, b₂) SAB film of $W_5$.*

In order to study the internal structure of the bumps, the cross section SEM images of SAB films have been provided in Figure 3. As enclosed in the red region in Figure 3a, many bumps could be observed. The cross-sectional view of the bump is clearly presented in Figure 3b. The void structure can be clearly observed, which proves bumps are actually bubbles.

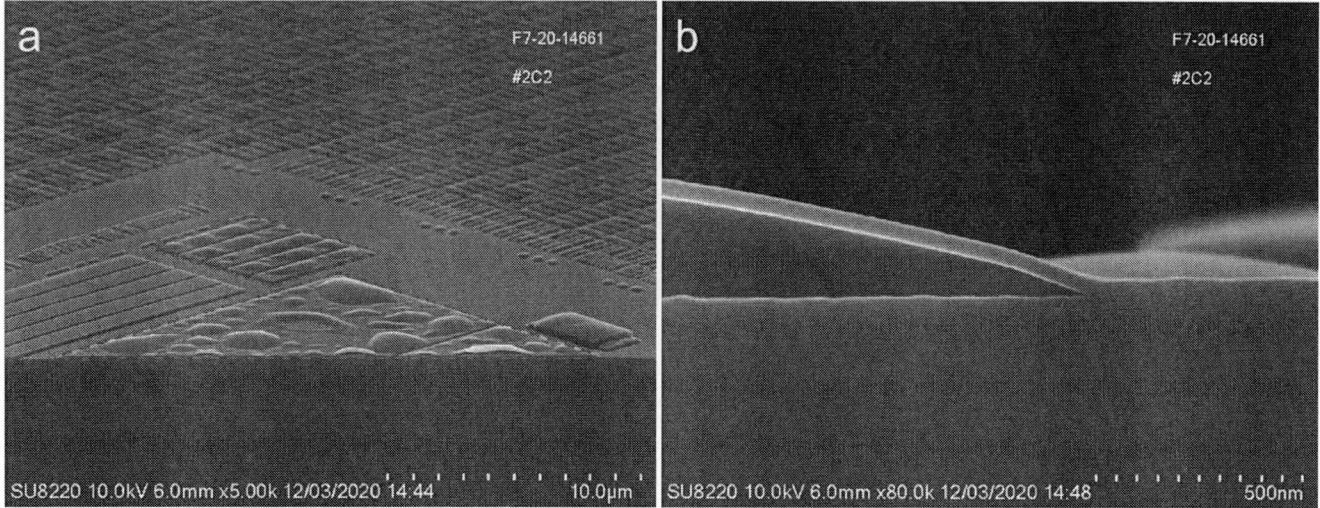

*Figure 3: The cross section SEM images of SAB films.*

**Analysis the forming of SAB bubbles.**

SAB bubbles appear due to the implant of $BF_2$ after RTA. In fact, a large amount of F escape from silicon to form the SAB bubbles. From the Figure 2, the size of SAB bubbles increase with the dose of $BF_2$ increasing. It has been improved reducing the dose could be a method of eliminating SAB bubbles.

## CONCLUSIONS

SAB bubbles can lead to failures on some devices due to their inherent defects and randomly distributed nature. In this paper, SAB bubbles have been observed clearly by SEM. We also find the relationship between bubble size and the dose of $BF_2$ by adjusting the dose of $BF_2$ implant. Moreover, the bubbles have been eliminated successfully. This study may help us understand the F implant better.

*Table 1: Implant conditions*

|  | F(dose/energy) | $BF_2$(dose/energy) | B+F(dose/energy) |
|---|---|---|---|
| $W_1$ | $2.0*10^{15}cm^{-2}/20keV$ | $2.0*10^{15}cm^{-2}/10keV$ | $1.5*10^{15}cm^{-2}/10keV(F)$ |
| $W_2$ | $2.0*10^{15}cm^{-2}/20keV$ | $2.5*10^{15}cm^{-2}/10keV$ | $8.0*10^{14}cm^{-2}/2keV(B)+1.5*10^{15}cm^{-2}/10keV(F)$ |
| $W_3$ | $2.0*10^{15}cm^{-2}/20keV$ | $2.5*10^{15}cm^{-2}/10keV$ | $1.0*10^{15}cm^{-2}/2keV(B)+1.5*10^{15}cm^{-2}/10keV(F)$ |
| $W_4$ | $2.0*10^{15}cm^{-2}/20keV$ | $3.0*10^{15}cm^{-2}/10keV$ | $1.5*10^{15}cm^{-2}/10keV(F)$ |
| $W_5$ | $2.0*10^{15}cm^{-2}/20keV$ | $3.5*10^{15}cm^{-2}/10keV$ | $1.5*10^{15}cm^{-2}/10keV(F)$ |

## REFERENCES

[1] G.S. Virdi, J. Lal, B.C. Pathak, W.S. Khokle, Some studies of $BF_2$ and B+F implanted Silicon, Radiation Effects, 106 (2006) 311-318.

[2] C.H. Poon, A. See, Effect of Fluorine Co-Implant on Boron Diffusion in Germanium Preamorphized Silicon During Post-LSA Rapid Thermal Annealing, IEEE Transactions on Semiconductor Manufacturing, 24 (2011) 333-337.

[3] J. Liu, D.F. Downey, K.S. Jones, E. Ishida, Fluorine effect on boron diffusion: chemical or damage? 1998 International Conference on Ion Implantation Technology. Proceedings (Cat. No.98EX144)1998, pp. 951-954 vol.952.

[4] S.Z.M. Saad, T.C. Lik, M.A. Othman, P. Holger, S.H. Herman, A study of fluorine implant in the formation of low leakage P+/N junction in BiCMOS technologies, 2012 International Conference on Enabling Science and Nanotechnology 2012, pp. 1-2.

[5] S.Z.M. Saad, T.C. Lik, M.A. Othman, P. Holger, S.H. Herman, An improved P+/N diode leakage current in BiCMOS technologies with fluorine co-implant, 2012 10th IEEE International Conference on Semiconductor Electronics (ICSE) 2012, pp. 690-693.

[6] C.L. Yang, C.I. Li, G.P. Lin, C.H. Tsai, Y.S. Huang, C. Fu, T.Y. Lu, H.Y. Wang, W.J. Chen, Y.L. Chin, M. Chan, J.Y. Wu, I.C. Chen, B. Colombeau, B.N. Guo, T. Wu, H. Gossmann, S. Lu, Improving device performance and variability for 28 nm and beyond low power SoC technology using advanced implant solutions, 2012 12th International Workshop on Junction Technology 2012, pp. 18-23.

# TAN BASED METAL-INSULATOR-METAL CAPACITOR WITH EXCELLENT LONG-TERM RELIABILITY

*Hualun Chen[1], Linlin Zhang[1,*], Hongxu Yang[1], Junwen Liu[1]*

[1] Huahong Semiconductor (WUXI) Limited, Wuxi 214028, China

*Corresponding Author's Email: Linlin1.Zhang@hhgrace.com

## ABSTRACT

Improving the long-term reliability of MIM capacitor is essential to tap its potential in IC application. At present, TiN is the mainstream material used for MIM capacitor metal plate with $Si_3N_4$ as insulator dielectric. However, the resistance characteristics of TiN are temperature unstable, and it has large lattice mismatch with $Si_3N_4$, which is detrimental to the device long-term reliability. Herein, we propose a strategy to employ TaN as MIM capacitor plate to solve these drawbacks. Experimental results reveal that the composition optimized TaN plate achieves a TCR near to zero. Finally, TDDB test reveals that the TaN based MIM capacitor can achieve more than $10^6$ years of long-term reliability, which is two orders of magnitude longer than TiN based device.

**Keywords:** *MIM capacitors; TaN; Reliability; TCR; TDDB*

## INTRODUCTION

With the rapid improvement of integrated circuit (IC) technology, more and more functional modules are integrated into a fingernail-sized chip. Among them, Metal-insulator-metal (MIM) capacitors, a typical passive component, have been widely used in silicon radio frequency (RF) and analog/mixed-signal product due to its advantages of high-conductance electrodes, low parasitic capacitance as well as depletion-free effect.[1-5] Dielectric layer and metal plate are the two major components of MIM capacitor. Aiming to obtaining high capacitance density, considerable efforts have been devoted to explore binary and multi-element high-*k* dielectric materials, such as $Ta_2O_5$, $HfO_2$, and $Al_2O_3$.[2-4] But there is relatively little research on metal plates. Actually, the metal plate and its matching with the dielectric layer material are also very important for the device long-term reliability. Hence, it is still necessary to conduct in-depth research on metal plate.

At present, titanium nitride (TiN) is the most widespread material used for MIM capacitor plate with silicon nitride ($Si_3N_4$) as insulator dielectric, especially when the back-end metal interconnection line is Al. But unfortunately, TiN has poor compatibility with more advanced low RC delay Cu interconnection process due to prone to defects. Certainly, as a common MIM capacitor plate material, TiN is meet the considerations of economy and easy to be integrated. However, there are two important potential drawbacks of TiN which should not be ignored with the continuous reduction of manufact-

TABLE 1. The comparison of lattice constant of TiN, $Si_3N_4$ and TaN

| Crystal System | Formula | a (Å) | b (Å) | c (Å) | α (°) | β (°) | γ (°) |
|---|---|---|---|---|---|---|---|
| Tetragonal | TiN | 4.22 | 4.22 | 4.22 | 90 | 90 | 90 |
| Hexagonal | $Si_3N_4$ | 7.60 | 7.60 | 2.91 | 90 | 90 | 120 |
| | TaN | 5.19 | 5.19 | 2.91 | 90 | 90 | 120 |

-uring costs and the progressively improvement of device performance requirements. One is that TiN is temperature unstable in term of electrical resistance. Its resistance will increase under continuous operation,[8] thus the device long-term reliability will be challenged. The other one is that the lattice mismatch between TiN and $Si_3N_4$ is large (Figure 1 and Table 1), which means defects are easily formed at the interface of plate and dielectric film. The large trap density will not only harm the device performance, but also the long-term reliability.

Tantalum nitride (TaN) is another very popular material used and suitable for IC manufacture. Moreover, TaN is commonly used as a barrier layer to suppress Cu diffusion. So, it is fully compatible with the Cu metal interconnection process.[5,6] According to the previous research, the resistance and temperature coefficient of resistance (TCR) of TaN is adjustable. And even a zero TCR can be obtained by regulating its chemical composition.[6-11] In addition, the lattice mismatch between TaN and $Si_3N_4$ is relatively small. The main phase both of TaN and $Si_3N_4$ in thin films are hexagonal (Figure 1). Therefore, TaN is a great candidate material for MIM capacitor metal plate fabrication to improve device reliability. However, there is very little systematic study on MIM capacitor device long-term reliability with TaN as metal plate until now.

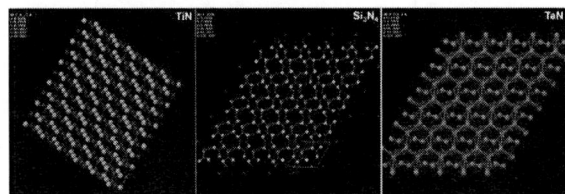

*Figure 1: The crystal structure of TiN, $Si_3N_4$ and TaN (unit cell structure is shown in the white frame at the bottom right corner)*

Here in, we replace TiN with TaN for MIM capacitor plate fabrication aiming to improve device long-term reliability. The TaN film was fabricated by physical vapor deposition (PVD) method, and its composition is controlled by tuning the $N_2$ flow rate. Experimental results revealed that the TaN based MIM capacitor with excellent electrical properties. In addition, the composition-optimized TaN achieves a TCR close to zero. Finally, constant voltage stress time dependent dielectric breakdown (CVS-TDDB) test indicates that the TaN based MIM capacitor can achieve more than $10^6$ years of reliability (Under JEDEC Criteria), which is two orders of magnitude longer than that of TiN based device.

## EXPERIMENT SECTION

### Fabrication of TaN Based MIM Capacitor

The TaN based MIM capacitor was inserted between top metal and sub-top metal layer of a standard CMOS back-end Cu interconnection process. A simplified process flow of MIM loop can be descript as follows. First, the bottom plate is formed by 1000A TaN via PVD method. A thin $Si_3N_4$ dielectric layer is deposited by chemical vapour deposition (CVD) technique on top of the bottom plate. Then, an additional 1000A TaN is deposited as top plate. Next, the top plate is patterned and etched with a selective chemistry that stop on the dielectric film. Finally, the dielectric together with bottom plate are patterned and etched in sequence. It is worth noting that depositing a thin $SiO_2$ buffer layer before bottom plate fabrication is very necessary to suppress the possible peeling issue.

### Measurements and Characterizations

Transmission electron microscopy (TEM) images were obtained on a JEOL JEM-2100 instrument. Wafer acceptance test (WAT) were carried out via an Agilent automated DC Parametric tester under atmosphere and room temperature. The testkey structure used for MIM capacitor WAT consists of 3 arrays, each array contains 9 devices. The single device area is $9 \times 9$ um$^2$. In order to analysis the temperature effect on resistance of TaN plate deposition with different $N_2$ flow rate, the plate resistance was recorded by a four-point probe at different temperature. The TDDB measurement was performed at 125°C using a constant voltage stress. According to the TDDB data, the long-term reliability of the device operation at normal condition is extracted through Weibull distribution along with an acceleration model.[12,13]

## RESULTS AND DISCUSSION

### Device Structure and WAT Result

The sandwich (Bottom plate/Dielectric/Top plate) structure MIM capacitor was fabricated according to the method described in experimental section. Figure 2a depicts the schematic diagram of the device built on sub-top metal. A thin $Si_3N_4$ insulator dielectric layer separate two TaN plate layers. An additional $Si_3N_4$ as spacer to

encapsulates the capacitor. And the top plate is connected to top metal by vias. Figure 2b shows the cross section TEM image of the TaN based MIM capacitor, it can be seen that the device keep excellent profile. The thickness of TaN plate film is very uniform. In addition, the film is also very flatness, which can effectively avoid the dielectric breakdown caused by possible tip discharge. These results indicate that the process is reliable.

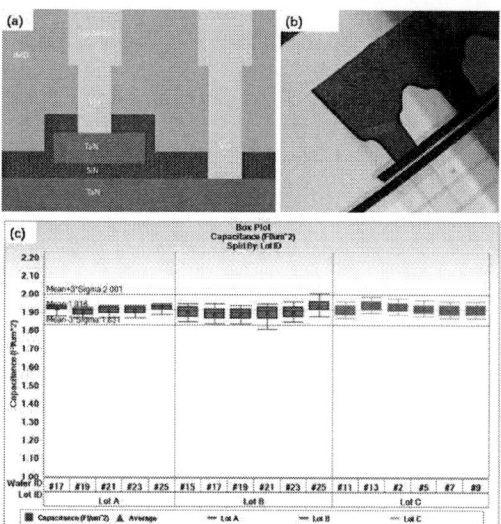

Figure 2: (a) The schematic structure of the MIM capacitor device, (b) the cross section TEM image of TaN based device, (c) the capacitance density box plot of TaN-based device

Capacitance density is a very important parameter to measure a MIM capacitor, which is determined by thickness and permittivity of the dielectric. Wafer acceptance test (WAT) data box plot of the measured MIM capacitor samples from 3 engineering lots are illustrated in Figure 2c. Each lot contained 5-6 wafers and each wafer consisted of 154 sites. It can be seen that the means value of capacitance density is about 1.92 fF/um$^2$, which is match the design specification ($2.0 \pm 0.15$ fF/um$^2$). More important is that the data distribution is very uniformity whether within wafers or within lots (range of $\pm 3$ sigma is below 0.2).

Breakdown voltage (BV) is another crucial parameter for MIM capacitor, which provides an upper limit for the application of the device. According to the electrical design rules in this experiment, once the leakage current density reaches 10 mA/cm$^2$, the device is judged to breakdown failure. Figure 3 illustrates the leakage current density variation of TaN based MIM capacitor under stress from 0 to 30 V. It can be seen that the voltage is approximately 23 V at fail criteria point, which is far higher than the design specification (15 V). Therefore, the TaN based MIM capacitor is fully meet the design requirement. To sum up, WAT results indicate that using

TaN as MIM capacitor plate for industry production is feasible.

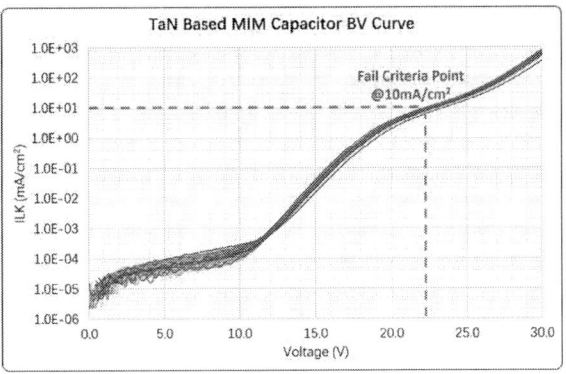

Figure 3: The leakage current density variation of TaN based MIM capacitor under stress from 0 to 30 V (total 154 samples)

**Resistance Temperature Effect Analysis**

As an important part of the MIM capacitor, the plate has a great influence on device performance. In addition, the metal plate is also used as resistor module in some products, so its inherent electrical resistance proprieties are also noteworthy, especially the resistance temperature characteristics. As mentioned earlier, the TCR of TaN is related to its chemical composition, and a zero TCR can be obtained by regulating its Ta/N molar ratio.[6,7,9-11] Here, in order to achieve the composition control, we varied the amount of the N source and fixed the amount of Ta ion during the preparation of TaN film. Specifically, the N source is regulated by control the $N_2$ flow rate during the PVD process. The sheet resistance of TaN plate deposited with different $N_2$ flow rate at different temperature are shown in Figure 4(a-c) (Plot). It can be found that the resistance value increase with the increase of $N_2$ flow rate, which is consistent with the literature report.[6,10] Others, first order and second order temperature coefficients, $TCR_1$ and $TCR_2$ are obtained by binominal fitting to the measurement result. The corresponding equation as following (1):

$$R=b_2(T-25)^2+b_1(T-25)+b_0 \qquad (1)$$

Where $T$ represent the temperature. The $TCR_1$ and $TCR_2$ can be obtained by $b_1/b_0$ and $b_2/b_0$.

The fitting curves of resistance versus temperature are shown in Figure 4(a-c) (Dotted line). It can be found that the absolute value of $TCR_1$ of TaN plate decreases first and then increases with the continuous increase of $N_2$ flow rate from 35 to 45 sccm (standard cubic centimeter per minute). While in case of $N_2$ flow rate is about 36 sccm, the $TCR_1$ is close to zero (Figure 4d). This result provide a powerful support that it is feasible to obtain a TaN plate with near-zero TCR by regulating $N_2$ flow rate in Fab fabrication.

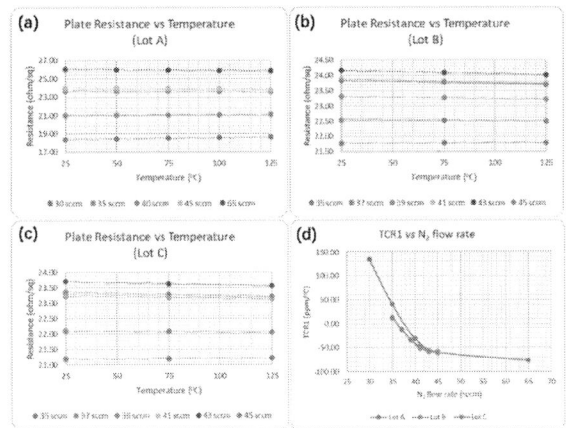

Figure 4: (a),(b) and (c)The Rs of TaN plate fabricated with different $N_2$ flow rate under different temperature (Plot) and the corresponding fitting curve (Dotted line);(d)The $TCR_1$ of TaN plate (Plot), and the fitting curve of correlation between $TCR_1$ and $N_2$ flow rate (Line)

**Device Long-term Reliability Assessment**

Dielectric breakdown is one of the potential reliability issues in MIM capacitor device. When a bias is applied to the metal plate, electrons and holes tunnel through the dielectric. With time increasing of device operation and device aging, the defect in the insulator layer will generate and accumulate. It is no doubt that the defects will act as electron traps. When a critical amount of defects is reached, the dielectric will lose its intrinsic properties, thus cause device failure. TDDB test and extrapolating breakdown time to mild operation conditions have been known widely as a method to determine the dielectric reliability.[13] In this work, the TDDB test was performed both for TiN and TaN based device at 125°C under a constant voltage stress.[14] The leakage current was continuously monitored during test. Once the current had significant increased, the measurement was stopped immediately and the time-to-breakdown ($t_{BD}$) was recorded. The identical $t_{BD}$ value is preferentially transformed into a Weibull distribution (2) [14]:

$$F(t)=1-\exp[-(\frac{t-\gamma}{\eta})^{\beta}] \qquad (2)$$

Where, the $\beta$ represents the Weibull distribution slope, $\eta$ is the 63% value, and $\gamma$ is the time delay or burn-in time. Addition, Weibull distribution are typically presented by plotting the Weibit $z$ versus the logarithm of time. The Weibit $z$ was calculated as following (3)[13]:

$$z = \ln[-\ln(1 - F(t))] \qquad (3)$$

Here, to simplify the calculation of $F(t)$, Benard approximation was used (4) [13]:

$$F(i)=\frac{i-0.3}{n+0.4} \qquad (4)$$

TABLE 2. The comparison of TDDB test result between TiN- and TaN-based MIM capacitor

| Metal Plate | Stress (V) | T63.2% (yrs) | Beta | T0.01% (yrs) | Fail Criteria |
|---|---|---|---|---|---|
| TiN | 19.5 | 2.63E+06 | 3.69 | 2.15E+04 | T0.01% > 10 years @1.1Vcc, 125℃, 10mm² |
| TaN | 19.5 | 2.13E+07 | 7.27 | 1.86E+06 | |

Where, $i$ is the serial number of the failed device (sort by $t_{BD}$), and $n$ is the total number of samples. Figure 5a illustrate the curve of Weibit $z$ versus the logarithm of time of MIM capacitor sample under 19.5V bias. Then, the distribution curve was fitted to determine the parameter $\beta$ and $\eta$. Finally, the lifetime down to operation conditions was further extrapolated depend on acceleration model.[13] According to JEDEC standard, supposing device area is 10 mm² and with 0.01% failure under 1.1 Vcc stress during extrapolation. Figure 5b depicts the Weibull distribution of MIM capacitor breakdown lifetime extrapolating to mild operation conditions. As shown in Table 2, the long-term reliability of TaN-based MIM capacitor has been significantly improved, reaching to 10⁶ years, an increase of almost two orders of magnitude than TiN based device. As mentioned before, we attribute this improvement mainly to the small lattice mismatch between TaN and dielectric layer, as well as the stable resistance temperature characteristic of TaN. Small lattice mismatch means fewer intrinsic defects and slower defect generation rate under stress. The near-zero TCR of TaN indicates that the plate are stable, which is beneficial to suppress more defects generation. In short, the synergy of the small lattice mismatch and peculiar TCR nature of TaN promote the improvement of the long-term reliability of the resultant devices.

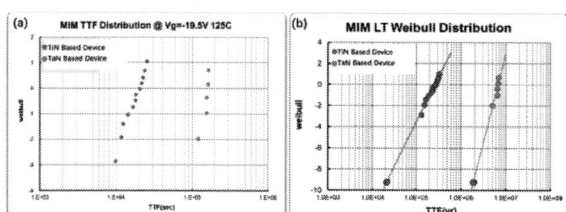

Figure 5: The TTF distribution of TiN and TaN-based MIM capacitor (a) under stress of Vg=-19.5 V and 125 °C, (b) extrapolating to mild operation conditions

## CONCLUSION

In summary, we replace traditional TiN with TaN as MIM capacitor plate. TaN overcomes the drawbacks of TiN resistance temperature instability and large lattice mismatch with Si₃N₄ dielectric layer. In addition, TaN is fully compatible with all-Cu back-end process with low RC delay. WAT results indicate the electrical performance of TaN based MIM capacitor is fully meet design rules. In addition, temperature resistance of plate measurement

indicates a TCR closing to zero can be obtained by control the N₂ flow rate during PVD process. Finally, CVS-TDDB indicates that the TaN based MIM capacitor achieve more than 10⁶ years of long-term reliability (@1.1 Vcc, 10 mm², and 0.01% failure, JEDEC Standards), which is two order of magnitude longer than that of TiN based device. This work paves a new way to prepare high quality MIM capacitor and further promotes the development of IC manufacturing.

## ACKNOWLEDGEMENTS

L.L. Zhang and H.L. Chen contributed equally to this work and should be regarded as co-first author. This work is supported by Huahong Semiconductor (WUXI) Limited. The authors would like to thank the contributions from process fabrication, failure analysis, wafer acceptance test, as well as reliability test department.

## REFERENCES

[1] J. L. Mu, X. J. Chou, Z. M. Ma, J. He and J. J. Xiong, *Micromachines*, vol. 9, 2018, pp. 69.

[2] A. Vilaret, E. Ebrard, N. Casanova, S. Guillaumet and T. Skotnicki, *European Solid State Device Research Conference* IEEE, 2007.

[3] T. H. Pergn, C. H. Chien, C. W. Chen, P. Lehnen and C. Y. Chang, *Thin Solid Films*, 2004.

[4] S. B. Chen, C. H. Lai, A. Chin, J. C. Hsieh and J. Liu, *IEEE Electron Device Letters*, vol. 23, 2002, pp. 185-187.

[5] S. J. Ding, Y. J. Huang, Y. Huang, S. H. Pan, W. Zhang and L. K. Wang, *Chinese Physics*, vol.16, 2007, pp. 2803-2808.

[6] H. C. Jiang, C. J. Wang, W. L. Zhang and X. Si, *Modern Physics Letters B*, vol.24, 2010, pp. 905-910.

[7] B. Wang, Z. G. Song and Q. T. Cao, *Advanced Materials Research*, vol. 10, 2015, pp. 34-37.

[8] F. F. Liu, Y. Tang, W. L. Zhang, H. C. Jiang and X. Si, *Electronic Components and Materials*, vol. 30, 2011, pp. 47-49.

[9] S. M. Kang, S. G. Yoon, S. J. Suh and D. H. Yoon, *Thin Solid Films*, vol. 516, 2008, pp. 3568-3571.

[10] S. M. Na, I. S. Park, S. Y. Park, G. H. Jeong and S. J. Suh, *Thin Solid Films*, vol. 516, 2008, pp. 5465-5469.

[11] T. Lee, K. Watson, F. Chen, J. Gill, D. Harmon, T. Sullivan and B. Z. Li, *IEEE International Reliability Physics Symposium*, 2004.

[12] W. Weibull, *Journal of Applied Mechanics*, vol. 18, 1951, pp. 293-297.

[13] T. Kauerauf, *Ph.D dissertation, Katholieke Universit Leuven, Belgium*, 2007.

[14] B. Kaczer, R. Degraeve, R. O'Connor, P. Roussel and G. Groeseneken, *International Electron Devices Meeting*, 2004, pp. 713-716.

# A STRATEGY OF ELIMINATING SILICON DISLOCATION IN 55 NM NODE TECHNOLOGY

*Hualun Chen[1], Chao Bao[1,*], Jiawei Gu[1], Junwen Liu[1], Zhongcai Niu[1]*
[1]Huahong Semiconductor (WUXI) Limited, Wuxi 214028, China
*Corresponding Author's Email: Chao Bao@hhgrace.com

## ABSTRACT

Dislocations are crystalline defects that can significantly change the electrical properties of semiconductor materials. In this paper, the lattice dislocation formed during 55 nm node high voltage (55HV) process have been studied in detail. By adjusting the conditions of ion implantation and temperature of thermal process, we found that thermal stress is the key process that inducing the lattice dislocation defects. Moreover, the dislocation defects were eliminated by reducing temperature in thermal process of HV and medium voltage (MV) gate oxide process. The improvement of the process not only improves the yield of 55HV product from about 1% to about 96%, but also brings new insights in eliminating similar defects in the future.

***Keywords—NMOS; crystal defect; dislocation; SRAM***

## INTRODUCTION

Dislocation are crystalline defects, which can obviously change the electrical properties of materials. The semiconductor materials with lattice dislocation defects present several electronic properties such as naturally formed PN junctions, anisotropic electrical conductivity, confinement for charge carriers and localization of implanted.[1] Moreover, if the dislocations locate in the depletion region of PN junction or Metal-Oxide-Semiconductor field effect transistor (MOSFET), undesirable current leakage may generate.[2] The current leakage is generated due to enhanced dopant diffusion along dislocations, as dislocations act as tubes of easy diffusion paths.[3-7] Dislocations induce diffusion spikes in the PN junction region and these defects then become a leakage path between the source and the drain.[8-11] there are many reasons for the formation of lattice dislocations in semiconductor manufacturing, but most of the dislocation defects in semiconductor devices are generated in the process of film growth, plasma etching, ion implantation and high temperature treatment process.[12-15] These manufacturing process steps have to be optimized in order to prevent the generation of dislocation defects.

Ion implantation is a common process to introduce dopants into semiconductor materials in semiconductor manufacturing. The ions enter the substrate and come to rest in typically $10^{-13}$s by nuclear and electronic stopping mechanisms.[16] The part of ions will occupy the position of silicon host atoms. Several defects called the primary damage may be formed after the implant. Annealing of primary damage results in the formation of dislocation defects. The primary damage consists only crystal damage. However several primary damage may be formed dislocation defects during thermal treatment. This formation of dislocation is closely associated with ions implantation and annealing process.

55HV is a new 55 nm node high voltage technology platform to implement the company's instructions of "12-inch characteristic process route" and follow the market development trend. There are three special types of MOSFET in 55HV product, as Figure 1 shown. They have different driving voltages of 32V, 6V and 1.2V, respectively. It provides solutions for liquid crystal display (LCD) drive products, organic light-emitting diode (OLED) drive products and panel touch products. The rapid establishment of 55HV platform can not only meet the customer's capacity requirements, but also can help introduce more competitive products in the future. The successful completion of 55HV platform will extremely enrich the characteristic technology.

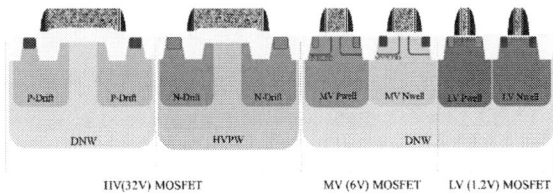

*Figure 1 schematic structures of three special types of MOSFET in 55HV product.*

Herein, we report the discovery of current leakage in the area of SRAM in 55HV technology. As the figure 2 shown, the yield of wafer is about 1% due to the current leakage. By electrical failure analysis (EFA) and TEM, lattice dislocation defects have been found. On this basis, the key process, which lead to the current leakage, has been screened out by controlling variable method. Finally, the low yield issue has been solved successfully through choosing proper processes.

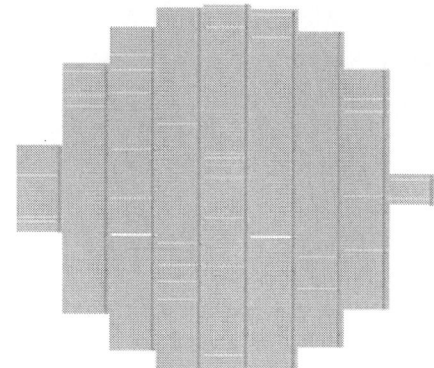

*Figure 2 Wafer map of SRAM test. The red chips represent chip probing (CP) failed chips.*

## EXPERIMENTAL PROCEDURE

### Wafers with different ion implantation

Six Si (110) wafers, named $a_i$, $b_i$, $c_i$, $d_i$, $e_i$ and $f_i$ were used in this paper in order to study the relationship between ion implantation and dislocation defects. These wafers had the same preparation steps except ions implantation. All the wafers have the different conditions of ion implantation, as the table 1 shown. After interlayer dielectric (ILD) process, each wafer was observed by TEM.

### Wafers with different temperature processes

Ten Si (110) wafers, named $a_t$, $b_t$, $c_t$, $d_t$, $e_t$, $f_t$, $g_t$, $h_t$, $i_t$ and $j_t$ were used in this paper in order to study the relationship between thermal steps and dislocation defects. These wafers had the same preparation steps except several thermal processes. All the wafers have the different conditions of thermal processes, as the table 2 shown. After ILD process, each wafer was observed by TEM.

### Measurements and Characterizations

TEM observations were performed using a JEOL JEM-2100 microscope that was equipped with a $LaB_6$ gun operated at 200 kV (Cs = 1.0 mm, point resolution of 2.3 Å). Images were recorded using a TENGRA CCD camera (2304×2304 pixels with a 2:1 fiber-optical taper and an effective pixel size of 18 $\mu m^2$).

*TABLE 1 Wafers with different conditions for ion implantation.*

| wafer | $a_i$ | $b_i$ | $c_i$ | $d_i$ | $e_i$ | $f_i$ |
|---|---|---|---|---|---|---|
| GE Blanket implantation | √ | skip | √ | √ | √ | √ |
| Core NLDD implantation | skip | √ | skip | √ | skip | √ |
| Core PLDD implantation | skip | √ | √ | skip | skip | √ |
| S/D implantation | skip | √ | √ | √ | √ | √ |
| Core PW implantation | √ | √ | √ | √ | √ | skip |

*TABLE 2 Wafers with different conditions of temperature in HV and MV gate oxide process.*

| Wafer | HV gate oxide | MV gate oxide |
|---|---|---|
| $a_t$ | --- | 700°C |
| $b_t$ | --- | 780°C |
| $c_t$ | --- | 850°C |
| $d_t$ | 900°C + anneal | 850°C |
| $e_t$ | 900°C + anneal | 700°C |
| $f_t$ | 950°C + anneal | 700°C |
| $g_t$ | 780°C | 700°C |
| $h_t$ | 780°C | 800°C |
| $i_t$ | 900°C | 800°C |
| $j_t$ | 900°C | 700°C |

## RESULTS AND DISCUSSION

The 55HV product suffers low yield issue caused by current leakage. In order to find out the reason of current leakage, we used electrical failure analysis to locate the hot spot. Figure 3 shows the GDS and TEM images of the hot spot area. Figure 3b and 3c show the TEM images of hot spot area, revealing the defects are silicon dislocation. Moreover, the lattice dislocation defects look like cracks, which locate about 200 nm far from the surface of silicon. Combined with Figure 3a, it can be seen that the lattice dislocation formed in the AA region and always locate between two poly-silicon. The specific orientation and location indicate that the surrounding environment has great influence on the formation of silicon dislocation.

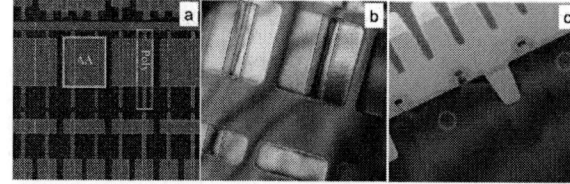

*Figure 3 (a) GDS image of the silicon dislocation in NMOS; (b) TEM image of vertical view of the silicon dislocation in NMOS; (c) TEM image of the transection of the silicon dislocation in NMOS.*

As we all known, most of the dislocation defects in semiconductor devices are formed in the process of film growth, plasma etching, ion implantation and high temperature treatment. In order to find out the key process, which lead to the silicon dislocation defects, in 55HV technology, the control variable method was adopted.

**Optimization of Ion Implant Processes:** During ion implantation, the crystal structure is damaged due to the collision of incident ions. Lattice damage occurs when an incident ion collides with an atom of the original matter and takes its lattice position. Crystal defects occur when

978-1-6654-9759-6/22 $31.00 © 2022 IEEE

atoms of the original matter are knocked out of their original positions by incident ions and remain in non-lattice position. Most of the defects formed by ion implantation are concentrated on the surface and are mainly point defects. However, TEM results showed that the lattice defects only appeared in the NMOS area, and never exist in the PMOS area. The types and concentrations of ions implantation in NMOS and PMOS region are different, so it is particularly important to study the relationship between ion implantation and silicon dislocation defects. In order to verify whether the generation of lattice dislocation is related to ion implantation, we selectively skipped several steps of ion implantation, as the table 1 shown.

All wafers in table 2 were sliced and observed by TEM. Figure 4 shows TEM images of silicon dislocation with different ion implantation conditions. All the wafers exist dislocation defects, indicating that ion implantation is not the critical process in the formation of dislocation defects. However, the position of the dislocation defects and the distance between the defects and silicon surface are different when skipping different ion implantation process. For example, in figure 4b and 4d, the lattice dislocations are adjacent to the bottom of shallow trench isolation (STI), while the other dislocation defects are located between two poly-silicon. This result strongly proves the formation of dislocation defects is closely associated with ions implantation. Unfortunately, the mechanism is still unclear and need further study.

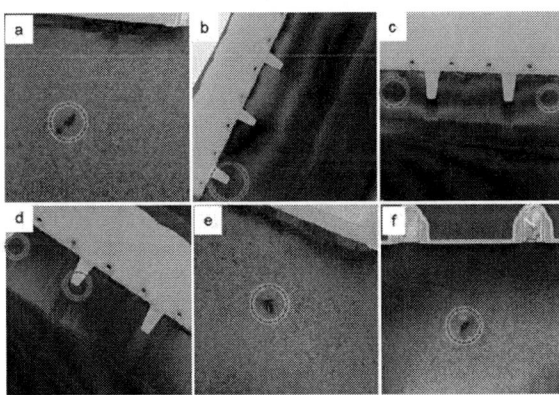

*Figure 4 TEM images of wafers with different conditions for ion implantation. The red circles represent silicon dislocation; a, b, c, d, e and f represent wafer of $a_i$, $b_i$, $c_i$, $d_i$, $e_i$ and $f_i$ in table 1, respectively.*

**Optimization of High Temperature Processes:** When the temperature of a solid material rises, the volume of its structure increases. This phenomenon is called thermal expansion. Different materials have different thermal expansion coefficients. In the semiconductor manufacturing, silicon, silicon dioxide and silicon nitride have different thermal expansion coefficients. The thermal stress will be generated from the interface among different materials in the process of temperature rising due to they have different expansion degree. With the temperature increasing, the defects will be generated in order to release the stress when the thermal stress reaching the limit. In order to verify whether the generation of lattice dislocation is related to the temperature in the semiconductor manufacturing, the temperature of HV and MV gate oxide process in 55HV technology have been adjusted, as the table 2 shown. Wafers (a, b, c) only run MV gate oxide process to reduce part of thermal in HV gate oxide process. Other wafers have different temperatures in HV and MV gate oxide process.

All wafers in table 2 were sliced and observed by TEM. Figure 5 shows TEM images of silicon dislocation in different temperature conditions in HV and MV gate oxide process. The results show no silicon dislocation defects in Figure (a, b, g, j) have been observed. However, obvious silicon dislocation defects can be observed in Figure (c-f, h, i) as the area of red circles shown. Combining table 2 and Figure 5, reducing thermal can suppress dislocation obviously.

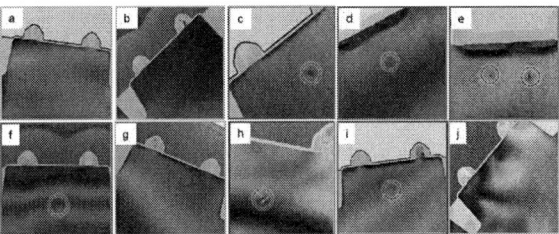

*Figure 5 TEM images of Wafers with different conditions of temperature in HV and MV gate oxide process. The red circles represent silicon dislocation. a, b, c, d, e, f, g, h, i and j represent wafer of $a_t$, $b_t$, $c_t$, $d_t$, $e_t$, $f_t$, $g_t$, $h_t$, $i_t$ and $j_t$ in table 2, respectively.*

## CONCLUSIONS

In summary, silicon dislocation defects, which lead to current leakage in 55HV technology, had been observed by TEM. Moreover, the dislocation defects were eliminated by reducing temperature in thermal process of HV and medium voltage (MV) gate oxide process. The improvement of the process improved the yield of 55HV product from about 1% to about 96%. We hope the study may bring new insights in eliminating the similar defects in semiconductor manufacturing.

## ACKNOWLEDGEMENT

Chao Bao and Hualuan Chen contributed equally to this work and should be regarded as co-first author. This work is supported by Huahong Semiconductor (WUXI) Limited. The authors would like to thank the contributions from process fabrication and failure analysis.

# REFERENCES

[1] S. Mil'shtein, *Physica Status Solidi (A)*, vol. 171, no. 1, pp. 371–376, Jan. 1999.

[2] D. Ha, C. Cho, D. Shin, G.-H. Koh, T.-Y. Chung, and K. Kim, *IEEE Trans. Electron Devices*, vol. 46, no. 5, pp. 940–946, May 1999.

[3] M. N. Shetty, *Delhi, India: PHI Learning*, 2013, pp. 314–315.

[4] J. G. Fiorenza et al., *IEEE Int. Rel. Phys. Symp.*, Apr. 2004, pp. 493–497.

[5] A. Qiang, J. Liew, W. Entalai, and K. Choong, *Asia Symp. Quality Electron. Des., Kuala Lumpur*, Malaysia, Jul. 2011, pp. 221–224.

[6] J. Xu and M. E. Law, *Proc. 11th Int. Conf. Ion Implantation Technol.*, Austin, TX, USA, Jun. 1996, pp. 630–633.

[7] H. Belgal, *Proc. 32nd Int. Annu. Rel. Phys. Symp.*, San Jose, CA, USA, Apr. 1994, pp. 399–404.

[8] F. Shimura, *Semiconductor Silicon Crystal Technology.* San Diego, CA, USA: Academic, 1989, pp. 343–344.

[9] C. T. Wang, H. Haddad, P. Berndt, B. -S. Yeh, and B. Connors, *Proc. 30th Int. Annu. Rel. Phys. Symp.*, San Diego, CA, USA, Apr. 1992, pp. 85–90.

[10] A. Owens, *Compound Semiconductor Radiation Detectors.* Boca Raton, FL, USA: CRC Press, 2013, pp. 72–74.

[11] H. Mehrer, *Diffusion in Solids. Berlin, Germany: Springer-Verlag*, 2007, pp. 584–589.

[12] C. Y. Nakakura et al., *Electrical and Electrochemical Phenomena at the Nanoscale*, S. Kalinin and A. Gruverman, Eds. New York, NY, USA: Springer-Verlag, 2007, pp. 639–641.

[13] B. El-Kareh, *Fundamentals of Semiconductor Processing Technology.* New York, NY, USA: Springer-Verlag, 1995, pp. 335–337.

[14] M. Yang and K. Nakata, *G. S. Mathad et al., Eds. Pennington, NJ, USA:Electrochemical Society*, 1998, pp. 29–35.

[15] M. Weling and C. Gabriel, *Diagnostic Techniques for Semiconductor Materials and Devices 1994*, D. K. Schroeder et al., Eds. Pennington, NJ, USA: Electrochemical Society, 1994, pp. 143–150.

[16] E E Morehead and B. L. Crowder, Radiat. Eft., 6 (1970) 27.

# 12-inch 90-nm BCD Process Optimization: Reducing Wafer Edge LDMOS Leakage

Hualun Chen, Li Wang, Li Xiao, Yong Chen, Tian Chen, Guangyuan Lu*, Jian Wang, Fan Ji

Huahong Semiconductor (WUXI) Limited, Wuxi, 214029, China

*Corresponding Author's Email: Guangyuan.Lu@hhgrace.com

### Abstract

12-inch 90-nm Bipolar-CMOS-DMOS (BCD) process has attracted significant attention because of better device performance and lower cost. As in 12-inch process, wafer edge process control is more challenging, yield loss in BCD products is frequently detected. With layer-by-layer failure analysis, one fail mode is found that the wafer edge yield loss is due to Metal-1 (M1) to gate leakage in laterally double-diffused metal–oxide–semiconductor (LDMOS) field-effect transistors. The influences of inter-layer-dielectric (ILD) thickness, M1 etch recipe, and LDMOS layout on the leakage were systematically studied. Based on the effort, a sequence of process optimizations have been applied, which help to improve the total production yield by 5-10%.

*Keywords— 12-inch 90-nm BCD Process; LDMOS Leakage; Yield Improve*

### Introduction

Bipolar-CMOS-DMOS (BCD) process has been widely used variable areas including display chip, power on ehernet (POE) and storage controller chip [1-5]. It combines the advantage of the high trans-conductance and strong load-drive capability of BJT device, high integration density and low power of CMOS devices and high power output of DMOS power devices [6, 7]. Recently, it has been paid significant attention to develop BCD chips using sub-130-nm processes, which helps to provide better device performance and lower cost. Following Moore's law, several foundries have announced successful mass-production of 12-inch 90-nm BCD products. Different from mature 8-inch 130-nm BCD products, some low yield cases due to the combination of BCD layout to 12-inch 90-nm process occurred during the earlier rick production period. In this work, we report our investigations on a wafer edge yield loss case, which is induced by Metal-1 (M1) to gate leakage in laterally double-diffused metal–oxide–semiconductor (LDMOS) field-effect transistors. Based on the effort, a series of process optimizations are introduced and help to improve the total production yield by 5-10%.

### Failure Analysis

Figure 1(a) shows the typical chip probing (CP) map tested in earlier 12-inch 90-nm BCD wafer. Chips in wafer edge 3-13 mm area usually showed bad performance, which leaded to a total 5-10% yield loss. Bin fail and OBIRCH analysis show that it is related to the abnormal performance of LDMOS array, as marked in Fig 1(b). Further study shows that at off-state, the bad chip exhibits obvious high side (HS) to low side (LS) leakage, as shown in Fig 1(c).

With layer-to-layer failure analysis, it is found that the leakage is caused by M1 to gate breakdown due to insufficient dielectric, as shown in Fig 2. M1 Cu lines in wafer edge bad

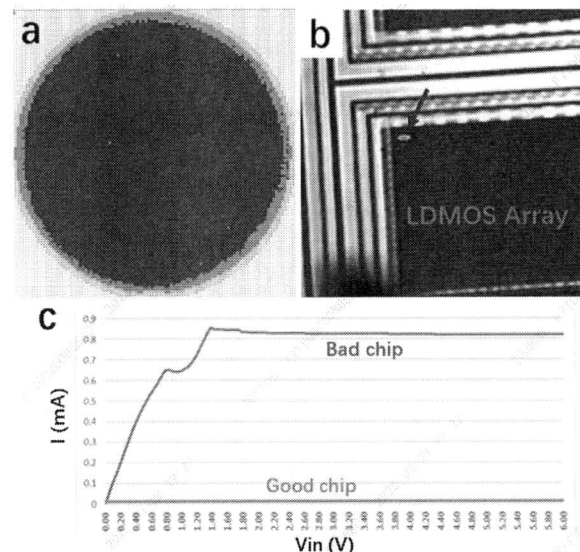

Fig.1. Wafer edge yield loss of 12-inch 90-nm BCD product. (a) CP map. (b) OBIRCH map highlighting leakage area in LDMOS array. (c) Comparison of off-state I-V curves between good and failed chips.

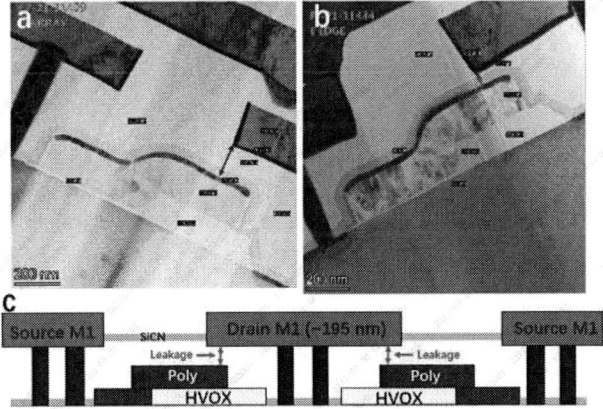

Fig.2. Failure analysis of LDMOS leakage. (a,b) Cross sectional images of LDMOS M1-gate layers obtained in (a) good and (b) bad chips, respectively. (c) Schematic illustration of LDMOS structure.

dies are found to be much thicker than those in good dies. For LDMOS devices, HVOX is applied to make poly higher in drift area to get better drain-source breakdown voltage (BV). As a result, the total inter-layer-dielectric (ILD) thickness is designed higher than logic products to make a certain distance between M1 lines and poly on HVOX. In wafer edge bad dies, the M1 thickness is ~50% higher than normal ties, making M1 to poly/HVOX distance only ~20% of target value. As the M1 line above gate/HVOX is connected to drain side, the bias voltage of drain to gate can easily break such thin barrier, inducing higher gate voltage. The LDMOS cannot be turned off, which finally leads to wafer edge yield loss.

**Investigation and Optimization**

## 1. ILD thickness control

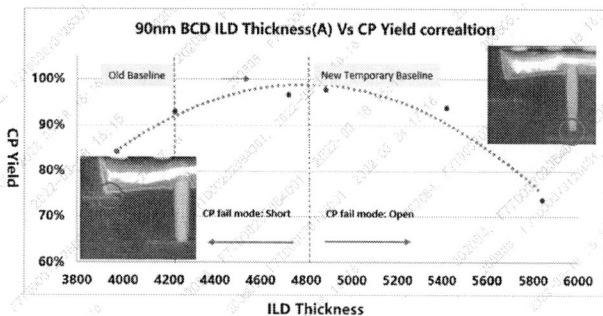

Fig.3. Relationship between CP yield and ILD thickness.

To obtain the best M1 to gate distance window, the ILD thickness split experiment was first conducted. As shown in Fig 3, by increasing ILD thickness by ~15%, wafer yield can be increased of ~10%. This is because that the thicker ILD keeps M1 lines away from gate in wafer edge, which helps to provide higher M1 to gate breakdown voltage, and lower LDMOS leakage. However, further increasing ILD thickness leads to severe yield reduction of 10-30%. It is found that this yield loss is owing to CT open, since the CT etch window cannot cover such high ILD thickness, as shown in the inset in Fig 3. Based on the investigation, as a temporary control plan, the ILD thickness was increased by 15% to get the best yield and reliability performance before M1 etch recipe tuning and LDMOS layout optimization.

## 2. M1 etch recipe optimization

Further investigation shows that the root cause of leakage is the excessive ILD loss introduced in M1 etch step. To obtain efficient M1-CT connect, by over etching SiCN stop layer, the ILD loss should be controlled at 5-15% during M1 etch. However, in the edge area of 3-13 mm to wafer bevel, the M1 main etch (ME) condition exhibits dramatically higher etch rate, which easily breaks SiCN layer before over etch (OE) step. As a result, chip dies around wafer bevel shows notably higher ILD loss after M1 etch step, as shown in Fig.4. In the outermost chip, the ILD loss is found to be more than 30%, which makes gate poly on HVOX almost exposed. High ILD loss means less space between M1 and gate, which finally induces severe LDMOS leakage.

Therefore, M1 etch body was tuned to optimize the device structure. Since the loading effect is mainly induced in M1 ME step, by adjusting etch rate and ME/OE time ratio, it is found that the excessive ILD loss can be dramatically depressed. Fig 5 displays the typical etch structures and SiCN OE structures obtained via the optimized etch recipe, with the calculated data summarized in Tab 1. Chip dies in wafer center, middle area as well as the outermost area possess uniform ILD loss of ~10%, and enough SiCN OE of above 30%. The optimized recipe ensures that both M1-gate isolation and M1-CT connect in all wafer dies are in good conditions, which helps to get lower LDMOS leakage, higher wafer yield as well as better reliability performance.

## 3. LDMOS layout optimization

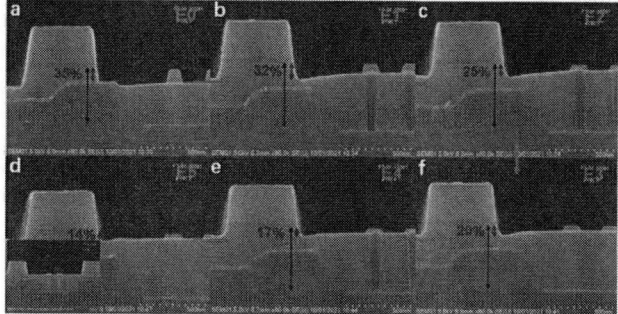

Fig.4. Wafer edge ILD loss induced by initial M1 etch recipe. Chip dies counted from wafer bevel (from E0 to E5, with the distance to wafer bevel ranges from ~3 to 18 mm) are checked after M1 etch step.

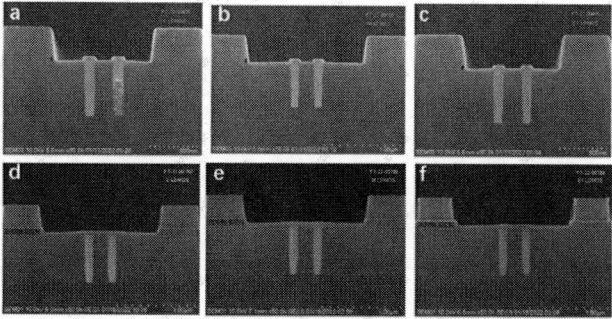

Fig.5. M1 etch recipe optimization. (a-c) Structures obtained in (a) wafer center, (b) wafer middle area and (c) wafer outermost edge via the optimized M1 etch recipe. (d-f) SiCN etch stop layer OE check via the optimized M1 etch recipe.

| Item | Spec | Center | Middle | Outermost |
|------|------|--------|--------|-----------|
| ILD Loss | 5-15% | 10% | 10% | 9% |
| NDC OE | >30% | 70% | 90% | 85% |

Tab. 1. ILD loss and SiCN OE obtained via the optimized M1 etch recipe.

As Cu Damascene technology is applied in 12-inch 90-nm BCD Process, it provides lower resistance and much higher current density than the traditional Al back-end-of-line (BEOL) process. Thus in LDMOS array, the M1 lines used in connecting drain side can be shrunk to smaller size without influencing on-resistance (Rsp) or reliability. As a result, the LDMOS layout is optimized as shown in Fig 6. M1 overlap poly above HVOX is no longer allowed. A lateral space of at least 250 nm from M1 to poly above HVOX is requested and added into topological layout rules (TLR). A much safer distance from M1 to gate/HVOX is designed, which finally eliminates the M1-gate leakage in LDMOS.

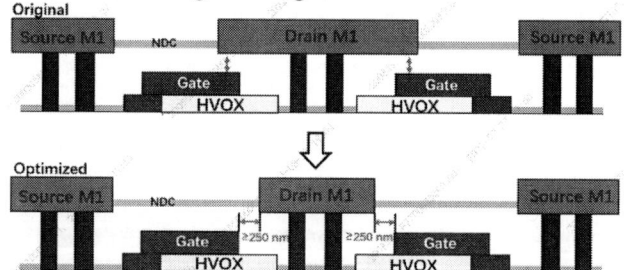

Fig.6. Schematic illustration of the layout optimization to reduce LDMOS leakage.

## Summary

As the development of semiconductor technology, many products are transferred from 8-inch to 12-inch process. Since the wafer size improves while the chip size reduces, the optimization of wafer edge process uniformity is often crucial to yield improvement. Here, we report our investigations on the wafer edge yield loss case appeared in 12-inch 90-nm BCD process, which is due to M1-gate leakage in LDMOS. Influences of ILD thickness, M1 etch recipe, and LDMOS layout were systematically studied. A series of optimization efforts have been applied, helping to improve the total production yield by 5-10%.

## References

[1] D. Riccardi et al., "BCD8 from 7V to 70V: a new 0.18 um technology platform to address the evolution of applications towards smart power ICs with high logic contents," Proc. of ISPSD, pp. 73-76, 2007.

[2] T. Uhlig et al., "A18 – a novel 0.18 um smart power SOC IC technology for automotive applications," Proc. of ISPSD, pp. 237-240, 2007.

[3] Sameer Pendharkar et. al., "7 to 30V state-of-art power device implementation in 0.25 um LBC7 BiCMOS-DMOS process technology," Proc.of ISPSD, pp. 419-422, 2004.

[4] A. Moscatelli et al., "LDMOS implementation in a 0.35 um BCD technology (BCD6)," Proc. of ISPSD, pp. 323-326, 2000.

[5] V. Parthasarathy et al., "A 0.25 um CMOS based 70V smart power technology with deep trench for high-voltage isolation," IEDM, pp.459-462, 2002.

[6] Zhengyuan Z et al., "A new method to reduce VDMOS on resistance in BCD process," 10th IEEE International Conference on Solid-State and Integrated Circuit Technology (ICSICT), pp. 117-119, 2010.

[7] Chang-Tzu W et al., "ESD Protection Design with Lateral DMOS Transistor in 40-V BCD Technology," IEEE TRANSACTIONS ON ELECTRON DEVICES, pp. 3395-3396, 2010.

# Enhanced Fill of Tungsten in 3D NAND Wordline: A View of Molecular Diffusion

## Xin Gan

Lam Research (Shanghai) Co., Ltd.
177 Bibo Rd, Pudong New District, Shanghai 201203, China
E-mail: Xin.Gan@lamresearch.com

## ABSTRACT

The increasing number of stacking layers in 3D NAND memory is challenging the tungsten fill process used in wordline (WL) structures. Herein, molecular diffusion in thermal CVD/ALD is studied to facilitate conformal coverage of tungsten WL fill. A revision of Knudsen diffusivity is proposed based on the geometry of 3D WL.

*Keywords — 3D NAND; tungsten; atomic layer deposition; diffusion*

## INTRODUCTION

Three-dimensional (3D) NAND has emerged and dominated the flash memory market for its high bit density and low cost per bit. Scaling of 3D NAND technology has overcome the challenge of 2D feature shrinkage by increasing the number of cells along the vertical channel. Each level of cells in charge-trap type 3D NAND operates by using metal wordline (WL) electrode to control electrical charge in the silicon nitride trap layer (Fig. 1). The 3D NAND structure requires a WL metal replacement process to fill the stack of high-aspect-ratio (HAR) horizontal gaps with highly conformal coverage. Therefore, chemical vapor deposition (CVD) and atomic layer deposition (ALD) of tungsten have been used in the 3D WL metallization. In pursuit of higher density of memory cells, advanced 3D NAND is designed with more WLs, smaller WL width and wider horizontal dimension (accommodating more of memory holes between two vertical slits). [1] Therefore, the challenge of tungsten CVD/ALD gap-fill is being raised to address the increasing complexity of the structure.

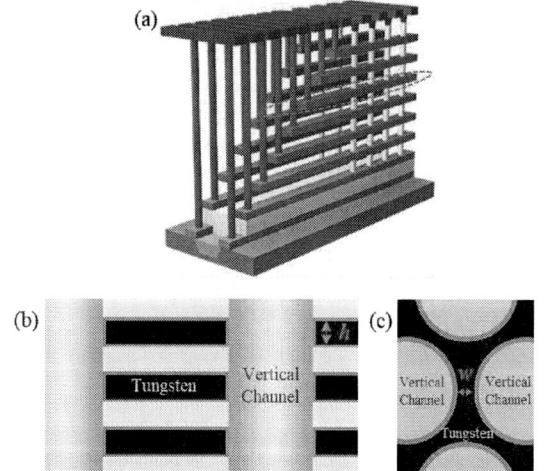

Fig. 1. Schematic illustrations of (a) 3D NAND, (b) side view of 3D WL and (c) top view of 3D WL. The WL is filled by replacing mold sacrificial film with tungsten metal in charge-trap memory technology.

ALD process uses alternating doses of different gas molecules achieve surface-limited reaction. The deposition conformality (coverage) mainly depends on the diffusion of gas molecules into the feature to fill. Herein, diffusion model of tungsten ALD gases into 3D WL feature will be discussed. Due to the confined space in WL structure, Knudsen diffusion is more favorable than Fickian diffusion. In this paper, a specific Knudsen diffusivity inside the actual WL structure is derived from the conventional one.

## RESULTS AND DISCUSSION

Molecular diffusion behavior is described by Fick's laws as below, where J is diffusion flux, D is diffusivity and $\phi$ is concentration of molecule and t is time.

$$\vec{J} = -D\nabla\varphi \tag{1}$$

$$\frac{\partial \varphi}{\partial t} = \nabla \cdot (D\nabla\varphi) \tag{2}$$

In typical of tungsten deposition conditions, different gas molecules have weak interaction in gas phase. Thus, the diffusivity is a product of molecular mean free path (MFP, $\lambda$) and average velocity u. The u is a result of gas constant R, temperature T and molecular weight M. [2]

$$D = \frac{\lambda u}{3} = \frac{\lambda}{3}\sqrt{\frac{8RT}{\pi M}} \tag{3}$$

Knudsen's model describes molecular diffusion in a pore (capillary) with a diameter of d $\ll \lambda$. By replacing $\lambda$ with d in the equation (3), Knudsen diffusivity is derived as below [2].

$$D = \frac{d}{3}\sqrt{\frac{8RT}{\pi M}} \tag{4}$$

For general tungsten CVD, $H_2$ and $WF_6$ are used as reaction gas, and Ar as carrier gas. Table 1 compares the MFP of $H_2$, $WF_6$ and Ar at 1333 Pa and 400 °C, with typical range of gap dimension in 3D WL.

978-1-6654-9759-6/22 $31.00 © 2022 IEEE

**TABLE 1**
Collision path of molecules used in tungsten ALD.

| Spacing | Molecule | Collision Path (nm) | Diffusivity ($m^2$/s) |
|---|---|---|---|
| Free space @ 1333 Pa and 400 °C | $H_2$ | $1.96 \times 10^4$ | $1.74 \times 10^{-2}$ |
| | $WF_6$ | $5.78 \times 10^3$ | $4.21 \times 10^{-4}$ |
| | Ar | $1.25 \times 10^4$ | $2.49 \times 10^{-3}$ |
| Circular pore d = 25 nm | $H_2$, $WF_6$, Ar | 25 | $4.98 \times 10^{-6}$ (Ar @ 400 °C) |
| Circular pore d = 35 nm | $H_2$, $WF_6$, Ar | 35 | $6.97 \times 10^{-6}$ (Ar @ 400 °C) |
| Rectangular pore of 25 × 35nm | $H_2$, $WF_6$, Ar | 32.8 | $6.53 \times 10^{-6}$ (Ar @ 400 °C) |

Because the mean free path is shorter than the physical dimension of 3D WL, Knudsen diffusion model can represent the spatial constraint to the diffusion in tungsten ALD. Nevertheless, original Knudsen model consider only cylindrical pore. In 3D NAND WL, the cross-section of molecular diffusion is a rectangle. The equivalent collision path of a rectangle pore can be calculated as followed [3].

$$d_{eq} = \frac{3}{4}\left[w \, sinh^{-1}\frac{h}{w} + h \, sinh^{-1}\frac{w}{h} - \frac{(w^2+h^2)^{\frac{3}{2}}}{3wh} + \frac{w^3+h^3}{3wh}\right]$$

(5)

As shown in Table 1, diffusivity of Ar with Knudsen model is much lower than Fickian model. The dimension in Table 1 is not real value of 3D NAND structure, but close to real ones [4]. Based on the revised Knudsen model, smaller WL gap leads to less diffusivity of gas molecules, thus enhanced diffusion is needed for gas molecules to reach inner space of 3D WL. The diffusion model is useful for optimizing tungsten gap-fill. For example, the diffusion-enhanced processes can result in excellent step coverage (>90%) [5]. Besides diffusion of gas molecules, the reaction chemistry and physics of tungsten are also utilized in the process development for WL fill of 100- to 300-layer 3D NAND.

The partial differential equation of diffusion model (Equation 2) is relatively simple as there is analytical solution in some cases. Adding reaction and convection into diffusion model [6] can yield more precise result, while extensive computation and experiment are needed.

## CONCLUSION

In this paper, the diffusion model for ALD tungsten fill in 3D NAND WL is refined. Different diffusion scenarios are compared, showing that the confined dimension of WL structure impedes gas molecules reaching the innermost space. The diffusivity is even worse when the WL tungsten fill is close to pinch-off. By using diffusion model combined with deposition chemistry and physical properties of tungsten, the ALD gap-fill of ~300-layer 3D WL structure can be improved.

## REFERENCES

[1] H. P. W. Hey, W. A. Mazak, R. K. Aggarwal, J. H. Curtin, "High throughput multi station processor for multiple single wafers," U.S. Patent No. 4 987 856, January 29, 1991.

[2] J. R. Welty, G. L. Rorrer, and D. G. Foster, Fundamentals of momentum, heat and mass transfer. Hoboken, NJ: John Wiley & Sons, 2015.

[3] S. L. Matson and J. A. Quinn, "Knudsen diffusion through noncircular pores: Textbook errors," AIChE Journal, vol. 23, no. 5, pp. 768–770, September 1977.

[4] 2020 Annual Memory Process Seminar, TechInsights, January 2021.

[5] L. Schloss and X. Ba, "Tungsten Films Having Low Fluorine Content," U.S. Patent No. 9 754 824, September 05, 2017.

[6] C. E. Baukal, Jr., V. Gershtein, and X. J. Li, Computational Fluid Dynamics In Industrial Combustion. S.L.: CRC Press, 2019.

# EFFECTS OF STACKED AL$_2$O$_3$ / HFO$_2$ FILMS DEPOSITED IN DEEP TRENCH ISOLATION ON WHITE PIXELS NOISE

*Qixin Wu, Hebao Liu, Haitao Jin, Felix Li, Enjoy Yin, Horse Ma, Xinhe Zheng\**
University of Science and Technology Beijing
Semiconductor Manufacturing International Corporation
Beijing, P. R. China
18801001577, 18801001577@163.com
*Correspondence author: xinhezheng@ustb.edu.cn

## ABSTRACT

In the stacked-BSI CMOS image sensor process, deposition of high-k thin films in deep trench isolation (DTI) to further reduce white pixels noise has attracted lots of attention. In this paper, we explore the influence of stacked two films of Al$_2$O$_3$/HfO$_2$ on the white pixel properties of the CMOS image sensor. Here, the two oxide layers are deposited using PEALD. We found that the as-deposited Al$_2$O$_3$ / HfO$_2$ (TH1/TH1) films show a suppression of the white pixels up to 175 PPM(P95). This achievement could open a possible way to remarkably further reduce the white pixels noise.

## I. INTRODUCTION

In CMOS image sensor (CIS) technology, the isolation of pixel area only by ion implantation cannot meet the requirements of dark current and white pixel. Therefore, physical isolation on the back of the structural wafer is needed. The importance of deep trench isolation technology (DTI) in backside illumination (BSI) is self-evident. It makes a great contribution to reducing the crosstalk and dark current phenomenon caused by recollection with incident light due to scattering to adjacent pixels or back interface. [1]

In this process, we have two directions to improve the leakage current. One is to reduce the ion bombardment during etching on the damaged area at the bottom of the deep groove or repair the damaged area. The depth of the groove can reach two-thirds of the wafer thickness, and the depth width ratio is more than ten to one. In order to meet the demand of this ratio, we use the Bosch process to etch deep grooves. Reducing bias voltage and increasing soft landing can reduce etching damage. [2,3]

The other is to protect the sidewall of the deep groove by depositing high dielectric electrostatic film. In recent years, the research on using high-k materials to reduce the leakage current between interfaces or substrates has become much hotter. Among them, Al$_2$O$_3$ is widely used because of its excellent passivation performance. In this paper, Al$_2$O$_3$ is combined with the traditional high-k material HfO$_2$ to study the effects of film thickness and stacking on leakage current.[4] When the HfO$_2$ layer is placed at different positions in the Al$_2$O$_3$ layer, it strongly

enhances electron capture. It is known that the interface of Al$_2$O$_3$ and HfO$_2$ has a high density of trap sites. This increased interface trapping could explain the better performance of the Al$_2$O$_3$/HfO$_2$ stacks. [5,6]

## II. EXPERIMENTAL

Twelve-inch p-type raw material wafers were selected for the experiments. In order to avoid the influence of process instability, we set the control group as the basic condition of the experiment. The fabrication began with a p-type bulk silicon wafer which has a thin layer of buried oxide and a lightly doped p-type Si active layer. The main structure is shown in figure 1.

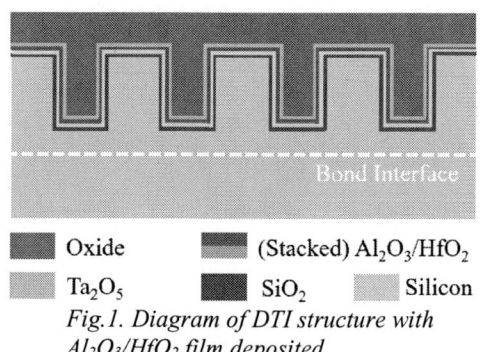

*Fig.1. Diagram of DTI structure with Al$_2$O$_3$/HfO$_2$ film deposited*

An oxide film is grown on the Si surface as a glue layer

*Tab. 1. Different stacking modes of high dielectric films*

| Group | Before High-K | Al$_2$O$_3$ | HfO$_2$ | Plies | After High-K |
|-------|---------------|-------------|---------|-------|--------------|
| A | same | TH1 | none | 1 | same |
| B | same | TH1 | TH1 | 1 | same |
| C | same | TH2 | TH2 | 1 | same |
| D | same | TH1 | TH1 | 2 | same |
| E | same | TH2 | TH2 | 2 | same |
| F | same | TH3 | TH3 | 2 | same |
| G | same | TH4 | TH4 | 3 | same |
| H | same | TH5 | TH5 | 6 | same |

between Al$_2$O$_3$ / HfO$_2$ and the silicon wafer. Al$_2$O$_3$ thin films were grown by PEALD at a certain temperature using $O_3$ as oxygen sources and TMA as Al sources. Then HfO$_2$ thin films were grown at the same temperature using $O_3$ as oxygen sources and TDMAH as Hf source. $Ta_2O_5$ has a high refractive index and light transmittance.[7] In order to increase the amount of light entering the PD area and reduce the light hitting the sidewall, a $Ta_2O_5$ film was grown after high-K film deposition. Finally, fill the deep groove up with oxide. Table 1 gives different conditions for different wafer experiments. The specific value of film thickness is replaced by the symbol TH. From TH1 to TH6, the films are getting thinner.

## III. RESULTS AND DISCUSSION

Figure 2 shows the cross-section of the deep groove and the interface of the deposited films. In this figure, we can see the stacked films clearly. The excellent passivation performance of $Al_2O_3$ High-K film is related to its negative fixed charges with a high density, which forms a hole-rich accumulated layer nearby the DTI-Si surface. It pushes the photon electron far away from the Si surface.

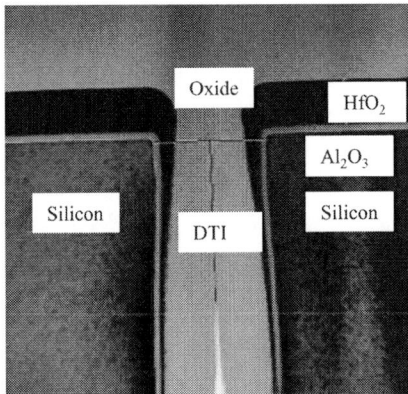

*Fig.2. TEM of the DTI cross section after films deposited.*

Figure 3 shows the white pixel properties of the experimental wafers. Unfortunately, the WP properties of wafer C with $Al_2O_3$ / HfO$_2$ thickness in TH2 are very poor due to process abnormalities. Therefore, subsequent discussions are carried out on the basis of excluding C. Figure (a) shows the probability of the WP properties for each wafer. The performance of groups G and H is much worse. Among the remaining wafers, groups B, D, and F have better WP properties than control group A. It can be determined that the WP properties can be improved by using two film stacking methods. In this experiment, group D with $Al_2O_3$ / HfO$_2$ thickness in TH1 shows the best WP properties, which reduces WP properties from 289 ppm(P95) of group A to 175 ppm(P95), improving 39%.

From figure 4 we can understand how this process happens. There are defect capture sites at the interface between alumina and hafnium oxide, so a negative fixed charge layer will be formed. Under the action of electric

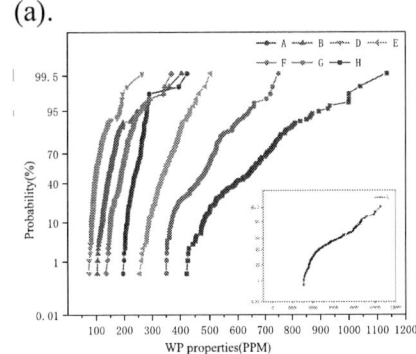

(a).

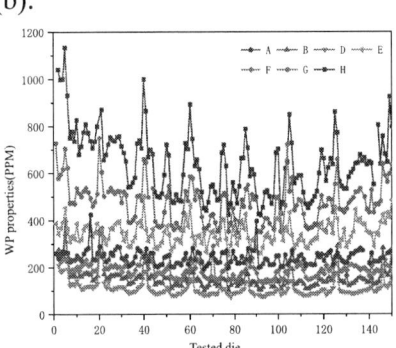

(b).

*Fig.3. (a) WP properties of different film stacking modes;(b) WP properties for each tested die.*

field force, the photogenerated carriers will be far away from the DTI sidewall to reduce the white pixel noise.

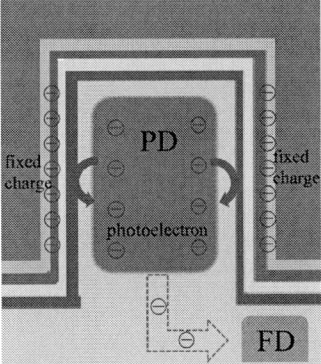

*Fig.4. Fixed charges at the interface reduces leakage*

Although group G and H have more stacking layers, the film of each layer is not thick enough, which leads to the insufficient negative fixed charge density at the interface.

Figure (b) shows the detail of the WP properties for every tested die in each wafer. For the experimental wafers with high white pixel noise, the variance of results obtained at different test positions is also large. Another important conclusion is that for group F, the white pixel noise of some measured die has been reduced to 80ppm. This provides an idea for our next work to further reduce white pixel noise.

The experimental results show that the stacked $Al_2O_3$ / $HfO_2$ films will produce more fixed negative charges at the interface layer. The interfacial charge density can be controlled by depositing $Al_2O_3$ / $HfO_2$ films with different thicknesses. The different combinations of the two films will also affect the interfacial charge density. In our research, the combination of $Al_2O_3$ / $HfO_2$ /$Al_2O_3$ / $HfO_2$ with thickness in TH2 for each layer produces more trapping sites among interfaces. Therefore, it can push more photoelectrons far away from the silicon surface, so as to reduce the leakage current and improve the properties of white pixels.

In addition, the stacked high-k films will induce an electric field on the sidewall of the deep groove. In the process of etching a deep groove, the sidewall will inevitably be affected by etching. The damaged area will adversely affect the properties of white pixels. Thus, the induced electric field will also drive the photoelectron away from the etching damaged area of the sidewall.

## IV. CONCLUSION

By using stacked $Al_2O_3$ / $HfO_2$ films, we can get better white pixel properties. Reducing the thickness of each layer and increasing the number of stacked layers cannot effectively improve the interface charge density, so the properties of white pixels will become worse. In this experiment, we did find a better condition. A layer of $Al_2O_3$ with TH1 thickness and a layer of $HfO_2$ with the same thickness can improve the properties of white pixels by 39%. Our next goal is to combine stacked high-k films with deep trench isolation at different depths in order to further reduce leakage current and white pixels.

## REFERENCES

[1] El, Gamal, A, et al. CMOS image sensors[J]. Circuits & Devices Magazine IEEE, 2005.

[2] Ertl G. The Development of the Haber-Bosch Process. 2013.

[3] Kissel D E. The historical development and significance of the Haber Bosch process. [J]. better crops with plant food, 2014.

[4] Simon D K, Jordan P M, Mikolajick T, et al. On the Control of the Fixed Charge Densities in Al2O3-Based Silicon Surface Passivation Schemes. [J]. Acs Applied Materials & Interfaces, 2015, 7(51):28215.

[5] Guo Y, Zhang X. Simulation of silicon solar cells with atomic layer deposited Al2O3 as passivation layers[C]// 2018 5th International Conference on Electrical and Electronic Engineering (ICEEE). 2018.

[6] Lai S C, Chen C P, Du P Y, et al. A study of barrier engineered Al2O3 and HfO2 high-K charge trapping devices (BE-MAONOS and BE-MHONOS) with optimal high-K thickness[C]// Memory Workshop. IEEE, 2010.

[7] Guo P, Xue Y, Huang C, et al. Optical Properties and Elemental Composition of Ta2O5 Thin Films. IEEE, 2009.

# AN ACHIEVEMENT OF LOW RESISTANCE NON-SALICIDE CT ON ACTIVE AREA

*Bingquan Wang, Simmons Zhang, Zhigao Wang*

Semiconductor Manufacturing International Corp., Beijing, China

BingQuan_Wang@smics.com

## ABSTRACT

Metal ions' contamination could degenerate the detection accuracy and increase the noise of Si-based photoelectric devices. Non-salicide contact technique has been developing to avoid metal ions' contamination during manufacturing process as much as possible. In this paper, several adjustments of process have been attempted to reduce the contact resistance between CT and Si without salicide loop. By using of proper process conditions including IMP, glue layer, and thermal, the contact resistance has been reduced from 916.6 ohms/CT to 85.5 ohms/CT, which shows significance for Si-based photoelectric and relevant whole circuits to reach a good performance.

*Keywords— photoelectric device; non-salicide CT; low M-S contact resistance*

## INTRODUCTION

With rapidly development of consumer electronics, intelligent driving and the internet of things through the last decade, the demand and industrial scale of photoelectric chips are becoming larger and larger. Si-based photoelectric sensor, such as CMOS image sensor (CIS) and time of flight (TOF) sensor has been widely used as pixel device, since it's easily to be integrated with other logic devices on wafer level package.

With the development of Si-based photoelectric sensor in sub-90μm node design, metal ions' contamination is more and more important to be reduced to improve the optical performance, thus silicide materials in pixel devices are considered to be reduced as much as possible since it's the largest metal source in FEOL process and contact with Si area directly. However, if salicide process loop is removed from mature sub-90μm node process, the contact resistance between CT and Si becomes very large (even can reach to MΩ level every single CT), which is not conducive to relevant logic circuits design and function achievement. On the other hand, to achieve proper optimized design of pixel devices, there's usually not too much layout area for CT design, thus the CD (critical dimension) of CT can't be too large and should match with the pixel/logic process node.

In a word, it's necessary to remove salicide loop from mature process and find a method to achieve a low level CT resistance. If we consider in retrospect the initial process of the above micron level node without salicide loop, silicide is formed just under the local large CT area. Thus, the similar local silicide idea can be transplant to the sub-90nm node, which is supposed to achieve the balance of little metal ions contamination and low M-S contact resistance.

In this paper, local Ti-silicide is formed from CT gule layer Ti/TiN to reduce the CT-Si contact resistance. Furthermore, by adjusting the process relevant parameters, resistance has been reduced to an acceptance level.

## CONTACT PROPERTIES

As known well, the contact type between glue layer metal W and Si is more likely as Schottky diode [1], which shows a barrier for carrier transportation on the interface. Silicide plays a role that make the contact more likely to ohmic contact, since silicide acts like both metal and Si [2].

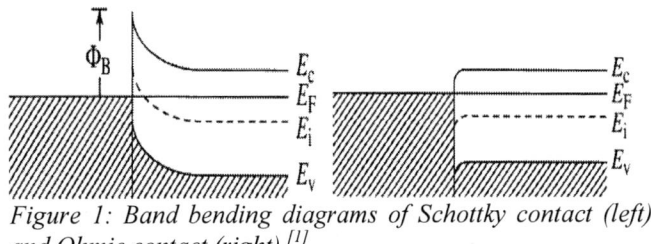

*Figure 1: Band bending diagrams of Schottky contact (left) and Ohmic contact (right)* [1]

Ti silicide can be formed by Ti and Si contacting and chemically reacting in high temperature ambience [3], thus it was proposed to add a thermal process after CT loop to make local Ti silicide formed by glue layer Ti and Si. An RTA thermal process was added, and N+/P+ IMP dosages have been adjusted to the same level, and final CT resistances (Rc) have been measured by wafer acceptance test.

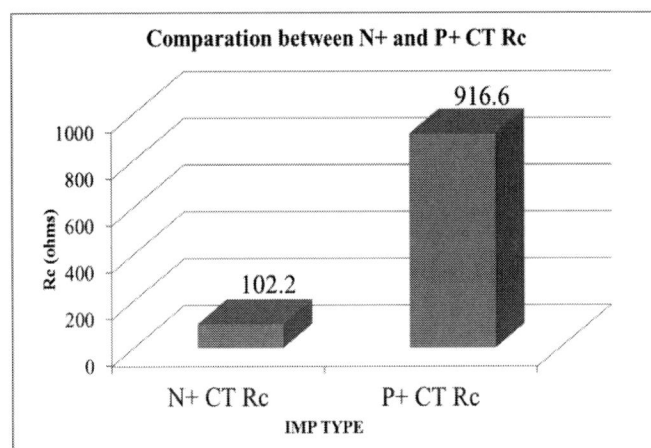

*Fig 2: Large size CT resistance on P+ and N+ active area*

According to Fig.2, It was found that CT resistance on the P+ active area was much larger than that on the N+ active area. Then we tested same small size CT resistance with currents in both forward and reverse directions and with different magnitudes, as shown in Fig.3.

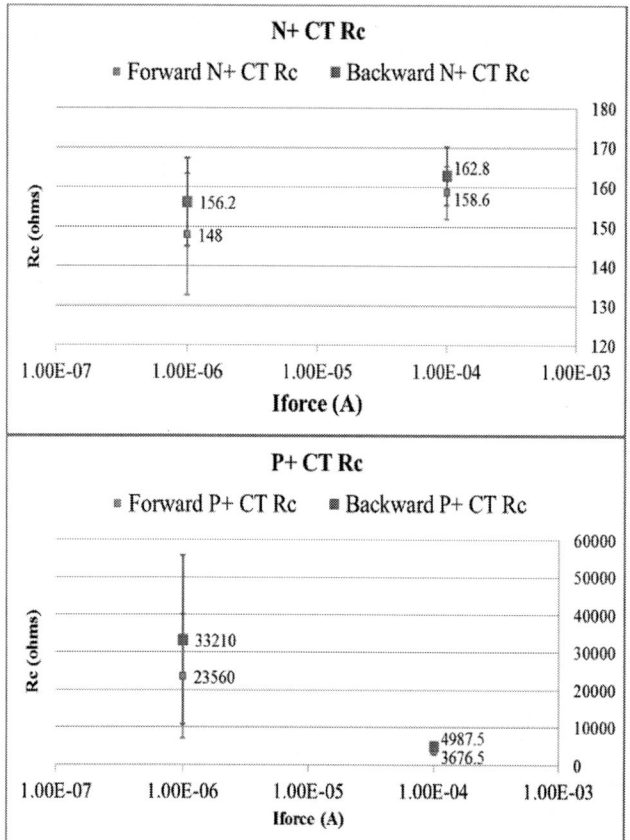

*Figure 3: Small size N+ and P+ CT Rc tested with currents in both forward and reverse directions and with different magnitudes*

It was found that N+ CT Rc values are closer, which indicated that contact between N+ active area and CT was very closer to ohmic contact with simply adjustment. However, as to P+ CT Rc values, there was large gap between different current magnitudes condition, and even larger gap between forward and reverse current directions conditions. Thus, it was believed that contact between P+ active area and CT showed little ohmic properties and was highly influenced by " Schottky" like non-ohmic effect. By comparing N+ and P+ CT Rc values, it was believed that if the contact type between P+ active area and CT could be improved to ohmic contact, the P+ CT Rc can be greatly reduced.

## P+ CT OHMIC CONTACT IMPROVEMENT

To improve the contact between P+ active area and CT, research of ohmic contact achievement and Ti-silicide properties have been done.

According to Fig.1, to improve the contact type from " Schottky" to ohmic, one method is to reduce the width of the barrier to makes it easier for carriers to tunnel through the barrier (as shown in Fig.4), which means the IMP dosage of Si active area under CT should be heavily p-type doped, as also the doped depth should be proper considering the position of CT bottom and silicide; the other method is to lower the barrier between metal and silicon, which means the local silicide

properties should be improved.

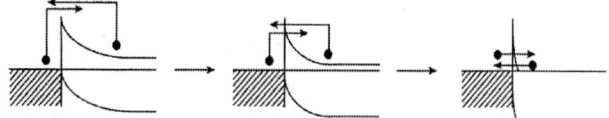

*Figure 4: Illustration of heavily doping in Si active area under CT to improve carriers tunneling and ohmic contact [1]*

When Ti is much less than Si, the composition of stable Ti-silicide is $TiSi_2$ [3]. The transforming steps follow [4]: Ti and Si firstly forms $Ti_5Si_3$, then C49 $TiSi_2$ forms, and finally C49 transforms to C54. C54 is a low sheet resistance phase compared with C49 phase. The transformation from C54 to C49 can be more radical with smaller grain size [5]. By properly adjusting the formation of local Ti-silicide to make it transformation more radical, the contact resistance between CT and P+ Si active area can be reduced obviously. The relevant process includes: pre-clean process, glue layer thickness, thermal temperature, thermal duration, etc.

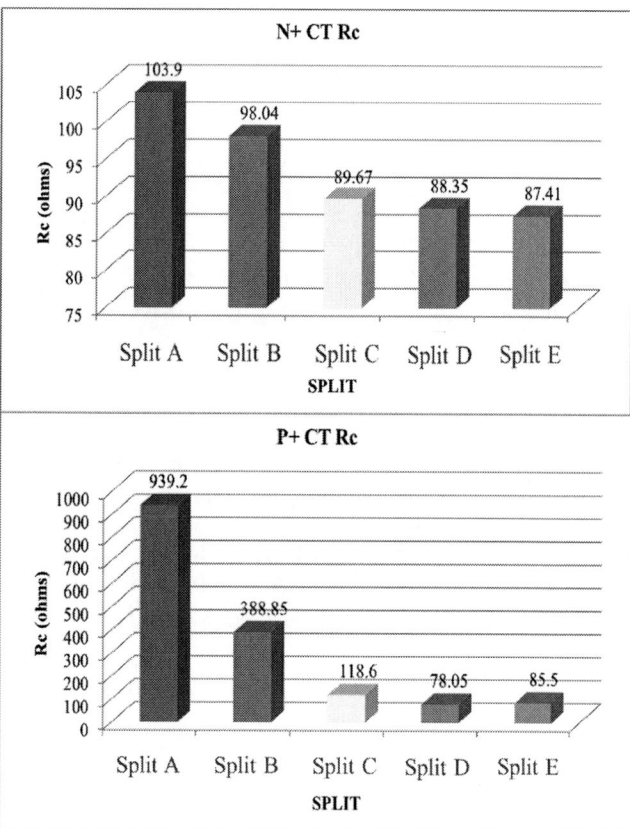

*Figure 5: N+ CT Rc and P+ CT Rc values in different process split.*

As shown in Fig.5, by adjusting the above-mentioned silicide forming relevant process parameters, and adjusting the IMP dosage and energy, the large size P+ CT resistance value was greatly reduced from 916.6 ohms/CT to 85.5 ohms/CT. Meanwhile, N+ CT resistance value also achieved a little reduction.

## CONCLUSION

978-1-6654-9759-6/22 $31.00 © 2022 IEEE

In this paper, on the basis of mature sub-90nm node process, to reduce the ions' contamination Si-based photoelectric device, non-salicide CT has been investigated with local silicide formation. By adjusting the relevant process, low N+/P+ resistance non-salicide CT on active area has been achieved.

## REFERENCES

[1] Robert F. Pierret, *Semiconductor Device Fundamentals. Addison Wesley,* April 1996.

[2] Gambino, J.P., E.G. Colgan, *Materials Chemistry and Physics,* vol. 52, 2, 1998, pp:99-146.

[3] Maex K., *Materials Science and Engineering: R: Reports*, vol. 11, 53, 1993.

[4] Z.B. Zhang, S.L. Zhang, D.Z. Zhu, H.J. Xu, Y. Chen, J*ournal of Materials Research*, vol. 17, 2002, pp:784.

[5] H. J. W. van Houtum, I. J. M. M. Raaijmakers, T.J. M. Menting, *Journal of Applied Physics*, vol 61, 8, 1987, pp: 3116-3118.

# DEVELOPING OF CMP HEAD-TO-HEAD COMPENSATION FUNCTION FOR GATE HEIGHT UNIFORMITY CONTROL

*Yurong Que[1]\*, Xing Ma[1], Shuxiang Wang[1], Jian Zhang[1], Haifeng Zhou[1], Jingxu Fang[2], Yu Zhang[2]*
*Advanced Module Technology Dept., Shanghai Huali Integrated Circuit Corp., Shanghai 201314,*
*China*
\*Corresponding Author's Email: queyurong@hlmc.cn

## ABSTRACT

The performance and yield of advanced devices directly depend on the control of within wafer uniformity during each chemical mechanical polishing (CMP) step, as multiple CMP steps have been developed to define the structure to meet the criteria of High-k metal gate (HKMG) devices. As a consequence, to meet the increasingly stringent requirements, advanced process control (APC) has become a common process regulating approach in the semiconductor industry, and the needs of APC function for CMP is increasing greatly along with the technology nodes advancing to 28nm and below. So far, APC has been used for within wafer uniformity (WiW) control based on consumable lifetime, but it cannot cover consumable variation. In this paper, a novel and useful APC model is built to achieve head-to-head (HtH) compensation function to control within wafer uniformity, especially for wafer extreme edge (WEE), during each period maintain (PM) cycle of machine. This function is developed for gate height uniformity control of inter-layer dielectric level zero (ILD0) CMP.

## INTRODUCTION

As the feature size of integrated circuit (IC) gradually decrease, the process control and stability enhancement of chemical mechanical polishing (CMP) are of great importance for surface planarization and within wafer uniformity[1-3]. To meet the surface planarization requirement of advanced node, many advanced in-situ process control techniques have been explored and applied in CMP process in succession, such as the advanced process control (APC) system[4-6] and the end-point (EP) system of a process machine[7]. However, many small changes of polishing parameters (like polish time, polish downforce, pad conditioning, condition cutting rate, condition downforce slurry flow rate, platen rotation speed, etc.) and small variations of consumables (like pad hardness, compressibility, pore size, porosity, grooving geometry, head retain ring thickness, etc.) always bring interference to the control of CMP process[8-12]. Thus, any considerably small variation needs to be counted in and compensated appropriately.

Inter-layer dielectric level zero (ILD0) CMP undertakes the polish task of dielectric film and the open task of dummy polysilicon (Poly-Si) gate which will be subsequently removed in the following steps. Then, the work function and aluminum metal will be deposited in the gate space to form high-k metal gate (HKMG). Therefore, ILD0 CMP plays an important role in the gate-last scheme of the typical random numeral generator (RMG) process flow. It should be noted that ILD0 CMP is also critical for the control of gate height. Good process control capability for minimizing within wafer gate height variation is necessary to achieve good device performance and reliability. In the worldwide market share, ILD0 CMP almost 100% employs the AMAT LK tool which includes three polish platens (Figure 1). On the first platen (P1), the dielectric oxide above the contact etch stop layer (CESL) SIN is removed with high selectivity slurry through EP system. On the second platen (P2), nitride on the gate is removed to expose the Poly-Si gate and control Poly-Si gate height. On the third platen (P3), a chemical clean agent is introduced to modify the hydrophobic to hydrophilic surface for defect improvement. It should be noted that the P1 process puts forward a higher requirement for the within wafer (WiW) uniformity, since it masters the start position of P2 process. The thicker local position may confront the risk of suffering SIN residue which can lead to integration issues like polysilicon residue, while the thinner local position may eventually present a lower gate height. However, the wafer extreme edge (WEE) (141mm and above) profile control is a great challenge for a long time[13]. Polishing head with retaining ring is a critical consumable in CMP process, which contacts with wafer directly and affects the removal rate (RR) as well as WiW uniformity. The different retaining ring thickness, and topography have different influence for polish performance, especially for WEE. In addition, numerous consumables like head, pad, disk, brush, machine need to do PM nearly every day for lifetime concern, in which the newly PM head is an important factor for the irregular shift of WEE RR profile. At present, in order to accurately control the WiW uniformity of gate height, the mainstream approach adopts the double-head operation after break-in, which will certainly sacrifice the capacity.

To meet this challenge, we have not only carefully studied the relationship between WiW uniformity with the pressure of different zone in the contour head, but also developed a novel APC model based on the rational retaining ring pressure feedback mechanism. This technique aims to solve the unpredictable variation of consumables in controlling WiW uniformity of gate height.

## EXPERIMENT

The Head-to-Head compensation function was setup on an Applied Material Reflexion LK APC system with 5-zone Titan Contour polishing heads for silicon oxide (OX) polishing with ceria slurry. This oxide polishing step is performed on Platen 1 of the Reflexion LK polisher and best known method (BKM) consumable set was used for these experiments. For OX CMP process characterization, 300mm silicon wafer blanket wafer was deposited by a thickness of 3000Å CVD (chemical vapor deposition) HARP (high aspect ratio process) OX. These blanket wafer is used to characterize process performance, such as RR and RR uniformity (U%), which were measured by NOVA i500. The function was online for ILD0 CMP of 28HKMG WiW uniformity control of gate height, a gate test structure was used to evaluate the performance of different processes, and the gate height were measured by NOVA T650.

## RESULT AND DISCUSSION

### Correlation of P1 blanket wafer RR with pattern wafer gate height in different use stage of polish head

As for the ILD0 CMP scheme in 28 nm technique, the bulk OX above CESL SIN should be adequately removed by the high selective slurry in P1, which is critical for P2 to achieve the polishing target of removing nitride and controlling Poly-Si gate height. Nevertheless, it is found that the WiW uniformity is always influenced by the use stage of P1 polish head during the same head PM cycle. In the initial use stage of new polish head, the WEE in P1 blanket RR profile shows irregular fluctuation then other zones. Such irregular fluctuation of WEE RR leads to the inadequate or excessive OX removal of pattern wafer in P1 polishing, which will directly affect the P2 polishing and transmit to the final gate height profile. As shown in Figure 1, when the P1 WEE shows a higher RR than other zones (Figure 1a), the final gate height at WEE is lower than wafer center (WC) (Figure 1b); on the other hand, the gate height at WEE is extremely higher than WC (Figure 1d) while P1 WEE exhibits a lower RR (Figure 1c). Although the polishing behavior of head tends to be stable and WiW uniformity gets better with enough usage of head (Figure 1e,f), it is urgent to develop an universal method for solving the fluctuation problem of WEE RR at each head PM cycle, since it is of great importance for capacity demand and economic benefits.

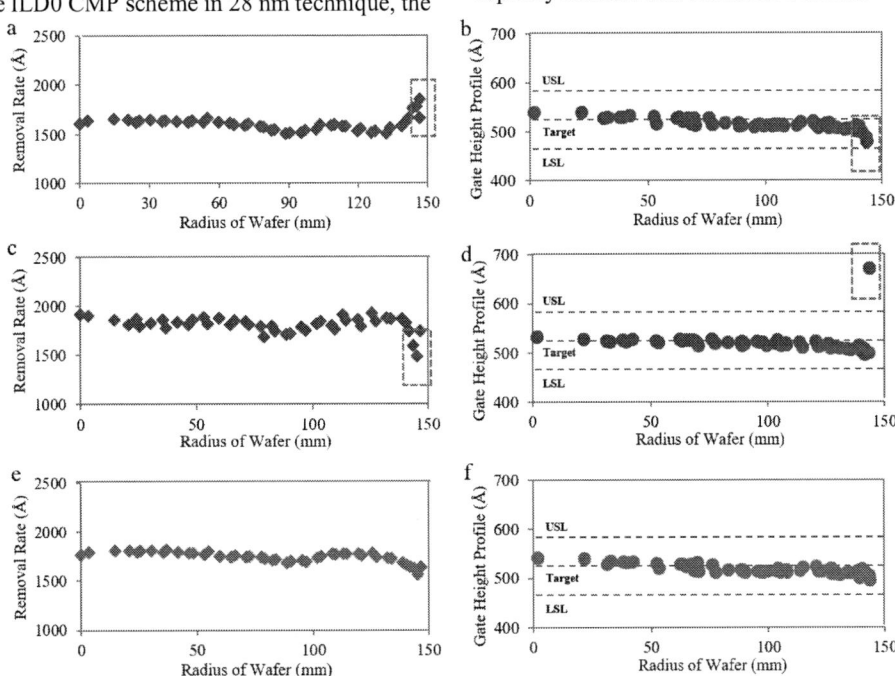

*Figure 1: a, c) The unhealthy P1 RR profile of blanket wafer in the initial use stage of new polish head, b, d) the unhealthy gate height profile of pattern wafer in the initial use stage of new polish head (USL: upper specification limit, LCL: lower specification limit), e) the expected P1 RR profile of blanket wafer with enough usage of polish head, f) the expected gate height profile of pattern wafer with enough usage of polish head.*

### Accurate control of WEE RR and gate height by building a novel and useful APC model

In advance CMP technique, the contour head is designed as multiple zone for exerting different pressure to control the RR profile. To meet the challenge of WEE RR controlling, it is expected to establish an efficient pressure feedback mechanism which can form a continuously improve process (CIP) of pressure to improve the WiW uniformity based on the RR of WEE in the previous polishing round. As we known, although the tuning of Z1 and Z2 pressure can influence the RR profile between 130mm to 145mm, the WEE RR profile is mainly decided by retaining ring pressure.

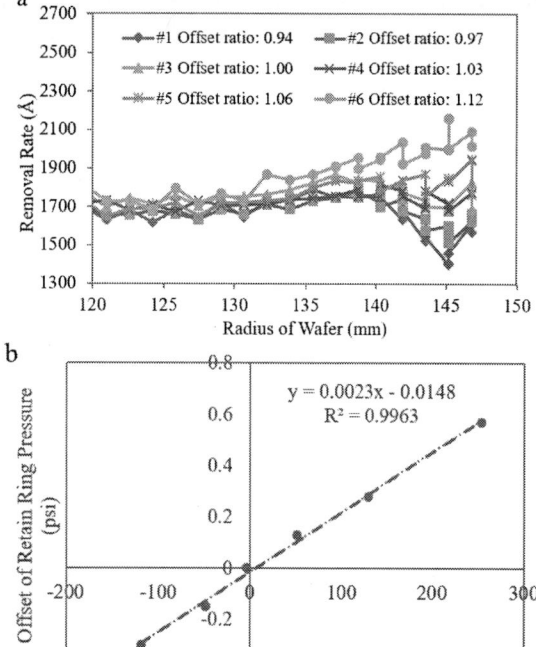

*Figure 2. a) The WE (wafer edge) RR profile with the retaining ring pressure possess different offset ratio, b) the linear relationship with the RR gap and the offset value of the retaining ring pressure.*

Table 1. DOE of different retaining ring pressure and the resulting RR gap.[a]

| Item | #1 | #2 | #3 | #4 | #5 | #6 |
|---|---|---|---|---|---|---|
| Offset ratio | 0.94 | 0.97 | 1.00 | 1.03 | 1.06 | 1.12 |
| Offset value (psi) | -0.30 | -0.15 | 0.00 | 0.13 | 0.28 | 0.57 |
| DOE (psi) | 4.75 | 4.90 | 5.05 | 5.18 | 5.33 | 5.62 |
| RR gap (Å) | -118.5 | -48.0 | -3.2 | 51.5 | 129.0 | 253.3 |

a: The $P_{BSL}$ of retaining ring is 5.05.

To gain detailed relationship of retaining ring pressure and WEE RR profile, the design of experiment (DOE) with different offset ratio of retaining ring pressure was carried out. As is shown in Figure 2 and Table 1, the gap between WEE RR and the average RR (RR gap) exhibits a linear relationship with the offset value of retaining ring pressure, which possess a linear fitting slope

of 0.0023 and $R^2$ of 0.9963. The research result indicated that this strong linear correlation can be utilized to build the efficient retaining ring pressure feedback mechanism for controlling WEE RR profile.

Based on the above mentioned linear formula, a novel APC model was built (Figure 3a) and then verified by blanket wafer. Firstly, based on the PBSL of 5.33 psi (BSL, baseline), the blanket wafer was polished on P1 by head 1-4, respectively. View from the WE RR profile in Figure 3b, the WEE RR present large fluctuations, and which were transported to the APC computer. Subsequently, APC system calculated the RR gap (44.2 Å, 33.1 Å, 93.9 Å and 50.2 Å) and the PCIP of retaining ring (5.23 psi, 5.25 psi, 5.11 psi and 5.21 psi) which were supplied to the tool for next round of work (Table 2). As clearly shown in Table 3, the RR gaps all get obvious decrease based on the PCIP, which is of great importance to the improvement of WiW uniformity and the realization of HtH compensation (Figure 3c).

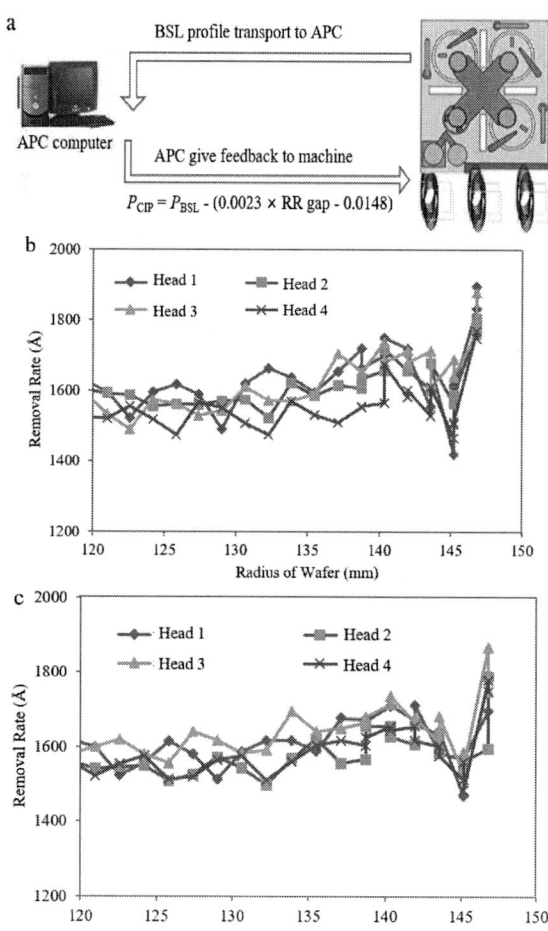

*Figure 3. a) The efficient retaining ring pressure feedback mechanism, b) the WE RR profile based on $P_{BSL}$, c) the WE RR profile based on $P_{CIP}$.*

In order to verify the feasibility of this method on pattern wafers, a series of pattern wafer was subsequently polished (Figure 4). As expected, the gate heights of pattern wafer edge are all on target with the average of 531.4 Å and a small standard deviation of 5.4 Å, suggesting the superiority of this APC model in optimizing WiW uniformity based on the retaining ring pressure feedback mechanism.

*Table 2. Verification of the retaining ring pressure feedback mechanism on blanket wafer.*

| Condition | Item | Head 1 | Head 2 | Head 3 | Head 4 |
|---|---|---|---|---|---|
| BSL | $P_{BSL}$ (psi) | 5.33 | | | |
| | RR gap (Å) | 44.2 | 33.1 | 93.9 | 50.2 |
| CIP | $P_{CIP}$ (psi) | 5.23 | 5.25 | 5.11 | 5.21 |
| | RR gap (Å) | 17.5 | 25.8 | 20.9 | 28.5 |

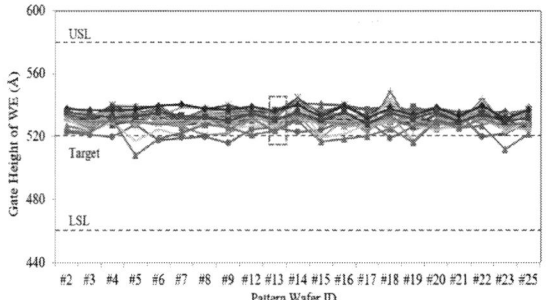

*Figure 4. The gate height of pattern wafer edge (R > 125mm).*

## CONCLUSION

In summary, we have developed a novel APC model based on the rational retaining ring pressure feedback mechanism, and applied in the gate height uniformity control of pattern wafer. The greatly optimized WEE RR and gate height indicates that this useful APC model can not only enhance the WiW uniformity and decrease the risk of residue, but also can realize the HtH compensation and boost economic benefits. This work provides significant inspiration for the exploration of other new APC function to optimize CMP process.

## ACKNOWLEDGEMENTS

All the experiment data was collected by Y. Que in HLMC. Y. Que, J. Zhang, H. Zhou, J. Fang and Y. Zhang have well discussed and analyzed these data to come out this article.

## REFERENCES

[1] C.-C. A. Chen, H.-T. Young, C.-H. Chiou, M.-Y. Xue, C.-L. Pan, Study on CMP Process of Glass Wafers with SiO$_2$ Based Slurry for Trench-Glass-Via Interposer, CSTIC, 2016.

[2] H.M. Wang, I. Kobata, T. Ishibashi, G. Stapf, D. França, Advanced CMP Processes for 450mm Applications, ASMC, 2016.

[3] J.G. Pan, S. Hung, Q.F. Zhang, C. Lu, Y. Cao, B. Qian, D. Chu, E. Su, W. Yang, Innovative CMP Solution for Advanced STI Process, ICPT, 2017.

[4] A. Tsen, Y.-W. Cheng, Y.-P. Hsieh, A novel multiple resolution APC on CMP and Litho-Etching, eMDC, 2013.

[5] Y. Sun, J. Reichelt, T. Bormann, A. Gondorf, A multi-step wafer-level run-to-run controller with sampled measurements for furnace deposition and CMP process flows: APC: Advanced process control, ASMC, 2016.

[6] T. Morisawa, H. Kobayashi, Y. Takeda, Non-Linear Process Model for CMP-APC, ICMSS, 2011.

[7] T.K. Das, R. Ganesan, A.K. Sikder, A. Kumar, Online End Point Detection in CMP Using SPRT of Wavelet Decomposed Sensor Data, IEEE TRANSACTIONS ON SEMICONDUCTOR MANUFACTURING, 2005, 18, 440-447.

[8] H. Doi, M. Suzuki, K. Kinuta, Effect of Ce$^{3+}$ on removal rate of ceria slurries in chemical mechanical polishing for SiO$_2$, ICPT, 2014.

[9] D. Lim, H. Kim, B. Jang, H. Cho, J. Kim, H. Hwang, A novel pad conditioner and pad roughness effect on Tunsten CMP, ICPT, 2014.

[10] F.C. Meyer, C.H. Kuo, C. Rudolph, P. Faustmam, Pad roughness effect on removal rate and selectivity in a STI ceria CMP process, ICPT, 2007.

[11] Y.L. Liu, W.S. Sie, C.L. Chen, P.C. Huang, Y.T. Li, R.G. Lin, Y.M. Lin, H.K. Hsu, O. Wang, J.F. Lin, J.Y. Wu, Defect reduction with CMP pad dressing optimzation, ICPT, 2014.

[12] C.C. Yang, K.E. lin, W.N. Fang, J.S. Chen, Y.C. Wu, Y.C. Kuo, H.B. Lu, Planarization improvement using non-porous polishing pad in ILD CMP, ICPT, 2015.

# SEVERAL STRATEGIES FOR ALUMINUM METAL GATE CHEMICAL MECHANICAL PLANARIZATION SCRATCH REDUCTION

Q. X. Hong, P. L. Zhu, F. Luo, D. C. Liu, J. Zhang, H. F. Zhou, J. X. Fang, Y. Zhang

Shanghai Huali Integrated Circuit Corporation

NO.6, Liangteng Rd. Pudong Shanghai, 201314, China

*Corresponding Author's Email: hongqingxuan@hlmc.cn

## ABSTRACT

Scratch has always been an important issue for Al metal gate chemical mechanical planarization (Al MG CMP). A scratch across gates will cause the short between gate and gate or gate and contact W (CTW), which will finally induce the failure of a chip and decrease yield; those scratches within a gate will be the noise for killer defect capture although they have little effect on yield loss. Several strategies for scratch decrease are proposed in this paper from two aspects---hardware and recipe. Filter with smaller pore size could reduce the amount of large size abrasive flowing to platen and decrease macro scratch caused by large particle; Slowing down the rotate speed(RS) of wafer and platen or reducing down force(DF) are effective methods for scratch performance improvement cause in that way wafer will suffer less mechanical effect which is the main factor for most scratches' formation. Balancing polish time is another method for scratch decrease because the risk of scratch will raise as polish time on one platen lengthens. The modification on recipe will usually cause some side-effect like selection ratio shift, low removal rate and et al. which will increase the risk of Al residue. The balance between scratch and Al residue is the final aim of polish recipe tuning.

## INTRODUCTION

Replacement metal gate is a well-known solution for gate leakage current and further MOS capability tuning as MOSFET dimensions shrink to 28nm [1-4]. Al CMP, as a key step of high-k metal gate(HKMG) process, has always been faced with defect issue, of which scratch is the most common one[5-8]. The form of scratch is on the one hand due to the born-with low mohs' hardness of AL and one the other hand due to the consumable this process chooses. Besides, process recipe also has some effect on Al scratch. Some conclusions have been put up by predecessors: using soft pad, decreasing down-force and adding pad clean after polish all will do good to scratch reduction[9]. In this paper, several strategies on abrasive size, polish recipe parameter modification and balance removal amount between P2 and P3 will be discussed.

## EXPERIMENTAL

12 inches pattern wafers are constructed with normal HKMG process flow to get gate height and defect data, and blanket wafers are prepared with CVD oxide and PVD Al film on bare silicon to collect Al offline removal rate. Al CMP process was carried on LK-lp3B5.0, defect data was collected on KLA-Tencor, Al film thickness was measured with KLA RS-200 4 points probe, gate height of pattern wafers was measured with Nova T650.

## RESULTS AND DISCUSSION

Al scratch will be classified into micro-scratch and macro-scratch in this paper based on the number of gates a scratch covers: scratches within one gate are defined as micro and those across more than one gate are macro.

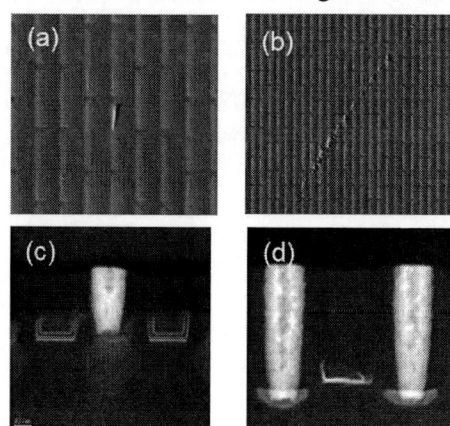

*Figure 1: (a) micro-scratch, (b)macro-scratch,(c) gate deformation,(d) gate damage*

Macro-scratches are of higher kill-ratio because the damage and deformation of metal gate may cause the short between gate and CTW [10]. Although micro-scratch has little effect on yield loss itself, it will be a noise for killer defect capture, like Al missing, Al residue and macro-scratch. So our aim is to decrease both of these two kinds of scratch.

Proper consumable and stable hardware are basic requirements of a healthy process. Slurry abrasive size may grow due to crystallization or agglomeration happened occasionally in its lifetime and large-size abrasive may induce scratch. Using filter is an effective way to control abrasive size. Two kind of filtration was compared in this paper: one single filter with 0.3 μm pore size and tandem 0.3 μm+0.1 μm filters. Scratch count decreased effectively when filter with 0.1μm pore size was added and the more the scratch count is, the more obvious the decrease is. Defect performance of 0.3μm+0.1 μm is more stable than single filter as well when more data was seen.

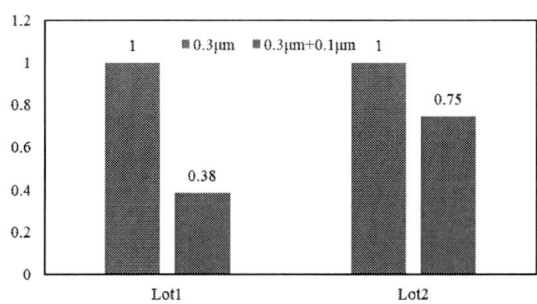

*Figure 2: Scratch performance compare between 0.3μm and 0.3+0.1μm filtration*

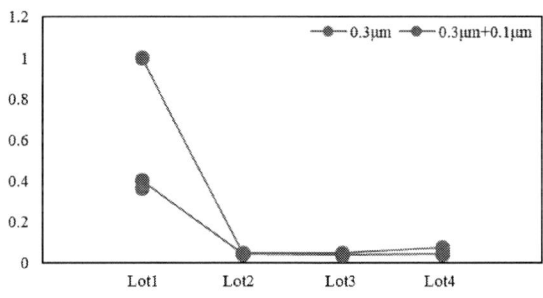

*Figure 3. Scratch performance compare of more lots*

To reveal the root cause of an issue, partition check is usually an effective way. According to partition result in Figure 4, we could come up with the conclusion that most of scratches have been formed before the polish on 3rd platen(P3), and P3 plays a repairing role in the whole process. That could be the result of hardness difference among these three polish pad. Polish pad used on P3 is the softest, and the softer the pad is, and the less scratch it will cause [9]. Based on the conclusion of partition check we did a split of different P2 over-polish(OP) time and the result was shown in Figure 5. Scratch count increasingly decreased as P2 OP time minishes which further proved that P2 produces most of the scratches. The ratio of

macro-scratch keeps in a comparable level between BSL and P2 OP- condition which means that macro-scratch count decreases at the same time. As P2 OP time further minishes, macro-scratch ratio increases, because scratch count could continuedly decreases, but the formation and repairing process of macro-scratch has met a dynamic balance.

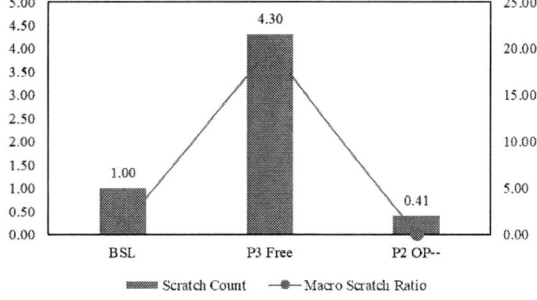

*Figure 4: Scratch result of partition check*

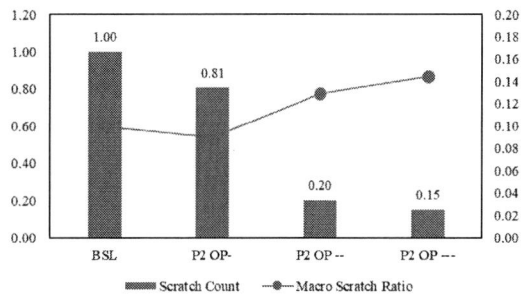

*Figure 5: Scratch performance of different P2 over-polish time*

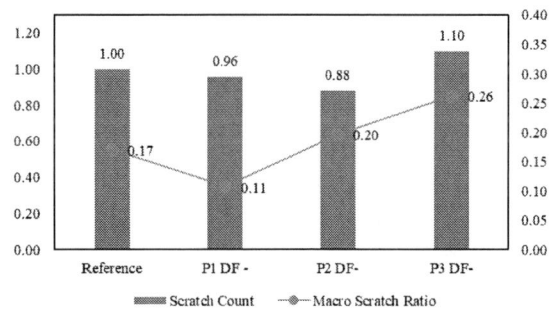

*Figure 6: Scratch performance of different down force on platen 1/2/3*

We chose P2 OP-- condition as reference condition for the concern of scratch and Al residue balance. Based on the result above, we further study the effect of down force(DF) and rotate speed(RS) on scratch. DF split was done on different platen and the result was shown in Figure 6. DF decrease on P1 does little help to scratch total count reduction, but could decrease macro-scratch ratio about 40% when compared with reference; P2 DF- could reduce scratch count about 12% but has little effect on macro-scratch count. P3 DF- makes scratch performance worse because the decrease of DF may weaken the repairing ability of P3. The effect of rotate speed(RS) on scratch performance was shown in Figure 7. Reducing RS could minish scratch count to some extent, but will soon met its limit. Macro-scratch counts of these split are always in a comparable level.

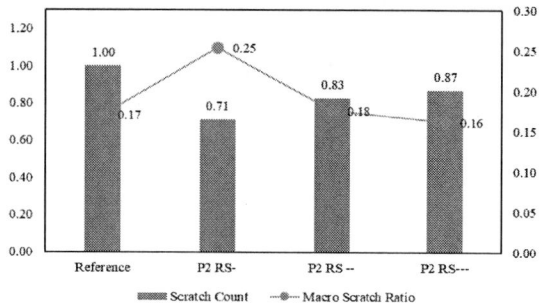

Figure 7: Scratch performance of different platen2 rotate rate

When process recipe got changed, we usually need to check dishing, gate height, profile and pattern loading besides defect [11]. After the BKM condition was decided, we collected gate height and profile, which are both comparable to BSL condition(Figure 9). But that conclusion does not suite every product. We tried BKM condition in another product(prod B) and gate profile of this two condition do not match each other(Figure 10). Then we usually need to fine tune polish pressure. Dishing was characterized by AFM on a Al bond pad, and compared with BSL, BMK condition dishing performance has a 66% improvement(Figure 8), that should be the result of P2 OP time decrease. The minishing of current-stage dishing sometimes implies the risk of metal residue, so Al residue should be another item which needs to be check in defect scan.

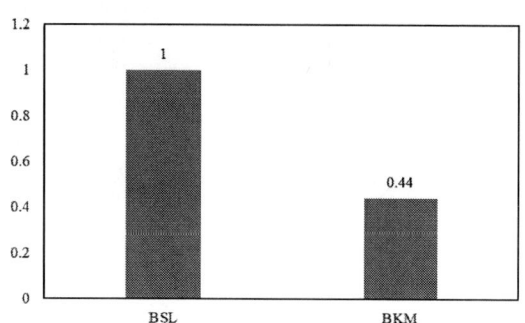

Figure 8: Al bond pad dishing compare between BSL and BKM condition

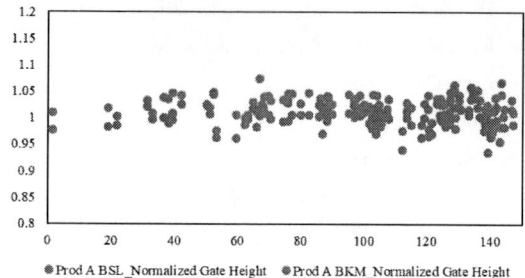

Figure 9: Gate height profile of product A

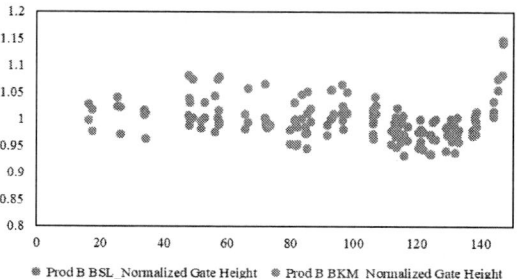

Figure 10:. Gate height profile of product B

## CONCLUSION

Several strategies were discussed in this paper from two aspects: tool hardware and process recipe. Adding a filter with smaller pore size could reduce scratch count and make scratch performance of the process more stable; decrease P2 OP time, P1 or P2 DF, P2 RS all could decrease scratch count to some extent. Besides, a whole procedure of process tuning for an issue, from partition check for trouble shooting to each recipe item optimizing, was shown in this paper. Process recipe tuning will always cause some side-effect, and the side-effect of P2 OP time decrease is less removal amount which may finally induce gate height abnormal and Al residue. The duty of process engineer is to find the balance point between scratch and al residue.

## REFERENCES

[1] S. Guha et al., Annu. Rev. Mater. Res., Vol. 39, 2009, p181-202.

[2] K. Mistry et al., IEDM Tech. Dig., 2007, pp. 247-25.

[3] C. Auth et al., Symp. on VLSI Tech., 2008, pp. 128-129.

[4] P. Packan et al., IEDM Tech. Dig., 2009, pp. 659-662.

[5] J. M. Steigerwald, IEDM Tech. Dig., 2008, pp. 37-40.

[6] J. M. Steigerwald, Proceedings of ICPT, 2009.

[7] P. Feeney, CMP for metal-gate integration in advanced CMOS transistors, Solid State Technology, Nov. 2010.

[8] Y. H. Hsieh et al., Proceedings of AMC, 2010.

[9] C. W. Hsu et al. 2012 IEEE International Interconnect

Technology Conference, 2012, pp. 1-3

[10] H. K. Hsu et al., Proceedings of IITC, 2011.

[11] J.-H. Han et al., Journal of The Electrochemical Society, Vol. 154,2007, pp H525-H529.

# MARK DAMAGE PHENOMENON CAUSED BY SUPERIMPOSED CMP DISHING ON LARGE-AREA STI REGIONS

*Xu WenSheng[12]\*, Ba You[1], Chen YongBo[1], Zhu YuJie[1], Li RunLing[1], Ding ShiJin[2], Ye JiongHan[1]*

[1] Shanghai Huali Microelectronics Corporation, Pudong New District, Shanghai, China
[2] School of Microelectronics, Fudan University, Shanghai 200433, China
\*Corresponding Author's Email: xuwensheng1990@126.com

## ABSTRACT

The phenomenon of mark damage observed at the large-area STI regions of scribe line will cause the overlay shift between different layers during the photo process, and it will further cause amount of residue/scratch defects on the wafer surface during the W CMP process. It has been confirmed that the decease of CMP dishing at the ILD0 CMP1 process may be the root cause. The hard mask (SiN) residue on-top of the dummy poly will block the dummy poly remove process. Thus, only partial of the marks can be filled with HKMG and W, causing the mark damage. The partial filled mark will further be the peeling source of tungsten residue at W CMP process, and introduce the scratches on wafer surface.

## INTRODUCTION

Despite the planarization technique advancing in modern semiconductor manufacturing technologies, the efficient chemical mechanical polishing (CMP) has become the primary choice of planarization for CMOS integrated circuit fabrication [1-2]. It has been widely used in the front end of line (FEOL) for the formation of the shallow trench isolation (STI), poly deposition planarization, and poly gate interlayer dielectric (ILD0) planarization to build the advanced CMOS device structures [3-5]. As the critical dimensions of CMOS devices shrink to *nm* level in modern semiconductor fabrication, the CMP of different structures with different film stacks has become more and more important [6-8]. The simplified process flow in an advanced FinFET technology includes several different CMP processes, such as STI CMP, Poly CMP, ILD0 CMP and HKMG W CMP. Here, the ILD0 CMP is divided into two parts. The ILD0 CMP1 is applied to open hard mask (HM) SiN of dummy poly, and the ILD0 CMP2 is applied to open the dummy poly, followed by dummy poly remove process.

Pattern density loading of different structures is one of the most important challenge of the CMP processes. The surface topography after CMP polishing such as the dishing and erosion effect can increase the variation of device performance. Besides, the poor surface topography and the accompanied CMP defects will bring some challenges in the following lithography and etch processes. Fig.1 shows the schematic demos of the key CMP steps in the FEOL of the advanced FinFET technology, and the figure (a), (b), (c) are corresponding to the STI CMP, Poly

CMP, and ILD0 CMP process respectively.

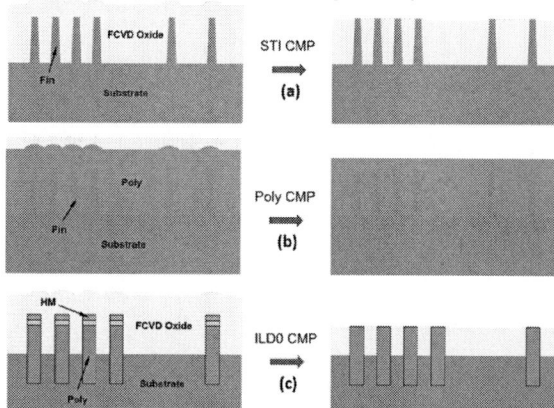

*Figure 1: The Schematic Demos of CMP Steps in FEOL, (a) STI CMP, (b) Poly CMP, (c) ILD0 CMP.*

In this paper, the superposed dishing of STI/Poly/ILD0 CMP on the scribe line has been observed and identified. Two kind of systematic defects caused by the worse CMP dishing on this large STI region has been found. A reliable model revealing the root cause of these defects has been raised and interpreted.

TABLE I.    THE SPLIT TABLE OF EXPERIMENTAL CONDITIONS

| Splits | CMP Process | | | |
|---|---|---|---|---|
| | *STI* | *Poly* | *ILD0* | *Defects* |
| 1 | BSL | BSL | BSL | 4 |
| 2 | DTE1 | BSL | BSL | 7 |
| 3 | DTE2 | BSL | BSL | 7 |
| 4 | BSL | CIP | BSL | 6 |
| 5 | BSL | BSL | DET1 | 23 |
| 6 | BSL | BSL | DET2 | 223 |

## EXPERIMENTS

Different split experiments have been applied on a batch of wafers to verify the influence of different CMP parameters on defect counts, as shown in Table 1. The split experiments mainly deteriorate and improve the dishing of oxide and dummy poly on the STI regions. Here, BSL mean normal condition. We established two deterioration conditions of STI CMP (DTE1/2, increase dishing), one improvement condition of Poly CMP (CIP, decrease dishing), and two deterioration conditions of ILD0 CMP1 (DET1/2, decrease dishing). The scanning electron microscope (SEM) are applied to confirm the status of mark damages. The focus ion beam (FIB) is applied to prepare samples from the damaged mark, and the transmission election microscope (TEM) is applied to explore the details of mark damages.

## RESULTS AND DISCUSSION

We have found many overlay flyers between current layer and poly layer with values out of specification for the BSL condition. Statistical data showed that the flyers are always located on wafer edge. The SEM review showed that the marks at the locations of flyers are damaged, as shown in Fig. 2. There are two types of overlay mark design for the poly layer: large poly pad with STI marks and large STI pad with poly marks, as shown in Fig. 2(a) and 2(b). It can be found that the damaged component are always the poly parts. The overlapped wafer map of the damaged marks at BSL conditions are all located on the wafer edge, consistent with the location of overlay flyers. It explains why the latter photo processes after dummy poly remove process have the overlay shift issue.

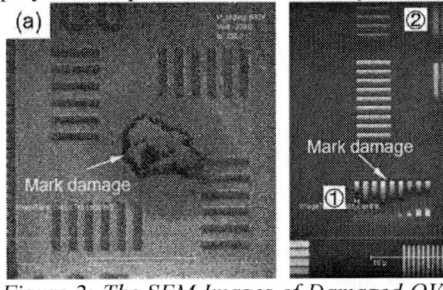

Figure 2: The SEM Images of Damaged OVL Marks, (a) Positive tone: Poly Pad with STI Marks, (b) Negative tone: STI Pad with Poly Marks.

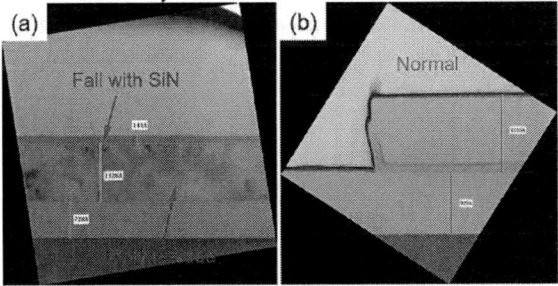

Figure 3: The TEM Images of Damaged Overlay Marks from the Locations in Figure 4(b), (a) FIB X cut from ①, (b) Reference of Good Mark of (a).

It has been confirmed that all the overlay marks are located on the scribe line with a very large area of STI regions. The TEM samples from the mark damage ① in Fig. 2(b) with FIB are analyzed in details, as shown in Fig. 3. Fig. 3(a) is the TEM images of the damaged OVL marks. Fig. 3(b) is the TEM image of a good mark for the references. It can be found that a layer of nitride (SiN) residue are caped on-top the poly parts of damaged mark. The SiN is the hard mask of dummy poly patterning process. The SiN residue will cause failure of dummy poly remove process after the ILD0 CMP2 step. Thus, the overlay measurement based on light diffraction will be fail due to the wrong film stack at the damaged mark.

After the deposition of HKMG film and contact metal W, the W CMP process is applied for the planarization to create separated metal gates for the devices. Another two types of defects (*surface residue and corresponding scratch*) are observed on the wafer surface, as shown in Fig. 4. In-situ EDS of SEM showed that the full-map random distributed residue are mainly W. Statistical data of the Brightfield Inspection from a lot of wafers showed that the locations of the scratch are close to the alignment and overlay mark.

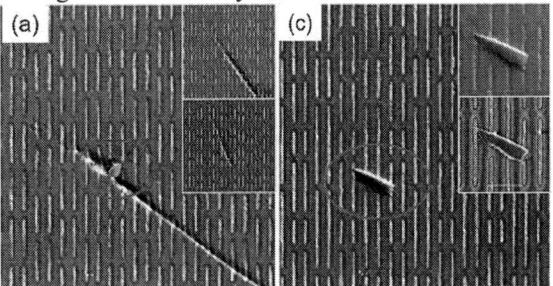

Figure 4: The Reviewed Defects after HKMG W CMP, (a) is Macro and Micro Scratches with Tungsten Residues, (b) is a clear image of Tungsten Residues.

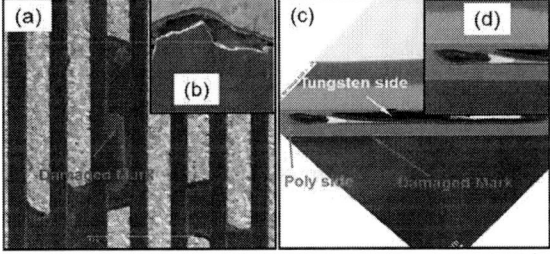

Figure 5: The SEM and TEM Images of Tungsten Residue Peeling Sources, (a) SEM Image of Damaged Mark, (b) Magnified Image of (a), (c) TEM image of Damaged Mark, (d) Magnified Image of (c).

The image of SEM review further confirmed that the peeling source are possibly the damaged marks on the scribe line, as shown in Fig. 5. Fig. 5(a) show a clear break off along the lines inside the mark. Fig. 5(b) is the magnified image of Fig. 5(a). Only part of the marks has been filled with W, but part of the mark is still the dummy poly. Fig. 5(c) and 5(d) are the TEM images by FIB along vertical (Y) direction of the damaged mark. It can be found that the partial filled W is very weak and fragile. The weak interface of the poly residue and W will be the peeling source during the W CMP process. The exfoliated W particle acted as the residue will cause many macro and micro scratch on the wafer surface.

The most possible reason of the mark damage is accumulated CMP dishing with the hard mask SiN residue before the dummy poly remove process. The three CMP processes before dummy poly remove has been suspected, that is STI CMP, Poly CMP, and ILD0 CMP1. From the summarized defect counts in the last row of Table 1, the conclusion can be obtained. The deteriorated dishing of STI CMP and the dishing CIP of Poly CMP do not have significant effect on the counts of the mark damage. However, the deterioration of dishing of ILD0 CMP1 has obviously significant influence on the mark damage, that is the decrease of ILD0 CMP1 dishing will introduce more mark damage. Because the decrease of ILD0 CMP1 dishing will go against with the HM SiN Remove process, thus it will block the dummy poly remove process. This will cause the mark damage effect.

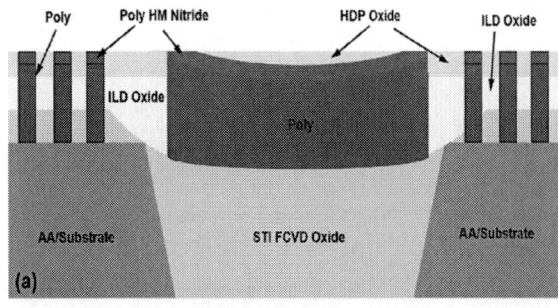

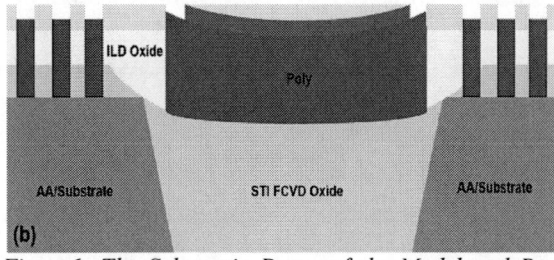

*Figure6: The Schematic Demo of the Model and Root Cause for Mark Damage.*

The potential model of the mark damage has been established, as shown in Fig. 6. The STI CMP process will cause a degree of dishing on the large-area STI region, because the remove rate of oxide is faster than the silicon substrate according to the choice of CMP slurry. The Poly CMP will also cause amount of poly dishing on the STI regions, because the large area of poly is removed faster. The ILD0 CMP1 need to open the HDP oxide and stop on the SiN on-top the poly. Then the SiN can be removed by a wet etch process, followed by the dummy poly remove process. Thus, the dishing of ILD0 CMP1 process should be increased on the STI region to make sure the HDP oxide is completely removed. If the dishing is decreased, the HDP oxide is not fully removed and the SiN residue will be introduced, causing the mark damage effect.

## CONCULSIONS

The mark damage phenomenon has been observed on the large-area STI regions. It will cause the overlay shift during the photo process and further it will cause the W residue and CMP scratch on the wafer surface during the W CMP process. The SiN residue on-top the dummy poly may be the root cause for mark damage. The decrease of ILD0 CMP1 dishing will introduce HDP oxide residue, and it will block the SiN remove process. Finally, it will further block the dummy poly remove process. Thus, only partial of the mark can be filled with HKMG and W, and the partial filled mark will be the peeling source of tungsten residue.

## REFERENCES

[1] Wu C. F., Chen C. H., Chen M. H. et al, "Application of advnced process control on mixed-product chemical mechanical polishing process," EMEIT, 2011, pp. 2116-2119

[2] V. N. Khmelev, A. V. Shalunov and E. S. Smerdina, "Increasing Efficiency of a Chemical-Mechanical Polishing of the Silicon Wafer," EDM, 2006, pp. 263-269

[3] C. Kuo, F. C. Meyer, M. Hollatz, et al, "Study of STI CMP Process Control on High Aspect Ratio Gap-fill Topographies by Motor Current EPD," ICPT, 2007, pp. 1-7.

[4] C. -T. Yeh et al., "A novel two-step poly CMP to improve dishing and erosion effect on self-aligned floating gate process," eMDC & ISSM, 2011, pp. 1-10.

[5] R. Ghulghazaryan, J. Wilson and A. Abouzeid, "FEOL CMP modeling: Progress and challenges," ICPT, 2015, pp. 1-4

[6] H. Huang et al., "New CMP processes development and challenges for 7nm and beyond," CSTIC, 2018, pp. 1-5

[7] R. F. Hafer, H. Lin and B. Y. -L. Hsieh, "Electron Beam Inspection: Within Die and Within- Wafer monitoring of RMG CMP," ASMC, 2019, pp. 1-4

[8] M. D. Wedlake, Adrian Santos Lopez, S. Trigno and P. Aniekwu, "Simplification of Replacement Metal Gate CMP metrology for FinFET," ICPT, 2015, pp. 1-5.

# COMPONENT OPTIMIZATION OF SAPPHIRE SLURRY BASED ON RESPONSE SURFACE METHODOLOGY FOR CHEMICAL MECHANICAL POLISHING

*Minghui Qu[1,2*], Xinhuan Niu[1,2*], Yanan Lu[1,2], Ziyang Hou[1,2], Han Yan[1,2], Fu Luo[1,2]*

[1]School of Electronics and Information Engineering, Hebei University of Technology, Tianjin 300130, The People's Republic of China

[2]Tianjin Key Laboratory of Electronic Materials and Devices, Tianjin 300130, The People's Republic of China

*qmha731@126.com, xhniu@hebut.edu.com

## ABSTRACT

For sapphire wafer, chemical mechanical polishing (CMP) is widely applied to obtain super-smooth and non-damaged wafer surface. To solve the problem of components accurate optimization of sapphire slurry with different crystal planes, the response surface methodology (RSM) is proposed, which can realize the distribution of slurry component ratio and predict material removal rate (MRR) of different crystal planes. Comparing the results, it can be seen that the RSM has high accuracy and good model fitting effect. The optimal ratio of the slurry components was obtained, which can make the MRR of sapphire with three crystal planes reach a higher level simultaneously.

***Keywords—sapphire with different crystal planes; chemical mechanical polishing (CMP); slurry; material removal rate (MRR); response surface methodology (RSM)***

## INTRODUCTION

Sapphire is widely used in high-tech fields such as microelectronics and optoelectronics because of its excellent optical, chemical and mechanical properties[1]. With the increasing requirement for sapphire surface quality, the challenge of realizing global planarization becomes more and more serious[2]. Chemical mechanical polishing (CMP) is the most important processing technology that can realize global planarization[3]. With the development of mathematical methods, people tend to predict or explain the polishing process and polishing results by establishing the relationship between the model and the influencing factors in CMP, so as to save time and avoid the waste of resources. J.K. Zhou et al[4] proposed a data prediction method based on response surface methodology (RSM) and neural network, and found that such less time-consuming optimization method can achieve more ideal CMP parameters and the proportion of slurry components. W.T. Wang et al[5] investigated the effects of pH, persulfate concentration and $TiO_2$ dosage on CMP by RSM, determined the optimal value, carried out repeated polishing experiments according to the optimized scheme, and finally confirmed that the RSM

experimental design method was accurate and reliable. In our previous research, Y.N. Lu et al[6] found that adding potassium persulfate ($K_2S_2O_8$) to the slurry containing fatty alcohol polyoxyethylene ether (JFCE) as surfactant can improve the material removal rate (MRR) of three different crystal planes. The polishing effects is closely related to the concentration of each component. However, when the concentration gradient of each component is crossed, a large number of experiments need to be done to achieve the component optimal ratio, which is time-consuming and labor-consuming.

Therefore, in order to determine the optimal ratio of each component more quickly and efficiently, it is necessary to use RSM for component optimization. The effects of additives on MRR of sapphire wafers were studied by RSM method. Then the optimal ratio of each component of slurry is determined by variance analysis and model analysis. Finally, it is compared with the measured experimental data.

## EXPERIMENTAL

**Actual experiments---** In slurry, silica sol (mean particle size of 80-90 nm) was used as abrasive, whose concentration was 40 wt%. Sodium ethylenediamine tetraacetic acid (EDTA-2Na) and JFCE were served as complexing agent and surfactant, respectively. $K_2S_2O_8$ was added to the sapphire slurry to enhance chemical action. KOH was used to adjust the pH value of the slurry to 10.5.

The MRRs were obtained by polishing two-inch single crystal commercial c-plane sapphire wafers using X62S82×305-D-S single-side polisher. The weight of the sapphire wafers before and after polishing was measured by professional electronic balance (AUY120) to calculate MRR by (1). Where MRR (μm/h) is the material removal rate, $\Delta m$ (g) is the mass loss, t (h) is the polishing time, $\rho$ (g/cm$^3$) is the sapphire density and r (cm) is the sapphire radius. In the test, t=1/2 h, $\rho$=3.98 g/cm$^3$, r=2.54 cm. Each group of experiments was repeated at least three times to take the average.

$$MRR = \frac{\Delta m \times 10^4}{\rho \pi r^2 t} \qquad (1)$$

**Optimization methods**--- In order to optimize the compositions of the slurry, three factors affecting MRR were selected for the experimental design in the research, containing A: $K_2S_2O_8$ (wt%), B: JFCE (wt%), and C: EDTA-2Na (wt%). Each factor included three levels, and MRR was the response. Table I shows the level and code of each variable. Box-Behnken design (BBD) was applied to obtain the experimental design. After the experimental results were obtained, the quadratic regression equation was obtained by RSM with three factors and three levels, and the best process parameters were found. According to the design principle of RSM, the 3-factor, 3-level response surface design consisted of 12 analytic points and 3 centroids. The experimental design and results are shown in Table II. Minitab was used to perform RSM to analyze the response surface data.

## RESULTS AND DISCUSSION

According to Table II, the response values of different factors were fitted by regression, and the fitting model is shown as (2).

$$MRR = 7.965 + 0.412A + 0.212B - 0.307C - 0.147A^2 - 0.406B^2$$
$$-0.743C^2 - 0.000AB + 0.000AC - 0.205BC \qquad (2)$$

To verify the adequacy of the fitted model for MRR, an analysis of variance was performed on the (2), and the results of the analysis are shown in Table III. The factors that have significant influence on the results are marked with "a". It was revealed that the P value check for the significance of each coefficient ($p < 0.05$ represents significant).Therefore, the effects of A and $C^2$ on MRR are significant, indicating that the effect of the A on the MRR is mainly a linear effect, and the effect of the C on the results is mainly a quadratic effect. The related statistical criteria values of the model are shown in Table IV. In the table IV, the S assesses the extent of the model describing the response values. $R^2$ value also plays a great role. The lower the S, the higher the $R^2$, and the better the model fits the data. It can be seen from Table IV that the S is less than 0.5, which is in the lower range. The $R^2$ is higher than 85%, which means that the model can explain more than 85% of the variation in the data. This proves that the model has high enough accuracy in reflecting the experimental results accurately. Figure 1 shows the normal probability plot of the residuals. It can be seen that the residuals are distributed around the diagonal, satisfying normality, and the model is reasonably usable. To further verify the reliability of the model, a backward prediction was performed based on the established model, and the results are shown in Figure 2. It can be seen that the predicted values are very close to the actual values. Both the actual and predicted scatter points are visible, and the scatter points are closely distributed on the diagonal, and the predicted and actual values fit well.

TABLE I. LEVELS AND CODES OF EXPERIMENTAL VARIABLES BASED ON RSM

| Codes | Variables | Levels | | |
|---|---|---|---|---|
| | | -1 | 0 | 1 |
| A | $K_2S_2O_8$ (wt%) | 0.1 | 0.2 | 0.3 |
| B | JFCE (wt%) | 0.2 | 0.3 | 0.4 |
| C | EDTA-2Na (wt%) | 0.1 | 0.2 | 0.3 |

TABLE II. EXPERIMENTAL DESIGN AND RESULTS

| Number | A(wt%) | B(wt%) | C(wt%) | MRR(μm/h) |
|---|---|---|---|---|
| 1 | -1 | -1 | 0 | 6.685 |
| 2 | 1 | -1 | 0 | 7.509 |
| 3 | -1 | 1 | 0 | 7.314 |
| 4 | 1 | 1 | 0 | 8.138 |
| 5 | -1 | 0 | -1 | 7.072 |
| 6 | 1 | 0 | -1 | 7.896 |
| 7 | -1 | 0 | 1 | 6.254 |
| 8 | 1 | 0 | 1 | 7.078 |
| 9 | 0 | -1 | -1 | 6.700 |
| 10 | 0 | 1 | -1 | 7.335 |
| 11 | 0 | -1 | 1 | 6.705 |
| 12 | 0 | 1 | 1 | 6.520 |
| 13 | 0 | 0 | 0 | 7.960 |
| 14 | 0 | 0 | 0 | 7.424 |
| 15 | 0 | 0 | 0 | 8.505 |

TABLE III. VARIANCE ANALYSIS OF THE EXPERIMENTAL RESULTS OF SAPPHIRE

| Source | Freedom | Sum of square | Mean square | F | P |
|---|---|---|---|---|---|
| Model | 9 | 5.1467 | 0.5719 | 3.81 | 0.078 |
| A | 1 | 1.3580 | 1.3580 | 9.04 | 0.030 |
| B | 1 | 0.3608 | 0.3608 | 2.40 | 0.182 |
| C | 1 | 0.7534 | 0.7534 | 5.01 | 0.075 |
| $A^2$ | 1 | 0.0795 | 0.0795 | 0.53 | 0.500 |
| $B^2$ | 1 | 0.6100 | 0.6100 | 4.06 | 0.100 |
| $C^2$ | 1 | 2.0381 | 2.0381 | 13.57 | 0.014 |
| AB | 1 | 0.0000 | 0.0000 | 0.00 | 1.000 |
| AC | 1 | 0.0000 | 0.0000 | 0.00 | 1.000 |
| BC | 1 | 0.1677 | 0.1677 | 1.12 | 0.339 |
| Residual | 5 | 0.7512 | 0.1502 | | |
| Lack of it | 3 | 0.1669 | 0.0556 | 0.190 | 0.895 |
| Pure error | 2 | 0.5843 | 0.2921 | | |
| Total | 14 | 5.8979 | | | |

*TABLE IV. MODEL SUMMARY TABLE*

| Response | S | $R^2$ | $R^2$(Adj) | $R^2$(Pre) |
|---|---|---|---|---|
| MRR | 0.3876 | 87.26% | 64.34% | 32.44% |

*TABLE V. THE PREDICTION RESULTS AND EXPERIMENTAL RESULTS*

| Response | A | B | C | fitted value | actual value |
|---|---|---|---|---|---|
| | (wt%) | | | ($\mu$m/h) | ($\mu$m/h) |
| MRR | 0.20 | 0.33 | 0.18 | 8.3029 | 8.235 |

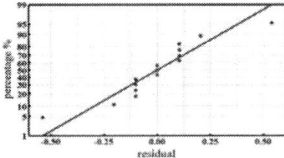

*Figure 1: Residual probability normal graph of MRR.*

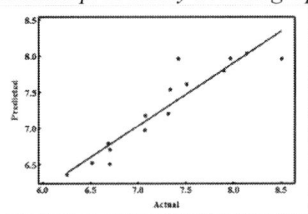

*Figure 2: Scatter diagram of predicted value and actual value of MRR.*

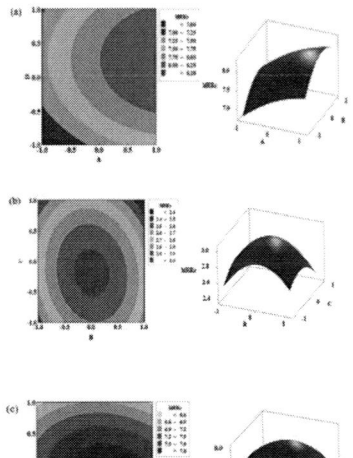

*Figure 3: Trend plots of different factors on the MRR by the RSM. (a) effect of A and B. (b) effect of A and C. (c) effect t of B and C.*

The response surface analysis plot is a three-dimensional surface plot , which consists of the response values and each experimental factor. It shows the effect of the remaining two factors on MRR when any one of the three factors A, B and C is taken to zero level. Fig. 3 (a) shows the AB contour plot and response surface plot, where MRR increases with increasing A factor levels over a range. As the level of B factor increases, MRR increases and then decreases. From Figure 3 (b), it can be seen that within a certain range, MRR increases with the increase of factor A level, and as increasing the level of C, the MRR tends to increase and then decrease. From Figure 3 (c), it can be seen that with the increase of B and C factors level, MRR shows an increasing trend, and when B and C factors reach a certain value, MRR shows a decreasing trend with the growth of B and C factors level.

The slurry ratios for reaching the maximum MRR was predicted, and the results are shown in Table V. The optimal composition ratio is about 0.30 wt% for $K_2S_2O_8$, 0.33 wt% for JFCE and 0.18 wt% for EDTA-2Na. The obtained MRR is extremely close to the theoretical value, indicating that the composition ratio of the slurry is reasonable and feasible.

## CONCLUSIONS

The prediction of the different component ratios of the slurry was carried out by RSM. It was found that the optimized response MRR could be significantly improved, which was verified by practical experiments, and the optimized parameters of the concentration of $K_2S_2O_8$, JFCE, and EDTA-2Na are 0.30 wt%, 0.33 wt% and 0.18 wt%, respectively. The concentrations of each component were more refined than the actual values. RSM can help to find an optimized set of parameters with the highest efficiency and lowest cost, which possesses excellent potential for application in the field of sapphire CMP.

## ACKNOWLEDGMENTS

This work was supported by the Major National Science and Technology Special Projects (No. 2016ZX02301003-004-007), Natural Science Foundation of Hebei Province (No. F2021202009), The authors also thank the teachers and classmates for their helpful suggestions.

## REFERENCES

[1] Khattak C P, Shetty R, Schwerdtfeger C R, Ullal S. *Journal of Crystal Growth, vol. 452, 2015, pp. 44-48.*

[2] Y.C. Xu, J. Lu, X.P. Xu. *Applied Surface Science, vol. 389, 2016, pp. 731-720.*

[3] L. Xu, H. Lei, T. X Wang, Y. Dong, S.W. Dai. *Ceramics International, vol. 45, 2019, pp. 8471-8476.*

[4] J.K. Zhou, X.H. Niu, T.L. Zhang, H. Wang, C.H. Yang, Y.C. Zhang, W.T. Wang, Z. Wang, Y.B. Zhu, Z.Y. Hou, R. Wang. *Nano Select, 2021.*

[5] W.T. Wang, B.G. Zhang, Y.H. Shi, T.D. Ma, J.K. Zhou, R. Wang, H.X. Wang, N.Y. Zeng. *Journal of Materials Processing Technology, vol. 295, 2021, pp.*

117150.

[6] Y.N. Lu, X.H. Niu, C.H. Yang, Z.Q. Huo, Y.Q. Cui, J.K. Zhou, Z. Wang. ECS Journal of Solid State Science and Technology, vol. 9, pp. 064006.

# DEFECT LAW OF CU/CO PATTERNED WAFERS AFTER USING A NOVEL BULK/BARRIER SLURRY AND CLEANING SOLUTION

*Lifei Zhang[1], Tongqing Wang[1], and Xinchun Lu[1]\**

[1]State Key Laboratory of Tribology, Tsinghua University, Beijing 100084, China
*Corresponding Author's Email: xclu@tsinghua.edu.cn

## ABSTRACT

As feature size shrinks to 7 nm and beyond, cobalt (Co) has been applied as one of the most promising alternatives of diffusion barrier layers to be employed in copper (Cu) interconnects. The present work describes the defect law of Cu/Co patterned wafers after using a novel bulk/barrier slurry and cleaning solution during chemical mechanical polishing (CMP) and post-CMP cleaning process. The novel slurry and cleaning solution were mainly composed of potassium persulfate (KPS) as an oxidizer and citric acid as a complexing agent, respectively. With the changes of different line width, line space and pattern density, the defects including copper dishing, dielectric erosion, fang as well as particle residues were investigated.

## INTRODUCTION

As device dimensions continue to decrease, Co has been emerged as one of the most promising candidates to be employed in Cu interconnects as a diffusion barrier layer[1]. Compared with the traditional Ta/TaN diffusion bilayers, Co shows an excellent gap-fill attribute, the capability of Cu directly electroplating as well as the conformal adhesion property with its relatively low resistivity. However, the integration of Cu and Co demands the compatibility with various wet process treatments, such as CMP and post-CMP cleaning process.

In our previous work, we have proposed a relatively stable KPS based barrier slurry and compared this KPS slurry with the conventional hydrogen peroxide based one[2]. It was found that 10 mM KPS with 3 vol.% colloidal silica could produce a material removal rate (MRR) selectivity of ~1 between Cu and Co films. On this basis, the coordination of different complexing agents and corrosion inhibitors with KPS has been studied in this research. Furthermore, it is remarkably that the polished Cu/Co surfaces could become inevitably contaminated by a large number of particles. It was demonstrated that citric acid could be used to remove silica particles from Cu and Co wafers[3], [4]. So, in this study, the citric acid cleaning solution has also been employed in Cu/Co patterned wafers to verify its cleaning ability.

Thus, a comprehensive study has been carried out to explore not only for the CMP process of Cu/Co patterned wafers, but also for the post CMP cleaning process. The objective of our research is to investigate the defect law of Cu/Co patterned wafers after using a novel KPS bulk/barrier slurry and citric acid cleaning solution.

## EXPERIMENTAL

Cu films (~1.5 μm) and Co films (90 nm) were deposited on 50 mm silicon wafers as blanket wafers, respectively. A specific kind of patterned wafers which has different line width and space was used in this study. The mask floor plan and cross-sectional view of this kind of patterned wafers are shown in Figure 1. The selection of complexing agents mainly included citric acid (CA), glycine (GLY), and ethylenediamine tetraacetic acid (EDTA). Besides, 1,2,4- triazole (TAZ), potassium oleate (PO) as well as 1-Phenyl-1H-tetrazole-5-thiol (PMTA) were chosen to be the corrosion inhibitors.

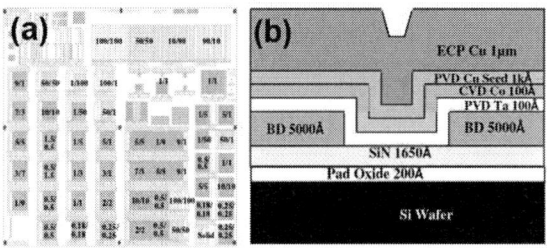

*Figure 1: (a)SKW 5-3.18 mask floor plan, (b) cross-sectional view of the Cu/Co patterned wafers*

## EFFECTS OF COMPLEXING AGENTS AND INHIBITORS ON MRRS

The MRRs of Cu and Co wafers in the presence of 10 mM KPS at pH 10 with different complexing agents (5 mM) were shown in Figure 2(a). A relatively high Cu MRR (1924.5 Å/min) can be observed when GLY works as a complexing agent with KPS. Since the superfluous Cu is supposed to be removed rapidly during the first step, GLY was selected to be the complexing agent in the KPS bulk slurry. On the other hand, to reduce copper dishing and other defects, the barrier slurry requires to make sure the removal selectivity between Cu and Co, which means the MRR of Cu couldn't be too fast. As a result, CA is the most suitable complexing agent in the KPS barrier slurry.

On the foundation of KPS and CA, the effect of different inhibitors (5 mM) on Cu/Co MRRs in the barrier slurry is presented in Figure 2(b). TAZ and PO have a certain inhibition effect on the MRRs of Cu and Co, while the MRR of Cu is still higher than that of Co. On the contrary, when PMTA is applied as the inhibitor, the MRR of Cu (113.6 Å/min) is hindered, along with a promoted removal rate of Co (475.0 Å/min), which could effectively reduce the risk of copper dishing defects.

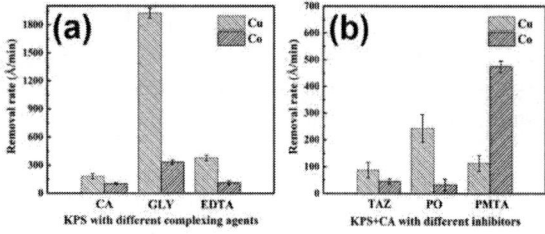

Figure 2: The effect of different (a) complexing agents and (b) inhibitors on Cu/Co MRRs in the KPS based slurry

## INHIBITION EFFICIENCY OF DIFFERENT CORROSION INHIBITORS

A set of potentiodynamic polarization plots of Co with different inhibitors and concentrations was presented in Figure 3. With the concentration of TAZ and PO raising from 1 to 7 mM, the corrosion current density (Icorr) of Co continuously dropped. While with the increasing addition of PMTA, the Icorr of Co is much higher than that without PMTA, which explains the reason of enhanced Co MRR. The calculation of inhibition efficiency ($\theta$) of each inhibitor was performed by the below equation:

$$\theta = 1 - \frac{I_{corr,with\ inhibitors}}{I_{corr,without\ inhibitors}} \times 100\% \qquad (1)$$

Aided by the computed $\theta$, the adsorption behaviors of inhibitors on Cu and Co surfaces can be analyzed with the adsorption isotherm. For example, a liner relationship between Co surface coverage (equals inhibition efficiency, $\theta$) and the concentration of inhibitor PO ($c_i$) can be obtained by using Langmuir adsorption isotherm, as shown in Figure 4. The Langmuir adsorption isotherm can be written as[5]:

$$\frac{c_i}{\theta} = c_i + \frac{1}{K} \qquad (2)$$

$$K = \frac{1}{55.5} \exp(-\frac{\Delta G^0_{ads}}{R_0 T}) \qquad (3)$$

where K is the adsorption equilibrium constant, $\Delta G^0_{ads}$ is the Gibbs energy of adsorption, $R_0$ is the universal gas constant and T is the absolute temperature. When the PO actioned as the corrosion inhibitor for Co wafers, the calculated $\Delta G^0_{ads}$ is equal to -27.23 KJ/mol, which indicates that PO is spontaneously adsorbed on the Co surfaces. Besides, the value of $\Delta G^0_{ads}$ on the order of -20 KJ/mol or higher represents physisorption behavior, while the value of $\Delta G^0_{ads}$ is around -40 KJ/mol or lower expresses that the adsorption is via the form of chemisorption. Based on the calculation, the adsorption of PO on Co wafers can be analyzed, which is the combination of physisorption (major) and chemisorption.

The adsorption behaviors of the other two inhibitors on Cu and Co surfaces can be achieved by the same method.

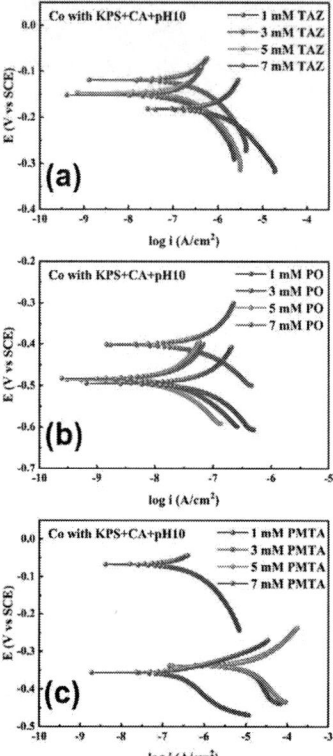

Figure 3: Potentiodynamic polarization plots of Co in the solution containing 10 mM KPS and 5 mM CA with different concentrations of (a) TAZ, (b) PO, (c) PMTA

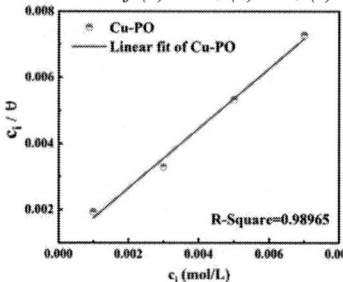

Figure 4: Langmuir isotherm plot for Co with inhibitor PO

To further explore and verify the adsorption behaviors, the XPS analysis of Co treated in different inhibitors was carried out, as shown in Figure 5. The reference curve presents the XPS spectra of the original Co surface. The XPS peaks which located at 61.1, 103.2, 780.8, 796.7, 927.9 eV are attributed to Co 3p, Co 3s, Co 2p$_{3/2}$, Co 2p$_{1/2}$, and Co 2s, respectively. With the addition of CA, the peaks of Co 2p and Co LMM increase, indicating the native oxide of Co was removed by complexation reaction. On this basis, N 1s appears when

TAZ and PMTA were added, while the enhanced C 1s comes into view with the addition of PO. These results proved that the three kinds of inhibitors are indeed absorbed or reacted on Co surfaces.

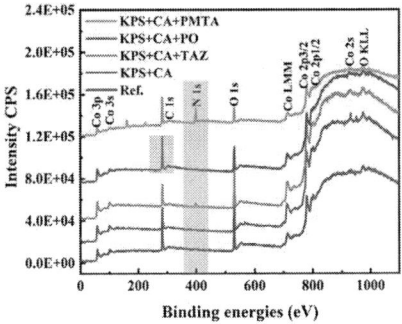

*Figure 5: XPS spectra of Co surfaces treated by different solution*

## DEFECT LAW OF CU/CO PATTERNED WAFERS AFTER POLISHING

TABLE I.    THE COMPONENTS OF BULK/BARRIER SLURRY AND CLEANING SOLUTION

| Solution | Components |
|---|---|
| **Bulk slurry** | Silica, KPS, GLY, pH 10 |
| **Barrier slurry A** | Silica, KPS, CA, pH 10 |
| **Barrier slurry B** | Silica, KPS, CA, PMTA, pH 10 |
| **Cleaning solution A** | CA, pH 11 |
| **Cleaning solution B** | CA, PMTA, pH 11 |

TABLE II.    DIFFERENT COMBINATIONS OF CMP AND POST CMP CLEANING PROCESS

| Experiment number | Combinations of CMP and post CMP cleaning process |
|---|---|
| *#1* | Bulk slurry |
| *#2* | Bulk slurry + Barrier slurry A |
| *#3* | Bulk slurry + Barrier slurry B |
| *#4* | Bulk slurry + Barrier slurry B + Cleaning solution A |
| *#5* | Bulk slurry + Barrier slurry B + Cleaning solution B |

Table I and II shown below summarize the specific components of the employed solution as well as different combinations of CMP and post CMP cleaning procedure. The SEM images of x/x arrays in Cu/Co patterned wafers after using the KPS bulk slurry are shown in Figure 6. It can be clearly seen that a large amount of silica particles has remained on the patterned surfaces after polishing.

Besides, a severe corrosion phenomenon can be observed at the junction between Cu line, Co barrier layer and the dielectric material. The reason is the standard equilibrium potentials difference between Cu and Co, which will cause galvanic corrosion in the presence of multiple slurry components.

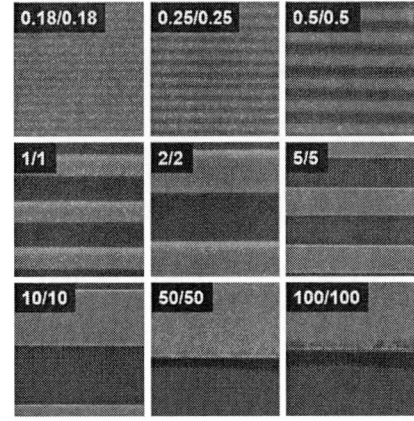

*Figure 6: SEM images of x/x arrays after polishing using the KPS bulk slurry (line width $\leq 2\ \mu m$, 40 kX magnification; line width $\geq 5\ \mu m$, 40 kX magnification)*

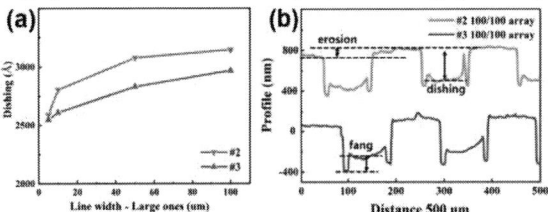

*Figure 7: (a) changes of x/x arrays dishing with large line width, (b) defect comparison of 100/100 μm array after using different barrier slurry*

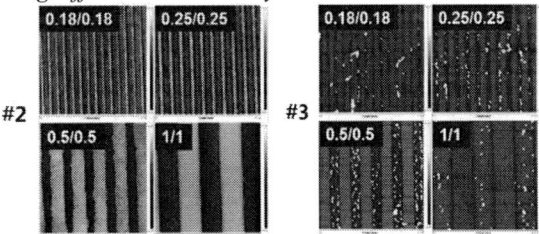

*Figure 8: Topographic AFM images of x/x arrays with small line width after using different barrier slurry*

The copper dishing results of x/x arrays with large line width after using two kinds of barrier slurry are presented in Figure 7(a). Copper dishing deteriorates gradually as the line width becomes larger. The main reason is that the polishing pad will deform a certain amount during the CMP process, and it is easier to bend towards the softer material, which is copper interconnect line. Meanwhile, the copper line with larger line width allows greater deformation, causing a bigger dishing defect. Besides, the comparison between #2 and #3 shows that the addition of PMTA in barrier slurry can improve the dishing defects. Figure 7(b) shows the line profiles of 100/100 μm array in #2 and #3. The barrier slurry without PMTA could give rise to the more serious erosion (935.0 Å), owing to the greater dishing and deformation. On the contrary, the barrier slurry containing PMTA leads to the deeper fang defect (1177.1 Å). The AFM images of the copper patterned structures with small line width after experiment #2 and #3 are shown in Figure 8. Obviously, the defect law of copper dishing is as same as the large line width arrays. However, compared with the barrier slurry without PMTA, the apparent particle residues after using the slurry containing PMTA indicate that it will weaken the ability of KPS and CA to impede the particle adsorption.

Figure 9 shows the dishing defect values of pattern density (PD). Taking #2 experiment as an example, when the conditions are PD<50% and consistent line width (such as 1/3 μm, 1/5 μm, 1/9 μm arrays), the copper dishing remains basically unchanged at 210.0 Å. Besides, the copper dishing values rise by degrees with the conditions of PD<50% and increasing line width. On the other hand, when PD is greater than 50%, regardless of the change of line space, the dishing values show an upward trend with the increment of line width. Also, the magnitude of dishing values of the PD>50% case is much greater than that of the array with PD<50%.

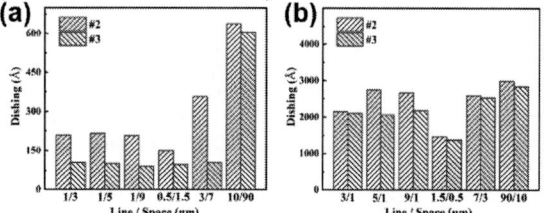

Figure 9: Changes of copper dishing with (a) PD<50%, (b) PD>50% arrays after using different barrier slurry

## DEFECT LAW OF CU/CO PATTERNED WAFERS AFTER CLEANING

The defect law of Cu/Co patterned wafers after two steps polishing and post cleaning process is shown in Figure 10(a). From the changes of dishing values, it can be found that the copper dishing becomes severer with the addition procedure of post cleaning. Similarly, the larger the line width, the more obvious the rising range of the copper dishing, and the greater the defect value caused by cleaning solution without PMTA. Figure 10(b) presents the defect comparison of 100/100 μm array after employing two kinds of CA cleaning solution. The CA cleaning solution can not eliminate the influence of polishing slurry on dishing and fang defect. Meanwhile, the addition of PMTA in CA cleaning solution displays the same effect on patterned wafers, which can increase the fang value slightly.

Figure 11 shows the AFM images of x/x arrays with small line width after using two kinds of CA cleaning solution. Compared with the polished surfaces, the adhesive silica particles on the patterned structures are greatly reduced by the cleaning effect of CA solution. As a result, CA solution can effectively remove the silica particles which residue on the Cu/Co patterned surfaces after polishing. And the addition of PMTA can reduce the risk of copper dishing on the premise of meeting the same cleaning effect.

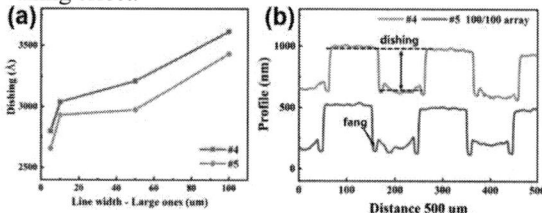

Figure 10: (a) changes of x/x arrays dishing with large line width, (b) defect comparison of 100/100 μm array after using different barrier slurry

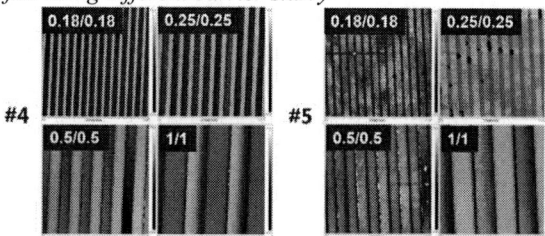

Figure 11: Topographic AFM images of x/x arrays with small line width after using different cleaning solution

## CONCLUSIONS

The defect law of Cu/Co patterned wafers with different line width, space and pattern density was studied. Regardless of the line space and PD, the values of dishing and erosion defect show an upward trend with the increment of line width. Also, the growth magnitude of the PD>50% arrays is much greater than that of PD<50%. Furthermore, the addition of PMTA in the KPS barrier slurry as well as CA cleaning solution could reduce the risk of copper dishing and dielectric erosion, while it will slightly lead to the fang defect. It can also be concluded that citric acid could remove a certain number of silica particles on Cu/Co patterned wafers.

## REFERENCES

[1] C. Yang, P. Flaitz, B. Li, F. Chen, C. Christiansen, S. Lee, P. Ma, D. Edelstein. *Microelectron. Eng.*, vol. 12, 2012, pp. 79-82.

[2] L. Zhang, T. Wang, X. Lu. *J. Mater. Sci.*, vol. 55, 2020, pp. 8992-9002.

[3] L. Zhang, T. Wang, X. Lu. *Microelectron. Eng.*, vol. 216, 2019, pp. 111090.

[4] L. Zhang, T. Wang, X. Lu. *Mater. Chem. Phys.*, vol.

275, 2022, pp. 125199.

[5] M. Hosseini, S. Mertens, M. Ghorbani, M. Arshadi.
*Mater. Chem. Phys.*, vol. 78, 2003, pp. 800–808.

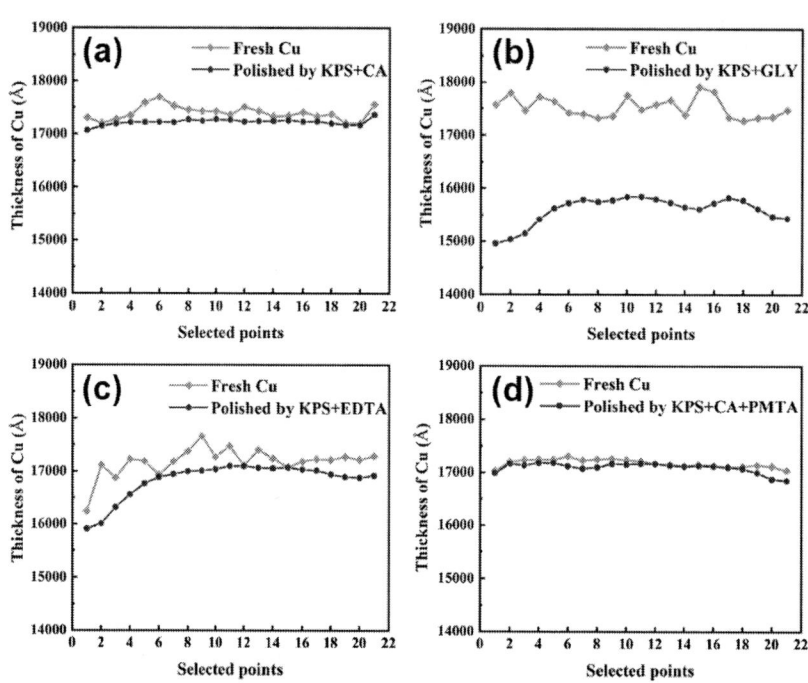

*Figure 1: The morphology and WIWNU of Cu wafers before and after polishing using different slurry*

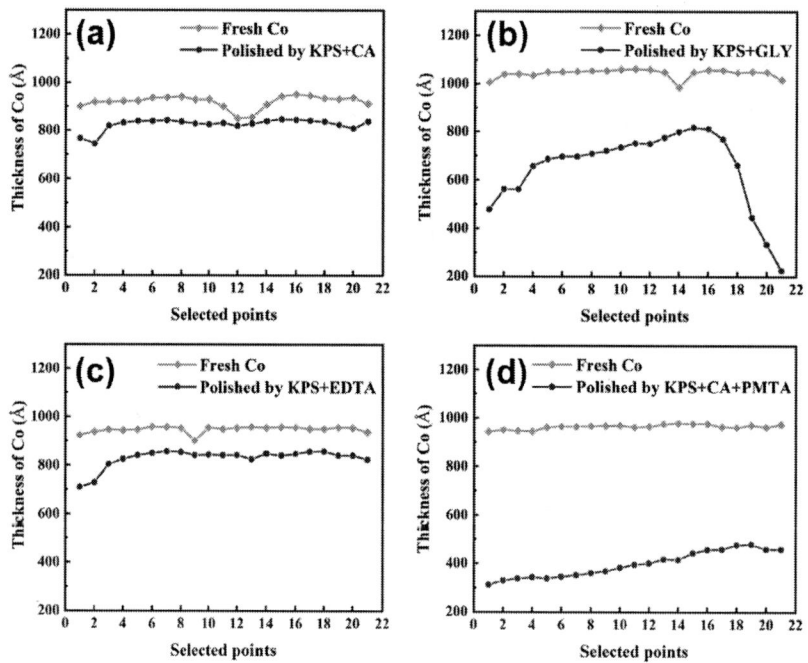

*Figure 2: The morphology and WIWNU of Co wafers before and after polishing using different slurry*

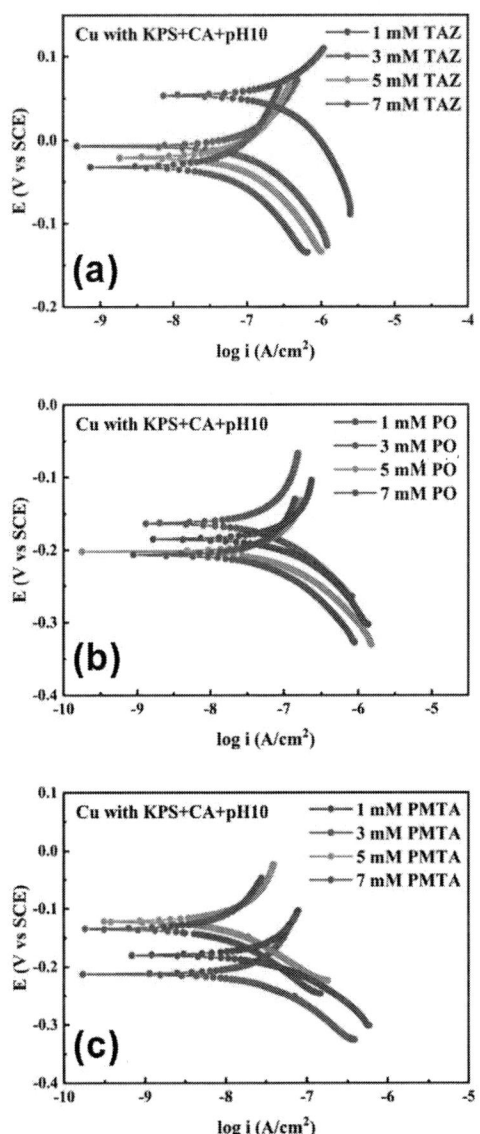

*Figure 3: Potentiodynamic polarization plots of Cu in the solution containing 10 mM KPS and 5 mM CA with different concentrations of (a) TAZ, (b) PO, (c) PMTA at pH 10*

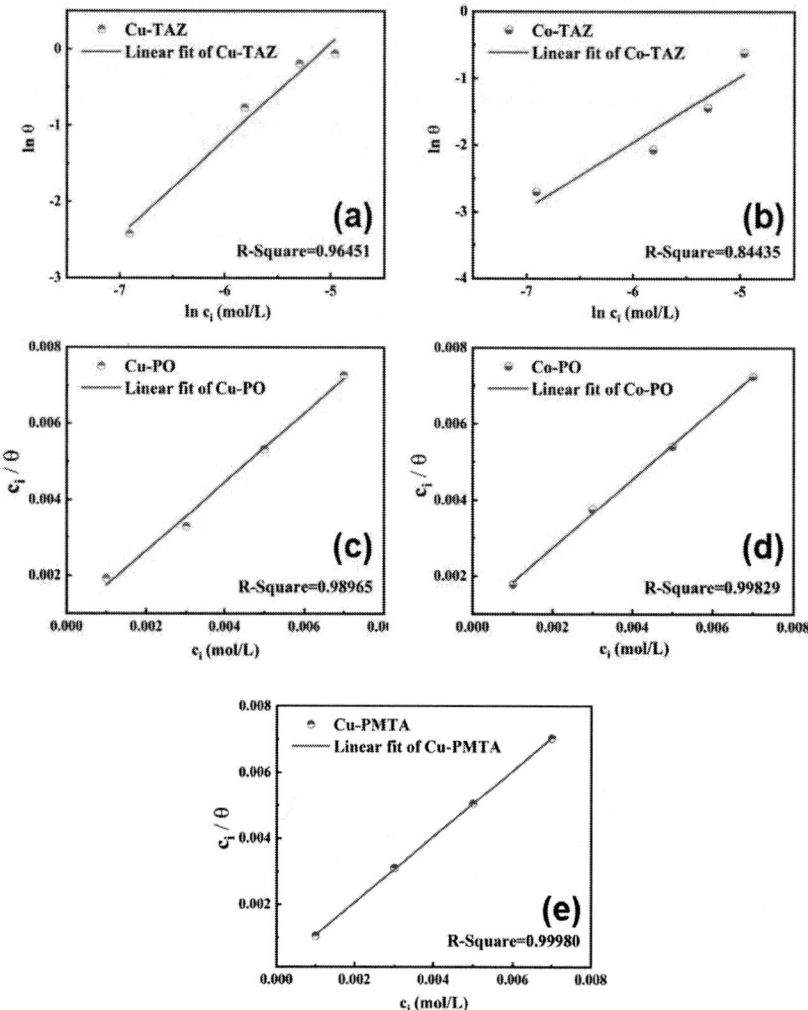

Figure 4: Isotherm plots for Cu and Co surfaces with three kinds of inhibitors

# IMPROVEMENT OF CU-CMP EDP CURVES FOR DIFFERENT PATTERN DENSITY

Yi Xian*, Yuanyuan Meng, Lei Zhang, Jian Zhang, Haifeng Zhou, Jingxun Fang

Advanced Module Technology Development, Shanghai Huali Integrated Circuit Corporation,
Shanghai 201314, China

*Corresponding Author's Email: xianyi@hlmc.cn

## ABSTRACT

Transmission rate and pattern density become the key factors affecting CU CMP EDP Curve at advanced node. Different CU line pattern density determines the difference of CMP incoming topography. This incoming pattern loading requires CMP to leave enough Process Window to fix. This article provides the direction of CU CMP EDP optimization for different pattern density products, and focuses on RTPC Remain thickness, so as to improve the CMP process window.

## INTRODUCTION

As the critical size of semiconductor devices shrinks，the requirements for the precision of semiconductor manufacturing processes are gradually increasing. IBM introduced chemical mechanical polishing（CMP）in 0.35 micron nodes to ensure the accuracy of lithographic exposure by optimizing the wafer surface flatness, and at the same time, it can also reduce the film thickness by polishing[1].

In the BEOL interconnection process, controlling the flatness to realize the correct connection and final conduction of copper lines, so as to avoid the occurrence of short circuit and open circuit. In addition, with the reduction of metal wire size, small changes in copper wire height have a greater impact on the resistance value. Controlling copper wire height to ensure the standard of resistance value has become one of the challenges of advanced node CU CMP[2].

The wafer surface is affected by pattern density, and the roughness of wafer surface is different in each region. The solution to within die uniformity problem is to optimize the thickness of ECP. [3]When the plating copper is very thick, the step height between the different patterns and the dishing within the pattering are optimized. CMP can also optimize the process and improve uniformity by optimizing the EDP curve, in addition to the more expensive method of thickening copper plating

CU CMP process consists of three steps. The first step is High Down Force (HDF) polishing large copper on the surface of the wafer, and using RTPC to control the thickness of copper to a set value. In the second step, Low Down Force (LDF) is used to fine polishing copper in contact with the barrier layer with a lower polishing rate and stop at the barrier layer through the endpoint detection technology. In the third step, the barrier layer and part of the dielectric oxide are buffed by the barrier layer abrasive solution.

Real time profile control(RTPC) based on Faraday and Lenz's Laws. Using electromagnetic sensor constantly reads CU profile during polishing ,and further feedback to the system to adjust the polish head pressure to improve wafer profile, so as to achieve closed-loop control. Therefore, the thickness of RTPC Remain is particularly important, which determines whether LDF has enough process windows to repair the incoming loading.[4]

This paper mainly studied the influence of Pattern density on the endpoint of Top Metal CU CMP. The effect of the thickness of HDF RTPC remain is optimized to effectively restore the incoming topography. In this way, LDF process window is improved and endpoint curve is further optimized to ensure CU clear.

## EXPERIMENTAL

As shown in Figure 1, the surface of copper before CMP is rough. In the first step, HDF removes most of the Cu through RTPC to compensate for changes in the incoming and CMP removal profile. The realization of HDF process is to stop at the set voltage according to the received electromagnetic signal, and the voltage value corresponds to the remaining copper thickness. In the LDF process, the window to capture the feature points of optical signals is optimized through the endpoint to make sure the copper is polishing clean.

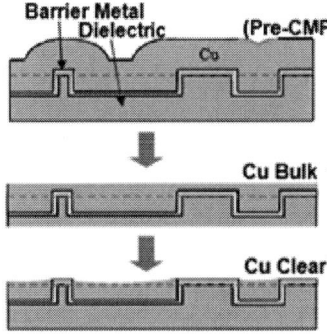

*Figure 1: Cu CMP HDF&LDF polishing process diagram*

When copper lines are closely aligned, ECP electroplating CU has good uniformity, while large areas are easy to form dishing. as shown in the figure 2, when the HDF polishing bulk copper stops at the set thickness, the sparse region may contact the barrier, part of oxide on the wafer surface has been exposed, this will interfere with the optical signals in the following LDF process, causing the signal to be incorrectly caught in advance and resulting in copper residue, or EDP fail.

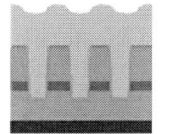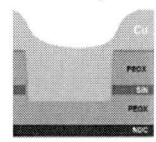

*Figure 2: Different pattern Density*

In Figure 3, the x-axis represents the diameter of wafer, and the y-axis is Cu Thickness. Sensor regularly collects signals to monitor the thickness of copper on the wafer surface. Dense lines indicate that the copper thickness does not decrease significantly during this time, that is, the polishing rate slows

978-1-6654-9759-6/22 $31.00 © 2022 IEEE

down. From HDF to LDF, there will be an obvious process of rate reduction. The thickness of CU ECP is 16K angstrom In the figure, the remaining thickness of RTPC changes from 4K→8K→10K. Obviously, when the remaining thickness of RTPC is 4K, the inner line of the red box is very dense，which indicates that when the HDF is finished, it directly to the barrier, LDF not polishing the barrier subsequently, and the signal thinks that copper has been worn clean, and the LDF process is directly ended, which is very bad for pattern loading repair. However, CU thickness decreases evenly when RTPC remains at 8K, and the optimization effect is most obvious when RTPC remains at 10K.

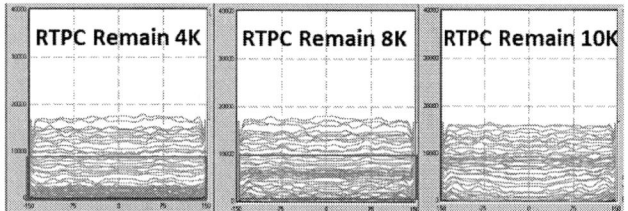

*Figure 3: Real time Cu thickness with different RTPC remain*

Figure 4 shows the EDP curve when the remaining thickness of RTPC is different. When the signal line goes out left and down in the red box, it is called window out, indicating that the signal descent slope reaches the set standard and the Cu thickness becomes significantly thinner. The signal in the red box is called Window in when left in and right out, it indicates that the signal flattens and CU is clear. FIG. 4 corresponds to FIG. 3. When RTPC remain 4K, EDP fail, and no EDP signal is caught and directly polishing to the maximum time，When RTPC Remain is 8K, window in signal is not caught, and EDP fails，Only the remaining thickness of RTPC is 10K, the signal changes significantly and Window is captured successfully.

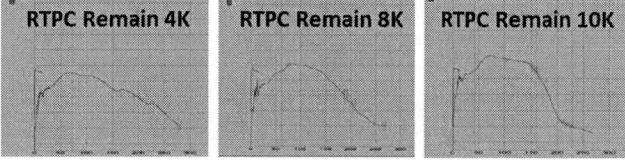

*Figure 4: Endpoint curve with different RTPC remain*

## RESULT AND DISCUSSION

The experimental results show that improving the thickness of RTPC Remain can effectively improve the EDP curve. Based on a large number of experimental data, it is summarized in Table 1 that when the thickness of CU ECP is 16K angstrom and the window setting is the same, wafer of different density corresponds to reasonable thickness of RTPC Remain.

When the chip is evenly distributed, it can be seen that the thickness of RTPC Remain gradually increases with the increase of pattern density and profile roughness, which is consistent with our expectation. The special case is when different chips are included and the pattern density varies greatly from chip to chip. There will be a step height difference between pattern, which requires the remaining thickness of RTPC to be higher, providing LDF with a certain process window for repair

TABLE 1: RTPC REMAIN FOR DIFFERENT PATTERN DENSITY

| Chip | | | | |
|---|---|---|---|---|
| Layer | Top metal | | | |
| Pattern density% | 71 | 78 | 80 | 81/50 |
| RTPC remain | 4k | 5k | 6k | 10k |

## CONCLUSION

With the decrease of the size of advanced node devices, CU CMP faces increasing challenges. The pattern density of different products varies. With the trench narrowing and copper wire height decreasing, only the smoothness of wafer surface cannot meet the requirements. It is necessary to further improve the uniformity of within die and solve the height difference between different patterns.

In this paper, the thickness of HDF RTPC Remain is adjusted according to different pattern density to ensure the accuracy and effectiveness of EDP curve. It is concluded that when the pattern density increases and step high exists between the patterns, the EDP curve can be optimized by increasing the remaining thickness of RTPC.

## REFERENCES

[1]. Y.Y Bai, et al. ECS Transactions.2004
[2]. J.X Fan, et al. *Ecs Journal of Solid Stateence & Technology*2019
[3]. W. Lixiao, et al. *IEEE Transactions on Semiconductor Manufacturing*.2015
[4]. L. Zhang, et al. *CSTIC*.2020

# EFFECT OF TAZ AS AN INHIBITOR ON ELECTROCHEMICAL AND CMP CHARACTERISTICS OF MOLYBDENUM

*Pengfei Wu [1,2], Baoguo Zhang[1,2*], Ye Wang[1,2], Mengchen Xie[1,2], Ye Li[1,2], Haoran Li[1,2]*

[1]School of Electronics and Information Engineering, Hebei University of Technology

Tianjin 300130, People's Republic of China

[2]Tianjin Key Laboratory of Electronic Materials and Devices, Tianjin 300130,

People's Republic of China

*Corresponding Author's Email: bgzhang2000@yahoo.com

## ABSTRACT

The effect of 1, 2, 4 – trizazole (TAZ) on CMP and electrochemical of Mo is investigated. It is found that removal rate (RR) decrease with the addition of TAZ into the alkaline $H_2O_2$-based slurry. The potentiodynamic polarization experiment shows that after adding TAZ into the slurry, the anti-corrosion ability of Mo increases. The mechanism of TAZ to CMP of Mo is probably due to TAZ adsorbs on the Mo surface.

*Keywords—TAZ; Alkaline slurry; Barrier CMP; Removal rate; Static etch rate*

## INTRODUCTION

With the development of integrated circuit, the feature sizes need to be reduced. RC delay is an important factor, which causes the signal delay of integrated circuit [1]. The decrease of groove gap leads the aspect ratio increases correspondingly. The traditional barrier structure of Ta/TaN bilayer is no longer applicable. Finding another metal as a barrier layer material is one of the solutions.

Mo has low electric resistivity to other barrier materials (Ta 14 μΩ·cm, Co 6.3 μΩ·cm, Ru 7 μΩ·cm, Mo 5.34 μΩ·cm) [2]. At the same time, the adhesion of Mo to Cu is stronger than that of Ta to Cu [3]. Therefore, Mo can be a best candidate of next generation of barrier material.

H Feng et al. investigated the effect of pH value on Mo CMP process with $H_2O_2$ as oxidizer [4]. In alkaline slurry, Mo transferred to $MoO_3$, which is soluble and form $MoO_4^{2-}$. The overall reaction can be written as:

$$Mo + 3H_2O_2 + 2OH^- = MoO_4^{2-} + 4H_2O \quad (1)$$

The basic CMP properties of Mo in $H_2O_2$ based slurry had already studied [5]. It is reported that greater static etch rate (SER) and removal rate (RR) can be achieved in alkaline $H_2O_2$ based slurry than in the acidic slurry [6].

Y Guang et al. investigated the effect of glycine on CMP of Mo [6]. Glycine can form complex with $MoO_3$ and promote dissolution of surface oxide. However, glycine can also adsorb on Mo and prevent Mo from contacting with $H_2O_2$. The slowed oxidation reaction dominates the whole process, resulting inhibited Mo corrosion.

TAZ is widely used as corrosion inhibitor on Cu CMP because the excellent inhibitory effect and be friendly to environment. Therefore, the inhibitory effect of TAZ on Mo has been investigated in this work. Polishing experiment and potentiodynamic polarization experiment was performed to study the inhibitory effect on Mo CMP and corrosion, respectively.

## EXPERIMENTAL

*Slurries preparation.* – Colloidal silica (mean particle size of 60 nm) was used for abrasive, 1, 2, 4-triazole (TAZ, Tianjin Sailboat Chemical Reagent Technology Co., Ltd, analytical grade) was used for corrosion inhibitor, hydrogen peroxide ($H_2O_2$, mass fraction of 30%, Tianjin Sailboat Chemical Reagent Technology Co., Ltd, analytical grade) was used for oxidant, KOH and $HNO_3$ (Tianjin Damao Chemical Reagent Technology Co., Ltd, analytical grade) were used as pH regulator.

*Polishing experiment.* – Mo disk (3 inch diameter, 99.99% purity, thickness of 2 mm) was polished on Alpsitec inc E460 polisher made in France. Politex^TM Reg polishing pad was used for all polishing experiments. After polishing, the disk was rinsed by deionized (DI) water and dried by high purity nitrogen. The polishing pressure was set up as 1.5 psi (10.3 kPa); the flow rate of the slurry was 300 ml/min; carrier/platen speeds were 87/93 rpm; and polishing time was set at 180 s.

The removal rates of Mo are determined by measuring the difference of mass before and after polishing by electronic analytical balance with a high precision of 0.1 mg (Mettler Toledo AB204-N). It is calculated by Equation 2.

$$RR = \frac{\Delta m}{\rho \pi r^2 t} \quad (2)$$

where $\Delta m$ is the reduction in the weight of the Mo disk after polishing. $\rho$ and $t$ is density of Mo ($\rho_{Mo} = 10.2$ g/cm$^3$) and polishing time, respectively.

*Electrochemical experiment.* – Electrochemical workstation (CHI660E Shanghai Chenhua Co., Ltd) is employed for experiments. The scanning rate is 5 mV/s for potentiodynamic polarization tests. Mo coupon (99.99% pure, $10 \times 20 \times 2$ mm) is used as working electrode and sealed with epoxy resin to expose a working area of $1 \times 1$ cm$^2$. Platinum electrode (99.99% pure, $10 \times 10 \times 0.2$ mm) is worked as counter electrode and placed in parallel to the working electrode. Saturated calomel electrode (SCE) is employed as reference electrode. Before each tests, Mo coupon is pretreated to remove the oxide film on the metal surface, cleaned with DI water and dried with $N_2$ gas.

## RESULTS AND DISCUSSION

As shown in Figure 1, the effect of different concentrations of $H_2O_2$ on Mo properties was investigated. The slurries containing 0.5% $H_2O_2$, vary concentrations of TAZ (0, 200, 400 and 600 ppm) and without abrasive. The corresponding data are shown in the inset table. With the addition of $H_2O_2$, the corrosion potential of Mo ($E_{corr}$) increased from -42 mV to 109 mV drastically. With more $H_2O_2$ adding into the slurry, $E_{corr}$ increased from 109 mV to 132 mV and then down to 129 mV. The corrosion current density ($i_{corr}$) increased from 3.494 μA·cm$^{-2}$ to 72.86 μA·cm$^{-2}$. A large amount of hydroxide ions was generated from cathodic side, as shown in Equation. 3 .

$$O_2 + 2H_2O + 4e^- = 4OH^- \qquad (3)$$

On the anodic side, $MoO_3$ was formed by the reaction of $H_2O_2$ and molybdenum. The chemical reaction equation as followed:

$$Mo + 2H_2O_2 = MoO_3 + H_2O + H_2 \qquad (4)$$

It indicates that $H_2O_2$ can react with molybdenum and form an oxidation film to protect it from corrosion. However, $MoO_3$ is not stable enough in alkaline solution.

$$Mo + 2OH^- = MoO_4^{2-} + 2H_2O \qquad (5)$$

The more $H_2O_2$ adding into the slurry, the more $MoO_4^{2-}$ emerging and accelerate the dissolution rate of $MoO_3$. When the slurry containing 0.5% $H_2O_2$, the highest $E_{corr}$ was gained and its $I_{corr}$ was not large. Therefore, 0.5% $H_2O_2$ is the most optimistic.

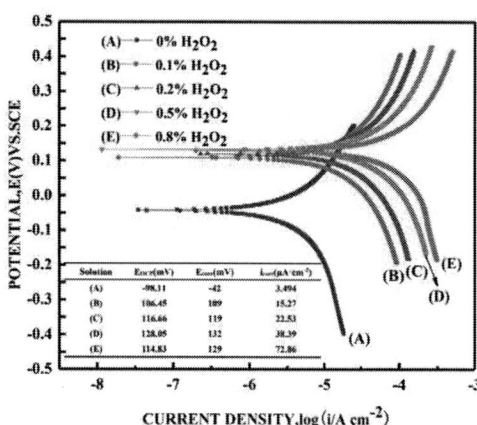

*Figure 1: Potentiodynamic plots of Mo electrode in different concentration of $H_2O_2$ at pH 10*

The effect of TAZ on electrochemical of Mo was investigated, as shown in Figure 2. The corresponding data are shown in the inset table. The corrosion potential of Mo increased dramatically from 102 mV to 184 mV

then to 203 mV slightly. At the same time, the corrosion current density increased from 63.18 μA·cm$^{-2}$ to 110.5 μA·cm$^{-2}$, which reveals that TAZ react with $MoO_4^{2-}$ and form a passivation film. The film was formed completely when adding 200 ppm TAZ. It can let hydroxide pass through and get in touch with the metal and its oxidation. That's the reason of corrosion current density increased with the augment of TAZ concentration. Therefore, 200 ppm TAZ is efficient for prevent Mo from corrosion.

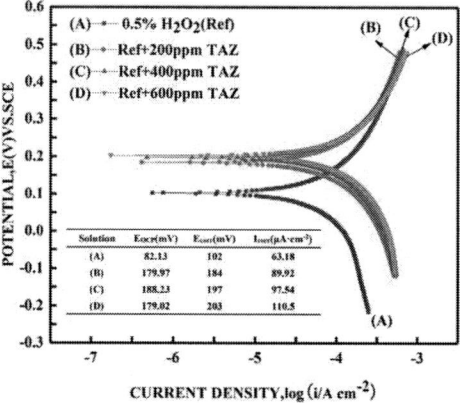

*Figure 2: Potentiodynamic plots of Mo electrode in different concentration of TAZ at pH 10*

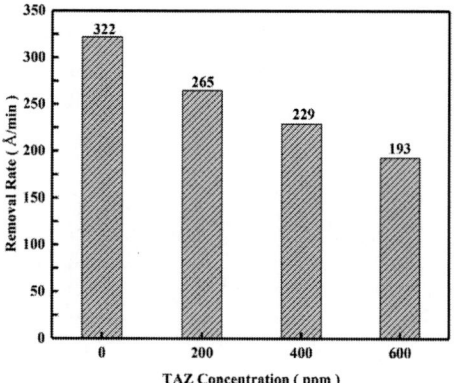

*Figure 3: Removal rate of Mo in different concentration of TAZ at pH 10*

Figure 3 exhibits the removal rate of Mo with the addition of vary concentrations of TAZ. The slurries were consisted with 5% $SiO_2$, 0.5% $H_2O_2$ and different concentrations of TAZ (0, 200, 400 and 600 ppm). With the introduction of TAZ from 0 to 600 ppm, the removal rate of Mo decreased from 322 Å/min to 193 Å/min. The whole CMP process is complicated, which concludes chemical and mechanical work. The chemical product formed by TAZ and $MoO_4^{2-}$ is more stable than oxide film of Mo, so that the removal rate of the absence of TAZ is higher than the presence of TAZ. It was continuously removed by mechanical effect of silica abrasives. Considering the polishing rate of disk is slightly slower than wafer's polishing rate. Therefore, the slurry added 200 ppm TAZ, its removal rate attained the requirements of barrier layer CMP.

## CONCLUSIONS

The addition of $H_2O_2$ increases the corrosion potential of Mo by forming $MoO_3$ on the metal surface. When adding 0.5% $H_2O_2$ into the slurry, the highest corrosion potential attained. In $H_2O_2$-based alkaline slurry, the addition of 200 ppm TAZ increases the corrosion potential of Mo from 102 mV to 184 mV, which means the anti-corrosion ability of Mo is improved. The RR of Mo is decreased from 322 Å/min to 193 Å/min by the introduction of appropriate amount of TAZ (200 ppm). The results show that TAZ has inhibitory effect on electrochemical and chemical mechanical polishing of Mo.

## ACKNOWLEDGEMENTS

This work was supported by the Major National Science and Technology Special Projects (No. 2016ZX02301003-004-007), Natural Science Foundation of Tianjin, China (16JCYBJC16100, 18JCTPJC57000). The authors also thank the teachers and classmates for their helpful suggestions.

## REFERENCES

[1] R. G. Grodon, H. Kim, Y. Au, H. Wang, H. B. Bhandari, Y. Liu, D. K. Lee, Y. Li. *Advanced Metallization Conference*, 2008, pp. 1-9.

[2] F. Chen, X, Zeng, J. B. Xu, H. Lu, X. P. Qu. *International Conference on Planarization/CMP Technology*, Grenoble, October 15-17, 2012, pp. 377-382.

[3] H. Feng, L. Cao, X. P. Qu. *2014 12th IEEE International Conference on Solid-State and Integrated Circuit Technology*, Guilin, October 28-31, 2014, pp. 148-155.

[4] H. Feng, L. Cao, J. Feng, X. P. Qu. *International Conference on Planarization/CMP Technology*, Kobe, November 19-21, 2014, pp. 66-69.

[5] X. P. Qu, G. Yang, P. He, H. Feng. *J. ECS Journal of Solid State Science and Technology*. vol. 6, 2017, pp. 470-476.

[6] G. Yang, P. He, X. P. Qu. J. *J. Applied Surface Science*, vol. 427, 2015, pp. 148-155.

# EFFECT OF OA AND JFCE AS SURFACTANTS ON THE STABILITY OF COPPER INTERCONNECTION CMP SLURRY

*Han Yan[1,2], Xinhuan Niu[1,2*], Fu Luo[1,2], Minghui Qu[1,2], Yinchan Zhang[1,2]*

[1]School of Electronics and Information Engineering, Hebei University of Technology, Tianjin 300130, China

[2]Tianjin Key Laboratory of Electronic Materials and Devices, Tianjin 300130, China

*xhniu@hebut.edu.cn

## ABSTRACT

In the slurry of copper interconnection chemical mechanical polishing (CMP), surfactants are beneficial to achieve a more consistent material removal rate (MRR). They also have a significant effect on the stability of the slurry. In this paper, the effects of cationic surfactant amine oxide (OA) and nonionic surfactant fatty alcohol polyethylene ether (JFCE) on copper MRR and the stability of slurry are discussed. In the FA/OII-based weakly alkaline slurry, the copper MRR tended to increase and then decrease with the increase concentrations of OA or JFCE. However, the slurry containing OA was less stable and the slurry with JFCE could be stable for a longer time.

*Keywords—Chemical mechanical polishing; OA surfactant; JFCE surfactant; material removal rate; slurry stability*

## INTRODUCTION

Since the mid-1990s, IBM, Intel, AMD and other integrated circuit (IC) manufacturers have decided to replace aluminum with copper as interconnection material because of its low resistivity and high electromigration resistance[1]-[2]. For conventional copper CMP, the slurry usually contains abrasive, oxidizer, pH adjuster, complexing agent, surfactant and corrosion inhibitor, among which surfactant plays a key role in material removal rate (MRR) and the stability of slurry due to the basic properties such as detergency, solubilization, foaming, defoaming, wettability, dispersion and emulsification[3][4]. Xu et al.[5] investigated the role of nonionic surfactant AEO in the planarization performance of copper wafers and achieved the good stability, high MRR, low within-wafer non-uniformity and good surface by optimal selection of concentration of AEO. Yin et al.[6] calculated the interaction parameters of the anionic surfactant LABSA and the nonionic surfactant JFCE. When the ratio of LABSA/JFCE was 2:3, they showed the strongest synergistic effect and the maximum particle removal of copper surface was 97.08 %. Luo et al.[7] investigated that the synergistic effect of DBSA and iAEO can reduce more than 86% scratches and improve particle dispersion. Zhang et al.[8] demonstrated the nonionic surfactant FA/O I can reduce copper depression and dielectric erosion.

In fact, the stability of slurry during the copper CMP has a crucial role in the production of industrial products.

Therefore, in this paper, OA and JFCE as surfactants were chosen for the study, and the influence of them on MRR and the stability of slurry was confirmed by dynamic polishing experiments and the changing of particle size and zeta potential of slurry.

## EXPERIMENTAL

The slurry consisted of colloidal silica (mean particle size about 80 nm-90 nm), complexing agent FA/OII, oxidizer $H_2O_2$ (the mass fraction of 30 wt%), surfactant OA and JFCE, pH was adjusted by KOH.

3-inch diameter Cu disk (purity of 99.99%) was polished on Alpsitecin-E460E polisher made in France. IC 1000 polishing pad was used in the all experiments. The polishing parameters are shown in Table I. Before polishing experiment, the polishing pad was conditioned by using diamond conditioner. After polishing, copper was taken out, washed in purified de-ionized (DI) water, and dried in $N_2$ air steam.

The weighing method was used to calculate the copper MRR by measuring the weight difference before and after polishing, which was determined by Eq. (1). A high-precision electronic balance (Mettler Toledo AB204-N, precision to 0.1 mg) was used to weigh. Mean particle size and zeta potential of particles were measured by NICOMP 380ZLS.

$$MRR = \frac{\Delta M}{\rho \pi R^2 t} \qquad (1)$$

## RESULTS AND DISCUSSION

**Effect of OA on copper MRR.** - The slurries consisted of 0.5 wt% $SiO_2$, 3 wt% FA/OII, 0.5 vol% $H_2O_2$, 0.004 wt% ADS, and different concentrations of OA at pH 9. Fig. 1 showed the effects of different concentrations of OA on the copper MRR. It can be observed that copper MRR increased and then decreased as the concentration of OA increased from 0 vol% to 0.5 vol%. At 0.1 vol% OA, copper MRR was the highest, about 5323 Å/min. When

*TABLE I. PROCESS PARAMETERS OF CMP*

| Parameters | Conditions |
|---|---|
| Down force (psi) | 1.5 |
| Back pressure (psi) | 0 |
| Polishing head speed (r/min) | 87 |
| Platen rotation speed (r/min) | 93 |
| Slurry flow rate (ml/min) | 300 |

the concentration of OA increased to 0.3 vol%, the copper MRR remained stable at about 5100 Å/min, the overall trend tended to level off. However, the MRR decreased significantly and the polishing stability was poor after three days.

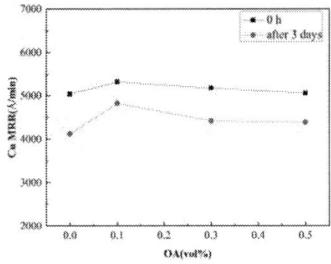

*Figure 1: Copper MRR with different concentrations of surfactant OA.*

**Effect of OA on particle size and zeta potential of the slurry.** - An increase in particle size or a decrease in the absolute value of zeta potential can cause agglomeration of silica sol, which in turn leads to gel delamination. The increase in the particle size of silica sol makes the collision probability of silica sol particles increase. The decrease in the absolute value of zeta potential makes the repulsive force between silica sol particles decrease. Both of these phenomena are not conducive to the stability of silica sol. The stability of the particles can be characterized by zeta potential. The particles are stable when the absolute value of zeta potential is greater than 30 mV, and particles will be unstable when the absolute value of zeta potential is less than 30 mV. The larger the absolute value of zeta potential is, the more stable the particles are.

As shown in Fig. 2, particle size of the slurry started at 77.4 nm. With the addition of different concentrations of OA, the particle size changed insignificantly and tended to be a straight line. After three days, the particle size decreased slightly. The particle size of the slurry varied about 3 nm. In terms of particle size, the three slurries containing different concentrations of OA were still relatively stable. After the addition of OA in the slurry, the absolute value of zeta potential was lower than the original slurry. Only when the concentration of OA was 0.1 vol%, the absolute value of zeta potential of the slurry was higher than 30 mV, about 40.19 mV at this time.

OA is a cationic surfactant which is not conducive to the stability of the slurry. The slurry became more and more unstable as time went by. Fig. 3 showed the slurry with different concentrations of OA after three days. It can be seen that the volume fraction at 0.1 vol% was better, and the volume fractions at 0.3 vol% and 0.5 vol% already showed the delamination. Therefore, the stability of the slurry is relatively good when the volume fraction of OA is 0.1 vol%.

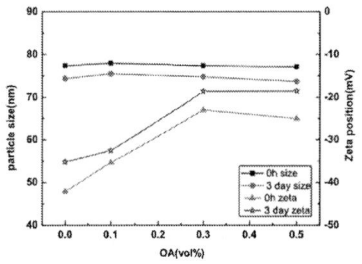

*Figure 2: Particle size and zeta potential of copper with different concentrations of surfactant OA.*

*Figure 3: Different concentrations of OA slurry after storage for three days.*

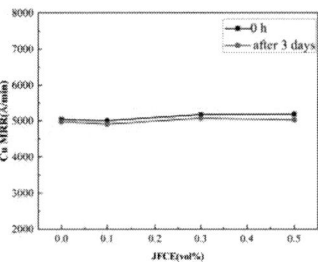

*Figure 4: Copper MRR with different concentrations of surfactant JFCE.*

**Effect of JFCE on copper MRR.** - The slurries consisted of 0.5 wt% $SiO_2$, 3 wt% FA/OII, 0.5 vol% $H_2O_2$, 0.004 wt% ADS and different concentrations of JFCE at pH 9. As shown in Fig. 4, the copper MRR increased with the increase of volume fraction of JFCE and then decreased, the MRR was about 5000 Å/min. In general, the MRR curve tended to be a straight line, and the MRR was the highest when the volume fraction of JFCE was 0.3 vol%, about 5176Å/min. After three days, the MRR was 5074 Å/min and the change of MRR was not much. This indicated that the slurry can be stable for a long time, and the introduction of JFCE as a surfactant can maintain the stability of the slurry well.

**Effect of JFCE on particle size and zeta potential of the slurry.** - As shown in Fig. 5, the particle size of the slurry was mainly in the range of 75-78 nm. The particle size basically did not change which was in the range of 74-77 nm after three days. Compared with the cationic surfactant OA, the particle size of the slurry with the introduction of JFCE remained basically unchanged, indicating that the stability of JFCE to the slurry was better than that of OA. The zeta potential plot revealed that the potential change increased relative to the original slurry after the addition of the nonionic surfactant JFCE to the slurry. It was also observed that at a volume fraction of 0.3 vol% of JFCE, the potential showed the maximum value of 47 mV. After three days, the zeta potential not only did not decrease, but also increased to 50 mV.

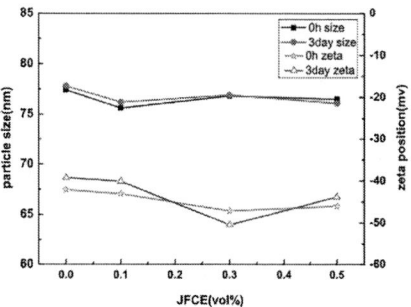

*Figure 5: Particle size and zeta potential of copper with different concentrations of surfactant JFCE.*

JFCE is a nonionic surfactant that can significantly improve the stability of the slurry, and the potential values showed an increase after the introduction of JFCE. The value of the potential did not decrease and changed significantly as time went by. This indicated that the stability of the slurry was very good. Fig. 6. showed the slurries with different concentrations of JFCE after three days. It was found that the slurry at a volume fraction of 0.3 vol% was better. Combined with the above analysis, the results showed that the stability of the slurry was relatively good when the volume fraction of JFCE was 0.3 vol%.

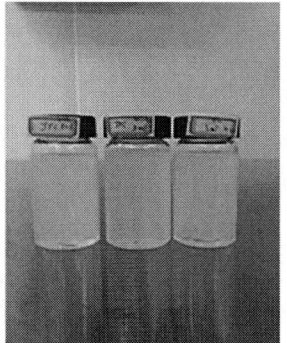

*Figure 6: Different concentrations of JFCE slurry after storage for three days.*

## CONCLUSIONS

In this paper, the effect of OA and JFCE in FA/OII-based weakly alkaline slurry on the copper MRR and the stability of the slurry are discussed in detail. The MRR, average particle size and the zeta potential of the slurry are used as reference indicators for analysis. The results showed that OA was not favorable to the stability of the slurry. However, JFCE has a catalytic effect on the stability of the slurry and the stability was the best when the volume fraction of JFCE was 0.3 vol%.

## ACKNOWLEDGMENTS

This work was supported by the Major National Science and Technology Special Projects (No. 2016ZX02301003-004-007), Natural Science Foundation of Hebei Province (No. F2021202009), The authors also thank the teachers and classmates for their helpful suggestions.

## REFERENCES

[1] T. Ohmi, T. Hoshi, S. Saka, S. Sakia. *Journal of The Electrochemical Society, vol. 104, 1993, pp. 1131-1137.*

[2] J.M. Steigerwald, S.P. Murarka, R.J. Gutmann, D.J. Duquette. *Materials Chemistry and Physics, vol. 41, 1995, pp. 217-228.*

[3] T.B. Du, Y. Luo, V. Desai. *Microelectronic Engineering, vol. 71, 2003, pp. 90-97.*

[4] E. Abelev, A.J. Smith, A.W. Hassel, Y. Ein-Eli. *Electrochimica Acta, vol. 52, 2007, pp. 5150-5158.*

[5] Q.Z. Xu, F. Yang, L. Chen, H. Cao. *International Journal of Precision Engineering and Manufacturing, vol. 19, 2018, pp. 1585-1595.*

[6] D. Yin, S.Y. Tian, N.N. Zhang, Q. Wang, X.Q. Sun, M.R. Liu, S.H. Zhang, B.M. Tan. *Materials Chemistry and Physics, vol. 257, 2020, 123841.*

[7] C. Luo, Y. Xu, N.Y. Zeng, T.D. Ma, C.W. Wang, Y.L. Liu. *Tribology International, vol. 152, 2020, 106576.*

[8] W.Q. Zhang, Y.L. Liu, C.W. Wang, J.J. Gao, X.H. Niu, J. Wang. *ECS Journal of Solid State Science and Technology, vol. 6, 2017, pp. 270-275.*

# Effect of different complexing agents on chemical mechanical polishing of copper film

Fu Luo[1,2*], Xinhuan Niu[1,2*], Han Yan[1,2], Minghui Qu[1,2]

[1]School of Electronics and Information Engineering, Hebei University of Technology, Tianjin 300130, China
[2]Tianjin Key Laboratory of Electronic Materials and Devices, Tianjin 300130, China

*15302038735@163.com, xhniu@hebut.edu.cn

## Biography

Fu Luo, a native of Tianjin, China, graduated from Hebei University of Technology. My main research direction is chemical mechanical polishing of materials.

## Abstract

As the main additive, complexing agent can effectively improve the removal rate of copper film during chemical mechanical polishing(CMP) process, so the basic research of the effects of FA/OII chelating agent, ethylenediamine(EDA) and glycine(GLY) as complexing agent on removal rate and surface quality of copper and cobalt for low tech-node were performed. The experimental results showed that all three complexing agents can achieve higher copper removal rates and lower cobalt removal rate. EDA has the highest removal rate of copper and serious surface corrosion. It was found the slurry with glycine can be stable for three days, which can meet the needs of industrial production.

*Keywords— Complexing agent; Copper slurry; Removal rate; Surface quality; Stability*

## Introduction

As feature size of integrated circuits (IC) gradually dropping down to 14 nm and even lower, copper (Cu), as one kind of low resistivity and high electromigration resistance metal, has been widely selected as an interconnecting metal[1-2]. A new type of ultra-thin barrier material with lower resistivity and better performance than traditional barrier layer tantalum (Ta) is needed. Cobalt (Co), with high stability, good conductivity and lower resistivity, has good step coverage and can be directly electroplated Cu wiring under seedless Cu crystals. Therefore, it is used as one of the barrier layer materials to replace tantalum[3]. For 14 nm and even lower Cu film it was changed from traditional two-step polishing to one-step polishing, the purpose is to remove the Cu completely and stop on the Co layer. And the Cu pattern wafer schematic is shown in Fig. 1.

J. K. Zhou et al[4] investigated the application of chitosan (CTS, a natural macromolecule organic compound) as a complexing agent in low-tech node Cu CMP, and the results showed that the removal rate and stability of slurry could be improved. J. Jun et al[5] investigated the polishing and electrochemical behavior of Co in an alkaline slurry containing a chelator based on a derivative of ethylenediamine tetraacetic acid (EDTA) in the absence of any oxidant. J. C. Wang et al[6] explored potassium nitrate to enhanced ionic strength in baseline slurry, and the removal mechanism of Cu and TEOS (tetraethylorthosilicate) were investigated. W. Q. Zhang et al[3] studied the effect of non-ionic surfactant on Cu dishing and dielectric erosion correction in alkaline barrier CMP solution.

However, there was still lack of systematic research on the role of removal rate of Cu and removal rate selection ratio of Cu/Co with different complexing agents using alkaline Cu CMP slurries for Cu film polishing.

In this paper, the effects of three complexting agents on Cu/Co removal rate and surface quality were studied. At the same time, the stability of three kind of slurries with different complexting agents was also investigated. Fig. 2 shows the structure formula of FA/OII chelating agent, ethylenediamine and Glycine.

## Illustrations

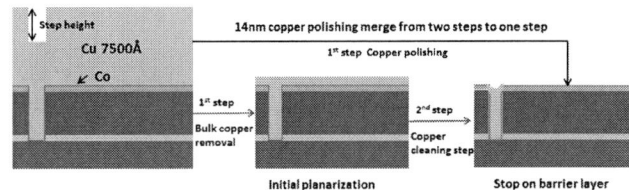

Fig 1 Schematic illustration of Cu CMP

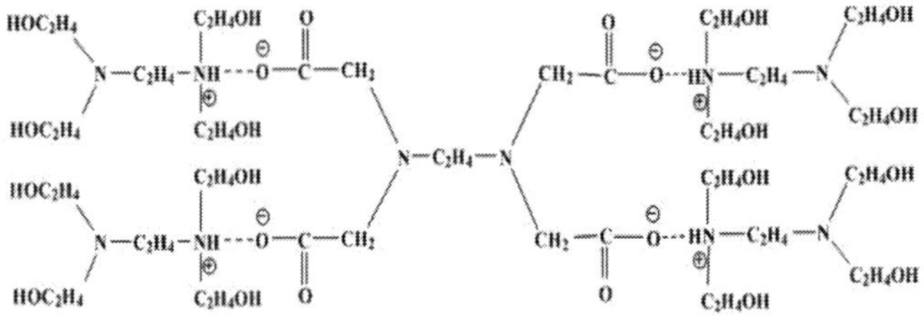

（a）FA/OII chelating agent

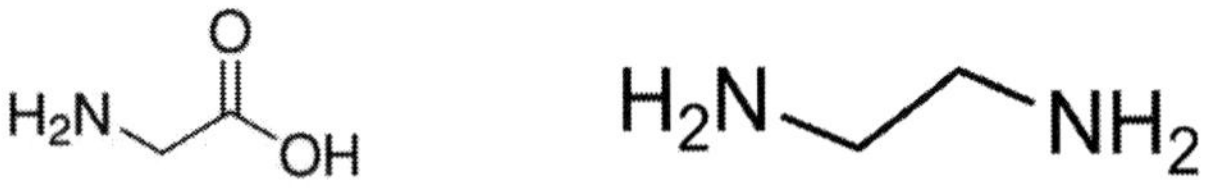

（b）Ethylenediamine     （c）Glycine

Fig 2 Structure formula of three chelating agents

## Experiment

**Polishing experiments**—3 inch diameter Cu and Co (purity of 99.99%) was polished using France-E460E polisher and IC 1000 polishing pad produced by Rohm and Haas Electronic Materials. After a large number of optimization experiments, the optimized process parameters were shown in Table 1. The used slurry with pH 10.5 was composed of colloidal silica (particle size between 80 nm-90 nm), FA/OII chelating agent, ethylenediamine, glycine, surfactant and hydrogen peroxide ($H_2O_2$, the mass fraction of 30 wt%, semiconductor grade). All the reagents were diluted by the purified de-ionized water. Before polishing experiment, the polishing pad was conditioned by using diamond conditioner. After CMP, Cu and Co was dried off by $N_2$ air stream. Removal rate (RR) was calculated by equation .

$$RR = \frac{\Delta m}{\pi \times \rho \times t \times r^2} \qquad [1]$$

$\Delta$ m is removal weight of Cu wafer measured by professional electronic balance (Mettler Toledo AB204-N) which resolution is 0.1 mg, r is the radius of Cu wafer, $\rho$ is the density of Cu and Co ($\rho_{Cu} = \rho_{Co} = 8.9$ g/cm$^3$) and t is polishing time ($t_{Cu} = 180$ s, $t_{Co} = 360$ s).

Table 1 Process parameters of CMP.

| Parameters | Conditions |
|---|---|
| Down force(psi) | 1.5 |
| Back pressure(psi) | 0 |
| Polishing head speed(r/min) | 87 |
| Platen rotation speeds(r/min) | 93 |
| Slurry flow rate(ml/min) | 300 |

**CMP performance measurement.**—Agilent 5600LS Atom Force Microscope(AFM) was used for measuring the surface morphology and roughness of Cu and Co after adding three chelating agents.

## Results and discussion

In general, CMP slurry stability is very important to the yield of industrial products. Chelating agent in slurry is the key factor of affecting the stability of silica sol, and the unstable silica sol will cause some problems. The change of Cu removal rate with the storage time of slurry with three complexing agents at 1 to 3 days was shown in fig 3. With the storage time from 1 to 3 days, the removal rate of Cu with 3 vol% FA/OII chelating agent dramatic decreased from 6960 Å/min to 1603 Å/min. It was indicated that the highly alkaline FA/OII chelating agent should be oxidized, so $H_2O_2$ was reduced and its decomposition was promoted. The lower the oxidant concentration, the faster it was consumed and decomposed, and the faster the polishing rate decreased as time went by. FA/OII chelating agent was the main factor affecting the stability of $H_2O_2$. With the storage time from 1 to 3 days, the removal rate of Cu with 0.3 vol% ethylenediamine dramatic decreased from 9294 Å/min to 2567 Å/min, that means the stability of removal rate was not good. While the change of Cu removal rate was the least in the slurry with 2 wt% glycine, which decreased from 6828 Å/min to 6328 Å/min. From the change of removal rate with storage time, glycine was significantly better than the other two complexants.

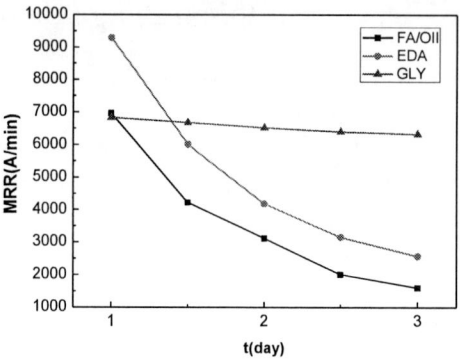

Fig 3 The Cu removal rate varies with storage time

If the removal rate selectivity between different materials cannot be well controlled, it will affect the flatness of wafer surface and even lead to device failure. Therefore, the removal rate of Co should not be too high. As the concentration of chelating agent increased, the Cu removal rate increased show in fig 4, while the Co removal rate increased slightly. When the chelating agent concentration was 3 vol%, the Co polishing rate was 83 Å/min, and the removal selection ratio of Cu/Co reached 84:1. At 0.3 vol% of ethylenediamine, Co removal rate was 67 Å/min and the removal selection ratio of Cu/Co reached 139:1. It can be found that when the concentration of ethylenediamine was low, the removal rate for Cu increased obviously, while Co increased slowly. As a chelating agent, ethylenediamine can achieve high removal rate of Cu film and high selection ratio of Cu and Co removal rate. With the increase of glycine, Cu removal rate increased and Co rate increased slightly. When the glycine concentration was 2 wt%, Cu and Co obtained the best rate ratio 131:1, and the removal rate of Co was 52 Å/min. The change of Cu removal rate was the least in the slurry with glycine, which decreased from 6828 Å/min to 6328 Å/min, when it reached the third day. At the same time, the removal rate of Co went from 52 Å/min to 69 Å/min. From the change of removal rate with storage time, glycine was significantly better than the other two complexants.

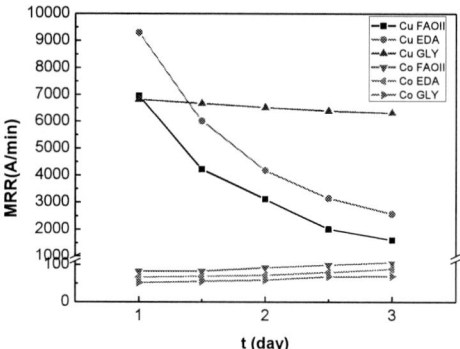

Fig 4 The removal rate varies with storage time

Chemical reaction will increase the removal rate of Cu surface and also affect the surface roughness to some extent. Therefore, surface analysis of Cu after CMP is needed to select complexing agent which can produce the lower surface roughness. Fig 5 (a), (b) and (c) showed the roughness of Cu surface under the polishing system of three chelating agents. It can be found that under the glycine system, the surface roughness of Cu was the lowest, at 1.71 nm(10 μm×10 μm), while the FA/OII and ethylenediamine systems were relatively higher, at 2.69 nm and 10.08 nm, respectively. Therefore, it can be concluded that glycine as a chelating agent can achieve lower Cu metal surface roughness. The reason is ethylenediamine and FA/OII are more corrosive to the metal surface, have stronger chemical effects, and have higher removal rate.

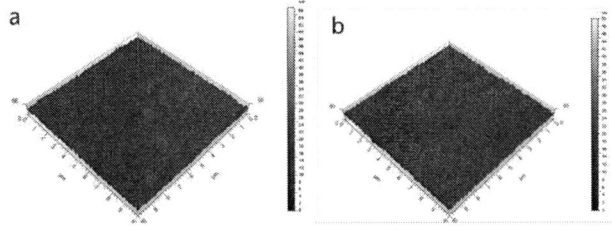

（a）Surface topography of Cu in 3 vol% FAOII slurry. Sq=2.69 nm
（b）Surface topography of Cu in 0.3 vol% EDA slurry. Sq=10.8 nm

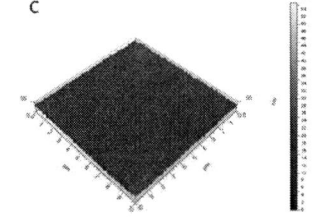

（c）Surface topography of Cu in 2 wt% Gly slurry. Sq=1.71 nm
Fig 5 The AFM images of Cu blanket wafer.

Figure 6 (a), (b) and (c) showed the surface roughness of Co under the polishing system of three complexing agents. Similar to Cu, the roughness was lower in the glycine system, which was 0.783 nm, and higher in the FA/OII and ethylenediamine system, which was 1.05 nm and 1.85 nm. Therefore, it can be concluded that glycine as a chelating agent can achieve lower metal surface roughness. It can be seen from the above results that the glycine system is more suitable for CMP of Cu film from the aspects of removal rate, polishing slurry stability and removal rate selection ratio.

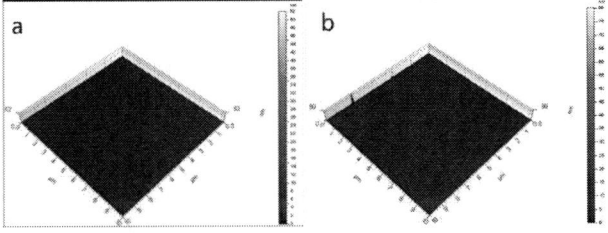

（a）Surface topography of Co in 3 vol% FAOII slurry.Sq=1.05 nm
（b）Surface topography of Co in 0.3 vol% EDA slurry. Sq=1.85 nm.

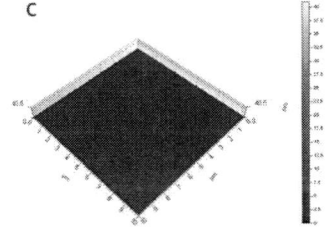

（c）Surface topography of Co in 2 wt% Gly slurry. Sq=0.783 nm
Fig 6  The AFM images of Co blanket wafer.

## Conclusions

The effects of three chelating agents on the removal rate and surface quality of Cu and the stability of slurry were studied. At the same time, Co validation test was carried out. The results showed that FA/OII can obtain a high removal rate of Cu film, but the low removal rate of Co was not obtained. Ethylenediamine can obtain higher removal rate of Cu film, low removal rate of Co and higher removal rate selection ratio of Cu and Co, but the slurry stability still existed insufficiency. In the glycine system, the high removal rate of Cu film, low removal rate of Co and the high removal rate selection ratio of Cu were obtained. At the same time, better surface quality was obtained. The glycine system slurry could be stable for 3 days which can meet the industrial demand. Under such basic research, we will continue to optimize the slurry composition to achieve better surface quality, higher Cu and lower Co removal rates and greater selection ratio.

## Acknowledgments

This work was supported by the Major National Science and Technology Special Projects (No. 2016ZX02301003-004 -007), Natural Science Foundation of Hebei Province (No. F2021202009), The authors also thank the teachers and classmates for their helpful suggestions.

## References

[1]  K. Zhang, X. H. Niu, C. W. Wang, J. C. Wang, D. Yin, and R. Wang. Effect of Chelating Agent and Ammonium Dodecyl Sulfate on the Interfacial Behavior of Copper CMP for GLSI [J]. ECS Journal of Solid State Science and Technology, vol. 7, 2018, PP. 509-517.

[2]  J. Hong, X. H. Niu, J. Wang, C. Wang, B. Zhang, R. Wang, S. Ming, and L. Y. Ling. Research on Si (100) crystal substrate CMP based on FA/O alkaline slurry[J]. Applied Surface Science,  vol. 420, 2017, PP. 483-488.

[3]  W. Q. Zhang, Y. L. Liu, C. W. Wang, X. H. Niu, J. Ji, Y. C. Du, and L. Han. Role of 1,2,4-Triazole in Co/Cu Removal Rate Selectivity and Galvanic Corrosion during Barrier CMP[J]. ECS Journal of Solid State Science and Technology, vol. 6, 2017, pp. 786-793.

[4]  J. K. Zhou, X. H. Niu, Z. Wang, Y. Q. Cui, J. C. Wang, C. H. Yang, Z. Q. Huo, and R. Wang. "Roles and mechanism analysis of chitosan as a green additive in low-tech node copper film chemical mechanical polishing," Colloids and Surfaces A: Physicochemical and Engineering Aspects, vol. 586, 2020, pp. 124293–124305.

[5]  J. Ji, G. F. Pan, W. Q. Zhang, Y. C. Du, and C. W. Wang. Role of Additive in Alkaline Slurries for Co CMP[J]. ECS Journal of Solid State Science and Technology, vol. 6, 2017, pp. 813-818.

[6]  J. C. Wang, X. H. Niu, Y. L. Liu, C. W. Wang, S. H. Yang, K. Zhang, Y. Xu, and T. D. Ma, and W. Q. Zhang. Improvement of Barrier CMP Performance with Alkaline Slurry: Role of Ionic Strength[J]. Ecs Journal of Solid State Science & Technology, vol. 7, 2018, pp. 462-467.

# IMPROVEMENT OF SENSITIVITY OF EDDY CURRENT THICKNESS SENSORS WITH FERRITE CORE FOR CMP PROCESS

*Chengxin Wang, Tongqin Wang, Fangxin Tian, Bangxu Liu, Xinchun Lu\**
State Key Laboratory of Tribology,
Tsinghua University, Beijing 100084, China
Corresponding Author's Email: xclu@tsinghua.edu.cn

## ABSTRACT

Accurate and real-time thickness measurement of metal film on Si-based wafer improves the global planarization by adjusting the polishing pressure and avoids over-polishing and under-polishing during the chemical mechanical polishing (CMP) process. Nevertheless, the large lift-off distance between the sensor and the target film with low conductivity, such as the tungsten, bring great challenges to measure the metal film thickness precisely in the metal CMP process. In this paper, a novel eddy current sensor with Mn-Zn ferrite core is designed to improve sensitivity of eddy current thickness sensors at the large lift-off distance. And the measured results are compared with an air core coil.

## INTRODUCTION

Chemical mechanical polishing (CMP) is a key technique used in large-scale integrated circuit manufacture. The metal CMP process has been widely used in Cu-interconnection processes and gate last approaches to provide global planarization [1]. The real-time thickness distribution of the metal film provides feedback information that can be used to control the polishing pressure, which is adjusted by the carrier head to decrease the range of global planarization [2][3]. And eddy-current thickness measurement is the most extensively used method to in the metal CMP process. However, the sensitivity of eddy-current sensors is limited by the lift-off distance [4], which is between the senor coil and the film surface in the range of 1.5 mm to 4 mm. Hence, it is very important to improve the sensitivity of the eddy-current thickness sensors [5].

In this paper, the principles of eddy-current thickness measurement are presented. Then two sensor systems with an air core coil and a Mn-Zn ferrite core coil are proposed to study their sensitivities when thicknesses of copper films and tungsten films, and the lift-off distances are different. Finally, the results are discussed to elucidate the improvement of the ferrite core coil sensitivity.

## METHOD

A description of the eddy-current method for nanoscale metal film thickness measurements is shown in Figure 1. When the time-varying current is fed into the coil, a time-varying magnetic field is generated around it, and eddy currents are induced in the metal film. The coil

resistance then increases, owing to the induced current, whereas the inductance decreases. The coil impedance is influenced by several factors, and the relationships between these factors are very complicated. Thus, other physical quantities are generally fixed when measuring the given metal film thickness, such as the lift-off distance $l_o$, the excitation frequency f, the outer diameter of a coil $2r_o$, the inner diameter of a coil $2r_i$, the number of the coil turns and etc.

The equivalent transformer model of a coil and induced eddy current loop in target metal film is developed and widely used to analyze many characteristics for eddy-current sensor, shown in Figure 2. The coil can be modeled as a primary circuit, which has an inductor $L_c$ and a resistor r in series. The metal film can be modeled as a secondary circuit, which has an inductor $L_t$ and a resistor $r_t$ in series. M ranging from 0 and 1, is the mutual inductance between the coil and the metal film, which only depends on the lift-off distance x.

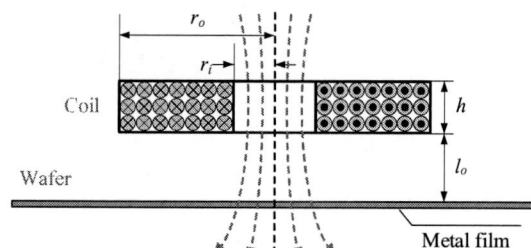

*Figure 1: Figure example*

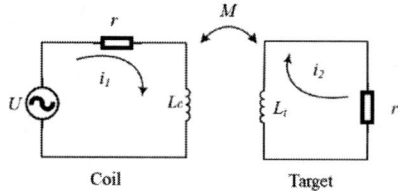

*Figure 2: Transformer model for eddy-current sensor*

According to Kirchhoff's law, the impedance of a eddy-current coil can be expressed as,

$$R = r + \frac{\omega^2 M^2}{r_t^2 + (\omega L_t)^2} R_t \tag{1}$$

$$L = L_c - \frac{\omega^2 M^2}{r_t^2 + (\omega L_t)^2} L_t \tag{2}$$

Where $\omega = 2\pi f$ is the angular frequency of the

excitation voltage, $R_t$ and $L_t$ are the equivalent resistance and inductance respectively of the metal film.

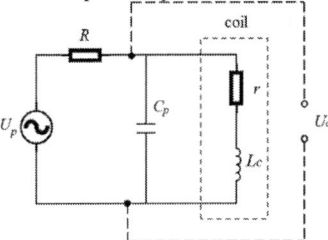

*Figure 3: A LC-resonance circuit of eddy-current sensor*

A typical LC resonant circuit is generally adopted to convert the slight changes in the impedance of the coil sensor due to the variations in the copper-film thickness to a voltage for output, as shown in Fig. 3. In this circuit, the capacitance of the parallel capacitor $C_p$ is $1/(\omega^2 L_c)$. The resistance R is in series connected with the coil. The output voltage $U_o$ corresponds to the metal film thickness.

## EXPERIMENTAL RESULTS

Two coils with air core were wound with same parameters, such as the outer diameter of 6mm, the inner diameter of 3mm, the coil turns number of 50. And then a air core was embedded on a Mn-Zn ferrite core. The resistance of the air core coil was 3.75 ohms and the inductance was 15.1 uH. The resistance of the ferrite core coil was 3.62 ohms and the inductance was 34.2 uH. Then, the capacitors of 180 pF and 420 pF were in parallel connected with them respectively. Their impedances changed with the excitation frequency changing from 1.5 MHz to 2.5 MHz by the test of impedance analyzer, as shown in Figure 4 and Figure 5. Their resonant frequencies were about 2 MHz and the resonance impedance of the ferrite core coil was greater than that of the air core coil.

Due to parasitic parameters in the test circuit, the actual excitation frequencies were set to 1.83 MHz and 1.66 MHz, respectively for the air core coil and the ferrite core coil, so that they were in LC resonant states. And the peak value of the excitation voltage $U_p$ was 3 V.

Copper films of different thicknesses ranging from 500 Å to 12000 Å and tungsten films of different thicknesses ranging from 500 Å to 8000 Å were prepared using the physical vapor deposition method. And all films were measured using the ResMap 273 diameter line model to get the precise thicknesses. Two sensor systems were used to measure these metal films at different lift-off distances of 1.6 mm, 2.6 mm and 3.6 mm. Experimental results are shown in Figure 6 and Figure 7. In order to compare the sensitivities more conveniently, all values of the output signals were normalized. The normalized value is the difference of the output signal divided by the standard value. The difference of the output signal is the

difference between the output signal amplitude of the sensor system and the air value signal. In addition, the standard value is defined as the difference of the output signal when the lift-off distance was 1.6 mm and the thickest copper film used in Figure 6(b) was measured by the ferrite core coil. The sensitivity is defined as the difference of the normalized value divided by the difference of film thickness, which represents the ability of the sensor system to measure metal thickness. Obviously, the sensitivity of the ferrite core coil was higher than that of the air core under any same experimental conditions. when the lift-off distance was 3.6 mm and the coil parameters were same, the sensitivity of a coil with Mn-Zn ferrite core was 4.5 to 5 times higher than that of a coil with air core.

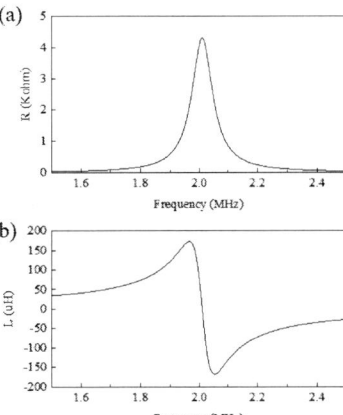

*Figure 4: Impedance change of a coil with air core while the excitation frequency changes from 1.5 MHz to 2.5 MHz*

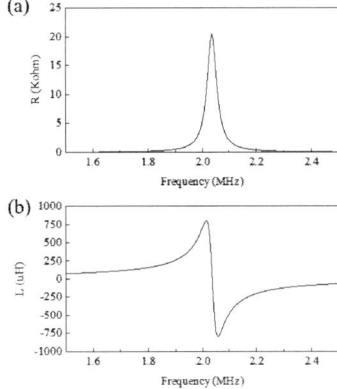

*Figure 5: Impedance change of a coil with Mn-Zn ferrite core while the excitation frequency changes from 1.5 MHz to 2.5 MHz*

Moreover, the sensitivity of both systems decreased,

with the increase of lift-off distance or the reduction of metal film conductivity. Comparing the measurement results of copper films and films, the lower the conductivity of the target film, the better the linearity of the obtained curve. And in both sensor systems, the larger the lift-off distance, the smaller the linear range of the curve. The difference of the output signal didn't change monotonically with the film thickness increasing for the air core coil, when the copper film thickness is between 7000 Å to 8000 Å at the lift-off distance of 3.6 mm, as shown in Figure 6(a). But the difference of the output signal changed monotonically for the ferrite coil as shown in Figure 6(b). This shows that when the coil parameters are the same, the thickness measurement range of ferrite core coil at a large lift-off distance is larger than that of air core coil. Hence, the ferrite core coil has better resolution and measurement range than the air core coil.

the sensitivity of a coil with Mn-Zn ferrite core was 4.5 to 5 times higher than that of a coil with air core. Moreover, the film thickness measurement range of a coil with Mn-Zn ferrite core was also larger than that of a coil with air core at the lift-off distance of 3.6mm. Hence, the ferrite core coil has better resolution and measurement range than the air core coil.

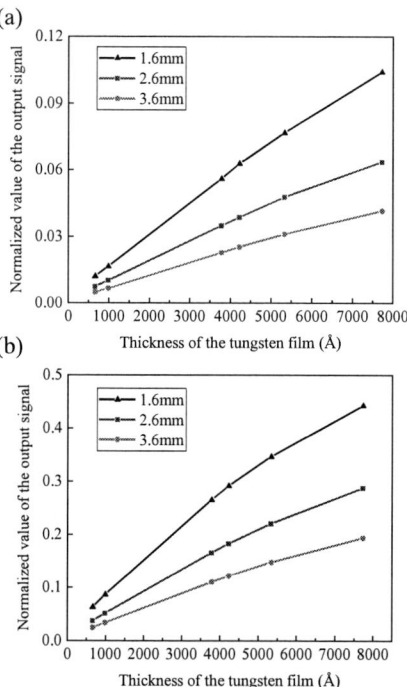

*Figure 7: Normalized value of the output signal when the thickness of the tungsten film changes at different lift-off distances: (a) a coil with air core, (b) a coil with Mn-Zn ferrite core.*

## ACKNOWLEDGEMENTS

This work was financially supported by the National Natural Science Foundation of China (Grant No. 51991374). The authors would like to thank HWATSING Co., Ltd. for providing a series of copper films and tungsten films with different thicknesses.

## REFERENCES

[1] K. C. Cadien and L. Nolan. Handb. *Thin Film Depos. Fourth Ed.*, 2018, pp. 317–357.

[2] T. Bibby and K. Holland. J. *Electron. Mater.*, vol. 27, 1998, pp. 1073–1081.

[3] C. Tan, W. Zhang, O. Huang, H. Gao, R. Zhao, and Z. Zhu. *ECS Trans.*, vol. 44, 2019, pp. 553–557.

[4] Z. Qu, Q. Zhao, and Y. Meng. *NDT E Int.*, vol. 61, 2014, pp. 53–57.

[5] C. Wang, T. Wang, T. Fang, and X. Lu. Unpublished.

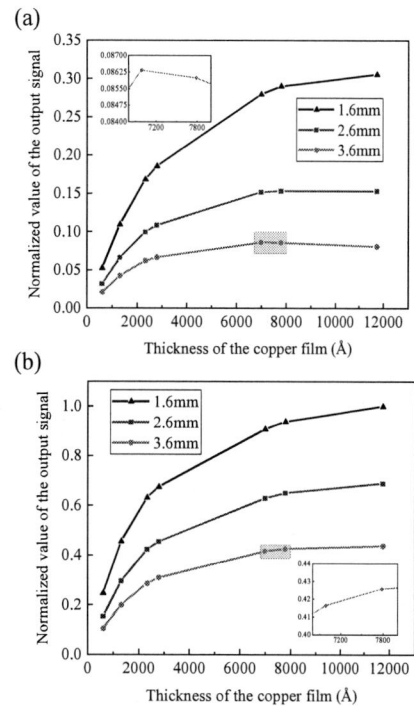

*Figure 6: Normalized value of the output signal when the thickness of the copper film changes at different lift-off distances: (a) an air core coil, (b) a magnetic core coil.*

## CONCLUSION

In this paper, we present two nanoscale thickness measurement sensor systems based on the eddy-current technique, which can be used in the CMP process. The experimental results show that the sensitivity of both systems decreased, with the increase of lift-off distance or the reduction of metal film conductivity. when the lift-off distance was 3.6 mm and the coil parameters were same,

# DISHING IMPROVE IN ADVANCED TECHNOLOGY NODES

*Yu Yang\*, Wei Zhao, Runcai Xiao, Hu Li, Jian Zhang, Haifeng Zhou, Jingxun Fang, Yu Zhang*

Advanced Module Technology Development, Shanghai Huali Integrated Circuit Corporation, Shanghai
201314, China
\*Corresponding Author's Email: yangyu@hlmc.cn

## ABSTRACT

As the semiconductor manufacturing process enters the FinFET node, the process window becomes narrower, and the requirements for the uniformity of chemical mechanical polishing (CMP) become higher，by which the negative effects brought are also unrestricted reduced. As one of the main side effect of CMP, the request of dishing is much stricter in advanced technology. The paper analyzes the causes of dishing, and introduces the control methods of dishing from three aspects: control before dishing generates, optimization during dishing and improvement after dishing exists. Finally, it should be pointed out that the successful stringing of advanced technology requires the coordination and integration of all modules. Only through joint cooperation can the process window be broadened and a win-win situation is achieved.

*Keywords—FinFET; Step height, Under polish; Buffer chemical，Design rule*

## INTRODUCTION

Dishing is increasingly seen as a critical performance parameter in CMP (Chemical Mechanical Polishing). It is mainly caused by the high selective ratio of slurry. Take Shallow trench isolation chemical mechanical polishing (STICMP) as an example, the selective ratio of slurry of silicon nitride (SIN) and silicon Oxide (Oxide) is about 15: 1. Once STICMP touches the SIN barrier layer, the oxide remove rate is higher than SIN as the polish continues, resulting in dishing (Fig1.). In addition, the deformation of polish pad caused by mechanical force during CMP will further aggravate dishing. The formation of dishing is one of the main reasons for CMP loading. It has a series of effects on subsequent processes and greatly impacts the electrical characteristics of the device.[1][2]

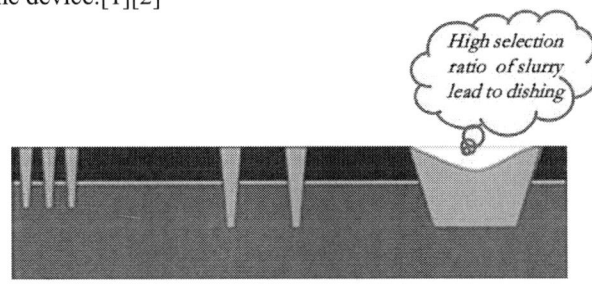

Fig1. Dishing schematic diagram

## RESULTS AND DISCUSSION

As the modern integrated circuit (lC) technology enters the Finfet node, it has put forward a higher demand for the dishing performance. Generally, the improvement of dishing can generally start from three aspects: control before dishing generates, optimization during dishing and improvement after dishing exists.

### 1. Thicken incoming deposition layer

The reason why the thickness of the incoming layer is placed first in the dishing improvement is that when CMP remove amount is insufficient to eliminate the step height of the incoming layer, the dishing is almost impossible to be improved by CMP. In general, step height could be effectively eliminated only when the CMP remove amount (b) is more than 2 times of the step difference (a) of the incoming layer (b≥2a). STICMP is taken as an example, when b≥2a, dishing is caused by the high selective ratio of SIN to oxide by over polish step after touch SIN stop layer, which is the dishing in the conventional sense . However, when b<2a, even polish to the SIN stop layer, the step height cannot be completely eliminated. At the moment, the shape shows like "dishing", but cannot be attributed to the dishing category as dishing is triggered by high selection ratio and over polish (Fig2.).

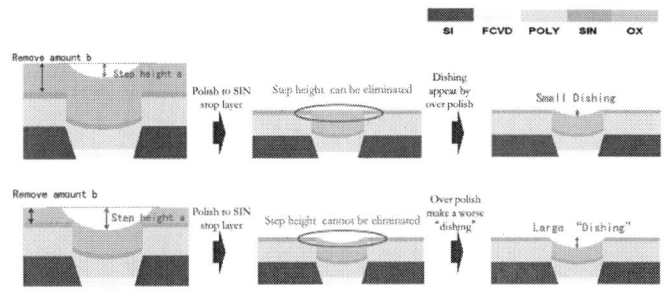

Fig2. A graphical representation of the relationship between remove amount (b) and step height (a)

## 2. Introduce a reference CMP layer

Normally, the more isolate area, the higher CMP remove rate. Also taken STICMP as an example, the remove rate of a isolated STI area is usually higher than that of a dense STI area, which is called as CMP iso/dense loading. In order to narrow the remove rate gap, A layer of reference film can be deposited before CMP. The rate of reference layer is lower than the target polish layer, therefore, it protects the large STI area in the initial polish period. The specific principle is shown in the Fig3, and the key factor is the remove rate difference between the target polish layer and the reference polish layer.[3]

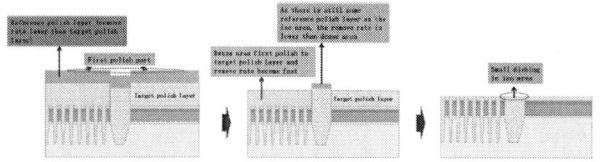

Fig3. Dishing improve by introducing a reference polish layer whose remove rate is lower than target polish layer

## 3. Under polish

As the dishing mainly comes from over polish, the most effective improvement is to reduce polish time, i.e. under polish. [4]An extreme case, stop polish before touch the stop layer, which almost cannot introduce dishing (Fig4.). Regretfully, reduce polish time has some disadvantages. Firstly, it will lessen CMP window and maybe resulted in residue after CMP. Secondly, it needs subsequent process such as wet etch to cover the reduce amount caused by under polish.

Fig4. Under polish for dishing improvement

## 4. Low pressure/low speed/low flow rate

For dishing and defect control, in advanced technology node, CMP is developing towards the direction of low pressure, low speed and low flow. I.e. in advanced process，AMAT tool, the polish pressure almost 1.0psi, which is close to the lower pressure limit of the machine, and the rotate speed also changes from 93/87 rpm to 67/53 rpm and even lower. Low pressure/rotational speed/flow rate significantly improved dishing, uniformity and scratch, but also suffer long polish time, low throughput, and endpoint signal difficulty to control etc. Therefore, the selection of pressure, speed and flow rate needs to comprehensively balance the above factors and select the optimal collocation.

## 5. Introduce buffer chemical

The above methods are basically improved before or during the generation of dishing. Furthermore, dishing can also be repaired in a special way . A series of new chemicals were introduced into the advanced technology for dishing improvement. The basic principle is much the same: with the dishing repair chemical, the remove rate of stop layer is nearly or even higher than that of polish layer. After the wafer trigger the endpoint, chemical is introduced to simultaneous removal of the stop layer and polish layer, thus repairing dishing (Fig5.). This method has an excellent improvement on dishing, but as it will remove some amount of stop layer and polish layer, it is necessary to comprehensively balance the deposition thickness and remove amount of the stop layer. In addition, as chemical consumes both the stop layer and the polish layer, there should not be residue before chemical buffer. Otherwise, chemical buffer will result in worse step height (Fig6.).

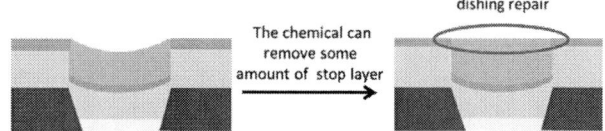

Fig5.Dishing repair by buffer chemical

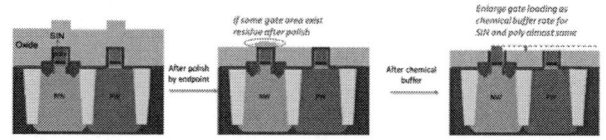

Fig6. Buffer chemical will enlarge gate loading if some gate area exist residue after polish as the remove rate for SIN and poly gate is almost same

## 6. Design rule revise

In advanced technology, especially FinFet node, the process window becomes narrower and narrower. The successful stringing of advanced technology requires the coordination and integration of modules, and only through joint cooperation can the process window be enlarged. As mentioned above, if the pattern density is too low, the CMP iso /dense loading will be too large, and the isolated area will likely suffer large dishing. At this point, relying only on CMP process cannot significantly improve dishing. It is necessary to revise pattern design, such as adding dummy (Fig7.) to avoid the existence of a single island and decrease iso/dense loading. [5] The means can obviously improve CMP loading, but mask revision requires verification and evaluation, and the time cycle is relatively long. In addition, dummy insertion must comply with the design rule before it is added, which means that this method has certain limitations, and cannot be used in very broad terms.

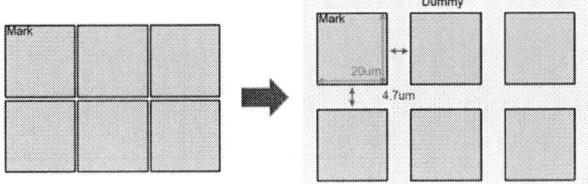

Fig7. Design rule revised for inserting dummy in isolate area

## CONCLUSIONS

In summary, there are roughly the above methods for dishing improvement. One method or a combination of several methods can be used to jointly realize dishing improvement according to the requirements of process specifications. The advantages and disadvantages are summarized as follows:

| Improve Action | Cross moduel (Y/N) | Advantage | Disadvantage |
| --- | --- | --- | --- |
| 1. Thicken incoming dep layer | Y | Least process modification | WIW/WTW uniformity difficult to control |
| 2. Introduce a reference CMP layer | Y | Effectively improve pattern loading within die | Complicated process , rely on the rate difference between the reference and target CMP layer |
| 3. Under polish | Y | Process simplicity,dishing improve significantly | CMP window margin, process stability difficult to control |
| 4.Low pressure/low rotate speed/low flow rate | N | WIW/WTW uniformity and scratch also can be improved | Lower tool throughput, CMP endpoint difficult to control |
| 5. Introduce buffer chemical | Y | Repair scratch while dishing improve | CMP Window margin |
| 6. Design rule revise | Y | Effectively improve pattern loading within die | Long evaluation cycle, not universal application |

## ACKNOWLEDGEMENTS

The authors would like to thank all the members of CMP group.

## REFERENCES

[1] Runzi Chang and C. J. Spanos, "Dishing-radius model of copper CMP dishing effects," in *IEEE Transactions on Semiconductor Manufacturing*, vol. 18, no. 2, pp. 297-303, May 2005, doi: 10.1109/TSM.2005.845110.

[2] Byoung-Ho Kwon, Jong-Hyup Lee, Hee-Jeen Kim, Seoung Soo Kweon, Young-Gyoon Ryu and Jeong-Gun Lee, "Dishing and erosion in STI CMP," ICVC '99. 6th International Conference on VLSI and CAD (Cat. No.99EX361), 1999, pp. 456-458, doi: 10.1109/ICVC.1999.820963.

[3] W. Shin, S. Park, H. Kim, K. Park, S. Joo and H. Jeong, "Reduction of Dishing in Polysilicon CMP for MEMS Application by Using protective Layer and High Selectivity," International Conference on Planarization / CMP Technology, 2007, pp. 1-5.

[4] S. W. Liew, L. S. Leong, R. Whsu and Y. S. Chong, "Poly CMP Process Challenges for Embedded Array Devices," *2021 32nd Annual SEMI Advanced Semiconductor Manufacturing Conference (ASMC),* 2021, pp. 1-5, doi: 10.1109/ASMC51741.2021.9435714.

[5] S. Lakshminarayanan, P. J. Wright and J. Pallinti, "Electrical characterization of the copper CMP process and derivation of metal layout rules," in IEEE Transactions on Semiconductor Manufacturing, vol. 16, no. 4, pp. 668-676, Nov. 2003, doi: 10.1109/TSM.2003.818956.

# AN OPTIMIZED METHOD FOR CU CMP DISHING IMPROVEMENT

*Lei Zhang[1]\*, Yuanyuan Meng[1], Yi Xian[1], Jian Zhang[1], Haifeng Zhou[1], Jingxun Fang[1], and Yu Zhang[1]*

[1]Advanced Module Technology Development, Shanghai Huali Integrated Circuit Corporation,
Shanghai 201314, China
*Corresponding Author's Email: zhanglei_td2@hlmc.cn

## ABSTRACT

With the development of advanced semiconductor device features shrinking, chemical mechanical polishing (CMP) is becoming an enabling technology to meet the precise machining in various applications. At the same time, the topography of wafer surface shows more and more serious challenges in advanced technology and beyond technology node. Tighten post CMP dishing, pattern loading and uniformity performance control has been evaluated during Cu CMP process. However, it is difficult to improve the Cu dishing and oxide dishing of the interconnect metals for copper CMP process at the same time, because of the selectivity of CMP slurry is fixed. This paper will present how to improve both Cu dishing and oxide dishing control with optimized NDC polish methodology. Experiment results shown that with the NDC CMP process, both Cu and oxide area show better dishing performance.

## INTRODUCTION

In order to improve the device performance of multilevel IC manufacturing, it is becoming more and more important to precisely control the sheet resistance (RS) of Cu interconnect lines. With continuous shrinking chip design rules in each technology node, copper interconnect structures shows more importance effect because of the higher electrical conductivity and electro migration resistance of copper, and CMP process becomes more challengeable to achieve planarization. Since the Cu line resistance is related to the Cu thickness, dishing control of the buck copper and erosion of line array area during Cu CMP is critical. On the other side, the topography after copper CMP polishing such as dishing and erosion causes the increasing or variation of wiring resistance, and the sort between the wirings in upper layer. Poor surface topography and defect will bring some challenges in lithography and wiring for the next layer. [1-2]

The planarization performance is influenced by slurry chemistry, pattern density, line/space width, topography characteristics, pad/disk/pressure/rotation speed, etc. This pressing need has driven development of optimized methods for controlling Cu dishing and erosion, which often occur at different pattern densities and line widths.

In Cu CMP, dishing is defined as the copper loss relative to the level of the neighboring oxide space, and erosion refers to the cu loss relative to the cu level of the neighboring area [3-4]. Wide trenches or open structures

usually enhance the dishing issue, while dense trenches lead to more erosion. Erosion exposes the underlying active devices will lead to device failure; On the other hand, dishing results in poor isolation. [5-6]

There are many solutions have been made toward more complete modeling and better supporting method of the CMP process. For Cu and oxide dishing and erosion defects, balance the Cu bulk and barrier layer polish process can have a bit of improvement [7]. However, with continuous shrinking chip design rules in advanced technology node, the limited improvement was far from enough. Optimize the selectivity of both copper slurry and barrier slurry can obvious improve the dishing and erosion performance. Slurry changing should collect profile, defect, WAT, CP and window check data, which need a long cycle time and resource [8-9]. This paper provides an optimized method for copper CMP dishing and erosion improvement, which shown minor different change for back end of the line (BEOL) metal loop. In general, post metal CMP was NDC and next metal layer films deposition. Our current work focus on the physical effects of metal dishing, and the design for manufacturability considerations for yield improvement, by adding a NDC planarization step to reducing the dishing and erosion after NDC deposition and achieve defect free. With fast technology scaling and increasingly tighter time, it is crucial to develop a simple and effective method to meet the objectives at advanced technology generation and beyond.

## EXPERIMENTAL

CMP polishing equipment mainly consists of three steps in the process, as shown in figure 1. Step 1 for preliminary polishing stage, with high Cu remove rate, the main role for preliminary polishing to remove most surface Cu, real time profile closed loop control for Cu process, well control surface uniformity; Step 2 is given priority to polishing, remove the surface part of the Cu and blocking layer, with low cu remove rate, and high selectivity for endpoint signal pressure, barrier layer as polishing end layer; Step 3 as barrier polish, the oxide has high selectivity, thus forming Cu pattern.

For advanced technology node, a robust copper CMP process with better post CMP polishing defect performance, smooth copper surface, pattern loading control, tighten metal line sheet resistance and profile has been evaluated during the copper CMP process. Especially defect control performance is more critical.

In Cu planarization, the bulk copper is removed first using a high removal rate process with high selectivity slurry, followed by reducing polish down force with lower removal rate process and finished with barrier removal process. The topography after polishing such as dishing and erosion causes the increasing or variation of wiring resistance, and the sort between the wirings in upper layer. And the barrier removal process as the last polish process in Cu CMP decided the final topography and defectively performance.

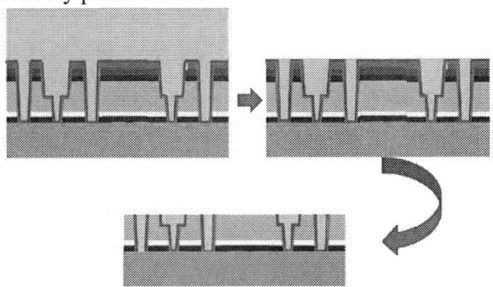

Figure 1: Cu CMP polishing process diagram

An advanced technology chip with various line/space CD and pattern density is used in this study. The design rule of the test-keys covers different technology nodes, and the range of pattern densities is from 5% to 50%. The surface profile measurement of dishing or erosion amount after Cu CMP is verified by an atomic force microscope (AFM). Furthermore, cross-sectional thickness verification was carried out on transmission electron microscope (TEM) tool.

The pattern wafers of study chip in Cu CMP operation was performed on a 300 mm Ebara system. The silicon slurry polishing is applied to provide a substantially planar topology, and following HSS (high selectivity slurry) polishing with end-point system is executed for copper bulk step.

TABLE I.　　PATTERN WAFER CONDITION & SPLITS

| Condition | Cu CMP Dishing Improvement Split Table | | | |
|---|---|---|---|---|
| | Split 1 | Split 2 | Split 3 | Split 4 |
| BKM recipe | V | | | |
| Balance bulk and barrier polish step | | V | | |
| Add NDC deposition | | | V | |
| Add NDC CMP | | | | V |
| Profile & Dishing | V | V | V | V |

A lot of pattern wafers was created with four splits to collect dishing performance as shown in Table 1. The wafers were polished on Platen A of an Ebara tool to clear the Cu and then further polished on Platen B. One wafer group used Cu CMP only and the other group was polished with NDC CMP step. Incoming as well as post polish dielectric thickness was measured using Nova.

## RESULT AND DISCUSSION

Baseline condition with only Cu CMP step, TEM result shows oxide dishing increase fast at large pitch area as shows in figure 2.　ISO area much thinner. Oxide dishing may cause Cu residue at next layer.

Figure 2: TEM result of post Cu CMP with oxide dishing

Another side effect of dishing defect is wide metal area dishing, which cause copper thickness not stable and worse WAT result, WAT for different width not follow ideal trend, thickness match with WAT, wide metal much thinner, as shown in figure 3.

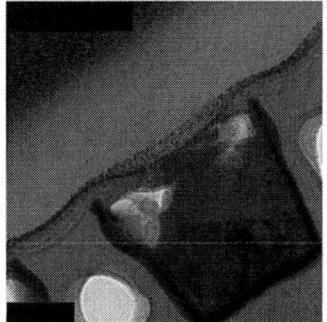

Figure 3: TEM result of post Cu CMP with copper dishing

Defect scan result shows only Cu CMP step suffer Cu residue defect issue, check layout pre layer is non-pattern oxide area. Suspect pre layer CMP oxide dishing induce Cu residue defect, as shown in figure 4.

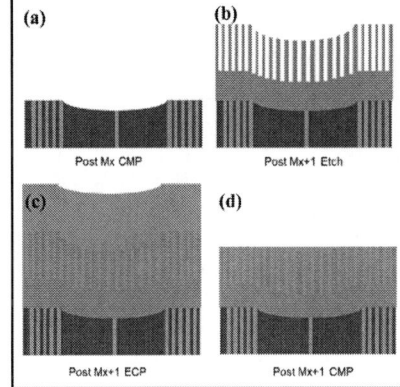

Figure 4: Pre layer Cu CMP include Cu residue diagram

In this paper an optimized polish methodology is presented that reduces the impact on dishing while improving the pattern loading through add a NDC polish step. As shown in figure 5, after Cu CMP (Fig.5a) add a deposition of NDC process (Fig.5b), then remove the surface of NDC layer(Fig.5c). In order to repair dishing effect during Cu CMP, series of experiments were carried with various NDC CMP split.

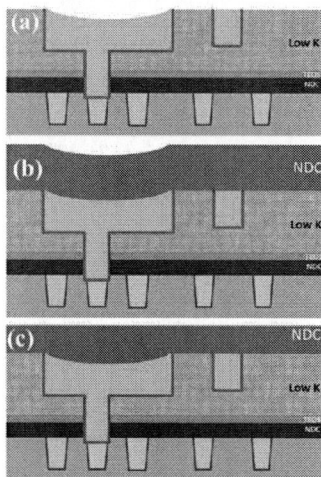

*Figure 5: NDC CMP process diagram*

According to the results, the performance of dishing and Cu remaining thickness after CMP process is highly dependent on the pattern density characteristics. In this case, optimized method can be obvious improve dishing, as shown in figure 6. From overall experimental results, under new CMP condition, dishing performances were better and more stable and had better uniformity. And narrower structures had much lower dishing.

These data clearly show that optimized condition is a promising candidate for the dishing improvement of the oxide dishing. Currently we are continuing the CMP optimization, further development with NDC CMP and new post-CMP cleaning to further improve the electrical performance for advanced technology node.

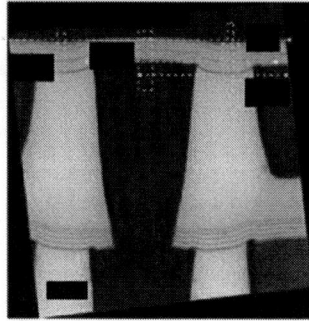

*Figure 6: TEM result post NDC CMP process*

## CONCLUSION

The continuous scaling in the logic device landscape towards advanced technology node dimensions has presented numerous engineering and manufacturing challenges in BEOL Cu interconnection. CMP engineers always struggle with dishing or erosion performance in critical layers during new product pilot run.

In this paper, the performance of new CMP condition not only meet the difficult criteria of pattern loading but also present much more steady performance about oxide and copper dishing.

Optimization CMP process to add a NDC polishing process is the primary source of pattern loading and dishing in Cu interconnects. Experiment result shown the dishing both dense trench line area and oxide area are quite obvious improvement. Trench area requests big top opening, while SRAM area requests more Cu step coverage on sidewall. Optimized parameters were finally presented and the defect level was significantly reduced.

## REFERENCES

[1] H.F. Kao, et al. Dishing and Erosion Amount Prediction According Pattern Density Calculation Algorithm in 3D Design Layout. Joint Symposium. 2015

[2] H. Nancy, et al. CMP process optimization for improved compatibility with advanced metal liners. IEEE Conference Publications. 2010

[3] L. Zhang, et al. Research and Solution of STI CMP Dishing and Uniformity Improvement for 28LP. IEEE Conference Publications. 2017

[4] Y.L. Hsieh, et al. Effects of BEOL Copper CMP Process on TDDB for Direct Polishing Ultra-Low K Dielectric Cu Interconnects at 28nm Technology Node and Beyond. Crown. 2013

[5] L. Zhang, et al. Study and Improvement on Tungsten Recess in CMP Process. IEEE Conference Publications. 2019

[6] X.Y. Hu, et al. Improve Cu Line Rs Control Using Feed-forward Information for CMP End pointing. IEEE Conference Publications. 2010

[7] H.F. Kao, et al. Dishing and Erosion Amount Prediction According Pattern Density Calculation Algorithm in 3D Design Layout. Joint Symposium 2015 - eMDC and ISSM

[8] L. Zhang, et al. Pattern Loading Effect Optimization of BEOL Cu CMP in 14nm Technology Node. IEEE Conference Publications. 2020

[9] L. Zhang, et al. Copper corrosion issue analysis and study on advanced CMP process. IEEE Conference Publications. 2021

# STUDY ON THE CORRELATION BETWEEN CMP CU LOADING AND EDP CURVE

*Yuanyuan Meng\*, Yixian, Lei Zhang, Jian Zhang, Haifeng Zhou, Jingxun Fang*

Advanced Module Technology Development, Shanghai Huali Integrated Circuit Corporation,
Shanghai 201314, China
\*Corresponding Author's Email: mengyuanyuan@hlmc.cn

## ABSTRACT

Endpoint detection technology plays an important role in chemical mechanical polishing (CMP) technology，and is affected by many aspects in the actual production process. Because the Endpoint curve can directly reflect the process information, in-depth study of the endpoint curve can predict the health of the process in advance. Taking the AMAT machine as an example, this paper focuses on the influence mechanism of Cu CMP incoming profile such as CU loading on Endpoint, and discusses how to improve endpoint stability to reduce the probability of endpoint capture failure.

## INTRODUCTION

With the increase in the number of metal layers and wafer size in the integrated circuit (IC) manufacturing, the need for global planarity across the wafer has become increasingly crucial [1]. Chemical mechanical planarization (CMP) [2–4] is presently the most widely used planarization technique. Due to the demand of metal wiring transformed from aluminum to copper, copper CMP in dual Damascene process was indispensable. The damascene structure of copper is characterized by the formation of through-holes and interconnect grooves in the inter metal dielectric (IMD) layer by etching technique, followed by the deposition based barrier layer tantalum and tantalum nitride. A large amount of copper is deposited through ionized metal plasma precipitation of copper seed crystal and electro chemical plating (ECP). Finally, the copper outside the trough is removed by CMP technology to achieve leveling and prevent short circuit between copper interconnects. Because of the planarization performance of a CMP process depends substantially on the properties of the incoming wafers, as the copper CMP pre layer, the depth of the IMD trench and the thickness of the ECP have a significant impact on the CMP process.

Loading caused by CMP also degrade the interconnect planarity and make it difficult to fabricate high density multilevel interconnect with a high yield. Copper loading effects are serious problems to be solved during copper CMP because of they reduce the final thickness of the copper line and copper loading leads to non-planarity of the surface, resulting in complications when adding multiple levels of metal. In addition, any Cu over polishing strongly enhances dishing and erosion, which will result in a decrease of the Cu line and the insulator thickness as well as in a deviation from the planarity. Those undesirable behaviors can be minimized using an optimized endpoint detection system. At present, various EPD methods [5] have been proposed, such as optical reflectance, frictional force, pad temperature, Cu-ion concentration, and eddy current. In the self-developed CMP system, the optical reflectance method has been used to develop independently an in-situ measurement module for detecting the Cu loading and layer thickness variation. For the same trench depth, the increase of trench width will reduce the amount of copper on the top that needs to be polished reflecting in CMP end point curve. Thus IMD trench and the thickness of the ECP have a significant impact on the CMP process.

During the complex chip manufacturing processes, the CMP process often needs to optimize and adjust the endpoint recipe to cope with different process changes based on previous layer changes. However, as we all know, the endpoint recipe setting is very complex, and frequent adjustments to the recipe are not conducive to production security and stability. This paper studies the change mechanism of endpoint curve based on different ECP deposition copper thickness, focuses on analyzing the influence of previous layer on CU CMP process, and avoids potential problems that may occur in CMP process through simple adjustment.

## EXPERIMENT

All CMP experiments were conducted on 300mm diameter wafers using dual damascene test structures with varying line widths. This experiment takes 28nm top metal as an example, to fill the 13KÅ Cu trenches, three thickness approximately 16KÅ, 20KÅ, 24KÅ copper overburden were deposited using electrochemical plating. The Cu CMP was performed using exactly same consumables at Applied Materials LKP polishing tool. Optical reflectance method end point detector was used to detect copper remove process, the endpoint curve and corresponding process schematic diagram are shown in Figure 1.

978-1-6654-9759-6/22 $31.00 © 2022 IEEE

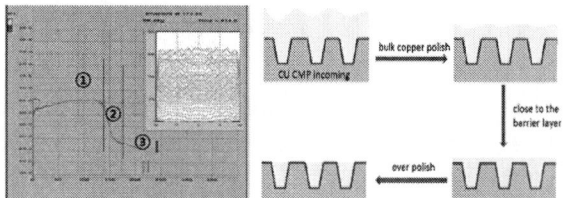

*Figure 1: CU CMP end point curve in AMAT and schematic diagram.*

Section ① refers to the bulk copper polish, with high Cu remove rate, the main role for preliminary polishing to remove most surface Cu, real time profile closed loop control for Cu process, well control surface uniformity. Section ② refers to copper CMP close to the barrier layer, with low Cu remove rate, and high selectivity for endpoint signal pressure, barrier layer as polishing end layer. Section ③ refers to the over polish stage to ensure remove the surface part of the Cu and blocking layer clearly. Line I refers to the signal of laser reflection intensity. Line II refers to the signal of Cu layer thickness. Take the line I in detail, during the initial stage of the bulk copper polish process, the signal of laser curve gradually rises because of the CU incoming surface is uneven and has diffuse reflection at the initial stage of grinding. As the polish process gradually tends to be flat, the signal intensity is strengthened, and the signal remains unchanged when the CU surface remains flat and continues to loss. When copper CMP close to the barrier layer, the CU reflected signal intensity decreases sharply because the barrier material is exposed on the surface. During the over polish process, the CU signal remains stable because the CU surface area remains constant. The small figure in the upper right corner of Figure 1 is the real time process control (RTPC) curve which uses the electromagnetic sensor to monitor (in real time) the copper thickness has been improved to provide better wafer profile information, especially near the edge of the copper film.

## RESULT AND DISCUSSION

For the copper thickness approximately 16KÅ experiment, as shown in Figure 2, the endpoint detection system was failed. During initial stage of the bulk copper polish process, the signal of laser reflection intensity was lower than the other two groups and risen sharply because of the copper surface that are too rough and quickly rubbed off. RTPC curves are read at the same interval at a fixed time. Sparse curves represent high copper remove rate, and dense curve intervals represent low copper remove rate. As shown in figure 3, when trench thickness is 13 KÅ, ECP copper thickness is 16 KÅ, increase of dishing in width trenches will occur due to the copper thickness does not sufficiently repair the step height witch caused by previous structures. During the bulk copper polish process, trench CD width area will remove the surface of the copper clean and touch barrier layer faster than CD narrow area witch still has copper residue on the barrier surface. When copper and barrier interfaces co-exist in wafer surface, because of the copper slurry has high selectivity between copper and oxide，the copper remove rate will be greatly reduced under the influence of barrier interface, resulting in the ambiguity of copper and oxide interface, which repeatedly interferes with optical signals and leads to the failure of endpoint capture mechanism.

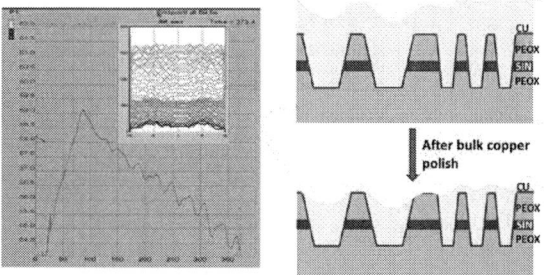

*Figure 2: ECP 16KÅ CU CMP end point curve in AMAT and schematic diagram*

In order to solve the above problems, ECP copper thickness was changed to 20KÅ, the endpoint curve was captured successfully. As shown in figure 3, with the increase of copper thickness, the dishing in width trenches has a significant improvement over the ECP 16KÅ. However, it is clear from the endpoint curve that it took 75s from the close of copper CMP to the barrier layer to the complete removal of copper, which is unhealthy polish process. Too long polish time will increase the loading between copper lines of different pitch sizes.

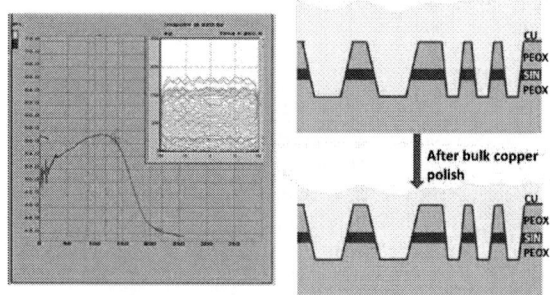

*Figure 3: ECP 20KÅ CU CMP end point curve in AMAT and schematic diagram*

To further explore the influence of ECP copper thickness on CMP process and endpoint curve, the ECP copper thickness was increased to 24 KÅ. As shown in figure 4, during initial stage of the bulk copper polish process, the signal of laser reflection intensity much flatter than the ECP 20 KÅ curve, which means the incoming topography was flatter than ECP 20 KÅ. The endpoint curve only took 30s from the close of copper CMP to the barrier layer to the complete removal of copper，this means that the loading of copper is very small. The TEM result in figure 5 also shows that ECP 24 KÅ loading improves approximately 3 KÅ compared with ECP 16 KÅ, which greatly reduces the variety of defect issues caused by copper loading.

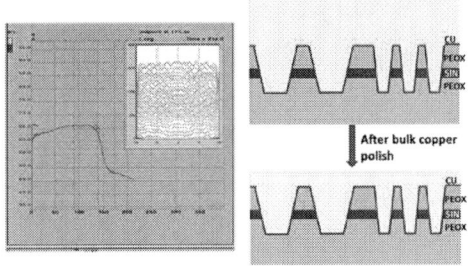

*Figure 4: ECP 24KÅ CU CMP end point curve in AMAT and schematic diagram*

| Condition | ECP16KA | ECP20KA | ECP24KA |
|---|---|---|---|
| | Center(0,0) | Center(0,0) | Center(0,0) |
| narrow trenches pattern | 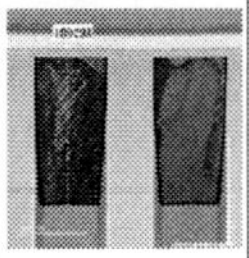 | 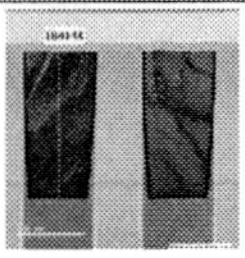 | 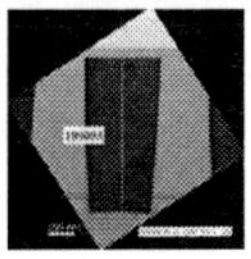 |
| Cu Height/A | 10929 | 10814 | 10909 |
| width trenches pattern |  | 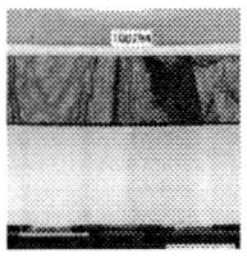 |  |
| Cu Height/A | 7671 | 10079 | 10972 |
| Loading/A | 3258 | 735 | 63 |

*Figure 5: Post CMP TEM images of ECP 16 KÅ, 20 KÅ, 24 KÅ CU thickness with different trenches width*

## CONCLUSION

Because of the post copper CMP performance is strongly influenced by previous processes, this paper mainly analysis of the copper CMP endpoint curve changed with the ECP deposition copper thickness. A process model is introduced to judge whether Cu incoming surface profile is reasonable by the expression of endpoint curve and to reduce post CMP copper loading by increasing the thickness of ECP copper thickness. Under the same process conditions, the test results show the thicker the copper thickness the lower the copper loading. Experimental results showed that this method effectively reduced the copper loading between narrow and width trenches from 3258Å to 63Å, significantly improved the stability of the product, and made an effective contribution to the improvement of product yield.

At the same time, the process is simple and low coast, which can be effectively popularized and applied in actual mass production.

## REFERENCES

[1] H. C. Hong, Y. L. Huang, *Int. J. Mater. Product. Technol.* 2003, 18, 469-486.

[2] Z. Stavreva, D. Zeidler, M. Plotner, K. Drescher, *Microelectron. Eng.* 1997, 37–38 (Suppl. 1–4), 143–149.

[3] P. B. Zantye, A. Kumar, A. K. Sikder, *Mater. Sci. Eng. R-Rep.* 2004, 45, 89–220.

[4] M. Krishnan, J. W. Nalaskowski, L. M. Cook, *Chem. Rev.* 2010, 110, 178-204.

[5] N. Ohashi, Y. Yamada, N. Konishi, H. Maruyama, T. Oshima, H.Yamaguchi, *Proceedings of the IEEE 2001 International Interconnect Technology Conference.* pp. 140 -142.

# TUNGSTEN CMP CONSUMABLE LOCALIZATION STUDY AT 28NM TECHNOLOGY NODE

*Shaojia Zhu*, Mingfei Yu, Hongming Pan, Yibin Li, Lei Zhang, Jian Zhang, Haifeng Zhou, Jingxun Fang, and Yu Zhang*

Advanced Module Technology Development, Shanghai Huali Integrated Circuit Corporation, Shanghai 201314, China *Corresponding Author's Email: zhushaojia@hlmc.cn

## ABSTRACT

With the rapid development of semiconductor in China, the localization of semiconductor materials has become a long-term development trend. The main consumables of chemical mechanical planarization (CMP) are pad and slurry, and the tungsten (W) CMP at 28nm node mainly used the imported pad and slurry which fabricated by American corporations. In this paper, the feasibility of replacing the pad and slurry by domestic consumable manufacturers is studied. The process parameters are fine-tuned according to the characteristics of the domestic pad and slurry. Offline remove rate (RR) and profile show comparable performance with baseline by using blanket wafer. Furthermore, the uniformity of film thickness, surface topography, defects, transmission electron microscope (TEM) were studied by using pattern wafer. The experimental results show that the domestic pad and slurry can be applied to W CMP at 28nm node, but the W protrusion control need fine tune by process parameters.

## INTRODUCTION

Since its emergence in the 1980s, CMP technology has become indispensable in ultra-large-scale integrated circuits, and its applications have also increased along with the progress of scaling. In the CMP process, a wafer is pressed down and rotated against a rotating polyurethane pad that is saturated with abrasive slurry particles and a chemical solution to achieve both local and global planarization. Further, the main CMP consumables pad and slurry which fabricated by foreign manufacturers play an important roles in CMP process. There are few reports on the application of domestic materials include slurry and pad [1].

Generally, W CMP process mainly includes two parts: the first part is a bulk polishing process, to remove most of the surface of W and barriers layer; the second part is the oxide buffing polish to control the local plug protrusion [2]. W CMP mainly through the combination of electrochemical corrosion and mechanical polishing surface effect, achieve the goal of removing W and oxide on the surface of the polishing. The surface recess on W plug or defects such as corrosion is a common problem in polishing process, this kind of defect will lead to higher contact resistance, caused by devices and metal wire connection fails, will seriously affect the product yield.

Oxide buffing polishing process with the reversed selectivity to W CMP process, which has lower W RR than oxide film, helps to form W plug, reduce the risk of W recess [3]. However, with the development of semiconductor process node, the key of the contact hole size has been reduced to below 28 nm, contact hole depth-to-width ratio more than 4:1. Contact between the higher level of integration, W plug may be higher, so precise control of oxide polishing technology of W plug height is also becoming more demanding. As device shrinkage, the buffing step may make the dense W plugs over-protruded and lead to defects trapped at W plugs easily. Therefore, the control of W plug protrusion is important to reduce the abrasive particles and organic residues. Further, the chemical clean process after polish is also a necessary process to reduce the defect [4-5].

In this study, we evaluated domestic slurry and pad in W CMP process at 28nm node to prevent the key material controlled by foreign manufacturers. Throughout the evaluation, Offline RR and profile show comparable performance with baseline by using blanket wafer. Furthermore, the uniformity of film thickness, surface topography, defects, transmission electron microscope (TEM) and process window were studied by using pattern wafer.

## EXPERIMENTAL

Two types of 300mm wafers were used in this study, none-pattern wafers for offline RR, profile, selectivity evaluation and 28nm pattern wafers for inline thickness, defect, topography evaluation. As shown in figure 1 CMP processes were carried out by EBARA F-REX 300 polisher using two platens process. First platen process was used for W removal and W clearing using eddy current end-point detection (R-ECM) on hard pad. Second platen was dedicated to barrier removal using a soft pad.

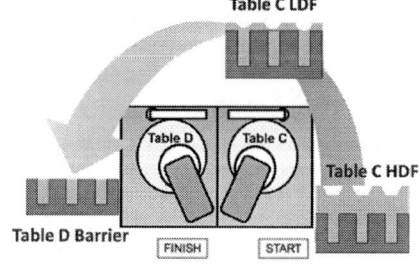

*Figure 1: CMP equipment polishing process flow diagram*

Oxide thickness was measured in patterned test box using a KLA-Tencor Aleris8510. Offline particle was measured using a KLA SP3. Inline defect was measured using KLA PUMA9980.W resistivity measurements for thickness calculation were done using NAPSON WS3000. W Recess on pattern wafer characterization was conducted using cross-sectional transmission microscopy (TEM). Post CMP topography dishing and erosion measurements have been realized on Atomic Force Microscope (AFM) BRUKER Nanoscope.

## RESULT AND DISCUSSION

In the W CMP Process, W is oxidized by etchant, and then removed by abrasives. Progressively, this process is cycled again until all bulk W is removed. Hydrogen peroxide (H2O2) and silica nanoparticles are commonly used as etchant and abrasive, respectively, to remove the material [6]. Slurry and pad are the most important factors in this process. As we know, RR, RR profile and particle are very important parameters in CMP process. So we first verified the feasibility of the domestic slurry and pad (NC) by using none-pattern wafer. As shown in figure 2, compared with the foreign leading materials (BSL), the domestic slurry and pad can reach the comparable performance at High down force(HDF), Low down force(LDF), oxide buffing RR and particle defect. But further validation on the pattern wafer is required.

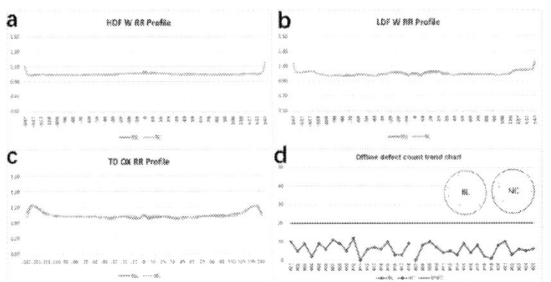

Figure 2: (a)HDF W RR Profile; (b) LDF W RR Profile; (c) TD OX RR Profile;(d) Particle trend chart (size 0.16um)

The pattern wafer results (Figure 3) show that based on domestic slurry only, the wafer thickness and uniformity can achieve the same performance compared with the imported slurry(BSL), but total defect count is higher than BSL, the defect image shows worse recess defect than BSL which will increase the contact resistance, effectively impact the product yield and reliability. The TEM shows baseline (BSL) has 25A protrusion meanwhile new condition (NC) has 80A recess. Furthermore, the wafer OCD test key topography(erosion) was studied by using Atomic Force Microscope(AFM). NC erosion shows 120A higher than BSL(Figure 4), supporting the defect conclusion above.

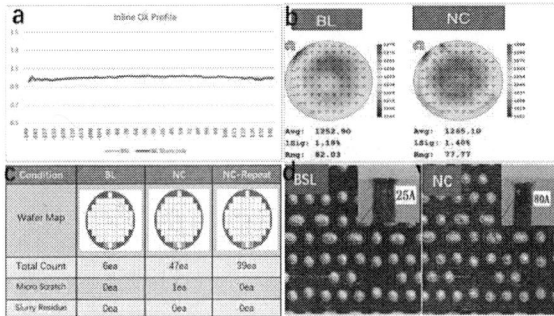

Figure 3: (a)Inline thickness profile; (b) Inline map; (c) Defect summary;(d) Defect image and TEM;

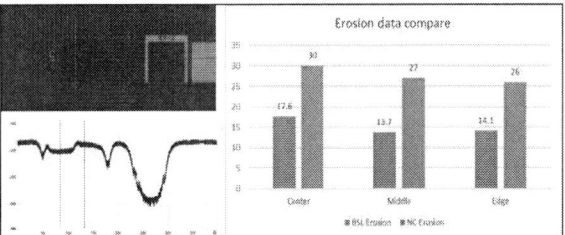

Figure 4: Erosion data by AFM(nm)

Additionally, based on erosion and recess issue talked above, a new domestic pad which has higher hardness than BSL was qualified as a solution, meanwhile the domestic slurry was used also. As figure 5a shows that the NC which uses new domestic slurry and pad can achieve comparable thickness and uniformity performance compared with BSL. As we all know, the harder pad can get better topography performance, figure 5b result shows NC (slurry + pad) has really better erosion, confirmed our prediction model. On the other hand, the defect was also verified by using full loop pattern wafer. The defect result (figure 5c) shows nearly total defect count, and the reviewed image found no micro_scratch or slurry residue defect which is easier to happen in CMP process. Furthermore, the reviewed image and TEM (Figure 5d) shows NC (Slurry + pad) combination can improve the recess defect exactly.

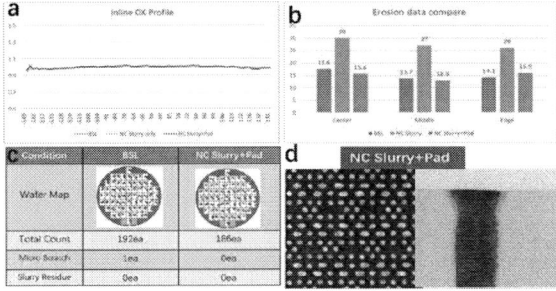

Figure 5: (a)Inline thickness profile; (b) Erosion data by AFM; (c) Defect summary;(d) Defect image and TEM;

## CONCLUSION

In this study, we evaluated domestic slurry and pad in W CMP process at 28nm node to prevent the key material control by foreign manufacturers. Throughout the evaluation, Offline RR and profile show comparable with baseline by using blanket wafer. Furthermore, the uniformity of film thickness, surface topography, defects, transmission electron microscope (TEM) were studied by using pattern wafer. We found that only replacing the slurry leads to worse recess defect which will increase the contact resistance, effectively impact the product yield and reliability. With the continuous development of semiconductor fabrication, particularly in the 28 nm and below nodes, in the process of W CMP, the impact of W recess on the product yield is more and more obvious, through the optimization of process, reducing the number of degree and the recess is helpful to improve product reliability and yield. In this paper, by combing with a new domestic harder pad will solve this problem. So, the experimental results show that the domestic pad and slurry can be applied to W CMP at 28nm node, but the W protrusion control need fine tune by process parameters. We hope that our research can contribute to the development of semiconductor in China.

## REFERENCES

[1] M. Tsujimura, et al. Cherishing Old Knowledge, Acquiring New - Past, Present and Future of CMP Technology. ICPT 2017, October, 11-13, 2017, Leuven, Belgium

[2] R. DeJule, et al. CMP Challenges below a quarter micron. Semiconductor International. 1997.

[3] L. Zhang, et al. STUDY AND IMPROVEMENT ON TUNGSTEN RECESS IN CMP PROCESS. IEEE Conference Publications. 2021

[4] K. Hwang, et al. Accelerated Corossion and Oxidation of Tungsten Plug During Tungsten CMP. IEEE Conference Publications. 2015

[5] Y. Li, et al. Study and Improvement on Tungsten Plug Corrosion in CMP Process for PCRAM. IEEE Transactions on Semiconductor Manufacturing. Feb., Vol.27, 2014, pp.38-42.

[6] D. Lim, et al. A Novel Pad Conditioner and Pad Roughness Effects on Tungsten CMP. IEEE Conference Publications. 2014

# AN OPTIMIZED MONITORING METHOD FOR 28HK ILDCMP

*Jingjing Li, Hongming Pan, Mingfei Yu, Shaojia Zhu, Feng Shi, Jian Zhang, Haifeng Zhou, Jingxun Fang, Yu Zhang*

Advanced Module Technology Development, Shanghai Huali Integrated Circuit Corporation, Shanghai 201314, China
*Corresponding Author's Email: lijingjing@hlmc.cn

## ABSTRACT

In this paper, according to the problems encountered in the research and development of the new generation product of 28HK project of our company, the problems existing in the ILD CMP measurement of the product were improved. The full Map profile and TEM data show that ILDb bond pad cannot reflect the real film thickness of the device area, while OCD pad has a pattern similar to the device area, so the measured value is closer to the actual value. This paper analyzes the reason why OCD pad measurement is closer to the actual value from the perspective of structural difference and CMP dishing. The analysis shows that OCD pad should be selected for subsequent ILD CMP film thickness measurement, and reasonable selection of pad will accelerate the development and mass production efficiency of 28HK products.

## INTRODUCTION

With the development of CMP technology, higher requirements are put forward for CMP measurement [1]. Take the CUCMP as an example, its measurement pad has two types, OCD pad and Bond pad [2]. The CUCMP measurement methods mainly include Nova measurement and Rudolph measurement. The principle of Rudolph measurement is to generate spectrum through laser and acoustic reflection, and determine a measured characteristic peak position through DOE. During measurement, the thickness of measurement point is deduced according to the measured peak position (acoustic delay time). The principle of Nova measurement is to generate a spectrum (diffraction reflection) through white light irradiation. In the measurement, the thickness of the measurement point is calculated by comparing the actual spectrum with the standard spectrum. Comparing the two measurement methods, Nova measurement is better than Rudolph measurement. CMP measurement techniques are also evolving. Masahiro HOME et al. [3] developed a fast and stable spectral film thickness measurement system and verified its effectiveness through tests. The spectroscopy measurement system can measure multilayer films with high precision and easy operation, and the measurement data has high repeatability even when the film surface conditions are not good enough.

With the development of 28HK technology of Shanghai Huali Integrated Circuit Corporation, the accuracy of film thickness measurement after CMP is also put forward higher requirements. How to select the appropriate pad for film thickness monitoring becomes the key point. This paper is to explore the difference of measurement results of different pad after ILD CMP under the current technological conditions of our company. The OCD pad was selected to monitor ILDCMP film thickness by contrast with the front film thickness profile and TEM result. In addition, by comparing the chip structure, the reason for the difference between different pad measurements was analyzed.

## EXPERIMENTAL

Tool used for CMP process in the experiment is AMAT's Reflexion LK machine type, mainly BCCOXA03 and so on. Tool used for ILDb pad measurement post CMP is KLA-Tencor's Aleris 8510 machine type, mainly BTTHKK01 and so on. Tool used for OCD pad measurement post CMP is Nova T650 machine type, mainly BEOCDN06 and so on.

## RESULT AND DISCUSSION

The industry for ILD CMP film thickness measuremen usually choose ILDb pad, Figure 1 shows the ILD CMP Full Map profile measured with ILDb pad, it can be seen that the edge region has a significant THK jump with a THK range of 267. In general, we would think that the ILD THK of the high jump area is thicker. In order to achieve good film thickness uniformity, it is necessary to increase the grinding pressure in these areas for thinning to flatten the entire profile. However, there is a possible situation, that is, in fact, the ILD film thickness is uniform, but because there are profiles of similar shape in the front layer, the profile after the ILD CMP is inherited, so as to present the image of high jump at the edge region. In this case, blindly flattening the profile will inevitably lead to the wafer edge area being thinner, resulting in serious uneven film thickness and affecting chip performance.

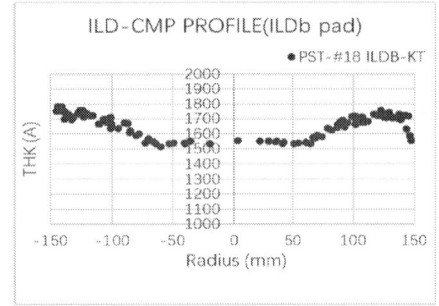

*Figure 1 ILDb pad measurement profile of ILDCMP*

In order to analyze the relationship between ILDCMP and the front layer, we collected the Full Map profile measured by ILDb pad, including ILD front layer ILD DEP, MG CMP and ILD0 CMP. As can be seen from the figure2, the profile of ILDCMP is very similar to that of ILD0CMP, MG CMP and ILDDEP, and they have an approximate thickness range of 200. It can be inferred that ILDCMP's profile is indeed inherited from the previous CMP.

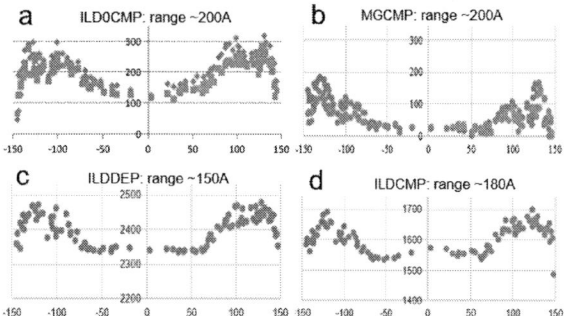

Figure 2: ILDb pad measurement profile at each layer: a. ILD0 CMP  b. MG CMP  c. ILD DEP  d. ILDCMP

Figure 3 shows the amount of loss in different areas of wafer from ILDDEP to ILDCMP measurement with ILDb pad. It can be seen that from ILDDEP to ILDCMP, the amount of oxide loss of wafer in different areas is relatively close, with a range of about 100 and no special map, that is to say, the thickness of the remaining ILD is relatively uniform. Obviously, using the ILDb pad measurement will lead us to believe that the ILD thickness fluctuates greatly, which will lead to the misadjustment of the profile.

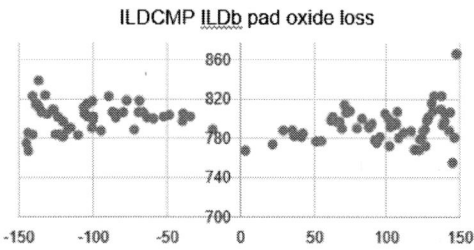

Figure 3: ILDCMP ILDb pad oxide loss

In fact, MGCMP is usually measured with OCD pad in our company. The OCD measurement data of MGCMP, ILDDEP and ILD CMP are shown as figure 4. As can be seen from the profile, The profile measured by MG CMP, ILDDEP and ILDCMP OCD was relatively flat. The THK range of MG CMP was 28, the THK range of ILDDEP was 32, and the THK range of ILDCMP was 157, much smaller than ILDb pad measurement range of 267. As can be seen from the analysis of oxide loss in Figure 3, the actual ILD film thickness is relatively uniform, so OCD pad measurement can better reflect the real film thickness.

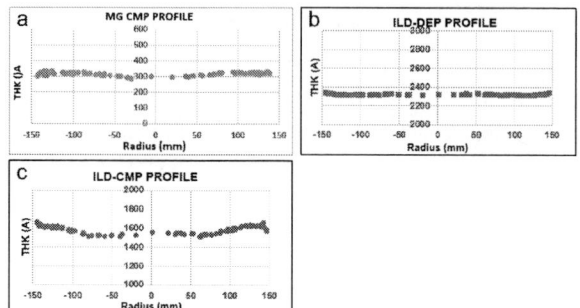

Figure 4: OCD pad measurement profile: a. MG CMP b. ILD DEP c. ILD CMP

In order to verify the above speculation, TEM was used to verify the effect of OCD pad measurement on the film thickness of real device region. In this experiment, we used OCD P110 pad as monitor pad to measure the MG film thickness of chip area. And we compared the actual film thickness of three different positions of OCD P110 pad and chip area, including SRAM-PD, SRAM-PU and Device-PFET. As shown in Figure 5, the three TEM images on the left are device region results, and the one on the right is OCD P110 pad results. It can be seen that the pattern of P110 pad is similar to that of device region, and the actual thickness of P110 pad is 347A. Compared with chip area thickness of 363/365/376, the gap is less than 30, indicating that OCD P110 pad can accurately reflect the true film thickness of device area. At the same time, due to the large gap between ILDb pad and OCD pad measurement profile, the measurement results of ILDb PAD are inferred to be deviated from the actual situation. Therefore, we choose OCD pad as the measurement pad of MGCMP.

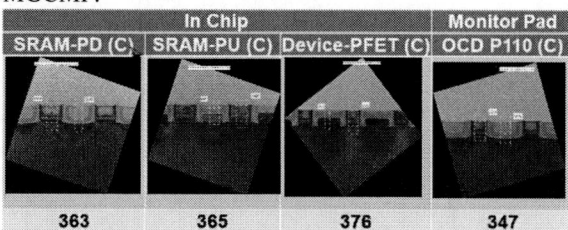

Figure 5: TEM results of OCD P110 pad and chip region

The following is to analyze why OCD pad measurement is more consistent with the fact while ILDb pad deviates from the fact. Firstly, analyze the structure: figure 6a and 6b are schematic diagrams of ILDb pad and OCD pad respectively. As shown in the figure 6a, ILDb pads are stacked with large block layers, and the top $SiO_2$ thickness is used for monitor the MG height. As shown in the figure 6b, MG height was monitored by the strip $SiO_2$ thickness at the OCD pad. It indicates that the structure of OCD pad is more close to that of real device region than ILDb pad. Secondly, the dishing generated in the grinding process: SiO2 on the top of ILDb pad is a large block structure, which is easy to produce dishing in the CMP process, and the dishing level of wafer may be different in different areas. On the one hand, the light spot may not be able to hit the center point of the pad, the more serious the dishing, the thinner the thickness measured. On the other hand, both SiN and NiSi under bulk $SiO_2$ are bulk structures, which are prone to dishing, leading to the accumulation of dishing at the top $SiO_2$, resulting in a larger gap between the film thickness and the actual device area. It is not easy to generate dishing at OCD pad in the CMP process due to the strip structure of $SiO_2$, which is close to the fact in the device area. According to the above analysis, OCD pad should be selected for MGCMP measurement

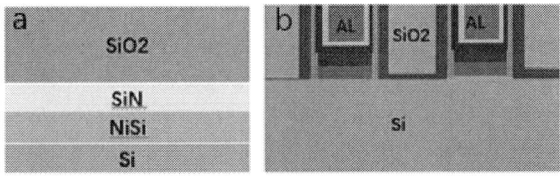

*Figure 6: structure diagram. a. ILDb pad b. OCD pad*

## CONCLUSION

According to the problems encountered in the research and development of the new generation product of 28HK of our company, the problems existing in the ILD CMP measurement were investigated. Firstly, the full map profile of ILDCMP and ILD DEP, MG CMP and ILD0 CMP ILDb pad or OCD pad was collected, and it was speculated that the measurement result of OCD pad could better reflect the film thickness of ILDCMP. Secondly, the MGCMP TEM experiment was developed to verify that the measured thickness of MGCMP on OCD pad was close to the actual chip area. ILDCMP has a uniform thickness, so the profile should be flat. Thirdly, the structure of ILDb pad and OCD pad is compared and analyzed, and the reason for the profile difference between ILDb pad and OCD pad is analyzed from the perspective of pad structure. The analysis shows that OCD pad should be selected for subsequent ILD CMP film thickness measurement, and reasonable selection of pad will accelerate the development and mass production efficiency of 28HK products.

## REFERENCES

[1] C. Kao, J. Li, Y. Wang, H. Xing and C. R. Liu, "Measurement of Layer Thickness and Permittivity Using a New Multilayer Model From GPR Data," in IEEE Transactions on Geoscience and Remote Sensing, vol. 45, no. 8, pp. 2463-2470, Aug. 2007, doi: 10.1109/TGRS.2007.900980.

[2] Yi Ding, Yefang Zhu, Junhua Yan, Conggang Wang, Wenbin Fan and A. Pang, "Application of measurement method on Cu-CMP process," 2015 China Semiconductor Technology International Conference, 2015, pp. 1-3, doi: 10.1109/CSTIC.2015.7153423.

[3] M. Horie, N. Fujiwara, M. Kokubo and N. Kondo, "Spectroscopic thin film thickness measurement system for semiconductor industries," Conference Proceedings. 10th Anniversary. IMTC/94. Advanced Technologies in I & M. 1994 IEEE Instrumentation and Measurement Technolgy Conference (Cat. No.94CH3424-9), 1994, pp. 677-682 vol.2, doi: 10.1109/IMTC.1994.352008

# CMP SCRATCH IMPROVE IN ADVANCED TECHNOLOGY NODES

Weiran Sun*, Yu Yang, Hu Li, Wei Zhao, Runcai Xiao, Jian Zhang, Haifeng Zhou, Jingxun Fang, Yu Zhang

Advanced Module Technology Development, Shanghai Huali Integrated Circuit Corporation, Shanghai 201314, China
*Corresponding Author's Email: sunweiran@hlmc.cn

## ABSTRACT

In the advanced-tech node's shallow trench insulation (STI) chemical mechanical planarization (CMP) process development, by doing design of experiments (DOEs) and repeat runs, the scratch defect count was successfully reduced and kept at 5 count per wafer. In the beginning of development, the process, consumables and tool were used 28nm node's. The scratch defect count in begin was >200eas. The following development improved defect count to 100, 10 at different phases, and finally kept it at 5 count per wafer. The experiment data showed that the updated consumables, improved process parameters, condition type and optimized rinse steps all three factors had significant benefits to scratch defect reduction.
*Keywords—STI; DOE; Scratch; Defect count; Consumables; Downforce; Rinse step*

## INTRODUCTION

The Moore's Law is pushing semiconductor industry shrinking the technology node size to half every two years. For catching the leading FAB's step, Huali is starting to develop the advanced node, beyond 20nm technology node. We use the familiar and matured platform, consumable, and process to develop the advanced node STI CMP as starting point. These are mainly based on 28nm technology's knowledge and hardware.

Because of the concern for electric property and trench filling on advanced node application, the flowable CVD (FCVD) method is used to deposit oxide film to fill the trench and form overburden. The positive about FCVD is that it has ability to fill trench under advanced node size with less void or seam. The negative is that the hardness is low, the oxide film is softer than other oxide film deposition methods [1]. The side effect is that there are much more defects generated on post STI CMP wafer surface, the most of them are scratch. This paper is focus on the way to find solution to minimize the defect count on post STI CMP with advanced node wafer.

## PROBLEM ANALYSIS

### STI CMP Process Analysis

The CMP process uses chemical and mechanical methods to planarize wafer surface uniformly. The consumables and hardware used in STI CMP are slurry, pad, head, platen and pad conditioner. Slurry contains chemicals to hydrate oxide surface, inhibitors to intentionally slow down planarization/removal rate on specific materials, abrasive particles to intentionally target oxide film to speed up planarization/removal rate, and other components to maintain stability for long pot life. Pad contacts wafer surface directly, it will respond to downforce from head and planarize wafer with slurry. Head carries the wafer, applies downforce to wafer and presses wafer on pad, and planarize/remove wafer film while rotating. Platen is where the pad attached on, it will rotate during the planarization. Pad conditioner is used to refresh the pad surface to maintain its performance during planarization process. Fig1 demonstrates CMP tool's components.

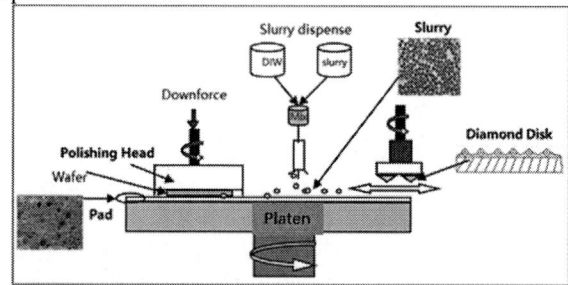

*Figure 1: The schematic of CMP tool components*

For STI CMP, the FCVD oxide overburden, bulk film of FCVD oxide and SiN, and small amount of SiN on the fin will be planarized off from wafer. The STI CMP process will stop on Ox/SiN interface, and the fin structure will be exposed [2]. Fig2 shows STI CMP incoming and post wafer schematic.

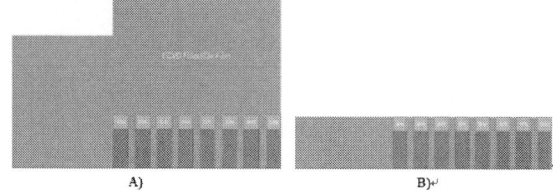

*Fig2. A). The incoming wafer schematic. On the fin there are SiN pad and FCVD oxide film. B). The post wafer schematic. The fin exposed*

### Defect Source Analysis

The defect is micro-scratch dominated, it is generated by mechanical force from solid particles. Fig3 is the defect image. There are two main mechanical force components, downforce and rotation speed. The solid particles are main come from slurry abrasive and process residue.

The possible way for scratch generation is that the solid tiny particle is embedded in oxide film by downforce then scratch is generated by rotation. The ways to minimize defect count are decrease downforce to wafer, slow down rotation speed, minimize residue on pad, and optimize slurry abrasive shape and size.

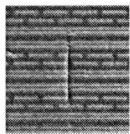 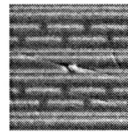 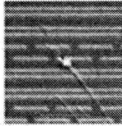

*Fig3. The defect images.*

## EXPERIMENT

The DOEs were based and focused on process parameters which include downforce and rotation speed, pad conditioning type, wafer rinse step, and upgraded slurry.

### Downforce

In STI CMP platen 2 process is the key for defect reduction, because the Ox/SiN interface is exposed at this step. There are two polishing steps require downforce applied on wafer, main-polish and over-polish steps. The main-polish time is controlled by end-point system that the system calls end-point, the main-polish step ends. The baseline settings of downforce of two steps are same. For downforce DOE, the downforce is decreased on over-polish step.

The over-polish step starts when Ox/SiN interface exposed theoretically, so that lower downforce can provide less damage to wafer surface.

### Rotation Speed

The head and platen are rotating in same direction during polishing, higher revolutions per minute (RPM) give higher removal rate. On the other hand, the higher RPM will give chance to generate more scratches compare to lower RPM. For RPM DOE, decrease platen 2 process head and platen RPM for both main-polish and over-polish step.

If the scratches are generated during polish steps, less count of revolutions can give less count of scratches.

### Pad Conditioning Type

There are two types of pad conditioning, ex-situ and in-situ. The process is using in-situ, the pad conditioner is down and apply downforce to do pad conditioning during wafer polishing. The ex-situ is that the conditioner is down and apply downforce to do pad conditioning before wafer polishing. The conditioning process is using small diamond tips on the conditioner to scrub pad with downforce. This process is possible to damage the pad and pad debris may be peeled off. Not only the pad debris, the residues stuck or trap in pad pores may be dug out. For pad conditioning DOE, the ex-situ conditioning is applied [3].

If the pad debris or residues are the defect source, the ex-situ conditioning can minimize this factor since there is rinse step between conditioning and polish steps.

### Wafer Rinse Step

There are wafer rinse steps after main-polish on each platen, and they are critical for wafer defect performance. In rinse step, there will be no slurry flow, the DIW will take place. The DIW will be ejected from a nozzle on pad with high flow, few liters per minute, to rinse wafer surface. The purpose is to clean out particles on wafer. The particles can be slurry residue, debris from pad, peeled chunk from wafer or any possible particle generated from polishing process. The side effect is the high flow of DIW can dig out particles which are buried in pad to pad surface and be potential defect source for scratch.

The modified wafer rinse step minimizes the particles on pad and therefor minimizes defect potential risk by adjusting rinse step sequence and parameters.

### Slurry Abrasive Screening

The STI CMP slurry is using ceria as abrasive. The used slurry abrasive size is around 500nm long with bar shape. The new test slurry abrasive is 100nm with ball shape for DOE. Fig4 shows two different slurry abrasives.

The abrasive is the potential defect source, especially for scratch. The bar shape ceria is possible to stick in oxide film or the uneven boundary of oxide in trench and SiN on fin.

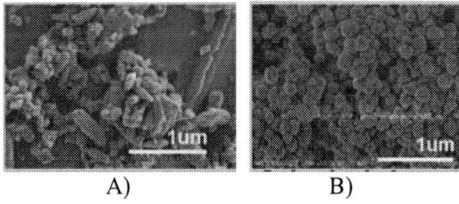

*Fig4. The 2 different slurry abrasives images. A) the bar shape abrasive has around 500nm long dimension. B) the ball shape abrasive's dimension is around 100nm.*

## RESULT & DISCUSSION

The results are followed DOE's expectation. The lower downforce, lower rotation speed, and ex-situ pad condition selection reduced defect count by 15%, 9%, and 5%, respectively. The new type of slurry with optimized abrasive shape and size provided thrilled defect reduction performance, the defect count improved from 200s to 10s.

The polishing parameters are the verified factors to reduce scratch defect, and the next generation of consumable, slurry, leads defect count to the next level. The advanced technology node process needs advanced consumable. The structure transform from 2-D planar shape to 3-D fin shape. There are times more patterns in same area, so that there are times more chances to generate defect. The bar shape slurry abrasive has corner which is easy to pierce in oxide film or any uneven surface, especially the interface of Ox/SiN since SiN pad on fin always protrudes from oxide film. Once the abrasive stuck in oxide film or Ox/SiN interface or SiN pad, the rotating wafer and pad will force to roll the abrasive, then scratch will be generated. The abrasive will not only stuck on wafer, but also it can be captured by pad micro pore. The captured abrasive will be defect source to scratch wafer.

The ball shape abrasive is next generation of ceria slurry, the most advantage compare to normal is defect reduction. The ball abrasive can slide on wafer surface without stuck in. The pad is hard to hold ball since the ball can easily roll away from pore.

By combining upgraded slurry and tunable polishing parameters, downforce, rotation speed, and conditioning type, all factors together to develop an optimized STI CMP process. The optimized process reduced total defect count less than 10 and it is repeatable.

## CONCLUSION

The CMP process now is consumable driven. The process parameters can improve performance at current consumable set, but it is possible to keep optimizing wafer performance by screening updated consumable since different consumable set will give a different performance. This study is a good learning that it is important to work with consumable vendor, advanced technology node needs advanced upgraded consumable.

## REFERENCES

[1]: H. Deng *et al.*, "New applications and challenges of dielectric films at 14nm FinFET technology and beyond," *2016 China Semiconductor Technology International Conference (CSTIC)*, 2016, pp. 1-4, doi: 10.1109/CSTIC.2016.7464017.

[2]: Lei Zhang, Junhua Yan, Kun Chen, Wenbin Fan, Yefang Zhu and Jingxun Fang, "Research and solution of STI CMP dishing and uniformity improve for 28LP," *2017 China Semiconductor Technology International Conference (CSTIC)*, 2017, pp. 1-2, doi: 10.1109/CSTIC.2017.7919818.

[3]: D. Lim, H. Kim, B. Jang, H. Cho, J. Kim and H. Hwang, "A novel pad conditioner and pad roughness effects on tungsten CMP," *Proceedings of International Conference on Planarization/CMP Technology 2014*, 2014, pp. 352-355, doi: 10.1109/ICPT.2014.7017318.

# STUDY ON PREPARATION AND POLISHING PERFORMANCE OF CERIA SLURRY

*Ye Wang [1,2], Baoguo Zhang[1,2,\*], Pengfei Wu[1,2], Mengchen Xie[1,2], Ye Li[1,2], Haoran Li[1,2]*

[1] School of Electronics and Information Engineering, Hebei University of Technology Tianjin 300130, People's Republic of China

[2] Tianjin Key Laboratory of Electronic Materials and Devices, Tianjin 300130, People's Republic of China

*Corresponding Author's Email: bgzhang2000@yahoo.com

## ABSTRACT

In this paper, propionic acid, oxalic acid and phytic acid were used as pH regulators in wet grinding, respectively. Propionic acid was found as the best pH regulator according to the result of particle distribution and zeta potential. The ceria slurry was prepared for chemical mechanical polishing with average particle size (D50) of $CeO_2$ is 225.4 nm. With the ceria slurry prepared, the material removal rate (MRR) of $SiO_2$ as high as 403 nm/min; at meantime, the $SiO_2$ roughness is smaller than 0.10 nm (Ra, 0.02 nm).

*Keywords—$CeO_2$; $SiO_2$; chemical mechanical polishing; MRR; Roughness*

## INTRODUCTION

$CeO_2$ polishing slurry is a high-quality polishing material, commonly used in the polishing of aviation glass, integrated circuit substrates, spectacle lenses, crystal displays, optical glass and various gems[1]. At present, the commonly used slurries for $SiO_2$ polishing are made up of $SiO_2$ and $CeO_2$ abrasive. The development direction of the polishing slurry for polishing the silica dielectric should be based on chemical action and choose small particle size abrasives, so as to achieve high material removal rate and low scratches. Compared to $SiO_2$, $CeO_2$ has a lower Mohs hardness ($CeO_2$ has a Mohs hardness of 6, $SiO_2$ has a Mohs hardness of 7). Its high polishing rate is not only due to mechanical action but also chemical action, so the use of nano $CeO_2$ particles as abrasives is in line with the future development trend of polishing fluids [2-4]. Now, relevant domestic enterprises have committed to getting rid of the dependence on imported slurry and realizing the research of high-quality and efficient slurry with independent intellectual property rights. The main research focus is on the improvement of the quality and efficiency of slurry.

Based on the difficulty of uniform dispersion of $CeO_2$ in water medium, organic acid is used as pH regulator and dispersant, and the method of wet grinding is adopted to study the particle size distribution of $CeO_2$ polishing solution and its chemical mechanical polishing (CMP) of $SiO_2$. In this experiment, the dispersion performance of the self-made $CeO_2$ suspension in an acidic system and its influence on the chemical mechanical polishing of $SiO_2$ were studied, which can provide reference for the research and development of ceria polishing slurry.

## EXPERIMENTAL

$CeO_2$ powder was obtained by high-temperature baking cerium carbonate ($Ce_2(CO_3)_3$) powder. $CeO_2$ suspension was prepared by wet ball milling. The $SiO_2$ dielectric used in this paper is 4-inch JGS2 with a thickness of 700 μm. All chemicals were used without further purification.

The chemical mechanical polishing experiment was performed on Ruixuan SSP-500 grinding and polishing machine. The polishing process conditions are: polishing pressure 4 psi, polishing head rotation speed 50 r/min, polishing pad rotation speed 50 r/min, and the volume flow rate of the polishing liquid is 100 ml/min. The polishing time is 300 s.

The surface roughness of characteristic region (5 μm × 5 μm) of $SiO_2$ dielectric before and after polishing was analyzed by atomic force microscope (AFM) made by Agilent Technologies. The particle size distribution of the ceria slurry was obtained from Laser nanometer (PSS 380). Material removal rate (MRR) is determined by the following Equation (1):

$$MRR = \frac{\Delta m}{\rho ts} \qquad (1)$$

where $\Delta m$ (g) is the mass loss, t is the polishing time (in the test, t = 5 min), $\rho$ is the $SiO_2$ dielectric density ($\rho$= 2.2 g/$cm^3$), s is the area of the quartz glass.

## RESULTS AND DISCUSSION

**The effect of adding different organic acids on the particle size distribution of $CeO_2$ suspension during ball milling.**—$CeO_2$ suspension was prepared by wet ball milling. During the ball milling, the pH value was adjusted to 3, the material-to-ball ratio was 1:4, and the total time of ball milling was 6 h. The effects of different organic acids (propionic acid, oxalic acid and phytic acid) as regulators on the particle size distribution and zeta potential of $CeO_2$ have been studied. The particle size distribution and zeta potential of ceria slurry used in the experiment was measured by the Nicomp 380 laser nanoparticle size analyzer produced by the US PSS company. The particle size distribution of ceria slurry obtained by adding different organic acids during ball milling is shown in Fig. 1. When propionic acid ($CH_3CH_2COOH$) is used as the regulator, the suspension after ball milling is uniformly dispersed, with a D50 of 225.4 nm. At this time, the zeta potential of the suspension is 52.19 mV, which indicated the dispersion stability of the suspension is relatively good. When oxalic acid ($C_2H_2O_4$) is used as the regulator, the D50 is 267.4 nm, the average particle size becomes larger, and the particle size distribution becomes significantly wider. At this time, the suspension zeta potential of 31.71 mV, and the dispersion of the suspension is poor. When phytic acid ($C_6H_{18}O_{24}P_6$) is used as the regulator, the D50 is 283 nm, the average particle size becomes larger and the particle size distribution becomes wider. At this time, the zeta potential of the suspension is 40.2 mV. After the suspension was placed for 3 h, the suspension began to stratify at a distance of 0.5 cm from the liquid surface. The particles in the suspension were easy to agglomerate and the suspension was unevenly dispersed.

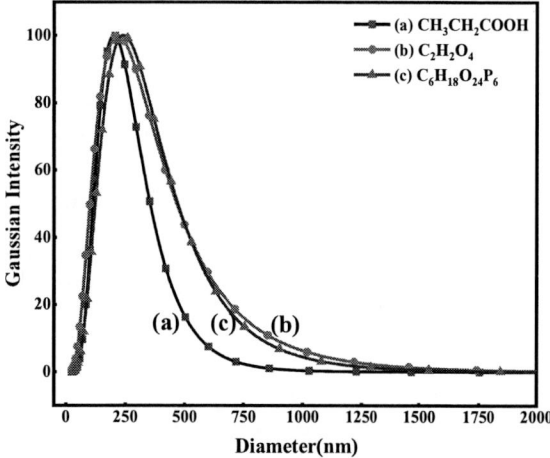

*Figure 1: Particle size distribution of $CeO_2$ suspension after ball milling of $CH_3CH_2COOH$, $C_2H_2O_4$ and $C_6H_{18}O_{24}P_6$*

The zero point potential (PZC) of $CeO_2$ is 7.9[5]. When the pH value of the slurry is below 7.9, the $CeO_2$ surface is positively charged, and under the effect of electrostatic force, the $CeO_2$ particles repel each other. When propionic acid is added, propionate will be adsorbed on the surface of $CeO_2$ particles, so that $CeO_2$ particles repel each other. Phytic acid has a large molecular weight and strong chelating ability, and can produce insoluble compounds with metal ions. Therefore, after adding phytic acid, the abrasive particles in the suspension agglomerate, the average particle size is larger and the particle size distribution is wider, and the suspension is unevenly

dispersed. Propionic acid has better performance than oxalic acid, so propionic acid is used as a regulator.

**The effect of pH on the removal rate of $SiO_2$ dielectric.**—The effect of pH on MRR of $SiO_2$ dielectric was studied in the slurry containing 1% $CeO_2$. As shown in Fig. 2, with the increase of pH, the MRR of $SiO_2$ dielectric firstly increases and then decreases. When pH is 6, the MRR of $SiO_2$ dielectric is 403 nm/min, which reaches the highest.

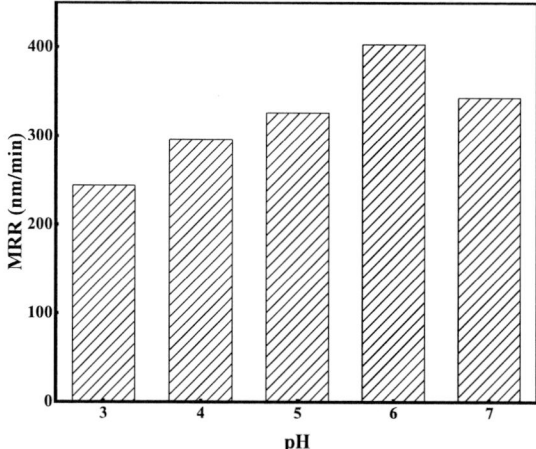

*Figure 2: The effect of pH on the polishing rate of $SiO_2$*

The pH value mainly affects the electrostatic force between particles in the slurry. The PZC of $SiO_2$ is 2.2[6], so when the pH is greater than 2.2, the surface of $SiO_2$ is negatively charged. For $CeO_2$, when the pH is below 7.9, the surface of $CeO_2$ particles is positively charged. When the electrostatic force is strong, the formation of the Si-O-Ce soft layer on the surface of $SiO_2$ dielectric can be accelerated. Therefore, the removal rate increases when the electrostatic force is strong, the abrasive particles in the slurry are evenly distributed, and the surface quality after polishing is improved[7]. It can be seen from Fig. 2 that when the pH is 6, the material removal rate reaches the maximum. This may be due to the electrostatic force between $CeO_2$ and $SiO_2$ is the strongest at pH 6, which accelerates the progress of the chemical action and improves the $SiO_2$ removal rate. The reduced electrostatic force will cause particles in the suspension to agglomerate, and larger particles may be further retracted into the substrate, causing greater surface damage.

**The effect of $CeO_2$ slurry on the surface roughness of $SiO_2$ dielectric.**—The surface roughness of $SiO_2$ dielectric before polishing is tested, and the results are shown in Fig. 3. When pH is 6, the polishing rate of $SiO_2$ dielectric reached the maximum, about 403 nm/min. The surface roughness result of the $SiO_2$ dielectric is shown in Fig. 4.

Ra=0.08 nm

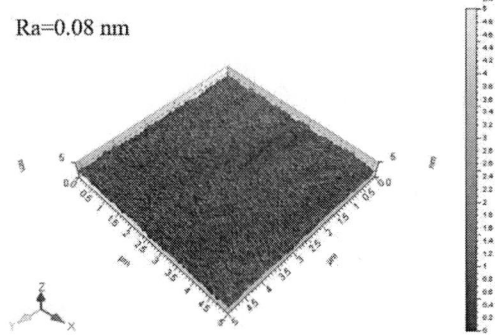

*Figure 3: AFM image of SiO₂ before polishing*

Ra=0.02 nm

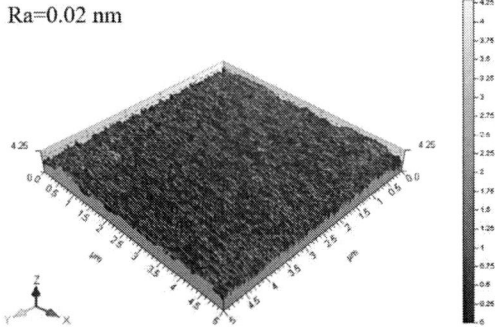

*Figure 4: AFM image of SiO₂ after polishing*

Fig. 4 shows that the use of ceria slurry with uniform dispersion and appropriate particle size can greatly improve the surface quality of silica when there is a higher removal rate.

## CONCLUSION

In this paper, the CMP of $SiO_2$ wafer was studied with the ceria slurry. The effect of adding different organic acids on the ceria suspension during ball milling was studied. When propionic acid is added during ball milling, the dispersion stability of the ceria suspension is best among the three pH regulators. At meantime, the average particle size(D50) is 225.4 nm, and the zeta potential is 52.19 mV. When the concentration of ceria is 1 wt% and the pH of slurry is 6, the highest MRR of $SiO_2$ dielectric is 403 nm/min. At the same time, the surface quality of $SiO_2$ dielectric has been greatly improved, and Ra is only 0.02 nm.

## ACKNOWLEDGEMENTS

This work was supported by the Major National Science and Technology Special Projects (No.2016ZX023 01003-004-007), Natural Science Foundation of Tianjin, China (16JCYBJC16100, 18JCTPJC57000). The authors also thank the teachers and classmates for their helpful suggestions.

## REFERENCES

[1] H. Tetsuya, K. Yasushi, T. Yuuki and K. Susa. *J.* Joumal of noncrystalline solids, vol. 283, 2001, pp. 129-134.

[2] R. Jairath, M. De Sai, M. Stell. *J.* MRS Proceedings, 1994, pp. 337.

[3] L. Cook. *J.* Journal of Non Crystalline Solids, vol. 120, 1990, pp. 152-171.

[4] T. Abiade, W. Choi, K. Singh. *J.* Journal of Materials Research, vol. 20, 2005, pp. 1139-1145.

[5] M. Nabavi, O. Spalla, B. Cabane. *J.* Journal of Colloid & Interface Science, vol. 160, 1993, pp. 459-471.

[6] S. Ramarajan, Y. Li, M. Hariharaputhiran. *J.* Electrochemical and Solid-State Letters, vol. 3, 2000, pp. 232-234.

[7] Y. Zhaoxia, B. Zhang, Y. Xiaofan and L. Ye. *J.* ECS Journal of Solid State Science and Technology, vol. 10, 2021, pp. 9.

# Galvanic corrosion caused by device structure in chemical mechanical planarization

*Lei Wang[1,2]\*, Guilin Chen[1], Luping Liu[1], Xinchun Lu[2]\**

[1]Zhejiang Hikstor Technology Co. Ltd., Hangzhou 311300, China
[2]State Key Laboratory of Tribology, Tsinghua University, Beijing 100084, China
\* Email: wang_lei@hikstor.com, xclu@tsinghua.edu.cn

## Abstract

Chemical mechanical planarization (CMP) has emerged as the most viable method to planarize thin films during fabrication of Spin-Transfer Torque-MRAM (STT-MRAM). Some metals when exposed to slurry in CMP process are subject to galvanic corrosion. This paper explores the influence of galvanic corrosion on the thickening of the hard mask oxidation layer at the top of MTJ module.

*Keywords—STT-MRAM; CMP; Galvanic corrosion*

## Introduction

Many new memory technologies are developed to meet the long-standing scaling challenges. STT-MRAM (Spin-Transfer Torque-MRAM) with fast random access capability, remarkable endurance and non-volatility, could possibly overshadow other memory technologies like static RAM, dynamic RAM and embedded ones [1]. It is known that STT-MRAM has at its core MTJ (magnetic tunnel junction) module to achieve better performance in signal strength and data retention (coercivity) [2].

In fabricating MTJ module, an encapsulation layer was deposited after MTJ etch to protect MTJ properties. And then a dielectric layer was deposited as dielectric insulation. Oxide CMP is introduced later to stop on top of MTJ module.

CMP process, however is a complicated mix of chemical action and physical effects with some causing unexpected results. For example, in the process of oxide CMP mentioned above, the galvanic corrosion [3] might lead to MTJ module's unnatural oxidation. The oxide coating then results in broken circuit.

Galvanic corrosion can be trigged by many factors, but this paper explores how the different bottom structures and hard mask (HM) materials of MTJ module in CMP process can cause different oxidation layer thickness due to galvanic corrosion effect.

## Experimental

### A. MTJ FABRICATION

Wafers without CMOS logic circuit were used as substrate in the fabrication process for the convenience of research. The procedures are as follows:

First, dielectric layer was deposited on the 300mm blanket wafer as substrate;

Second, back-end of line (BEOL) process was used to form bottom metal (BM) and bottom via (BV);

Third, bottom electrode (BE) and MTJ module were later deposited on the bottom via and patterned. MTJ module is composed of MTJ films and metal hard mask;

Fourth, a dielectric insulation film was then filled for further oxide CMP process.

### B. MTJ Array Structure

MTJ with three different bottom structures was designed to study the galvanic corrosion effect in the oxide CMP process, as shown by Figure 1. Figure 1(a) shows the MTJ bottom electrode with no circuit structure below it. While MTJ bottom electrode in both Figure 1(b) and 1(c) had bottom via and bottom metal, the bottom metal in 1(b) is complete and that in 1(c) is separate.

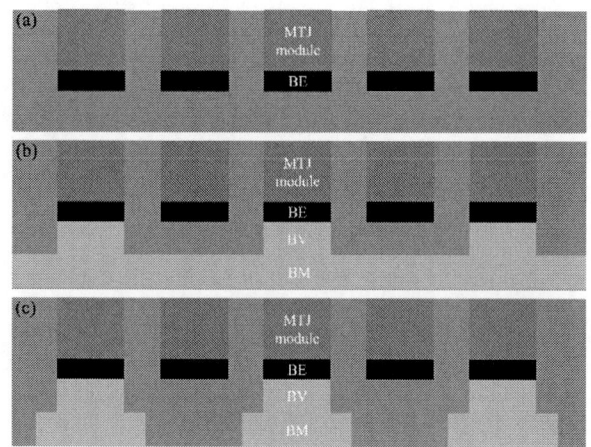

*Figure.1. MTJ Array structures: (a) shows the MTJ bottom electrode with no circuit structure below it, while in 1(b) and 1(c), MTJ bottom electrode had BM and BV; the BM in (b) is complete but that in (c) is separate.*

### C. CMP Polisher and Metrology Tool

CMP process was carried out on Reflexion LK prime (Applied Materials, Santa Clara, CA, USA). The measurement of MTJ module oxidation thickness can be obtained by TEM.

## Results and Discussion

In the first experiment, MTJ module as shown by Figure 1(a) was formed and its HM was made of metal A. Following the oxide CMP process, 500Å nitride was first deposited to prevent oxidation and TEM was then performed to measure how much the HM remains and if it was oxidized. The TEM results suggest that the HM's remains fulfilled the requirement for MTJ module fabrication, and HM had on top of it an oxidation layer with a thickness of $H_1$, as shown by Figure 2(a). This layer was considered to be a natural oxidation layer because metal A could generate passivation during oxidation which would prevent further oxidation from happening.

In the second experiment, MTJ module with metal A as HM material both in Figure 1(b) and Figure 1(c) were fabricated on a single complete wafer. After the oxide CMP process, 500Å nitride was also deposited to prevent oxidation and TEM was performed. The results that can be found in Figure 2(b) & 2(c) indicate that HM at the top of both MTJ modules had oxidation layer. The relationship of these two structures' HM oxidation thickness was given by $H_1 = 0.4H_2 \approx H_3$, where $H_2$ denoted the thickness of Figure 1(b) structure and $H_3$ meant the thickness of Figure 1(c) structure. HM oxidation layer of Figure 1(b) structure was basically twice as thick as that caused by the natural oxidation, and this accelerated oxidation was likely to result from galvanic corrosion in the CMP process. Meanwhile, there was an interesting phenomenon. When the dummy without BM and BV were added to structures both in Figure 1(b) and 1(c) to alleviate the patterning density effect in the CMP process, it is observed that the HM oxidation thickness at the top of the dummy (namely, H) was nearly the same as that in the first experiment ($H_1$).

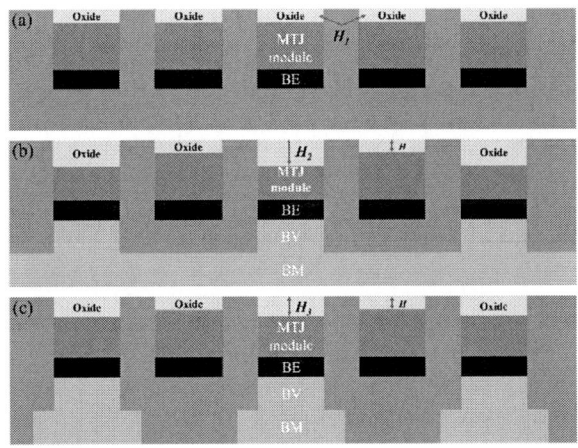

*Figure.2. HM oxidation thickness.*

The third experiment aimed to explore the influence of time on the HM oxidation thickness. On one complete wafer, MTJ module still with metal A as HM material shown both in Figure 1(b) and 1(c) were fabricated. The difference was that the wafers after CMP process were exposed to the atmosphere for certain time periods (2h, 4h, 6h) before nitride deposition and TEM. The final TEM results suggest that despite different exposure periods, the HM oxidation thickness relationship remained the same: $H_1 = 0.4H_2 \approx H_3$. In other words, HM oxidation thickness was irrelevant to the exposure time.

The fourth experiment intended to find a way to solve the problem of greater HM oxidation thickness due to different bottom structures. HM material was changed from metal A to metal B and metal C. There were two identical wafers with both MTJ structures in Fig 1(b) and Fig 1(c). One

wafer used metal B as HM material but the other used metal C as HM material. After CMP processes, the wafers were exposed to the atmosphere for two hours. The nitride deposition and TEM were later performed. It was observed that for Figure 1(b) structure, the HM oxidation thickness relationship was $H_1 = 0.4H_A = 0.6H_B > H_C$, but for Figure 1(c) structure, the thickness relationship was $H_1 \approx H_A \approx H_B > H_C$. The galvanic corrosion effect, therefore seemed to be restrained with certain HM materials.

## Conclusion

In fabricating MTJ module, the difference in bottom structures induced galvanic corrosion, which then accelerated the MTJ module oxidation in the CMP process. To protect wafer from the impact of the galvanic corrosion, two measures can be taken, that is, either optimizing the bottom circuit structure of MTJ or using stable metal materials as hard mask.

## References

[1] S. Chung et al., "Fully integrated 54nm STT-RAM with the smallest bit cell dimension for high density memory application," *2010 International Electron Devices Meeting,* San Francisco, CA, 2010, pp. 12.7.1-12.7.4.

[2] S. Hassan, L. Xue, J. Anh, M. Pakala, G. Sin and M. Okazaki, "STT-RAM device performance improvement using CMP process," *2017 28th Annual SEMI Advanced Semiconductor Manufacturing Conference (ASMC), Saratoga Springs,* NY, USA, 2017, pp. 209-211.

[3] G. Banerjee, R. L. Rhoades, "Chemical Mechanical Planarization Historical Review and Future Direction," *ECS Transactions*, 13 (4) 1-19 (2008)

# Post CMP Cleaning Study Of Ceria Slurry

Lei Wang[1,2*], Luping Liu[1], Zhen Li[1], Changzhen Jia[3], Shoutian Li[3], Xinchun Lu[2*]

[1]Zhejiang Hikstor Technology Co. Ltd., Hangzhou 311300, China
[2]State Key Laboratory of Tribology, Tsinghua University, Beijing 100084, China
[3]Anji Microelectronics Technology (Shanghai) Co. Ltd. Pudong New District, Shanghai 201203, China
* Email: wang_lei@hikstor.com, xclu@tsinghua.edu.cn

## Abstract

Post CMP cleaning is a very important step to remove the CMP slurry abrasive from wafer surface and get rid of the CMP particles defects. In this paper, we studied the relationship between the ceria particle surface charge and effective cleaning process of oxide wafers. We also developed an effective on-platen cleaning process to remove ceria particles from TEOS (tetraethyl orthosilicate) wafer surface when the ceria slurry had positive zeta potential. In this process, the cleaning solution was introduced on polishing platen but not into cleaner. This process offers advantage over traditional ceria cleaning process.

Keywords—Ceria CMP; surface particle defects; on-platen cleaning

## Introduction

The generation of surface defects during CMP processes can profoundly affect chip manufacture process and degrade device performance [1]. One of surface defects during CMP is the slurry abrasive particles left on wafer surface. CMP cleaning is an important process to get rid of these so-called surface particles on wafers.

Ceria-based CMP slurry can be used in ILD (Interlayer Dielectric) and STI (Shallow Trench Isolation) CMP processes [2]. Ceria slurry can be formulated with either negative or positive zeta potential [3]. Ceria particles with positive zeta potential have strong affinity to adhere on oxide wafer surface and these particles can be difficult to be cleaned off. CMP polishers like Mirra and Reflexion manufactured by AMAT have integrated cleaners and three polishing platens. First platen is used for ILD CMP process, second platen is applied for STI CMP process and third platen is reserved for buffing process to get rid of the surface particles of wafers. Without the platen 3 buffing step, surface particle counts are typically high even after the wafers are cleaned again in the cleaner. In many practical applications, laboratory or fabs, platen 3 is not always available. Thus, developing an on-platen cleaning processes is necessary to remove CMP slurry particles from wafer surface.

In this paper, we studied the relationship between the ceria slurry surface charge and effective cleaning process. Ceria slurries can be formulated with positive charges (i.e., positive zeta potential) and negative charges at the applied pH. The ceria abrasive particles with positive charges are more difficult to be cleaned off than the one with negative charges. For the ceria slurry with positive zeta potential, we develop an effective on-platen cleaning process to remove ceria particles from TEOS (tetraethyl orthosilicate) wafer surface. In this process, the cleaning solution is introduced on polishing platen but not into cleaner. This process offers advantage over traditional ceria cleaning process.

## Experimental

### A. Ceria Slurries

To study the relationship between ceria slurry properties and cleaning process, we chose ceria slurries with opposite charges. Slurry A was formulated with positive charges, i.e., positive zeta potential; Slurry B was formulated with negative charges. At the slurry's pH, silicon oxide wafer (PETEOS) has negative zeta potential. The zeta potentials of the two slurries and PETOS are listed in Table 1.

TABLE 1
PHYSICAL PROPERTIES OF CERIA SLURRIES

| Slurry | pH | Zeta potential |
|--------|-----|----------------|
| Slurry A | 4.5 | +35 mV |
| Slurry B | 6.7 | -58mV |
| PETEOS | 4.5 | -33 mV |
| PETEOS | 5.0 | -40 mV |

### B. Cleaning Solutions

Several cleaning solutions were used in this study. These solutions included citric acid-based acidic cleaning solution and different alkaline cleaning solutions. The cleaning solutions were either added into cleaner's cleaning solution tank or dispersed on polishing platen through a 2nd line.

### C. CMP Polisher and Metrology Tool

CMP processes were carried out on Reflexion (Applied Materials, Santa Clara, CA, USA). Wafer surface particles counts were detected by SP2 (KLA-Tencor, Corp., Milpitas, CA, USA). The threshold for the SP2 recipe was 0.12um. Wafers were 300 mm plasma-enhanced tetraethyl-orthosilicate (PETEOS) (Hikstor, China).

## Results and Discussion

In the first experiment, we compared the SP2 defect counts of PETEOS wafers polished by Slurry A and Slurry B

978-1-6654-9759-6/22 $31.00 © 2022 IEEE

and cleaned with citric acid based cleaning solution without buffing step. The citric acid based cleaning solution was added into the Reflexion polisher's cleaning solution tank and dispersed on the cleaner's brushes. PETEOS wafers polished by either Slurry A or Slurry B were sent through the cleaner and the wafers were scanned by SP2 for defectivity. This cleaning procedure is called baseline procedure in this experiment. Table 2 displays the SP2 defect maps of slurry A and B with citric acid based cleaning solution.

TABLE 2
SP2 DEFECT PERFORMANCE OF SLURRY A AND B WITH CITRIC ACID BASED CLEANDING SOLUTION

| Condition | slurry | A | B |
|---|---|---|---|
| | Clean chemical | Citric Acid Based | |
| SP2 PRE (µm) | 0.12-0.16 | 1 | 1 |
| | 0.16-0.2 | 0 | 0 |
| | 0.2-0.5 | 5 | 7 |
| | 0.5-1 | 0 | 1 |
| | >1 | 3 | 1 |
| | Map | | |
| SP2 PST (µm) | 0.12-0.16 | 625 | 37 |
| | 0.16-0.2 | 2 | 3 |
| | 0.2-0.5 | 18297 | 35 |
| | 0.5-1 | 2 | 0 |
| | >1 | 0 | 0 |
| | Map | | |
| Adder (ea) | @0.16µm | 18293 | 29 |

Table 2 clearly shows that citric acid-based cleaning solution can clean the PETEOS wafers polished by Slurry B well but massive slurry particles left on wafers surface after polished by Slurry A. The results can be explained by charge repulsion and attraction. The citric acid-based cleaning solution is in acidic pH regime. In acidic pH regime, the ceria particle surface of the Slurry B has negative charges and PETEOS wafer surface has negative charges as well. The charge repulsion between ceria particles and PETEOS wafer surface makes the ceria particles easy to be brushed off from the wafer surface. On the other hand, the ceria particle surface of the Slurry A has positive charges and PETEOS wafer

surface has negative charges. There is strong charge attraction between ceria particles and PETEOS surface and such attraction force makes the ceria particles adhere to PETEOS surface. In addition, the citric anions can bridge positive ceria particles together to form agglomerations which have strong affinity to adhere to negative PETEOS wafer surface and PVA brushes. All these factors contribute to the high SP2 counts.

In the second experiment, we compared the SP2 defect counts of PETEOS wafers polished by Slurry A and Slurry B and cleaned with trimethylamine-based commercial cleaning solution without buffing step. The results are similar to the first experiment. The PETEOS wafers polished with Slurry B has very low SP2 defects (26ea@0.16µm)and the PETEOS wafers polished with Slurry A has very high SP2 defect counts (2047ea@0.16µm). The results are shown in Table 3.

TABLE 3
SP2 DEFECT PERFORMANCE OF SLURRY A AND B WITH TRIMETHYLAMINE BASED CLEANDING SOLUTION

| Condition | slurry | A | B |
|---|---|---|---|
| | Clean chemical | Trimethylamine Based | |
| SP2 PRE (µm) | 0.12-0.16 | 0 | 0 |
| | 0.16-0.2 | 0 | 0 |
| | 0.2-0.5 | 1 | 2 |
| | 0.5-1 | 2 | 1 |
| | >1 | 3 | 2 |
| | Map | | |
| SP2 PST (µm) | 0.12-0.16 | 1040 | 42 |
| | 0.16-0.2 | 0 | 12 |
| | 0.2-0.5 | 2053 | 19 |
| | 0.5-1 | 0 | 0 |
| | >1 | 0 | 0 |
| | Map | | |
| Adder (ea) | @0.16µm | 2047 | 26 |

The results can be also explained by surface charge repulsion or attraction. The trimethylamine-based commercial cleaning is formulated at alkaline pH regime. At polished pH which is weak acidic for both Slurry A and Slurry B, ceria particles in Slurry A have positive charges but negative in Slurry B. The charge force between Slurry B and PETEOS is repulsion but attraction in Slurry A. So the ceria particles in Slurry A have much stronger affinity to adhere PETEOS surface during the polishing than the Slurry B. Although during the cleaning step, the trimethylamine-based cleaning solution is in alkaline pH regime, the surface charges of ceria particles from both Slurry A and Slurry B are negative and both repulse to the negative PETEOS surface. However, such repulsion force is not enough to lift the ceria abrasive particles that already adhere to PETOES surface. The results in the 2nd experiment also indicate the force between brushes and wafers is not enough to sweep away the ceria particles once they stick on the wafers surface.

To further test the effect of cleaning solution chemistry, we conducted another experiment to compare the cleaning solution chemistries and clearer brush/chemistry setup. The cleaning solutions in this experiment include: DIW, ammonia, potassium hydroxide (KOH), tetramethylammonium hydroxide (TMAH), organic amine A and B. The partition of each clean solution in brush 1 and brush 2 boxes of the Reflexion cleaner is shown in Table 4, as well as the SP2 results. The PETEOS wafers were polished with Slurry A. The use of high pressure rinse (HRP) after main polishing is indicated in Table 4. The results in Table 4 indicate the high surface particle counts by SP2 regardless of the nature of cleaning solutions. The force between brushes and wafers is not enough to sweep away the ceria particles on the wafers surface. To reduce the SP2 defect counts, we need to prevent ceria particles from adhering on PETEOS wafer surface after polishing. In the following experiments, we focused on how to remove the surface particles of Slurry A.

TABLE 4
SP2 DEFECT COUNTS OF DIFFERENT CLEANING SOLUTIONS

| Procedure | HRP | Solution in brush 1 box | Solution in brush 2 box | SP2 counts |
|---|---|---|---|---|
| 1 | Y | TMAH | TMAH | 64,313 |
| 2 | N | NH4OH | NH4OH | 64,447 |
| 3 | N | NH4OH | Open | 66,230 |
| 4 | N | Amine A | Open | 65,753 |
| 5 | N | Amine B | Open | 68,104 |
| 6 | N | Amine B | Open | 68,261 |
| 7 | N | Amine A | Amine B | 67,063 |
| 8 | Y | KOH | Open | 66,188 |
| 9 | N | DIW | DIW | 64,356 |
| 10 | Y | DIW | DIW | 64,123 |

In the third experiment, we added a buffing step in a separate platen. The experiment was conducted in the following procedure using AMAT's Reflexion polisher: a PETEOS wafer was polished on Platen 2 using Slurry A. Then the wafer was moved to Platen 3 and buffed with

de-ionized water (DIW) at 2.0psi for 45 seconds. Polytex pad was used on Platen 3. Cleaner setup was not changed. The cleaning solution was ammonia. SP2 threshold in this experiment was set at 0.16um. In this experiment, after added buffing step on Platen 3, the surface ceria particles on the PETEOS wafers were removed resulting in low SP2 defect counts, as seen from Figure 1. For comparison, the new wafers SP2 counts are included. The surface particles on the new wafers detected by SP2 come from environment since the wafers were not processed in CMP step. The experiment indicates the buffing step with 2psi 45second on Polytex pad generates enough mechanical force to sweep away the ceria particles from the wafer surface.

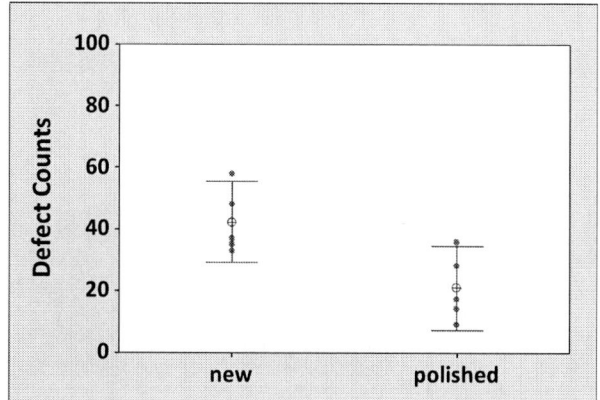

Fig.1: SP2 defect counts comparison between new PETEOS wafers and polished PETEOS wafers that were buffed on Platen 3

TABLE 5
SP2 DEFECT COUNTS OF ON-PLATEN CLEANING

| Wafer # | On-platen buff solution | Brush 1 Chem | Brush 2 Chem | SP2 Counts |
|---|---|---|---|---|
| 1 | TMAH | DIW | DIW | 313 |
| 2 | TMAH | DIW | DIW | 548 |
| 3 | TMAH | DIW | DIW | 282 |
| 4 | TMAH | DIW | DIW | 390 |
| 5 | TMAH | DIW | DIW | 367 |
| 6 | TMAH | DIW | DIW | 670 |
| 7 | Amine B | DIW | DIW | 271 |

In the fourth experiment, we moved the buffing step into the same polishing platen utilized Slurry A. The polishing pad was hard pad – IC1010 (Dow Chemical, Midland, MI, USA). The buffing started right after the main polishing. The TMAH buffing solution was introduced the polishing platen via 2nd line. The buffing down force was 2.0psi and buffing time was 30 seconds. In this experiment, we also tested if cleaning solution was needed in cleaner. The SP2 measurement threshold was set at 0.12um. The SP2 results are shown in Table 5. The results show when the buffing step was moved to the polishing platen, DIW is sufficient in cleaner's brush 1 and brush 2 boxes. Tetramethylammonia hydroxide solution was effective way to get rid of ceria particles from PETEOS wafers surface. This procedure offers advantage over the third experiment since the third platen is not always available due to platen configuration in our laboratory.

**Conclusions**

The negative ceria particles have less adhesion force on the negative PETEOS surface during CMP and these ceria particles can be effectively cleaned off using normal cleaner setup. However, the positive ceria particles have stronger adhesion force on the negative PETEOS surface. Buffing step was necessary to clean off these positive ceria particles. When the separate platen for buffing is not available, we developed an on-platen buffing procedure to clean off the ceria particles. In this on-platen buffing processes, we found the DIW, ammonia and KOH cleaning solution is less effective but hydroxylamine and organic amine solutions are more effective.

# REFERENCE

[1] P. Feeney, C. Hawes, C. Baker, C. Schmidt, H. Chou, J. Hawkins, K. Moeggenborg, D. Patin, D. Mateja, and K. Harris, *Defect Issues in CMP*, 2002 Proceedings of seventh international chemical-mechanical planarization for ULSI multilevel interconnection conference (2002 CMP-MIC conference), p175. Feb.27 – March 1, 2002.

[2] P. B. Zantye, A. Kumar, and A.K. Sikder, *Chemical mechanical planarization for microelectronics applications*, Materials Science and Engineering, R 45, pp. 89–220, 2004

[3] A. Dietrich, A. Neubrand, and Y. Hirata, *Filtration behavior of nanoparticulate ceria slurries*, J. Am. Ceram. Soc., 85 [11] pp.2719–24, 2002

# STUDY ON WEIGHTED BINARY CLASSIFIER WITH IMBALANCED SEM DEFECT DATA

*Hairong Lei, Cho Teh, Lingling Pu, Qian Dong*
ASML, 80 W Tasman, San Jose, CA 95131 USA
hairong.lei@asml.com; hairong_lei@yahoo.com;

## ABSTRACT

Class imbalance classifier is a common and hard topic in semiconductor industry field because of the limited killing defect counts. Most of the recent research are focusing on data level methods. This paper presents some interesting observations using algorithm level method – the weighted binary (for simplicity) classifier. We would want to have the binary classifier heavily weight the few samples (for example hole bridge defects) that are available. The imbalanced data skew rate is in different ranges from 27% to 1% and to extreme ratio of 0.1%. Our binary classifier performance is evaluated using metrics of confusion matrix and the receiver operating characteristic curve (ROC) instead of only the accuracy.

## INTRODUCTION

Class imbalance is a common problem that has been comprehensively studied in classical machine learning, yet only limited systematic research is available in the context of deep learning, especially for SEM defect classification. Methods for addressing class imbalance can be divided into two main categories. The first category is data level methods that operate on training set ("Synthetic Minority Over-sampling Technique"- SMOTE or Over-sampling using GAN network) [1]. SMOTE's procedure or Over- sampling using GAN can be troublesome since it blindly generalizes the minority class without regard to the majority class. It is particularly problematic in the case of highly skewed class distributions since, in such cases, the minority class is very sparse with respect to the majority class, thus resulting in a greater chance of class mixture.

The other category covers classifier (algorithmic) level methods, which keeps the training dataset unchanged and adjust training or inference algorithms. In our study, we use the algorithm level method – the weighted binary classifier for simplicity (for example: Hole Bridge vs. Hole Normal).

## WEIGHTED IMBALANCED BINARY CLASSIFIER

### Problems and Solutions of Machine Learning Approach

In our case we don't have many positive samples to work with and we would want to have the binary classifier heavily weight the few samples that are available. We do this by passing CNN Keras weights for each class through a parameter for our SEM defect dataset. These will cause the model to "pay more attention" to the minority class.

Predictive accuracy, a popular choice for evaluating the performance of a binary classifier, might not be appropriate when the data is imbalanced. The imbalanced data skew rate of our experiments is in different ranges from 27% to 1% and even 0.1%. Our binary classifier performance is evaluated using metrics of the confusion matrix and the receiver operating characteristic curve (ROC) instead of only the accuracy.

### Class Weight Calculation and Apply

The goal is to identify SEM inspection killing defects, but we don't have many positive samples to work with, so we would want to have the classifier heavily weight the few samples that are available. We do this by passing Keras weights for each class through a parameter. These will cause the model to "pay more attention" to samples from an under-represented class. Using class_weights changes the range of the loss. This may affect the stability of the training depending on the optimizer. Optimizers whose step size is dependent on the magnitude of the gradient, like tf.keras.optimizers, SGD, may fail in this case. The optimizer used here, tf.keras.optimizers, Adam, is unaffected by the scaling change. Also note that because of the weighting, the total losses are not comparable between the two models (no weighted vs. weighted model)

```
# Scaling by total/2 helps keep the loss to a similar magnitude.
# The sum of the weights of all examples stays the same.
weight_for_0 = (1 / neg) * (total / 2.0)
weight_for_1 = (1 / pos) * (total / 2.0)

class_weight = {0: weight_for_0, 1: weight_for_1}

print('Weight for class 0: {:.2f}'.format(weight_for_0))
print('Weight for class 1: {:.2f}'.format(weight_for_1))
```
```
Weight for class 0: 0.50
Weight for class 1: 289.44
```

*Figure 1: Example of weight calculation for imbalanced binary classifier.*

### Evaluate Metrics

False negatives and false positives are samples that were incorrectly classified. True negatives and true positives are samples that were correctly classified. Accuracy is the percentage of examples correctly classified (true samples/total samples) while Precision is the percentage of predicted positives that were correctly classified (true positives/(true positives + false positives)). Recall is the percentage of actual positives that were correctly classified (true positives/(true positives + false negatives)). AUC refers to the Area Under the Curve of a Receiver Operating Characteristic curve (ROC-AUC). This metric is equal to the probability that a classifier will rank a random positive sample higher than a random negative sample, although some argue that it has nothing to do with classification.

Finally, AUPRC refers to Area Under the Curve of the Precision-Recall Curve. This metric computes precision-recall pairs for different probability thresholds. Apparently, accuracy is not a very helpful metric for imbalanced case. You can 99.9%+ accuracy on this task by predicting False all the time. We use a confusion matrix to summarize the actual vs. predicted labels, where the X axis is the predicted label and the Y axis is the actual label. Evaluate the model on the test dataset and display the results for the metrics we created. If the model had predicted everything perfectly, this would be a diagonal matrix where values off the main diagonal, indicating incorrect predictions, would be zero.

Besides confusion matrix we also use the ROC. This plot is useful because it shows the range of performance the model can reach just by tuning the output threshold. The AUPRC is the Area under the interpolated precision-recall curve, obtained by plotting (recall, precision) points for different values of the classification threshold. Depending on how it's calculated, AUPRC may be equivalent to the average precision of the model. We will not use this matric in our result analysis.

## EXPERIMENT RESULTS AND ANALYSIS
### Two Datasets and Weighted Model

For simplicity reason, we adopted the binary classifier for weighted class impact analysis. We use two sets of dataset for our experiments: one is called difficult data, which has hole-bridge defect and normal types (non-defect); the other is called easy data, which has Nu (nuisance type) and Type 7 (it is a bridge defect type). The later is relatively easy to classify as shown in Figure 2.

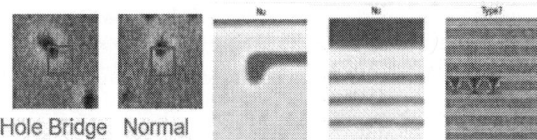

*Figure 2: Sample images for difficult data (Hole-Bridge vs. Normal) and easy data (Nu vs. Type7)*

As we mentioned previously, in our case we don't have as many of positive samples to work with. We would want to have the binary classifier heavily weight the few samples that are available. We force the model to "pay more attention" to the minority class using different class weight.

Our model is fit using a larger than default batch size of 8, this is important to ensure that each batch has a decent chance of containing a few positive samples. If the batch size was too small, they would likely have no critical defect to learn from. This model will not handle the class imbalance well. We know the dataset is imbalanced. One good suggestion is to use output layer's bias to reflect that. It supposed to help with initial convergence. The correct bias to set can be derived from: $b0= log_e(pos/neg)$. However we did not obverse the obvious benefit to set the bias and we do not set the bias in our experiments.

### Imbalanced Binary Classifier Result for Difficult Data:

The difficult data has 250 hole bridge defect type and 10000 normal type. We use 50/50 for training and validation. With weighted classes during model training, the classifier identifies 122 of 125 hole bridge defects and 5000 normal types while without weighted classes model only identifies 121 of 125 hole bridge defects and 5000 normal types as shown in Figure 3.

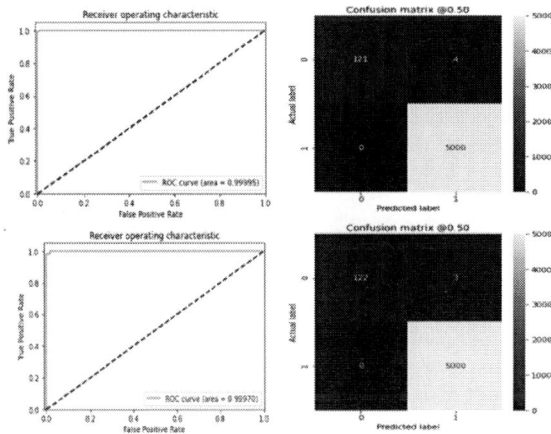

*Figure 3: Binary classifier without/with class weight ({0: 1, 1: 0.025}) for hole-bridge vs. Normal.*

**Imbalanced Binary Classifier Result for Easy Data:**
For easy data, we also use 50/50 for training and validation. We have three cases: 1) imbalance ratio is 27% (2000 Nuisance type and 542 Type7 defects); 2) imbalance ration is 1% (8000 Nu and 80 Type7 defects); and 3) imbalanced ratio is 0.1% (8000 Nu and 8 Type7 defects). Without class weight during model training, the case 1 (imbalance ratio of 27%) can handle the imbalanced easy dataset easily (1000/1000 Nu and 271/271 Type7). Both accuracy and purity are 100%.
For case 2 (imbalanced ratio of 1%) without class weight, the result is (3999/4000 Nu and 40/40 Type7).
For case 3 (imbalanced ratio of 0.1%) without class weight, the result is (4000/4000 Nu and 3/4 Type7).

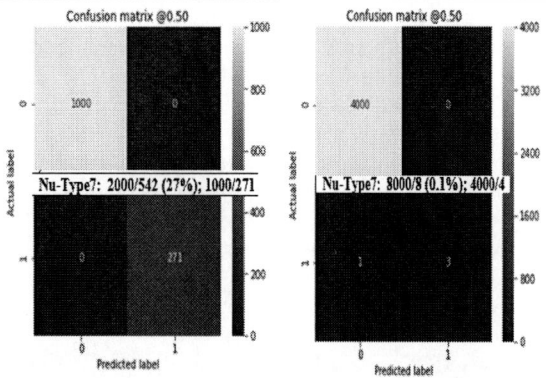

*Figure 4: Confusion matrix comparison for easy data with different imbalanced ratio (27% vs. 0.1%) no class weight.*

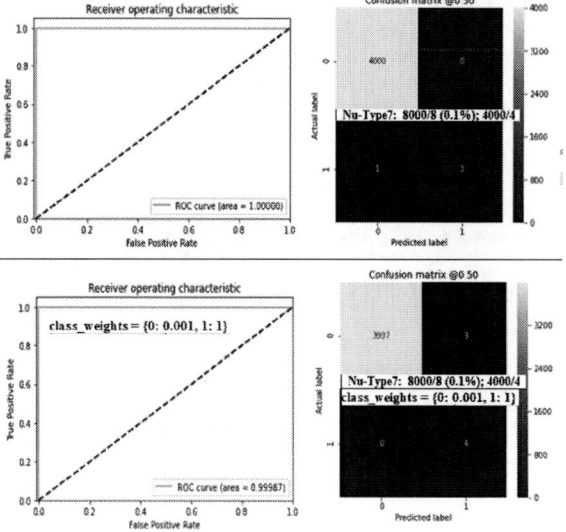

*Figure 5: Extremely imbalanced (0.1%) case: no class weight vs. class weight*

With class weight for case 3, the result is shown in Figure 5. We have 100% of the critical defect – Type7 while only have 3997/4000 Nu. Here you can see that with class weights, the ROC curve area is slightly smaller (0.99987 vs. 1.00000; the accuracy and precision are lower) because there are more false positives. Conversely the recall and AUC are higher because the model also found more true positives. Despite having lower accuracy, this model has higher recall (and identifies more critical defect). Of course, there is a cost to both types of error. Carefully consider the trade-offs between these different types of errors is important in real application. Figure 6 shows the accuracy, loss, and AUC for the case 3 (skew rate of 0.1%).

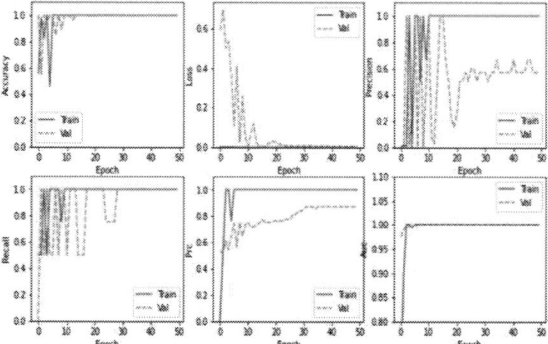

*Figure 6: Accuracy, Loss, and AUC for skew rate 0.1%*

## CONCLUSION

Imbalanced data classification is an inherently difficult task since there are so few samples to learn from. We should always start with the data first and do our best to collect as many samples as possible.

We did not compare the result with Over-sampling method. We showed that the weighted class binary classifier does help in finding more critical defects (type7 or hole bridge defect) without the hassle of manipulating data too much. Weighted class is observed to have more impact in extreme imbalanced case and difficult data case.

At some point the model for extreme skew rate (0.1%) may struggle to improve and yield the results we want, so it is important to keep in mind the context of problem and the tradeoffs between different types of errors.

## REFERENCES

[1] Hairong Lei, Cho Teh et al., "CONVOLUTIONAL NEURAL NETWORK (CNN) BASED AUTOMATED DEFECT CLASSIFICATION (ADC) WITH IMBALANCED DATA", CSTIC 2021, Shanghai, China.

# Advanced Modeling Techniques
# Expand the Applications of Picosecond Laser Acoustics for RF Process Monitoring

*Johnny Dai[1], Cheolkyu Kim[2], Robin Mair[1], Priya Mukundhan[1]*

[1]Onto Innovation, 550 Clark Drive, Budd Lake, NJ 07828, USA

[2]Onto Innovation, 16-6, Sunae-dong, Bundang-gu, Sungnam-si,Gyunggi-do, 3965 Korea

*Corresponding Author's Email: johnny.dai@ontoinnovation.com

## ABSTRACT

Picosecond Ultrasonics (PULSE[TM] technology) [1] is a first principle, rapid non-contact, non-destructive, technology, for the measurement of single layer and multilayer metal films in semiconductor process control. It has a strong footprint and is uniquely positioned as a tool-of-record for metal film thickness metrology in RF filter monitoring. In addition to thickness measurements, the technique can be used to characterize acoustic velocity values for dielectric and piezoelectric materials, which is a critical parameter for process control. We have previously reported on the advantages of PULSE technology for RF applications and its excellent performance to meet the stringent requirements for process monitoring and control [2].

Most of the RF applications involve multilayer metal stacks or films on oxide that are more intuitive and are easier to measure and model using our standard modeling algorithms. However, at the measurement wavelength of 522nm, measuring thinner metal films directly on Si substrate is challenging and often requires films on oxide or other films as the signal is complex and dominated by Si oscillations. Another situation in which modeling is challenging is when an $SiO_2$ film is present as a capping layer. The oxide film is included as part of the device stack to obtain a low thermal coefficient of the acoustic device. In a typical full stack, when the oxide capping layer is present over a multilayer stack of oxide/electrode/piezoelectric layer/electrode/Si, PULSE measurement signal is a convolution of $SiO_2$ oscillations and echoes from the multilayer stacks (most often three or more layers).

To expand the application space of the PULSE technology and improve its performance, we have developed advanced modeling techniques for the deconvolution of different components of the signal to reliably model parameters of interest. In this paper, we review one of the approaches we have successfully used to improve sensitivity, accuracy and robustness without impacting the repeatability needed for high-volume manufacturing.

## INTRODUCTION

A recent study shows that RF filter market is going to grow steadily by nearly $16 billion from 2020 to 2024 at a compound annual growth rate (CAGR) of approximately 20% [3]. The increased adoption of 5G technology in IoT requires more complex, smaller, higher-frequency RF filters. Multilayer metal/dielectric/metal stacks are common in RF manufacturing. The ability to measure multilayer stacks simultaneously eliminates the need to measure on monitor wafers and provides direct feedback for process monitoring

and control. Metrology systems used for in-line process monitoring and control are required to measure the full stack (top electrode/piezo layer/bottom electrode) with excellent repeatability and long-term stability. A typical sampling includes several hundred points across the wafer, which requires a high throughput system to make it production worthy.

PULSE technology is a non-contact, non-destructive first principle technique. It is capable of measuring metal films from 50Å to 20µm with the option to extend to 35µm. The laser beam is focused to a tight spot (8µm x10µm) on the wafer, enabling direct measurements on devices (15µm). Measurements take a few seconds per site; the high throughput allows mapping of the whole wafer in minutes. Measurements are used to feed forward to the trimming process to adjust the center frequency and improve across-wafer and wafer-to-wafer variability.

We have previously discussed [2] the benefits of PULSE technology for the measurements of multilayer metal thickness simultaneously. The capability to measure acoustic velocity of dielectric and piezoelectric films offers significant insight and is directly correlated to device performance. In order to measure sound velocity and thickness, films need to be >5,000Å and have a good metal transducer film on top or below the dielectric films to generate a strong signal.

It is challenging when the films are directly on Si, especially at the current measurement wavelength of 522nm. While measuring select applications, such as thin metal films on Si, or a $SiO_2$ cap layer on a full stack of surface acoustic wave (SAW) or film bulk acoustic resonator (FBAR) device, the signal is quite complex and difficult to model using conventional modeling algorithms.

## APPLICATION OF ADVANCED MODELING TECHNIQUE TO RF APPLICATIONS

In this paper, we demonstrate how we have developed and applied advanced modeling techniques to overcome some of the limitations we identified in the preceding section. We discuss a few specific use cases where the technique has proven to be promising.

### Thin metal film measurements on Si substrate

At a measurement wavelength of 522nm, a thin single-layer metal film (200Å or less) on silicon substrate is dominated by Si oscillations and cannot be modeled using currently available techniques. Fig. 1 shows the measurement signal (change in reflectivity vs. delay time in picoseconds) from Mo films on Si substrate ranging in thickness from 50Å to 300Å. The measurement signal is dominated by Si

978-1-6654-9759-6/22 $31.00 © 2022 IEEE

oscillations, with a period of ~7.5ps. In this case, it is hard to use traditional modeling techniques to report the thickness of Mo films. Using advanced filtering techniques, the data has been reprocessed to eliminate Si oscillations. Fig. 2(a) and Fig. 2(b) are plots of the post-processed data; the maxima in the data correspond to the acoustic echoes that are used in the determination of film thickness.

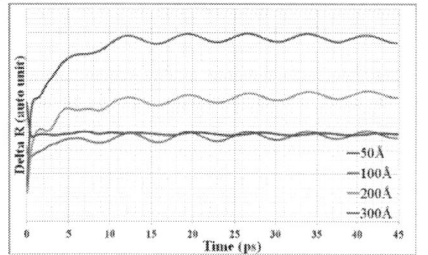

*Fig. 1. PULSE measurement for Mo films with thickness of 50Å, 100Å, 200Å, and 300Å on Si substrate.*

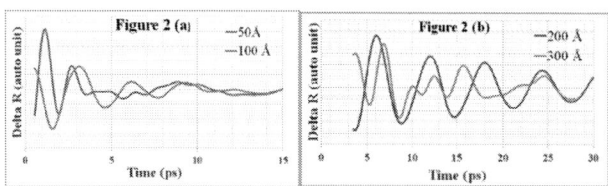

*Fig. 2 (a). Signal from 50Å and 100Å Mo films post-processing, (b). Signal from 200Å and 300Å Mo films post-processing*

To test the sensitivity of this approach, a design-of-experiments (DOE) skew of 12 thin Mo wafers was created from 50Å to 300Å. Fig. 3 is a correlation plot between thickness measured using the picosecond technique vs. reference data. We see excellent linear correlation with respect to the reference data and the correlation coefficient, $R^2 \sim 1.0$. We tested the repeatability performance of the films by performing load/unload measurements across 13 sites. Typical repeatability performance ($3\sigma$) is better than 1% for the thinner films and 0.5% for >150Å films.

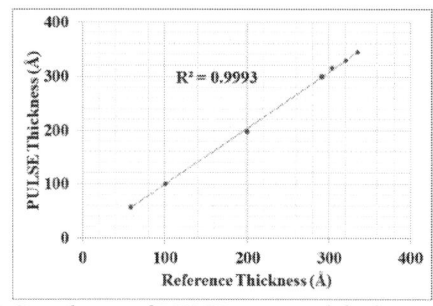

*Fig. 3. Correlation of PULSE measured thickness with the reference value*

## Multilayer film stack with SiO₂ cap layer

$SiO_2$ has been widely used to obtain low thermal coefficients in the manufacturing of Bulk Acoustic Wave (BAW)/Film Bulk Acoustic Resonator (FBAR)/Surface Acoustic Wave devices (SAW). The accurate control of thickness for the full stack, including the $SiO_2$ cap is the key to frequency control for RF filters, and the center frequency relies on tighter thickness control for every layer in the stack, although the piezoelectric layer plays the most critical role.

Fig. 4 shows overlay plots of measurement from a FBAR device with $SiO_2$ capping film. The blue curve represents the raw measurement signal. Green and red curves represent the background subtracted and post-processed signals, respectively. The measurement signal and background subtracted signals are both complex; it is challenging to model using the existing method with a theoretical model to fit the echo locations and model all the layers. However, after applying some of the advanced filter techniques, we were able to decouple the relevant features in the signal from the oscillations, and the subsequent data can then be modeled. Using this approach, we report the thickness of all the four layers in the stack: top $SiO_2$/top electrode A/dielectric AlN/bottom electrode B.

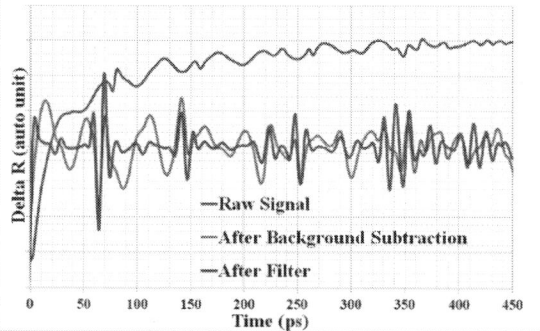

*Fig. 4. Measurement signal from a FBAR device with $SiO_2$ film using picosecond ultrasonics.*

In our previous report [2], we demonstrated excellent repeatability for multilayer measurement using PULSE technology. In Fig. 5, we show the correlation of thickness from PULSE technology to reference thickness. Since top layer $SiO_2$ is constant, we only show three layers, top electrode A (Fig 5a), piezoelectric layer (Fig 5b) and bottom electrode (Fig 5c). The plots show the excellent linear correlation of the PULSE technique to the reference data.

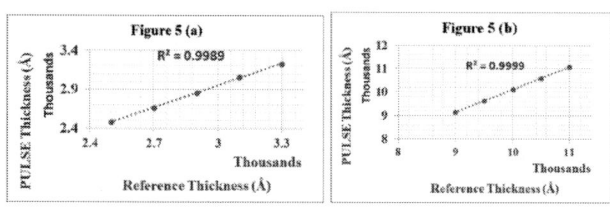

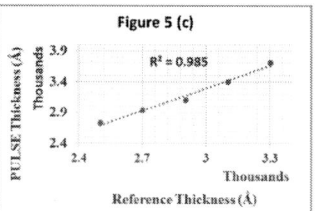

*Fig. 5. Thickness correlation between PULSE and the reference thickness for a full stack of (a). Top electrode film; (b). Piezoelectric film; (c). Bottom electrode film*

## Simultaneous measurement of thickness and sound velocity for challenging applications

978-1-6654-9759-6/22 $31.00 © 2022 IEEE

Simultaneous measurement of both thickness and sound velocity of dielectric/piezoelectric layers is one of the key advantages of PULSE technology. For films, such as oxide or AlN films on silicon or other metal transducers, the opaque substrate absorbs energy from the pump pulse, launching a sound wave that travels up through the transparent film at the speed of sound. The strain causes a local change in the index of refraction of the film. The partial reflection of the probe beam from the moving sound wave, combined with the partial reflection from the film surface, leads to destructive and constructive interference at the detector [4]. As a result of this time-dependent interference, the measured signal oscillates with a period, $\tau$, from which the sound velocity ($V$) in the material can be determined by

$$V = \frac{\lambda}{2n\tau cos\varphi}$$

where $n$ is the index of refraction, $\lambda$ is the wavelength and $\varphi$ is the angle of refraction. We have shown typical examples in our earlier paper [2, 5] on PULSE measurement for thickness and sound velocity that relies on Brillouin oscillations to simultaneous measure sound velocity and thickness.

In some cases, we do not see a strong oscillation from oxide or piezoelectric films either because the layer is too thin (<5,000 Å) or sound waves generated from under layer is low. For example, in the case of $SiO_2/Si$, the measurement signal is weak, and as explained earlier, the signal is clearly dominated by the oscillations from the underlying Si substrate. Oscillations from $SiO_2$ are not obvious for sound velocity measurement.

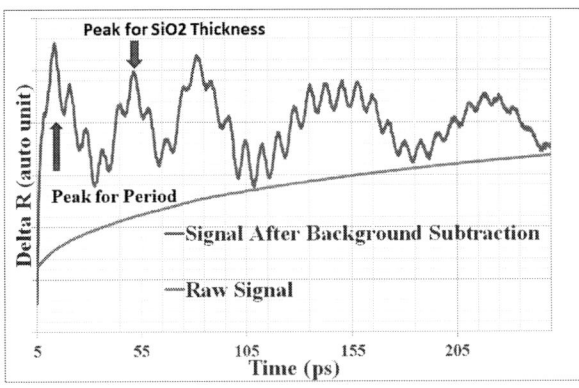

*Fig. 6. Raw measurement signal (green) and signal after background subtraction (blue).*

Fig. 6 shows raw measurement signal and signal after background subtraction. The green curve represents the as-acquired data; prior to background subtraction, it is impossible to see any of the relevant features in the signal. After background subtraction, shown as the blue curve, the signal is dominated by Si oscillation. As shown in Fig. 6, we use the first arrow to identify the peak to calculate the period of the oscillations; this is used for sound velocity measurement, and the second arrow indicates the peak for the echo from $SiO_2$ and Si interface used for $SiO_2$ thickness measurement.

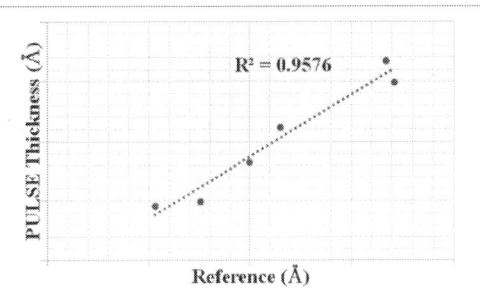

*Fig. 7. Correlation of thickness from PULSE measurement to the reference value.*

Fig. 7 shows the correlation of thickness from PULSE measurement to the reference thickness. We have excellent correlation, with a correlation coefficient of $R^2 \sim 0.96$. The thickness in the above plot is over a very narrow range of 12Å, and we expect the correlation to improve over a wider range. The algorithms developed and demonstrated in this paper have shown that sensitivity and repeatability are more than adequate for applications in production.

## CONCLUSIONS

RF filter process control requires stringent metrology due to tight process tolerances. PULSE technology is a workhorse and has carved out a unique space in the RF process monitoring space due to the numerous technical advantages. PULSE technology can measure both thickness and sound velocity for $SiO_2$ and AlN with excellent gage capable of repeatability and reproducibility. With the implementation of these new advanced modeling techniques, we have addressed some of the limitations of the existing modeling approaches and further expanded the application's capability. The advanced modeling techniques developed here have expanded the capability of the technique to other device segments as well and is especially beneficial during process chamber qualifications.

## REFERENCES

[1] C. Thomsen, H. T. Grahn, H. J. Maris, J. Tauc, Phys. Rev. B, vol. 34, 1986, pp. 4129-4138

[2] J. Dai, Johnny Mu, Cheolkyu Kim, Priya Mukundhan, March 14-15, 2021 CSTIC, Shanghai, China

[3]. https://www.everythingrf.com/news/details/11259-rf-filter-market-expected-to-grow-steadily-to-reach-15-billion-by-2024

[4] J. L. Arlein, S. E. M. Palaich, B. C. Daly, P. Subramonium, and G. A. Antonelli, J. of Appl. Physics, vol 104, 2008, pp 033508 1-6

[5] J. Dai, P. Mukundhan, R. Mair, M. Mehendale, C. Wang, E. Wang, C. Kim, March 14-15, 2020 CSTIC, Shanghai, China

# ALL SIDE OPTICAL CHIPPING INSPECTION OF DICE IN A DIE ATTACH MACHINE AT HIGH THROUGHPUT

*Norbert Ackerl\*, Anton Rigner, Andreas Wiedmer, Rudolf Grüter, and Katharina Schmeing*
Besi Switzerland AG, Hinterbergstrasse 32a, 6312 Steinhausen, Switzerland
*Corresponding Author's Email: norbert.ackerl@besi.com

## ABSTRACT

A full optical inspection of dice for electronic components with high reliability needs is proposed. All sides are assessed checking for defects possibly trashing bad dice to mitigate fatal error in operation. The inspection sequence is optimized using an adjustable lens for focusing, thus allowing more than 5'000 units per hour output at high optical resolution better than 6μm/pixel. This study presents detected defects originating from the dicing process using cutting, laser full ablation, and stealth dicing. Distinct features at the sides captured with the up-looking 6 side module are correlated with micrographs proving the capability of this inspection routine.

## INTRODUCTION

In recent years, the interest in a full inspection of all dice in the backend processing chain increased. An automated optical inspection (AOI) with high speed is a viable non-destructive technique used [1]. Singulation processes like wafer sawing leading to mechanical stress might introduce defects. Thin dice are especially prone to cracking, and a sawing process can result in additional chipping [2]. Using die attach film (DAF) substituting the liquid epoxy introduces additional challenges, where the DAF can overlay defects at the bottom [3]. While there are solutions to assess front and backside chipping [4], a fast AOI of all 6 sides for die attach machines is lacking. First concepts have been proposed [5] [6], however, the need in resolution, throughput, and flexibility for varying die sizes is not satisfied.

Here, the latest up-looking vision 6 side (UVI6S) module from Besi [7] is presented in figure 1 with the

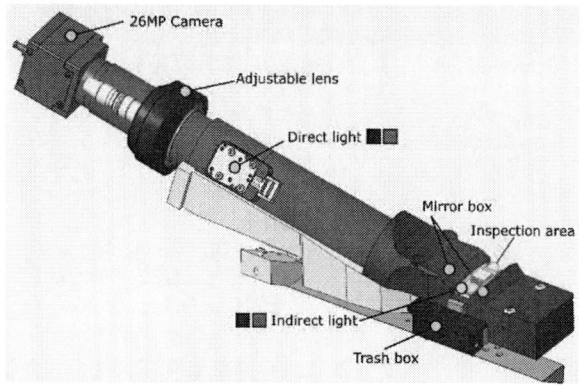

*Figure 1: UVI6S module using AOI available for the Besi Esec 2100 sD advanced [i] die attach machines.*

associated process sequence shown in figure 2. The concept uses one pair of 45° tilted mirrors in a 9x9mm² box for the side inspection combined with various illumination options. An adjustable lens enables rapid refocusing and the die can rotate 90° as shown in the figure allowing an inspection of the other two sides depicted in figure 2B. The module is mounted between the pick and bond position facilitating the AOI and bad dies can be trashed.

This study focuses on the defects introduced by different dicing methods on bare silicon wafers investigating characteristic features on all sides. The capability of the UVI6S module in terms of resolution is assessed and compared with microscopy. In addition, the throughput given in units per hour (UPH) is investigated and an estimation for different die sizes provided.

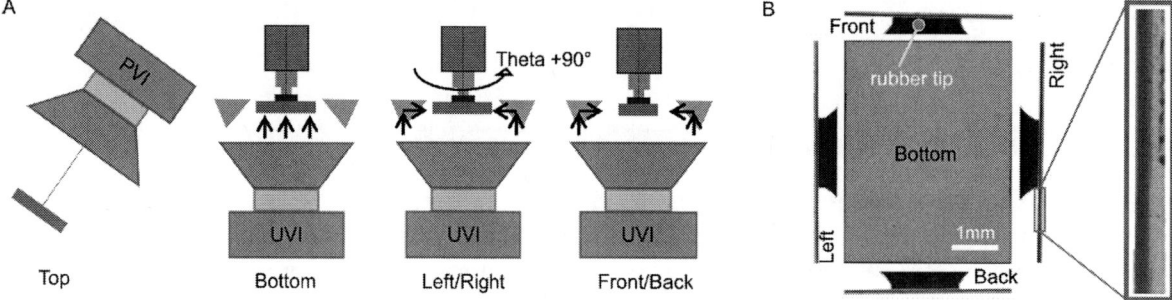

*Figure 2: A: Sequence for inspecting all sides of a die. The topside is captured before pick with the pick vision (PVI) and the die transported to the UVI6S inspection site. The bottom is aligned to be centered within the mirror box and the first two sides assessed before rotating the die by 90°, refocusing and inspecting the second sides. B: Images captured with the UVI 6S module and zoom on chipping defects on the right side of a 100μm thin die.*

## SETUP, PROCESS, AND MATERIAL

A Besi Esec 2100 sD *advanced* [i] die attach machine with the UVI6S inspection module serves as experimental testbed. The module is equipped with individually adjustable dual wavelength illumination of blue and red for the detection shining direct and indirect light on the bottom and sides of the die. Before pickup, the top inspection assesses edge and corner defects, see figure 2A. A defect-free die is transported to the UVI6S site. The bottom is examined and the die position is aligned for later correction at the bond site. Following, the die is centered in the module. Left and right images are recorded and the die is rotated by 90°. Next, the front and back side (not to be confused with bottom and top) is evaluated by eventually refocusing the adjustable lens for non-square dies. If one edge is bigger than 6mm, a lift-up z-move rotates the die above the module mitigating crashes at the mirrors.

### Detection algorithm

The detection algorithm consists of a sequence of image processing steps. In short, the grabbed image from the UVI6S undergoes feature detection, adaptive thresholding and segmentation steps, summed up in figure 3. Within the set region of interest (ROI) a Gaussian filter is applied to smoothen out high frequency noise in the image. A Sobel gradient filter emphasizes the edges of interest, removing possible cutting marks at a selectable angle (3B). Following, the greyscale image with applied threshold is segmented and a blob detection is carried out after exclusion of selected regions like the DAF. A final clustering step groups the detections together spatially to represent the defect and compare its length and depth against the specifications. All algorithmic steps are controlled by a set of selectable parameters defining the inspection area plus depth and regions to neglect. Especially, the Sobel gradient angle, scale factor, threshold intensity, minimal chipping area, and the clustering distance allow setting the detection to the

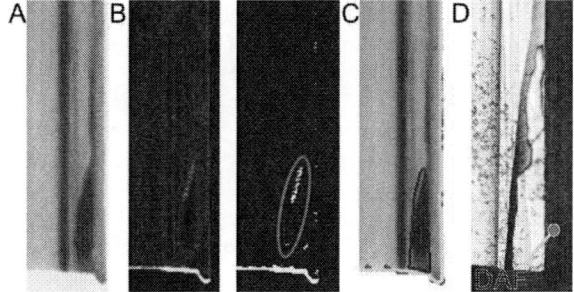

*Figure 3: Detection steps on an exemplary MEMS die with DAF (A). A filtering, masking, edge and blob detection with clustering in red (B) step resolves chipping (C). The detected size correlates well with the micrograph (D).*

specification and material properties. Exemplarily, a detected chipping stuck on the DAF of a MEMS die is shown in figure 3C compared with a microscopy image in D unraveling the detection capability.

### Material

A 100μm thick 8" Silicon Prime Wafer from DISCO HI-TEC EUROPE GmbH with ELEGRIP tape UHP-0805MC is used. The dice are singulated to 3.6 x 4.68mm² using Disco machines by means of diamond blade cutting with a ZH05-SD3500- N1-110 CC and NBC-ZH 127F-SE KO2 for single and step cut accordingly. In addition, laser full ablation and laser stealth dicing (SD) are inspected with details given in table 1.

*Table 1: Singulation process details of the used dicing step.*

| Process | Machine | Settings |
|---|---|---|
| Single cut | DFD6340 | Z1: 60k rpm, 30mm/s |
| Step cut | DFD6340 | Z2: 35k rpm, 30mm/s |
| Laser full cut | DFL7340 | UV laser, 300mm/s |
| Laser SD | DFL7340 | NIR laser, 300mm/s |

## RESULTS AND DISCUSSION

### Detected defects and comparison with microscopy

The differently singulated dice show typical marks on the sides with an overview provided in figure 4.

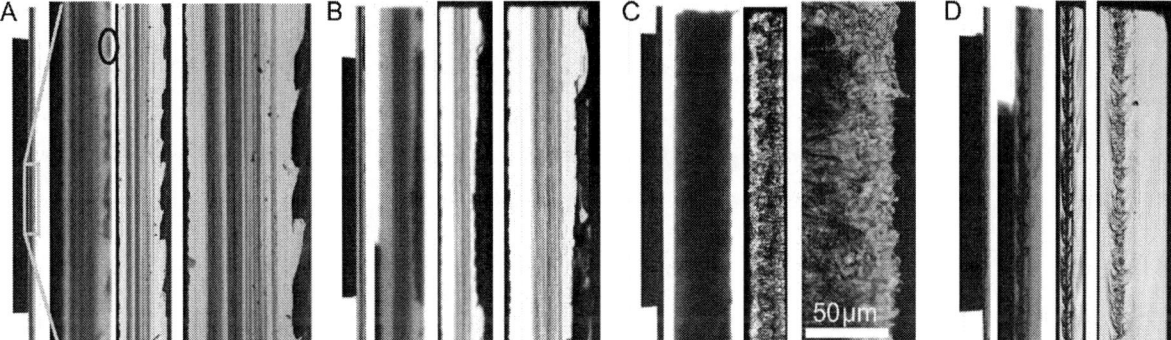

*Figure 4: Side view comparing the UVI6S images (first 2 graphs) with microscopy of 100μm thick silicon dies singulated by A: single, B: step cut, C: laser full cut, and D: laser stealth dicing. Defects like chipping occur for single and step cut and the laser cut reveals high roughness and melt ejection. SD shows no classic defects, but the edges can cleave tilted*

Especially, for standard disc cutting, a higher amount of chipping defects for the last cutting path occurs. The die can move more freely as seen in the samples of figure 4A and B. In contrast, the laser processes shown in figure 4C/D reveal a different topography. The full laser cut, as a thermal process, leads to a molten surface with ejected material at the bottom of high roughness. These surfaces revealed a high absorption not showing much detail at the sides. Nevertheless, the roughness of the inspected edges could be used as criterion if necessary, to define a maximal deviation. The SD edges are well defined and the typical subsurface structures from the focused NIR laser beam are visible forming the modification zone for defined cleaving. However, at the edge of the wafer the cleavage can lead to defects and tilted edges originating from the non-homogeneous tape expansion. The proposed AOI UVI6S module resolves different defects and enables to sort out bad dies. Defects smaller than 15µm are reproducibly detected with an example of high contrast marked blue in figure 4A. The micrographs point to a reliable detection of below 50µm thin dies. Depending on the defect and dicing process, the biggest challenge is a good illumination and robust settings for the detection algorithm. With the combination of direct and indirect light a multitude of settings are feasible.

**Production speed at full optical inspection**

Besides the capability of detecting defects, the AOI must be fast to keep the productivity as high as possible. Therefore, the inspection cycle time and movements to the inspection site are optimized for different scenarios like alternate 0/180° dies on the wafer or changing bonding positions. Generally, the full process time and reachable UPH is governed by the necessary pick and bond times, which are process dependent. The sprint throughput gives a measure on machine capability and in figure 5 the loss in

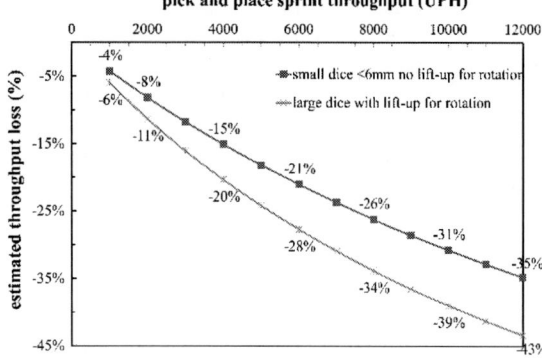

*Figure 5: Sprint UPH loss from the pick and place system for AOI of all dice with and without z lift-up move.*

terms of UPH points to a theoretical maximum of 7800. Two cases are calculated, where a die size below 6mm rotates inside the unit and bigger dice above. This adds

mechanical axes moves leading to a UPH loss. Within the first trials 5100 UPH have been reached at a pick and bond time of 50ms pointing to a fast real production speed. However, the inspection time can vary strongly depending on the necessary illumination to resolve the defects and the die size.

## CONCLUSIONS

The UVI6S module enables an AOI of every die on a die attach machine enhancing the reliability of the full package. Detection of defects down to 15um proved to be feasible with a high UPH. Four dicing techniques and the resulting sides have been investigated, where chipping occurred for sawing majorly from the last cut. Laser full cut and stealth dicing show characteristic marks, which can be used to define a defect specification if needed.

## ACKNOWLEDGEMENTS

N.A. wants to thank the whole UVI6S team, without their effort, this contribution would not have happened. Many thanks to Ernst Barmettler, Marco Hug, Enda McCague, Ralf Weise, Pascal Bazelli, Andre Knecht, Axel Schiers, and Bruno Hausheer. The support of Charles Galea to invest time on this topic is highly appreciated.

## REFERENCES

[1] A. Ebayyeh and A. Mousavi, "A Review and Analysis of Automatic Optical Inspection and Quality Monitoring Methods in Electronics Industry," *IEEE Access*, vol. 8, pp. 183192–183271, 2020.

[2] W. H. Teh, D. S. Boning, and R. E. Welsch, "Multi-Strata Stealth Dicing Before Grinding for Singulation-Defects Elimination and Die Strength Enhancement: Experiment and Simulation," *IEEE Trans. Semicond. Manuf.*, vol. 28, no. 3, pp. 408–423, Aug. 2015.

[3] H. Baharuddin, "Step cut for dicing laminated wafer in a QFN package," *Solid State Sci. Technol.*, vol. 16, no. 2, pp. 198–206, 2008.

[4] V. Perminov, V. Putrolaynen, M. Belyaev, E. Pasko, and K. Balashkov, "Automated image analysis for evaluation of wafer backside chipping," *Int. J. Adv. Manuf. Technol.*, vol. 99, no. 5–8, pp. 2015–2023, Nov. 2018.

[5] A. Salvi, "Dispositif optique et module d'inspection," EP1602001B1, May 27, 2009.

[6] M.-F. Chen, C.-W. Chen, C.-Y. Chen, C.-H. Hwang, and L.-Y. Hwang, "An AOI system development for inspecting defects on 6 surfaces of chips," in *2016 IEEE International Instrumentation and Measurement Technology Conference Proceedings*, Taipei, Taiwan, May 2016, pp. 1–6.

[7] R. Weise, "Apparatus and method for optical inspecting three or more sides of a component", patent pending.

# REVIEW OF MICRO- AND NANOPROBE METROLOGY FOR DIRECT ELECTRICAL MEASUREMENTS ON PRODUCT WAFERS

*Benny Guralnik[1,2*], Peter F. Nielsen[1], Dirch H. Petersen[2], Ole Hansen[2], Lior Shiv[1],*
*Wilson Wei[1], Thomas A. Marangoni[2], Jonas D. Buron[1], Frederik W. Østerberg[1],*
*Rong Lin[1], Henrik H. Henrichsen[1], Mikkel F. Hansen[1]*

[1]CAPRES – a KLA company, Diplomvej 373B, 2800 Kgs. Lyngby, Denmark
[2]Technical University of Denmark, 2800 Kgs. Lyngby, Denmark
*Corresponding Author's Email: benny.guralnik@gmail.com

## ABSTRACT

At the nanoscale, the electrical resistivity of solids is strongly and nonlinearly affected by their chemistry, crystallography, and geometry (e.g., critical dimensions). To achieve on-spec performance of semiconductor devices, an exceptional process control is thus essential. Four-terminal sensing is a well-established electric metrology, where resistivity is obtained from applying a known current across one pair of electrodes in contact with the sample, while measuring the voltage drop across another. Thanks to microfabrication, the downscaled Micro Four-Point Probes (M4PP) are characterized by (sub-)μm inter-electrode spacing, which enables to accurately determine resistivity on comparable length scales, while reducing the risk of current leakage through adjacent layers/devices. In addition to electrical resistivity (typically determined at a <0.1% precision), other key transport parameters can often be concurrently quantified (e.g., Hall carrier density and mobility, the temperature coefficient of resistance, and certain thermoelectric parameters). Here, we review milestones in M4PP development, showcase its characteristic use for in-line process monitoring of product wafers, and flag recent methodological improvements and advances.

## IN-SITU FOUR TERMINAL SENSING ON THE MICROSCALE

Four-terminal measurements based on the van der Pauw theorem [1] are widely used in the semiconductor industry to monitor the deposition and annealing processes of metal and semiconductor thin films on the blanket wafer level. However, with a typical pitch of ~1 mm, and a contact force of ~1 N per electrode, most macroscopic four-point probes [e.g., 2] are rarely suitable for measurements on product wafers due to their large footprint, and the ensuing surface damage to the sample. In comparison, M4PPs have a typical pitch of 8 μm, and contact force of ~10 μN per electrode. When such probes are landed obliquely (at 30° to the probed surface), they exert minimal impact (if at all) on the probed surface, making M4PPs suitable for landing on metrology test pads located in scribe lines of product wafers. Another benefit of M4PP miniature size and gentle landing is that it reduces the risk of current leakage in thin multilayer samples through conductive layers below the layer of interest [3]. Key milestones in the miniaturization and automation of four-point probing at the microscale by CAPRES include:

a)  the original straight cantilever design [4] (Fig. 1a),
b)  vibration-tolerant L-shaped cantilevers [5] (Fig. 1b),
c)  integrated surface detection (Fig. 1b-d) and automated probe change [6],
d)  multi-cantilever (Figs. 1b-c) and non-equidistant (Fig. 1c) probes [7] down to sub-μm pitch [8], utilizing multi-electrode probing algorithms that eliminate geometric uncertainties [8-10], and
e)  custom probe designs enabling the nondestructive probing of particularly fragile 3D nanostructures such as fins (Fig. 1d).

Current mainstream M4PP applications include the process control of ultra-shallow junctions [11], and in-line production monitoring of Magnetic Random-Access Memory (MRAM) devices [12].

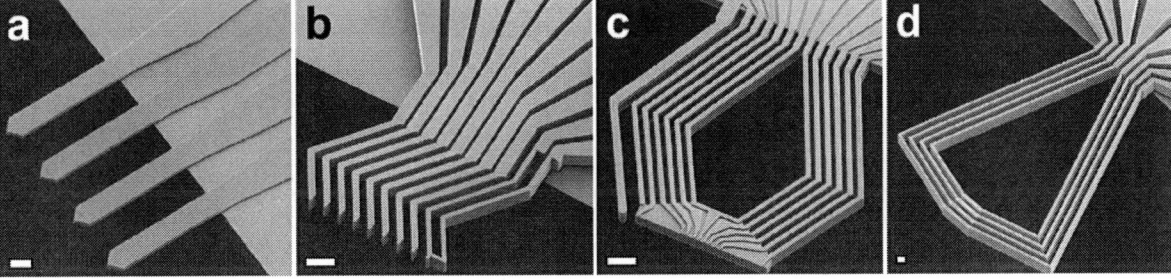

*Figure 1: Evolution of CAPRES M4PPs during 2000–2020 (SEM images of the probes' contact surface facing up; bars are ~5 μm each): (a) the original, straight-cantilever, four-point probe design, (b) multi-cantilever, vibration-tolerant design, featuring a strain gauge for surface detection (the rightmost loop cantilever), (c) extreme small-pitch, non-equidistant, multi-cantilever probe (notice electrodes on top of a common, electrically-isolated, supporting plate), and (d) loop probe, for perpendicular engage on fragile elongated arrays of interconnects or fins.*

978-1-6654-9759-6/22 $31.00 © 2022 IEEE

## SHEET RESISTANCE AND HALL CARRIER DENSITY AND MOBILITY

By applying a magnetic flux density normal to the surface of a conducting thin film, M4PP can be used to determine three key electromagnetic parameters of the thin film [13], namely:

a) Sheet resistance, $R_{sq}$,
b) Hall sheet carrier density, $N_{HS}$, and
c) Hall carrier mobility, $\mu_H$.

In the seminal study, the isolation of the magnetoresistive component required measurements at multiple locations approaching a straight boundary to a nonconductive medium [13]. Later, the measurement time was dramatically reduced via a multi-cantilever sampling at a single landing point in the vicinity (a few µm) of the half-plane boundary [9–10]. Most recently, and by utilizing the conformal mapping technique, accurate Hall effect measurements were demonstrated on rectangular pads of arbitrary dimensions, some as small as $70\times70$ µm² [14]; the latter technique has been accordingly termed by CAPRES as the Micro Hall Effect (MHE).

Here, we present the next generation of the CAPRES Micro Hall Effect module (MHE2; red symbols in Fig. 2a-c), which significantly improves on the original methodology (MHE; blue symbols in Fig. 2a-c) both in terms of better reproducibility (Fig. 2a), as well as reduced test pad dimensions (Fig. 2b). The first aspect of MHE2 has to do with a new design of experiment and of the data regression algorithm, which adopts the multi-cantilever geometric correction developed for M4PP Current-In-Plane Tunneling measurements [8]. Fig. 2a shows 25-point repeatability measurements performed using the L7PP probe [9] on a 25 nm thick $Si_{0.6}Ge_{0.4}$ thin film, benchmarking MHE against its successor MHE2. Both approaches yield matching (within their respective uncertainties) estimates of the sheet resistance (116.54 $\Omega/\square$), Hall sheet carrier density ($3.32\times10^{15}$ cm$^{-2}$) and Hall carrier mobility (16.15 cm V$^{-1}$ s$^{-1}$), whilst the relative standard errors of the updated algorithm (MHE2) are two to five times lower, close or below the 0.1% limit.

The second improvement is due to a redesigned and twice as small probe (Fig. 1b), that can obtain reproducible measurements on significantly smaller test pads. Fig. 2b

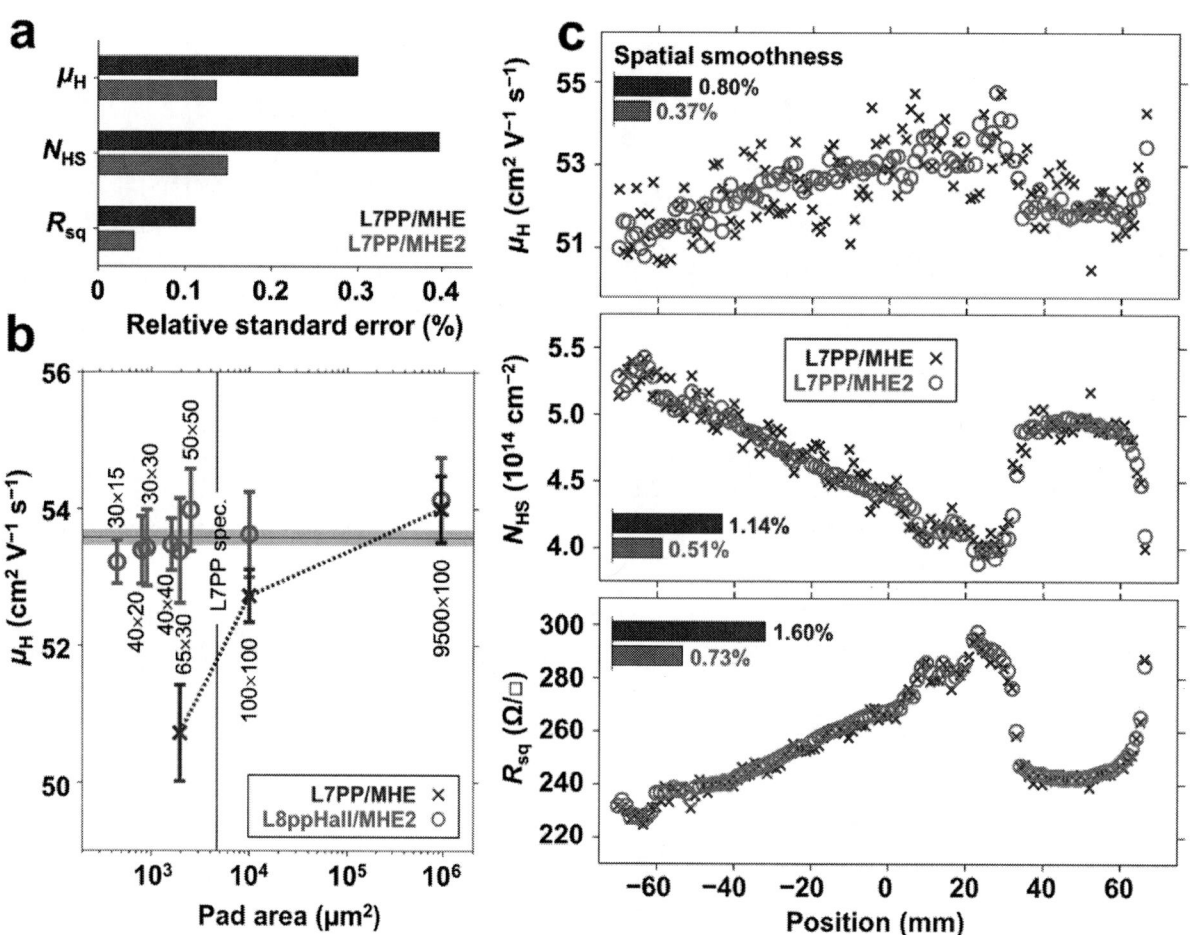

*Figure 2: Next generation M4PP Micro Hall Effect measurements (MHE2), compared with its predecessor (MHE): (a) a two- to fivefold increase in reproducibility (N=25) via algorithm improvement, (b) elimination of pad size dependence, alongside a tenfold reduction in measurable pad size (pad dimensions in µm²), and (c) a 150 mm patterned wafer transect, showing a twofold increase in spatial smoothness (reduction of unaccounted spatial variance).*

compares Micro Hall Effect measurements ($N$=5 for each data point; error bars correspond to 1 standard error) of rectangular pads (ranging from 30×15 to 9450×100 μm² in size) of a patterned Boron-doped Si wafer (B(11)⁺ implanted at 2 keV to $10^{15}$ cm⁻²; annealed at 1100 °C for 30 s). One subset of the measurements (crosses in Fig. 2b) was carried out using the larger L7PP probe [9] and analyzed using the old MHE algorithm; the other subset (circles in Fig. 2b) was collected using the new L8ppHall probe (Fig. 1b) and analyzed via MHE2. In combination, the new probe and algorithm (L8ppHall/MHE2) demonstrate reproducible measurements and narrow uncertainties over 4 orders of magnitude of test pad area, exhibiting no pad size dependence (as does the L7PP/MHE combination beyond the probe's spec limits, cf. vertical line in Fig. 2b), and enabling accurate measurements of ca. tenfold smaller structures than previously possible.

To give an idea how Micro Hall Effect measurements on ~200−300 μm² test pads located in the scribe lines of a product wafer, could be used for in-line process control, Fig. 2c shows a line scan of sheet resistance and Micro Hall Effect measurements along the diameter of a 150 mm wafer (same sample as in Fig. 2b). The algorithm performance (MHE vs. MHE2) is evaluated through spatial smoothness, which we arbitrarily define here as the median of the relative differences across all pairs of neighboring sites ($N$=125 sites spaced 1 mm apart), i.e. median $[2|y_i - y_{i+1}|/(y_i + y_{i+1})]$, $i \in \{1, ..., N-1\}$. The MHE2 algorithm is not only visually smoother but yields a quantitative improvement by a factor of ~2.2 in all the regressed parameters (inset bar charts in Fig. 2c). Together with better reproducibility (Fig. 2a), we believe that this adds up to an order-of-magnitude increase in the method's spatial sensitivity (spatial resolving power), thus opening up new opportunities for in-line monitoring of process inhomogeneities on a tenfold finer scale than previously attainable.

## CRITICAL DIMENSIONS, THERMAL EFFECTS, AND COMPLEMENTING CONTACTLESS THZ METROLOGY

Four recent M4PP applications (Fig. 3a-d), which extend the familiar domain of four-point electrical metrology, include:

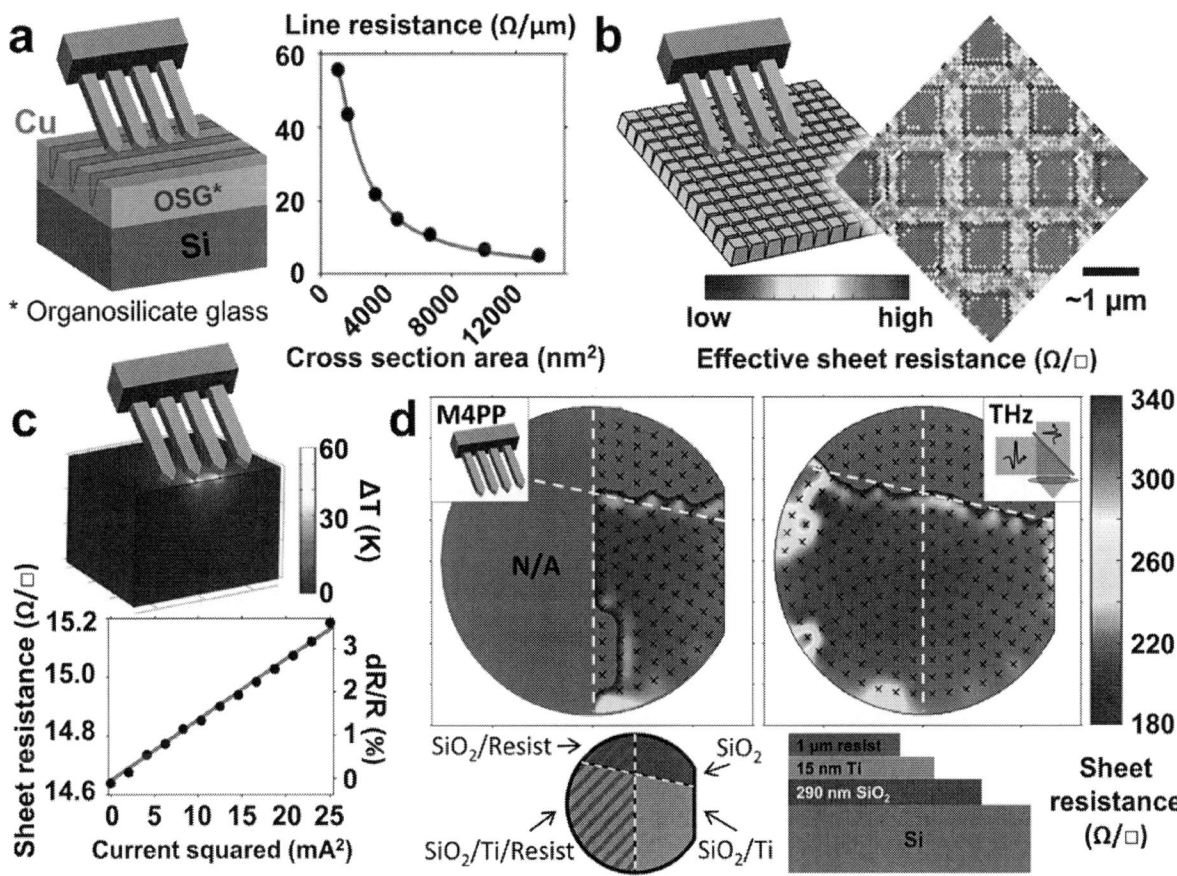

*Figure 3: Recent CAPRES M4PP applications to measure on (a) nanowires, (b) doubly-periodic USJ, (c) TCR, and (d) buried metal layers (left: M4PP reference on exposed metal only; right: non-contact terahertz metrology seamlessly probing both exposed and buried metal; bottom: a color-matching top-view and cross-section of the wafer).*

a) Non-destructive characterization of line resistance of nanoscale interconnects (Fig. 3a) as small as ~1000 $nm^2$ in cross section (~16 nm width). Small pads (50×50 µm) of tightly pitched interconnects, placed in the scribe lines of product wafers, can be used for performing in-line quality control measurements immediately after each successive metallization layer for process review and control, ultimately eliminating the need for monitor wafers.

b) Sheet resistance metrology of doubly-periodic square nanocomposites whose constituents have contrasting resistivities [15] can be achieved via small-area (e.g. ~6×6 µm in Fig. 3b) and densely-sampled (e.g. 100 nm steps in Fig. 3b) maps. These maps resolve the sub-µm spatial variation in resistance of 3D-patterned ultra-shallow junctions (USJ), which can be useful in cases where the resistivity gradient across cells and trenches is critical for device operation.

c) At sufficiently high sampling currents (~ mA range), M4PP exerts a considerable degree of Joule heating, which can be quantified to determine the Temperature Coefficient of Resistance (TCR) (Fig. 3c) [16]. During electrical probing, heating is localized to the M4PP volume and thus avoids the need to heat large areas, or to microfabricate and intercalibrate microheaters and thermocouples. Using a $3\omega$ technique, the TCR contribution to resistance (the so-called "self-heating") can be corrected for [17], yielding more accurate (up to a percent level) estimates of the electromagnetic properties. The sensing of thermoelectric properties through the $2\omega$ technique is currently under development [18].

d) To provide an alternative metrology for situations where no Ohmic contact can be established between the material or device of interest and the M4PP, CAPRES has recently integrated a non-contact electrical metrology module on its A301 tool, based on terahertz (THz) spectroscopy [19]. Such THz metrology is capable of sensing electrical characteristics (e.g., resistivity and carrier mobility) of buried conductive layers where traditional M4PP may fail (Fig. 3d) and is also capable of extracting carrier mobility from unpatterned films.

## REFERENCES

[1] van der Pauw, L. J. *Philips Res. Rep.*, vol. 13(1), 1958, pp. 1–9.

[2] Ye, Q., Cui, J., Yu, L., Shapoval, T., Flach, F., Haupt, R., Heider, F., Petersmann, W. and Haberjahn, M. *Proceedings of the 25th IEEE Annual SEMI Advanced Semiconductor Manufacturing Conference (ASMC 2014)*, pp. 169–171.

[3] Petersen, C.L., Lin, R., Petersen, D.H. and Nielsen, P.F. *Proceedings of the 14th IEEE International Conference on Advanced Thermal Processing of Semiconductors*, 2006, pp. 153–158.

[4] Petersen, C.L., Hansen, T.M., Bøggild, P., Boisen, A., Hansen, O., Hassenkam, T. and Grey, F. *Sensors and Actuators A*, vol. 96(1), 2002, pp.53–58.

[5] Petersen, D.H., Hansen, O., Hansen, T.M., Petersen, P.R. and Bøggild, P. *Microelectronic engineering*, vol. 85(5–6), 2008, pp.1092–1095.

[6] Nielsen, P.F., Petersen, D.H., Lin, R., Jensen, A., Henrichsen, H.H., Gammelgaard, L., Kjær, D. and Hansen, O. *Proceedings of the 12th IEEE International Workshop on Junction Technology*, 2012, pp. 100–105.

[7] Worledge, D.C. and Trouilloud, P.L. *Applied Physics Letters,* vol. 83(1), 2003, pp. 84–86.

[8] Cagliani, A., Østerberg, F.W., Hansen, O., Shiv, L., Nielsen, P.F. and Petersen, D.H. *Review of Scientific Instruments*, vol. 88(9), 2017, p. 095005.

[9] Henrichsen, H.H., Hansen, O., Kjaer, D., Nielsen, P.F., Wang, F. and Petersen, D.H. *Proceedings of the 2014 IEEE International Workshop on Junction Technology (IWJT),* pp. 1–4.

[10] Witthøft, M.L., Østerberg, F.W., Bogdanowicz, J., Lin, R., Henrichsen, H.H., Hansen, O. and Petersen, D.H. *Beilstein Journal of Nanotechnology*, vol. 9(1), 2018, pp. 2032–2039.

[11] Petersen, D.H., Hansen, O., Hansen, T.M., Bøggild, P., Lin, R., Kjær, D., Nielsen, P.F., Clarysse, T., Vandervorst, W., Rosseel, E. and Bennett, N.S. *Journal of Vacuum Science and Technology B,* vol. 28(1), 2010, pp. C1C27–C1C33.

[12] Cagliani, A., Østerberg, F.W., Hansen, O., Petersen, D.H., Shiv, L. and Nielsen, P.F. *Proceedings of the 2017 IEEE International Memory Workshop (IMW)*, pp. 1–3.

[13] Petersen, D.H., Hansen, O., Lin, R. and Nielsen, P.F. *Journal of Applied Physics*, vol. 104(1), 2008, p. 013710.

[14] Østerberg, F. W., Petersen, D. H., Nielsen, P. F., Rosseel, E., Vandervorst, W., and Hansen, O., *Journal of Applied Physics*, vol. 110(3), 2011, p. 033707.

[15] Guralnik, B., Hansen, O., Henrichsen, H.H., Caridad, J.M., Wei, W., Hansen, M.F., Nielsen, P.F. and Petersen, D.H. *Nanotechnology*, vol. 32(18), 2021, p. 185706.

[16] Marangoni, T.A., Guralnik, B., Borup, K.A., Hansen, O. and Petersen, D.H. *Journal of Applied Physics*, vol. 129(16), 2021, p. 165105.

[17] Guralnik, B., Hansen, O., Henrichsen, H.H., Beltrán-Pitarch, B., Østerberg, F.W., Shiv, L., Marangoni, T.A., Stilling-Andersen, A.R., Cagliani, A., Hansen, M.F. and Nielsen, P.F. *Review of Scientific Instruments*, vol. 92(9), 2021, p. 094711.

[18] Witthøft M.-L., Folkersma, S., Marangoni, T. M., Bogdanowicz, J., Schulze, A., Henrichsen, H. H., Østerberg, F. W., Hansen, O., Vandervorst, W., and Petersen, D. H. *Proceedings of the 2019 International Conference on Frontiers of Characterization and Metrology for Nanoelectronics (FCMN 2019)*, pp. 1–3.

[19] Buron, J.D., Mackenzie, D.M., Petersen, D.H., Pesquera, A., Centeno, A., Bøggild, P., Zurutuza, A. and Jepsen, P.U. *Optics express* 23(24), 2015, pp. 30721–30729.

# AN INDUSTRY EXAMPLE TO REDUCE THE TEST TIME BY OPTIMIZING DATA EXTRATION METHOD

*Xiaofeng Liang, Deguang Zheng*
Wafer Test Department, NXP semiconductor (China) Ltd., Tianjin 300385, China

## ABSTRACT

Test time reduction(TTR) is one of critical steps for test engineers to reduce the running cost in Semiconductor Probe. The test time is highly influenced by the test program, such as the program logic and methodology. This paper presents an industry example to reduce the test time by optimizing the data extraction method through changing the VB code on a volume automotive IC product with stringent quality requirements. Based on the experiment result, there is a tremendous test time reduction opportunity available through the data extract method.

## INTRODUCTION

The semiconductor process is far from perfect, therefore wafer-level probe is needed to distinguish between functionally correct and reliable devices and devices that are defective or non-reliable due to production abnormalities. And test cost may rate more than 50% of the overall production cost [1]. So in order to reduce the test cost, usually two directions are adopted, one method is to increase the parallelism of probe card Device Under Test (DUT), which is expensive and increasingly hard pressed to keep up with memory capacity growth[2]. Furthermore, higher parallelism will be lead to lower operation efficiency, so this method is running out of gas and novel directions [3]. The other direction is test program development, such as using pattern overlapping[4] , full SCAN to selective SCAN [5] , or an asynchronous clock test methodology[6] etc. Also of course, along with development of the design rule, these potential issue or dummy pattern were already considered and have been optimized. However, some logic aspects or program code aspects will not be easy to be found during device development stage, but rather, be identified until mass production phase. This paper addresses this in-process challenge with an algorithm that provides significant TTR through data extract method.

In this Paper, we examine the test time (TT) starting from the real production environment, monitor and investigate the TT trend and distribution. Once the a significant TT gap is observed, we believe that a designed program didn't fully consider the independent accessibility. The imperfect program architecture may offer even larger test time reduction opportunity than design phase. By narrowing down the root cause step by step, in the paper, we will focus on the code aspect to understand the reason of longer test time, the purpose is to optimize the test flow to reduce the test time and benefit the output.

## METHODOLOGY TO OPTIMIZE DATA EXTRACTION METHOD

### Description of the problem

A test program, with the higher integration and the resulting increase in the number and complexity of functions, was created in the conventional manner by a group of persons[7]. However, as [8] discussed, the conventional method lead to a skeleton program designed for idealistic test conditions with low efficiency.

To study the TT impact of the combinational program, four automotive IC products from Teradyne J750 platform were selected from the similar family. As shown in Figure 1, TT data of 100 touch downs were studied for these four products, there is a huge TT variation observed from one touchdown to another one for P1 product. So, it's necessary to investigate the reason to cause this, furthermore, seek for the opportunity to improve it.

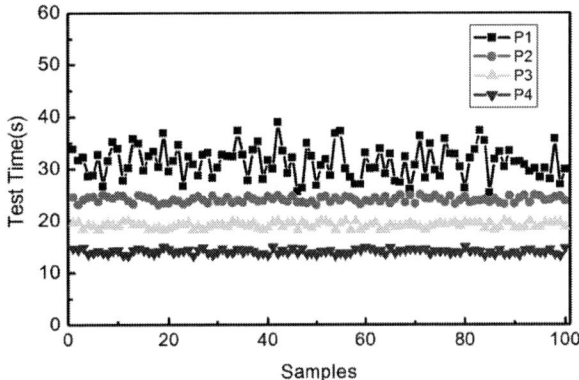

*Figure 1: Test time distribution for several products*

### TTR Opportunities

Basically, the wafer-level test time can be described by equation (1) which is composed of multiple test instances in the test flow. A test flow is the collection of all subsequent test instances, so if one or several test instances have test time variation, the total test time would be possible to show an accumulated variation. For production P1, we can arrange an experiment to repeat several touch-downs, and collect the test time of each test instance. In Figure 2, the distribution of the TT of each test group is produced, the test flow was repeated 4 times, and it is clearly seen that 6 of 7 test groups do not show much difference among each repeat. However, the test group Func_PMC data is dispersed between repeats, the range is from around 12s to 18s. It can be concluded that the test instance of Func_PMC would be the source of the TT variation, and which is an important direction of in-depth study.

$$Wafer\ Test\ Time = \sum_{i=1}^{n} \sum_{j=1}^{m} (TT_{ij} + TI_{ij}) \qquad (1)$$

Where

Wafer test time is the time to test (probe) a whole wafer;

$TT_{ij}$ is the test time of No. i touchdown/No. j test instance in the test flow;

$TI_{ij}$ is the probe index time from one touch-down to next touch-down

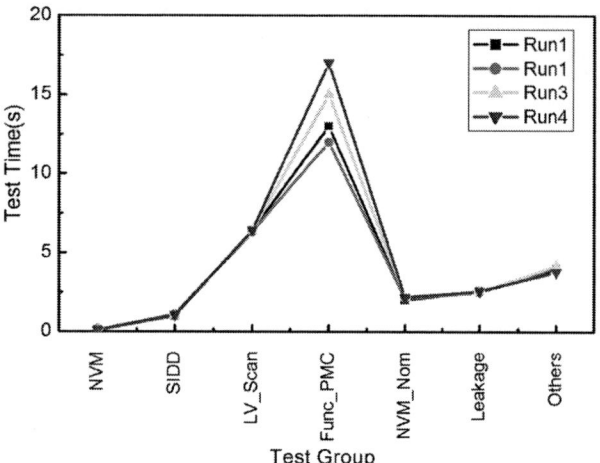

*Figure 2: Test time variation for different test groups*

Since the product is tested on Teradyne J750 tester, the test program for this tester is a excel-VBA based software, also, Teradyne integrated all needed add-ins in the test environment to ease the program and debug[9]. As a leading test solution developer that Teradyne provided several standard templates to execute the test, such as PPMU template, function template and memory test template etc. General speaking, these templates can meet the basic needs of programmers, but if users have additional or special test requirements, such as, users want to do some parameters measurement during pattern running, they need to program the VB code to realize the customized requirement, and use the standard template to call these customized function. Back to this case, the func_PMC test used the Teradyne function template, and also used the customized VB code to realize the whole test requirements. So we need to identify where the test time variation produced, is it in Teradyne function template or in the customized VB code? In order to confirm which section of the test caused the test time variation and consumption, we listed the test time with the VB code section and without the VB code section into Table 1. It is very clear to see that once the VB code was removed from the test instance, the test time would be greatly reduced from 3.7s-4.2s to 0.1s. In meanwhile, there is almost no test time variation existing among 4 test instances without VB code. So we can conclude that the longer test time >3.7s and large variation of >0.4s should be caused by the VB code section.

| Flow | Item | Step | Total | PreBody | Body | PostBody |
|------|------|------|-------|---------|------|----------|
| func_pmc_lvd1_taps | With VB code | 112 | 4.2 | 0 | 4.2 | 0 |
| | Without VB code | 112 | 0.1 | 0 | 0.1 | 0 |
| func_pmc_lvd2_taps | With VB code | 115 | 3.7 | 0 | 3.7 | 0 |
| | Without VB code | 115 | 0.1 | 0 | 0.1 | 0 |
| func_pmc_lvd3_taps | With VB code | 120 | 3.7 | 0 | 3.7 | 0 |
| | Without VB code | 120 | 0.1 | 0 | 0.1 | 0 |
| func_pmc_lvd4_taps | With VB code | 123 | 4.1 | 0 | 4.1 | 0 |
| | Without VB code | 123 | 0.1 | 0 | 0.1 | 0 |

TABLE I. EXECUTION TIME

**New Data Extraction Method**

Looking into the VB code section of the funct_PMU test, the purpose of this VB code is to read the VDD value and get the DC spec from the worksheet, then use these data to judge pass or fail for a DUT. The flow would be:

A. Define parameters for Bandgap trim test--------- (12ms)

B. Define parameters for LVD test------------------ (15ms)

C. Get VDD value--------------------------------------(8ms)

D. Get the number of cycles captured by HRAM--- (30ms)

E. Get limits from worksheet DC spec---------------- **(4.1s)**

F. Execute pass or fail judgment and log the result- (34ms)

A-F flow was developed using sequential execution order, so we can easily set a timer to record execution time of each step. Based on the recorded time, we observed that E step consumed 4.1s test time, in contrast, the test time of other steps only consumed a millisecond-level of TT. Continue to investigate the E step, which is to get limits from DC spec worksheets.

For a standard function test, the precondition parameters would be loaded into LVM (Large Vector Memory) during program validation, so the test execution should be very fast and did not need to re-load the timing parameters and voltage parameters from worksheet each time. However, for a customized section, it will need to execute the VB code each time. Below code (2) is for the parameter load section in E step.

```
… …
If t_type = 1 Then
    If i = 3 Then
        Vbg_min = TheExec.VariableValue("_bandgap_min")
        Vbg_max = TheExec.VariableValue("_bandgap_max")
    End If
End If
If t_type = 2 Then
    If i = 1 Then                                              (2)
        Vbg_min = TheExec.VariableValue("_bandgap_min")
        Vbg_max = TheExec.VariableValue("_bandgap_max")
    End If
    If i = 2 Then
        Vlvdlvw_min = TheExec.VariableValue("_lvd0Trip_min")
        Vlvdlvw_max = TheExec.VariableValue("_lvd0Trip_max")
    End If
… …
```

For Teradyne J750 program, the variables are saved in certain worksheet, such as DC spec, AC spec and Global spec worksheets. When VB area need to get the value of the variables, just need to use "_" + variable name to query.

With respect to this test, the desired variables are saved in the DC spec worksheet. The program used the function "theexec.variablevalue" to search and get the value from the worksheet. From the J750 IGXL manual, we can learn that the property of "theexec.variablevalue" is used to get the value of a variable (spec) in the context of a test[9]. This property is widely used in the J750 test program since it is very convenient and flexible to get the data from the worksheet, whereas test time consumption will accumulate over frequent value extraction from work worksheet by each execution. So the optimized data extraction method is requested to minimize the number of times data is transferred between excel and the VB code area instead of looping through cells frequently [10][11]. But this point usually is ignored by the program designer.

For this case, The E step used a dual-nested loop to get all needed limits, the Figure 3 showed the process to extract limit from the worksheet with the function of "theexec.variablevalure". From the loop, we can conclude that the data extraction will need to execute 13*32=416 times for each LVD test, 13 means 13 pins and 32 means the test parallelism is a 32 DUTs test for each touch-down, so for 4 LVD tests, the function will need to execute 416*4*2=3328 times.

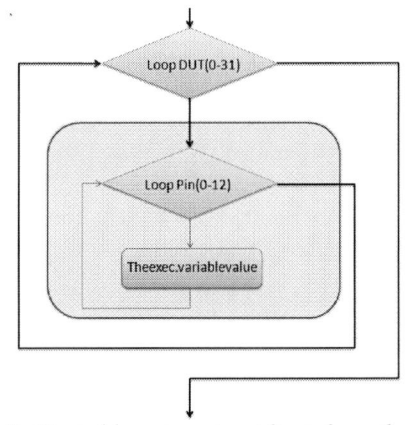

*Figure 3: Nested loop to extract limit from the DC spec*

We definitely cannot remove the limit extraction action for test time reduction purpose since the limits are necessary for the pass and fail judgment. However, we can change the data extract method to speed up the execution. One possible method is, when the test program firstly runs the VB code, we can still keep the current data extraction method to search all the necessary limits from the target worksheet. Once all the data are available, then we save these data into a static array. For the sequential runs, when there is any data is needed, it just need to extract the limit from the static array. The associated code (3) is listed below.

$$
\begin{aligned}
&... ...\\
&If\ TheExec.ExecutionCount = 0\ Then\\
&\quad ... ...\\
&\quad If\ t\_type = 1\ Then\\
&\quad\quad If\ i = 3\ Then\\
&\quad\quad\quad Vbg\_min = TheExec.VariableValue("\_bandgap\_min")\\
&\quad\quad\quad Vbg\_max = TheExec.VariableValue("\_bandgap\_max")\quad (3)\\
&\quad\quad End\ If\\
&\quad End\\
&\quad ... ...\\
&End\ If\\
&... ...
\end{aligned}
$$

The property of "TheExec.ExecutionCount" is used to count the number of times that program has been executed. This is automatically reset whenever the test program is loaded, the test program is validated, the tester is initialized, or the DataTool is started. So when the test program is initial executed, the value of variables will be loaded into memory, however, with the subsequent tests, these values will keep as static variables. Reset conditions of these static variables are same as the property of "TheExec.ExecutionCount".

The other method to ease the data extraction is to use the constant value method. The variables are saved in the DC spec worksheet, we can check and calculate the values of these variables used here, then record these limits data into the VB code section as code(4) shown. With this method, when program need to run the VB code any time, it will not need to extract any data from DC spec worksheet any more, just to use the available data value as the limit for the pass and fail judgment. Compared to the static array method, this method is clearer for the user to know what limits are used here, that will be more beneficial for further program debug. However, the static array method require fewer program changes and will still use the existing variable values from the DC spec worksheet. In addition to the first run, other subsequent tests will not consume test time since there is no data extraction action any more. To verify which solution could be the better from the test time aspect point of view, Table 2 shows the execution time using each proposal. We can come to a conclusion that two novel methods only consume one thousandth execution time in contrast with existing "theexec.variablevalue" method.

$$
\begin{aligned}
&... ...\\
&If\ t\_type = 1\ Then\\
&\quad If\ i = 3\ Then\\
&\quad\quad Vbg\_min = 1.1901\\
&\quad\quad Vbg\_max = 1.2099 \quad\quad\quad (4)\\
&\quad End\ If\\
&End\ If\\
&... ...
\end{aligned}
$$

TABLE 2.   EXECUTION TIME WITH DIFFERENT METHOD

| Method | Theexec.variablevalue | Array | Constant |
|--------|----------------------|-------|----------|
| Time | 12-26ms | 10-28μs | 10-28μs |

Compared to the significative variation between executions for P1, the new data extraction methods have a better repeatability and lower test time.

## EXPERIMENTAL RESULTS

Since there is no noticeable difference between the static array method and constant method, eventually we picked up the constant method to verify the real test time impact. Figure 4 shows the test time difference between the constant method and current "theexec.variablevalue" method. Obviously, the test time variation with the constant method was reduced to a very low level compared to the existing "theexec.variablevalue" method. Meanwhile, the average test time was also reduced from current TT=32s to 19s per touch-down, the TTR rate is about 40%. The wafer-level test time (55min) was reduced to 36min, the TTR rate is around 35%.

The consistent of the test time from touch-down to touch-down and from wafer to wafer is also improved, known from Figure 4, only slight TT variation observed with constant method.

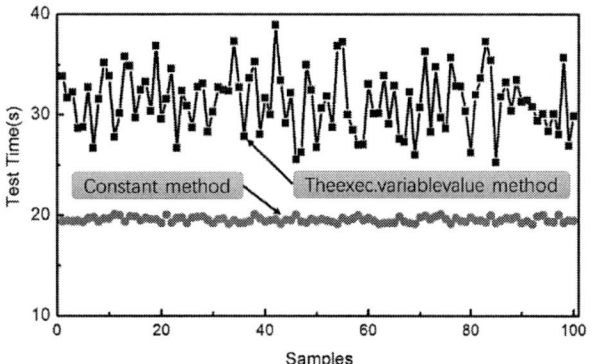

*Figure 4: Test time difference with different data extraction method*

## CONCLUSION

Actual test environment generally has restriction with regard to designed test program, so the demand for TTR methodologies that reduce the test cost of semiconductor manufacturing continues. In this paper, we studied the P1 product with large test time variation, and the experiment proved that the repetitive execution of VB code sentence of "theexec.variablevalue" from the DC spec worksheet will consume excessive test time, also cause a wide TT variation from one touchdown to another. The static array method and constant methods were proposed in this paper to fix this issue, the corroborative experiment has demonstrated the effectiveness of new methods and enabled us to obtain a test time reduction amounting to 35%.

Generally speaking, the corresponding data extraction method is widely used in the J750 test program since it can query the variable value which is saved in the worksheet. But excessive test time consumption is most commonly seen for the similar program with presetting or preconditioning, so TTR needs to be taken into account and, ultimately, reduce the Test Time caused by data extraction method. In the future, we intend to categorize and optimize the data extraction methods as a reference working instruction for programmer to use.

## REFERENCES

[1] F. Poehl, F. Demmerle, J. Alt, and H. Obermeir, "Production Test Challenges for Highly Integrated Mobile Phone SoCs - A Case Study", Proc. of European Test Symp., 2010 pp. 17-22

[2] J. -C. Yeh, S. -F. Kuo, C. -H. Chen and C. -W. Wu, "A Systematic Approach to Memory Test Time Reduction", IEEE Design & Test of Computers, vol. 25, no. 6, pp. 560-570

[3] H. Hashempour, J. Dohmen, B. Tasić, B. Kruseman, C. Hora, M. van Beurden, and Yizi Xing "Test time Reduction in Analogue/Mixed-Signal Devices by Defect Oriented Testing: An Industrial Example", Design, Automation & Test in Europe Conference & Exhibition (DATE), 2011, pp. 1-6

[4] M. Chloupek, O. Novak, and J. Jenicek, "On Test Time Reduction Using Pattern Overlapping, Broadcasting and On-Chip Decompression", 2012 IEEE 15th International Symposium on Design and Diagnostics of Electronic Circuits & Systems (DDECS), Apr. 2012, pp. 300-305

[5] Wen-Joung Lai, Chen-Pin Kung and Chen-Shang Lin, "Test time Reduction in Scan Designed Circuits", 1993 European Conference on Design Automation with the European Event in ASIC Design, 1993, pp. 489-493

[6] P. Venkataramani and V. D. Agrawal, "ATE test time reduction using asynchronous clock period", 2013 IEEE International Test Conference (ITC), 2013, pp. 1-10

[7] J. Hirase, "Test Time Reduction through Minimum Execution of Tester-Hardware Setting Instructions", Proceedings 10th Asian Test Symposium, 2001, pp. 173-178

[8] Bo Zhong, Deguang Zheng, Xina Dai, and Yu Hu, "A Novel Automatic Probe-to-Pad Alignment Error Correction Approach", IEEE Transactions on Semiconductor Manufacturing, vol. 35, no. 1, 2022, pp.146-148

[9] J750 basic programming V3.4 student training manual PN553-405-50, Rev August 2002

[10] Diego Oppenheimer, https://blogs.office.com/2009/03/12/excel-vba-performance-coding-best-practices/, on March 12, 2009

[11] http://msdn.microsoft.com/en-us/library/dd721892.aspx, Visual Basic for Applications.

# HIGH INHERITABILITY AND FLEXIBILITY SMARTTEST8 TESTMETHOD LIBRARY ON ULTRA-HIGH-SPEED SERDES MEASUREMENT

*Jiaying Xiang, Yichen Xiao, Juyang Sun, Qingqing Xia and Yanfen Fang*

Advantest, Shanghai, China
E-mail Address: hunter.xiang@advantest.com

## ABSTRACT

Compatibility with new test head cards, diversification of program development supported by open source of native Code, likewise, highly efficient test time reduction (TTR) and cost cutting, the Smartest8 is a growing trend for automatic test. SerDes, the high-speed wideband digital communication with stable and efficient program is participated in this reformation. The traditional Smartest8 Testmethod is re-organize into procedure-oriented coding style called Test Method Runner (TMR) for test engineering.

However, more complicated measurement scenarios of SerDes are concerned in released products, for instance, management for at least ten kinds of registers with plenty Read-Modify-Write (RMW) operations, seasonable Control and States Register(CSR) verification for each configuration and requirement of optional protocols. The Common Building Block (CBB) of SerDes on Smartest8 provides solutions for all cases above with Register Management and Register Access (RA) implemented. As the CBB is presented with introductions and rules, for further projects which inherit similar framework, the difficulty and time cost will be significantly reduced by 80% for new engineer, some specified codes can even be reused directly from Smartest7 to Smartest8. Additionally, we provide tools for improved functions with concise Graphical user interface (GUI). This CBB has achieved the promotion of the SmarTest8 SerDes Test in kiku and been uniquely qualified for further SerDes involved products. Here we will illustrate the problems correspond to representative scenarios and related CBB solutions.

*Keywords—smartest8; serdes; register manager; RMW;protocool;*

## INTRODUCTION

With the increasing speed and quantity of data transmission, the shipment of SerDes chips is also growing rapidly. As the transmission rate of SerDes increases, there are strict requirements on the number of IO, so SerDes will deploy a large number of registers to control the chip. The demo device is an 8T8R high-speed SerDes, the highest speed can reach 30gbps, can be widely integrated in RF, network, consumer equipment for high-speed data transmission.

Figure 1 shows the basic architecture of the SerDes device. A macro is usually composed of Service Slice(SS), Clock Slice(CS) including PLL and MCU, DS lanes and control plane. SS receives external 100 MHz reference clock and provides

clock routing to PLL. High precision PLL will lock the received clock to a 9-10Ghz high-speed clock with low jitter, usually in femtosecond level.

DS consists of TX and Rx. TX channel has testability function design such as PRBS code generator, jitter generation and ppm generation, and the driver end has equalization functions such as fir adjustment. RX channel is designed with testability functions such as PRBS code type verifier and eye pattern detector, and the receiver end is equipped with equalization functions such as CTLE and DFE. SerDes macro has a common on-chip JTAG TAP controller interface and supports 1149.1 and 1149.6 standards. The registers of each module can be configured through JTAG interface. A MCU 8051 is integrated in the macro. Firmware in hex format can be loaded through the JTAG interface and reset the MCU. The on-chip MCU performs corresponding operations to complete SerDes calibration, CTLE adaptation or ABIST (built-in self-test).

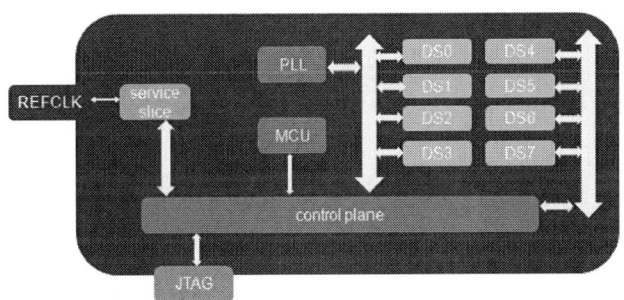

*Figure 1: A macro of SerDes chip structure*

The rate of ultra-high-speed SerDes can generally reach more than 30Gbps. It is difficult for the current test board to reach such a high rate. Therefore, ABIST (built-in self-test) is used to test ultra-high-speed SerDes. In CP phase, the inner loopback test method will be used, while in FT phase, the cross outer loopback test method will be used. Before these tests, we need to configure the chip state to the specified state through registers.

Figure 2 shows the register architecture of the SerDes device.

Each module of SerDes has one or more types of registers. For a specific class of registers, there are dozens of CSR (control and states register). A CSR has 16bit or 32bit wide data, which are allocated to the so-called field registers. Therefore, an 8t8r macro usually has more than ten kinds of

CSR and 10000+ field registers. CSR is divided into per macro, per lane, and API CSR. Per macro CSR means that there is only a single CSR in each macro. Per lane CSR means that there are multiple CSR with the same name in each macro, and its number is consistent with that of DS. Per lane CSR can be configured by broadcast. For example, when the CSR of DS0 is configured and the broadcast is turned on, the CSR of DS1-DS7 will also be configured together. The API CSR is unique. It is also divided into per macro and per lane CSR. However, for per lane API CSR, it cannot be configured by broadcast, but should be configured one by one.

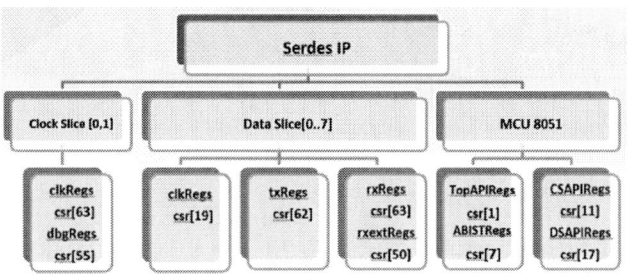

*Figure 2: Various CSR in SerDes module*

In the face of complex and numerous register management, another challenge that needs to be paid attention to is RMW(Read-Modify-Write). The register of chip is read and written with 16 bit data as a whole. However, the specific configuration only needs to modify one of the 16 bits. When the program is executed, it will only give the modification information on the relevant bits, and the values of the bits that are not involved are unknown. In order to ensure that the value of the unconfigured bit is not modified, it is necessary to read back the 16 bit data from the chip, modify the corresponding bit data, and then write it to the chip. Figure 3 is an example of RMW operation. If we need to set the minimum 4 bits to 0xC, the values of other bits are unknown except for the minimum 4 bits. First read back to get the current status of the chip is 0x5A5A. Because the first 12 bits are not the bits to be modified, the read back data is reserved, and the lowest 4 bits are replaced with our set value. Finally, we get the data to be written: 0x5A5C.

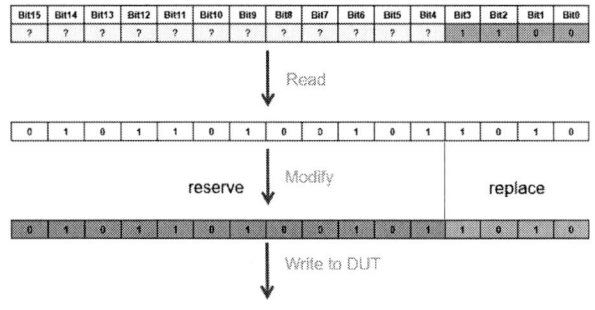

*Figure 3: Example of RMW operation*

In addition, we also need to pay attention to the multi protocol switching of complex chips, as well as the measurement management on smartest8 and the subsequent increase in test

time. Facing the challenges of these complex and flexible register management on smartest8, the high inheritability and flexibility SMT8 test method library provides a one-stop solution for register management, RMW, protocol management and making test program structured.

**Registers management with plenty RMW operations**

As is stated above, for the Bit Field configuration of the register, we need the RMW operation to keep its unchanged bits unaffected. The TML provides learning mode and production mode to solve the problem of RMW. In the setup phase under learning mode, TML will ergodic the measurement with register configuration in the current test suite and collect the registers with RMW actions. Then TML generates an auxiliary measurement, which is placed before all other measurements. The content of auxiliary measurement is to read back all previously collected registers. In the execute phase under learning mode, the auxiliary measurement will read back all the previously collected registers and write the registers and corresponding values to the RCF (register cache file) file with the same name as the current test suite. At the same time, the names and read back values of these registers are stored in a class called registers cache to prepare for subsequent modification. Finally, when the measurement of register configuration is executed, TML will take the read back value of the current configuration registers from the register cache, modify the bit we need, and then write it to the device. While in the setup phase under production mode, the TML will fetch the saved register read back value from the RCF file and modify the bit we need, then generate the operation sequence and pattern based on modified data value. In the execute phase under production mode, the solidified vector written to the register will be executed, which will greatly reduce the test time.

We have discussed the case where there is only one measurement in a test suite. So how does TML work when there are multiple measurements in a test suite? In one test suite, after the first measurement operates on the register, the value of the register in the second measurement may change. This will cause the register value read back at the beginning of the test suite to deviate from the register value when the second measurement is executed. Therefore, we introduce the sync function, which can do an additional readback between two measurements. This ensures that the value in the register cache is consistent with the DUT register in real time.

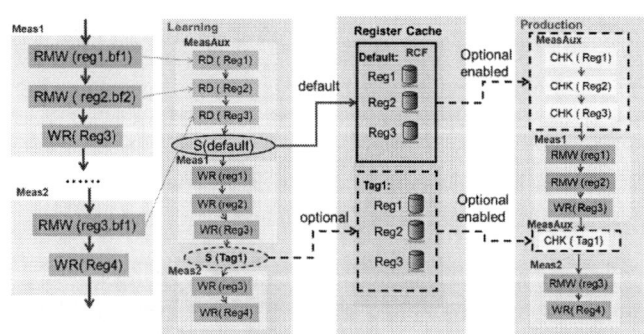

*Figure 4: Multiple RMWs over Multiple Measurements*

Figure 4 shows a scenario where multiple RMWs over multiple measurements in a test suite. The default sync is required and is arranged between the secondary measurement and the first measurement. The sync used by the user between meas1 and meas2 is optional. It can be given a user-defined tag, and then the read back value of the registers is stored in the RCF file with the tag as the ID. Before executing meas2, TML will get the latest read back value from the RCF file and refresh the written pattern. The optional sync costs performance, so user may think about to ignore the additional sync, and add them as needs while is it out of sync.

Another function of RMW is CSR check, which is only used in production mode. When the CSR check function is not turned on, TML will not execute auxiliary measurement in production mode. When CSR check is turned on, TML will execute auxiliary measurement in production mode and compare the read back value with the value in register cache. When there is a deviation, TML will return the result of check fail. CSR check will prompt the engineer that the current measurement register has changed, and additional sync needs to be added.

**Flexible Register Manager Structure**

There are more than ten kinds of CSR and 10000+ field registers in an 8t8r macro. TML designs an advanced register management structure to accommodate these complex and numerous registers. The register structure is divided into three levels: Peripheral, RegBase and Bitfield. The peripheral level represents a collection of registers. For example, peripheral_ CS_CLK contains the registers which control the CLK configuration in CS. The regbase level represents a single CSR for a class of registers. There are multiple CSR in CS_CLK, and each CSR is a regbase. Bitfield is the smallest unit of register structure. Each CSR is composed of 16 bits of data. Some of these 16 bits can form a bitfield to implement a specific configuration of the device. The absolute address of a register consists of a base address and an offset address. A peripheral corresponds to a unique base address. A regbase in the peripheral also corresponds to a unique offset address. When we need to calculate the absolute address of a CSR, we only need to obtain the base address of its peripheral plus the offset address corresponding to the CSR. If the absolute address of per lane CSR is calculated, the offset address of each lane needs to be added.

The advanced register structure is not only very flexible and efficient, but also can adapt to a variety of devices through programmable base address and offset address. In addition, it is perfectly fit to RMW action and check action. When we operate on a bitfield, TML will record the regbase corresponding to the bitfield and read it back in the auxiliary measurement. When we directly operate the regbase, TML will not read it back, which will greatly reduce the test time. Furthermore, the advanced register structure is compatible with any protocol because users do not need to pay attention to the underlying protocol logic of register reading and writing.

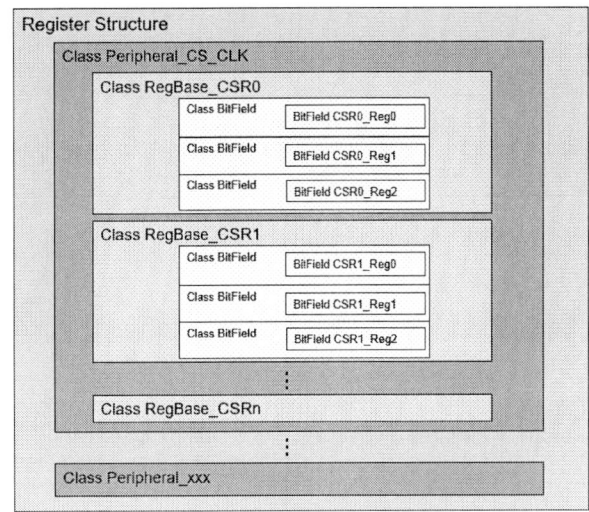

*Figure 5: Register structure of CS_CLK*

**The Common Building Block of SerDes Test**

The common-building-block(CBB) integrates macro-lane-control class, protocol management, customized data structure and common functions. User only need to fill in protocol information and SerDes information and use universal tools to generate register structure and test methods to develop the test program.

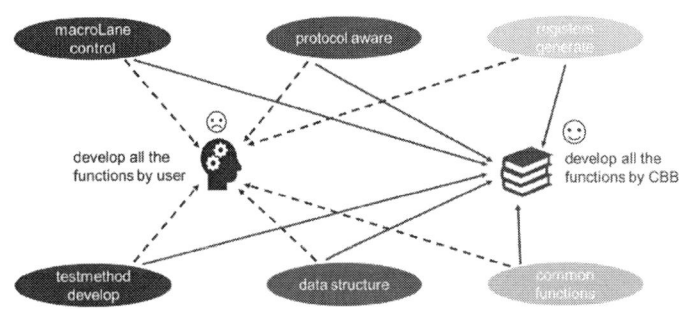

*Figure 6: Developing the Serdes test method library by CBB*

For the development of SerDes test program, test engineers need to make protocol declaration, register information injection, macro lane control, common function development, data structure development and test method development. The CBB provides a flexible and general macro lane control scheme. CBB decouple register operation from protocol, and connect register action with protocol through register access library (RA). For multi-protocol devices, users only need to declare the required protocols and switch the protocol through RA, and they can use different protocols to configure the register. In terms of program architecture, CBB also provides some general functions. For example, DumpAllCSR, this function can read back all registers and print, which brings great convenience to engineers' debugging. CBB also has some well-defined data structure classes, such as MACRO_INT. This class can store MultisiteLong data distinguished by macro, provides corresponding set and get methods, and overloads the toSting method to print results quickly. Finally, CBB also provides some tools to automatically generate registers and testmethod, which can greatly reduce the time of program

978-1-6654-9759-6/22 $31.00 © 2022 IEEE        503

development.

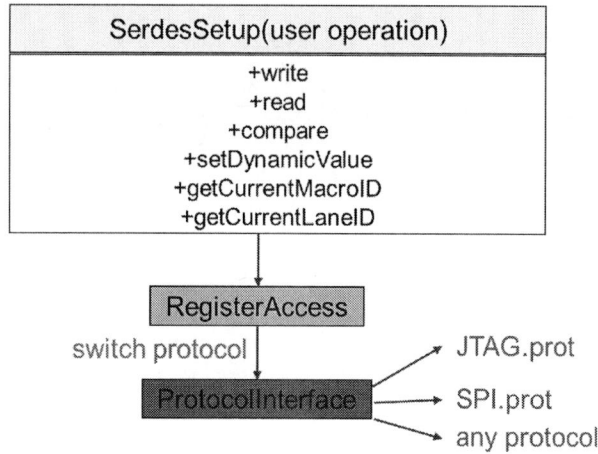

*Figure 7: Flexible protocols configuration and switch*

## SUMMARY

Facing the challenge of ultra-high speed SerDes testing, the high inheritability and flexibility SMT8 test method library provides a one-stop solution for register management, RMW, protocol management and making test program structured. As a result, the TML not only provides a complete test solution for high-speed SerDes, but also significantly reduce the difficulty and time cost by 80% for new engineer. The TML is used for SerDes IP now. But in the future, it can also be used in any device that needs register management, RMW or protocol management.

## REFERENCE

[1] S. Sunter, A. Roy and J. -. Cote, "*An automated, complete, structural test solution for SERDES,*" 2004 International Conferce on Test, 2004, pp. 95-104, doi: 10.1109/TEST.2004.1386941.

[2] S. Abdennadher, K. Tripician and S. Singaravelu, "*At Speed Testing Challenges and Solutions for 56Gbps and 112Gbps PAM4 SerDes,*" 2020 IEEE Latin-American Test Symposium (LATS), 2020, pp. 1-5, doi: 10.1109/LATS49555.2020.9093685.

[3] Robertson, Iain, et al. "*Testing high-speed, large scale implementation of SerDes I/Os on chips used in throughput computing systems.*" IEEE International Conference on Test, 2005.. IEEE, 2005.

[4] Evans, Andrew C. "*The new ATE: Protocol aware.*" 2007 IEEE International Test Conference. IEEE, 2007.

[5] Li, Baohu, Bei Zhang, and Vishwani D. Agrawal. "*Testing with reduced ATE channels.*" 23rd IEEE North Atlantic test workshop. 2014.

# THE FLEXIBLE AND INHERITABLE SMT7 SOLUTION FOR HIGH-SPEED DDR PHY IP WITH IJTAG

Yichen Xiao[1]*, Hao Wu[2], Xin Song[3]

[1]Advantest (China) Co. Ltd, Shanghai 201203, China

*Corresponding Author's Email: shaw.xiao@advantest.com

## ABSTRACT

The high-speed Double Data Rate Interface intellectual property (DDR PHY IP) integrated into system on chip (SOC) is part of a memory subsystem, supporting the frequency up to 3200Mbps data rate at 1.2V for communication with external memories. To enhance the capabilities of embedded IPs, especially for multiple level architecture, the IEEE P1687 Internal JTAG (IJTAG) is applied. The flexible access is established by the selectable gateway Segment-Insertion-Bit (SIB). Nevertheless, the adjustive scan path signify plenty combinations of SIBs which results in a more complex interface protocol. In this paper, a general solution for DDR PHY validation with IJTAG is presented. Moreover, by testing the packaged products on Automated Test Equipment (ATE) platform, the correlation between traditional design for test (DFT) test and IJTAG protocol-based instruction vector test is accomplished.

*Keywords – Memory Interface Test, ATE, IEEE P1687 IJTAG, DC measurement*

## INTRODUCTION

The DDR interface plays an important role in data transmission between the external memory modules and embedded memory controller, especially the application for the High-Performance Computing (HPC) with large volumes of data from consumer. The product consists of calibration module for initialization and a slice-based architecture (Address /Control, Data and IO pads as shown in figure1) to connect to the memories as shown in figure.1.

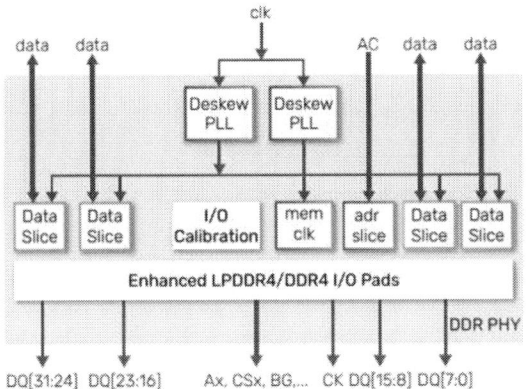

Figure 1: The IP-level PHY diagram with main components [1]

The DC measurement is implemented for the input and output impedance in enhanced I/O pads. As extensive application scenarios are concerned, the accompanied Build-in self-test (BIST) covers three routes: slices, slices with IO part, and thirdly slices with the IO pads connected memory. The third loopback with Dynamic Random-Access Memory

(DRAM) on load board simulates a real work condition of the DDR interface for improving test coverage. As the number and type of integrated IPs on chip are increased, the access protocol is optimized by the IJTAG. With the participation of designed and programed SIBs, the scan path and test sequence are dynamically configured.

In protocol framework section, the industry concerns and benefits of utilizing IJTAG are described. It is also indicated that the basic architecture of IEEE P1687 standard and the solution for variable protocol format. The third part covers the BIST and DC measurement, and the last presents correlation trouble shooting between DFT and IJTAG generated pattern on the ATE Smartest 7 software (SMT7).

## PROTOCOL FRAMEWORK

The traditional test equipment and the legacy verification generally rely on external physical probes. However, for several reasons, such as heterogeneous multi-chip integration has no direct port for top core at substrate and the priority of constructed intrusive instrumentation limits the access point, the efficiency and test coverage of legacy validation are diminishing. As the Moore's law continues to drive the chip-manufacture process, at the chip level, the size, work frequency and more complex package technology are also driving the industry toward software-based embedded instrumentation system [2]. Moreover, the period of electronic products launch into market is significantly shrinking which affects the development cycle of the chip verification and test.

To enhance the capacities of embedded instruments for above industry trend and problem, the IEEE P1687 IJTA standard was designed.

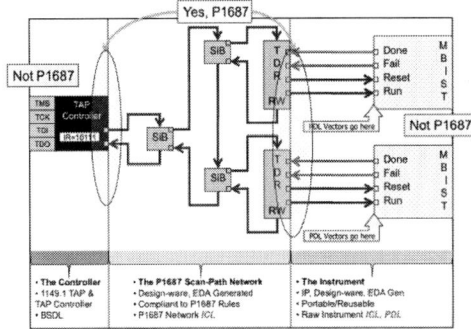

Figure 2: The architecture with IEEE P1678 IJTAG standard applied at the chip level [2].

The basic concept of the IEEE 1149.1 boundary-scan standard Test Access Port (TAP) and Test Data Register (TDR) are respectively illustrated with controller and the potential IPs in figure 2. The IEEE P1687 IJTAG Network, especially the SIBs and Procedural Description Language (PDL) are indicated to simply show how the access between TAP and embedded IP established. For a chain of instruction or data for each side, different TDRs should be allowed to

978-1-6654-9759-6/22 $31.00 © 2022 IEEE

access. Thus, the routes constructed by variable SIBs are designed and managed which result in plenty formats of bit stream.

The ATE vectors are generally created by the specific protocol-based test program or the vector configuration file (WGL) provided by DFT engineer. Nevertheless, the reconfiguration of ATE SMT7 pattern for the adjustive SIBs path leads to the uncertainty of protocol program and WGL file. The complex protocol construct system (Fig3), which is developed based on Dynamic to Static (D2S) and support the ATE pattern creation by programming protocol, is applied to solve the pattern iteration cause of variable SIBs format [3][4]. The information for configuring the inheritable framework on SMT7 platform are described below and the setting for TDRs is recorded in CSV file.

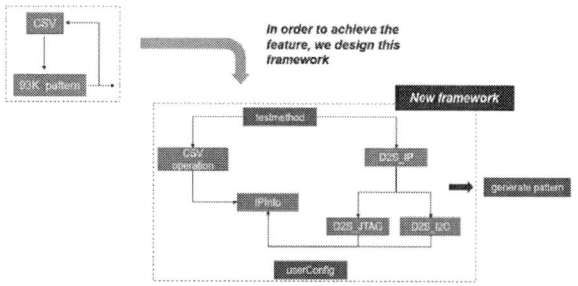

*Figure 3: The framework design of "complex protocol construct system" [3].*

There are several kinds of instruction formats for scan paths and the basic fragments are demonstrated in figure 4. The SIBs are managed based on the actual hardware design. For some, as the IEEE P1687 IJTAG generally work with IEEE 1149.1 standard, it is assigned after the instruction register (IR) shift [5]. Others are loaded at the end of one complete action. Then there are short SIBs configured inside the vector of data register (DR).

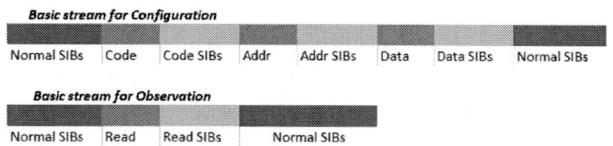

*Figure 4: The basic stream should be confirmed for writing and reading TDRs (the length of block does not stand for the real bit length)*

In this experiment, a write action consists of two of first fragments for DR and read with one more second stream. To correspond to the IJTAG algorithm in DFT system, the application scenarios "write – read" and "read – write" on ATE platform also add pre-config for SIBs circuit open/close for next action.

Protocol verification procedure:

- Confirm the SIBs scan path requirements of the TDRs observation and modification for the product, to ensure that the "complex protocol construct system" can cover these scenarios.
- Check all conditional statements of the framework. The bit stream of protocol consists of address, data, specialized code and SIBs configuration, Portion of management for vector instruction are accomplished

by the conditional statements.

- Verification of write pattern for register setting, compare with the lines of DFT vector.
- For TDR observation, Error Map and Capture are the conventional methods to harvest the of information for test device. Then the binary data obtained by firmware command is converted into the real value of addressed register.

Nevertheless, for the SIBs with extremely long, unique and infrequently applied format, the DFT pattern is used directly at the designed position.

## DC MEASUREMENT AND BIST
### *Write leveling and clock calibration*

As the topology technology based Dual Inline Memory Module (DIMM) enhance the capabilities of memory, the offset between the control signal (or clock) and data becomes a generally concerned issue in transmission. Within the DIMM, the route layout for command and data stream are decided by the number of integrated DDR Synchronous Dynamic Random-Access Memory (SDRAM) and applied type of topology. To reduce the lines for Address and Command (AC), the fly-by topology prefer one specific route for it and link the DRAM inside the DIMM [6]. However, the clock from AC signal has timing skew and is mismatched to the data signal due to the varying length of routes. Write leveling, the method is implemented to calibrate the edge of data and differential clock with aid of a dedicated strobe signal (DQS).

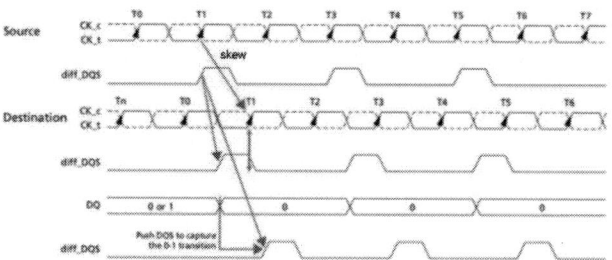

*Figure 5: The edge calibration by write leveling procedure [6].*

The transmission with write leveling is demonstrated in figure5, which has a synchronous rising of differential CK and DQS at source. The skew of clock occurs at destination. The calibration for solving the delay is to capture the state of clock on DQS firstly. Secondly, the information is sent back to memory controller though data line (DQ) of DRAM under training. Then loop above until the low to high change of clock is detected Finally, the delay message from DQ is applied to DQ accompanied DQS to accomplish the synchronization of differential clock and data.

As the delay time is a crucial factor in memory system for data transmission, once the basic function of test is accomplished, the continuously optimization is focus on calibrating the skew to obtain optimum eye diagram. The configuration for memory controller (MC) embedded DDR interface is designed and shift in through the IJTAG chain. Accordingly, on ATE platform, the quality of received data is evaluated inside and observed by return value of specific TDR.

### *Impedance measurement and BIST*

The Scatter (S factor) which includes the reflection and consumption is induced by different material and characteristic of transmission line also affect the height and width of eye diagram. The DC measurement is implemented firstly for the input and output impedance in enhanced I/O pads, especially the dynamic on die termination (ODT) which optimize termination value for the varying loading conditions. By applying one of I or V from ATE instrument, the resistance is calculated and judged. It should be mentioned that, before resistance test, the optional degree for output driver impedance (RON) and ODT resistance are configured by IJTAG based ATE pattern.

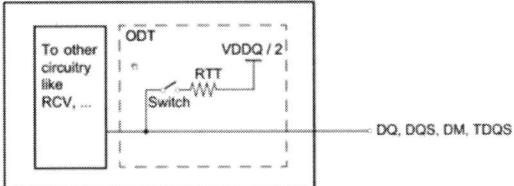

*Figure 6: The basic view of ODT in DDR IO pads [7].*

The BIST is implemented for comprehensive application scenarios, for instance, the external loopback improves test coverage for actual work. During these validations, the instruction vectors are distinct due to different layout of data and AC slices, as shown in Fig.1. Additionally, the definition for bi-directional IO pins on ATE platform allow the excitation from instrument driver loop back into DDR interface. All the error count before optimum signal is read back via mentioned IJTAG standard.

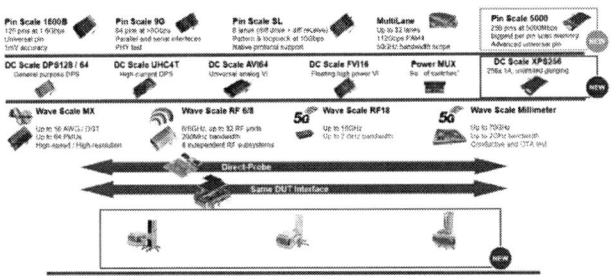

*Figure 7: The ATE product road map of the Advantest [8].*

Several instruments are implemented to support the validation. The DC scale cards from the ADVANTEST, such as AVI64, UHC4 and DPS128/64 (Fig.7), feature high voltage, high current or analogy capabilities. The Digital cards including PinScale1600, PS9G and PSSL address test requirements, for instance, the new IPs with increasing number of cores and higher speed serial and parallel interface. Consequently, driving high into DDR interface to simulate the function of external memories can be achieved by using PS1600 test card.

## CORRELATION ANALYSIS

The IJTAG based instruction vector is utilized for clock skew optimization, DC measurement preparation and the BIST. However, the convenience of procedure "ATE basic setup – configuration – read TDR result" leads to an abstract validation. The action is not clearly presented as traditional intrusive instrumentation. Thus, when the result is unexpected,

the schedule probably gets effect from a lack of debug method.

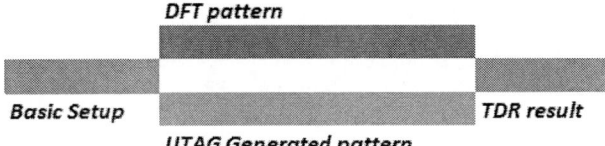

*Figure 8: The correlation between DFT and ATE pattern.*

To address the issue, the correlation with flexible protocol system is applied to route variously but achieve same function. For example, the voltage at input is detected and recognized as logic high or low for test chip. After inside DFT process, the result is shift out for comparison. Once the DFT result in error, the partial validated but with similar function setting from BIST can be implemented for constructing instrument vector to replace the suspicious lines in DFT pattern without time cost for iteration. The "basic setup" can be verified by intrusive instrument on ATE platform or the reasonable reacted "TDR result" which also indicates the function of "configuration" for trouble shooting.

## CONCLUSION

Advanced integration technology and protocol standard add new possibility for design and test. The validation of an IJTAG based DDR PHY IP presents the enhancement in protocol structure for SIBs configuration. According to the flexible framework, instrument vector can be implemented to achieve the calibration of clock skew, impedance measurement and BIST on ATE platform. The solution is inheritable for IPs with variable protocol. The correlation between DFT and generated instruction can be effectively exploited to address the mistake in designed test plan which allows time cost reduction about sixty percent for the reliable DDR interface test program development before products launch into market.

## REFERENCES

[1] Cadence, "Denali DDR PHY IP for TSMC", Design IP Datasheet, 2020.
[2] Asset, "IEEE P1687 Internal JTAG (IJTAG) Tutorial", AS SET InterTech, Inc,2011.
[3] Xin Song, Man Cao, "Complex protocol construct system on ATE platform", IEEE and CSTIC China, 2021, in press.
[4] ADVANTEST. *Test Documentation Center d2s*, Rev.7.6.1, topic. 122756, unpublished.
[5] Sam Gallagher, "The JTAG Test Access Port (TAP) state machine", ALL ABOUT CIRCUITS, Technical Article, Nov 2020.
[6] Logic Jitter Gibbs, "DDR learning Part B-3: Write Leveling", OpenIC SIG, Jan 2021.
[7] Leo Fun, "Chat about ODT in DDR3", CSDN, FC, Jul 2018, unpublished.
[8] Kevin Yan "Test Evolution for Heterogeneous Multi-chip Integration", Advantest, 2021, unpublished.

# AN ATE SOLUTION FOR HIGH RESOLUTION ADC/DAC

*Feng Qin, Mitch Royals*

Advantest (China) Co., Ltd
*Corresponding Author's Email: qin.feng@advantest.com

## ABSTRACT

We introduce a test solution for high resolution ADC/DAC. Including the works on both hardware and software. from the instrument selection, board design, to software coding rules. we explain the reason of grounding, signal isolation and power filtering, their contributions to the solution. In the end, figures will illustrate the impact of an inaccurate synchronization to the measurement result.

## INTRODUCTION

ADC and DAC are the most typical mixed signal devices. In this kind of testing, analog stimulus signal is generated by an arbitrary waveform generator (AWG) which employs a D / A converter inside, and analog signal is measured by a digitizer or a sampler which employs an A/D converter inside. Though the stimulus signal itself is created with mathematical method, and the measured signal is also processed with mathematical method, extracting various parameters. These instruments are installed in the tester, its parameter should be better than DUT, otherwise onboard component will be needed to cover the gap. A good tester should be capable to provide a stable and noise clean electrical environment. The complicate part is to keep it clean when you connect them to the DUT, which involve the load board design. In Figure.1, it's the function diagram of an A / D convertor with its instruments, each part has its own corresponding connections to the tester side, and all of them are going through load board. The isolation among is the key.

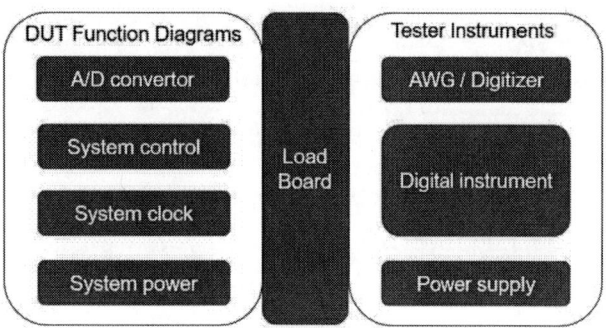

*Figure.1 A/D convertor function diagram and connections*

## ANALOG SIGNAL ISOLATION

A high resolution convertor is capable to reach a resolution up to 24 bits, which left them a very low immunity against noise. Therefore it's very easy to have an interference during board transmission, so called crosstalk. It often happens in two cases. One is caused by the same analog signal located on the same layer stayed too close. They interference with each other when running in parallel. The other could be caused by a high speed signal runs not in the same layer, but just on the top or bottom.

The principle to suppress the crosstalk is keep those signals far away from each other. It is recommended to keep all digital signals in the same layer, in the same area, then wrap them up by a ground layer on both top and bottom. A full-sized ground layer will act as a good isolation between them. This works for analog signal too, if the space is limited in one layer, multiple layers is acceptable. Just keep the signals in the same parallel run away from each other.

## DIGITAL SIGNAL ISOLATION

Unlike the analog signal, the digital signal has a stronger noise immunity by its nature. So no extra work for them. except for the reference signal. The voltage reference signal works as the conversion reference value, and a common-mode bias voltage is for the use with any external input circuitry. The stability of these two is a key to the performance of the device. An ESR de-coupling capacitors is necessary, and place them as close as possible to the device pin is necessary too. Because the trace itself will form a parasitic noise by nature. For parallel test, an isolation circuit applied on each reference signal pin will be a plus.

all the sigma-delta convertor requires an external clock, from which the modulator oversampling and digital sub-system clk are derived. A clock signal with jitter as low as 100 ps or less is necessary for an optimal performance. In order to keep the signal away from other interferences, place it on the top layer and keep it as the only signal on it is recommended.

## GROUNDING

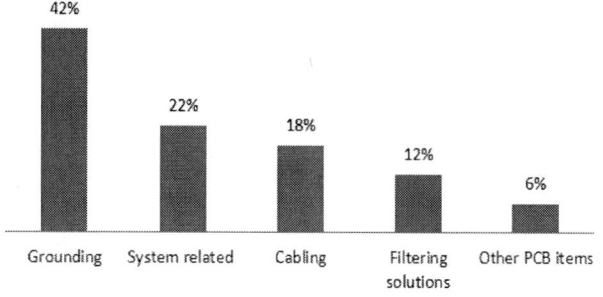

*Figure.2 the importance of grounding compared to other compliance solutions.*

Grounding is the most import part in design. (Figure 2) Not only multiple ground layers presented, but there're multiple type of ground need to be linked together. By the each one's noise cleanness level, they're categorize to 3 degrees.

The cleanest ground is the analog ground, which connected to nothing else but analog grounds of each on board device. As the connection to the tester, only the analog ground is linked to it too. No need to worry about the noise from the tester side, since it has its own way to consume it. Just make sure neither high voltage nor large current go through this plane, and each small volume is stable enough.

Then the medium degree ground is digital ground, it includes all the digital ground from the device. The operate current of a digital circuit contains harmonics, which is a noise source by itself. it flows through not only signal lines but also the power supply and ground causing common mode noise. It has much bigger impact to the analog signal than to themselves. Which is the reason don't mix with analog ground. It connects to the digital card in tester, and since its power supply related plus the digital signal has a quite immunity by nature, so power supply ground can be connected to it safely.

At last is the noisiest ground called utility ground. Almost all ate tester has a special block for utility line control. Utility line allows you to control or read the states of external devices as relays, EEPROMS and other indicators on board. Considering the unknown electrical environment of different applications, it's necessary to separate it from other blocks in the tester. On the board level, the ground pins can be connected to the digital ground planes, however you need to insert a galvanic separate circuitry between the two ground domains. This is to prevent a negative impact on level accuracy and introduction of noise to your measurements, due to ground loops. A simpler approach is applying a ferrite bead.

If there are multiple ground planes of the same type across the board which is often the case, they will be connected through vias at many different places on the board. They allow you to have access to the ground plane from anywhere you can fit in a via. Using vias can help you to avoid ground loops. They connect the components directly to the ground points, which connect through low impedance to all other ground points. They also help to keep the length of return loops short.

## POWER SUPPLY FILTERING

Power supply noise suppression is much simpler, since power source from the tester is already filtered and smoothed before it arrives to the board. A common mode filter or an EMI filter should be enough for most of the case. plus, a few decoupling capacitors have become a must for any power pin.

## TEST POINTS APPLYING

The last thing you must not forget is add multiple test points to the board. Though for the safety reason, most digital pins are asked to add one for things like spike check. Yet test point for analog signal is also necessary. It gives you an easy access to see the original signal at different stages of the circuit. link it to the tester resource is another option as well. This will serve a great help during debugging and board condition monitoring in production.

## SOFTWARE CODING RULES

Unlike the hardware part, software has much less contribution to the noise suppression. once the environment is set, what left to do is make the measurement as accurate as possible. But on the good side, each measurement run, you will have a detailed waveform, and by advantage of the tester software, any need to be improved point is showed.

Measurements on a high resolution A/D convertor also have challenges to the tester as well.

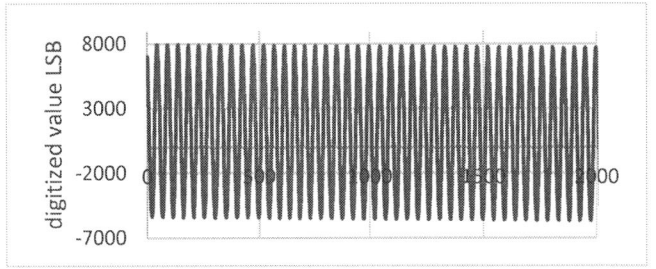

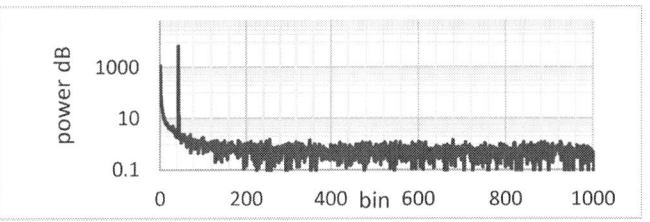

*Figure.3 the waveform and spectrum of a capture without initial discard.*

In order to capture a waveform with high AC performance of the digitizer, start capturing the waveform after discarding the initial samples is necessary. Without this action (showed as figure 3), the noise increases at the frequencies near the input signal frequency. This noise is caused by the drift of the A/D convertor inside the digitizer, and disappears when the above time elapses, then the digitizer starts samplings. By discarding the initial samples from the beginning, the drift ends, and you can use the digitizer with high AC performance.

In mixed signal tests, test signal waveforms are digitized with a waveform digitizer/sampler or an A/D converter, and the captured signal is processed with DFT/FFT basically. DFT / FFT is the main tool to analyze signals. The Fourier transform is based on the infinitely continuous waveform. So the final data point of the waveform should be smoothly continuous to its start point. If there is discrepancy between them, spectrum would be smeared. And If you would like to have a beautiful spectrum, you must set up your test condition strict coherent. A whole number of signal cycles should exactly fit the UTP. This indicates the digitizer or sampler has to work exactly on the same frequency with the convertor. But often limited by hardware specs or some other reasons, they won't be able to reach it. Which leads to a synchronize accuracy issue.(showed in Figure. 4)

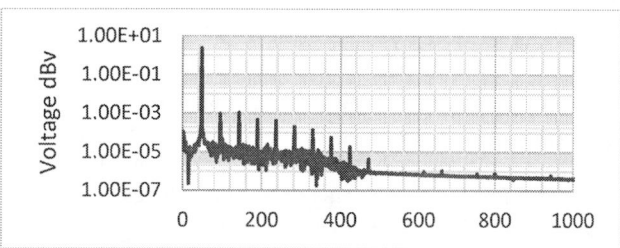

*Figure.4 spectrum of a capture with an dis-accurate synchronization.*

Then we need to investigate on the period / clock system of the tester.

The master clock in the V93000 test system is generated from the master clock generator on the clock board. it's distributed to different hardware. Thus all the clock in tester is generated via the divider from the master clock.

Specifying a period resolution means to decide a master clock rate, leading all the test system period equals to a multiple of the period resolution value. This will help greatly on the synchronize accuracy problem, which improve the result both for big signal and small signal level.(figure 5)

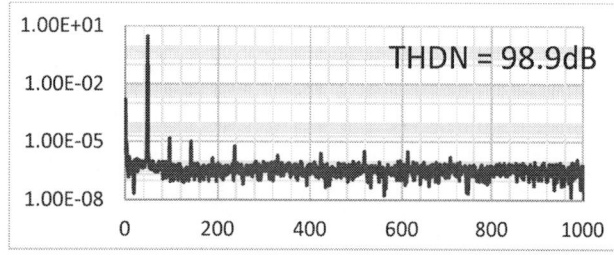

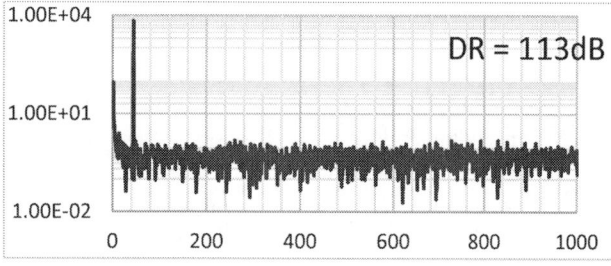

*Figure.5 spectrum of an accurate synchronization capture, both 0dB input and -60dB input.*

## CONCLUSUSION

The paper introduced a way to developpe an ATE solution for high resolution A/D convertor on both hardware and software level. Listed some typical noise sources, and the way to suppress them; showed some key points to take extra attention on the software coding. The figures showed in this paper are the result extracted from a real case, collected on the V93000 test system.

## ACKNOWLEDGEMENTS

We would like to acknowledge the support from company for providing us of this development opportunity. We also would like to acknowledge the colleague from all the different sites who have helped us a lot to debugging.

## REFEERENCE

1] Advantest Co., Ltd. "Advantest Technical Documentation Center"

2] Hideo Okawara "DSP-Based Testing Fundamentals"

# TOWARDS LEAN FRONT END IC MANUFACTURING
## (with AMHS implants)
### *George W Horn** Middlesex Industries SA, <u>gwhorn@midsx.ch</u>

## ABSTRACT

Just-in-time arrivals of WIP to process tools in the recursive and highly variant front-end manufacturing is not possible. On the other hand, the current degree of process decoupling via large buffers and storage systems is an overreaction to the above variability. This creates long cycle times and a manufacturing process loaded with up to three times more wafers in idle then those being processed. Contradicting lean efforts. Reviewing the fundamentals of arrival variability, it is shown that a major contributor to it is the AMHS system's inability to closely couple process steps due to its own variance. In fact, the AMHS multiplies process variance, making things worse. Current AMHS imitates the solution used in old manual fabs, stuffing excess WIP between a majority of processes, resulting in low fab efficiency. Instead, a method of closer process coupling is shown via reduction of WIP through the use of AMHS implants.

## THE INTER PROCESS GAP
### Exponentially Distributed Probabilities

If the wafer lot output of a process is a stochastic Poisson process, estimated as an exponential distribution of the time intervals between output events, and $1/\lambda_x$ mean and $1/\lambda_x^2$ as variance, and the transport of an output lot similarly being an independent Poisson process (discrete vehicles), estimated with exponential distribution of its service time, and $1/\lambda_y$ mean and $1/\lambda_y^2$ as variance, then, the variance of lot arrivals to the next step will be the combined variance of the two independent variables. One: the variable of wafer lot output intervals from the first process and the variable of hat lot's service time by the AMHS. The probability distribution of arrivals to the next process is computed via multiplication of the two independent variables.

In general, the outcome of two independent random variables $X$ and $Y$ acting as a system is a probability distribution obtained by the multiplication of the two, $XY = Z$. [1]

$$P(X \cap Y) = P(X) P(Y) = P(Z) \qquad (1)$$

If $X$ and $Y$ are independent, and continuous, described by probability density functions $f_X(x)$ and $f_Y(y)$, then the probability density function $f_Z(z)$ of $XY$ is [1]

$$f_Z(z) = \int_{-\infty}^{\infty} f_X(x) f_Y(z/x) \frac{1}{|x|} \qquad (2)$$

And the variance of the product $XY$ in general is [1]

$$\text{VAR}(XY) = (\sigma_x^2 + \mu_x^2)(\sigma_y^2 + \mu_y^2) - \mu_x^2\mu_y^2 \qquad (3)$$

Where $\sigma^2$ is variance and $\mu$ is the mean of the distribution. With the symbolism of exponential distributions,

$$\text{Var}(XY) = [1/(\lambda_x)^2 + (\lambda_x)^2][1/(\lambda_y)^2 + (\lambda_y)^2] - (\lambda_x)^2(\lambda_y)$$

And solving for the ratio:

$$\text{Var}(XY)/\text{Var}(X) = 3\text{Var}(Y) = 3/(\lambda_y)^2 \qquad (4)$$

Meaning that a multiplication of discharge variance from upstream tools happens, resulting in arrival variance to downstream tools, as a function of the AMHS variance.

The resultant of multiplication of the two random and independent variables is a bivariate density function of $X$ & $Y$, $f_Z(z)$. This bivariate density function, and its variance describe the wafer lot arrivals to a typical process in the fab. Let $f_X(x)$ be the density of time intervals between wafer lots exiting a process and let $f_Y(y)$ be the density of AMHS service times in moving those wafer lots to the next process, with the following characteristics,

$$\{X(a \leq x \leq b) \cap Y(c \leq y \leq d)\}$$

where the height of space above the area (a-b) · (c-d) is bounded by the upper surface $f_Z(z)$. In other words, the probability of the area in the $x$-$y$ plane given by the multiplicand of $XY$ as a resultant variate $Z$.

If X=Y, then the AMHS service is synchronized with the evolution of wafer lots from a process. In such a case Z is fixed and defined only by the common frequency of X & Y. Wafer lot arrivals to a process are determined by wafer lot evolution from the previous process, i.e. $\lambda_x = \lambda_y = \lambda_z$, and $1/(\lambda_x)^2 = 1/(\lambda_y)^2 = 1/(\lambda_z)^2$. Therefore, the existing variance of a previous process has not been multiplied, or if considering the global fab, inherent variance in the process has held steady. This case would define an idealized AMHS, and close coupling of the processes. Such idealization with the OHT type of AMHS (i.e. discrete vehicle principle) is not possible. While at the same time, the conveyor (hybrid) type of AMHS would better approximate it.

In case X is fixed to be constant (i.e. $f_X(x)$ is invariate) the wafer lot evolution from a process is fixed and steady state, the arrivals to the next process will be determined by the AMHS only. This will be exponential in Y, with a variance of $1/(\lambda_y)^2$ modifying arrivals around $\lambda_x$ mean. The normal industry procedure to reduce or eliminate this introduced variance into the process by the OHT type of AMHS is by creating a queue at the destination process. Of course, this will then increase fab cycle time.

In case Y is fixed to be constant (i.e. $f_Y(y)$ is invariate), the AMHS does not increase the exit variance of a process, and so that variance gets transferred unchanged to the destination process. Variance for the fab is then created only in a process. This scenario is generally assumed by fab management and measures implemented to deal with, as tools in a toolbox to increase competitiveness of the fab. These tools are the release and dispatch algorithms adopted by the fab. The current methodology to achieve this state of the fab is by making the AMHS irrelevant in manufacturing. Such irrelevance is achieved by creating high WIP content, which in turn assures high content of queues between process steps. To state this simply, fab cycle time is increased to quiet the instability that would be created by inter process moves through the use of variance multiplying AMHS types.

In a hybrid AMHS fab a moderate number of process steps should be closely coupled (a degree of synchronization) via AMHS systems without variance, i.e. direct and continuous flow transports (hybrid systems – conveyors + local OHT). This process is similar to using cluster tools, except on a macro scale of integration. A common belief has been that of scale. Meaning that the individual process variances are far larger than the AMHS variance in the process gap, therefore the latter is of little concern. However, the above calculations show that the significance of the AMHS introduced variance is in its multiplying the process variance.

## IMPLICATIONS OF THE OC CURVES

Controlling queue contents [2] by regulating flow rates in the process gap (via the availability of parallel tools or via holding WIP idle at upstream tools) implies a degree of underutilization of process tools. This directly contradicts common sense and presents us with a paradox. The problem is highlighted by the OC curve of the fab, which relates throughput to cycle time.

The Pollaczek-Khinchin result from queuing theory for Markovian arrivals to a single server M/G/1 queue is the basis for generating OC curves[6].

$$\frac{T}{\left(\frac{1}{\mu}\right)} = 1 + \frac{\rho(1 + C^2)}{2(1 - \rho)}$$

T- average time in queue and in service
$\mu$ - average service rate (throughput)
$\rho$ - server utilization $(1/\mu)$
$C^2 = \sigma^2/(1/\mu)^2$ coefficient of service variance

Combined with Little's Law (by the IBM Consulting Group),[6]

W (WIP) = (service time)·(arrival rate)

And further normalizing Cycle Time with Pure Process Time, as well as throughput (Service rate) with design capacity, the current OC relationship is used for factories:

$$CT = \alpha \frac{U}{1-U} + 1 \qquad (5)$$

$CT$ is (cycle time)/(pure process time), also known as FF (flow factor)
$\alpha$ is Coefficient of variance $(C_1^2 + C_2^2)/2$,
$U$ is throughput/capacity,
$C_1$ is coefficient of variance for inter process arrivals,
$C_2$ is coefficient of variance for the value-add processes,[3]

Note, however, that the coefficient of variance for inter process arrivals is a result of multiplying the discharge variable of the upstream process with the independent AMHS variable that brings the WIP to the arrival queue of the next process. Therefore the variance of the AMHS becomes a factor in the variance of the arrivals.

It is also important to recognize that the above derivation with queuing theory and little's law does not consider the physics of inter process moves. It merely looks at a system from the outside, may that have any types of physical components. Thus the Pollaczek-Khinchin Queueing model assumes Markovian arrivals with a general service rate distribution. Meaning a service rate with variance unspecified (M/G/1). Consequently, the derivative formula for OC curves is also devoid of any specific probability distribution of service (i.e. AMHS type). Yet, specifying AMHS will alter arrival rates and hence $\alpha$, the coefficient of the OC. It can be estimated that in practice the discrete vehicle AMHS will move carriers to destination via a stocker stop in 60% of cases (two inter process moves per wafer lot). And in these moves the arrival variance will be purely the AMHS variance (the variance of the second move). At the same time, the 40% direct inter-process moves will have the AMHS multiplying the variance of the upstream process discharge.

According to the probabilistic OC relation of Throughput and Cycle Time of a given factory, and due to the NP-Hardness[4] to create algorithms used to improve its inherent process variabilities (dispatch), the tool remaining to achieve a more favorable OC is via the revision of its logistics.

The focus on Dispatch to minimize the inherent process variance is justified[2]. However in these the regulation of overall WIP content of the Queues is overlooked. The proposal is to reduce Queue sizes to the point, where the logistics executed by the AMHS becomes relevant. And then, at that point to apply the hybrid AMHS, with is near zero variance multiplier in the gap for single step direct inter process moves.

## OVERALL WIP AND THROUGHPUT

Considering the relation of Fab throughput and WIP content the OC curve is converted by bringing WIP into the expression. *W= wip/ideal wip*.

$$W = \alpha \frac{U^2}{1-U} + U \qquad (6)$$

There is an idealized basic WIP content in the fab [5] which resides in process tools, i.e. none in the transport gap between processes. Or summing over each process tool group :

$$\sum_i (ct)_i (cap)_i \text{ which may be } \sum_i (ct)_i (cap)_{bneck}$$

Considering Fig 1, the tangential to the initial rise of the WIP curve quantifies the ideal WIP in the system as W= 1. Further on fig 1 it can be seen that at 80% capacity the Fab WIP content is W=4 ($\alpha$=1). This means that 3 times more WIP is in the process gaps (i.e. idle or in transit between processes) than in the process itself. Clearly, an inefficient manufacturing system.

## FAB EFFICIENCY AND WIP

To improve fab performance the wip content in waiting needs be reduced. This gets us closer to the distant ideal of Just-in-time. As we reduce WIP content while maintaining 80% utilization, we would violate the OC curve. Therefore we need to step outside of the current OC curve on which it is not possible to maintain the utilization at 80% with less WIP. And hence generate a new and better OC (fig 1). On this new OC curve at a WIP content of W= 2.4 we achieve a coefficient of overall variance of 0.5 (as opposed to the original =1). By reducing WIP content incrementally a whole family of OC curves may be generated. Decreasing the WIP will incrementally create curves with more efficiency. While theoretically such a reduction of WIP content is possible, it would prove to be impossible with current AMHS, which is not able to maintain tool utilizations without large WIP. Therefore, we need to create a new kind of AMHS to enable the WIP reduction. This is the rationalization of the hybrid AMHS, where we introduce a transport mechanism into the OHT system (conveyors) with less inherent variance to help the OHT maintain tool utilizations with less WIP.

In fig. 1 it is clearly seen that increasing the WIP in production beyond some value, will yield diminishing returns for throughput. When should one stop the pumping of more WIP into the system? To help with this

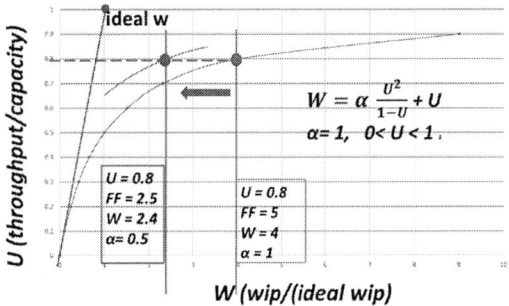

*Figure 1: WIP vs. throughput of a fab at 80% capacity. Reducing WIP of a fab a new family of curves can be generated. Example: exiting the curve with α = 1 will result in a new OC with α=0.5, requiring a wip content of only 2.4 times the ideal WIP.*

decision the value for WIP is incrementally reduced to the point where Utilization, or throughput starts to fall. It is currently estimated that this will happen at the normalized WIP value of W = 2.5, where the AMHS emerges from irrelevance. Expanding equation (6) it is seen that the exponential component of the curve does not intersect the

$$W= (\alpha\text{-}1)U^2 + (W+1)U \qquad (7)$$

linear part, therefore, it remains dominant. This means that for each increment in utilization we need to increase the WIP near exponentially. This diminishing return on WIP investment into the process is caused by the lack of close coupling between process tools preventing the ultimate lean process. However, the current inefficiency can substantially be improved by lending assistance to the OHT type AMHS with conveyor implants at logic flows between bays which are overburdened by WIP.

## REFERENCES

[1] An Introduction to Probability and Statistics, Second Edition, *Wiley*, Vijay K. Rohatgi, A. K. MD. & Ehsanes Saleh.

[2] An Integrated Release and Dispatch Policy for Semiconductor Wafer Fabrication, *International Journal of Production Research, 2014*, You LI & Zhibin Liang.

[3] Friedrich Böbel, PhD, Stefan Halmel, Siemens Semiconductors, *ASMC 1998*

[4] NP hard, *Wikipedia*

[5] Development and Simulation of Semiconductor production System Enhancement for Fast Cycle Time, *Technical University, Dresden, Killian Stubbe.*

[6] The Operating Curve, Steven S Aurand Ph.D., IBM Consulting Group, Peter Miller, IBM Microelectronics, *ASMC 1997.*

# STUDY ON STRESS MIGRATION OF FCQFN PACKAGE WITH UNBALANCED ARRANGEMENT OF BUMPS

*Shuanshe Chao[2]\*, Xinyi Lin[2]\*, Dan Yang[2], Na Mei[2], Tuobei Sun[2], Keqing Ouyang[1]*

[1] State Key Laboratory of Mobile Network and Mobile Multimedia Technology, ZTE Corporation, Shenzhen, 518055, P.R.China

[2] Department of Packaging and Testing, ZTE Corporation, Shenzhen, 518055, P.R.China

Corresponding Author's Email:10205916@zte.com.cn

## ABSTRACT

Stress migration is a common failure of Copper process chips, which is often related to the packaging structure and temperature stress. This paper mainly studies the stress migration of FCQFN package with unbalanced arrangment of bumps, which confirms unbalanced bump structure will cause greater stress risk and even lead to stress migration of Cu interconnection structure. In addition, we analyzed the relationship between plastic sealing materials, die thickness, bump size, temperature cycle and stress migration, It is found that the appropriate plastic packaging material and die thickness can effectively improve the problem of excessive stress. In the future, we will also perform this BKM for our development production.

*Keywords — stress migration; unbalanced bump structure; FCQFN; temperature stress;*

## INTRODUCTION

We have entered the era of 5nm process. At the same time, more and more 2.5D and 3D packaging technologies help us beyond Moore. However, for some vehicle gauge or analog power chips, the most advanced is not exactly the best. The requirement of high reliability is the most critical of the products, less is more[1]. Stress-induced migration failure is a common issue of copper process chips, which is often related to the packaging structure and temperature stress in Cu/Low-k Interconnects[2]. This paper mainly studies stress migration of FCQFN package with unbalanced bumps, as shown in *Figure 1*. In order to evaluate the stress capability of the chip in practical application, such as high temperature reflow soldering, we try to estimate stress risk of reflow and temperature cycle from the conditions of plastic packaging material and temperature cycling conditions[3].

Under unbalanced design of bump, select the plastic sealing material and die thickness that meets the requirements of high reliability of chip and the highest strength temperature stress can be borne. Especially for the new design, there will be unrecongnized variables in the stress simulation model. This paper try to identify design risk in development phase through reliability verification test.

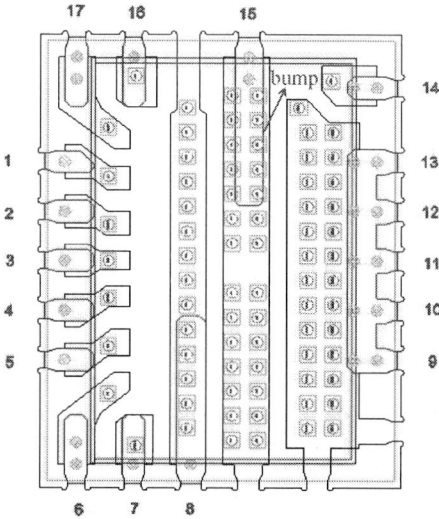

*Figure 1: Schematic diagram of packaging structure*

For packaging reliability test of QFN packaged chips, TCC(-65°C~150°C) condition is generally selected in temperature cycling test according to JESD22-A104E.Meantime, and stress simulation is carried out for the chip at the beginning of the project, as shown in TABLE I. However, during the reliability verification test, it is found that the chip has problem with varying degrees of over stress in the middle of chip, as shown in *Figure 2*. At the same time, it is found that electrical parameter drift related to bangap circuit units in precondition, HAST(Highly Accelerated Stress Test) and HTST(High Temperature Storge Test).

Due to the requirements of current carrying capacity, there are more but not less bumps in the bump dense area shown in Figure 1, which will lead to failure caused by stress imbalance in the stress environment related to temperature stress, such as the reduction of chip yield and the drift of electrical parameters in precondition,HAST and HTST with low temperature stress, such as reference voltage parameters drift.Large temperature stress in TCC test will causecracks from the plastic package to passivation layer, and even the problem of short circuit caused by stress migration in the metal layerof die.Without changing the chip function design and bump distribution, this paper reduces the stress impact of unbalanced bump design on the chip by optimizing the thickness of plastic packaging material and die. On this basis, it verifies the reliability of the chip to ensure that it can meet the application requirements of the chip.

TABLE I.     STRESS SIMULATION RESULTS

| Die thickness | Max Equivalent Stress/MPa | | Loading condition |
| --- | --- | --- | --- |
| | pin | bump | PMC |
| 250um | 231.9 | 481.5 | |

## STRESS MIGRATION

The integrated circuit chip is stored at a certain temperature for a certain time without applying current. In some cases, holes or even disconnection can be observed on some metal wires. Stress migration is a diffusion process caused by mechanical stress. The generation of mechanical stress mainly has two aspects: on the one hand, the thermal expansion coefficient is inconsistent between the metal process of integrated circuit and the process of protective insulation layer; Another aspect is the stress of the chip caused by the unbalanced bump distribution under the temperature stress. No obvious abnormality is found after precondition, HAST and HTST, but the chip reference voltage tend to drift up within the specification range. There is no obvious crack on the surface of the package, but delamination occurs at the interface between bump and plastic sealing material. However there is obvious crack on the surface corresponding to bump dense area in TCC, and crack extends from the plastic surface to bump, passivation, as shown in *Figure 2* to *Figure 4*, which indicates under temperature cycle conditions, the stress in the area with the most dense bump distribution is the largest, and this problem is obviously exacerbated in TCC 1000 cycles.

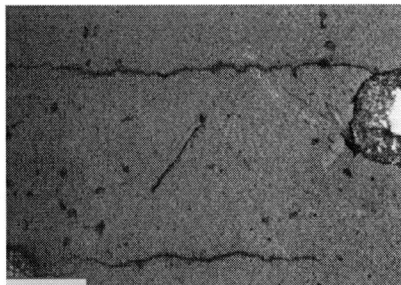

*Figure 2: Cracks on the surface of plastic package*

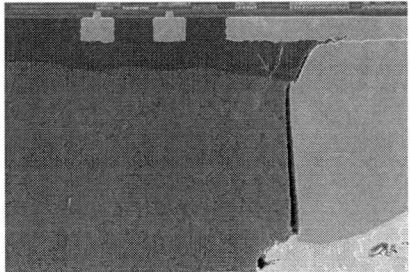

*Figure 3: Crack propagation*

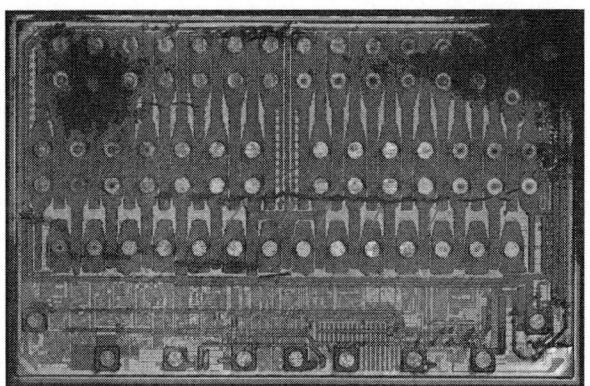

*Figure 4: Crack of passivation layer*

*Figure 5: Deformed morphology of TiN layer*

According to the failure phenomenon of TiN layeras shown in *Figure 5*, it is suspected that mechanical stress caused stress migration in TC test. TC test is conducted to determine the ability of components and solder interconnects to withstand mechanical stresses induced by alternating high- and low-temperature extremes. Permanent changes in electrical and/or physical characteristics can result from these mechanical stresses. The hot spot results are shown in *Figure 6*,When 0.01V voltage is applied to OBIRCH machine, 2.1mA leakage occurs in the deformation area of TiN layer. Physical failure analysis shows that top metal has obvious cracks and discoloration, as shown in *Figure 7*. After FIB confirmation of the abnormal position, it is found that continuous metal holes appear in the metal layer, and the metals of different circuits are bridged,as shown in *Figure 8*.

Through the above analysis, temperature cycling is a critical root causes of chip stress migration.The stress simulation, as shown in *Figure 9*, and TCC 1000 cycles test results show that there is a large stress in the bump imbalance area.

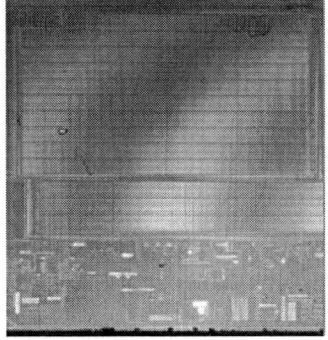

*Figure 6: Hot spot of failure sample by OBIRCH*

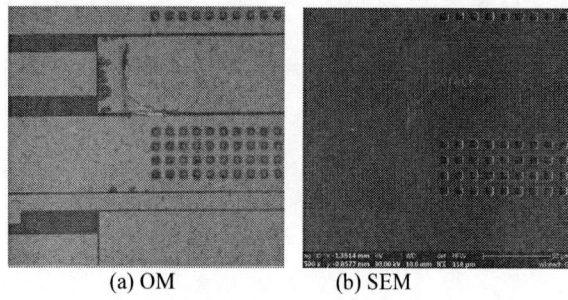

| (a) OM | (b) SEM |

Figure 7: Abnormal morphology of top metal layer

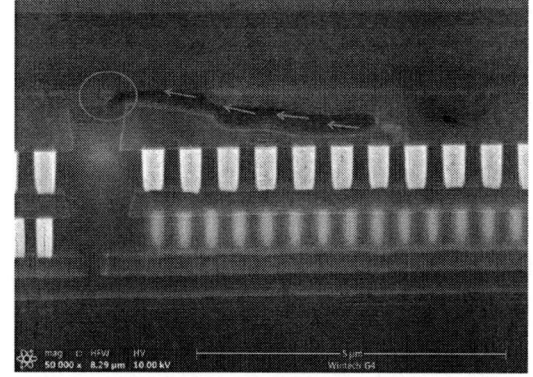

Figure 8: Stress migration of metal layer

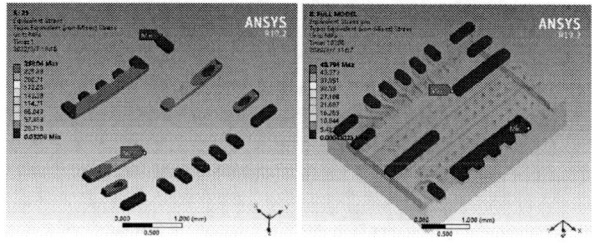

Figure 9: Stress simulation

## DISCUSSION ON OPTIMIZATION

The problem of stress migration appears in the reliability verification test, which proves that the unbalanced bumps arrangement will cause the problem of excessive local stress of the chip. First of all, we confirmed that the stress migration problem has been significantly improved after reducing the TC conditions. At the same time, the TCB conditions meet the needs of customers and industry standards. In addition, a variety of methods were used to opimize the chip design to meet higher reliability requirements and solve the stress migration issue.

### Optimize Bump

The demand of chip current carrying capacity can not be reduced in the dense area of bump distribution, but the stress can be balanced by adjusting the diameter size of the bump in the dense area. However, the simulation results show that the stress increases with changing the size of the bump, indicating that modifying the diameter size of the bump in the dense area can not balance the stress,as shown in TABLE II.

TABLE II.    STRESS SIMULATION RESULTS

| Type | Max Equivalent Stress/MPa | | Remark |
| --- | --- | --- | --- |
| | Pin | Bump | |
| round bump | 231.99 | 481.5 | |
| Elliptical bump horizontal | 232.33 | 503.4 | 250um |
| Elliptical bump vertical | 232.17 | 487.8 | |

### Optimize Chip Structure

The chip area accounts for a large proportion, and there is no space for optimization. We choose to optimize the die thickness, and the stress of the bump is reduced after thinning. As shown in the TABLE III, 200um can be selected as the optimized die thickness.

TABLE III.    STRESS SIMULATION RESULTS OF THICKNESS

| Die thickness/um | Max Equivalent Stress/MPa | | Loading condition |
| --- | --- | --- | --- |
| | pin | bump | |
| 200 | 249.5 | 470 | |
| 250 | 231.9 | 481.5 | PMC |
| 300 | 226.1 | 484.4 | |

### Optimize Plastic Sealing Materials

The short problem caused by stress migration was significantly reduced after reducing the TC condition and

978-1-6654-9759-6/22 $31.00 © 2022 IEEE

optimizing bump and chip structure, but it was not completely solved; After selecting the plastic sealing material with lower Young's modulus, there was no short circuit after TCB 1000 cycles test, the problem of Vref parameter drift has been solved.

By optimizing the chip design, we solved the problem of stress migration caused by unbalanced arrangement of bumps, and also optimized the problem of parameter offset.

# CONCLUSION

FCQFN package with unbalanced arrangement of bumps has stress risks. we analyzed the relationships between plastic sealing materials, die thickness, bump size, temperature cycles and stress migration. It is found that the appropriate plastic packaging material and die thickness can effectively improve the problem of excessive stress. In the future, we will also perform this BKM for our development production.

# REFERENCES

[1] J. M. Passage, N. Azhari and J. R. Lloyd, "Stress Migration Followed by Electromigration Reliability Testing," 2019 IEEE International Reliability Physics Symposium (IRPS), 2019, pp. 1-5.

[2] C. Lee and A. S. Oates, "A New Stress Migration Failure Mode in Highly Scaled Cu/Low-k Interconnects," in IEEE Transactions on Device and Materials Reliability, 2012, pp. 529-531.

[3] A. Tezaki, T. Mineta, H. Egawa and T. Noguchi, "Measurement of three dimensional stress and modeling of stress induced migration failure in aluminium interconnects," 28th Annual Proceedings on Reliability Physics Symposium, 1990, pp. 221-229.

# MACHINE LEARNING BASED AGING CRITICAL PATH SELECTION

*Keqing Ouyang[1,2]\*, Jiadong Yao[2], Qi Wei[1,2], Chenfei Wu[1,2], Guohua Zhou[1,2]*

[1]State Key Laboratory of Mobile Network and Mobile Multimedia Technology, China
[2]Sanechips Technology Co., Ltd, Shenzhen, Guangdong, China
\*Corresponding Author's Email: ouyangkeqing@sanechips.com.cn

## ABSTRACT

Under the current advanced semiconductor technology node, the aging of transistors becomes one of the main factors which affect the reliability of circuits. In order to cover the aging effect, the proper timing margins are usually added in the design according to the aging-aware timing analysis of the timing paths. However, the popular solution built on aging-aware SPICE simulation could hardly cover all the potential aging-critical paths due to its low operation speed. In this work, we developed a machine learning based classification system which could quickly complete critical path selection from a large number of paths. It extracts the path-level statistical features from the STA timing reports and invokes the integrated classifier which contains Random Forest Classifier, Support Vector Classifier and K-Nearest Neighbors Classifier to implement critical path (identified based on specific benchmarking threshold) selection. The verification results showed that the proposed system could classify about 1000 paths within 10 s, and the recall scores reach over 90% for various thresholds and process nodes. It is thought that the critical path selection scheme proposed in our work is well compatible with the current mainstream design flow and improves the aging analysis efficiency in the large design.

## INTRODUCTION

With the rapid development of the technologies about automatic drive, medical devices, high-performance computing and so on, demand for high-reliability systems becomes more and more urgent. And it raises a greater challenge to the advanced digital circuit design. Therefore, reliable analysis methods and corresponding solutions about various reliability issues must be established to meet the application requirements of the market [1].

Under the current advanced semiconductor technology node, the aging of transistors is one of the main factors affecting circuit reliability. With the decrease of its size, the effective electric field and junction temperature in the transistors become higher. As a result, the aging effects (such as bias temperature instability and hot carrier injection [2]) in the standard cells are enhanced, which results in performance degradation and even function invalidation. In order to solve this problem, the most popular engineering strategy

is to analyze the aging influence of the timing path based on SPICE simulation, and then add the timing margin in the circuits design according to the simulation result to cover the timing decline caused by the aging effect. However, the simulation efficiency of SPICE is very low. And so, in the large design, it is almost impossible to cover as many potential aging critical paths as possible through the simulation of a huge number of paths. Therefore, aging critical path selection becomes a key issue that affects the analysis efficiency and results accuracy.

Critical path selection could be attributed to a classification problem, which is exactly the area that the machine learning technology is good at. Therefore, in this work, we defined the path as critical path where the path delay variation percentage before and after aging exceeds a certain threshold, and then developed a machine learning based classification system which could quickly complete critical path selection in a large amount of path data. The following measurement results showed that the proposed critical path selection scheme is well compatible with the current circuit design flow and improves the aging analysis efficiency in the large design.

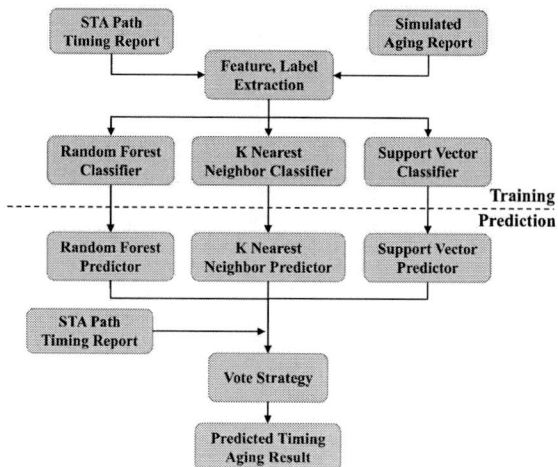

*Figure 1: Schematic Diagram of Critical Path Selection Process Based on Machine Learning Models.*

## PROPOSED APPROACH

Figure 1 shows the schematic diagram of the process of the machine learning based critical path selection. It consists of two stage, model training and critical path

*Table I: Path Level Features for Model Training.*

| Feature Type | Feature Description | Feature Counts |
|---|---|---|
| cell delay | information like total and max delay | 3 |
| VT type | cell voltage threshold type like LVT | 3 |
| cell drive | information like total and max drive | 3 |
| cell type | types like buffer and ND gate | 16 |
| output arc | cell timing arc type like rise to rise | 8 |
| other | aspects like derate and si | 10 |

prediction, which are described in detail below.

**Model training**

First, path-level statistical features were extracted from STA timing report for model training. We selected a series of path features from many aspects, such as cell delay, VT type, cell drive and so on. Table 1 lists the specific feature types, the corresponding feature description and total number of features. These selected features basically cover the factors that may be related to the aging effect and can be extracted easily from the STA timing report. On the other hand, the classification label of training was obtained from the aging report of SPICE simulation. Considering the fact that those paths with a greater aging delay variation are more possible to cause timing violations, we marked the paths as critical path where the aging delay variation percentage exceeded a certain threshold.

Then, with the features and labels extracted from thousands of data paths, three machine learning models based on random forest algorithm, support vector machine and K nearest neighbor algorithm [3-5] were trained independently for critical path prediction.

**Critical Path Prediction**

Since the critical paths should be selected as many as possible to reduce the risk, we adopted a biased vote strategy to improve the critical path selection ratio further on the basis of three independent prediction results. The detailed prediction procedure is as follows: first, as same as model training, the path features were extracted from the STA timing report; then, the three prediction models were used to predict critical path independently; finally, vote to determine the final prediction result. The specific vote strategy (referred as Vote: Any) used is that as long as any of the models predicts the path as critical path, we finally treated it as critical path.

## EXPERIMENTAL RESULTS

Figure 2 shows the prediction results of each model.

As shown in the figure, the classification accuracy, precision and recall scores of the three independent prediction models based on random forest, support vector machine and K nearest neighbor algorithms are all above 0.9. This indicates that the three machine learning models adopted by us have good applicability for aging critical path selection, thus ensuring the accuracy of the final result with vote strategy. On this basis, we adopted the aforementioned biased vote strategy (Vote: Any) as the final prediction result in order to select as many high-risk critical paths as possible. It can be seen from the comparison with three independent prediction models and the unbiased vote strategy (referred as Vote: More, the principle that the minority obeys the majority), although the precision score of Vote: Any strategy was reduced (this cost is acceptable in engineering), the recall score was much improved and so more critical paths can be selected to reduce aging risk.

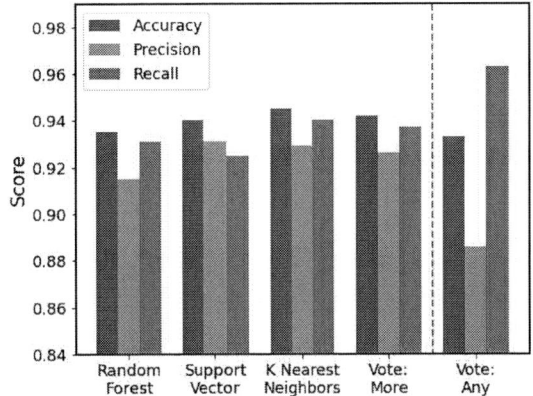

*Figure 2: Prediction Scores of Different Models for Critical Path Selection.*

In addition, we further explored the accuracy of prediction results under different classification thresholds considering that it is needed to select critical paths with different aging degrees in various cases. The prediction results at different aging thresholds are shown in Figure 3, where the thresholds are chosen within the standard deviation range near the mean value (mean: 3.93, standard deviation: 0.50). It can be seen from the figure that the accuracy and recall scores at different thresholds were all above 0.9 and precision scores were also above 0.8, which indicates that our classification system has good versatility under different thresholds and can be used for selection of critical paths with different aging degrees.

Furthermore, we studied the versatility of this classification systems in different process nodes. We applied it to multiple blocks from different designs, trained the models and predicted the critical paths respectively. The prediction results are listed in Table 2.

We can see that all the recall scores at different processes and blocks could reach over 0.9, which indicates that the system can be applied to different processes and blocks and implement critical path selection very well.

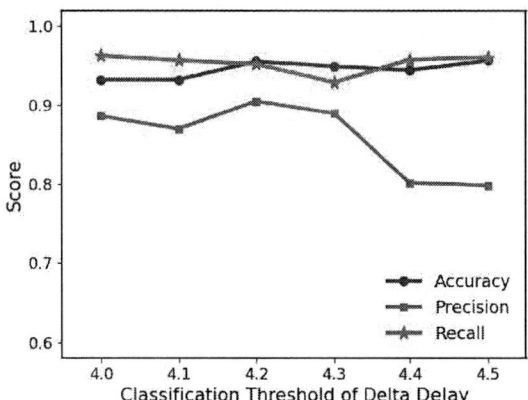

*Figure 3: Prediction Results at Different Classification Thresholds.*

Finally, the operation speed of the classification system was measured. The result showed that the system could complete the prediction of 1000 paths within 10 s, while the aging simulation with SPICE takes dozens of hours. Therefore, the proposed machine learning based system could quickly select critical path with high aging risk from a large number of timing paths, and so improve the efficiency of aging analysis in the large-scale chip design.

## CONCLUSIONS

In this work, we proposed a machine learning based classification system for aging critical path selection. The system extracts path-level features, uses random forests, support vector machine, and K nearest neighbor algorithms for model training and prediction, and finally adopts a biased vote strategy to further improve the

*Table 2: Prediction Results (Recall/Precision) of Different Processes and Blocks.*

| Process | Block 1 | Block 2 | Block 3 |
|---|---|---|---|
| Process A | 0.941 / 0.808 | 0.912 / 0.786 | 0.972 / 0.983 |
| Adv Process B | 0.921 / 0.785 | 0.903 / 0.731 | 0.951 / 0.911 |
| Adv Process C | 0.952 / 0.783 | 0.960 / 0.776 | 0.913 / 0.889 |

selection ratio. And all the recall scores could reach over 0.9 in the prediction results of different thresholds, processes and blocks. Meanwhile, the prediction of 1000 paths could be completed within 10 seconds, which showed a much better operation speed than SPICE simulation. The above results show that the proposed scheme for aging critical path selection has good performance and versatility and improve the efficiency of aging analysis in the large-scale chip design. It's also thought that this work could provide certain reference to handle critical path selection problems in other fields.

## ACKNOWLEDGEMENTS

This work is supported by the department of back-end design of Sanechips Technology Co., Ltd. Thank all the team members for help in preparing and analyzing of STA timing reports.

## REFERENCES

[1] K. Huang, X. Q. Zhang and N. Karimi. *IEEE T. Instrum. Mean.*, Vol. 68, No. 12, 2019, pp. 4756-4764.

[2] X. F. Guo and M. R. Stan, *Circadian Rhythms for Future Resilient Electronic Systems*, Springer, 2020.

[3] T. Cover and P. Hart, *IEEE Transactions on Information Theory*, Vol. 13, No. 1, 1967, pp. 21-27.

[4] L. Breiman. *Machine Learning*, 45, 2001, pp. 5-32.

[5] J. Qin and Z. S. He, *International Conference on Machine Learning and Cybernetics*, Aug 18-21, 2005, pp. 5144-5149, Vol. 8.

# Digital Defect Traceability across Sapphire Processing: Case Study on Micro-LED Chain

Dr. Ivan ORLOV [1], Dr. Gourav SEN [2] and Frédéric FALISE [1]

[1] Scientific Visual, Lausanne, Switzerland | Ivan.Orlov@scientificvisual.ch
[2] Fametec-Ebner GmbH, Leonding, Austria | Sgo@fametec.co

## ABSTRACT

We present preliminary results of a case study carried out by four major market players. An automated crystal scanner recorded defects in sapphire boules as they passed through consecutive production stages: *raw crystals (boules)* ➔ *cores* ➔ *EPI-ready wafers*. In parallel, the defectiveness was assessed by experts using manual methods. The experiment design allowed tracing individual defects through all stages.

We also show how digital quality control increases wafer yield by 'intelligent wafering', i.e. by correctly positioning the sapphire core in a wafering system.

## INTRODUCTION

LEDs and micro-LEDs industry highly depend on the sapphire substrate. According to IHS LED Intelligence Service, 96.7% of the global LED production in 2020 was achieved on sapphire wafers. Material defects in the substrates - such as micro-bubbles, clouds, and structures - cause rejection of finished wafers and may also disrupt large production lines for weeks.

Manufacturers mitigate this risk by imposing increasingly stringent quality requirements on their sapphire core suppliers. The latter adapt by rejecting all suspicious areas in crystals, even if they are clean but located near a defect. As a result, a considerable part of fresh-grown sapphire becomes scrap, even though it can be used if adequately evaluated.

This study is to help sapphire manufacturers to transition to digital quality control that leads to material-saving by:

- tracing how defects in raw sapphire crystals affect the quality of micro-LED wafers
- identifying critical defect features that decrease the yield for the wafer manufacturer
- coding such indicators into the raw crystal scanner software
- increasing productivity by proactively improving core and wafer processing through computer-aided optimisation.

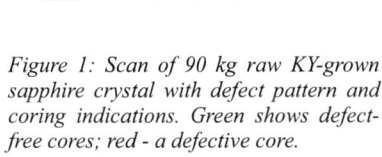

*Figure 1: Scan of 90 kg raw KY-grown sapphire crystal with defect pattern and coring indications. Green shows defect-free cores; red - a defective core.*

The project is led by Scientific Visual and Fametec-Ebner and involves third-party manufacturing companies and materials experts worldwide.

*Scientific Visual* operates the world's largest raw industrial crystal scanner, TotalScan™. Equipped with a fully automated 4-axis scanning head, it can examine crystals up to 350 kg and accurately detect internal defects down to 8 µm in sapphire, LT/LN, LBO, ZnSe, and semiconductor crystals. This service facility, located in Switzerland, was used as the analytical center for the study.

*Fametec-Ebner* is the market leader in heat treatment furnaces, running its own sapphire growing facility. The company meets the industry's needs for high-quality wafers by growing several 7- or 9-inch crystals in a single furnace. The crystals grown with a proprietary technique yield ultra-low bow and warp wafers that are specifically suited for micro-LED manufacturing.

The authors emphasise that this research is ongoing, and the paper shows only its interim results.

## RESEARCH METHODOLOGY

**1. Raw material.** For this study, we used *c*-axis ⌀7" HEM crystals subsequently cored to ⌀6" cylinders, which is a standard size in the industry. The material was grown in several furnaces and at different times to reduce the effect of random fluctuations that may arise during the crystal growth. To select crystals suitable for the study, the authors scanned 96 raw crystals with a total weight of 1'480 kg. The crystals had to present a substantial variability of defect morphology and size. Five 7" rough crystals, representing the full range of quality from clean to very defective, were retained for processing into 6-inch cores.

The authors underline that the highly defective crystals chosen for this project do not represent the quality of Fametec-Ebner production.

**2. Crystal scanning.** A Scientific Visual TotalScan™ scanner was used for defect inspection. The scanner automatically detects bubbles, structures, and clouds in raw ("as is") crystals of any shape and determines defect-free zones suitable for wafer fabrication. For illustration purposes, For illustration purposes, Fig. 1 shows an example of such defect-conscious coring on a sapphire crystal grown by Kyropolous method. The 3D defect patterns obtained at this stage served as a starting point for defect tracing.

In addition to the automated measurement, experts from the industry evaluated the crystals manually to compare the reliability of human and TotalScan™ defect identification.

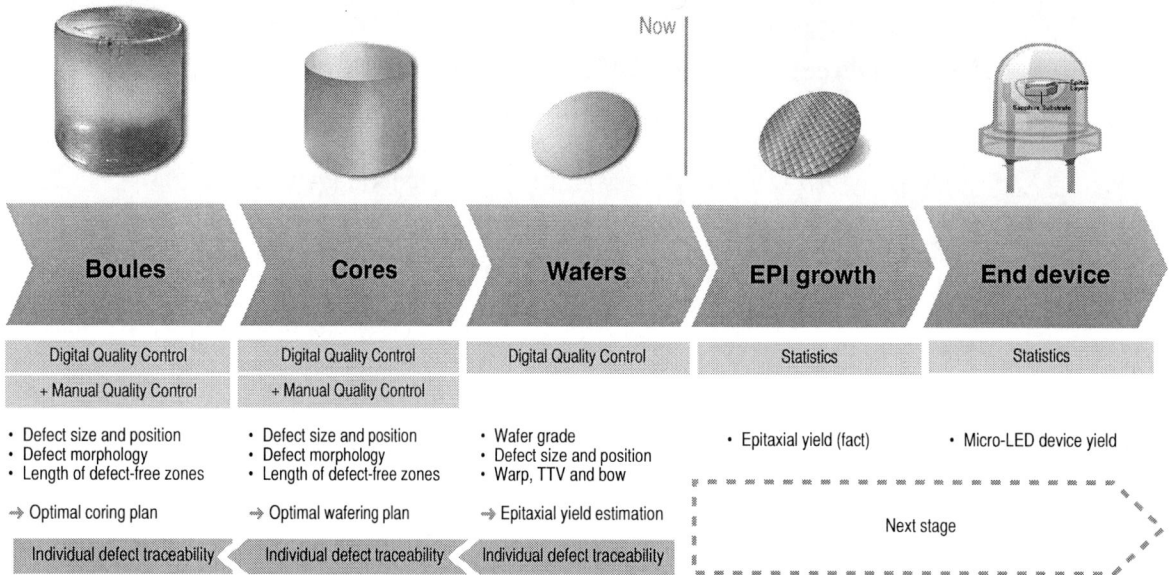

*Figure 2: End-to-End study design*

**3. Coring.** A reputable tier-one core maker processed the selected crystals into ⌀6" cores. It had the references that allowed to unambiguously locate cores in the volume of the corresponding parental crystal. As in the previous stage, a fabrication expert independently evaluated the defect-free zones before passing the cores to the next wafering stage.

**4. Core scanning.** The five cores were (re-)scanned in TotalScan™ with the same settings as the original crystals, ensuring that their orientation was identical to one of the parent boules. The software output included defect pattern and length and position of the defect-free zones suitable for wafer processing. The 3D defect fingerprints obtained at this stage served as the second reference point in defect tracing.

**5. Wafering.** The cores were wafered with 1.75 mm pitch and consequently polished to 0.9 ± 0.025 mm thickness by a leading LED wafer manufacturer. They were marked with reference lines and points so that their orientation and position within the parental core could be unambiguously reconstructed. Thus, the authors obtained a mutual orientation of wafers, cores, and the original raw crystals. By now, we have received the complete end-to-end dataset for the first crystal; the others are in progress.

**6. Wafer inspection.** Each wafer was analysed using an automated KLA-Tencor Candela® system, which gave the indication of whether it complains to micro-LED specification and, if not, the reason for it: material defects, polishing/lapping defect, or AFM.

Warp, TTV, and bow values were also measured at this stage. Fametec-Ebner wafers exhibited average values that are 2 to 15 times ahead of the industry standards (shown in brackets): average bow 12.55um (0±20 um), average TTV 2.53 (≤40 um), and average warp 13.52 (≤31 um). This high structural quality directly impacts the uniformity of the LED wavelength, maximising the output of LED devices and increasing the price of specification-compliant devices.

**7. Integration of results.** 3D defect patterns collected at various stages were integrated using software *Yield Pro v4.4* by Scientific Visual for sapphire quality analysis. The integration took into account their initial orientation inside each other through reference points. It allows to trace the evolution of each defect individually and draw conclusions about its specific impact on the wafer yield.

## RESULTS AND CONCLUSIONS

### 1. Eye control largely overestimates defectiveness in raw crystals

Fig 3 shows an example of waferable area identified by the scanner and a human expert in raw crystals and in the same volume after coring.

The scanning was set to a zero-tolerance profile: any defect within the scanner's detection range (down to 8 um) rejects the corresponding wafer.

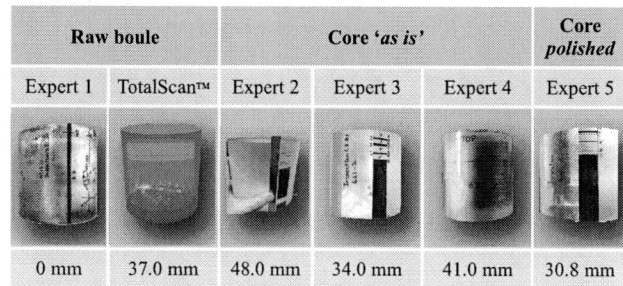

| Raw boule | | Core 'as is' | | | Core *polished* |
|---|---|---|---|---|---|
| Expert 1 | TotalScan™ | Expert 2 | Expert 3 | Expert 4 | Expert 5 |
| 0 mm | 37.0 mm | 48.0 mm | 34.0 mm | 41.0 mm | 30.8 mm |

*Figure 3: Example of crystal evolution: sample 1621-2 (h165 mm, ⌀170 mm). The bottom raw shows waferable areas identified by five human experts and TotalScan™ at various stages.*

Fig. 4 shows how such evaluation correlates with the crystal defectiveness.

It can be clearly seen that:

a) visual inspectors repeatedly underestimate waferable length. It is due to objective obstacles to visualising defects in raw crystals and the psychological tendency to overestimate defects in order to mitigate risks of defective material entering the production chain.

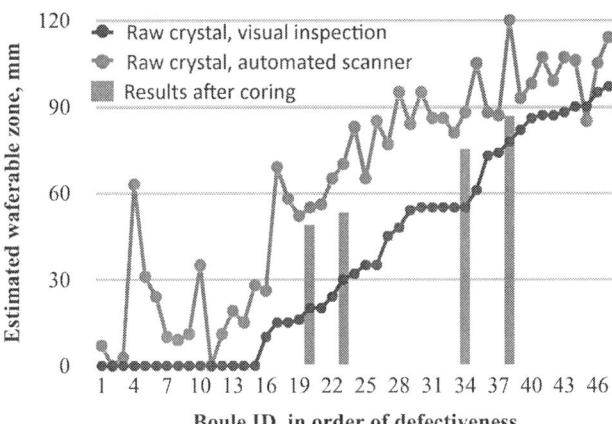

Figure 4: Waferable areas identified by a human expert and TotalScan™ in 47 raw crystals before coring. Grey bars indicate results after coring.

(b) the human error is higher in crystals with high defectiveness. It might be related to "bias trap": when finding significant defective areas, the human expert tends to consider the whole crystal defective, despite the presence of commercially significant waferable zones.

The overall statistics show that human expert underestimates waferable areas by ~32% in raw crystals and ~27% in cores. With manual quality control, most of these erroneously defective zones would be sent to scrap.

## 2. Defect-conscious wafering enables up to 7% more wafers

When there are few defects in the core, their positioning to the cutting planes matters. Knowing the precise defect coordinates allows calculating a core offset to position more defects into sawing gaps and out of future wafers.

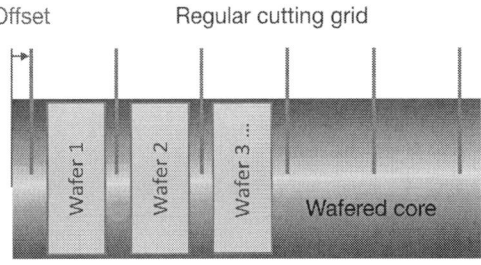

Figure 5: The principle of Intelligent Wafering: pre-computed offset of cutting grid increases the number of good wafers (patent pending).

Table 1 shows an example of such gain. The only parameter to control is the offset of the first cutting plane to the edge of the core.

Out of all possible offset values, the table shows the theoretical worst and the theoretical best, yielding 44 or 49 good wafers out of 57, respectively. That makes 7% yield difference. In practice, the cutting wire or blade is positioned with imprecision of ±0.25mm, so that the first cut is unlikely to be exactly at the best offset (1.4 mm in this case). The correction for the imprecision reduces number of good wafers to 48, therefore ensuring a gain of 4.5% over uncontrolled "blind" core positioning, which averages at 46 quality wafers.

Table 1. Intelligent wafering: wafer yield as function of core offset, as sketched in Figure 5.

|  | Offset, mm | Good wafers | Bad wafers |
|---|---|---|---|
| **Worst cut** | 0.6 | 44 | 13 |
| **Best cut (theory)** | 1.4 | 49 | 8 |
| **Best cut (practice)** | 1.4 ± 0.25 | 48 | 9 |

## 3. The study allows to calibrate yield-impacting defect threshold for LED wafers

The cores chosen for the end-to-end test have typically a conical distribution of defects in the seed area. In other words, the defect density in wafers decreases from the bottom upwards and was zeroed at a certain height.

As the scanner has a higher defect sensitivity than required by LED standards, it raises the question where to set the threshold separating yield-impacting defects from tolerable ones. Combining data from TotalScan™ and Candela® let us derive the cut-off: defects below it are not diagnosed at wafer quality control and, thus, can pass into production. Fig. 6 shows the distribution of defects and the wafers' status.

*Disclaimer:* This article presents first available data, which is not statistically valid. The information will be refined with the oncoming end-to-end statistics. For example, the compliance of an individual defect-containing wafer may be influenced by the defect location depth (on the surface or in the volume). This random fluctuation will be averaged out with the statistically valid dataset.

Figure 6: Wafer compliance in the crystal 3121-4. Green - good wafers, yellow - defects detected by TotalScan™ in raw crystal, red - defects detected by TotalScan™ in raw crystal and Candela® in wafers.

## SUMMARY

Thanks to end-to-end defect tracing, the authors correlated defects yield-impacting defects in polished wafers with the ones identified by TotalScan™ in the raw crystals.

The consortium will continue to gather more statistics. The complete dataset and metrics are available to project participants.

The obtained correlations confirm that digitalisation of crystal quality control offers tangible opportunities to improve profitability. Processing companies could extract from 5 to 20% more quality wafers with intelligent processing.

## ACKNOWLEDGEMENTS

The consortium is proud to be financially supported by Eurostars program of the European Union, as well as the Swiss innovation agency – Innosuisse.

# RESEARCH ON RELIABILITY OPTIMIZATION MECHANISM OF 28HKMG TECHNOLOGY

*Weiwei Ma[1]\*, Ran Huang[1], Yamin Cao[1], Wei Zhou[1]*

[1]Shanghai Huali Integrated Circuit Corporation, Shanghai, China
Corresponding Author's Email: maweiwei@hlmc.cn

## ABSTRACT

With the aggressive scaling down of the gate dielectric, reliability issues especially NBTI (Negative Bias Temperature Instability) and HCI (Hot Carrier Injection) become serious challenges. In this study, the critical roles of both gate dielectric thickness and DPN process are investigated in 28 nm HKMG technology. It's found that under a commercial DPN process, a certain thickness of gate dielectric is required. When this requirement is fulfilled, qualitative changes of NBTI performance could be made through DPN power optimization. And thus, a nitrogen profile modulation model is proposed to explain this significant improvement of NBTI performance after choosing a proper gate dielectric thickness and DPN process power.

## INTRODUCTION

With the continuously scaling down of transistors, high-k dielectrics and metal gates are introduced as new gate stacks for solving intolerable tunnel leakage problems [1-4]. Meanwhile, reliability problems such as NBTI and HCI have become major concerns when realizing highly reliable integrated CMOS devices [5-6]. Various NBTI models have been proposed, and Reaction–Diffusion (R–D) model is the most prevalent one, which is highly related to hydrogen terminated silicon dangling bonds [7]. However, the roles played by nitrogen in NBTI performance still have controversies.

In this study, the critical roles of both gate dielectric thickness and DPN process were investigated in 28 nm HKMG technology. And a nitrogen profile modulation model was proposed to explain the internal mechanism of the improvement of NBTI performance.

## EXPERIMENT

Wafers of different gate dielectric thicknesses and various DPN powers were prepared. In this experiment DPN power was used to adjust N concentration. All wafers analyzed in this study were manufactured based on HLIC 28 nm HKMG technology. And the NBTI lifetimes of samples were tested by common practice in the industry.

## RESULTS AND DISCUSSION

### 1. The Thickness of Gate Dielectric was Critical

As shown in table 1, only with a certain increase of gate dielectric thickness then a significant improvement of NBTI lifetime could be achieved. The benefit of adjusting the gate dielectric thickness or DPN power separately was low.

| DPN Power | Split  | IL BSL | IL+x | IL+2x | IL+3x |
|-----------|--------|--------|------|-------|-------|
| BSL       | HK BSL | 1X     | 6X   | 7X    | 70X   |
| DPN-w     | HK BSL | 5X     |      |       |       |
| BSL       | HK+y   |        |      | 10X   |       |
| DPN-w     | HK+2y  |        |      | 200X  |       |
| DPN-3w    | HK+6y  |        | 100X |       |       |

*Table 1. The Extent of NBTI Lifetime Improvement under Different Conditions*

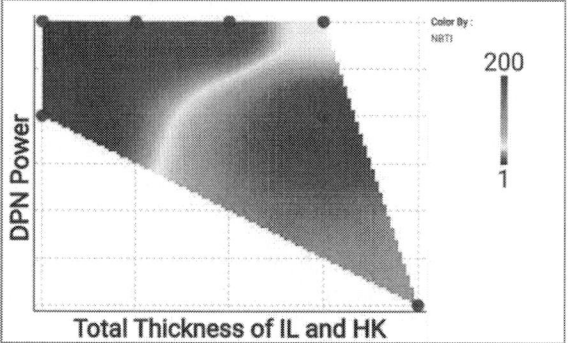

*Figure 1. The Extent of NBTI Lifetime Improvement under Different Conditions*

To make table 1 more easily be understood, figure 1 was made to illustrate the relationship between dielectric thickness, DPN power and NBTI lifetime. There were mainly three color regions, namely red, yellow and green, which stood for different NBTI lifetime performances. The dots in the red region meant under these combinations of gate dielectric thickness and DPN power the NBTI lifetime performance was not good enough. The results pass a certain criteria would distribute in the yellow or green region. To be noted that the darker the green region, the better the NBTI lifetime performance.

The top four dots in figure 1 illustrated that it's hard to improve NBTI lifetime performance by merely increasing the total thickness of gate dielectric and without lowering the DPN power. Similarly, DPN power alone could hardly make a difference to the reliability performance. However, combine these two factors together significant improvement was made, which was as high as 200 times. Moreover, it was found that to increase same thickness IL got better improvement than HK.

In a word, a certain thickness of gate dielectric was crucial to reliability of NBTI. And it should be related to both electric field intensity of gate dielectric and N distribution of commercial DPN process.

978-1-6654-9759-6/22 $31.00 © 2022 IEEE

## 2. DPN Power Could Make a Difference

After the fundamental dielectric thickness was settled, further DPN power splits were carried out and the reliability results were impressive.

With a thicker interface layer (IL), IL+2x, DPN power could make a significant change of NBTI lifetime. Even 2 orders of improvements were demonstrated.

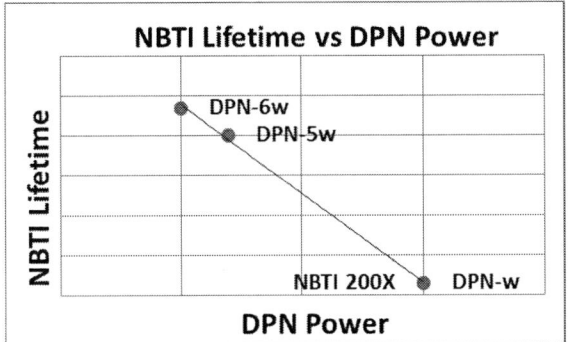

Figure 2. The Extent of NBTI Lifetime Improvement under Different DPN Power, The Interface Layer Thickness was IL+2x

N concentrations under two different DPN powers with same film stack were measured by SIMS as shown in figure 3. On one hand N concentration could be modulated by DPN power efficiently. On the other hand, there was still considerable N concentration in the region of IL even under lower DPN power. And this may be the reason why DPN power alone could not make a major difference to NBTI lifetime.

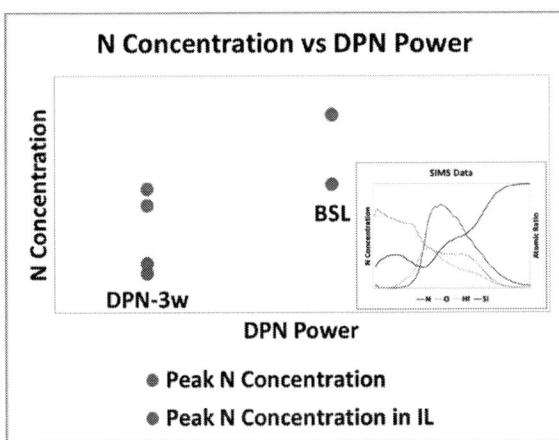

Figure 3. Nitrogen Concentration and Distribution

## 3. Proposed Model

Based on these results, we proposed a nitrogen profile modulation model to explain the internal mechanism of NBTI improvement, as illustrated in figure 4 and figure 5.

The main idea of this model was that limited by the minimum penetration depth of nitrogen of commercial DPN process, the gate dielectric required a minimum thickness in order to push nitrogen profile away from the interface of IL and silicon(channel). Because nitrogen terminated silicon dangling bonds trend to attract positive hydrogen ions and become stable positive charge centers, which were detrimental to NBTI. Only

with enough thickness of gate dielectric, DPN power modulation could be efficient of improving NBTI lifetime.

$$V_T = \phi_{MS} - \frac{Q_{ox}}{C_{ox}} - \frac{Q_{it}}{C_{ox}} + 2\phi_F + \frac{Q_S}{C_{ox}} \tag{1}$$

$$\Delta V_T = -\frac{q(\Delta N_{ox} + \Delta N_{it})}{K_{ox}\varepsilon_0} t_{ox} \tag{2}$$

$$\Delta N_{it}(t) \approx k_F N_0 t \tag{3}$$

$$\Delta N_{it}(t) \approx \sqrt{\frac{k_F N_0}{2 k_R}} (D_H t)^{1/4} \tag{4}$$

Formula (1) and (2) were about threshold voltage (Vt) and its shift. In order to minimize the shift of Vt during NBTI test initial interfacial trap density was important according to formula (3) and (4), representing the reaction and diffusion dominated regions respectively according to R-D theory. Further, gate dielectric thickness was also important. Not only because it would affect the electric field intensity directly, but also the N distribution that discuss in this paper.

N in the interface of IL and silicon channel would form Si-N bonds, which had strong attraction to H+ that released from hydrogen terminated silicon dangling bonds. And thus seriously affect NBTI lifetime.

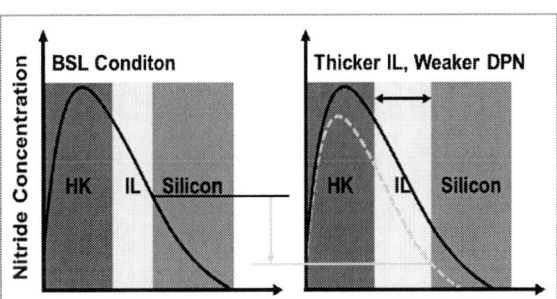

Figure 4. Nitrogen Profile Modulation Model of NBTI Improvement

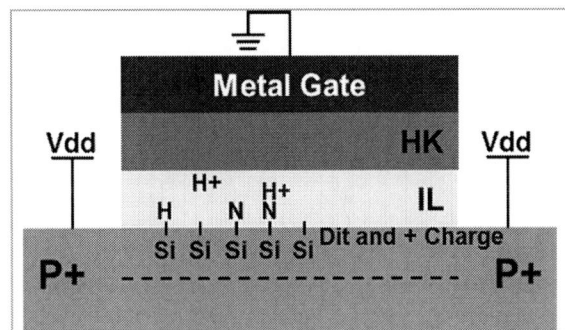

Figure 5. Nitrogen and Dangling Bonds in the Interface

## CONCLUSION

We proposed a nitrogen profile modulation model to explain the internal mechanism of NBTI improvement in 28 nm HKMG technology. Limited by the minimum penetration depth of nitrogen of commercial DPN process, the gate dielectric required a minimum thickness in order to push nitrogen profile away from the interface of IL and silicon. With enough thickness of gate dielectric, DPN power modulation could be efficient of improving NBTI lifetime.

## ACKNOWLEDGEMENTS

I would like to appreciate senior engineer Ran Huang, Yamin Cao, Wei Zhou for their support and advice on this experiment. At the same time, I am grateful to the reliability team of HLIC for their testing and discussion.

## REFERENCES

[1] "Work Function Setting in High-k Metal Gate Devices," Complementary Metal Oxide Semiconductor Chaper 3, 2018.

[2] "Challenges for The Integration of Metal Gate Electrodes," International Electron Devices Meeting, 2004.

[3] "Characteristics and Mechanism of Tunable Work Function Gate Electrodes Using a Bilayer Metal Structure on $SiO_2$ and $HfO_2$," Electron Device Letters, 2005.

[4] "Dipole-induced Modulation of Effective Work Function of Metal Gate in Junctionless FETs," AIP Advances, 2020.

[5] "NBTI enhancement by nitrogen incorporation into ultrathin gate oxide for 0.10-pm gate CMOS generation," IEEE VLSI Symp, 2000.

[6] "Negative bias temperature instability: What do we understand?," Microelectronic Reliability, 2007.

[7] "Negative bias temperature instability: Road to cross in deep submicron silicon semiconductor manufacturing," Journal of Applied Physics, 2003.

# EFFECTIVE USAGE OF ATTENUATORS ON PRODUCTION TEST

*Hao Chen[1]\*, XueQuan Chen[1], Qin Feng[1]*
Advantest (China) Co., Ltd, Shanghai 201203, China
\*Corresponding Author's Email: hao.chen@advantest.com

## ABSTRACT

In this paper, we will discuss how to effectively use attenuators together with ATE instruments for best COT. Two test cases will be shared. The first one is a DigRF transceiver using v93K WSRF splitter & attenuator to avoid RF reflection from mismatch. Splitter performance and splitter results with attenuators will be shown by mass experiment data. Through that, we reduce from 2 WSRF card to only 1WSRF card and support Octal site MIMO test.

The Second case is a UWB device which uses on board programmable attenuator together with WSRF. This makes per-site power sweeping test from serial to parallel, which saved overall test time for more than 50%.

Both cases have been released to and help reduce COT significantly. We will also give some tips and suggestions about attenuator selection, as well as limitations and restrictions.

*Keywords—Attenuator; ATE; Cost-of-Test*

## INTRODUCTION

With today's 5G & IOT trend, high end RF device's ports & complexity has been significantly increased. For production test, those devices typically will be tested with high multisite up to x8 or x16. In order to achieve a reasonable cost of test, RF resource sharing between sites is a common solution.

When talking about ATE RF resource sharing, there's limitations which might be over-looked due to RF instrument structure. In some cases, attenuators should be added to avoid critical failure or performance degradation. On the other hand, adding proper attenuator can not only save tester resources and test time, but also improve performance.

### Attenuator Introduction

An attenuator is a two-port resistive network designed to weaken or "attenuate" the power supplied by a source to a level that is suitable for the connected load, without distorting the signal waveform,

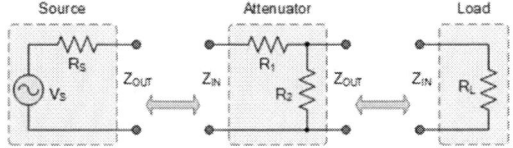

*Figure 1: Attenuator circuit*

Attenuators are used in a wide variety of applications and can satisfy almost any requirement where a reduction in power is needed. Attenuators are typically used as below. Extend the dynamic range of devices such as power meters and amplifiers. Reduce signal levels to detectors. Match circuits and balance out transmission lines that otherwise would have unequal signal levels. Provide isolation between a source and a troublesome load

Usually, attenuators are classified into three types in terms of attenuation level.

Fixed Attenuators: They are passive attenuators that provide a fixed attenuation level. Most of these products are available in a series, which have several attenuation levels available.

Variable Attenuators: These attenuators provide an attenuation and they can be regulated to provide a certain attenuation level. Typically, the attenuation level can be modified by changing the voltage applied to an input control line.

Step Attenuators: These are attenuators where the attenuation level can be modified in fixed steps.

Regarding production test, Advantest Wave Scale RF card is the world most integrated card. Each card has 4 independent RF subsystems providing a total of: 32 RF ports (8 RF ports per subsystem), 4 independent CW or modulated stims and 4 independent measurements.

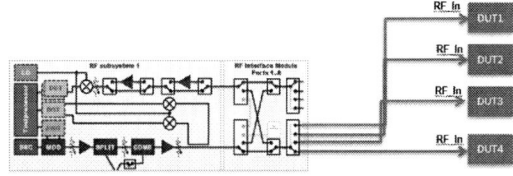

*Figure 2: WSRF Stimulus Switch or Splitter*

As shown in Fig 2, its unique splitter function able to Split the RF Stim signal into 1, 2, 3 or 4 ports for both inter-site and intra-site use cases, thus increasing testing parallelism multisite efficiency

However, it still has some limitations including maximum measurement power, even higher multi-site requirements. Eventually ATE always want to get the best COT through less instrument, more DUTs and less test time.

In some special cases, attenuator can be used with WSRF in order to: reduce output Power of PA to meet the Measurement spec of WSRF, match circuit and reduce reflection from shared resource, and adjust per site power from shared resource etc.

### RF Transceiver with Fixed Attenuator

This device under test is a highly integrated, single-die radio transceiver chip that supports 5G NR, 4G LTE with LTE-A for FDD-LTE, TDD-LTE, 3G WCDMA, HSDPA, HSUPA, GSM/EDGE as well as TD-SCDMA operation. The device has over 40 single-ended transmit and receive ports.

Its original tester configuration for Octal sites use 2x PS1600,1x Pure Clock (LPN) ,1x DPS128/64, 2x WSRF

In order to reduce cost, we try to reduce one WSRF card by using WSRF interface module's splitter function. Which can split the RF Stim signal into 4 ports.

However, splitter itself has limitations for un-matching cases, which will cause different site results are dependent. Reflection occurs impacting RX tests when using RFIM splitter to stimulus 4 RF signals simultaneously

Even without splitter, DUT mismatch can cause measurement uncertainty. This measurement uncertainty displayed in Fig 3 is mostly observed when there is a change in setup during the different measurement.

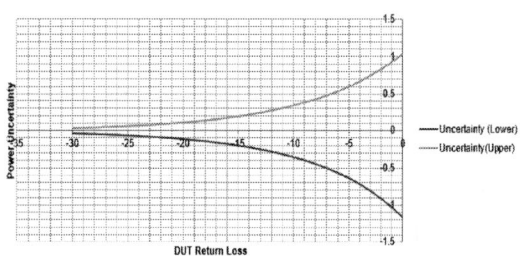

*Figure 3: DUT Return Loss vs Power Uncertainty*

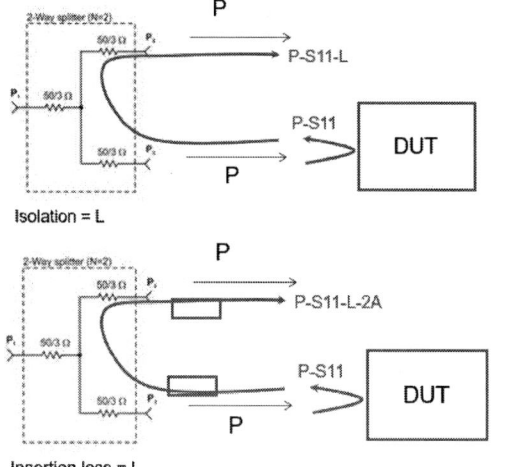

*Figure 4: Reflected Power in Splitter Mode with/without Attenuator*

Actual Power will depend on the phase of the reflected power relative to P(Fig3). Worse case when they are in conjugate. Max delta P due to reflect powers from other sites = 10(P-S11-L) *(num_site-1). If add extra attenuator (Fig4), it becomes 10(P-S11-2A-L) *(num_site-1).

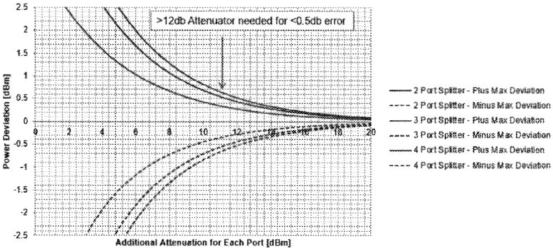

*Figure 5: Power Deviation vs Attenuation at unmatching case*

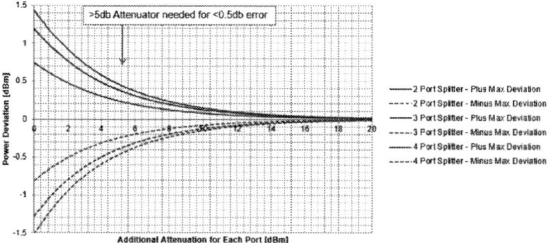

*Figure 6: Power Deviation vs Attenuation at matching case*

As shown in Fig5 and 6, We can see stim error greatly reduced with match DUT ports. For unmatched cases, we need more than 12 dB attenuator for less than 0.5dB error. But we only need 5 dB attenuator for matched cases.

In many cases, the cable loss, trace loss and board switch loss would make up more than 3 dB attenuation.

*Figure 7: Test LoadBoard with Attenuator BW-S10W2+*

We verified different attenuator level. Eventually, we select 10dB fixed attenuator BW-S10W2+[1] to help matching circuit & reducing reflection from shared resource.

This splitter & attenuator solution has been accepted with better COT. It saves one WSRF card with about 25%

total tester cost and almost no test tine penalty. Attenuator test result matches with old solution, with even better distribution.

**Ultra-Wideband Module with Step Attenuator**

This device under test is a UWB module, UWB is a radio technology that can use a very low energy level for short-range, high-bandwidth communications over a large portion of the radio spectrum [2].

Since the wide bandwidth (>500 MHz) is beyond WSRF's bandwidth, external scope and waveform generator is used together with v93K as illustrated in Fig8.

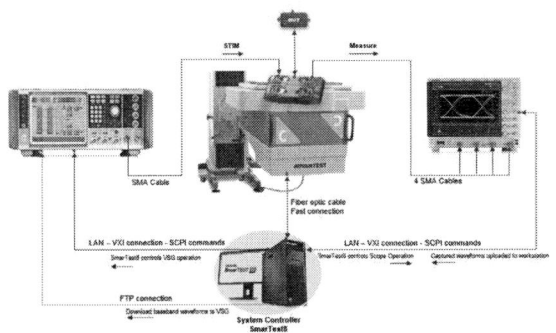

*Figure 8: SmarTest8 Central and Fully-Automated Solution with external instrument*

For this device, a special Gain Step test is required in order to find best RF stim input power with modulated signal. Over 50 Different condition & waveform need to be tested. There are 16 input power steps per condition and each step test takes longer than 100ms. 16 site RF Stim sharing the same RF source from external waveform generator (2 source).

In order to get best test time, we performed three different approaches as Fig9. a). Binary Search: 16 sites perform test semi- parallelly, with 4 steps; b). Linear Search: 16 sites perform test parallelly with 16 steps; c). Step Attenuator with binary search: 16 sites parallel with 4 steps. With the help of step attenuator, we successfully reduce test time to 13% of the original way.

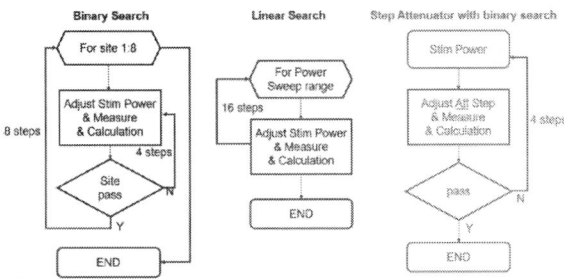

*Figure 9: Test Procedures*

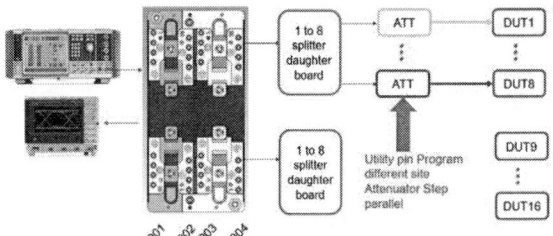

*Figure 10: Hardware View when Testing with Step Attenuator*

Programable step attenuator and daughter board is designed to adjust the input power per site from shared resources. We also developed a special attenuator control class to use utility 256 of v93K, which helps reduce PS1600 resource. Meanwhile, we achieved better and faster control through dynamically (sequencer-controlled) controlling utility to program attenuator to different attenuate level per site.

With the application of step attenuator, we minimize the test steps from 32 steps per site binary test to 4 steps parallel test. Test time is only 13% compare to the original. The hardware view for 16 sites gain step test is shown as Fig 10.

**CONCLUSION**

For RF production test on ATE, we usually suffer from limited resource. Adding attenuator together with RF hardware helps reduce COT.

This paper introduced two cases when applying different types of attenuator which helps reduce overall cost significantly. Both two projects have been released to production with good test results.

The hardware and software can be borrowed by further applications. Several other IOT applications with attenuator solution have already been used and achieve production release.

**ACKNOWLEDGEMENTS**

Authors would like to take this chance to thank Advantest for the support on these projects. We also want to acknowledge everyone who has taken part in the project but cannot be listed here.

**REFERENCES**

[1]    Minicircuit    BW-S10W2+    datasheet. "https://www.minicircuits.com/WebStore/dashboard. html?model=BW-S10W2%2B"

[2]    Zhou, Yuan; Law, Choi Look; Xia, Jingjing. *"Ultra low-power UWB-RFID system for precise location-aware applications". 2012 IEEE Wireless Communications and Networking Conference Workshops.* 2012, pp. 154–158.

# MCU+ TEST SOLUTION ON V93000 SMARTTEST8

Tianyu Zhang[1*], Lei Zhu[2], Wenyu Zhang[3]

[2]Advantest, Shanghai 201210, China

[3] Advantest, Shanghai 201210, China

*Corresponding Author's Email: tianyu.zhang@ advantest.com.cn

## ABSTRACT

With the development of semiconductor technology, MCU is becoming more and more integrated. The traditional MCU includes processor module, memory module, IO module, power gain amplifier module and other analog modules. However, with the development of IOT technology, many new MCU integrates ADC, DAC and Bluetooth communication modules, this type of MCU is referred to MCU+ in this article. MCU+ test integrates digital test, dc test, analog test, RF test compared with traditional MCU test only requiring digital test and dc test. This paper introduces the test solution of MCU+ chip based on V93K platform, including the hardware of Wave Scale series and the software solution of the new generation platform SmartTest8. Meanwhile, this paper designed a set of MCU+ test program library based on SmartTest8 API, which includes the common function of dc action, analog waveform generator action, digitizer action, rfStim action, rfMeas action. It can help users improve the test program development efficiency by more than 30% and improve the quality of test program development and reduce the maintenance cost.

*Keywords— MCU+ tests; SmartTest8; test program library; V93000.*

## INTRODUCTION

### MCU+ Brief Introduction

MCU is a chip level computer that properly reduces CPU frequency and specifications, it integrates Memory, Timer, AD\DA, UART, DMA and other peripheral interfaces and even LCD driver circuits into a single chip. MCU+ integrates the communication module based on the MCU as shown in figure1.

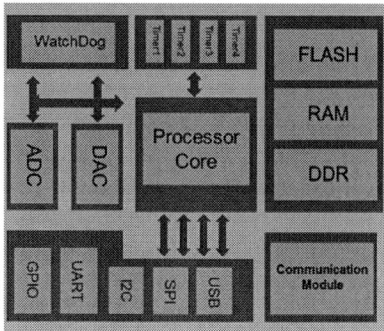

*Figure 1:MCU+ Block*

The core component of MCU is the processor core, which includes two main parts, arithmetic unit and controller. Figure2 shows the category of MCU, according to the processed data bits it can be classified into 4bits, 8bits, 16bits, 32bits and 64 bits MCU, it can also be classified into CISC and RISC according to the instruction structure. MCU can be divided into Harvard architecture and Von Neumann architecture according to the classification of memory architecture, and it can also be divided into general type microcontroller and special type microcontroller according to the classification of use.

*Figure 2: Classification of MCU*

### Application scenarios

The MCU market can be divided into IOT, consumer electronics, autonomous drivin g and industrial control as shown in figure3.

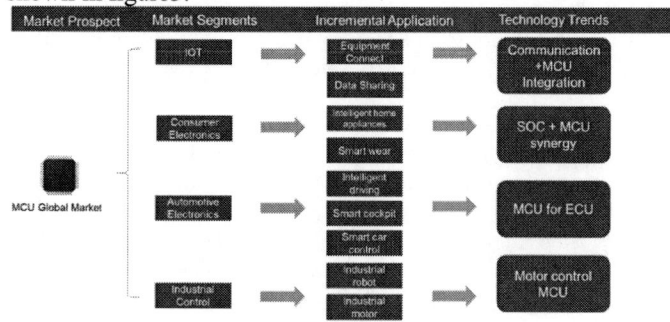

*Figure 3: Application Scenarios*

This paper focues on MCU+ chip, while this type of chip mainly targets the Internet of Things and consumer electronics. After integrating the communication module, MCU+ chip can reducing the cost of system integration and also reduce the system PCB area, thus lower the costs and accelerating customer product design, which widely used in smart watches, remote controls, high-end toys and other applications.

## V93000 TEST SOLUTION

### Hardware Solution

This paper presents a hardware configuration scheme based on PS1600, DPS128, WSMX and WSRF to test MCU+ chip. As a powerful digital card, PS1600 can be used for DFT

978-1-6654-9759-6/22 $31.00 © 2022 IEEE

test such as scan and mbist, and also voltage and current testing can be carried out with its high accuracy per-pin parametric measurement unit。 The MCU integrated with communication module usually needs test gain, IMD3, p1db and other spec. The WSRF card paried with RFIM module covers the frequency range 0.01-8GHz with good phase noise performance, which can well cover the test of bluetooth, WeFi and other communication modules. The Wave Scale MX consists of 16 source and 16 measure units that operate independently and can be used in parallel[1]. With the 500Mbps sampling rate and 16bit resolution, this card can completely cover the AD\DA tests inside MCU+. Figure4 lists the test items for the MCU+ chip, including DC, DFT,Digital, analog, RF.

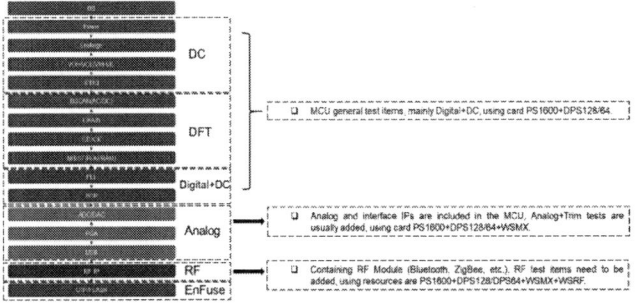

*Figure 4:MCU+ Chip Test Items*

The combination of PS1600+DPS128+WSMX+WSRF can well cover the requirements of the above test items, and multi-site execution can be achieved due to high density of pin count.

**Software Solution**

In order to be compatible with hardware scheme WSMX and WSRF, this paper uses SMT8 as the software solution foundation. Instruments is a very import concept in SMT8, which provide you with a unified way to set up the tester hardware resources. The instruments reflect the electronics of the hardware resources, and give access to the capabilities and features of the test head cards[2]. PS1600 mainly uses DigInOut and DcVI two instruments as shown in figure5, mainly for the control of driver, comparator, activeload, pmu, and badc, which can well meet the digital and dc test in MCU+.

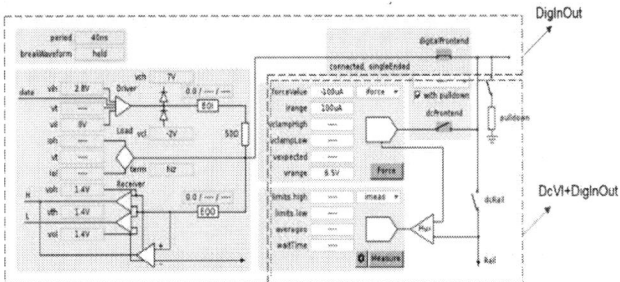

*Figure 5: DigInOut & DCVI for PS1600*

DigInOut is used to operate digital drivers and receivers, usually used in Function test, and can also be used to control PMU and BADC. DigInOut instrument is required when execute DFT test items. DcVI is used to operate PMU and BADC. It is recommended to use in some DC test items, such as force voltage measure current, force current and measure voltage. In one scenario, digInOut instrument must be used when both pattern run and dc actions need to be done in one test suite. The reason is that DCVI does not support the use of the driver, and instrumnet switching within one test suite is also not supported. There are two things user need to pay attention to when using PS1600. The first point is the status of the relay, the digitalFrontend is still closed by default when using DigInOut Instrument to measure DC data, that means the driver and pmu are connected in the same time, which affecting the accuracy of the measurement results. To solve this problem, user needs to disconnect the relay by setting digitalFronted to disable in test program. The second point is the IO pin state when channel disconnect. The IO channel will drop to the ground by default when PS1600 is disconnect[3]. In some scenarios, users need to set the pin to the high resistance state when they disconnect. Therefore, the default mode can not take effect, user must manually set disconnectMode to high resistance. DPS128 is mainly used with DCVI instrument, there is no need to distinguish between SIG mode and DPS mode, which brings convenience. Note that DPS128 is required to set the connect mode to HIZ mode when use as volt meter. IsetupAwg instrument is used to control the waveform sent by AWG, it supporting ramp, sine, multitone, noise and other waveforms. During AD\DA test, ramp and sine wave are commonly used. IsetupDigitizer instrument is used to control DGT waveform sampling, which can be started directly according to the operating sequencer instruction without trigger signal. It supports both single-end and differential connections and internal loopback connections. There are 3 tips to recommended when using WSMX, the first one is when set the AWG send waveform continuously, the sampling condition of DAC must be follow the coherent sampling theorem, otherwise, periodic waveform splicing will be a non-integer periodic waveform as shown in figure6. In the spectrum, it is reflected as spectrum leakage. The second one is when using AWG differential mode, the actual output differential voltage is twice of the set amplitude, while the

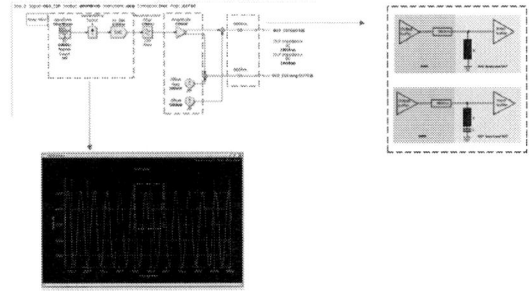

*Figure 6: WSMX Using Tips*

offset value of the differential signal is the specified value. The third one is that user need to consider the impedance setting, the impedance setting will affect the actual level value of the WSMX output signal to the IO terminal of the chip. If the resistance of DUT pin is high resistance, user do not need to set the parameter of dutImpedanceAC and dutImpedanceDC, due to they have been set to high impedance by default. If the DUT IO have a 50Ω pull-down resistance, set dutImpedanceDC to

$50\Omega$, dutImpedanceAC is not required in this case. When the chip impedance matching mode is AC termination, set dutImpedanceAC to $50\Omega$ and dutImpedanceDC need to set to $1M\Omega$.

This paper develops a software development kit based on the real project from the perspectives of program quality, development efficiency and program maintainability. From the quality point of view, building software development kit and validate in real projects, using the SDK can reduces the risk of bugs from a complete redesign when a new project is being developed. From the perspective of efficiency, SDK encapsulates the operations of various cards such as PS1600,DPS128,WSMX,WSRF. Users only need to fill in relevant parameters and do not need to spend too much time to do the test method programming with SMT8 API, thus greatly improving the efficiency of test program development. From the program maintainability point of view, the common actions for the card operations are abstracted and realized in the SDK. User-written test programs can call functions from the SDK, which greatly improves the convenience of program update of the project when the operations need update. For testing programs of different projects, the use of SDK has greatly increased the convenience of reviewing programs or updating programs. The entire SDK framework is shown in Figure 7, which includes three broad categories.

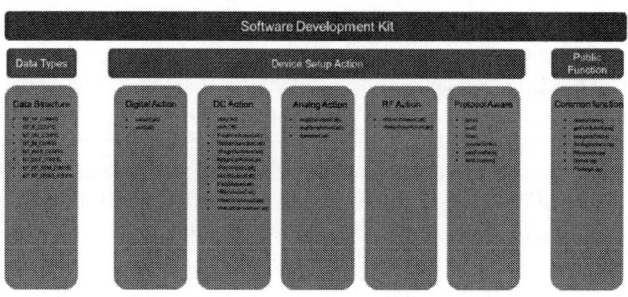

*Figure7: SDK Framework*

In the category of Data Types, commonly used data structures are encapsulated to increase the convenience of function parameter transmission. In the category of Device Setup Action, five subclasses are defined, namely digital Action, DC Action, Analog Action, RF Action and Protocol Aware. In subcategory of digital action, the main functions are pattern-related, including pattern execution, execution wait, result acquisition, etc. In subcategory of DC action, the main functions include force voltage, force current, measure voltage, measure current, and these actions are decoupled from the card. Meanwhile, utility control bit operations are also supported in this DC action. In subcategory of Analog Action, the main functions include controlling the AWG of WSMX to send sine waveform,Ramp waveform, etc, which is used to test the static and dynamic parameters of ADC. At the same time, also supports the control of WSMX digitizer, to sample the waveform from DAC or mixer and achieve the power measurement, gain measurement, SNR, etc. In subcategory of RF action, the main function includes controlling WSRF board to send or receive RF single tone signal and dual tone signal, which is necessary for MCU+ communication module test. In the subcategory of protocol Aware, the functions implemented include registers static write, dynamic write, registers read and

wait. The static writing is mainly used for chip configuration with fixed registers values, it has the characteristics of fast write and short execution time. Dynamic write mainly applies to the scenario where register values need to be dynamically modified or different sites configured with different register values. Compared with static write, dynamic write can realize the real-time configuration of different register values for each chip according to its own chip state. However, the problem is to increase the test execution time and reduce the test efficiency. When configuring the register, the program can automatically read back the register value and compare the auto-read value with the written value. If there is any difference, the user will be provided with warning information. This function helps to find out whether the register value has been written to the chip, with this feature, the time for this common debugging approach is greatly reduced. Meanwhile, this function can be turned off during in mass production, directly written the registers value without reading back, thus test time will not be affected. In common function category, features implemented include release tester, information print control, etc. Based on multi-thread technology, release tester function can realize the synchronization of board execution and test data processing, this helps to reduce test time effectively. The hierarchical control function of printing information can be useful in the program debugging, users do not need to add printing statesments everywhere in the program, just turn on the corresponding flag.

## CONCLUSION

This paper proposes an ATE test solution based on MCU+ chip, the hardware solution includes PS1600/DPS128/WSMX/WSRF, the configuration combination can well meet the chip testing requirements. At the same time, the test cost can be reduced when apply the multisite solution. The software solution is developed based on SMT8, the concept of instrument allows users to focus on test itself rather than operations of the hardware cards. The software development kit has been developed, it can effectively help customers reduce the difficulty of SMT8 program development, improve the efficiency of program development, and improve the maintainability of the program. The application results show that the software development kit can improve the efficiency of program development by at least 30%.

## ACKNOWLEDGEMENTS

I would like to extend my sincere gratitude to Hu Jiaming, for his instructive advice and useful suggestions on my thesis. I am deeply grateful of his help in the completion of this thesis.

## References

[1] SmarTest Documentation Center, Topic 250804.
[2] SmarTest Documentation Center, Topic 248936.
[3] SmarTest Documentation Center, Topic 350586.

# CONDITIONALLY EXECUTED TESTS, BRANCHING AND ALGORITHMIC BINNING – GETTING IT RIGHT

*Lai-Choon Chan*

Applications Development Centre, Teradyne Asia Pte Ltd, Singapore, Singapore

Email: lai-choon.chan@teradyne.com

## ABSTRACT

Test programs for large SOCs contain complex flow and binning logic. To perform tasks such as memory repair, trimming and speed binning, flows are non-deterministic and incorporate branching, loops and conditional test execution. Binning decisions are made based on elaborate algorithms. Testing multiple devices in parallel adds another dimension of complexity, which increases the possibility of coding errors. To eliminate such errors, the test program needs to be checked. Current methods of verification have drawbacks. This paper introduces a tool that addresses the deficiencies using speed binning as an example.

## INTRODUCTION

### Test programs are getting more complex

Today, test programs are required to do much more than take a measurement, execute a test pattern or acquire and analyze a signal. For example, embedded memory has to be tested to determine if it is prime. If it isn't, it has to be checked if it can be repaired. If repair is unsuccessful, another attempt may be made. Searching and looping are needed to trim oscillators, regulators or sensors. As a result, the program has to execute different tests depending on the state of the device.

### Challenges test engineers face

To ensure there are no coding errors that lead to incorrect branching or test omission, every branch of the test flow should be exercised when verifying the flow logic. This may require execution of a large number of different test combinations.

To do this, some options available include:
1. Use real devices that, when tested, will produce responses that lead to the target combinations of tests getting executed
2. Modify the test code to simulate different results that cause the flow to branch or loop in the desired manner.

Disadvantages with using these methods are:
1. Regressive testing is inefficient. If changes were made to the test program, there is no easy way to run through all the test combinations again
2. To simulate different test results, the test program has to be repeatedly modified and reverted. The process is not only tedious, but can introduce

errors.
3. Cherry-picking devices that provide a required set of responses is difficult. In a wafer probe situation, trying to find 2 adjacent dice with the required responses to verify independent branching when the 2 devices are tested in parallel is even more challenging.

### A tool for testing the test program

Testing a complex test program for bugs takes effort, as in any software testing exercise [1]. A tool with the following features and capabilities has been developed to overcome the drawbacks of existing methods:
1. DUT (device-under-test) responses are stored in a database. Any combination of responses can be used to enable verification of all branches of the test flow. A particular combination of responses creates a "scenario". Real devices are no longer required.
2. Because the tool can inject the desired DUT responses, the test code no longer has to be modified to simulate failures or other results.
3. The responses and scenarios constructed from the responses can be stored and re-applied. This speeds-up regressive testing.
4. When testing multiple devices in parallel, any scenario can be applied to any device.
5. Responses of a real DUT can be recorded and stored. These responses can be modified if necessary to create different responses
6. Results generated by the test program when executed using a scenario can be recorded and stored as "expected results". These can then be checked to ensure the flow and binning logic worked as intended. Once verified, the stored results can be used in regressive testing and automatically compared against results produced after any test program update.
7. A list of scenarios with expected results can be saved to a "test suite". Using the test suite, the test program can be automatically executed by the tool to run through all the scenarios so that different branches of the test flow are exercised and the results compared.

Figure 1 shows the components of the tool.

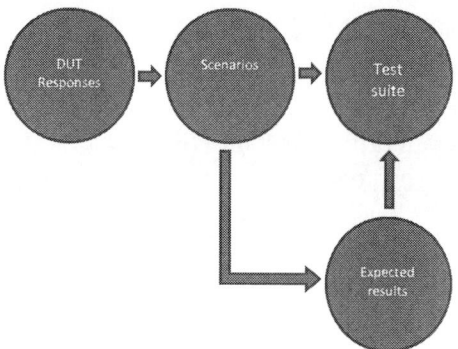

*Figure 1: Components of flow/binning verification tool*

## IMPLEMENTATION

### Binning algorithm

Speed binning is a means to improve yield by classifying devices that perform at lower than optimum levels as good but lower grade devices rather than as rejects, so they retain some economic value. Ways in which it is implemented varies depending on the device and manufacturer.

The example used is an application processor with 4 blocks broadly categorized as CPU, peripherals, connectivity and DSP, each of which is split into smaller cores, as illustrated in Figure 2.

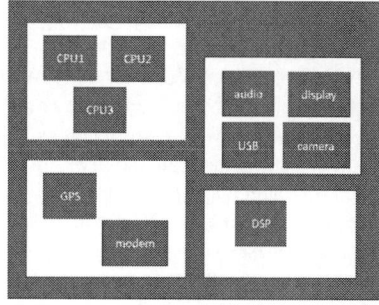

*Figure 2: Application processor block diagram*

Each core is tested at 4 performance levels and if every core passes at every level, the device is assigned to bin-1.

If all critical tests pass but there are tests failing at some performance levels, the device is downgraded or rejected based on the criteria in Figure 3.

The test program is structured into 10 sub-flows based on the cores. The criteria for rejection in Figure 3 specify that if any test failed at performance level 3 or 4 in any of the sub-flows in the CPU, peripherals or connectivity blocks and tests failed at performance level 1 or 2 in the DSP block, the device is rejected. The device is also rejected if any test failed at performance level 3 or 4 in the CPU, peripherals or connectivity blocks and there are failing tests in multiple blocks. Finally, if performance level 3 or 4 tests failed in the DSP block, the device is rejected.

| | P3 | P4 | DSP P1 | DSP.P2 | >1 block | |
|---|---|---|---|---|---|---|
| **CPU** | fail | | fail | | | |
| | fail | | | fail | | |
| | fail | | | | fail | |
| | | fail | fail | | | |
| | | fail | | fail | | |
| | | fail | | | fail | |
| **Peripheral** | fail | | fail | | | |
| | fail | | | fail | | |
| | fail | | | | fail | |
| | | fail | fail | | | |
| | | fail | | fail | | OR |
| | | fail | | | fail | |
| **Connectivity** | fail | | fail | | | |
| | fail | | | fail | | |
| | fail | | | | fail | |
| | | fail | fail | | | |
| | | fail | | fail | | |
| | | fail | | | fail | |
| **DSP** | fail | | | | | |
| | | fail | | | | |
| Reject | | | | | | |

| | P1 | P2 | P3 | P4 | |
|---|---|---|---|---|---|
| **CPU** | | | pass | pass | |
| **Peripheral** | | | pass | pass | AND |
| **Connectivity** | | | pass | pass | |
| **DSP** | | | pass | pass | |
| Performance | | | | | |

| Peripheral | Connectivity | CPU1 | CPU2 | CPU3 | |
|---|---|---|---|---|---|
| Pass | | Pass | Pass | Pass | |
| | Pass | Pass | Pass | Pass | |
| Pass | Pass | Pass | | | OR |
| Pass | Pass | | Pass | | |
| Pass | Pass | | | Pass | |
| Core binning | | | | | |

*Figure 3: Binning criteria*

If the device is not rejected, it is evaluated for downgrade. If all performance level 3 and 4 tests pass, it qualifies for performance binning. The downgrade bin assigned depends on results of the remaining tests. Otherwise, it is evaluated for downgrade based on which cores passed all tests as specified in the core binning criteria. Failures must be confined to only one of the 3 blocks (CPU, peripherals and connectivity) and at least one of the 3 CPU cores must pass all tests.

The logic for determining the final binning is not trivial, so coding errors can occur.

To check that different test result combinations are correctly binned, 26 different scenarios have to be constructed and the test program has to be executed using these scenarios. Dual-site verification, where 2 devices are tested in parallel, requires even more scenarios.

If coding errors are found or there are changes in the binning criteria, the process has to be repeated, which is very time consuming if the scenarios have to be re-created each time and the test program execution is not automated.

### Using the tool

Instead of verifying the binning logic manually, the new tool is used, starting with the creation of the response database.

| IGSIM RESPONSE DATA V1 | | | | | | | |
|---|---|---|---|---|---|---|---|
| Response Key | Flow Table | Flow Line | Test Instance | Default | Fail | I_32mA | I_52mA |
| Iddq.0.vdd1 | Flow_DC | | 7 IddqTest | 1.23E-03 | | 0.03162 | 0.052 |
| CPU1Perf1.0 | Flow_CPU | | 8 CPU1_Perf1_Test | TRUE | FALSE | | |
| Display1.0 | Flow_Peripherals | | 6 Display_Perf1_Test | TRUE | FALSE | | |

*Figure 4: Response database*

| TEST SUITE ID | SITE | RESULT | Notes |
|---|---|---|---|
| Test_Reject1 | 0 | FAILED | Value Mismatch For: Bin Expected: 4 Actual: 14; Value Mismatch Fo |
| Test_Reject2 | 0 | FAILED | Value Mismatch For: Bin Expected: 5 Actual: 14; Value Mismatch Fo |
| Test_Reject3 | 0 | FAILED | Value Mismatch For: Bin Expected: 5 Actual: 14; Value Mismatch Fo |
| Test_Reject4 | 0 | PASSED | |
| Test_Reject5 | 0 | PASSED | |

*Figure 7: Verification results*

Figure 4 shows a few entries in the response database. The "default" column specifies the default response, which is usually a passing value. In the case of functional tests, there are only 2 possible responses, either true (pass) or false (fail). Parametric tests can have as many responses as needed. Responses for mixed signal tests can be a link to a file containing digitized samples of a waveform.

| IGSIM TESTSCENARIOS V1 | | |
|---|---|---|
| Scenario ID | Response ID | Response Set |
| Sc_Reject1 | CPU2Perf4.0 | Fail |
| Sc_Reject1 | DSP1.0 | Fail |
| Sc_Reject2 | Display3.0 | Fail |
| Sc_Reject2 | Display4.0 | Fail |
| Sc_Reject2 | DSP2.0 | Fail |

*Figure 5: Scenarios*

Figure 5 shows how scenarios are constructed from the responses. Tests that pass do not have to be included in the scenarios since the passing response is the default. So, when scenario "Sc_Reject1" is executed, the tool will provide failing responses for the CPU2 performance level 4 and DSP performance level 1 tests and passing responses for all other tests.

Finally, the scenarios are used to set up tests in the Test Suite as shown in Figure 6.

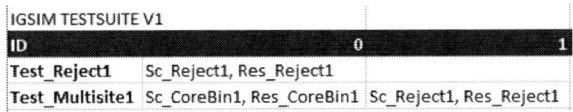

| IGSIM TESTSUITE V1 | | |
|---|---|---|
| ID | 0 | 1 |
| Test_Reject1 | Sc_Reject1, Res_Reject1 | |
| Test_Multisite1 | Sc_CoreBin1, Res_CoreBin1 | Sc_Reject1, Res_Reject1 |

*Figure 6: Test suite*

In this example, there are 2 tests listed in the test suite. The first test will execute the test program using responses from scenario "Sc_Reject1" on site #0 in a dual-site test program, ie, 2 devices can be tested in parallel. The corresponding expected result for that scenario, "Res_Reject1" is specified for site #0. The next test executes the test program using responses from 2 different scenarios on the 2 sites, with each having a different result. Columns can be added to accommodate as many sites as needed, with any combination of scenarios.

The test suite can contain all the tests needed to verify the binning logic. So, in this case, there are 26 single site tests and additional tests for multi-site verification. The tests in the test suite can then be selected and executed, and the tool automatically executes one run of the test program for each test selected and reports the results.

As shown in Figure 7, three of the tests for the reject binning logic failed during the initial verification because the results from the datalog when the test program was executed did not match the expected results.

The test program implements the binning logic using "bin tables", which sets a reject flag using flags that are set when tests fail, as shown in Figure 8.

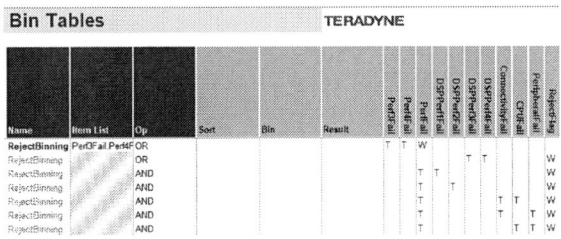

*Figure 8: Reject bin table used in test program*

The failures in the verification tests were traced to a bug in the bin table which was subsequently fixed. The same tests in the test suite were then easily re-executed to verify the fix.

## CONCLUSION

The example described shows how verification of complex binning algorithms can be aided by a tool that provides storage for DUT responses and for scenarios that are constructed from these responses. Different scenarios can then be simulated without any modification to the test program. If the test program is updated to resolve any bugs found, the same scenarios can be simulated to re-verify the binning logic without additional effort. The tool can be applied to the verification of complex flow logic in a similar manner.

## ACKNOWLEDGEMENTS

Thanks to Dwayne Dohmann and Roberto Rotili for checking the manuscript and to Danny Liew and Randy Kramer for their guidance in the writing of this paper.

## REFERENCES

[1] J. Whittaker. "What is software testing? And why is it so hard?", *IEEE Software*, Jan/Feb 2000, pp. 70-79.

# Study of Negative Charge Accumulation Mechanism and Removal Method on the Wafer Surface in Via Photo Development Process

*Haihua Chen, Shaojian Hu, Xiaobing Liu, Heguang Shi, Manhua Shen, and Yushu Yang*

Shanghai IC R&D Center

497 Gaosi Road, Zhangjiang Hi-Tech Park, Shanghai 201210, PR China

Email: hhch1984@126.com

## ABSTRACT

Influence of the wafer surface charging effect on the precision of measuring ADI features during photo development process using CD-SEM was reported. An initial mechanism has been proposed that the surface charge generation may be related to previous hard mark opening and via photo development, but the essential generation mechanism has not been clarified clearly. In this paper we will report some further studies for the mechanism analysis and removal of the generated surface charge. By measuring the wafer surface potential with partition check, we found that deionized water (DIW) used in via photo development process would induce to negative charge accumulation on the wafer surface. Based on this mechanism, a functional rinse containing surfactant molecules was used after DIW rinse to absorb the surface charge and take it away. Finally, the removal approaches of the generated surface charge was discussed by optimizing the dispense time and spin speed of functional rinse.

## INTRODUCTION

As semiconductor process technology scales down, the precision of critical dimension (CD) measurement becomes more and more important, and any measurement error will induce to bad device performance. In our previous work, blurred image from scanning electron microscope measurement (SEM) was observed after via photo development during dual damascene, which would bring big measurement error [1]. By partition check, an initial mechanism was proposed that negative electric charge will accumulate on the wafer surface after via photo development, inducing the formation of asymmetric electric field on the wafer surface [2]. When the e-beam of SEM arrives the wafer surface, incidence direction will be changed by the asymmetric electric field, which will impact the auto focus.

In this work, we focused on the root cause study of negative charge accumulation on the wafer surface after via photo development, and a detailed mechanism was proposed. Then, a kind of function rinse was employed to eliminate the negative charge.

## EXPERIMENT

Spin-coating of bottom anti-reflectivity coating (Barc) and photoresist (PR) was performed on 300mm diameter Si wafers using a TEL Clean TRACK, and exposure was carried on ASML 193nm Scanner, following by a post exposure bake (PEB). Development process and function rinse treatment were also conducted in TEL Clean TRACK. SEM image was obtained on Hitachi CDSEM tools. And the negative potential of wafer surface was measured to demonstrate wafer surface charge on Wafer Surface Charge Analyzer.

## RESULT AND DISCUSSION

*Table 1: Litho conditions of five wafers for partition check*

|     | Barc coating | PR coating | Exposure | TMAH DEV | DIW rinse |
| --- | --- | --- | --- | --- | --- |
| #1 | √ | -- | -- | -- | -- |
| #2 | √ | √ | -- | -- | -- |
| #3 | √ | √ | √ | -- | -- |
| #4 | √ | √ | √ | √ | -- |
| #5 | √ | √ | √ | √ | √ |

To determine the process stage when negative charge accumulates, five wafers with different litho conditions were prepared as below table for partition check, and the wafer surface potential was measured on Wafer Surface Charge Analyzer, as shown in Figure 1. A significant phenomenon was observed that the wafer surface charge appeared after DIW rinse during development, indicating that DIW rinse will induce negative charge accumulation on the wafer surface.

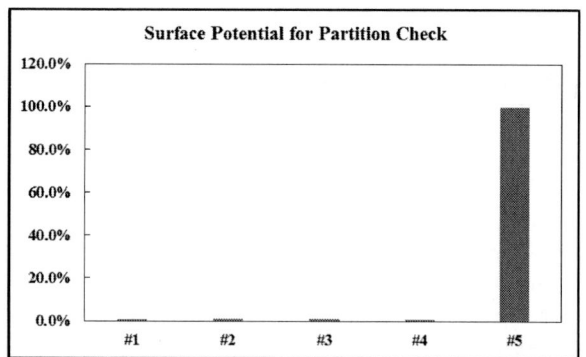

*Figure 1: Surface potential of five different litho conditions for partition check (the surface potential of condition #5 was defined as 100%, and other values were all normalized to condition #5)*

To further confirm the key impact factor of DIW rinse on surface potential, more patterned wafers with various DIW rinse time (10s, 20s, 30s, 40s, and 50s) and various DIW spin speed (1000rpm, 1500rpm and 2000rpm) were prepared, followed by surface potential measurement, as shown in Figure 2 and Figure 3, from which we can see that wafer surface potential increases with the increase of DIW rinse time and spin speed obviously. As known, DIW is used to remove the reactant of TMAH and PR post exposure. After DIW was dispensed onto wafer surface, friction will occur between DIW and PR surface under high spin speed, which was supposed to be the root cause of negative charge generation. And the negative charge continued accumulating on wafer surface with the DIW rinse time.

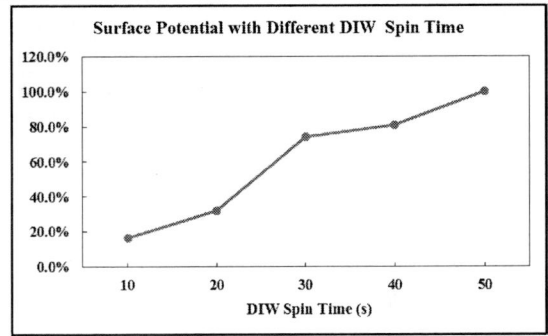

*Figure 2: Surface potential with different DIW rinse time (the surface potential with DIW rinse time 50s was defined as 100%, and other values were all normalized to it)*

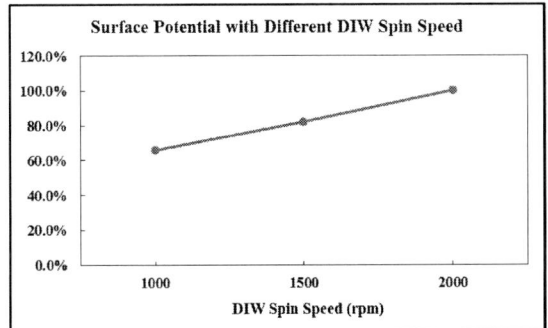

*Figure 3: Surface potential with different DIW spin speed (the surface potential with DIW spin speed 2000rpm was defined as 100%, and other values were all normalized to it)*

Combination with our previous conclusion that pre-layer metal hardmask (HM) opening is also an indispensable precondition of negative charge accumulation on wafer surface, a reasonable mechanism of wafer surface charge and SEM image blur after via photo development was proposed, as illustrated in Scheme 1. When DIW is dispensed onto patterned wafer, friction between PR and DIW under high spin speed will induce negative charge generation. Because of its insulation, DIW cannot take the negative charge away from wafer surface during spin-dry process. Consequently, more and more negative charge accumulates on wafer surface as DI-

*image blur*

-W rinse time increases, resulting in a so-called capacitor formation, where the patterned PR could be considered as upper electrode, and the pre-layer patterned HM could be considered as lower electrode. As the potential increases of upper electrode, negative charge will move to the lower electrode, namely, the interface of Barc and patterned HM, until a dynamic equilibrium status is achieved. Due to the existence of negative charge, an asymmetrical electric field will form near wafer surface, and the asymmetry depends on the pre-layer HM pattern density. During ADI measurement, the e-beam of SEM will deviate its original orientation caused by the asymmetrical electric field, resulting in astigmatism, which is the root cause of SEM image blur.

A temporary method to obtain clear SEM image is to use auto-stigma function developed by tool vendor [3, 4], where the orientation of e-beam could be corrected during measurement. Although the image blur could be avoided, the wafer surface charge still exists. In this work, a kind of functional rinse containing surfactant was employed to eliminate the surface charge. Based on our proposed mechanism, DIW is insulated and cannot take the generated negative charge away, while the surfactant, amphipathic molecule, in the functional rinse is supposed to be able to adsorb the negative charge and take it away during spin-dry process. To validate this hypothesis, three wafers after development were performed with different rinse process, as shown in table 2. Wafer #1 was rinsed only with DIW; and wafer #2 firstly was rinsed with DIW, and then treated with functional rinse; while wafer #3 was firstly rinsed with DIW, and then with functional rinse, then with DIW. And their surface potential was measured as shown in Figure 4. A significant improvement is observed on wafer #2 with 32% of surface potential relative to wafer #1, indicating the functional rinse works well on the elimination of surface negative charge. The residual negative charge should be attributed to un-optimized spin recipe of functional rinse. Additionally, a high surface potential appeared again when DIW was used again after functional rinse treatment, as denoted by wafer #3, further demonstrating that our proposed mechanism of negative charge generation is reasonable.

*Table 2: Conditions of three wafers with functional rinse*

| Wafer | DIW | Functional rinse | DIW |
|-------|-----|------------------|-----|
| #1 | √ | -- | -- |
| #2 | √ | √ | -- |
| #3 | √ | √ | √ |

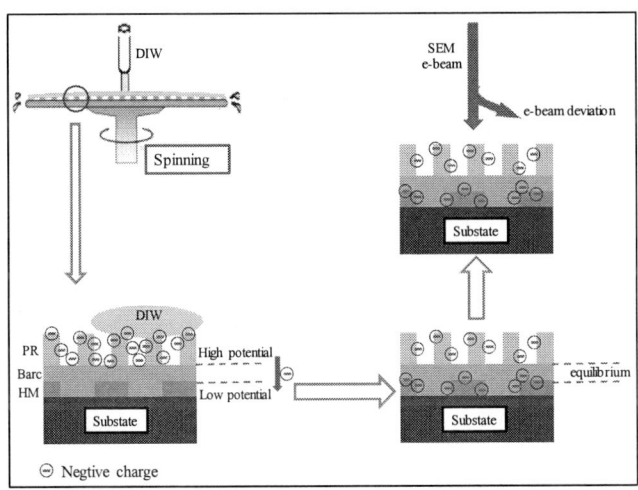

*Scheme 1: Mechanism of surface charge generation and SEM*

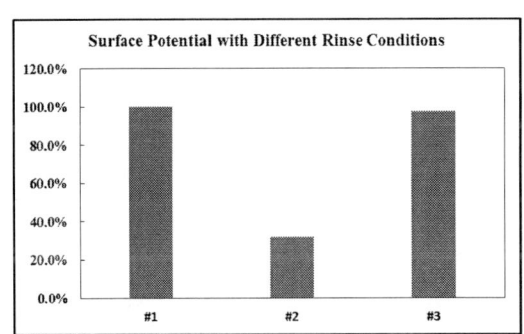

*Figure 4: Surface potential with different rinse conditions (surface potential of wafer #1 only rinsed with DIW was*

978-1-6654-9759-6/22 $31.00 © 2022 IEEE

*defined as 100%, and other values were all normalized to it)*

To entirely eliminate the surface negative charge, the function rinse treatment recipe was optimized based on two key parameters: dispense time and spin speed, as exhibited in Figure 4 and Figure 5. It is observed that surface potential decreases with dispense time and spin speed increase, indicating that the removal of negative charge should be a gradual process. As the dispense time increases, more and more negative charge on upper electrode is absorbed by the functional rinse, resulting in the break of the electric equilibrium. So the potential of lower electrode becomes higher than upper electrode, and the negative charge will gradually move from lower electrode to upper electrode, and then was removed along with functional rinse under high spin speed. With the optimized condition (10s of dispense time and 1000rpm of spin speed), only 0.03% of surface potential is obtained indicating that the surface negative charge has been eliminated entirely.

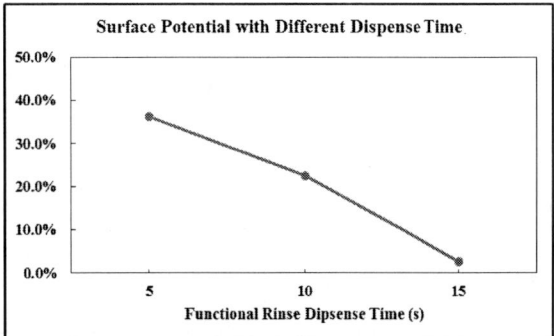

*Figure 5: Surface potential of different functional rinse dispense time (surface potential of the wafer without functional rinse treatment was defined as 100%, and the other values were all normalized to it).*

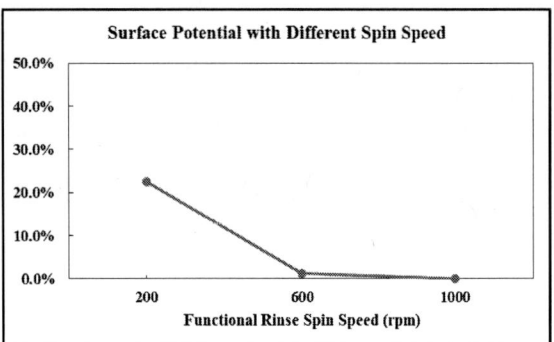

*Figure 6: Surface potential of different functional rinse spin speed (surface potential of the wafer without functional rinse treatment was defined as 100%, and the other values were all normalized to it).*

Finally, the SEM image post functional rinse treatment with optimized condition is obtained, as shown in Figure 7 (b), and the left SEM image (a) is obtained on the patterned wafer without functional rinse treatment as control. The clear SEM image post functional rinse treatment demonstrates that wafer surface charge has been removed, and astigmatism of e-beam could be eliminated successfully.

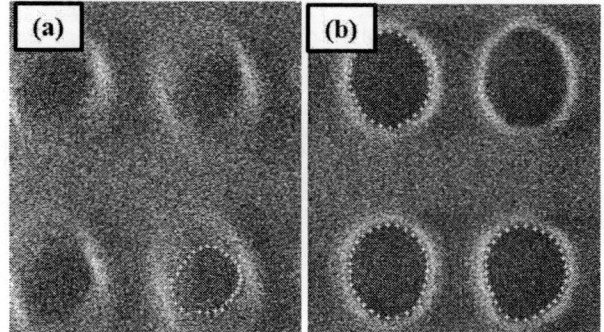

*Figure 7: SEM image (a) without functional rinse treatment; (b) with functional rinse treatment*

## CONCLUSIONS

In this work, the mechanism of wafer surface charge during dual damascene process has been clarified. It is demonstrated that the friction between PR and DIW under high spin speed will induce negative charge generation. As the accumulation of negative charge on wafer surface, the surface potential becomes higher, and more and more negative charge will move to the interface between HM and Barc until an electric equilibrium is achieved. During ADI measurement, the SEM e-beam will deviate its original orientation induced by the asymmetric electric field, resulting in astigmatism, which is the root cause of SEM image blur. Based on this mechanism, a functional rinse was used after DIW rinse to remove the negative charge, because its amphipathic molecules could absorb the negative charge. After optimizing the dispense time and spin speed of functional rinse, only 0.03% of surface potential was obtained indicating that the negative charge has been eliminated entirely.

## ACKNOWLEDGEMENTS

The authors would like to thank the higher management team from Shanghai IC R&D Company for the support of this work.

## REFERENCES

[1] Q. Zhang, G. G. Deng, B. Xing, J. G. Hao, Q. Wu and Y. S. Lin, "Study of CDSEM Measurement Issue Caused by Wafer Charging," 2015 China Semiconductor Technology International Conference, 2015, pp. 1-3.

[2] D. S. L. Mui, E. H. Lenz, C. Cyterski, K. Venkataraman and M. Kawaguchi, "Wafer Surface Charging Model for Single-Wafe Wet-Spin Processes," IEEE Transctions on Semiconductor Manufacturing, 2011, vol. 24, pp. 552-558.

[3] B. Choo, S. Punjabi, C. Morales, B. Singh, M. K. Templeton and M.P. Davidson. Proceedings of SPIE Vol, 2000, pp. 57-64.

[4] S. Dupuis, T. Hayes, C. Archie and E. Solecky, Proceedings of SPIE, 2001, pp. 344-354.

# 14 BITS CURRENT DAC TESTING WITH 16 BITS INSTRUMENTS

*Pio Marcozzi[1*] and Luffy Jin[2*]*

[1]Teradyne SEG, Milan, Italy

[2] Teradyne SEG, Hefei, China

* Authors' Email: pio.marcozzi@teradyne.com luffy.jin@teradyne.com

## ABSTRACT

In this paper the authors describe a pedestal-based method coupled with sub-ranging VI and current summing nodes for measuring source/sink current generators driven by 14 bits DACs. We will show how to implement it using 16 bits DACs/ADCs VI sources, they can be used for testing current DACs linearity parameters like DNL/INL. The modern instruments have other techniques like oversampling and dithering can obliterate the discrete nature of the underlying ADCs & DACs, but in this paper, the authors will focus on elaborating "pedestal-based" method test concept. The pedestal source doesn't perform any measure and its effect is calibrated directly by the sub-ranging VI. It can be shared across several source/sink current DACs to avoid multiple pedestals. The sub-ranging VI source measures the DAC's current with a small range to guarantee accuracy. The equivalent achievable accuracy is -at least- 3 bits higher than the 14 bits DACs under test.

## BACKGROUND

We worked with a device containing several 14-bit current DACs. Each current DAC has an open-drain structure and handles several current ranges: from few milliamperes to tens of milliamperes. The target instrument is a 16-bit DACs/ADCs VI instrument. However, the 16-bit ammeter of the instrument is not capable enough in terms of resolution because the sign bit is lost (due to the open drain structure of the device's DACs) then the ammeter becomes a 15-bit equivalent instrument.

Referring to the following table1 "used VI measure current specification", say that DAC's 5mA range would show nominal 305nA LSB while the best matching measure range of the VI ammeter, the 10mA range, would show 305nA resolution as well. Obviously, this would make the 5mA range untestable. Things get even worse when the device overtakes the 10mA cross-point and the VI must forcefully switch to 50mA/100mA range which shows up 3.05uA resolution for current ammeter. The 1mA range would be perfect for our purpose as the resolution is very promising being around 30nA, but it would allow to test the device only in the 0mA-1mA interval.

Can we add up an offset and still use the 1mA range accuracy for current measurements? The 1mA VI may be called "sub-ranging" while the other one at higher range, even 100mA, may be referred as "pedestal".

TABLE I. USED VI MEASURE CURRENT 1X SPECIFICATION

| Current Range | Resolution, nominal | Offset | Gain |
|---|---|---|---|
| ±100 uA | 3.05 nA | ±100 nA | ±0.05% rdg |
| ±1 mA | 30.5 nA | ±300 nA | ±0.05% rdg |
| ±10 mA | 305 nA | ±2 uA | ±0.05% rdg |
| ±50 mA | 3.05 uA | ±12.5 uA | ±0.05% rdg |
| ±100 mA | 3.05 uA | ±12.5 uA | ±0.05% rdg |

## PEDESTAL-BASED METHOD ON SINGLE DAC CHANNEL

Let us think of the device as a current limiter MOSFET with on top an ON/OFF MOSFET in such a way to reach an open drain current sink generator.

The sub-ranging source (VI1 in figure1) is directly connected to the device pin and its purpose is to force the required drain voltage and measure the device's currents. Remember that VI1 resolution is 30nA, 10 times smaller than the resolution of the device's 5mA range.

The pedestal source (VI2 in figure1) is connected to the device through a simple current adding node. The two VIs are decoupled through a series resistor which should be in the range 0Ω-1kΩ.

The purpose of the decoupling resistor is twofold: allow the pedestal source (VI2) to be able to work as FV or FI source and try to reduce the inevitable noise introduced by the VI2.

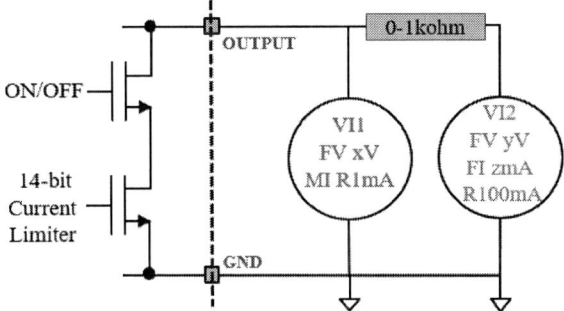

*Figure 1: Pedestal-based method concept*

### Technique description

The following examples explain the logical current sub-ranging procedure.

At first step it is necessary to "calibrate" the "nominal 1 mA device current", assuming previous codes are tested:
- VI1 forces the requested output/drain voltage in 1mA

current range (30.5nA resolution) and measures $I_{OUPUT}$.
- VI1 measures $I_{OUTPUT}$ = 1.003mA. Don't care about VI1 over-range for now. Later in the paper we will explain how to overcome the over-range problem in the real world.

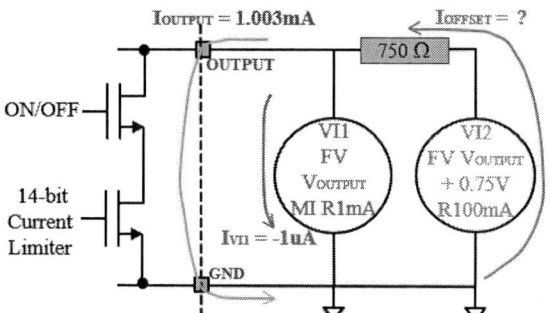

*Figure 2: Nominal 1 mA calibration*

While the device is still generating the 1.003mA insert the pedestal VI2 which should generate a nominal 1mA offset (in the example VI2 forces $V_{OUTPUT}$ + 1mA * 750Ω) and measure again with VI1. VI1 measures -1uA so the "VI2 introduced offset" is $I_{OFFSET}$ = 1.003mA + 1uA = 1.004 mA.

If device's output current is below 2mA we can measure any current using VI2 as pedestal (VI2 is not measuring anything as its "pedestal effect" is grabbed by VI1 using 1mA range) and we can prove it in the next example. Program the device to nominally sink 1.5 mA and then:
- VI1 measures +498uA
- The "real" device's current is $I_{OUPUT} = I_{OFFSET} + I_{VI1} =$ 1.004mA + 498uA = 1.502mA

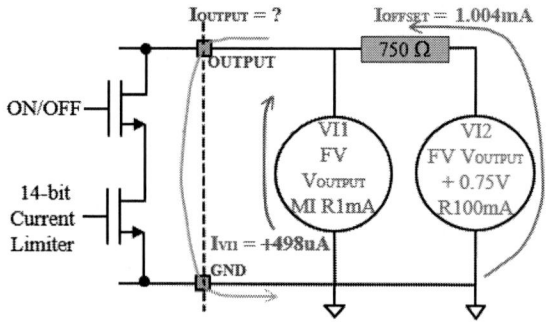

*Figure 3: Currents between 1 mA ~2 mA*

At some point the device's current will exceed the 2mA value and the VI1 will get into overcurrent problems. To overcome this problem, we must increase the pedestal offset and at the same time accurately calibrate it by means of the 1mA range ammeter of the VI1. The process can be done in two steps using the device as a "calibration vehicle". Program the device to sink nominal 2mA and then:
- VI1 measures +999uA
- VI2 is still generating the 1.004mA offset

- Obviously, the calibrated device current now is $I_{OUPUT}$ = 1.004mA + 999uA = 2.003mA
Again, $I_{OFFSET}$ is the value calibrated by VI1 in 1mA current range in the previous calibration step.

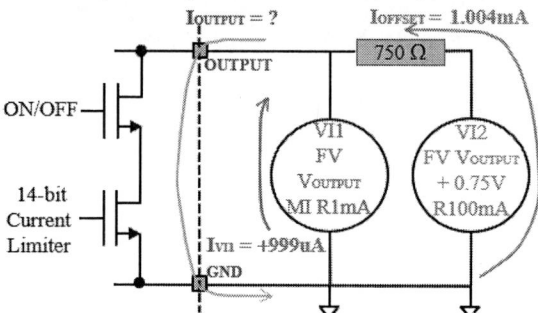

*Figure 4: Nominal 2 mA calibration*

While the device is still generating $I_{OUPUT}$ = 2.003mA (nominal 2mA as programed as before), we increase the voltage of VI2 to generate a nominal 2mA offset (VI2 adds up 1mA step forcing $V_{OUPUT}$ + 1mA * 750 Ω * 2) in 100mA current range and measure again with VI1. VI1 measures +5uA so the calibrated current generated by VI2 now is $I_{OFFSET}$ = 2.003mA - 5uA = 1.998mA.

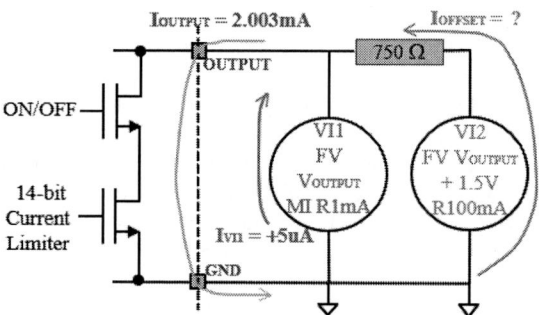

*Figure 5: Nominal 2 mA calibration*

Let's measure a nominal devices' current sinking of 2.5mA:
- VI1 measures +501uA
- VI2 is still generating 1.998mA offset
- Apparently, the calibrated device current now is $I_{OUTPUT}$ = 1.998mA + 501uA = 2.499mA

Same procedure for currents until the maximum DAC current. VI2 increase the pedestal offset step by step and every time VI1 calibrates it in 1 mA range.

**More efficient technique**
At this point the rationale of this method should be clear, below also find a graphical representation:
- VI2 (pedestal source) provides support current in such a way that VI1 never gets out of 1mA range
- VI1 (sub-ranging resource) keeps the needed device's drain voltage and measures/calibrates everything staying always in the 1mA range.

978-1-6654-9759-6/22 $31.00 © 2022 IEEE

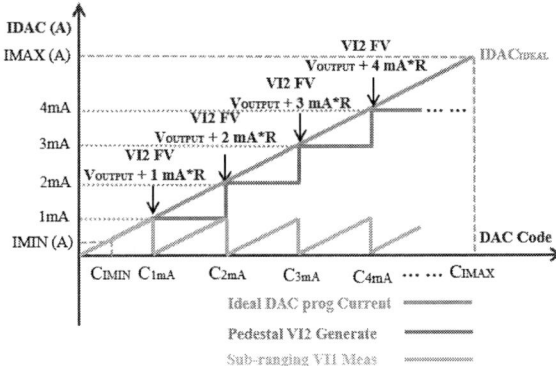

*Figure 6: Half range simulation graph*

After the first raw explanation let us think about some improvements:
- Sub-ranging resource VI1 should be used also for negative currents
- Pedestal resource VI2 jumps on bigger steps if VI1 takes advantage of the +/-1mA range
- Pedestal resources jumps are in the order of $V_{OUTPUT}$ + 2mA * 750Ω * N, N = 1, 2, 3 … etcetera

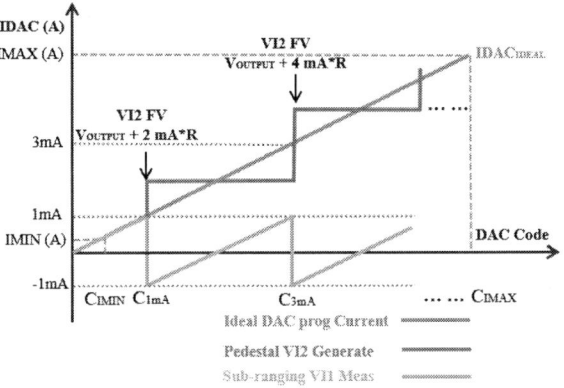

*Figure 7: Full range simulation graph*

### Differences between ideal case and realistic case

Up to now things are based on the ideal behavior from the device and perfect value of the decoupling resistor. However, experience shows that stray resistances can play badly and device itself disrupts the "perfect plan" especially when the code-to-current characteristic is slightly different from the desired one. The effect is shown by the sub-ranging source VI1 running into over-range/ overcurrent alarms. We cannot rely anymore on VI1 working in the full +/-1mA range. Instead, we can reduce the usage of VI1 range to +/-800uA, for example. Obviously, it is highly dependent on the specific device. So, it is up to the reader to find the best range usage of the sub-ranging VI1.

Below equations tell each switch point of pedestal source programing with usage of +/-800uA range for a theoretical 16mA range DAC (N ≤ 10).

$$SubRangingVI = V_{OUTPUT} \qquad (1)$$

$$IDeviceNom = 0.8 * (N - 1) * 1.6mA \qquad (2)$$

$$PedestalVI = V_{OUTPUT} + 1.6mA * 750 * N \qquad (3)$$

## PEDESTAL-BASED METHOD ON MULTIPLE DAC CHANNEL

When testing multiple current DACs, one sub-ranging VI per pin/DAC is mandatory, however, a unique pedestal VI can be used to provide the necessary pedestal currents, the pedestal VI has a decoupling resistor vs each sub-ranging VI works in 1mA range, the pedestal VI can still use the 100mA range.

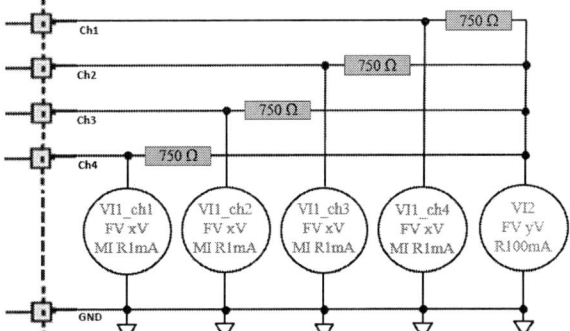

*Figure 8: Pedestal-based method on 4 DAC channels*

## SUMMARY

The pedestal and sub-ranging resources (or coarse and fine resources) are widely used in the ATE world. We adapted the general method and minimized application circuitry through easily manageable VIs existing in many test platforms.

This method provides a solution to use 16 bits VI to test 14 bits DACs. The Sub-ranging VI can measure device's DAC maximum current in a much smaller range to guarantee precise current measurements meanwhile the pedestal VI runs a synchronized staircase whose steps are runtime calibrated by the sub-ranging VI. All the codes can be tested when DAC can continuously source from one code to another, the pedestal VI source is stable especially after every switch point to eliminate dynamic effect, and the decoupling resistors do not drift.

Potentially higher bits DACs would also be testable with this technique if the sub-ranging VI scales down to smaller range from 1mA to 100uA for example. We hope that the reader has found some useful hints from this current DAC testing strategy.

## REFERENCES

[1] *An Introduction to Mixed-Signal IC Test and Measurement, second edition*, G. Roberts , F. Taenzler, M. Burns

# A NEURAL NETWORK APPROACH TO ANALYZE FDC DATA IN SEMICONDUCTOR MANUFACTURE

*Wei Yu[1]\*, Xu Chen[2], Guiyun Mao[3], Yong Wang[4], Zhengying Wei[5]*

[1, 2, 3, 4, 5] E1, HLMC, Zhangjiang Hi-Tech Park, Shanghai 200120, People's Republic of China
\*Corresponding Author's Email: yuwei_e1@hlmc.cn

## ABSTRACT

Fault Detection and Classification (FDC) Data reflects machine real-time status that provide information for diagnosing processing quality. In practice, most semiconductor engineers analyze FDC data by experience. Usually, they only use parameters which they thought are important and resolve data in an inherent subjective method. However, with the development of sensor and communication technology, there are increasing numbers of variables collected in milliseconds. Therefore, analyzing by human experience becomes more complicated, inconsistent, and unreliable. To solve this problem, we proposed a system to transform FDC data into summary statistics that can assist engineer to trace root causes and provide forecasted inline value for corresponding process. The system consists of two parts: a Convolutional Neural Network (CNN) model that pre-process FDC Data and a neural network model based on transformer framework to predict inline value. Numerical results show that the mixed model outperform simple transformer or CNN model and nearly a hundred times faster than pure Long Short-Term Memory (LSTM) model.

## INTRODUCTION

Sensors are utilized in a large number of industries for automation of systems. Along with the development of sensor technology, a great deal of sensor data which reflect the actual manufacturing statuses can be continuously produced and accumulated. These data are high-dimension and large-scale, and often used in Fault Detection and Classification (FDC). As a result, we also called these data as FDC data in order to distinguish from other data collected in production. Nowadays, the novel and critical problem has become how to make full use of FDC data [1]. Virtual Metrology (VM) is then proposed.

In semiconductor manufacturing, metrology tools are utilized to ensure production quality. However, with time and cost limitations involved, inline metrology cannot be applied after every processing. To overcome it, VM systems are developed to forecast metrology variables based on sensor data mentioned above [2].

A high variety of VM methodologies have been addressed in open literature. Traditional VM methods can be summarized as follow: calculating characteristic values (such as mean, median, minimum and maximum value) from sensor data and applying partial least squares (PLS)

[3], [4], principal component regression [5] and Gaussian process regression (GPR) [6]. These methods consider different types of VM scenarios which can occur realistically in practice. However, FDC data are composed of many status variable identifications (SVIDs) collected as time series at high sampling rates [7]. Using characteristic values (e.g. mean, median, minimum and maximum value) often lead to the ignorance of hidden information behind time series data. Recently, neural networks (NNs) have been proposed [8]. Reference [9] considers four methods, including MLR, GPR and NN, and evaluates these models with a CVD dataset. Although NN was not the one that performed best, it did reach similar level.

In this paper, we propose a hybrid system that combines CNN and a neural network model based on transformer framework. Instead of exploring data by characteristic values, a CNN model is applied to filter redundant data while transformer based NN model is then utilized to predict corresponding inline values.

## METHODS

### Convolutional Neural Network (CNN)

To avoid overfitting caused by excessive parameters in the second part of this system, we established a simple CNN to filter useless information and reduce noise. In this part, we only adopted one convolutional layer and trained it based on the principle of auto-encoder (Figure 1) [9].

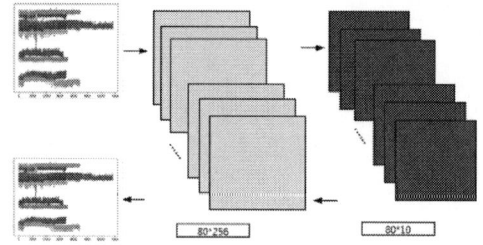

*Figure 1: CNN structure used in our study*

To emphasize, in this model, we do not reduce the dimensions of variables. The purpose is to integrate information on time steps. We used Adam optimizer setting the learning rate to 0.01 and the decay rate to 0.9. The model was trained with a batch size of 64 for 500 epochs. The hyper-parameters were optimized by evaluating reconstruction performance. The measurement of reconstruction performance is determined by Dynamic

Time Wrapping (DTW) [10]. For two time series $Q = (q_1, q_2, ...q_i, ...q_n)^T$ and $C = (c_1, c_2, ...c_j, ...c_m)^T$, the purpose of is finding $DTW(Q,C)$ that minimizes the sum of total distance in (eq.1), where $\|\cdot\|_P$ represents the $l_p$ norm.

$$DTW(Q,C) = \min \sum_{i=1}^{m} \sum_{j=1}^{n} d(q_i, c_j) \qquad (1a)$$

$$d(q_i, c_j) = \|q_i - c_j\|_P \qquad (1b)$$

### Transformer

Transformer [11] is first raised in the Natural Language Processing (NLP) task and achieves dominant performance [12], [13]. It contains four main parts: embedding, positional encoding, encoder and decoder. The key parts of encoder and decoder are scaled dot-product attention and multi-head attention. Considering the similarity between natural language and sensor data, we adopted the basic structure of positional encoding and encoder, and a two-layer perceptron on top of the encoder to carry out the final result (Figure 2). The hyper-parameters were set according to the instructions in [11] while we reduced $N$ to 4 due to the limited amount of data. The model was trained by minimizing the mean squared error (MSE) between predicted and actual value.

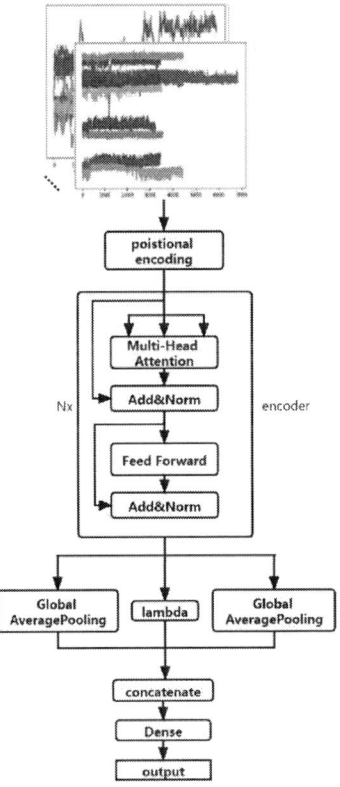

*Figure 2: NN structure based on transformer applied in our study*

### EXPERIMENT

We used data from CVD sensors to train and evaluate the proposed models. To balance between accuracy and generality, a total of 2000 sensor data were collected in half a year, among which nearly 200 data were randomly selected for testing while the rest of it were put for training.

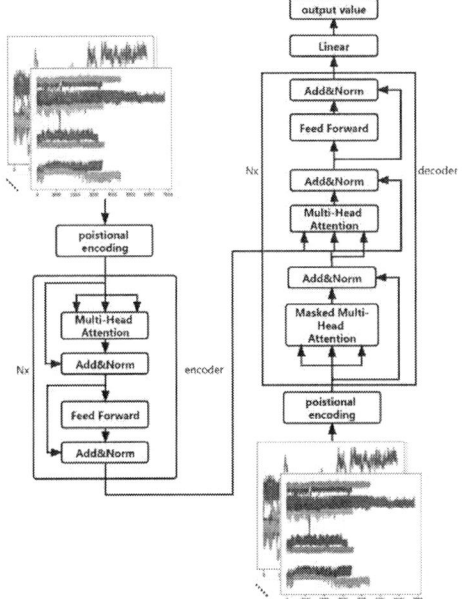

*Figure 3: model structure used to train encoder*

In actual production, the sampling rate of physic metrology is less than 20%. That is to say, 80% of the data could not be utilized if we just use model in Figure 2. To take maximum use of the data, we actually expanded the transformer-based NN (Figure 2) into a larger model (Figure 3) and trained the model by minimizing reconstruction error. The error was calculated as accumulative MSE (eq.2), where $m$ represents the number of variables and $n$ shows the time steps. After pre-training, we adopted the predict-model illustrated in Figure 2 and trained it with parameters fixed in encoder.

$$\text{error} = \frac{\sum_{i=1}^{m} \sum_{j=1}^{n} (value_{predict} - value_{actual})}{m*n} \qquad (2)$$

### RESULTS

Table1 manifests the results of prediction by different methods. It is obvious that the system we mentioned above achieves the highest level of performance. The system without pre-training shows its limitation in forecasting extreme values, which may be caused by overfitting in training data set. The comparison between model with and without pre-training indicates that it would be better to apply a full transformer-based model (as shown in Fig.3) to train the encoder part and only use limited inline values training the two-layer perceptron. Besides, we experimented on the model without CNN. The results showed that this part of system plays a good role in noise reduction. The numeric decline, from 1.59 in LSTM to 1.37 in CNN+LSTM, confirmed this point. As can be noticed, traditional method, GPR, cannot predict well, even failed in forecasting the trend sometimes. The reason for this phenomenon could be the variety and complexity of semiconductor FDC Data. Only applying characteristic values will omit key factors which have significant influence on final results.

TABLE I:RESULTS FOR DIFFERENT MODELS

|  | MSE |
| --- | --- |
| Model | 0.76 |
| Model without pre-training | 0.97 |
| Model without CNN | 1.03 |
| GPR | 2.64 |
| LSTM | 1.59 |
| CNN+LSTM | 1.37 |

## CONCLUSION

In general, we proposed a hybrid stochastic algorithm that combines CNN and transformer-based neural network. CNN is functioned as a filter to reduce time steps and to reduce noise while the deformed transformer is then connected to forecast the corresponding metrology value. Using real data, this approach has been proved effective, and showed better performance compared to competing method.

## REFERENCES

[1] H. Liu, J. Chen, F. Huang and H. Li, "An Electric Power Sensor Data Oriented Data Cleaning Solution," *2017 14th International Symposium on Pervasive Systems, Algorithms and Networks & 2017 11th International Conference on Frontier of Computer Science and Technology & 2017 Third International Symposium of Creative Computing (ISPAN-FCST-ISCC)*, 2017, pp. 430-435, doi: 10.1109/ISPAN-FCST-ISCC.2017.29..

[2] R. Clain, V. Borodin, M. Juge and A. Roussy, "Virtual metrology for semiconductor manufacturing: Focus on transfer learning," *2021 IEEE 17th International Conference on Automation Science and Engineering (CASE)*, 2021, pp. 1621-1626, doi: 10.1109/CASE49439.2021.9551567.

[3] T. Hirai and M. Kano, "Adaptive Virtual Metrology Design for Semiconductor Dry Etching Process Through Locally Weighted Partial Least Squares," in I*EEE Transactions on Semiconductor Manufacturing, vol. 28, no. 2, pp. 137-144, May 2015*, doi: 10.1109/TSM.2015.2409299.

[4] H. Roh et al., "Development of the Virtual Metrology for the Nitride Thickness in Multi-Layer Plasma-Enhanced Chemical Vapor Deposition Using Plasma-Information Variables," in *IEEE Transactions on Semiconductor Manufacturing, vol. 31, no. 2, pp. 232-241, May 2018*, doi: 10.1109/TSM.2018.2824314.

[5] S. Park, S. Jeong, Y. Jang, S. Ryu, H. -J. Roh and G. -H. Kim, "Enhancement of the Virtual Metrology Performance for Plasma-Assisted Oxide Etching Processes by Using Plasma Information (PI) Parameters," in *IEEE Transactions on Semiconductor Manufacturing, vol. 28, no. 3, pp. 241-246, Aug. 2015*, doi: 10.1109/TSM.2015.2432576.

[6] J. Wan and S. McLoone, "Gaussian Process Regression for Virtual Metrology-Enabled Run-to-Run Control in Semiconductor Manufacturing," in *IEEE Transactions on Semiconductor Manufacturing, vol. 31, no. 1, pp. 12-21, Feb. 2018*, doi: 10.1109/TSM.2017.2768241.

[7] W. Yang, J. Blue, A. Roussy, J. Pinaton and M. S. Reis, "A Structure Data-Driven Framework for Virtual Metrology Modeling," in *IEEE Transactions on Automation Science and Engineering, vol. 17, no. 3, pp. 1297-1306, July 2020*, doi: 10.1109/TASE.2019.2941047..

[8] 14. R. Asim, G. Sandeep and M. Raymond, "An algorithm to generate radial basis function (RBF)-like nets for classification problems", *Neural Netw.*, vol. 8, no. 2, pp. 179-201, 1995.

[9] D. E. Rumelhart, G. E. Hinton and R. J. Williams, "Learning representations by back-propagating errors", *Nature*, pp. 533-536, 1986.

[10] J. Zhao and L. Itti, "shapedtw: Shape dynamic time warping", *Pattern Recognition*, vol. 74, pp. 171-184, 2018.

[11] GAshish Vaswani, Noam Shazeer, Niki Parmar, Jakob Uszkoreit, Llion Jones, Aidan N Gomez, Łukasz Kaiser, and Illia Polosukhin. Attention is all you need. In *NeurIPS*, 2017.

[12] Tom B Brown, Benjamin Mann, Nick Ryder, Melanie Subbiah, Jared Kaplan, Prafulla Dhariwal, Arvind Neelakantan, Pranav Shyam, Girish Sastry, Amanda Askell, et al. Language models are few-shot learners. In *NeurIPS*, 2020.

[13] Jacob Devlin, Ming-Wei Chang, Kenton Lee, and Kristina Toutanova. Bert: Pre-training of deep bidirectional transformers for language understanding. In *NAACL*, 2019.

# A VOLTAGE SCREEN MODEL AND METHOD FOR EARLY FAILURE SCREENING

*Wen Ying\*, Canny Chen, Kelly Yang*

Q&R, Semiconductor Manufacturing International Corporation (SMIC) Shanghai 201203, China
\*Corresponding Author's Email: Wen_Ying@smics.com

## ABSTRACT

Bathtub curve is widely used in semiconductor reliability, which divides the product lifecycle into three stages: infant mortality stage, useful life stage and wear out stage. Though, a batch of products will be monitored before delivery, some infant-mortality products may escape the monitor as it is sampling inspection, especially when cumulative failure rate is small. Thus, a voltage screening method, basing on the voltage screening model, was built to screen out the early failures. The voltage screening model is transformed from Weibull distribution, which can perfectly describe the whole Bathtub curve. The voltage screening method is high-efficiency as one stress is enough to detect, and the saturated cumulative failure rate can be sharply decreased after screening.

## INTRODUCTION

When it comes to reliability, Bathtub curve is a key which can't be opened around [1]. Bathtub curve divides product lifecycle into three stages: infant mortality, useful life and wear out, whose hazard rate is degressive, invariable and rapidly increasing, respectively, as shown in Figure 1. To quantitatively evaluate semiconductor device performance, we need a mathematical tool as Bathtub curve is just a physical and qualitative description, and Weibull distribution [2] is the qualified one. Depending on the shape factor, β, Weibull distribution can perfectly describe the above three stages. When β < 1, = 1 or > 1, it can fit infant mortality stage, useful life stage or wear out stage, respectively. As far as we know, Weibull distribution is mainly used to extrapolate device lifetime through wear out stage, and hardly used in infant mortality stage. For semiconductor devices, extrinsic defects determine the fail in infant mortality stage, while intrinsic property does in wear out stage.

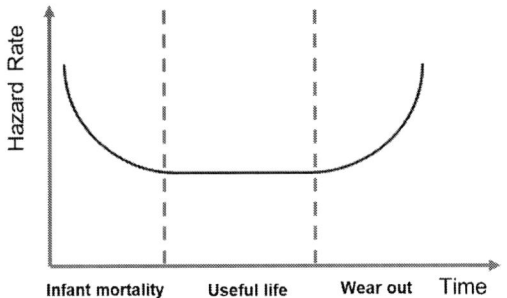

Figure 1: Bathtub curve

The intrinsic property in wear out stage can reflect device quality and thus catches much attention, while early failures caused by extrinsic defects in infant mortality stage is easily ignored, as WAT/RE monitor tests will filtrate it. However, limited by the sample size, some early failures may escape from monitor when cumulative failure rate is low (~ several hundred ppm).

In this work, we proposed a method to efficiently screen out these early failures during monitor. The method is based on voltage screening model (VSM), a simply transition from Weibull distribution. After that, the saturated cumulative failure rate will sharply decrease and improve the quality of final products.

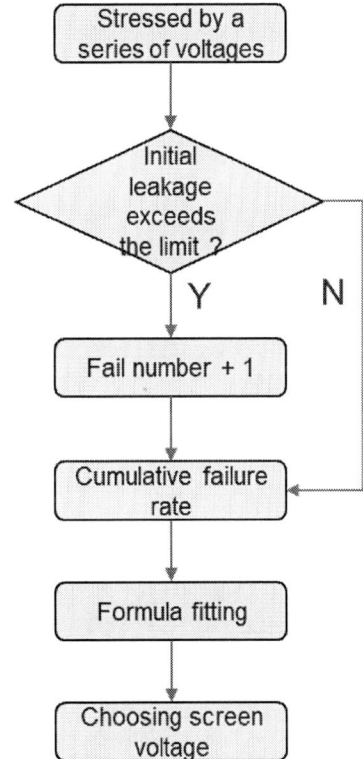

Figure 2: flow chart of voltage screen model building

## EXPERIMENTAL

The voltage screening method consists of two steps: model building and voltage screening. Process of VSM building was shown in Figure 2. We successively applied 2.8V, 3.3V, 4.5V, 5.0V, 7.5V and 10V on samples and

978-1-6654-9759-6/22 $31.00 © 2022 IEEE

checked their current. If the current was larger than the criteria, the corresponding sample was marked as a failure.

Cumulative failure rate was calculated by failure number/total samples. After we got a set of voltage and cumulative failure rate, VSM was fitted by Gauss-Newton method. More than 50 thousand samples were tested to build VSM. Based on VSM, we can choose a proper voltage to screen out those early failures.

## RESULT AND DISCUSSION

In Weibull distribution, Equation (1), the constant 1 represents the maximum cumulative failure rate is 100%. There is no doubt that all samples will fail if the stress is large or stress time is long enough. However, in infant mortality stage, the stress or stress time is limited, not all samples should fail. So, in VSM, Equation (2), we introduced a variate $\alpha$ to replace constant 1, and $0 <= \alpha <= 1$. This variate represents saturated cumulative failure rate, and could reflect product quality to some extent. If $\alpha$ is large, it means the product may be low quality with many early failures.

$$F = 1 - \exp(-(V/\tau)^{\wedge}\beta) \qquad (1)$$

$$F = \alpha - \exp(-(V/\tau)^{\wedge}\beta) \qquad (2)$$

Six voltages were used in this work, and the corresponding cumulative failure rate was listed in Table1, while the fitting result was shown in Figure 3. The equation in Figure 3 was the fitted VSM, and we could know that the saturated cumulative failure rate was ~850 ppm. If we want final product low cumulative failure rate, such as A ppm, then we could put A into the dependent variable F to calculate the independent variable V. The resulted voltage V will be used in voltage screening step. Theoretically, we can get extremely low-ppm product with appropriate screen voltage.

TABLE I.        VOLTAGE SCREEN MODEL FITTING DATA

| Voltage (V) | Cumulative Failure Rate |
|---|---|
| 2.8 | 0.000309 |
| 3.3 | 0.000328 |
| 4.5 | 0.000367 |
| 5 | 0.000386 |
| 7.5 | 0.000444 |
| 10 | 0.000463 |

As this method could be used in WAT/RE monitor, as well as CP test, the screen test should be nondestructive, and thus the screen voltage, as well as voltage used in VSM building, should be limited. The criteria is set by two rules: one is no more than mode B voltage, the ceiling voltage of early failure; the other one depends on the acceptable maximum lifetime loss by voltage screening,

such as 1%. The voltage corresponding to certain lifetime loss can be calculated through Equation (3), which is based on E model and derivate from AC lifetime calculation.

$$V_{target} = V_{max@sepc} + T_{ox}/\gamma * \ln(\varphi * t_{sepc}/t_{test}) \qquad (3)$$

In Equation (3), $V_{target}$ is the stressed voltage that device lifetime will be reduced by $\varphi$, and its value changes with the lifetime loss $\varphi$. $V_{max@sepc}$ is the maximum applied voltage with device lifetime just meeting spec. $T_{ox}$ is thickness of dielectric, and $\gamma$ is voltage acceleration factor. $t_{sepc}$ is the spec of lifetime, such as 10 yrs, and $t_{test}$ is the stress time during voltage screening, usually < 1s.

Compared with traditional monitor method, which is sampling and applies ramped voltage, our method is time-saving and unimpaired.

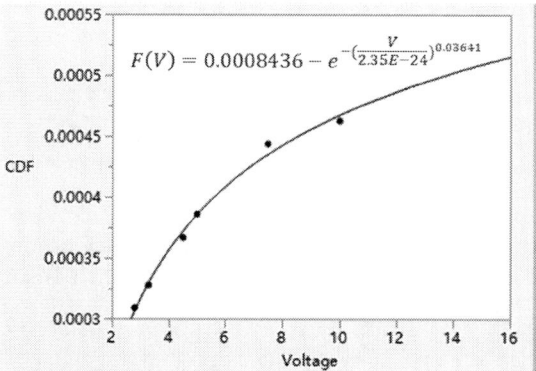

Figure 3: Voltage screen model fitting result

## CONCLUSION

WAT/RE monitor is an important method to monitor the product quality, but it is not competent when the cumulative failure rate is relatively low, because it is hardly to find early failure in limited samples. CP test is also insufficient to screen out most early failures efficiently. Herein, we proposed a voltage screening method to make up it. The voltage screening method, containing the voltage screening model, is based on Bathtub curve and Weibull distribution. As soon as voltage screening model is build, the method can be applied in all samples, as it is nondestructive, and test time is much shorter than traditional monitor tests. The method can be used not only for gate oxide, but also for dielectric in MOL or BEOL, and the cumulative failure rate of final product could be greatly reduced after voltage screening.

## ACKNOWLEDGEMENTS

Many thanks to Kaili Li for her toilsome tests.

## REFERENCES

[1]  R. W. Smith and D. L. Dietrich. *Proceedings of*

*Annual Reliability and Maintainability Symposium (RAMS)*, Anaheim, January 24-27, 1994, pp. 241-247.

[2] M. R. Gurvich, A. T. Dibenedetto and S. V. Rande. *Journal of Materials Science*, vol. 32, 1997, pp. 2559-2564.

# STUDY AND IMPROVEMENT ON MEASUREMENT ACCURACY OF IMAGE BASED OVERLAY

*Zhi-Feng Gan, Xiao-Chi Xu*
Technology R&D, Semiconductor Manufacturing International Corp.
Pudong New Area, Shanghai, P.R.China
+86(21)20817484, zf_gan@163.com

## ABSTRACT

The accuracy of overlay metrology becomes crucial when shrinking device dimensions to the 1X nm nodes and below. TIS (Tool Induced Shift) and WIS (Wafer Induced Shift) are the main issues for IBO (Image Based Overlay). Many factors could induce measured offset such as low signal contrast for opaque stack materials or mark profile degrade due to processes like CMP (chemical mechanical polishing). These optical measurement offsets will induce final device performance to degrade. In this paper, investigations of measured offset in IBO metrology induced by mark profile are presented, in the meanwhile, an analytical mechanics model is presented to provide insight into the connection between mark profiles and overlay measured offset. In the second part of the work, methods of compensation have been introduced via shift focus in the optical measurement system, and corresponding experimental results have been investigated as evidence to validate the IBO overlay results.

Keyword: Overlay, Accuracy, Image based, SEM, CMP, Mark profile, Focus, Target design

## INTRODUCTION

The future development of silicon manufacturing beyond the usual scaling of dimensions implies the introduction of new material but also the increase of process complexity. It means that metrology will need to enable the extraction of profile information, allow the increased sampling. In-line metrology must enable process control. To do so, the most important is measurement consistency, which means being process-robust to allow the reduction of metrology uncertainty [1].

Traditional overlay marks usually have feature widths that are significantly larger than the device feature size. The processing steps (CMP or etch) are optimized for the smaller device features. The resulting overlay structure might therefore differ from the ideal symmetrical pattern and introduce false overlay error not representative of the real device overlay error [2].

Under certain circumstances, inaccuracy due to such target imperfections can become the dominant contribution to the metrology uncertainty and cannot be quantified by the standard TMU contributors [3]. In this paper, we describe a calibration method that makes the overlay measurement robust to target imperfections. Methods of compensation have been introduced via shift focus in the optical measurement system, and corresponding experimental results have be investigated as evidence to validate the image based overlay results.

## RESULTS AND DISCUSSIONS

The analysis was done for the BEOL critical metal layer in the Copper Dual-Damascene process. The IBO measurements presented in this paper are performed using Archer500 metrology tool from KLA-Tencor. To address the question of the reliability of OVL measurements we used the Qmerit accuracy key and CD-SEM reference data coming from a CG6000 from Hitachi.

### IMAGE BASED OVERLAY

Image based overlay (IBO) uses a standard optical microscopy systems using white light (or a subset of it called filter or color) while auto-focus is performed via the interferometer. Each target is symmetric with respect to 90° rotation. This grating target is characterized by periodic series of lines and spaces. The center of inner and outer structures is found with the kernel (boxes to check the grating profile, see Figure 1). The overlay is the vector between the two centers [4]. In this study, Reference layer as M1 is copper layer, and current layer M2 is resist layer.

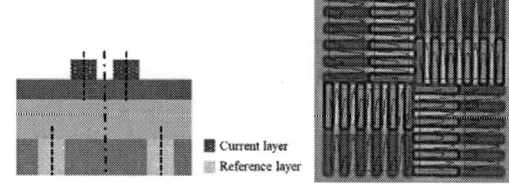

*Figure 1: Kernel (red boxes) working area and periodical signal processed for Image based overlay mark*

### WAFER INDUCED OVERLAY SHIFT

In Figure2, we show an overlay map example of mark damaged by process in the BEOL critical metal layer, the overlay map is wafer rotation vector map significantly both in raw data and modeled data set. The results indicate two possible, one is real overlay shift, and one is wafer induced overlay shift.

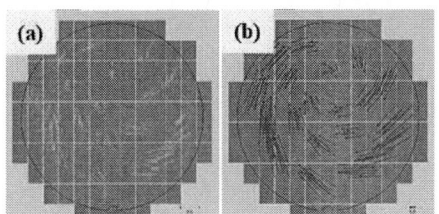

*Figure 2: Overlay map of M2 to M1 in one wafer (a) Raw Data, (b) Modeled Data*

The selection of the correct color filter is crucial to ensure good contrast and thus the performance and robustness of the recipes. The available in-place OVL tool generation, Archer500, is equipped with nine color filters.

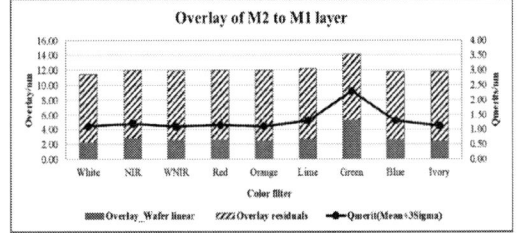

*Figure 3: Overlay and Qmerits of M2 to M1 layer by color filters*

We ran full map data collection on the ADI stage for one wafer and performed an offline accuracy analysis using the Qmerit metric. Looking at the Qmerit value (Figure 3), we can remove the green color, from our considerations as it is sensitive to the existing wafer-to-wafer and process variations. As Qmerit analyzes the target asymmetry separately for two parts of the target, we can conclude that the asymmetry of the resist layer and copper layer can be neglected.

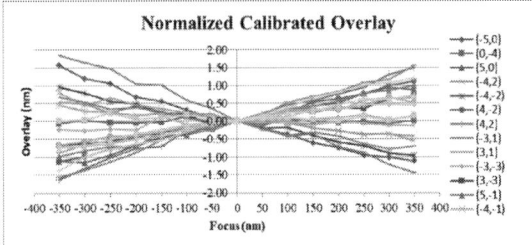

*Figure 4: Through Focus overlay dependence to overlay metrology results*

An asymmetric target profile will highly influence the results of overlay metrology. Another experiment was implemented to determine how asymmetry induced by the process affects the overlay metrology result, as shown in Figure 4. In this experiment, overlay error was measured at thirteen sites across a 300mm wafer. Each site was measured at fifteen different focal positions. Results analysis was then performed on each data set and compared. Each line in Figure 5 represents the one-site

overlay variance at fifteen focal planes. The conclusion could be: Focus shift induces overlay measurement error in metal layer. These data demonstrate that process induced target asymmetry is a linear, wafer-level contributor to overlay shift. In this study, the Qmerit accuracy flag cannot be used to quantify the measurement error.

## MECHANISM ANALYSIS FOR OVERLAY MEASURED SHIFT

To understand why overlay shifts occur, we analyze the measurement mechanism of an IBO metrology tool. Two basic measurement modes are used for measuring the overlay mark targets: Bright field microscopy and Coherence Probe Microscopy (CPM).

For the Bright Field Microscopy the tool uses the standard method of optical microscopy systems. The illumination in the Bright Field microscope is directed through the optical system. Measurement is performed using white light while auto-focus is performed via the interferometer. To obtain the overlay, it depends of target used.

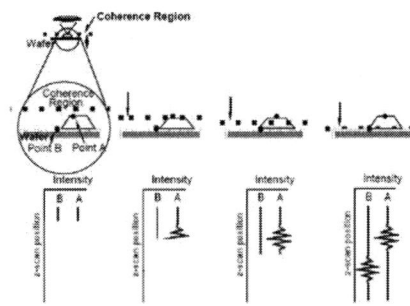

*Figure 5: CPM modes principle, illustration of coherence region scan to obtain a cross-section image of the target.*

The tool uses Coherence Probe Microscopy to enable 3-D measurement of pattern dimensions. CPM technology uses a Linnik interferometer to collect 3-D information, measuring both the amplitude and phase of an image. The output signal corresponds to the signal of the coherent region, by making a focus scan; the tool gives a 'cross-section' image of the target (see figure5). The overlay is deduced from this image.

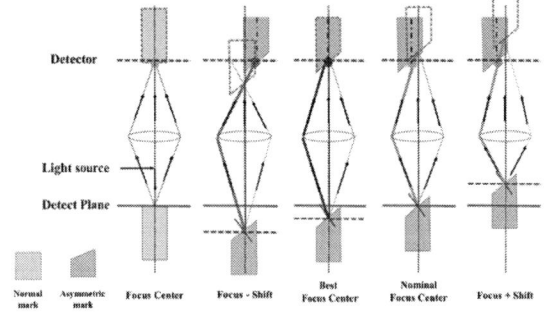

*Figure 6: Mark image at the different focus position*

978-1-6654-9759-6/22 $31.00 © 2022 IEEE

Linnik subassembly moves up and down on the axis direction of the lens to focus the objective lens on the target. The PIN Diode array measures the signal intensity as the focus position is moved, and determines the best focus position. As illustrated in Figure 6, for normal mark, maximum intensity equals best focus position; while for the asymmetric mark of reference layer, nominal best focus position decided by Linnik system will induce mark image position to have a bias to real mark's position, which induce false rotation vector map have been output. Whether single grab mode or double grab mode cannot eliminate the effect in measurement system since reference layer best focus bias occurs.

Asymmetry of the metrology target at copper level is suspected. If the overlay is different with the focus, it means that the target at reference layer might suffer surface asymmetric profile. This is confirmed with the AIM cross-section performed on the test wafer.

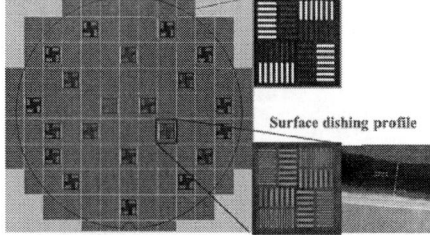

*Figure 7: Cross section on BEOL AIM target*

CMP dishing have effects on overlay in the following ways: Causes overlay mark asymmetry that results in overlay errors and causes depth variation across the wafer. Figure 7 shows the top view image of reference copper layer AIM mark target, the optical image clearly shows the damage on the target, which indicates the wafer suffers CMP dishing effect across the wafer, and cross section of the target mark suffer worse dishing profile.

## SOLUTION AND CORRECTION

One solution is target design optimization. The segmented AIM reduces the marginality of the loading of CMP process by getting closer to transistor dimensions.

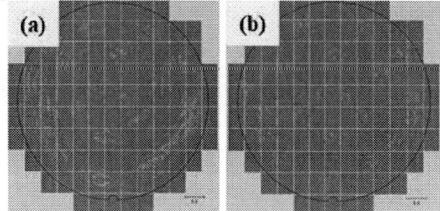

*Figure 8: Overlay map of M2 to M1 in one wafer (a) IBO Overlay data, (b) SEM Overlay data after etch stage*

We have tested in scribe-line, a target much design rule compliant and minimized CMP dishing effect in mark region significantly. But the trend of segmentation and

dummification, is decreasing the contrast of the target.

The second solution explored is complementary reference (High Voltage SEM) to calibrate optical measurement. Figure 8 shows Overlay map of M2 to M1 in one wafer, Figure 8(a) is IBO overlay at ADI stage, Figure 8(b) is SEM overlay data after the etching stage, the bias results from optical measurement shift, lithography stage and other non-lithography (CMP, etc.) stage processes variation.

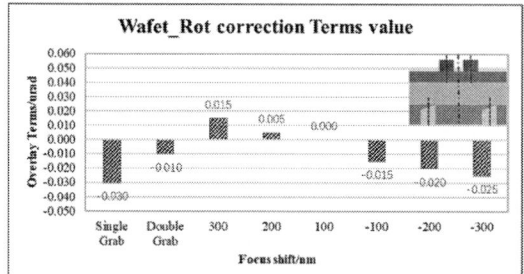

*Figure 9: Through Focus overlay terms of Wafer Rotation*

We can conclude wafer rotation-related vector bias "value A" via SEM and IBO overlay bias vector modeling calculation. At the same time, the related parameters between wafer rotation correction terms and focus position in the IBO measurement system can obtain via through focus overlay measurement. As shown in Figure 9, each wafer was measured at seven different focal positions, and corresponding wafer rotation correction "terms Bx" is obtained. The best focus position in the IBO measurement recipe is to be corrected via finding process introduces "terms Bx" reference to SEM calibrated "value A", and, then process introduce overlay measurement error can be eliminated.

## CONCLUSION

In this work we discussed the connection between mark profiles and overlay measured offset. If the overlay is different with the focus, it means that mark target might suffer surface asymmetric profile introduced by non-lithography stage processes. Combining image based overlay measurement methods of correction via shift focus in the optical measurement system and SEM overlay calibration results, makes the overlay measurement robust to target imperfections.

## ACKNOWLEDGEMENTS

The authors would like to thank vendors' useful discussions and help from KLA-Tencor. We also thank TD teams for providing valuable support.

## REFERENCES

[1] Jens T. Neumann et al. , Proc. SPIE. Vol 8326, 832602.

[2] Philippe Leray., Proc. of SPIE Vol. 10145 1014503.

[3] Eran Amit et al.，Proc. of SPIE Vol. 8681 86811G.

[4] Yoann Blancquaert et al., Proc. of SPIE Vol. 8681 86812O.

# TINY SADP DEFECT DETECTION AND REDUCTION FOR 19NM NAND FLASH TECHNOLOGY SEMICONDUCTOR MANUFACTURING ENGINEERING

*QiuFeng Cao [*], Qiliang Ni , and Jianye Song*

Shanghai Huali Microelectronics Corporation, Shanghai 201210,China

*Corresponding Author's Email: caoqiufeng@hlmc.cn

## ABSTRACT

In this paper, the method of combining electron beam scanning detection machine and optical bright field detection machine was explored to detect the extreme tiny defects in the 19nm NAND Flash self-aligned dual imaging process and the defect monitoring index was established. Combined with the principle of electron beam and bright field detection machine, the principle of defect detection was analyzed and a short process was established to improve the defect. According to the experimental results, the process conditions of lithography were optimized to reduce the defect. The mechanism of the defect formation was verified on this basis. Electron beam scanning detection makes up for the limitations of bright field detection machine with low resolution and material constraints. The combination of the two can understand the causes of defects more quickly and establish monitoring indicators so as to speed up the development of advanced manufacturing products.

**Keywords: 19nm, NAND Flash, defect detection, lithography, electron beam scanning**

## INTRODUCTION

Along with the feature size of semiconductor devices becoming smaller and smaller, more and more new processes, new materials and new equipment are introduced into semiconductor integrated circuit manufacturing, which leads to new defects and becomes an important factor affecting yield. With the change of the international situation, the domestic IC industry is in urgent need of rapid technological accumulation and breakthrough. In this context, the research and development of 19nm NAND Flash platform and yield breakthrough are of vital importance to the future development of the company and the realization of national strategic goals. Nand- Flash memory is a kind of Flash memory, which uses nonlinear macro - cell mode, and provides a cheap and effective solution for the implementation of solid state large - capacity memory. Nand-flash memory has the advantages of large capacity, fast rewriting speed and so on. It is suitable for the storage of a large amount of data, so it has been more and more widely used in the industry. For example, integrated circuits such as digital cameras, MP3 walkman memory cards and compact U disks in embedded products continue to develop towards high density.[1][2]

As the density of components in the integrated circuit becomes higher and higher, more and more components are integrated on a smaller and smaller chip, which makes the chip less and less tolerant of defects. More and more tiny size defect gradually become the yield killer, and it is more and more difficult to detect defects. The 19nm Nand Flash project is a 2D storage design based on a minimum line width of 19nm. For integrated circuit manufacturing, 193nm infiltration lithography machine has been unable to meet the production requirements when the line width less than 30nm. In order to fill the gap between infiltrating lithography and EUV lithography, dual graphics technology was invented. Among them, self-aligned dual imaging process (SADP), as shown in Figure 1, is widely used in the manufacturing process of memory devices for its advantage of defining high-density array graphics without alignment error.

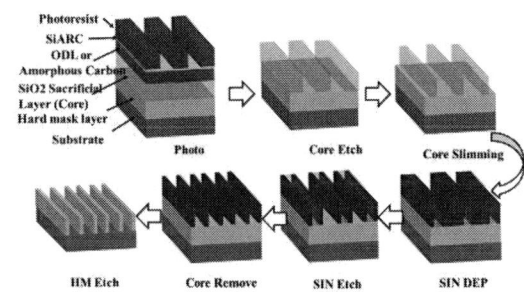

*Figure 1: Schematic diagram of self-aligned dual imaging process (SADP)*

Many new and advanced inspection tools have been developed to meet the requirements of challenging defect detection, including Bright Field (BF), Dark Field (DF) and e-beam Inspection (EBI). Optical detection tool (BF&DF) is widely used in on-line defect monitoring due to its high detection speed. Optical detection uses lasers on the scanned area of the wafer and collects scattered or reflected light to detect physical defects. Compared with optical detection, EBI adopts scanning electron microscope technology to capture defect signals by detecting secondary electronic signals with high-energy electrons as incident beam, which has been widely used in electrical defect detection. EBI is also used to inspect surface defects due to its superior resolution and ability to classify defects. However, the main disadvantage of EBI is low throughput for slower scan speed and smaller pixel size. The effective speed of 5nm EBI detection may be 5000 times slower than that of 50nm bright field detection.[3]

With the decrease of defects' dimensions in advanced manufacturing processes, electron beam scanning has been widely used in defect detection and monitoring as a supplement to optical inspection.[4] This study explored a method to detect and improve the silicon oxide defect caused by lithography in the 19nm Nand Flash process. The reason why this extreme tiny defect could not be monitored by optical detection equipment was analyzed in this paper, and a brand new defect monitoring method was carried out, which based on marking the position of defect by electron beam scanning. Through the experimental analysis, the formation mechanism of defects was further deduced, and the photolithography process was optimized to solve tiny SADP defect.

## DETECTION AND MECHANISM OF TINY LITHOGRAPHY DEFECTS

The bright field scanning machines involved in this study include KLA29XX produced by KLA-Tencor, and Escan3XX series produced by Hermes Microvision, Inc. The defect observation equipment is the KLA-Tencor EDR series observation machine.

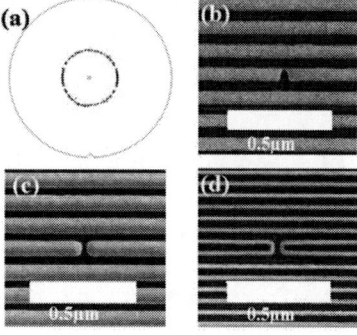

*Figure 2: (a) Distribution diagram of line missing defects; (b) SEM images of line missing defects after Core Etch; (c) SEM images of line missing defects after silicon nitride deposition (SIN DEP); (d) SEM images of line missing defects after Core Remove.*

In the development of 19nm Nand Flash self-aligned dual imaging process (SADP), the line missing defects with circular map was found, as shown in Figure 2(a). At the beginning, bright field defect scanning first discovered the defect after silicon nitride deposition(SIN DEP), as shown in Figure 2(c). This defect would cause AA and Poly line bridging, which resulting in abnormal data read/write storage, as shown in Figure 2(d). The detected

defect size was around 30nm. At first, bright-field defect scanning did not find defects after Core etching, thus electron beam scanning was used to detect tiny SADP defects. Electron beam scanning results showed that defects showed same distribution after Core etching, and scanning electron microscope (SEM) image showed that defect was core lines missing, as shown in Figure 2(b). When the defect coordinates were found by EBI, BF tool was used to defect scanning for its high-throughput. Different scanning parameters were combined to find the optimal condition with maximum SNR.

There are various factors affecting the defect signal in the optical detection，including pixel size, spectral wavelength, focal length, incident light direction and so on of the scan tool, and the material, size, shape and other factors of defects. The defect as shown in Figure 2(b) presented weak signal-to-noise ratios with different spectral mode at the minimum pixel, as shown in Figure 3(a).However, after silicon nitride deposition (SIN DEP), the SNR of defects at the same position increased significantly under the same optical detection conditions, and showed the strongest SNR with medium wavelength (Mode5), as shown in Figure 3(b).Because the silicon oxide film is used as core material in SADP process, the transmittance of silicon oxide is strong in the range of ultraviolet and visible light, and the reflected light signal of line missing defect is weak, so a high signal-to-noise ratio cannot be obtained. While silicon nitride deposition uses silicon nitride material, which has weak transmittance under ultraviolet and visible light and strong defect signal in reflected light, so as to obtain a higher signal-to-noise ratio.

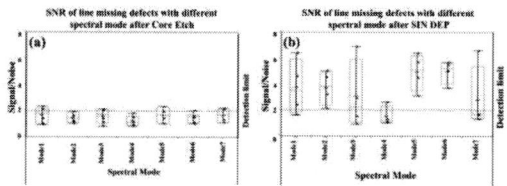

*Figure 3: (a) SNR of line missing defects with different spectral mode after Core Etch; (b) SNR of line missing defects with different spectral mode after SIN DEP.*

As mentioned before, the line missing defect cannot be detected by the optical bright field tool at the first time but can be captured by electron beam scanning tool. However, the disadvantage of EBI is also obvious for its small scanning area and slow scanning speed. Thus a short-process defect detection process was established for this defect, as shown in Figure 4. This process can shorten the experimental verification cycle and provide a fast and effective method for defect improvement.

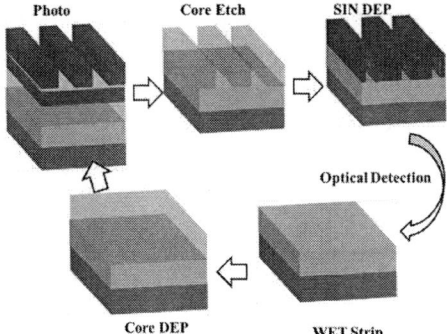

*Figure 4: Short process defect detection process*

## FAILURE MECHANISM AND IMPROVEMENT METHOD OF TINY LITHOGRAPHY DEFECTS

In the 19nm Nand Flash self-aligning dual imaging process, in order to improve line uniformity ODL (Organic Dielectric Layer) is used instead of Amorphous Carbon as hard mask before photography. The experiment results that the defects can be completely improved by enhancing the affinity between ODL and SiARC materials, as shown in Table 1. According to the experimental results, it is highly speculated that the abnormal affinity between ODL material and SiARC material results in the formation of bubbles during the coating of SiARC material. Bubbles broke out in the subsequent process, resulting in the loss of photoresistance, which further leads to the formation of missing lines in the etching process, as shown in Figure 5.[5]

TABLE I.    SUMMARY OF IMPROVEMENT EXPERIMENTS OF LINE MISSING DEFECTS

| NO. | Experimental conditions | Result |
|---|---|---|
| Test1 | Baseline | bad |
| Test2 | Optimization RRC dispense/casting | slight |
| Test3 | Add WET pre-clean | bad |
| Test4 | Change develop rinse | bad |
| Test5 | Change HMDS process | bad |
| Test6 | Improvement of the affinity between ODL and SiARC materials | better |

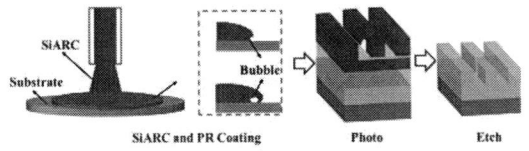

*Figure 5: Schematic diagram of defect formation*

## CONCLUSION

Based on the line missing defect found in the 19nm NAND Flash self-aligned dual imaging process, the failure mechanism and signal intensity under different detection machines were analyzed. The defect monitoring index was established by combining the defect detection with electron beam scanning and the bright field monitoring machine on different material substrates. On this basis, the short process defect detection is established to shorten the cycle for process development and defect improvement. Based on a series of experiments for the defect failure model, the line missing defects were completely improved by optimizing photolithography process.

## ACKNOWLEDGEMENTS

Thanks for the great efforts and help of all our colleagues payed in this thesis.

## REFERENCE

[1] Lin Y Y , Chen C C , Li C Y , et al. *Pattern wiggling investigation of self-aligned double patterning for 2x nm node NAND Flash and beyond[C]*.Advances in Resist Materials and Processing Technology XXX. International Society for Optics and Photonics, 2013.

[2] Chang Y S, Tsai M F, CC Lin, et al. *Pattern decomposition and process integration of self-aligned double patterning for 30nm node NAND FLASH process and beyond[J]*. Proceedings of SPIE - The International Society for Optical Engineering, 2009, 7274(5):72743E-72743E-8.

[3] Patterson, O. D., Lee, J., Lei, C., Salvador, D., *E-Beam Inspectionfor Detection of Sub-Design Rule Physical Defects*, Proceedings of ASMC, May 2012, pp. 383-387.

[4] HY Chien, CH Hsu, YY Yen, TH Ying, *A Case Study on Inline Defect Diagnosis by Applying E-beam Inspection System*, 2016 27th Annual SEMI Advanced Semiconductor Manufacturing Conference (ASMC), ISSN: 2376-6697.

[5] Yang X, Zhu X Z, Cai S . *Wafer surface pre-treatment study for micro bubble free of lithography process*. Metrology, *Inspection, & Process Control for Microlithography XXVIII. Metrology, Inspection, and Process Control for Microlithography XXVIII*, 2014.

# OPTIMIZATION OF ESD DIODE DESIGN FOR RF APPLICATIONS IN FINFET TECHNOLOGY

*Mingxing Zhou[1], TongQing Zhu[1], Lifei Zhang[1]\**

[1]Semiconductor Manufacturing International (Shanghai) Corporation

Zhangjiang Road 18, Pudong New Area, Shanghai 201203, CHINA

*Corresponding Author's Email: Cindy_Zhang@smics.com

## ABSTRACT

Electrostatic Discharge (ESD) protection for FinFET technologies is a challenge because introduction of new processes could induce weak ESD robustness. Compared to other ESD protection devices, shallow trench isolation (STI) diode is one of the most popularly used components for I/O ESD protection due to its simple layout design and high ESD protection capability. For RF applications, not only the ESD robustness but also the parasitic capacitance should be considered in parallel. The figure-of-merits It2/Capacitance and It2/Layout Area are used as an important factor to evaluate the comprehensive performance of RF ESD diodes. In this paper, ESD STI diodes with different perimeter, active area length/width, array geometry and back-end routing style were designed and characterized. With layout splits DOE, an optimized ESD diode design was proposed to meet ESD HBM 2kV spec. and RF application requirements.

## INTRODUCTION

With the continuous scaling of semiconductor technologies, ESD protection becomes more and more complicated and challenging. The dual-diode circuit has been found to be a suitable ESD protection circuit for high frequency CMOS applications, such as RF I/O pins [1], [2].

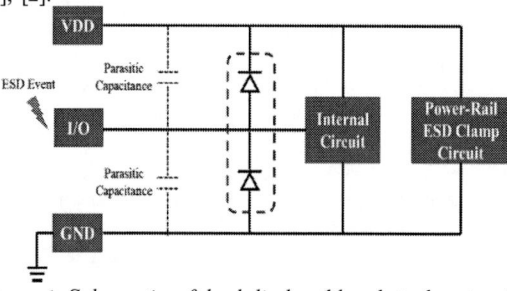

*Figure 1: Schematic of dual diodes (blue dotted rectangle) in a circuit used for ESD protection.*

Under forward-bias conditions, current conduction occurs primarily along the perimeter of the PN junction rather than on the bottom plate of the junction [3], [4]. Away from the edges, the PN junction does little to enhance the ESD current carrying capability of the diode and mainly adds parasitic capacitance. Therefore, maximizing the PN junction perimeter and minimizing the bottom junction area should serve to increase the ratio of failure current to parasitic capacitance .The failure current and parasitic capacitance are dependent on the P-N junction perimeter and junction area, respectively. The ratio of junction perimeter to junction area can by expressed by:

$$\frac{Junction\ Perimeter}{Junction\ Area} = \frac{2 \times (L+W)}{L \times W} \approx \frac{2}{W} \qquad (1)$$

$L$ is junction length, $W$ is junction width. The ratio of

junction perimeter to junction area can be mathematically simplified as $2/W$ since the junction length is normally much larger than junction width.

For very high-frequency I/Os, e.g. ultra wide band (UWB), parasitic capacitance of ESD devices savings on the order of 10fF can be important. Such specification has led to the proposal of figures of merit (FOM) such as $V_{HBM}$/Capacitance and It2/Capacitance. Layout optimization for ESD diodes is one of the most effective and significant methods to provide a high protection level per unit capacitance (C) and reduce the parasitic capacitance induced performance degradation on high frequency I/O pins.

The STI-based diodes have been proven to have a relatively high ratio of failure current to parasitic capacitance compared to other ESD protection devices, and are widely used in low capacitance I/O protection applications in planner technology node [5-7]. In this paper, we will focus on the design and optimization of ESD STI-based diode in advanced bulk FinFET technology. The key layout design parameters (e.g., active area length and diffusion width) are studied in detail to figure out their effects on device ESD protection performance and to achieve further optimization.

## MEASUREMENT AND SIMULATION

The STI-based diode is fabricated on a baseline FinFET CMOS technology. The failure current was measured by a commercial TLP test system (BARTH Model 4002 HBM TLP Pulse Curve Tracer). The rise time and pulse width of the TLP are 10 ns and 100 ns, respectively. The parasitic capacitance was simulated by SPICE model.

## RESULTS AND DISCUSSIONS

### STI-based Diode $W_N$ and $W_P$ Optimization

Firstly, the effect of N+ and P+ diffusion widths ($W_N$ and $W_P$) on the TLP performance of N+-diffusion/P-well STI-based diode is studied. Fig. 2(a) displays the structure schematic diagram of N+-diffusion/P-well STI-based diode. The fin direction is shown as an arrow. The N+-diffusion/P-well STI-based diode direction is perpendicular to the fin direction. The corresponding cross-section view of the N+-diffusion/P-well STI-based diode is shown in Fig. 2(b). It should be pointed that the layout structure of P+-diffusion/N-well STI-based diode is as same as the layout structure of N+-diffusion/P-well STI-based diode.

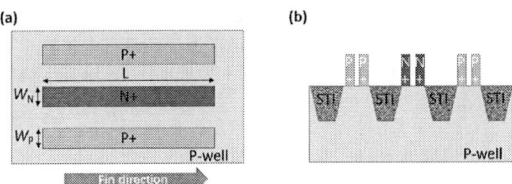

*Figure 2: (a) Structure schematic diagram for optimization study for N+-diffusion/P-well stripe diode diffusion width. $W_N$ represents the width of the N+ diffusion. $W_P$ represents the width of the P-well Pick up P+ diffusion. (b) Cross-section view of the N+-diffusion/P-well stripe diode.*

To obtain the optimum $W_N$, $W_P$ is fixed as 0.504 μm, while $W_N$ varies with values of 0.315, 0.504, and 0.567 μm, and the P-N junction area will be changed simultaneously. Compared to the relationship between P-N junction area and $W_N$, the It2 is not a linearly increasing function of $W_N$. When the $W_N$ larger than 0.504 μm, the value of $W_N$ become saturated. It2 saturates for large $W_N$ values, indicating that current flow is dominated by the perimeter p-n junction, not the bottom-plate junction. The similar results can also be found in P+-diffusion/N-well stripe diode, as shown in Fig. 3(b). These results show that the increases of $W_N$ does little to improve the ESD current handle capability of the diode but mainly increases layout area and junction parasitic capacitance. This phenomenon can also be observed in planner technology node [6].

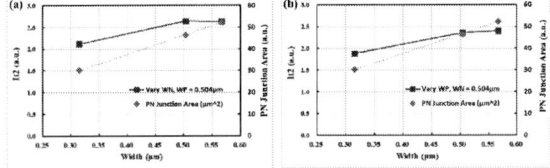

*Figure 3: (a) $I_{t2}$ for N+-diffusion/P-well STI-based diode in the $W_N$ optimization experiment. (b) $I_{t2}$ for P+-diffusion/N-well STI-based diode in the $W_P$ optimization experiment. $I_{t2}$ saturates for large $W_N$ and $W_P$ values, indicating that current flow is dominated by the PN junction perimeter, not by the PN junction area.*

The drop-off in It2 at small values of $W_N$ means that current conduction is not purely perimeter-based because all STI-based diodes have the close perimeter (202.82~205.09 μm). As mentioned before, if current conduction is limited by the junction perimeter, the It2/Junction area should be linearly increasing function of $1/W_N$. However, the relationship between It2/Junction and $1/W_N$ shows a nonlinear trend in N+-diffusion/Pwell (a) and P+-diffusion/Nwell (b) STI-based diodes. In another circumstance, if the current is uniformly distributed across the entire P-N junction, It2/Junction

area would be a constant, independent of $W_N$, which is not consistent with the result of the Fig. 4. We can found that when $W_N > 0.504$ μm, the value of It2/Junction area comes to saturate. This result indicates that current conduction is mostly confined by the P-N junction perimeter within a particular region, such as $W_N > 0.504$ μm in this paper.

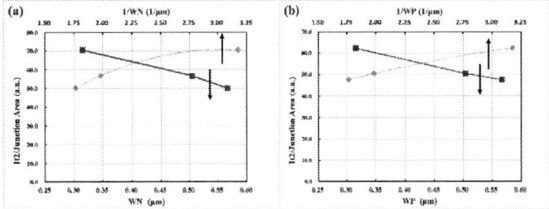

*Figure 4: Normalized failure current per junction area for N+-diffusion/Pwell (a) and P+-diffusion/Nwell (b) STI-based diode. As $W_N$ increases, It2/junction area decreases because current-handle capability is dominated by the P-N junction perimeter, not the junction area.*

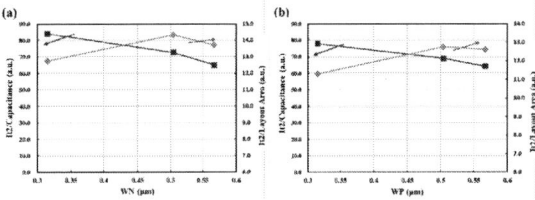

*Figure 5: The It2/Capacitance and It2/Layout area for N+-diffusion/P-well (a) and P+-diffusion/N-well (b) diode at different diffusion width.*

**Diffusion length Optimization**

To enlarge the current handle capacity of the STI-based diode, multiple stripes can be connected parallel to form an array configuration in practical applications. After $W_N$ and $W_P$ optimization, the $W_N$ and $W_P$ are fixed to 0.504 μm in N+-diffusion/Pwell and P+-diffusion/Nwell diode, respectively. Therefore, array number or diffusion length is another important layout parameter for STI-based diode structure design.

To obtain the optimum diffusion length, junction perimeter is fixed as ~305 μm, while diffusion length varies with values of 5.13, 7.98, 24.89, 30.02, 37.62, 50.35 μm, and the corresponding array number is 27, 18, 6, 5, 4, and 3. The diffusion length dependence of It2 and junction area for N+-diffusion/P-well and P+-diffusion/N-well diode are depicted in Fig. 6. It can be seen that the junction area slightly increases with the increasing of diffusion length. The value of It2 increases as the diffusion length decreases initially and slightly increases afterwards. This result indicates that the values of It2 saturate at large diffusion length values. Therefore, current handle capacity dose not benefit from further increasing the diffusion length due to the weak heat dissipate capacity in center diffusion region (self-heating and temperature rise) [8].

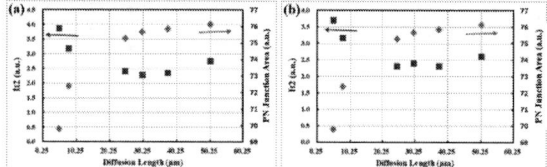

*Figure 6: (a) $I_{t2}$ for N+-diffusion/P-well STI-based diode in the diffusion length optimization experiment. (b) $I_{t2}$ for P+-diffusion/N-well STI-based diode in the diffusion length optimization experiment.*

### Layout Area Optimization

The diffusion length dependence of It2/Capacitance for STI-based diode is displayed in Fig. 7. The diodes with large diffusion length ($\geqslant 24.89$ μm) show ~45% reduction in It2/Capacitance compared to the diodes with relatively small diffusion length due to the reduced It2. Furthermore, an optimum ESD STI-based diode layout design should also take layout area into consideration. It should be pointed out that the layout area not only includes the junction area, but also includes well pick-up area [9]. The diffusion length dependence of It2/Layout area is not absolutely the same as the dependence of It2/Capacitance. The It2/Layout area increases with the diffusion length and peaks at the diffusion length of 7.98 μm. Therefore, considering these two figures of merit, it is safe to conclude that the optimum ESD STI-based diode design should have appropriate diffusion width (0.504 μm) and diffusion length (5.13 μm or 7.98 μm) layout parameters, as shown in (red dotted rectangle in Fug. 7. These two STI-based diode not only can pass 2 kV HBM specification, but also display relatively high It2/Capacitance and It2/Layout area.

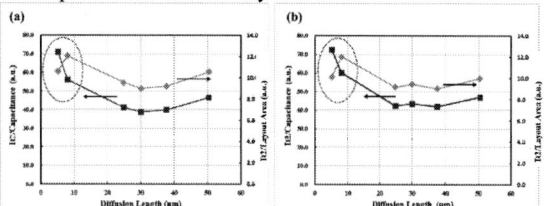

*Figure 7: The It2/Capacitance and It2/Layout area for N+-diffusion/P-well (a) and P+-diffusion/N-well (b) diode at different diffusion length.*

## CONCLUSIONS

In this paper, the diffusion width and diffusion length are optimized to achieve the best ESD protection performance in advanced FinFET technology. The best comprehensive ESD protection performance, including high current-handle capacity, It2/Capacitance, and It2/Layout area, can be obtained at specified layout configuration. These results indicate that the optimum STI-based diode can be used as ESD protection device in RF applications.

## ACKNOWLEDGEMENTS

This work was performed at the SMIC ESD/Latch-up team. The authors would like to thank Amanda Chen and Kalk Zhang for designing layout and collecting silicon data.

## REFERENCES

[1] C. Richier, P. Salome, G. Mabboux, I. Zaza, A. Juge and P. Mortini. *Electrical Overstress/Electrostatic Discharge Symposium Proceedings 2000*, 2000, pp. 251-259.

[2] ZH. Gan, A. Zhang, W. Wong, LF. Zhang, HH. Ye and CL. Tseng. *2015 China Semiconductor Technology International Conference*, 2015, pp. 1-3.

[3] E. R. Worley and A. Bakulin. *2002 Electrical Overstress/Electrostatic Discharge Symposium*, 2002, pp. 62-72.

[4] Ben Streetman and Sanjay Banerjee, *Solid State Electronic Devices*, 5th ed., Prentice Hall, 2000.

[5] S. -H. Chen et al., *2014 IEEE International Electron Devices Meeting*, 2014, pp. 20.4.1-20.4.4.

[6] K. Bhatia, N. Jack and E. Rosenbaum. *IEEE Transactions on Device and Materials Reliability*, 2009, pp. 465-475.

[8] Y. Li, M. Miao and R. Gauthier. *2018 40th Electrical Overstress/Electrostatic Discharge Symposium (EOS/ESD)*, 2018, pp. 1-7.

[9] A. Dong, J. Xiong, S. Mitra, W. Liang, R. Gauthier and A. Loiseau. *2018 40th Electrical Overstress/Electrostatic Discharge Symposium (EOS/ESD)*, 2018, pp. 1-8.

# DIFFERENT MODULE'S PROCESS AFFECT TO POLY PATTERN ETCH STICK PARTICLE

*Jiayi Fu, Qiliang Ni, Chao Han*

Shanghai Huali Microelectronics Corporation, Shanghai 201210, China

*Corresponding Author's Email: fujiayi@hlmc.cn

## ABSTRACT

In the semiconductor manufacturing process, because of the backside films' characteristic, some process will produce special defects. Particularly, stick particle is one of the special defects in Poly etch process. This paper will discuss the formation and improvement measures of stick particle through the film materials, process integration and other aspects.

*Keywords—Stick Particle; Process; Poly etch; Improvement*

## INTRODUCTION

With the rapid development of semiconductor industry in the recent years, the current semiconductor chip manufacturing process needs more rigorous requirements. With the continuous improvement, the line width can be smaller and smaller even to the extent that the current line width has been able to achieve mass production of 28nm, and the tolerance of defect also be difficult [1]. Poly process is one of the most important parts in the front-end process, and its process stability plays a critical role in connecting the past and the future process, so the demand for defect improvement is increasingly urgent.

CMOS image sensors (CIS) products tend to suffer the long strip particle with strong adhesion after poly etch (Stick Particle) [2], and the scan results of defect detection show that there are obvious traces of wet cleaning, and which can't be removed after multiple wet process [3]. It will make the final product yield by a certain degree of loss.

In this paper, we will analyze the source of defects and the impact of improvement of each module's process on defects from multiple perspectives.

## DEFECT SOURCE
### Defect Map and PFA Analysis

As shown in Fig.1(a) and (b) , there are obvious traces of water scoring in the map, and in the poly etch station, the stick particle means on the surface. It was preliminarily judged that the distribution of such defects was formed after the etching was finished and the defects fell out of the wet tank. According to Fig.1(c), abnormalities also occurred when the process was made to CT. Fig.1(d) and (e) shows the PFA result: cross section also found spacer film abnormal between the node of CT and Poly. And the elements are not special components, mainly Si and O.

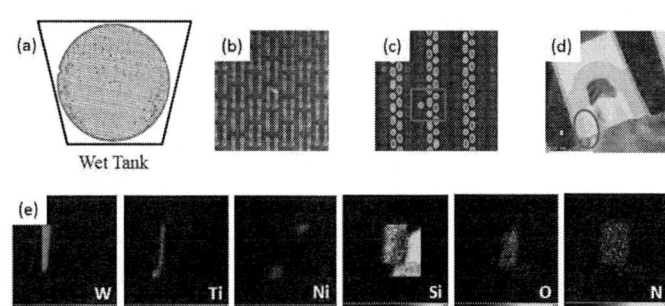

*Fig.1(a) wafer in wet tank; (b)stick particle image; (c)process to CT; (d)PFA;(e) components of stick particle*

## IMPACT OF DIFFERENT MODULE PROCESS
### Impact of DEP Film

The thicker medium on the backside of wafer will store charge in the chuck state for a long time, and the static electricity will absorb the polymer on the E-chuck (ESC) [4]. After the wafer operation is completed, with the gradual release of charge on the backside of the wafer, the polymer particles adsorbed on the backside will drop down. The thicker dielectric layer will cause the more charge storage, finally the more particles [5]. The single variable experiment adopts wafer with different wafer backsides. After the process of etch, the CIS wafer placed in the reservoir vessel (FOUP) made more particles the next wafer under it, which is significantly higher than bare wafer, and the suffered map image formed shows a ring shape. The shape is consistent with etch ESC, shown in Fig.2.

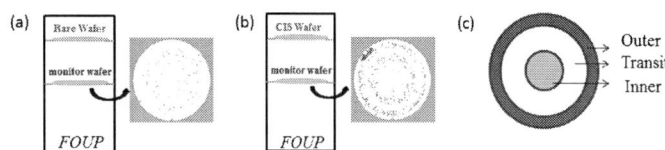

*Fig.2(a) bare wafer idle 15min in FOUP; (b) CIS wafer idle 15min in FOUP; (c)etch E-chuck mode*

### Impact of Etch's Process

It's known from 2.2 that the plane of wafer upper will affect the backside of wafer lower. Related experiments show that it is strongly correlated with ESC. But from another set of experiments, it is unable to draw conclusions relevant to its own jobs.

| Test | Test1 | Test2 | Test3 | Test4 |
|---|---|---|---|---|
| Test Type | chuck on/off (30times) | chuck on/off (30times) | pin up/down (30times) | chuck on/off+ pin up/down (30times) |
| Wafer Type | Product Wafer | Bare Wafer | Bare Wafer | Bare Wafer |
| Map | | | | |

*Fig.3 four types test for ESC work step*

As shown in Fig.3, 1) chuck on or off, 2) pin up or down, both of the change can't make stick particle. But will cause particle higher.

In view of the different technological process, etch will be targeted to appropriate preprocessing of the chamber [6]. Including AA and poly, after the pre-clean, subsequent to a series of chemical reactions, in order to make sure the chamber in a stable environment, mainly some silicide combined with oxygen to produce $SiO_2$ for protection, as in (1). In the actual process, if the chemical reaction is inappropriate, it will lead to the fall of $SiO_2$ and cause particle instead.

$$SiCl_4 + O_2 \rightarrow SiO_2 + Chloride \qquad (1)$$

On the contrary, spacer etch does not have the similar pretreatment before the process, the representation is different, its performance is better than AA and poly, as shown in Fig.4.

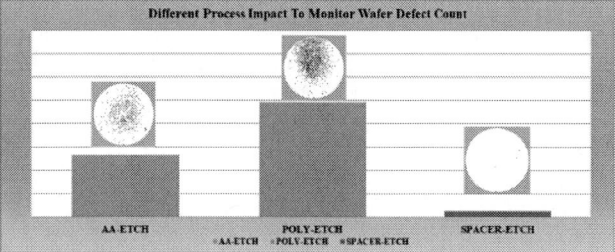

*Fig.4 different process impact to monitor wafer*

### Impact of Wet's Process

We get the information from the previous chapter that stick particle has already formed after etching, so the performance of wet clean is serious of removing the defect [7]. At present, the online process sequence is HF-H2SO4:H2O2-NH4OH:H2O2, but from a practical point of view, this process sequence can Slight reduce stick particle [8]. Fig.5 shows the degree of decline, which can't be completely removed, can be found in particle distributions of different sizes.

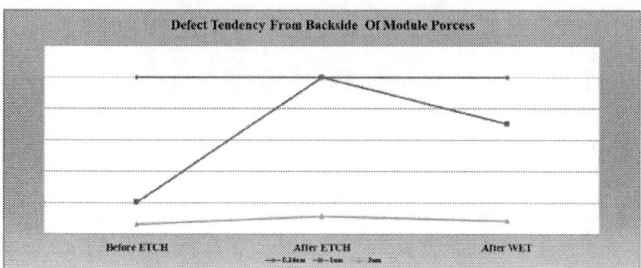

*Fig.5 different process suffer defect count of wafer backside*

In the experiment, for exchanging the sequence of the three types cleaning chemicals of wet recipe:

*HF-H2SO4:H2O2-NH4OH:H2O2*
*→(Test1) H2SO4:H2O2- NH4OH:H2O2*
*→(Test 2) HF-H2SO4:H2O2-NH4OH:H2O2*
*-HF-NH4OH:H2O2*
*→(Test 3) H2SO4: H2O2-HF-NH4OH: H2O2*

And exchange the project（3）HF clean time:

*Baseline: X(s)*
*→ [Test 1] X+10(s)*
*→ [Test 2] X+20(s)*
*→ [Test 3] X+30(s)*

With different proportions in the experiments, we can find the optimal result in Fig.6(a) and (b).

| | (a)Chemical Sequence Experiment | | | (b)HF Process Time Experiment | | | |
|---|---|---|---|---|---|---|---|
| Chemical | Stick Particle (ea) | CD (nm) | CD image | HF Process Time | Stick Particle (ea) | CD (nm) | CD image |
| Baseline | 995 | 72.0 | | Baseline | 0 | 71.7 | |
| (Test 1) | 0 | 69.9 | | [Test 1] | 0 | 72.0 | |
| (Test 2) | 0 | 76.3 | | [Test 2] | 0 | 71.6 | |
| (Test 3) | 0 | 73.2 | | [Test 3] | 0 | 72.4 | |

*Fig.6(a) chemical sequence experiment; (b) HF process time experiment*

According to the data, the particle can be removed by changing the operation sequence of chemicals, but the result of each sequence has different effects on the critical dimension (CD) of poly, which is obvious. The steps of *(Test 3)* can approach the original line width to the maximum extent on the premise of improving the defects. Sequence of HF-H2SO4:H2O2-NH4OH:H2O2 is the best chemical operation, when the time is increased by 10 seconds, the optimal condition is obtained.

## CONCLUSION AND DISCUSSION

The investigation aims at proving the influence of different processes on stick particle and analyze it from the perspectives of crystal backside material, etch and wet process respectively. 1) Wafer backside material: the CIS wafer placed in the reservoir vessel (FOUP) made more particles the next wafer under it because of its material (Silicon Nitride); 2) Etch process: the reason for need of erosion about AA and poly profile that the chamber needs to do pre-deposit, which becomes the source of particle. 3) Wet process: to dispense the sequence of chemical reagents and cleaning time, the optimal solution can be found without affecting the critical dimension, recipe sequence: $H_2SO_4:H_2O_2-HF-NH_4OH:H_2O_2$, time: baseline add 10seconds. According to the actual process situation of the assembly line, 1) and 2) are not adjustable, and the stick particle can be removed by 3) changing the process to a minimum degree.

The analysis and research of this paper have very important reference significance for the improvement of stick particle and other defect research of similar problems.

## REFERENCES

[1] L. Sheng, P. Porath, E. Glines, On the Generation and Elimination of Lonely Poly-Silicon Crater-Defects and Their Impacts on Gate Oxide Integrity (GOI) in Dual-Gate Technology, *IEEE, ASMC* 2013, pp. 378-384.

[2] X. Liu, K. Lim, Z. Wu, Z. Xiong, Y. Ding, H. Nong and Y. Wu, "A Study of Inverse Narrow Width Effect of 65nm Low Power CMOS Technology," *Incorporating Knowledge in Genetic Algorithms for Device Synthesis*, 2008, pp. 1138-1141.

[3] A. A. Pande, D. S. L. Mui, D. W. Hess, SiO2 Etching With Aqueous HF: Design and Development of a Laboratory-Scale Integrated Wet Etch/Dry Reactor, *IEEE Transactions on Semiconductor Manufacturing*, 2011, Vol.11, pp.104-116.

[4] Y. Kondo, K. Ishikawa, T. Hayashi, Y. Miyawaki, K. Takeda, H. Kondo, et al., Silicon nitride etching performance of CH2F2 plasma diluted with argon or krypton, *J. Appl. Phys.*, 2015, vol. 54 040303.

[5] M. Zawierta, M. Martyniuk, R. D. Jeffery, G. Putrino, A. Keating, K. K. M. B. D. Silva, L. Faraone, Control of Sidewall Profile in Dry Plasma Etching of Polyimide, *Journal of Microelectromechanical Systems*, 2017, Vol.26, pp.593-600.

[6] J. Charmet, J. Bitterli, O. Sereda, M. Liley; P. Renaud; H. Keppner, Optimizing Parylene C Adhesion for MEMS Processes: Potassium Hydroxide Wet Etching, *Journal of Microelectromechanical Systems*, 2013, Vol.22, pp.855-864.

[7] S. Guillemin, P. Mumbauer, H. Radtke, M. Fimberger, S. Fink, J. Kraxner, A. Faes, J. Siegert, Etching Mechanisms of SiO2 and SiNx:H Thin Films in HF/Ethanol Vapor Phase: Toward High Selectivity Batch Release Processes, *IEEE, Journal of Microelectromechanical Systems*, 2019, Vol.3, pp. 717–723.

[8] S.M. GadelRab, A.M. Miri, S.G. Chamberlain, A comparison of the performance and reliability of wet-etched and dry-etched a-Si:H TFTs, *IEEE Transactions on Electron Devices*, 1998, Vol.45, pp.560-563.

# CLASSIFICATION OF WAFER BACKSIDE IMAGES VIA FASTERRCNN-BASED NEURAL NETWORK

*Junjun Zhuang [1]\*, Guiyun Mao [1], Yong Wang [1], Xu Chen [1]\*, Yansheng Wang [2] and Zhengying Wei [1]*

[1] Shanghai Huali Microelectronics Corporation, Shanghai, China
[2] Shanghai Huali Integrated Circuit Corporation, Shanghai, China

\*Corresponding Author's Email: chenxu@hlmc.cn

## ABSTRACT

Automatic classification of defect patterns in wafer backside image stands as a challenging problem in semiconductor manufacturing industry, especially different wafer backside defects start to impact the front side pattern defined and cause yield loss because of the design rule shrinkage. Deep neural network has recently shown decent progress in image classification, including wafer bin maps, but has rarely been applied in wafer backside image identification as far as we are aware. A Faster-RCNN based neural network model has been proposed for defect pattern detection and classification in wafer backside images. As a prepositive strategy, customized data augmentation has been proposed to diminish the expense of explicit manual supervision. Meanwhile, Class Activation Map (CAM) of each image has been generated to guide the training process. The results of this practical study showed the validity and practical viability of the model even at little expense of manual-labeled images.

Keywords: wafer backside image; defect pattern; neural network; classification; Gradient-based Class Activation Map (Grad-CAM)

## INTRODUCTION

In dealing with emerging challenges of advanced wafer processing, a shrinking in Design-Rule, and a dramatically increasing process/measurement steps in order to enable new functionality and to meet the ever-stringent performance requirements, backside defects can no longer be tolerated. It has been reported that backside defects affect device yield, scrap, and production cost. It has recognized that backside defectivity is one of the key defect reduction requirements for front-end processes.

Backside defects are often caused by many sources and experientially be classified as particles, residues, scratches and so on. Backside defects can contribute to photolithography issues by distorting wafer flatness during exposure and causing photolithography hot spots [1]. If the hot spots are detected prior to the etch, wafers can be rework; otherwise, the wafers must be scrapped or will have significant yield loss. Therefore, classifying the different types of defect patterns on the backside of wafers after each process in time seems extremely important to judge whether the wafer can be repaired by rework or

proceed to the next fabrication step. In practice, most semiconductor engineers use subjective and time-consuming eyeball analysis backside images of wafers to assess defect types, which is complicated, inconsistent, and unreliable.

Deep learning (DL) has been exploratory applied in recognizing and classifying some types of images in semiconductor manufacture, such as wafer bin maps and the scanning electron microscope (SEM) images of defects on the wafer front-side [2]. An auto-encoder based deep model has been used to extract features of input data, and proved distinguish performance, which is benefit to the following tasks, just like regression or classification [3]. For identifying a case by backside images or frontside images, the location of the defects also a key factor that need to be taken into consideration to judge the impact wafers and root causes. The DL models for object detection have developed rapidly in recent years, such as Fast-RCNN, Yolo and so on, which is appropriate for the defect location detection. In the literature so far, the use of object detection models for identifying the pattern types and locations on the wafer backside is not explored.

In this article, we propose an effective method for defect pattern detection on wafer backside, which improves the feature extraction part based on the original Faster-RCNN. In the feature extraction part, we combine CNN and FPN [4] in order to use multiple layers of strong semantic information to extract feature maps of various scales. Meanwhile, we utilize Class Activation Mapping (CAM) techniques to obtain a view of a deep network's perception of aerial images and identify salient local regions, gaining insights on how CNN's interpret these images for classification and be a guidance for training. The Methodology part of the paper introduces the complete algorithm flow and the improved feature extraction part of the Faster-RCNN. The second part gives the comparison, testing, verification and analysis of the algorithm performance. Finally, the conclusion is given.

## METHODOLOGY

This section describes the methods that used in this study, including the preprocess of the images, model architecture and training scheme.

### Image Denoise

we use contour defection algorithm which refers to the method proposed by SUZUKI [5] for getting rid of noise, which is the main inference factor for defect pattern location.

**Image Augmentation**

Apart from the traditional method for image augmentation, such as rotation, vertical flip, horizontal flip and zoom, image combination, which refers to a new strategy [6] for augmentation, is also used in this article. As described in the article, the goal of this image combination is to generate a new training sample by combining two training samples $(xA; yA)$ and $(xB; yB)$. The generated training sample $(x; y)$ is used to train the model with its original loss function. We define the combining operation and y denote a training image and its label, respectively.

**Feature Pyramid Network (FPN)**

The original Faster-RCNN structure is complex and can be divided into four parts, namely Feature Extraction, Region Proposal Network (RPN), RoI Pooling, and Classification. The traditional feature extraction network only uses the high-level strong semantic information, and discards the detailed information in the low-level network. However, these low-level networks contain a large number of object location information. Discarding this information will greatly increase the error rate of defect detection. Therefore, this paper applies the feature pyramid network to the feature extraction part to accurately locate the object and improve the accuracy of small defect detection. The Feature Pyramid Network (FPN) is mainly used to solve small target detection. It adds layer-to-layer connections based on the original network and generates feature pyramids. with strong semantic information on all scales. The FPN itself is not a defect detector, but a feature detector that cooperates with the defect detector. FPN doesn't increase the amount of calculation and the occupied memory significantly, but greatly improves the accuracy of detecting small defects. FPN is mainly divided into two steps: first, build a bottom-up pathway, and secondly build a top-down pathway with lateral connections. Due to the huge memory footprint of conv1, it was not included in the pyramid.

The feature extraction networks widely used currently are VGG16 and ResNet. Since the ResNet has been put forward, more and more algorithms use it as feature extraction network because of its superior performance. Here we adopt the ResNet-50 with FPN for our scheme's backbone. The details about the combining of ResNte-50 and FPN is shown in the Fig. 1.

**Gradient-based Class Activation Map (Grad-CAM)**

Zhou et al. [7] generates the class activation maps (CAM) using the global average pooling (GAP) in CNNs. utilizing a specific network structure that replaces the fully-connected layer of the image classifier with the global average pooling layer. Later, Grad-CAM enhances

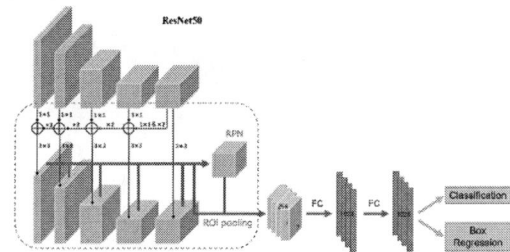

*Figure 1: The structure of Faster-RCNN with FPN*

the generalization ability of this technique, which enables generating class activation maps for any off-the-shelf CNN-based image classifier possible. Grad-CAM utilizes the average gradients of a feature map to represent its importance to the target category. Although these methods can effectively locate the target objects, a common issue among them is that they all rely on the final convolutional layer of CNN to generate class activation maps. Class activation map for a particular class indicated the regions in the image used by the CNN to identify that category. Same technique can also be applied in wafer defects detections.

While Grad-CAM visualizations are class-discriminative and localize relevant image regions well, they lack the ability to show fine-grained importance like pixel-space gradient visualization methods (Guided Backpropagation and Deconvolution), it is unclear from the low-resolutions of the heat-map why the network predicts this particular instance as its category. In order to combine the best aspects of both, we fuse Guided Backpropagation and Grad-CAM visualizations via pointwise multiplication, which is named as Guided Grad-CAM [8].

## EXPERIMENT

We used real backside images to train and evaluate the proposed model. A total of real 500 backside images from different recipes with the size of 128*128 contains different types of defect patterns and are labeled manually. Fig. 2 showed the typical images with different defect patterns obtained after different recipes scan. Also, we utilized some transformations such as rotation, vertical flip, horizontal flip, mirror flip and image combination (Fig.3), which was described in METHODOLOTY, for the data augmentation and enlarged the database to 10000, which tended to make the model more robust for the different cases.

*Figure 2: The typical backside images with different defect patterns.*

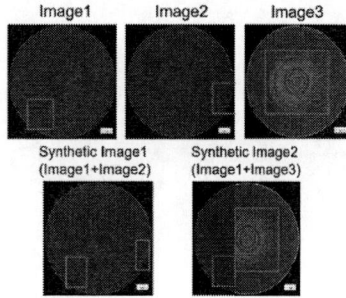

*Figure 3: The image combination strategy of two different images.*

As we could see in the Fig. 2, the defect patterns in the most defective images were just a tiny ratio of the total pixels, especially in images that were labeled as 'dot' and 'scratch', which was very obscure to be judged just by naked eyes. Therefore, some approaches were applied to get rid of the background noise for making the patterns more obviously. In concern of the area difference between background noise and the defect patterns that need to be detected, we employed contour detection to inspect the noise by setting the appropriate threshold, which was just under the area value of the defect patterns. After contour detection, the noise areas were filled with the pixel value as the background to vanish the noise influence in the coming model training. As shown in Fig. 4, the raw images in first column were dotted with random noise, which would inference the object detection especially when the object was tiny. After contour detection with the threshold under the area value of defect patterns, the noises were detected and marked in red (shown in second column), followed by filling the noise area with the pixel of background, the denoised images were obtained in the last column, which was revealed that the contour detection was an applicable method to get rid of the random noise in the backside images.

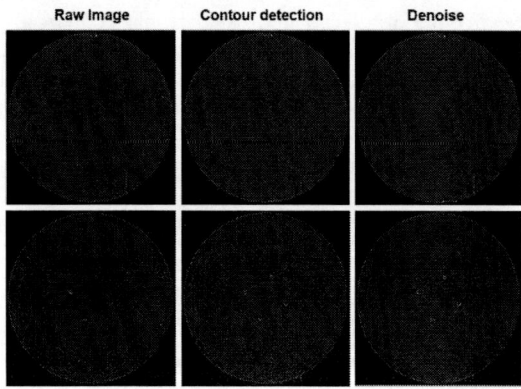

*Figure 4: Contour detection for noise abolish, raw images were in first column, images after contour detection were in second column and denoised images were in last column*

The object detection results were shown in Fig. 5, the different patterns in images were all detected by the bounding boxes with the accurate classes on the top. What was worth to notice that the last image was combined with two different images and it was also detected accurately, which demonstrated that the image combination made the model more robust for the complex background. Table 1 showed the numerical results of classification, compared with the precision and accuracy, we would pay more attention to the recall in term of the defective image was rarely appear in the real world, but when it did, it could have a serious impact to the production. As shown in Table 1, only the value of 'scratch' was under 0.8 because the number of defective images was a tiny ratio of total images. The F1 scores of all classes were above 0.85, which revealed that the model had a good capability to classify different defect patterns.

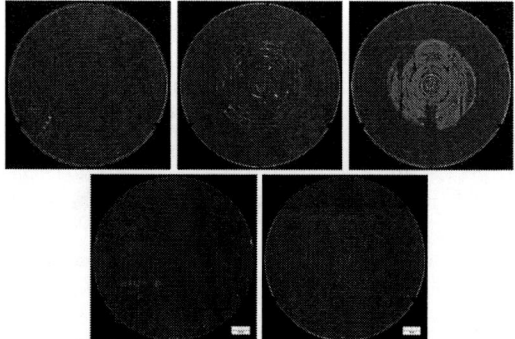

*Figure 5: The detection result of different typical images. The defect patterns were bounded in the bounding boxes with corresponding classes.*

Fig. 6 shows the Grad-CAM and Guided Grad-CAM results of single pattern detection, these images are of single category. It shows that explanation results of the proposed method are less noisy and can locate target objects precisely in these cases. As shown in Fig.6, these images contain objects of two categories. When specifying the category, it shows that our explanation results are more class-discriminative and less noisy. The results of Fig.6 and Fig.7 demonstrate our model has an excellent capability for defect pattern detection of backside images because it can always locate the pattern accurately and classify it to the corresponding category according to the features that are bounded.

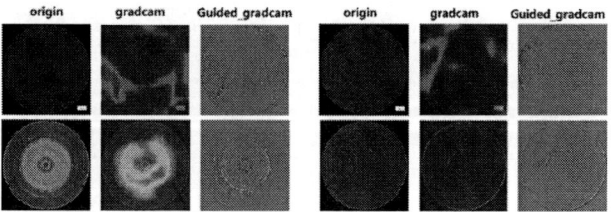

*Figure 6: The CAM result of the multi pattern images*

*Table 1: The classification indicator for every defect pattern*

| Category | TP | FP | FN | TN | Accuracy | Precision | Recall | False | F1 |
|---|---|---|---|---|---|---|---|---|---|
| center_slight | 6 | 0 | 1 | 39 | 0.96 | 1 | 0.85 | 0 | 0.92 |
| center_medium | 17 | 1 | 0 | 28 | 0.97 | 0.94 | 1 | 0.05 | 0.97 |
| center_serious | 6 | 0 | 1 | 39 | 0.97 | 1 | 0.86 | 0 | 0.92 |
| scratch | 3 | 0 | 1 | 42 | 0.97 | 1 | 0.75 | 0 | 0.85 |
| dot | 17 | 3 | 0 | 26 | 0.93 | 0.85 | 1 | 0.15 | 0.91 |

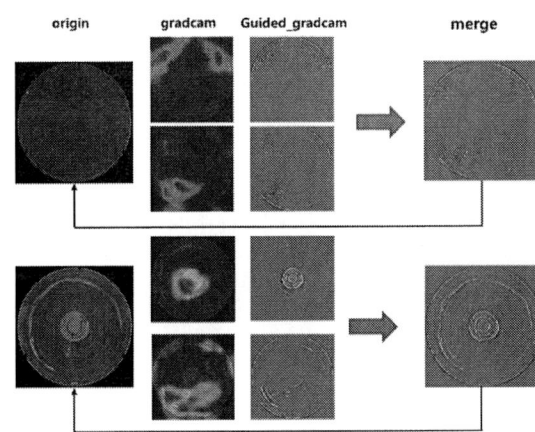

*Figure 7: The CAM result of the multi pattern images*

## CONCLUSIONS

A Faster-RCNN based model is constructed for the defect pattern detection in backside images and FPN is incorporated into the base model to adapt the different size and shape of defect patterns in the backside images. Image augmentation is an easy way to improve the robust of the model to complex background. An empirical study shows that the proposed method can effectively improve the efficiency and accuracy of defect patterns detection in backside images. Specifically, even though the number of images with defect pattern is extremely small, the model also presents a satisfied performance of abnormal classification. Due to the promising high classification accuracy of the networks, it can be considered as powerful tools for implementing object defection in the semiconductor manufacturing.

## REFERENCES

[1] MLABalu, Elango, et al. "Wafer Backside Cleaning for Defect Reduction and Litho Hot Spots Mitigation DI: Defect Inspection and Reduction." 2018:216-221.

[2] Yuan, T, W. Kuo , and S. J. Bae . "Detection of Spatial Defect Patterns Generated in Semiconductor Fabrication Processes." IEEE Transactions on Semiconductor Manufacturing 24.3(2011):392-403.

[3] Hong, Chaoqun , et al. "Multimodal Deep Autoencoder for Human Pose Recovery." IEEE Transactions on Image Processing A Publication of the IEEE Signal Processing Society 24.12(2015):5659-5670.

[4] Lin, T. Y., et al. "Feature Pyramid Networks for Object Detection." 2017 IEEE Conference on Computer Vision and Pattern Recognition (CVPR) IEEE Computer Society, 2017.

[5] Fuchs, M. B.. "Topological structural analysis." Structural optimization 13.2(1997):104-111.

[6] Yun, S., et al. "CutMix: Regularization Strategy to Train Strong Classifiers with Localizable Features." (2019).

[7] B. Zhou, A. Khosla, L.A., A. Oliva and A. Torralba, "Learning Deep Features for Discriminative Localization", CVPR, 2016.

[8] Selvaraju, R. R., et al. "Grad-CAM: Visual Explanations from Deep Networks via Gradient-based Localization." International Journal of Computer Vision 128.2(2020):336-359.

# WAT THROUGHPUT IMPROVED BY ALGORITHM OPTIMIZATION

*Tong Chen[1], Liang Wang[1], Guiyun Mao[1], Xun Chen[1]\**

[1] Shanghai Huali Microelectronics Corporation,

Shanghai, China

*Corresponding Author's Email: chenxu@hlmc.cn

## ABSTRACT

WAT (Wafer Acceptance Test) is an important stage before wafers fab out. The goal of WAT is to be both fast and accurate. It is well known that algorithm is a key factor to determine data accuracy and test speed. In this paper, the throughput of WAT has been improved by optimizing the VTI, CIS_VTH_LIN and CIS_VPIN_ VSW algorithms. After the new algorithm has been implemented on LP and CIS platforms, the test speed of wafer has been improved 13.52% ~ 18.87%. A total of 12,146.7 hours of WAT test capacity were saved between April and September 2021.

## INTRODUCTION

The rapid development of IC technology requires continuous improvement of its testing technology to improve the production efficiency of semiconductor industry. Precise control and evaluation of wafer manufacturing is carried out throughout the manufacturing process to ensure wafer products meet specifications. In order to achieve this, in addition to accurately monitoring the critical dimensions and thickness of the deposited film in each production step during the production process, WAT [1] is also required to ensure that the electrical parameters of the key components of the chip conform to the electrical design rules before the wafer is shipped [2-5].

In the process of WAT, it is necessary to establish corresponding algorithms for different electrical parameters, such as IDSAT algorithm for testing MOS saturation current, VTI algorithm for threshold voltage, RFV algorithm for resistance testing by two-terminal method, RKV algorithm for resistance testing by four-terminal method, etc. The advantages and disadvantages of the testing algorithm on the one hand determine the accuracy of parameter testing data, affect the accuracy of process monitoring, on the other hand determine the wafer testing time, affect the test throughput. We want to minimize test time and increase WAT throughput while ensuring the accuracy of parameter test data. The following three algorithms involved in this paper are introduced:

VTI algorithm measures the threshold voltage Vt of MOS in constant current mode. The device connection diagram during the test is shown in Figure 1. The test port of this algorithm has five terminals: Drain is connected to PIN1, Gate is connected to PIN2, Source is connected to

PIN3, and Bulk is connected to PIN4. If there is Deep Nwell (Dnwell), PIN5 is connected. The algorithm is executed as follows: 1) Connect PIN1-49 to GNDU, wait 0.01 seconds, release accumulated charge on Pinboard, and then disconnect all; Drain, Source, Gate and Bulk are connected to GNDU. After discharge, Drain is connected to SMU1, Gate is connected to SMU2, and Bulk is connected to SMU4; 2) Set the parameters of SMU test accuracy at Drain, and apply Bulk voltage of 0 V in general; 3) Apply the corresponding voltage (0.05V at Vtlin and VDD at Vtsat) at the Drain, sweep the voltage from 0 to 0.8VDD at the Gate, while measure the current at the Drain. When the current reaches the target value, the voltage at the Gate is Vt. The algorithm adopts the Search mode of Linear Search, in which one of the setting variables is VGDELAYTIME, which is the time interval between the output of the Search unit (SMU2) and the test unit (SMU1).

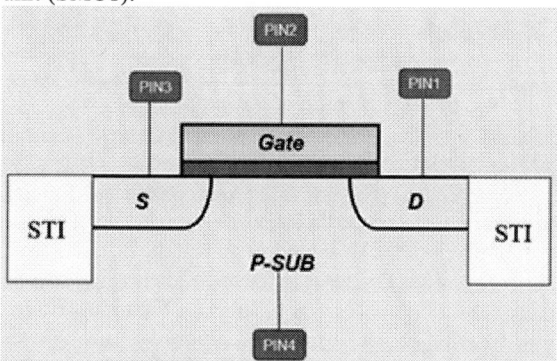

*Figure 1: The device connection diagram of MOS threshold voltage measurement.*

CIS_VTH_LIN algorithm is used to measure the threshold voltage Vthi of each MOS in pixel area of CIS

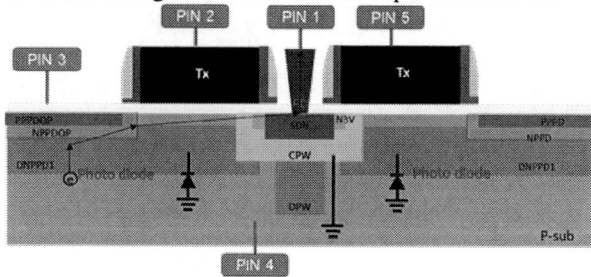

*Figure 2: The device connection diagram of MOS threshold voltage measurement in the pixel area of CIS product.*

978-1-6654-9759-6/22 $31.00 © 2022 IEEE

product under constant current mode. Taking TX1 device as an example, its device connection diagram is shown in Figure 2. FD, the Gate of TX1, Photodiode and Bulk are connected to Pin1-4 respectively, and the Gate of TX2 is connected to PIN5. The algorithm adopts Binary Search mode and iterates the Gate voltage several times. When the current at Drain (FD in this case) reach the target value, and the Gate output voltage at this time, i.e. Vthi. The algorithm also contains a Delaytime variable.

CIS_VPIN_VSW algorithm is used to measure the clamp voltage of device of CIS products, which is characterized by the threshold voltage when the substrate is subjected to negative voltage and the leakage current from Photodiode to the substrate is greater than the normal N-type region leakage current. The voltage represents the isolation effect between Photodiode and the substrate. The connection diagram of the test device is shown in Figure 3. This test port has three terminals. One of the $N^+$ zones on both sides of Photodiode is connected to PIN1 for Drain, PIN2 for Source and PIN3 for base Substrate. Binary Search is used twice in this algorithm to output substrate voltages Vpin1na and Vpind1na when the Source current is 1 nA and 0.1 nA, respectively. This algorithm also contains the Delaytime variable.

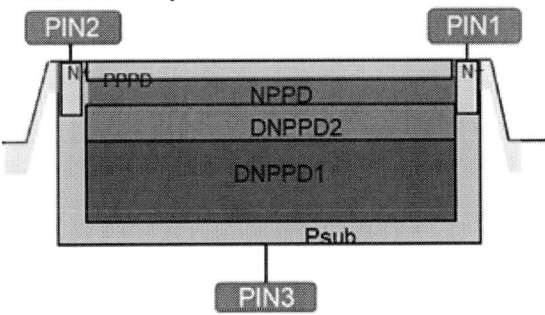

*Figure 3: The device connection diagram of the device clamp voltage of CIS products.*

After careful observation, it can found that all these algorithms are needed to set up time interval between data output and data detection. This interval length depends on the response speed of the device. It is well known that the response speed of MOS device and PN junction is very fast, therefore, we can improve the testing speed by reducing this interval time, and improve the test throughput. By shortening the time to 1/10 of the initial, we greatly improved the test speed of some parameters on the premise of ensuring the accuracy of the test data. The test speed increase rate of Vt parameters of LP product reached 75.54%; The improvement rate of Vt parameters in the logic area of CIS product reached 76.74%, that of Vt parameters in the pixel area reached 23.08%, and that of VPIN parameters reached 37.57%. So far, the new algorithm has been applied to 15 products on LP and CIS platform, and the overall wafer test speed is improved

from 8.42% to 48.41%, saving a total of 12146.7 hours of WAT capacity.

## EXPERIMENT

Take 0043 product as an example:
**Set up an experiment: 0043_WATDB**
**Set up 0043_WATDB TXT file**

Open the TST file of product 0043 and select Vt parameters of MOS with different types and sizes. Taking NRV_Vtlin_D27_D054_M as an example. Copy it as NRV_Vtlin_D27_D054_M_1, NRV_Vtlin_D27_D054_M_2, NRV_Vtlin_D27_D054_M_9, and VGDELAYTIME of each parameter is set to 1/10, 2/10 and ⋯ 9/10. Vt parameters of other MOS devices are added as the same way. The new TST file is named 0043_WATDB.txt, as shown in Figure 4.

*Figure 4: The txt file of 0043_WAT*

**Set up 0043_WATDB LIMIT file**

Add the new parameters in the 0043_WATDB to the 0043 limit file-0043_WAT.lim, set Keyflag to 2, and keep the values of Spec, Valid, Ctr, and target the same as the original parameters. The new limit file is named 0043_WATDB.lim.

**Set up the recipe: 0043_WATDB**

Select the appropriate WAF FILE, DIE FILE, and PRB FILE along with the new 0043_WATDB.tst and 0043_WATDB.lim to form the recipe: 0043_WATDB.

**Collect data**

Select 6 wafers (#2/5/8/10/15/18/20/25) of lot AP34321, and set up a Run Card using 0043_WATDB to collect data. As for CIS platform, product 0186 and 0224 is chosen to test. The Vt parameters of logic were split into 10 groups and the Vt and VPIN parameters of pixel were split into 3 groups (1/10, 1/2 and 1 of the initial values) according to VGDELAYTIME

## RESULT AND DISCUSSION

Table 1 shows the correlation results between experimental data and the Baseline value of product 0043 Vt parameters of LP platform. It can be seen that shortening VGDELAYTIME has no effect on Vt

*Table I: Correlation results between experimental data and the Baseline value of Vt parameters of product 0043*

| Parameter | BL | | NEW | | Mean Shift | STD Extend |
|---|---|---|---|---|---|---|
| | MEAN | STD | Mean | STD | | |
| NRV Vtlin D27 D054 M 1 | 0.357 | 0.025 | 0.357 | 0.025 | 0.00 | 1.00 |
| NRV Vtlin D27 D054 M 2 | 0.357 | 0.025 | 0.357 | 0.025 | 0.00 | 1.00 |
| NRV Vtlin D27 D054 M 3 | 0.357 | 0.025 | 0.357 | 0.025 | 0.00 | 1.00 |
| NRV Vtlin D27 D054 M 4 | 0.357 | 0.025 | 0.358 | 0.026 | 0.01 | 1.00 |
| NRV Vtlin D27 D054 M 5 | 0.357 | 0.025 | 0.357 | 0.025 | 0.00 | 1.00 |
| NRV Vtlin D27 D054 M 6 | 0.357 | 0.025 | 0.357 | 0.025 | 0.00 | 1.00 |
| NRV Vtlin D27 D054 M 7 | 0.357 | 0.025 | 0.357 | 0.025 | 0.00 | 1.00 |
| NRV Vtlin D27 D054 M 8 | 0.357 | 0.025 | 0.358 | 0.026 | 0.01 | 1.00 |
| NRV Vtlin D27 D054 M 9 | 0.357 | 0.025 | 0.357 | 0.025 | 0.00 | 1.00 |
| NRV Vtlin D27 D054 M | 0.357 | 0.025 | 0.357 | 0.025 | 0.00 | 1.00 |
| NIO25 Vtlin 9 D252 M 1 | 0.519 | 0.008 | 0.519 | 0.008 | 0.02 | 1.01 |
| NIO25 Vtlin 9 D252 M 2 | 0.519 | 0.008 | 0.519 | 0.008 | 0.02 | 1.01 |
| NIO25 Vtlin 9 D252 M 3 | 0.519 | 0.008 | 0.519 | 0.008 | 0.00 | 1.00 |
| NIO25 Vtlin 9 D252 M 4 | 0.519 | 0.008 | 0.519 | 0.008 | 0.00 | 1.00 |
| NIO25 Vtlin 9 D252 M 5 | 0.519 | 0.008 | 0.519 | 0.008 | 0.02 | 1.01 |
| NIO25 Vtlin 9 D252 M 6 | 0.519 | 0.008 | 0.519 | 0.008 | 0.00 | 1.00 |
| NIO25 Vtlin 9 D252 M 7 | 0.519 | 0.008 | 0.519 | 0.008 | 0.00 | 1.00 |
| NIO25 Vtlin 9 D252 M 8 | 0.519 | 0.008 | 0.519 | 0.008 | 0.02 | 1.01 |
| NIO25 Vtlin 9 D252 M 9 | 0.519 | 0.008 | 0.519 | 0.008 | 0.00 | 1.00 |
| NIO25 Vtlin 9 D252 M | 0.519 | 0.008 | 0.519 | 0.008 | 0.00 | 1.00 |

parameters of LP products. Compared with the Baseline data, the MEAN shift value is between 0 and 0.02 while STD Extend is between 1.0 and 1.01. As can be seen, MEAN and STD value don't shift. Table 2 shows the test time and the improvement of test speed when testing Vt parameters of product 0043. It can be seen that with the shortening of the VGDELAYTIME, the average test time of the parameters is also shortened, from the initial 400-500ms to about 100ms. When the VGDELAYTIME is 1/10 of the initial, the test speed increased the most, reaching 75.03%-78.95%.

*Table II: The test time and the improvement of test speed of Vt parameters of product 0043*

| Parameter | | |
|---|---|---|
| NRV_Vtlin_D27_D054_M_3 | 169.89 | 58.36% |
| NRV_Vtlin_D27_D054_M_4 | 203.92 | 50.02% |
| NRV_Vtlin_D27_D054_M_5 | 237.58 | 41.77% |
| NRV_Vtlin_D27_D054_M_6 | 271.65 | 33.42% |
| NRV_Vtlin_D27_D054_M_7 | 306.21 | 24.95% |
| NRV_Vtlin_D27_D054_M_8 | 340.30 | 16.59% |
| NRV_Vtlin_D27_D054_M_9 | 374.36 | 8.24% |
| NRV_Vtlin_D27_D054_M | 407.98 | 0.00% |
| NIO25_Vtlin_9_D252_M_1 | 113.94 | 78.95% |
| NIO25_Vtlin_9_D252_M_2 | 160.57 | 70.33% |
| NIO25_Vtlin_9_D252_M_3 | 208.28 | 61.51% |
| NIO25_Vtlin_9_D252_M_4 | 255.70 | 52.75% |
| NIO25_Vtlin_9_D252_M_5 | 303.77 | 43.87% |
| NIO25_Vtlin_9_D252_M_6 | 351.10 | 35.12% |
| NIO25_Vtlin_9_D252_M_7 | 398.99 | 26.27% |
| NIO25_Vtlin_9_D252_M_8 | 446.33 | 17.52% |
| NIO25_Vtlin_9_D252_M_9 | 493.85 | 8.74% |
| NIO25_Vtlin_9_D252_M | 541.16 | 0.00% |

*Table III: The correlation results between experimental data and the Baseline value of Vt parameters in the logic area of product 0186*

| Parameter | BL | | NEW | | Mean Shift | STD Extend |
|---|---|---|---|---|---|---|
| | MEAN | STD | Mean | STD | | |
| NRV Vtlin D9 D054 S 1 | 0.434 | 0.018 | 0.434 | 0.018 | 0.01 | 0.99 |
| NRV Vtlin D9 D054 S 2 | 0.434 | 0.018 | 0.434 | 0.018 | 0.01 | 0.99 |
| NRV Vtlin D9 D054 S 3 | 0.434 | 0.018 | 0.433 | 0.018 | -0.01 | 0.99 |
| NRV Vtlin D9 D054 S 4 | 0.434 | 0.018 | 0.434 | 0.018 | 0.00 | 1.00 |
| NRV Vtlin D9 D054 S 5 | 0.434 | 0.018 | 0.434 | 0.018 | 0.00 | 1.00 |
| NRV Vtlin D9 D054 S 6 | 0.434 | 0.018 | 0.434 | 0.018 | 0.01 | 0.99 |
| NRV Vtlin D9 D054 S 7 | 0.434 | 0.018 | 0.434 | 0.018 | 0.00 | 1.00 |
| NRV Vtlin D9 D054 S 8 | 0.434 | 0.018 | 0.434 | 0.018 | 0.00 | 1.00 |
| NRV Vtlin D9 D054 S 9 | 0.434 | 0.018 | 0.434 | 0.018 | 0.01 | 0.99 |
| NRV Vtlin D9 D054 S | 0.434 | 0.018 | 0.434 | 0.018 | 0.00 | 1.00 |
| HO33 Vtlin 9 9 S 1 | -0.739 | 0.002 | -0.739 | 0.002 | 0.00 | 1.00 |
| HO33 Vtlin 9 9 S 2 | -0.739 | 0.002 | -0.739 | 0.002 | 0.00 | 1.00 |
| HO33 Vtlin 9 9 S 3 | -0.739 | 0.002 | -0.739 | 0.002 | 0.00 | 1.00 |
| HO33 Vtlin 9 9 S 4 | -0.739 | 0.002 | -0.739 | 0.002 | 0.00 | 1.00 |
| HO33 Vtlin 9 9 S 5 | -0.739 | 0.002 | -0.739 | 0.002 | 0.00 | 1.00 |
| HO33 Vtlin 9 9 S 6 | -0.739 | 0.002 | -0.739 | 0.002 | 0.00 | 1.00 |
| HO33 Vtlin 9 9 S 7 | -0.739 | 0.002 | -0.739 | 0.002 | 0.00 | 1.00 |
| HO33 Vtlin 9 9 S 8 | -0.739 | 0.002 | -0.739 | 0.002 | 0.00 | 1.00 |
| HO33 Vtlin 9 9 S 9 | -0.739 | 0.002 | -0.739 | 0.002 | 0.00 | 1.00 |
| PIO33 Vtlin 9 9 S | -0.739 | 0.002 | -0.739 | 0.002 | 0.00 | 1.00 |

Table 3 shows the correlation results between experimental data and the Baseline value of Vt parameters in the logic area of product 0186 of CIS platform. It can be seen that shortening VGDELAYTIME has no effect on Vt parameters in the logic area of CIS product. Compared with the Baseline data, the MEAN shift ranged from 0 to

*Table IV: The test time and the improvement of test speed of Vt parameters in the logic area of product 0186*

| Parameter | Time(ms) | Increase |
|---|---|---|
| NRV_Vtlin_D9_D054_S_1 | 100.07 | 78.46% |
| NRV_Vtlin_D9_D054_S_2 | 140.38 | 69.79% |
| NRV_Vtlin_D9_D054_S_3 | 180.66 | 61.12% |
| NRV_Vtlin_D9_D054_S_4 | 221.36 | 52.36% |
| NRV_Vtlin_D9_D054_S_5 | 261.93 | 43.63% |
| NRV_Vtlin_D9_D054_S_6 | 302.77 | 34.84% |
| NRV_Vtlin_D9_D054_S_7 | 343.04 | 26.18% |
| NRV_Vtlin_D9_D054_S_8 | 383.58 | 17.45% |
| NRV_Vtlin_D9_D054_S_9 | 424.51 | 8.65% |
| NRV_Vtlin_D9_D054_S | 464.69 | 0.00% |
| PIO33_Vtlin_9_9_S_1 | 128.21 | 82.11% |
| PIO33_Vtlin_9_9_S_2 | 191.22 | 73.33% |
| PIO33_Vtlin_9_9_S_3 | 256.89 | 64.16% |
| PIO33_Vtlin_9_9_S_4 | 322.66 | 54.99% |
| PIO33_Vtlin_9_9_S_5 | 388.35 | 45.83% |
| PIO33_Vtlin_9_9_S_6 | 454.05 | 36.66% |
| PIO33_Vtlin_9_9_S_7 | 519.76 | 27.50% |
| PIO33_Vtlin_9_9_S_8 | 585.41 | 18.34% |
| PIO33_Vtlin_9_9_S_9 | 651.12 | 9.17% |
| PIO33_Vtlin_9_9_S | 716.87 | 0.00% |

*Table V: The correlation results of experimental data of Vt parameters of four MOS devices and VPIN parameters locate in the pixel area of product 0186*

| Parameter | BL | | NEW | | Mean Shift | STD Extend |
|---|---|---|---|---|---|---|
| | MEAN | STD | Mean | STD | | |
| VT1_A1_TX1F_1 | -0.145 | 0.017 | -0.145 | 0.017 | 0.00 | 1.00 |
| VT1_A1_TX1F_2 | -0.145 | 0.017 | -0.146 | 0.018 | -0.03 | 1.01 |
| VT1_A1_TX1F | -0.145 | 0.017 | -0.145 | 0.017 | 0.00 | 1.00 |
| VT1_A2_RST_1 | 0.510 | 0.048 | 0.509 | 0.048 | -0.01 | 1.00 |
| VT1_A2_RST_2 | 0.510 | 0.048 | 0.509 | 0.048 | 0.00 | 1.00 |
| VT1_A2_RST | 0.510 | 0.048 | 0.510 | 0.048 | 0.00 | 1.00 |
| VT1_A3_SF_1 | 0.268 | 0.058 | 0.268 | 0.059 | 0.00 | 1.00 |
| VT1_A3_SF_2 | 0.268 | 0.058 | 0.268 | 0.058 | 0.00 | 1.00 |
| VT1_A3_SF | 0.268 | 0.058 | 0.268 | 0.058 | 0.00 | 1.00 |
| VT1_D3_RS_1 | 0.401 | 0.032 | 0.401 | 0.032 | 0.00 | 1.00 |
| VT1_D3_RS_2 | 0.401 | 0.032 | 0.401 | 0.032 | 0.00 | 1.00 |
| VT1_D3_RS | 0.401 | 0.032 | 0.401 | 0.032 | 0.00 | 1.00 |
| VPIN1NA_VDD1_B1_PIN_1 | 1.701 | 0.073 | 1.701 | 0.073 | 0.00 | 1.00 |
| VPIN1NA_VDD1_B1_PIN_2 | 1.701 | 0.073 | 1.701 | 0.073 | 0.00 | 1.00 |
| VPIN1NA_VDD1_B1_PIN | 1.701 | 0.073 | 1.701 | 0.073 | 0.00 | 1.00 |
| VPING1E_2D5_F1_PIN_1 | 2.483 | 0.022 | 2.483 | 0.022 | 0.00 | 1.00 |
| VPING1E_2D5_F1_PIN_2 | 2.483 | 0.022 | 2.483 | 0.022 | 0.00 | 1.00 |
| VPING1E_2D5_F1_PIN | 2.483 | 0.022 | 2.483 | 0.022 | 0.00 | 1.00 |

0.04, STD Extend ranged from 0.96 to 1.0, indicating that both MEAN and STD value don't shift. Table 4 shows test time and the improvement of test speed when testing product 0186 logic Vt parameters. It can be seen that with the shortening of VGDELAYTIME, the average test time of parameters is also shortened accordingly. RV device is shortened from the initial 400ms to 100ms. IO devices are shortened from the initial 716ms to 128ms.When the VGDELAYTIME is 1/10 of the original, the test speed increases the most, for RV devices, the improvement rate

*Table VI: the test time and the improvement of test speed of the Vt parameters of the four MOS devices and VPIN parameters in the pixel area of product 0186*

| Parameter | Time(ms) | Increase |
|---|---|---|
| VT1_A1_TX1F_1 | 327.50 | 20.35% |
| VT1_A1_TX1F_2 | 365.70 | 11.06% |
| VT1_A1_TX1F | 411.17 | 0.00% |
| VTI_A2_RST_1 | 454.99 | 23.47% |
| VTI_A2_RST_2 | 523.60 | 11.93% |
| VTI_A2_RST | 594.50 | 0.00% |
| VTI_A3_SF_1 | 445.80 | 25.03% |
| VTI_A3_SF_2 | 518.38 | 12.83% |
| VTI_A3_SF | 594.66 | 0.00% |
| VTI_D3_RS_1 | 456.31 | 24.41% |
| VTI_D3_RS_2 | 512.99 | 15.03% |
| VTI_D3_RS | 603.69 | 0.00% |
| VPIN1NA_VDD1_B1_PIN_1 | 782.33 | 33.01% |
| VPIN1NA_VDD1_B1_PIN_2 | 959.77 | 17.82% |
| VPIN1NA_VDD1_B1_PIN | 1167.89 | 0.00% |
| VPING1E_2D5_F1_PIN_1 | 6225.04 | 37.57% |
| VPING1E_2D5_F1_PIN_2 | 8021.38 | 19.56% |
| VPING1E_2D5_F1_PIN | 9971.63 | 0.00% |

is 78.46%, for IO devices, the improvement rate is about 82.11%.

Table 5 shows the correlation results of experimental data of Vt parameters of four MOS devices and VPIN parameters in the pixel area of product 0186 of CIS platform. It can be seen that shortening VGDELAYTIME has no effect on Vt parameters and VPIN parameters in pixel area of CIS product. Comparing with the Baseline value, MEAN shift ranges from 0 to 0.09 and STD Extend ranges from 1.0 to 1.02. Obviously, neither MEAN nor STD shifts. Table 6 shows the test time and the improvement of test speed of the Vt parameters of the four MOS devices and VPIN parameters in the pixel area of product 0186. It can be seen that with the shortening of VGDELA YTIME, the average test time of each parameter also decreases, but the improvement rate of Vt parameters of the four MOS devices in the pixel area is not as obvious as that in the logic area. The reason may be that the algorithm used for testing the Vt parameter in the pixel area adopts Binary Search, while for the Vt parameter in the logic area it adopts Liner Search. DELAYTIME takes a relatively lower proportion in the Binary Search than in the Liner Search. Therefore, the improvement rate of test speed is lower after reducing DELAYTIME. The test speed improvement rate of VPIN parameter is 33.01%-37.57%.

According to the experimental data of LP and CIS products, shortening the DELAYTIME in the algorithm to 1/10 of the initial value can significantly improve the test speed of Vt/VPIN parameters without affecting the test data, thus improving the test speed of the whole wafer. Therefore, we gradually test LP and CIS products using new testing conditions under monitoring. Figure 5 shows the test results of Vt parameters product 0043 and in logical area of product 0186 one week after they went online. Figure 6 shows the test results of Vt and VPIN parameters in pixel area of product 0186 one week after they went online. It can be seen that the test results of the LP and CIS platforms are stable, with MEAN values

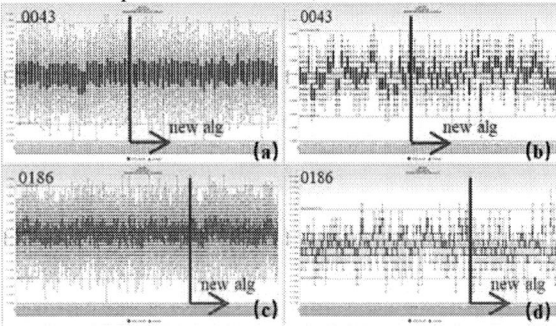

*Figure 5: The box charts of Vt parameters product 0043 and in logical area of product 0186 one week after they went online: (a) NRV_Vtlin_D27_D054_M, (b) NIO25 _Vtlin_9_D252_M, (c) NRV_Vtlin_D9_D054_S, and (d) PIO33_Vtlin_9_9_S.*

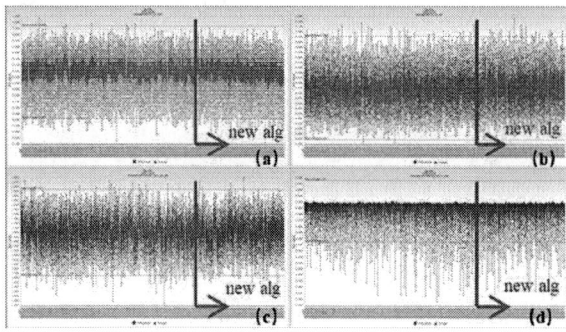

*Figure.6. The box charts of Vt and VPIN parameters in pixel area of product 0186 one week after they went online: (a) VT1_A1_TX1F, (b) VT1_A2_RST, (c)VPIN1NA_VDD1_B1_PIN, (d)VPING1E_2D5_F1_PIN.*

*Table VII: The test time of each wafer of product 0043 and product 0186 before and after the launch of the new algorithm*

| Product | Test time old (min) | Test time new (min) | decrease rate |
|---------|---------------------|---------------------|---------------|
| 0043 | 17.94 | 14.60 | 18.87% |
| 0186 | 21.17 | 18.30 | 13.52% |

fluctuating within 3sigma, and STD values of each lot are almost the same. Therefore, appropriately reducing the DELAYTIME value in the algorithm will not affect the test results of parameters.

Table 7 shows the test time of each wafer of product 0043 and product 0186 before and after the launch of the new algorithm. Obviously, the test time of each wafer of the two products is shortened to varying degrees. Among them, the test time of product 0043 is shortened from 17.84 minutes to 14.60 minutes, which is 18.87%, and the test time of product 0186 is shortened from 21.17 minutes to 18.30 minutes, which is 13.52%. Figure 7 shows the number of wafers tested using the new algorithm and test time used every month after the new algorithm was launched. A total of 12,146.7 hours of WAT capacity were saved between April and September 2021.

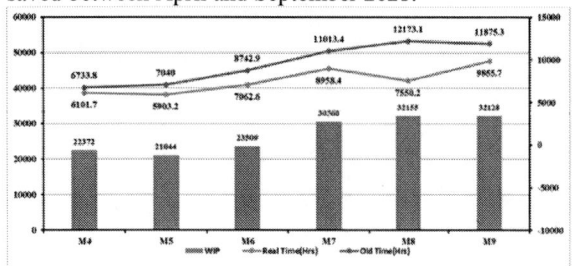

*Figure 7: The number of wafers tested using the new algorithm and the test time used every month after the new algorithm was launched.*

## CONCLUSION

In this paper, we greatly improve the test speed and shorten the test time of wafer by optimizing the algorithm on the premise of ensuring the accuracy of data. The test speed increase rate of Vt parameters of LP product reached 75.54%; The improvement rate of Vt parameters in the logic area of CIS product reached 76.74%, that of Vt parameters in the pixel area reached 23.08%, and that of VPIN parameters reached 37.57%. So far, the new algorithm has been applied to 15 products on LP and CIS platform, and the overall wafer test speed is improved from 13.52% to 18.87%, saving a total of 12146.7 hours of WAT capacity.

## REFERENCES

[1] P. C. Chen, WAT Method of Plasma Damage in CMOS Process, Electronic & Packaging, 16(6), 2016, pp: 31-35.

[2] D. T. Wen, Integrated Circuit Manufacturing Technology and Engineering Application, China Machine Press, 2018.

[3] H. B. Zhuo, A Review of Wafer Probe Testing for Integrated Circuits, Engineering Technology, 7, 2016, pp: 236-238.

[4] J. D. Peng, Design and Implementation of Acceptability Test for a Memory Chip Wafer, University of Electronic Science and Technology of China, Chengdu, 2011.

[5] X. X. Chen, Study on Wafer Acceptance Test in Foundry, Fudan University, Shanghai, 2007.

# ANOMALY DETECTION OF SEMICONDUCTOR PROCESSING DATA BASED ON DTW-LOF ALGORITHM

Wang Yong, Mao Guiyun, Chen Xu, Wei Zhengying
Shanghai Huali Microelectronics Corporation, Shanghai, China
(86)21-61871212, wangyong@hlmc.cn
*Corresponding Author's Email: chenxu@hlmc.cn

## ABSTRACT

Fault detection and classification (FDC) of equipment sensor data is of great significance in semiconductor wafer manufacturing to monitor equipment and the processing wafer condition. The early detections of anomalies of sensor data facilitate later process adjustments and avoid further economic loss. Sensor data varies greatly among different processing wafer lots because of machine status changes during the maintenance cycles. This paper combines the Dynamic Time Warping (DTW) and Local Outlier Factor (LOF) algorithms to achieve stable anomaly detection of sensor data under unstable machine conditions. The DTW-LOF model shows good anomaly detection accuracy in different chip processing technologies with a small amount of data. Since the anomalies of semiconductor processing were usually reflected as comprehensive abnormality of multi-sensor in same time ranges, the normalized DTW distance of these sensors could effectively identify the processing abnormal stage and sensors.

*Keywords—DTW; FDC; Dynamic Time Wrapping*

## INTRODUCTION

In the processing of modern semiconductor chips, hundreds of sensors were used to monitor the machine conditions and changes of the chip wafer. The observation of the topography of the semi-finished chip and the control of defects were expensive because of increasing the production costs and reduce the processing efficiency. The collection and analysis of sensor data during processing can realize early detection of abnormalities to improve product quality without increasing costs due to excessive testing.

There were many different studies discussing the anomaly detection of sensor data in semiconductor chip processing using different machine learning algorithms, neural networks, automatic encoders[1], GAN network[2], etc. However, these analyses regard the sensor data in wafer processing as conventional multi-dimensional time series data, and not fully considering the characteristics of these sensor data. When the traditional machine learning model learns features from training data, it i presupposed that there will the samples in training set is not related, so as to learn the common features between samples. However, semiconductor processing is usually processed in batch that the machine conditions of the same product vary greatly between different batches but almost the same among the wafers in a single batch. Therefore, when the anomaly detection model is trained with sensor data from semiconductor wafer processing, the number of sample with different features was far less than the number of samples.

At the same time, due to changes in machine conditions between different batches, if strict criterions were set in the abnormal detection model of semiconductor processing, a large number of false alarms would be generated after the change in processing conditions. In order to adapt to changes in operating conditions and set loose criterions, it is difficult for the anomaly detection model to detect weak abnormalities in a single batch.

This paper proposes an anomaly detection algorithm that combines DTW distance and LOF algorithm. Taking into account the characteristics of semiconductor processing that is usually carried out in batches, it pays full attention to the anomaly analysis in a single batches. The experiments shown that making full use of the similarities between different wafers within Lot may provide a higher anomaly detection effect.

## METHODOLOGY

### Data Set

There were hundreds of sensors in each machine in modern semiconductor wafer manufacturing. Typically, sensor data were gathered at a specific sampling rate of about 0.1-2 sites/second. The processing procedure of a single wafer could be divided into different stages, and characteristics of sensor data varied between the stages. It was discerned that the data of different Wafers in the same Lot could have differences like optional stages omitting or different sampling rates.

The data used in this paper was obtained from an advanced 300mm semiconductor foundry. Sensor Data was collected from etching process with training set of 1261pcs normal processing wafer and test set of 544pcs wafer with 61pcs having actual anomalies. The etching process was monitored by 50 different sensors and the processing time was about 300 seconds.

### Algorithm Description

The Local Outlier Factor algorithm[3] is a traditional density-based anomaly detection algorithm, which considers data points in lower density area as anomalies. This method evaluates the density of data points by calculating the reachability distance between the data points. The specific reachability distance from point $\mathbf{x}$ to $\mathbf{x}'$ is calculated as:

$$RD_k(x, x') = max(\|x - x^k\|, \|x - x'\|)$$

where $x^k$ is the $k$th nearest neighbor of $\mathbf{x}$. Base on reachability distance, the local reachability density of $\mathbf{x}$ is defined as:

$$LRD_k(x) = \left( \frac{1}{k} \sum_{i=1}^{k} RD_k(x^i, x) \right)^{-1}$$

Local reachability density is the reverse of average RD from $x^i$ to $x$. And the local outlier factor of $x$ is defined as:

$$LOF_k(x) = \frac{\frac{1}{k} \sum_{i=1}^{k} LRD_k(x^i)}{LRD_k(x)}$$

Local outlier factor is the ratio of average local reachability density of $x^i$ and the local reachability density of $x$. Data point x is regarded as an anomaly when the local outlier factor has as large value that $x^i$ in a high density area while $x$ in a low density area.

Since sensors behave differently in different stages in semiconductor processing, and actual anomalies usually manifests on multiple sensors in a specific stage. Therefore, this paper divides the sensor data according to stages, and performs LOF anomaly detection on the same sensor at the same stage between wafers. In order to compare the anomalies of different sensors, the LOF score of the training set is standardized and the same scaling methods were applied to test data. So the abnormal score of stage k in defined as the average abnormal score of all the sensors of same stage:

$$S_k = \frac{1}{n} \sum_{i=1}^{n} S_{k,i}$$

The abnormal score of a single wafer was defined as the maximum of the abnormal scores of all stage:

$$S = Max(S_1, S_2, \dots S_n)$$

Because two time series data might differ in amplitude and time direction. Two completely identical time series data will produce a big Euclidean distance shift only one second. The sensor data collected in semiconductor processing is accompanied by different start times due to sampling or other issues. Unlike the Euclidean distance, DTW algorithm[4] allowing similar shapes to match even if they are out of phase in time dimension. So DTW algorithm can compare time series of variable size and is robust to shifts or dilatations across the time dimension. Therefore, this paper uses DTW distance to minimize the influence of time differences to more accurately compare the differences in sensor values among different wafer processing.

## RESULTS AND DISCUSSION

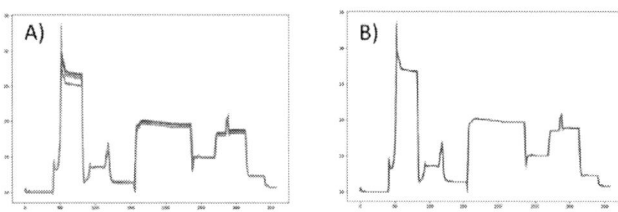

Figure 1. Sensor data from etching process: (A) different lot (B) same lot

The processing of semiconductor wafers is usually processed in units of Lots. Even if the processing is based on wafers, the wafers in the same lot is usually carried out under the same machine conditions in adjacent times. The processing of different batches was often separated by days, during which the machine may process a large number of other products or have undergone maintenance. As shown in Figure 1, the sensor data of different lot were quite different, and the sensor data difference of wafers in the same lot is very small. Therefore, if a fixed numerical standard is set for all sensor data, a large number of alarms may be triggered when the machine status changes. When a loose standard is set, the sensor data offset caused by abnormal processing of some wafers in the same batch will not be detected.

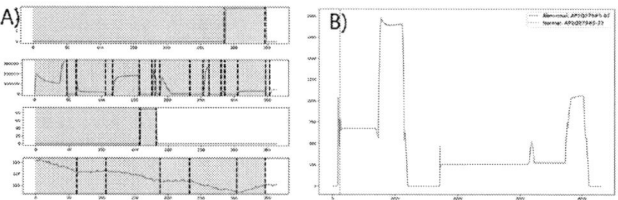

Figure 2. semiconductor processing sensor data. A) steps B) actual anomalies

Each step of the semiconductor wafer processing process is usually divided into multiple stages, and different sensors have large differences between these stages. Therefore, the sensor data was divided into stable phases and unstable phases. In the unstable phase, the sensor data may fluctuates greatly, and the abnormal fluctuation of the key sensor in stable phases is extremely small. As depicted in Figure 2A, the time points of the changes of different sensors are not the same, which brings further difficulties to the actual anomaly detection algorithm. The actual anomaly shown in Figure 2B only exists in 0.3-0.4 seconds of all the processing processes of about 300 seconds.

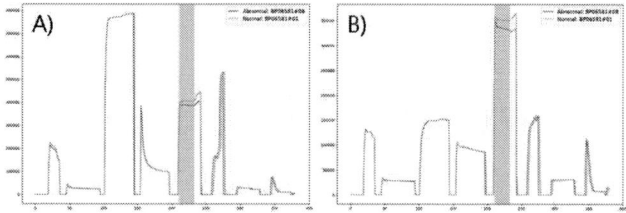

Figure 3. Anomaly of processing reflected by different sensors

Large number of sensors are used to monitor the semiconductor processing process. In fact, abnormal processing conditions often cause abnormal fluctuations of multiple different sensors. As shown in Figure 3, the actual sensor fluctuation is caused by sensor No. 88 and 89 at the same time. Therefore, when evaluating the abnormal processing stage, this article will score the abnormal scores of each sensor at this stage, avoiding the isolated judgment of a certain sensor abnormality to improve the efficiency and accuracy of abnormal detection. At the same time, since there is no direct correlation between non-neighboring processing stages, it is not necessary to consider all stages when judging wafer processing abnormalities, and the influence of unstable stages in the processing process is also avoided.

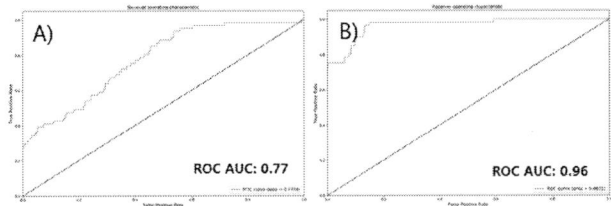

*Figure 4. Receiver operating characteristic curve. (A) Euclidean distance (B) DTW distance*

*Figure 4* (A) is receiver operating characteristic curve of test set with model trained with DTW distance and Euclidean distance. It can be seen from t that when using the Euclidean distance in the LOF algorithm to evaluate the difference in semiconductor sensor data can achieve the ROC AUC score at 0.77. And change the distance function to DTW distance can greatly improve the effect of this anomaly detection, which achieved ROC score 0.96. As shown in *Figure 5*, the LOF algorithm with Euclidean distance takes the time shift as the sensor outlier, but in fact, the waiting of a certain process to be ready and starts processing often does not affect wafer quality.

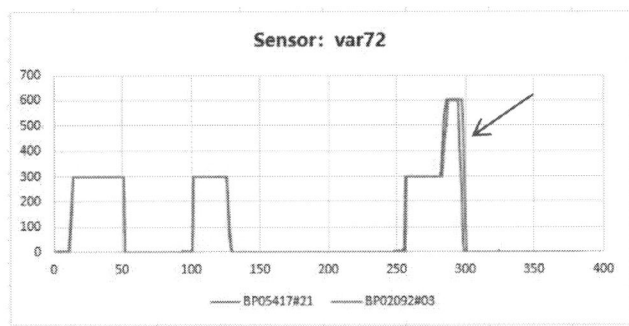

*Figure 5. Detailed sample of LOF algorithm with Euclidean distance*

In fact, semiconductor processing is carried out in batches, and it is rare that all wafers in the entire batch have the anomalies of same pattern, because this can only cause from systemic reasons such as machine failures or changes in production procedures. In the complex wafer production process, machine failures are rarely found to cause quality problems for the entire batch, and wafer level abnormalities caused by accidental factors are more likely to occur. If the entire batch has similar sensor data fluctuations, the local density of different wafer data is higher. The different performance of a small number of wafers in the entire batch of wafers makes these wafers have a lower local reachability density, that is, a higher local outlier factor. If the hyper parameter of the number of neighbors in the LOF algorithm is less than the number of wafers in the wafer processing batch, the overall fluctuation between batches will not cause an increase in LOF scores, only abnormality between the wafers within a batch will cause the rise of LOF score. Therefore, the algorithm can effectively avoid the false alarm of the model caused by the change of the machine state and raw materials,

and improve the accuracy of anomaly detection.

## CONCLUTION

This paper proposed a local outlier algorithm with dynamic time warping distance to identify anomaly sensor data in semiconductor processing. Experiments shows that, this method can achieve a good anomaly detection result in semiconductor processing. The LOF algorithm can effectively use the characteristics of wafer batch processing, accurately detect the small anomalies of the sensors within the batch, and can ignore the sensor fluctuations caused by the differences in materials and machines between batches.

## REFERENCES

1. W. Yong, C. Xu, and W. Zhengying, in *2020 China Semiconductor Technology International Conference (CSTIC)* (2020), pp. 1–3.
2. M. Hashimoto, Y. Ide, and M. Aritsugi, Procedia Computer Science **192**, 873 (2021).
3. M. Breunig, H.-P. Kriegel, R. T. Ng, and J. Sander, in *Proceedings of the 2000 Acm Sigmod International Conference on Management of Data* (ACM, 2000), pp. 93–104.
4. M. Müller, editor , in *Information Retrieval for Music and Motion* (Springer, Berlin, Heidelberg, 2007), pp. 69–84.

# REASON FORECAST WITH BERT ON TEST RESULT ALARM IN SEMICONDUCTOR FAB

*Meng Xue[1], Xu Chen[2], Guiyun Mao[3], Yong Wang[4], Zhengying Wei[5]*

[1,2,3,4,5]Shanghai Huali Microelectronics Corporation, Shanghai 200433, China

Author's Email: xuemeng@hlmc.cn

## ABSTRACT

The development of artificial intelligence in fab raised the opportunity of improving the product yield and quantity by predictive analytics. That improving the keen awareness of the knowledge of relationship between anomalous data and product yield can help engineers to find potential risk in wafer processing, which contributes to improving product yields. In this paper, we extracted a set of alarm records with confirmed reasons from engineers as data set. And we used an attention-based convolutional neural network with BERT as the embedding layer and added some more attentive features to find the alarm reason and the alarm influence. The result indicated that this resolution could help engineer find alarm reason timely.

## INTRODUCTION

The rapid pace of technology innovation like Artificial Intelligence, 5G, etc., it's no surprise that today's wafer manufacturing execution became more complicated and competitive. Further, fabs still face a sobering reality: complexity and variability make it a constant challenge for fabs to balance critical priorities: keeping manufacturing operating costs low, processes integrated, throughput high, quality uncompromised, and cycle time minimized [1]. To keep pace with the conventional growth in demand, the quality assurance has an integral part of each process step in a fab. Identify the root cause that possibly led to a failure, with detailed components of the asset involved, along with failure reasons with probabilities helps provide automated early detection of quality deviations and faults. Multi-label text classification is one of the most common tasks in Natural Language Processing (NLP) [2]. With neural networks' capacity of learning representation from data without complex feature engineering, deep learning becomes the hottest research model in this area as well as many others [3]. Modern Transformer-based models make use of pre-training on vast amounts of text data that makes fine-tuning faster, use fewer resources and more accurate on smaller datasets. Here, we used BERT which is one of the top pre-trained deep language models has achieved impressed results in various NLP tasks [4]. In order to improve the accuracy of judgement on processing alarm in Fab. We employed Bert to classify the alarm with specific labels and find reason.

## METHOD

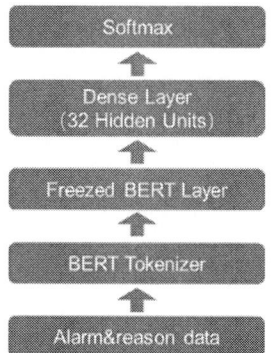

*Figure 1: Fine-tuned BERT model for alarm classification.*

BERT is a masked language model that uses transformer architecture which needs to use a special tokenizer. Fig. 1 shows that the network architecture contains a special tokenizer, frozen BERT layer and a new trainable dense layer with 32 hidden units. This architecture is useful to fine-tuned BERT layer with a new dense layer to categorize alarms.

The dataset consists of 13000 alarm data and has 9 different categories. Figure 2 presents the distribution of alarm in the dataset.

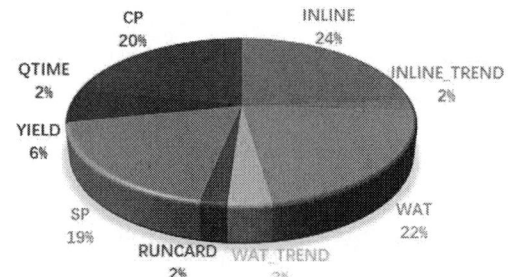

*Figure 2: Fine-tuned BERT model for alarm classification.*

## EXPERIMENTS

We used a BERT-base-uncased pre-trained model which has 12-layer, 768-hidden, 12-heads, 110M parameters. And we set the maximum sequence length as 128, training batch size as 16, learning rate as 2e-5 and training epochs as 1, 2, 3, 4, respectively. The dataset was divided for 3 parts, and the proportion of training set, development set, and test set is 70%, 20%, 10%, respectively.

Before using the pre-trained Bert, we modify the codes. We create a new processor method and processor dictionary to satisfy our dataset. Then we set the above-mentioned parameters to run with the pre-trained model. Then we adjust the corresponding parameters and using the test data and the model after fine-tuning to predict the categories of alarms.

## RESULTS

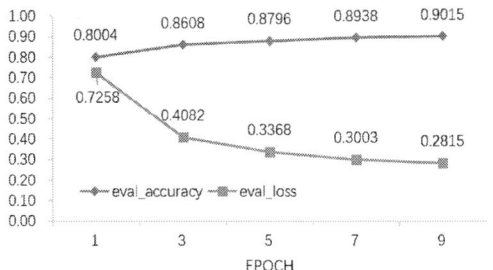

*Figure 3: Trends of accuracy and loss with the epochs changes.*

As show in Fig.3, the classification accuracy of development set of this model increased with the increasing training epochs, and the growth rate of accuracy becoming less rapid. Increasing the number of epochs will bring the overfitting and this process takes time. So there is no need to add epochs to train the model. To save time, we chose 7 as the training epoch as the saved model to test the testing set.

We used confusion matrix to evaluate the test result.

$$precison= tp / (tp + fp) \qquad (1)$$

$$recall= tp / (tp + fn) \qquad (2)$$

$$acuuray= (tp+tn)/total \qquad (3)$$

The number of true positives is tp, the number of false positives is fp, the number of false negatives is fn, and total is the total number of data.

The precision is intuitively the ability of the classifier not to label as positive a sample that is negative. And the recall is intuitively the ability of the classifier to find all the positive samples. The F-beta score weights recall more than precision by a factor of beta. beta = 1.0 means recall and precision are equally important [5].

As shown in Fig4, we get the final accuracy is 0.8815. Experimental results show that the characters of data itself vary, their precision, recall and f1-score also vary. The precision of each type is dissimilar. Like the type of "INLINE" and "INLINE_TREND", they are very similar. It is hard to distinguish these two types, so the precision of these two types are relatively lower than other types. For recall and f1-score, they are in the same condition. For the type of "SP", it is properly that the discrimination of "SP" alarm between the other is large. So the result looks like very good. All in all, we can find the model is very effective in the accuracy of alarm classification.

|  | precision | recall | f1-score |
|---|---|---|---|
| YIELD | 0.7636 | 0.9835 | 0.8597 |
| CP | 0.9416 | 0.9699 | 0.9555 |
| INLINE | 0.6538 | 0.4857 | 0.5574 |
| SP | 1.0000 | 1.0000 | 1.0000 |
| WAT | 0.9355 | 0.9667 | 0.9508 |
| WAT_TREND | 0.9920 | 1.0000 | 0.9960 |
| RUNCARD | 0.8576 | 0.9893 | 0.9187 |
| INLINE_TREND | 0.0000 | 0.0000 | 0.0000 |
| QTIME | 0.7692 | 0.1176 | 0.2041 |
|  |  |  |  |
| accuracy |  |  | 0.8815 |
| macro avg | 0.7681 | 0.7236 | 0.7158 |
| weighted avg | 0.8504 | 0.8815 | 0.8463 |

*Figure 4: Prediction result of trained model.*

## CONCLUSION

We used a pre-trained model named BERT to make multi-label classifications with alarm data from Fabs. As we all know, BERT has a fantastic ability on specific tasks by fine-tuning, with further adjust the BERT model, the accuracy of the model can be 88.15%. We can get the alarm reason automatic and help the engineer to find the alarm reason timely.

## ACKNOWLEDGMENT

I would like to express my deepest appreciation to my colleagues who helped me during learning ML, especially my leader. Many thanks to my colleagues, they are kind, funny, and generous.

## REFERENCES

[1] A. Ahmadi, H. -G. Stratigopoulos, A. Nahar, B. Orr, M. Pas and Y. Makris, 2015 IEEE/ACM International Conference on Computer-Aided Design (ICCAD), 2015, pp. 9-14.

[2] D. W. Otter, J. R. Medina and J. K. Kalita, *IEEE Transactions on Neural Networks and Learning Systems*, vol. 32, no. 2, Feb. 2021, pp. 604-624.

[3] T. Young, D. Hazarika, S. Poria and E. Cambria, *IEEE Computational Intelligence Magazine*, vol. 13, no. 3, Aug. 2018, pp. 55-75.

[4] J. Devlin, M.W. Chang, K. Lee and K. Toutanova, 2018, arXiv:1810.04805v2.

[5] Scikit-learn: Machine Learning in Python, Pedregosa et al., JMLR 12, 2011，pp. 2825-2830.

# VIRTUAL METROLOGY OF WAT VALUE WITH MACHINE LEARNING BASED METHOD

*Tao Zhou[1], Xuling Diao[1], Yiyi Jiang[2], Shiyuan Wen[2], Xuelong Shi[1], Quan Jing[2], and Chen Li,[1]\**

[1]Shanghai Integrated Circuits R&D Center Co., Ltd., Shanghai, China
[2]Shanghai Huali Microelectronics Corporation, Shanghai, China
*Corresponding Author's Email: lichen@icrd.com.cn

## ABSTRACT

Plasma etch is a key process during semiconductor manufacturing, materials are removed from the surface of semiconducting, typically silicon. The etch process is highly complicated because of chemical and electrical interactions, which makes the etch process notoriously difficult to model and control. Traditional metrology depends on expensive equipments and exhausts a lot of time, which has become new bottleneck of high-volume and high-throughput manufacturing at advanced nodes. In this work, we focus on the virtual metrology technique which could be used for prediction of WAT value based on the key parameters in etch process. By using random forest (RF) and stepwise regression (SR) method, the predicted values of NRV_Idsat reached a high R2 compared with true measurements. It reveals high efficiency of the proposed method and great potentials in the applications of virtual metrology in real manufacturing.

*Keywords—virtual metrology (VM); etch process; random forest (RF); stepwise regression (SR); wafer acceptance test (WAT);*

## INTRODUCTION

The plasma etch process, as the critical dimension (CD) keeps shrinking at advanced nodes, has become more and more challenging due to the complexity of chemical and electrical interactions. Traditionally, the process and yield engineers add some measurement sites to monitor and ensure the quality etch operation. However, the metrology often takes a lot of time and reduces the manufacturing throughput. In addition, the metrology equipments are often expensive. Therefore, if there are some highly reliable models to reduce/replace the actual measurements, it will be very attractive for semiconductor manufacturing both in efficiency and cost. Fortunately, some previous works on VM have been reported and gained impressive results [1-3], with future optimization, the VM methods will become popular in full-auto Fabs.

In this work, a machine leaning based method of random forest combined with stepwise regression strategy has been designed to train the prediction model of WAT value (NRV_Idsat) for etch process. This method could absorb the experience easily by adding new feature (leaf) in the decision tree of forest, and on the other hand, the useless/high-similar features are filtered to lower the tree structure redundant.

Furthermore, with specified training dataset, the prediction model gains precise results with sorted high-correlated sensors.

## METHOD

### A. Distribution of metrology data

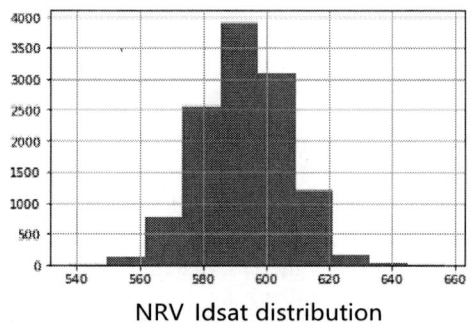

Fig. 1. The distribution of metrology data of NRV_Idsat.

Before model training, the overall status of metrology data should be checked. Traditionally, the distribution could well reflect the data status, to the central limit theorem, in nature and production, many phenomena are affected by amount independent random factors, if the influence of each factor is very small, the total influence can be regarded as a normal distribution. Under the assumption, the model trained on a sampled dataset then could be trusted to be used on other new data in the same environment.

The overall status of the NRV_Idsat dataset is as follows: available wafer number is 11856, and the NRV_Idsat value ranges from 537.7 to 657.2. The distribution of data is shown in Figure 1, as it indicates, the data is normal distribution and satisfies the modeling assumptions.

### B. Schematic of Algorithm

The framework of model training is based on random forest technique, which is constructed by many decision trees, and each tree is an independent model to prediction. The leaves of each tree are the features/properties to describe the problem. Then how to choose the feature to represent is very important. However, the dimensions of actual problems are usually very large, dimension reduction should be done first. There are two methods are applied, one is to use the statistic values instead of raw data for model training, the other is to compute the mutual information to judge the similarity of data for filtering. After dimension reduction, the new features are added into the RF structure for model training, if the validation value of R2_adj is lower than the previous, then the added features are proved useless, otherwise, the features are kept for next iteration. Finally, if R2_adj is stable during iterations, then we stop the updates and output the trained model for prediction.

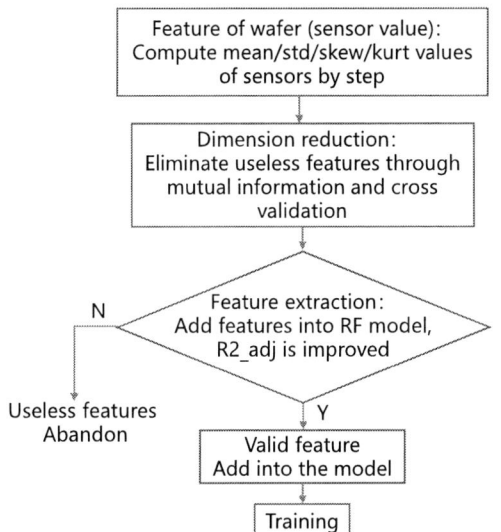

Fig. 2. Schematic of designed algorithm for training of WAT prediction with etch process parameters.

## RESULTS AND DISCUSSION

After filtering and training iterations, the number of features remained is 8, which means based on the 8 sensors/statistic values, the prediction could reach a satisfied level.

The regression results are shown in Figure 3 and Table 1, Figure 3 shows the prediction results of model on train and test datasets, respectively. The red lines are the fitting curves, which indicate that the model reaches good results because of slope is very close to 1 for both lines. It could could be found that the prediction of training dataset is better than test, which is normal for machine learning methodology, at the same time, several popular quantitative indicators have been calculated for quality check of regression model, such as the correlation coefficient R2 (R2_adj in this work), MAE, RMSE and MAPE, etc. The R2 is widely used in model evaluation, the ideal value is 1, which means that the model could exactly predict the result. In general, actual production and engineering fields, this value is often lower than 0.6, however, our model reaches over 0.7 on the test dataset. The MAPE is also reveal that the averaged error is less than 3 percent. All the calculations parameters show that the designed model could get a good prediction of NRV_Idsat parameter without doing a real metrology.

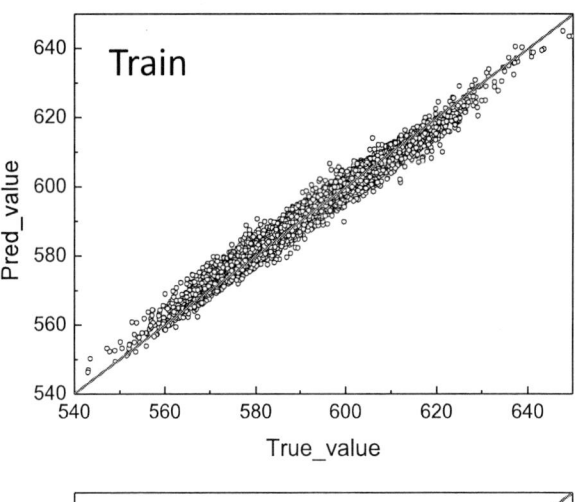

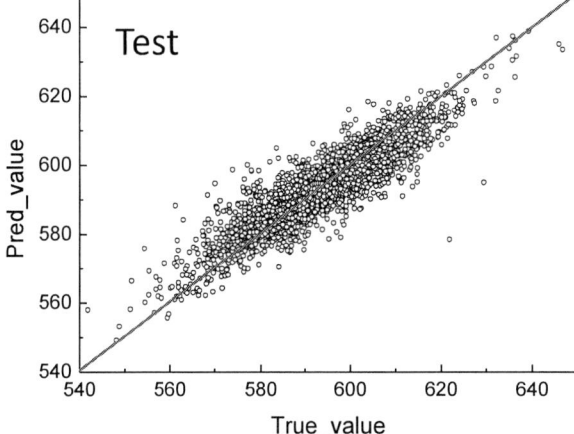

Fig. 3. Demonstration of some test patterns for SEM images.

TABLE I
SUMMARY OF THE REGRESSION MODEL RESULTS

|       | R2    | R2_adj | MAE   | RMSE  | MAPE  |
|-------|-------|--------|-------|-------|-------|
| Train | 0.969 | 0.968  | 1.859 | 2.448 | 0.025 |
| Test  | 0.789 | 0.789  | 4.947 | 6.405 | 0.025 |

## CONCLUSION

To summarize, a prediction model has been designed for WAT value (NRV_Idsat) prediction of etch process, the model uses random forest structure and combines stepwise regression strategy to absorb the effective features. The prediction results shows that the model possesses a very attractive ability which provides precise results without actual metrology and proves the effectiveness of our method.

## REFERENCES

[1] J. –W Kwon, S. Ryu, J. Park, et al., "Development of Virtual Metrology Using Plasma Information Variables to Predict Si Etch Profile Processed by SF 6 /O 2 /Ar Capacitively Coupled Plasma", *AEC/APC Symposium Asia*, 2021.

[2] S. A. Lynn, N. MacGearailt, J. V. Ringwood, "Real-time virtual metrology and control for plasma etch", *Journal of Process Control*, vol. 22, no 4, pp. 666-676, 2012.

[3] S. A. Lynn, J. Ringwood, N. MacGearailt, "Global and Local Virtual Metrology Models for a Plasma Etch Process", IEEE Transactions on Semiconductor Manufacturing, vol. 25, pp. 94-103, 2012.

# HYBRID SOLUTIONS FOR ROOT CAUSE TRACING OF RANDOM ALARM ON ETCH TOOL

*Xuling Diao[1]\*, Yiyi Jiang[2], Shiyuan Wen[2], Xuelong Shi[1], Quan Jing[2], Chen Li, and Tao Zhou[1]\**

[1]Shanghai Integrated Circuits R&D Center Co., Ltd., Shanghai, China
[2]Shanghai Huali Microelectronics Corporation, Shanghai, China
*Corresponding Author's Email: diaoxuling@icrd.com.cn

## ABSTRACT

Root cause tracing of random alarm is one of the most critical challenges at advanced nodes in manufacturing. As device geometries shrink, process become increasingly sensitive to small differences in chamber environment and performance. The worth of finding root cause for alarm cases is effectively controlling sources of variability that impact on-wafer results, more importantly, bringing the payoff of higher device yield and improved tool stability. In this work, a hybrid solution is proposed for random alarm of etch tool, our method merges supervised and unsupervised way to hand the high dimensional problem, dimension reduction and model training are employed for chamber matching and analysis, convinced results are obtained compared with the traditional methods.

Keywords: root cause tracing; chamber matching (CM); machine learning; principal component analysis (PCA); random forest (RF);

## INTRODUCTION

The manufacture of each semiconductor components products requires hundreds of processes. After sorting, the entire manufacturing process is divided into eight steps: Wafer Processing, Oxidation, Photography, Etching, Film Deposition, Interconnection, Test, and Package. The etching process is one of the most important processes. After the photolithography of the circuit diagram is completed on the wafer, an etching process is used to remove any excess oxide film and only the semiconductor circuit diagram is left. To do this, liquid, gas or plasma is used to remove the unselected parts. There are two main etching methods, depending on the material used: wet etching that uses a specific chemical solution for chemical reaction to remove the oxide film, and dry etching that uses gas or plasma.

Because of the semiconductor wafer fabrication's long process and hundreds of complex chemical steps, it needs to monitor a large number of process parameters [1]. With the advancement of digitization in the semiconductor industry, Artificial Intelligence (AI) can use algorithms such as machine learning and neural network for wafer defect detection and classification[2,3], optical measurement, chip manufacturing and modeling, photoresist contour prediction[4], semiconductor production result prediction[5,6], wafer process control and monitoring[7], root cause tracing, so as to optimize the manufacturing process and improve the production efficiency.

Root cause tracing is a popular and commonly used technique that helps people answer why a problem occurs first. It tries to use a specific set of steps and related tools to determine the root cause of the problem to find the main cause of the problem so that you can determine what happened, determine why it happened, and figure out how to reduce the likelihood of recurrence.

Root cause tracing of alarm cases at advanced nodes in 300mm Fab is very important, such as the fault detection and classification (FDC) function module has been taken as a necessary part in world-class engineering and manufacturing, field responsiveness, dynamic control loops and automatic monitoring are becoming more essential as tolerances shrink. FDC transforms sensor data into summary statistics and models that can be analyzed against user defined limits to identify process excursions.

In the field of semiconductor manufacturing, process and equipment engineers are often asked to quickly find the root cause of yield problem or a chamber deterioration with a combination of experience, instinct and a mass of sensor data. However, the memory of the existing FDC is too small to monitor all processes, and its data analysis method is relatively simple to meet the needs of engineers. In order to solve this problem, scholars in the industry are using machine learning methods to explore more effective, comprehensive and fast solutions.

In this work, a hybrid method is proposed for root cause tracing of random alarm cases on etch tool. The method adopts an unsupervised way to reduce dimension of sensor data and to find the mismatches between chambers or wafers, and then drill down the key sensor which cause the deviation. On the other hand, based on the golden wafer information, a supervised method of random forest could be applied for sorting sensors, and directly find out the key sensors of alarm cases.

## METHOD

In this section, we present the details of the proposed model for root cause tracing of random alarm cases on etch tool. First, the sensor data from FDC is cleaned and the statistical features are extracted for the subsequent model analysis. Second, we provide a detailed account of the proposed model and its ability to effectively use both supervised and unsupervised method.

### Data preprocessing

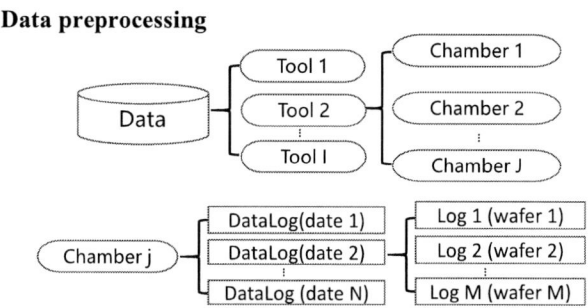

*Fig 1: Data cleaning of etch tool.*

The processing data of etch tool is stored by fault

detection and classification module, as well as the tool itself. The data include tool ID, chamber ID, recipes, lot ID, wafer ID, wafer's sensor data such as chamber's temperature, pressure, gas flow in etching process, and alarm wafer ID from FDC.

The enormous data should be cleaned and extract useful feature before feed to algorithm models. The clean flow is shown in Figure 1, the data has been recorded according to tool type, chamber ID. The main work is to arrange each datalog to each wafer based on different recipes, then the problem shooting is based on wafer level. At the same time, due to the large number of sensors and the long time series data it receives, we use the statistical features [8] such as mean, median, minimum, maximum, standard deviation (SD), kurtosis and skewness to reduce the dimension.

**The framework of hybrid solution**

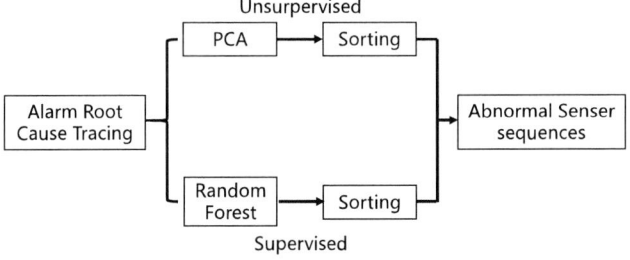

*Fig 2: The schematic of hybrid solution.*

To trace the root cause of random alarm of etch tool, the target is to fix the problematic sensor, or sort them when there are several suspicious sensors. The traditional way is to check the FDC data or depend on intrinsic and experience, however, at advanced nodes, the number of sensors continuously increases as well as the complicity of alarm cases, which poses severe challenges for the solutions, engineers and Fabs.

The proposed hybrid solution has been shown in Figure 2. Two methods have been combined for root cause tracing, both of which are suitable for high dimension data processing. One is principal component analysis (PCA), which is used for reducing the dimension of data, and finding out the main features to distinguish the 'good' and 'bad' wafer; the other is random forest (RF), which is constructed by decision trees, and after trained with labeled dataset, the importance of the leaves (sensors) could be indicated by their weights.

**Unsupervised PCA**

PCA [9], also known as principal component analysis or principal component regression analysis, is an unsupervised data dimensionality reduction method. Firstly, the data is transformed into a new coordinate system by linear transformation; Then, using the idea of dimension reduction, the first variance of any data projection is on the first coordinate (called the first principal component, PC1) and the second variance is on the second coordinate (the second principal component, PC2). This idea of dimensionality reduction first reduces the dimension of the data set, while maintaining the feature that the opposite difference of the data set contributes the most, and finally makes the data appear intuitively in the two-dimensional coordinate system.

As mentioned earlier, the data monitored by FDC is limited and the monitoring method is very simple, so it may not cover all process exceptions. We can find the potential abnormal wafer through PCA visualization. The figure of PC1 and PC2 can 'tell' us the 'bad' wafers' wafer ID. Then use RF to quickly find the abnormal sensor.

**Supervised RF**

RF [10] is an algorithm that integrates multiple trees through the idea of ensemble learning. Its basic unit is the decision tree, and its essence belongs to a major branch of machine learning- ensemble learning method. From an intuitive point of view, each decision tree is a classifier (assuming that it is aimed at the classification problem now), then for an input sample, n trees will have n classification results. While the RF integrates all classification voting results and specifies the category with the most votes as the final output.

In the RF model, the importance ranking is calculated based on the degree of chaos (impulse / Gini coefficient), that is, the standard to measure the importance of a feature is to see how much the degree of chaos is reduced in the process of building a random forest through the decision tree. After integrating all trees, the greater the average reduction is, the more important the feature is determined to be.

Based on the known FDC alarm information or abnormal phenomena found through PCA, we train an RF model with good classification accuracy, obtain the importance ranking of features through such a model, and then calculate the importance of sensors based on features, that is, the anomaly degree of sensors.

## EXPERIMENT

We use actual data from a semiconductor factory to verify our method, the test results are shown in Figure 3. Some experimental details should be elaborated. First, the alarm case chosen is a random alarm case, and abnormal data could be clearly identified. The PCA method is used to compare all wafers' performance, and then RF method is used to trace the key sensors.

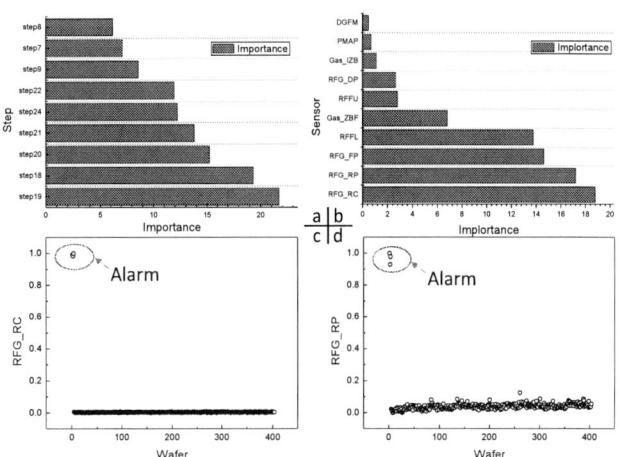

*Fig 3: The results of root cause tracing are shown in (a) and (b)，the (c) and (d) reveal the sensor performance on all the test wafers.*

As shown in Figure 3 (a) and (b), the 'importance' sensors are sorted, and the the most key step and sensor are 'step19' and 'RFG_RC'. And the other sensors are also meaningful, and because of the complicity of the etching chambers, the most key sensor may not have a direct correlation with the alarm, or it is an indirect value (affected by many status), even it is not a precise value due to the sensor's shift of responsivity. And then the other parameters could offer move choices for equipment engineers.

To verify the output of our method, the performance of the sensors 'RFG_RC' and 'RFG_RP' are checked on each test wafer, as shown in Figure 3 (c) and (d). From the plots, the breakpoints could be easily found out, which are consistent with the (a) and (b), more importantly, comparing (c) and (d), we find the that the sensor 'RFG_RC' is more representative because the its baseline is much flatter than (d), and this point is also very consistent with the sensor sorting shown in (b).

## CONCLUSION

To summarize, the designed hybrid method for root cause tracing has gained convinced results, which not only sorts the sensors, but also give a comprehensive view for analysis. The results present all the potential correlated sensors together, which makes the trouble shooting easier. However, the results are data and model based, and traditionally should be verified by equipment engineer and pi-run process, etc., and then put it into production use.

## ACKNOWLEDGEMENTS

This paper and the research behind it would not have been possible without the exceptional support of Shanghai integrated circuits research & design Center (ICRD). Thanks for my mentor, Tao Zhou. His enthusiasm, knowledge and exacting attention to detail have been an inspiration and kept my work on track. In addition, I would also like to thank Chen Li, Yiyi Jiang, Shiyuan Wen, Xuelong Shi and Quan Jing for providing technical scenarios and data support for my research. The generosity and expertise of one and all have improved this study in innumerable ways and saved me from many errors; those that inevitably remain are entirely my own responsibility.

## REFERENCE

[1] Chien, Chen-Fu, Kuo-Hao Chang, and Wen-Chih Wang. *An empirical study of design-of-experiment data mining for yield-loss diagnosis for semiconductor manufacturing.* Journal of Intelligent Manufacturing 25.5 (2014): 961-972.

[2] Jin, Cheng Hao, et al. *Wafer map defect pattern classification based on convolutional neural network features and error-correcting output codes.* Journal of Intelligent Manufacturing 31.8 (2020): 1861-1875.

[3] Jin, Cheng Hao, et al. *A novel DBSCAN-based defect pattern detection and classification framework for wafer bin map.* IEEE Transactions on Semiconductor Manufacturing 32.3 (2019): 286-292.

[4] Wei, Chih-I., et al. *Better prediction on patterning failure mode with hotspot aware OPC modeling.* Metrology, Inspection, and Process Control for Semiconductor Manufacturing XXXV. Vol. 11611. International Society for Optics and Photonics, 2021.

[5] Kim, Dasol, Mintae Kim, and Wooju Kim. *Wafer Edge Yield Prediction Using a Combined Long Short-Term Memory and Feed-Forward Neural Network Model for Semiconductor Manufacturing.* IEEE Access 8 (2020): 215125-215132.

[6] D. Sharma, H. Armer, and J. Moyne, *A comparison of data mining methods for yield modeling, chamber matching and virtual metrology applications,* 2012 SEMI Advanced Semiconductor Manufacturing Conference.

[7] F. Dennis, *Enhanced process metrology using plasma parameters in FDC,* Solid State Technology, vol.51, no 6, 2008.

[8] T. H. Pan, D. S. –H. Wong, S. –S. Jiang, *Chamber Matching of Semiconductor Manufacturing Process Using Statistical Analysis,* IEEE Transactions on Systems, Man, and Cybernetics—Part C: Applications and Reviews, vol. 42, no 4, 2012.

[9] Wold, Svante, Kim Esbensen, and Paul Geladi. *Principal component analysis.* Chemometrics and intelligent laboratory systems 2.1-3 (1987): 37-52.

[10] Svetnik, Vladimir, et al. *Random forest: a classification and regression tool for compound classification and QSAR modeling.* Journal of chemical information and computer sciences 43.6 (2003): 1947-1958.

# THE TYPE AND SOLUTION OF INLINE POLY RESIDUE DEFECT FOR 28 HK PROCESS IMPROVEMENT

*Min Wang [1], Hunglin Chen [1], Kai Wang[1], Yin Long[1], and Hao Guo [1]*

[1] Shanghai Huali Microelectronics Corporation

Shanghai, China

Author's Email: Wangmin_E1@hlmc.cn

## ABSTRACT

In the high-k advanced semiconductor process, we always use new metals for example AL to replace the NMOS and PMOS poly silicon gate electrodes. The systematic defect of poly residue is a key defect with higher killer ratio in HK process, which will cause chip yield loss or reliability issue. We found that the poly residue closely related to a key parameter of horn height, which is effected by EB1 and EB2 process. More interesting, these processes are related to each other during PREB~ILD0 loop, the EB1 and EB2 process determined the height of horn, and the CMP process is mainly influenced by the height of the horn. Whether the horn is too high or too low will all cause CMP process loading, and finally result in poly residue defect. Moreover, the PR loading on the different poly CD, leading to the large patch or high poly density place is easy to suffer poly residue defect. Meanwhile, the weak point of CMP process is another factor, which also can easily cause poly residue defect. On the basis of this reason, the root cause of these defects was very intricately. In this study, we discussed every fail model of poly residue defect in depth, and provide effective improvement actions.

**Keywords:** 28 HK technology node, defect inspection, poly residue, PREB~ILD0 CMP, yield improvement

## INTRODUCTION

In the high-K advanced semiconductor process, we always use the new materials for example AL to replace NMOS and PMOS poly silicon gate electrodes[1-2]. Nevertheless, with the development of semiconductor process, the critical dimension (CD) of the poly gate is shrinking continually, and the aspect ratio (AR) of the poly gate is still increasing[3-4]. Thus, it is a big challenge for removing the poly silicon in the poly trench completely, and then depositing AL fully in the poly gate in the 28 or 22 nm technology-node. The whole poly gate remove process contains five main steps: PREB photo, EB1 etch, EB2 etch, ILD0 CMP, Dummy poly remove (etch and wet). In this process flow, the major killer defect is poly residue, which will block the AL deposition and finally caused AL missing in the poly gate.

We found that the poly residue defect closely related to the key parameters of horn height and uniformity, which is effected by EB1 and EB2 process. Whether the horn is too low or too high, it will all cause the poly residue defect.

Meanwhile, the ILD0 CMP process is also affected by the uniformity of the horn. Therefore, each process step is critical and can effect or limit the window of the next step, thus causing a chain reaction during this loop and forming serious defect problems after process integration. What's more, the ILD0 CMP process itself also can generated poly residue defect, due to the weak stability and uniformity of the remove rate in wafer edge [3-4].

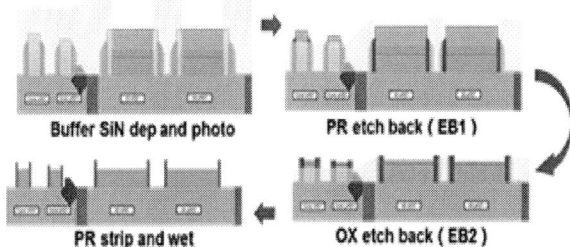

Figure1.The process flow of EB1 and EB2.

In this work, we investigated the affect factors of the poly residue defect comprehensively. The result showed that the key parameter of horn height is mainly affected by the EB1 and EB2 steps, and the uniformity of the next ILD0 CMP process is also influenced by the height of horn. Thus, the four processes of PREB~ILD0 CMP loop are not independent of each other, but closely interrelated to each other, we make a comprehensive and systematic adjustment for this loop. Moreover, the poly residue defects' failure mode and formation mechanism were discussed deeply to optimize the process conditions.

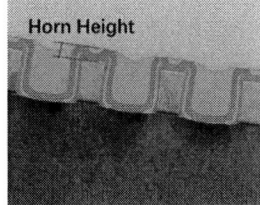

*Figure2. The image of horn height.*

## EXPERIMENTS AND RESULTS

### The EB1 and EB2 process

The EB1 and EB2 steps are crucial step in this loop, and the height of horn is primarily affected by these two processes. Meanwhile, the process window of PREB loop

978-1-6654-9759-6/22 $31.00 © 2022 IEEE

is very narrow. As we known, the EB1 only etch the PR, and EB2 can etch PR, oxide and SIN. As shown in Figure 4, the thickness of PR is thickened gradually with the increasing of the poly CD and poly density, and the thick PR will block etch, then formed Poly residue after dummy poly remove. So, the region of large poly CD or high poly density is the weak point that will result in the poly residue defect. Meanwhile, the PR thickness is thin in the region of ISO, which is easily causing NISI damage. Whether the poly residue or NISI damage is all key defect, and thus the process window is especially narrow during the PREB loop.

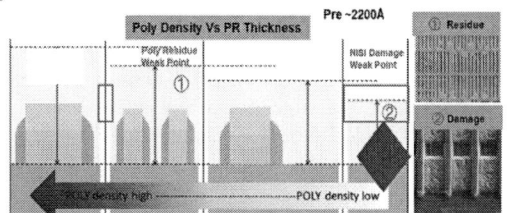

*Figure 3: the schematic diagram of the PREB process window*

### The height of horn

When the height of horn is too low, it is easy to suffer poly residue. The gate high of PMOS is lower than NMOS because of the SIEG process, which etch the partial poly of PMOS. Moreover, the EB2 process not only can etch PR, SIN and oxide, but also can etch poly. When the process time of EB2 is too long, it will etch the poly, and while the etching rate of SIN and oxide is higher than poly. Thus, the profile of poly gate is like a little hill, as shown in Figure 4, and the horn is negative. Furthermore, if the gate high of PMOS poly is less than the spec of ILD0 CMP (Figure 4 C), the SIN above the PMOS poly will not be polished by CMP process. Therefore, the SIN on the PMOS poly will block the dummy poly etch, which inevitably result in the poly residue defect. Because of the special formation mechanism of defect, this type poly residue have an obvious characteristic that the defect only occurred on PMOS, and the poly width of PMOS is always narrower than NMOS.

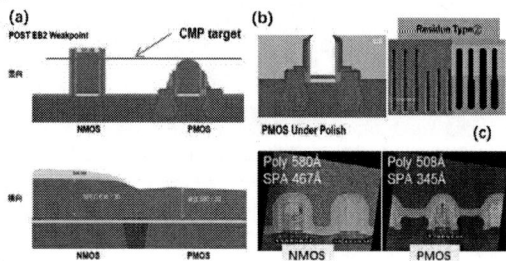

*Figure 4:(a)The schematic diagram of PMOS and NMOS after EB2 over etching. (b) The defect image (c)The TEM image of PMOS and NMOS profile after EB2.*

When the height of horn is out of spec, it is also easy to suffer poly residue. As shown in Figure 5, the inline CESL OCD is 40 A higher than baseline, there is oxide residue or SIN residue remaining on poly gate after ILD0 CMP, which will block dummy poly remove, and thus forming poly residue. The higher inline horn always caused by the lower SIN etching rate after EB2 tool PM, as displayer in Figure 4c. Thus, it is critical to keep inline horn height stable.

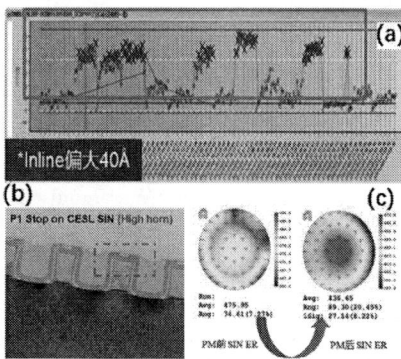

*Figure 5:(a)The inline chart of CESL OCD, (b)The TEM image of over high horn, (c) The EB2 SIN etch rate before and after PM.*

### The PR loading poly residue defect

The PR loading defect is caused by the different PR thickness on the different Poly CD and Poly density. The region of large poly CD have more thicker PR, and thus it is easy to suffer SIN or oxide remaining with the same EB1 and EB2 time. Based on this reason, the desige rule stipulate that the PR on the region with poly CD more than 330 nm will open by PREB photo. However, some products have the poly CD around 330 nm, and these region are easy to suffer poly residue due to the thicker PR. As shown in Figure 6, the region of poly CD 329 nm suffer poly residue, and all the defect in the same die area. This type poly residue defect can be solved by extending EB1 time, and reducing PR remeaing thinkness after EB1 (Figure 7).

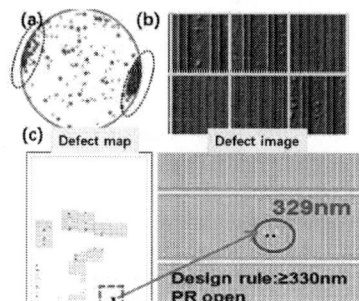

*Figure 6: (a) The defect map (b) The defect image of poly residue (c) The stack die and JDV of poly residue defect.*

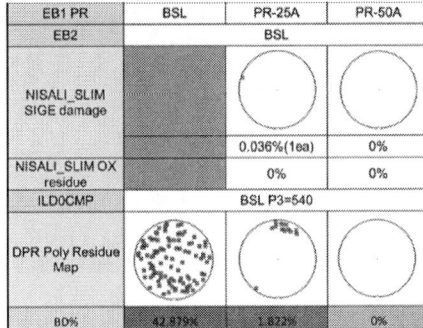

| EB1 PR | BSL | PR-25A | PR-50A |
|---|---|---|---|
| EB2 | BSL | | |
| NISALI_SLIM SIGE damage | | ◯ | ◯ |
| | | 0.036%(1ea) | 0% |
| NISALI_SLIM OX residue | | 0% | 0% |
| ILD0CMP | BSL P3=540 | | |
| DPR Poly Residue Map | | ◯ | ◯ |
| BD% | 42.879% | 1.822% | 0% |

*Figure 7: The EB1experiment split defect data*

**ILD0 CMP process condition**

The ILD0 CMP process is also a crucial step in the loop, it must polish oxide and SIN on the poly. As displayed in Figure7, if the wafer edge remove rate of the head is low, the offline thickness will be high in the edge. Meanwhile, because of the no polished SIN on the poly, the inline poly gate high will also out of spec, and finally, the wafer edge suffer poly residue after dummy poly remove. The remove rate of the polish head always decrease, especially the new head. Thus, for the new head of CMP tool, we usually polish 200 PCS dummy wafer before using in the product wafer. Meanwhile, in order to cover the new head remove rate low problem, we also use a look up table system to put more pressure in the new head.

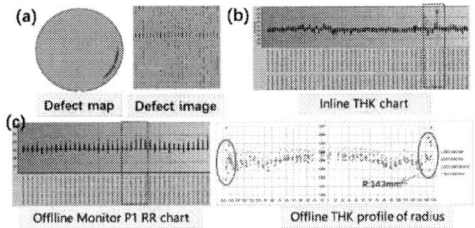

*Figure 7: (a) The defect map and defect image of poly residue (b) The inline THK chart (c) The offline monitor P1 RR chart and offline THK profile of radius*

## CONCLUSION

In this work, the systematic poly residue defects were comprehensively investigated, which could be monitored by the novel BFI inspection tool.

The various failure mode and the corresponding formation mechanism of poly residue are discussed in detail, and the resulted showed that each process step is critical and can effect or limit the window of the next step during the PREB~DPR loop. The key parameter of horn height is determined by the EB1 etch and EB2 etch, and whether the horn is too low or too high, it will all cause the poly residue defect. The uniformity of horn height in the wafer also influences the ILD0 CMP profile. What's more, The ILD0 CMP only can control one PAD poly gate high,

and thus the uniformity of horn height in different poly CD is also very important to keep stable. Finally, the corresponding improvement actions to removal poly residue were executed.

## REFERENCES

[1] W.C. Shen, C. Y. Mei, Y.D. Chih, High-K metal gate contact RRAM (CRRAM) in pure 28nm CMOS logic process. Electron Devices Meeting. IEEE, 2012:31.6.1 - 31.6.4.,

[2] P.Lan, J.Y.Yang, G. Chen, High-K metal gate contact process optimization for yield improvement via innovative defect inspection technique. 2014, pp: 405-407.

[3] M.S Kuo, H. Chen, R.W. Fan, *Materials Science in Semiconductor Processing,* vol. 20, 2014, pp. 17-22.

[4] X.P. Wang, Y. Huang, Q.H. Han, Dry Etching Solutions to Contact Hole Profile Optimization for Advanced Logic Technologies. 2012, pp: 351-355.

[5] B. Egan, H.J. Kim, *ECS Journal of Solid State Science and Technology,* vol. 18, 2019, pp: 3206-3211.

[6] H. Lee. D. Lee. H. Jeong, International journal of precision engineering and manufacturing, vol. 17, 2016, pp: 525`-536.

# STUDY OF THE DEGRADATION IN LDMOS WITH STI TECHNOLOGY AND IMPROVE THE RELIABILITY WITH SEVERAL METHODS

*Xiaoming Zhang[1,2]\*, Donghua Liu[2], and Wensheng Qian[2]*

[1]School of Information Science and Technology, ShanghaiTech University, Shanghai 201210, China
[2] HuaHong Grace Semiconductor Manufacturing Corporation, Shanghai 201206, China
*Corresponding Author's Email: Xiaoming.Zhang@hhgrace.com

## ABSTRACT

The degradation in lateral double-diffused MOS (LDMOS) is increased obviously with the higher operation drain voltage and the shrink of technology node. Therefore, it is important to study the damage of degradation, which is caused by the long- term hot carrier stress (HCS). Recently, the TCAD tools have already used the degradation model with some additional parameters in device electronic simulation. In this paper, we will determine the location of impact ionization and discuss the mechanism of degradation in LDMOS with STI technology. According to different gate voltage and drain voltage stress conditions, the key degradation region influenced the reliability of devices is deduced. This paper also discusses several methods to improve the degradation of the device. These methods are useful to improve the degradation of the device, which are proved by TCAD simulation results.

## INTRODUCTION

Typical mixed-signal applications, which are in automotive, power management and etc. require high voltage capability and fast switch velocity. High voltage lateral double-diffused MOS (LDMOS) are widely used in these applications to meet requirements. Many studies have already focused on the optimization of breakdown voltage and on-resistance. However, with the scaling of technology node and increase of operation voltage, the electrical fields and impact ionization in LDMOS are obviously increased. Hence, it is necessary to analyze the hot–carrier stress (HCS) and optimize the degradation in LDMOS. The HCS model is based on the numerical solution of the full-band-structure Boltzmann transport equation (BTE) [1-2]. The essential physical effects have been concerned in the model, which were the single- and multiple-particle mechanisms in Si-H bond dissociation [3], the self-heating in lattice thermally driven increase of the degradation [4] and the localization of damage in the interface of Silicon and Oxide [5]. In this paper, the Synopsys TCAD tool has been used to simulate the degradation in LDMOS with the lucky-electron model. The lucky-electron model has been applied for many years in standard CMOS to describe the degradation problems [6]. In recent years, several advanced models [7] have proposed to adapt the low voltage operating region for standard CMOS. However, the lucky-electron can be suitable in LDMOS device with the high voltage operation region.

This paper is organized as follows. In Section II, the device structure and its stress degradation under different conditions are presented. In Section III, several methods to improve the degradation and the simulation and compared results are presented.

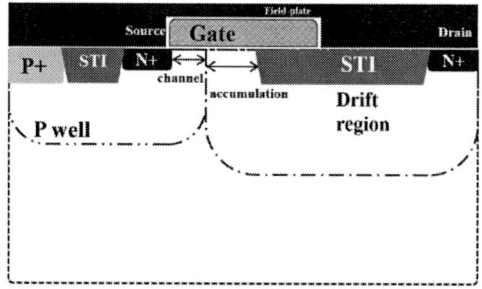

*Figure 1: Cross-section of the STI based LDMOS device. The main geometrical features are reported.*

## DEVICE STRUCTURE AND DEGRADATION RESULTS

Fig. 1 shows a 2-D sketch of the n channel LDMOS device, which is processed in a 0.18μm CMOS technology. With the use of the RESURF and the STI technology, the device can protect very large drain voltages. The gate oxide thickness is 20 nm. The whole device is embedded in a low doped layer to connect with other devices. The off-state breakdown voltage $V_{BV}$= 95V, the threshold voltage $V_t$= 0.7V, the on-resistance $R_{on}$= 98.8 mOhm $*mm^2$, the above being the basic transistor parameters. The channel, accumulation and drift regions are indicated in Fig. 1.

Base on the device, the degradation could be analyzed. In switching applications, devices are always on a state with high $V_{DS}$ / low $V_{GS}$ and a state with high $V_{GS}$ / low $V_{DS}$. Therefore, the simulation would be carried out with the attention to high $V_{DS}$ and $V_{GS}$, respectively. When the device is stressed in the on-state, the hot carriers flow through the drift region from drain to source and accelerate at the high electric field regions. These hot carriers can cause impact ionization to generate hot holes and hot electrons which can damage the device. In the meanwhile, the carriers can be trapped into the oxide and have enough energy to damage the Si-Oxide layer and

978-1-6654-9759-6/22 $31.00 © 2022 IEEE

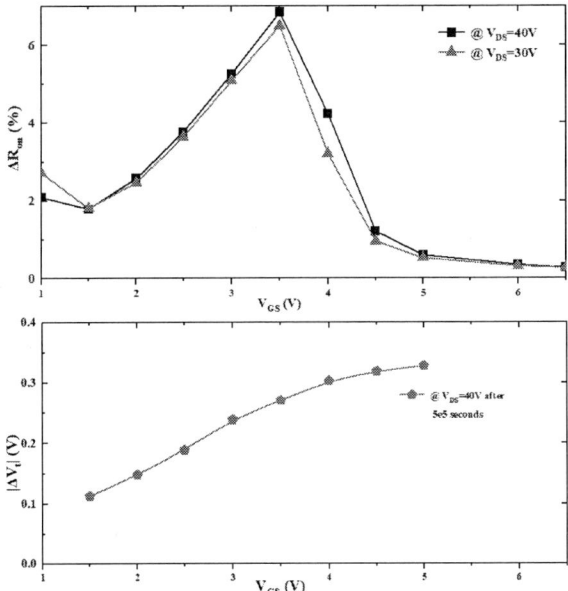

*Figure 2: (Top)$\Delta R_{on}$ and (bottom) $\Delta V_t$ absolute value measured in the device as functions of the stress $V_{GS}$ vs. $V_{DS}$=20V/30V/40V with long stress time.*

generate new interface states. As a result, the TCAD tools have been carried out using the lucky-electron model coupled with the hole equation, electron equation and the lattice temperature equation to monitor the on-resistance, which the most relevant quantities to reflect the degradation influence, and the threshold voltage. The measured quantities are defined as

$$\Delta R_{on} = \frac{R_{on} - R_{on,ref}}{R_{on,ref}} \quad (1)$$

where $R_{on,ref}$ is the on-resistance of the fresh device and $R_{on}$ is the on-resistance measured at stress time, while the shift of the threshold voltage is $\Delta V_t = V_t - V_t'$.

In Fig. 2, $\Delta R_{on}$ and $\Delta V_t$ measured on the above device are indicated as function of the stress $V_{GS}$ with a fixed stress $V_{DS}$=20V/30V/40V and time t=5e5 seconds. The high electric field nearby the source side of STI corner is the main cause of the degradation. The energy which supplied by electric field is enough for hot carriers to generate large density electrons and holes. The large density of highly energetic electrons flow through the accumulation region leads to the electron trapping. The trapping of electrons leads to the large shifts of the drain current in the on-state condition. There is a lower on-resistance degradation taking place at $V_{GS} \approx 1.5V$. It is main cause that the electric field is a bit lower than those at $V_{GS}$=1V/2V. With the further increasing of the stress $V_{GS}$, the electric field at the accumulation region is reduced obviously. On the contrary, it is increased at the drift

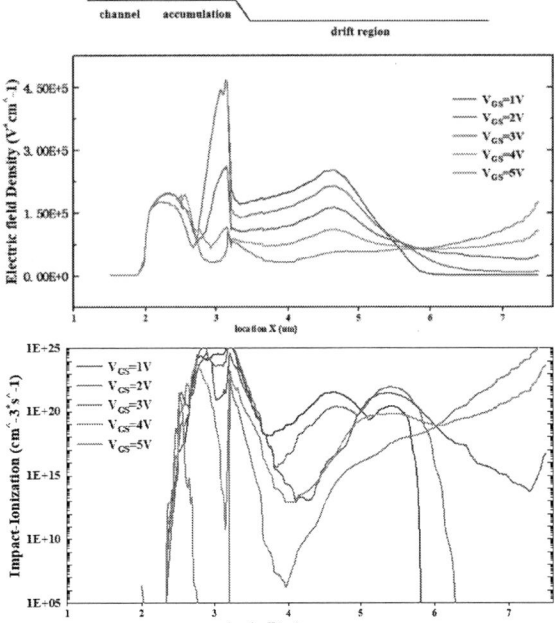

*Figure 3: Cutline of (top) electric field density and (bottom) impact ionization along the Si-Oxide interface at $V_{DS}$=40V and different $V_{GS}$ after long stress time.*

region indicated in Fig. 3, leading to a higher impact ionization taken place below the STI interface. It causes a less severe degradation of on-resistance at $V_{GS} \approx 5V$. When $V_{GS} \geq 5V$, the degradation is not only ascribed to the high electric field at the STI source side corner, but also the trapping along the drift region nearby drain contact. In the meanwhile, the trapping nearby drain contact has little effect on drain current. Therefore, the degradation will be at a lower value when the $V_{GS} \geq 5V$. It is clear indication that the degradation within channel region is not obvious which measures $\Delta V_t$ at different $V_{GS}$ in Fig. 2. Although the electric field is high which is due to high $V_{GS}$, the longitudinal electric field in the channel is low. It is crucial to prevent the hot-carrier induced in channel region.

## IMPROVE THE DEGRADATION

From the analysis of the above simulation results, two different methods can improve the degradation in LDMOS. The first on is to change the current path. It will effectively improve the state to make hot-carriers no longer flow next to the Si-Oxide interface. The second one is to reduce the electric field or longitudinal electric field at the accumulation region.

### Modulate doping concentration to change current path

Aim to the implement the first method, it is useful to change the doping in the drift region. In 0.18μm LDMOS technology, there are multiple implantations in drift region

978-1-6654-9759-6/22 $31.00 © 2022 IEEE

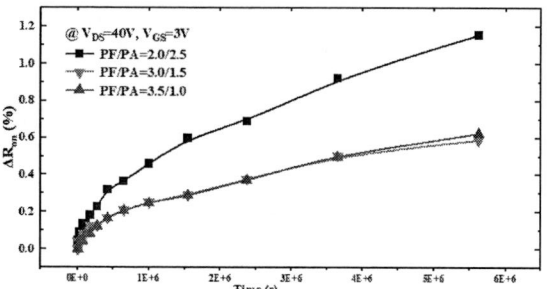

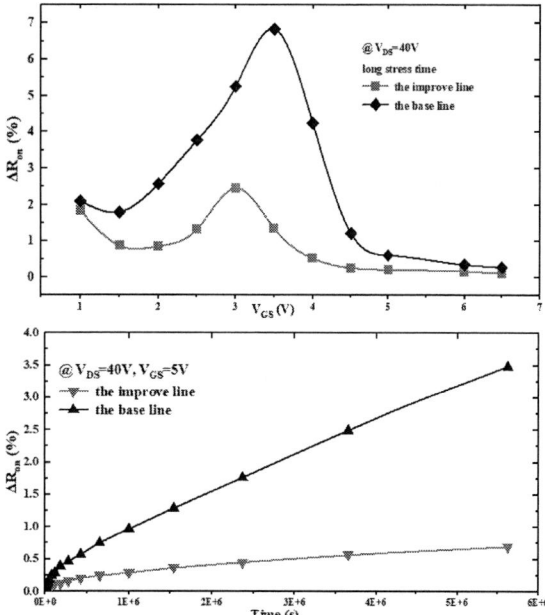

*Figure 5: $\Delta R_{on}$ measured in the device with different ratio of PF and PA as function of the stress time with $V_{DS}=40V$, $V_{GS}=3V$.*

When the field plate terminal is close to drain, the effect of balancing the electric field is weakened gradually. So, the degradation would not decrease when the width of PF is large shown in Fig .5 that the blue line and the red line almost coincide. Obviously, it has good effect at lower $V_{GS}$ regime due to the change at the accumulation region.

## CONCLUSION

The proposed analysis of the degradation in LDMOS with STI technology is based on the use of Synopsys TCAD tools. It was shown that the lucky-electron model can be applied to high voltage switching devices. In the meanwhile, the determinants of degradation are different at various $V_{GS}$. Aim to different $V_{GS}$ states, two methods are provided to improve the degradation at the drift region and accumulation region, respectively. Based on the simulation and comparison results, it has great effect to get a less severe degradation.

## REFERENCES

[1] S. Jin, A. Wettstein, W. Choi, F. Bufler, and E. Lyumkis, in Proc. Int. Conf. Simul. Semicond. Process. Devices, 2009, pp. 1–4.

[2] S. Tyaginov, I. Starkov, C. Jungemann, H. Enichlmair, J. M. Park, and T. Grasser, in Proc. ESSDERC, 2011, pp. 151–154.

[3] A. Bravaix and C. Guerin, in Proc. 47th Annu. Int. Rel. Phys. Symp., 2009, pp. 531–548.

[4] J. W. McPherson, R. B. Khamankar, and A. Shanware, J. Appl. Phys., vol. 88, no. 9, pp. 5351–5359, Nov. 2000

[5] S. Reggiani, S. Poli, E. Gnani, A. Gnudi, G. Baccarani, M. Denison, S. Pendharkar, R. Wise, and S. Seetharaman, in Proc. 48th Annu. Int. Rel. Phys. Symp., 2010, pp. 881–886.

[6] C. Hu, S.C. Tam, F-C. Hsu, P-K. Ko, T-Y. Chan, K.W. Terril, IEEE. Trans. El. Dev., vol. ED-32, pp. 375-385, Feb. 1985.

[7] C. Guerin, V. Huard, A. Bravaix, IEEE. Trans. Dev. Mat. Rel., vol.7, pp. 225-235, Jun 2007.

*Figure 4: The comparison of $\Delta R_{on}$ measured in the device as functions of the stress (top) $V_{GS}$ with $V_{DS}=40V$ and the stress (bottom) Time with $V_{DS}=40V$, $V_{GS}=5V$.*

to make the doping well-distributed. Therefore, it makes higher doping concentration near the Si-Oxide interface to form the current path. No additional mask is required, a slight boron implantation can be carried out to modulate the doping concentration below the interface after P well ion implantation and before filling the shallow trench. The simulation results which compare with the initial device are indicated in Fig. 4. It is shown that the method can improve the degradation. Obviously, it has good effect at high $V_{GS}$ regime due to the change in drift region.

### Modulate the electric field at the accumulation region

When the device is on the low $V_{GS}$ state, the main cause of the degradation is the electric field at the accumulation region. Aim to this operation regime, it is useful to modulate the electric field to improve the degradation. In 0.18μm technology, the field plate is used to balance the electric field peak to obtain the higher breakdown voltage. Therefore, it can be used to modulate the electric field at the accumulation region as well. In Fig. 1, the width of the STI region covered with field plate is defined as PF and the width of the STI region without field plate is defined as PA. The whole width of PF and PA is fixed. After modulating, the electric field peak will appear below the field plate terminal. Therefore, the electric field at the accumulation region will be decreased with the increase of PF width. The simulation results are shown in Fig. 5. With the increase of PF width, the electric field has been modulated, leading to a lower sever degradation.

# INTEGRATION STRATEGY ON LOW-COST CHIP-FIRST FAN-OUT PANEL LEVEL PACKAGING

*Cheng-Tar Wu\*, Junbo Jiang, Mengqiang Li, Chen Xiang, and Minghao Shen*

Chengdu ESWIN System IC Co., Ltd.

Chengdu, Sichuan Province, China

\*Corresponding Author's Email: terry.wu@eswin.com

## ABSTRACT

In the fan-out packaging (FOP), a higher lithography cost induced by the reconstituted die shift has yet been overcome, especially in the gradually increased carrier size in recent advance, e.g. panel-level-packaging (PLP). Our previous study demonstrates that with proper die shift compensation, the post-molding/grinding 3-sigma die shift value in X and Y for the whole panel with size of 510mm × 515mm are over 200um. The die shift data is collected and input into a software algorithm to synthesis a 99.8% photolithography yield within ±12um lithography overlay specification under an exposure field size that contains 4 × 4 die array (28.4mm x 28.4mm).

By confining the die shift induced by the mold flow and pre-shifting the die in consideration of the effective coefficient of thermal expansion of FOPLP, the post-molding/grinding die shift 3-sigma value in X and Y is then improved to 109um and 108um, respectively, rendering the exposure field size increasing to greater than 10 times (99.4mm x 99.4mm) than that of 4 × 4 die array, meanwhile maintain the lithography yield at a comparable level.

*Keywords—FOPLP; Die shift; Lithography yield; Exposure field.*

## INTRODUCTION

It is well known that fan-out packaging (FOP) has better electrical and thermal performance, as well as smaller form factor. Typical FOP can be categorized into chip-first face-down (CFFD), chip-first face-up (CFFU) and chip-last face-down (CLFD). Most of commercial packages, e.g. fan-in wafer level packaging (WLP) and PBGA, can be adopted to FOP, however, the incentive in production cost is not quite obvious, especially in CF approach.

In a CF process, the critical challenge is the die shift resulted from molding process and carrier expansion [1]. Therefore, a costly step-by-step exposure correction is required to remedy the shifted die, providing a good patterning yield. Though with proper die shift compensation ease the loading of the exposure tool, the site-by-site correction is inevitable.

In this study, CFFU approach (Fig. 1) is used to reconstitute chip on a 510mm × 515mm panel followed by redistribution layer (RDL) process. Fan-out panel-level packaging (FOPLP) has a larger carrier usage ratio in comparison with FOWLP. The larger the effective area, the longer the lithography process time is anticipated. Thus, an intention to maximize the exposure field size to increase the throughput for low-cost consideration.

Here, attempts to improve the post-molding/grinding die shift during die attach process are investigated. Furthermore, a stepper is used for die position mapping followed by data transferring to a software algorithm to synthesize the atop patterning overlay and yield at different exposure field size. With the integration of the improvement in the die attach process and overlay/yield prediction, offering a low-cost strategy that is suitable for FOPLP.

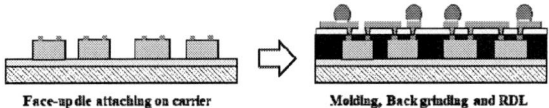

*Figure 1: Schematic of chip-first face-up process*

## EXPERIMENTAL

### Test Vehicle Chip

A bumped test vehicle (TV) chip of a package size of 7.1mm × 7.1mm was designed for the experiment. Fig. 2 (a) shows the pillar pattern of the TV. Fig. 2 (b) and 2 (c) are the alignment marks for the position mapping and overlay test, respectively.

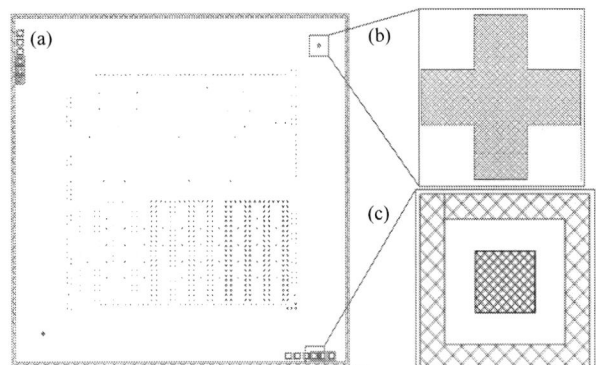

*Figure 2: Test vechicle layout of (a) die pattern, (b) alignment mark for mapping, and (c) pad (green mesh)/via (grey mesh) overlay test pattern*

## Reconstituted Panel Fabrication

CFFU with global alignment was initially adopted for the reconstitution process. Bumped dies were transferred from a diced wafer to a bonding adhesive pre-applied glass panel with a size of 510mm × 515mm followed by a compression molding process. The reconstituted panel was consisted of 4,100 units of TV chip. A grinding process were then applied to expose the pillar pattern. Die position was then collected by measuring the alignment marks via a corrected mapping system. Since the main contribution of the die shift comes from the shrinkage of the EMC, pre-shift and enhanced local alignment were then introduced for the die shift compensation.

## Mapping system for die shift measurement

A reliable mapping system is needed to provide a clear guidance for the following work on die shift compensation. To prevent tool basis, the stepper was the only mapping tool for the offset measuring.

Calibration method of mapping system was used the same approach as we reported elsewhere [2]. As seen from Fig. 3, the overlay error 3-sigma is within ±1.5μm, which is at a good level and consistent with our previous study.

The grinded panels were then fully characterized by the stepper of the die positions in each step.

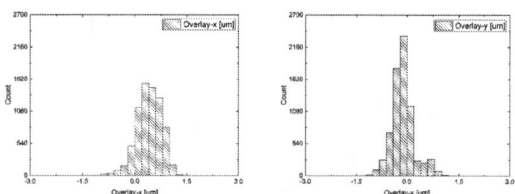

*Figure 3: Overlay error with mapping and compensation from the stepper*

## Simulation on Exposure Field Size

A universal lithography solution for overcoming the dramatic die shift in the FOP is to expose the atop pattern by die-by-die exposure, which has been proven for a high yield, but less productivity surely. For high-volume-manufacturing (HVM), throughput and lithography overlay yield need to be considered simultaneously. Fig. 4a shows the die layout of a whole panel. The simulation of the software algorithm was carried out to guide a reasonable design for the exposure field size that matches with HVM boundaries [3]. Fig. 4b shows the schematics of three different field sizes of the 2nd layer via pattern we applied to the software algorithm, covering 1 (1 × 1), 16 (4 × 4), and 196 (14 × 14) TV chip, as shown in Fig. 4b. Site-by-site alignment was used for the simulation, synthesizing the corresponding overlay. The software algorithm was then worked out the yield in each field size.

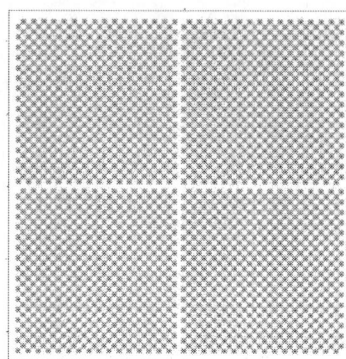

*Figure 4a: Full panel die layout*

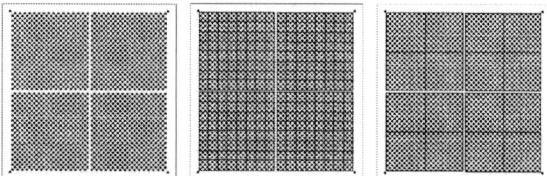

*1x1 die array    4x4 die array    14x14 die array*
*Figure 4b:  Die layouts with designated exposure field size*

## RESULTS AND DISCUSSION

### Die Attach Improvement in Reconstituted Process

Die shift compensation is a straightforward way to enhance the backend patterning yield in the FOP [4]. Here, few attempts were made to control the post-molding die shift in a good manner:

a) BKM: the baseline as described in the experimental that a global alignment with respect to the corner of the panel is used for die positioning.

b) Phase 1: on the basis of BKM, the effects of shrinkage/expansion of involved materials were calculated and compensated during the die attach process.

c) Phase 2: on top of the work in the phase1, a precise local alignment was added.

Table I summarizes the 3-sigma die shift in each panel at different stages. The die position was shifted dramatically away from its designated position over ±600μm in both X- and Y-direction and the θ rotation can up to 5mrad. Offset improvements to below ±190um was achieved while the mismatched of involved materials is considered. With the local alignment enhancement introduced in the phase 2, the within panel die shift was then improved to ca. ±100um, downing to one sixth of the value obtained from the BKM.

TABLE I.    COMPARISON OF DIE SHIFT 3-SIGMA UNDER BKM, PHASE 1, AND PHASE 2 CONDITIONS

| Die Attach Condition | Die Shit Performance | | |
|---|---|---|---|
| | Δx (um) | Δy (um) | (mrad) |
| BKM | 627 | 701 | 5.0 |
| Phase 1 | 186 | 187 | 3.8 |
| Phase 2 | 109 | 98 | 2.2 |

The full panel die shift in each phase is illustrated in Fig. 5. Fig. 5a, the BKM, shows the largest shift, concentrating at outer zone of the panel. This is because the center of the coordinate system for the reconstitution process was referred by the corner of a panel. The farther from the center, the larger shift impacted by the expansion of the carrier and the shrinkage of the molding compound was observed. Phase 1 has offset the contribution of material expansion/shrinkage for the process, reducing most shift, nevertheless near edge die gives a comparatively large shift, as shown in Fig. 5b. Local alignment further confined die shift, giving evenly distributed dies on the panel in the phase 2 (Fig 5c).

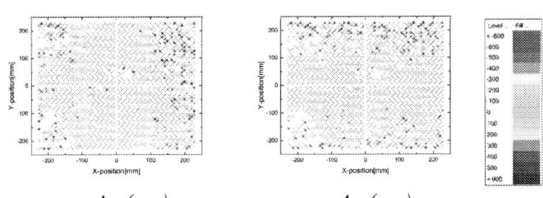

*Δx (um)*          *Δy (um)*

*Figure 5a: Die shift performance without compensation*

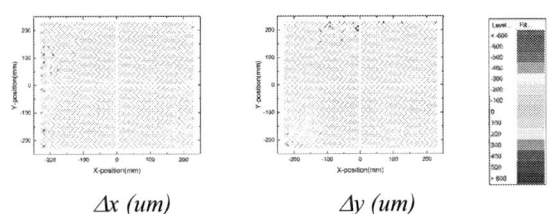

*Δx (um)*          *Δy (um)*

*Figure 5b: Die shift with compensation in phase 1*

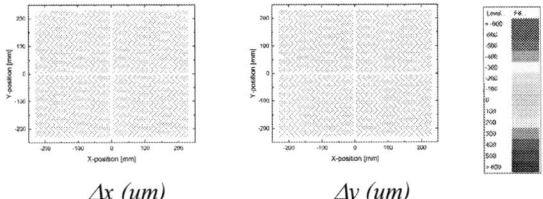

*Δx (um)*          *Δy (um)*

*Figure 5c: Die shift with compensation in phase 2*

**Lithography Overlay Prediction**

Table II summarizes the 3-sigma overlay at different phases while different exposure field sizes were implemented for yield prediction [5]. Both phase 1 and phase 2 did a great improvement in respect of overlay on a greater die array.

Comparing with the 4 × 4 die array exposing, the overlay obtained over phase 2 was improved to ca. 1.5 times than that of phase 1. Even on a larger exposing field size of 14 × 14 die array at the phase 2, the overlay in X and Y are 7.82um and 7.24um, respectively, which can be controlled at the same level of phase 1.

TABLE II.    COMPARISON OF OVERLAY ERROR 3-SIGMA (MICRON) OVER 1 × 1, 4 × 4, AND 14 × 14 DIE ARRAY ON STEPPER, WITH CONTINUOUS IMPROVEMENT IN DIE ATTACH PROCESS

| Overlay Error- Standard Deviation (um) | | | | | | |
|---|---|---|---|---|---|---|
| Die Array | BKM | | Phase 1 | | Phase 2 | |
| | X | Y | X | Y | X | Y |
| 1 × 1 | 1.49 | 1.52 | 1.52 | 1.49 | 1.51 | 1.51 |
| 4 × 4 | 27.68 | 23.61 | **7.52** | **7.24** | 4.80 | 4.79 |
| 14 × 14 | 37.25 | 36.35 | 12.35 | 13.47 | **7.82** | **7.24** |

We further analyze the combination of different field sizes in phase 2 as shown in Fig. 6, expressing that the larger the exposure field size, the less stable of lithography overlay is obtained. It is, however, a reasonable overlay standard needs to be set for evaluating the yield of lithography overlay efficiently.

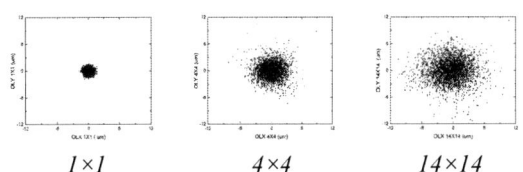

*1×1*          *4×4*          *14×14*

*Figure 6: Scatter maps of overlay error with different exposure field sizes in phase 2*

**Lithography Yield and Throughput Improvement**

Overlay yields are highly correlated with overlay specification as well as the exposure field size. Table III summarizes the prediction of the lithography overlay yields corresponding to different die array exposure fields and overlay specifications.

For HVM, the industrial overlay specification and yield are ±15μm and 99.5%, respectively. A tighten spec of smaller than ±12μm was used in this paper. It is obvious that die-by-die exposure can provide 100% yield even the tightest overlay spec is applied, which has the greatest number of exposure shots and is also time-consuming.

A 99.8% yield can be synthesized under a larger exposure field size of 4 × 4 die array when materials effect was well considered for die shift compensation, providing a practical result for HVM, but again, costly. A favorable result is obtained from the phase 2, the field size is increased to 14 × 14 die array (99.4mm × 99.4mm), meanwhile maintain the yield at 99.9% under the ±12μm overlay spec. This promising result suggests that we can expose with the fewest shots. Consequently, the throughput can be greatly improved to 12.2 times than that of 1 × 1 exposing, as seen in Fig. 7.

TABLE III.    OVERLAY YIELD PREDICTION UNDER DIFFERENT EXPOSURE FIELDS

| Overlay Yield with Various Exposure Field (%) | | | | |
|---|---|---|---|---|
| Conditions | Die Array | ±3µm | ±6µm | ±12µm |
| BKM | 1 × 1 | **100.0%** | **100.0%** | **100.0%** |
| | 4 × 4 | 5.7% | 29.0% | 73.6% |
| | 14 × 14 | 1.4% | 11.4% | 42.9% |
| Phase 1 | 1 × 1 | **100.0%** | **100.0%** | **100.0%** |
| | 4 × 4 | 52.5% | 92.7% | **99.8%** |
| | 14 × 14 | 21.8% | 64.5% | 97.7% |
| Phase 2 | 1 × 1 | **100.0%** | **100.0%** | **100.0%** |
| | 4 × 4 | 82.8% | 99.3% | **100.0%** |
| | 14 × 14 | 53.7% | 93.0% | **99.9%** |

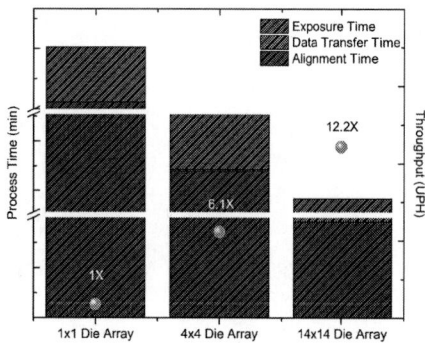

*Figure 7: Throughput under different exposure field sizes, whole process time is considered*

## CONCLUSION

An integration of die reconstitution, die position mapping and exposure field size simulation on CF FOPLP has been demonstrated. The continuous improvement in the reconstitution process has successfully compensated the die shift that induced by compression molding, giving evenly distributed dies over the entire panel. Moreover, the post-molding die shift is reduced significantly to ca. a hundred microns level within a 510mm × 515mm panel.

The software algorithm was also presented to simulate the lithography overlay and yield over corresponding exposure field sizes, preventing massively rework in multilayers lithography process. In addition to these, an optimized field size can be predicted and implemented that helps to improve the throughput greatly. This approach has the potential for lower cost of ownership and superimpose the inherent advantage of higher carrier usage ratio of panel, offering a promising strategy on low-cost FOPLP production.

## REFERENCES

[1] T. Braun et al., *"Challenges and opportunities for fan-out panel level packing"*, 2014 9th International Microsystems, Packaging, Assembly and Circuits Technology Conference (IMPACT), 2014, pp. 154-157

[2] J. Jiang, C-T. Wu et al., *"High Yield and High Throughput Lithography Solution for Emerging High Density Fan-Out Panel Level Packaging"* 2021 22nd International Conference on Electronic Packaging Technology (ICEPT), 2021, pp. 1-5.

[3] Hsieh, R. et. al., *"Lithography Challenges and Considerations for Emerging Fan-Out Wafer-level Packaging Applications"*, International Wafer-level Packaging Conference, 2009, pp. 1-6

[4] K. Nishido et al., *"Study of the die position accuracy in the fabrication process of a die first type FO-PLP"* 2018 IEEE 20th Electronics Packaging Technology Conference (EPTC), 2018, pp. 509-513.

[5] S-C Horng, *"Compensating Modeling Overlay Errors Using the Weighted Least-Squares Estimation"*, Institute of Electrical and Electronics Engineers Transactions on Semiconductor Manufacturing (IEEE), 2014, pp. 60-70.

# DEVELOPMENT AND APPLICATION OF ELECTROPLATING COPPER PRODUCTS WITH LOW OR ZERO INTERNAL STRESS

*Yun Zhang[1\*], Jing Wang[1\*], Peipei Dong[1], Xingxing Zhang[1], Wei Zhao[1]*
*Michael Herkommer[2], Klaus Leyendecker[2], Volker Wohlfarth[2]*
[1]Suzhou Shinhao Materials LLC, Suzhou, Jiangsu, China
[2]Umicore Galvanotechnik GmbH, 73525 Schwaebisch Gmuend, Germany
*Corresponding Author's Email: yunzhang@shinhaomaterials.com; wj@shinhaomaterials.com

## ABSTRACT

Shinhao Material LLC has developed several kinds of stress-free products, such as SC-5 and SC-6, which can meet the requirements of matte copper and bright copper respectively. The internal stress after electroplating is less than 3 MPa. After anneal or aging, the internal stress is approximately zero. This is mainly due to the unique role of electroplating additives developed by Suzhou Shinhao Material LLC, which makes the electroplated copper form a distinctive microstructure, so that the internal stress of electroplated copper is very small and can be competent in the field with high internal stress requirements, such as panel level fan out packaging, LED panel, wafer electroplating, etc.

*Keywords—stress free; low stress; electroplating; additive; deformation*

## INTRODUCTION

Low internal stress copper electroplating is very important for large electroplated area. For example, in PLP (Panel Level Packaging), LED (Light Emitting Diode) panel, copper foil electroplating and other fields, low internal stress means that deformation is not easy to occur before and after electroplating. In addition, in the fields of thick copper electroplating and electroforming, internal crack is not easy to appear because of low internal stress, which is more convenient for subsequent process. Shinhao Material LLC. has developed low internal stress and zero internal stress electroplating additive products for many years. Its internal stress is much lower than that of other products in the market, and there are bright copper and matte copper to choose from. At present, it has been widely used in relevant industrial fields.

The internal stress of the copper is mainly composed of the interface stress and the growth stress of the copper itself. The interface stress is generated by the growth of heterogeneous materials at the interface due to the discontinuous electron density at the interface. The microcrystals produced by electrodeposition form epitaxial nucleation on the surface adjacent to the base metal, and their lattice type is the same as that of the base metal. At the same time, with the increase of thickness, it gradually transits to the inherent lattice type of coated metal. In this thin layer, lattice distortion and deformation are generated, resulting in internal stress. For the crystal produced by electrodeposition, the base metal exerts attraction on it to produce compression. At the same time, the surface tension also plays a compression role during crystal growth, resulting in compressive stress in the coating. When the adjacent crystals grow and tend to contact, the mutual attraction will gather them together, resulting in tensile stress in the coating. The effect of crystal compression and mutual attraction is particularly obvious in the thin layer adjacent to the base metal, and the internal stress of the coating is the result of their joint action.

The internal stress of the copper is the embodiment of the resultant force of interfacial stress and growth stress. In addition, the internal stress is also related to the microstructure of copper, such as grain size, orientation and impurities in the copper layer, and is also closely related to the external environment.

## EXPERIMENTAL

The electroplating solution adopts copper sulfate sulfuric acid system, the composition is shown in Table 1, and the current density is 5ASD and 10ASD.

The anode was adopted phosphor copper anode, and the cathode was adopted blank wafer and stress tape respectively. Blank wafer was used for XRD and FIB-SEM respectively, and the stress tape was mainly used to test the internal stress. We use Shinhao low internal stress copper electroplating products to electroplate whole wafers, measure the warpage of whole wafer respectively, and regularly observe the warpage to judge whether the warpage will change. The whole wafer was electroplated on the equipment. After electroplating, the warpage was measured by 3D laser scanning microscope.

The annealing condition is that the samples were placed in a tubular furnace, protected by nitrogen atmosphere, and annealed at 120 ℃ for 2 hours.

Table 1 Bath composition of SC-6 system

| Bath Composition | | Value | Unit |
|---|---|---|---|
| VMS | $Cu^{2+}$ | 50 | g/L |
| | $H_2SO_4$ | 100 | g/L |
| | Cl- | 50 | mg/L |
| Additives | A | 4 | mL/L |
| | S | 10 | mL/L |
| | L | 15 | mL/L |

## Results and discussion

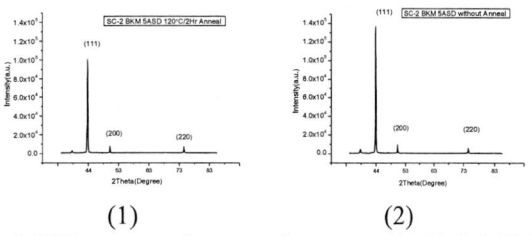

(1)          (2)

*Fig. 1 XRD pattern of copper electroplated with SC-6 BKM at 5 ASD before (1) and after (2) anneal at 120 °C for 2 hours*

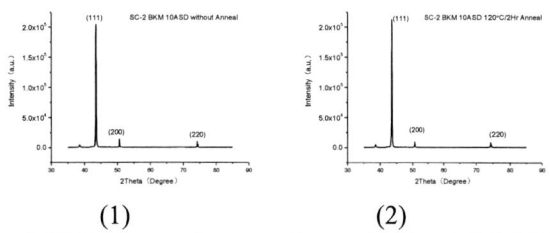

(1)          (2)

*Fig. 2 XRD pattern of copper electroplated with SC-6 BKM at 10 ASD before (1) and after (2) anneal at 120 °C for 2 hours*

It can be seen from Fig. 1 and Fig. 2 that the XRD patterns of the electroplated copper layer prepared by SC-6 system, whether 5 ASD or 10 ASD, have not changed significantly before and after anneal

Table 2 Proportional relationship of XRD peak height

| Experimental Condition | | Peak height normalization | | |
|---|---|---|---|---|
| | CD | ( 111 ) | ( 200 ) | ( 220 ) |
| AS Electroplated | 5 ASD | 100 | 7 | 4 |
| 120 °C/2hr Anneal | 5 ASD | 100 | 8 | 8 |
| As Electroplated | 10 ASD | 100 | 7 | 5 |
| 120 °C/2hr Anneal | 10 ASD | 100 | 5 | 4 |
| The Powder Diffraction File of Copper ( PDF of Copper ) | | 100 | 46 | 20 |

Table 2 is normalized according to the XRD data in Fig. 1 and Fig. 2. Compared with the standard PDF card of copper, it can be seen from the table that the diffraction peak intensity corresponding to the (111) crystal plane of SC-6 is much stronger than that corresponding to the standard PDF card, that is, the (111) crystal plane of copper electroplated with SC-6 has a certain degree of preferred orientation.

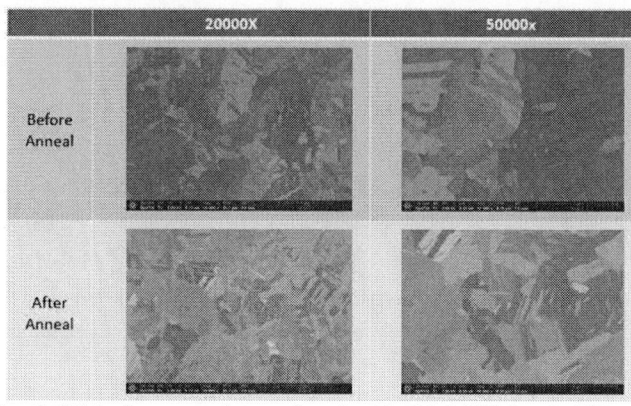

*Fig. 3 SEM of copper electroplated with SC-6 BKM at 5 ASD before (1) and after (2) anneal at 120 °C for 2 hours, the sample was cut by FIB.*

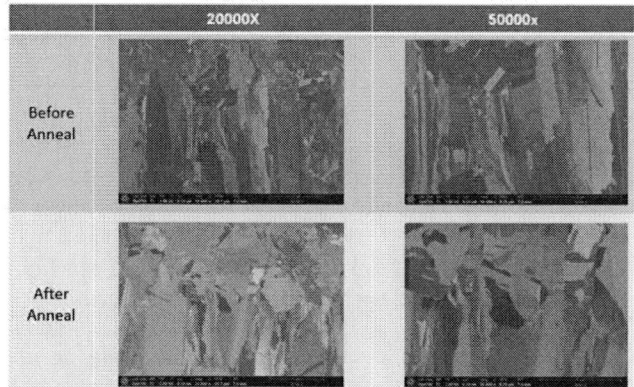

*Fig. 4 SEM of copper electroplated with SC-6 BKM at 10 ASD before (1) and after (2) anneal at 120 °C for 2 hours, the sample was cut by FIB.*

It can be seen from the Fig. 3 and Fig. 4 that no matter 5 ASD or 10 ASD, the grain size of electroplated copper has no obvious change before and after anneal. This result is also consistent with the XRD results.

Generally speaking, the specific surface energy of the crystal surface plays an important role in the shape and state of the crystal. The shape of any crystal under equilibrium conditions must meet the condition that the total surface energy is the minimum. When a small metal grain of simple cubic crystal system is formed on the matrix, that is, when one of its crystal planes contacts with an external object, its equilibrium shape will be different from the cube, its size in the axial direction is reduced, and the shape of the growing actual grain is different from that of the equilibrium grain. When additives are added, its growth speed on the axis is particularly fast. If the volume of this small grain is the same as simple cubic crystal, it is obvious that its total surface energy will be greater than the total surface energy of equilibrium grain, that is to say, under the same volume, the surface area of a simple cube is smaller than that of other cuboids, so it is a non-equilibrium grain. Due to the existence of excess surface energy, the surface force will shorten the grain along the *z* axis and extend the bottom along the *x* and *y* axis, resulting in compressive stress. Then, if the (111) crystal plane occupies a dominant position in the grains of the copper electroplating layer, the lattice structure of the electrodeposited copper can be regarded as an equilibrium grain state, showing a state of low internal stress or even zero internal stress.

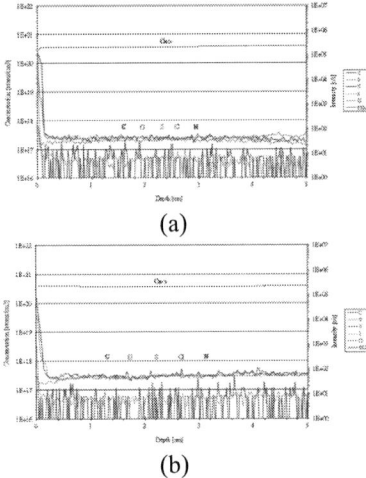

(a)

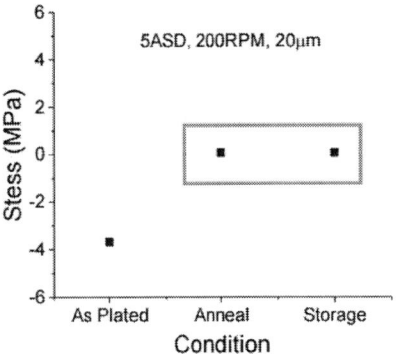

Fig. 6 Stress of SC-6 system

(b)

*Fig. 5 C, O, N, S and Cl elements are quantified based on Cu material, and the depth is converted by the etching rate of the standard. (a) 5 ASD, (b) 10 ASD*

Fig. 5 shows the data of secondary ion mass spectrometry, where (a) is the sample after 5 ASD electroplating, and (b) is the sample after 10 ASD electroplating. The detection depth of secondary ion mass spectrometry is from the sample surface to the place 5 μm away from the surface. It can be seen from the Fig.5 that the contents of C, O, S, Cl and N in the sample are very low whether using 5 ASD or 10 ASD, that is, the purity of electroplated copper is very high, and the degree of lattice distortion caused by impurities in the copper lattice is also very low.

In fact, in the electroplating process, the electroplating additives in the electroplating solution are usually indispensable, and many electroplating additives are various organic molecules. They participate in the electrochemical reaction and are easy to be mixed in the electroplated copper layer to form various defects, which will lead to lattice distortion and internal stress. With the continuous growth of crystal and the gradual movement of electric double-layer interface to electrolyte, the effect of field decreases gradually until it stops. Organic molecules are no longer constrained in a certain direction. The influence and restriction of thermal movement lead to the disorder of molecular arrangement to a certain extent, which further increases the volume of the coating and produces compressive stress at the same time.

Based on the above analysis, the electroplating additive developed by Shinhao is relatively strong in the (111) direction, (111) is also the densest crystal surface of copper ions, and the impurity content in copper is also very low. Therefore, using this electroplating additive can theoretically obtain better low internal stress copper.

Fig. 6 shows the change of internal stress of stress strip after electroplating, anneal and anneal standing. It can be seen from the figure that after electroplating, the stress strip shows slight compressive stress. After anneal, the stress is basically zero. After anneal standing, the stress continues to remain zero without change, that is the internal stress of the coating remains constant and close to zero or equal to zero.

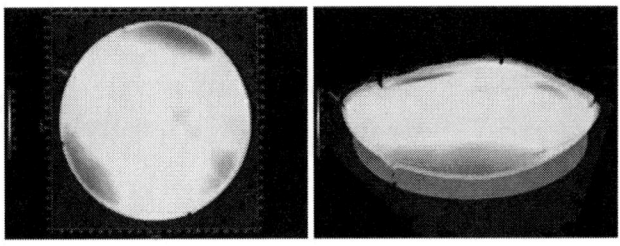

2D view          3D view

*Fig. 7  20μm copper was electrodeposited on blank wafer, almost zero warpage for 24 hours after electroplating*

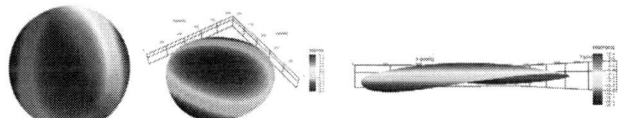

*Fig.8 20μm copper was electrodeposited on blank wafer, 72 hours after electroplating*

Fig. 7 and Fig. 8 are the warpage data of the electroplated wafer with SC-6 system, which is the warpage data distribution of 24 hours and 72 hours respectively. It can be seen from the Fig. 8 that the warpage degree is very small.

*Fig. 9 No obvious warpage was observed after electroplating, and after 14 days*

Fig. 9 shows that the electroplated wafer with SC-6 system is placed on the stage after 14 days, and no warpage can be seen by the naked eye. For the SC-6 system of Shinhao, if used in the semiconductor field, the wafer will not warp or warp very slightly after electroplating, so as to improve the yield of products and reduce the complexity of relevant processes.

## CONCLUSIONS

Shinhao Materials LLC has developed a kind of low/zero stress copper electroplating systems. The electroplated wafers can basically achieve zero warpage and significantly improve the yield of products. This is mainly due to Shinhao's electroplating additives, which realize the continuity of electron density at the heterostructure interface in electrochemical deposition, and modulate the grain size and lattice orientation to realize the stress relaxation.

## REFERENCES

[1] Philippe Nivelle, John A. Tsanakas, Jef Poortmans, Micha"el Daenen. Stress and strain within photovoltaic modules using the finite element method: A critical review. Renewable and Sustainable Energy Reviews 145 (2021) 111022, p11-p18.

[2] Zihan Dong, Yuanwei Lin. Ultra-thin wafer technology and applications: A review. Materials Science in Semiconductor Processing 105 (2020) 104681, p1-p8.

[3] Omar Bchir, Houssam Jomaa, Chin Kwan Kim, Layal Rouhana, Kuiwon Kang, Milind Shah, Steve Bezuk. Improvement of Substrate and Package Warpage by Copper Plating Process Optimization. 2014 Electronic Components & Technology Conference, p1396-p1400.

[4] Vincenzo Vinciguerra, Antonio Landi. On the Way to understand the Warpage in 8" Taiko Semiconductor Wafers for Power Electronics Applications (Si and SiC). 2021 22nd International Conference on Thermal, Mechanical and Multi-Physics Simulation and Experiments in Microelectronics and Microsystems (EuroSimE), p1-p14.

# DESIGN AND FABRICATION OF HIGH-Q IPDS FOR PROCESS DESIGN KITS ON GLASS SUBSTRATE

*Haozhe Ma[1,2], Zhihui Hu[1,2], Qing Zhou[1,2], Jiabin Chen[2] and Daquan Yu[1,2*]*

[1]School of Electronic science and Engineering, Xiamen University, Xiamen 361005, China

[2]Xiamen Sky Semiconductor Co., Ltd, Xiamen, China

*Corresponding Author's Email: yudaquan@xmu.edu.cn

## ABSTRACT

Based on glass substrate, a series of high-Q passive devices such as capacitors and inductors with different specifications are designed and fabricated for the application in PDK (Process Design Kits). In addition, three-dimensional full-wave electromagnetic field simulation tools were employed to analyze the electrical properties of the capacitors, inductors on glass substrate.

***Keywords—IPD ((Integrated Passive Device); PDK (Process Design Kits); capacitors; inductors; Q factor; glass substrate***

## INTRODUCTION

In recent years, with the rapid development of 5G communication, IPDs (Integrated Passive Device) are used in RF with its advantages of high frequency and miniaturization [1][2]. At the same time, glass becomes a superior alternative substrate material with the characteristics of low dielectric loss, high barrier performance, simple process flow and low cost [3]. With TGV (Through Glass Via) technology, high performance 2D/3D IPDs can be manufactured on glass substrate [4]. Compared with substrate materials such as high resistance silicon and gallium arsenide, passive devices based on glass substrate have higher Q factor, especially at high frequency field.

PDK (Process Design Kits) is a bridge between the design company, the foundry and EDA manufacturers. It is used to reflect the documentation of the foundry process. It is not only the cornerstone used by the design company for physical verification, but also the key factor for the process flow.

In this work, devices structure and layout was introduced, electrical properties of capacitor, inductor on glass substrate were simulated, the process flow of IPD fabrication on glass substrate is shown.

## DESIGN OF DIFFERENT STRUCTURES

Based on the glass substrate, 48 different structures were designed in this work. It contains 23 kinds of capacitors, 22 kinds of inductors and 3 kinds of de-embedded structures.

**Design of Capacitors**

In this paper, MIM (metal-insulator-metal) capacitor structure is used in our design. MIM capacitor based on glass substrate has the advantages of high Q factor and high precision. In the design, the capacitance can be estimated by the area of the upper electrode plate and the dielectric layer. By controlling the plate area, dielectric layer thickness and dielectric type, 10 different capacitance values are realized: 6pF, 5pF, 4.1pF, 3.5pF, 2.3pF, 1pF, 0.55 pF, 0.5 pF, 0.1pF and 0.01pF.

|  (a)  |  (b)  |  (c)  |

*Figure 1: (a) Silicon nitride dielectric capacitor. (b)PI dielectric capacitor. (c) series connection capacitor structure.*

In our design, the plate material is copper and the dielectric layer material is silicon nitride. In addition, in order to realize small capacitance capacitor, we designed the structure of capacitor series connection and the structure of using PI layer as dielectric material.

**Design of Inductors**

Our design includes two different inductor structures: 2-D inductor and 3-D inductor. The structure of 2-D inductor adopts plane spiral coil and 3-D inductor structure uses TGV to form a spatial spiral coil. Compared with two-dimensional inductor, three-dimensional inductor has higher Q factor but smaller inductance.

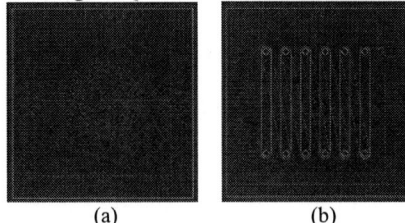

|  (a)  |  (b)  |

*Figure 2: (a) 2-D inductor. (b) 3-D inductor.*

The inductance is related to line width, line distance and the number of spiral turns. By changing these parameters, we achieve three different kinds of inductance :2.5nH, 5nH and 10nH.

**Design of De-embedding Structures**

The influence of embedding on device measurement needs to be considered in the design process. The test structure, via and interconnect will introduce unwanted parasitic resistance, parasitic capacitance and parasitic inductance. In order to eliminate this effect, we designed "Open" structure and "Short" structure for de-embedding.

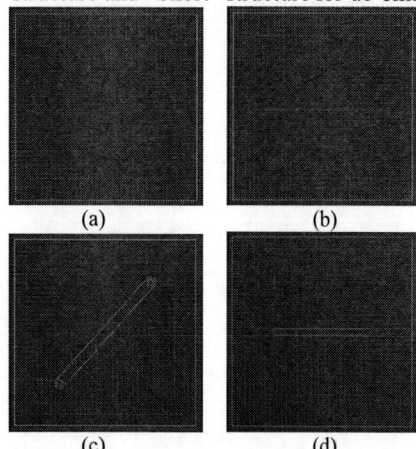

(a)　　　　　(b)

(c)　　　　　(d)

*Figure 3: (a) "Open" structure. (b) "Short" structure for 2-D inductor. (c) "Short" structure for 3-D inductor. (d) "Short" structure for capacitor.*

The "Open" structure only preserves the test structure, corresponding to the parallel parasitic factors such as *Y1, Y2, Y3* shown in Figure 4. The "Short" structure is a short-circuit structure based on the "Open" structure that short-connects **Port1** to **Port2**, corresponding to the direct parallel connection of Parallel Parasitic Factor (*Y1, Y2, Y3*) and Serial Parasitic Factor (*Z1, Z2, Z3*) in Figure 4.

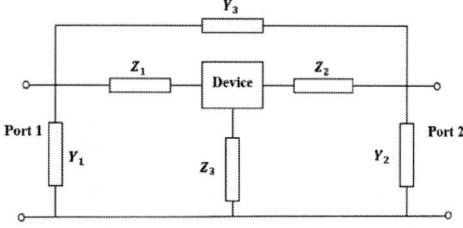

*Figure 4: Open-Short test structure equivalent circuit.*

The S-parameters of the device, the "Open" structure and the "Short" structure are obtained and converted to the Y-parameter ($Y_{dut}$, $Y_{open}$, $Y_{short}$), then the Y parameters of the de-embedded intrinsic device are calculated as follows:

$$Y = [(Y_{dut} - Y_{open})^{-1} - (Y_{short} - Y_{open})^{-1}]^{-1} \quad (1)$$

## SIMULATION OF DESIGN

Three-dimensional full-wave electromagnetic field simulation tools were employed to verify the design. We build de-embedded models for simulation to eliminate the impact of embedding. The simulation results include capacitance/inductance value and Q factor. The design target frequency is 2GHz. The simulation results of capacitance and inductance are as follows, which match our design well.

**Simulation of Capacitor**

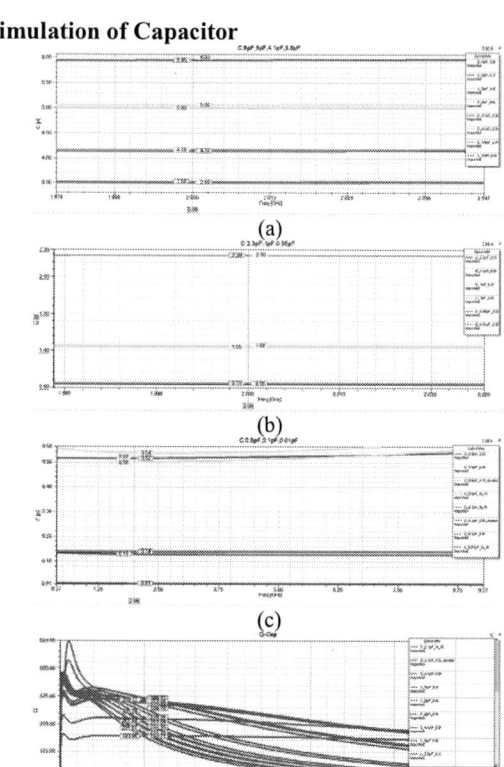

(a)

(b)

(c)

(d)

*Figure 5: (a) C curves: 6pF, 5pF, 4.1pF, 3.5pF. (b) C curves: 2.5pF, 1pF, 0.55pF. (c) C curves: 0.5pF, 0.1pF, 0.01pF. (d) Q curves.*

The simulation results of capacitor are basically consistent with the design. The Q factor of silicon nitride dielectric layer is significantly higher than that of PI dielectric layer. The capacitors' Q value of silicon nitride dielectric layer is more than 190, while the PI dielectric layer is only about 25.

**Simulation of Inductor**

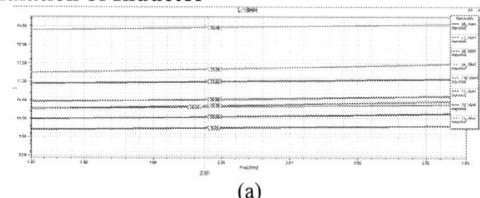

(a)

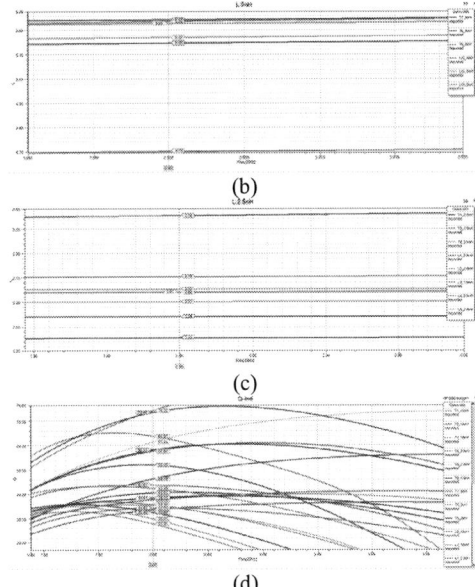

(b)

(c)

(d)

Figure 6: (a) L curves:10nH. (b) L curves: 5nH. (c) L curves:2.5nH. (d) Q curves.

The simulation results of inductance are in fairly agreement with the design. Under the same inductance, the Q factor of 3-D inductance is higher than that of 2-D inductance. The Q value of 3-D inductance is above 50, while the Q value of 2-D inductance is only about 35.

## FABRICATION OF DESIGN

As shown in Figure 7, 230um thickness glass wafer was prepared. After wafer-cleaning, we used laser drilling to form TGV cavity. Then double side PVD process was done to form Ti/Cu seed layer which covers the both sides of the glass and the wall side of the TGV cavity. Next, double side full fill electroplating was done to fill the TGV cavity with copper and form the top side metal layer M2 and bottom side metal layer Ml. After that, the Si3N4 layer and a metal layer M3 was formed on the top side of M2 to formulate MIM capacitors. And the upper metal layers M4 were fabricated to generate spiral coil inductor and GSG (Ground-Signal-Ground) test structure. Pads were finally connected to the top metal, which were used for package.

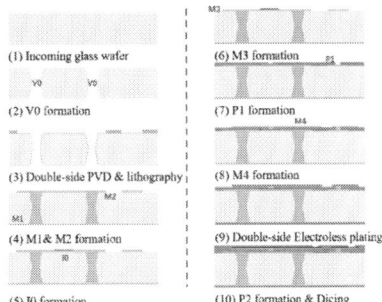

Figure 7: Process flow about fabrication.

At present, the fabrication of the wafer has been completed, and the structures are shown in the figure below.

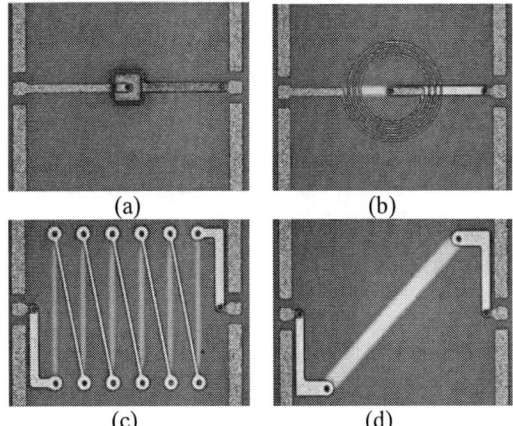

Figure 8: (a) capacitor. (b) 2-D inductor. (c) 3-D inductor. (d) De-embedding structure.

## SUMMARY

In this work, a series of high-Q passive devices on glass substrate were designed and simulation verification was carried out. Then, the designed structure was manufactured and the process flow applied to PDK fabrication has been verified. In addition, the simulation results are in good agreement with our design, and show the advantage of high Q factor of IPD based on glass substrate.

## ACKNOWLEDGEMENTS

Thanks Xiamen Sky Semiconductor Co., Ltd for support and assistance in this work.

## REFERENCES

[1] P. J. Turner, B. Garcia, V. Yantchev, G. Dyer, S. Yandrapalli, L. G. Villanueva, et al., "5 GHz band n79 wideband microacoustic filter using thin lithium niobate membrane", Electronics letters, vol. 55, no. 17, pp. 942-944, August 2019.

[2] R. Xiaoli, W. Junying and Y. Daquan, "Design and Fabrication of High Density 3D IPDs (Integrated Passive Devices) on Glass Substrate," 2020 IEEE International Conference on Integrated Circuits, Technologies and Applications (ICTA), 2020, pp. 127-129, doi: 10.1109/ICTA50426.2020.9332031.

[3] Michael Topper et al., "3-D Thin Film Interposer Based on TGV (Through Glass Vias): An Alternative to Si Interposer", In Proc of 60th Int. Conf on ECTC, pp. 66-73, 2010.

[4] S. Takahashi et al., "Development of through glass via technology for 3D packaging," 2013 Eurpoean Microelectronics Packaging Conference (EMPC), 2013, pp. 1-4.

# INVESTIGATION ON YIELD IMPROVEMENT OF FAN-OUT WAFER-LEVEL PACKAGING

*Kai Zhu2*,Shuanshe Chao2*,Yang Chen2,Na Mei2,Jianmin Fang2,Tuobei Sun2,Keqing Ouyang1*
*1, State Key Laboratory of Mobile Network and Mobile Multimedia Technology*
*2, Department of Packaging and Testing*
*ZTE Corporation,Shenzhen, China*
*\*Corresponding Author's E-mail: Kai.zhu2@sanechips.com.cn*

## ABSTRACT

This paper mainly studies the problem of yield improvement of fan-out wafer-level packaging(FOWLP). To solve the problem of low yield, firstly, we analyze and confirm the root cause of open/short failure. It is confirmed that the open is due to the misjudgment caused by the test scheme. Short is the weak ESD protection circuit design of relevant pins, and the gate oxygen breakdown is caused by environmental static electricity during the installation and test; Then the chip warpage is reduced by optimizing the chip packaging design to solve the problem of chip parameter drift. Finally, after modifying the ESD protection circuit and test scheme, the problem of O/S is solved, the problem of chip warpage and parameter drift after DOE is also improved, and the yield of the chip is significantly improved.

*Keywords — fan-out wafer-level packaging;open/short;warpage;parameter drift;yield improvement*

## INTRODUCTION

In recent years, advanced packaging has gradually become a focal point on surpassing Moore's Law persistening. Fan-out wafer-level packaging as shown in Figure 1 is a prominent representative of advanced packaging. As shown in Figure 2,Fan-out wafer-level packaging breaks through the limitation of the number of I/O terminals, and increases the area of a single package through wafer reconstruction.Then, it uses advanced wafer-level package making process to complete multi-level redistribution layer(RDL) and bump preparation. After dicing and separating, the package body that can be directly mounted on the PCB and interconnected electrically with the outside is obtained, or the package body that can be interconnected with the package carrier to form a secondary package.Since fan-out technology did not require an interposer, it has been cheaper than 2.5D/3D packaged devices. Compared with packages such as traditional wire bonding, fan-out wafer-level packageing has obvious advantages in package volume, product performance,

package cost, and package efficiency. But at the same time it will also bring new challenges. The main challenge of fan-out packaging is warpage/wafer bow. In addition, chip placement will also affect parameter drift.This paper mainly discusses the fanout packaging O/S problem and the parameter drift problem. By solving these two problems, we can improve the fanout packaging yield.

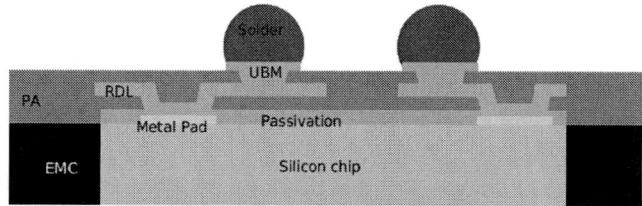

*Figure 1:Structure diagram*

*(a)Top view of fan-out wafer-level packaging*

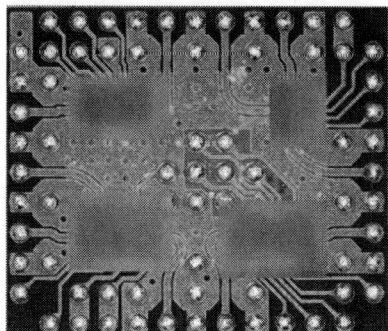

*(b)buttom view of fan-out wafer-level packaging*
*Figure 2:Appearance drawing*

## O/S FAILURE ROOT CAUSE

After confirming that the test contact is good, there is no abnormality in the Scanning Acoustic Tomography(SAT), as shown in Figure 3.And there is no obvious abnormality in the package RDL circuit.The IV curve of the short failure pin to the ground is shown in Figure 4, and the IV curve of the good pin to the ground is shown in Figure 5. The short-circuit fault pin does not appear obvious diode characteristics, which proves that its interior has been breakdown.

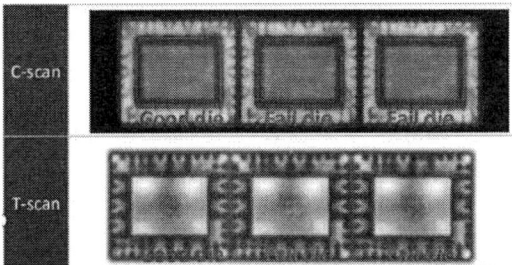

*Figure 3:Scanning Acoustic Tomography*

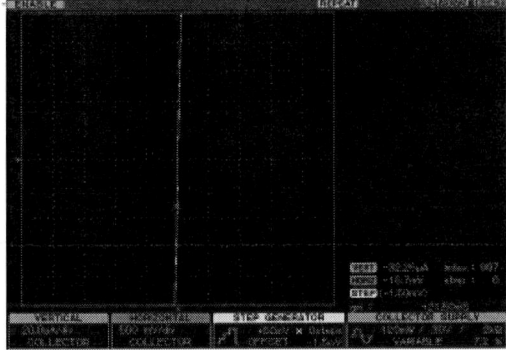

*Figure 4:IV Curve of Short Pin to Ground*

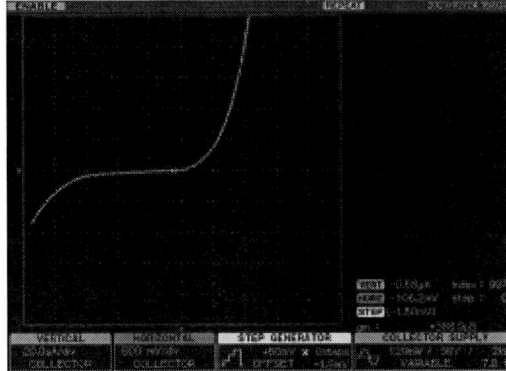

*Figure 5:IV Curve of Good Pin to Ground*

As shown in Figure 6,through the thermal EMMI of the short-circuit failure samples, 1mA current and 301mV

voltage are applied between the failure pin and GND, and we found that the short-circuit heating area is located inside the chip. Through confirmation with the wafer design, the pin ESD protection design has defects, which caused the ESD to be short-circuited during the packaging and testing process[1].

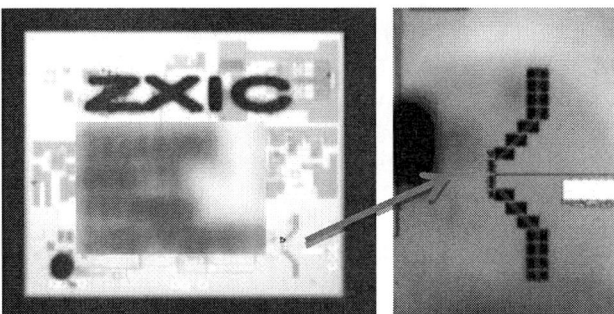

*Figure 6:Location of failed thermal EMMI*

In subsequent versions, the ESD protection of the wafer was optimized and the short problem was solved. It passed the HBM1000V and CDM250V tests.

Comparing the good die and the open die's IV curve of the failed pin to ground, it is found that both have diode characteristics and there is no significant difference.At the same time, the tested open failed chip passed the electrical performance test on the EVB board.

There is no golden sample because this device is worked for the first time. The limit setting of the first version of the test program was not reasonable enough, resulting in a false test. After grabbing the data of 1000 samples, removing the absolute open test(about -3V) and absolute short test(within -0.050V) part, manually calculate according to the average value +/-6sigma, and optimize the new test limit.The new limit is used to arrange the FT test, which eliminates the open false test caused by the unreasonable limit. The test yield has been increased from 83% to 98%.

## WARPAGE AND PARAMETER DRIFT

The warpage problem of fanout package is a common problem. For the problem of parameter drift, we find that the problem of parameter drift of the chip with small warpage will be reduced and disappear. Select the packaging process with the smallest warpage through DOE of the packaging process, At the same time, the problem of parameter drift is improved.

During the engineering lot stage, we found that the warpage of the recon wafer after the high temperature process would have a relatively large change and affect parameters of die, so we designed 4 sets of DOE(Design of Experiment), and in the case of determining the thickness of the Si die, the thickness of the recon wafer was adjusted by different mold cap thicknesses[2].Collecting the warpage data after recon, post PA1 cure, post RDL1 etch, post PA2 cure, post RDL2 etch and reflow, it determined the optimal parameters.Figure 7 shows the warpage data of each key process measured through different mold thickness combinations. When the mold thickness is equal to 670um, the warpage is the smallest.

Through the DOE, we have improved the parameter drift of the chip, and at the same time, we have not found parameter problems and obvious warpage risks of the chip in the temperature cycle and high temperature storage test.

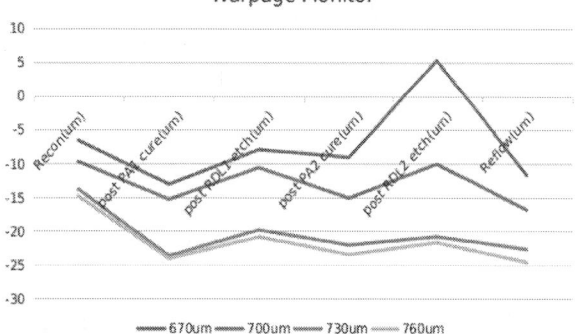

Figure 7:Warpage Monitor

# CONCLUSION

Fan-out wafer-level packaging is a popular advanced packaging widely used in communications, artificial intelligence, mobile terminals and other fields. First, we through the analysis of the chip design and the test plan, we find the root cause of the open circuit and the short circuit, and make corresponding improvements.Then we have solved the problems of wafer warpage and parameter drift through the DOE.Finally, improve the yield of fan-out wafer-level packaging.

# ACKNOWLEDGEMENTS

The author would like to acknowledge the support ofthe department of packaging and testing of ZTE corporation.

# REFERENCES

[1] S. s. Chao, D. Yang, N. Mei, T. b. Sun and L. kai Yuan, "Investigation on ESD failures of RF IC," 2020 21st International Conference on Electronic Packaging Technology (ICEPT), 2020, pp. 1-3.

[2] T. Yu, X. Ren, D. Yu, X. Zhang and F. Jiang, "Developing of Wafer Level Fan-Out Packaging Technology for Millimeter-Wave Chip Using Different Carriers," 2019 IEEE International Conference on Integrated Circuits, Technologies and Applications (ICTA), 2019, pp. 129-133.

# ASSEMBLY REFLOW PROCESS EFFECT ON INTERLAYER DIELECTRIC CRACK AND NEW FINDING OF FAILURE ANALYSIS

*Suming Wang, Zhidan He, Wenyuan Chen, Honghu Ji*
TF-AMD, Suzhou 215000, China
Simon.Wang@tf-amd.com

## ABSTRACT

With requirement of superior electrical performance for today's semiconductor devices, ultra-low k (ULK) interlayer dielectric (ILD) materials are used. During back-end flip chip assembly process, due to thermal expansion coefficient (CTE) difference between die and substrate, reflow conducted thermal stress causes ILD crack. In this paper, process reflow condition is investigated to minimize failure rate of ILD crack. The result shows large temperature gradient would cause high failure rate. Longer cooling time is also an effective method to reduced ILD crack failure rate.

ILD crack failure is also known as "white spots" which can be detected by C-mode Scanning Acoustic Microscope (CSAM) and confirmed by decapsulation and Focused Ion Beam (FIB). After reflow profile optimization, a new type of CSAM failure "black spots" is observed with the same ILD crack failure mode.

## INTRODUCTION

IC package dimension has been shrinking during past decades. To achieve better electrical performance, ultralow-k dielectrics were introduced since 1990s [1]. By the year of 2019, dielectric constant k has been reduced to less than 2.50 and gradually shorten interconnect delay time. The reduced k is accomplished by non-conformal deposition [2] and removal of sacrificial materials after multi-level interconnects [3]. The introduction of porosity into dielectrics would affect mechanical properties. Both elastic modulus and hardness decrease with increasing porosity [4].

*Figure 1 Scheme side view of copper pillar solder bump flip chip*

Reduced mechanical strength of ILD, large die size, CTE mismatch between die and substrate, the use of copper pillar bump all contribute to initiation of ILD crack. Figure 1 shows side view of flip chip and ILD is embedded in ELK area. Other studies show those ILD

crack mostly occur at corner or edge of the die. This is because the greater distance from neutral axis, the greater stress is experienced [5]. Those ILD defect is circular shape can be observed by CSAM. In the past, it is often referred as "white spots". Normally those white bumps are observed immediately after die attach reflow and before underfilling. ILD crack may cause pin to pin short or leakage fail which affect final yield rate. Sometimes ILD crack may not cause direct failure. However, it still has potential reliability problem at customer end. Thus, it is necessary to eliminate ILD crack issue as much as possible during flip chip assembly process.

## THERMO STRESS INDUCED ILD CRACK

A lot of work has been done by other researchers and process engineers [6] [7] to simulates thermal mechanical stress induced during process assembly. During reflow, since CTE of substrate is greater than CTE of die, different displacement magnitude of two materials is generated and lead to warpage. Then the magnitude of stress on corner bumps is much greater than stress on center bumps. Both longitudinal and horizontal stress would cause ILD crack around corner bumps. In principle, coefficient of thermal expansion for most solid can be expressed by

$$ 1/V * dV/dT = \alpha_v \qquad (1) $$

Where $\alpha_v$ is volume expansion coefficient, $V$ is volume, $dV/dT$ is rate of change between volume and temperature. By this definition, CTE is strongly temperature dependent. Normally substrate undergoes larger displacement than die and lead to concave shape package profile. On the other hand, strain is defined as

$$ ( L_{final} - L_{initial} ) / L_{initial} = \varepsilon \qquad (2) $$

And for most solid material, thermal energy induced strain is proportional to temperature change, thus thermal strain can be expressed as

$$ \Delta T * \alpha = \varepsilon_{thermal} \qquad (3) $$

Based on virtual crack closure technique [8], strain energy release rate $G$ is given by

$$ (-1/2 * R_{x,y} * \Delta u_{x,y}) / \Delta A = G \qquad (4) $$

where $\Delta A$ is change of surface area. $R_{x,y}$ is force applied on crack for x and y directions, in other word, stress. $u_{x,y}$ is displacement for x and y directions, which refers to strain. Combining equation (1) to (4), total energy release rate can be expressed by

$$(-1/2 * R * \Delta T * \alpha) / \Delta A = G \qquad (5)$$

With controlled variables during reflow process, large temperature gradient would cause high energy release rate and lead to fast crack propagation. A specific package type FCLGA product suffers high failure rate due to ILD crack. In order to reduce total amount of ILD crack affected units, reflow profile optimization experiments are designed. Pre-heat temperature, peak temperature, exit temperature, heat time, cooling time are considered as key factors contributing to ILD crack. The die and substrate dimensions, bump size and pitch, metal layout are all identical to exclude noise during reflow process. Failure rate is calculated based on 100% CSAM inspection for "white spots" observation. For all experiments, D represents default condition. First experiment is to study peak temperature and exit temperature effect.

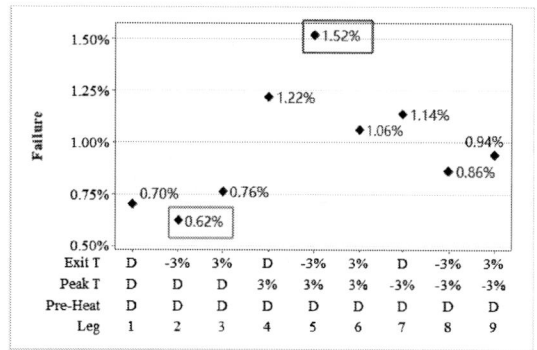

*Figure [2] Failure Rate vs Reflow Condition 1*

The result shows leg 5 suffers highest failure rate and leg 2 has lowest failure rate. This correlates with previous assumption that large thermal gradient would cause ILD crack issue. For leg 7 to leg 9, the overall failure rate is slightly higher than leg 1 to leg 3. This indicates with the same total temperature difference range; stress level differs with boundary condition. By excluding the best and worst legs, there is no obvious regularity of failure rate with respect to different temperature settings. This indicates temperature gradient is not the only factor affecting ILD crack failure rate. Based on this experiment, leg 2 condition is used as default condition for next experiment. The next variable is heating time and the goal is to investigate heating slope effect on stress accumulation.

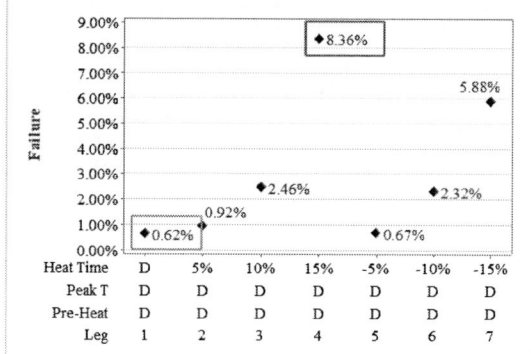

*Figure [3] Failure Rate vs Reflow Condition 2*

Second experiment shows Leg 4 has highest failure rate. Under the same peak temperature, smoother slop would not help to improve ILD crack issue. Longer heating time allows more space for crack propagation. It is interesting that leg 7 with shortest heat time also suffers high failure rate. This indicates fast heating time would also accelerate stress accumulation. With this experiment result, it is able to conclude smooth or steep heat slope is not efficient way to revolve ILD crack issue. As leg 1 still has lowest failure rate, this condition is used as default for next set experiment. Since heating time cannot be changed based on current result, cooling time is considered as variable for next experiment.

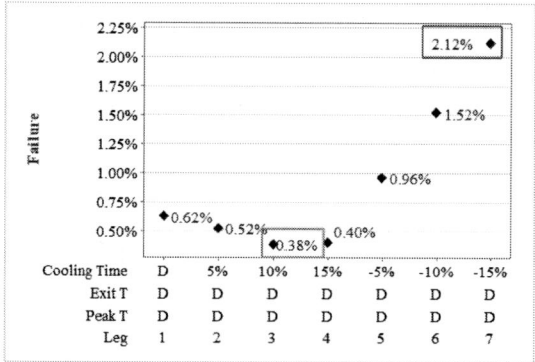

*Figure [4] Failure Rate vs Reflow Condition 3*

Figure 4 result shows longer cooling time reduces ILD failure rate. Thus, shorter cooling time is very efficient for failure reduction. To summarize, large temperature gradient caused high energy release rate would increase ILD crack failure rate. On the other hand, heating time change also contributes to thermal stress accumulation.

In flip chip assembly, reflow process is the most important step since this involves direct contact between substrate pre-solder and wafer bump. Other process such as underfill dispense, ball mount, lid attach is considered as low risk for ILD crack. This is because after reflow process, bumps are formed and will be further surrounded by other material. For front-end parameters such as PI opening size, Aluminum pad location, ELK and metal line structure, those design helps to create the foundation of ILD crack because all those materials have different CTE. Sometimes, even all process is controlled within specifications at back-end, ILD crack may still occur and affect final yield. By above three sets of experiments, ILD crack failure rate is reduced from 0.70% to 0.38% which meets expectation. While during CSAM inspection, failure analysts find noteworthy phenomenon and will be discussed in the next section.

## FAILURE ANALYSIS OF ILD CRACK

Acoustic microscopy has been widely used in semiconductor field, especially C-mode scanning acoustic microscope. The ultrasonic transducer scans over flip chip and thousands of pulses enter sample each second. Each pulse will be scattered or absorbed through whole device. As shown in figure 1 from previous section, flip chip has different material interfaces. When those pulses meet material interface, some pulses reflected back to the transducer. Those signal counts and amplitude will be recorded. The software will process those signals into 2D phase image. CSAM is an efficient tool for non-destructive analysis to detect major defects for flip chip such as void, crack, and delamination. This is because high frequency acoustic energy cannot transmit in air. Any kind of gap will cause 100% signal refection. Figure 5 is a traditional "white spots" CSAM image and refers to ILD crack failure. As described at previous section, those crack occurs near die edge.

*Figure [5] CSAM "white spots" for flip chip*

In previous section, different reflow condition is studied to solve ILD failure. Figure 6.a is a typical unit with many bumps of "white spots" on upper corner before reflow optimization. This is very typical and share similar shape with traditional "white spots" image. Figure 6.b and 6.c are CSAM images of units after optimization. It is obvious that no typical "white spots" was observed. While "black spots" take place around die corner. In order to further identifying this CSAM anomaly, destructive analysis such as parallel lapping from substrate side, mechanical fine polishing and FIB X-sectioning are performed. The same procedure is performed for all three units shown in Figure 6 to avoid artificial defect and ensure the result is credible.

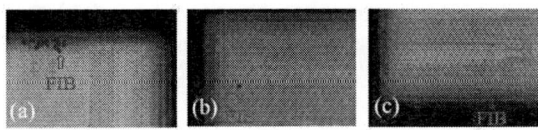

*Figure [6] CSAM image before (a) and after (b),(c) optimization*

Figure 7 shows FIB result with respect to three units in Figure 6. Three units all share the same failure mode ILD crack. Figure 7.d with respect to Figure 6.a shows traditional "white spots" has longest crack length. Figure 7.e and 7.f "black spots" suffer relatively shorter crack length. Combining those result, crack is seen to take place near 3rd and 4th ULK layer and propagate through ULK layers horizontally. There is no obvious sign of vertical growth. Traditional "white spots" has circular shape ILD crack with maximum width larger than bump diameter. And "black spots" also has circular shape ILD crack with maximum width smaller than bump diameter.

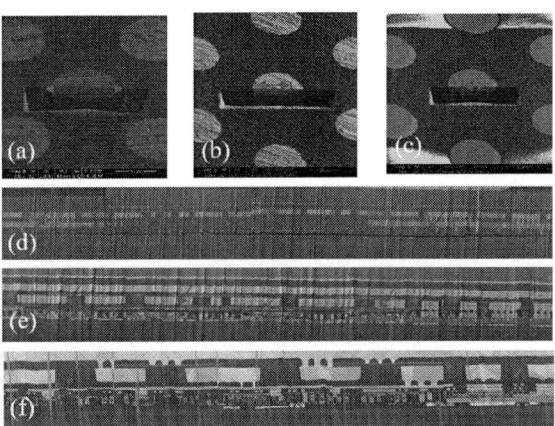

*Figure [7] image (a) and (d) FIB cross section of Figure 6.a. (b) and (e) FIB cross section of Figure 6.b. (c) and (f) FIB cross section of Figure 6.c.*

Back to CSAM images, Figure6.c has less darkness intensity compared to Figure6.b, which indicating failure should be less severe. This aligned with FIB result, crack length is much shorter. For traditional "white spots", crack width is greater than bump diameter, with enough ILD crack area, one observes black edge and white center circular shape anomaly. However, for "black spots", total crack area is too small for CSAM to restore tiny circular shape morphology due to its detection limit. If CSAM detection has high enough resolution, with local zoom in, it should be able to observe small size circular shape ILD crack which shall have the same appearance as traditional "white spots".

Another common flip chip failure mode is delamination between two interfaces. This also appears as "black spots" on CSAM image. Figure.5 shows an example of CSAM "black spots" and its failure analysis result. By doing destructive analysis including mechanical cross-section, polishing and SEM, failure mode is confirmed to be delamination between PI and underfill interface near top bump edge. The delamination width is very small, less than ILD crack.

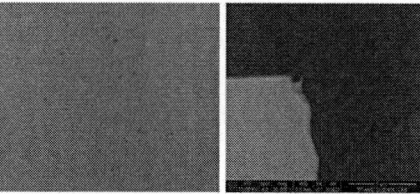

*Figure 5 CSAM "black spots" and related PI/UF interface delamination*

To isolate ILD crack "black spots" and delamination caused "black spots", first is to observe failure location. ILD crack only occurs around die corner, while other failure may take place elsewhere. Next is to compare waveform. Since ILD crack has longer width, its peak magnitude value should be larger than small delamination because most signals are reflected back. This also contributes to contrast difference. ILD crack "black spots" shall have darker contrast. Another method is to compare anomality shape. ILD crack is circular shape, thus on CSAM image, the failure anomaly also tends to be circular shape. For small delamination in Figure 5, it appears on side of bump. So CSAM anomaly has irregular non-rounded shape. For failure analyst, it is critical to identify different failure modes before any further destructive analysis. With wrong judgment, the final analysis result may be unsatisfying and affect process parameter improvement.

## SUMMARY

Flip chip interlayer dielectric crack is one of the most common failure in semiconductor industry. In this paper, thermal effect study is presented and illustrates large temperature gradient caused high energy release rate would lead to high ILD crack failure rate. With optimized reflow profile, ILD failure rate is reduced effectively. For failure analysis, ILD crack induced CSAM "white spots" is also studied. With smaller and narrower ILD crack size, typical "white spots" turns to "black spots". Failure location, waveform peak magnitude, anomality shape are ley factors to identify ILD crack induced "black spots" and other CSAM failure mode.

## REFERENCE

[1] Edelstein, Dan, et al. "Full copper wiring in a sub-0.25/spl mu/m CMOS ULSI technology." *International Electron Devices Meeting. IEDM Technical Digest*. IEEE, 1997.

[2] Yoo, H. J., et al. "Demonstration of a reliable high-performance and yielding Air gap interconnect process." *2010 IEEE International Interconnect Technology Conference*. IEEE, 2010.

[3] Daamen, Roel, et al. "Multi-level air gap integration for 32/22nm nodes using a spin-on thermal degradable polymer and a SiOC CVD hard mask." *2007 IEEE International Interconnect Technology Conferencee*. IEEE, 2007.

[4] Iacopi, Francesca, et al. "Challenges for structural stability of ultra-low-k-based interconnects." *Microelectronic engineering* 75.1 (2004): 54-62.

[5] Garner, Luke, et al. "Finding Solutions to the Challenges in Package Interconnect Reliability." *Intel Technology Journal* 9.4 (2005).

[6] Raghavan, Sathyanarayanan, et al. "Study of chip–package interaction parameters on interlayer dielectric crack propagation." *IEEE Transactions on Device and Materials Reliability* 14.1 (2013): 57-65.

[7] Iacopi, Francesca, et al. "Challenges for structural stability of ultra-low-k-based interconnects." *Microelectronic engineering* 75.1 (2004): 54-62.

# THE STUDY ON WARPAGE OF EPOXY MOLDING COMPOUND

*Yangyang duan[1]\*, Wei Tan[1], Hongjie Liu[1], Lanxia Li[1], Dandan Fan[1], Lingling Liu[1], Xingming cheng[1]*
Jiangsu Hua Hai Cheng Ke Advanced Material Co. Ltd., Lianyungang 222047, China
\*Corresponding Author's Email: yangyang.duan@hhck-em.com

## ABSTRACT

There are many factors have effects on the warpage control of epoxy molding compound (EMC). Some key factors such as epoxy resin, filler content and stress reliever were selected to design different formulations of EMC to study the effect on warpage. Epoxy resin A with the stucture in Fig.1a and epoxy resin B with the stucture in Fig.1b were used to the formula. Varying the weight ratio of epoxy resin A/epoxy resin B resulted in products with different warping, ranging from short-edged smiling to long-edged crying. The same transformation were also observed with the filler content increased. In addition, stress reliever are used to sovle the problem of warpage varies greatly for different thickness of epoxy molding compound.

Keywords：epoxy molding compound; warpage; epoxy resin; filler content; stress reliever

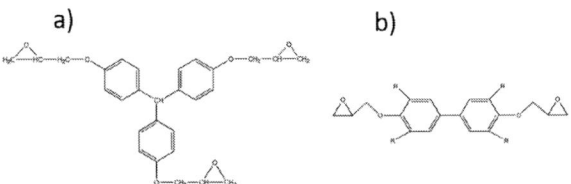

*Figure 1:a）Stucture of epoxy resin A ;b) Stucture of epoxy resin B*

## INTRODUCTION

With the rapid development of the advance integrated circuit (IC) packaging, more and more challenges have been put forward to the packaging materials. For example, one sided asymmetric package such as QFN, BGA and FBP[1] is easy to generate warpage. The warpage performance of one sided asymmetric package is very critical to the board SMT and later quality control.

There are many factors have effects on the warpage control, include assembly materials of epoxy molding compound, substrates, package dimension, etc. The epoxy molding compound is one of the most important factor to control the warpage in the packaging process. The EMC consists of epoxy resin, Phenol resin, filler, catalyst, coupling agent, release agent, stress release agent, coloring agent, flame retardant, ion trapping agent and so on. There are multiple factors which will impact warpage behavior, such as chemical shrinkage due to polymerization conversion during post molding cure of EMC and physical shrinkage due to expansion coefficient mis-matching[2-4]. Some key factors such as different types of epoxy resin, filler content and content of stress reliver

were selected to design different formulations to study the effect on warpage.

## EXPERIMENTAL

Epoxy resin A was provided by Huntsman Advanced Materials. Epoxy resin B was got from Dainippon Ink & Chemicals Incorporated. Silicon dioxide as filler was produced by Novoray Corporation. Fig.2 shows the morphology of silicon dioxide and it has spherical shape. Phenolic resin as harder with the multiaromatic (MAR) structure was bought from Meiwa Corporation. The catalyst is a latent catalyst. Stress reliever was provided by Shin-Etsu. Coupling agent and wax were received from Momentive and Clariant respectively.

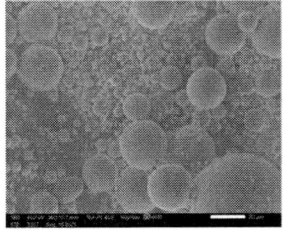 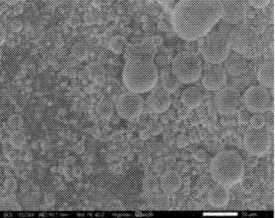

*Figure 2: SEM images of silicon dioxide with spherical shape.*

Epoxy molding compound were prepared as described in reference[5]. For a typical process, all the ingredients were weighted up according to the formulation A and mixed in the high speed mixing machine. After this, the mixture was feeding to double screw extruder under the heating condition for extrusion, and then the discharged material go through sheet formation, cooling, pre-braker, granulation, post blend.

For basic formula, the weight ratio of silicon dioxide/epoxy resin /phenolic resin/ catalyst/ coupling agent/wax/Stress reliever/other addictives is 88:5.86:2.65:0.24:0.6:0.28:0.4:1.97. Other addictives including colorants, silicone oil and ion trapping agents etc. By varying the weight ratio of epoxy resin A/epoxy resin B, the silicon dioxide content and the content of stress reliever, epoxy molding compound with different warpage and thermodynamic/dynamic mechanical properties can be readily fabricated.

The specimens for warpage test with plane bump package (FBP)[1]package were prepared through transfer molding method with the temperature as 175℃ and cure time as 140s followed by a six-hour post mold cure at the same temperature. The warpage after post mold cure (PMC) were measured. For different packing form of FBP package, thickness of EMC was difference. The thickness of EMC corresponding to 0.55mm, 0.35mm and 0.27mm

978-1-6654-9759-6/22 $31.00 © 2022 IEEE

while the long side of 120mm and the short side of 64mm were labeled as FBP-1, FBP-2 and FBP-3, respectively. As shown in Fig. 3, there are two different warping types the smiling warpage and the crying warpage. The warping of the short side is generally represented by the smiling face which the short side curves up when the epoxy molding compound is on top. while the long side is crying which the long side curves down when the epoxy molding compound is on top. Other test specimens were prepared with mold cure time of 120sec at 175℃ followed by a six-hour post mold cure at the same temperature. For thermodynamic properties, Tg and coefficient of thermal expansion were measured by TMA. For dynamic mechanical properties, storage modulus were measured by DMA. TA instrument model, from 25℃ to 280℃ at 3℃ per minute. Further more, the gel time (GT), spiral flow (SF) and hot hardness (HH) of molding 90 s of the EMC samples were tested according to SJ/T 11197-1999.

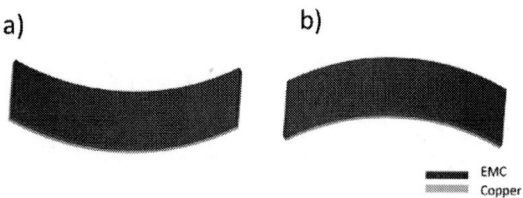

*Figure 3:a) the smiling warpage and b) the crying warpage.*

## RESULT AND DISCUSSION
**Effect of the weight ratio of epoxy resin A/epoxy resin B**

The warpage of FBP-2 package with weight ratio of epoxy resin A/epoxy resin B at 0.5,1,2,4 (labeled as S1, S2, S3, S4) are shown in table 1. With the epoxy A/epoxy B weight ratio of 0.5, FBP package with smile warpage in short side direction were formed, which the warpage value was 3.6mm. When the ratio was increased to 1 and further to 2, the warpage value in short side direction was gradually decreased. There is almost no long side warping with the weight ratio of epoxy resin A/epoxy resin B from 0.5 to 2. The further increasing of the weight ratio of epoxy A/epoxy B to 2 led to the smile warping in short side direction with the value of 0.55mm. Meanwhile, the warpage of the long side direction was appeared with the value of 1.50mm. These results demonstrates that as the proportion of epoxy resin A increased, the shrinkage becomes smaller. These changes may be explained by the Tg covert.

Fig.4 is a typical Tg transition curve measured by TMA. It is clear that the rate of dimension change is smaller before Tg (part A) and increasing after Tg (part B). Higher Tg led to the region with a higher rate of dimension change get smaller, which is conducive to control shrinkage. As shown in table 1, with weight ratio of epoxy resin A/epoxy resin B from 0.5 to 4, the Tg increased from

105℃ to 135℃. That's why the warping gets smaller in short side direction. At the same time, with weight ratio from 0.5 to 4, the spiral flow decreased from 162cm to 128cm. The drop in liquidity also need attention.

TABLE Ⅰ. EFFECT OF THE WEIGHT RATIO OF EPOXY RESIN A/EPOXY RESIN B

| Formula | S1 | S2 | S3 | S4 |
|---|---|---|---|---|
| Epoxy resin A/Epoxy resin B | 0.5 | 1 | 2 | 4 |
| GT(s) | 38 | 37 | 36 | 36 |
| SF(cm) | 162 | 145 | 135 | 128 |
| 90s HH | 83 | 85 | 83 | 80 |
| Tg(℃) | 105 | 110 | 120 | 135 |
| α1(×10-6/℃) | 8.08 | 7.86 | 8.12 | 7.96 |
| α2(×10-6/℃) | 35.52 | 35.80 | 35.45 | 35.54 |
| The warpage value in short side (mm) | 3.7 | 1.95 | 1.10 | 0.55 |
| The warpage value in long side (mm) | <0.5 | <0.5 | <0.5 | 1.5 |

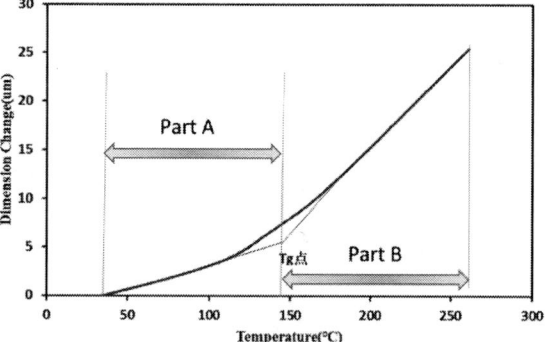

*Figure 4:A typical Tg transition curve measured by TMA*

**Influence of the silicon dioxide content**

The impact of the silicon dioxide content have also investigated, with epoxy resin A/epoxy resin B at 2 and other materials fixed, on warpage of FBP package. The warpage of FBP-2 package with silicon dioxide content at 87%, 88%, 89% (labeled as C1, C2, C3) are shown in table 2. When silicon dioxide content increased from 87% to 88%, the warpage value in short side direction was decreased from 1.95mm to 1.1mm. The further increasing of the silicon dioxide content to 89%, the warpage value in short side direction went down to less than 0.5mm. But the warping become serious for the long side warpage getting worse. The warpage decrease in short side direction cause by increasing filler content indicates a smaller shrinkage.

Fig. 5 shows the TMA curves of C1, C2, and C3. The higher silicon dioxide content led to the lower rate of dimension change after Tg. As shown in table 2, with silicon dioxide content increased from 87% to 89%, the α2 decreased from 37.13×10-6/℃ to 33.62×10-6/℃. Increasing the filler content is benefit to reduce physical shrinkage and improvements in CTE mismatch to control the short direction warping[2-4]. But if adding too much, the shrinkage is too small and it is easy to produce long side warping.

TABLE Ⅱ. EFFECT OF THE SILICON DIOXIDE CONTENT OF 87%, 88%, 89%

| Formula | C1 | C2 | C3 |
|---|---|---|---|
| silicon dioxide content(%) | 87 | 88 | 89 |
| GT(s) | 35 | 36 | 32 |
| SF(cm) | 146 | 135 | 105 |
| 90s HH | 81 | 83 | 84 |
| Tg(℃) | 120 | 119 | 121 |
| α1(×10-6/℃) | 8.0 | 8.7 | 8.7 |
| α2(×10-6/℃) | 37.13 | 35.15 | 33.62 |
| The warpage value in short side (mm) | 1.95 | 1.1 | <0.5 |
| The warpage value in long side (mm) | <0.5 | <0.5 | 4.5 |

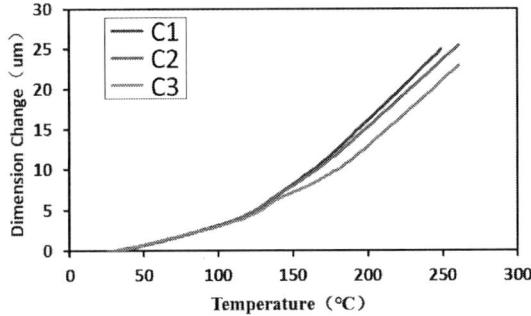

Figure 5: TMA curves of C1, C2 and C3 with the silicon dioxide content of 87%, 88%, 89%, respectively.

### Influence of stress reliever

The warpage with different thickness of EMC corresponding to 0.55mm, 0.35mm and 0.27mm labeled as FBP-1, FBP-2 and FBP-3 were shown in table3. Formula S4 were used to this experiment. As shown in table 3, the thinner of EMC, the weaker to resist warpage. It is more pronounce in FBP-3 with the thickness of EMC 0.27mm.

TABLE Ⅲ. WARPING PERFORMANCE WITH DIFFERENT THINKNESS OF EMC

| Different thickness of EMC | FBP-1 | FBP-2 | FBP-3 |
|---|---|---|---|
| The warpage value in short side (mm) | <0.5 | 0.55 | 0.6 |
| The warpage value in long side (mm) | 1.2 | 1.5 | 4.6 |

In order to get a low-warpage product for different thickness of EMC, stress reliever are essential. Functional silicone rubber with linear structure and hydroxyl equivalent of 2294 as a stress release agent used to study the effects of warping. When the silicon rubber were added to 1.1, different thickness of EMC were all low-warpage with the warpage value less than 2.0mm for

both short side and long side. The warpage change described above may be attributed to the change of Storage Modulus.

Fig. 6 shows Storage Modulus - Temperature patterns with silicon rubber of 0.4 and 1.1. With silicon rubber of 1.1, Storage Modulus is higher at room temperature and form 25℃ to 175℃ the change is much greater. Storage Modulus is actually young's Modulus, which is the index of material rebound after deformation, indicating the ability of material to store elastic deformation energy. High Storage Modulus at room temperature indicates that it has strong resistance to warpage. A greater amount of variation form 25℃ to 175℃ indicates a more complete deformation of the release. That's why stress relievers work for warping.

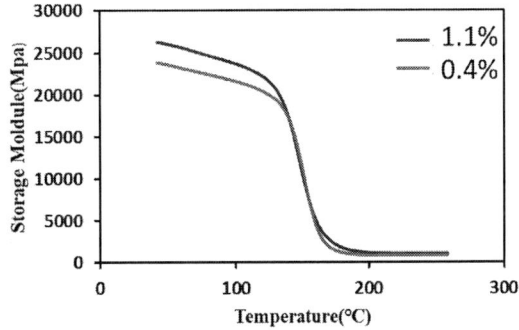

Figure 6:DMA curves with silicon rubber of 1.1 and 0.4

## CONCLUSION

Varying the weight ratio of epoxy resin A/epoxy resin B from 0.5 to 4 resulted in products with different warping, ranging from short-edged smiling to long-edged crying. The Tg increased from 105℃ to 135℃ , which is conducive to control chemical shrinkage. The decrease of shrinkage is benefit to reduce the smiling-warpage .But if the shrinkage is too small it will bring out the crying-warpage.

The same transformation were also observed with the filler content increased. with silicon dioxide content increased from 87% to 89%, the α2 decreased from 37.13×10-6/℃ to 33.62×10-6/℃. Increasing the filler content is benefit to reduce physical shrinkage and improvements in CTE mismatch to control the short direction warping. But if adding too much, the shrinkage is too small and it is easy to produce long side warping.

In addition, stress reliever are used to sovle the problem of warpage varies greatly for different thickness of epoxy molding compound. Finally, low-warpage produce were manufactured for different thickness of EMC with the warpage value less than 2.0mm for both short side and long side.

## REFERENCES

[1] Z. Liang, Y. Tao. *Flat bump package* . Electronics&Packing. 2006, 34, pp. 5-8.

[2] P. Liu, L. Tong, et al. *Warpage analysis of FBGA*

*packaging during assembling process.* Electronic component and Materials. 2014, 33, pp. 48-51.

[3] G. Yu, L. Tan, et al. *Investigation on Block Warpage of FBGA Package.* Equipment for Electronic Products Manufacturing.2014, 232, pp.19-23.

[4] W. Tan, et al. *Requirements and solutions of epoxy molding compound for high-end packaging.* Equipment for Electronic Products Manufacturing. 2015, 241,pp.40-42..

[5] W. Tan, H. Liu, Y. Duan. *The study on the mold ability and reliability of epoxy molding compound.* CSTIC, 2017.

# STATE OF THE ART METAL DEPOSITION SYSTEM FOR ADVANCED UBM, RDL AND FAN-OUT WAFER LEVEL PACKAGING

*Clinton Goh[*], Junqi Wei, Tuck Foong Koh, Kang Zhang, Kelvin Boh, Hannah Tang, Bridger Hoerner*
Applied Materials Singapore Technology Pte Ltd.
*Corresponding Author's Email: Clinton_Goh@amat.com

## ABSTRACT

As we enter the fourth wave of computing, advanced packaging becomes a key enabler of Moore's law scaling. Two of the key building blocks: 1D Under Bump Metallization and 2D Re-Distribution Layer (RDL) form critical interconnects in Fan-Out Wafer Level Packaging (FOWLP) which deliver benefits in power, performance, area, cost and time to market (PPACt). The use of polymeric materials in FOWLP give rise to several challenges for PVD, such as low temperature processing, warpage control and contact resistance. This work presents a state-of-the-art multi-chamber PVD platform capable of achieving >45% contact resistance reduction, and >30% productivity improvement over legacy PVD solutions without exceeding thermal budgets required.

## INTRODUCTION

In 1965, Gordon Moore predicted not only doubling of transistor density but also the important role which advanced packaging would have to play in the future: *"it may prove to be more economical to build large systems out of smaller functions, which are separately packaged and interconnected"* [1]. In today's AI era, designers are no longer able to get all the circuits they want into a single die anymore. And with the high cost of shrinking, the cost reductions can no longer keep up with Moore's law.

Current state-of-the-art advanced packaging using wafer fab equipment to stack and connect multiple logic, memory and specialty chiplets to optimize the system for higher performance, lower power consumption, smaller area and lower costs, while accelerating time to market. As a result, designers are looking to heterogeneous design to create an alternative way of continuing Moore's law.

The architecture shown in Figure 1 represents how heterogeneous design allows us to combine logic, memory, and specialty devices to deliver optimized system performance. Here, multiple DRAMs are stacked and connected to a logic die stack. The 3D memory stacks, and logic stacks are connected using dense 1D Under Bump Metallization (UBM) and 2D Re-Distribution Layer (RDL) connections incorporated into advanced substrates such as High-Density Fan-Out Polymer RDL and Silicon Interposers. This approach delivers high data bandwidth and much lower power consumption. It also allows us to pack more transistors in a unit area, delivering clear Power, Performance, Area, Cost and Time to market (PPACt®) benefits [2].

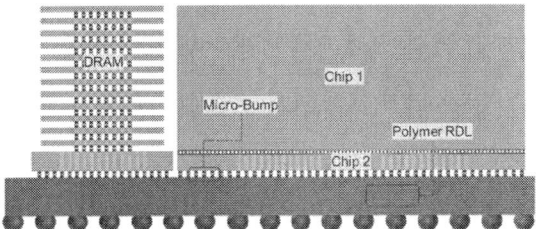

*Figure 1: Advanced Packaging Enabled Multi-Chip Heterogenous System (adapted from [2])*

## PVD CHALLENGES IN FOWLP

The Fan-Out RDL-first wafer level packaging process involves creating Cu interconnects on polymeric materials. Figure 2 describes the typical process flow [3].

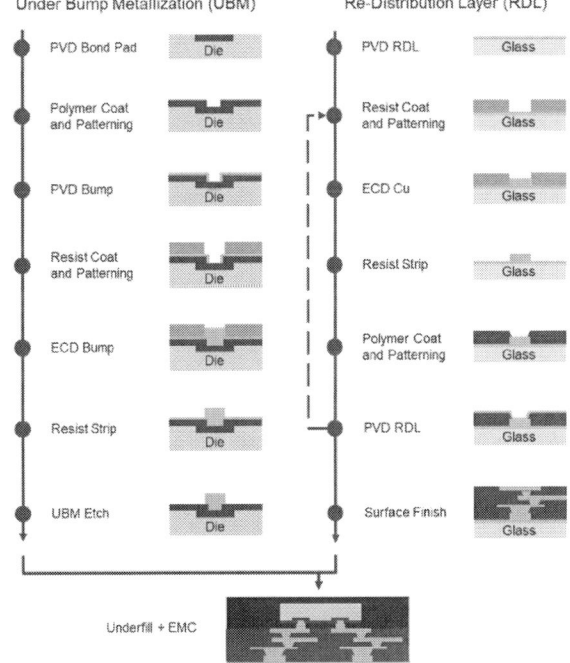

*Figure 2: Process flow of Fan-Out RDL-first WLP*

The use of polymeric materials such as Polyimide, PBO and EMC present several challenges to PVD as shown in Figure 3. Firstly, thick polymers retain a lot of moisture, and if substrates are not degassed sufficiently, this may lead to poor adhesion and PVD delamination. Next, contact resistance challenges are faced during bumping as polymer outgassing contaminates the metal bond pad and creates a poor ohmic interface between bump and bond pad. These challenges if not addressed sufficiently result in poor mechanical reliability and electrical performance.

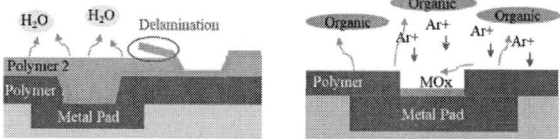

*Figure 3: Moisture (H2O) outgassing resulting in poor PVD adhesion (left) and organic outgassing resulting in poor contact resistance (right)*

## 2-Stage Advanced Degas vs. Traditional Degas

To enable a fast degas process for high moisture retaining polymers, we introduce a 2-stage degas approach which combines a multi-wafer pre-heat step with a second stage high efficiency single wafer degas. Pre-heating the wafers enables drier wafers to enter the degas chamber. The second stage uses advanced volumetric heating technology to degas single wafers quickly and efficiently.

Firstly, we benchmark the performance of a single stage, single wafer traditional heater degas against a 2-stage multi-wafer pre-heat followed by heater degas. Substrates used were Fan-Out RDL wafers consisting of reconstituted Si dies embedded in EMC. Degas efficiency is represented by post degas vacuum chamber pressure. Figure 4 shows that duration needed for the single wafer heater degas to achieve similar residual moisture outgassing was reduced by 10x by the addition of the pre-heat step. This enhances the productivity of the PVD system when processing Fan-Out RDL substrates.

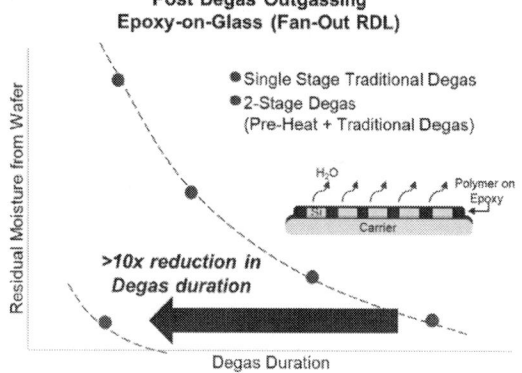

*Figure 4: Performance of 2-Stage Degas using a multi-wafer Pre-Heat with Single Wafer Heater Degas*

Next, we benchmark the performance of advanced volumetric heating technology against traditional heater degas. Figure 5 shows that the advanced degas reduces post degas residual moisture by 4x as compared to traditional heater. This highly effective method of degassing wafers also enables process integration to be simplified. Typical Fan-Out process requires wafers to be pre-baked in large batch ovens before entering the PVD system. These ovens take up clean room space and queue time control is needed for batch-to-batch repeatability. Figure 6 shows that even without a pre-bake step, advanced degas can achieve lower post degas outgassing versus traditional heater degas with pre-bake.

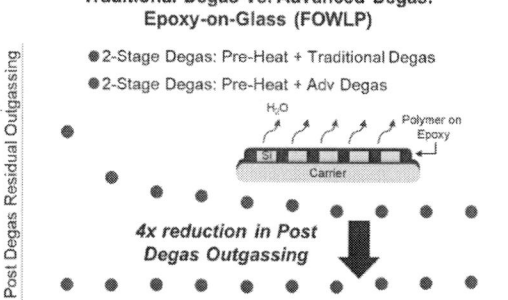

*Figure 5: Performance of Traditional Degas vs. Advanced Degas for Fan-Out RDL-first wafers*

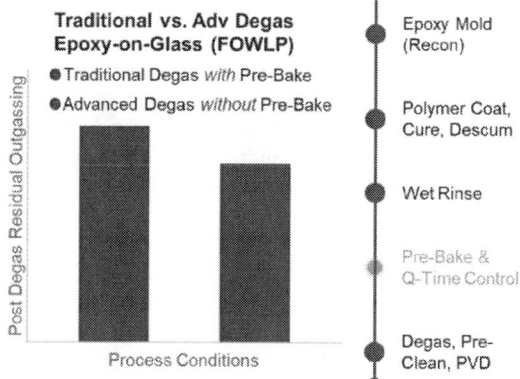

*Figure 6: Elimination of pre-bake and queue time control by advanced volumetric degas technology*

## Advanced Pre-Clean

Organic outgassing from polymers take place during sputter etch due to the presence of high vacuum UV energy and weak polymeric bonds. If not controlled, carbon and oxygen containing organic by-products can contaminate the underlying metal bond pad, resulting in a non-ohmic contact between the metal barrier and metal bond pad. TEM EDX analysis of the Bump and Bond Pad interface shown in Figure 7 indicates the presence of Carbon, Oxygen and Fluorine elements in a standard pre-clean sputter etch chamber which was not optimized for treating substrates with polymer passivation [6].

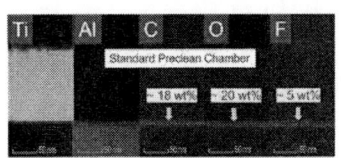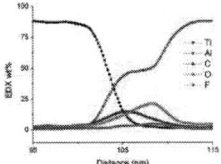

*Figure 7: Presence of C, O and F elements found at the Ti/Al interface using standard pre-clean technology (adapter from [6])*

To ensure minimal organic contamination of the metal bond pad, we introduce an advanced pre-clean sputter etch chamber based on capacitively coupled plasma (CCP) technology. The advanced CCP sputter etch chamber is capable of actively cooling the substrate using an electrostatic chuck, thus minimizing the amount of organic outgassing. Next, the chamber's unique ultra-high conductance design ensures that any organic outgassing present in the chamber is quickly pumped away. Lastly, the CCP design avoids using RF coils unlike alternative inductively coupled plasma (ICP) technology [4, 5]. This enables the use of a metal shield which enables in-situ metal pasting. This minimizes defects and reduces any secondary outgassing from re-sputtered polymers.

We benchmark the electrical performance of the advanced pre-clean technology against standard pre-clean using standard Kelvin probe contact resistance test vehicles as shown in Figure 8. These test vehicles were fabricated with PBO passivation and an underlying Aluminum bond pad on 300mm Silicon wafers [6].

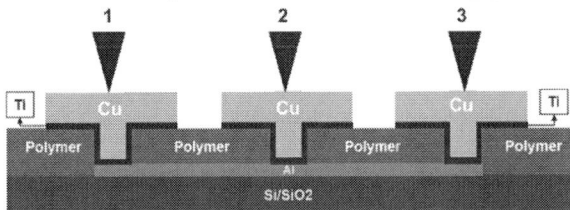

*Figure 8: Kelvin probe contact resistance test vehicle*

By minimizing organic outgassing and effectively removing it from within the process cavity, the advanced pre-clean chamber achieves up to 48% reduction in contact resistance for the smallest vias down to 10um opening, as shown in Figure 9, thus enabling low power consumption for the most advanced devices integrated in a high density Fan-Out RDL-first wafer level package.

## CONCLUSIONS

The ability to efficiently degas and minimize metal pad contamination during PVD processing is key for ensuring cost effectiveness, mechanical and electrical reliability of Fan-Out RDL-first wafer level packaging. The work presented here highlights the state-of-the-art technologies developed by Applied Materials to effectively deal with moisture and organic outgassing challenges faced during Fan-Out PVD. The 2-stage degas

approach combining a pre-heat step with an advanced volumetric degas offers the shortest degas time with the ability to eliminate pre-bake and queue-time control steps prior to PVD. And the advanced CCP pre-clean technology achieves best-in-class contact resistance performance for the most advanced devices. When integrated on Applied Materials' Charger® PVD system, these technologies make it an ideal product solution for UBM, RDL and Fan-Out RDL volume production.

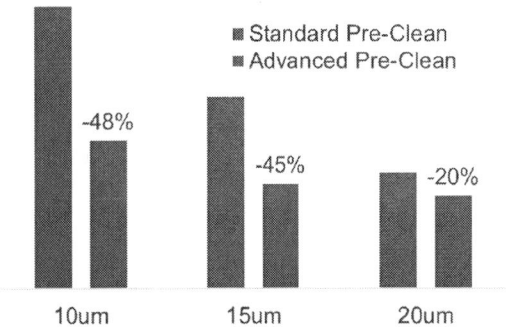

*Figure 9: Contact resistance reduction achieved by Advanced CCP Pre-Clean technology*

## REFERENCES

[1] G. E. Moore, "Cramming more components onto integrated circuits," *Electronics,* vol. 38, no. 8, pp. 114-117, 1965.

[2] Applied Materials Inc., "2021 ICAPS and Packaging Master Class," 2021. [Online]. Available: https://investors.appliedmaterials.com/events/event-details/2021-icaps-and-packaging-master-class.

[3] J. H. Lau, "FOWLP: Chip-Last or RDL-First," in *Fan-Out Wafer-Level Packaging*, Singapore, Springer, 2018.

[4] P. Carazzetti et. al., "Impact of Process Control on UBM/RDL Contact Resistance for Next-Generation Fan-Out Devices," in *IEEE 67th Electronic Components and Technology Conference*, 2017.

[5] C. Jones et. al., "UBM/RDL Deposition by PVD for FOWLP in High Volume Production," in *IEEE 68th Electronic Components and Technology Conference*, 2018.

[6] K. Zhang e. al., "Advanced Preclean Chamber for UBM/RDL Contact Resistance Improvement in Advanced Node Packaging Applications," in *International Wafer Level Packaging Conference*, 2020.

# GLASS CARRIERS FOR ADVANCED PACKAGING

*James Li and Jay Zhang*
Corning Incorporated
Corning, New York, USA
zhangjj@corning.com

## ABSTRACT

Glass carriers play an important role in today's advanced packaging fields. We describe two major industrial applications: device wafer thinning and fan-out wafer level packaging. In wafer thinning, glass offers CTE matching, low TTV and superior flatness, and supports laser and mechanical debonding. For fan-out, glass can deliver tailorable CTE in the range of 3.4-12.6 ppm/C, high stiffness through engineered high Young's modulus, and flexible thickness to control in-process warp in both chip-first and RDL-first scenarios. For laser debonding using UV light, glass can be further engineered to provide high UV transmission to increase throughput and reduce cost.

## INTRODUCTION

### Why glass

Glass is an amorphous material of unlimited compositional possibilities with a wide a range of properties. Modern finishing technology can make flat wafers of glass at low cost, making glass wafers that can meet semiconductor packaging needs, specifically as a carrier to support wafer processing. As a carrier, the glass is not in the final product stage. Proper handling (e.g., avoidance of sharp contact with hard materials) makes glass carrier reusable, thus cost efficient.

### Light-based debonding

The transparent nature of glass makes light-based debonding a natural fit. Compared to alternative debonding methods, such as thermal and mechanical, light-based debonding offers low stress, and is attractive in cases where the supported device wafer is prone to breakage due to being very thin or structurally weak. High throughput is another benefit of light-based debonding.

### Mechanical debonding

Evolution of advanced packaging technologies started with mechanical debonding using non-glass carrier wafers made of Si, ceramic, or metal. Mechanical debonders tend to be less costly as well. Working with our partner, Suss Microtec, we will demonstrate fully automated mechanical debonding with glass carriers. This makes current and future owners of mechanical debonders suitable users of glass wafers that deliver these attractive attributes.

## GLASS CARRIERS FOR WAFER THINNING

A wide range of device types, device performances, or form factor requirements drive the need for wafer thinning. Prime examples include high bandwidth memories (HBMs) and power devices such as IGBTs. This wafer thinning trend continues for the foreseeable future [1].

### Si thinning

Si-based semiconductor chips represent a great majority of all chips, with more than of all Si chips thinned during chip packaging. Traditional thinning is supported with a tape and vacuum chucking. Consensus is that as final thickness becomes much less than 100um rigid backing, such as a carrier wafer, must be used to ensure high thinning yields. Post-thinning processing, such as adding RDLs, also requires rigid carrier support. Corning's SG3.4 carrier wafers is the industry's workhorse product that provides superior carrier attributes such as low TTV (total thickness variation), low bow and warp, excellent CTE match to Si (see Figure 1), and availability in 0.3-1.1mm thicknesses.

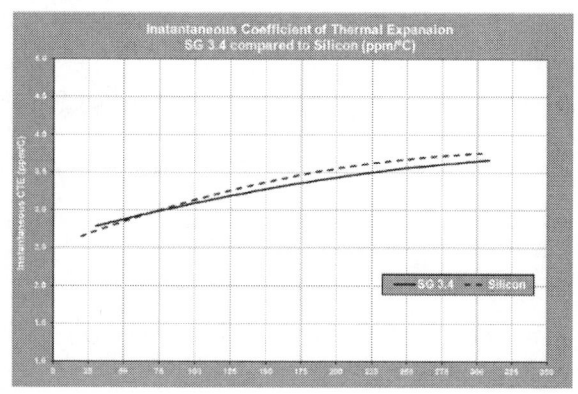

*Figure 1: Corning's SG3.4 glass matches Si CTE in a wide temperature range, supporting Si thinning and post-thinning processing.*

For users of UV light sources, Corning also offers a higher UV transmission version of SG3.4, called SG3.4HT, which enables more efficient use of UV photons to bring benefits such as higher debonding throughput, lower equipment maintenance and consumables, and less heating of the carrier during use. Figure 2 shows UV transmission characteristics and its dependence on carrier thickness. Given the glass type, thickness is another lever to effect UV transmission.

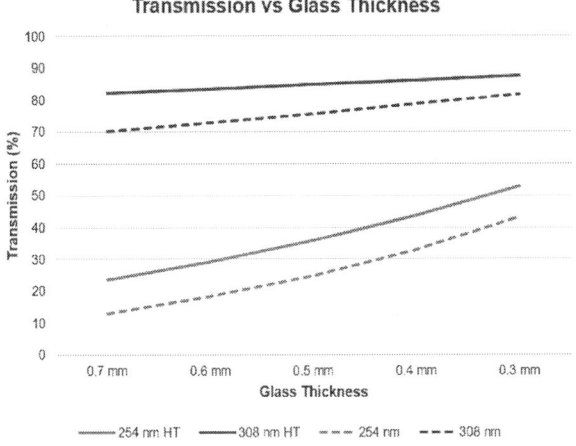

*Figure 2: Corning's SG3.4HT (solid lines) offers higher UV transmission than regular SG3.4 (dashed lines).*

**Mechanical debonding**

While glass enables efficient debonding using light, its mechanical properties render it suitable for mechanical debonding. Figure 3 illustrates initiation of debonding using Suss Microtec's fully automated debonder and a summary of the debonding results. Both materials A and B in the table are commercially available adhesives. Some adhesive residues were found at the initiator position on the glass for material A, and small adhesive traces were found on the device wafer using material B. These residues can be easily removed, and glass carriers are in excellent condition and can be readily reused.

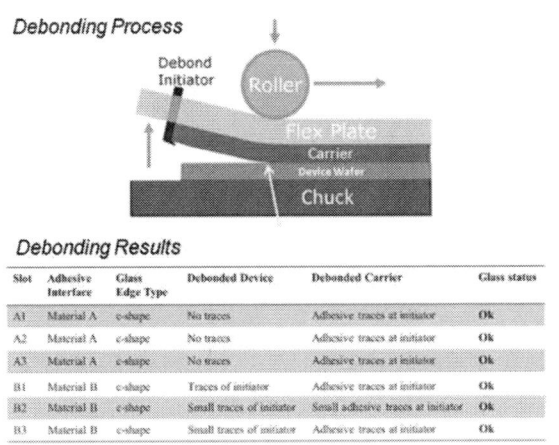

*Figure 3: Glass is not only suitable for light-based debonding, it can also support mechanical debonding.*

**Non-Si thinning**

While Si is the highest volume semiconductor material, compound semiconductor materials and piezoelectric materials have become more important in meeting the demanding requirements of power, RF, light generation, and display applications. The table lists some examples with their respective thermal expansion coefficient (CTE, ppm/K). Those showing a range have strong crystalline orientational dependency.

| Type | Si | SiC | GaN | GaAs | sapphire | LiTaO3 |
|------|-----|------|--------|--------|----------|--------|
| CTE | ~3 | ~4 | 3-5.6 | ~6 | 6-7 | 4-16 |

Glass is an ideal carrier wafer material because it offers CTE that matches the wafer type to be thinned, its Young's modulus engineered to be relatively high, and its thickness can be optimized within the constraints of the application. Transparency of glass makes light-based debonding possible and enables simple bonding quality inspection and control.

Corning developed a new series of products called Advanced Packaging Carriers (APC). Figure 4 shows the range and increment the CTE APC glasses cover and the Corning standard glass types that are already used in semiconductor packaging. The APC group not only expands the CTE range but shows higher Young's modulus by glass design. APC wafers are more flexible in thickness and are not limited by typical sheet forming process limitations.

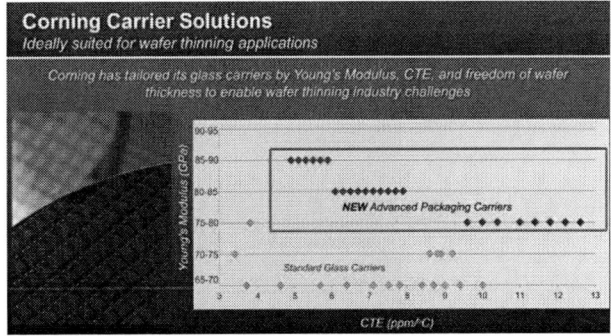

*Figure 4: Corning offers a wide range of CTEs with fine granularity to enable thinning of any compound semiconductor wafers.*

Using GaAs CTE in the temporary bonding temperature range, we identified the most suitable APC glass with CTE of 6.9, product code HS6.9. This ensures that the GaAs/glass stack has close to zero warp after cooling to room temperature from the adhesive curing temperature where we expect stress-free between the wafer pair. Figure 5 plots modeling and experimental results to compare HS6.9 APC glass performance with sapphire and sodalime glass (SLG); CTE-matched APC wafer clearly outperforms the alternatives.

Figure 6 shows an image of HS6.9 supported GaAs that has successfully undergone back-grinding to reduce GaAs thickness from 580um to 125um with excellent performance in total thickness variation (TTV).

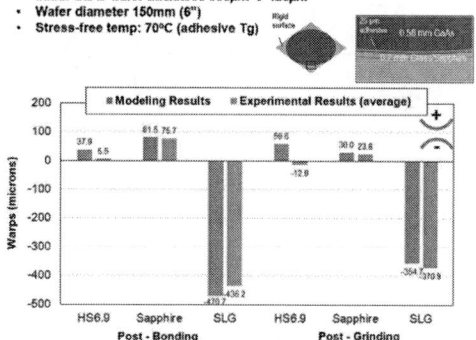

**In-process warp of bonded GaAs wafers: Modeling and experimental data**

Modelling assumption:
- High temperature GaAs wafer bonding with different substrate
- Substrate thickness 0.7mm and adhesive thickness 25um
- Initial GaAs wafer thickness 580μm → 125μm
- Wafer diameter 150mm (6")
- Stress-free temp: 70ºC (adhesive Tg)

*Figure 5: APC glass HS6.9 with close CTE match to GaAs at adhesive curing temperature outperforms sapphire and SLG.*

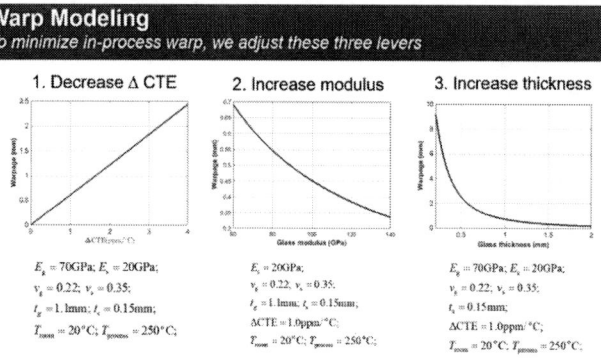

**Bonded GaAs on HS6.9 after back grinding**

| Carrier | | Org. | Final Thickness (um) (Measurement points) | | | | | | |
|---|---|---|---|---|---|---|---|---|---|
| Material | Thickness (um) | thickness | 1 | 2 | 3 | 4 | 5 | Ave | TTV |
| HS6.9 | 700 | 1383 | 824.8 | 825.3 | 824.9 | 824.7 | 825.0 | 824.9 | 0.6 |

*Figure 6: APC glass HS6.9 as carrier successfully supported back grinding of GaAs from 580um to 125um.*

## GLASS CARRIERS FOR FANOUT PACKAGING

Fan-out wafer level packaging (FOWLP) has seen growing adoption by foundries as well as OSATs in recent years. There is industry consensus that glass is the material of choice for the carrier in high density scenarios where epoxy molding, multi-layer RDLs, and other processing steps make warp control challenging.

*Figure 7: Warp modeling points to three levers to control warp in FOWLP: minimize CTE mis-match, increase Young's modulus, and maximize carrier thickness to the extent possible.*

APC wafers are ideal to support high density fan-out. Figure 7 shows warp modeling results highlighting three key levers to minimize in-process warp:

1. CTE match to the extent possible. When multiple CTE materials are used, there will be an optimal carrier CTE to minimize the highest warp
2. Increase Young's modulus of the carrier
3. Use the highest possible thickness

Corning's APC wafers offer wide CTE coverage plus fine CTE granularity to allow the user to choose the most ideal CTE, high stiffness because of high Young's modulus, and all wafers are available in flexible thicknesses up to 3mm. In addition, these wafers have excellent flatness and surface smoothness. Furthermore, the users will receive the same quality wafers during R&D phase as well as mass production, ensuring smooth transition in product development. Figure 8 summarizes the benefits of using glass carriers for fan-out.

| | | Benefits |
|---|---|---|
| **Properties of Glass** | Range of CTEs in fine granularity | ↓ ΔCTE → ↓warp |
| | Optical transparency | Enables light based bonding/debonding |
| | Flatness & smoothness | Enables thinner device wafer <<50 μm |
| | High stiffness | ↓warp |
| | Size of glass substrates | ↓cost + ↓handling damage |
| | Variety of thicknesses | ↓warp + ↓handling damage |
| | Manufacturing scalability | Enables R&D + ↓ cost |

*Figure 8: Corning's Advanced Packaging Carrier (APC) products are serving the top FO companies worldwide due to its technical performance and support for both R&D and MP.*

## CONCLUSION

Glass carriers are ideal for wafer thinning and high-density fan-out applications. Glass wafers support both light-based and mechanical debonding. The APC wafers add to the existing glass portfolio to cover a wider range of CTE with finer granularity to enable effective product development.

## REFERENCES

[1] Yole Développement, *Thinning Equipment Technology & Market Trends for Semiconductor Devices.*

## ACKNOWLEDGEMENTS

The authors acknowledge the mechanical debonding work performed at Suss Microtec. Corning contributors to this work include Indy Dutta, Andy Teng, Yu Xiao, Steven Lin, Erica Chang, and Julia Brueckner.

# DIELECTRIC PROPERTY DESIGN BASED ON BaTi$_2$O$_5$ NANORODS AND BaTiO$_3$ NANOPARTICLES COUPLE AND ITS APPLICATION IN EMBEDDED CAPACITOR

*Wenzhong Zou[1*], Guoyun Zhou[1,2], Wei He[1,2*], Shouxu Wang[1,2], Baoliang Ren[3], Zesheng Wen[4], Quanyong Wang[4], Yongqiang Xu[4]*

[1] School of Materials and Energy, University of Electronic Science and Technology of China, Chengdu 610054, China

[2] Jiangxi Electronic Circuit Research Center of UESTC, Pingxiang 337011, China

[3] Pingxiang Fengdaxing Circuits Co., Ltd., Pingxiang 337011, China

[4] Ganzhou Sun&Lynn Circuits Co., Ltd., Ganzhou 341000, China

[*]Corresponding Author's Email: 202011030514@std.uestc.edu.cn, he_wei@uestc.edu.cn

## ABSTRACT

In this work, a composite structure based on BaTi$_2$O$_5$ (BT2) nanorods and BaTiO$_3$ (BT) nanoparticles couple was designed to explore the impact of the nanorods and nanoparticles on composite properties. For this purpose, huge numbers of high-purity BT2 nanorods and BT nanoparticles have been obtained respectively by molten-salt synthesis. The mixture of prepared BT2 and BT was filled into the epoxy resin as a whole to fabricate the embedded capacitor copper clad laminate (ECCCL) with a filler mass fraction of 60 %. As the content of BT2 increases, the relative permittivity of the ECCCLs shows an increasing trend, and the relative permittivity of ECCCL filled with 100 wt% BT2 nanorods reaches a maximum of 47@1 MHz.

**Keywords**—BaTi$_2$O$_5$ nanorods, BaTiO$_3$ nanoparticles, molten-salt synthesis, embedded capacitor copper clad laminate, relative permittivity

## INTRODUCTION

BaTiO$_3$ (BT) is a representative dielectric material with high relative permittivity, and lots of works have made to fabricate the dielectric composites by using different morphologies BT nanomaterials as fillers [1-3].

BaTi$_2$O$_5$ (BT2) is a very promising dielectric material. BT2 possesses a high Curie temperature ($T_C$ is 430 °C), and the relative permittivity of needle-like BT2 reaches 30000 in its b-axis direction at $T_C$ [4]. For this property, Zou et al. synthesized BT2 nanorods by the molten-salt synthesis (MSS), then the high dielectric nanocomposites with a dielectric constant of 62 @1 MHz were fabricated by using the BT2 nanorods as filler [5].

Nanorods with a high aspect ratio can effectively improve the dielectric properties of composites. However, when the content of nanorods is too high, the disorganized connection between the nanorods will produce more voids, which will greatly reduce the dielectric properties of composites. Small-sized high dielectric nanoparticles are introduced between nanorods to fill the voids between

them, this will significantly improve the dielectric properties of the composites. Therefore, we designed a composite structure based on BT2 nanorods and BT nanoparticles couple to study the composition of fillers in composites on the dielectric properties of embedded capacitor copper clad laminate (ECCCL). Finally, 100 wt% BT2 nanorods were filled into the composite in ECCCL, which made the relative permittivity of ECCCL attain the maximum of 47@1 MHz.

## EXPERIMENT

### Preparation of BT2 nanorods and BT nanoparticles

BT2 nanorods and BT nanoparticles can be synthesized by a similar method to reference [5]. To prepare uniform and high-purity BT2 nanorods, Ba(NO$_3$)$_2$ and anatase TiO$_2$ were weighed at the molar ratio of 1:2, then KCl was introduced to above mixture at a mass ratio of 10:1. After the same ball milling, drying, grinding, sintering and other steps as in [5], the expected nanorods products will be obtained. Unlike the reference [5], the temperature in the box furnace rises by 5 ℃/min. As for the synthesis of BT nanoparticles, most of steps are similar to above-mentioned process of BT2 nanorods, except for the ratio of raw materials (Ba(NO$_3$)$_2$, TiO$_2$ and KCl were mixed at a molar ratio of 1:1:20) and the holding time.

### Preparation of BT2-BT based ECCCL

The BT2 nanorods and BT nanoparticles used to make ECCCLs were calcined at 950 ℃ for 40 min and 30 min, respectively, and the matrix is epoxy resin (EP). BT2-BT based ECCCLs were fabricated by the same process as the composite preparation in reference [5]. The weight fractions of BT2 nanorods in the mixture of BT2 and BT were 0, 30, 50, 70, 100 wt%, and the weight fractions of the mixture in their respective composite are fixed at 60 %.

### Measurements

X-ray diffraction (XRD, Model MiniFlex 600) were applied to the as-synthesized nanorods and nanoparticles to investigate their crystal structures. The scanning electron microscope (SEM, Model Inspect F) was used to observe the morphologies of as-prepared powder and the distribution of nanorods and nanoparticles in ECCCLs. To study the differences of the relative permittivity and dielectric loss in different ECCCLs, impedance analyzer (IM7580A, HIOKI) was carried out to ECCCLs for testing their dielectric properties.

## RESULTS AND DISCUSSION

The SEM and XRD of as-prepared BT2 nanorods and BT nanoparticles are shown in Fig. 1. It can be seen that there are many uniform nanorods with over 10 μm in length (Fig. 1a) and there are hardly the diffraction peaks of impure phase except for the BT2 peaks (Fig. 1b). From Fig. 1c, large amounts of angular nanoparticles can be obviously found in the picture. The XRD in Fig. 1d confirm that the nanoparticles are cubic BT and no clear peaks of other phases can be observed. Consequently, the BT2 nanorods and BT nanoparticles synthesized by the molten-salt synthesis are uniform in size and possess a very high purity, respectively.

*Figure 1: (a) SEM and (b) XRD of BT2 nanorods synthesized at 950 °C for 40 min, (c) SEM and (d) XRD of BT nanoparticles synthesized at 950 °C for 30 min.*

In order to evaluate the dielectric properties of ECCCL, the relative permittivity and dielectric loss of the composites in ECCCL with various contents of BT2 nanorods were measured from 1 to 100 MHz, and the results are shown in Fig. 2. As frequency increase, all relative permittivity from composites with different BT2 nanorods levels continuously decreased and the value under higher filler contents changed faster from 1 to 20 MHz (Fig. 2a). At higher frequencies, the dipole mobility of the EP and interfacial polarization between EP and ceramic fillers will lead to this behavior [6-8]. At 1 MHz, the relative permittivity constantly increases with the increasing BT2 contents, and reaches the maximum of 47.3 at 100 wt%, which is 13 times to the pure EP. For the dielectric loss (Fig. 2c), the ECCCLs with higher BT2

contents possess much bigger dielectric loss than others at low frequency (1-20 MHz). As Fig. 2d shown, the dielectric loss exhibits the same variation trend as the permittivity in Fig. 2b. Apparently, there must be a close connection between dielectric loss, relative permittivity and interface areas.

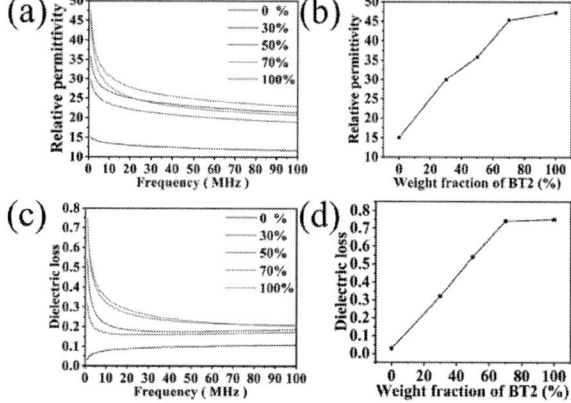

*Figure 2: (a, b) Relative permittivity and (c, d) dielectric loss of composites in ECCCL.*

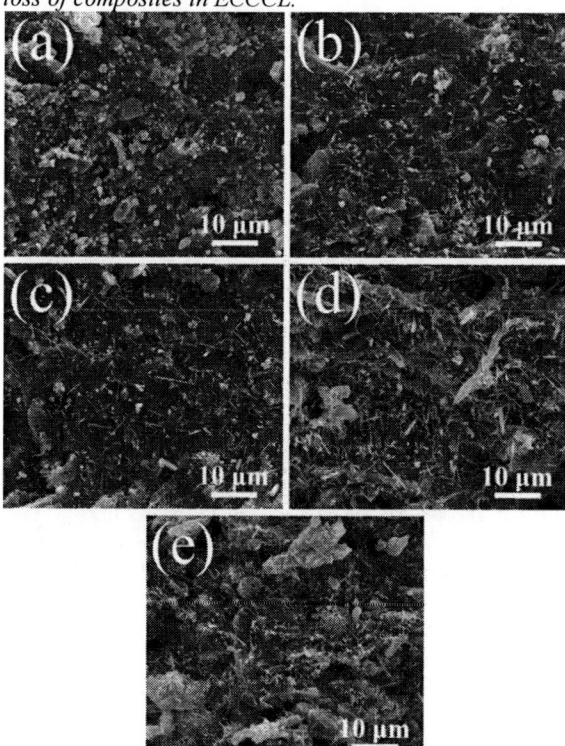

*Figure 3: The SEM of the section of composites filled with (a) 0%, (b) 30%, (c) 50%, (d) 70% and (e) 100%.*

To further investigate the reasons of the variation of the relative permittivity and dielectric loss, the SEM of the section of the composites in ECCCLs filled different contents of BT2 nanorods are shown in Fig. 3. It can be seen that the ECCCLs with less BT2 nanorods show a lower filler concentration, and more EP can be found in Fig. 3a-c. Therefore, the contents of BT2 nanorods play an important role in the improvement of the dielectric constant and dielectric loss of ECCCLs.

## CONCLUSION

The ECCCLs filled with different contents of BT2 nanorods and BT nanoparticles demonstrate an obvious improvement in relative permittivity and dielectric loss while the concentrations of BT2 nanorods increase, and the relative permittivity in the contents of 100 wt% BT2 nanorods is up to the maximum of 47@1 MHz, which reaches 13 times as high as the pure EP. The BT2 nanorods play a greater role than BT nanoparticles in improving the dielectric properties for ECCCLs under the same filler content.

## ACKNOWLEDGEMENTS

This work was supported by the National Natural Science Foundation of China (No. 61904026) and Innovation Team Project of Zhuhai City (No. ZH0405190005PWC).

## REFERENCES

[1] J. Fu, Y. Hou, M. Zheng, Q. Wei, M. Zhu, H. Yan, "Improving Dielectric Properties of PVDF Composites by Employing Surface Modified Strong Polarized BaTiO$_3$ Particles Derived by Molten Salt Method", *ACS Applied Materials & Interfaces*, vol. 7, 2015, pp. 24480-24491.

[2] J. Fu, Y. Hou, M. Zheng, M. Zhu, "Comparative study of dielectric properties of the PVDF composites filled with spherical and rod-like BaTiO$_3$ derived by molten salt synthesis method", *Journal of Materials Science*, vol. 53, 2018, pp. 7233-7248.

[3] H. Luo, Z. Wu, X. Zhou, Z. Yan, K. Zhou, D. Zhang, "Enhanced performance of P(VDF-HFP) composites using two-dimensional BaTiO$_3$ platelets and graphene hybrids", *Composites Science and Technology*, vol. 160, 2018, pp. 237-244.

[4] Y. Akishige, K. Fukano, H. Shigematsu, "New ferroelectric BaTi$_2$O$_5$", *Japanese Journal of Applied Physics Part 2-Letters & Express Letters*, vol. 42, 2003, pp. L946-L948.

[5] W. Zou, G. Zhou, W. He, S. Wang, Y. Hong, Y. Chen, C. Wang, C. Ma, S. Guo, H. Miao, J. Zhou, "Process, fundamental and application of one-step molten-salt synthezed BaTi$_2$O$_5$ nanorods", *Journal of Alloys and Compounds*, vol. 826, 2020, 154064.

[6] Z.-M. Dang, H.-P. Xu, H.-Y. Wang, "Significantly enhanced low-frequency dielectric permittivity in the BaTiO$_3$/poly(vinylidene fluoride) nanocomposite", *Applied Physics Letters*, vol. 90, 2007, 012901.

[7] X. Huang, L. Xie, P. Jiang, G. Wang, F. Liu, "Electrical, thermophysical and micromechanical properties of ethylene-vinyl acetate elastomer composites with surface modified BaTiO$_3$ nanoparticles", *Journal of Physics D-Applied Physics*, vol. 42, 2009, 245407.

[8] L. Xie, X. Huang, C. Wu, P. Jiang, "Core-shell structured poly(methyl methacrylate)/BaTiO$_3$ nanocomposites prepared by in situ atom transfer radical polymerization: a route to high dielectric constant materials with the inherent low loss of the base polymer", *Journal of Materials Chemistry*, vol. 21, 2011, pp. 5897-5906.

# OPTIMIZATION OF WAFER DICING-SAW TO REDUCE THE CHIPPING DEFECT BY USING THE RESPONSE SURFACE METHODOLOGY

*Hong Zhang[1*], WeiFeng Wang[2], and Po-Yu Huang[2]*

[1] School of Microelectronics, Fudan University, Shanghai 200433, China

[2] Zing Semiconductor Corporation, Shanghai 201306, China

*Corresponding Author's Email: hzhang18@fudan.edu.cn

## ABSTRACT

With the development of IC integration scaling, chip after packaging is also sizing down. Dicing-Saw induces the high risk of wafer chipping by physical process characteristic especially when final thickness of chip is less than 100um. The process condition of Dicing-Saw is the key to control the physical damage during the Dicing-Saw process and maintain the high quality of chip. In this paper, we applied the Design of Experiment (DoE) to collect the result of process optimization and used Response Surface Methodology（RSM）to analyze the result to find out the best conditions to improve the yield loss by top-side chipping for the ultra-thin wafer sawing. Based on the result of DoE, matrix of feed speed and Z1 cutting height are the significant factors to reduce the risk of chipping and achieve high yield. The research contribute not only to knowing the best condition of wafer dicing saw process to reduce the wafer top-side chipping but also showing the DoE methodology performed well to improve the trouble solving efficiency.

## INTRODUCTION

With the development of IC integration scaling, chip after packaging also need to be sizing down. As the target to provide the smaller and lighter chip, process optimization by front-end and back-end attracted engineers' focused. Dicing-Saw, as the key process of small chip packaging, it plays important role before pick-and-place because the risk of low yield during the physical process for the thinner wafer. Current mainstream of chip thickness is no longer as 150um but going thinner than 100um. How to control the physical damage during the dicing-saw process became the key to maintain the high quality chip.

The chipping is the fragmentation or cracking during dicing-saw process. The chipping can occur on wafer topside and backside and both of chipping lead to the risk of yield and reliability performance. The previous researches showed the poor efficiency of cooling during sawing process would lead to the blade overloaded, and the serious backside chipping [1]. To increase the dice counts, the test patterns is designed in the scribe lane on the chips, which lead to debris of test-key and affects the spalling, and regeneration of the blade. The phenomenon will result in the bluntness of the blade and lead to the chip cracking and the low yield. As Figure 1 and Figure 2.

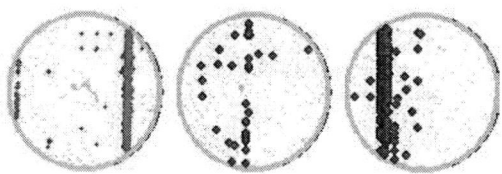

Figure 1 : Defect Map

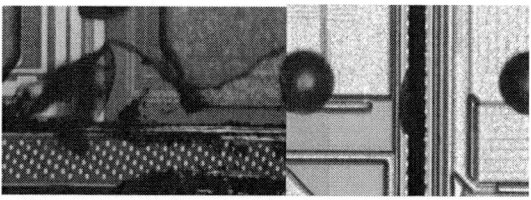

*Figure 2 Picture of defect. Defect position is along the wafer row direction, which is some different test key pattern.*

In this paper, we discuss the method of experiment design and find out the best conditions to improve the yield loss by topside chipping for the ultra-thin chip design. The DoE method (design of experiment) is applied to find the best condition of process.

## Experiment Design and Methodology

### Opportunities Analysis and Experiment Design

After the blade type and the cutting method, the temperature effect is considered at first. The blade is made by bonding of diamond and the exposure of diamond at the edge lead to blade serrated and related to performance of sawing. The more exposure of diamond appeared, the blade was sharper. During the sawing process, the diamond particles on the blade edge will be worn, peeled and regenerated continuously. If the regeneration was not in time, the blade will be blunt and the cutting load increased which lead to the cutting temperature too high, that's so-called the blade overload [1, 2].

When overload of blade occurred, the quality of sawing is affected and lead to wafer cracked on topside or backside. In order to prevent the chip cracking, optimize the process condition after the appropriate blade and the selected step-cut method starts from the cooling effect.

The DOE is designed by two majors considerations including cooling during the dicing-saw and sawing condition. Cooling factor is composited as backside temperature and cooling water efficiency. Sawing process condition is concluded as composition of factors, feed speed (CH2), Z1 Cut Height (CH2) [3, 4]. RSM (Response Surface Methodology) is applied to analyze the result of experiments.

**Response Surface Methodology**

Response Surface Methodology (RSM) is a collection of mathematical and statistical techniques which are useful for the modeling and analysis of problems in which a response which might interest in the influences by several variables and objective is to optimize the response. In most RSM application, the relationship between the response and variables is not clear. Thus, the first step in RMS is to find a suitable approximation for the true functional relationship between Y and the set of the variables. Then the nonlinear relationship between the variables and the fastest rising curve on the variance surface had been dug out. [5, 6]

# Experiment Result and Discussion
### Cooling Effect Clarification

There are two factors applied to two simulation tests to find the appropriate cooling condition of the sawing process, which are the chuck temperature and cooling shower. The first simulation is to fix process condition exclude the chuck temperature as room temperature versus 40C to check the correlation. Base on the simulation result (as figure 3), there is no obvious different by the yield of pick-and-place (PnP) between the chuck temperature set as room temperature 25℃ or 40℃.

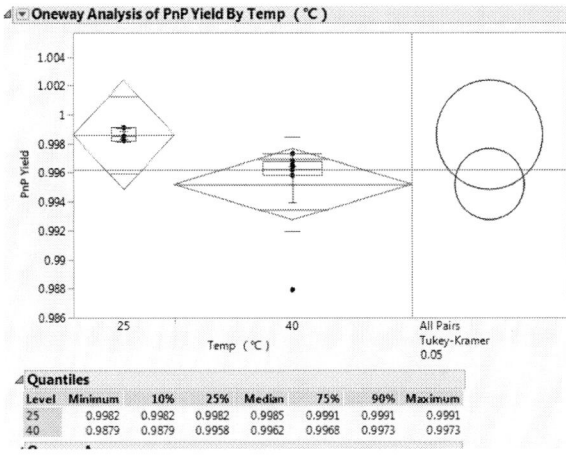

*Figure 3: Graph Simulation 1 result for temperature*

The wafer rise is for the topside fragment clean during the dicing-sawing process. There are two steps clean including the shower and Spray. The better efficiency of the fragment clean will lead to the less over-loading phenomena of blade. In Simulation 2, we try to find the correlation between the water pressures of this step with topside chipping performance. The experiment condition is set as baseline, +20% -20% matrix, as table 1. The result of simulation 2 showed there is no obvious different performance in each condition, as Figure 4.

*Table 1: Process matrix of simulation 2*

| Leg | Wafer ID | Shower (LPM) | Spray (LPM) | AOI Yield |
|---|---|---|---|---|
| Baseline | #02 | baseline | baseline | 99.79% |
| | #03 | baseline | baseline | 99.81% |
| 1 | #04 | -20% | -20% | 99.87% |
| | #06 | -20% | -20% | 99.83% |
| 2 | #07 | 20% | -20% | 99.79% |
| | #08 | -20% | -20% | 99.87% |
| 3 | #09 | -20% | 20% | 99.81% |
| | #11 | -20% | 20% | 99.83% |
| 4 | #12 | 20% | 20% | 99.92% |
| | #13 | 20% | 20% | 99.68% |

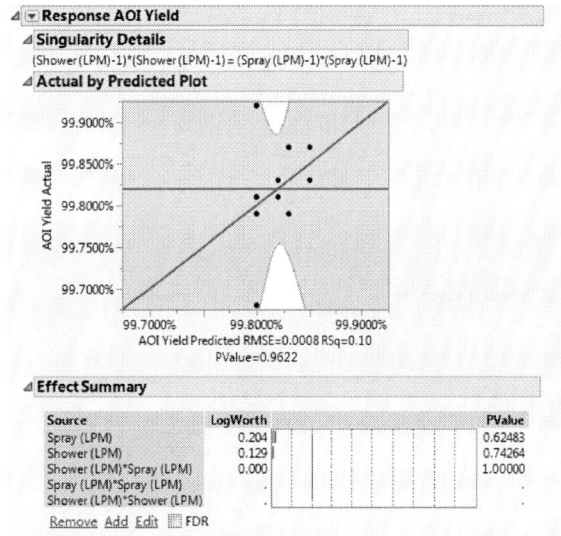

*Figure 4: Result of Simulation 2*

### Cooling Effect Clarification

The requirement of package thickness reduction lad to

the new challenges of thin wafer process. Back-grinding, DAF (die attach film) lamination are established to meet the requirement of process especially as wafer thinner than 3 mils, sawing process meet the significant challenge to prevent or reduce the side effect as wafer chipping. Consider about the leverage of cost and defect density, an optimized process condition by saw parameters including feed speed, RPM and blade types are also elucidated [5]. Based on the result of 2 simulation, other factors -Feed speed (CH2) and Z1 Cut Height (CH2) is designed in DoE for the effective condition finding.

In DoE of this study, the control factors are set as Z2 BG wheel/Temp/Time/Mount type and keep as baseline. The matrix combined by Feed Speed (CH2) and Z1 Cut Height (CH2) is performed in the experiment. We perform the DOE of RSM (Response Surface Methodology ) CCD and set 13pcs of full metal stack wafer by the matrix of feed speed (CH2)(15/20/25) and Z1 Cut Height (CH2)(0.5/0.65/0.7/0.8) and the result of split is AOI yield. The design matrix table, prediction variance surface and correlations map as Table 2. Figure 5 shows the prediction of Variance Surface.

*Table 1 DoE Matrix Table*

| | Pattern | Block | Feed speed (CH2) (mm/s) | Z1 Cut Height (CH2) | AOI Yield (%) |
|---|---|---|---|---|---|
| 1 | 00 | 1 | 20 | 0.7 | • |
| 2 | ++ | 1 | 25 | 0.8 | • |
| 3 | 00 | 1 | 20 | 0.7 | • |
| 4 | -+ | 1 | 15 | 0.8 | • |
| 5 | -- | 1 | 15 | 0.6 | • |
| 6 | +- | 1 | 25 | 0.6 | • |
| 7 | 00 | 1 | 20 | 0.7 | • |
| 8 | 00 | 2 | 20 | 0.7 | • |
| 9 | 00 | 2 | 20 | 0.7 | • |
| 10 | 0a | 2 | 20 | 0.5396432549 | • |
| 11 | 00 | 2 | 20 | 0.7 | • |
| 12 | A0 | 2 | 28.017837257 | 0.7 | • |
| 13 | 0A | 2 | 20 | 0.8603567451 | • |
| 14 | a0 | 2 | 11.982162743 | 0.7 | • |

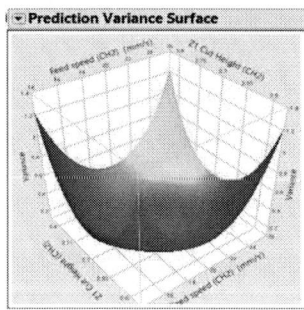

*Figure 5: Prediction of Variance surface.*

The RSM is applied to obtain the best condition of experiment and showed in Figure 6. The DoE result showed the optimal parameter is Feed speed 20/Z1& Cut height 0.66% and the optimal AOI yield is 99.96%, which meet the criteria of experiment.

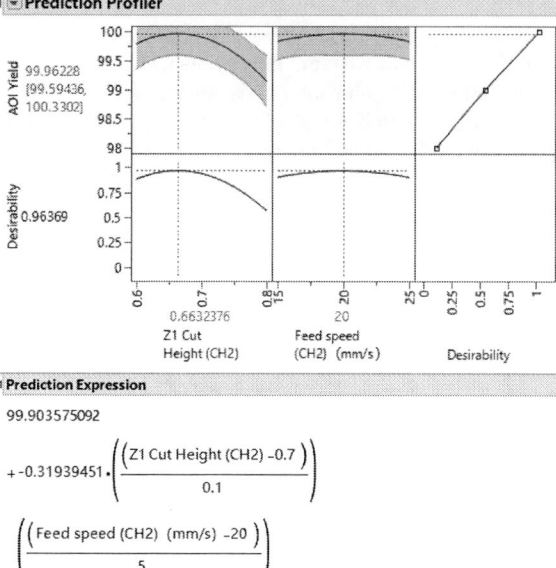

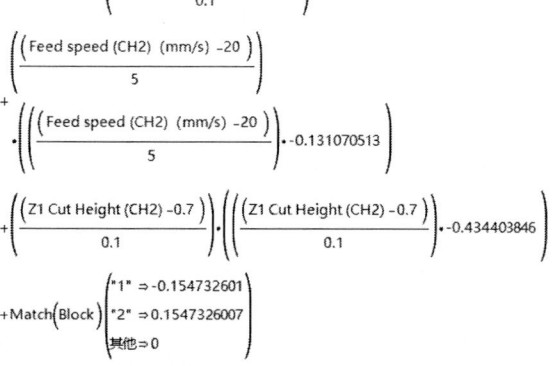

*Figure 6 : Graph DoE result and Prediction Expression.*

## CONCLUSION

In this paper, the research of factors to reduce the wafer chipping during wafer sawing process is evaluated. From the simulations of cooling effect, there is no obvious different in the condition of backside temperature and water rise pressure. In the DoE by matrix of Feed speed and Z1 cutting height, 2 conditions to achieve high yield are achieved. After the analysis of RSM, the best condition is obtained.

The research contribute not only to know the best condition of wafer sawing process to reduce the wafer top-side chipping but also the DoE methodology which well performed to improve the trouble solving efficiency.

## REFERENCES

[1] Jiang J., and Z.L. Zhang,"Study on the Technology of the Ultrathin Wafer Sawing, *China integrated Circuit*, 2009, pp. 66-69.

[2] Pang Ling, "Improvement of Dicing Process of Silicon Chip for Precaution Against its Disintegration in Package", *J. Suzhou Vocational University*, 2011, pp. 36-39.

[3] C.E.Tan & F.K.Chin, "Saw Chipping Improvement to Achieve Defect Free Bare Die Products", *36th International Electronic Manufacturing Technology Conference*, 2014.

[4] Amri M S, Liew D, Harun F, "Chipping free process for combination of narrow saw street (60um) and thick wafer (600um) sawing process", *Electronic Manufacturing Technology Symposium. IEEE,* 2010.

[5] IR Durán, Profili J, Stafford L, et al. "Response surface methodology as a predictive tool for the fabrication of coatings with optimal anti-fogging performance", *J. Thin Solid Films,* 2021, 718:138482.

[6] Umrp A , Nsr B , Sc A , et al. "Estimation of coating thickness in electrostatic spray deposition by machine learning and response surface methodology", *J. Surface and Coatings Technology,* 2021.

[7] Paydenkar C, Poddar A, Chandra H, et al. "Wafer sawing process characterization for thin die (75micron) applications", *International Electronics Manufacturing Technology Symposium. IEEE*, 2004.

# THE RESEARCH ON SMALL "DEAD ZONES" PACKAGING TECHNOLOGY FOR MASS PRODUCTION OF SILICON PHOTOMULTIPLIER

*Yuxiao Liu, Xingan Zhang, Yang Shao, Yongqiang Yan, Heng Yi, Ru Yang\*, Kun Liang, Dejun Han*
College of Nuclear Science and Technology, Beijing Normal University, Beijing 100875, China

*Corresponding Author's Email: yangru@bnu.edu.cn

## ABSTRACT

Silicon photomultiplier (SiPM) is one of the best choices for Positron Emission Computed Tomography (PET) detectors due to its excellent properties. In order to achieve high fill factor of SiPM tile for PET detector, we have developed a small "dead zones" packaging technology based on lithography, SU-8 photoresist and conductive silver paste rather than Au wire bonding. It was simple, cost-effective to form a small "dead" zone as small as 60 μm of packaging SiPMs for mass production, the ratio of chip area to packaged device area is as high as 0.9037. The preliminary yield reached 49.8%.

*Keywords—Lithography, Packaging, SU-8 photoresist*

## INTRODUCTION

PET scanner is one of the most advanced nuclear medical imaging equipment at present. SiPM has been the best choices for PET detector due to its advantages of high photon detection efficiency, low bias voltage, compact and etc. Since PET detectors require the integration of individual devices into large arrays, known as SiPM tiles[2],However, there are various problems with the current electrical connection that hinder the packaging density. For example, TSV packaging has the disadvantages of high cost and technical difficulty, while traditional lead bonding has the disadvantages of large dead zone area, which seriously hinders the increase of packaging density. The smaller the dead zone area, the higher the integration of the device, the smaller the area of a single pixel, that means it will have higher resolution in practical applications. Thus, the dead zone area has a significant impact on detection efficiency.

## MATERIALS AND METHODS

### Materials

SU-8 negative epoxy photoresist has advantages include excellent imaging properties, high photosensitivity, low optical absorption in the near UV-range, good adhesion with different substrate materials as well as superior chemical, thermal, andmechanical stability. Because of these advantages, SU-8 photoresist is best for high aspect ratio graphics,

used as UV patterning material in many microfluidics and micro-electro Mechanical Systems (MEMS) applications since it was first emplyed in the 1990s by the EPFL-Institute of Microsystems and IBM-Zurich[3].

The material we use is SU-8 3000 series, compared to 2000 series, it improved adhesion and reduced coating stress. Thus, the resulting graphics have a better aspect ratio.

### Methods

The new packaging technique innovation point lies in the way of using lithography that can reduce the size of Insulation wall to micro-level (only if the materials have good insulativity ) . By the mean time, it replaced the wire bonding, greatly reduces the package area required to accommodate pressure solder joints.

After the whole wafer is cut into a single device, a group of 196 devices are attached to a PCB, called a Block (BK).Then a BK unified photolithography, development, to construct an insulating side wall. In order to place the edges and corners exposed, the whole device is coated twice. After the first coating, we should have a short soft-bake, to solidify the adhesive layer that has been formed. The temperature is 65℃ and 95℃,after second coating ,the time of soft-baking need to be extend to ensure that the solvent is fully volatile. After exposure, we still use 65℃ and 95℃ to post-bake. Because we still need to use 100℃to solid the silver glue and 105℃ to dry the device after cutting, after developing, we use 135℃ to hard-bake. Then use conductive silver glue to connect. Finally, we use SU-8 to protect the surface. The specific process steps are shown in Table I and Figure.1.

*Table I*

| Process | |
|---|---|
| Insulating structure forming | Coating |
| | Soft-bake |
| | Exposure |
| | Post- Bake |
| | Development |
| | Hard-bake |

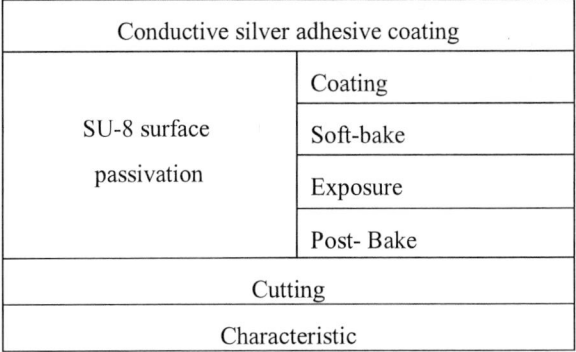

| Conductive silver adhesive coating | | |
|---|---|---|
| SU-8 surface passivation | Coating | |
| | Soft-bake | |
| | Exposure | |
| | Post- Bake | |
| Cutting | | |
| Characteristic | | |

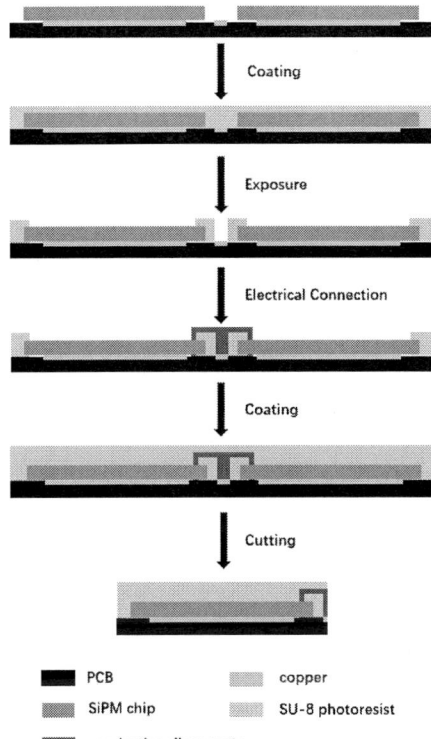

PCB
copper
SiPM chip
SU-8 photoresist
conductive silver paste

*Figure.2. Schematic process flow*

## CHARACTERIZATION

### Surface Appearance

As shown in Figure 2, the minimum area of the photodetector (Chip area 3.2mm×3.2mm) successfully packaged was 11.28mm2 (3.36mm×3.36mm), the ratio of chip area to packaged device area is as high as 0.9037, compared with the area of 12.6mm2 (3.75mm×3.36mm) by wire bonding, it is greatly reduced. It's only 60μm between the edge of chip and the edge of packaged device. The ratio of chip area to packaged device area is as high as 0.9037.Compared to the through silicon via(TSV), its has the same duty cycle(which is almost 2%)[3]-[5].

Compared with hole wire bonding(HWB) encapsulation, the dead zone area is smaller.

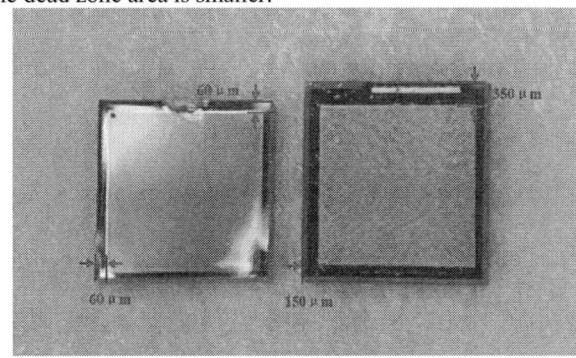

*Figure.2. Planer-picture of the packaged devices (SiPM): the new packaging technique(left) and the regular wire bonding(right).*

### Current-Voltage characteristics

The dark current level of new package wayis No difference compared to the wafer level

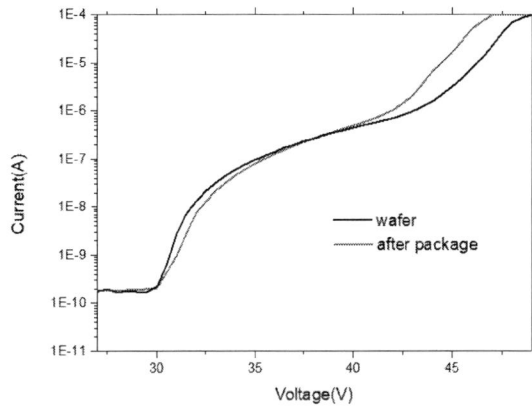

*Figure.3. Current-Voltage characteristics of the photodetectors packaged with the method presented in this study and wafer.*

### Dark count rate(DCR)

Compared to surface mount type, there is only a slight difference,but the difference may because of the different source of wafer.

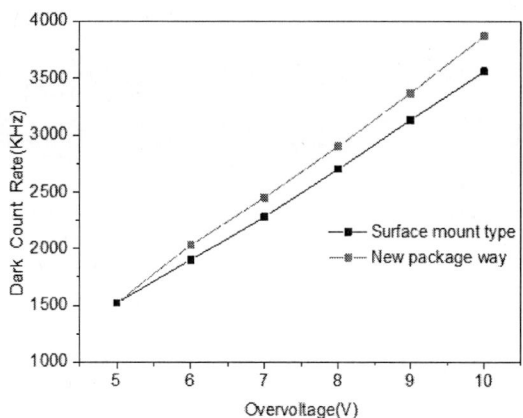

*Figure.4.DCR of the photodetectors packaged with the method presented in this study and the wire bonding way.*

### Photon Detection Efficiency(PDE)

The protection method studied in this paper is to use SU-8 to cover the surface. Therefore, the surface flatness will be improved accordingly. We select the wavelength of 420nm to characteristic the device ,the result shown as Figure.5.

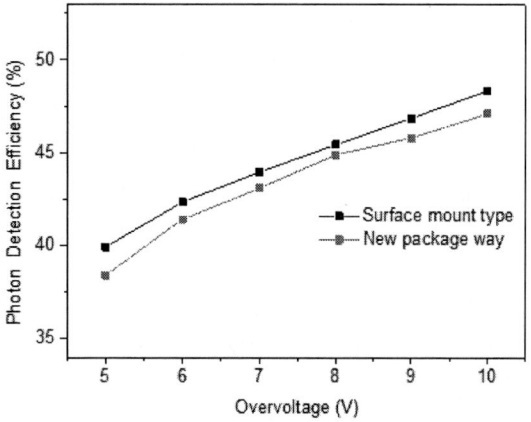

*Figure.5. PDE of the photodetectors packaged with the method presented in this study and the wire bonding way.*

### Yield

The yield reached 49.8%. But in fact, after every package ,the disconnected devices and short-circuit devices accounted for 22.14% and high leakage devices accounted for 28.06%. The high leakage of current might caused by the problem of the wafer and other reasons. So the yield of single chip can reach 80% Theoretically.

The mostly reason of disconnection is that the silver glue isn't covered surface electrode. Because the area of silver glue should be control only cover a few avalanche photodiode(APD) , so that the amount of silver

glue should be as little as possible, if it can be control, the number of the disconnected devices can down to 0. The mostly reason of short-circuit is that at the bottom of the insulation wall has void , which leads to infiltration of conductive silver glue. It can be improve by dry thoroughly after cleaning (to increase the adhesion of SU-8 to the bottom) and post-baking be cooled slowly (to reduce separation due to the different stress).

Because of the alignment errors in various process steps, the thickness of the insulating side wall is not consistent, so the size cannot be further reduced after packaging. If the minimum feature size of the process can be reduced and the error can be reduced, the insulating side wall can be down to the several micron scales .

## CONCLUSION

This paper successfully tested the application of small "dead zones" packaging in the mass production of SiPM. This method has low cost and simple operation, and achieved a yield of 49.8%, which can be applied to the production of SiPM tile.The photoelectric characteristics of the packed device were unaffected.

## ACKNOWLEDGEMENT

This work was supported by the National Key R&D Program of China (Grant No. 2020YFC01220001).

## REFERENCES

[1] T. Tekin, Review of packaging of optoelectronic, photonic, and MEMS components, IEEE J. Select. Top. Quant. Electron. 17 (2011) 704‑719.

[2] Alessandro Ferri,Fabio Acerbi, Alberto Gola,Claudio Piemonte, Giovanni Paternoster, Nicola Zorzi. Performance of a 64-channel, 3.2×3.2cm 2 SiPM tile for TOF-PET application[J]. Nuclear Inst. and Methods in Physics Research, A,2016,824:

[3] A. Ayoib, U. Hashim, M. K. M. Arshad and V. Thivina, "Soft lithography of microfluidics channels using SU-8 mould on glass substrate for low cost fabrication," 2016 IEEE EMBS Conference on Biomedical Engineering and Sciences (IECBES), Kuala Lumpur, Malaysia, 2016, pp. 226-229.

[4] Z.K. Esfahani, H.W. van Zeijl, Kouchi Zhang, High aspect ratio lithography for lithodefined wire bonding, In Proceedings - 64th Electronic Components and Technology Conference, 2014, pp. 1556‑1561.

[5] X. Sun et al., "Investigation of TSV noise coupling in 3D-ICs using an experimental validated 3D TSV circuit model including Si substrate effects and TSV capacitance inversion behavior after wafer thinning," 2016 IEEE MTT-S International Microwave Symposium (IMS), San Francisco, CA,

USA, 2016, pp. 1-4.

[6] Ting Kang,Zhaowen Yan,Wei Zhang,Jianwei Wang. Research on the Transmission and Coupling Issue of Multilayer TSV and Multilayer TSV Array[A]. IEEE Beijing Section 、 Beijing Jiaotong University. Proceedings of 2015 IEEE 6th International Symposium on Microwave, Antenna, Propagation,and EMC Technologies(MAPE 2015)[C].IEEE Beijing Section、Beijing Jiaotong University:IEEE BEIJING SECTION,2015:4.

# FINE - PITCH CU - SN TRANSIENT - LIQUID - PHASE BONDING BASED ON REFLOW AND PRE - BONDING

*Yunfan Shi[1], Zilin Wang[1], Zheyao Wang[1,2]\**

[1]Institute of Microelectronics, Tsinghua University, Beijing 100084, China
[2] Beijing Innovation Center for Future Chips, Tsinghua University, Beijing 100084, China
*Corresponding Author's Email: z.wang@tsinghua.edu.cn

## ABSTRACT

Among the various metal bonding techniques, Cu-Sn transient-liquid-phase (TLP) bonding has received extensive attention in recent years because the melting of Sn allows low bonding temperature and ease of operation. In this paper, a Cu-Sn TLP bonding method is proposed to achieve fine-pitch Cu-Sn bonding to avoid the problem of Sn extending by using reflow and pre-bonding. By elaborating control of reflow, the remaining Sn in the bonding interface is sufficient for wettability and compensation for the height variations of Cu bumps, but is insufficient for lateral expansion. By avoiding Sn extrusion, Cu-Sn bumps with a diameter of 5μm and a pitch of 25μm can be bonded successfully with a bonding strength higher than the adhesion strength of sputtering.

## INTRODUCTION

Metal bonding is an important technology for three-dimensional (3-D) integration as it realizes electrical interconnection between neighboring chips. Among the various metal bonding techniques, Cu-Cu direct bonding is prevalent due to its excellent electrical conductivity, low morphology, and good reliability [1, 2]. However, direct Cu bonding relies on tight surface contact and therefore chemical mechanical polishing (CMP) is highly required to achieve extremely flat surfaces. In contrast, metal transient-liquid-phase (TLP) bonding by using a low-melting-point metal (such as Sn, Ag, etc) as an intermediate layer between two high-melting-point metal (such as Cu) bumps avoids the use of CMP because the melted Sn can compensate the rough surface and height variations of Cu bumps [3-5]. Through the chemical reaction between the melted Sn and the solid Cu, inter-metallic compound (IMC) are formed to join the two Cu bumps. However, the melted liquid Sn may expand laterally under the bonding pressure, causing the problem of size extending of the bumps and the difficulty in obtaining high-density and fine pitch bumps.

As TLP bonding relies on the liquid state of Sn to accelerate the diffusion of Cu into Sn to form IMC, temperature higher than the melting-point of Sn is needed during bonding. When the temperature reaches the melting-point, Sn has a certain degree of flowing and will be easily squeezed out of the bumps when bonding pressure is applied. The squeezed Sn may extrude laterally and bridge neighboring bumps in case of small pitches.

In this paper, we report a method to achieve fine-pitch bonding by exploring Sn reflow and pre-bonding. Through elaborate reflow and pre-bonding, the amount of Sn can be probably consumed such that the remaining Sn sufficient for surface wetting but insufficient to cause extrusion. The experimental results show that the Sn amount can be well controlled and the bonding strength is strong for use.

## EXPERIMENTAL APPROACH

Fig.1 shows the bonding structure. It includes a top chip and a bottom chip before bonding. The top chip has an Cu bump array that are electroplated on a 20 nm TiW adhesion layer and a 100 nm Cu seed layer, and the bottom chip is sputtered with a 20nm TiW adhesion layer and a 2μm thick Cu pad . The bumps are covered with a thin Sn layer deposited using electroplating. The orignal bump diameter is 5.4μm, height 4μm, and pitch 25μm.

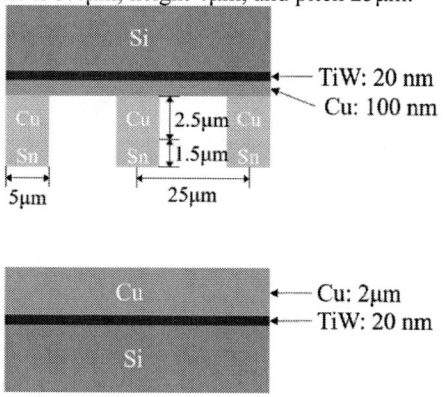

*Figure 1: Schematic diagram of the structure before bonding.*

Reflow process is performed in vacuum environment before bonding. During reflow, the temperature is firstly raised from 200°C to 260°C in 3 minutes and held for 3 minutes. Then, it is cooled down to 200°C in 3 minutes. After reflow, the bonding pair is tranferred into a bonding facility, FC150 (Smart Equipment Technology), for pre-bonding and final bonding. Pre-bonding at 200°C for 10 minutes makes the bumps and the pads fully contacted.The final bonding is performed at 260°C for 30min. A certain pressure is applied during pre-bonding and bonding. To reduce the residual stress, the bonding pair is slowly cooled down to room temperature within 10 minutes after bonding.

## RESULTS AND DISCUSSIONS
### Cu-Sn bumps after reflow process

Fig. 2 shows the Cu-Sn bumps after reflow. It can be seen that the bumps are spherical-like. The cross-section of the Cu bumps shows that they have three distinct layers, i.e., Sn, Cu-Sn IMC, and Cu. The intermediate areas with transitional gray indicate that Cu-Sn IMCs have been formed. The Sn left on the surface plays a vital role in bonding process. It can maintain wettability and compensate for the height variations of Cu bumps. The bump diameter after reflow is 5.7μm, about 0.3μm larger than that before reflow.

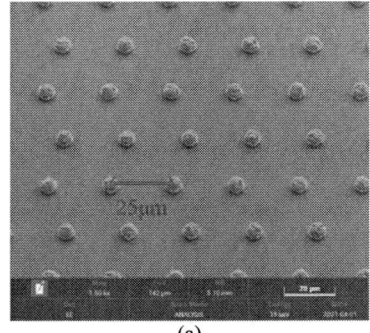

(a)

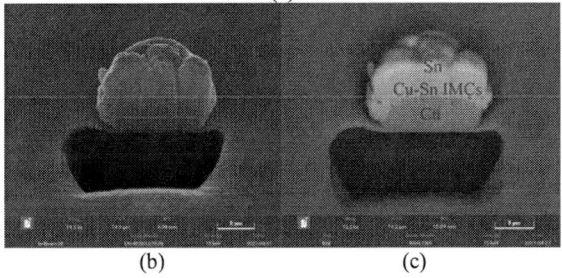

(b)                    (c)

*Figure 2: SEM images of Cu-Sn bumps after reflow. (a) Array of Cu-Sn bumps after reflow, (b) Cross-sectional image of a single bump, (c) Backscattered electrons image of a single bump.*

### Bonding Results

After bonding, the bonded chip pair is separated by applying a large shear force on the sidewall of the top chip. It is destructive, such that the bump array can be checked and the bonding strength can be measured, as shown in Fig.3.

It can be seen that the two bonded chips are pulled apart, with the breaking point appearing at the interface between the TiW adhesion layer and the substrate instead of at the bonded interface, indicating that the bonding strength is stronger than the adhesion strength of the sputtered adhesion layer. The measured shear strength of the bonded chips is 20.13MPa. It should be noted that this measured strength is the adhesion strength of Cu seed layer sputtered on the chip, and the actual bonding strength between the Cu bumps and the Cu pads is higher than the measured value.

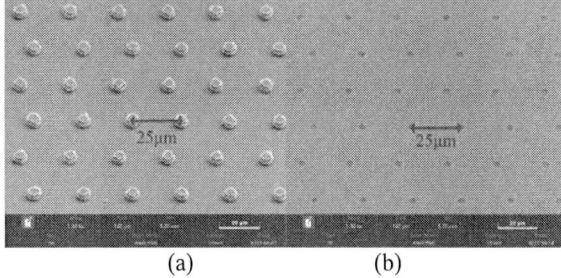

(a)                    (b)

*Figure 3: The surfaces of the two bonded chips after shear test. (a) Bumps transferred onto the bottom chip, (b) The adhesion TiW and Cu seed layer on the top chip peeled off with the bumps.*

The cross sectional photo in Fig.4 shows that all the remaining Sn after reflow is consumed after pre-bonding and final bonding by reaction with Cu, and an IMC $Cu_3Sn$ is produced while the excessive Cu in the bumps remains below the IMC. There is no distinguishable voids appeared at the bonding interfaces, indicating a good bonding quality. It also shows that after bonding, there is no significant Sn extrusion occurred, and the diameter of the bumps is around 6.05μm, about 12% wider than that before bonding. The slight diameter extension implied that the pitch of the bumps can be further reduced to achieve high density.

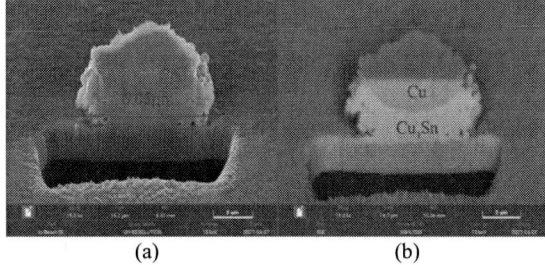

(a)                    (b)

*Figure 4:. Cu-Sn bumps after separating the bonded chips. (a) Cross-sectional image of a transferred bump, (b) Backscattered electrons image of a transferred bump.*

## CONCLUSION

A Cu-Sn TLP bonding method based on reflow and pre-bonding has been developed to achieve good bonding quality and high bump density. By applying reflow and pre-bonding, the extrusion of liquid Sn is avoided and high density Cu-Sn bonding is realized by the interface wettability of the remaining Sn. The results show that the

978-1-6654-9759-6/22 $31.00 © 2022 IEEE

bonding strength is higher than 20.13 MPa.

## ACKNOWLEDGEMENTS

We acknowledge Rutian Huang, Hanlin Wang, and Yao Zheng help for experiment.

## REFERENCES

[1] S. Senturia. *Proceedings of Transducers2003*, Boston, June 8-12, 2003, pp. 10-15.

[2] T. Tsuchiya, O. Tabata, J. Sakata and Y. Taga. *J. Microelectromech. Syst.*, vol. 7, 1998, pp. 106-113.

[3] R. P. Feynman, *Lectures on Physics*, Addison Wesley, 1989.

[1] H. Park, H. Seo, Y. Kim, S. Park, and S. E. Kim. *IEEE Transactions on Components Packaging and Manufacturing Technology*, vol. 11, no. 4, 2021, pp. 565-572.

[2] S. YewChung Wu, L. Meiyi, L. Tung-Yen, L. Tsan-Feng, W. Yu Hsiang, and C. Jiun-Wei. *ECS Journal of Solid State Science and Technology*, vol. 10, no. 4, 2021, pp. 044004.

[3] J. Cai, J. Q. Wang, and Q. Wang, *Microelectronic Engineering*, vol. 236, 2021, pp. 10.

[4] H. K. Kannojia, and P. Dixit, *Journal of Materials Science-Materials in Electronics*, vol. 32, no. 6, 2021, pp. 6742-6777

[5] L. Sun, L. Zhang, Y. Zhang, M.-h. Chen, and C.-p. Chen, *Journal of Manufacturing Processes*, vol. 68, 2021pp, 1672-1682.

# A SIMPLE METHOD FOR FINE VERTICAL INTERCONNECTION BY STENCIL PRINTED VIAS ON FLEXIBLE PRINTED CIRCUIT BOARD WITH LOW TEMPERATURE SINTERING NANO-SILVER PASTE

*Xun Xiang [1], Zibai Li [1], Hongjian Zeng [2], Wei Zheng [1], Chuan Hu [1] and Yao Wang [1]\**

[1]Institute of Semiconductors, Guangdong Academy of Sciences, Guangzhou, China.
[2]Electrical and Electronic Engineering, School of Engineering, Merz Court, Newcastle University, Newcastle upon Tyne, United Kingdom.
*Corresponding Author's Email: yaowang530@foxmail.com

## ABSTRACT

The demand of packaging technology with smaller size, higher density and more diversity is growing strong. It is essential to develop more advanced interconnection technology to connect different redistribution-layers (RDLs). In this paper, a nano silver paste is transferred into blind vias with 50 μm diameter on two layer flexible printed circuit (FPC) like board by stencil printing, and then the filled vias are sintered at 180 °C for one hour to complete electrical via formation. With the modest process temperature and fine via filling processes, a vertical interconnection technology is developed and can be utilized for buildup of multiple RDLs. Daisy chains are designed to test electrical resistances of the filled vias. An average resistance of vias is found to be 0.50 ± 0.03 Ω, which indicates that a good electrical connection is obtained between the two copper layers. Thus, we have demonstrated a simple, fast and inexpensive method for vertical interconnection and this method is expandable to multiple layer of buildups.

## INTRODUCTION

The development of advanced packaging, especially 3D packaging, put forward higher requirements for vertical interconnection technology. Vertical interconnections between different layers of encapsulation structure are commonly realized by electroplating metal or printing solder paste [1]. Fine vertical structures are mainly fabricated by electroplating

method. However, electroplating technology is a complexed and environmentally unfriendly process. Moreover, electroplating is slow and costly when fine structures are fabricated. Printing solder paste is a fast and inexpensive method, but fine structure is not achieved in this way [2].

Nano silver particles with diameter less than 10 nm have significantly low melting temperature less than 300 °C [3]. A paste with such particles can be used in many packaging processes as inter layer vias of multiple RDLs, which typically requires a process temperature lower than 250 °C. In this study, a two layer FPC board is used to demonstrate the vertical interconnect of multiple RDLs. FPC board acts as the substrate of fine structure and fine interconnection experiments are carried out on them. A simple, fast and inexpensive new technology for vertical interconnection with nano silver is investigated and tested.

## MATERIALS & EXPERIMENTS

Nano silver paste is prepared with nano silver particle (about 5 nm), high polymer resin, organic solvent and other additives. FPC board consists of two copper layers (12 μm respectively) and a middle polyimide (PI) layer (25 μm). 50 μm vias are pre-prepared on FPC board by laser drilling. Nano silver paste is transferred into the vias by stencil printing. Vacuum and pressure are adopted to assist the flowing of nano silver paste into the vias. Nano silver paste is sintered in an oxygen-free oven with a protective gas of $N_2$. Figure 1 shows the TEM picture of nano silver particle (left) and the picture of stencil printer (right).

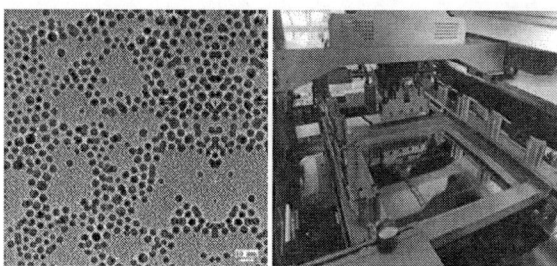

Figure 1: The TEM picture of nano silver particle (left)(the scale bar is 10 nm) and picture of stencil printer (right).

A test block (6x6 mm) with 5 daisy chains is designed. Every daisy chain contains 120 vias and corresponding 60 test pads. The preparation process of daisy chains is shown in figure 2. After laser drilling and vias filling, the pattern of daisy chains is prepared by exposure, development, etching and stripping processes.

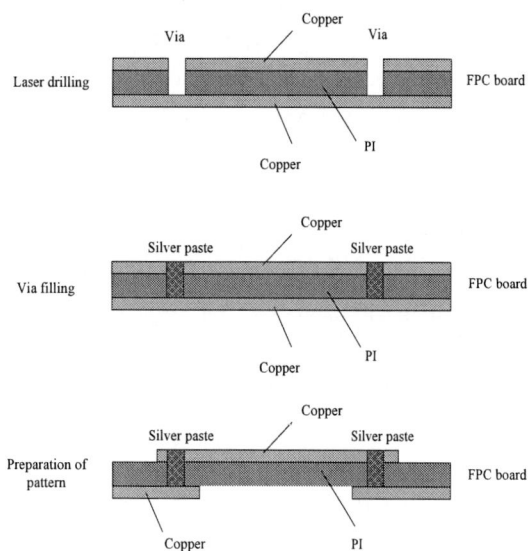

Figure 2: The preparation process of daisy chains.

The cross section of nano silver paste filled samples is conducted on a grinding and polishing machine. The morphology of the cross section is observed by an optical microscope. A digital multimeter is used to check the resistance of daisy chains. The resistance of single via as well as multiple vias on one daisy chain can be measured at once.

## RESULTS AND DISCUSSIONS

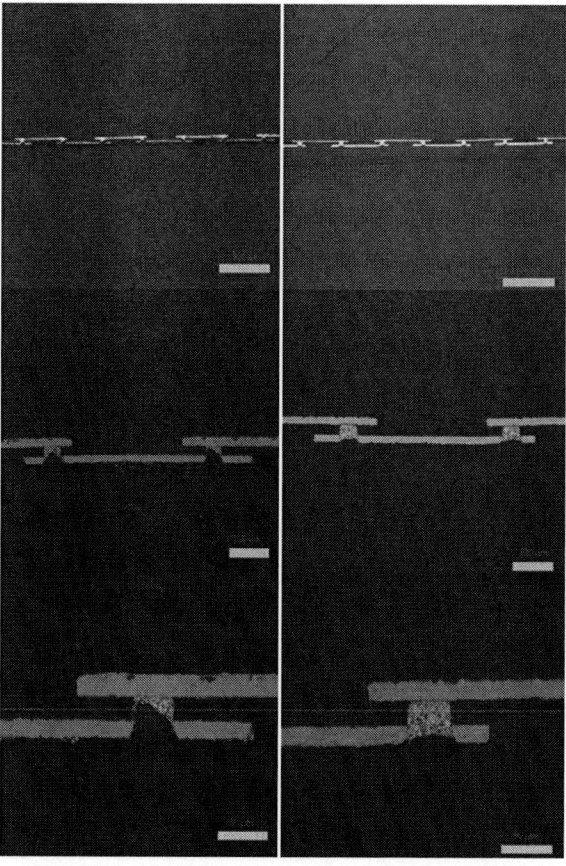

Figure 3: Cross section of filled vias with different magnification. Left: single filling. Right: multiple filling.

The filling process is studied first. Nano silver paste is transferred into vias and then sintered at 180 °C for 1 h for first time. Due to the volatilization of organic solvent in paste and fusion of nanoparticles, the volume of paste will shrink after sintering. Large number of big voids can be observed after first filling. The result is shown in figure 3 (left). The vias is then re-filled and re-sintered under the same condition for several times. In this case, only very small voids in the top of vias are observed, but basically all the vias are well filled. The result is shown in figure 3 (right). In this paper, a sintering temperature of 180 °C is selected, which is friendly for semiconductor processing. The cross-section result shows that the nano silver paste is well sintered at 180 °C. A good conductive network is formed between the silver particles after sintering, and an effective connecting channel is formed between the two copper layers by the sintered nano silver paste (multiple filling).

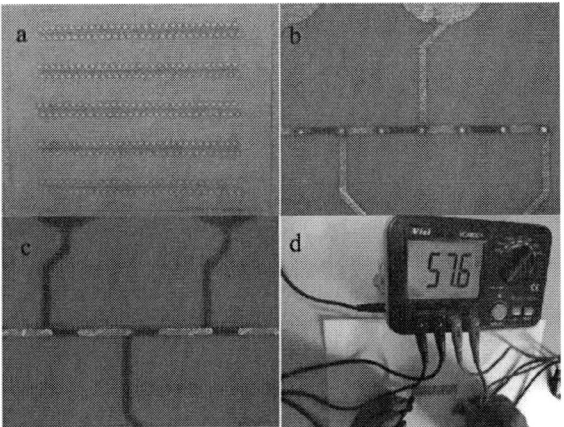

Figure 4: a. The whole test block with 5 daisy chains. b. The top pattern of daisy chain. c. The bottom pattern of daisy chain. d. The resistance test of daisy chain.

To further investigate filling result, daisy chains are designed and prepared. The as prepared daisy chains are shown in figure 4(a-c). Top copper circuits are connected with bottom copper circuits through vias. Filled vias can be observed from the optical micrograph. The resistances of daisy chains are checked, as shown in figure 4(d). An average test result of 59.9 ± 3.9 Ω for five whole daisy chains is obtained, as shown in table 1. The average resistance for per single via is 0.50 ± 0.03 Ω (120 vias for every daisy chain). This result shows that the top copper layer has a good connection with the bottom copper layer through the vias. This result is consistent with the previous cross section result. The resistance of filling vias consist of bulk resistance of sintered silver and contact resistance between silver and copper. We have found the resistance of filling vias decreases with higher sintering

TABLE 1. The Resistance of Daisy Chains.

| Daisy chain No. | 1 | 2 | 3 | 4 | 5 |
|---|---|---|---|---|---|
| Resistance (Ω) | 57.6 | 54.5 | 60.7 | 64.2 | 62.4 |
| Average (Ω) | 59.9 | | | | |
| Standard deviation | 3.9 | | | | |

temperature. In our future work, we will further explore ways to reduce the resistance of filling vias, and study the reliability of filling vias.

Stencil printing is a fast, simple and cheap technology. Vertical connection by stencil printing is a competitive approach compared with the traditional electroplating way. The method for vertical connection fabrication proposed in this paper is very potential in 3D and other advanced packaging.

## ACKNOWLEDGMENT

Thanks for the financial support of Department of Science and Technology of Guangdong Province (2020B0101320001), Guangdong Academy of Sciences (2021GDASYL-20210103078, 2021GDASYL-20210103079, 2020GDASYL-20200302012).

## REFERENCE

[1] T. H. Hoang, K.W. Lee, D. Ando, Y. Sutou, M. Koyanagi and J. Koike. *IEEE International Interconnect Technology Conference 2017*, Hsinchu, May 16-18, 2017, pp. 1-3.

[2] T. Iwai, T. Sakai, D. Mizutani, S. Sakuyama, K. Iida, T. Inaba, H. Fujisaki, A. Tamura and Y. Miyazawa. *IEEE CPMT Symposium Japan 2018*, Kyoto, Nov 19-21, 2018, pp.105-108.

[3] S. Kang, S. Cho, R. Shanker, H. Lee, J. Park, D.S. Um, Y. Lee and H. Ko. *Science Advances*, vol. 4, 2018, eaas8772.

# HIGHLY PROCESSABLE WTSV MODULAR MANUFACTURING FOR NEXT GENERATION 3D MEMS/NEMS INTEGRATED SYSTEM

*Simian Zhang[1], Xiaonan Deng[1], Yuqi Wang[1], Yifei Wu[1], Shengxian Ke[1], Linsen Li[2], Chaoyang Xing[3], Zhengcao Li[1*] and Chen Wang[1*]*

[1] State Key Laboratory of New Ceramics and Fine Processing, Key Laboratory of Advanced Materials of Ministry of Education, School of Materials Science and Engineering, Tsinghua University, Beijing, China, 100084

[2] Anhui Province Key Laboratory of Microsystem, The 43$^{rd}$ Research Institute of CETC, Hefei, Anhui, China, 230088

[3] Beijing Institute of Aerospace Control Devices, Beijing，100094

*Corresponding Author's Email: zcli@tsinghua.edu.cn; chenwang0101@tsinghua.edu.cn

## ABSTRACT

As the basis of three-dimensional (3D) integration for scaling in the field of MEMS/NEMS, wide through-silicon via can transform planar interconnections into vertical ones, realizing higher interconnection density, shorter line size, and shorter signal delay, together with a much smaller device size and higher integration density. In this paper, a comprehensive modular manufacturing strategy for 3D MEMS/NEMS wTSVs with aspect ratio up to 11, consisting of 3 etching and 2 filling designs and process innovation, meeting all the above requirements, together with extra advantages such as high processability and integrability, is proposed. An organic liner layer process can be applied to low-temperature bio-chips, inorganic liner process will be essential for high reliability and high load 3D power devices. The filling quality of two processes are further analyzed, exhibiting a low leakage current of 20 pA under 30V bias and a low parasitic capacitance lower than 100 fF, paving a clear pathway for the next-generation 3D MEMS/NEMS integrated system.

## INTRODUCTION

During the past decades, Moore's law became the driving force of the revolution of IC technology. However, the most advanced chips are more and more difficult to scale down and integrate with traditional semiconductor technology and planar integration architecture. Thus, a new strategy to break through the technical limitations and continue the hyper Moore's law is to employ the 3D architecture based on TSV technology to realize high-density integration.

In principle, 3D packaging technology can bring a new dimension to MEMS/NEMS fields. By vertically stacking subsystems with different functions, multifunctional 3D systems with smaller sizes and high performance are centrally researched and in urgent industrial needs. Figure 1 shows a typical 3D MEMS/NEMS integrated system architecture.

For 3D MEMS/NEMS pursuing high power and high current, traditional narrow TSV for chips may not suitable for its low reliability. Here we define wide-TSV (wTSV) as TSV with a via diameter larger than 30 μm. The current density in copper interconnection is generally controlled below $10^5$ A/cm$^2$ to avoid failure caused by serious electro-migration [1]. Under such circumstances, TSV with a typical diameter <10 μm [2] can only bear a maximum current of 0.02 A, while the wTSV with a diameter of 30 μm can load current 35t times up to 0.7 A, which is a basic requirement of high power devices.

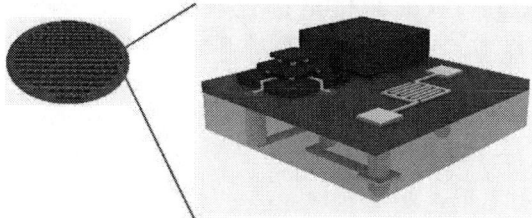

*Figure 1: Schematic of a typical 3D MEMS/NEMS integrated architecture and its wafer-level integration.*

However, in the process of wTSV, there will be many emerging technical challenges. Cu electro-deposition process may form voids in Cu pillars [3-5], which will seriously reduce the withstand voltage of the device. Another common issue of the wTSV filling process with a high aspect ratio (AR) is that the barrier layer and seed layer tend to be discontinuous when prepared by the traditional PVD process [6, 7]. The discontinuity of the barrier layer will lead to the diffusion of Cu into the liner layer and silicon substrate, while the discontinuity of the seed layer will lead to the failure of electroplating. To avoid such problems, some researchers adopt conical hole structure [8] or very large size vias (larger than 100 μm) [9], which is hard to be directly used in 3D MEMS/NEMS integrated system.

In this work, we propose a comprehensive strategy for wTSV process, consisting of three etching methods for forming high-aspect-ratio vias, and two filling methods of high aspect ratio wTSV. First one consists of fabrication of organic liner layer, electroless Ni barrier/seed layer, and Cu electroplating process for low-temperature and low-cost processes. The second is composed by an inorganic silicon oxide/nitride liner layer together with an ALD TiN barrier layer, electroless Cu seed layer, and Cu electroplating filling process with higher stability and higher reliability. Both methods can achieve wafer-level filling and are well compatible with existing semiconductor processes.

## ETCHING OF HIGH AR WTSV
### Single-side Etching

For the whole wTSV structure process, the first and essential step is etching silicon wafer vertically from the surface to form holes. Specially designed deep reactive ion etching process is used to fabricate vertical TSVs with a high aspect ratio and smooth sidewall. By appropriately adjusting the recipe, a high-quality wTSV with an ultra-high 11.3 aspect ratio (i.e. diameter of 40.70 μm, depth of 460.56 μm in Figure 2(a)) can be demonstrated. These results testify that the wTSV structure with a large diameter, high aspect ratio, and the smooth sidewall is highly feasible. With a relatively severe etching condition, through-wafer via (TWV) without back thinning can be formed, which will avoid uneven thinning or fracture due to excessive wafer bowing. However, etching through vias in wafers with single-side deep etching requires a longer process time, more violent process conditions, and worse quality controllability as etching efficiency decreases significantly as via depth beyond 100μm.

### Double-side etching

Based on single-side etching, we developed another effective method of directly forming wTSV by the double-sided etching, i. e., after the half vias are fabricated by frontside etching, the backside of the wafer is patterned by double-sided exposure and etched.

*Figure 2: SEM side view images of the overall morphology of (a) single-side etched wTSV and (b)(c)(d) double-side etched wTSVs with different depths.*

Fig. 2(b)(c)(d) shows the double-side etching dynamics. This method is much more effective for high aspect ratio wTSV since both etching processes can be completed with soft parameters. It's noted, the additional lithography process arises the requirement of alignment precision. .

## FILLING OF HIGH AR WTSV
### Highly Processable, Low-cost, and Low-Temperature wTSV Technology

Based on the wTSV structure, the low-cost, highly processable, and low-temperature organic liner (OL) layer process is first adopted. The OL layer is formed by vacuum-spin-coating [10]. High vacuum can release air bubbles in the via and assist solution to form void-free filling, while the high-speed spinning removes the excess solution and form a uniform layer attached to the sidewall. Then the wafer is annealed to finally form an OL layer. Ni barrier/seed layer is deposited by the electroless plating method which is occurred at a much lower temperature (< 70 ℃) compared to ALD or CVD process [11]. The barrier property of Ni to Cu diffusion is enough to meet the application in wTSV filling[12].

High-density Cu electroplating is further carried out on the structure to realize the complete void-free filling of wTSV. Figure 3 shows the filling result. The overall structure of a filled wTSV with a width of 32.01 μm, depth of 242.05 μm showed in Figure 3(a), and no void and crack can be seen in the Cu pillar, indicating that the Ni seed layer is continuous on the side wall of the whole hole and has good conductivity. In the top region (Figure 3(b)) and middle region (Figure 3(c)), the organic and Ni layer remain continuous and uniform, not affected by the electroplating process. The Energy Dispersive Spectrum (EDS) analysis in Figure 3(c) demonstrate clearly well defined polymer, Ni and Cu regions (Figure 3 (d)(e)(f)). No any Cu signal appears in the OL liner and Si region, proving the excellent property of Ni as a barrier layer.

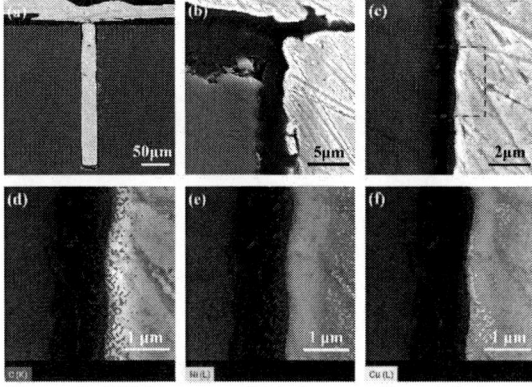

*Figure 3: SEM images of (a) the overall morphology, (b) top region, and (c)middle region of filled wTSV by the low-temperature process. (d)(e)(f) are the EDS analysis result of the organic liner, Ni, and Cu regions.*

**Highly Reliable and Highly Integrable wTSV Technology**

In the high stability and reliability wTSV process, $SiO_2$ formed by thermal oxidation serves as the inorganic liner (IL) layer, which is compatible with subsequent high-temperature processes and high-temperature applications, since the temperatures of thermal oxidation is around 1000℃, much higher than the glass transition temperature of most organic bio-chips or other temperature sensitive materials. [11].

ALD TiN film is used as a barrier layer in this structure. Compared with the Ni barrier layer, the TiN layer has excellent insulation performance with a sheet resistance of around 20.63 kΩ, and can more effectively realize the insulation between Cu pillar and silicon substrate. What's more, the melting point of TiN can reach as high as 2930℃ [13], while that of Ni is around 1460℃ [14], which implies that TiN can be used in high load cases of high-temperature 3D power chips.

To ensure the smooth progress of electroplating, the seed layer is a Cu layer formed by electroless plating [15], which shows a low sheet resistance of 12.18 mΩ. The Cu pillar after electroplating on Cu seed layer is observed by SEM in Figure 4. Cu pillar is fully filled in a through-wafer via with a diameter of 39.91 μm, depth of 226.13 μm, and aspect ratio of around 5.7 (Figure 4(a)). Figure 4(b) and (c) show the top and middle region of the via, respectively, indicating the good uniformity of filling films. EDS results at the middle region (red box in Figure 4(c)) show the integrity of each layer and its good barrier effect on Cu (Figure 4(d)(e)(f)).

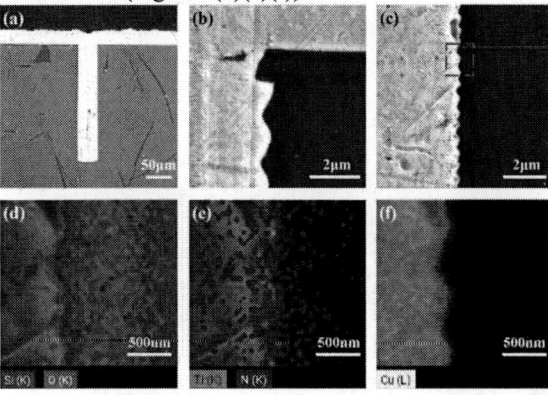

Figure 4: SEM images of (a) the overall morphology, (b) top region, and (c)middle region of filled wTSV by the high-reliability process. (d)(e)(f) are the EDS analysis result of the SiO₂, TiN, and Cu regions.

## ELECTRICAL CHARACTERIZATION

The copper pillar formed after electroplating is evidenced to have an ultra-low resistivity of $3.36 \cdot 10^{-8}$ Ω·m. Consider a prepared copper pillar with a typical size of 38 μm in diameter and 240 μm in length, the resistance

of it is 7.11 mΩ, much smaller than same AR small-size TSV reported in the previous works [16-18].

The leakage current of IL layer under room temperature is measured by Keysight B1500A. Figure 5(a) shows the comparable leakage current in the order of 20pA under a voltage less than 30 V, showing the excellent insulation properties of the wTSV liner layer prepared by the technology.

The *C-V* characteristic of the trench capacitors with IL layer is evaluated also by Keysight B1500A at 1MHz. With voltage swept from -20V to 20V. As shown in Figure 5(b), when the bias voltage switches from -20 V to 20V, the sample capacitance of the IL layer increases from 49.6 fF to 85.3 fF, indicating the change from depletion region to accumulation region as a feature of n-type silicon.

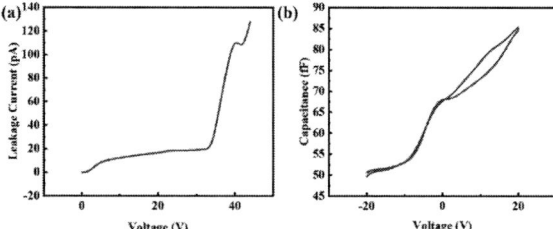

Figure 5: The (a) leakage current and (b) capacitance of wTSV structures with IL layer.

## CONCLUSION

In this work, two unique and widely applicable wTSV technologies for 3D MEMS/NEMS device and integrated system applications are demonstrated. The highly processable, low-cost, and low-temperature wTSV process adopts spin-coated OL layer and electroless plated Ni barrier/seed layer, thus providing a lower thermal budget and better compatibility with other process steps requiring low temperature. The highly reliable and highly integrable wTSV technologies make use of thermally grown IL layer, ALD TiN barrier layer and electroless plated Cu seed layer. The structure can withstand a much higher temperature and is compatible with subsequent high-temperature processes; In addition, the structure has a much higher breakdown voltage useful for high-power devices. Overall, wTSV pillar shows a very low resistivity, lower electro-migration, and low parasitic capacitance which together with the aforementioned advantage of wTSV provides a totally new possibility for device applications with high voltage, high current density, high/low temperature and large load. Thus, the developed wTSV with different specifications can be a excellent candidate for next-generation 3D MEMS/NEMS and high power 3D devices with highly processable, highly integrable, high current density, low cost, high-reliability applications, enlightening a clear road for next generation 3D integrated system.

## ACKNOWLEDGMENTS

This work was financially supported by National Key R&D Program of China (2020YFB2008704), and Tsinghua Foshan Innovation Special Fund (2021THFS0215).

## REFERENCES

[1] M. Seehra and A. Bristow, *Noble and Precious Metals: Properties, Nanoscale Effects and Applications*, BoD–Books on Demand, 2018.

[2] J. U. Knickerbocker, P. S. Andry, L. P. Buchwalter, A. Deutsch, R. R. Horton, K. A. Jenkins, Y. H. Kwark, G. McVicker, C. S. Patel, R. J. Polastre, C. D. Schuster, A. Sharma, S. M. Sri-Jayantha, C. W. Surovic, C. K. Tsang, B. C. Webb, S. L. Wright, S. R. McKnight, E. J. Sprogis and B. Dang. *IBM Journal of Research and Development*, vol. 49, 2005, pp. 725-753.

[3] H. Liu, F. Geng and P. Sun. *2020 21st International Conference on Electronic Packaging Technology (ICEPT)*, Guangzhou, Aug. 12-15, 2020, pp. 1-3.

[4] C. Shang-Chun, C. Yiu-Hsiang, H. Yu-Chen, C. Chien-Chou, C. Jui-Chin, C. Po-chih, T. Pei-Jer, L. Sue-Chen, C. Shin-Chiang, W. Chung-Chih, H. Tzu-Chien, L. Yu-Ming, C. Erh-Hao, L. Cha-Hsin and K. Tzu-Kun. *2013 International Symposium on VLSI Technology, Systems and Application (VLSI-TSA)*, Hsinchu, April 22-24, 2013, pp. 1-2.

[5] Y. Li and D. Goyal, *3D Microelectronic Packaging: From Fundamentals to Applications*, 2017.

[6] S. V. Huylenbroeck, Y. Li, N. Heylen, K. Croes, G. Beyer, E. Beyne, M. Brouri, S. Gopinath, P. Nalla, M. Thorum, P. Meshram, D. M. Anjos and J. Yu. *2015 IEEE 65th Electronic Components and Technology Conference (ECTC)*, San Diego, May 26-29, 2015, pp. 66-72.

[7] S. Killge, I. Bartusseck, M. Junige, V. Neumann, J. Reif, C. Wenzel, M. Böttcher, M. Albert, M. Jürgen Wolf and J. W. Bartha. *Microelectronic Engineering*, vol. 205, 2019, pp. 20-25.

[8] N. Khan, V. S. Rao, S. Lim, H. S. We, V. Lee, X. Zhang, E. B. Liao, R. Nagarajan, T. C. Chai, V. Kripesh and J. H. Lau. *IEEE Transactions on Components and Packaging Technologies*, vol. 33, 2010, pp. 3-9.

[9] Y. Suzuki, H. Fukushi, M. Muroyama, Y. Hata, T. Nakayama, R. Chand, H. Hirano, Y. Nonomura, H. Funabashi and S. Tanaka. *2017 IEEE 30th International Conference on Micro Electro Mechanical Systems (MEMS)*, Las Vegas, Jan. 22-26, 2017, pp. 744-748.

[10] Y. Yan, Y. Ding, T. Fukushima, K. Lee and M. Koyanagi. *IEEE Transactions on Components, Packaging and Manufacturing Technology*, vol. 6, 2016, pp. 501-509.

[11] M. Xiong, Z. Chen, Y. Ding, H. Kino, T. Fukushima and T. Tanaka. *IEEE Electron Device Letters*, vol. 40, 2019, pp. 95-98.

[12] K. W. Lee, C. Nagai, A. Nakamura, J. C. Bea, M. Murugesan, T. Fukushima, T. Tanaka and M. Koyanagi. *2014 International 3D Systems Integration Conference (3DIC)*, Kinsdale, Dec. 1-3, 2014, pp. 1-4.

[13] W. Li, U. Guler, N. Kinsey, G. V. Naik, A. Boltasseva, J. Guan, V. M. Shalaev and A. V. Kildishev. *Advanced Materials*, vol. 26, 2014, pp. 7959-7965.

[14] E. T. Chen, R. N. Barnett and U. Landman. *Physical Review B*, vol. 41, 1990, pp. 439-450.

[15] Z. Zhang, Y. Ding, L. Xiao, Z. Cai, B. Yang, Z. Chen and H. Xie. *IEEE Electron Device Letters*, vol. 42, 2021, pp. 1520-1523.

[16] P. Ramm, M. J. Wolf, A. Klumpp, R. Wieland, B. Wunderle, B. Michel and H. Reichl. *2008 58th Electronic Components and Technology Conference*, Lake Buena Vista, May 27-30, 2008, pp. 841-846.

[17] D. Chen, W. Chiou, M. Chen, T. Wang, K. Ching, H. Tu, W. Wu, C. Yu, K. Yang and H. Chang. *2009 IEEE International Electron Devices Meeting (IEDM)*, Baltimore, Dec. 7-9, 2009, pp. 1-4.

[18] M. Stucchi, D. Perry, G. Katti, W. Dehaene and D. Velenis. *IEEE Transactions on Semiconductor Manufacturing*, vol. 25, 2012, pp. 355-364.

# EVALUATION OF CAPACITIVE HUMIDITY SENSOR WITH POLYIMIDE AS SENSING MATERIAL *

Xiaoxu Kang[1*], Xiaolan Zhong[1], Zhenghui Chu[1], Min Zhang[1], Xiaoqiang Zhou[1], Jiming Qi[1], Qi Jia[2], Weifa Zhong[2], Huajian Liang[2], Xiaozhi Kang[3], Qingqing Sun[3]

[1] Process Technology Department, Shanghai IC R&D Center, Shanghai, 201210, China

[2] Zhuhai SiSensor Technology Co. Ltd., Guangdong, 519080, China

[3] State Key Laboratory of ASIC and System, School of Microelectronics, Fudan University, Shanghai 200433, China

*Corresponding Author's Email: kangxiaoxv@icrd.com.cn

## ABSTRACT

In this paper, capacitive humidity sensor was fabricated on 200mm CMOS Al Back End of Line (BEOL). Al interconnect layer with metal FORK structure was used as electrode of the capacitor, and polyimide was used as sensing material which was filled into metal FORK electrode structure. After fabrication, its humidity sensing performance was evaluated and compared with standard sensor, including IV curve, capacitance variation with different relative humidity (RH%), recovery time, etc. Based on the measured data, the humidity sensor device based on polyimide material can well meet the humidity measurement requirements.

## INTRODUCTION

With the technology developing and market growing, requirement for high performance and low cost sensor product is increasing rapidly, especially in application of the Internet of Things, sensor networks, biosensor and consumer electronics.

Humidity sensor is one of the widely used sensors. Its application including environment monitoring, industry control, agriculture, automobile defoggers, etc. And with the trend of MEMS/Sensor "Consumerization", humidity sensor market will increase dramatically, especially for the application in consumer electronics products, such as mobile phone, etc. [1-5]

In this paper, capacitive humidity sensor was fabricated on 200mm CMOS Al BEOL. Interconnect metal layer with FORK structure was used to fabricate the capacitor structure, and polyimide was used as sensing material which was filled into metal FORK structure. After sensor fabrication, dedicated readout circuit module on PCB level was designed and fabricated, and humidity sensing performance was then evaluated.

## EXPERIMENTAL DETIALS

All the process was developed on 200mm wafer Al BEOL. Metal interconnect layer was used to fabricate the capacitor with FORK structure, and a series of structures with different line/space and different capacitor area were designed. Schematic sensor device structure was shown in Fig.1.

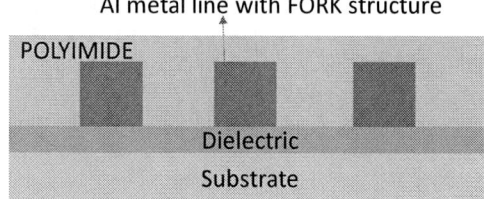

*Figure. 1 Schematic sensor device*

After metal fork structure was fabricated, polyimide was then spin-coating and filling into metal FORK structure. The thickness of polyimide after coating was about 10um, and after baking its thickness was shrunk to about 5um. As polyimide thickness was larger than metal thickness, and there was polyimide covering the entire capacitor structure surface which was schematically in Fig.1. In order to simplify the process and achieve high performance and better process control, photosensitive polyimide was used, and no further etch and strip process was needed which can achieve better sensing material performance.

After device fabrication, its humidity performance was evaluated in test environment. All the relative humidity test environment was built by saturated salt solution, the detailed information can be seen elsewhere [6].

## RESULTS AND DISCUSSION

### 1. Physical evaluation of the humidity sensor

After device fabrication, cross-sectional SEM was done to check its physical performance, which was shown in Fig.2. As shown in the SEM photo, polyimide was about ~5um thick which was measured from the bottom of the FORK structure to top of the surface. There was no void in the gap between the metal lines, which indicated good filling performance. And very flat surface can be achieved after polyimide filling and baking process, which means good planarization performance.

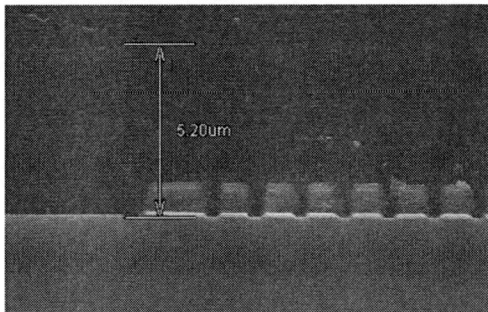

Figure. 2 X-SEM photo of the Sensor device

### 2. Electrical Evaluation of the humidity sensor

Capacitance was measured for the sensor structures by HP4284A, which was shown in Fig.3. In order to guarantee the capacitance measurement data accuracy, leakage was measured by HP4156C for the structure with largest capacitance (on the right in Fig.3). As shown in Fig.4, leakage can be controlled to less than 1E-10A for IV sweeping mode from 0V to 5V, which indicated a good leakage control.

**Capacitance (Units: pF)**

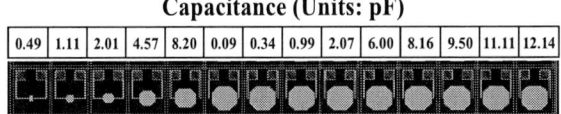

Figure.3 Measured capacitance of different structures

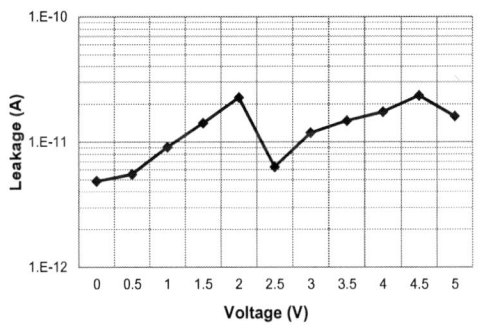

Figure.4 Measured IV Curve of the sensor device

### 3. System level evaluation of the humidity sensor

In order to evaluate humidity performance of the sensor device, readout circuit was designed and fabricated on PCB level to extract the device capacitance. The sensor under test was firstly bonded to the PCB with readout circuit, and then attached to a cover of a sealed container together with a standard humidity sensor. And humidity sensor of Rotronic HC2A-S was chosen as the standard sensor to measure the RH% in the test environment. The sealed container can provide environment with fixed

relative humidity by saturated salt solution. Measurement system was shown in Fig.5.

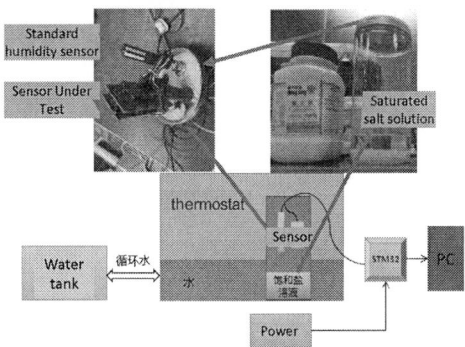

Figure. 5 Schematically designed humidity measurement system and building system photo.

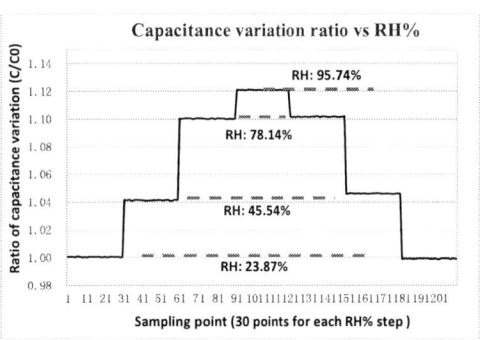

Figure.6 Sensor Capacitance variation ratio (C/C0) with RH%

Sensor device with largest capacitance was selected to evaluate its humidity response in order to reduce parasitic capacitance influence. In each relative humidity environment, the device was measured by sampling mode after about ~10min stabilization. At each RH% step, there was 30 points was sampled with fixed time interval between each point. The device was measured from low humidity to high humidity and then back to low humidity, which can form a humidity testing loop and can be used to check its hysteresis performance.

Taking capacitance at lowest RH% as C0 and C as the capacitance at current RH% environment, C/C0 was plotted in Fig.6. As shown in the figure, the sensor capacitance increased about 12% from ~ RH 24% to ~ RH 96%, and there was no obviously humidity hysteresis problem for sensor under test, which was compatible with standard device.

Recovery time was then evaluated by taking sensor from high RH% environment to indoor environment, and standard sensor was measured together to compare their performance difference. As can be seen in Fig.7, performance of sensor in this work was comparable with standard humidity sensor.

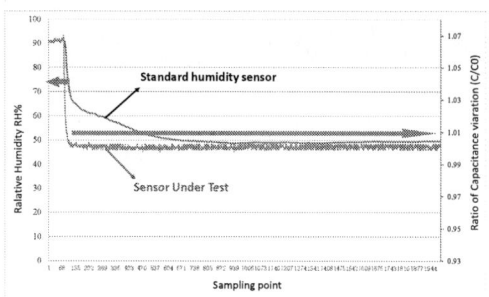

*Figure.7 Comparison of Recovery time between fabricated sensor device and standard sensor*

Based on above data, performance of humidity sensor device in this work was comparable with standard device, can well match the humidity sensor requirements.

## CONCLUSION

In this paper, capacitive humidity sensor was developed with polyimide as sensing material. After sensor fabrication, dedicated readout circuit module was designed and fabricated, and humidity sensing performance was evaluated and compared with standard sensor. Based on the measured data, the humidity sensor device can well meet the humidity measurement requirements.

## ACKNOWLEDGEMENTS

The author would like to thank R&D department of Shanghai IC R&D Center for the support in this work. Special thanks to Qingyun Zuo, Weijun Wang, Ruoxi Shen, Ming Li and Bin Jiang for their strong support in this work.

## REFERENCES

[1] Vigna, B. More than Moore: Micro-machined products enable new applications and open new markets. In Proceedings of IEEE International Electron Devices Meeting, Washington, DC, USA, 1-8 December 2005.

[2] X. Y. Li, X. D. Chen, X. Yu, X. P. Chen, X. Ding, and X. Zhao, A High-Sensitive Humidity Sensor Based on Water-Soluble Composite Material of Fullerene and Graphene Oxide, IEEE Sensors Journal, 18(3)(2018)962-966.

[3] Kang X, Zuo Q, Yuan C, et al. Humidity sensor with graphene oxide as sensing material, IEEE, International Conference on Asic. IEEE, 2016:1-3.

[4] Chen Z, Lu C, Humidity Sensors: A Review of Materials and Mechanisms [J]. Sensor Letters, 2005, volume 3(4):274-295(22).

[5] Shengxue, Yang, Chengbao, Jiang, Su-huai, & Wei. Gas sensing in 2d materials. Applied Physics Reviews, 4(2), 1-34, 2017.

[6] Feng, Jinfeng, Kang, et al. Sensors, Vol. 16, Pages 314: Fabrication and Evaluation of a Graphene Oxide-Based Capacitive Humidity Sensor [J]. Sensors, 2016, 16(3):314.

# QUANTUM EFFICIENCY ENHANCEMENT BY OPTIMIZING BST DESIGN FOR NIR CMOS SENSOR

Chunshan Zhao[1]*, Wuzhi Zhang[1], Yamin Cao[1], Wei Zhou[1]

[1]Shanghai Huali Integrated Circuit Corporation

Shanghai, China

*Corresponding Author's Email: zhaochunshan@hlmc.cn

## ABSTRACT

We demonstrate the quantum efficiency enhancement by using the optimized BST (backside scattered technology) structures for NIR (near-infrared) BSI (backside-illuminated CMOS image sensor). Two type BSTs with different structures are developed for NIR BSI with a cell size of 2.0 um. By using a field-shaped BST, a quantum efficiency of 36.6% at 850 nm and 23.0% at 940 nm was obtained, and using a meter-shaped BST, the quantum efficiency improved to 47.2% at 850 nm and 28.2% at 940 nm.

*Keywords—BSI; NIR; DTI; BST; Quantum efficiency*

## INTRODUCTION

Nowadays, near-infrared (NIR) image sensors have been more and more employed into iris certification, face recognition and motion sensing [1-4]. The NIR sensitivity or quantum efficiency is the key for its applications. Therefore, NIR sensors are based on BSI (backside-illuminated CMOS image sensor), which have higher quantum efficiency than FSI (frontside-illuminated CMOS image sensor). A thicker Si layer is usually needed for more photo absorption and higher photoelectric conversion efficiency, which results in high cost such as high-energy ion implanters [2]. Another may to increase NIR quantum efficiency is to enhance the diffraction and scattering effect with Si, which extend the optical path length with Si. It is reported previously that using the BST (backside scattered technology) structure, NIR sensitivity can be improved significantly [2, 5]. In this study, we prepare NIR BSI pixels with a BST structure and cell size of 2.0 um, and investigate the influence of the BST design on the enhancement of NIR quantum efficiency.

Fig.1 shows the schematic diagram of the BSI pixels with BST and DTI (depth trench isolation). The BST structure is expected to enhance the quantum efficiency by extending the optical path length within C-Si, and the DTI is used to suppress crosstalk between pixels [6]. The depth of the DTI is 2~4 times larger than that of the BST. To maximize the fill factor and sensor sensitivity, each pixel has an on-chip microlens (OCL). In this study, the BST and DTI are applied in the NIR BSI pixels with a cell size of 2.0 um and the process flow is shown in Fig.2. After wafer bonding and thinning, the BST and DTI is performed. To reduce crosstalk, high-K film is necessary for better isolation between pixels. ALD (atomic layer deposition) technology is applied to fill in the BST and DTI with oxide for good quality. Good BST and DTI filling quality is necessary for NIR sensor, which has a significant influence on the pixel noise level, such as white pixel and random telegraph signal noise.

Two types of the BSTs with different structures were developed as shown in Fig.3, one is the field-shaped BST, and another is the meter-shaped BST. For comparison, the NIR BSI without BST was also prepared. Fig.4 shows the top view photographs of the two types BSTs.

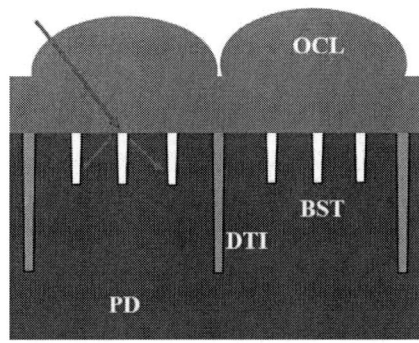

*Fig.1. Schematic diagram of NIR BSI pixels with a backside scattered technology (BST) structure and deep trench isolation (DTI). Each pixel had an on-chip micro lens (OCL).*

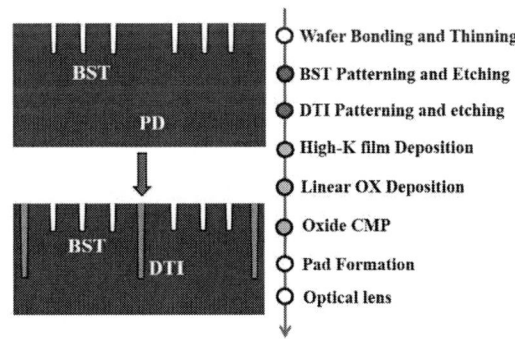

○ Wafer Bonding and Thinning
● BST Patterning and Etching
● DTI Patterning and etching
● High-K film Deposition
● Linear OX Deposition
● Oxide CMP
○ Pad Formation
○ Optical lens

*Fig.2. Process flow for the BST and DTI structure applied in NIR BSI pixels.*

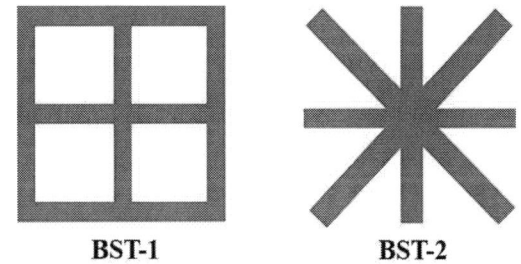

*Fig.3. Two types of the BST structures with different designs*

978-1-6654-9759-6/22 $31.00 © 2022 IEEE

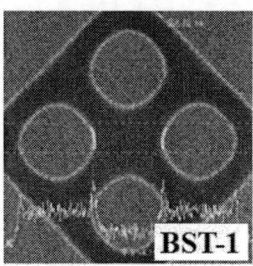

*Fig.4. Top view photographs of the two types BST structures with a cell size of 2.0 um*

Fig.5 shows the quantum efficiencies of 850 nm and 950 nm wavelength of the NIR BSI pixels with a cell size of 2.0 um. Without BST structure, the NIR quantum efficiency was at a very low level, which was only 19.8% at 850 nm and 7.9% at 940 nm. By using the BST structure, significant improvement of quantum efficiency was achieved. With a field-shaped BST, a quantum efficiency of 36.6% at 850 nm and 23.0% at 940 nm was obtained, and with a meter-shaped BST, the quantum efficiency improved to 47.2% at 850 nm and 28.2% at 940 nm. Compared to the filed-shaped BST, the meter-shaped BST exhibited a better performance, 29% higher at 85 0nm and 23% higher at 940 nm wavelength. It suggests that the higher quantum efficiency of the meter-shaped BST is due to its more complicated structure, which should result in a stronger diffraction and scattering effect, leading to a higher photoelectric conversion efficiency.

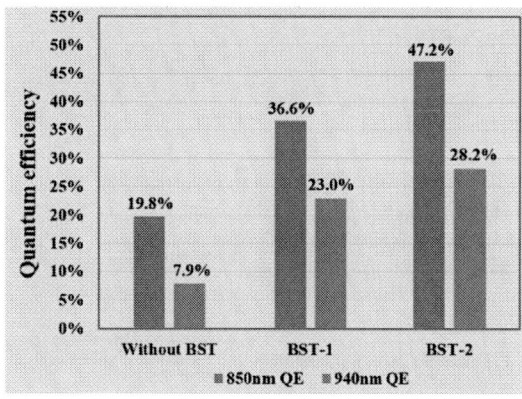

*Fig.5. Quantum efficiencies of 850 nm and 950 nm wavelength of the NIR BSI pixels with a cell size of 2.0 um.*

Although the meter-shaped BST had a preferred performance than the field-shaped one, its complicated structure and larger CD bias between space and cross positions leads to a narrow process window. Therefore, more carefulness is needed for the BST process, especially for BST etching and filling. For example, it is shown in Fig.6 that with inadequate ALD oxide filling, the cross positon shows obvious dishing. In our study, it is found this phenomenon seriously increased the pixel noise level. Increasing the thickness of ALD oxide can fix this problem, which results in a higher cost and reduced capacity.

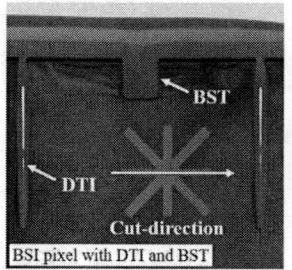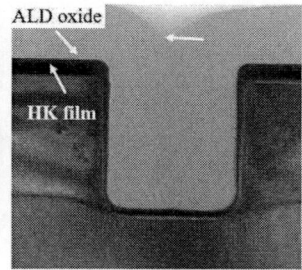

*Fig.6. The micrograph of the cross section of the BSI pixel with DTI and meter-shaped BST structure. With inadequate ALD oxide filling, the cross positon shows obvious dishing.*

## CONCLUSION

NIR BSI without and with different structure BSTs were fabricated. By using the BST structure, significant improvement of quantum efficiency was achieved. With a field-shaped BST, a quantum efficiency of 36.6% at 850 nm and 23.0% at 940 nm was obtained, and with a meter-shaped BST, the quantum efficiency improved to 47.2% at 850 nm and 28.2% at 940 nm. It suggests that the higher quantum efficiency of the meter-shaped BST is due to its more complicated structure, which should result in a stronger diffraction and scattering effect, leading to a higher photoelectric conversion efficiency.

## REFERENCES

[1] H. Sumi et al., Next-generation Fundus Camera with Full Color Image Acquisition in 0-lx Visible Light by 1.12-micron Square Pixel, 4K, 30-fps BSI CMOS Image Sensor with Advanced NIR Multi-spectral Imaging System. IEEE, 2018.

[2] Oshiyama et al., "Near-infrared sensitivity enhancement of a back-illuminated complementary metal oxide semiconductor image sensor with a pyramid surface for diffraction structure," 2017 IEEE International Electron Devices Meeting (IEDM), 2017, pp. 16.4.1-16.4.4

[3] C. Cao et al., "A Dual NIR-Band Lock-In Pixel CMOS Image Sensor With Device Optimizations for Remote Physiological Monitoring," in IEEE Transactions on Electron Devices, vol. 68, no. 4, pp. 1688-1693, April 2021, doi: 10.1109/TED.2021.3057035.

[4] Y. Monno, H. Teranaka, K. Yoshizaki, M. Tanaka and M. Okutomi, "Single-Sensor RGB-NIR Imaging: High-Quality System Design and Prototype Implementation," in IEEE Sensors Journal, vol. 19, no. 2, pp. 497-507, 15 Jan.15, 2019, doi: 10.1109/JSEN.2018.2876774.

[5] Yokogawa, S., Oshiyama, I., Ikeda, H. et al. IR sensitivity enhancement of CMOS Image Sensor with diffractive light trapping pixels. Sci Rep 7, 3832 (2017).

[6] Kitamura Y, Aikawa H, Kakehi K, et al. Suppression of crosstalk by using backside deep trench isolation for 1.12 um backside illuminated CMOS image sensor[J]. 2012, 24.2.1-.2.4.

# DEVICE MODELING AND SIMULATION OF FERROELECTRIC TUNNEL JUNCTION FOR COMPUTING-IN-MEMORY APPLICATION

*Yuyao Lu[†], Linpu Zhai[†], Bin Gao\*, Jianshi Tang, Feng Xu, Yue Xi, Qingtian Zhang,*
*Zhigang Zhang, He Qian and Huaqiang Wu*

School of Integrated Circuits (SIC),
Beijing Innovation Center for Future Chips (ICFC), Tsinghua University, Beijing, China.
[†]Authors with equal contributions. \*Corresponding Author's Email: gaob1@tsinghua.edu.cn

## ABSTRACT

For computing-in-memory applications implemented by ferroelectric tunnel junction (FTJ), a multi-pulse FTJ switching model is required. Here, based on the single-pulse nucleation-limited switching (NLS) model, a multi-pulse model capable of calculating the change of ferroelectric polarization under a series of arbitrary waveform pulses at different frequencies is proposed, which shows good agreement with literature-reported experimental results. In addition, the multi-pulse model was adopted in the simulation of an FTJ-based neural network, where it was found that the programming scheme with increasing pulse amplitude could achieve higher recognition accuracy and better FTJ conductance fluctuation tolerance than those with identical pulse or increasing pulse width. This work provides a useful model for further optimization and application of FTJ in neuromorphic computing.

## INTRODUCTION

Various applications based on deep learning have raised higher and higher demand for computing power and data processing capabilities. In the conventional von Neumann architecture, the data computing and storage units are separated [1], and hence most of the energy consumption and delay of the system are spent in memory accessing, which severely limits the data processing speed and leads to the so-called "memory wall" issue. The computing-in-memory (CIM) architecture based on emerging memories such as memristor emerges as a promising computing paradigm to solve this problem [2], [3]. As the fundamental computing unit, memristor with analog resistive switching characteristics plays a vital role in the performance of CIM systems [4]. Thanks to the excellent characteristics of ferroelectric materials, including fast polarization switching speed and stable polarization state, ferroelectric memory has the potential advantages of high speed, low power consumption and non-volatility [5]. In particular, as one type of ferroelectric memory devices, ferroelectric tunnel junction (FTJ) has attracted wide attention because of its ability to realize continuous conductance modulation [6].

Based on the analog resistive switching characteristics of FTJ, an FTJ crossbar array can realize the vector-matrix multiplication operation in one step by taking the advantage of CIM, which is highly desired for accelerating deep learning algorithms. For the implementation of neural network training and inference, the linearity, symmetry, number of conductance levels, fluctuation and other resistive switching characteristics of the FTJ devices can directly affect the final recognition accuracy [7]. For this purpose, establishing a multi-pulse switching model of FTJ, and studying the influence of its non-ideal characteristics on the array performance are critical for the CIM application of FTJ devices.

The key issue of the FTJ multi-pulse switching model is the polarization reversal of the ferroelectric film. So far, there are two mainstream models which describe the polarization reversal process: Preisach model and nucleation-limited switching (NLS) model [8], [9]. However, neither Preisach model nor NLS model could accurately describe the change of ferroelectric polarization under a series of arbitrary pulses.

In this article, a method is proposed to expand the NLS model to a new multi-pulse model by calculating the cumulative effect of previous pulses. This new multi-pulse model could accurately calculate the polarization changes of ferroelectric materials under arbitrary voltage pulses by dividing an arbitrary voltage waveform into a series of rectangular pulses based on the idea of differentiation. The multi-pulse model could also predict the frequency-dependent characteristics of ferroelectric materials very well. Furthermore, the FTJ multi-pulse switching model was applied in artificial neural network simulation, and the influence of different programming schemes on the classification accuracy was studied.

## DEVICE MODELING

Here it is assumed that n pulses with the same amplitude $V$ and a summation of total pulse width $T_n$ are equivalent to one single pulse with pulse amplitude $V$ and pulse width $T_n$. According to this assumption, the switched fraction of the ferroelectric film under n voltage pulses with the same amplitude $V$ is given by:

$$p(n) = p_{NLS}(V, T_n) = p_{NLS}\left(V, \sum_1^n t_i\right) \#(1)$$

where $p_{NLS}$ denotes the switched fraction of the ferroelectric film calculated by NLS model, $t_i$ is the pulse width of the $i^{th}$ pulse and $T_n = \sum_1^n t_i$.

Similar to Equation (1), the switched fraction of the

ferroelectric film under n+1 voltage pulses namely $p(n+1)$ can be derived as follows:

$$p(n+1) = p_{NLS}(V, T_{n+1}) = p_{NLS}(V, T_n + t_{n+1}) \#(2)$$

where $T_{n+1}$ denotes the summation of total pulse width of n+1 pulses, $t_{n+1}$ is the pulse width of the $(n+1)^{th}$ pulse and $T_{n+1} = \sum_1^{n+1} t_i = T_n + t_{n+1}$.

Note that this equation of calculating the accumulative effect is only suitable for pulses with same amplitude.

For pulses with different amplitudes, suppose that the value of $p(n)$ has been obtained. The amplitude of the $(n+1)^{th}$ pulse is $V^*$ and the pulse width is $t_{n+1}^*$. In order to calculate $p(n+1)$, the accumulative effect of previous pulses namely $p(n)$ is assumed to be equivalently induced by one single pulse with constant amplitude $V^*$ and width $T_n^*$, thus converting the different amplitudes condition to same amplitude condition. $T_n^*$ can be derived by solving Equation (1):

$$T_n^* = solve[p(n) = p_{NLS}(V^*, T_n^*)] \#(3)$$

Then by substituting $T_n^*$ into Equation (2), $p(n+1)$ can be calculated.

$$p(n+1) = p_{NLS}(V^*, T_n^* + t_{n+1}^*) \#(4)$$

In this way, once $p(1)$ is known, the evolution of ferroelectric polarization under the pulse sequence can be calculated. Here $p(1)$ is the fraction of switched volume under the first pulse, which can be directly calculated by NLS model:

$$p(1) = p_{NLS}(V_1, t_1) \#(5)$$

The overall calculation process is summarized as follows:

$$p(1) = p_{NLS}(V_1, t_1)$$

$$\vdots$$

$$p(n) = p_{NLS}(V_n, T_{n-1}^* + t_n)$$

$$T_n^* = solve[p(n) = p_{NLS}(V_{n+1}, T_n^*)]$$

$$p(n+1) = p_{NLS}(V_{n+1}, T_n^* + t_{n+1})$$

$$\vdots$$

where $V_i$ is the amplitude of the $i^{th}$ pulse.

Figure 1(a) shows the fitting results of experimental data (Cu/P(VDF-TrFE)/Cu ferroelectric capacitor) with the NLS model under different electric field strengths [10]. By dividing an arbitrary voltage pulse into short rectangular pulses, each with an approximately constant amplitude, the multi-pulse switching model described above could be utilized, and the simulation results of the polarization change with the external electric field at different frequencies are shown in Figure 1(b). The good fitting between the multi-pulse switching model and the experimental data indicates that the multi-pulse switching model can accurately describe the FTJ switching process.

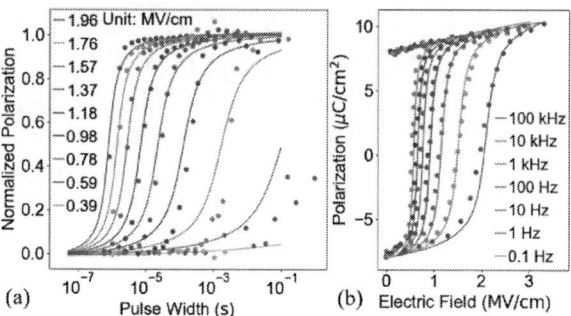

*Figure 1: Points denote experimental data and lines denote simulation results. (a) Fitting results of the experimental data with NLS model under different electric field strengths. (b) Polarization vs. electric field at different frequencies calculated by the multi-pulse switching model. The experimental data are extracted from Reference [10].*

## SIMULATION RESULTS

### Programming Scheme Simulation Results

Simulations of three different FTJ programming schemes were implemented, using the multi-pulse switching model. The parameters of set and reset pulses and the calculation results of FTJ current density are shown in Figure 2. It can be seen that the programming scheme with increasing pulse amplitude yields the best linearity and symmetry.

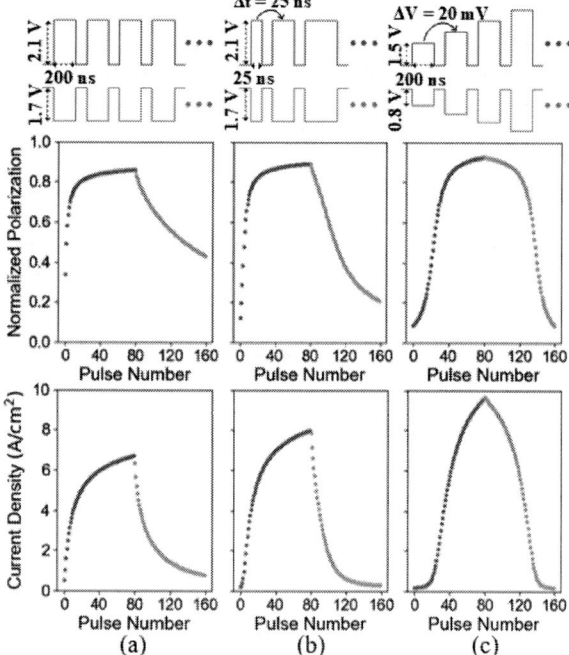

*Figure 2: Polarization and current density vs. ordinal number of pulses with different programming schemes: (a) identical pulses; (b) increasing pulse width; (c) increasing pulse amplitude. Pulse amplitudes are reduced in reset*

*process to improve the linearity and symmetry of FTJ.*

### Neural Network Simulation Results

Utilizing the analog switching characteristics of FTJ under different programming schemes in Figure 2, neural network simulations were performed for handwriting digits recognition on the widely used Mixed National Institute of Standards and Technology (MNIST) database. Figure 3(a) illustrates the structure of the simulated multi-layer perceptron of 784×200×10. Differential pairs consisting of two FTJ cells were adopted to realize both positive and negative synaptic weights, and the hidden layer was binarized to simplify the hardware implementation [11]. Device fluctuations were introduced in the training process, by renewing the conductance with a stochastic increment of $\Delta G \times N(1, \sigma)$ at each time when the weight was updated. The simulation results of recognition accuracy with different programming schemes are shown in Figure 3(b). It can be seen that compared with the other two programming schemes, the programming scheme of increasing amplitude pulses gives the highest recognition accuracy of about 90% and the highest tolerance to conductance fluctuation.

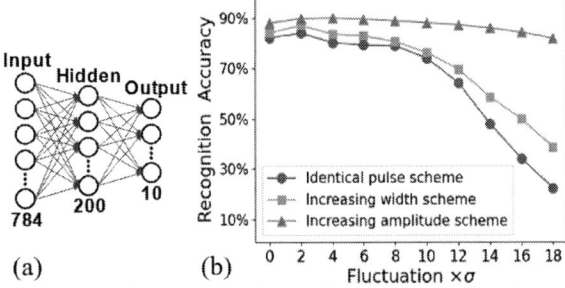

(a)  (b)

*Figure 3: (a) Structure of neural network. (b) Simulation results of recognition accuracy using different pulse programming schemes. σ is the standard deviation, which represents the degree of dispersion of the conductance value drift caused by fluctuation.*

## CONCLUSION

To sum up, a comprehensive multi-pulse switching model for FTJ is established, extending the single-pulse NLS model to calculate the change of ferroelectric polarization under a series of arbitrary waveform pulses.

The simulation results of the model were in good agreement with the experimental data at different frequencies. The multi-pulse switching model was further applied to simulations of FTJ-based artificial neural network for MNIST dataset recognition. It was found that using pulses with increasing amplitude achieved the best recognition accuracy up to 90% and the highest tolerance to conductance fluctuation compared with the other two schemes. The developed multi-pulse switching model of FTJ provides a useful guidance for device optimization and also neural network application in the future.

## ACKNOWLEDGEMENTS

This work was in part supported by the National Science and Technology Major Project of China (2017ZX02315001-005) and Natural Science Foundation of China (61974081, 62025111, 61851404).

## REFERENCES

[1] J. Von Neumann, "First Draft of a Report on the EDVAC," *IEEE Annals of the History of Computing,* vol. 15, no. 4, pp. 27-75, 1993, doi: 10.1109/85.238389.

[2] Y. Li *et al.,* "Oscillation neuron based on a low-variability threshold switching device for high-performance neuromorphic computing," *Journal of Semiconductors,* vol. 42, no. 6, p. 6, 2021, doi: 10.1088/1674-4926/42/6/064101.

[3] P. Yao *et al.,* "Fully hardware-implemented memristor convolutional neural network," *Nature,* vol. 577, no. 7792, pp. 641-646, 2020, doi: 10.1038/s41586-020-1942-4.

[4] M. A. Zidan, J. P. Strachan, and W. D. Lu, "The future of electronics based on memristive systems," *Nature electronics,* vol. 1, no. 1, pp. 22-29, 2018, doi: 10.1038/s41928-017-0006-8.

[5] R. Guo, W. Lin, X. Yan, T. Venkatesan, and J. Chen, "Ferroic tunnel junctions and their application in neuromorphic networks," *Applied Physics Reviews,* vol. 7, no. 1, p. 011304, 2020, doi: 10.1063/1.5120565.

[6] A. Chanthbouala *et al.,* "A ferroelectric memristor," *Nature materials,* vol. 11, no. 10, pp. 860-864, 2012, doi: 10.1038/nmat3415.

[7] Y. Xi *et al.,* "In-Memory Learning With Analog Resistive Switching Memory: A Review and Perspective," *Proceedings of the IEEE,* vol. 109, no. 1, pp. 14-42, 2020, doi: 10.1109/JPROC.2020.3004543.

[8] G. Bertotti and I. D. Mayergoyz, *The Science of Hysteresis: Mathematical modeling and applications.* Academic Press, 2006.

[9] A. K. Tagantsev, I. Stolichnov, N. Setter, J. S. Cross, and M. Tsukada, "Non-Kolmogorov-Avrami switching kinetics in ferroelectric thin films," *Physical Review B,* vol. 66, no. 21, p. 214109, 2002, doi: 10.1103/PhysRevB.66.214109.

[10] W. J. Hu *et al.,* "Universal ferroelectric switching dynamics of vinylidene fluoride-trifluoroethylene copolymer films," *Scientific reports,* vol. 4, no. 1, pp. 1-8, 2014, doi: 10.1038/srep04772.

[11] H. Wu *et al.,* "Device and circuit optimization of RRAM for neuromorphic computing," in *2017 IEEE International Electron Devices Meeting (IEDM),* 2017: IEEE, pp. 11.5. 1-11.5. 4, doi: 10.1109/IEDM.2017.8268372.

# POLARITY-CONTROLLABLE WSE_2 TRANSISTOR ENABLED BY INSERTING A H-BN LAYER

*Zheng Bian[1,2], Jialei Miao[1,2,3], Tianjiao Zhang[1,2] and Yuda Zhao[1,2]\**

[1] School of Micro-Nano Electronics, Zhejiang University, Hang Zhou, China
[2] ZJU-Hangzhou Global Scientific and Technological Innovation Center, Hang Zhou, China
[3] Faculty of Electrical Engineering and Computer Science, Ningbo University, Ningbo, China
*Corresponding Author's Email: yudazhao@zju.edu.cn

## ABSTRACT

The polarity control is highly imperative for the development of complementary metal-oxide-semiconductor (CMOS) technology based on two-dimensional (2D) layered materials. The conventional doping strategy by ion implantation is not suitable for atomically thin 2D materials and the widely adopted chemical doping method on 2D materials is not industry-friendly. In this paper, we demonstrated the polarity modulation of WSe_2 field-effect transistors (FETs) from p-type to n-type transport by inserting a thin hexagonal boron nitride (h-BN) as the interfacial dielectric layer. Based on this, a WSe_2 CMOS inverter was successfully achieved. This air-stable and process-friendly strategy promotes the integration of transistors based on 2D materials in silicon CMOS techniques.

## INTRODUCTION

Complementary n-channel and p-channel MOSFETs are the basic building blocks to construct logic gates in digital circuits, which requires an effective method to dope semiconductors. Owing to the atomic thickness, 2D semiconductors show the ability to overcome the scaling limit and exhibit high immunity to short channel effects. [1] However, it is a major challenge to effectively modulate the carrier type in 2D channels and the polarity of 2D FETs by a CMOS compatible method. Among semiconducting transition metal dichalcogenides (TMDs), the intrinsic ambipolar character of WSe_2 makes it a suitable channel material for complementary FETs. [2], [3] Several methods have been demonstrated to modulate the carrier type of WSe_2, including variation of channel thickness [4], chemical doping [5], contact engineering [6], and tuning of metal work function [7]. But these methods suffer from the low stability and the weak tunability. In this work, we inserted a thin layer of h-BN between the WSe_2 channel and the SiO_2 dielectric layer, acting as the interfacial dielectric layer. The perfect interface provided by 2D insulating h-BN can reduce the random scattering from the dielectric disorders [8] and improve the transistor performance. Furthermore, by selecting an appropriate thickness of h-BN layer, the WSe_2 FETs were modulated from p-type transport to n-type transport. A CMOS inverter composed of a p-type WSe_2 FET and an n-type WSe_2 FET was successfully

constructed. Our results not only demonstrate a robust and CMOS-compatible strategy to control the polarity of WSe_2 FETs, but also provide a viable solution to build CMOS logic gates based on single 2D semiconductor, shedding light to high-performance 2D electronics and CMOS design.

## RESULTS AND DISCUSSION

In the reported works, [9] few layers WSe_2 FETs mostly exhibit a p-type transfer characteristic on SiO_2 substrate. In our work, we firstly investigated the contribution of substrate on the carrier type of the WSe_2 channel layer by fabricating WSe_2 FETs on SiO_2 and h-BN/SiO_2 dielectric layer, respectively. Fig. 1 shows the schematic of the device structure of the few-layer WSe_2 FETs on the h-BN/SiO_2 dielectric. Few-layer h-BN was first exfoliated on the 285nm SiO_2/Si substrate followed by the dry transfer of WSe_2 flake onto the h-BN layer area. Metal contacts (5nm/50nm Cr/Au) were patterned by mask free photolithography and deposited by thermal evaporation.

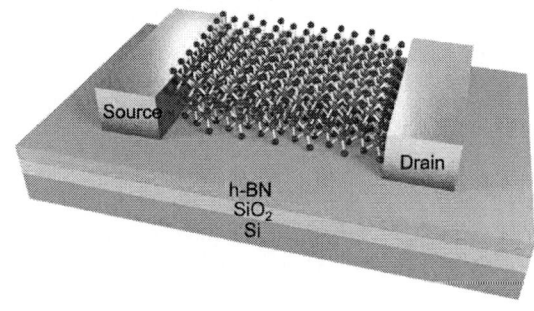

*Figure 1: Schematical illustration of polarity reversed WSe_2 FET by inserting the h-BN layer.*

As shown in Fig. 2, WSe_2 transistors on SiO_2 and h-BN/SiO_2 dielectric layer exhibited distinct electrical transport properties. The transfer curve of WSe_2 FET on SiO_2 showed a p-type transport characteristic while an n-type transport behavior was observed on h-BN/SiO_2 substrate. Both devices showed stable polarity through

978-1-6654-9759-6/22 $31.00 © 2022 IEEE

several cycles of test and this polarity control method can be repeated in different batches of devices, showing little cycle-to-cycle and device-to-device variation. Here, chromium was selected as the contact material due to the appropriate alignment of metal work function (4.5 eV) [10] with the Fermi level of few layer WSe$_2$ (intrinsic layer: 4.3 eV) [4].

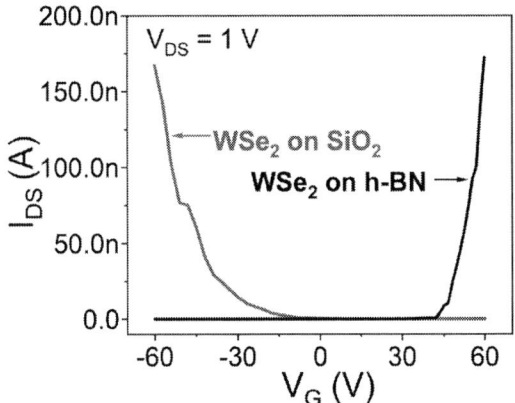

Figure 2: Transfer curves of the WSe$_2$ FETs demonstrate p-type (red line) and n-type (black line) characteristics on SiO$_2$ and h-BN/ SiO$_2$ substrate, respectively.

Fig. 3 showed the transfer characteristics of WSe$_2$ FETs with different thickness of inserting h-BN layer. The device on bare SiO$_2$ substrate acted as the control sample. The thickness of h-BN was characterized by atomic force microscope (AFM), which is 17 nm, 23 nm, and 30 nm, respectively. WSe$_2$ layer with a thickness of 5 nm (about 6-7 layers) was selected as the channel material. As we can see, the transfer curves of WSe$_2$ FETs changed from p-type transport on bare SiO$_2$ substrate to a p-dominant ambipolar transport by inserting a 17 nm h-BN layer. When the h-BN thickness increased to 23 nm, the transfer curve becomes an n-dominant ambipolar transport. And a fully n-type transfer characteristic was achieved with a 30 nm h-BN inserting layer. These results demonstrated that the WSe$_2$ layer was gradually n-type doped with the increasing thickness of h-BN layer and the majority carrier type of WSe$_2$ layer was switched from p-type to n-type, revealing that our technique has the capability to dramatically modulate the carrier density in 2D channels.

Fig. 4 reveals the polarity switching mechanism of WSe$_2$ FETs on different substrates. The thin SiO$_2$ layer is deposited by the thermal oxidation of Si and this process inevitably introduces positively charged impurities on the surface of SiO$_2$/Si substrates. These positively charged impurities induce the p-type doping on the adjacent WSe$_2$ layer, resulting in the p-type transport on the SiO$_2$ substrate. When a thin h-BN layer is inserted between

WSe$_2$ and SiO$_2$, it acts as a blocking layer and reduces the charged impurities scattering originated from the SiO$_2$ surface. Meanwhile, the positively charged impurities underneath the thin h-BN layer induce the electrostatic doping effect, leading to the n-type transport behavior in WSe$_2$ FETs. When the h-BN layer is very thin, the electrons in WSe$_2$ layer are attracted by the positively charged impurities on the SiO$_2$ surface and have the possibility to tunnel through the h-BN layer. With the increase of the h-BN thickness, the electrostatic doping becomes dominant and the device shows strong n-doping effect (Figure 3, green curve). When the h-BN thickness further increases to over 50 nm, WSe$_2$ FETs restored to the ambipolar transport behavior, which is the intrinsic character of WSe$_2$.

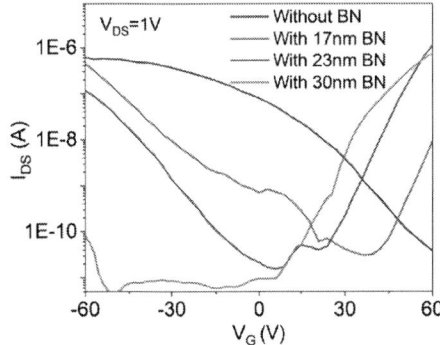

Figure 3: Transfer curves of WSe$_2$ FETs with different thickness of the h-BN inserting layer.

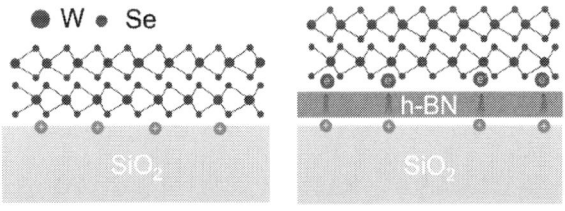

Figure 4: Polarity switching mechanism of WSe$_2$ FET by inserting a h-BN layer between WSe$_2$ and SiO$_2$.

On this basis, we fabricated a CMOS inverter by using p-type (on SiO$_2$ substrate) and n-type (on h-BN/SiO$_2$ substrate) WSe$_2$ devices as pull-up and pull-down transistors, respectively. The h-BN layer was only placed beneath part of the WSe$_2$ layer, and hence the two FETs (p-type and n-type) use the same WSe$_2$ flake as the channel layer, which is shown in Fig. 5. A 30 nm h-BN layer was used as the interfacial layer to achieve a fully n-type WSe$_2$ transistor.

Fig. 6 shows the voltage transfer curves of the WSe$_2$ CMOS inverter with input voltage V$_{dd}$ from 1V to 5V. It shows a complete inverter operation function and repeatable property. And the switching threshold voltage shifts along with the increasing of V$_{dd}$. Our results demonstrate that WSe$_2$ layer can be used to fabricated both n-type and p-type transistors via an industry-friendly method by inserting a h-BN layer. This method is more reliable than chemical doping methods.

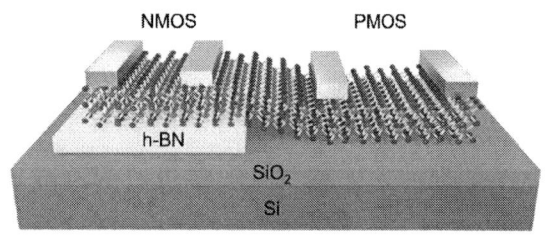

*Figure 5: Schematical illustration of WSe$_2$ inverter. The WSe$_2$ transistor on h-BN/SiO$_2$ substrate operates as pull-down transistor while the WSe$_2$ transistor on SiO$_2$ substrate operates as pull-up transistor.*

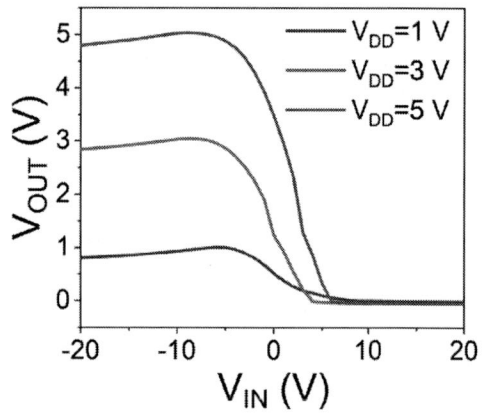

*Figure.6: Voltage transfer characteristic of the WSe$_2$ inverter with different input voltage V$_{dd}$.*

## CONCLUSION

In summary, we demonstrated a novel approach by inserting a thin h-BN layer to modulate the polarity of WSe$_2$ FETs and successfully constructed a CMOS inverter. The polarity of WSe$_2$ FETs was fully switched from p-type to n-type by inserting a 30 nm h-BN. The electrostatic doping and the reduction of charged impurity scattering synergistically contribute to the switching of the transistor polarity. With the development of wafer-scale growth of TMDs/h-BN heterostructures, this method will enable the fabrication of high-performance 2D CMOS devices and systems on a large scale.

## ACKNOWLEDGEMENTS

We thank ZJU Micro-Nano Fabrication Center and Southeast University for the supports. The project was primarily supported by the National Natural Science Foundation of China (62090030, 62090034, 62104214).

## REFERENCES

[1] T. D. Ngo et al., "Fermi-Level Pinning Free High-Performance 2D CMOS Inverter Fabricated with Van Der Waals Bottom Contacts," Adv. Electron. Mater., vol. 7, no. 5, pp. 1–7, 2021, Art. no. 2001212.

[2] Y. Zhao *et al.*, "Molecular Functionalization of Chemically Active Defects in WSe$_2$ for Enhanced Opto-Electronics," *Adv. Funct. Mater.*, vol. 30, no. 45 , 2020，Art. no. 2005045.

[3] F. Ali et al., "Traps at the hBN/WSe$_2$ interface and their impact on polarity transition in WSe$_2$," 2D Mater., vol. 8, no. 3, 2021, Art. no. 035027

[4] A. A. and A. Kis, "Electron and Hole Mobilities in Single- Layer WSe$_2$," ACS Nano, vol. 8, no. 7, pp. 7180–7185, 2014.

[5] H. G. Ji et al., "Chemically Tuned p- and n-Type WSe$_2$ Monolayers with High Carrier Mobility for Advanced Electronics," Adv. Mater., vol. 31, no. 42, 2019, Art. no. 1903613.

[6] L. Kong et al., "Doping-free complementary WSe$_2$ circuit via van der Waals metal integration," Nat. Commun., vol. 11, no. 1, 2020, Art. no. 1866.

[7] T. D. Ngo et al., "Fermi-Level Pinning Free High-Performance 2D CMOS Inverter Fabricated with Van Der Waals Bottom Contacts," Adv. Electron. Mater., vol. 7, no. 5, pp. 1–7, 2021, Art. no. 2001212.

[8] A. Raja et al., "Dielectric disorder in two-dimensional materials," Nat. Nanotechnol., vol. 14, no. 9, pp. 832–837, 2019.

[9] C. Zhou et al., "Carrier Type Control of WSe$_2$ Field-Effect Transistors by Thickness Modulation and MoO$_3$ Layer Doping," Adv. Funct. Mater., vol. 26, no. 23, pp. 4223–4230, 2016.

[10] H. B. Michaelson, "The work function of the elements and its periodicity," J. Appl. Phys., vol. 48, no. 11, pp. 4729–4733, 1977.

# STUDY ON SMALL DEAD AREA HIGH-VOLTAGE SILICON PIN DETECTORS

*Tiesong Li[1], Xin He[1], and Min Yu[1*]*

[1]National Key Laboratory of Nano/Micro Fabrication Technology, Institute of Microelectronics,
Peking University, Beijing 100871, China
*Corresponding Author's Email: yum@pku.edu.cn

## ABSTRACT

The purpose of this work is to investigate the effect of dicing damages on breakdown voltage of silicon PIN detectors to achieve small dead area high-voltage detectors. A numerical simulation study has been carried out by introducing traps at detector die edge to simulate the dicing defects and cracks. The experimental results from the electrical characterization of the high-voltage silicon PIN detectors before and after different dicing methods, are presented. It's concluded based on the agreement between simulation and experimental results that the edge width of about 1mm is relatively safe to avoid the electrical breakdown due to defects at die edge, which ensures that the breakdown voltage of the detectors can reach 1000V.

## INTRODUCTION

In nuclear radiation detection and aerospace, detectors with high breakdown voltage and small dead area are highly desired. Various edge termination approaches, such as Field Plate and Multiple Floating Guards in different configurations[1-3], have been designed and produced on high resistivity silicon wafer in order to obtain high breakdown voltage. Most studies focus on obtaining the high breakdown voltage. However, few studies have looked at detectors with smaller dead area and high breakdown voltage as well. Usually, there exists a gap of at least some hundreds of micrometers between the edge of main junction and the physical edge of detector die, which is called dead area here. The detectors are typically separated from the wafer by using a diamond blade and the dicing procedure inevitably introduces defects and cracks along the edge of the detectors which act as generation–recombination centers. In case the depletion area spreads from the active area and reaches these defects, a high leakage current can be drawn by the detectors, which limits its work under high voltage[4]. In order to reduce the dead area as much as possible without affecting the breakdown performance of the detectors, we focus on finding the optimum edge dead area for the high-voltage silicon PIN detectors.

In this work, we have fabricated the high-voltage silicon PIN detectors. In particular, making full use of TCAD simulations, we introduced the trap model to simulate the defects and cracks at detector die edge. We designed several silicon PIN detector structures with different dead areas. The structures are simulated and fabricated as well to compare the theoretical and experimental results. Different dicing methods, i.e. diamond blade and laser dicing, are applied and compared to investigate the best compromise between dead area and breakdown performance.

## DEVICE SIMULATIONS

We first use Sentaurus TCAD tools to simulate and analyze the effect of dicing damages on the detectors. The defects at detector die edge are modeled as carrier traps which can act as generation-recombination centers and contribute to device leakage current. By changing the gap, i.e. the distance from the physical die edge to the edge of main junction, we simulate detectors with different dead areas. Figure 1 shows a schematic cross section of the basic detector structure. The detector is made on a 300μm thick, n-type silicon substrate with a bulk doping concentration of $2 \times 10^{11} cm^{-3}$ and a relatively low concentration of oxide layer fixed charge ($10^{10} cm^{-2}$) is applied for simulations. An acceptor level in the middle of the band gap with a concentration of $10^{17} cm^{-3}$ and capture cross sections of $10^{-14} cm^2$ are supposed to describe the traps region at the physical edge of the detector die. In the physical model part, models including SRH, surfaceSRH and Avalanche[5] are included in order to further consider the effect of traps at the edge.

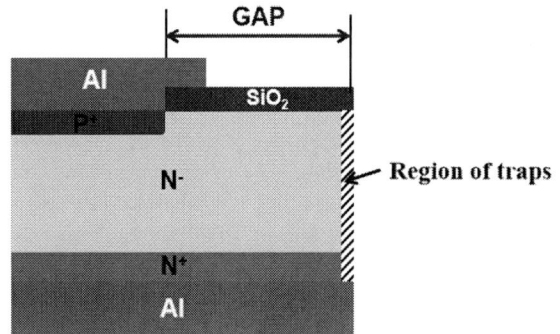

Figure 1: schematic cross section of silicon PIN radiation detector for simulations

Then, negative bias voltage is applied to the aluminum electrode above the $P^+$ region and the current is limited to 200nA. The simulation results of device I-V characterization are obtained as shown in Figure 2. When the gap is 1mm or more, traps at the edge have almost no effect on the breakdown performance of the detectors, and the breakdown voltage of the detectors can reach 1200V. However, when the gap is reduced to 700μm, the leakage current of the detector increases obviously under the bias voltage of 1000V~1200V. If the gap is reduced to only 500μm or 400μm, the leakage current of the detectors increases rapidly when the voltage is over 400V and 200V respectively, which indicates the breakdown of the detectors. A dramatic decrease of breakdown voltage is observed.

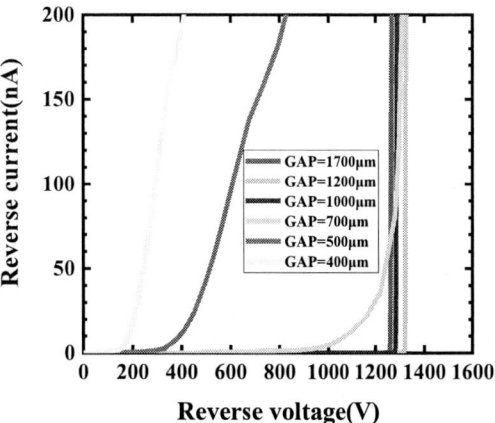

Figure 2: Simulated I-V curves for different gaps

To gain insight into the above result, we extract the electrostatic potential distribution of the detectors at breakdown from the simulations. If the gap is 1mm, as shown in Figure 3(a), the depletion region is still hundreds micrometers far from the die edge, which indicates that the traps at edge has no effect on the leakage current. On the other hand, as can be seen from the electrostatic potential distribution in Figure 3 (b), the transverse depletion region of the detector shows that the depletion region can be extended to a distance about 500μm from the edge of main junction and reaches the edge trap region. Thus a high leakage current can be drawn by the detector, which damages the high-voltage performance. In fact, the range of defects and cracks damages caused by the dicing can be larger than the traps range set by the simulations and may cause more serious problems. Additionally, we can also see a larger carrier current density inside the detector from the simulation when the gap is 500μm compared to 1mm.

## EXPERIMENTAL

We have fabricated high-voltage silicon PIN detectors on 300μm thick, n-type high resistivity silicon substrate that is also used in the previous simulations. Boron and phosphorus are doped in the front and back side of the wafer to form the $P^+$ and $N^+$ region with the active area of 50mm$^2$. Aglient Technologies B1505A Power

Device Analyzer are used to test the high voltage electrical performance of the detectors. Seven detectors (number as F7-01~F7-07) are selected for further dicing experiments.

(a)
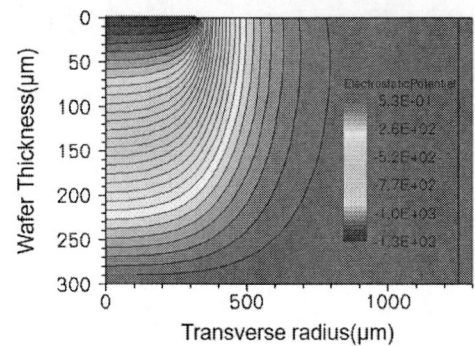

(b)
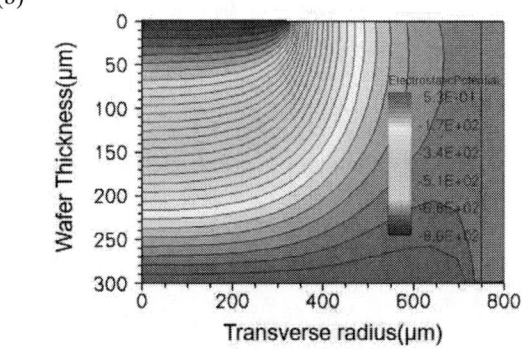

Figure 3: Electrostatic potential distribution in detectors with gap value of (a) 1mm; (b) 500μm

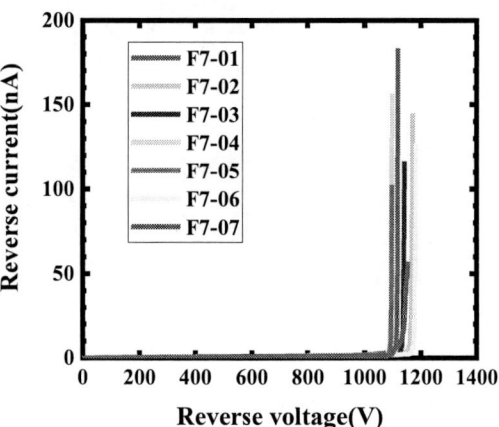

Figure 4: Electrical characterization results of the detectors before dicing

The test results are shown in Figure 4. The breakdown voltage of all the detectors are between 1000~1200V, and the leakage current of the detectors are all about 2nA when the reverse bias voltage is below breakdown voltage. Then the edge width, i.e. the gap, of

1mm, 700μm, 500μm are realized by diamond blade dicing on detector samples of F7-01, F7-02 and F7-03. Figure 5 shows electrical characterization results for the detectors after diamond blade dicing. The breakdown voltage of all the detectors decrease. Although the sample F7-01 and F7-02 still maintain breakdown voltage higher than 800V, F7-03 breaks down at about 200V.

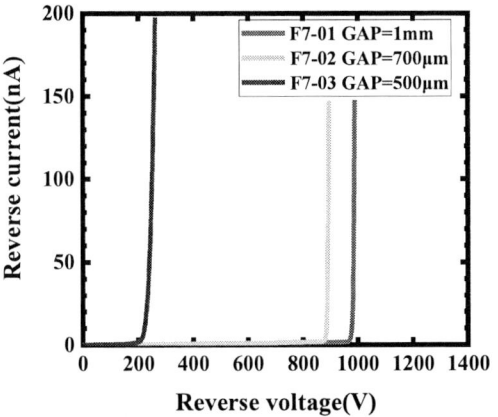

Figure 5: Electrical characterization results for the detectors after diamond blade dicing

In order to further reduce the dicing damages, the laser dicing method is applied. The sample F7-04 and F7-05 are diced with Laser Ablation Dicing, F7-06 and F7-07 with Laser Stealth Dicing. The electrical characterization results for the detectors after laser dicing (Figure 6) show that the breakdown voltage of the detectors diced with gap of 1mm by Laser Ablation Dicing and Laser Ablation Dicing both reach nearly 1200V. And the breakdown voltage is some higher than 200V when the gap is 500μm. This two methods of laser dicing result in very similar breakdown voltages. The breakdown voltages are some higher than that of blade dicing shown in Figure 5, which can be explained as the less dicing damages in laser dicing. It is obvious for gap of 500μm, both laser and blade dicing show breakdown voltage around 200V which are comparable with the simulation prediction shown in Figure 2. It is obvious that all experimental results and simulations show that the critical edge width or dead area is around 500μm, below which the breakdown voltage can decrease seriously. The agreement between dicing experiments and simulations also confirms that it is the carrier trap effect induced by dicing damages at the die edge of the detectors that can lead to the increase of leakage current and thus reduce the breakdown voltage of the detectors. On the other hand, although the breakdown voltage can be as low as 200V due to dicing, it is still large enough for normal application of detectors. Only if the detector is required to work at very high voltage, the edge width or dead area should be large enough.

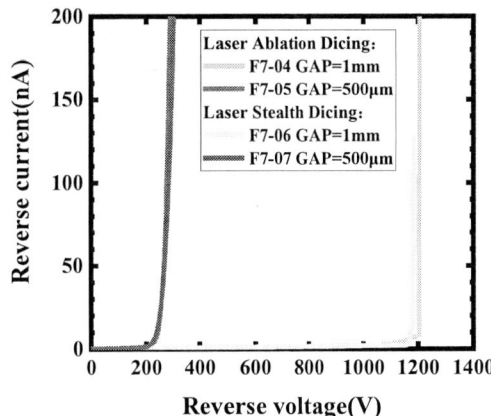

Figure 6: Electrical characterization results for the detectors after laser dicing

## SUMMARY

The results of numerical simulations as well as diamond blade dicing and laser dicing of high-voltage silicon PIN detectors are presented in this work. The agreement between simulations and dicing experimental results indicate that the carrier trap effect induced by dicing damages can severely decrease the breakdown voltage of detectors. The edge width, i.e. dead area, of 1mm is a relatively safe value that ensures the breakdown voltage of detectors can reach 1000V. Whereas, 500μm edge width is also possible if 200V breakdown voltage is acceptable in the specific applications.

## ACKNOWLEDGEMENTS

The silicon PIN detectors in this paper were fabricated in the National Key Laboratory of Nano/Micro Fabrication Technology, Institute of Microelectronics, Peking University. We sincerely thank the engineers for their technical support and assistance.

## REFERENCES

[1] M. D. Rold, N. Bacchetta, et al. IEEE Transactions on Nuclear Science, USA, August, 1999, pp. 1215-1223.

[2] M. A. Benkechkache, S.Latreche, et al. IEEE Transactions on Nuclear Science, USA, August, 2017, pp. 1062-1070.

[3] B. J. Baliga. Fundamentals of Power Semiconductor Devices, Springer, 2008

[4] M. Povoli, A. Bagolini, et al. Nuclear Instruments and Methods in Physics Research A, vol. 658, 2011, pp. 103-107.

[5] I. Synopsys. Sentaurus Device User Guide.Version A-2008.09, 2008.

# ULTRA-FAST ON-CHIP SENSING FOR PCR DETECTION WITH MACHINE LEARNING-BASED IMAGE PROCESSING ALGORITHM

*Jingxuan Shen[1], Lilei Hu[1,2*], Yichen Zhang[2], Zhu Chen[1], Liying Liu[5], Yang Liu[1], Yuqian Ma[1], and Chang Chen[1, 2, 3, 4*]*

[1]School of Microelectronics, Shanghai University, Shanghai 200444, China
[2]Shanghai Industrial µTechnology Research Institute, Shanghai, China
[3]State Key Laboratory of Transducer Technology, Shanghai Institute of Microsystem and Information Technology Chinese Academy of Sciences, Shanghai, China
[4]Shanghai Academy of Experimental Medicine, Shanghai, China
[5]Shanghai Si-Gene Biotech Co., Ltd, Shanghai, China

*Corresponding Authors' Email: hulilei@shu.edu.cn, chang.chen@mail.sim.ac.cn

## ABSTRACT

This work developed a highly integrated and ultra-fast optoelectronic on-chip Polymerase Chain Reaction (PCR) system, in which multiple Si-based optoelectronic on-chip sensing techniques have been used, i.e., CMOS imaging sensor, novel snake-shaped microfluidic chip fabricated using MEMS technology, and a thermoelectric cooler chip module. A machine learning-based algorithm has also been developed to improve the PCR testing speed and accuracy. This design exhibits an ultra-fast PCR testing process of fewer than 5 minutes, and a PCR testing accuracy as high as 91.1% has been achieved.

## INTRODUCTION

Polymerase Chain Reaction (PCR) is a technique widely used in molecular biology. It's possible to amplify a single or few copies of a piece of DNA across several orders of magnitude, generating millions or more copies of the DNA piece. Therefore, PCR is capable of testing even very small numbers of bacteria. However, this conventional PCR approach generally needs 2-3 hours to complete a test. A faster testing technique is strongly demanded of quick disease screening [1, 2], especially for a pandemic such as Covid-19. Therefore, in this work, we developed a highly integrated and ultra-fast photoelectronic on-chip PCR system and applied machine learning algorithms to improve the PCR testing speed and accuracy, providing a new approach for ultra-fast disease screening in point-of-care testing applications.

## ON-CHIP PCR DETECTION

From Fig. 1(a), the main components of the system include a CMOS imaging sensor (CIS), snake-shaped microfluidic chip, and a thermoelectric cooler (TEC) chip [3]. Based on this technique, commercial PCR instruments have been manufactured by Shanghai Si-Gene Biotech Co., Ltd, as shown in Fig.1(b). PCR nucleic acid amplification reaction occurs in a microfluidic chip. PCR nucleic acid amplification reaction includes 30-50 heating cycles and for each one includes denaturation, annealing, and extension in order. The heating is controlled with temperature monitored by a TEC module, as shown in Fig.1(a). A CIS is used to collect the fluorescence light emitted from the microfluidic chip, and finally, the generated image is processed and analyzed. The CIS image is taken at the amplification stage during each PCR heating cycle. PCR testing uses a standard Covid-19 testing kit and was prepared following a general procedure including mixing the to-be-tested sample and the testing kit and centrifuging. Finally, the precursor solution is injected into the snake-shaped microfluidic chip for PCR testing.

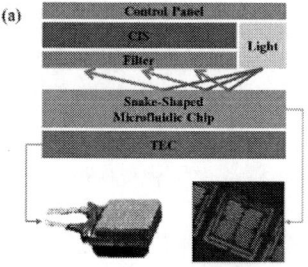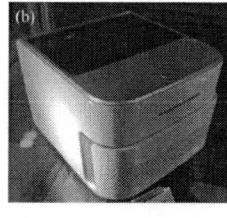

*Figure 1: (a) Schematic showing the module of the on-chip PCR detection system. (b) Commercial PCR instrument based on the technique shown in (a).*

## MACHINE LEARNING-BASED IMAGING PROCESSING ALGORITHM

Conventional PCR testing requires a complete PCR nucleic acid amplification reaction, i.e., 45 heating cycles, and generates a standard PCR curve [4]. Then a threshold value is calculated from the statistics of a large amount of experimental data. The threshold is applied to determine whether the sample being tested is positive or negative, as shown in Fig. 2. The accuracy of the PCR test depends on the accuracy of the threshold value [5]. Compared to the conventional method, the machine learning-based method requires only simple statistical values of the original data, such as its mean and standard deviation, and uses it as a feature in the machine learning algorithm to classify each tested sample into two categories: positive, negative. Since there is no need to run a standard PCR curve fitting, the machine learning-based method does not require a whole-process (45 cycles) PCR reaction while still can obtain the required features and classify the sample. In this paper, the random forest algorithm is used which is essentially a collection of decision trees, where each tree is slightly different from the others, and each tree might have a good chance to improve the predicting [6, 7]. Besides, Bayesian Optimization (BO) is chosen as an alternative to the traditional grid search for hyper-parameters [8]. BO is set to solve for an acceptable maximum value (optimal value of hyper-parameters) when the objective function is a black box (i.e., the structure of the function is not yet known). Compared with the grid search method, BO requires fewer iterations, however, the granularity can be too small, making it harder to find the global optimal solution.

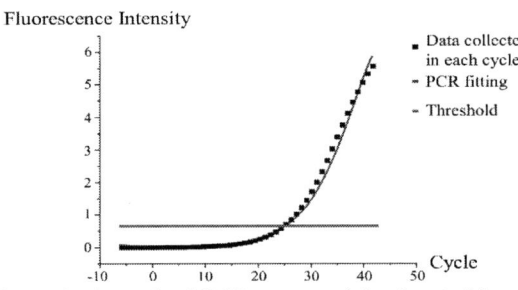

Figure 2: A standard PCR curve and the threshold.

## EXPERIMENTS

### Experimental setups

A brief flow of preparing PCR data sets is shown in Fig. 3. After one PCR reaction, 45 images are collected. Each image is divided into 3 images, i.e., the R, G, and B color channels, and the green channel is retained. Then the images of the interested PCR reaction chamber are chopped. The average fluorescence intensity in the snake-shaped reaction channel is calculated. This process is repeated 45 times to obtain one set of PCR reaction data.

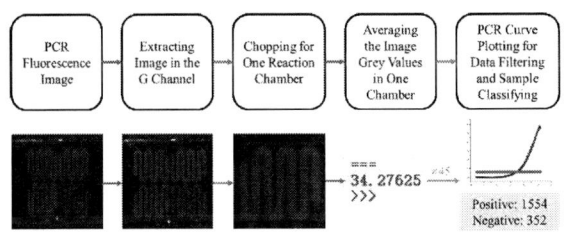

Figure 3: Flowchart for preparing PCR data sets for training.

Following the above procedure, 3000 sets of data were experimentally collected in 1 month from Shanghai Customs in the first half of 2021, and each set of data is generated from one PCR testing. Eventually, 1906 sets of data were prepared for training, including 1554 sets for positive samples and 352 groups for negative samples. Then, the obtained data were trained using the machine learning-based models, the data processing flow is shown in Fig. 4.

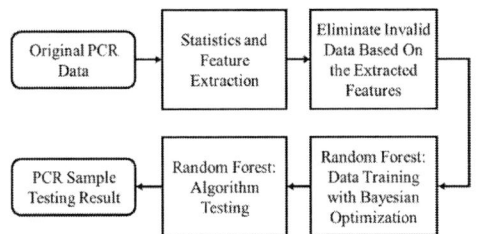

Figure 4: Flowchart for the random forest algorithm–based training and data processing.

Regarding feature extraction, as shown in Fig. 4, first, the mean value $Ave_{15}$ for the fluorescence intensities of the first 15 PCR heating cycles was chosen as a data feature due to its indication of the precursor solution concentration and the baseline of the testing system. With the PCR reaction forwarding, the PCR curve for positive samples grows exponentially, while it remains flat or needs more heating cycles for the exponential growth to appear for negative samples. The way a PCR curve is evolving can be quantitatively expressed by the relative difference $Diff_n$, which is defined as:

$$Diff_n = \left| \frac{Value_n - Ave_{15}}{Ave_{15}} \right| \qquad (1)$$

Where $n$ denotes the n-th heating cycle in a reaction and $Value_n$ denotes the fluorescence intensity value collected in the n-th cycle. As shown in Fig. 5, $Diff_{30}$ and $Ave_{15}$ can clearly distinguish the positive samples from negative samples by a feature correlation comparison. In this way, fewer PCR heating cycles are required, and the correlation between the two characteristics is poor. Therefore, $Diff_n$ is chosen as the second data feature.

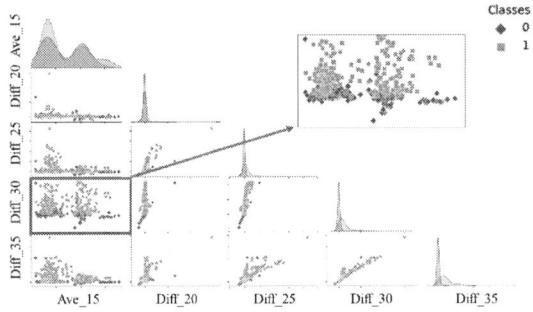

978-1-6654-9759-6/22 $31.00 © 2022 IEEE

*Figure 5: Pair plot showing the data features of PCR, and the correlation between $Diff_{30}$ and $Ave_{15}$.*

To ensure a qualified generalization performance of the machine learning-based algorithm, the ratio of the numbers of positive and negative samples used for training needs to be as close as to 1:1. As shown in Table I, using $Diff_{30}$ as a criterion, 298 positive samples are grouped into 7 ranges, with the inclusion of 298 negative samples, the final dataset for training has been obtained. The results of the training were evaluated by precision, recall, and f-score.

TABLE I.     DISTRIBUTION OF PCR TESTING DATA USED FOR ALGORITHM TRANING

| $Diff_{30}$ | | $Amount$ |
|---|---|---|
| **Negative Samples** | | 298 |
| **Positive Samples** | 0-0.02 | 50 |
| | 0.02-0.05 | 50 |
| | 0.05-0.08 | 50 |
| | 0.08-0.12 | 50 |
| | 0.12-0.26 | 49 |
| | >0.26 | 49 |

**Experimental results**

After preparing the data according to the above method, the random forest was applied for training. Specifically, k-fold cross-validation was used to form the training and test datasets, where the value of k is taken as 10. First, a grid search method is used to optimize the fitting parameters for each algorithm. A comparison of the evaluation of training results between the traditional machine learning models, such as k-NN, MLPs, SVM, and decision trees, and our methods are shown in Table II. The results show that the random forest has higher recall and precision scores, but its f-score is lower, indicating an overfitting problem, and are prone to obtain false-negative results.

TABLE II.     COMPARISON OF THE PERFORMANCE OF DIFFERENT ALGORITHMS.

| Model | Negative Samples | | | Test Accuracy |
|---|---|---|---|---|
| | *Precision* | *Recall* | *f-score* | |
| k-NN | 0.80 | 0.92 | 0.86 | 0.850 |
| MLPs | 0.83 | 0.92 | 0.87 | 0.900 |
| SVM | 0.83 | 0.96 | 0.89 | 0.885 |
| Decision Trees | 0.89 | 0.92 | 0.91 | 0.904 |
| Random Forests | **0.83** | **0.96** | **0.89** | **0.885** |

| Model | Negative Samples | | | Test Accuracy |
|---|---|---|---|---|
| | *Precision* | *Recall* | *f-score* | |
| Random Forests with BO | **0.87** | **0.97** | **0.92** | **0.911** |

The BO method was then used to replace the original grid search method for the hyper-parameter optimization, and the results are shown in Table II and were compared with the previous random forest model. The f-score and the test accuracy of the BO-based random forest were improved and show better performance than other traditional machine learning models, indicating that the BO-based random forest model has a better generalization performance and can obtain better training results.

## CONCLUSION

This paper combines an improved machine learning-based image processing algorithm with Si-based photoelectronic on-chip sensing techniques. The system exhibits an ultra-fast PCR testing of fewer than 5 minutes. Compared with the traditional approach that uses a threshold value, the as-developed BO-adjusted random forest-based algorithm makes it unnecessary to run through a whole-process PCR test and significantly reduces the testing time, meanwhile achieving high accuracy of 91.1%, showing a better generalization performance of the random forest model.

## ACKNOWLEDGEMENTS

The authors acknowledge the Shanghai Rising-Star Program (21QB1405700), the Research Instrument and Equipment Development Project of the CAS (GJJSTD20210006), the National Key R&D Program of China (2020YFC0847600), and the Program of Shanghai Technology Innovation Center of IVD Chip (20DZ2220500).

## REFERENCES

[1] H. Sang, P. Seung-min, K. Brian, K. Oh, R. Won-Yep, and J. Bong-Hyun. *Biosensors and Bioelectronics*, vol. 141, 2019.

[2] Y. Minli, L. Zedong, F. Shangsheng, G. Bin, Y. Chunyan, H. Jie, and X. Feng. *Trends in Biotechnology*, vol. 38, 2020, pp. 637-649.

[3] Shanghai Industrial μTechnology Research Institute. *A PCR fluorescent nucleic acid detection device*: CN202021199467.0[P]. 2021-04-06.

[4] A. Larionov, A. Krause, and W. Miller. *Bmc Bioinformatics*, vol. 6, 2005, pp. 1-16.

[5] M. Gunay, E. Goceri, and R. *2016 15th IEEE International Conference on Machine Learning and Applications (ICMLA)*, 2016, pp. 588-592.

[6] Y. Jianfeng, Q. Peirui, L. Yongmei, and W. Ning. *Statistics & Decision*, vol. 6, 2019, pp. 36-40.

[7] L. Breiman. *Machine Learning*, vol. 45, 2001, pp. 5-32.

[8] D. Hongyao, W. Yidan, and L. Lihong. *China Computer & Communication*, vol.33(17), 2021, pp. 34-37.

# BASED ON DEEP LEARNING CD-SEM IMAGE DEFECT DETECTION SYSTEM

*Shijia Yan[1], Shenglan Ding[1]\*, Sen Wang[1], Cong Luo[1], Lei Li[1], Juan Ai[1], Qiang Shen[1], Qing Xia[1], Zhi Li[1], Qilin Cheng[1], Shilin Li, Hongwei Dai[1], Xiangang Hu[1]*

[1] Wuhan Xinxin Semiconductor Manufacturing Co., Ltd.(XMC), Wuhan, China

*Corresponding Author's Email: SHENGLAN_DING@XMCWH.COM

## ABSTRACT

In the semiconductor manufacturing, defects are inevitably created on the wafer. When CD SEM tool is used to measure the CD of wafer, it will take pictures of the line profile on the surface of wafer. Based on the deep learning algorithm, we use CD SEM image to detect wafer defect. We also developed a system to detect defects in CD SEM images online in real time, and the detection accuracy is as high as 100%. It not only reduces the need for engineers to manually review each image, but also expands the ability of inline defect inspection, so as to improve the economic benefits for the factory.

## INTRODUCTION

In semiconductor manufacturing, it is very important to improve yield and product quality. Finding defects in time is an important means of monitor inline process, which can reduce the number of impacted products in time. It is an important guarantee for maintaining and improving yield.

The latest developments in semiconductor process technology have greatly increased the density of transistors. Therefore, more and more defects will inevitably be introduced, and we need a more accurate and effective detection method to manage them. At the same time, as the output continues to increase, the defect inspect resources will become tight, and the throughput capacity of the scanning electron microscope (SEM) machine is limited. Therefore, we built a defect detection model and used the SEM image of Critical Dimension SEM (CD-SEM) to automatically detect the defect in a specific area on the wafer. This also expands the defect detection capabilities of the CD -EM machine, and at the same time improves the defect detection capabilities in inline wafer production, which can help engineers find defects in time.

With the development of deep learning algorithms, people began to try to use deep convolutional neural networks to automatically detect defects on images acquired by SEM or automatic optical inspection tools. The automatic detection of defects by the system can prevent engineers from being unable to conform to the standards of defect detection due to differences in their respective experiences. It can also reduce the need for manpower.

Jaehoon Kim[1] et al proposed a defect detection model based on deep learning to detect and classify defects by using an adversarial network architecture of conditional GAN. The model was validated on industrial data sets to defect hard and sorft defects. Marco Bellini[2] et al They proposed a neural network model that can identify defective dies on the labeled image dataset, and then used DCNN's feature maps to generate unsupervised classification of wafer die defects. They used t-Distributed Stochastic Neighbor Embedding (t-SNE) to map defect-related 200-dimension feature maps to 2-dimentional space. This can realize unsupervised clustering of defects and refine the classification labels of defects. Sejune Cheon[3] et al. proposed a model based on a single convolutional neural network (CNN) model for the classification of surface defects on scanning electron microscope images. The model can extract effective features for defect classification without applying additional feature extraction algorithms. In addition, it can also compare the features of the unseen defect class and the seen defect class in the CNN, so as to realize the correct classification of the unseen defect class.

However, the above-mentioned researches are based on the detection effect on the datasets, and have not been further deployed on the inline wafer real-time production system. And they are all aimed at machines that specifically detect defects on wafers, while the grey image on CD-SEM machines has not been studied. But we are performing defect detection on the SEM image of the CD-SEM tools and have developed a real-time wafer defect detection system.

The rest of this paper is organized as follows. The Section II introduces the wafer defect detection model based on deep learning and the real-time detection system we developed. The Section III introduces datasets preparation and data preprocessing and experimental results, followed by the conclusion in Section IV.

## DEFECT DETECTION MODEL AND REAL-TIME DETECTION SYSTEM

### Defect Detection Model

The defect detection model we built uses YOLOv2[4,5] as the base model. YOLOv2 is based on regression for target detection, through several convolutional layers and pooling layers. It turns the input image into a three-dimensional tensor and uses the 448*448 input image fine-tune classifier to improve the resolution. YOLOv2 uses

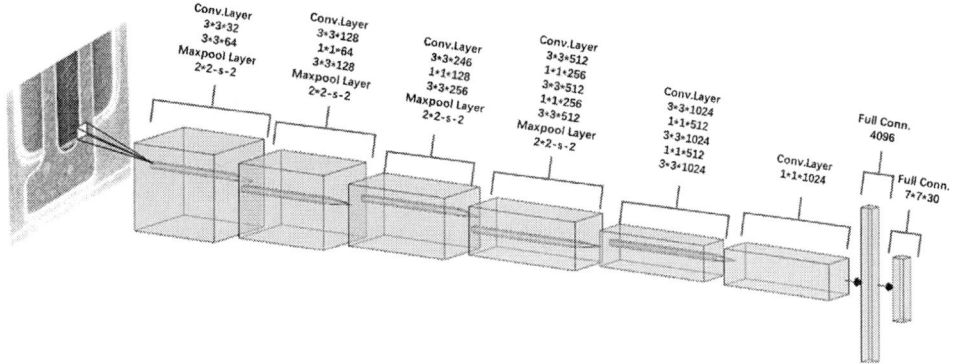

*Figure 1: The structure diagram of the CD-SEM defect detection model with YOLO-V2 as the base model.*

anchor as a predefined candidate area to detect whether there is a target in the neural network. It uses the k-means clustering method to cluster the bounding boxes in the training set and predict the location based on the coordinates of the grid cell. It uses logistic activation to control the IOU of the prediction box and ground truth.

YOLOv2 uses Darknet-19 as a feature extractor with 19 convolutional layers and 5 max pooling layers, as shown in Fig.1. It mainly uses a 3*3 filter, the number of double channels after each pooling. This further improves the accuracy and training speed of the model. YOLOv2 uses a mechanism for jointly training on classification and detection. It learns the specific features in the detection in the labeled datasets and classifies common objects and predicts the bounding box coordinate.

The author of YOLOv2 constructs a hierarchical tree to represent the affiliation between objects and makes softmax on objects of the same category. When predicting, the model starts from the root node of the tree and searches downward. It selects the child node with the highest predicted score each time until the predicted scores of all selected nodes are multiplied and are less than a certain threshold and stop searching.

In order to reduce the model's initial learning of defects on the CD-SEM gray image, the cost is very high. We use transfer learning and use existing knowledge to assist in learning new knowledge as soon as possible, so that defect detection can be quickly implemented on mini-datasets. The core of transfer learning is to find the similarities between old and new knowledge.

Based on the transform learning mechanism, we end-to-end fine-tuning the pre-trained YOLOv2 on the data set marked by CD-SEM images, which is the process of updating all the weights (parameters) for the defect detection task in the CD-SEM images. The loss of the model consists of three parts: position error, confidence error and classification error, which are balanced by using sum-squared error loss.

**Real-time CD-SEM Defect Detection system**

We deploy the defect detection model of offline training on our GPU server. The system monitors the results of wafer cd measurement by fab inline CD SEM Tools, and performs real-time defect detection on CD-SME gray image. The architecture diagram of the system is shown in the Fig.2 below.

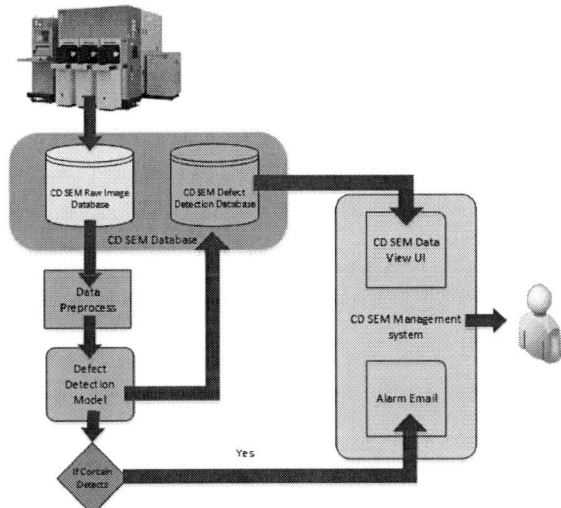

*Figure 2: Real time CD-SEM defect detection system architecture diagram.*

When CD-SEM Tools completes the CD measurement of a wafer, it will output the CD measurement data to the CD-SEM server database. Then, the system will automatically extract the SEM gray image, preprocess the data, and enhance the features. It then uses the defect detection model to perform defect detection on it. Then save the test result as a review image and output it to the CD-SEM server database. If the detection result finds that there is a defect, it will trigger an alarm email to be sent to the engineer. Engineers can also view the detailed CD-SEM data and review image in the CD SEM Data View UI developed by us.

Based on the real-time CD-SEM defect detection system developed by us, it can save engineers from manually viewing hundreds of CD-SEM gray images. At the same time, avoid the defect miss due to different experiences. In addition, the real-time detection system can detect defective wafers in time and notify engineers to take corresponding actions. This can effectively reduce the impact of products, which is very important for the improvement of yield.

# EXPERIMENT RESULT & ANALYSIS

## Data Preprocess & Datasets

In order to further enhance the features on the CD-SEM gray image, we have dynamically adjusted the contrast and brightness of the image. Thereby, the defect features on the defect wafer can be highlighted. We both use image linear blending technology for contrast and brightness enhancement. The original image and the processed transparent channel image are linearly fused with corresponding weights. For the two input images $I_0$ and $I_1$, linearly add the pixel values at the same position, as shown in formula 1.

$$g(x) = \lambda I_0(x) + (1 - \lambda)I_1(x) \qquad (1)$$

where, $\lambda$ is the weight, which controls the weight of the fusion of the pixels at the corresponding positions of the two images. In this way, different contrast and intensity image enhancement effects are achieved, as shown in the Fig.3 below.

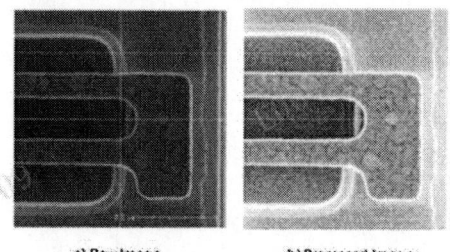

a) Raw Image          b) Processed Image

*Figure 3: a) CD-SEM raw image, b) CD-SEM Processed image, in which features are enhanced.*

We collect CD-SEM gray image from the actual fab production of Wuhan Xinxin Integrated Circuit Manufacturing Co., Ltd. After enhancing the features, we mark them as defects and use the open software label-me software, as shown in Fig.4. Since transfer learning can be used for training on small data sets, we have marked a total of 88 gray images. This is the CD-SEM defect image mini-dataset that we will use in the next training model. In order to optimize the problem of unbalanced sample distribution, we try to make the number of samples of defective images and non-defect images as close as possible to prevent the model from overfitting.

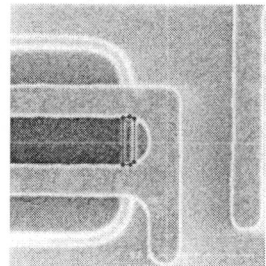

*Figure 4: Labeling CD-SEM processed image based on Labelme.*

## Model Train & Test

The defect detection model training steps are all carried out on a machine with the following configuration: Inter(R) Xeon(R) Gold 6140 CPU @2.30GHz, image processor model is Tesla M60, memory is 125G, operating system is Centos 7.6, hard disk capacity It is 2T. The experimental environment is python3.6.0 and the open source deep learning frameworks Keras and Turi Create. The code used in this article adopts python language for experiment.

In the model training process, set the batch size to 32 and the max iteration to 2000. The training results are shown in the Table I below. The real-time test results of real-time defect detection inline pi-run mode are shown in Table II.

The defect detection accuracy on the test set is 100%, and some test results are shown in the Fig.5 below.

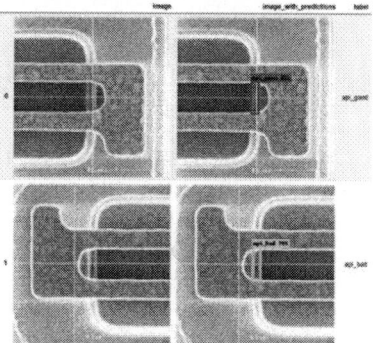

*Figure 5: The effect of defect detection on test dataset.*

TABLE I.     CD-SEM DEFECT DETECTION MODEL
TRARING RESULT

| Iteration | Loss | Elapsed Time |
|-----------|----------|--------------|
| 1 | 60.3221 | 18.52s |
| 2 | 60.2291 | 20.22s |
| 1995 | 0.340144 | 55m 15s |
| 2000 | 0.346497 | 55m 22s |

TABLE II.    THE TEST OF REAL-TIME DEFECT DETECTION INLINE PI-RUN MODE

| Image Count | 1 | 188 |
|---|---|---|
| Time | 17s | 23s |
| Accuracy | 100% | 100% |

**Real-Time Defect Detection**

When a CD-SEM Tool in the fab measures a wafer CD, a maximum of 188 gray images will be generated. We perform defect detection of batch CD-SEM images by wafer. We conducted batch inspection tests on 188 CD-SEM gray images, and the test results are shown in the table II below. After our detection efficiency testing, it is found that the deep learning framework is more efficient than single sheet in batch target detection. Therefore, we adopt batch defect detection strategy on CD-SEM gray image on the official version of inline to improve detection efficiency.

The real-time inspection results are displayed to engineers through the CD SEM Management System software developed by us. As shown in the figure below, due to experimental needs, the fig.6 shows the test results of wafers and abnormal products. When a defect is detected, it will be marked with a red box, if there is no defect, it will be marked with a green box. At the same time, detailed defect detection results will be displayed, including the predicted bounding box position, defect category and confidence value.

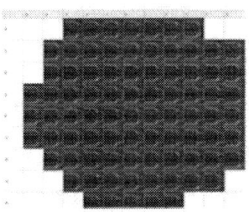 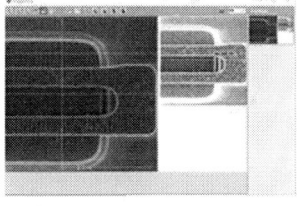

a) 1 wafer defect detection result          b) 1 CD-SEM gray image defect detection result

*Figure 6: The Real-Time Defect Detection Visual System display defect detection result.*

## CONCLUSION

With the rapid increase in the output of fabs, the defect detection resources in the fab have become tight. In addition, the peculiarities of some processes cannot be effectively detected by defect inspection tools caused by process issues. However, CD-SEM Tools itself measures Line CD based on SEM gray image. In other words, the SEM image of the corresponding position will be generated while measuring the CD. Therefore, based on deep learning and migration learning, we established a real defect mini-dataset and passed the end-to-end fine-tuning YOLOv2 model. Therefore, the defect detection model is constructed and deployed on the CD-SEM server, which can perform defect detection on the SEM image of the CD-SEM on the production line in real time. At the same time, the CD-SEM management system software developed by us will show engineers the detailed results of CD-SEM auto defect detection, including the predicted defect box bounding position, defect class, and confidence level.

With the rapid increase in the output of fabs, the defect detection resources in the fab have become tight. In addition, the peculiarities of some processes cannot be effectively detected by defect inspection tools caused by process issues. However, CD-SEM Tools itself measures Line CD based on SEM gray image. In other words, the SEM image of the corresponding position will be generated while measuring the CD. Therefore, based on deep learning and migration learning, we established a real defect mini-dataset and passed the end-to-end fine-tuning YOLOv2 model. Therefore, the defect detection model is constructed and deployed on the CD-SEM server, which can perform defect detection on the SEM image of the CD-SEM on the production line in real time. At the same time, the CD-SEM management system software developed by us will show engineers the detailed results of CD-SEM auto defect detection, including the predicted defect box bounding position, defect class, and confidence level.

## ACKNOWLEGEMENTS

We thank the higher management team from Wuhan Xinxin Semiconductor Manufacturing Company for the support of this work.

## REFERENCES

[1] J. Kim et al., "Adversarial Defect Detection in Semiconductor Manufacturing Process," in IEEE Transactions on Semiconductor Manufacturing, vol. 34, no. 3, 2021, pp. 365-371.

[2] M. Bellini, G. Pantalos, P. Kaspar, L. Knoll and L. De-Michielis, "An Active Deep Learning Method for the Detection of Defects in Power Semiconductors," 2021 32nd Annual SEMI Advanced Semiconductor Manufacturing Conference (ASMC), 2021, pp. 1-5.

[3] S. Cheon, H. Lee, C. O. Kim and S. H. Lee, "Convolutional Neural Network for Wafer Surface Defect Classification and the Detection of Unknown Defect Class," in IEEE Transactions on Semiconductor Manufacturing, vol. 32, no. 2, 2019, pp. 163-170.

[4] J. Redmon, S. Divvala, R. Girshick and A. Farhadi, "You Only Look Once: Unified, Real-Time Object Detection," 2016 IEEE Conference on Computer Vision and Pattern Recognition (CVPR), 2016, pp. 779-788.

[5] J. Redmon and A. Farhadi, "YOLO9000: Better, Faster, Stronger," 2017 IEEE Conference on Computer Vision and Pattern Recognition (CVPR), 2017, pp. 6517-6525.

# IMPROVING WHITE PIXEL THROUGH THE OPTIMIZATION OF STRUCTURE AND IMPLANTATION IN CIS DEVICE

*Lu Wang\*, Cuiyu Mei, Jiong Xu, Chang Sun*

Shanghai Huali Microelectronics Corporation, Shanghai, China

*\*Corresponding Author's Email: wanglu@hlmc.cn*

## ABSTRACT

White pixel is one of the most important indicators of the CMOS image sensor, which represents the sensitivity of a photodiode. In this work, we propose a viable method to improve the white pixel performance through the optimization of surface implantation. Experimentally, white pixel could be improved dramatically, while other pixel capabilities still keep in a high level. Meanwhile, the physical model is proposed, including the illustration of parallel and perpendicular diffusion of implanted ions, which would further benefit the design and development of high-performance image sensor.

## INTRODUCTION

As the rapid improvements achieved for information technology, the capture and transmission of messages demands for high performance. CMOS image sensor (CIS) has drawn intense interests, which is the key device for image capture in the applications of phone, camera, and surveillance system [1-3]. The image sensors are developed to meet higher demands in sensitivity, resolution, dynamic range, and power consumption [4, 5]. White pixel is one of the most important indicators, representing the sensitivity of a photodiode. It is of great significance to improve the white pixel performance of CIS device [6, 7].

In this work, we demonstrate the correlation between the surface P+ implantation condition and white pixel, which would help researchers and engineers optimizing the sensitivity of CIS device during design and fabrication. To clarify the mechanism of surface P+ implantation, the cross-section view of photodiode is shown in Figure 1. Here, Boron implantation, as the surface isolation for ultra shallow PN junction, is proposed. Furthermore, the model of the diffusion of implanted Boron is illustrated, matching with experimental results of variable source, dose, and energy. Experimentally, white pixel is improved up to 20% without degradation of other pixel properties, such as full well capacity. Thus, it gives a hint of the design of next-generation high-performance CIS device.

A wild and diverse variety of pixel design of CIS chips has been achieved to meet targeted demands. As for mobile phones and other applications which request for high resolution, smaller pixel size is utilized usually, thus realizing high-resolution capabilities up to 100 and 200 million pixels on a small chip. While for products aiming at surveillance system, it is important to keep the high sensitivity whether during day or night. Thus, the increased pixel size is adopted and developed to enhance the quantum efficiency. However, as the increase of pixel size, the sensitivity towards metal contamination is grown exponentially, resulting in the difficulty during fabrication. In summary, improving the resistivity of metal contamination is the key point, especially for big pixel-size products.

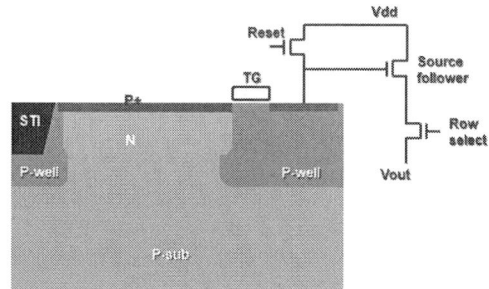

*Figure 1. Schematics for the photodiode in CIS device.*

As the increasing pixel size, the ultra shallow PN junction in the pixel region become more and more significant. We experimentally adopted Boron and BF2, two kinds of normal implantation sources, as the P-type source, respectively. As shown in Figure 2, it can be seen from the probability chart that the white pixel performance is highly improved with a magnitude of up to 20%. Besides, for the total ~3000 chip from the 12-inch wafer in this work, the better tailing performance indicates the improved uniformity of the wafer under the condition of Boron implantation.

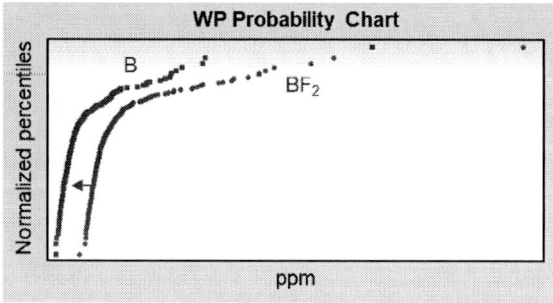

*Figure 2. WP (White pixel) probability chart of ~3000 chips from a 12-inch wafer under the condition of B and BF2 ion implantation.*

The model is schematically shown in Figure 3, exhibiting

the mechanism of ultra shallow PN junction. Since Boron has smaller atomic weight than BF2, the diffusion behavior of P+ type elements is strengthened [8]. Thus, the solid ultra shallow PN junction of Boron implantation can better protect the photodiode from the surface contamination and defect. Here, both the in-plane and out-of-plane diffusions are observed and demonstrated during experiments of various dosages and energies of implanted Boron and BF2. Also, we noticed the surface film uniformity of a 12-inch wafer. It can be deduced that the diffusion of Boron can make up for the lack of film uniformity, matching with the better tailing performance shown in the P-chart of Figure 2.

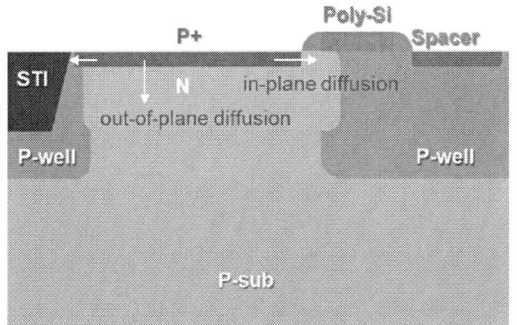

Figure 3. Schematics of the in-plane diffusion of P+ ions.

Moreover, we compared the resistivity of metal contamination between the Boron and BF2 implantation, using the magnitude of white pixel shift. The experiments are adopted under the known and consistent environment contaminated by metal. As shown in Figure 4, the normalized x-axis stands for the time the wafer put in the contaminated environment, while the normalized y-axis stands for the magnitude of white pixel shift compared with the baseline performance. Firstly, as the longer the wafer stays in the contaminated environment, the white pixel performance becomes more and more serious, the shift magnitude gets stronger, which matches with the all-known relationship between white pixel and metal contamination. Secondly, the Boron, used as the P+ type source for the ultra shallow PN junction, can extremely decrease the metal influence to white pixel. The magnitude of white pixel shift is dramatically decreased to 1/5, using Boron instead of the BF2 source.

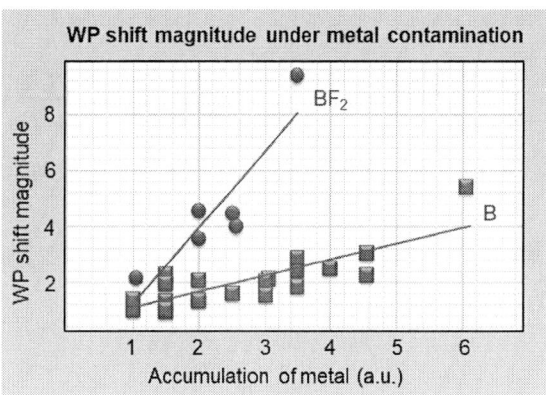

Figure 4. Magnitude of WP (White pixel) shift under various metal accumulation.

In conclusion, CMOS image sensor is one of the key devices for information technology that captures and transmits optical images of the real world. This paper developed a viable method to optimize the CIS device performance-- white pixel. We firstly demonstrate the correlation between surface P+ implantation condition and white pixel performance, which is the basics of developing the physical model for further study. Additionally, the physical model is illustrated to help the data analysis, optimization, and theoretical study of CIS devices. What is more, a series experiments are adopted to demonstrate the improved resistivity of metal contamination based on both Boron and BF2 implantation, further indicating the meaning of the optimized ultra shallow PN junction during mass production. In this work, results of white pixel performance with variable implantation conditions are analyzed experimentally and theoretically, achieving improvement of white pixel performance up to 20%, and a dramatic shift decrease down to 1/5 of baseline condition when the environment is contaminated. It is significant to note that the physical model we proposed and experiment result we demonstrated can lead to development for the optimization in pixel performance, making high-performance image sensors come true.

## ACKNOWLEDGEMENTS

The author acknowledges the support from colleagues in both CIS team and other departments from Shanghai Huali Microelectronics Corporation. We thank Jinhua Jiang and Shuntian Liu for their help in the experiment part and thank Zhengying Wei for his help in experiment resource and mechanism study.

## REFERENCES

[1] S. Park and H. Uh, *Microelectron. J.*, vol. 40, no. 1, pp. 137–140, Jan. 2009.

[2] B. Mheen, Y. Song and A. J. P. Theuwissen, *IEEE Electron Device Letters*, vol. 29, no. 4, pp. 347-349, April 2008.

[3] T. Shinohara et al., *2013 IEEE International Electron Devices Meeting*, 2013, pp. 27.4.1-27.4.4.

[4] M. Sarkar, B. Buttgen, and A. Theuwissen, *IEEE Trans. Electron Devices,* vol. 60, no. 3, pp. 1154–1161, Mar. 2013.

[5] J. Tan, B. Buttgen, and A. Theuwissen, *IEEE Sensors J.,* vol. 12, no. 6, pp. 2278–2286, Jun. 2012.

[6] C. Y.-P. Chao, Y.-C. Chen, K.-Y. Chou, J.-J. Sze, F.-L. Hsueh, and S.-G. Wuu, *IEEE J. Electron Devices Soc.,* vol. 2, no. 4, pp. 59–64, Jul. 2014.

[7] Z. Gao, S. Yao, C. Yang, and J. Xu, *IEEE Sensors J.,* vol. 15, no. 6, pp. 3265–3273, Jun. 2015.

[8] H. L. Liu, S. S. Gearhart, J. H. Booske and Wei Wang, *IEEE Conference Record - Abstracts. 1997 IEEE International Conference on Plasma Science*, 1997, pp. 212.

# FABRICATION AND PERFORMANCE RESEARCH OF SILICON NANOSHEET FIELD EFFECT TRANSISTOR BIOSENSOR

*Enyi Xiong[1], Zhaohao Zhang[2], Qingzhu Zhang[2], Shuhua Wei[1]\*, Qianhui Wei[3], Jing Zhang[1], and Jiang Yan[1]*

[1]School of Information Science and Technology, North China University of Technology, Beijing 100144, China

[2] Key Laboratory of Microelectronic Devices and Integration Technology, Institute of Microelectronics, Chinese Academy of Sciences, Beijing 100029, China

[3]State Key Laboratory of Advanced Materials for Smart Sensing GRINM Group Co. Ltd., Beijing 100088, China

*Corresponding Author's Email: weishuhua@ncut.edu.cn

## ABSTRACT

Silicon nanowire (SiNW) FET biosensor has made great contributions to chemical and bio-molecular sensing, and is considered to be one of the best candidates for biosensors. However, due to high sensitivity to pH changes of the buffer solution when detecting biomolecules and low binding rate to biomolecules for small area, SiNW FET biosensor still has the problems of high variation and long response time. To solve the problems, In this paper, we investigate the sensing performances of silicon nanosheets (SiNS) FET with different areas, and found that SiNS FET biosensor with large areas can reduce the sensitivity to pH effectively, thus reducing the detection error caused by pH changes and improving sensing accuracy. In addition, by COMSOL simulating, we also found that devices with wide SiNS show higher bonding rate than that of SiNW devices at low analyte concentrations, thereby decreasing the responding time for high-speed detecting.

## INTRODUCTION

Since 1970s, FET-based pH sensors have begun to be popularized and successfully commercialized [1]. Benefiting from large surface area-to-volume ratio, high sensitivity and good biocompatibility, SiNW FET biosensor is currently considered to be one of the best promising candidates sensors for biosensors [2].

However, there are still some problems with SiNW FET sensors. According to the work of Serge Ismael Zida et al. [3], when detecting biomolecules, the sensing performances of small-sized SiNW may be interfered by the pH of the solution, resulting in large variations. In addition, Paul E. Sheehan et al. [4] also stated that the small area of SiNW may lead to the problem of long specific binding time when the SiNW FET biosensor detects biomolecules. In this paper, SiNS FET biosensors with different sensing areas were fabricated. Firstly, we used the device to detect single stranded DNA(ssDNA),

showing a good concentration gradient curve in the range of 10 nM to 100 μM. Meanwhile, we found that pH is a very critical factor in ssDNA detection, and the change of pH affects the detection accuracy. Subsequently, we conducted pH sensitivity dectects on SiNS FET biosensors with different widths, and found that the SiNS FET biosensor with a larger sensing area achieved lower pH sensitivity. Therefore, the detection error caused by pH changes can be reduced by using the wide SiNS FET biosensor. In addition, through COMSOL simulation, we also found that the large-size SiNS can bind the analyte faster and more at a low concentration of 50 nM analyte than SiNW FET biosensor.

## EXPERIMENTAL

The SiNS FET biosensors were fabricated on 200 mm (100) silicon-on-insulator (SOI) wafers. The wafers were featured with a 145 nm thick buried oxide layer (BOX) and a 20 nm top silicon layer. SiNS was formed using a conventional photolithography process (i-line) and dry etching process. Subsequently, nickel-platinum alloy ($Ni_{0.95}Pt_{0.05}$) were used to form metal silicides at the source and drain. After opening the channel, a 5 nm $HfO_2$ layer was deposited by ALD. Finally, dry etching was used to expose the source and drain contacts. We have fabricated SiNS devices with widths of 0.5 μm, 1 μm and 1.5 μm, and channel lengths of 0.5 μm, 1 μm and 1.5 μm, respectively.

To improve the sensitivity and specific detection of DNA, the surface of the SiNS biosensors needs to be functionalized by surface modification. To form amino groups (-$NH_2$) on the $HfO_2$ surface, 1% v/v 3-aminopropyltriethoxysilane(APTES) in absolute ethanol was used as a crosslinker solution. After immersion in APTES solution for 30 min, the SiNS biosensors were rinsed with ethanol three times and annealed at 110 ℃ for 30 min to achieve a firmly silanization. And then the unreacted surface -$NH_2$ was blocked with 10% trimethoxy-n-propylsilane(PTMS) in absolute ethanol solution for 30 min. After the same 110 ℃ annealing for 30 min to complete the blocking. After the surface functionalization procedure, 100 μM ssDNA(1 × PBS, GTG TTG) was added onto the fabricated SiNS device and incubated for 20min(later called B-DNA). Finally, the SiNS biosensor can be used for real-time and label-free detection of ssDNA. The sequence of ssDNA used for detection is CAC AAC(later called G-DNA), and the short-stranded DNA is also used to effectively reduce the Debye screening effect and improve the reliability of the results.

## RESULTS AND DISCUSSION

### Device basic parameter characterization

Figure 1a shows the structure schematic diagram of the SiNS FET biosensor and the connection of the source, drain, and gate. Figure 1b shows the scanning electron microscope (SEM) image of a SiNS FET biosensor with a channel width of 0.5 μm and a channel length of 1.5 μm.

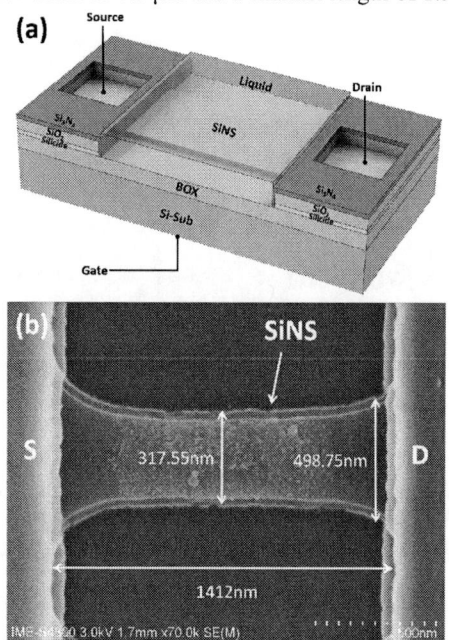

*Figure 1: (a) Device structure diagram and source-drain gate connection; (b) SEM image of SiNS FET biosensor in top view.*

The $I_d$-$V_g$ and $I_d$-$V_d$ curves of the back gate bias voltage of the SiNS FET biosensor are shown in Figure 2. It can be seen from Figure 2a and 2b that the device has bipolar characteristics. The bipolar characteristics of the device may be due to the inversion layer in both source and drain with the gate voltage changing. In addition, Figure 2c and 2d show that $|I_d|$ increases as $|V_g|$ increases, which is also in line with the general law of transistors.

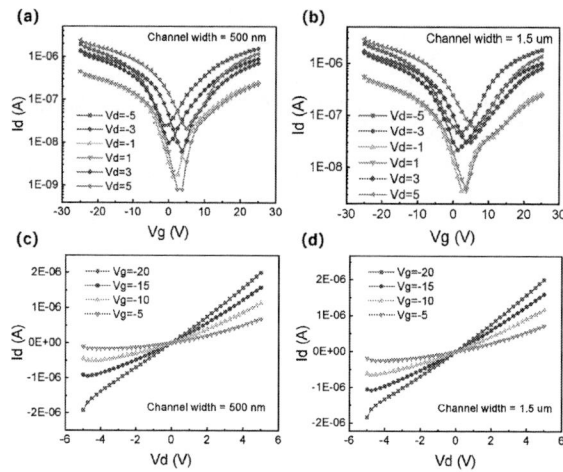

*Figure 2: (a,b) The transfer characteristic curve of 500nm and 1.5 μm wide SiNS ; (c,d) The output characteristic curve of 500 nm and 1.5 μm wide SiNS.*

### PH and DNA sensing detection

We performed a sensor detection of G-DNA on SiNS FET biosensor with a width of 1.5 μm in 1 × PBS (Figure 3). It can be seen that a good concentration gradient curve from 10 nM to 100 μM is shown, which is also in line with DNA is negatively charged in the solution (DNA has a low isoelectric point), so the current tends to increase.

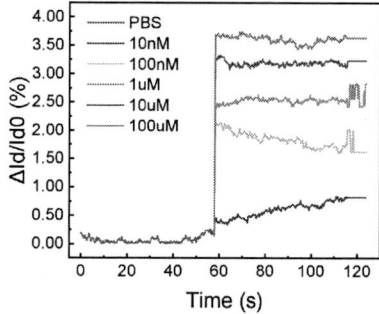

*Figure 3: Plots of normalized current change versus time with G-DNA at a series of concentrations (10 nM, 100 nM, 1 μM, 10 μM, and 100 μM) for B-DNA modified 1.5 μm wide SiNS device.*

In the G-DNA detection, we found that pH has a great influence on biological detection.To prove the important impact of pH changes in biological detection, G-DNA with different pH was detected. We modified 100 μM B-DNA on the surface of the 1.5 μm wide SiNS and added 1 μM G-DNA with different pH. It can be seen from Figure 4 that when adding 1 μM G-DNA in the order of pH=9, 7, 5, the current did not increase with the increase of absolute value of G-DNA concentration as expected, but gradually decreases with the decrease of pH. This result can prove that even for a 1.5 μm wide SiNS, the response to two pH values change is far exceed the response to 1 μM G-DNA. This indicates that the change of pH has a great influence on the results of biological detection.

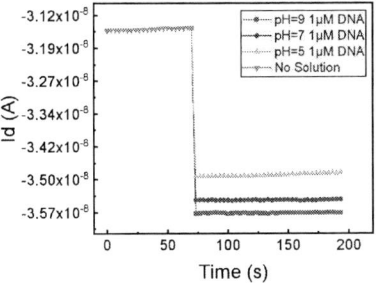

*Figure 4: The time curve of 1 μM G-DNA solution on a 1.5 μm SiNS device at different pH(pH is added in the order of 9, 7, 5).*

Then, pH sensitivity dectection on SiNS FET biosensors with different widths was performed to investigate the influence of sensing area on pH sensitivity Figure 5a and 5b show the $I_d$-$V_g$ curves with fifferent pH solutions for 500 nm and 1.5 μm channel length, respectively. Figure 5c and 5d show the threshold voltage ($\Delta V_{th}$) of Figure 5a and 5b, respectively, which are extracted by the constant current method. It can be seen from Figure 5 that the average sensitivity of 1.5 μm SiNS FET biosensor is $\Delta V_{th}$=29.39mV/pH, while the sensitivity of 500nm SiNS is $\Delta V_{th}$=39.09mV/pH, which indicated that the wider SiNS had lower sensitivity to pH. This is in line with the theory proposed in the literature [3][5]. For this result, people first think about whether the sensitivity of pH will affect the sensitivity to biomolecules. This problem is mentioned in literature [6], and the conclusion is that there is no obvious connection. Therefore, combined with DNA detection results at different pH and the influence of SiNS with different widths on pH sensitivity, we believe that the low pH sensitivity of wide SiNS can effectively reduce detection errors caused by pH changes in biological detection.

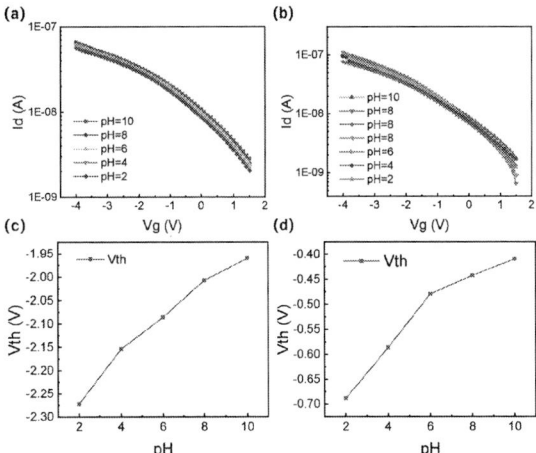

*Figure 5: (a) The transfer characteristic curve of 500 nm wide SiNS at different pH values; (b) The transfer characteristic curve of 1.5 μm wide SiNS at different pH values; (c) and (d) The threshold voltage extracted by the constant current method from (a) and (b).*

**Analyte binding simulation of different width SiNS**

Figure 6 shows the simulated difference in protein adsorption capacity and adsorption rate between 20 nm wide SiNW and 1.5 μm wide SiNS FET biosensor by COMSOL. The simulation model(Figure 6a and 6b) is to set up SiNW and SiNS in the center of a microchannel (5 μm length, 1 μm width, 0.5 μm height), and then pass into 50 nM analyte for 10 minutes. The amount of analyte bound on the surface of SiNW and SiNS can be seen in Figure 6c. Due to the large reaction area of the nanosheet, the 1.5 μm SiNS adsorbed about 2 pM of the analyte within 10 minutes, while the 20 nm nanowire adsorbed less than 0.3 pM of the analyte. Figure 6d is the first derivative extraction of the fitted curve, which shows that at an analyte concentration of 50 nM, SiNS with a width of 1.5 μm has a significant advantage in binding rate compared to SiNW with a width of 20 nm.

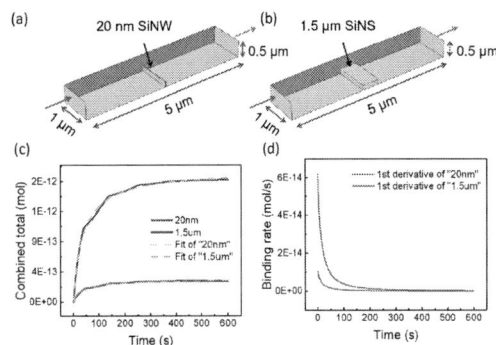

*Figure 6: (a,b) Microchannels, nanowires and nanosheets*

*modeled by COMSOL. The red arrow is the analysis inlet and the green arrow is the analysis outlet; (c)The total amount of analyte bound on the surface of SiNW and SiNS; (d) The first derivative calculation of (a) indicates the binding rate of SiNW and SiNS to the analyte.*

## CONCLUSION

SiNS FET biosensors with different widths were fabricated, which showed bipolar characteristics and good electrical properties. Then we detected ssDNA and got a good electrical response in the range of 10 nM to 100 μM. But in the process, it was found that pH has a great influence on ssDNA detection. Next we detected the 1 μM G-DNA with different pH, and we found that the current change was mainly caused by pH, rather than G-DNA concentration, which proved that the pH had great influence on biological detection. Furthermore, pH detection was performed on SiNS with different sensing areas, and it was found that wide SiNS can effectively reduce the pH sensitivity and thus reduce the impact of pH changes on biosensing. Therefore, wide SiNS is valuable to reduce the detection error caused by pH changes to improve sensing accuracy. Finally, by simulating with COMSOL, we found that SiNS FET biosensor is more conducive to the binding of analytes at low analyte concentrations than SiNW FET biosensor, which reduced the response time of high-speed detection.

## ACKNOWLEDGEMENTS

This work was supported by the National Natural Science Foundation of China (No. 61874002). We thank the Integrated Circuit Advanced Process Center (ICAC) at the Institute of Microelectronics of the Chinese Academy of Sciences for the devices fabricated on their advanced 200 mm CMOS platform.

## REFERENCES

[1] Oelssner W, Zosel J, Guth U, Pechstein T, Babel W, Connery JG, et al. Encapsulation of ISFET sensor chips[J]. *Sensors & Actuators B Chemical*, 2005, 105(1):104-117.

[2] Namdari P, Daraee H, Eatemadi A. Recent Advances in Silicon Nanowire Biosensors: Synthesis Methods, Properties, and Applications[J]. *Nanoscale Research Letters*, 2016, 11(1).

[3] Zida S I, Yang C C, Khung Y L, Lin Y D. Fabrication and Characterization of an Aptamer-Based N-type Silicon Nanowire FET Biosensor for VEGF Detection[J]. *Journal of Medical and Biological Engineering*, 2020, 40(12): 601-609.

[4] Sheehan P E, Whitman L J. Detection limits for nanoscale biosensors[J]. *Nano Letters*, 2005, 5(4):803-807.

[5] Elfström N, Juhasz R, Sychugov I, Engfeldt T, Karlström A E, Linnros J. Surface Charge Sensitivity of Silicon Nanowires: Size Dependence[J]. *Nano Letters*, 2007, 7(9):2608-2612.

[6] Lowe B M, Sun K, Zeimpekis I, Skylaris CK, Green NG. Field-effect sensors – from pH sensing to biosensing: sensitivity enhancement using streptavidin – biotin as a model system[J]. *Analyst*, 2017, 142(22):4173-4200.

# STUDY ON PERFORMANCE OF SI₃N₄ ENHANCED MICROBOLOMETER WITH SALICIDED POLYSILICON THERMISTOR IN CMOS TECHNOLOGY

*Haolan Ma, Yaozu Guo, Ke Wang, Haoyu Zhu, Jiabing Liu and Xiaoli Ji*[*]

School of the Electronic Science and Engineering, Nanjing University, Nanjing 210023,China
*Corresponding Author's Email: xji@nju.edu.cn

## ABSTRACT

In this paper, we studied the enhancement of additional $Si_3N_4$ layer on the voltage responsivity and detectivity of Si-based microbolometer in the standard CMOS process, which is composed of a $SiO_2$ absorber coupled with a salicided poly-Si thermistor The FDTD simulation results show that the growth of $Si_3N_4$ can improve the absorptivity of the microbolometer at 7-13 μm wavelength and optimal thickness is 500 nm. The experimental results show that the responsivity and detectivity can be improved after deposition of $Si_3N_4$ significantly. Especially, the enhancement of the detectivity can reach the maximum of 122% and 29.6% at 11.5 μm and 8.5μm, respectively.

## INTRODUCTION

Microbolometers are widely used in various fields of civil and military due to low cost, long life and wide spectral response and so on [1,2]. The microbolometer using salicided polysilicon as the thermistor is a new type of uncooled infrared detector based on standard CMOS process. Compared with the microbolometer of MEMS technology, it has the advantages of high integration. However, the responsivity and the detectivity of Poly-Si microbolometer are still lower than main $VO_x$ and a-Si microbolometers [3, 4] due to its poor infrared absorptivity. Here, an additional $Si_3N_4$ layer is grown on the surface of this type microbolometer to increase the infrared absorptivity and enhance responsivity and detectivity further. The experimental results reveal that the performance of microbolometer with salicided poly-Si thermistor in CMOS technology can be improved effectively by simply depositing $Si_3N_4$ layer on surface of the $SiO_2$ absorber.

## DESIGN AND FABRICATION

The microbolometer is designed with 0.18μm CMOS standard technology. Figure 1 shows the three-dimensional structure of Si microbolometer. Its thermistor material is salicided poly-Si, which is designed in a serpentine shape. The surrounding silica dielectric filling layer serves as an infrared absorbing material. The thermally isolated cavity under absorber and micro-bridge are formed by etching the substrate Si through the Post-CMOS process.

The main manufacturing process flow of the microbolometer is shown in Figure 2, in which the standard 0.18μm CMOS process step is marked with a hollow circle, and the post-CMOS process step is marked with a solid circle.

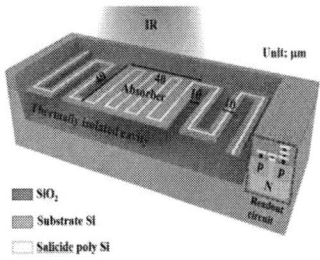

*Figure 1: 3D schematic illustration of the polysilicon microbolometer and design dimensions*

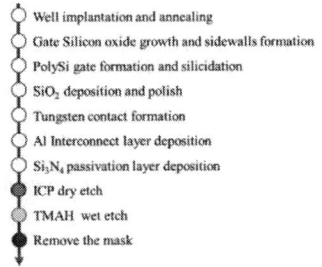

*Figure 2: Process flow for the fabrication of the microbolometer*

Firstly, N-well implantation and thermal annealing are performed sequentially. Then, a gate oxide is grown. The first aluminum interconnection layer is grown after $SiO_2$ is deposited and the subsequent $Si_3N_4$ passivation layer is deposited. Then, the post-CMOS process step is performed to fabricate the thermally isolated cavity. The inductively coupled plasma (ICP) is used to etch on the material without a mask cover until the silicon substrate is exposed. After this step is completed, the chip is immersed in TMAH solution and the Si substrate is etched; then, the mask is dissolved in the $H_3PO_4$ solution. After the etching is completed, the microbolometer structure with thermally isolated cavity is accomplished. Finally, the $Si_3N_4$ layer is deposited on the surface of the detector.

The influence of $Si_3N_4$ layer thickness grown on $SiO_2$ absorber on the infrared absorptivity was studied by the Lumerical simulation software of the finite difference time domain method (Lumerical FDTD Solutions). In the FDTD simulation, the size of the absorber is set to 40 μm ×40 μm, and the thickness of $SiO_2$ is set to 1.1 um. The plane wave source and perfect match layer conditions were used in the simulation and the thickness of $Si_3N_4$ is set to 0 nm, 100 nm, 200 nm, 300 nm, 400 nm and 500 nm, respectively. The simulation results of range from 7 μm to 13 μm are shown in Figure 3 and it is found that the absorptivity increases with the increase of the thickness of $Si_3N_4$, and the absorptivity reaches a saturation value at a wavelength of 500μm. It can be seen that the absorption peak is located at 8.7 μm which is nearly 87%.

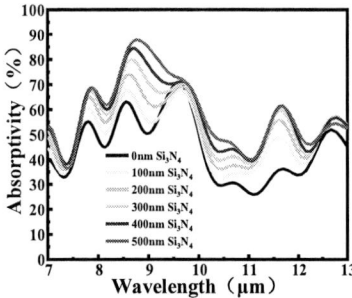

Figure 3: The simulated absorptivity at $Si_3N_4$ thicknesses of 0, 100, 200, 300, 400, and 500 μm

## RESULTS AND DISCUSSIONS

The temperature coefficient of resistance (TCR), and effective thermal conductivity ($G_{eff}$) have been obtained through previous experiments results, which are 0.35%/K and $3.08×10^{-7}$ W/K, respectively [5]. The voltage responsivity ($R_v$), detectivity ($D^*$) and noise equivalent power ($NEP$) of the microbolometer are characterized by the schematic diagram shown in Figure 4. First, the test chip is packaged in a dual-in-line ceramic package (DIP), and placed in a vacuum chamber, which pressure is kept less than 10 Pa. A mid-infrared quantum cascade laser (QCL) is used to generate infrared light with wavelength of 7-13 μm. Then the SR830 Lock-In amplifier generates and collects the signal voltage $\Delta V$.

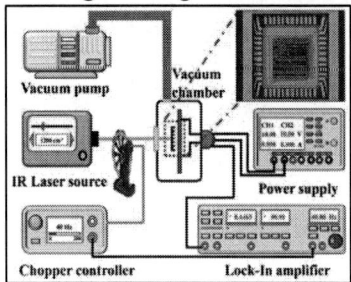

Figure 4: The schematic of the experimental set-up

The voltage responsivity is then calculated under different bias currents by the following formula：

$$R_v = \frac{\Delta V}{P_{in}} \qquad (1)$$

Here $R_v$ is the voltage responsivity, $\Delta V$ is the output voltage, and the $P_{in}$ is the input power of the IR illumination. Figure 5 shows the relationship between $\Delta V$ and input power at different currents. It is obvious that all data fall on the linear fitting curve, indicating that the $\Delta V$ and input power present linear relationship. In addition, the $\Delta V$ increases with increasing current under the same input power.

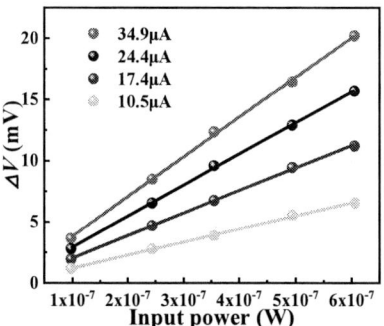

Figure 5: $\Delta V$ as a function of input power and bias current

Figure 6 shows the change of responsivity with wavelength range of 7-13 μm before and after deposition of 500 nm $Si_3N_4$. It can be found that the responsivity is improved as a whole after 500 nm $Si_3N_4$ is deposited. The responsivity increases from $1.67×10^4$ V/W to $2.09×10^4$ V/W at 8.5 μm and from $7.12×10^3$ V/W to $2×10^4$ V/W at 11.5 μm. Especially, the maximum of responsivity enhancement reaches 108.5% on average in the wavelength of 10-13 μm.

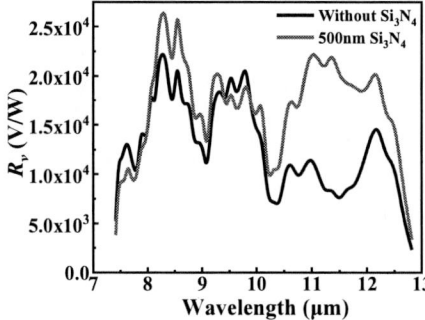

Figure 6: Voltage response of the microbolometer vs wavelength

The $D^*$ and $NEP$ of the CMOS microbolometer under various bias currents at 8.5 μm and 11.5 μm before and after the growth of 500 nm $Si_3N_4$ is shown in Figure 7 (a) and (b), respectively, which are obtained according to the following formulas[6]:

$$D^* = \frac{R_v \sqrt{A_d \Delta f}}{V_n} \qquad (2)$$

$$NEP = \frac{V_n}{R_v} \qquad (3)$$

Where $D^*$ is the detectivity, $A_d$ is the area of the absorber, $\Delta f$ is the electrical bandwidth and $V_n$ is the noise voltage of 5.1 μV obtained from previous study [4]. It can be seen from Figure 7 (a) that the $D^*$ improves after deposition of 500 nm $Si_3N_4$ at 8.5 μm. The detectivity increases from $1.25 \times 10^9$ cmHz$^{1/2}$/W to $1.62 \times 10^9$ cmHz$^{1/2}$/W when the bias current is 33 μA and the efficiency of improvement can reach about 29.6%. The NEP decreases from $3.18 \times 10^{-10}$ W to $2.47 \times 10^{-10}$ W. Moreover, the increasing efficiency of $D^*$ is more significant at 11.5μm, which can reach 122%, when the detectivity increases from $6.49 \times 10^8$ cmHz$^{1/2}$/W to $1.44 \times 10^9$ cmHz$^{1/2}$/W under the situation that bias current is 33 μA, as shown in Figure 7 (b). The NEP decreases from $6.16 \times 10^{-10}$ cmHz$^{1/2}$/W to $3.87 \times 10^{-10}$ cmHz$^{1/2}$/W, and the decrease rate is as high as 37%.

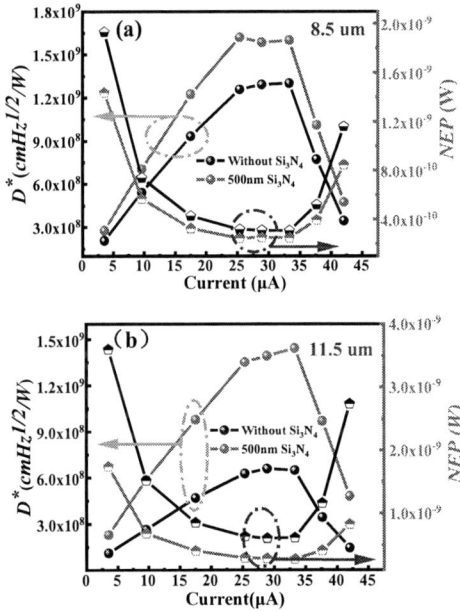

Figure 7 : $D^*$ and NEP of the microbolometer under various bias currents at 8.5 μm (a) and 11.5 μm (b) before and after the growth of 500 nm $Si_3N_4$

## CONCLUSION

The performance enhanced by $Si_3N_4$ of microbolometer with salicided polysilicon thermistor in standard CMOS technology is studied. The FDTD simulation results show that the optimal thickness of deposited $Si_3N_4$ is 500 nm. The experimental results show that the responsivity of the microbolometer at wavelength of 8-9 μm and 10-13μm improve markedly. Detectivity can be improved from $1.25 \times 10^9$ cmHz$^{1/2}$/W to $1.62 \times 10^9$ cmHz$^{1/2}$/W at 8.5 μm and from $6.49 \times 10^8$ cmHz$^{1/2}$/W to $1.44 \times 10^9$ cmHz$^{1/2}$/W at 11.5 μm, the enhancement efficiency can reach 29.6% and 122%, respectively.

## REFERENCES

[1] L. Yu, Y. Z. Guo, H. Y. Zhu, M. C. Luo, P. Han, and X. L. Ji, "Low-Cost microbolometer type infrared detectors," *Micromachines*, vol. 11, September 2020. pp. 800.

[2] B. Du, Y. He, Y. He, and C. L. Zhang, "Progress and trends in fault diagnosis for renewable and sustainable energy system based on infrared thermography: A review," *Infrared Phys. Technol*, vol. 109, January 2020, pp. 025007.

[3] N. Shen, Z. A. Tang, J. Yu, and Z. Huang, "A low-cost infrared absorbing structure for an uncooled infrared detector in a standard CMOS process," *Journal of Semiconductors*, vol. 35, March 2014,pp. 034014.

[4] P. S. Lin, T. W. Shen, K. C. Chan, and W. Fang, "CMOS MEMS thermoelectric infrared sensor with plasmonic metamaterial absorber for selective wavelength absorption and responsivity enhancement," *IEEE Sens. J.*, vol. 20, October 2020, pp. 11105-11114.

[5] Y. Z. Guo, M. C. Luo, H. L. Ma, H. Y. Zhu, L. Yu, F. Yan , P. Han, and X. L. Ji, "Microbolometer with a salicided polysilicon thermistor in CMOS technology," *Optical Express*, vol. 29, November 2021, pp. 37787-37796.

[6] T. W. Shen, K. C. Chang, C. M. Sun, and W. Fang, "Performance enhance of CMOS-MEMS thermoelectric infrared sensor by using sensing material and structure design," *J. Micromech. Microeng*, vol. 29, September 2020, pp. 103383.

# INFLUENCE OF B IONS DOPING ON THE PERFORMANCE OF *P*-TYPE SILICON NANOWIRE FIELD EFFECT TRANSISTOR BIOSENSOR

*Jiawei Hu[1,2], Qingzhu Zhang[2*], Shuhua Wei[1*], Jing Zhang[1], Zhaohao Zhang[2], Jin biao Liu[2] and Jiang Yan[1]*

1. School of Information Science and Technology, North China University of Technology, Beijing 100144, China;
2. Key Laboratory of Microelectronic Devices and Integration Technology，Institute of Microelectronics，Chinese Academy of Sciences，Beijing 100029，China;
E-mail: zhangqingzhu@ime.ac.cn; weishuhua@ncut.edu.cn

## ABSTRACT

In this paper, the influence of different doping concentrations on the performance of *p*-type polycrystalline silicon nanowire (SiNW) field effect transistor biosensors was investigated. Silicon nanowire biosensors with different doping concentrations were botained by implanting B ions with various concentrations of 0, 1E13, 5E13, and 1E14 into the poly-silicon film. With the increase of the doping concentrations of B ions, the threshold voltage ($V_{th}$) of the SiNW biosensor moves to the positive direction. In addition, the switching ratios ($I_{on}/I_{off}$) and the subthreshold swing (SS) is also significantly improved, the biosensor working in the subthreshold region has the best sensing sensitivity.

*Keywords—Silicon nanowire; field effect transistor, biosensor; B ions doping; spacer image transfer*

## INTRODUCTION

In recent years, the application of semiconductor field effect transistors (FET) biosensors have attracted a lot of attentions because of their ability to convert the interaction between FET surface and target molecules into electrical signal directly [1-4]. Silicon nanowires (SiNW) biosensors have been considered as one of the most promising candidates for biochemical sensors [2-4], due to their large surface area to volume (S/V) ratio, high sensitivity and good biocompatibility. Generally, the field effect transistor biosensor has three operating regions, the off region, subthreshold region and saturation region, and the SiNW biosensor could achieve the largest sensitivity at the subthreshold region. The approaches of adjusting the threshold voltage ($V_{th}$) of the biosensor include channel doping and changing the work function.

In this paper, the influence of different doping concentrations on the performance of *p*-type polycrystalline SiNW field effect transistor biosensors was investigated. With the increase of the doping concentrations of B ions, the threshold voltage ($V_{th}$) of the SiNW biosensor moves to the positive direction[5]. In addition, the switching ratios and the subthreshold swing (SS) is also significantly improved.

## EXPERIMENTAL

The SiNW biosensor was fabricated on 8-inch *p*-type Si (100) wafers with a resistivity of 0.1-100Ω•cm (see Fig. 1). Firstly, 145 nm thermal oxide ($SiO_2$) and 40 nm amorphous silicon were formed on the silicon surface, and then different doses of B ions were implanted with an energy of 5 keV. The flowing fabrication steps of NW formation, source and drain doping, self-allied silicide, gate trench and contacted formation were similar to our previous reports [2-3].

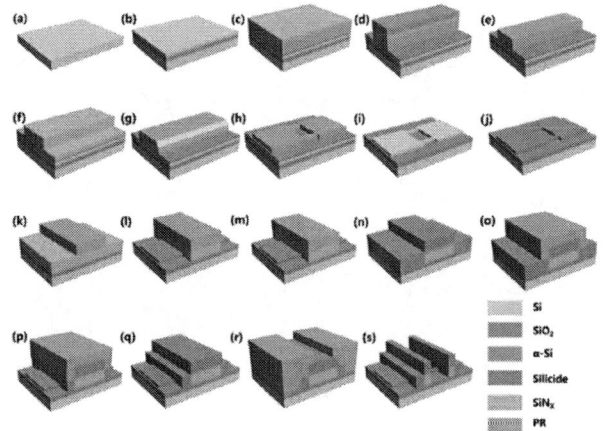

Fig.1. Fabrication flow of SiNW biosensors.

## RESULTS AND DISCUSSION

Fig. 2 (a), (b) and (c) show the transfer curves ($I_d$-$V_g$) of the SiNW biosensor by bias gate voltages with the B ions doses of 1E13, 5E13 and 1E14 for 5μm gate length ($L_g$), respectively. As can be seen from the images, the smooth and uniform *p*-type MOSFET curves were achieved for the biosensors. Due to the incensements of different doping concentrations, the threshold voltage ($V_{th}$) of the SiNW biosensors shifts to the right. Fig. 2(d) depicts the output curves ($I_d$-$V_d$) of the SiNW biosensor with the B ions dose of 1E13. The drain current has a linearly incensement with the

978-1-6654-9759-6/22 $31.00 © 2022 IEEE

increase of $V_g$, which means that the carrier concentration in SiNW channel can be adjusted linearly and the parasitic resistance ($R_{para}$) is very small.

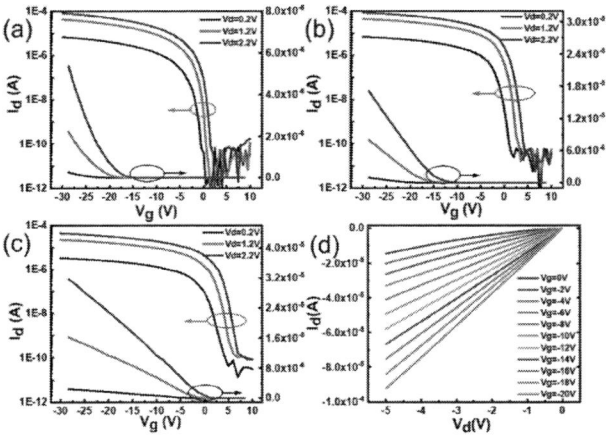

Fig.2. (a-c) The $I_d$-$V_g$ curves of the SiNW biosensor with the B ions doses of 1E13, 5E13 and 1E14, respectively. (d) The $I_d$-$V_d$ curves of the SiNW biosensor with the B ions doses of 1E13.

The extracted values of $V_{th}$, SS, $I_{on}$ and $I_{on}/I_{off}$ of the SiNW biosensors with the B ions doses of 0, 1E13, 5E13 and 1E14 are shown in Fig. 3. As can be seen from the images, the $V_{th}$ of the SiNW biosensors with doping concentrations of 0, 1E13, 5E13, and 1E14 are -6.1 V, 0.1 V, and 2.7 V and 5.5V, respectively. With the increase of doping dose, the values of $V_{th}$ shift to the positive direction. The value of $V_{th}$ of the SiNW biosensor is about 0V after the implantation of 1E13 B ions, which is conducive to achieve higher sensitivity. In addition, the values of SS of the doped SiNW biosensors are smaller than that of the undoped SiNW biosensor. Furthermore, the switching ratio and the open-state current are also improved with the increase of the doping doses.

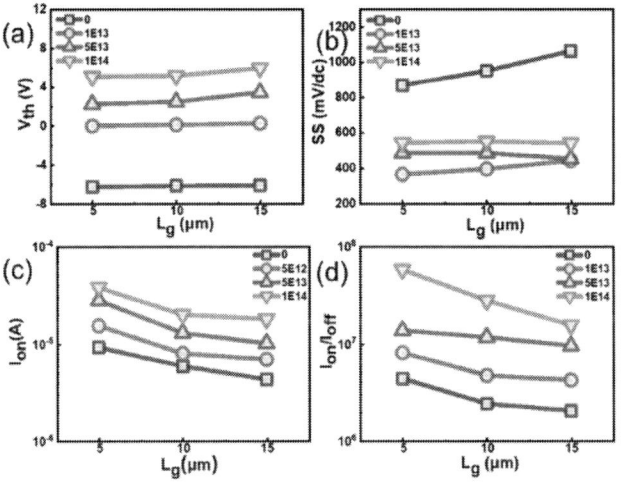

Fig.3. Comparison of the fabricated SiNW biosensors with different doses of B ions. (a) $V_{th}$, (b) SS, (c) $I_{on}$ and (d) $I_{on}/I_{off}$ ratio.

The extracted values of $R_{para}$, mobility ($\mu$), transconductance ($g_m$) and Drain Induced Barrier Lowering (DIBL) of the SiNW biosensors with the B ion doses of 0, 1E13, 5E13 and 1E14 are shown in Fig. 4. The parasitic resistance is obtained by fitting the resistance of devices with different gate lengths, the $R_{para}$ of the SiNW devices with injection amount of 0 is about 35450$\Omega$, and the $R_{para}$ of doped devices is around 31606$\Omega$ , $R_{para}$ declined slightly. The $\mu$ and $g_m$ are also improved with the increase of B ion implantation. Usually the doping concentration will lead to DIBL. However the DIBL is not obvious in our devices, and the DIBL values change little.

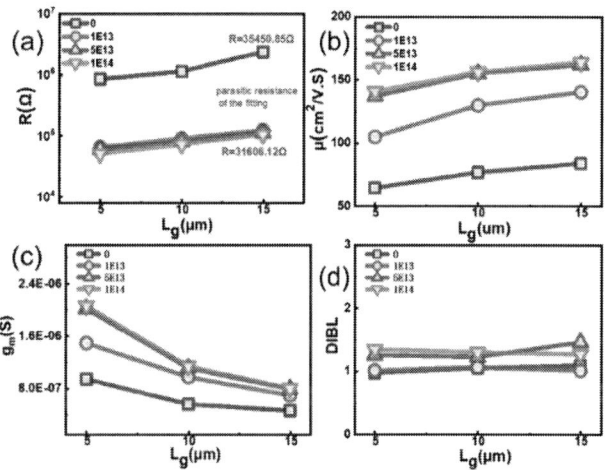

Fig.4. Comparison of the fabricated SiNW biosensors with different doses of B ions. (a) R, (b) $\mu$, (c) $g_m$ and (d) DIBL.

Fig. 5 shows the working state of the SiNW biosensor implanted with 0, 1E13, 5E13 and 1E14 B ions when $V_d$ = 1.2V and $V_g$= 0V, and the change of the current of the biosensor with pH under this working state. The biosensor with injection amount of 0 works in the off region and unable to carry out pH sensing detection. When the SiNW injection amount is 1E13, the biosensor works in the subthreshold region, and the current increases significantly with the increase of pH. When the SiNW injection amounts are 5E13 and 1E14, the biosensors work in the saturation region, and the current increases slightly with the increase of pH.

978-1-6654-9759-6/22 $31.00 © 2022 IEEE

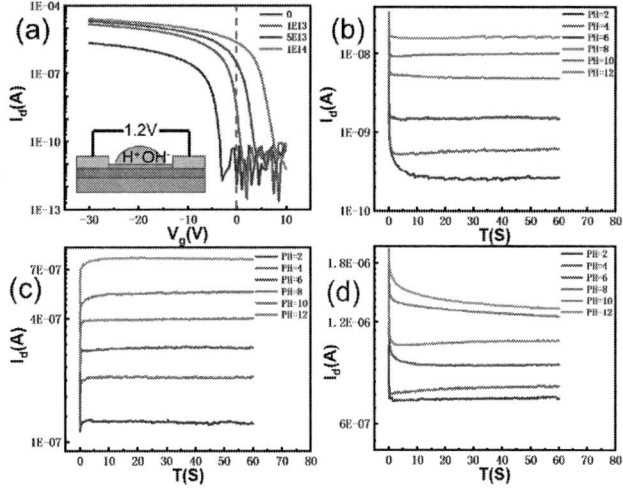

Fig .5. Sensing results of pH from SiNW biosensors of different doses of B ions (a) The transfer characteristic curve of the biosensor injected with different doses at $V_d$= 1.2V, (b), (c), and (d) are the pH sensing results of SiNW biosensors with injection amounts of 1E13, 5E13, and 1E14, respectively.

The current variation of SiNW biosensors with different injection amounts was calculated to obtain the sensitivity of each biosensor to pH, as shown in Fig. 6. As can be seen in the figure, the biosensor current with injection amount of 1E13 increases exponentially with the change of pH. The biosensor currents with injection amounts of 5E13 and 1E14 increase linearly with the change of pH.

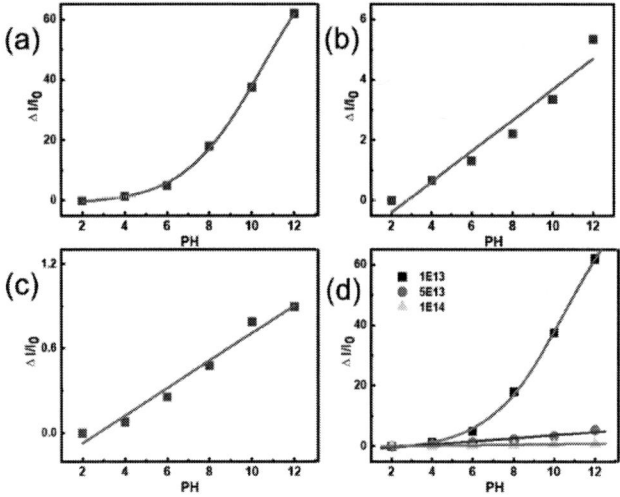

Fig .6. Using the current at pH = 2 as $I_0$, calculate the current change rate of SiNW biosensors with different injection amounts for each pH value, (a) 1E13, (b) 5E13, (c) 1E14 and (d) the current change rate of SiNW biosensors with different injection amounts.

## CONCLUSION

By implanting different concentrations of B ions into silicon nanowires, a SiNW biosensor with a threshold voltage near 0V is obtained. And the SS and $I_{on}/I_{off}$ of SiNW biosensor have been improved to some extent. The $R_{para}$ of the biosensor is reduced, μ and $g_m$ have been improved to a certain extent. Finally, the pH sensing detection of SiNW biosensor is carried out, and the biosensor current working at subthreshold increases exponentially with the increase of pH, which has the best sensing sensitivity.

## ACKNOWLEDGEMENTS

This work was supported by the National Natural Science Foundation of China,(Grants No. 61874002), We thank the Integrated Circuit Advanced Process Center (ICAC) at the Institute of Microelectronics of the Chinese Academy of Sciences for the devices fabricated on their advanced 200 mm CMOS platform.

## REFERENCES

[1] Yi, Cui, Qingqiao, Wei, Hongkun, & Park, et al. (2001). Nanowire nanosensors for highly sensitive and selective detection of biological and chemical. Science, 293(5533), 1289-1289.

[2] Tang, S. , Yan, J. , Zhang, J. , Wei, S. , & Tu, H. . (2020). Fabrication of low cost and low temperature poly-silicon nanowire sensor arrays for monolithic three-dimensional integrated circuits applications. Nanomaterials, 10(2488), 1.

[3] Qingzhu, Z.; Hailing, T.; Huaxiang, Y .; Feng, W.; Hongbin, Z.; Qianhui, W.; Zhaohao, Z.; Wenwu, W. Si Nanowire Biosensors Using a FinFET Fabrication Process for Real Time Monitoring Cellular Ion Actitivies. IEDM 2018. ,18, 679–682

[4] Zhang, N.; Zhang, Z.; Zhang, Q.; Wei, Q.; Zhang, J.; Tang, S.; Lv , C.; Wang, Y .; Zhao, H.; Wei, F. $O_2$ plasma treated biosensor for enhancing detection sensitivity of sulfadiazine in a high-k $HfO_2$ coated silicon nanowire array .Sens. Actuators B Chem.2020,306.

[5] Gao, X. , Zheng, G. , & Lieber, C. M. . (2010). Subthreshold regime has the optimal sensitivity for nanowire fet biosensors. Nano Letters, 10(2), 547-552.

# AN SOI-BASED SIGE HETEROJUNCTION PHOTOTRANSISTOR WITH LARGE OPERATION CURRENT RANGE

*Fu Zhu [1], Hong-Yun Xie[1]\*, Zi-Hang Wang[1], Rui-Lang Ji[1], Yang Xiang[1], Yin Sha[1], Dong-Yue Jin[1], Wan-Rong Zhang[1]*

[1] Faculty of Information Technology, Beijing University of Technology, Beijing 100124, china
\*Corresponding Author's Email: xiehongyun@bjut.edu.cn

## ABSTRACT

In this paper, the incident light distribution in device and photo-generated holes concentration of SOI-based and Si-based SiGe phototransistor (HPT) are simulated and analyzed. The parameters of the photocurrent gain, optical characteristic frequency, optical output characteristic curve and total capacitance change curve of two devices are compared. Compared with Si-based SiGe HPT, the maximal optical characteristic frequency and maximal optical gain of SOI-based SiGe HPT are increased by 24% and 49%. The kirk current density of SOI-based SiGe HPT for the maximal photo-charateristic frequency is 0.53 $mA/um^2$, a 51% improvement than the value obtained from Si-based SiGe HPT.

## INTRODUCTION

The high-performance photodetector with low cost and low power consumption is the demand for the development of optical communication technology and silicon-based photoelectric integration[1-2]. SiGe HBT devices have great advantages in low cost, low power consumption, large-scale monolithic integration circuits and their compatibility with complementary metal-oxide-semiconductor (CMOS) technology. Therefore, SiGe HPT detector, usually adopting the same structure with SiGe HBT and inheriting the advantages of SiGe HBT, has become one of the optoelectronic devices recently proposed for direct integration with the high speed SiGe technologies[3]. However, the vertical illumination SiGe HPT has the problem of kirk effect when collector current become large enough, which always make gain and working speed reducing and narrow the device operation amplitude. Thus, researchers increasingly interest in the development of SiGe HPT with wide operation range, high response and high working speed[4].

In this paper, we report a vertical illumination SiGe HPT based on SOI substrate. The influence of the SOI substrate on the operating current amplitude of SiGe HPT is studied through analyzing photocurrent gain and photo characteristic frequency, relevant explanations are provided. It is verified that the use of SOI substrate has great advantages in improving the operation current and other performance of the device

## DEVICE STRUCTURE

The detailed structure of SiGe HPT is shown in Figure 1, SOI-based SiGe HPT is shown in Figure 1(a) and Si-based SiGe HPT is shown in Figure 1(b). Both are NPN transistor in which only the base is made from the $Si_{1-x}Ge_x$ alloy, where x is the relative Germanium concentration. The heavily $P^+$-doped ($\approx 1\times10^{19}$ $cm^{-3}$) base is realized by a thin 100nm strained SiGe/Si film with a Ge content about 20%. The collector is 600nm thick Si layer with low doping. The sub-collector is 300nm thick Si layer with high $N^+$ doping. The emitter is made of 300nm poly-Si with high $N^+$ doping in the range of $2\times10^{20}$ $cm^{-3}$. The 300nm buried oxide layer locates between sub-collector and Si substrate for SOI-based SiGe HPT, which is the only difference between the two devices. As shown in Figure 1, both total emitter mesas are $8\times8\mu m^2$, each optical window is set in emitter mesa with the window area of $6\times6\mu m^2$.

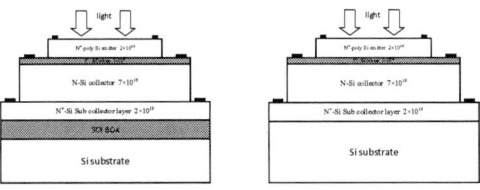

*(a) SOI-based          (b) Si-based*
*Figure 1: Device structures of SiGe HPT*

## RESULT AND DISCUSSION

### Photo-Current Gain

The output collector current of both SiGe HPTs are shown in Figure 2. Under the same incident light power and bias voltage of $V_{ce}$, the SOI-based SiGe HPT obtains a larger collector current of 74uA with incident power of 1.2mW, which is larger than the collector current of Si-based SiGe HPT under same conditions.

Figure 3 shows the light power distribution inside the device under the condition of 10mW and 940nm simulated by TCAD software. SOI-based SiGe HPT is shown in Figure 3(a) and Si-based SiGe HPT is shown in Figure 3(b). It is obvious that the interference of light occurs in the buried oxide layer, the sub-collection layer and collector layer in Figure 3(a), but there is no such phenomenon in Figure 3(b). The reason for it is there exists a large refractive difference between the Si-layer and $SiO_2$ BOX layer and part of the incident light are reflected at the $Si/SiO_2$ interface and then result in the light interference between incident light and reflected light.

Therefore, the reflected light existing in SOI-based SiGe HPT will be re-absorbed and its absorption efficiency increase correspondingly, which increasing the collector output current.

Figure 4 presents the photo-current gain of SOI-based and Si-based SiGe HPTs with different collector current density at the reverse bias $V_{ce}$ of 2V. The results show that when the collector current density is 1.37 mA/um$^2$, the photo-current gain $G_{opt}$ of SOI-based SiGe HPT reaches the maximum of 52. Obviously, the collector current density of achieving maximal $G_{opt}$ promote by 52% when comparing SOI-based and Si-based SiGe HPTs.

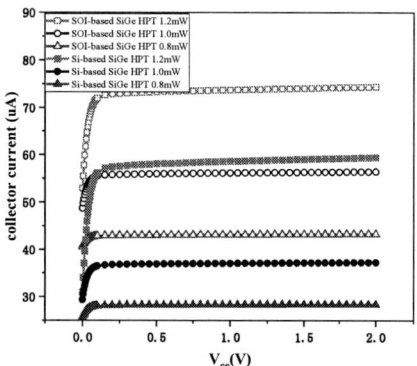

*Figure 2: Optical output characteristic curves of SiGe HPT with different optical powers*

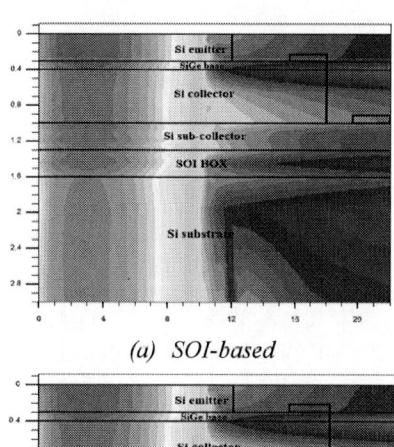

*(a) SOI-based*

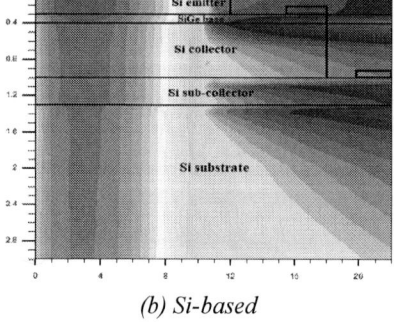

*(b) Si-based*

*Figure 3: Light power distribution in SiGe HPT*

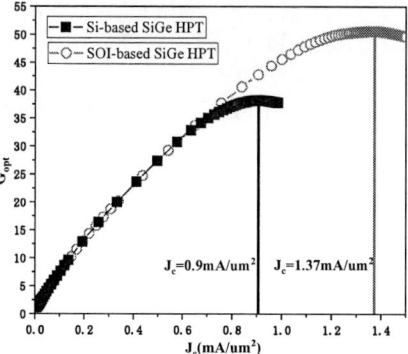

*Figure 4: Photo-current gain of SiGe HPT with different collector current density*

**Photo-Characteristic Frequency**

Figure 5 shows the curves for photo-characteristic frequency $f_{t,opt}$ of SiGe HPT under different collector current density at the reverse bias $V_{ce}$ of 2V. The maximal $f_{t,opt}$ of SOI-based SiGe HPT is 24.7GHz, which is 24% higher than the maximal $f_{t,opt}$ of Si-based SiGe HPT. The total capacitance $C_{total}$ of SiGe HPT with different collector current density are shown in Figure 6. The total capacitance $C_{total}$ is the sum of emitter diffusion capacitance $C_{DE}$, emitter barrier capacitance $C_{TE}$ and collector barrier capacitance $C_{TC}$ of the SiGe HPT. It is concluded that the maximal capacitance of SOI-based SiGe HPT is 0.66pF, which is 31% smaller than the maximal capacitance of Si-based SiGe HPT. The decrease in $C_{total}$ benefits the reduction of RC delay time and the total transit time of carriers and results the promotion of the optical characteristic frequency.

Figure 5 also shows clearly that the collector current density for achieving maximal $f_{t,opt}$ is higher when compare the value of SOI-based SiGe HPT with Si-based device. Generally, collector current density at which the characteristic frequency reaches the maximum is considered as the kirk current density. So, the corresponding kirk current density of the SOI-based SiGe HPT $J_{kirk}$ is 0.52 mA/um$^2$, improved by 51% by compared to the Si-based SiGe HPT of 0.35 mA/um$^2$, which is consistent with the analysis of photocurrent gain above.

For the HPT, the photo-absorption generate lots of photo-generated carriers in the absorption area which is always composed of base, collector and BC junction depletion layer. When the photo-generated electrons are extracted by collector, the photo-generated hole also diffuse/drift to emitter. However, the mobility of hole is usually low and then there will be large number of photo-generated holes accumulated in the BC junction depletion layer, especially when the incident power is high. As the SOI layer increases the light absorption rate, the photo-generated carriers generated in the BC junction depletion layer are higher, the holes accumulated here are also higher, as shown in Figure 7.

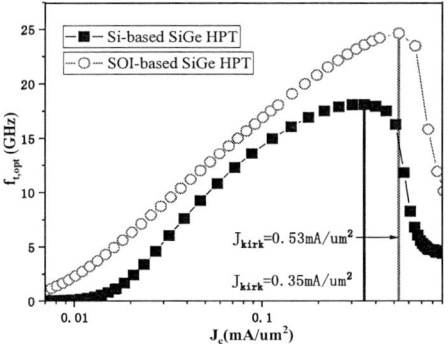

*Figure 5: Optical characteristic frequency of SiGe HPT with different collector current density*

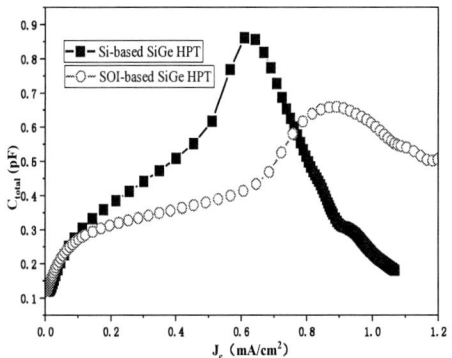

*Figure 6: Total capacitance of SiGe HPT with different collector current density*

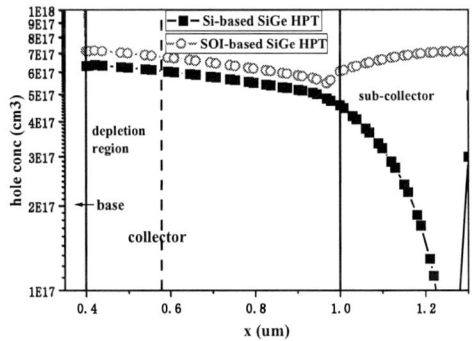

*Figure 7: Hole concentration at different locations of SiGe HPT*

When the incident light power is large, the kirk effect would be caused because large amounts of electrons are injected into base and collector and part of them would stay in BC junction depletion layer. As illustrated above, there generally exits accumulated holes in BC junction depletion layer of HPT, the stayed electrons will be recombined in certain degree. For SOI-based SiGe HPT, the improved the light absorption efficiency, the more accumulated holes in BC junction depletion layer, the more recombination with the stayed electrons under large incident light power, the larger kirk current density. Therefore, the normal operation current amplitude of SOI-based SiGe HPT is enlarged with 51% compared with the Si-based SiGe HPT.

## CONCLUSION

An innovative wide operation amplitude SiGe HPT with SOI substrate is demonstrated in this paper. The use of SOI substrate causes the reabsorption of incident light and reduces the junction capacitance. The photocurrent gain and optical characteristic frequency of SiGe HPT have increased by 49% and 24% respective. At the same time, the kirk current density of the SOI-based SiGe HPT is as high as 0.53 mA/um$^2$ when the optical characteristic frequency reaches its maximum, which is an increase of 51% compared to the kirk current density of the Si-based SiGe HPT. The reason for it that the reabsorption in SOI-based SiGe HPT make higher accumulated photo-generated holes in collector depletion layer and more recombination with the stayed electrons there and finally delay the occur of the kirk effect.

## ACKNOWLEDGEMENTS

Project supported by the National Natural Science Foundation of China (Grant Nos. 61604106, 61774012, and 61901010), Beijing Municipal Natural Science Foundation, China (Grant Nos. 4192014, 4204092) and the Shandong Province Natural Science Foundation (Grant No. ZR2021MF077).

## REFERENCES

[1] B. Sha. *Optical Materials*, vol. 106, Aug. 2020, p. 109 959.

[2] Liu Y L, Yu C C and Lin K T. *ACS nano*, vol. 9, 2015, pp. 5093-5103.

[3] Z. G. Tegegne, C. Viana, J.-L. Polleux, M. Grzeskowi ak, and E. Richalot, *IEEE Journal of Quantum Electr onics*, vol. 54, 2018, pp. 1–9.

[4] P. Ma, Hong-Yun Xie, Quan-Xiu Chen and Rui Liu, *I EEE International Conference on Solid-State and Int egrated Circuit Technology (ICSICT)* , 2018, pp. 1–3.

# Using Mixed Logic Synthesis Tools in Open-Source FPGA Design Framework

Liangtao Shi*, Yong Xiao[†], Yun Shao[†] and Zhufei Chu*[‡]

*EECS, Ningbo University, Ningbo 315211, China
[†]Giga Design Automation Co. Ltd., Shenzhen 518055, China
[‡]Email: chuzhufei@nbu.edu.cn

## ABSTRACT

In recent years, several open-source logic synthesis tools and design frameworks have emerged. However, there is limited research on the interactive combinations of these logic synthesis tools. To exploit the optimization capabilities of these tools, in this paper, we mixed three logic synthesis tools to construct a synthesis toolchain. The toolchain was integrated into an open-source FPGA design framework (OpenFPGA) to demonstrate the results in the full design flow. We evaluate Yosys, ABC, and the in-house synthesis tool ALSO for comparison. According to experimental results over EPFL benchmark suites, the combination of these three tools results in an average improvement of 47% in critical path delay with 22% overhead in the logic block area.

## INTRODUCTION

Logic synthesis is a crucial step in the *electronic design automation* (EDA) design flow. Generally, EDA flow can be divided into front-end logic design and back-end physical design stages. Logic synthesis converts the designs from *register-transfer level* (RTL) to an efficient gate-level implementation in front-end, which is critical to the performance of subsequent back-end tools [1].

Translation, technology-independent optimization, and technology mapping constitute the three steps of logic synthesis. The logic representation method involves the three steps for efficient manipulation of Boolean functions. Currently, *directed acyclic graph* (DAG) based logic representation is widely used for large-scale circuit synthesis. The typical example is *and-inverter graph* (AIG), in which the circuit is represented by only two-input AND and inverters.

Despite the mentioned AIG logic representation method, there are several other tools based on *majority-inverter graph* (MIG) and *exclusive-OR (XOR)-majority graph* (XMG) representations. These tools are distinct in their optimization algorithms and advantages. It is insufficient to evaluate a design framework by only using one logic synthesis tool. For example, most design frameworks only use Yosys, where ABC is used for gate-level optimization, to synthesize, thereby ignoring optimization possibilities. Optimizing with a toolchain that does not depend on a single tool could be an alternative.

In this paper, we propose a mixed logic synthesis

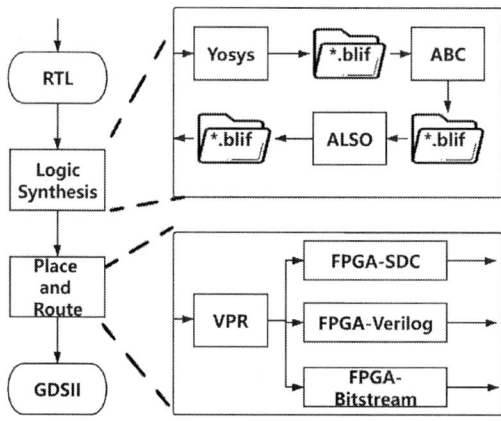

Figure 1: The flow diagram based on OpenFPGA [2].

toolchain that uses three open-source logic synthesis tools to optimize delay. Through DAG transformations, each tool is capable of optimizing a netlist through its command, which will significantly enhance the synthesis optimization performance. By tuning the order and comparing the performance of various scripts, the toolchain with the optimal delay can be created. This method optimizes the circuit delay by optimizing the depth as much as possible at the expense of small amounts of area.

## BACKGROUND

In this section, we will give backgrounds of open-source logic synthesis tools and logic representation methods for Boolean function.

### Logic Synthesis Tools

1) Yosys[1]: Yosys [3] supports industrial-strength front-ends by reading a design written in *hardware design language* (HDL), such as Verilog-2005. It is a framework for RTL synthesis tools, in which ABC is embedded for gate-level logic optimization.

2) ABC[2]: ABC is a set of tools for sequential synthesis and verification systems. It was developed by the Berkeley logic synthesis and verification group. ABC

---

[1]https://github.com/YosysHQ/yosys
[2]https://github.com/berkeley-abc/abc

TABLE I. Technology dependent results for EPFL benchmarks.

| Benchmarks | Yosys | | | Yosys-ABC | | | Yosys-ABC-ALSO | | |
|---|---|---|---|---|---|---|---|---|---|
| | critical_path $e^{-9}$ | logic_block_area $e^6$ | net_length | critical_path $e^{-9}$ | logic_block_area $e^6$ | net_length | critical_path $e^{-9}$ | logic_block_area $e^6$ | net_length |
| ctrl | 4.30 | 0.38 | 8.64 | 4.19 | 0.32 | 9.18 | 3.60 | 0.27 | 9.28 |
| int2float | 7.89 | 0.70 | 11.93 | 7.41 | 0.59 | 11.84 | 5.06 | 0.65 | 9.29 |
| router | 23.2 | 0.65 | 7.56 | 9.87 | 0.49 | 6.08 | 6.37 | 0.65 | 9.71 |
| dec | 2.82 | 0.86 | 16.28 | 3.15 | 0.92 | 18.42 | 2.87 | 0.86 | 16.34 |
| cavlc | 8.45 | 1.99 | 16.75 | 7.84 | 1.83 | 14.96 | 7.86 | 1.94 | 14.84 |
| priority | 53.6 | 1.72 | 11.52 | 46.2 | 1.46 | 10.39 | 33.0 | 1.67 | 10.43 |
| adder | 110 | 2.16 | 10.44 | 110 | 2.48 | 10.28 | 8.42 | 4.74 | 12.93 |
| i2c | 8.91 | 3.61 | 15.26 | 6.58 | 3.13 | 15.18 | 5.92 | 3.18 | 13.82 |
| max | 126 | 5.93 | 18.77 | 94.1 | 7.98 | 15.96 | 16.6 | 15.80 | 17.01 |
| bar | 6.80 | 7.87 | 20.21 | 7.70 | 8.57 | 19.39 | 7.57 | 8.68 | 20.54 |
| sin | 95.9 | 12.60 | 21.14 | 80.1 | 13.90 | 20.34 | 54.8 | 18.80 | 21.12 |
| arbiter | 39.8 | 34.00 | 50.15 | 39.3 | 34.00 | 50.32 | 8.05 | 20.00 | 22.25 |
| voter | 33.8 | 36.50 | 13.81 | 29.0 | 22.70 | 11.48 | 28.4 | 24.10 | 13.03 |
| square | 109 | 42.90 | 22.86 | 110 | 43.90 | 16.39 | 23.1 | 50.80 | 15.42 |
| sqrt | 2440 | 69.60 | 17.81 | 2330 | 53.20 | 17.56 | 1920 | 63.60 | 16.63 |
| multiplier | 118 | 53.90 | 25.42 | 121 | 67.00 | 22.86 | 56.9 | 98.50 | 24.84 |
| log2 | 179 | 71.00 | 30.28 | 171 | 80.50 | 28.14 | 111 | 110.00 | 25.62 |
| div | 1940 | 162.00 | 18.58 | 2040 | 111.00 | 16.04 | 514 | 195.00 | 17.93 |
| Avg. | 294.86 | 28.24 | 18.74 | 289.86 | 25.22 | 17.49 | 156.31 | 34.40 | 16.17 |
| Ratio | 1 | 1 | 1 | 0.98 | 0.89 | 0.93 | 0.53 | 1.22 | 0.86 |

Avg. : Average, net_length : average_net_length.

is also an industrial-strength synthesis and verification tool and is widely used and integrated by other design flows.

3) ALSO[3]: ALSO is based on the EPFL logic synthesis library [4] and is designed to develop advanced logic synthesis tools for modern FPGA and emerging nanotechnologies. ALSO is mainly researched by using *Majority-of-three* (MAJ) as the logic primitive and combining it with *Exclusive-OR* (XOR) operator [5].

## LOGIC REPRESENTATION

AIG is the main logic representation data structure used in ABC. Compared with a truth table, sum-of-products, and *binary decision diagrams* (BDDs), AIG is supposed to be more efficient for large-scale problem-solving. MIG is analogous to AIG. Instead of using two-input AND as logic primitive, a three-input MAJ node is used as an alternative. Note that MAJ can be reduced to two-input AND or OR if one of the inputs is set to constant zero or one, respectively. XMG is extended by MIG by adding XOR operations in order to offer better support for exact synthesis, arithmetic circuit representation, and quantum circuit synthesis. Each logic representation has its strengths and weaknesses. An open problem remains how to select an appropriate method for general logic synthesis.

[3] https://github.com/nbulsi/also

## OPENFPGA INTEGRATION

In this section, we demonstrate the integration of several open-source tools to create a synthesis-enhanced FPGA design flow, the flow diagram is shown in Fig 1. The backbone of the flow is OpenFPGA [2], which is an open-source suite for FPGA synthesis from RTL to GDSII. Experimental evaluation using EPFL benchmark suites is conducted to compare critical path delay, logic block area, and average net length after placement and routing.

The ALSO tool was integrated into the OpenFPGA flow to improve logic synthesis performance to demonstrate interoperability with other open-source tools. With OpenFPGA default flow, Yosys is used for trivial optimizations and ABC is used exclusively for technology mapping in Yosys. The benefits of using more sophisticated optimization methodologies were demonstrated by using the OpenFPGA flow modified to use an ABC optimization script (resyn2rs) with an ALSO script (xmgrw) applied after Yosys generic synthesis.

Performance is the focus of the proposed flow. Typically, academic synthesis tools stop at technology mapping, with a synthesized netlist for the physical design process. However, OpenFPGA has performed technology-dependent logic synthesis as well as full consideration of physical information. Because ABC, Yosys, and ALSO have different focuses, they are tested in serial order. Yosys performs high-level logic synthesis, while ABC and ALSO are still used for low-level logic

978-1-6654-9759-6/22 $31.00 © 2022 IEEE

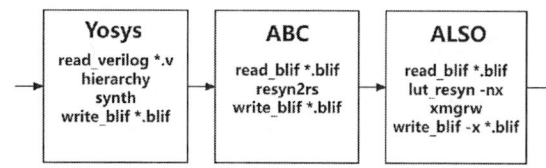

*Figure 2: Synthesis tools flow.*

synthesis.

**Our OpenFPGA Flow**

The logic synthesis flow is shown in Fig 2, which is one step in the OpenFPGA flow diagram. To be specific, the proposed flow works as follows.

1) Yosys read in RTL design described in Verilog. After executing several RTL optimization and mapping commands, the *berkeley logic interchange format* (BLIF) file is exported.
2) ABC read in BLIF file for AIG-based optimization, then ALSO is further adopted for logic optimization based on XMG. The optimized BLIF file is written for subsequent processing.
3) The logic netlists are verified to check the functional equivalence.
4) VPR is used to perform placement and routing for physical design. The design statistics are then reported.

## EXPERIMENTAL RESULT

Table I shows the experimental results over the EPFL benchmark suites after FPGA placement and routing. The critical path delay improved by 47% and the average net length decreased by 14% in comparison with the original flow. When compared to the ABC optimized design, both critical path delay and average net length were improved by 46% and 7.5%, respectively. Nevertheless, the method results in a 22% and 37% tradeoff in the logic block area, respectively.

Additionally, a technology-independent performance analysis was conducted for each design, resulting in the information in Table II. Through ABC optimization, AIG was able to optimize network size by up to 20% and depth by 3%. ALSO improved the logic depth by 53% with increased network size by 66%. The sacrifice of logic block area is less significant despite the greater increase in network size. Reduced network depth correlates with reduced critical path delays.

## SUMMARY

In this paper, an open-source logic synthesis toolchain for delay optimization is proposed. The advantages of each tool are utilized to bring a variety of scopes to improve the circuit area and performance. As part of

TABLE II. Technology independent results for EPFL benchmarks.

| Benchmarks | Yosys | | Yosys-ABC | | Yosys-ABC-ALSO | |
|---|---|---|---|---|---|---|
| | gates | level | gates | level | gates | level |
| ctrl | 132 | 9 | 85 | 8 | 93 | 6 |
| int2float | 254 | 16 | 210 | 15 | 218 | 9 |
| router | 219 | 53 | 169 | 21 | 221 | 11 |
| dec | 304 | 3 | 304 | 3 | 304 | 3 |
| cavlc | 681 | 16 | 629 | 16 | 651 | 14 |
| priority | 557 | 125 | 496 | 105 | 589 | 68 |
| adder | 764 | 255 | 893 | 255 | 1709 | 15 |
| i2c | 1219 | 19 | 1059 | 14 | 1087 | 12 |
| max | 2121 | 285 | 2832 | 206 | 5653 | 31 |
| bar | 2833 | 11 | 3141 | 12 | 3141 | 12 |
| sin | 4492 | 211 | 5015 | 172 | 6743 | 103 |
| arbiter | 11839 | 87 | 11839 | 87 | 6454 | 10 |
| voter | 12675 | 69 | 7954 | 59 | 87899 | 55 |
| square | 15505 | 250 | 16028 | 249 | 18390 | 38 |
| sqrt | 24774 | 5056 | 19105 | 4967 | 22942 | 3859 |
| multiplier | 19497 | 259 | 24389 | 263 | 35515 | 106 |
| log2 | 25243 | 402 | 29208 | 376 | 39724 | 210 |
| div | 57420 | 4372 | 20842 | 4364 | 69221 | 870 |
| Avg. | 10029.39 | 638.78 | 8011.00 | 621.78 | 16697.44 | 301.78 |
| Ratio. | 1 | 1 | 0.80 | 0.97 | 1.66 | 0.47 |

Avg. : Average.

the logical design process, critical paths are minimized to improve circuit performance. When compared with traditional logic synthesis techniques, the toolchain in this paper enables us to investigate synthesis-enhanced FPGA design flow. Experimental results show that the method can achieve a 47% critical path improvement over the EPFL benchmark suites compared with the default synthesis flow A future study will address the area aspect after addressing the delay aspect in this paper.

## ACKNOWLEDGEMENT

This work was supported in part by the NSFC under Grant 61871242 and in part by the State Key Laboratory of ASIC & System under Grant 2021KF008.

## REFERENCES

[1] S. Temple, W. L. Neto, M. Austin, X. Tang, and P.-E. Gaillardon, "Lsoracle: Using mixed logic synthesis in an open source asic design flow," in *Workshop on Open-Source EDA Technology*, 2021.
[2] X. Tang, E. Giacomin, B. Chauviere, A. Alacchi, and P.-E. Gaillardon, "Openfpga: An open-source framework for agile prototyping customizable fpgas," *IEEE Micro*, vol. 40, no. 4, pp. 41–48, 2020.
[3] C. Wolf, "Yosys open synthesis suite," 2016.
[4] L. Amarú, P.-E. Gaillardon, and G. De Micheli, "The epfl combinational benchmark suite," in *Proceedings of the 24th International Workshop on Logic & Synthesis (IWLS)*, no. CONF, 2015.
[5] Z. Chu, M. Soeken, Y. Xia, L. Wang, and G. De Micheli, "Structural rewriting in xor-majority graphs," in *Proceedings of the 24th Asia and South Pacific Design Automation Conference*, 2019, pp. 663–668.

# Design-for-Recovery Techniques for Combating Chip Aging Issues

Xinfei Guo

University of Michigan – Shanghai Jiao Tong University Joint Institute
Shanghai Jiao Tong University, Shanghai, China
Corresponding author email: xinfei.guo@sjtu.edu.cn

## BIOGRAPHY

Xinfei Guo is an Assistant Professor at Shanghai Jiao Tong University. He is also a Senior Member of the IEEE. He received his PhD from the University of Virginia, where he worked on improving chip lifetime through various design techniques. Before joining academia, he spent over 3 years working at Nvidia as a senior design engineer in US. His current research interests include EDA, VLSI, Reliability and Low power design. More details can be found on his website https://sites.ji.sjtu.edu.cn/xinfei-guo/.

## ABSTRACT

Nowadays, on a computer chip with size of a fingernail, there are billions of transistors that serve as the smallest computing units. Just as in the biological systems, these transistors and their interconnects will age with time, the degradation over time leads slowly but surely to decreased switching speeds, and it can even result in outright circuit failures. It is also unfortunate that aging is becoming much more troublesome for design teams at 10nm and below. The reduction in PPA, coupled with a demand for longer lifetime from the application spaces, has pushed the designers to examine alternative approaches to deal with aging issues instead of assigning conservative static margins as what has been done for a long time. In this work, we propose a new design dimension that explores the unique recovery properties several aging mechanisms offered. A set of design for recovery techniques will be introduced. Implementation details and results are presented.

## I. AGING AND RECOVERY

Chip aging is a growing issue at advanced technology including 5nm and below. The primary root cause of these effects is the stress effect of high electric fields across the dielectric, as geometries have got smaller, but voltages have not scaled at the same rate, the electric field across gates has increased, resulting in worse aging behavior [1], [2]. One top of this, the emerging markets such as industrial IoT, automotive, cloud computing, healthcare have proved that the chip will function as expected over time is becoming more and more difficult [3], [4]. At the transistor level, the dominant aging mechanism is bias temperature instability (BTI), that is characterized by the increase of the absolute value of threshold voltage $V_{th}$; at the interconnect level, electromigration (EM)

is the dominant aging threat that increases the wire resistance R over time. Traditionally, transistor aging in the design has been addressed by margining, which is to either add static margin to the clock uncertainty or add derates, computing the different performance of this aged cell versus the original cell, and using that ratio (old/new) to derate all cells of that type in the design. While EM was addressed by margining in a different format, e.g. conservative design rules or sizing. The margining technique has fundamental limitations and can lead to very pessimistic performance and power target. A good sign is that many of these aging mechanisms can be recovered slowly, but the existing research only "scratch a surface" on understanding the underlying mechanism due to the difficulties of setting up experiments and capturing the degradations within a reasonable time frame. To fulfill the need for understanding recovery behaviors thoroughly, in our previous work, we designed a set of novel experiments to collect measured results from actual chips (FPGA for transistor aging [5], test chips for interconnect aging [6]), each set of measurement has been carefully designed by considering different combinations of recovery conditions and lasts for more than 3 days. Through extensive experiments, we discovered that recovery can be made active, and the irreversible components can be completely avoided. It was also the first time that we found aging mechanisms follow a *"circadian rhythm"* like pattern - the whole process of chip aging and recovery can be compared directly to the biological world, in which scheduled recovery periods (or healing) are necessary after extensive workouts (or stress), with their athletic performance actually getting even better after the rest periods [7]. We borrowed the idea and successfully applied to the electronic world, thus the equivalent circadian rhythm for an operating chip would start with the active status until the irreversible aging kicks in, then an active recovery period is followed so that the irreversible aging becomes almost unobservable after active recovery even under extreme stress cases. These experimental results provided brand new insights on recovery from aging, such as frequency dependency [8], accelerated and active recovery and long-term vs. short recovery behaviors [7]. Such insights contributed as new experimental evidences for reliability community to create better and more accurate device models for emerging aging effects, they also offered strong implications of designing reliable systems by considering recovery in the early phase.

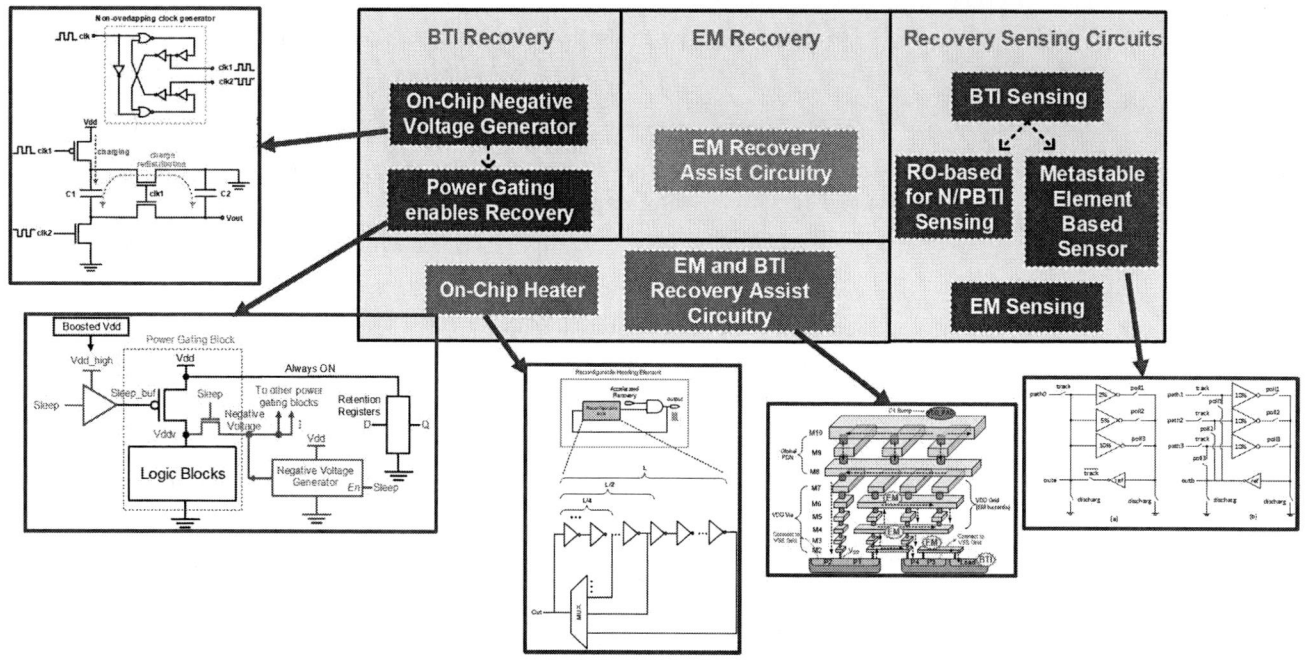

Fig. 1. Circuit components as the foundations for the design-for-recovery techniques

TABLE I
SUMMARY OF PPA METRICS FOR DIFFERENT DFR CIRCUIT IP COMPONENTS (IN 28NM TECHNOLOGY UNLESS SPECIFIED)

| Type | Design Name | Leakage Power | Dynamic Power | Area | Performance |
|---|---|---|---|---|---|
| BTI Accelerated 2*& Active Recovery Circuits | Neg. Voltage Generator[1] | $68.85nW$ | $64.47\mu W$ | $4300\mu m^2$ | $> 66.7MHz$ |
| | On-chip Heater[2] | $16.8nW$ | $75\mu W$ | $16\mu m^2$ | - |
| EM Accelerated & Active Recovery Circuits | Multi-mode Assist Circuit[3] | - | - | $58.24\mu m^2$ | Wake-up time~ $170ns$ |
| 2*BTI Sensors | RO-Based P/NBTI Sensors | $19nW$ | - | $92\mu m^2$ | - |
| | Metastable Element -Based Sensors[4] | $40.64nW$ | - | $54.1\mu m^2$ | Acq. Time $12ns$ |
| EM Sensors | Metal-line Based Sensors[5] | - | - | $100 - 500\mu m^2$ | - |

[1] Corresponds to a negative voltage generator designed for generating $-0.3V$ for BTI active recovery.
[2] Corresponds to one on-chip heater (41-stage RO) that can generate temperature of more than 80, in the real use case, multiple copies of this heater will be distributed.
[3] PPA metrics of this circuit depends on the load size and application, here we list the examples for 8 ring oscillators running in parallel.
[4] The numbers presented in this table shows metrics for a 2-path version of the sensor.
[5] Data is reported by [9] with $180nm$ technology node.

978-1-6654-9759-6/22 $31.00 © 2022 IEEE

## II. Design for Recovery (DFR)

Based on our encouraging experimental results, we investigated on how to bring recovery into the context of design processes. We propose a new design technique called "design-for-recovery", in which recovery is treated as a tunable design knob for dealing with aging issues. The fundamental idea is to develop design infrastructure that facilitate designers to consider recovery in the initial design phase and potentially improve aging not unlike improving power or performance. As the first step, we propose a set of circuit level recovery assist components that can be instantiated as an embedded IP, similar to the process monitor or temperature sensors in the current SoC designs. They are summarized in Fig. 1. On-chip negative voltage is designed for generating the negative voltage for activating the BTI recovery; on-chip tunable heater is a modified version of configurable ring oscillator which can be deployed for providing high temperature when necessary. A multi-mode EM/BTI recovery assist circuit is able to support both BTI and EM recovery simultaneously, and it can be designed based on the existing power gating infrastructure on chip, thus less design effort and overhead are required. As with the recovery solutions, sensor components that are designed to track the ongoing aging behavior are required so that any recovery solutions can be asserted or de-asserted. Table I summarizes the power, performance, and area metrics for these circuit IP components discussed above. It shows that most of these components are small in terms of area and fast in terms of response time.

With the IP foundation built, the designers are able to take the recovery circuit IPs and plug into their large-scale designs. Aging, especially transistor aging, is usually not part of the EDA optimization problems, due to the fact that it highly depends on the workload and environment. While the design for recovery techniques are able to engage upfront aging considerations and dynamic solutions. Figure 2 illustrates an example design-for-recovery methodology which starts from from the design specifications (SPEC) and applications, defining the expected lifetime, clock frequency, power and area budget, and so on. This flow assumes that high-level architecture-level decisions based on the SPEC have been made; these decisions can be type of ISA, number of cores, and so on. This information can be used as the input for pre-RTL and architecture-level simulation tools to estimate the power, thermal, and performance behaviors. Reliability tools such as OldSpot proposed in [10] that were based on aging models can take the reported behaviors from such tools and determine the aging behavior at a higher level, such as which units could be potentially aging-critical and what would be the possible operating schedules for the proactive recovery. These decisions can guide the frontend design phase when designing the microarchitecture; for example, more aging sensors can be instrumented for the aging-critical units, multiple copies of these units can be integrated in the design, and so on. The control logic for closing the aging and recovery loop is also implemented in this stage of the design process. During the

backend design stage, core/resource allocation solutions that are beneficial to recovery and PDN design against aging can be part of the floorplan decisions. Similarly, recovery circuit components discussed above can be integrated into the aging-critical units. During the rest of the physical design steps, circuit knobs that affect the recovery levels also need to be considered; examples of such knobs are the logic depth or power gating styles (for low power designs). This skeleton methodology offers a high level perspective of taking recovery as a proactive design decision to ensure the reliability of the chip.

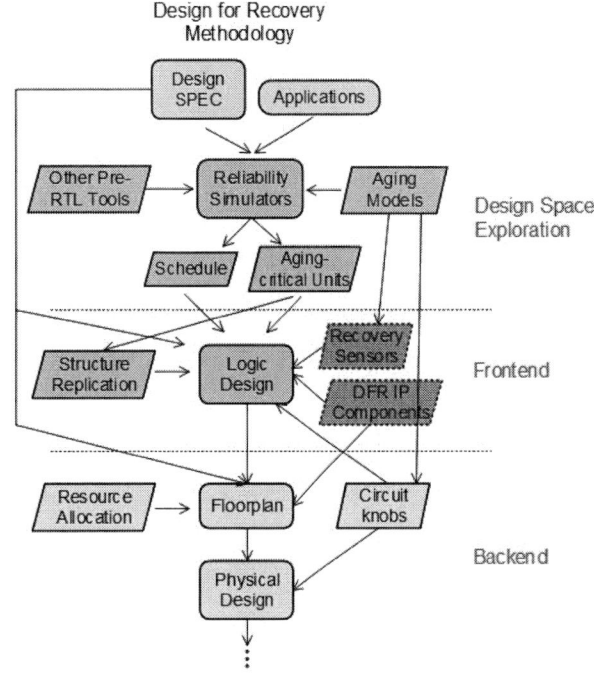

Fig. 2. An example design-for-recovery methology from concept to implementation

## III. Conclusions

In this paper, we introduced a new design dimension against the chip aging issues that are dominated by BTI and EM. As the conventional margining solution introduces significant pessimism to the performance and requires more design effort to close the design, we discovered that recovering from aging can potentially relaxing the design margin requirement. We presented a series of circuit IP components which lay the foundation for a design-for-recovery approach. An example of such approach was also discussed. In summary, with further advancement of semiconductor manufacturing nodes, the issues associated with chip aging will only intensify, we oversee that design-for-recovery can be a cost-effective and portable design solution to ensure a long chip lifetime and better overall PPA.

## Acknowledgement

This work was partially funded by a SJTU Explore-X Research Grant and the startup funding from Shanghai Jiao

978-1-6654-9759-6/22 $31.00 © 2022 IEEE

Tong University.

## REFERENCES

[1] I. Meric, S. Ramey, S. Novak, S. Gupta, S. Mudanai, and J. Hicks, "Modeling framework for transistor aging playback in advanced technology nodes," in *2020 IEEE International Reliability Physics Symposium (IRPS)*. IEEE, 2020, pp. 1–6.

[2] O. Prakash, C. K. Dabhi, Y. S. Chauhan, and H. Amrouch, "Transistor self-heating: The rising challenge for semiconductor testing," in *2021 IEEE 39th VLSI Test Symposium (VTS)*. IEEE, 2021, pp. 1–7.

[3] A. Shamshiri, M. Ghaznavi-Ghoushchi, and A. Kariman, "Ml-based aging monitoring and lifetime prediction of iot devices with cost-effective embedded tags for edge and cloud operability," *IEEE Internet of Things Journal*, 2021.

[4] G. Tshagharyan, G. Harutyunyan, Y. Zorian, A. Gebregiorgis, M. S. Golanbari, R. Bishnoi, and M. B. Tahoori, "Modeling and testing of aging faults in finfet memories for automotive applications," in *2018 IEEE International Test Conference (ITC)*. IEEE, 2018, pp. 1–10.

[5] X. Guo, W. Burleson, and M. Stan, "Modeling and experimental demonstration of accelerated self-healing techniques," in *2014 51st ACM/EDAC/IEEE Design Automation Conference (DAC)*. IEEE, 2014, pp. 1–6.

[6] X. Guo and M. R. Stan, "Deep Healing: Ease the BTI and EM Wearout Crisis by Activating Recovery," in *2017 47th Annual IEEE/IFIP International Conference on Dependable Systems and Networks (DSN)*. IEEE, 2017, pp. 184–191.

[7] ——, *Circadian Rhythms for Future Resilient Electronic Systems - Accelerated Active Self-Healing for Integrated Circuits*. Springer, 2020.

[8] ——, "Work hard, sleep well-avoid irreversible ic wearout with proactive rejuvenation," in *2016 21st Asia and South Pacific Design Automation Conference (ASP-DAC)*. IEEE, 2016, pp. 649–654.

[9] K. He, "Parallel cad algorithms and hardware security for vlsi systems," Ph.D. dissertation, University of California, Riverside, 2016.

[10] A. Roelke, X. Guo, and M. R. Stan, "OldSpot: A Pre-RTL Model for Fine-grained Aging and Lifetime Optimization," in *Computer Design (ICCD), 2018 IEEE International Conference on*. IEEE, 2018.

# Analytical Optimization Method for VLSI Global Placement

Weijie Chen[1], Haishan Huang[1], Zhipeng Huang[2], and Jianli Chen[2,3]

[1]College of Computer and Data Science, Fuzhou University, Fuzhou 350108, China
[2]Center for Discrete Mathematics and Theoretical Computer Science, Fuzhou University, Fuzhou 350108, China
[3]State Key Lab of ASIC & System, Fudan University, Shanghai, China
{200320066, M185410003}@fzu.edu.cn, chenjianli@fudan.edu.cn

## Abstract

Placement is one of the critical stages in the physical design of very large scale integrated circuits (VLSI), which has a significant impact on the performance of subsequent stages. Modern placement algorithms need to handle large-scale designs with millions of cells and complex placement constraints. Further, placement often includes three steps: global placement, legalization, and detailed placement. To a great extent, global placement affects the quality of the final placement and thus is considered a crucial part of the placement. Based on the recent research progress, this paper briefly introduces related optimization models and algorithms commonly used in VLSI global placement and discusses possible research directions.

*Keywords—VLSI; global placement; optimization algorithm*

## Introduction

Placement is the central part of the physical design of VLSI and a typical non-deterministic polynomial-time hard (NP-hard) problem that plays a significant effect on IC performance metrics like timing delay, routability, power and circuit reliability. Placement refers to the process where cells are placed with some measurement metrics (e.g., wirelength, routability) optimized on the condition that there is no overlap between different cells. Although decades worth of research on the placement problem, the most recent studies indicate that the current placement tools still fail to generate near-optimal placement results. Therefore, how to design high-performance placement has drawn extensive attention.

Presently, there are mainly three categories of algorithms solving the VLSI placement problems: algorithms based on simulated annealing, partitioning and analytical. The SA-based algorithm simulates the annealing mechanism to optimize placement by changing the positions of units in placement regions. The partitioning-based placement algorithm is to recursively partition circuit and placement regions, and then assign sub-circuits into sub-regions in a top-down manner until sub-circuits are smaller enough. Analytical-based algorithms transform the placement problem into a mathematical programming problem that consists of the objective function of wirelength and non-overlapping constraints of cells, adopt different smooth techniques to smooth the objects while considering constraints, construct a variety of optimization algorithms or heuristic methods to solve the placement optimization model. Recent literature and placement contests show that analytical-based placement algorithms can obtain solutions with better quality.

In analytical-based placement, it often includes three steps: global placement, legalization and detailed placement. The global placement ignores some placement constraints (e.g., cell overlap) and finds the best position for each cell.Legalization eliminates all cell overlaps while reserving the "fairly good" solution obtained in the global placement stage as much as possible. The detailed placement further improves the solution after legalization. To a great extent, global placement affects the quality of the final placement outcome and thus is considered a crucial part of the placement.

## Problem Statement

VLSI global placement is a multi-objective, multi-constraint optimization problem, making it extremely difficult to find the optimal solutions. The most commonly used method is to construct an algorithm that only optimizes the wirelength in the global placement problem first to calculate reasonable solutions with great extensibility within an acceptable time, allowing other objectives and constraints

to be simultaneously considered or processed in the next step.

In the VLSI global placement, the cells in the circuit are regarded as vertices, and the nets are regarded as hyperedges. Given a circuit with $n$ cells and $m$ nets, it can be transformed into a hypergraph $H(V, E)$, where $V = \{v_1, v_2, ..., v_n\}$ is the set of all vertices, and $E = \{e_1, e_2, ..., e_m\}$ is the set of all hyperedges. The center coordinates, the width (height), and the area of the cell $v_i$ are represented by $(x_i, y_i)$, $w_i$, $h_i$, and $A_i$, respectively. Due to a close correlation between the wirelength and circuit performance, the most common indicator in global placement is minimized the total wirelength, which can be described as the following problem of constraint minimization.

$$\min W(x, y)$$
$$\text{s. t. no cell ovelap} \tag{1}$$

where $W(x, y)$ is the wirelength function, which is usually calculated by the sum of the half-perimeter wirelength (HPWL) of all nets, and smooth by some smoothing wirelength model.

## Analytical Global Placement

This section introduces representative optimization models and algorithms for the VLSI analytical global placement.

### A. Quadratic Programming

The quadratic programming based global placement generally uses two methods to make the cell distribution uniform. In these methods, even if there are additional constraints/forces, the quadratic wirelength minimization problem can still be modeled as convex quadratic programming, which can be solved effectively. The first method is to physically divide the placement region, or add linear constraints to change the center of gravity of some elements to prevent cells from clustering together. Another integrated wirelength model and overlap processing technology is the fixed point method. Obtain the target position of each cell using overlap processing technology, create a fixed point on the position and establish a virtual net between the cell and the fixed point. Then, solve the placement problem on the modified netlist and repeat the diffusion process until the distribution of placement density satisfies requirements.

### B. Penalty Method/Lagrangian Multiplier Method

Placement tools based on non-convex optimization usually approximate HPWL by non-quadratic wirelength functions, control cell diffusion by non-quadratic density constraints, and then use quadratic penalty methods to convert the optimization problem consisting of wirelength objectives and density constraints to the unconstrained minimization problem. Most of the non-quadratic placement tools (e.g., NTUplace3 [1] and ePlace [2]) use LSE or WA to smooth the wirelength model. In these methods, NTUplace3 use the bell-shaped density model, and ePlace use the electric field energy model.

In NTUplace3 [1], to fully understand the placement information and effectively use nonlinear programming techniques, the placement region is evenly divided into non-overlapping grids bin. Then the global placement problem is transformed into an unconstrained minimization problem by using the quadratic penalty method:

$$\min W(x, y) + \lambda_p \sum_b (\widehat{D}_b(x, y) - M_b)^2, \tag{6}$$

where $\lambda_p$ is the penalty parameter that used to balance the weight of the wirelength and the density penalty, and it is constantly updated in the algorithm iteration (e.g., increasing to meet the density constraint). $M_b$ is the maximum allowable area of movable blocks in bin $b$. $M_b$ can be computed by

$$M_b = t_{density}(w_b h_b - P_b), \tag{7}$$

where $t_{density}$ is a user-specified target density value for each bin, $w_b(h_b)$ is the width(height) of bin b, and $P_b$ is the base potential that equals the preplaced block area in bin b. Original overlap functions can be computed by

$$D_b(x, y) = \sum_{v \in V} P_x(b, v) P_y(b, v) \qquad (8)$$

where $P_x$ and $P_y$ are the overlap functions of bin b and block v along x - and y -directions. Where NTUplace3 [1] uses a bell-shaped potential function $p_x$ to smooth $P_x$. $p_x$ is defined by

$$p_x(b, v) = \begin{cases} 1 - ad_x^2, & 0 \le d_x \le \frac{w_v}{2} + w_b \\ b(d_x - \frac{w_v}{2} - 2w_b)^2, & \frac{w_v}{2} + w_b \le d_x \le \frac{w_v}{2} + 2w_b \\ 0, & \frac{w_v}{2} + 2w_b \le d_x \end{cases} \quad (9)$$

where

$$a = \frac{4}{(w_v + 2w_b)(w_v + 4w_b)}$$
$$b = \frac{2}{w_b(w_v + 4w_b)}, \qquad (10)$$

so, the nonsmooth function $D_b(x, y)$ can be replaced by a smooth one

$$\widehat{D}_b(x, y) = \sum_{v \in V}^{n} c_v p_x(b, v) p_y(b, v), \qquad (11)$$

where $c_v$ is a normalization factor so that the total potential of a block equals its area. The conjugate gradient method is usually used to solve the above global placement problem.

Solving the above-mentioned very-large-scale nonlinear programming problems in a direct manner can be time-consuming. In that case, a multi-level clustering framework is typically used to reduce the calculation time of the placement algorithm. Specifically, the algorithm first clusters the circuit cells in a multi-level manner to obtain coarse netlists with diminishing scale. It solves the nonlinear programming problem on the coarse netlist to quickly generate a coarse placement. During de-clustering, the nonlinear programming problem for placement optimization is solved with the position of the cluster in the previous as the initial position. In the global placement, commonly used clustering algorithms include Best-Choice and First-Choice.

ePlace [2] treat the cell as a positive charge in the global placement, and regard the density constraint of the cell distribution as a total potential energy constraint N(v). Then according to the electrostatic balance, the global placement problem is transformed into the following equation by using a Lagrangian multiplier $\lambda_l$:

$$\min W(v) + \lambda_l N(v). \qquad (12)$$

Due to the electric force generated by the two charges, the sum of the potential energies of the system is equal to the sum of the potential energies generated between all charge pairs, so $N(v) = \frac{1}{2} \sum_{i \in V} q_i \psi_i$, where $q_i = A_i$ represents the electric power of cell i. Different from Equation (6), multiple constraints are transformed into a single zero potential energy constraint N(v).

*C. Augmented Lagrangian Method*

Since the augmented Lagrangian method can reduce the morbid possibility of minimization problem, the work [3] proposes to solve the global placement problem of the integrated circuit using the augmented Lagrangian method rather than the penalty method. At the same time, the reduction coefficient and damping coefficient are added to modify the augmented Lagrangian method according to the characteristics of the actual problem. Besides, the wirelength objective and density constraint are balanced by using a cautious dynamic density weighting strategy, and a wide range of searches in the initial state is enabled after using an adaptive step size strategy in the conjugate gradient method. Also, the step size is reduced when the optimal position is possibly approaching. Note that the solution obtained from the augmented Lagrangian method will not be a sufficient approach to the stable point of the original problem without the precise line search [3]. The work [4] analyzes the strengths and weaknesses of four methods of handling constraints: the quadratic penalty method, the Lagrangian multiplier method and two extended augmented Lagrangian methods. Next, the author proposes the generalized augmented Lagrangian method (GALM) to solve the global placement of the integrated circuit as follows:

$$\min \widehat{W}(x, y) + \frac{1}{w} \sum_b \mu_b^k (\widehat{D}_b(x, y) - M_b) + \frac{\rho}{2} \sum_b (\widehat{D}_b(x, y) - M_b)^2 \quad (13)$$

By appropriately updating the value of $\omega$ and the Lagrangian multiplier, GALM not only has the advantages of the rapid diffusion cell of the penalty method in the initial stage, but also retains the advantages of the augmented Lagrangian method in the later iteration and avoid its shortcomings. This work proves that GALM can converge to the KKT point of the original problem under certain conditions.

The work [5] extends the ePlace model to the FPGA global placement to deal with heterogeneous modules in FPGA design. Different from the Lagrangian multiplier method used in ePlace to deal with density constraints, this work uses the augmented Lagrangian approach to assign different weights to the modules of each resource type:

$$\min \widehat{W}(x, y) + \sum_{s \in S} \lambda_s (\Psi_s(x, y) + \frac{c_s}{2} \Psi_s(x, y)^2), \qquad (14)$$

where $s \in S = \{LUT, FF, DSP, RAM\}$ corresponds to different resource types, and $c_s$ is the coefficient that controls the quadratic term. To better control the overall behavior of density constraints and reduce the sensitivity of the algorithm to the initial placement, the quadratic term is also controlled by $\lambda_s$. Setting the parameters correctly, when a specific resource type s has high potential energy $\psi_s(x, y)$ (a lot of overlap), the quadratic penalty term is dominant to enhance the convexity of the objective function and improve convergence. When a certain resource type s has relatively small potential energy, the primary term will dominate. It can be optimized by using the Lagrangian multiplier method without encountering the ill-condition problems related to the penalty method.

*D. Alternating Direction Multiplier Method*

The alternating direction multiplier method (ADMM) is an effective algorithm for optimization problems with separable structures, particularly the separable linearly constrained convex optimization problems. ADMM decomposes a large global problem into multiple local sub-problems that are small in size and easily solved, and then converges to a high-quality (feasible) solution to the global problem through coordinating the solutions of the sub-problems. The method has also been applied in the VLSI global placement.

In the work [6], the traditional VLSI global placement problem with wirelength optimization is converted into a particular non-convex minimization problem. More precisely, the objective function is the sum of two convex wirelength functions, and the function in the constraint condition is a non-convex density function. Then, the above problem is solved by using the proximal point alternating direction multiplier method. Considering the fogging and proximity effect in integrated circuit manufacturing, the work [7] models the VLSI global placement as the following separable minimization problem with linear constraints to reduce the complexity of the problem:

$$\begin{cases} \min \ \theta_1(x, y) + \theta_2(x, y) \\ s.t. \ x - g = 0 \\ \quad\ \ y - h = 0, \end{cases} \qquad (15)$$

Where $\theta_1(x, y) = \lambda_1 \widehat{W}(x, y) + \lambda_2 \widehat{D}_b(x, y)$ represent wirelength and density functions, $\theta_2(g, h) = \lambda_3 S_f(g, h) + \lambda_4 S_p(g, h)$ is the value of fogging and proximity effects. A proximal group ADMM is proposed to solve the problem by adding a strongly convex proximal term to the objective function. The first sub-problem (mainly related to wirelength and density) is calculated by the steepest descent method, while the second one (mainly related to fogging and proximity effects) is to directly obtain the analytical solution using some approximation techniques.

For 2.5D FPGA placement problems, the work [8] considers optimizing the number of super long lines (SLL), a continuous cost function is proposed to solve the finite discrete SLL problem and relax the clock constraints. Then the 2.5D FPGA global placement problem considering SLL and clock constraints is expressed as a continuous differentiable minimization problem. To solve this problem with the ADMM algorithm, two sub-problems need to be considered. The first sub-problem includes wirelength, density function and SLL constraints, and the second sub-problem is clock constraints.

An ADMM-based timing-driven global placement algorithm is proposed in the work [9] for superconductive electronic circuits. By imposing hard constraints on the maximum interconnection length of all nets on the timing critical path of the circuit, the wirelength is minimized while seeking the placement of the target clock frequency. In the constructed ADMM, the timing-driven global placement problem breaks up into two sub-problems. One is to optimize the wirelength and reduce overlap, while another is to find a placement solution approaching the previous sub-problem and satisfying the path delay constraints.

### E. Non-Smooth Optimization

Different from other smoothing approximation methods, the work [10] proposes a model of $l_1$ norm wirelength that accurately calculates HPWL. When only 2 or 3 pins are connected to the net, the $l_1$ norm function is HPWL; In the case of $p_e$ ($p_e > 3$) pins connected, HPWL can also be more accurately approximated as a weighted $l_1$ norm function. $l_1$ norm wirelength model is represent as:

$$W_{l_1}(x,y) = \sum_{e \in E}(\|A_e x\|_1 + \|A_e y\|_1), \quad (16)$$

where $A_e = (\omega_{pg} a_{pg})$ is the weighted connection matrix of the hyperedge e. In addition, the overlap function is accurately calculated by the definition of the density control model, rather than a smooth approximation. For example, for bin b and cell i, let $d_x = |x_i - x_b|$, if $w_b > w_i$, then

$$\theta_x(b,i) = \begin{cases} w_i, & 0 \le d_x \le \frac{w_b - w_i}{2}; \\ \frac{w_b - 2d_x + w_i}{2}, & \frac{w_b - w_i}{2} < d_x < \frac{w_b + w_i}{2}; \\ 0, & \frac{w_b + w_i}{2} \le d_{x\circ} \end{cases} \quad (17)$$

Since the proposed model is non-smooth, the paper gives a subgradient that is easy to calculate.

### F. GPU Acceleration Method Based on Deep Learning

With the increasing scale of design, the complex constraints and objectives are two challenges faced by the semiconductor industry. The latest advance in GPU acceleration brings new opportunities for high-performance design automation. The work [11] equivalently converts the continuous optimization-based placement method of ePlace [2] into a training neural network, and proposes a GPU-accelerated placement framework DREAMPlace, in which the wirelength and density function are analogous to the prediction error and regular term of the neural network. In the neural network training, each data is identified by the feature vector $\tilde{x}_i$ and the label $\tilde{y}_i$, and a predicted label $\phi(\tilde{x}_i; w)$ would be obtained after being input into the neural network. The task of training is to adjust the weight w to minimize the total objective, where the objective is the sum of the prediction error of all training data and the regular term R(w). To compare the VLSI placement with neural network training, the work [11] integrates the cell location $(x, y)$ into w, each data $(\tilde{x}_i, \tilde{y}_i)$ is replaced with a network $e_i$ and a data instance $(e_i, 0)$ with a label of zero, and thus calculate the wirelength cost $WL(e_i; w)$. Since the wirelength is non-negative, the prediction error of the neural network is transformed into $\sum_{i=1}^{n} WL(e_i; w)$. Since the density cost D(w) has irrelevant with the network instance, it can correspond to the regular term R(w). In this way, the placement problem can be further solved according to the neural network training process, in which the forward propagation calculates the objective function, and the backward propagation calculates the gradient.

The widely used deep learning toolkits PyTorch and TensorFlow provide mature and effective GPU acceleration solutions. DREAMPlace [11] constructed a method to quickly calculate the wirelength and potential in forwarding propagation and quickly calculate the wirelength gradient and electric field in backward propagation, and implemented it on PyTorch. In the absence of quality degradation, DREAMPlace has speeded the global placement by more than 30 times, and the framework can complete one million cells in one minute. At the same time, the placement design of up to 10 million cells maintains almost linear scalability.

## Future Research Directions

### A. Multi-Objective Mixed Optimization Problem

The wirelength driven placement is a fundamental problem for the VLSI placement, of which each larger improvement for the algorithm can greatly advance the research on placement problem. Although it has been a mature problem, it is also necessary to consider many performance indexes such as routability, timing delay, and power. Therefore, when these factors are added to the wirelength driven placement, the placement problem is not only transformed into a multi-objective optimization problem, but also may mixed with continuous optimization, discrete optimization, and differential equation. Although many scholars have studied these important problems, from the view of the optimization theory and algorithm, they are still at the preliminary stage as a whole.

### B. Global Placement in Advanced Node Technologies

More complicated design rules and constraints are introduced in advanced node technologies to achieve higher performance, which makes the placement problem much more challenging. For example, in order to make a better trade-offs among routability, timing, power and performance, designs with mixed-row-height cells become popular in advanced node technologies. As feature size continues to shrink, placement with multi-VTs may violate the minimum-implant-area (MIA) layer rule arising from the limitations of patterning technologies. Further, a drain-to-drain abutment (DDA) constraint has emerged as a new challenge for the FinFET technology. In addition, leveraging the non-integer multiple-height (NIMH) cells entirely can give more room to optimize PPA, providing larger solution space and so better QoR. Therefore, developing an effective method to handle the global placement with several constraints (e.g., power-rail alignment, routability, and timing) in advanced node technologies is necessary.

### C. GPU Acceleration

In an analytical-based global placement, high-quality solutions can be obtained using optimization algorithms, but they are usually more time-consuming than heuristic methods. With the rapid increase in the scale of the placement problem, the runtime has also increased dramatically. GPU acceleration is a common choice for processing big data, and it also has some applications in global placement issues [11]. It is further necessary to consider the wide application of GPU acceleration in global placement algorithms.

## Conclusion

Many algorithms have been developed to solve the VLSI global placement problem, in which the continuous optimization-based algorithm is the current mainstream algorithm. However, many studies indicate that there is a large gap between the solution solved with the existing algorithm and the optimal solution. Therefore, there is a certain space for studying the placement problem of VLSI circuits.

## References

[1] Chen, T.-C., Jiang, Z.-W., Hsu, T.-C., Chen, H.-C., and Chang, Y.-W. NTUplace3: An analytical placer for large-scale mixed-size designs with preplaced blocks and density constraints. IEEE Transactions on Computer-Aided Design of Integrated Circuits and Systems, 2008, 27(7):1228-1240.

[2] Lu, J., Chen, P., Chang, C.-C., Sha, L., Huang, D. J. H., Teng, C.-C., and Cheng, C.-K. ePlace: Electrostatics-based placement using fast fourier transform and Nesterov's method. ACM Transactions on Design Automation of Electronic Systems, 2015, 20(2):1-34.

[3] Zhu, W., Chen, J., and Li, W. An augmented Lagrangian method for VLSI global placement. The Journal of Supercomputing, 2014, 69(2):714-738.

[4] Zhu, Z., Chen, J., Peng, Z., Zhu, W., and Chang, Y.-W. Generalized augmented lagrangian and its applications to VLSI global placement. In 2018 55th ACM/ESDA/IEEE Design Automation Conference, 2018, 1-6.

[5] Li, W., Lin, Y., and Pan, D. Z. elfPlace: Electrostatics-based placement for large-scale heterogeneous fpgas. In 2019 IEEE/ACM International Conference on Computer-Aided Design, 2019, 1-8.

[6] Peng, Z., Chen, J., and Zhu, W. A proximal alternating direction method of multipliers for a minimization problem with nonconvex constraints. Journal of Global Optimization, 2015, 62(4):711-728.

[7] Chen, J., Yang, L., Peng, Z., Zhu, W., and Chang, Y.-W. Novel proximal group ADMM for placement considering fogging and proximity effects. In 2018 IEEE/ACM International Conference on Computer-Aided Design, 2018, 1-7.

[8] Chen, J., Zhu, W., Yu, J., He, L., and Chang, Y.-W. Analytical Placement with 3D Poisson's Equation and ADMM Based Optimization for Large-Scale 2.5 D Heterogeneous FPGAs. In 2019 IEEE/ACM International Conference on Computer-Aided Design, 2019, 1-8.

[9] Shahsavani, S. N., and Pedram, M. TDP-ADMM: a timing driven placement approach for superconductive electronic circuits using alternating direction method of multipliers. In 2020 57th ACM/IEEE Design Automation Conference, 2020, 1-6.

[10] Zhu, W., Chen, J., Peng, Z., and Fan, G. Nonsmooth optimization method for VLSI global placement. IEEE Transactions on Computer-Aided Design of Integrated Circuits and Systems, 2015, 34(4):642-655.

978-1-6654-9759-6/22 $31.00 © 2022 IEEE

[11] Lin, Y., Jiang, Z., Gu, J., Li, W., Dhar, S., Ren, H., Khailany, B., and Pan, D. Z. Dreamplace: Deep learning toolkit-enabled gpu acceleration for modern vlsi placement. IEEE Transactions on Computer-Aided Design of Integrated Circuits and Systems, 2020, 40(4):748-761.

# Area-aware optimization of XOR-AND Graph based on Reed-Muller logic expansion

Hongwei Zhou[*], Yong Xiao[†], Yun Shao[†] and Zhufei Chu[*‡]

[*]EECS, Ningbo University, Ningbo 315211, China
[†]Giga Design Automation Co. Ltd., Shenzhen 518055, China
[‡]Email: chuzhufei@nbu.edu.cn

## ABSTRACT

XOR-AND Graph (XAG) is a logic representation method for Boolean functions. A major advantage of using XAG is that it is efficient for cryptography and security applications in which the number of AND gates relates to the multiplicative complexity. A Reed-Muller logic expansion is used in this paper to develop an area-aware optimization method for XAG. The logic functions can be represented in several ways by Reed-Muller logic based on different polarities. To find an area-efficient representation of the logic network, the cuts of the logic network are enumerated and then the mixed polarity conversion algorithm is utilized. When the representation is improved over the existing one, we replace the existing cut. Following the experimental results over EPFL benchmark suites, the proposed method can optimize XAG size. Compared with the best known results, we achieve a reduction of up to 50% and an average reduction of 14% in the number of XOR gates.

## INTRODUCTION

Logic synthesis is one step of the modern *electronics design automation* (EDA) flow that is considered one of the key steps in the realization of competitive and cutting-edge integrated circuits. Traditional logic synthesis data structures are designed for CMOS applications, which have logic abstraction based on primitives {AND, OR, NOT}. Reed-Muller (RM) logic, which only uses AND and exclusive-OR (XOR) operations, is another circuit expression that differs from Boolean logic. RM representation may be more efficient for circuits like arithmetic circuits, parity check circuits, and communication circuits in terms of power, area, speed, and testability than their Boolean representation [1]. In recent years, the work in [2] has extended logic synthesis into a new field addressing cryptography and security, which uses XOR-AND Graph (XAG) as the base for the optimization, because it efficiently abstracts cryptography circuits over {AND, XOR, NOT}.

Rewriting can reduce the number of logic nodes. As an example, in [3], a structural rewriting method was proposed for the XOR-Majority graph (XMG), which allows structural rewriting and optimization of XMG network area and logic depth. Because RM logic is based

on AND/XOR primitives, the RM logic optimization algorithm could be used for XAG rewriting. A fast minimization algorithm has been developed for fixed-polarity Reed-Muller (FPRM) expressions in [1], which can also be extended for mixed-polarity Reed-Muller (MPRM) expressions. However, the method is not applied to logic networks. In this paper, we propose a method for area-aware XAG optimization that uses RM logic expansion. The experimental results over EPFL benchmark suites show the number of XOR gate is reduced up to 50% compared with the best-known results.

## BACKGROUND

### Reed-Muller logical function expressions

For a logic function $f(x)$, we can use the Shannon expansion to obtain the RM logic expansion based on the XOR/AND form, as shown in (1).

$$
\begin{aligned}
f(x_1, x_2...x_n) &= \bar{x}_1 f(0, x_2, ..., x_n) + x_1 f(1, x_2, ..., x_n) \\
&= (1 \oplus x_1) f(0, x_2, ..., x_n) \oplus x_1 f(1, x_2, ..., x_n) \\
&= f(0, x_2, ..., x_n) \oplus x_1 [f(0, x_2, ..., x_n) \\
&\quad \oplus f(1, x_2, ..., x_n)] \\
&= f(0, x_2, ..., x_n) \oplus x_1 f_1(x_2, ..., x_n)
\end{aligned}
\tag{1}
$$

where $f_1(x_2...x_n)$ is defined as $f(0, x_2, ..., x_n) \oplus f(1, x_2, ..., x_n)$. By continuously iterating the Shannon expansion on the remaining functions, we can obtain (2).

$$
f(x) = \oplus \sum_{j=0}^{2^n - 1} b_j p_j
\tag{2}
$$

where the number of input variables of $f(x)$ is $n$, $b_j$ belongs to $\{0, 1\}$, and $p_j$ is a product term composed of variables.

### Polarities of Reed-Muller logic

RM logic expressions are divided into fixed polarity logic expressions and mixed polarity logic expressions.

The logic function $f(x_{n-1}, x_{n-2}, ..., x_0)$ with $n$ input variables has three RM expansion forms, as shown in (3), (4), (5), where substituting $\bar{x}_k = 1 \oplus x_k$ and $x_k = 1 \oplus \bar{x}_k$ into (3) yields (4) and (5), respectively.

If the logic functions are expanded according to (4) and (5), $2^n$ different FPRM expressions can be obtained.

978-1-6654-9759-6/22 $31.00 © 2022 IEEE

Similarly, if the logic functions are expanded according to (3), (4), and (5), $3^n$ different MPRM expressions can be obtained. Hence it follows that fixed polarity means that the variables in the expression can only appear in one of the original variables or the complemented variables. In contrast, mixed polarity means that the variables in the expression can appear in both the original and complemented variables. Therefore, the MPRM representation of the same logic function is more likely to yield the compact RM logic function compared to the FPRM representation.

$$f(x_{n-1}, x_{n-2}, ..., x_0) = \bar{x}_k f(x_{n-1}, ..., x_{k+1}, 0, x_{k-1}, ..., x_0)$$
$$\oplus x_k f(x_{n-1}, ..., x_{k+1}, 1, x_{k-1}, ..., x_0) \quad (3)$$

$$f(x_{n-1}, x_{n-2}, ... x_0) = f(x_{n-1}, ..., x_{k+1}, 0, x_{k-1}, ..., x_0)$$
$$\oplus x_k f(x_{n-1}, ..., x_{k+1}, 0, x_{k-1}, ..., x_0)$$
$$\oplus f(x_{n-1}, ..., x_{k+1}, 1, x_{k-1}, ..., x_0) \quad (4)$$

$$f(x_{n-1}, x_{n-2}, ..., x_0) = f(x_{n-1}, ..., x_{k+1}, 1, x_{k-1}, ..., x_0)$$
$$\oplus \bar{x}_k f(x_{n-1}, ..., x_{k+1}, 0, x_{k-1}, ..., x_0)$$
$$\oplus f(x_{n-1}, ..., x_{k+1}, 1, x_{k-1}, ..., x_0) \quad (5)$$

### Cuts of logic network

A large-scale logic network can be partitioned into cuts or windows. In this paper, we adopt the cuts and cut enumeration technique for RM logic rewriting. The cut of node $X$ is a sub-logical network of $n$ nodes, each input of this sub-logical network is called a leaf node, and every path from *primary input* (PI) of the logical network to node $X$ will pass through at least one leaf node. Node $X$ is called the root node of this cut set. If a cut set is itself, we call it a trivial cut set. A cut set is said to be $k$-viable if it has no more than $k$ leaf nodes. If all nodes of a cut set are contained within another cut set [4], we say the cut set is covered.

## PROPOSED DESIGNS

### Polarity conversion algorithm

It is possible to obtain RM expressions in two ways: one is to use a Boolean function through polarity conversion algorithm, the other is to use RM expressions with known polarity through polarity conversion algorithm. The tabular technique is the most widely used polarity conversion algorithm because it is simple to use and easy to implement using computer language [5]. Our paper also uses the tabular technique to perform the polarity transformation.

Polarities are indicated by numbers. The number '2' indicates the variable can appear as either the original or complemented one; while the number '1' indicates the variable can only appear as the complemented one

and the number '0' shows the variable must be shown as the original one. Given a logic expression represented in minterms

$$f = \bar{A}\bar{B}\bar{C} + \bar{A}B\bar{C} + A\bar{B}\bar{C} + ABC, \quad (6)$$

if we specify a polarity '210', which is corresponded to Boolean variables $A$, $B$, and $C$, respectively, one possible MPRM expression through tabular technique polarity conversion is

$$f = \bar{A} \oplus \bar{A}C \oplus AC \oplus A\bar{B}. \quad (7)$$

One can check both $A$ and $\bar{A}$ appeared in (7), and $B$ appears in complemented form while $C$ is shown in original form. Note that different polarity configuration leads to distinct MRPM logic expression and thus a varied cost in the number of logic gates.

### The proposed algorithm for area optimization

The logic function of $n$ input variables has $3^n$ different mixed polarities. Each polarity corresponds to a different MPRM expression, and the number of AND and XOR terms corresponding to each is also different. The area optimization of MPRM circuits is to minimize the number of ANDs and XORs by finding the optimal polarity.

Traversing all polarities will allow us to discover which MPRM logic circuit has the best area using the enumeration method. We propose an area optimization algorithm combining the tabular technique with the cuts technique, following the enumeration method to optimize mixed-polarity RM logic circuits. The proposed optimization algorithm involves finding an area-efficient representation of the logic network, enumerating the cuts of the logic network, and then using the tabular method to find the optimal representation. If the representation is improved over the existing one, we replace the existing cut.

The outline of the enumeration-based algorithm for area optimization is shown in Algorithm 1. Given a benchmark circuit, which is stored as a XAG network $N$ after a series of transformations, the algorithm will return a resynthesized logic network with a reduced number of logic nodes $N'$. We iterate through each cut to identify opportunities for optimization (line 3). We use the function to obtain truth tables $str$ of the cut network (line 4). The $minterm$ of this cut is obtained from the truth table (line 5). Then, we search for the optimal mixed polarity and obtain the optimal mixed polarity $optimal\_P$ (line 6). We can combine the minterm and the optimal mixed polarity to obtain the optimal RM product term set $product$ (line 7). Our next step is to calculate the number of gates in the optimal RM product term set obtained at line 8 ($node\_num$) and the number

---

978-1-6654-9759-6/22 $31.00 © 2022 IEEE

**Algorithm 1:** the enumeration-based algorithm for area optimization

---

**Input** : Logic network $N$
**Output** : Resynthesized logic network $N'$

1  $cuts \leftarrow$ cut_enumeration($N$);
2  **for** each node $n$ in $N$ **do**
3    **for** auto const& $cut$ : $cuts$ **do**
4      $str \leftarrow$ compute_truth_tables($cut$);
5      $minterm \leftarrow$ get_minterm($str$);
6      $optimal\_P \leftarrow$ search($minterm, polarity$)
7      $product \leftarrow$ conversion($minterm$,    $optimal\_P$)
8      $node\_num \leftarrow$ count($optimal\_P, product$)
9      $old\_node\_num \leftarrow$ cut_view($cut$)
10     **if** $node\_num < old\_node\_num$ **then**
11      $opt \leftarrow$ creat($optimal\_P, product$);
12      substitute_network($opt,n$);
13     **end**
14    **end**
15 **end**
16 **return** $N'$ = network-cleanup-and-sweeping($N$)

---

TABLE I. Experimental Results of XAG-Size Optimization

| Benchmark | PI/PO | Initial [7] | | Our Method | | |
|---|---|---|---|---|---|---|
| | | #AND | #XOR | #AND | #XOR | impr. |
| cavlc | 10/11 | 494 | 197 | 494 | 137 | 30.46% |
| ctrl | 7/26 | 85 | 8 | 85 | 4 | 50.00% |
| i2c | 147/142 | 623 | 502 | 623 | 302 | 39.84% |
| int2float | 11/7 | 100 | 101 | 100 | 66 | 34.65% |
| log2 | 32/32 | 19436 | 9371 | 19436 | 8214 | 12.35% |
| mem_ctrl | 1204/1231 | 5113 | 4168 | 5113 | 2949 | 29.25% |
| multiplier | 128/128 | 11940 | 8614 | 11941 | 8007 | 7.05% |
| sin | 24/25 | 4075 | 1770 | 4075 | 1428 | 19.32% |
| sqrt | 128/64 | 6244 | 9640 | 6244 | 9104 | 5.56% |
| voter | 1001/1 | 5651 | 6066 | 5651 | 4757 | 21.58% |
| Sum | | 53761 | 40437 | 53762 | 34968 | |
| Ratio | | 1 | 1 | 1 | 0.86 | |

of nodes in the original cut at line 9 ($old\_node\_num$). After the representation is improved, we will create a new subnetwork $opt$ based on the expression (line 11) and replace the existing cut (line 12). An optimized network $N'$ is obtained once all cuts have been traversed.

## EXPERIMENTAL RESULT

We implemented our approach in C++ as a command called `rm_mixed_polarity` on top of the logic synthesis framework ALSO[1]. The benchmarks considered are general combinational circuits from MCNC. Our results are verified by `cec` command in ABC[2] to ensure functional correctness [6].

The experimental results are shown in Table I, where the "Benchmarks" column lists the benchmark name, "PI/PO" indicates the number of primary inputs/outputs. As a baseline, we use the best-known results presented in [7], that are obtained applying the rewriting algorithm until convergence of the results is achieved.

In Table I, we separate AND and XOR for comparison. In this paper, we propose an algorithm that first optimizes the number of ANDs and then optimizes the number of XORs. On the basis of the experimental data in this paper, the algorithm proposed in this paper has no optimization for the number of ANDs, but a great optimization for the number of XORs, which improves the number of XOR gates up to 50%, with an average 0f 14% optimization.

[1] https://github.com/nbulsi/also
[2] https://github.com/berkeley-abc/abc

## SUMMARY

This paper proposes an approach to area optimization in XOR-AND Graphs using Reed-Muller logic, which uses Reed-Muller logic with the nature of the network to optimize the number of gate nodes in the network. The optimization potentials in local subcircuits are explored using a fully automatic approach utilizing cut enumeration algorithms. Using EPFL benchmarks, we improved the best-known numbers for EPFL by up to 50%, resulting in an average of 14% optimization.

## ACKNOWLEDGEMENT

This work was supported in part by the NSFC under Grant 61871242 and in part by the State Key Laboratory of ASIC & System under Grant 2021KF008.

## REFERENCES

[1] Z. He, L. Xiao, Z. Huo, T. Wang, and X. Wang, "Fast minimization of fixed polarity Reed-Muller expressions," *IEEE Access*, vol. 7, pp. 24 843–24 851, 2019.

[2] E. Testa, M. Soeken, H. Riener, L. Amaru, and G. D. Micheli, "A logic synthesis toolbox for reducing the multiplicative complexity in logic networks," in *2020 Design, Automation Test in Europe Conference Exhibition (DATE)*, 2020, pp. 568–573.

[3] Z. Chu, M. Soeken, Y. Xia, L. Wang, and G. De Micheli, "Structural rewriting in XOR-majority graphs," in *Proceedings of the 24th Asia and South Pacific Design Automation Conference*, 2019, pp. 663–668.

[4] A. Mishchenko, S. Chatterjee, and R. K. Brayton, "Improvements to technology mapping for LUT-based FPGAs," *IEEE Transactions on Computer-Aided Design of Integrated Circuits and Systems*, vol. 26, no. 2, pp. 240–253, 2007.

[5] A. Almaini, P. Thomson, and D. Hanson, "Tabular techniques for Reed—Muller logic," *International Journal of Electronics*, vol. 70, no. 1, pp. 23–34, 1991.

[6] R. Brayton and A. Mishchenko, "ABC: An academic industrial-strength verification tool," in *International Conference on Computer Aided Verification*. Springer, 2010, pp. 24–40.

[7] E. Testa, M. Soeken, L. Amarù, and G. D. Micheli, "Reducing the multiplicative complexity in logic networks for cryptography and security applications," in *2019 56th ACM/IEEE Design Automation Conference (DAC)*, 2019, pp. 1–6.

978-1-6654-9759-6/22 $31.00 © 2022 IEEE

# POLYNOMIAL FORMAL VERIFICATION OF GENERAL TREE-LIKE CIRCUITS

*Alireza Mahzoon[1] and Rolf Drechsler[1,2]*

[1] Institute of Computer Science, University of Bremen, Bremen, Germany
[2] Cyber-Physical Systems, DFKI GmbH, Bremen, Germany
{mahzoon,drechsle}@informatik.uni-bremen.de

## ABSTRACT

In recent years, the size and complexity of digital circuits have grown drastically. Consequently, bugs may appear in a circuit in different phases of design and synthesis. If they remain undetected, they propagate to the physical chip and cause huge financial loss for the designers and manufacturers. Formal verification is an important task after design to ensure the correctness of a circuit. Recently, polynomial formal verification has gained special attention. If we prove that the space and time complexity of a verification method is polynomial, we can ensure that it is always scalable.

Many digital circuits which are used in different applications have a tree-like structure. It has been proven that if a tree-like circuit is made of basic logic gates (AND, OR, NOT, NAND, and NOR), it can be verified polynomially using BDDs. However, these proofs cannot be extended to the general tree-like circuits containing XOR gates. In this paper, we propose a method to prove the correctness of a general tree-like circuit in polynomial time.

## INTRODUCTION

Bugs might appear in a circuit in different phases of design, optimization, or synthesis. They can even propagate to the physical chip in the lack of a proper method to detect them. Faulty chips cause huge financial losses for the designers, and using them might have disastrous consequences. Formal verification is an important task after the design to ensure the correctness of a digital circuit. Formal verification methods take advantage of rigorous mathematical reasoning to prove that a design meets all or parts of its specification [1].

In the last 30 years, formal verification methods have achieved many successes in proving the correctness of digital circuits. *Binary Decision Diagrams* (BDDs) [2,3] report very good results when it comes to the verification of adders and subtractors. Combinational equivalence checking using *Boolean Satisfiability* (SAT) [4,5] is used to verify adders as well as to prove the correctness of circuits after optimization. *Binary Moment Diagrams* (*BMD and K*BMD) and *Symbolic Computer Algebra* (SCA) are employed to verify multipliers and dividers [6,7,8,9].

Despite the success of formal verification methods in proving the correctness of a large variety of circuits, their space and time complexity is not fully investigated. Knowing the complexity of a formal verification method brings us several advantages: (1) before applying the method, we can realize whether it is scalable for a specific design, (2) we can estimate the required run-time and memory for the verification process, and (3) we can compare the complexity of two different methods for verifying a circuit and choose the best one.

BDD-based verification is used for many years to prove the correctness of digital circuits. The core concept of this verification method is based on symbolic simulation. First, the inputs of the circuit are encoded as BDDs. Then, starting from the *Primary Inputs* (PIs), the BDDs for the outputs of gates/components are computed. This process continues until the BDDs for the *Primary Outputs* (POs) are obtained. Finally, the output BDDs are evaluated to see whether they match the reference BDDs. If yes, the circuit is correct; otherwise, it is buggy.

Recently, some research works have focused on polynomial BDD-based verification. PolyAdd [10] proves that polynomial BDD-based verification of three adder architectures; i.e., ripple carry adder, conditional sum adder, and carry look-ahead adder is possible. The verification complexity bounds for the conditional sum adder and prefix adders are calculated in [11] and [12], respectively. The authors of [13] show that a simple ALU consisting of basic logic and arithmetic operations can be verified in polynomial time using BDD-based verification. The polynomial BDD construction of symmetric functions is investigated in [14].

A big group of digital circuits whose polynomial verification has not been yet fully investigated is tree-like circuits. These circuits do not have any fanouts. This special property is employed for the efficient representation and verification of the circuits. This paper focuses on the polynomial formal verification of general tree-like circuits. It has been proven in [15] that polynomial BDD-based verification of tree-like circuits is possible if they are only made of basic logic gates (i.e., AND, OR, NOT, NAND, and NOR). However, this proof cannot be extended to general tree-like circuits containing XOR gates. We first prove that the verification of a general tree-like circuit is exponential using only BDDs. Then, we come up with a method based on partitioning to prove the correctness in polynomial time using *Kronecker Functional Decision Diagrams* (KFDDs).

---

978-1-6654-9759-6/22 $31.00 © 2022 IEEE

## PRELIMINARIES

In this section, we first introduce tree-like circuits. Then, we give a brief overview of BDD and KFDD representations.

### Tree-like Circuits

Combinational circuits consist of many different designs with a wide variety of structures. In order to study these circuits, we usually categorize them based on their function or structure.

Tree-like circuits are a big group of designs with a common feature: they do not have any fanouts in their structure. In a general tree-like circuit, the logic gates, including AND, OR, NOT, NAND, NOR, and XOR are used.

### BDDs and KFDDs

A BDD is a directed, acyclic graph whose nodes have two edges associated with the values of the variables 0 and 1. A BDD contains two terminal nodes (leaves) that are associated with the values of the function 0 or 1. A BDD is called Reduced Ordered BDD (ROBDD) if the variables occur in the same order in each path from the root to a leaf, and it contains a minimum number of nodes for a given variable order. The ROBDD of a Boolean function is always unique. In the rest of the paper, we simply refer to ROBDDs as BDDs.

In a BDD, each node is expanded by Shannon decomposition:

$$f = \bar{x}_i f_i^0 + x_i f_i^1, \tag{1}$$

where $f_i^0$ ($f_i^1$) denotes the cofactor of $f$ with respect to $x_i = 1$ ($x_i = 0$). Moreover, $f_i^0$ ($f_i^1$) creates the low (high) branches of the node. The BDD representation of the function $f = x_1 \wedge (x_2 \vee x_3)$ is depicted in Fig. 1.

A BDD only takes advantage of Shannon decomposition to expand nodes. However, it is also possible to use other decompositions such as positive Davio and negative Davio:

$$f = f_i^0 \oplus x_i f_i^2, \tag{2}$$

$$f = f_i^1 \oplus \bar{x}_i f_i^2, \tag{3}$$

where $\oplus$ is XOR, and $f_i^2$ is defined as $f_i^2 = f_i^0 \oplus f_i^1$.

A KFDD takes advantage of three decompositions to represent a Boolean function, i.e., each node in a KFDD is expanded by *Shannon (S)*, *positive Davio (pD)*, or *negative Davio (nD)* [16]. It has been shown that Shannon decomposition results in smaller graphs for some Boolean functions such as $f = x_1 \wedge (x_2 \vee x_3)$. On the other hand, positive Davio and negative Davio are more efficient for representing functions like $g = x_1 \oplus x_2 \oplus x_3$. Fig. 2 shows the KFDD representations of function $g = x_1 \oplus x_2 \oplus x_3$. In Fig. 2(a) and Fig. 2(b), Shannon decomposition and

positive Davio are used to expand nodes, respectively. It is evident that KFDD with Shannon decomposition has fewer nodes for function $g$. The methods for the construction of small KFDDs are presented in [17]. The authors develop different heuristics to find good variable orderings and decomposition types.

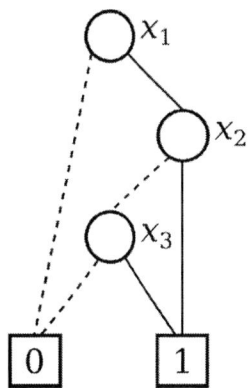

Fig. 1. BDD representation of $x_1 \wedge (x_2 \vee x_3)$

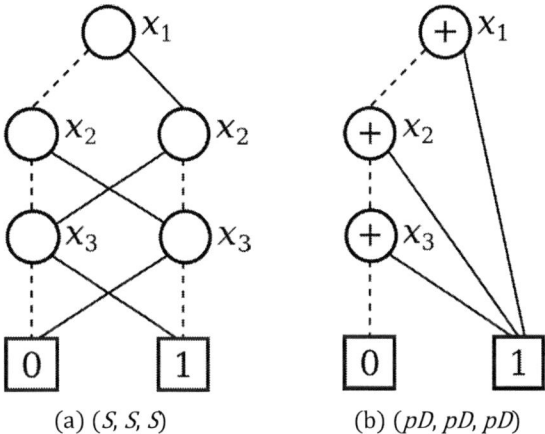

(a) (S, S, S)        (b) (pD, pD, pD)

Fig. 2. KFDD representation of $x_1 \oplus x_2 \oplus x_3$

Fig. 3 presents the algorithm for XOR operation between two KFDDs $F$ and $G$. The algorithm should be run recursively to obtain the result. It has a worst-case complexity of $|F| \cdot |G|$, where $|F|$ and $|G|$ are the sizes of the graphs. However, if one of the terminal cases happens, the algorithm is terminated immediately. An example of a terminal case is as follows:

$$\text{if } (F == 0) \text{ return } G. \tag{4}$$

It has been also shown that a KFDD might have an exponential worst-case complexity for some operations such as AND. It is possible to avoid the exponential complexities during symbolic simulation by confining the decompositions based on the logic gates in a circuit.

978-1-6654-9759-6/22 $31.00 © 2022 IEEE

**Input:** $F$, $G$ KFDDs

**Output:** $F \oplus G$ KFDD

**if** terminal case **then**

    **return** result

**else if** computed-table has entry $(F, G)$ **then**

    **return** result

**else**

    $v$ = top variable for $(F, G)$

    $low(v) = F_{low(v)} \oplus G_{low(v)}$

    $high(v) = F_{high(v)} \oplus G_{high(v)}$

    **if** $Shannon(v)$ **then**

        **if** $high(v) == low(v)$ **then**

            **return** $low(v)$

        **else**

            **if** $high(v) == 0$ **then**

                **return** $low(v)$

    $R = find\_or\_add\_unique\_table(v, low(v), high(v))$

    $insert\_computed\_table(F, G, R)$

    **return** $R$

Fig. 3. Algorithm for XOR operation

## POLYNOMIAL VERIFICATION

In this section, we first show that verification of a general tree-like circuit has a worst-case exponential complexity using BDDs. Then, we propose a method based on partitioning to prove the correctness of a general tree-like circuit in polynomial time.

### Verification Using BDDs

It has been proven in [15] that the time complexity of verifying a tree-like circuit consisting of only basic logic gates (i.e., AND, OR, NOT, NAND, and NOR) is polynomial. The proof is based on the fact that the size of the BDD representing an intermediate signal is always equal to the number of PIs that affect the value of the signal; i.e., there is only one node for each PI. As a result, since the BDDs of the internal signals are polynomially bounded, the symbolic simulation can be carried out in polynomial space and time.

Unfortunately, for a general tree-like circuit containing XOR gates, the story is different. In a general tree-like circuit, there are potentially multiple nodes for a PI. Based on the structure of the circuit, the size of BDDs can grow exponentially during the symbolic simulation. Since the complexity of each operation depends on the size of intermediate BDDs, the space and time complexity become exponential. We can conclude that using only BDDs (or equivalently using KFDDs with only Shannon decomposition) is not a scalable and efficient approach for the verification of general tree-like circuits.

### Polynomial Verification Using Partitioning

We now propose a method based on circuit partitioning to divide the general tree-like circuit into two smaller parts that are verifiable in polynomial time. Our proposed method consists of three main phases:

1. The circuit is partitioned based on the XOR gates. The XOR gates which are connected to each other are grouped as a partition. We collect all these groups and call them an XOR partition. The rest of the circuit which does not have any XOR gate is called a basic partition. Please note that the outputs of the XOR partition are the new PIs for the basic partition.
2. The basic partition which is only made of basic logic gates is verified using KFDDs with only Shannon decomposition (or equivalently BDDs).
3. The XOR partition is verified using KFDDs with positive Davio decomposition.

In our proposed method, in addition to the output KFDDs, the correct (reference) KFDDs on the boundaries of the partitions have to be available. Our method can be particularly used for the verification of the tree-like circuit in which the XOR gates are concentrated in a part of the circuit as e.g. chain of XOR gates.

In a tree-like circuit, a subset of the circuit is also a tree-like circuit. Thus, the basic and XOR partitions are tree-like circuits as well. In order to prove that the polynomial formal verification of a general tree-like circuit is possible, we have to prove that the basic and XOR partitions are verifiable in polynomial time.

**Theorem 1.** The symbolic simulation of a tree-like circuit consisting of basic logic gates has a polynomial time complexity if only Shannon decomposition is used for expanding KFDDs.

KFDDs whose nodes are expanded using Shannon decomposition are BDDs. It has been proven in [15] that symbolic simulation using BDDs has a polynomial time complexity for the tree-like circuits which are only made of basic logic gates. The proof is based on the complexity of the IF-Then-Else (ITE) operation and the structure of BDDs in a tree-like circuit.

We now investigate the time complexity of symbolic simulation for a tree-like circuit consisting of XOR gates: The size of a KFDD representing the output of an XOR tree-like circuit is equal to the number of PIs if only positive Davio decomposition is used. In other words, for each PI, there is only one node. However, it has to be also proven that the size of KFDDs is polynomially bounded during the construction.

978-1-6654-9759-6/22 $31.00 © 2022 IEEE

**Theorem 2.** Let $g$ be a KFDD, representing a circuit, and $x_i$ be a variable that does not occur in $g$. Then $f = x_i \oplus g$ can be constructed in $O(|g|)$, if $x_i$ is the top-variable in the ordering of $f$.

*proof.* In the positive Davio decomposition of $f$, the functions for the high and low branches are obtained based on the algorithm in Fig. 3 as follows:

$$f_{low} = (x_i = 0) \oplus g = g, \tag{5}$$

$$f_{high} = [(x_i = 0) \oplus (x_i = 1)] \oplus [g \oplus g] = 1. \tag{6}$$

Thus, for the case of $f_{high} = 1$, in the recursive call of the algorithm in Fig. 3, a terminal case is reached, immediately. For the low edge branching, the KFDD of $g$ has to be traversed but no new nodes are generated. Thus, the time complexity is $O(|g|)$.

The same argument also holds for the XOR gates with more than two inputs, e.g. 4-input XOR gates. The gates have to be decomposed into 2-input gates and then Theorem 2 is applied.

Theorem 2 indicates that the symbolic simulation of an XOR tree-like circuit has a polynomial time complexity if the positive Davio decomposition is employed for the expansion of the nodes in KFDD. Since symbolic simulation can be carried out polynomially for both basic and XOR partitions of a general tree-like circuit, the polynomial formal verification of the circuit becomes possible.

## CONCLUSION

In this paper, we investigated the polynomial formal verification of general tree-like circuits. We proved that pure BDD-based verification has exponential worst-case complexity. Then, we came up with a partitioning technique to divide a general tree-like circuit into polynomially verifiable parts. Thus, the correctness of each part is proved independently using KFDD in polynomial time.

In our future research, we focus on the polynomial formal verification of other architectures. Specifically, we plan to extend [18] for the polynomial formal verification of complex arithmetic circuits using Symbolic Computer Algebra (SCA).

## ACKNOWLEDGEMENTS

This work was supported by the German Research Foundation (DFG) within the Reinhart Koselleck Project *PolyVer: Polynomial Verification of Electronic Circuits* (DR 287/36-1).

## REFERENCES

[1] R. Drechsler. Advanced Formal Verification. Kluwer Academic Publishers, 2004.

[2] R.E. Bryant. Binary decision diagrams and beyond: enabling technologies for formal verification. ICCAD, 1995, pp. 236-243.

[3] R. Drechsler and D. Sieling. Binary decision diagrams in theory and practice. Int. J. Softw. Tools Technol. Transf. vol. 3(2), 2001, pp. 112–136.

[4] A. Kuehlmann and F. Krohm. Equivalence checking using cuts and heaps. DAC, 1997, pp. 263–268.

[5] A. Kuehlmann, V. Paruthi, F. Krohm, and M.K. Ganai. Robust boolean reasoning for equivalence checking and functional property verification. TCAD, vol. 21(12), 2002, pp. 1377–1394.

[6] A. Mahzoon, D. Große, and R. Drechsler. RevSCA-2.0: SCA-based formal verification of non-trivial multipliers using reverse engineering and local vanishing removal. TCAD, 2021.

[7] A. Mahzoon, D. Große, C. Scholl, and R. Drechsler. Towards formal verification of optimized and industrial multipliers. DATE, 2020, pp. 544–549.

[8] D. Kaufmann, A. Biere, and M. Kauers. Verifying large multipliers by combining SAT and computer algebra. FMCAD, 2019, pp. 28–36.

[9] C. Scholl, A. Konrad, A. Mahzoon, D. Große, and R. Drechsler. Verifying dividers using symbolic computer algebra and don't care optimization. DATE, 2021, pp. 1110-1115.

[10] R. Drechsler. PolyAdd: Polynomial formal verification of adder circuits. DDECS, 2021, pp. 99–104.

[11] A. Mahzoon and R. Drechsler. Late breaking results: Polynomial formal verification of fast adders. DAC, 2021, pp. 1376-1377.

[12] A. Mahzoon and R. Drechsler. Polynomial Formal Verification of Prefix Adders. ATS, 2021.

[13] R. Drechsler, A. Mahzoon, and L. Weingarten. Polynomial Formal Verification of Arithmetic Circuits. ICCIDE, 2021.

[14] R. Drechsler and C. Dominik. Edge verification: Ensuring correctness under resource constraints. SBCCI, 2021.

[15] R. Drechsler. Polynomial circuit verification using BDDs. arXiv:2104.03024, 2021.

[16] R. Drechsler and B. Becker. Ordered Kronecker functional decision diagrams - a data structure for representation and manipulation of Boolean functions. TCAD, vol. 17(10), 1998, pp. 965-973.

[17] R. Drechsler, B. Becker, and A. Jahnke. On Variable Ordering and Decomposition Type Choice in OKFDDs. TC, vol. 47(12), 1998, pp. 1398-1403.

[18] M. Barhoush, A. Mahzoon, and R. Drechsler. Polynomial word-level verification of arithmetic circuits. MEMOCODE, 2021.

978-1-6654-9759-6/22 $31.00 © 2022 IEEE

# AN FPGA-BASED VERIFICATION PLATFORM FOR HIGH-SPEED INTERFACE IPs

*C.-Z. Chen[1,2,*], Xuhui Liu[1], and Hanming Wu[1,2]*

[1]Peng Cheng Laboratory, Shenzhen 518000, China

[2] School of Micro-Nano Electronics, Zhejiang University, Hangzhou 311200, China

*Corresponding Author's Email: chenchzh@pcl.ac.cn, czchen126@126.com

## ABSTRACT

Multiple CMOS ICs for die-2-die (D2D, or chip-2-chip, C2C) integration, typically using 65nm to 7nm processes, are assembled in today's SoC designs. The interconnection between CPU and high-speed interface (I/F) IPs SerDes and memory (DDR or Flash memory), are through HBM, 2.5D and/or SiP following PCIe 4.0 or NVMe protocols. For high-data rate at 100 Gbps (25Gbps x 4 channels) or above communications, D2D is including silicon to photonics (SiPh) ICs, progressing to chiplet technology, to enable heterogeneous CMOS and SiPh integration possible. In either or both of D2D and/or CMOS (si-to-si) to SiPh substrate, data transmissions of the chiplet intra-connection are to be tested, verified and validated. An effective verification FPGA-based platform fit to serve this purpose. This study proposes a customized FPGA platform, for the verification of high-speed I/F IPs, at 100 Gbps and opt to 400 Gbps (100 Gbps x 4 channels), its output to a network interface card (NIC, Ethernet 4.0) is tenable to a photonics module via RX/receiver IC and Tx/driver IC of a PAM4-based SerDes. The performance and features of data transferring parameters, such as Bit Error Rate (BER), bandwidth and latency can be measured and validated for proposed applications.

**Keywords**: I/F IP, DDR4, SerDes, PCIe 4.0, D2D, SiP, chiplet, PAM-4, BER

## INTRODUCTION

### Background of Interface IP

Increased demand on interface IP (I/F IP) has shown it a fast growing sector due to big data and high data-rate applications in such as data center, automotive, and wireless communications. Study in recent years on I/F IP, has focused on each segments of system modeling, IC design, manufacturing to packaging and testing. For DDR (double data rate SDRAM), for example, for the high-speed memory such as the latest DDR4/DDR5 (Table 1), the focus is how to reduce the delay due to connectivity via various packaging technique 2.5D/SiP, 3D/HBM [1] (high bandwidth memory, see Fig. 1) as well as via chiplet package.

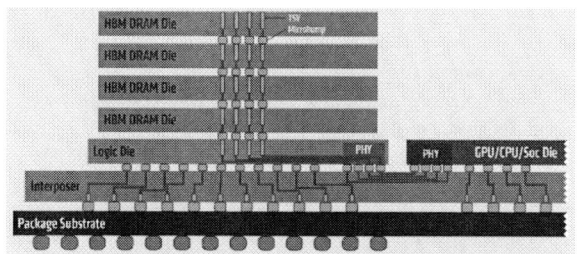

*Figure 1: Schematic of HBM Package [1]*

The PCIe (Peripheral Component Interconnect express) is a high-speed serial computer expansion bus, a standard designed as an interface for adapters, SSDs (solid-state drive) etc. The data rate and bandwidth of PCIe specifications are shown in Table 2. In an SoC or ASIC application, the PCIe can communicate to multiple number of I/F IP SerDes (serializer/deserializer), i.e. ×1, ×2, ×4, ×8 and ×16 based on the needs (see Fig. 2). It is noted that older PCIe uses LVDS (low-voltage differential signaling) instead of DDR; until PCIe 5.0 is used, can DDR5 be supported.

*Table 2: Data rate and bandwidth of PCIe versions*

| Year Released | PCIe Generation | Data Rate (Gb/s) (Encoding) | Bandwidth |
|---|---|---|---|
| 2003 | 1.0 | 2.5 (8b/10b) | 8 GB/s |
| 2007 | 2.0 | 5.0 (8b/10b) | 16 GB/s |
| 2010 | 3.0 | 8.0 (128b/130b) | 32 GB/s |
| 2017 | 4.0 | 16.0 (128b/130b) | 64 GB/s |
| 2019 | 5.0 | 32.0 (128b/130b) | 128 GB/s |
| 2021 | 6.0 | 64.0 (PAM-4) | 256 GB/s |

*Table 1: Data rate, bandwidth and process of DDR*

| Year | RAM | Data Rate | Bandwidth | PHY Process |
|---|---|---|---|---|
| 2002 | DDR1 | 400 MT/s | 3.2 GB/s | 65, 55 nm |
| 2004 | DDR2 | 800 MT/s | 6.4 GB/s | 55,40,28 nm |
| 2007 | DDR3 | 2,133 MT/s | 17.1 GB/s | 28,16,7 nm |
| 2013 | DDR4 | 3,200 MT/s | 25.2 GB/s | 16,12,7 nm |
| 2020 | DDR5 | 6,400 MT/s | 32.0 GB/s | 16,10,7 nm |
| Note: DDR1 is usually written as DDR | | | | |

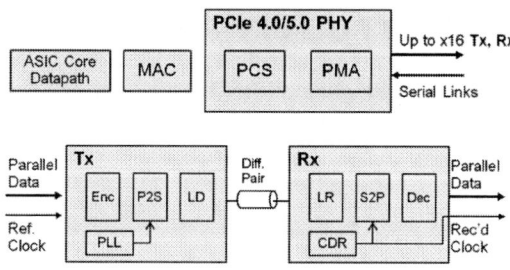

*Figure 2: Schematic of PCIe and SerDes (Tx, Rx)*

Then Ethernet IP establishes interconnect between devices and equipment. In less than 40 years, the link speed has evolved from 10 Mb/s Ethernet to the next level 1.6 TbE (100 GbE ×16 or 200 GbE ×8, see Fig. 3).

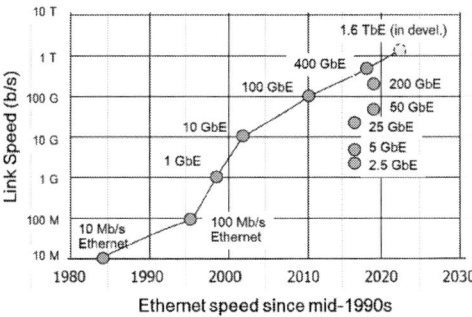

*Figure 3: Link Speed of Ethernet IP*

In recent years, on top of PCIe and Ethernet IP, to overcome the bottleneck of interface connections, various protocols are created, to keep the latency to a minimum, such as OpenCAPI, CCIX, Gen-Z, Nvlink (NVL), and a new standard CXL (Compute Express Link) [2] (Fig. 4), for the sake of interconnect between CPU and devices, it is an industry-supported Cache-Coherent Interconnect (CCI) protocol for Processors, Memory Expansion and Accelerators. Based on PCIe 5.0, CXL is a high-bandwidth, low-latency serial bus interconnect between host processors and devices such as accelerators, memory controllers/buffers, and I/O devices; thus it includes three components (I/O, memory, and cache).

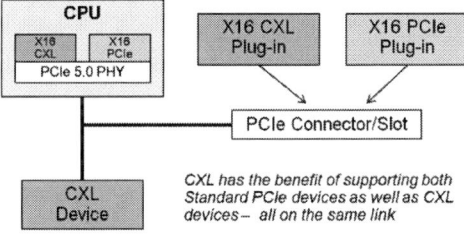

*Figure 4: Benefit of using CXL [2]*

### FPGA Verification for SmartNIC and TOR

Ethernet Card, USB Ethernet Card (such as a USB dongle), and Network Interface Card (NIC) have been widely used and are still in use today. An NIC is in fact a network interface controller. A regular NIC enables communication between computers on the same local area network (LAN) and large-scale networks through the internet protocol (IP).

By adding a compute layer with multi-core processors on the regular NIC, the SmartNIC can be used additionally to accelerate networking, storage, and security functions. The FPGA based SmartNIC, used in a data center, can also

have an option to accelerate workloads in machine learning/deep learning (ML/DL) data processing.

In high data rate application scenarios, typically in server racks at a data center, where a SmartNIC is essential for each server station, each SmartNIC is used to connect to Top-of-Rack (TOR) in these server racks.

## METHODOLOGY

### Selection of FPGA Board and Setup

An FPGA platform using Xilinx Alveo U50 is chosen for this study (Fig. 5). The U50 board is designed for data center acceleration. U50 has QSFP28 (Quad Small Form Pluggable transceiver) networking I/O at 100 GbE (25GbE×4 lanes or 100GbE×1), supporting PCIe Gen4 (and CCIX), and 8 GB of HBM2 for a bandwidth of 460GBps. U50 consumes typical power 50W or max. 75W. With HBM2 instead of DDR4, U50 offers a good form factor for the test.

*Figure 5: FPGA Platform using Alveo U50*

### Interconnect Reach and Application Distance

To verify high-speed I/F IP using SerDes and DDR4/DDR5, the interconnect reach and application distance of the proposed FPGA platform is collected for the analysis (Table 3); the test can start from a HBM2 coming along with the FPGA board.

*Table 3: Interconnect Reaches and Application Distances for CEI 56Gbps*

| Serial Reach | Reach Distance | Connection | Modulation |
|---|---|---|---|
| USR | 10 mm | D2D | NRZ |
| XSR | 50 mm | C2OE | PAM4 |
| VSR | 150 mm | C2M | PAM4 |
| MR | 500 mm | C2C | PAM4 |
| LR | 1000 mm | BP/CC | PAM4 |

The initial SerDes of 100 GbE can be first established and tested; while at later stage SerDes of 400 GbE, and above can be used. Table 3 lists the CEI (common electrical interface) roadmap of 56 Gbps in various applications, which follows the OIF (Optical Internetworking Forum) protocol [3]. In today's common USR (ultra-short reach) for D2D (2.5D/3D) packaging, the modulation uses NRZ (non-zero return); in XSR (extra-short reach) for C2OE (chip-to-optical engine), VSR (very short reach) for C2M (chip-to-module), MR (mid-reach) for C2C (chip-to-chip), and LR (long reach)

978-1-6654-9759-6/22 $31.00 © 2022 IEEE

for BP (backplane) or CC (copper cable), PAM-4 (pulse amplitude modulation) is commonly used.

**CMOS and Silicon Photonics**

To establish communications between CMOS based SerDes (Tx in, Rx out) and laser optical signals (ROSA in, TOSA out), a common CDR block (or DSP block by optical partners) is bridging between the heterogeneous materials (Fig. 6). Likewise, optical modules, such as TIA and LA are migrating away from SiGe process to CMOS silicon process (*e.g.* 180nm, 65nm, 40nm). Other optical details are not shown.

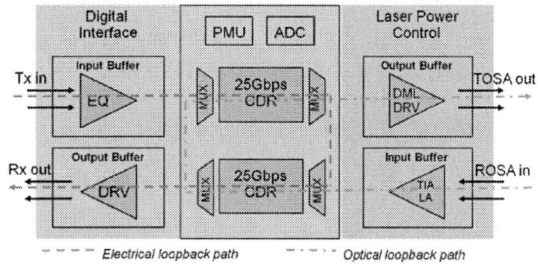

Figure 6: Silicon SerDes and photonic laser blocks. (SerDes Tx/Rx: transmitter/receiver, ROSA/TOSA: receiver/transmitter optical sub assembly, CDR: clock data recovery, TIA: transimpedance amplifier, LA: limit amplifier)

## RESULTS AND DISCUSSION

Current study of I/F IP, involving I/O IP is mainly focused on the whole IC product chain from system modeling of various IP modules, IC design, manufacturing to packaging and testing, which can extend the process technique from 180nm (such as TIA, LA) to 65/55nm (e.g. power management unit, PMU), down to 7nm (e.g. 112G SerDes, DDR5, Table 1). The analysis is guided with the principle of DTCO (design technology co-optimization), including the study of memory reliability [4], to establish an intuitive method for DTCO as well as for STCO (system technology co-optimization).

In all SoC designs, CPU and I/O communication bus has gone through PCI (clock 33MHz), followed by PCI-X (2nd generation, 133MHz) to today's PCIe (2.5GHz) as the performance of the required bandwidth of I/O pins has been tremendously increased. While PCIe deals with interconnect mainly inside a serve or internal components; Ethernet IP is for low latency communication between computers in a local area network (LAN). For the coverage of a long term strategy in between a local server computer and multiple computers in the LAN, both PCIe and Ethernet IP are continuously under development.

The FPGA platform is set up to evaluate the interconnects by dealing with I/F IP and including I/F I/O through SerDes (Tx, RX). For memory interface world beyond DDR4, Wide I/O (Samsung) interface standard for the memory bandwidth for 3D/2.5D is raised; secondly HBM (AMD and Nvidia) is used for ultra-low power bandwidth; and thirdly HMC (hybrid memory cube, Intel and Micron) for multi-core processing has been proposed. In addition, the RapidIO architecture has been developed for circuit board-to-board networking and communication, using a backplane.

The FPGA platform can also accommodate a setting for CMOS and silicon photonics, following HIR [5] (heterogeneous integration roadmap), extending 2.5D/3D IC SiP to chiplet packaging technology.

The methodology presented in Fig. 5 using U50 is for a 100 GbE suitable for studying the SmartNIC bandwidth with PCIe Gen 4; the technology is scalable by using, e.g. a FPGA Platform for EIC (electronic IC) and PIC (photonic IC) verification, e.g. for 400 GbE, suitable for studying the TOR bandwidth used in data centers (Fig. 7).

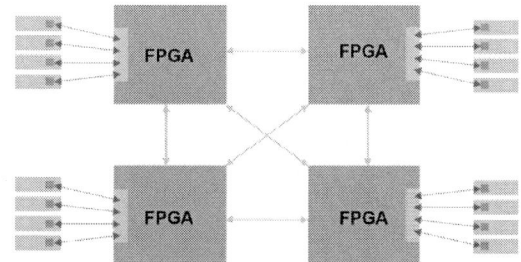

Figure 7: FPGA Platform for EIC+PIC Verification

## REFERENCES

[1] HBM by AMD, https://www.amd.com/system/files/documents/high-bandwidth-memory-hbm.pdf

[2] CXL 2.0, https://www.computeexpresslink.org/download-the-specification

[3] OIF-CEI. https://www.oiforum.com

[4] Chen, C.-Z., Hu, D.Y. and Hanming Wu. "Effective Radiation Damage to Floating Date of Flash Memory," CSTIC 2021, Mar 14-15, 2021, Shanghai), https://ieeexplore.ieee.org/document/9461408/

[5] HIR, https://eps.ieee.org/images/files/HIR_2021/ch19_security.pdf

# AUTOMATIC PLACEMENT ALGORITHM OF INTEGRATED CIRCUITS FOR WIRE BOND PACKAGING APPLICATION

*J. Wan [1]\*, HC. Wang [1]*

[1] School of Information Science and Engineering, Fudan University, Shanghai, China
\* Email: jingwan@fudan.edu.cn

## ABSTRACT

In this work, we have explored the automatic placement algorithm of chips during wire bond packaging based on multi-objective optimization technique. We use genetic algorithm to optimize the chip's position and angle during packaging. Pareto front is obtained indicating the best compromise with performances, such as the distance between the wires, wire length and so on. Optimization with up to four objectives has been successfully demonstrated.

## INTRODUCTION

Wire bond is an important packaging method commonly used in integrated circuits. This process uses gold wires to complete the connection between the chip and the package lead frame.

Digital circuits' wire bond is relatively easy, because when designing the digital circuits' layout, the pads are usually moved from their original position to the edge of the chip. Meanwhile, they are placed in the order of the pins information. The process of changing the position of the pads requires to use metal wires. In the analog circuits, the current of some branches is very large. If all the pads are still moved to the edge, maybe the wire resistance will bring a large error. Therefore, pads of some analog circuit chips will be placed in the original position, as shown in Fig.1.(a). The wire bond of such analog circuits usually relies on manual design, which greatly increases the working time.

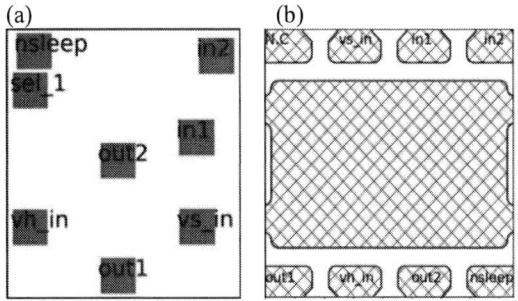

Fig.1. (a) An analog circuit chip structure.
(b) Package structure and pins information.

In recent years, there has been little research on the automatic placement of chips during wire bond packaging[1-2]. However, the optimization methods are all through manual adjustment of the layout, which do not fundamentally achieve the purpose of automatic design.

In terms of algorithms, genetic algorithm and particle swarm optimization are commonly used to search the co-optimization Pareto front[3-4]. Under such a background, we propose an automatic placement of chips technique, which realized the optimization of the position and angle parameters of a chip with eight pads.

## RESULTS AND DISCUSSIONS

### Chip and Package structure

Fig.1.(a) shows the chip used in this research. It has 8 pads. The connection relationship is that the pins and pads with the same name need to be wired. "N.C" means no connection. "sel_1" means selection which can be selected to be wired or not wired. In this research, we select not to wire it. Fig.1.(b) shows the corresponding package structure and pins information. The blue shaded area in the middle of the picture is the area where the chip can be placed, which is called the plating area.

We choose two indicators to determine the chip's location, one is the vertex coordinates (x, y) with the upper left corner of the chip as the origin. The other is the chip rotation angle. Therefore, a total of three parameters of the circuit are used as design variables. The range of vertex coordinates is determined by the area of the packaged plating area, and the specific value range of each variable is as follows:

TABLE I.   VARIABLE RANGE

|  | X($\mu$m) | Y($\mu$m) | Angle(degree) |
|---|---|---|---|
| **From** | 80 | 380 | 0 |
| **To** | 1640 | 1420 | 360 |

### Single-objective optimization

When designing the wire bond, the distance between the wires should be as large as possible to avoid short-circuit. At the same time, the wire length is also a parameter that needs to be considered to reduce the cost and signal transmission delay. We also need to add two constraint rules to ensure the feasibility of the design. First, the chip cannot be placed beyond the area of the plating area. Second, each wire cannot cross. If any of the above two constraint rules are not met, theoretically this design should not be considered.

We do two experiments. The optimization process is conducted in Python. We optimize the distance between the wires and the wire length respectively, which can be expressed as:

$$
\begin{cases}
minimize \quad wire\ distance\ normalized\ fitness \\
s.t. \quad wires\ can't\ cross \\
s.t. \quad chip\ can't\ go\ out\ of\ plating\ area
\end{cases}
\quad (1)
$$

$$
\begin{cases}
minimize \quad wire\ length\ normalized\ fitness \\
s.t. \quad wires\ can't\ cross \\
s.t. \quad Chip\ can't\ go\ out\ of\ plating\ area
\end{cases}
\quad (2)
$$

Taking the wire length optimization as an example. One of the initial solutions generated randomly is shown in Fig.2.(a). The normalized fitness varies from 0 to 100. 0 represents the best performance. The normalized fitness of each generation in the entire calculation process is shown in Fig.3. After optimizing for 20 generations, the fitness tends to be stable. One of the final solutions is shown in Fig.2.(b), which has reached the requirement of obtaining a short wire length and satisfying the two constraint rules.

(a)

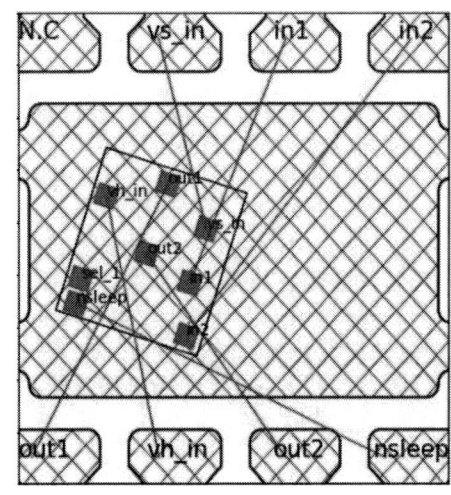

(b)

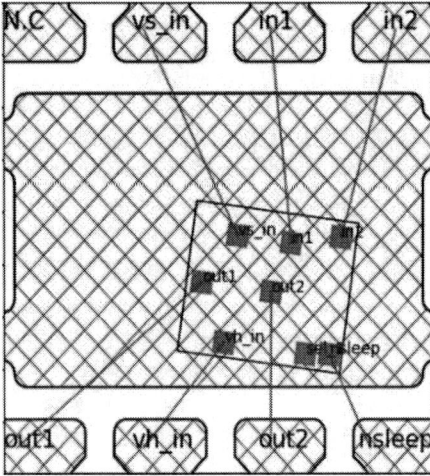

Fig.2. (a)  One of the initial solutions.
(b)  One of the final solutions.

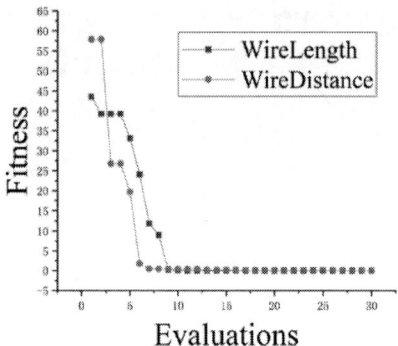

Fig.3. The normalized fitness of each generation.

**Multi-objective optimization**

When designing the wire bond, the position and the angle of the chip are also important factors to ensure the most correct placement of the chip and minimize the negative impact caused by the process deviation. The chip should be placed as close as possible to the center of the plating area. Meanwhile, it is best to place the chip at an angle close to 90 degrees or its integer multiples.

First, we choose the wire length and the distance from the center of the chip to the center of the plating area as the co-optimized targets, and the two constraint rules are the same with the two mentioned before, which can be expressed as:

$$
\begin{cases}
minimize \quad wire\ distance\ normalized\ fitness \\
minimize \quad chip\ position\ normalized\ fitness \\
s.t. \quad wires\ can't\ cross \\
s.t. \quad Chip\ can't\ go\ out\ of\ plating\ area
\end{cases}
\quad (3)
$$

The optimization problem is solved by the NSGA-II algorithm, which is a typical genetic algorithm used in a variety of multi-objective optimization fields [5-6]. The population size of the genetic algorithm is 50, and 50 generations of genetic algorithm are used in the optimization process. The entire optimization process takes about 3 minutes. The set of the optimization solutions is called the Pareto front, as shown in Fig.4.

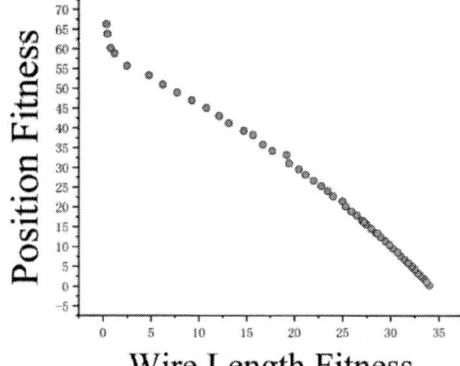

Fig.4. Wire Length-Position Pareto

These Pareto front points clearly form a boundary in the performance space. With a certain wire length, the best position can be found on the Pareto front.

Finally, we take the distance between wires, wire length, the position and the angle of the chip as the co-optimization targets for four-objective optimization, which can be expressed as:

$$
\begin{cases}
minimize & wire\ distance\ normalized\ fitness \\
minimize & wire\ length\ normalized\ fitness \\
minimize & chip\ position\ normalized\ fitness \\
minimize & chip\ angle\ normalized\ fitness \\
s.t. & wires\ can't\ cross \\
s.t. & Chip\ can't\ go\ out\ of\ plating\ area
\end{cases}
\quad (4)
$$

We have obtained a set of Pareto front solutions and the normalized fitness. We respectively show the extreme values and average values of the four normalized fitness as shown in Fig.5.

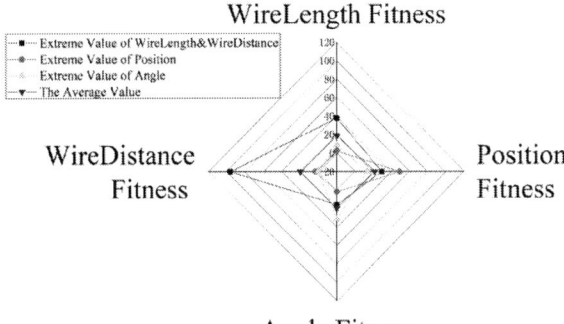

Fig.5. The extreme values and average values of the four normalized fitness.

Among all the solutions, we show a solution whose all the four optimization objectives perform well, as shown in Fig.6.

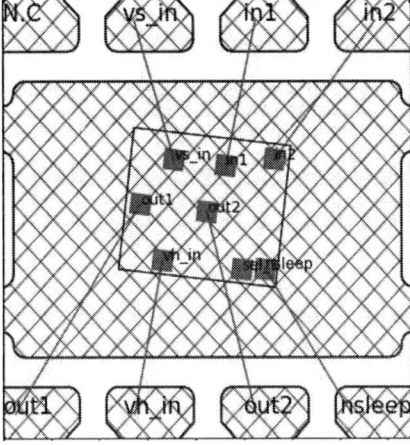

Fig.6. A solution whose all the four optimization objectives perform well.

## SUMMARY

This paper discusses the application of genetic algorithm in chip wire bond design. We use a chip with 8 pads and a package with 8 pins for demonstration. We use the NSGA-II algorithm for multi-objective optimization and finally found a wire bond design that all its four optimization objectives performed well.

## ACKNOWLEDGEMENTS

This work was supported by the Shanghai Science and Technology Commission "explorer project" (21TS1401300), National Key R&D Program of China (2021YFA1200500), and Pioneering Project of Academy for Engineering and Technology, Fudan University (gyy2021-001).

## REFERENCES

[1] Zhu Yue, Peng Ying, Zhang Jiyao, Wang Chao, Zhang Mingyu. *A high-power chip layout and its layout and packaging and wiring optimization method* [P]. Hubei Province, China: CN109585442B, 2021-02-12.

[2] Lu Jiangping, Liu Xia, Chen Yuanjin, Chen Chao, Wang Lili. *A chip pad layout design method that can adapt to a variety of different packaging requirements* [P]. Jiangsu Province, China: CN103903996A, 2014-07-02.

[3] Oltean G, Hintea S, Sipos E. *Knowledge-based & Intelligent Information & Engineering Systems*, International Conference, Kes, Santiago, Chile, September, Part II, (2009).

[4] Fakhfakh M, Cooren Y, Sallem A. *Analog Integrated Circuits & Signal Processing*, 63(1), p.71-82 , (2010).

[5] Wei Mao, Jia-Hao Wei, Jing Wan, 2020 IEEE 15th International Conference on Solid-State & Integrated Circuit Technology (ICSICT) , (2020).

[6] Huang H, Zeng Y, Liao J, 2018 IEEE Asia Pacific Conference on Circuits and Systems (APCCAS), (2018).

# INTEGRATED SUPERCONDUCTING ISOLATOR-CIRCULATOR-ISOLATOR DEVICE

*Rutian Huang[1], Genting Dai[1], Jianshe Liu[1], and Wei Chen[1,2,\*]*

[1] School of Integrated Circuits, Tsinghua University, Beijing, 100084, China
[2] Beijing Innovation Center for Future Chips, Tsinghua University, Beijing, 100084, China
[\*] Corresponding Author's Email: weichen@mail.tsinghua.edu.cn

## ABSTRACT

There needs a large quantity of microwave devices like circulators, isolators and quantum noise limited amplifier in the development of scalable superconducting quantum computers. Traditional isolators and circulators have huge size and could induce a high thermal dissipation. This paper designs a kind of on-chip Isolator-Circulator-Isolator (ICI), which is able to act as isolator and circulator synchronous. The ICI is composed of tunable inductor bridge which based on an array of superconducting quantum interference devices (SQUID). It works on the frequency between 4 and 8 GHz, and operation temperature at 20 mK, with isolation exceeds 40 dB and insertion loss less than 3 dB.

## INTRODUCTION

In superconducting quantum systems, there are a series of devices such as filters, isolators, circulators and quantum noise limited amplifier that are used to readout quantum information [1,2]. However, traditional isolators and circulators introduce thermal load for huge size and not suitable for the development of large-scale superconducting quantum circuit. This paper designs an on chip integrated Isolator-Circulator-Isolator (ICI) device. ICI is a nonreciprocal device, it has smaller size and could be used as an isolator and circulators simultaneously, which can be connected directly with quantum processor and quantum noise limited amplifier like Josephson parametric amplifier (JPAs) or impedance-transformed parametric amplifier (IMPAs). ICI will simplify quantum measurement circuit, protect qubits from external interference noise and is of great importance in the development of large-scale quantum circuits.

## PRINCIPLE AND ANALYSIS

The circuit symbols and ideal scatter matrixes of circulator, isolator and ICI are given in Table I. The ICI is a nonreciprocal device with three ports which marked as 1, 2, 3 in anti-clockwise direction respectively. The signal is able to transmitted from 1 to 2, 2 to 3, then out of 3, or transmitted from 1 to 2, then directly out of 2, while it is prohibitive to transmit from 3 to 1, while traditional circulator allows signal to go through from 1 to 2, 2 to 3, and then 3 to 1. To some extent, the ICI could act as isolator and circulator at the same time.

TABLE I. THE CIRCUIT SYMBOLS AND IDEAL SCATTER MATRIXES OF CIRCULATOR, ISOLATOR AND ICI.

| Device Name | Circuit Symbol | Ideal Scatter Matrix |
|---|---|---|
| Isolator | | $\begin{bmatrix} 0 & 0 \\ 1 & 0 \end{bmatrix}$ |
| Circulator | | $\begin{bmatrix} 0 & 0 & 1 \\ 1 & 0 & 0 \\ 0 & 1 & 0 \end{bmatrix}$ |
| ICI | | $\begin{bmatrix} 0 & 0 & 0 \\ 1 & 0 & 0 \\ 0 & 1 & 0 \end{bmatrix}$ |

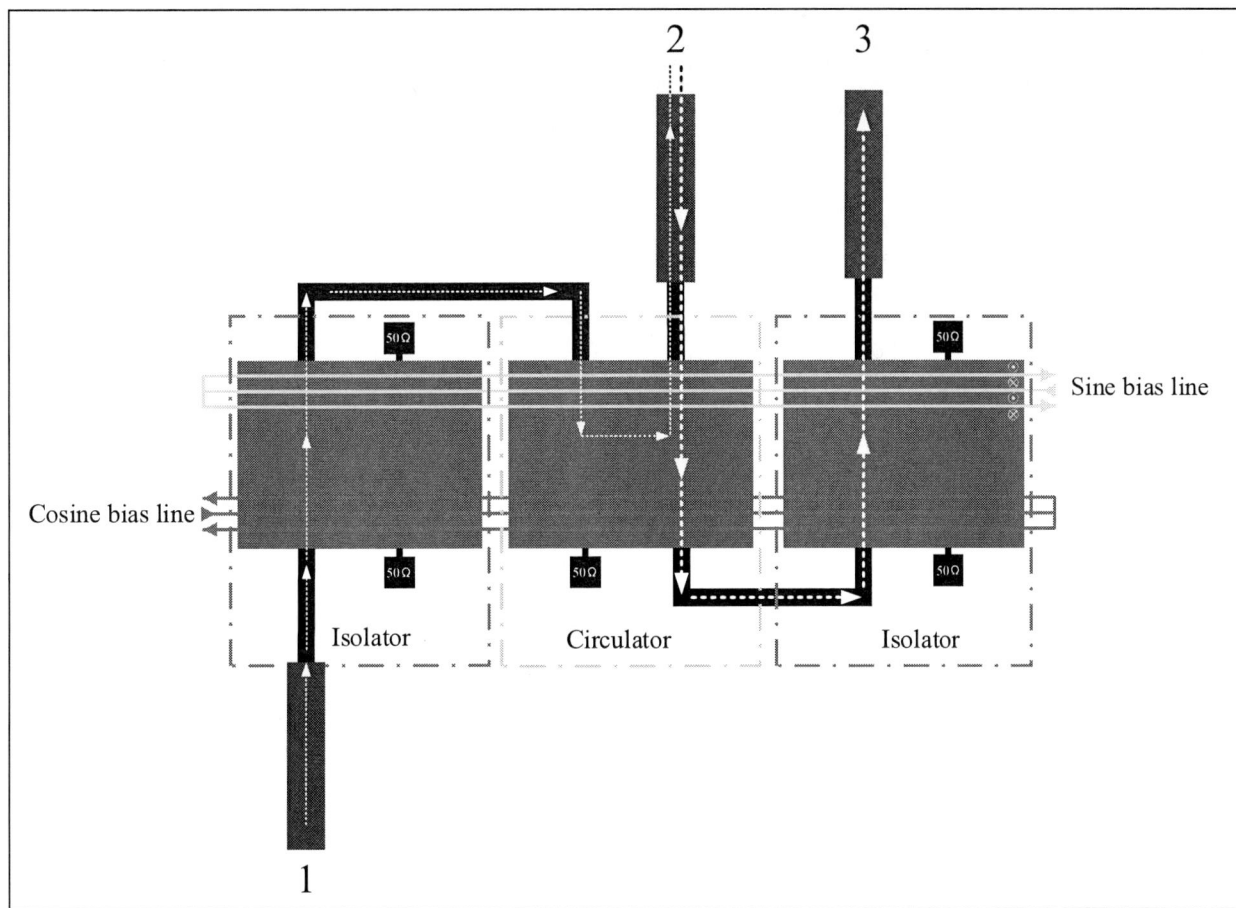

*Fig.1. Composition diagram of ICI. It includes one circulator (in the green dash line box) sandwiched by two isolators (in the red dash line box), with one sine bias line above and one cosine bias line below.*

Two isolators are formed by 4-port circulator [3-5] with 2 ports terminated by matching load. Bias lines are used to tune the magnetic field around ICI. According to Auster's principle, the magnetic flux through SQUIDs can be changed by adjusting the current in the bias line, so that the equivalent inductance of SQUIDs will also be changed. This operation tunes the center frequency of ICI. The transmission route of microwave signal in ICI is presented by white dashed lines.

The ICI is composed of tunable inductor bridge which based on an array of SQUIDs. Its working frequency is between 4 and 8 GHz, operating at 20 mK. The isolation ($\widetilde{IS}$) of isolator is around 20 dB the insertion loss ($\widetilde{IL}$) of circulator [5] is around 1 dB, so the $\widetilde{IS}$ of ICI is:

$$\widetilde{IS}_{\mathrm{ICI}} = 2\widetilde{IS}_{\mathrm{iso}} + \widetilde{IL}_{\mathrm{cir}},$$

$$\widetilde{IS}_{\mathrm{ICI}} = 41 \text{ dB}.$$

The insertion loss ($\widetilde{IL}$) of isolator is around 1 dB, the isolation ($\widetilde{IS}$) of circulator [5] is around 20 dB, so the $\widetilde{IL}$ of ICI is:

$$\widetilde{IL}_{\mathrm{ICI}} = 2\widetilde{IL}_{\mathrm{iso}} + \widetilde{IL}_{\mathrm{cir}},$$

$$\widetilde{IL}_{\mathrm{ICI}} = 3 \text{ dB}.$$

So, isolation of ICI exceeds 40 dB and insertion loss is less than 3 dB.

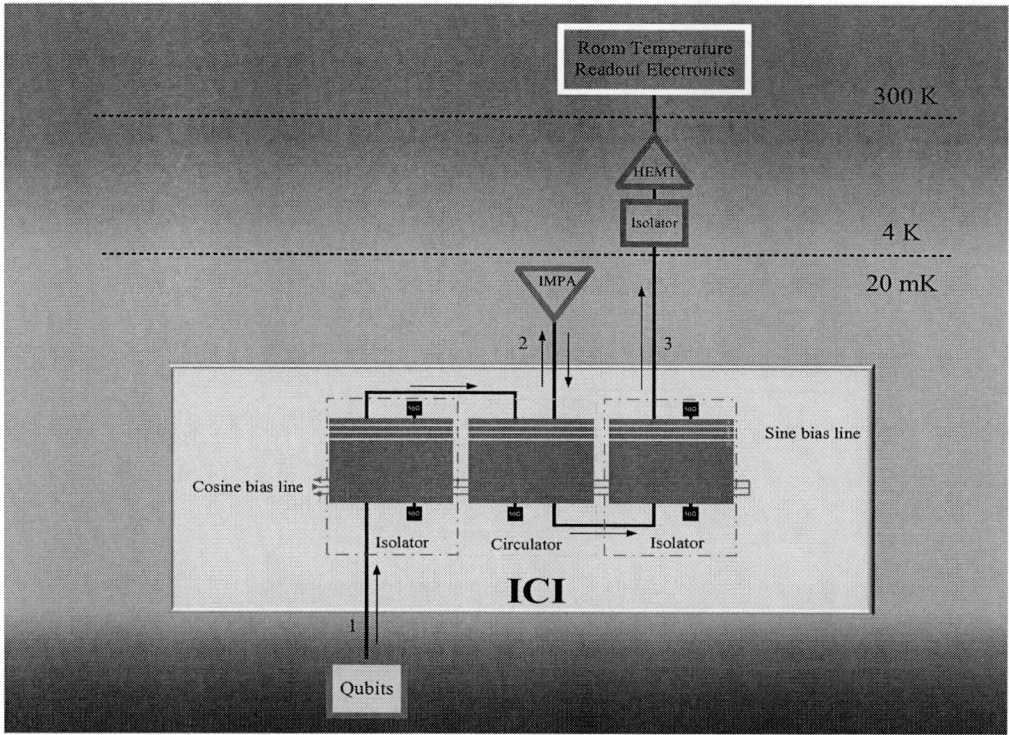

*Fig.2. Schematic of ICI used in superconducting quantum system.*

The concept of superconducting quantum test system which ICI is used in shown in Fig.2. Quantum information processed in qubits is readout through resonators. Signal is input ICI via port 1, output via port 2 to IMPA, and then reflected back to port 2 of ICI after being amplified in IMPA, output from port 3. The layout of ICI is about 6.8 mm * 6.8 mm. The fabrication of superconducting ICI is compatible with semiconductor technology.

## CONCLUSION

Compared with traditional isolators and circulators, the integrated ICI chip has the advantages of nonreciprocity, smaller size, lower heat load and can be integrated with the superconducting circuits. Its working frequency is between 4 and 8 GHz, and operating at 20 mK, with isolation exceeds 40 dB and insertion loss less than 3 dB. ICI will simplify quantum measuring system, protect qubits from external interference noise and is of great importance in the development of scalable quantum computers.

## ACKNOWLEDGMENTS

We acknowledge Changhao Zhao, Xiao Geng, Yunfan Shi, Qing Yu, Yongcheng He, Kaiyong He, Xinyu Wu, and Liangliang Yang help for stimulating and valuable discussions. This work is supported by the National Natural Science Foundation of China (Grant Nos. 11653001, 11653004, and 60836001).

## REFERENCES

[1] Arute F, Arya K, Babbush R, et al. Quantum supremacy using a programmable superconducting processor. *Nature* 2019; 574:50 5–10.

[2] Gong M, Wang S, Zha C, et al. Quantum walks on a programmable two-dimensional 62-qubit superconducting processor. *Science* 2021;372: 948–52.

[3] Chapman B J, Moores B A, Rosenthal E I, et al. General purpose multiplexing device for cryogenic microwave systems[J]. *Applied Physics Letters*, 2016, 108(22):1169-1174.

[4] Kerckhoff J, K Lalumière, Chapman B J, et al. On-chip superconducting microwave circulator from synthetic rotation[J]. *Physical Review Applied*, 2015, 4(3).

[5] Chapman B J, Rosenthal E I, Kerckhoff J, et al. Widely Tunable On-Chip Microwave Circulator for Superconducting Quantum Circuits[J]. *Physical Review X*, 2017, 7(4).

# A GAUSSIAN ROCESS AND MULTI-SWARM OPTIMIZER ASSISTED OPTIMIZATION APPROACH FOR ANALOG CIRCUIT DESIGN

*Xu Fu[1], Changhao Yan[1*], Dian Zhou[2] and Xuan Zeng[1*]*

[1]State Key Lab of ASIC & System, Fudan University, Shanghai 201203, China
[2]Department of Electrical Engineering, University of Texas at Dallas, TX, U.S.A
*Corresponding Author's Email: {yanch, xzeng}@fudan.edu.cn

## ABSTRACT

In this paper, we propose an analog circuit synthesis approach that consists of a Gaussian process and niching migratory multi-swarm optimizer assisted differential evolution algorithm. Instead of building an artificial neural network (ANN) to perform a local minimum search by sampling a large number of samples in the local area, we use samples simulated before to build a Gaussian process model and combine the Gaussian process model and niching migratory multi-swarm optimizer to exploit the local area. The weighted expected improvement function is selected as the criterion to judge which point is more worthy of simulation. Compared to the ANN assisted optimization system, our proposed method reduces the simulation time by up to $15\times$.

## INTRODUCTION

Nowadays, digital circuit design has been automated for a long time, but analog circuits still rely on experienced designers. Automated analog circuit design tools are greatly needed because of the growing demands for high-performance and short-to-market circuits.

Device sizing is one of the steps in analog circuit design. In this paper, we treat analog circuit sizing as a constrained optimization problem. Various approaches have been proposed to solve this optimization problem. These methods can be classified as simulation-based methods and model-based methods.

An artificial neural network assisted optimization system [1] uses GA for global optimization and constructs an ANN model for local minimum search. It can make full use of parallel simulation but still cannot solve the problem of GA that the sampling efficiency is low, and a large number of data points need to be simulated to build a local ANN model every iteration. Memetic machine learning-based differential evolution (MMLDE) [2] uses a surrogate model to choose the most worthy individual to simulate, populations are generated by differential evolution (DE) algorithms. It prevents parallel simulation due to iteratively search.

In this article, we proposed a framework to improve the efficiency of DE sampling and exploit the local area without additional sampling. The framework is GA-based global search [3], Gaussian Process (GP) model [4] is employed to help exploit the local area and choose points that are more worthy of simulation according to the value of the acquisition function. Weighted Expected Improvement (wEI) function is selected as the select function to handle constraints. Niching migratory multi-swarm optimizer (NMMSO) [5] is used with the GP model to find the separate local modes for exploiting and exploring.

## PROPOSED ALGORITHM

### Analog Sizing Problem Formulation

Analog sizing problem can be formulated as a general constrained nonlinear optimization problem, the formula can be written as below:

$$\text{minimize} \quad f(\boldsymbol{x})$$
$$\text{s.t.} \quad c_i(\boldsymbol{x}) < 0 \quad \forall i \in 1\ldots N_c \tag{1}$$

Where the $f(x)$ is an objective function and $c_i(x)$ denotes the ith constraint needs to be satisfied. $N_c$ is the number of constraints.

### Gaussian Process

In the literature, the GP model has been built as a surrogate model to approximate the behavior of the objective function and constrain function. GP not only provides a prediction but also can show uncertainty for the prediction.

To solve the constrained optimization problem, we use wEI [4] function as a criterion to select the individual being simulated, and its form is:

$$wEI(x) = EI(x)\prod_{i=1}^{N_c}PF_i(x)$$
$$PF_i(x) = \Pr\left(c_i(x) < 0\right) = \Phi\left(\frac{-\mu_i}{\sigma_i}\right) \tag{2}$$

Where $N_c$ is the number of constraints. PF is a function used to measure feasibility. EI is the expected value of the gain of the new measured value compared to the historical optimal value. By comparing the wEI value of different points, point with smaller objective function value and higher feasibility is more likely to be selected.

### Niching Migratory Multi-Swarm Optimize

Optimization problems are confined to identifying the optimized function. When this function contains multiple solutions it is called to be multimodal function. The multimodal optimization problem is formulated as:

$$\max_{XS \subset D}|XS| \tag{3}$$

978-1-6654-9759-6/22 $31.00 © 2022 IEEE

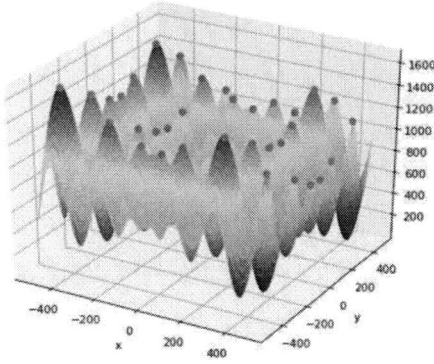

Fig. 1: An example of multimodal function

Where $[\ ]$ is the size of the set, $XS = \{x_i^{lo}, i = 1, 2, \cdots, n\}$,

and $x_i^{lo}$ is the local optimum.

Fig.1 shows the topography of the schwefe1, which is derived from deap.beachmarks. An efficient multi-modal optimization solver NMMSO [5] can locate all globally optimal and suboptimal solutions accurately. In this paper, this method is applied to generate offspring whose wEI value is local optimum.

In the actual circuit design process, we often need to make tradeoffs between different circuit specifications, such as the gain, unity gain frequency, and phase margin of the two-stage operational amplifier. Therefore, The optimal point is often near the boundary of the feasible region, even if the model error is small, the predicted optimal point meeting the constraints may not meet the constraints in the actual simulation. However, NMMSO generates multiple local optimums, All local optimal solutions can be the current best, which makes the optimization process more robust and the convergence speed faster.

### General Framework

The flow diagram of the framework is given in Fig.2. Latin hypercube sampling (LHS) [6] is used to initialize the datasets for sampling the design space more uniformly. At the beginning, we compare the previous generation of the population with the newly simulated individuals and select the $\gamma$ individuals as the new population. Through mutation and crossover, a new offspring population will be generated, we pick the best $\lambda$ data points in the new population to build the GP model. Then separate local peaks of wEI function are located by solving a multi-modal problem. These peaks are added to the new offspring population. All offsprings are sorted according to the wEI value and $\alpha$ individuals with the largest wEI value are simulated.

Compared with ANN modeling, a large number of additional points need to be sampled to ensure the

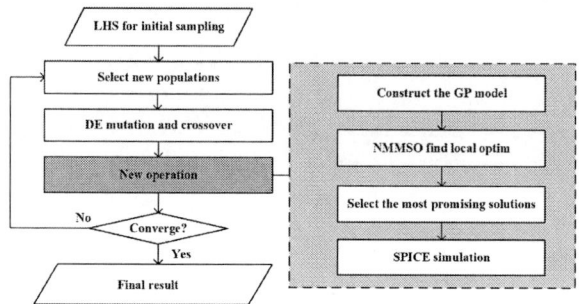

Fig. 2: Flow diagram of the algorithm

accuracy of the model, our method makes full use of the real points simulated before to build the GP model, which can reduce the simulation consumption. NMMSO combines GP to perform a local exploit to speed up the convergence speed. The selected points do not interfere with each other and can be simulated in parallel.

## EXPERIMENTAL RESULTS

In this paper, we apply our algorithm to two real-world analog circuits. Our proposed method is compared with three algorithms: DE[4], GA+ANN[1] and DE+GP[2]. The degree of parallelism in the experiment is 40, which means that 40 circuits can be simulated simultaneously in one simulation cycle. We record the equivalent simulation time consumption with the equivalent number of simulation cycles on average to achieve the final circuit design (Avg. \# Sim).

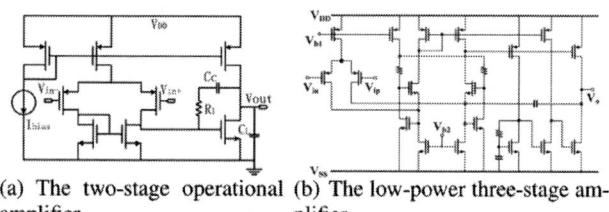

(a) The two-stage operational amplifier  (b) The low-power three-stage amplifier

Fig. 3: The schematic of the two circuits

### Two-Stage Operational Amplifier

The first circuit is a two-stage operational amplifier with 10 design variables and SMIC 180nm process. The schematic is shown in Figure3 (a). Design specifications of the circuit are formulated as the following form:

$$\text{maxmize GAIN}$$
$$\text{s.t.} \quad \text{UGF} \quad >40\text{MHz} \tag{4}$$
$$\text{PM} \quad >60°$$

For DE+GP and DE+GP+NMMSO, when the number of simulation cycles exceeds 50 or the gain is greater than 89 and all constraints are met, the program stops running. For DE and GA+ANN, the maximum number of simulation cycles is 150.

The experimental result is shown in table I. Compared with DE, GA+ANN and DE+GP, DE+GP+NMMSO

TABLE I: The optimization results of two-stage operational amplifier

| Algo | DE | GA+ANN | DE+GP | DE+GP+NMMSO |
|------|------|--------|-------|-------------|
| GAIN | 89.00 | 89.21 | 89.11 | **89.11** |
| UGF | 40.07 | 40.22 | 40.19 | **41.25** |
| PM | 61.42 | 60.74 | 61.62 | **60.55** |
| Avg. # Sim | 128 | 69 | 30 | **21** |
| # Success | 10/10 | 10/10 | 10/10 | **10/10** |

reduces the simulation time by up to $6\times, 3.3\times, 1.4\times$ The

TABLE II: optimization results of three-stage amplifier

| Algo | DE | GA+ANN | DE+GP | DE+GP+NMMSO |
|------|------|--------|-------|-------------|
| IQ | 29.90 | 29.71 | 29.88 | **29.61** |
| GAIN | 101.86 | 100.21 | 100.44 | **101.47** |
| UGF | 0.92 | 0.92 | 0.92 | **0.92** |
| PM | 52.52 | 52.81 | 52.93 | **52.71** |
| GM | 20.60 | 19.54 | 20.43 | **19.95** |
| SR+ | 0.20 | 0.22 | 0.20 | **0.21** |
| SR- | 0.44 | 0.54 | 0.50 | **0.50** |
| Avg. # Sim | 120 | 62 | 12.4 | **4** |
| # Success | 10/10 | 10/10 | 10/10 | **10/10** |

inefficient problem that ANN modeling requires a large amount of sampling is successfully solved by constructing gaussian model using simulated points. Meanwhile, the experimental results demonstrate that the optimization process can be sped up by introducing NMMSO into our algorithm framework.

**Low-Power Three-Stage Amplifier**

The three-stage amplifier is shown in Figure3 (b). There are 24 design variables, including transistor length and width, capacitance, resistance, and bias current. Specifications of the circuit are listed as below:

$$
\begin{aligned}
\text{Minimize} \quad & Iq \\
\text{s.t.} \quad & GAIN > 100dB \quad UGF > 0.92\text{MHz} \\
& PM > 52.50 \quad GM > 19.5dB \\
& SRR > 0.18\text{V}/\text{u.s} \quad SRF > 0.2\text{V}/\text{us}
\end{aligned}
\tag{5}
$$

The maximum simulation cycle is set to be 50, When Iq is less than 30 and no constraint is violated, the program converges. For the rest of the algorithms, the maximum simulation cycle is 150.

The experimental results are shown in Table II. Compared with DE, GA+ANN and DE+GP, our proposed algorithm reduces the simulation time consumption by up to $30\times$, $15\times$, $3\times$ respectively. This shows that DE+GP+NMMSO has a much higher efficiency than the state-of-the-art optimization algorithms.

## CONCLUSION

In this paper, we propose an algorithm for dealing with the analog circuit sizing problem. This method includes building a GP model to help select simulation samples so that the sampling efficiency can be improved and avoid falling into the local optimum. At the same time, there is a new strategy to produce children, using NMMSO to find the maximum point of wEI, so that the current promising area can be fully exploited. The experimental results on two analog circuits show that our proposed method improves the sampling efficiency and requires fewer simulation time to achieve the same goal.

## ACKNOWLEDGEMENTS

This research is supported partly by the National Key R&D Program of China 2019YFB2205000, 2019YFB2205002, partly by National Natural Science Foundation of China (NSFC) research projects 61974032, 61774045, and 61929102.

## REFERENCES

[1] Y. Li, Y. Wang, Y. Li, R. Zhou and Z. Lin, "An Artificial Neural Network Assisted Optimization System for Analog Design Space Exploration," IEEE Transactions on Computer-Aided Design of Integrated Circuits and Systems, vol. 39, no. 10, pp. 2640-2653, Oct. 2020

[2] B. Liu, D. Zhao, P. Reynaert and G. G. E. Gielen, "Synthesis of Integrated Passive Components for High-Frequency RF ICs Based on Evolutionary Computation and Machine Learning Techniques," IEEE Transactions on Computer-Aided Design of Integrated Circuits and Systems, vol. 30, no. 10, pp. 1458-1468, Oct.2011

[3] McKay, M.D.; Beckman, R.J.; Conover, W.J, "A Comparison of Three Methods for Selecting Values of Input Variables in the Analysis of Output from a Computer Code". Technometrics. American Statistical Association 21 (2): 239–245. May 1979

[4] K. Price, R. M. Storn and J. A. Lampinen, Differential Evolution. A Practical Approach to Global Optimization, Germany, Berlin:Springer, 2005.

[5] J. E. Fieldsend, "Running Up Those Hills: Multi-modal search with the niching migratory multi-swarm optimiser," 2014 IEEE Congress on Evolutionary Computation (CEC), 2014, pp. 2593-2600

[6] Y. Wang, M. Orshansky and C. Caramanis, "Enabling efficient analog synthesis by coupling sparse regression and polynomial optimization," 2014 51st ACM/EDAC/IEEE Design Automation Conference (DAC), 2014, pp. 1-6.

[7] Yan, Zushu & Mak, Pui-In & Law, Man-Kay & Martins, R.P.. (2013). A 0.016-mm2 144-μW three-stage amplifier capable of driving 1-to-15 nF capacitive load with 0.95-MHz GBW. IEEE Journal of Solid-State Circuits, 48(2), 527-540. IEEE Journal of Solid-State Circuits. vol. 48. pp. 527-540

[8] W. Lyu et al., "An Efficient Bayesian Optimization Approach for Automated Optimization of Analog Circuits," in IEEE Transactions on Circuits and Systems I: Regular Papers, vol. 65, no. 6, pp. 1954-1967, June 201

# An Approximating Twiddle Factor Coefficient Based Multiplier for Fixed-Point FFT

Songyu Sun[1], Zhicheng Xu[1], Xi Chen[1] and Xunzhao Yin[1,2*]

[1]College of Information Science & Electronic Engineering, Zhejiang University, Hangzhou, China
[2]Zhejiang Lab, China [*]Corresponding Author's Email: xzyin1@zju.edu.cn

## Abstract

Fast Fourier Transform (FFT) is playing an important role in signal processing. This paper proposes a FFT-specific approximate multiplier design to improve the energy efficiency. The approximate multiplier is based on the approximated twiddle factor and compressor to reduce the energy during partial product accumulation. Instead of complex arithmetic operations, the design can achieve reduced power consumption and hardware cost with limited error.

*Keywords—Approximate computing; Fast Fourier Transform; twiddle-factor-coefficient*

## Introduction

Fast Fourier Transform (FFT) has become ubiquitous in signal processing, which has attracted interest from both academia and industry [1-3]. One classical approach to fixed-point FFT implementation is Cooley-Tukey algorithm [2]. Recently, many digital signal processing (DSP) applications [4,5,6] have been integrated to portable devices, which often utilizes an embedded FFT module to reduce power overheads. Such devices need to handle large quantities of data almost real time with limited power, which allows researchers to trade off between accuracy and energy efficiency.

In an FFT architecture, multiplication is probably the most resource-expensive arithmetic operation. For example, there are 5,632 multiplications in a 1024-point Radix-2 FFT, which almost quadratically increases with the number of points. Since FFT often needs to be repeatedly invoked, the growing number of multiplications inevitably degrades the speed of FFT and hence impairs the overall performance of the devices. Recently, approximate multiplier is considered as a promising alternative to improve the energy efficiency [7,8,9]. For example, V. Ariyarathna *et al.* [7] proposed a multiplier-less design to reduce area and power, which suffered from high error rate. Reference [4] used two floating-point operations, fused dot product and fused add-subtract, to implement an approximate FFT.

However, it is noted that the multiplication in an FFT is not exactly the same as a regular multiplication, which has two unfixed multiplier and multiplicand. FFT employs pre-determined twiddle factor coefficients (TFC) to calculate a series of multiplications. TFC, which can be mathematically written as the real or imaginary part of

$e^{-j2\pi/n}$, is used to simplify the complexity of FFT from $O(n^2)$ to $O(n\log_2 n)$, where *n* is the number of points for FFT. Thus, instead of directly employing a regular approximate multiplier, it is necessary to explore the characteristics of TFC in FFT and design an FFT-specific approximate multiplier. Since TFCs in a fixed-point FFT are always constants, it is found that most power consumed in the TFC multiplication originates from the accumulation operations instead of partial product generation. It is then a natural idea to reduce the number of partial products in TFC multiplication by removing the nonzeros in the TFCs. Based on the approximated TFCs, we then redesign the compressor circuit in the multiplier to reduce the number of additions and hence significantly improves its power consumption with almost negligible accuracy loss.

In summary, we propose a fixed-point FFT-specific approximate multiplier by simultaneously exploring the approximation of TFC representation and compressor circuit. Experimental results show that with the proposed FFT-specific approximate multiplier, we are able to achieve 40% power saving with a mean relative error distance (MRED) of 0.10554 for a 1024-point radix-2 FFT.

## Background

The 1-D DFT of an input signal sequence $x(n)$ is given by

$$X(k) = \sum_{n=0}^{N-1} x(n) W_N^{nk}, n = 0,1, \dots, N-1$$

where $X(k)$ is the frequency domain representation of $x(n)$ with $k = 0,1, \dots, N-1$. $W_N^{nk} = \exp\left(-j\frac{2\pi nk}{N}\right)$ is the twiddle factor and denotes the $N^{th}$ primitive root of unity. Using Euler's rule to rewrite the twiddle factor, we have:

$$W_N^{nk} = exp(-j\frac{2\pi nk}{N}) = cos\frac{2\pi nk}{N} - jsin\frac{2\pi nk}{N}$$

where $cos\frac{2\pi nk}{N}$ and $sin\frac{2\pi nk}{N}$ are twiddle factor coefficients (TFCs). The most commonly used FFT algorithm is Cooley-Tukey decimation-in-time (DIT) FFT [2], which depends on the recursive decomposition of an $N$ point transform into two $(N/2)$-point transforms. Fig.1 shows the algorithm diagram of the 8-point DIT algorithm. Many hardware architectures have been proposed for the FFT implementation, from parallel, pipelined, to loop structure [10].

978-1-6654-9759-6/22 $31.00 © 2022 IEEE

In an FFT module, the multiplication with twiddle factor coefficient is repeatedly invoked during the entire computation, which is a critical and very power consuming component. A few prior works proposed to encode TFCs with canonical signed digit (CSD) to share the common part between the CSD codes [11]. A shared CSD complex constant multiplier can then be developed for parallel mixed radix FFT [11]. X. Han et al [12] proposed a common subexpression (CS) sharing algorithm to reduce the number of multiplications in FFT architecture and increase area efficiency. The truncated TFCs can be further approximated by adjusting some digits in the initial TFCs to decrease the hardware overhead for adders in partial product accumulation. In [13], by approximating TFCs, the common part of the TFC can be further increased under acceptable approximate threshold. Since much energy is consumed by partial product accumulation, it is then necessary to further exploit the characteristics of TFC and improve the approximation to the accumulation stage.

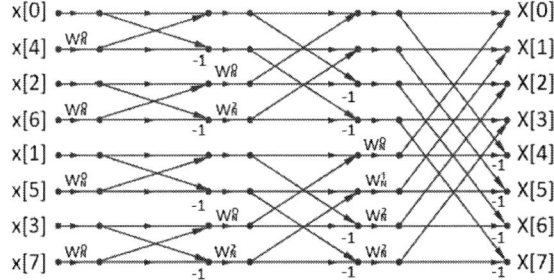

**Fig. 1. An example of 8-point FFT algorithm.**

### Proposed Approximate Multiplier Design

Since a TFC is a sine or a cosine function, its value is always in the range of $[-1, 1]$, with many nonzeros after the decimal point. In our approximate multiplier design, we would like to further reduce the non-critical nonzero bits in the TFC so as to improve the energy efficiency of the multiplication. Fig. 2 presents an example of TFC approximation, in which the number of partial products is reduced from 4 to 3 by removing the last nonzero in the TFC.

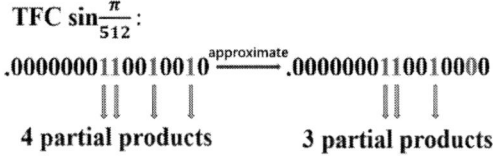

**Fig. 2. An example of approximating TFC $\sin\frac{\pi}{512}$.**

In order to control the nonzero bits in the TFCs with the desired accuracy, we here propose Algorithm 1 in Fig. 3 to repeatedly remove the non-critical nonzero bits from the TFC until it reaches the desired bit-width. In Algorithm 1, if the least significant nonzero bit is followed by another nonzero bit, the least significant bit is flipped to 0 with a carry propagated to the previous bit. This procedure is repeated until the previous bit is 0.

With Algorithm 1, we can approximate TFCs with fewer nonzeros. In left subfigure of Fig. 4, we illustrate the histogram of the nonzero bit reduction rate after the approximation for TFCs in a 1024-point FFT. It is

observed that, with *TFC_length*=16, *num_approximation*=1, we can already reduce quite significant number of nonzero bits, which later saves the energy of multiplication with those approximated TFCs. The right sub-figure of Fig. 4 shows the impact on accuracy with the proposed approximation, which has a maximum relative error less than 2%. Thus, this clearly demonstrates the benefit of the proposed TFC approximation technique with limited accuracy loss but non-trivial energy saving. Moreover, in order to confirm the scalability of the proposed approximation, we study the performance of FFT with different points, from 256 to 2048. As shown in Fig. 5, the average count of nonzero bit is reduced by almost 70% with almost negligible average relative error, which is 2.35e-4 for 2048-point FFT.

---

**Algorithm 1** Algorithm for Twiddle Factor Coefficient Approximation

**Input:** *Original_TFC, TFC_length, num_approximation*

**Output:** *approximated_TFC, error, count_of_bit_1*

$i = TFC\_length$;

**for** $j = 1 : num\_approximation$

    **while** $i > 0$

        **if** *Original_TFC* $* 2^i \% 2 = 1$

            **if** *Original_TFC* $* 2^{i-1} \% 2 = 1$

                *approximated_TFC* $=$ *Original_TFC* $+ 2^{-i}$;

            **else**

                *approximated_TFC* $=$ *Original_TFC* $- 2^{-i}$;

                break;

            **end if**

        **end if**

        $i = i - 1$;

    **end while**

    *Original_TFC* $=$ *approximated_TFC*;

**end for**

*count_of_bit_1* $=$ Number of Nonzeros in *approximated_TFC*;

*error* $=$ *approximated_TFC* $-$ *Original_TFC*;

---

**Fig. 3 Algorithm for TFC approximation.**

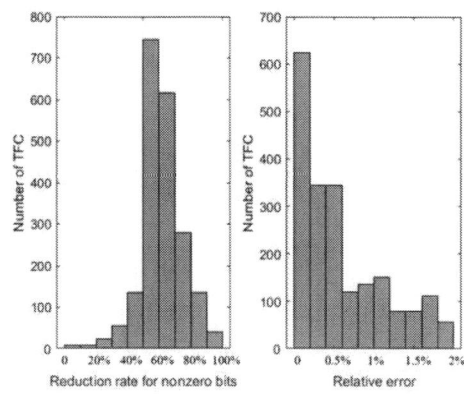

**Fig. 4 Histogram of the reduction rate of nonzero bits for TFCs in a 1024-point FFT (Left); and relative error introduced by the approximation (Right).**

With the proposed approximate TFCs, we further design a low power multiplier using an approximated 3-2 compression circuit. Instead of generating partial products and conducting exact compression, we here use the circuit in Fig. 5 to compute the approximated partial product accumulation, which can replace the 3-2 compressors in the first few stages of a Wallace Tree. The signals of y0, y1, and y2 are the flag

978-1-6654-9759-6/22 $31.00 © 2022 IEEE

bits, which are obtained from TFCs at the beginning and iteratively updated with the circuit in the dashed box. The flag signals are computed only once for every 3 partial product accumulation. The signals of in0, in1, and in2 represent the three bits from the three partial products to be compressed and generate the signals of out0 and out1. This procedure is repeated for every three bits from the three partial products. Instead of actually computing the sum and carry of the three inputs, we can directly drop a bit and output the other two bits when in0 to in2 are all '1'. Since after TFC approximation, there are very few consecutive 3-bit nonzeros, such approximation may induce very small error but speeds up the partial product generation and accumulation. For the last few stages of partial product accumulation, since it is closer to the most significant bit, we use accurate compressor instead.

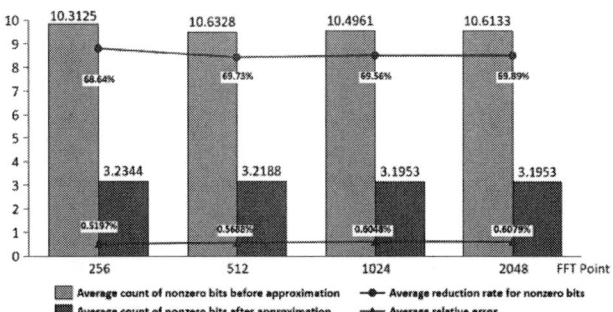

**Fig.5. Average count of nonzero bits before and after approximation (bar chart in light and dark green), average reduction rate for nonzero bits (red line with round dot), and average relative error (purple line with triangle dot) for all TFCs of FFT with different number of points.**

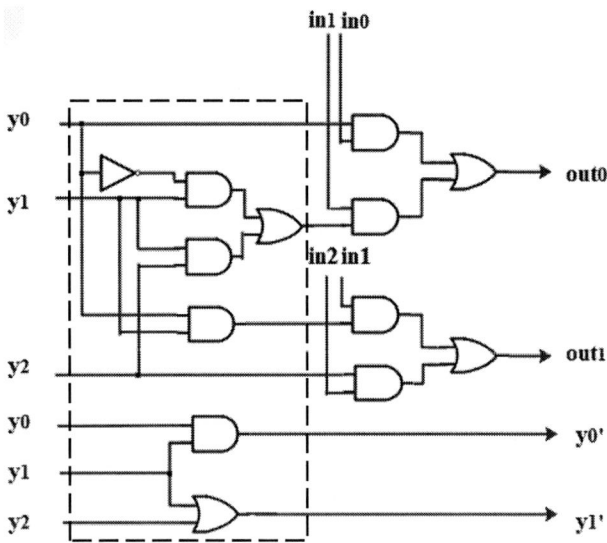

**Fig.5 Approximated 3-2 compressor circuit netlist.**

**TABLE I Comparison between the proposed approximate and accurate multipliers**

| | Power (mW) | Area (um²) | Error |
|---|---|---|---|
| Accurate | 4.5571 | 5886.0172 | |
| Proposed | 2.7171 | 3577.4900 | MRED:0.10554 |
| | | | NRED:0.000981 |
| Relative Saving | 40.4% | 39.2% | |

To evaluate the performance of the proposed approximate multiplier, we conducted experiments on a 16-bit multiplier for 1024-point FFT, which was synthesized using UMC 40nm process and commercial design flow. The approximated compressors were used for the first two stages of the Wallace Tree. We randomly generate 100,000 vectors and compare their multiplication results using the proposed multiplier and an exact multiplier. As shown in Table I, the proposed multiplier has very small error, with a mean relative error distance (MRED) of 0.10554 and normalized relative error distance (NMED) of 0.000981. The power and area are both reduced by 40.4% and 39.2%, respectively, when compared to the original accurate multiplier.

## Conclusion

In this paper, an approximate multiplier is proposed for the multiplication with TFCs in fixed-point FFT design. The approximations are introduced for both TFC representation and compressor design in the multiplier, which is found to save 40% power with very limited accuracy change.

## Acknowledgements

This work was supported in part by the National Key R&D Program of China (Grant No. 2018YFE0126300) and Zhejiang Lab (2021MD0AB02).

## References

[1] P. Duhamel and M. Vetterli, "*Fast Fourier transforms: A tutorial review and a state of the art,*" Signal Processing, vol. 19, pp. 259-299, 1990.

[2] J. Cooley and J. Tukey, "*An algorithm for machine computation of complex Fourier series,*" Math. Comput., vol. 9, pp. 297-301, 1965.

[3] S. Bouguezel, et al., "*Arithmetic complexity of the split-radix FFT algorithms,*" in Proc. IEEE ICASSP, pp. 137-140, 2005.

[4] Y. Hoon, et al., "*Energy efficient 8-bit microprococessor for wireless sensor network applications,*" in ICEAC, pp 147-151,2013.

[5] G. Ganjikunta and S. Sahoo, "*An area-efficient and low-power 64-point pipeline fast Fourier transform for OFDM applications,*" Elsevier Integration, vol. 57, pp. 125-131,2017

[6] J. Deng, et al., "*Energy Efficient Real-Time UAV Object Detection on Embedded Platforms,*" IEEE TCAD, vol. 39, issue 10, pp. 3123-3127, 2020.

[7] V. Ariyarathna, et al., "*Multibeam digital array receiver using a 16-point multiplierless DFT approximation,*" IEEE TAP, vol. 67, no. 2, pp. 925–933, 2019.

[8] C. Guo, et al., "*A Reconfigurable Multiplier for Signed Multiplications with Asymmetric Bit-Widths,*" ACM JETCS, vol. 17, issue 4, pp. 1-16, 2021.

[9] C. Chen, et al., "*Optimally Approximated and Unbiased Floating-Point Multiplier with Runtime Configurability,*" in Proc. IEEE/ACM ICCAD, pp. 1-9, 2020.

[10] B. Liu et al., "*Precision Adaptive MFCC Based on R2SDF-FFT and Approximate Computing for Low-Power Speech Keywords Recognition,*" in IEEE Circuits and Systems Magazine, vol. 21, no. 4, pp. 24-39, 2021.

[11] R. Kaur and T. Singh, "*Design of 32-point mixed radix Fft processor using CSD multiplier,*" in Proc. IEEE PDGC, pp. 538-543,2016.

[12] T. T. Bao Nguyen and H. Lee, *"Shared CSD complex constant multiplier for parallel FFT processors,"* in Proc. IEEE ISOCC, pp. 27-28,2015.

[13] X. Han, et al., *"A Novel Area-Power Efficient Design for Approximated Small-Point FFT Architecture,"* in IEEE TCAD, vol. 39, no. 12, pp.4816-4827,2020.

# A FAST TIMING ANALYSIS AND OPTIMIZATION FOR LATCH-BASED CIRCUITS

*Kaixiang Zhu, Xiao Di, Wai-Shing Luk, Lingli Wang[*] and Jun Tao*

School of Microelectronics, Fudan University, Shanghai 201203, China

*Corresponding Author's Email: luk@fudan.edu.cn

## ABSTRACT

The transparency of latches leads to complicated timing analysis because it allows critical long and short paths to be extended across multiple combinational stages. In this paper, we propose an algorithm to perform the timing analysis of latch-based circuit by modeling it as a minimum cost-to-time ratio problem. Traditionally, timing analysis of latch-based circuit usually uses linear programming (LP) or iterative methods, which may take huge runtime. By using this approach, we can greatly reduce the runtime and optimize the timing performance through the time borrowing on the critical path. Experimental results on the FPGA's VPR benchmarks by reconfiguring flip-flops as latches show that the runtime can be reduced by 136.69% on average with the same critical paths found by LP. Furthermore, our latch-based optimization provides 10.02% performance improvement on average over the original flip-flop-based designs, without changing the placement and routing results.

*Keywords— Latch-based design, timing analysis, timing optimization, minimum cost-to-time ratio*

## I. INTRODUCTION

A traditional flip-flop is composed of two latches which are connected in a master-slave fashion. Therefore, latches have smaller delay, area and clock load than flip-flops. A latch is a level-sensitive device. During the valid period of the clock signal, it is in a transparent state and can transmit data continuously. As a result, each combinational block in a latch-based circuit is not isolated from each other and timing analysis between latches cannot be considered independently, which makes timing analysis of latch-based circuit very difficult. On the other hand, due to the existence of this transparency, latches allow the combinational blocks to have a delay of more than one clock period, commonly known as time borrowing or cycle stealing. Time borrowing has been used in many designs, such as the two giants of the FPGA industry, Intel and Xilinx, both support time borrowing in their own designs 0[2]. Besides latches have higher tolerance to clock skew and clock jitter than flip-flops because the transparency window shifted by skew can still capture data.

Although Intel and Xilinx have already used latches in their products, there are few researches on how to model latch-based circuit timing analysis, and all related existing works have made some assumptions to simplify calculation. For example, [3] and [4] made an assumption that the earliest time data arrives at latch is on the rising edge of the clock to simplify the calculation. In our work, we give a detailed inference calculation process and convert the timing analysis problem into a more complete solution model. The main contributions of this paper are as follows.

1. In this work, we propose a method to model the timing analysis problem for latch-based circuit as a minimum cost-to-time ratio (MC2TR) problem.

2. In our modeling, the timing constraint derivation process is very detailed and complete, without ignoring any parameter to simplify the derivation

3. The corresponding timing analysis algorithm is proposed. In our approach, not only the solving efficiency is greatly improved, but also the critical path can be determined during the solution process.

The remainder of this paper is organized as follows. In section II and III, we review the timing constraints of the latch and formulate the latch timing constraints as a MC2TR problem. We propose our algorithm in section IV. Experimental results are presented in section V. Finally, conclusions of our work are drawn in section VI.

## II. TIMING CONSTRAINTS OF THE LATCH

The main difference between latch and flip-flop timing constraints is the ability of short and long paths to be extended through multiple latches [4]. We assume that all the latches are driven by a pulse with a width $W$ and consider a data path between latch $i$ and latch $j$. In a positive level-sensitive sequential circuit, data can arrive at any time before the falling edge of the clock, which is the setup constraint:

$$A_j \leq T + W - T_{su}. \qquad (1)$$

$A_j$ is the latest arrival time of latch $j$, which shows that the data should arrive at latch $j$ earlier than the setup time($T_{su}$) before the falling edge. $A_j$ can also be computed from the data arrival times of latches connected to latch j through the combinational paths:

$$A_j = \max(T_{cq}, A_i + T_{dq}) + D_{ij}. \qquad (2)$$

Combining equation $A_j \leq T + W - T_{su}.$ $\qquad (1)$ and $A_j = \max(T_{cq}, A_i + T_{dq}) + D_{ij}.$ $\qquad (2)$, we can obtain the setup time constraint between latches $i$ and $j$, which can be described by

$$\max(T_{cq}, A_i + T_{dq}) + D_{ij} \leq T + W - T_{su}. \qquad (3)$$

The earliest arrival time of latch $j$ shows that the data launched from latch $i$ should arrive at latch $j$ no earlier than the hold time($T_h$) after the falling edge of the clock:

$$a_j \geq W + T_h. \qquad (4)$$

Similarly, $a_j$ can also be computed by

$$a_j = \max(T_{cq}, a_i + T_{dq}) + d_{ij}. \qquad (5)$$

Therefore, the hold time constraint can be described by

$$\max(T_{cq}, a_i + T_{dq}) + d_{ij} \geq W + T_h. \qquad (6)$$

The equations and constraints defined in Section II can be solved by linear programming (LP). However, we choose to model it as a MC2TR problem and use Howard's algorithm to solve it more efficiently.

## III. PROBLEM FORMULATION OF

## LATCH TIMING CONSTRAINTS

In this section, we discuss how the constraints derived in Section II can be mapped into a graph problem.

### A. Minimum cost-to-time ratio problem

Let $G = (V, E, \omega, \tau)$ be a strongly connected graph, where $v \in V$ represents the vertices and $e \in E$ represents the edges. Let $c$ and $C$ denote a cycle and the set of all cycles in G, respectively. Let $\omega(c)$ be the sum of the costs of all edges on the path and $\tau(c)$ be the sum of the transit times of all edges on the path. $\rho$ is the ratio of the two sums:

$$\rho(c) = \frac{\omega(c)}{\tau(c)} = \frac{\sum_{e \in c} \omega(e)}{\sum_{e \in c} \tau(e)}. \tag{7}$$

For a strongly connected graph, which contains finite number of cycles, there must be a cycle with the smallest ratio. Finding the cycle with the smallest cycle ratio is what we called the MC2TR problem.

Let us now introduce a new variable $r$ to illustrate it more clearly. With $r$, the graph $G = (V, E, \omega, \tau)$ can be converted to $G = (V, E, \omega - r \cdot t, \tau)$ and $\omega(c)$ can be converted to $(\omega - rt)(c) = \omega(c) - r \cdot \tau(c)$. Let $\rho^*$ be the smallest cycle ratio. Now, let us introduce two theorems.

I.  *When $\rho^*(G) < 0$, there is a negative cycle in the graph G.*

II. *For any real number $r$, $\rho(c) - r = \frac{\omega(c) - r \cdot \tau(c)}{\tau(c)}$ is true.*

The combination of these two theorems leads to the following corollary:

*If $r > \rho^*(G)$, there is one or more negative cycles in $G_r$; if $r = \rho^*(G)$, there is one or more zero cycles in $G_r$; if $r < \rho^*(G)$, there is no negative cycle in $G_r$.*

This is the basis of many algorithms for solving the MC2TR problems. We can describe the MC2TR problem as its dual linear form:

$$\begin{aligned} \max \quad & r \\ \text{s.t.} \quad & d_v \le d_u + (\omega_{u,v} - r \cdot \tau_{u,v}), (u,v) \in E. \end{aligned} \tag{8}$$

### B. Timing analysis of latch-based circuit

We should formulate the problem of latch-based timing analysis as a MC2TR problem before proposing the algorithm. The timing constraints we discussed before can be summarized as follows :

$$\begin{cases} \max(T_{cq}, A_i + T_{dq}) + D_{ij} \le T + W - T_{su} \\ \max(T_{cq}, a_i + T_{dq}) + d_{ij} \ge W + T_h \end{cases} \tag{9}$$

To illustrate the problem and analysis, we introduce some new variable, $a_i$ and $A_i$ to represent the earliest and latest arrival time of latch $R_i$; $d_i$ and $D_i$ to represent the earliest and latest departure time of latch $R_i$. The path delays $d_i$ and $D_i$ can be described by

$$d_i = \max(a_i + T_{dq}, T - W + T_{cq}), \tag{10}$$

and

$$D_i = \max(A_i + T_{dq}, T - W + T_{cq}). \tag{11}$$

Similarly, the earliest and latest arrive time $a_j$ and $A_j$ of $R_j$ can be described by

$$a_j = \left(\min_{\forall i \sim j}[d_i + d_{ij}]\right) - T, \tag{12}$$

$$A_j = \left(\max_{\forall i \sim j}[D_i + D_{ij}]\right) - T. \tag{13}$$

After we introduce these new variables, we can formulate the setup constraint and hold constraint as follows:

$$T_h \le a_j, \tag{14}$$

$$A_j \le T - T_{su}. \tag{15}$$

To reduce the difficulty of solving, we still need to linearize these constraints. We will use linear relaxation to linearize the maximum and minimum functions. To transform the linear model into a MC2TR problem, we need a new variable $r$ to convert the optimization objective of the model from minimizing $T$ to maximizing $r$.

$$r = \frac{1}{2}(T_0 - T). \tag{16}$$

Also, we introduce a reference point t to ensure that $a$, $A$, $d$, $D$ are solved relative to the same time reference point. Now, we can express the equivalent MC2TR model as follows:

TABLE I MC2TR MODEL OF LATCH CIRCUIT TIMING ANALYSIS

| maximize | $r$ |
|---|---|
| s.t. | $(1)\, t - a_j \le -T_h$ |
| | $(2)\, A_j - t \le T_0 - 2r - T_{su}$ |
| | $(3)\, a_j - A_j \le 0$ |
| | $(4)\, d_j - D_j \le 0$ |
| | $(5)\, a_j - d_{i1} \le d_{i1 \sim j} - T_0 + 2r$ |
| | $\cdots\cdots$ |
| | $a_j - d_{in} \le d_{in \sim j} - T_0 + 2r$ |
| | $(6)\, D_{i1} - A_j \le -D_{i1 \sim j} + T_0 - 2r$ |
| | $\cdots\cdots$ |
| | $D_{in} - A_j \le -D_{in \sim j} + T_0 - 2r$ |
| | $(7)\, a_i - d_i \le -T_{dq}$ |
| | $t - d_i \le -T_0 + 2r + W - T_{cq}$ |

After we obtain the optimal solution $r$, the corresponding minimum clock period is

$$T_{min} = T_0 - 2r. \tag{17}$$

Let $G = (V, E, \omega, \tau)$ represent the graph of the MC2TR problem, each constraint corresponds to an edge in $G$, which represented by $E_{ta}$, $E_{At}$, $E_{aA}$, $E_{dD}$, $E_{a_jd}$, $E_{DA}$, $E_{a_id}$, $E_{td}$, $E_{AD}$ and $E_{tD}$. The four variables $a$, $A$, $d$, $D$ of each latch correspond to one vertex. Since the constant term does not affect the form of the problem and the generality of the solution process, in order to facilitate subsequent calculations and experiments, we pre-set the clock signal of 50% duty cycle and set the parameter $T_{su}$, $T_{cq}$, $T_h$, $T_{dq}$ to 0. Then, we can construct the following cost function and conversion time function:

$$\omega(u,v) = \begin{cases} -T_h, (u,v) \in E_{ta} \\ T_0 - T_{su}, (u,v) \in E_{At} \\ d_{uv} - T_0, (u,v) \in E_{a_jd} \\ -D_{uv} + T_0, (u,v) \in E_{At} \\ -\frac{1}{2}T_0, (u,v) \in E_{td} \cup E_{tD} \\ 0, (u,v) \in E_{aA} \cup E_{dD} \cup E_{a_id} \cup E_{AD} \end{cases} \tag{18}$$

$$\tau(u,v) = \begin{cases} 2, (u,v) \in E_{At} \cup E_{DA} \\ -2, (u,v) \in E_{a_jd} \\ -1, (u,v) \in E_{td} \cup E_{tD} \\ 0, (u,v) \in E_{ta} \cup E_{aA} \cup E_{dD} \cup E_{a_id} \cup E_{AD} \end{cases} \tag{19}$$

## IV.  THE PROPOSEDALGORITHM

The most famous and very efficient cycle-based algorithm for obtaining the optimal solution to the MC2TR problem is Howard's algorithm [5][5]. The method of the cycle-based algorithm is to start with a larger $r$ ($r \geq r^*$) and maintain an upper bound on the optimal solution. In each iteration, we will perform the shortest path judgment and relaxation operations and check for the presence of negative cycles in $G_r$. If there is a negative cycle, the current $r$ is reduced, otherwise the optimal solution is found. The input of this algorithm is $G = (V, E, \omega, \tau)$, which represents the set of constraints, and the accuracy $\varepsilon$. It maintains a policy graph $G_p$ in the running process, which is a simple subgraph of G and an out-degree of 1 for each node.

## V. EXPERIMENTAL RESULTS

Our approach has been implemented with the VPR 8.0 framework. In this work, we first restore the circuit timing information in the form of graph, in which each node corresponds to a sequential element and the edge represents the connection relationship of the element. Each edge is associated with two attributes, which are the maximum and minimum combinational logic delays of the data path corresponding to the edge. Then, the timing analysis problem of the latch-based circuit is constructed as a LP model or a MC2TR model and store the constraints in the form of graphs for the convenience of subsequent solution.

We carried our experiments on a set of sequential circuits taken from the VPR 8.0 benchmarks. In TABLE II, we compare the results of the MC2TR model and the LP model to verify the advantages of the MC2TR model. Column $|V|$ and $|E|$ respectively represent the number of vertices edges in the model, which can intuitively show the scale of the problem. $T_{init}$ is the longest path delay in the original flip-flop-based design circuit, which also the minimum cycle period of the circuit. $T_{opt}$ is the minimum cycle period of the latch-based design circuit solved by the MC2TR model. The next column compares the run-time between the LP model and the MC2TR model, from which we can find the run-time can be reduced on average by 136.69% with the same critical paths found by the LP. Furthermore, the last column lists the improvement of $T_{opt}$ relative to $T_{init}$, which represents the comparison of the minimum clock period of the latch-based circuit and the flip-flop-designed circuit. Our latch-based optimization provides a 10.02% performance improvement on average by time borrowing compared to original design based on flip-flops without changing the placement and routing results.

## VI. CONCLUSION

In this paper, we have carried out a complete and detailed derivation process to model latch-based circuit timing constraints into a MC2TR problem and the corresponding algorithm is proposed to solve the timing analysis of latch-based circuit. Results in table and table show that the run-time can be greatly reduced by using the MC2TR methods.

TABLE II BENCHMARK RESULTS OF HOWARD'S ALGORITHM

| Bench-mark | $T_{init}$ (ns) | $T_{opt}$ (ns) | Howard (ms) | GLPK (ms) | speed-up |
|---|---|---|---|---|---|
| s1 | 9.647 | 7.523 | 0.013009 | 0.699436 | 53.76 |
| s27 | 8.527 | 7.198 | 0.020442 | 1.08714 | 53.18 |
| s208.1 | 4.486 | 4.127 | 0.019696 | 0.892371 | 45.30 |
| s298 | 18.138 | 12.554 | 0.024379 | 0.981526 | 40.26 |
| s344 | 4.541 | 3.4815 | 0.045618 | 3.39776 | 74.48 |
| s349 | 4.747 | 3.806 | 0.023859 | 2.07112 | 86.80 |
| S400 | 3.785 | 3.24 | 0.03878 | 3.19796 | 82.46 |
| s420.1 | 4.752 | 4.4775 | 0.036439 | 2.163 | 59.36 |
| s444 | 3.566 | 3.32 | 0.027957 | 3.21331 | 114.94 |
| s510 | 4.346 | 3.8435 | 0.026118 | 0.980705 | 37.55 |
| s526n | 3.978 | 3.4835 | 0.035538 | 7.23576 | 203.60 |
| s526 | 3.722 | 3.371 | 0.064419 | 2.32209 | 36.04 |
| s641 | 10.053 | 10.053 | 0.021092 | 3.77171 | 178.82 |
| s713 | 9.065 | 9.065 | 0.025728 | 3.1443 | 122.21 |
| s832 | 4.564 | 4.533 | 0.008516 | 2.65581 | 311.86 |
| s838.1 | 9.072 | 8.532 | 0.06167 | 7.2687 | 117.86 |
| s953 | 4.38 | 4.212 | 0.091415 | 7.61123 | 83.26 |
| s1423 | 17.083 | 11.3887 | 0.127741 | 19.3181 | 151.23 |
| s1494 | 5.146 | 4.721 | 0.016296 | 0.871056 | 53.45 |
| s5378 | 8.037 | 7.82 | 0.839758 | 78.8612 | 93.91 |
| s38417 | 13.223 | 12.629 | 1.09306 | 917.946 | 839.79 |
| s38584 | 9.966 | 9.966 | 2.58147 | 1219.89 | 472.55 |
| sand | 5.301 | 4.9815 | 0.022768 | 0.644399 | 28.30 |
| diffeq | 6.842 | 6.483 | 0.499624 | 192.137 | 384.56 |

## REFERENCES

[1] Ganusov I, Devlin B. Time-borrowing platform in the Xilinx UltraScale+ family of FPGAs and MPSoCs[C]//2016 26th International Conference on Field Programmable Logic and Applications (FPL). IEEE, 2016: 1-9.

[2] Chromczak J, Wheeler M, Chiasson C, et al. Architectural Enhancements in Intel® Agilex™ FPGAs[C]//Proceedings of the 2020 ACM/SIGDA International Symposium on Field-Programmable Gate Arrays. 2020: 140-149.

[3] X. Di, W. -S. Luk and L. Wang, "Multi-parameter Timing Optimization for Pulsed-Latch Circuits," 2020 IEEE 15th International Conference on Solid-State & Integrated Circuit Technology (ICSICT), 2020, pp. 1-3, doi: 10.1109/ICSICT49897.2020.9278397.

[4] B. Teng and J. H. Anderson, "Latch-Based Performance Optimization for FPGAs," 2011 21st International Conference on Field Programmable Logic and Applications, 2011, pp. 58-63, doi: 10.1109/FPL.2011.21.

[5] Cochet-Terrasson J, Cohen G, Gaubert S, et al. Numerical computation of spectral elements in max-plus algebra[J]. IFAC Proceedings Volumes, 1998, 31(18): 667-674

# A TRANSIENT-IMPROVED SPIKE TIME REDUCTION CIRCUIT FOR LDO

*Zongyuan Zheng, Chen Zhang, Bo Wang[*] and Xinan Wang*

The Key lab of IMS, School of ECE Peking University Shenzhen Graduate School,
GuangDong, China

*Corresponding Author's Email: wangbo@pkusz.edu.cn

## ABSTRACT

In this paper, a spike time reduction circuit is proposed for improving the transient response of low-dropout (LDO) voltage regulator. Test results show that the circuit can significantly reduce the overshoot recovery time of output voltage after the sudden load current decrease. The main part of presented circuit includes an asymmetrical comparator and a discharging transistor. By setting a lower comparator input reference voltage value or increasing the size of the discharging transistor, the overshoot recovery time can be reduced from 240.5μs to 126.1μs during load current change from 1mA to 200mA. The prototype of the circuit is fabricated by 0.18um CMOS processes. Its consumption is only 102.4nA. The circuit has been verified on a commercial LDO production.

## INTRODUCTION

The low dropout nature of the regulator makes it suitable for the increasing demand of portable electronic devices. In addition, the increased level of integration in portable devices not only demands the LDO to deliver high load current, but also requires the low load current for its start and standby requirements to lower the consumption [1]-[3]. Previously, the works are mainly focused on reducing the magnitude of spike. [4] uses dynamic-replica LDO regulator to suppress the spike. [5] uses assisted pass-transistors and control circuit to smooth the spike. [6] realizes the comparator control and [7] increases the bias current. Above methods indeed reduce the amplitude of spike when the load change is about 50mA. However, while the load change is over 50mA, their effectiveness gets worse, and the overshoot time becomes too long when the load current changes from 200mA to 1mA, which is very common in practical applications.

Figure 1 illustrates a standard analog LDO structure diagram. Given the stability of output voltage, its loading condition is comprised of filter capacitor $C_{out}$, associated equivalent series resistance $R_{esr}$ and output load current.

When the load current suddenly increases, before the feedback loop responds, the increased load current is provided by filter capacitor [7]. At this time, the output voltage will be lower than the regulated value until the negative feedback loop response is established, and the power PMOS provides more current which restores the voltage of the filter capacitor and the output voltage to a

new regulated value corresponding to the new load current.

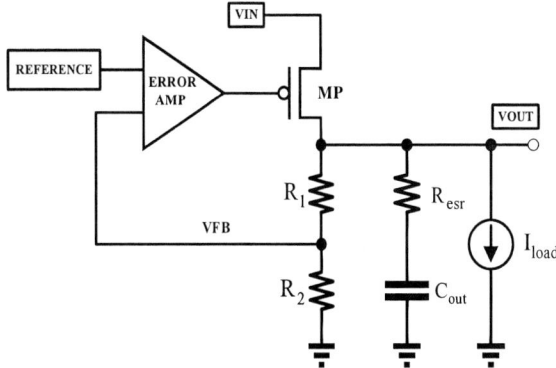

*Figure 1: Typical structure of LDO*

Similarly, when the load current suddenly decreases, the excessive current charges the output filter capacitor that makes the output voltage higher than the regulated value [7]. It takes more time to regulate with the larger change of load current.

This paper proposes some techniques to reduce the spike recovery time. Verified on a commercial LDO, with the output current of 200mA, the proposed circuit can reduce the spike time from 240.5μs. to 126.1μs.

## CIRCUIT IMPLEMENTATION

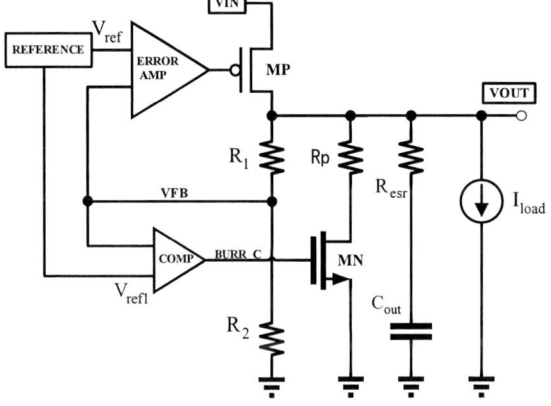

*Figure 2: Structure of proposed circuit*

The proposed circuit can speed up the time of output voltage returning to the target regulation value after the load resistance suddenly decreases. This circuit consists of an asymmetrical comparator, a discharge NMOS MN and a drain protection resistor $R_p$. Figure 2 shows the structure of the circuit, and the comparator circuit used is illustrated in Figure3.

The proposed circuit has been added to an existing LDO with output voltage of 3.3V. $V_{ref}$ =1.2V is the reference voltage of ERROR AMP(EA) and the voltage $V_{ref1}$ =1.25V is sent to the comparator also generated by the reference source circuit. VFB is the feedback voltage. When the regulator works normally, VFB≈1.2v, so the feedback coefficient β=1.2/3.3.

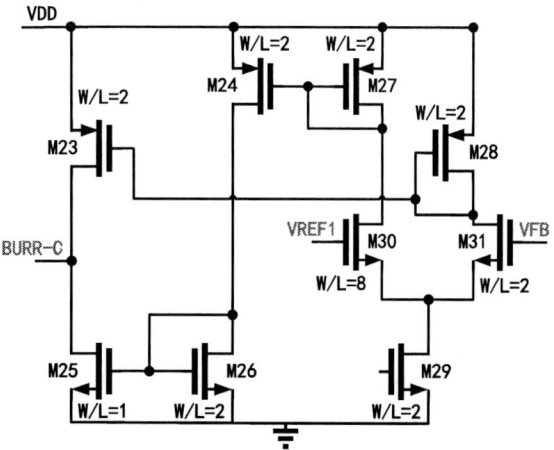

*Figure 3: Structure of asymmetrical comparator*

Now we calculate the flip threshold ΔTR of the comparator in Figure 3, assuming all the transistors of the comparator work in the saturation region. If $VFB < V_{ref1}$, the comparator outputs a low level, and as $VFB$ increases, the output voltage of the comparator begins to flip.

$$\frac{1}{2}kn * \frac{8W}{L} * (V_{ref1} - V_s - V_{th})^2 * \frac{1}{2}$$
$$= \frac{1}{2}kn * \frac{2W}{L} * (V_{FB} - V_s - V_{th})^2 \quad (1)$$
$$\Delta TR = V_{FB} - V_{ref1} = (\sqrt{2} - 1)(V_{ref1} - V_s - V_{th}) \quad (2)$$

where kn is the coefficient of NMOS and $V_s$ is the source voltage of M29. If the load current decreases suddenly, the output voltage is higher than the regulated voltage initially. Once $V_{out} * \beta - V_{ref1} > \Delta TR$, the comparator outputs a high level, and the MN is turned on. Currently, the time constant of the discharging circuit is dependent on $C_{out}$ and $(R_{on} + R_p)$. $R_{on}$ is the on resistance of MN. In this spike time reduction circuit, $R_p$=70Ω, $R_{on}$≈24Ω and at light load condition, the value of equivalent load resistance is at least a few hundred ohms. Therefore, after

adding about 94 ohms load resistance in parallel, the discharge time constant can be significantly reduced.

With continuously discharging, $V_{FB} - V_{ref1}$ will be smaller than ΔTR. Afterwards, MN has been turned off, and the time constant of the discharging loop returns to (5). According to the above analysis if $V_{FB} - V_{ref1} > \Delta TR$, the discharging loop will accelerate the discharge process, so the regulator can recover to the regulated voltage faster.

## SIMULATION RESULT

The proposed spike time reduction circuit is simulated by 0.18μm CMOS technology. Simulation environment sets $V_{in}$ =3.8V, $C_{out}$ =1 μF, $R_{esr}$ =100mΩ, the output is 3.3V, the equivalent load resistance changes from 11Ω to 3.3kΩ, and then back to 11Ω. The time interval is 500μs and the load change time is set to 1ns. First, we simulate the transient waveform of the LDO output voltage without the spike time reduction circuit. The simulation result is shown in Figure 4(a). $t_{ry1}$ is the recovery time when load current suddenly decrease and $t_{ry2}$ is the recovery time of the opposite situation.

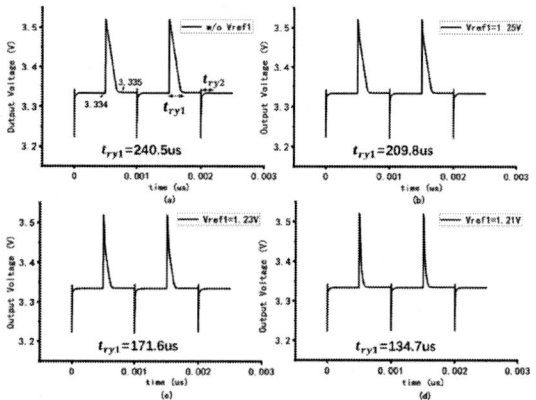

*Figure 4: Transient waveform of LDO Output voltage with different Vref1: (a)w/o (b)1.25V (c)1.23V (d)1.21V*

Figure 4(a) shows that load resistance is equal to 11Ω and the LDO output voltage is 3.334V. If the load resistance is equal to 3.3kΩ, LDO output voltage is 3.335V. $\Delta_{LDR}$ is only 1mV. As previously expected, $t_{ry1}$ is significantly longer than $t_{ry2}$ .Without the spike time reduction circuit, $t_{ry1}$ is 240.5μs and $t_{ry2}$ just only 86μs.

Adding the proposed circuit and setting $V_{ref1} = 1.25V$, repeat above load current change situation. The transient output voltage change is shown in Figure 4(b). Currently, $t_{ry1}$ is 209.8μs. Compared with Figure 4(a), the slope of the initially discharging procedure becomes larger. It is precisely because $V_{FB} - V_{ref1} > \Delta TR$, so that the comparator outputs a high level to turn on MN causing the discharging time constant to decrease, as shown in Figure 4(b).

Setting $V_{ref1} = 1.23V$ and repeating the above simulation. (2) illustrates that $\Delta TR$ will decrease with $V_{ref1}$ getting smaller. As a result, the turning on time of MN increased, and the discharging time extends. The simulation result is shown in Figure 4(c). $t_{ry1}$ reduced to 171.6μs. On the prerequisite of ensuring $\Delta TR > 0$, keeping lower the value of $V_{ref1}$, $t_{ry1}$ become shorter and shorter.

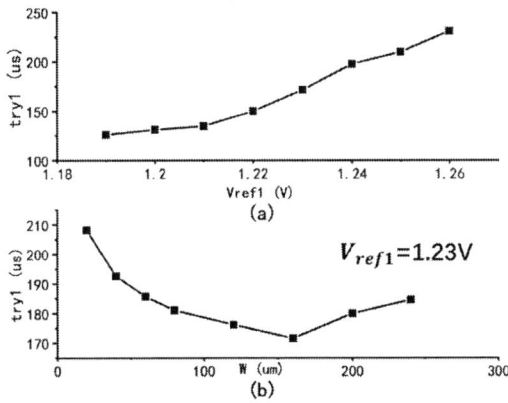

*Figure 5: (a) Relationship between $V_{ref1}$ and $t_{ry1}$*

*(b) Relationship between $t_{ry1}$ and width of MN*

Figure 5(a) shows the $t_{ry1}$ corresponding to different values of $V_{ref1}$. Obviously, under the guarantee of $\Delta TR > 0$, keeping lower the value of $V_{ref1}$, the turning on time of MN is also continuously increased, and this finally close to the discharging time of $C_{out}$. Therefore, the output voltage recovery time also depends on the value of $R_{on}$ and $R_p$.

Figure 5(b) shows the change of the recovery time $t_{ry1}$ corresponding to the varied sizes of width of MN at $V_{ref1}$ =1.23V. The width of MN is relatively small, increasing width can reduce the on-resistance and $t_{ry1}$ will reduce. While considering the width of MN is large enough, increasing width just gives a little contribution to reduce the on-resistance, but will increases the gate capacitance causing turn-on time of MN become longer which means $t_{ry1}$ will increase slightly.

## TEST RESULT

Figure 6(left) shows the measured waveform results without adding the proposed circuit during output load current jumps between 1mA and 200mA. When the load current suddenly becomes smaller, the overshoot recovery time $t_{ry1}$ is close to 275μs. Figure 5(right) shows the measured waveform results after adding the proposed circuit. Under the same test conditions, when $V_{ref1}$ is 1.2V,

the recovery time $t_{ry1}$ is reduced to nearly 138μs, which is close to the simulation value of 131.1μs.

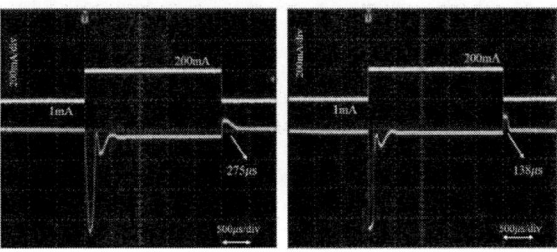

*Figure 6: Output waveform of LDO*

## CONCLUSION

In this paper, a spike time reduction circuit for improving LDO transient response is presented, mainly implemented by an asymmetric comparator. The overshoot recovery time reduces from 240.5μs to 126.1μs during load current change from 1mA to 200mA. It consumes 102.4nA and has been verified on a commercial LDO product.

## REFERENCES

[1] M. Al-Shyoukh, et al, "A Transient-Enhanced 20μA-Quiescent 200mA-Load Low-Dropout Regulator with Buffer Impedance Attenuation," IEEE Custom Integrated Circuits Conference, 2006, pp.615-618.

[2] P. Luo, Z. Liu, L. Huang and S. Zhen, "A Fast-Response NMOS-LDO Voltage Regulator without on-chip Compensated Capacitor," International Midwest Symposium on Circuits and Systems (MWSCAS), 2018, pp.611-614.

[3] C. Hsieh, et al, "A low-dropout regulator with smooth peak current control (SPCC) topology for over current protection," (ICECS), 2009, pp. 363-366.

[4] G. Ma, C. Zhan and Y. Zhang, "A Transient-Improved Dynamic-Replica LDO Regulator with Bulk Modulation," IEEE International Conference on Electron Devices and Solid-State Circuits (EDSSC), 2018, pp.1-2.

[5] X. Tong and K. Wei, "A Fully Integrated Fast-Response LDO Voltage Regulator with Adaptive Transient Current Distribution," IEEE Computer Society Annual Symposium on VLSI (ISVLSI),2017, pp. 651-654.

[6] M. Amayreh, J. Leicht and Y. Manoli, "A 200ns settling time fully integrated low power LDO regulator with comparators as transient enhancement," IEEE International Symposium on Circuits and Systems (ISCAS),2016, pp. 494-497.

[7] P. Y. Or and K. N. Leung, "An Output-Capacitorless Low-Dropout Regulator with Direct Voltage-Spike Detection," IEEE Journal of Solid-State Circuits,

2010, vol. 45, no. 2, pp. 458-466.

978-1-6654-9759-6/22 $31.00 © 2022 IEEE

# RULE CHECK OF PAD PLACEMENT IN IC LAYOUT WITH YOLO V3

*Chunxi Lin[1], Tao Su[2]*

[2] School of Electronics and Information Technology, Sun Yat-sen University
Guangzhou, Guangdong, 510006, China
*Corresponding Author's Email: sutao@mail.sysu.edu.cn

## ABSTRACT

Deep learning has been widely applied in the image processing. However, few works have been done on the layout check of integrated circuits. This paper tests the capability of Yolo v3 in identifying the rule violation of the IC layout. The test case is the VDD and VSS pad arrangement of a 44-pin die. The result is quite positive. All error spots in the test set have been identified. Even the layout is irregularly shaped, Yolo can still find all errors. This work shows that deep learning has great potential in integrated circuit layout inspection, which is worthy of further exploration.

*Keywords—integrated circuits; layout check; deep learning; Yolo; pad arrangement*

## INTRODUCTION

Layout check is critical for the success of type-out of integrated circuits. Calibre of mentor graphic is popular commercial tool to perform layout check. It explores the original design data of the IC layout and identifies spots of rule violation. Another possible way for the layout check is through the visual technology. Errors can be found by looking at the images of the chip layout. Visual inspection is friendly for human nature. It helps the project manager to understand more precisely the work of the design engineers and even to point out the mistakes in the work.

Yolo is developed in 2016. It has great success in the object recognition. Human faces, animals, plants and so on can be detected in images. However, how Yolo performs in layout check is still a question.

The motivation of the paper is to test the capability of Yolo in layout rule check. To build a visual inspector, we are also interested in tolerance of Yolo on irregularity of the images. As an initial study, we apply Yolo to detect the pad mis-arrangement of the VDD and VSS pins. Section 2 introduces the rule to be checked. Section 4 illustrates the experiment setup. Section 5 shows the experiment results. The last section concludes the paper.

## LAYOUT RULE TO BE CHECKED

This paper focuses on the pad arrangement of the power and ground pins. Fig.1 shows a layout of 20 pads. Pad 1 and pad 13 form a power and ground pair. The two pads are on the opposite sides of chip resulting a large loop in the supply network. That is bad arrangement. Pad 7 and pad 8 form another power and ground pair. The two pads are close to each other which brings a small inductance in

the supply network and generate low electromagnetic emission. Therefore is a good arrangement.

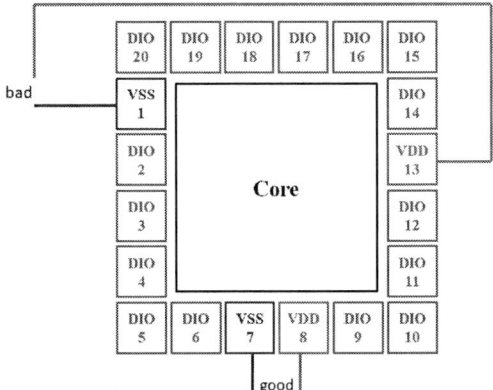

*Fig.1. Example of the pad arrangement of an IC layout*

The criterion for the layout check is following: the rule is passed if they are placed near each other while the rule is failed if they are separated by other pads.

## NEURAL NETWORKS

There are many models of deep convolutional neural networks, among which the main ones used for target detection and entered into practical use are Faster R-CNN, SSD and Yolo.[1] In this paper, the v3 version of Yolo [2] is used for version checking.

Yolo v3 shows superior performance in the detection of small targets. It is also more accurate and stable in target frame position detection and object localisation. The bounding box generated by Yolo v3 is almost identical in size and closer to the true size of the target for the same size sample detection.

## EXPERIMENT SETUP

A 44-pin layout was used as an experimental example, see Fig.2. 100 sample images were first generated using a self-written python script, which was subsequently labelled using labelImg and then a dataset was generated.

The targets were divided into two categories: 'right' and 'wrong'. For VDDs and VSSs that are adjacent to each other on the same edge we frame them and label them as right; for VDDs and VSS pads that do not conform to the layout specification we frame them together with the common pads or pads of the same polarity that are adjacent to them on the left and right and label them as wrong as shown in Fig.2.

*Fig.2. Marked graphic*

After the dataset was created, we randomly divide it into three parts: training set, test set and detect set, obtaining the file name and address of each part.

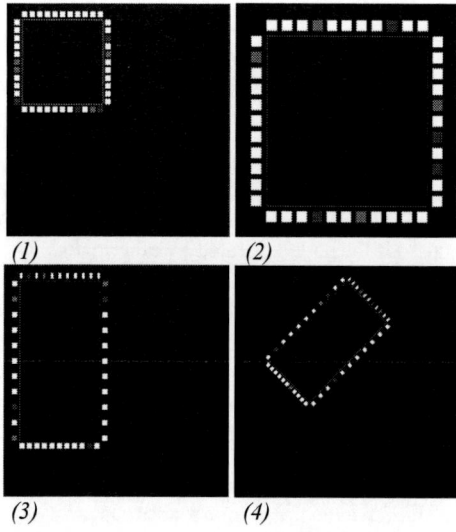

*Fig.3. Irregular sample: (1) shrunk, (2) enlarged, (3) stretched in one direction, (4) rotated.*

The trainer was invoked to train the training set. After two hundred iterations of the gradient descent algorithm, the best weight file weight1.pt was generated. The IC layout checker (hereinafter called checker) was then invoked to test the weights.

Additional rregular datasets were produced to test the generalisation capabilities of the checker: (1) shrunk, (2) enlarged, (3) stretched in one direction (hereinafter called stretched), (4) rotated, as shown in Fig.3.

A test set of irregular samples was tested and then retrained three times for the weaknesses exposed in the test. The characteristics of the four training sessions and the weights obtained are shown in Table 1.

*TABLE I. Marking of weights*

| Training | Dataset | Weight |
|---|---|---|
| First | regular | weight1.pt |
| Second | shrunk or enlarged | weight2.pt |
| Third | stretched | weight3.pt |
| Fourth | rotated | weight4.pt |

## EXPERIMENT RESULTS

First, we used the regular training set for training. The generated weights has superior performance. When the confidence lies in the interval about 0.02 to 0.75, the average effect of recall and accuracy (hereafter referred to as the "F1 value") increases nearly linearly from approximately 0.70 to a maximum value of 0.94 (where confidence=0.524), and then decreases to about 0.75 at the maximum confidence, as shown in Fig.4.

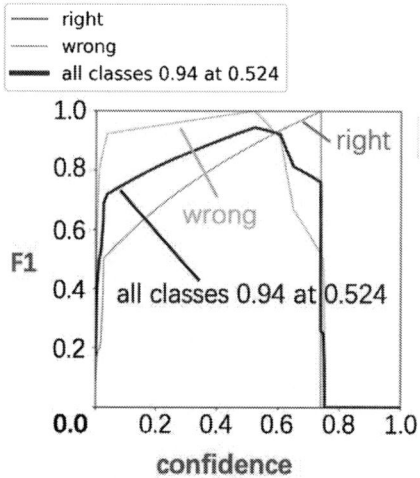

*Fig.4. F1 curve after the first training*

*Fig.5. The detection effect of the regular sample*

The results showed a 100% accuracy rate and a recall rate of over 99% for the round. This shows that the checker has demonstrated excellent performance in the detection of similar samples, as shown in Fig.5.

The second round of experiments is for testing the generalisation ability of the checker, still using the weight

weight1.pt for irregular samples, and the results are shown in Fig.6. and Table 2 (first time). The layout checker has a low detection accuracy for isometric reduced samples and a low check-all rate for gap changes and angle changes. The accuracy and recall rates of the layout checker for irregular samples are not high.

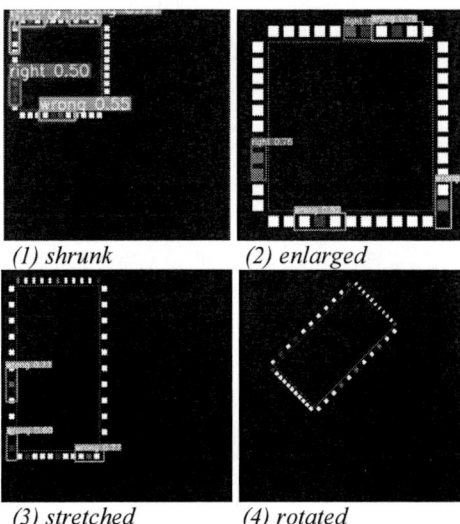

*(1) shrunk*          *(2) enlarged*

*(3) stretched*          *(4) rotated*

*Fig.6. The detection effect of the irregular sample after the first training.*

TABLE II.    *The detection effects of various samples*

| Training | Test Set | Accuracy | Recall Rate |
|---|---|---|---|
| weight1 | regular | 1.00000 | 0.99464 |
| | shrunk | 0.82979 | 0.59091 |
| | enlarged | 0.96970 | 0.90141 |
| | stretched | 0.85417 | 0.49400 |
| | rotated | 0.00000 | 0.00000 |
| weight2 | shrunk | 0.90164 | 0.85938 |
| | enlarged | 0.92105 | 0.98952 |
| | stretched | 0.91490 | 0.96629 |
| | rotated | 0.92308 | 0.80899 |
| weight3 | shrunk | 0.98438 | 0.90000 |
| | enlarged | 1.00000 | 0.87654 |
| | stretched | 0.97531 | 0.96341 |
| | rotated | 0.92593 | 0.90361 |
| weight4 | shrunk | 0.95455 | 0.88732 |
| | enlarged | 0.93056 | 0.93056 |
| | stretched | 0.94048 | 0.95181 |
| | rotated | 1.00000 | 0.96429 |

In the third round of experiments, scaling samples were added for training to obtain weight2.pt and then tested on irregular samples, and the results are shown in Table 2 (weight2). The detection performance of checker

for small-sized targets was substantially improved. In this round of detection, the bounding boxes of small targets fit the objects more closely and the detection accuracy rose.

In the fourth round of experiments, the datasets of samples stretched in one direction was added for training. The anchoring of the checker for small target borders was further optimised. It also has a higher recognition accuracy for irregular rectangular pads with irregular width gaps. A recall rate of over 90% has been achieved for rotated samples. However, as no training has yet been carried out for this sample, the bounding box is not framed in a regular way and the accuracy rate is poor.

In the fifth round of experiments, the rotated dataset was added for training. The generalisation performance of the checker was further improved. For the irregular samples, the accuracy of the checker met expectations. At the same time, the bounding box fit on the detection of small targets was reduced due to the improved generalisation capability. The accuracy of the checker has been fully achieved for both regular and irregular samples.

TABLE III.    *Performance of the plate checker*

| Training | Maximum F1 | Corresponding Confidence |
|---|---|---|
| weight1 | 0.94 | 0.524 |
| weight2 | 0.99 | 0.641 |
| weight3 | 1.00 | 0.577 |
| weight4 | 0.91 | 0.363 |

The first three training directions focused on improving the checker's detection accuracy for small samples and irregular samples. While the fourth training round required a more universal and inclusive labelling approach for the samples. After the fourth training round, the checker's detection results for all types of samples tended to be even, and the detection results for similar samples and non-skewed shaped samples decreased compared to the previous training rounds as shown in Table 3. After the targeted training, the plate checker showed a strong adaptability to changes in pad layout rules.

## CONCLUSION

Yolo v3 shows excellent performance in rule check of pad placement in IC layout and demonstrates strong generalization capabilities. This study shows that there is great potential in layout check in Yolo v3, which is worthy of continuous exploration and mining.

## REFERENCES

[1] Y. Zhao, Y. Rao, S. P. Dong, and J. Y. Zhang, A review of deep learning target detection methods, Chinese Journal of Image Graphics, 25(4), 629-654

[2] Redmon J and Farhadi A, Yolov3: an incremental improvement [EB/OL], [2019-06-20], https://arxiv.org/pdf/1804.02767.pdf

## A

Ackerl Norbert
Ai Juan

## B

Ba You
Bao Chao
Bao Rui
Bao Shuchao
Bao Yu
Bian Yu Yang
Bian Zheng
Boh Kelvin
Buron Jonas D

## C

Cai Shenqi
Cai Xinchen
Cai Yaguo
Cai Ying
Cao Lei
Cao QiuFeng
Cao Xiaoqing
Cao Yamin
Cao Yanpeng
Cao Zherui
Cao Zhijun
Chai Jingrui
Chan Lai-Choon
Chang Sean
Chao Shuanshe
Che Siyuan
Che Yongqiang
Chen Canny
Chen Chang
Chen Chen
Chen C-Z
Chen Guilin
Chen Haihua
Chen Hao
Chen Hualun
Chen Hui
Chen Hunglin

Chen Jiabin
Chen Jianli
Chen Jin
Chen Kecheng
Chen Kun
Chen Phil
Chen Shanshan
Chen Tian
Chen Tong
Chen Wei
Chen Weijie
Chen Wenyuan
Chen Wi-Lun
Chen Xi
Chen Xinyu
Chen Xu
Chen Xuandbo
Chen XueQuan
Chen Xun
Chen Yang
Chen Ying-Yi
Chen Ying-Yin
Chen Yong
Chen YongBo
Chen Yongyue
Chen Yu
Chen Yu De
Chen Zhu
Cheng Ming
Cheng Qilin
Cheng Weichi
Cheng Xingming
Cheng Xinhua
Cheng Yingbo
Chi Min-Hwa
Chi Yu Shan
Chi Yushan
Chou J J
Chu Zhenghui
Chu Zhufei
Claeys Cor
Condo Eric


# **D**
Dai Genting

Dai Hongwei
Dai  Johnny
Dan Yang
Dang Qi
Danli Gong
Deng Xiaonan
Di Xiao
Diao Xuling
Ding Huwen
Ding Li
Ding RongZheng
Ding Shenglan
Ding ShiJin
Dong Lisong
Dong Peipei
Dong Qian
Drechsler Rolf
Du Yide
Du Yihang
Duan Wenting
Duan Yangyang
Dube Belinda Langelihe Yolanda

## **E**
Ervin Joseph

## **F**
Falise Frédéric
Fan Dandan
Fang J X
Fang Jianmin
Fang Jingxu

Fang Jingxun

Fang Mingxu
Fang Robb
Fang Yanfen
Fang Ziquan
Feng Ji
Feng Long
Feng Peng
Feng Qin
Fu Xu
Fu Jiayi
Fuh Yiin-Kuen

Fujimori Toru

## G

Gan Xin
Gan Zhi-feng
Gao Bin
Ge Qiang
Goh Clinton
Gong Bin
Grüter Rudolf
Gu Jiawei
Gu Junwei
Gu Lin
Gu Ming
Guo Hao
Guo Xiaobo
Guo Xin
Guo Xinfei
Guo Yaozu
Guo Yuning
Guo Zhao
Guo Zhong Ning
Guralnik Benny

## H

Hansen Mikkel F
Hansen Ole
Hao Yanxia
Han Chao
Han Dejun
Han Jianglong
Han Tianyu
Han Xin-Ru
He Ting
He Wei
He Xin
He Yimin
He Zhidan
He Zhongyi
Hendric Bryan
Henrichsen Henrik H
Herkommer Michael
Hoemer Bridger
Hong Jiaqi
Hong Q X

Horiguchi Naoto
Horn George W
Hou Bin
Hou Jianqiu
Hou wenrong
Hou Ziyang
Hsu Chien-Pin Sherman
Hu Chuan
Hu Jiawei
Hu Jiuli
Hu Lilei
Hu Shaojian
Hu Xiangang
Hu Yongjie
Hu Zengwen
Hu Zhihui
Huang Haishan
Huang Huojun
Huang Jacky
Huang Jian
Huang Jun
Huang Po-Yu
Huang Qianqian
Huang Ran
Huang Ru
Huang Rutian
Huang Shan
Huang Xi
Huang Xin-Wen
Huang Yibin
Huang Ying
Huang Yuanhao
Huang Yunlong
Huang Zhipeng
Huang Ziqiang
Huo Jiali

# I

# J

Ji Fan
Ji Honghu
Ji Ruilang
Ji Shiliang
ji Xiaoli

Ji Yujie
Ji Zheng
Jia Changzhen
Jia Qi
Jiang Changcheng
Jiang Hao
Jiang Junbo
Jiang Lan
Jiang Linpeng
Jiang Xiping
Jiang Yiyi
Jiang Yu-Long
Jiang Zhong Wei
Jiang Zhongyuan
Jiao Jiahui
Jiao Shuang
Jiao Zhijie
Jin Dong-Yue
Jin Feng
Jin Haitao
Jin Luffy
Jing Quan

## K

Kang Xiaoxu
Kang Xiaozhi
Kang Yi
Kar Gouri Sankar
Ke Shengxian
Ke Xing
Khaydarov Sherzod
Kim Cheolkyu
Kim Gi
Kim Yujin
Kobayashi Hiroyuku
Koh Tuck Foong
Kong Weiran

## L

Lai Walter
Lan Jun
Lau W S
Lei Hairong
Leyendecker Klaus
Li Bo

Li Chen
Li Cheng
Li Fei
Li Felix
Li Fengmei
Li Haoran
Li Hu
Li Hua
Li James
Li Jiahe
Li Jingjing
Li Ke
Li Lanxia
Li Lianlian
Li Lei
Li Ling Feng
Li Lingling
Li Linsen
Li Mengqiang
Li Ming
Li Peipei
Li Perry
Li Quanbo
Li Run-Ling
Li Shilin
Li Shipu
Li Shoutian
Li Tiesong
Li Tomi T
Li Wenqiang
Li Wenxin
Li Xiaohui
Li Xiaoyu
Li Xingyu
Li Xinyi
Li Yajie
Li Yang
Li Yanli
Li Ye
Li Yibin
Li Yida
Li Yijun
Li Yimei
Li Yu
Li Yuan

Li Zhen
Li Zhengcao
Li Zhengning
Li Zhi
Li Zibai
Li Xiaoyu
Li Xingyu
Li Zhixiong
Liang Huajian
Liang Jinxuan
Liang Kun
Liang Xiaofeng
Lin Bing-Hui
Lin Chunxi
Lin Rong
Lin Song-Yu
Lin Xinyi
Lin Xuanyu
Lin Yangkui
Ling Haiyang
Litta Eugenio Dentoni
Liu Bangxu
Liu Biqiu
Liu Botong
Liu D C
Liu Donghua
Liu Fanyu
Liu Hailong
Liu Hao
Liu Hebao
Liu Hongjie
Liu Janifer
Liu Jiabing
Liu Jianrui
Liu Jianshe
Liu Jin Biao
Liu Junwen
Liu Lihong
Liu Jintao
Liu Jipeng
Liu Luping
Liu Junwen
Liu Living
Liu Mei-Hua
Liu Muyi

Liu Qingwei
Liu Qun
Liu Senter
Liu Shaoxiong
Liu Tao
Liu Xianhe
Liu Xiaobing
Liu Xiaoxi
Liu Xueqiang
Liu Xuhui
Liu Yali
Liu Yang
Liu Yuxiao
Liu Zhang
Liu Zhenghong
Lo Hsiao-Han
Long Yin
Lu Chia Lin
Lu Guangyuan
Lu Hanwei
Lu Jiqing
Lu Lian
Lu Wei
Lu Xinchun
Lu Yanan
Lu Yuyao
Luk Wai-Shing
Luo Cong
Luo F
Luo Fu
Luo Jin
Luo Kun
Luo Lianbo
Luo Wen
Luo Xiantao
Luo Yanna
Lyu Peng Fei
Lyu Pengfei
Lyu Qingpeng

## M

Ma Haolan
Ma Haozhe
Ma Horse
Ma Jing-Lun

Ma Li
Ma T P
Ma Weiwei
Ma Xing
Mahzoon Alireza
Mair Robin
Mao Guiyun
Marangoni Thomas A
Marcozzi Pio
Mei Cuiyu
Mei Na
Meng Fayan
Meng Fei
Meng Xiangguo
Meng Yuanyuan
Mertens Hans
Mi Qing
Miao Jialei
Mu Tian Lei
Mukundhan Priya

## N
Na Weicong
Ni Dong
Ni Qiliang
Nie Wangxin
Nielsen Peter F
Niu Xinhuan
Niu Zongcai
Nozawa Toshihisa

## O
Oh Hansu
O'Neill James A
Orlov Ivan
Østerberg Fredrik W
Ouyang Keqing

## P
Pan Hongming
Pan Junjie
Pan Zhecheng
Peng Li
Peng Yufei

Petersen Dirch H
Pu Jiaming
Pu Lingling

# Q

Qi Jiming
Qian He
Qian Jun
Qian Wensheng
Qian Wihong
Qian Yuanyuan
Qin Feng
Qiu Chenchen
Qu Minghui
Qu Xiaofeng
Que Yurong

# R

Rassoul Nouredine
Ren Baoliang
Ren Hongnong
Ren Xiaobing
Rigner Anton
Ritzenthaler Romain
Royals Mitch

# S

Schmein Katharina
Sen Gourav
Sha Yin
Shao Xiong
Shao Yang
Shaoy Yun
Shen Hongjie
Shen Jiangxuan
Shen Manhua
Shen Mei
Shen Minghao
Shen Qiang
Shen Yaoting
Shen Yijiang
Shi Feng
Shi Heguang

Shi Liangtao
Shi Xian-Yu
Shi Xuelong
Shi Yunfan
Shim Byung Sup
Shiv Lior
Simoen Eddy
Song Jianye
Song Wan
Song Xin
Su Bo
Su Chang
Su Tao
Su Wen
Su Xian Wen
Su Xinruo
Sui Zhenchao
Sun Chang
Sun Juyang
Sun Lei
Sun Lifei
Sun Qin
Sun Qingqing
Sun Songyu
Sun Tuobei
Sun Weiran
Sun Xiaoting
Sun Xin
Sun Zheng-Qing

## T
Tan Huaqing
Tan Jun
Tan Wei
Tan Xiaoyu
Tang Hannah
Tang Jianshi
Tang Yongjin
Tang Zhong-Hai
Tao Da-Wei
Tao Jun
Teh Cha
Teng Tianjiao
Tian Fangxin
Tian Guoliang

Tian Jiajia
Tian Tian
Tong Jing
Tu Huojin

## U

## V
Vincent Benjamin

## W
Wan J
Wan Jing
Wan Min
Wang Bingquan
Wang Bo
Wang Cheng
Wang Chengxin
Wang Dawei
Wang DeJin
Wang H C
Wang Haihua
Wang Hu
Wang Hui
Wang Huimei
Wang Jian
Wang Jiang
Wang Jing
Wang Jingang
Wang Jun
Wang Kai
Wang Ke
Wang Kui
Wang Lei
Wang Li
Wang Liang
Wang Lingli
Wang Lu
Wang Mingxiang
Wang Pan
Wang Peter J
Wang Qiang
Wang Qing Peng
Wang Qingpeng

Wang Quanyong
Wang Sen
Wang Shasha
Wang Shihao
Wang Shouxu
Wang Shuxiang
Wang Song
Wang Suming
Wang Tongqing
Wang Wei
Wang Weifeng
Wang Wenhui
Wang Xiaodong
Wang Xiaofang
Wang Xiaoyan
Wang Xinan
Wang Xinhe
Wang Xining
Wang Xue-Sheng
Wang Yan
Wang Yansheng
Wang Yao
Wang Yinshua
Wang Ye
Wang Yong
Wang Yuan
Wang Yubing
Wang Yong
Wang Yuqi
Wang Zhe
Wang Zheyao
Wang Zhigao
Wang Zi-Hang
Wang Zhuangzhuang
Wang Zhuo
Wang Zilin
Wei Hongxue
Wei Junqi
Wei Qi
Wei Qianhui
Wei Shuashua
Wei Wilson
Wei Yanzhao
Wei Yayi
Wei Zhengying

Wiedmer Andreas
Wen Shiyuan
Wen Zesheng
Wolhlfarth Volker
Wu Chenfei
wu Cheng-Tae
Wu Chunlei
Wu Cinan
Wu Haijing
Wu Hanming
Wu Hao
Wu Huaqiang
Wu Lisa
Wu Pengfei
Wu Qiang
Wu Qixin
Wu Yifei
Wu Zhenhua

# **X**
Xi Xiaoyang
Xi Yue
Xia Jun
Xia Qing
Xia Qingqing
Xiao Yichen
Xian Chen
Xian Yi
Xiang Jiaying
Xiang Xun
Xiang Yang
Xiao Haochun
Xiao Jiaxing
Xiao Kai
Xiao Li
Xiao Runcai
Xiao Zhiqiang
Xhiaoy Yong
Xie Hongyun
Xie Hong-Yun
Xie Mengchen
Xie Meng-Yu
Xie Qiushi
Xing Chaoyang
Xing Yangyang

Xiong Enyi
Xiong Shao
Xiu Chun Yu
Xu Feng
Xu Gaobo
Xu Haoqing
Xu Jiazhang
Xu Jiong
Xu Kaidong
Xu Li Tiang
Xu Lu
Xu Min
Xu Ran
Xu Saisheng
Xu Shaodi
Xu Weikai
Xu WenShang
Xu Xiaojun
Xu Xiao-Chi
Xu Yongqiang
Xu Youfeng
Xu Zhaozhao
Xu Zhicheng
Xu Zhiyuan
Xue Meng

## Y

Yamada Yoshiaki
Yan Changhao
Yan Gangping
Yan Haitao
Yan Han
Yan Jiang
Yan Shijia
Yan Yongqiang
Yan Yu
Yana Zhongwen
Yang Dan
Yang Guang
Yang Hong
Yang Jianguo
Yang Jingwen
Yang Kelly
Yang Qing
Yang Ru

Yang Shang Shian
Yang Yongsheng
Yang Yu
Yang Yu-Pu
Yang Yushu
Yang Yuxin
Yao Jiadong
Yao Jiaxin
Yao Xing-Jun
Ye JiongHan
Yin Enjoy
Yin Huaxiang
Yin Jun
Yin Xunzhao
Ying Wen
Yiwen He
Yixian
Yoo Abraham
Yu Daquan
Yu Jiahong
Yu Mingfei
Yu Shaofeng
Yu Shirui
Yu Wei
Yu Weiwei
Yu You-Quan
Yuan Jie
Yuan Ning-Hsiu
Yuan Zhiqi

# **Z**
Zaheer Muhammad
Zang Kaiyan
Zeng Hongjian
Zeng Jiangji
Zeng Xuan
Zeng Zhaqin
Zhai Dongmei
Zhai Linpu
Zhan Kangshu
Zhang Baichun
Zhang Baoguo
Zhang Bo
Zhang Chen
Zhang David Wei

Zhang Ge
Zhang Guobiao
Zhang Haiyang
Zhang Hong
Zhang J
Zhang Jan-Kun
Zhang Jay
Zhang Jian
Zhang Jian-Kun
Zhang Jiefeng
Zhang Jing
Zhang Kang
Zhang Lei
Zhang Li
Zhang Libin
Zhang Lifei
Zhang Linlin
Zhang Min
Zhang Mingchuan
Zhang Qingtian
Zhang Qingzhu
Zhang Qinrui
Zhang Shaw
Zhang Shi-ran
Zhang Shuai
Zhang Simian
Zhang Simmons
Zhang Song
Zhang Tianhao
Zhang Tianjiao
Zhang Tianyu
Zhang Wanrong
Zhang Wenyu
Zhang Wei
Zhang Wuzhi
Zhang Xiaoming
Zhang Xiaotong
Zhang Xin
Zhang Xingan
Zhang Xingxing
Zhang Xuexiang
Zhang Y
Zhang Yanyan
Zhang Yaxin
Zhang Yichen

Zhang Yijun
Zhang Yinchan
Zhang Yingchun
Zhang Yiyang

Zhang Yu

Zhang Yun
Zhang Zhaozhao
Zhang Zhigang
Zhao Chunshan
Zhao Peng
Zhao Wei
Zhao Yuda
Zhao Zhenyang
Zhao Zun-Hua
Zhen Chongchong
Zheng Deguang
Zheng Jun-Fei
Zheng Wei
Zheng Xinhe
Zheng Ze
Zheng Zhaqin
Zheng Zongyuan
Zhong Liang
Zhong Qinghua
Zhong Weifa
Zhong Xiaolan
Zhong Yi
Zhou Bing
Zhou Dian
Zhou Gouhua
Zhou Guoyun
Zhou H F

Zhou Haifeng

Zhou Hongwei
Zhou Kan
Zhou Mingxing
Zhou Qing
Zhou Qing Jun
Zhou Tao
Zhou Wei
Zhou Wei Yu
Zhou Wen Zhan
Zhou Xiaofeng

Zhou Xiaoqiang

Zhou Ya

Zhou Yaohui

Zhu Alexander

Zhu Fu

Zhu Haoyu

Zhu Kai

Zhu Kaixiang

Zhu Lei

Zhu Limin

Zhu P L

Zhu Quanzhou

Zhu Ruiping

Zhu Shaojia

Zhu TongQing

Zhu Xiaona

Zhu Yanping

Zhu YuLie

Zhuang Jian-Zhu

Zhuang Junjun

Zhuang Shiwei

Zhuo Fengguo

Zhuo Jiaxiang

Zou Wenzhong

**IEEE**
445 Hoes Lane
Piscataway, NJ 08854-4141

ISBN 978-1-6654-9759-6